# TMS 2012

141st Annual Meeting & Exhibition

# Magnesium Technology 2012

Proceedings of a symposium sponsored by the
Magnesium Committee of the Light Metals Division of
The Minerals, Metals & Materials Society (TMS)

Held during
TMS 2012 Annual Meeting & Exhibition
Orlando, Florida, USA
March 11-15, 2012

Edited by

Suveen N. Mathaudhu
Wim H. Sillekens
Neale R. Neelameggham
Norbert Hort

*Editors*
Suveen N. Mathaudhu
Wim H. Sillekens
Neale R. Neelameggham
Norbert Hort

ISBN 978-3-319-48571-3 ISBN 978-3-319-48203-3 (eBook)
DOI 10.1007/978-3-319-48203-3

Chemistry and Materials Science: Professional


Published by Springer International Publishers, Switzerland, 2016
Reprint of the original edition published by John Wiley & Sons, Inc., 2012, 978-1-11829-121-4



Printed on acid-free paper

This Springer imprint is published by Springer Nature
The registered company is Springer International Publishing AG
The registered company address is: Gewerbestrasse 11, 6330 Cham, Switzerland

# TABLE OF CONTENTS
# Magnesium Technology 2012



## Magnesium Technology 2012

### Plenary Session



### Primary Production

## Deformation Mechanisms



## Casting and Solidification

## Alloy and Microstructural Design

## Corrosion and Coating



## High Temperature Processing and Properties

## Processing-Microstructure-Property Relationships I



## Advanced Processing and Joining

## Processing-Microstructure-Property Relationships II

## Energy and Biomedical / Primary Production



## Poster Session

# PREFACE

*"The market for magnesium may be expected to develop along lines similar to those along which aluminum developed, but whether to anything like the same extent is entirely problematic. The market is somewhat unacquainted with many of the special qualities of the metal, and increased sales are largely a matter of education and research whereby a demand will be created and developed hand in hand with production. [...] The importance of a metal with properties similar to that of aluminum, but nearly half as light in these days of enormous automobile and aircraft expansion hardly needs much argument."*

- 1919 United States Bureau of Mines Bulletin

In 1920, the television was not invented yet, and radio was king, with Louis Armstrong and Duke Ellington playing their legendary jazz in living rooms everywhere. The prohibition of alcohol began in the U.S.A. ushering in an era of organized crime (most famously, Al Capone) and of homemade whisky. The average daily wage for a male was US$5.00 per day; the average car cost US$265.00; a gallon of milk was US$.56, and a gallon of gas was US$.22. It was clearly a different time, but as this quote illustrates, some things are jarringly as valid today as they were 90 years ago. While the length of time since the bulletin may indicate the resilience of the Mg community, it also points to the fact that our mission is far from done; we must continue to support education and research for Mg and its alloys to reach their full potential.

It is this mission that supports the continued propagation of the Magnesium Technology Symposium, which is held yearly under the auspices of the TMS Light Metals Division's Magnesium Committee at the TMS Annual Meeting and Exhibition. This year is the 13th edition of this meeting, and will be held in Orlando, FL (USA) from March 11-15, 2012. The symposium is arranged into an opening plenary session, a poster session, and a number of technical sessions which address the full scope of Mg research and development. These sessions include, deformation mechanisms, primary production, casting and solidification, alloy and microstructural design, corrosion and coating, high temperature processing and properties, processing-microstructure-property relationships, advanced processing and joining, and new applications in biomedical and energy.

This proceedings volume represents the proceedings of these symposia. They represent contributions from 19 countries spanning the globe. The largest contribution for the 2012 proceedings comes from the United States (30%), followed by Canada (19%), China (14%) and Japan (11%). All submissions were thoroughly peer-reviewed by volunteer referees acting on the behalf of the Magnesium Committee. These reviewers are critical to the success of this volume, and are commended in the following pages. Some of the symposium contributions related to Mg surface science will be published in an upcoming JOM special issue sponsored by the Magnesium Committee, titled "Magnesium: Engineering the Surface" in June 2012, and to be edited by Michele Manuel (University of Florida, USA), and Chamini Mendis (Helmholtz-Zentrum Geesthacht, Germany).

Additionally, a one-day, invited speaker mini-symposium will be held on the specific topic of "Phase Transformation and Deformation in Magnesium Alloys", and is organized by Jianfeng Nie (Monash University, Australia), Sean Agnew (University of Virginia, USA) and Suveen Mathaudhu (U.S. Army Research Office, USA). These sessions are co-sponsored by the TMS/ASM Phase Transformation Committee, and aim to examine the current understanding on phase transformations and roles of phase transformation products in controlling the deformation behavior of commercially and strategically important magnesium alloys, and to provide a platform for rational design of microstructures for better mechanical properties and formability. The submissions to this mini-symposium will be published in a forthcoming Special Issue of *Metallurgical and Materials Transactions A.*

Lastly, the organizing of these proceedings would not have been possible without the dedication and assistance of TMS staff. In particular, we would like to recognize Ms. Chris Wood for her tireless patience and hard work.

Over the past 12 years, the Magnesium Technology Proceedings, through its high quality technical content, has become the primary proceedings for the magnesium industry and technical community. We would hope that this year's volume will serve you just as well.

*Suveen N. Mathaudhu (U.S. Army Research Office, USA)*
*Wim H. Sillekens (TNO, Netherlands)*
*Neale R. Neelameggham (IND LLC, USA)*
*Norbert Hort (Helmholtz-Zentrum Geesthacht, Germany)*

# ABOUT THE EDITOR

## SUVEEN NIGEL MATHAUDHU
## MAGNESIUM TECHNOLOGY 2012 EDITOR

**Suveen Nigel Mathaudhu** (1975) serves as the Program Manager for Synthesis and Processing of Materials with the U.S. Army Research Office (ARO), Materials Science Division. He received his B.S.E. in Mechanical Engineering from Walla Walla University in 1998, and Ph.D. in Mechanical Engineering from Texas A&M University in 2006. In his current position, he manages programs which focus on the use of innovative approaches for processing high performance structural materials reliably and at lower costs. He also concurrently serves as an Adjunct Assistant Professor in the Department of Materials Science and Engineering at North Carolina State University, where his research interests include: ultrafine-grained and nanostructured materials by severe plastic deformation, consolidation of metastable particulate materials and processing-microstructure-property relationships of refractory metals and lightweight metals, and thermally stable nanocrystalline materials. He has co-authored over 50 technical publications in these areas.

# ABOUT THE ORGANIZERS

**Wim H. Sillekens** (1963) is a Senior Scientist at the Netherlands Organization for Applied Scientific Research (TNO), where he is involved in national and European research projects. He obtained his Ph.D. from Eindhoven University of Technology, Netherlands, on a subject relating to metal-forming technology. Since he has been engaged in light-metals research (aluminum and magnesium), amongst others on (hydro-mechanical) forming, (hydrostatic) extrusion, forging, recycling/refining, and more recently on biomedical applications. His professional career includes positions as post-doc researcher at his alma mater and as a research scientist / project leader at TNO. International working experience covers a placement as a research fellow at the Mechanical Engineering Laboratory (AIST-MITI) in Tsukuba, Japan, and – more recently – shorter stays as a visiting scientist at GKSS in Geesthacht, Germany, and at PNNL in Richland WA, USA. He has co-authored a variety of journal and conference papers (about 50 entries to date). Current research interests are in the physical and mechanical metallurgy of light metals in general and magnesium wrought alloys in particular.

**Neale R. Neelameggham** is 'Guru' at IND LLC, involved in Technology marketing and consulting in the field of light metals and associated chemicals [ boron, magnesium, titanium, lithium and alkali metals], rare earth elements, battery and energy technologies, etc. He has over 38 years of expertise in magnesium production technology from the Great Salt Lake brine in Utah, involved in Process Development of its startup company NL magnesium through the presently known US Magnesium LLC, and was its Technical Development Scientist from where he retired. Dr. Neelameggham's expertise includes all aspects of the magnesium process, from solar ponds through the cast house including solvent extraction, spray drying, molten salt chlorination, electrolytic cell and furnace designs, lithium ion battery chemicals and by-product chemical processing. In addition, he has an in-depth and detailed knowledge of alloy development as well as all competing technologies of magnesium production, both electrolytic and thermal processes worldwide. Dr. Neelameggham holds 13 patents and a pending patent on Boron Production, and has several technical papers to his credit.

As a member of TMS, AIChE, and a former member of American Ceramics Society he is well versed in energy engineering, bio-fuels, rare-earth minerals and metal processing and related processes. Dr. Neelameggham has served in the Magnesium Committee of LMD since its inception in 2000, chaired it in 2005, and has been a co-organizer of the Magnesium Symposium since 2004. In 2007 he was made a Permanent Co-organizer for the Magnesium Symposium. He has been a member of the Reactive Metals Committee, Recycling Committee and Programming

Committee Representative of LMD. In 2008, LMD and EPD created the Energy Committee following the symposium on $CO_2$ Reduction Metallurgy Symposium initiated by him. Dr. Neelameggham was selected as the inaugural Chair for the Energy Committee with a two-year term. He was a member of LMD council. He received the LMD Distinguished Service Award in 2010. He has been a co-editor of the Energy Technology symposium. Dr. Neelameggham holds a doctorate in extractive metallurgy from the University of Utah.

**Norbert Hort** (1964) is the head of the Magnesium Processing Department at the Magnesium Innovation Centre (MagIC) within the Helmholtz-Zentrum Geesthacht Zentrum für Material- und Küstenforschung, Geesthacht, Germany (formerly the GKSS Research Centre). Concurrently he is lecturer at the Leuphana University, Lüneburg, Germany. He studied Materials Sciences at the Clausthal University of Technology (CUT), Germany, where he already got involved in magnesium research since the early 1990ties. His diploma thesis (1994) was dealing with gas atomized, hot extruded magnesium alloys. During 1994-95 he worked as a researcher at the Institute of Materials Sciences (CUT) and he joined the HZG in 2000. He got his PhD degree in Materials Sciences in 2002 from the Clausthal University of Technology. In the MagIC he is responsible for the development of new creep resistant magnesium alloys, grain refinement and the castability of magnesium alloys (viscosity and density of melts, fluidity, and mould filling). A major focus of his research is additionally the development of new magnesium alloys for biodegradable implants in a close collaboration with biologists and clinicians. He is co-author of more than 60/120 peer reviewed journal papers and contributions to conference proceedings. In recent years he was involved in the organizing committees of the conference series "Magnesium Alloys and their Applications" and of the conference "Light Metal Technologies 2011". Since 2009 he is also the chairman of the technical committee "Magnesium" of the German Society of Materials (DGM).

## TMS Magnesium Technology 2012 – Session Chairs *(provisional)*

**Plenary Session** – Monday AM
- Suveen N. Mathaudhu (US Army Research Office, USA)
- Wim Sillekens (TNO, Netherlands)
- Norbert Hort (Helmholtz-Zentrum Geesthacht, Germany)

**Primary Production** – Monday PM
- Neale Neelameggham (IND LLC, USA)
- Adam Powell (Metal Oxygen Separation Technologies Inc., USA)

**Deformation Mechanisms** – Monday PM
- Bin Li (Mississippi State University, USA)
- Alok Singh (National Institute of Materials Science, Japan)

**Casting and Solidification** – Wednesday AM
- Norbert Hort (Helmholtz-Zentrum Geesthacht, Germany)
- Yongho Sohn (University of Central Florida, USA)

**Alloy and Microstructural Design** – Wednesday AM
- Jian-Feng Nie (Monash University, Australia)
- Nack Kim (POSTECH, South Korea)

**Corrosion and Coating** – Wednesday PM
- Guang-Ling Song (General Motors, USA)
- Michele Manuel (University of Florida, USA)

**High Temperature Processing and Properties** – Wednesday PM
- Paul E. Krajewski (General Motors, USA)
- Warren Poole (University of British Columbia, Canada)

**Processing-Microstructure-Properties I** – Thursday AM
- Hidetoshi Somekawa (National Institute of Materials Science, Japan)
- Kyu Cho (U.S. Army Research Laboratory, USA)

**Advanced Processing and Joining** – Thursday AM
- Tyrone Jones (U.S. Army Research Laboratory, USA)
- Brian Jordan (University of Alabama, USA)

**Processing-Microstructure-Properties II** – Thursday PM
- Alan Luo (General Motors, USA)
- Fabrizio D'Errico (Politecnico di Milano, Italy)

**Energy and Biomedical / Primary Production** – Thursday PM

- Wim Sillekens (TNO, Netherlands)
- Neale Neelameggham (IND LLC, USA)

**Poster Session**

- Eric A. Nyberg (Pacific Northwest National Laboratory, USA)

# TMS Magnesium Technology 2012 – Reviewer Pool

The completion of this volume would not have been possible without the volunteer assistance of these fine individuals from 11 different countries. Their commitment to the advancement of Mg research and science is to be commended. They are presented here in last-name alphabetical order:

John Allison
*University of Michigan (USA)*

Matthew Barnett
*Deakin University (Australia)*

Jan Bohlen
*Helmholtz-Zentrum Geesthacht (Germany)*

Carlos Caceres
*University of Queensland (Australia)*

Gokhan Demirci
*Aselsan Inc.(Turkey)*

Fabrizio D'Errico
*Polytechnico di Milano (Italy)*

Kevin Doherty
*U.S. Army Research Laboratory (USA)*

David Foley
*Texas A&M University (USA)*

Mark Gibson
*CSIRO (Australia)*

Louis Hector
*General Motors (USA)*

Daniel Höche
*Helmholtz-Zentrum Geesthacht (Germany)*

Norbert Hort
*Helmholtz-Zentrum Geesthacht (Germany)*

Yuanding Huang
*Helmholtz-Zentrum Geesthacht (Germany)*

Vipin Jain
*National Physical Laboratory (India)*

William Joost
*U.S. Department of Energy (USA)*

Brian Jordan
*University of Alabama (USA)*

Karl Kainer
*Helmholtz-Zentrum Geesthacht (Germany)*

Ishak Karakaya
*Middle East Technical University (Turkey)*

Paul Krajewski
*General Motors (USA)*

Zi-Kui Liu
*The Pennsylvania State University (USA)*

Alan Luo
*General Motors (USA)*

Luke Mackenzie
*Novelis (USA)*

Bilal Mansoor
*Masdar Institute of Technology (United Arab Emirates)*

Michele Manuel
*University of Florida (USA)*

Suveen Mathaudhu
*U.S. Army Research Office (USA)*

Chamini Mendis
*National Institute of Materials Science (Japan)*

Rajiv Mishra
*University of North Texas (USA)*

Neale Neelameggham
*IND LLC, (USA)*

Soobhankar Pati
*Metal Oxygen Separation Technologies, Inc. (USA)*

Warren Poole
*University of British Columbia (Canada)*

Adam Powell
*Metal Oxygen Separation Technologies, Inc. (USA)*

Wim Sillekens
*TNO (Netherlands)*

Alok Singh
*National Institute of Materials Science (Japan)*

Kiran Solanki
*Arizona State University (USA)*

A. Srinivasan
*Helmholtz-Zentrum Geesthacht (Germany)*

John Tartaglia
*Stork Climax Research Services (USA)*

Domonkos Tolnai
*Vienna University of Technology (Austria)*

Kishore Venkatesan
*CSIRO (Australia)*

Ravi Verma
*General Motors (USA)*

Lei Yang
*Helmholtz-Zentrum Geesthacht (Germany)*

Zhisheng Zhen
*Magontec (Germany)*

Yuntian Zhu
*North Carolina State University (USA)*

# Magnesium Technology 2012

## Plenary Session

### Monday AM

*Session Chairs:*

**Suveen N. Mathaudhu**
**(US Army Research Office, USA)**
**Wim Sillekens**
**(TNO, Netherlands)**
**Norbert Hort**
**(Helmholtz-Zentrum Geesthacht, Germany)**

**Magnesium Technology 2012**
*Edited by: Suveen N. Mathaudhu, Wim H. Sillekens, Neale R. Neelameggham, and Norbert Hort*
***TMS (The Minerals, Metals & Materials Society), 2012***

# Magnesium Alloy Development Using Phase Equilibria Computation and Microstructure Validation

A.A. Luo, R.K. Mishra, B.R. Powell, A.K. Sachdev
General Motors Global Research and Development, Warren, MI 48090, USA

Keywords: Magnesium alloys, alloy development, computational alloy design, microstructure

**Extended Abstract**

This paper summarizes the development of new cast and wrought magnesium alloys using computational thermodynamics and experimental methods. The work illustrates the role of calculated phase diagrams, solidification paths and phases in predicting and interpreting the final microstructure of Mg-Al-Ca and Mg-Al-Sn cast alloy systems and Mg-Al-Mn and Mg-Zn-Ce wrought alloy systems.

The Mg-Al-Ca alloys show excellent creep resistance due to the formation of high-temperature $(Mg,Al)_2Ca$ phase. Computational thermodynamics calculations and experimental investigation of the Mg-Al-Ca ternary system have validated the earlier development of creep-resistant AX52 (Mg-5Al-2Ca[1]) and AX53 (Mg-5Al-3Ca) alloys. The Scheil simulation of alloy solidification has suggested key guidelines of alloy composition design in promoting the thermally stable $(Mg,Al)_2Ca$ phase, while replacing the less stable $Mg_{17}Al_{12}$ in the microstructure. The suppression of the $Mg_{17}Al_{12}$ phase can increase the solidus temperature, reduce the freezing range, and increase the latent heat during solidification, all of which contribute to improved castability in the AX53 and AX52 alloys compared with the AX51 (Mg-5Al-1Ca) alloy. The quantitative phase equilibrium data, microstructure characterization, as well as the thermal physical properties of the Mg-Al-Ca alloy system generated from this study are important basis for further optimization of alloy composition and microstructure for elevated-temperature applications.

The solidification microstructure and mechanical properties of as-cast Mg-Al-Sn alloys have been investigated using computational thermodynamics and experiments. The as-cast microstructure of Mg-Al-Sn alloys consists of α-Mg, $Mg_{17}Al_{12}$ and $Mg_2Sn$ phases. The amount of $Mg_{17}Al_{12}$ and $Mg_2Sn$ phases formed increases with increasing Al and/or Sn content, and show good agreement between the experimental results and the Scheil solidification calculations. Generally, the yield strength of as-cast alloys increases with Al and/or Sn content, while the ductility decreases. This study has confirmed an early development of Mg-7Al-2Sn alloy for structural applications and led to a promising new Mg-7Al-6Sn alloy with significantly improved strength and ductility comparable to commercial AZ91 alloy.

AM30 (Mg-3Al-0.3Mn) alloy was developed to have significantly improved extrudability and slightly improved mechanical properties compared to AZ31 alloy. The limited extrudability of AZ31 (Mg-3Al-1Zn) alloy is reported to be caused by incipient melting of the $Mg_{17}Al_{12}$ binary phase with a eutectic temperature of 438°C. Computational phase equilibria and microstructural examination, however, show that Mg-Al-Zn ternary phases with eutectic temperatures as low as 338°C are actually present in the AZ31 alloy, which is indeed the real limitation to its extrudability. The significantly improved extrudability of AM30 alloy (Mg-3Al-Mn) is due to the absence of these zinc-containing eutectic phases. In the Mg-Zn-Ce (ZE) system, Zn provides strength through solid solution strengthening, while Ce increases the ductility by creating a favorable texture. Computational thermodynamics and experimental work confirm a final solidification structure of the ZE alloys: α(Mg) + $Mg_7Zn_3$ + MgZn. Ce was found not tied up by Zn and $Mg_7Zn_3$ exists at temperatures above 325°C and decomposes into α(Mg) and MgZn at below 325°C, a potentially age-hardening phase.

[1] All compositions in wt.%.

**Magnesium Technology 2012**
*Edited by: Suveen N. Mathaudhu, Wim H. Sillekens, Neale R. Neelameggham, and Norbert Hort*
***TMS (The Minerals, Metals & Materials Society), 2012***

# ATOMS-TO-GRAINS CORROSION MODELING FOR MAGNESIUM ALLOYS

Hyunwook Kwak[1], Jie Xiao, and Santanu Chaudhuri[1*]

[1]ISP/Applied Sciences Laboratory, Washington State University, WA 99210-1495

Keywords: Corrosion Protection, First-Principles Calculations, MgO, Rare-Earth Doping

## Abstract

Development of new Mg-alloys needs greater efforts in systematic evaluation of corrosion protection methods under service environments. A deeper understanding of microscale and nanoscale metallurgy is needed to slow down corrosion. Results from our multiscale modeling effort will be presented to demonstrate the power of first-principle theories in predicting the composition-dependent kinetics of corrosion reactions. Our recent results show that the corrosion prevention properties of Mg-alloys can be significantly enhanced by doping rare-earth elements such as Cerium that protects the oxide layer from rapid hydrolysis. Furthermore, we are developing kinetic Monte Carlo and finite-element analysis (FEA) based models to extend the predictions from the atomic length scales to nano- and microscale models that will include grains and alloy microstructure. The potential for incorporating mechanical and corrosion performance in continuum length scale combined with the insights from first-principles based surface chemical models can have powerful impact on the development of Mg-alloys.

## Introduction

Our initial multiscale simulation results on Mg alloys indicate that corrosion mechanisms in Mg alloys are unique.[1] Therefore it is important that corrosion prevention strategies need to include consideration for specific mechanistic pathways to prevent failures and find mature alloys for structural applications. From optimizing composition of promising alloys, increasing the stability of the oxide layer, and obtaining a uniform microstructure through carefully controlled processing conditions, all play a part in reducing corrosion vulnerabilities. Our simulation effort is geared towards understanding different galvanic and dissolution processes in hot and humid environments. Some theoretical development will be needed to address the role of electrochemical field inside the pits at the intermetallic particle boundaries. In alloys such as AZ31, AZ80 and AZ91D, there are different phases of intermetallic precipitates that are formed in the alloy matrix. Intermetallic impurities or inclusions such as $Al_2Mn$, $Mg_{17}Al_{12}$ etc. have the potential to act as a cathode and expedite the oxide film and alloy etching where the Mg-alloy matrix acts as anode. In this current work, we will discuss stability of MgO film without intermetallic impurities across different length scales and identify the links between elemental doping of the MgO surface and the correlation with metal loss from atomistic to microscale.

Figure 1 illustrates the framework we are developing for understanding and prescribing predictive pathways to slow-down corrosion. It consists of four layers of modeling blocks, which are bi-directionally coupled and closely connected with a block of experimental data. At the bottom layer, first-principles-based calculations are responsible for generating thermodynamic information, such as Gibbs free energy of activation for a chemical or electrochemical reaction. This information is critical for kinetic Monte Carlo and molecular dynamics (kMC/MD) simulation of corrosion phenomena on and nearby the metal-electrolyte interface at nanoscale, e.g., oxide layer formation, ion adsorption, desorption, and diffusion through oxide/hydroxide layer, and metal dissolution, etc. On the other hand, reaction kinetics derived from first-principles calculation and kMC/MD simulation can be fed into an upper layer. At this layer, finite element models describe the growth of a single pit, where metal microstructure, pit surface morphology, and corrosion environment are systematically taken into account. Detailed investigation on the growth of a single pit will eventually lead to a successful study of the evolution of an ensemble of pits at macroscopic time and length scales by using the damage function analysis (DFA) method. Finally, corrosion performance evaluation (such as life time quantification) will be conducted at the top layer.

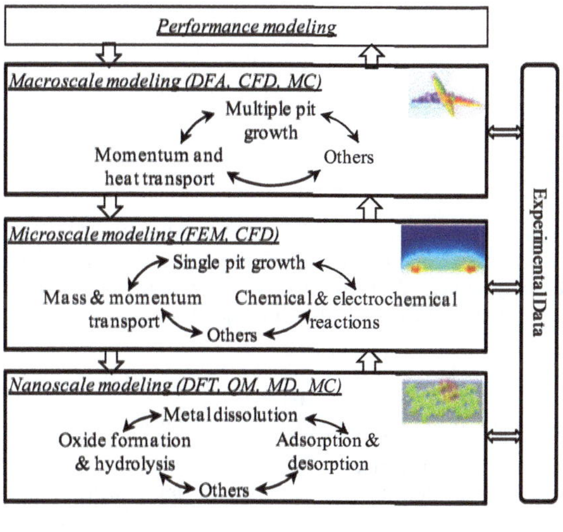


Figure 1. Multiscale modeling framework of corrosion

Oxide film on top of Mg alloys plays an important role on the corrosion protection of pure magnesium and Mg alloys. Presence of alloying compounds in the alloy is also important for the oxide films because the alloying species can transport to the metal surface and alter the composition of the oxide film. This observation is backed by numerous previous work where traces of excessive alloying compounds were detected at the protective films on top of Y-enriched and Ce-enriched Mg alloy systems. The primary emphasis of first-principles based *corrosion modeling effort* is to develop understanding of interfacial reactions as a function of surface composition of alloys under humid and corrosive conditions. Mg-alloys are particularly vulnerable in humid environment due to hydrolysis of MgO layers to Mg()H)2 as shown in figure 2 . Once this barrier layer is damaged, galvanic processes leads to rapid loss of Mg due to its position in the electrochemical series.

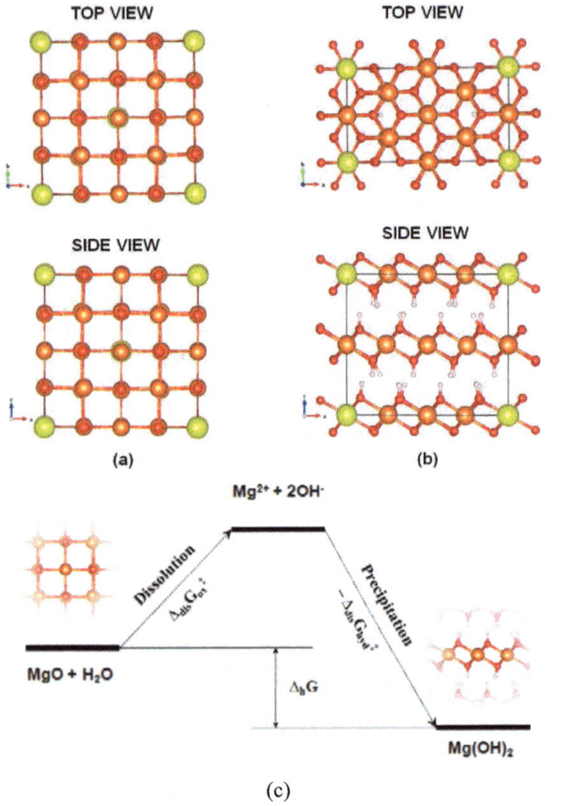


Figure 2 (a) The unit cell of MgO, and (b) $Mg(OH)_2$ from top and side-view of atomic arrangements. (c) We calculate the thermodynamics from a two-step reaction model for the hydroxylation of the oxide films on Mg alloy.

## Methods

<u>Thermochemical properties from first principles</u>

The key to our thermodynamic description of the hydroxylation process for the doped oxides is the use of first-principles methods to calculate the overall change in thermochemical potential during the following hydroxylation process,

$$\mathrm{MgO(periclase)} + \mathrm{H_2O_{(l)}} \longrightarrow \mathrm{Mg(OH)_2(brucite)} \tag{1}$$

For oxide films in contact with aqueous solution at room temperature, we consider a system that is exposed to a solution reservoir in which temperature, pressure, and activities of solute species are kept constant at standard condition. The appropriate thermodynamic potential to describe such a system is Gibbs free energy. Based on the definition of Gibbs free energy (G = U – TS + pV), the temperature-dependent free energy of a solid can be approximated as follows,

$$G \simeq U_{\mathrm{elec}} + F_{\mathrm{vib}}(T) + pV \tag{2}$$

The first term indicates the electronic total energy that corresponds to the ground state energy of a system at 0 K. The second term corresponds to the free energy contribution from the vibrational excitation of solids $(F_{\mathrm{vib}}(T) = U_{\mathrm{vib}}(T) - TS_{\mathrm{vib}}(T))$. For gas phase substances, similar expression can be given,

$$G \simeq U_{\mathrm{elec}} + F_{\mathrm{trans}}(T) + F_{\mathrm{rot}}(T) + F_{\mathrm{vib}}(T) + pV \tag{3}$$

where the translational and the rotational free energy of a gas molecule ($F_{\mathrm{trans}} + F_{\mathrm{rot}}$) is also considered. From our methodology, the first three terms of Eqn. (2) are calculated from electronic structure and phonon frequency calculations based on DFT. Change in the $pV$ term is neglected for solids since they are at least several orders of magnitude smaller than the change in the internal energy at standard temperature and pressure condition.

From the phonon frequency calculations of solids, vibrational contribution of the thermodynamic function ($F_{\mathrm{vib}}(T)$) can be calculated using a direct method. Direct method uses first-principles-based force constants to form up a dynamical matrix. The force constants are calculated from the Hellmann-Feynman forces induced by the displacement of individual atoms. In this formalism, dynamical matrix of a crystal is defined as,

$$D(k;\mu\nu) = \frac{1}{\sqrt{M_\mu M_\nu}} \sum_{\mathbf{m}} \Phi(0,\nu;\mathbf{m},\nu)\, e^{-2\pi i \mathbf{k}\cdot[\mathbf{R}(0,\mu)-\mathbf{R}(\mathbf{m},\nu)]} \tag{4}$$

where $M_\mu$, $M_\nu$ and $\mathbf{R}(0,\mu)$, $\mathbf{R}(\mathbf{m},\nu)$ are masses and positions of atoms $\mu$ and $\nu$, respectively. The summation index $\mathbf{m}$ runs over all atoms of the crystal. $\mathbf{k}$ is the wave vector in the first Brillouin zone and $\Phi(0,\nu;\mathbf{m},\nu)$ is the force constant matrix which can be constructed from the induced Hellman-Feynman forces by atomic displacements. Phonon frequencies $\omega(\mathbf{k},j)$ and the polarization vector $\mathbf{e}(\mathbf{k},j)$ are given from the eigenvalues and the eigenvectors of the dynamical matrix:

$$\omega^2(\mathbf{k},j)\mathbf{e}(\mathbf{k},j) = D(\mathbf{k})\mathbf{e}(\mathbf{k},j) \tag{5}$$

with the index $j$ for the normal modes. With this, total phonon density of states can be evaluated from $\omega(\mathbf{k},j)$ and $\mathbf{e}(\mathbf{k},j)$ as,

$$g(\omega) = \frac{1}{nd\Delta\omega} \sum_{\mathbf{k},j} \delta_{\Delta\omega}[\omega - \omega(\mathbf{k},j)] \tag{6}$$

where $n$ is the number of sampling wave vectors, $d$ is the dimension of the dynamical matrix, $\Delta\omega$ is the frequency interval, and the delta function $\delta_{\Delta\omega}$ is defined as

$$\delta_{\Delta\omega}(x) = \begin{cases} 1 & \text{if } -\frac{\Delta\omega}{2} < x < +\frac{\Delta\omega}{2} \\ 0 & \text{otherwise.} \end{cases} \tag{7}$$

For a given phonon density of states, Sternik and Parlinski showed that the vibrational contribution of the free energy, $F_{\mathrm{vib}}(T)$, can be calculated within the harmonic approximation as,

$$F_{\text{vib}}(T) = rk_B T \int_0^{\infty} d\omega\, g(\omega) \ln\left[2\sinh\left(\frac{\hbar\omega}{2k_B T}\right)\right] \quad (8)$$

Using Eqns (2) and (8), free energies of solid MgO and $Mg(OH)_2$ crystals at constant temperature can be calculated solely based on first-principles methods.

Effect of doping on MgO hydroxylation

The fundamental question is, how can we improve the stability of the MgO layer? In this work, we applied the cationic doping concentration of 1 out of 16 Mg sites (6.25 at%Mg) for both MgO and $Mg(OH)_2$. For the doped $Mg(OH)_2$ crystals, 4×2×2 orthorhombic supercell with $16(Mg(OH)_2)$ was built with one cationic doping site. Electronic energy, thermal energy, and entropy values for MgO and $Mg(OH)_2$ are all in close agreement with previous first-principles study by Duan and Sorescu. The calculated Gibbs free energy of hydroxylation (-26.5 kJ / mol) also agrees very well with the prediction from experimental database (-27.1 kJ / mol). The negative Gibbs free energy indicates that the oxide film degradation via hydroxylation occurs spontaneously. The results of the dissolution thermodynamics is summarized in Tables I and II.

**Table I** First-principles-calculated thermochemical properties of reactants and products from the overall hydroxide reaction shown in Eqn. (1). All units are in kJ / mol.

| $T = 298$ K, $p = 1$ bar | $U_{\text{elec}}$ | $F_{\text{vib}}(T)$ | $pV$ | $E_{\text{rot}}(T)$ | $G(T,p)$ |
|---|---|---|---|---|---|
| MgO (s) | -1156.1 | 10.82 | - | - | -1145.2 |
| $H_2O$ (g) | -1370.0 | 17.90 | 2.48 | 7.44 | -1352.1 |
| $Mg(OH)_2$ (s) | -2601.2 | 68.90 | - | - | -2532.3 |
| $\Delta_h G = G(Mg(OH)_2) - \left[G(MgO) + G(H_2O_{(g)}) + \delta G^{g\to l}_{H_2O}\right] = -26.5$ (kJ/mol) | | | | | |

In Table I, the first-principles calculated energy and entropy values for the oxide and the hydroxide crystals are shown for standard temperature and pressure condition (298 K, 1 bar). Electronic energy, thermal energy, and entropy values for MgO and $Mg(OH)_2$ are all in close agreement with previous first-principles studies. The calculated Gibbs free energy of hydroxylation at standard condition is -26.5 kJ / mol. This agrees very well with the prediction from the experimental database (-27.1 kJ / mol). The negative Gibbs free energy indicates that the oxide film degradation via hydroxylation occurs spontaneously. We also found that the thermodynamic driving force for the hydroxylation mostly attributes to the decrease in electronic energy ($\Delta_h U_{\text{elec}} < 0$) which compensates the increase in thermal energy ($\Delta_h F_{\text{vib}} > 0$).

Subsequently, we calculated hydroxylation free energies for the doped oxide and hydroxide crystals with six different types of cationic doping ($Zn^{2+}$, $Al^{3+}$, $Y^{3+}$, $Ce^{3+}$, $Ce^{4+}$, and $Zr^{4+}$). The results are summarized in Table II **Table** along with electronic ($\Delta_h U_{\text{elec}}$) and thermal ($\Delta_h F_{\text{vib}}$) contributions to $\Delta_h G$. When doped with cationic impurities, a clear trend was found for $\Delta_h G$ with respect to the charge state of the impurity atoms. While doping with isovalent $Zn^{2+}$ shows minimal impact, doping with higher charge state (3+ and 4+) elements leads to a significant increase in the hydroxylation free energy. For $Ce^{4+}$ and $Zr^{4+}$ doped oxides, the doping effect is significant enough to change the sign of $\Delta_h G$ from negative to positive. As a result, hydroxylation reaction becomes no longer spontaneous with the doping. The similar trend found in $\Delta_h U_{\text{elec}}$ suggests that such doping effect has the electronic origin. The trend is most apparent from a comparison between $Ce^{3+}$ doping and $Ce^{4+}$ doping, where significant difference between hydroxylation free energies (26.8 kJ / mol) can only attribute to the difference in their impurity charge state. The change in sign from a highly favorable reaction to unfavorable one is a potentially powerful information on use of doping in slowing down failure of MgO grains.

**Table II** First-principles calculated electronic, vibrational, and overall free energy changes during the hydroxylation of the doped MgO crystals.

| (298 K, 1 bar) | $\Delta_h U_{\text{elec}}$ (kJ/mol) | $\Delta_h F_{\text{vib}}$ (kJ/mol) | $\Delta_h G$ (kJ/mol) |
|---|---|---|---|
| Undoped | -75.2 | +40.2 | **-26.4** |
| $Zn^{2+}$-doped | -73.1 | +40.5 | **-24.0** |
| $Al^{3+}$-doped | -53.8 | +40.8 | **-4.5** |
| $Y^{3+}$-doped | -59.1 | +40.3 | **-10.2** |
| $Ce^{3+}$-doped | -61.1 | +41.0 | **-11.5** |
| $Ce^{4+}$-doped | -33.4 | +40.2 | **+15.3** |
| $Zr^{4+}$-doped | -31.1 | +40.6 | **+18.1** |

First-principles-assisted kinetic Monte Carlo (kMC) model

Let us examine first how similar information can be useful in understanding nanometer scale evolution of surface with doping elements such as Ce. The first model we are developing is for the alloy exposed to electrolyte without the MgO layer on the top. We will eventually describe the MgO grains and their hydrolysis. To analyze nanometer-scale corrosion of Mg surfaces, we employ lattice-based kinetic Monte Carlo (kMC) technique. While the dissolution parameters for Mg atoms on the surfaces that are in contact with aqueous solution will be set from experimental polarization curve of pure Mg, other parameters for atomistic diffusion process will be directly extracted from first-principles DFT calculations. For a given surface morphology and the alloying composition, the outcome of the simulation will provide the corrosion current density (*e.g.* dissolution rate) and the surface morphology of the alloy surface as a function of time. The model will also provide selective dissolution mechanism during the corrosion process and will give insights on the corrosion performance of a given alloying composition.

Simulation setup

In KMC simulation, atoms and molecules are modeled in a coarse-grained way which would represent averages of behaviors that arises from the transitions between different atomistic-scale configurations. We use lattice-based kMC module from SPPARKS code from Sandia National Laboratory to describe dissolution and diffusion of metal atoms at alloy surfaces. The same code will be used in future to describe grains.

For the lattice-based kMC simulation, we employ (100)-facing fcc structured slab surface model with lattice parameter of 4 Å. The 2D periodic surface unit cell is 8 nm × 8 nm wide (800 lattice sites per layer) and 34 atomic-layer thick. The rest of the 6 layers of atomic lattices are initially assigned as solution sites, *i.e.* non-occupied sites. The resulting slab is 6 nm thick with 27200 sites that are initially occupied with Mg or RE atoms. At the beginning of simulation, metal sites are assigned by randomly distributing Mg and RE atoms with a given alloying composition, *i.e.* the doping concentration. Both upward and downward facing surfaces in the slab are exposed to the solution. During the simulation, only diffusion and dissolution steps are considered.

Dissolution model

Selective dissolution of Mg alloy surface is simulated by allowing only Mg atoms to dissolve from the surface while more cathodic RE element remains at the surface. For each Mg atom, propensity of its dissolution is determined by Arrhenius-type formula:

$$r_{\text{dissolution}} = \nu_{\text{diss}} \exp\left[-\left(E_{\text{diss}}(n) - \phi\right)/\text{kT}\right] \tag{9}$$

where $r_{\text{dissolution}}$ is dissolution rate, $\nu_{\text{diss}}$ is frequency for the dissolution process, $\phi$ is overpotential, and $E_{\text{diss}}$ is dissolution barrier for each Mg atom as function of number of nearest occupied sites $n$. We employed the frequency value of $\nu_{\text{diss}} = 10^4$ $\text{sec}^{-1}$ from the similar kMC studies.

Diffusion model

Diffusion of Mg and RE atoms at the surface is modeled with hopping mechanism between the nearest-neighbor sites. Instead of using the conventional bond-breaking scheme for diffusion model, we use site-energy model in which final state atomic configuration as well as initial state atomic configuration is taken into account. In site-energy model, diffusion barrier is the sum of site-dependent energy $E_{\text{site}}(n)$ and the constant diffusion barrier $E^0_{\text{diff}}$. The interaction energy between Mg and RE atoms ($E^{\text{Mg-RE}}$) could also affect the diffusion barrier. The resulting expression for the rate of diffusion is as follows:

$$r_{\text{diffusion}} = \nu_{\text{diff}} \exp\left[-\left(E_{\text{site}}(n) + E^0_{\text{diff}} + E^{\text{Mg-RE}}\right)/\text{kT}\right] \tag{10}$$

We set $\nu_{\text{diff}} = 10^{13}$ $\text{sec}^{-1}$ based on the Debye frequency of solid magnesium at room temperature (~ $1.2 \times 10^{13}$ $\text{sec}^{-1}$). The site-dependent energy and the diffusion barrier will be determined by performing first-principles DFT calculations of Mg structures embedded with vacancies, surfaces, islands, and clusters.

Selective dissolution of RE-doped Mg surface

As preliminary analysis, we performed KMC simulation on the selective dissolution of 10 at% Ce-doped Mg surface. Based on first-principles transition-state calculation of surface vacancy diffusion and surface ad-atom diffusion on Mg (100) surface, we obtained the average diffusion barrier of 0.40 eV for hopping Mg atoms between the same atomic configurations. For Ce atoms, slightly higher diffusion barrier of 0.50 eV was applied. For both types of atoms, 0.25 eV of cohesive energy per bonding was applied for each occupied nearest neighbor site to assign the site-dependent energies. Diffusion barrier for Mg atoms was assumed to be the same as the diffusion barrier at the surface. To reflect the segregation of RE elements within the alloy matrix, a repulsive interaction potential of 0.05 eV / site was set between Mg and Ce atoms.

Figure illustrates the doping effect on the dissolution of alloy surfaces by plotting the surface current density and the surface composition for the first 50 *ms* of selective dissolution with a given overpotential of -0.3 *V*. First, increase of the dissolution rate becomes noticeably slower with the Ce doping when compared with pure Mg dissolution. It is also found that the surface composition of Mg gradually decreases from its bulk composition (90 *at*%) and converges to a much lower concentration (~75 *at*%). This indicates that the selective dissolution of Mg atoms leads to a surface segregation of Ce atoms near surface so that it would slow down the Mg dissolution. Nanometer-scale evolution of 10 *at*% Ce-doped Mg surface depicted in Figure confirms this argument. The figure clearly shows that the selective dissolution of Mg leads to higher concentration of segregated Ce clusters at the surface.

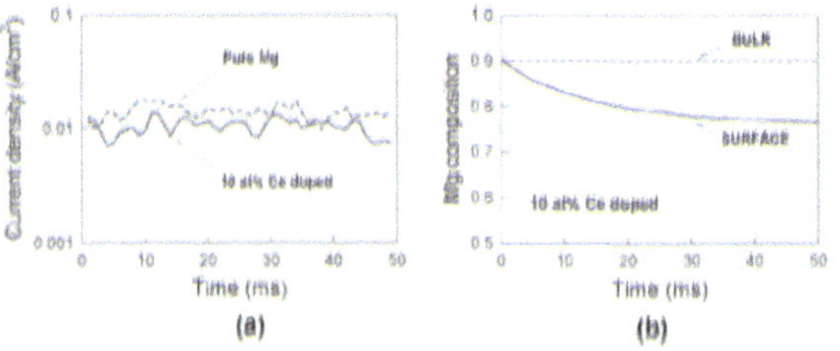


Figure 3. (a) Dissolution rate (in the unit of current density) of pure and 10 at% Ce-doped Mg surfaces, and (b) surface Mg composition of the Ce-doped surface. Dashed line in (b) indicates the bulk Mg composition of the alloy (10 at%). Both graphs are plotted for the first 50 ms of KMC simulation.

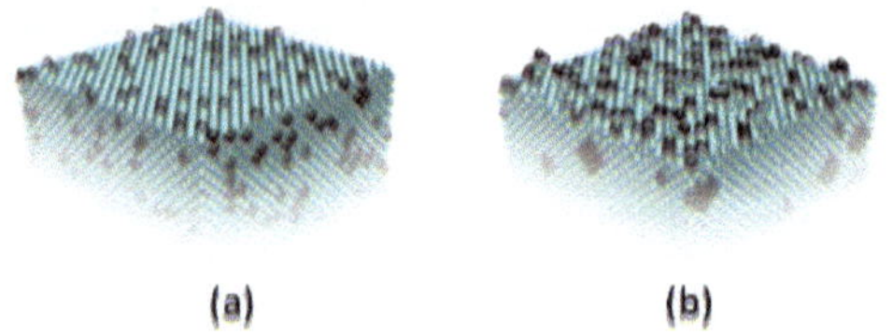


Figure 4. Snapshot from the dissolution of 10 at% Ce-doped Mg surface (8nm × 8nm) at (a) t = 0 ms and (b) t = 50 ms. Grey spheres represent Mg atoms and black spheres represent Ce atoms.

Before we describe the larger length-scale beyond kMC, it shows that elemental design of surfaces and understanding the fundamental science aspects of failure of MgO surface layer is critical for better design of alloy phase. In addition, we have results from reactive surfaces that are treated using coupled electrochemical models using computational fluid dynamics (CFD), both negative differential effect and effects of RE doping is currently being implemented in the microscale models. Long-term goal of our research is to improve the description of failure of grains and dissolution mechanics by coupling DFT-kMC and CFD methods.

Fluid-dynamics Simulation of Corrosion at a Pit

The calculated thermodynamic quantities from DFT can be fed into FEM model by resorting to an information integration method described in this section. Rate constants for chemical and electrochemical reactions can be calculated from Gibbs free energy of activation by:

$$k = \frac{k_B T}{h} \exp\left(-\frac{\Delta G^{*}}{RT}\right) \tag{11}$$

where $k_B$ is Boltzmann's constant and $h$ is Planck's constant. The surface coverage of reactive species ($\theta_j$) in electrochemical reactions can be related to the solution chemistry using a Langmuir model [4]:

$$\theta_j = \frac{r_j a_j}{1 + \sum_k r_k a_k} \tag{12}$$

$$r_j = \frac{k_B T}{h} \exp\left(-\frac{\Delta G_{ads,j}}{RT}\right) \tag{13}$$

where $a_j$ is the activity of the $j$-th reactive species at the metal surface and the coefficient $r_j$ is a scaled adsorption constant. Conventionally, both anodic metal dissolution and cathodic reduction reactions are assumed to follow an empirical Tafel's law and the parameters in Tafel's equation are fitted from experimental data or from first-principles calculated kinetics with corrections for potential. In this work, the general form of current density is:

$$i = i_a + i_c \tag{14}$$

$$i_a = \sum_k \sum_j n_m^k F k_{a,j}^k \left(\theta_{j,s}^k\right)^{\omega_j} \exp\left(\frac{\alpha_{a,j}^k n_m^k F(\psi - \phi)}{RT}\right) \tag{15}$$

$$i_c = -\sum_j n_j F k_{c,j} \left(\theta_{o,s}\right)^{\omega_j} \exp\left(-\frac{\alpha_{c,j} n_j F(\psi - \phi)}{RT}\right) \tag{16}$$

In this way, anodic and cathodic current densities are quantitatively correlated to the applied potential ($\psi$), electrostatic potential in the solution ($\phi$), temperature ($T$), and most importantly, the material and solution chemistry, which can be reflected by the reaction rate constant ($k$) and surface coverage of reactants ($\theta$). Note that the subscript $j$ represents the $j$-th aggressive ion in Eq. (15) and the $j$-th reduction reaction in Eq. (16). It is widely recognized that the negative difference effect (NDE) is a critical phenomenon in magnesium corrosion. Evolution of gas bubbles from a pit can perturb ion diffusion, and hence the growth of a pit. In order to quantitatively characterize such phenomenon, $H_2$ gas bubble generation as well as bubble redistribution due to multi-phase flow need to be fully taken into consideration. Figure 5 gives $H_2$ bubble distribution in systems with different electrolyte flow rates. Electrochemical and chemical reactions involved in this system were assumed to be the same as those in pure Mg corrosion.

$$Mg \rightarrow \equiv Mg^{+} + e^{-} \quad \text{and } Mg^{+} \rightarrow Mg^{2+} + e^{-}$$

$$Mg^{+} + H_2O \rightarrow Mg^{2+} + OH^{-} + 0.5H_2 \text{ and } 2H^{+} + 2e^{-} \rightarrow H_2$$

$$Mg^{2+} + 2OH^{-} \rightarrow Mg(OH)_2 \text{ and } MgO + H_2O \rightarrow Mg(OH)_2$$

In a stagnant electrolyte system (Figure 5a), gas bubbles arise along the peripheral lip of the pit, which was confirmed in experiments. As we increased the flow speed, the volume fraction of gas bubbles was reduced and the most prominent volume fraction appeared on one side of the pit surface facing towards the flow direction (see figure 5b). The NDE phenomenon was clearly identified (see figure 5d), where the hydrogen evolution rate increased rather than decreased with increasing potential. We have also developed a detailed pit growth and propagation model in the CFD scale using surface dissolution equations for intermetallic particles around Al-2024 alloy using a similar approach. [2, 5]

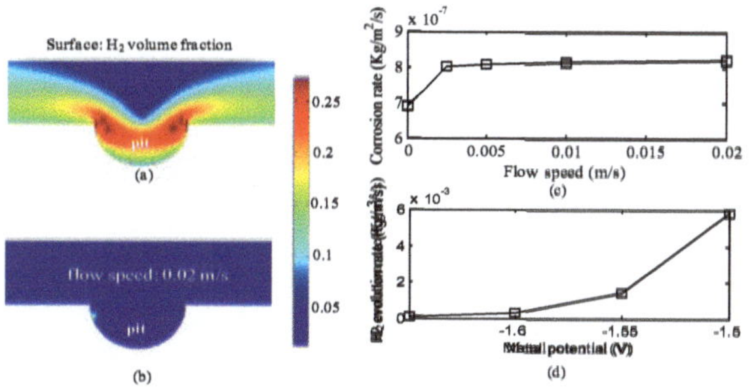


Figure 5. The effect of multiphase flow on corrosion: (a) $H_2$ bubble volume fraction in a stagnant system, (b) $H_2$ bubble volume fraction in a flowing system, (c) corrosion rate as a function of flow speed, and (d) $H_2$ evolution rate

In conclusion, robust mesoscale model of grain structure and incorporation of intermetallic phase across the length scale will be required for more realistic simulations that allow us to explore the effects of doping and alloy composition in a more comprehensive fashion. Such model will therefore be used to quantify the effect of microstructure and chemical environment in determining failure of surface film and the subsequent metal loss from damaged surface or pits.

## Acknowledgments

Research was sponsored in part by the Army Research Laboratory and was accomplished under Cooperative Agreement Number W911NF-10-2-0112. The views and conclusions contained in this document are those of the authors and should not be interpreted as representing the official policies, either expressed or implied, of the Army Research Laboratory or the U.S. Government. The U.S. Government is authorized to reproduce and distribute reprints for Government purposes notwithstanding any copyright notation herein. The rest of the funding is from the Office of Naval Research grant #s N00014-04-1-0688 and N00014-06-1-0315.

## References

1. H. Kwak and S Chaudhuri; "Cationic doping of MgO surfaces to build corrosion protection in Mg alloys." *Journal of Alloys and Compounds* 509, 32 (2011) 8189-8198.

2. J. Xiao, H. Kwak, S. Chaudhuri, "Understanding the Role of Intermetallic Particles in Localized Corrosion" *Proceedings of MS&T 2011, Columbus, OH*; 2011; p1374-1385.

3. H. Kwak, J. Xiao, S. Chaudhuri, "First-Principles-Based Analysis on the Role of Rare-earth Doping on Mg Alloy Corrosion" *Proceedings of MS&T 2011, Columbus, OH*; (2011) 1386-1396

4. A. Anderko, N. Sridhar, D.S. Dunn, A General Model for the Repassivation Potential as a Function of Multiple Aqueous Solution Species, *Corrosion Science*, 46(2004) 1583-1612

5. J. Xiao, S. Chaudhuri; "Predictive modeling of localized corrosion: An application to aluminum alloys"; *Electrochim Acta* 56, 16(2011), 5630-5641

# SOLID STATE JOINING OF MAGNESIUM TO STEEL

Saumyadeep Jana[1], *Yuri Hovanski*[1], Siva P Pilli[1], David P Field[2], Hao Yu[2], Tsung-Yu Pan[3], M L Santella[3]
[1]Pacific Northwest National laboratory, Richland, WA, 99354, USA
[2]Washington State University, Pullman, WA, 99164-2920, USA
[3]Oak Ridge National Laboratory, Oak Ridge, TN 37381-6096, USA

Keywords: Magnesium, Steel, Friction stir welding, Ultrasonic welding

## Abstract

Friction stir welding and ultrasonic welding techniques were applied to join automotive magnesium alloys to steel sheet. The effect of tooling and process parameters on the post-weld microstructure, texture and mechanical properties was investigated. Static and dynamic loading were utilized to investigate the joint strength of both cast and wrought magnesium alloys including their susceptibility and degradation under corrosive media. The conditions required to produce joint strengths in excess of 75% of the base metal strength were determined, and the effects of surface coatings, tooling and weld parameters on weld properties are presented.

## Introduction

The U.S. Department of Energy's (DOE) Vehicle Technologies Program (VTP) is leading the development of high-performance, cost-effective materials-and the processes needed to manufacture them-to make advanced vehicles more fuel efficient and affordable. One key goal is to validate a cost-effective 50% weight reduction in passenger-vehicle body and chassis system by 2015 [1]. Recently, a joint project between VTP and U.S. automakers overcame several hurdles to the introduction of lightweight magnesium components, resulting in a magnesium engine cradle for the Chevy Corvette Z06 that is nearly 60% lighter than a steel cradle [1]. A key combination of low density, and high specific strength makes magnesium alloys an ideal choice for automotive applications because of the associated effect of weight reduction on better fuel economy. Though currently there are no adequate wrought magnesium alloys available to meet the requirements of automotive body applications, extensive research is taking place to enhance the properties of magnesium sheet products [2]. Steel, on the other hand, is currently the automaker's material of choice. This is because cold rolled mild steel offers excellent ductility, consistent properties and low overall component production costs. However, in a modern multi-material vehicle, lightweight materials such as aluminum and magnesium alloys can be a challenge to join and attach to the underlying sub-structure, usually composed of steel. Joining methodologies available in the cost environment relevant to automotive manufacturing include resistance spot welding, adhesives, linear fusion welding, hemming, clinching, bolting, and riveting [3].

However, joining Magnesium to steel poses a specific challenge. The maximum solid solubility of Fe in (Mg) is 0.00043 at%, and solid solubility of Mg in (Fe) is nil [3]. Melting points of Mg and Fe are 922 K (649 °C) and 1812 K (1539 °C), respectively. This huge difference in melting points makes it very difficult to melt both metals at the same time as might be required for fusion welding processes. Moreover, both metals are immiscible in a liquid state, and they do not react to form any congruent melting phase(s). Therefore, joining Mg alloys to steels through conventional fusion welding is problematic at best. Recently, laser-GTA hybrid welding technique was tried to join AZ31B alloy to 304 steel in lap configuration with AZ31B on top [4]. Results indicate poor joining strength and Mg-Fe interfacial fracture. The authors attributed such low mechanical strength to severe oxidation at Mg/Fe interface. Joining methods that require a large amount of plastic strain in the magnesium component suffer from magnesium's poor ductility at room temperature.

A recent DOE funded project looked at the feasibility of joining magnesium to steel through solid state route. Two specific joining technologies, e.g., friction stir welding and ultrasonic welding, were examined in this program. Pacific Northwest National Laboratory (PNNL) played the lead role in developing the friction stir welding route. A few recent studies indicate that friction stir welding (FSW) has the ability to join Mg alloys to steel [5,6]. FSW, which is a solid state welding technique invented by Thomas et al. at TWI, UK [7], has the added advantage of minimal associated oxidation because of the solid state nature of the process. A short preliminary effort at PNNL indicated higher joint strength in Mg to steel lap welds when the bottom steel sheet was deformed by the FSW tool [8].

Ultrasonic welding (USW) of Mg alloys to steel sheet was developed at the Oak Ridge National Laboratory (ORNL). The work focused on identifying the factors critical to increasing joint strength while understanding the influence of sonotrode tip design, material stack-up, coatings, and process time.

## Summary of the FSW Initiative

Following our initial trial, an effort was made to optimize the FSW tool design in order to increase the Mg/steel joint strength in lap configurations. Two tool designs were tested in this study. The first tool (Tool 1) had a convex scrolled shoulder and a stepped spiral pin. The second tool (Tool 2) had identical features as the first tool plus a short hard WC insert on the tool pin bottom. Both the tools were made of H13 tool steel. A nominally 2.32 mm thick AZ31 magnesium alloy (O temper) was used as the top sheet. Two different steels were used, e.g., (i) HDG 1.5 mm thick HSLA steel and (ii) EG 0.8 mm thick mild steel (MS), as the bottom sheet material. Mechanical performance of the joints was determined through quasi-static unguided lap-shear tests, and fatigue tests. Further, joint interface microstructure was characterized in detail using optical microscope/SEM. Additionally, EBSD technique was employed to learn about the crystallographic information at the joint interface.

The welding was carried out in position controlled mode in a high stiffness, precision FSW machine manufactured by TTI, Inc. and located at PNNL. Complete details about the welding procedure has been published elsewhere [3,8]. Subsequently, three to five lap-shear specimens of width ~30 mm were obtained for each weld covering the entire weld length. Tensile tests were carried out at room temperature at a crosshead speed of 1 mm/min. Both ends of the lap-shear specimens were appropriately shimmed in order to minimize misalignment. Average failure load of a standard 30 mm wide lap-shear AZ31/MS weld coupon varied between 4.6 to 6.8 kN, as a function of process parameters. In the case of AZ31/HSLA weld coupons, the average failure load varied between 3.7 to 9.7 kN. Comparison of the lap-shear data revealed that the failure load was generally higher in the case of AZ31/HSLA welds, reflecting the higher strength of the bottom steel sheet used therein. Moreover, use of Tool 2 helped in lowering the scatter.

For graphical comparison purpose, weld efficiency of all the joints were calculated by comparing the failure load of the joint, with the maximum load carrying capacity of the weaker parent sheet section. Whichever parent sheet section shows a lower total load carrying capacity, is noted to be the weaker one. 0.8 mm steel is the weaker one for AZ31/MS welds, and AZ31 sheet is the weaker one for AZ31/HSLA welds. For calculating joint efficiency, the following equation was used:

$$\text{Joint efficiency} = (\text{Max. load of joint} / (A_w \times TS_{\text{parent sheet}})) \times 100$$

$A_w$ is the area of the weaker parent sheet section, and TS is the tensile strength. Weld efficiency data for all the welds are shown in Figure 1a and 1b. The average minimum joint efficiency is

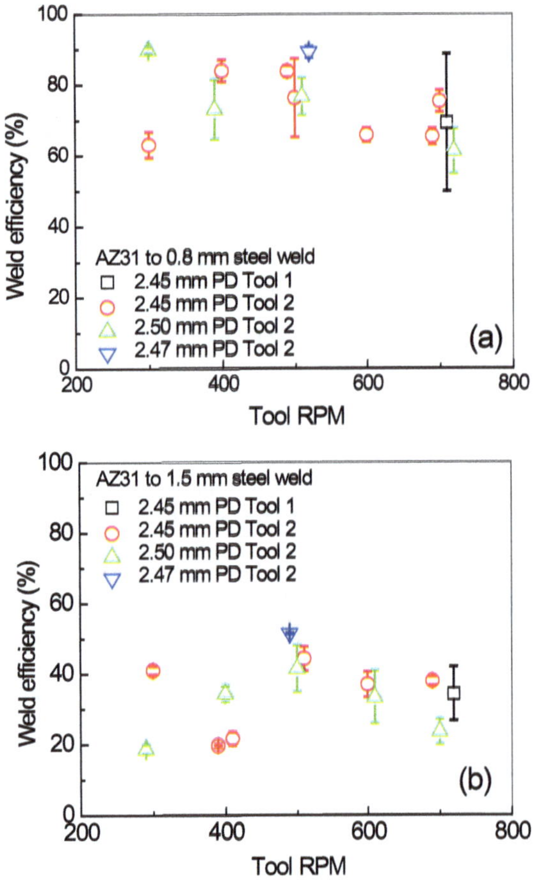

Figure 1. Weld efficiency plots: (a) AZ31/MS, (b) AZ31/HSLA.

>60 % in the case of AZ31/MS welds, and ~40% for AZ31/HSLA welds. Joint efficiency varied as a function of process parameters. Further, the benefit of using Tool 2 in order to obtain uniform strength is evident from the weld efficiency plots. Maximum joint efficiency is achieved at ~500 tool RPM for the AZ31/MS welds, with a tool plunge depth of 2.45 mm (Fig. 1a). For AZ31/HSLA welds, the highest joint efficiency is noted at 500 tool rotation rate, with the tool plunge depth being 2.50 mm (Fig. 1b). Based on the lap-shear test data, a single tool rotation rate of 500 rpm at an intermediate tool plunge depth of 2.47 mm was selected to carry out further welds. It should be noted that the tool travel speed was fixed at 100 mm/min for the entire study. The combination of 500 RPM, 100 mm/min travel speed, and 2.47 mm plunge depth resulted in >90% weld efficiency in the case of AZ31/MS welds. At same process parameters, weld efficiency of AZ31/HSLA weld was noted to ~52%, which was the highest in its class. Additionally, one of the AZ31/MS weld specimens at this particular combination of process parameters showed failure of the steel sheet, leaving the weld intact. The example of such a lap-shear specimen is shown in Figure 2.

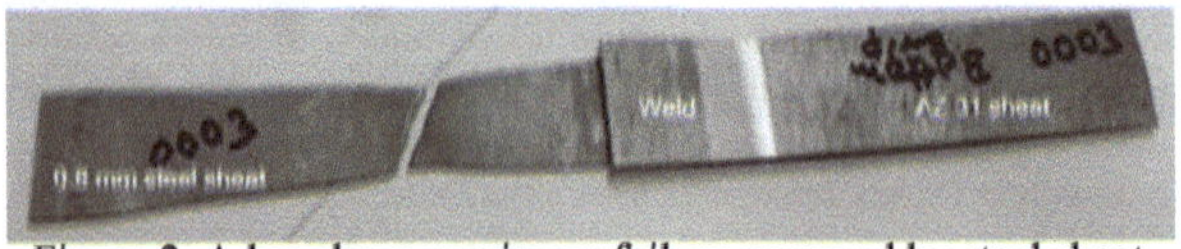

Figure 2. A lap-shear specimen: failure occurred by steel sheet fracture, leaving the weld intact.

Following quasi-static testing, performance of the lap joints at the best process parameters was determined under dynamic loading. A number of 30 mm wide lap-shear weld coupons were tested in a servo-hydraulic fatigue testing machine at a load ratio R of 0.1. The test frequency was 5 Hz. Fatigue tests were conducted in the load-control mode. S-N plot of the lap-shear fatigue specimens is represented in Fig. 3. At similar fatigue lives, the maximum

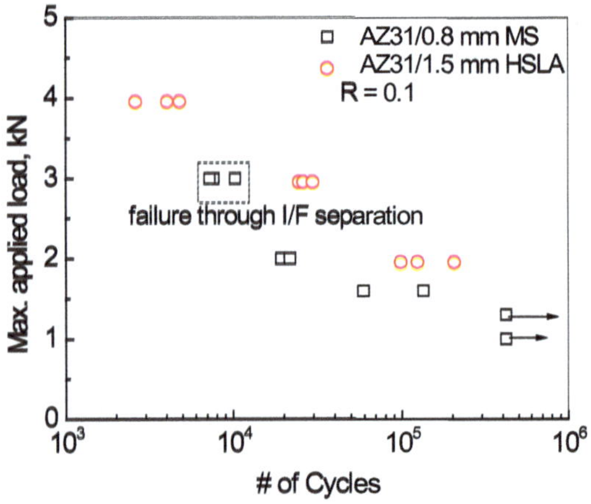

Figure 3. S-N plot for the Mg/steel friction stir lap welded specimens.

applied loads are found to be higher for the AZ31/HSLA welds than the AZ31/MS welds, reflecting the higher failure load of the AZ31/HSLA welds. However, when the maximum applied load is normalized against the respective failure load, as shown in Fig. 4, specimens from both material combinations are noted to have similar fatigue lives at any given applied load to failure load ratio. The major mode of failure was found to be top Mg sheet fracture.

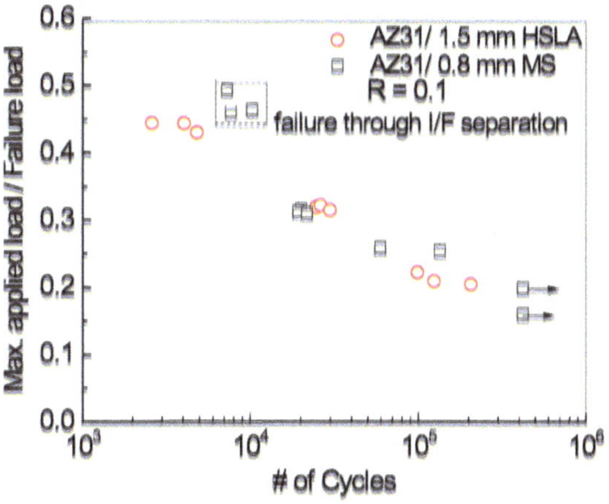


Figure 4. Normalized S-N plot for the Mg/steel friction stir lap welded specimens.

However, three AZ31/MS weld specimens that were tested at 3 kN, showed relatively longer life. Interfacial separation of the top and bottom sheet was the mode of failure for these three specimens. It can be further noted from Fig. 4 that two of the AZ31/MS weld specimens did not fail when applied load to failure load ratio was less than 0.2. Fatigue tests for these two specimens were stopped at ~$5 \times 10^5$ cycles.

A typical cross-section image of a joint fabricated by Tool 2 is shown in Figure 5a. The effect of the WC insert on the tool pin bottom is clearly evident on this image. Presence of the insert resulted in machining of the bottom steel sheet. Further, it led to the formation of two "dove-tail" like features on both ends, as noted by the arrows. Some of the higher magnification SEM images obtained from various locations along the joint interface are shown in Figure 5b and 5c. Microstructural features observed in these images, indicate formation of solidification products through eutectic reaction, due to the melting of the Zn coating present on the steel sheet. However, no such solidification product was found along the region where the steel sheet showed signs of deformation due to the penetrating action of the tool. As the FSW tool plunges through the top Mg sheet, the Zn coating on the bottom steel sheet melts. The liquid Zn layer dissolves Mg from the top Mg sheet, and, hence forms a liquid Zn-Mg alloy. However, this newly formed liquid alloy layer gets squeezed out laterally due to the forging action of the tool.

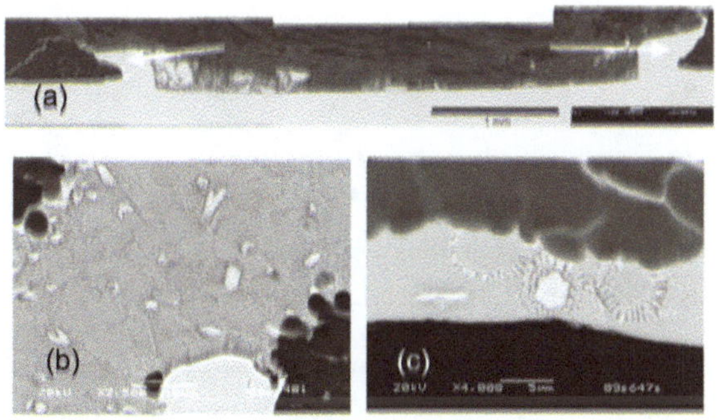


Figure 5. (a) Cross-section of a joint fabricated by Tool 2, (b),(c) Eutectic microstructure at Mg/steel interface.

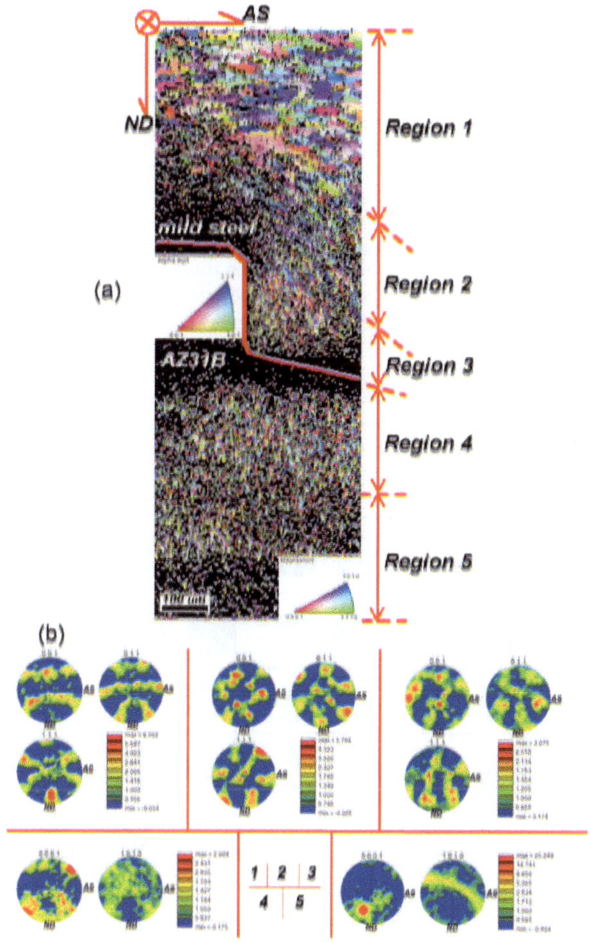


Figure 6. (a) Orientation map of the interface region at the advancing hook of FSW of AZ31B to mild steel, (b) Pole figures of the different regions in shown in (a).

A detailed EBSD study was carried out to characterize the weld nugget and the Mg/steel interface. As shown in Fig 6a, the area around the advancing hook in the FSW of AZ31B to mild steel could be divided into five regions. Region 1 is the base metal of the mild steel and there is no influence of the shear that happens in this region. Region 2 is the region near the interface which lies between the hook area and the base metal. The grains in this region are elongated in the shear direction. Due to the influence of the shear, the grains' thickness decreases as shear increases. Several shear textures form in this region. Region 3 is the region of hook near the weld nugget. The strain is heavy in this region. Grains in this region show a different orientation to the other region of the steel. Region 4 is the interface region of the Mg side. The texture differs to that of the steel side. Region 5 is the region near the weld nugget. The pole figures of each region are shown in Fig.6b.

Fig.6b-1 shows the texture of the base metal of the mild steel. The maximum intensity of the pole figure is 8.069. Fig.6b-2 shows the texture of the interface region of the steel side. The grains in that region rotated in favor of the shear and the strength of the texture decreased to 5.794. The texture of the interface near the weld nugget was influenced by the lower surface of the pin and the side face of the tip. The texture is randomized in this region and the maximum intensity decreases to 3.075. The interface of the Mg side shows a texture much like that of normal direction (ND)-advancing side. The two components of the texture are perpendicular to each other. The maximum strength of the texture is only about 2.904. The texture of the Mg side away from the interface shows a strong texture. The maximum intensity of the pole figures of that region is 25.249.

The pole figures of the weld nugget region are shown in Fig. 7. The texture changes little in the ND in the region far away from the shoulder and the interface. The rotation of shoulder and the pin forms a truncated cone and the crystals within the cone have their basal planes parallel to the cone surface. The direction of the crystalline alignment is due to the local change of the shear direction. The texture under the shoulder is influenced by the lower surface of the shoulder, and the texture changes little in this nugget region. The texture near the interface differs from that of the weld nugget, but the texture strength in this region is low.

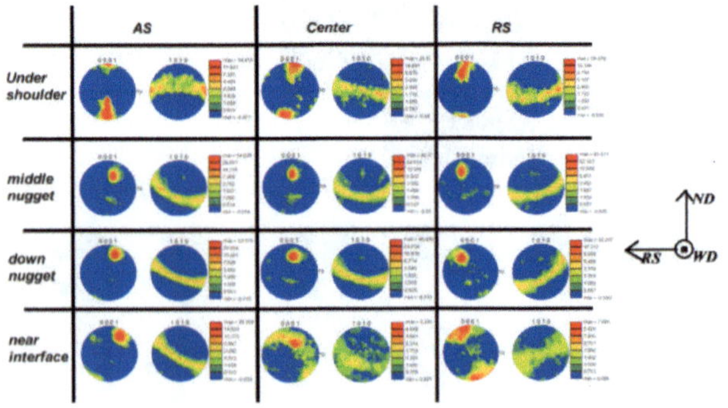


Figure 7. Pole figures of the nugget region of FSW of AZ31B to HSLA

**Summary of the USW Initiative**

The ultrasonic welds for this study were made with a Sonobond CLF2500 pedestal welding station. The rated conditions at the welding tip were 20 kHz frequency with 25 micrometers (μm) amplitude. The sonotrode tips were made from T1 steel. They typically have flat, rectangular faces with dimensions of either 5 mm x 7 mm or 7 mm x 7 mm. The line pressure to the tip clamping mechanism was adjusted to make the welds under constant nominal pressure of 39 MPa.

The materials used for the current experiments were sheets of 0.8-mm-thick hot-dip-galvanized (HDG) mild steel and 1.6-mm-thick AZ31B-H24. The zinc coating on the steel was about a 9 μm-thick. Prior to welding the surfaces of the AZ31B sheet were buffed with non-metallic abrasive pads to remove surface oxides and produce shiny surfaces. Both metals were also degreased with acetone followed by isopropyl alcohol to remove lubricants and surface debris.

Coupons of AZ31 nominally 30 mm wide x 100 mm long were welded to mild steel coupons of the same size to produce specimens for mechanical testing and metallographic analysis. A 25 mm overlap was used for making lap-welded coupons with spot welds centered in the overlap regions. Specimens were positioned for welding so that the primary vibration direction of the sonotrode was perpendicular to their long axis. Spot welding was typically performed using power of 1500-2500 W, and welding times ranged from 0.2-1.2 s.

The results of the tensile lap-shear testing are presented in Figure 8 where the variations of failure load with welding time are plotted. With both tips there was an initial rapid rise in lap-shear failure load from around 0.5 kN to near 2.9 kN as the welding times increased from 0.2-0.4 s. After the noted increase, failure loads continued increasing with welding times but at a much lower rate. From 0.4-1.2 s, failure loads increased to a maximum average value of 3.7 kN for the 5-mm x 7-mm tip and 4.2 kN for the 7-mm x 7-mm tip. Welds of the AZ31 to bare steel were unsuccessful. Weak bonding occurred in this case, but joints could be easily broken by hand.

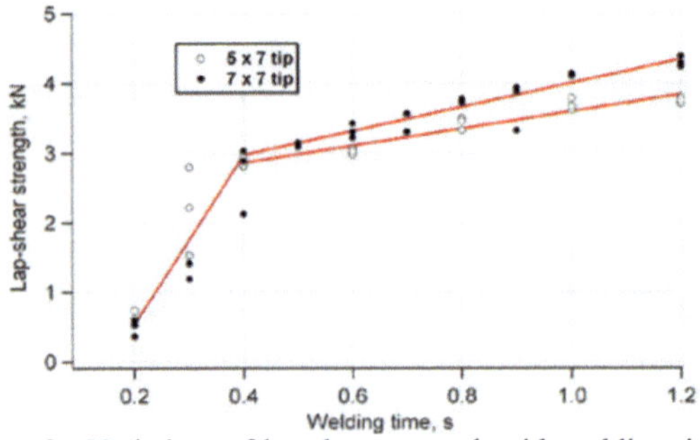


Figure 8. Variations of lap-shear strength with welding time and tip size

The lap-shear strength results can be placed in perspective by comparison with data from resistance spot welds and friction stir spot welds of AZ31. Published work [9] confirmed that AZ31 could be resistance spot welded with lap-shear strengths ranging to 5.7 kN. Lap-shear strengths increased with spot diameters and were about 4.0 kN for diameters with areas equivalent to the 7-mm x 7-mm USW tip. More recently, Lang and co-workers [10] measured lap-shear strengths up to 3 kN for spot welds of AZ31 with nugget diameters of about 7 mm. Another published study showed that AZ31 sheet could be friction stir welded and some joints had lap-shear strengths as high as 4.75 kN. No comparable published data could be found for magnesium-steel spot-welded joints made by these two processes. Nevertheless, strengths for ultrasonically-welded joints at times greater that about 0.5 s consistently exceeded 3 kN and so they are generally consistent with values found for magnesium-magnesium spot welds.

The results of the fatigue testing are shown as Figure 9 where the fatigue life expressed as cycles to failure is plotted against the load range for testing. The USW spot welds were tested at constant values of R = 0.1 and R = 0.2, where R = minimum load/maximum load.

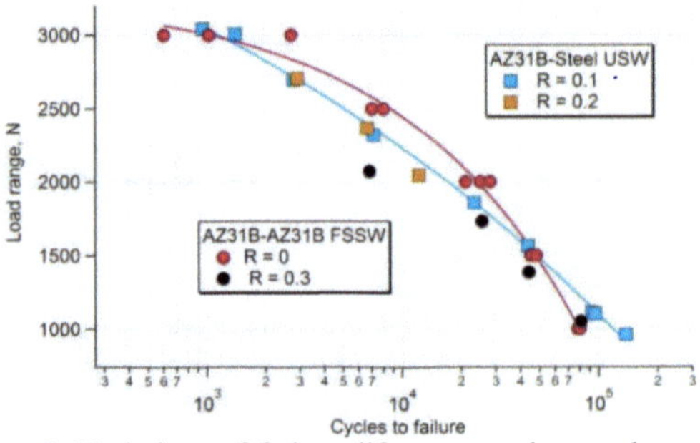


Figure 9. Variations of fatigue life expressed as cycles to failure with loading range

There are few published data for similar or dissimilar spot welds involving magnesium alloys, and Figure 9 compares the USW data to similar data from friction stir spot welds of AZ31-to-AZ31 lap welds. Regression analysis was used for the lines drawn through the USW data at R = 0.1 and the FSSW data of R = 0.0. However, a single line could have been fit through the entire combined data set. This suggests the fatigue properties of the AZ31-steel USW spots welds will not differ significantly from other types of spot welds made using AZ31. A more detailed examination of the fatigue performance for this work was previously documented [11, 12].

Additional results from tensile lap-shear testing are presented in Figure 10 where the variations of failure load with welding time are plotted for two arrangements: Mg-steel, where the sonotrode engaged the AZ31, and Steel-Mg, where the same engaged the hot-dip- galvanized (HDG) steel. The maximum lap-shear failure loads for Mg-steel spot welds are near 4 kN for 1.0 s of welding time. Those lap-shear strengths are in the range of the levels found for resistance spot welding and friction stir spot welding AZ31 to itself [9,10]. Given the disparities of properties between AZ31 and the mild steel, the lap-shear results are remarkable. Figure 10 indicates that similar maximum failure loads can be achieved with the either arrangement. This important result indicates that the USW has desirable flexibility in how it is presented to components in the manufacturing environment.

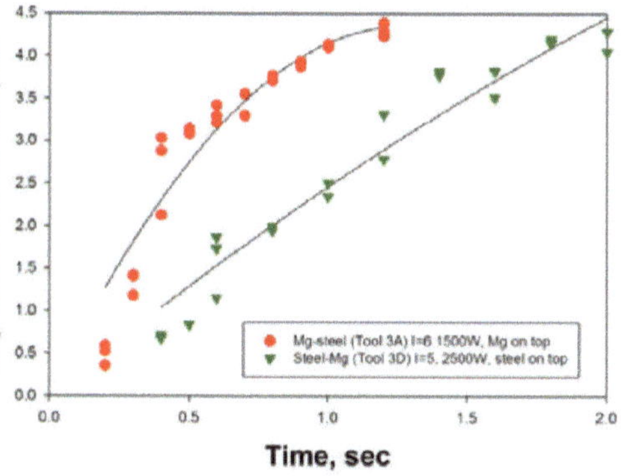


Figure 10. Variations of lap-shear strength with welding time and metal arrangement using a 7-mm x 7-mm sonotrode tip

One additional point of interest was the evaluation of hybid joining. Initial results from this evaluation are presented in Figure 11 where the Mg-steel results are replotted for comparison. Results are shown for three adhesive (Dow Betamate 73305) situations. First, USW spot welds were made through uncured adhesive as is commonly done with resistance spot welding. This approach caused a significant decrease of lap-shear strength. Subsequently, some spot welded coupons were cured. Curing produced lap-shear strengths in the range of 6 kN. As a reference, coupons that were only adhesive bonded and cured were also tested, and these had lap-shear strength of 4-5 kN. These results suggest that USW spot welding through adhesive is not advisable. Presumably, this is at least partial due to the adhesive modifying friction properties at the welded interface. Additional studies will be required to determine how much strength can be preserved in USW spot welds made through uncured adhesives. Once the adhesive is cured, lap-shear strengths can exceed those made by either processes alone. In all of the experiments, the area covered by adhesive was approximately 25 mm x 25 mm, a significantly larger area than that of the USW spot welds, approximately 7 mm x 7 mm.

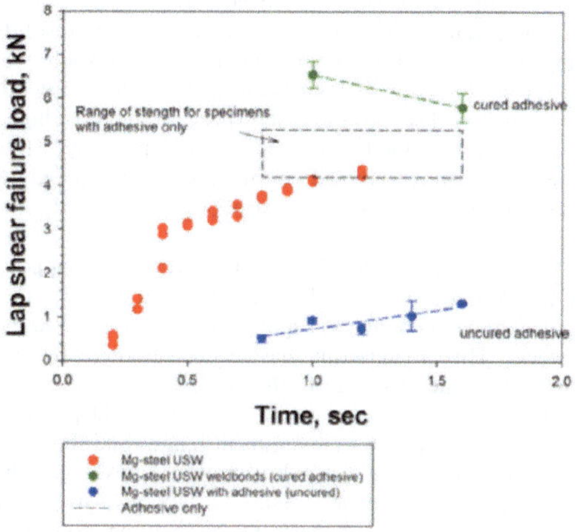


Figure 11. Variations of lap-shear strength with welding time and adhesive cure condition

Finally, results from an evaluation of the effect of corrosion on lap-shear strength are shown in Figure 12. A total of 20 Mg-steel specimens were made for exposure using the Ford Arizona Proving Ground Equivalent Corrosion Cycle (APGE) test, a widely accepted screening test for automotive industry. The exposure consisted of immersion in 5% NaCl bath at room temperature for 15 minutes followed by air-drying for 3 hours. Following this, the dried specimens were placed in a chamber where relative humidity was controlled to 80% and temperature was maintained at 50°C. Each cycle of exposure in the chamber lasted 20.75 hours. Starting at the fifth cycle of exposure, one specimen was tested for lap-shear strength after each cycle. Figure 3 shows that strength decreased linearly with each cycle of exposure up to 12 cycles. From that point on, strengths became very erratic, and many specimens could be broken by routine handling. A compound identified as $Mg(OH)_2$ was deposited in the crevice between the AZ31 and the steel. The deposits were observed after even one exposure, but they continued to thicken with each exposure cycle contributing to a separation force between the sheet specimens. Examination of fracture surfaces found that welded areas also decreased with exposure cycle, but even in the weakest joints there was evidence of metal-metal bonding.

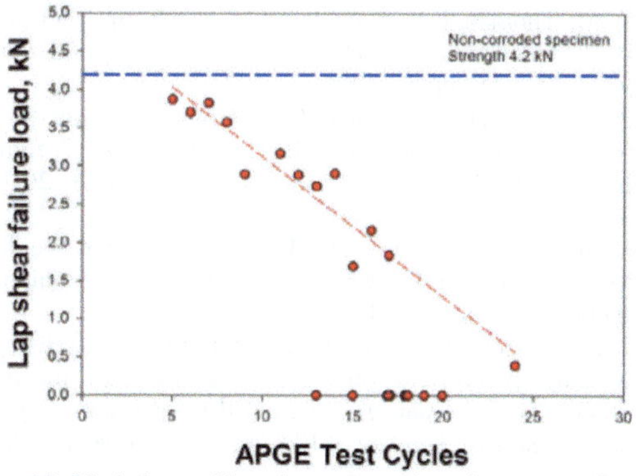


Figure 12. Variations of lap-shear strength and exposure times

## Conclusions

A comparative study has been conducted to evaluate the joining potential of Mg/steel couples through two solid state welding routes, e.g., friction stir welding and ultrasonic welding. Initial results in terms of both static and dynamic strengths are very promising. The observed lap-shear strength is higher for FSW route as the bonded area is larger than USW. However, FSW suffers from a restricted configuration of using only Mg as top sheet material. This is not the case for USW method.

The authors would like to acknowledge the support from Dr. Carol Schutte and Mr. William Joost at the US Dept. of Energy, Vehicle Technology, Light Weight Materials Program for this work. This manuscript has been authored by Battelle Memorial Institute, Pacific Northwest Division, under Contract No. DE-AC05-76RL01830 with the U.S. Department of Energy. 

## References


1. www.eere.energy.gov, Vehicle technologies program factsheet, Dec 2010.
2. R.S. Beals, C. Tissington, X. Zhang, K. Kainer, J. Petrillo, M. Verbrugge and M. Pekguleryuz, "Magnesium Global Development: Outcomes from the TMS 2007 Annual Meeting", *JOM*, August 2007, pp. 39-42.
3. S. Jana, Y. Hovanski, G. J. Grant, and K. Mattlin, "Effect of Tool Feature on the Joint Strength of Dissimilar Friction Stir Lap Welds", *Friction Stir Welding and Processing VI*, TMS 2011, pp. 205-211.
4. L.M. Liu, X. Zhao, "Study on the Weld Joint of Mg Alloy and Steel by Laser-GTA Hybrid Welding", *Material Characterization*, 2008, 59, pp. 1279-84.
5. T. Watanabe, K. Kagiya, A. Yanagisawa, and H. Tanabe, "Solid state welding of steel and magnesium alloy using a rotating pin – solid state welding of dissimilar metals using a rotating pin", *Quart J Japan welding soc.*, 2006, 24, pp. 108-15.
6. Y.C. Chen, K. Nakata, "Effect of tool geometry on microstructure and mechanical properties of friction stir lap welded magnesium alloy and steel", *Mat. design*, 2009, 30, pp. 3913-19.
7. W.M. Thomas, E.D. Nicholas, J.C. Needham, M.G. Murch, P. Templesmith, C.J. Dawes, *GB Patent application no. 9125978.8*, December 1991.
8. S. Jana, Y. Hovanski, and G. J. Grant, "Friction Stir Lap Welding of Magnesium Alloy to Steel: A Preliminary Investigation", *Metallurgical and Mat. Trans. A*, 2010, pp. 3173-3182.
9. Klain, P., Knight, D. L., and Thorne, J. P., "Spot Welding of Magnesium with Three-Phase Low Frequency Equipment," *Welding Journal* 32:7-18, 1953
10. Lang: B., Sun, D. Q., Li, G. Z., and Qin, X. F., "Effects of welding parameters on micro-structure and mechanical properties of resistance spot welded magnesium alloy joints *Science and Technology of Welding and Joining* 13:698-704, 2008
11. Teresa J. Franklin, Jwo Pan, Michael Santella, and Tsung-Yu Pan, "Fatigue Behavior of Dissimilar Ultrasonic Spot Welds in Lap Shear Specimens of Magnesium and Steel Sheets," SAE Paper No. 2011-01-0475.
12. Santella, M., T. Franklin, J. Pan, T. Pan, and E. Brown. 2010. Ultrasonic Spot Welding of AZ31B to Galvanized Mild Steel. SAE Paper 2010-01-0975.

**Magnesium Technology 2012**
*Edited by: Suveen N. Mathaudhu, Wim H. Sillekens, Neale R. Neelameggham, and Norbert Hort*
***TMS (The Minerals, Metals & Materials Society), 2012***

# GRAIN EVOLUTION DURING HIGH TEMPERATURE NECKING OF MAGNESIUM ALLOYS

P. E. Krajewski

General Motors Company, Warren, MI, 48090-9055, USA

Keywords: magnesium, microstructure, elevated temperature, necking

## Abstract

Magnesium sheet materials AZ31, AM50, ZM21, ZK10, and ZK30 were tested to failure at 450C at a strain rate of 0.001/s. Each of these materials exhibit significant changes in grain structure during deformation, especially in the necked regions of the failed samples. Grain size in the neck is shown to vary with thickness strain and strain rate. Some materials exhibit significant coarsening along the necked region while others show very little change. The results of this work can be used to help validate models for dynamic recrystallization in magnesium.

## Introduction

Magnesium sheet can provide significant lightweighting for automobile applications. A key limitation of magnesium sheet is its lower formability compared to steel and aluminum. This has necessitated the use of thermal assisted forming techniques to make parts [1, 2]. The combination of strain and temperature during forming can lead to significant microstructural changes, including grain growth and dynamic recrystallization. These changes in microstructure can affect both forming and the final performance of the formed part. Previous research on 5xxx series aluminum alloys at elevated temperature demonstrated that recrystallization in the necked region of tensile samples can occur both dynamically [3] and with post deformation static aging [4]. This can lead to significant changes in microstructure including dramatic increases in grain size. The present paper describes a preliminary look the microstructural evolution in five different magnesium alloys, AZ31, AM50, ZM21, ZK10, ZK30 after elevated temperature tensile testing at 450C, by characterizing the static vs. dynamic change in grain size, as well as the effect of strain.

### Experimental Procedure

Five magnesium materials were used in the present study: AZ31, AM50, ZM21, ZK10, and ZK30. The materials were provided by Alubin as extruded plates which were 175 mm x 175 mm x t mm where t is shown in Table 1 below. The extrusions were straight, smooth, and had no visible cracks. Thechemical composition of the materials is shown in Table 1.

Elevated temperature tensile samples were extracted using water jet cutting. All tensile specimens were aligned with the extrusion direction parallel to the testing direction. Specimen dimensions were 25.4 mm wide x 63.5 mm long with a 6.4-mm-wide x 25.4-mm-long gage section following ASTM E 2448-06. An Instron, model 5568, screw-driven, test frame, with an Instron 3119-007 furnace and Bluehill 2, version 2.4 data acquisition system were used for all testing. Samples were allowed to equilibrate in the furnace at 450C for 15 min. prior to testing. Testing was performed at a constant strain rate of 0.001/s. Total elongation was measured after fracture.

Metallographic sample preparation was performed using standard procedures, including the use of cold mounting rather than hot compression mounting to avoid creating artifact twins in the microstructure. Acetic-picric etchant (10 ml acetic acid, 4.2g picric acid, 70ml 95% ethanol, 10 ml water) was used to reveal grain boundaries.

## Results and Discussion

*Tensile Results* – Total elongation to failure for the five different magnesium materials at 450C and 0.001/s is summarized in Figure 1. All materials exhibited over 150% elongation which is very good considering they were in the as extruded condition with no additional thermal mechanical processing. The AZ31 material exhibited the lowest ductility of the five materials tested. The ZK30 material exhibited almost 300% elongation which is close to the General Motors specification for hot blow formed material [5]. Additional results on elevated temperature tensile testing of these materials can be found elsewhere [6].

*Microstructural Analysis* - The grain structure of each of the five materials tested is presented in Figures 2-6 below by showing a micrograph of the following four conditions: (a) as-extruded, (b) taken from the grip section after testing, therefore seeing thermal exposure without strain, (c) fracture tip, and (d) failed tensile sample away from fracture region. The grain sizes for the as-extruded and exposed conditions are shown in Table 2.

*AZ31*– The grain structure in the various conditions for AZ31 are shown in Figure 2. The as-extruded material has a relatively homogenous grain structure. After thermal exposure at 450C, very little change in structure is observed, with only a small increase in grain size from 12.8 µm to 15.1 µm. After the thermal exposure, the grain structure appears even more homogeneous. The failed samples, shown in Figure 2 (c) and (d) show significant changes in microstructure. At the fracture tip, the grains are extremely large, over 200 µm across. This has been observed previously in AZ31 [7]. Away from the fracture tip, the grain size is closer to the starting material, however the grains appear to be a bit more angular with a few local grains that are 2 -3 times the size of neighboring grains.

*AM50*– The grain structure in the various conditions for AM50 are shown in Figure 3. The as-extruded material is less homogeneous than the AZ31 exhibiting regions of fine grains and coarse grains in bands along the extrusion direction. After thermal exposure, the grain size has increased significantly from 8.7 µm to 12.9 µm and become much more homogeneous both in morphology and average size. At the fracture tip, very large grains are observed, but not greater than 100 µm, and they are much more uniform than in the AZ31. The "large" grains are also smaller than those observed in AZ31. Away from the fracture edge, the grain structure is more uniform without a significant bimodal distribution.

*ZM21*– The grain structure in the various conditions for AM50 are shown in Figure 4. The as-extruded grain structure is very uniform in size and grain morphology. After thermal exposure, the grain size has increased modestly from 18.1 µm to 22.8 µm, and the grain boundaries are a bit straighter than in the as-extruded material. In the region near the fracture tip, a very large grain is present which runs the full width of the fracture surface which is almost 250 µm across. Remote from the fracture surface, the grains are slightly coarser than in the exposed material.

*ZK10*– The grain structure in the various conditions for AM50 are shown in Figure 5. The as-extruded microstructure of the ZK10 material is very uniform in both size and morphology. Very little change in structure is observed after thermal exposure, with only slight coarsening from 12.8 µm to 14.4 µm. In the fracture region, the grain structure is very similar to the starting and exposed materials. There are two very small regions where a coarse grain is observed right at the tip of the failure zone. Everyplace else, the grain structure remains uniform and relatively fine across the necked region.

*ZK30*– The grain structure in the various conditions for AM50 are shown in Figure 6. The grain structure of the as-extruded material is very bimodal with a network of sub 10 µm grains surrounding some 20-30 µm grains. After thermal exposure, the grain structure is coarser and much more uniform almost doubling from an average size of 9.4 µm to 16.4 µm. In the necked region, very little change in grain structure is observed compared to the exposed material, and there are no extremely large grains present.

*Discussion*- The grain structure of the five materials tested showed varied response in the necked region after testing at 450C and 0.001/s. The materials which exhibited the largest grains in the neck after testing, AZ31 and ZM21, exhibited the lowest total elongation values. The material which showed no coarse grains in the necked region and likely had the most stable microstructure, ZK30, exhibited the highest total elongation. Figure 8 shows the effect of both material and strain in the neck on grain size after testing. The data is limited, but clearly shows that the grain structure in the ZK materials was very stable, while it varied significantly in the AZ31, AM50 and ZM21 materials. These preliminary results demonstrate that the microstructural evolution in the necked region may correlate with tensile ductility. However, additional work is required to characterize the structure including evaluating different temperatures and strain rates, as well using electron backscatter diffraction (EBSD) to interrogate texture evolution.

## Conclusions

The elevated temperature tensile behavior and necking behavior of five magnesium extruded sheets was characterized at 450°C and 0.001/s. Specific conclusions are given below:

1. The as-extruded materials exhibited very good elongation, with almost 300% for the ZK materials at strain rates of 0.001/s.
2. All materials exhibit some level of static grain coarsening during testing.

3. AZ31, AM50, and ZM21 materials exhibited significant grain coarsening in the necked region with grains over 100 microns.
4. The ZK materials exhibit very stable microstructures in the necked region without the presence of large grains.

## References


1. P.E. Krajewski, S. Kim, J.T. Carter, R. Verma, "Magnesium Sheet: Automotive Applications and Future Opportunities", Korean Institute of Metals and Materials: Trends in Metals & Materials Engineering, vol. 20, no. 5, (2007) 60.
2. J.T.Carter, P.E. Krajewski, and R. Verma, "The Hot Blow Forming of AZ31 Mg Sheet: Formability Assessment and Application Development", JOM, vol 60 no 11, (2008)77.
3. S. Agarwal, P.E. Krajewski, and C.L. Briant, "Dynamic Recrystallization of AA5083 at 450C: The Effects of Strain Rate and Particle Size," Met. Mat. Trans. A., Vol. 39, no. 6, (2008) 1277.
4. J-K Chang, K. Takata, K. Ichitani, and E.M. Taleff, "Abnormal Grain Growth and Recrystallization in Al-Mg Alloy AA5182 Following Hot Deformation", Met. Mat. Trans. A, Vol 41A, (2010) 1942.
5. J.R. Bradley. "Bulge Testing of Superplastic AA5083 Aluminum Sheet", in *Advances in Superplasticity and Superplastic Forming*, Ed. E.M. Taleff, P.A. Friedman, P.E. Krajewski, R.S. Mishra, and J.G. Schroth, TMS, (2004) 109.
6. P.E. Krajewski and A. Ben-Artzy, "Elevated Temperature Tensile Behavior of Extruded Magnesium Plates," Magnesium Technology 2010: Proceedings of a symposium sponsored by the Magnesium Committee of the Light Metals Division of The Minerals, Metals, & Materials Society, (2010) 221.
7. P.E. Krajewski, "Elevated Temperature Behavior of Sheet Magnesium Alloys," *Magnesium Technology 2002*, Ed. H.I. Kaplan, The Minerals, Metals, and Materials Society, (2002)175.


**Table 1: Chemical composition of extruded magnesium sheets used in the present study.**

| Alloy | *t* (mm) | Al | Mn | Si | Zn | Zr | Fe | Mg |
|---|---|---|---|---|---|---|---|---|
| AZ31 | 1.65 | 2.9 | 0.34 | <0.5 | 0.97 | <0.01 | <0.005 | Bal. |
| AM50 | 1.75 | 4.9 | 0.33 | <0.5 | <0.01 | <0.01 | <0.005 | Bal. |
| ZM21 | 1.43 | 0.03 | 1.0 | <0.03 | 2.10 | <0.01 | 0.03 | Bal. |
| ZK10 | 2.93 | <0.01 | <0.003 | <0.03 | 1.10 | 0.6 | 0.003 | Bal. |
| ZK30 | 2.93 | <0.01 | <0.003 | <0.03 | 3.10 | 0.5 | 0.003 | Bal. |

**Table 2: Grain size of magnesium materials in as-received and thermally exposed condition.**

| Alloy | *Grain Size As-received (μm)* | *Grain Size Exposed (μm)* |
|---|---|---|
| AZ31 | 12.8 | 15.1 |
| AM50 | 8.7 | 12.9 |
| ZM21 | 18.1 | 22.8 |
| ZK10 | 12.8 | 14.4 |
| ZK30 | 9.4 | 16.4 |

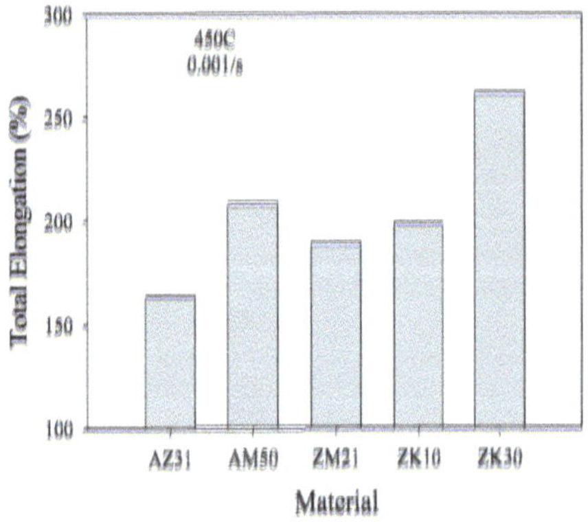


**Fig 1:** As-extruded microstructure of (a) AZ31, (b) AM50, (c) ZM21, (d) ZK10, and (e) ZK30.

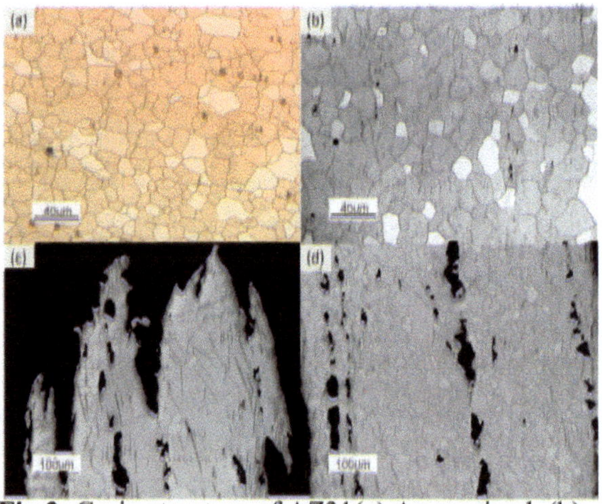


**Fig 2**: Grain structure of AZ31(a) As-received, (b) From grip section – thermal exposure without strain, (c) fracture tip, and (d) failed tensile sample away from fracture region.

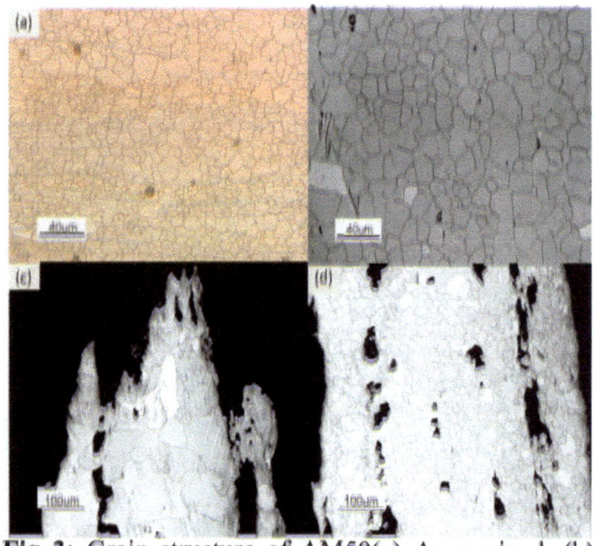


**Fig 3:** Grain structure of AM50(a) As-received, (b) From grip section – thermal exposure without strain, (c) fracture tip, and (d) failed tensile sample away from fracture region.

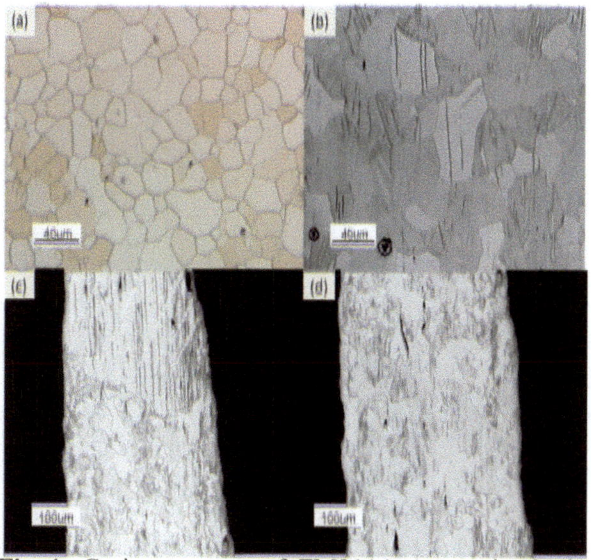


**Fig 4:** Grain structure of ZM21(a) As-received, (b) From grip section – thermal exposure without strain, (c) fracture tip, and (d) failed tensile sample away from fracture region.

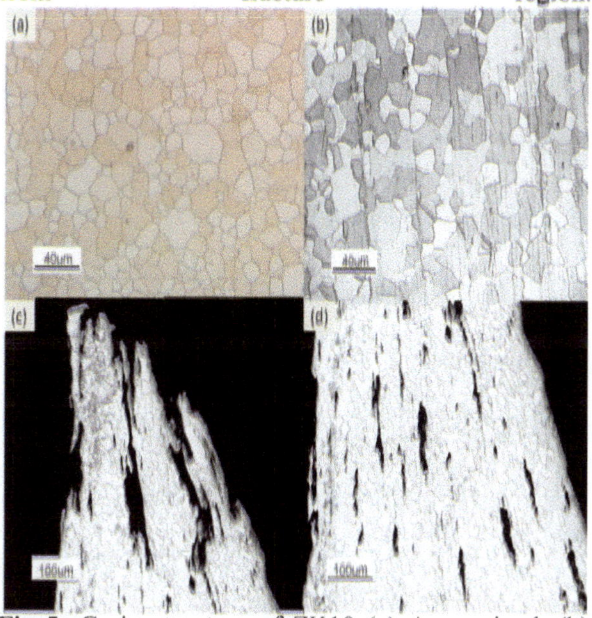


**Fig 5:** Grain structure of ZK10 (a) As-received, (b) From grip section – thermal exposure without strain, (c) fracture tip, and (d) failed tensile sample away from fracture region.

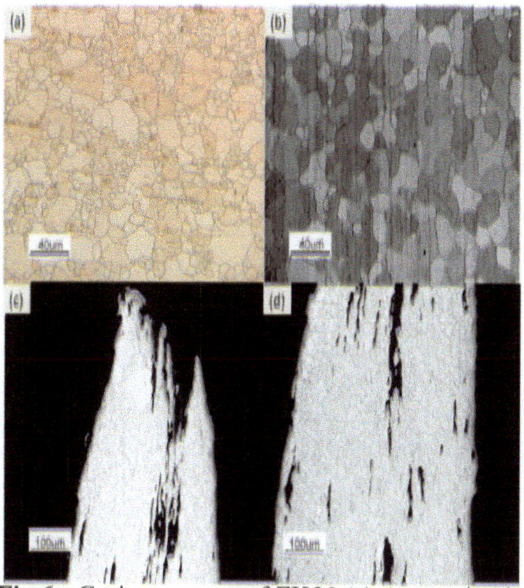


**Fig 6:** Grain structure of ZK30 (a) As-received, (b) From grip section – thermal exposure without strain, (c) fracture tip, and (d) failed tensile sample away from fracture region.

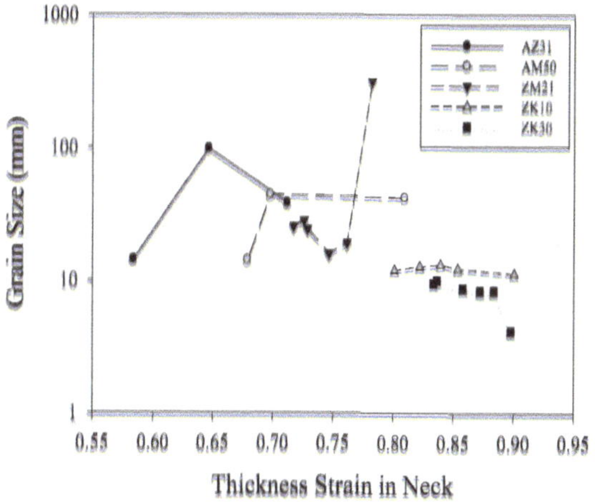


**Fig 7:** The effect of alloy composition and thickness strain in the necked region on grain size.

**Magnesium Technology 2012**
*Edited by: Suveen N. Mathaudhu, Wim H. Sillekens, Neale R. Neelameggham, and Norbert Hort*
***TMS (The Minerals, Metals & Materials Society), 2012***

# Production of wide shear-rolled magnesium sheet for part forming

D Randman[1], B Davis[1], ML Alderman[1], G Muralidharan[2], TR Muth[2], WH Peter[2], TR Watkins[2]
[1]Magnesium Elektron North America; 1001 College St.; P.O. Box 258; Madison; Illinois 62060; USA
[2]Materials Science and Technology Division; Oak Ridge National Laboratory; 1 Bethel Valley Rd; Oak Ridge; Tennessee 37831; USA

Keywords: Magnesium, Asymmetric Rolling, Texture, Formability

## Abstract

In recent years the process of shear rolling has seen considerable study, particularly for heavily textured materials such as magnesium. The goal of this work has been to produce sheet with greater formability for industries such as automotive and aerospace. To date, almost all work on asymmetric rolling has been carried out on small strips that are not large enough to produce parts. The current work will discuss scaling-up of the shear rolling process to generate wider sheet. A mill at the Magnesium Elektron North America plant was modified to allow shear rolling at a ratio of 1:1.35 on sheets up to 36" wide. AZ31B and ZEK100 sheets of size 36"x72" were shear rolled and demonstration automotive parts have been formed by Superform USA.

## Introduction

Magnesium alloys are strong and lightweight, making them very attractive to the automotive and aerospace industries, with a density approximately two thirds of that of aluminum and one fifth of steel. It could therefore provide significant weight savings in applications for these industries. Magnesium alloys are utilized in various applications in cast form, but sheet material has not gained wide acceptance in modern engineering solutions. A significant hindrance to the greater use of magnesium sheet is the poor low temperature formability. In general magnesium sheet is either superplastically formed at elevated temperatures or hot deep drawn. Both processes are more expensive compared to room temperature matched die deep drawing, as is currently employed for steel sheet. Forming of sheet components at temperatures below 200°C simplifies the lubrication system needed for deep drawing, as oil can be used. Above this temperature, solid state lubricants such as boron nitride and graphite are used that require specialized cleaning routes to remove them. Oils can be removed with a simple degreasant such as an alkali dip, which saves time and energy costs.

Production of more formable magnesium sheet has been approached in a variety of ways. Examples include Equal Channel Angular Processing (ECAP)[1, 2], Equal Channel Angular Rolling (ECAR)[3, 4], 'cross-roll rolling'[5], and asymmetric or shear rolling[6-8]. Asymmetric rolling is considered to be the only viable method to produce sheet on a commercial scale and therefore has been the focus of the current work.

Asymmetric, or shear, rolling introduces shear into the sheet during the rolling process by having one roll moving faster than the other. This can be achieved either by having independently driven rolls, or by having differently sized rolls on a single drive. Shear rolling has been shown to provide more homogeneous deformation through the thickness of the sheet, along with grain refinement and texture changes[9-14]. These factors are expected to increase the formability of the sheet.

Two alloys were chosen for this study: AZ31B (Mg-3wt%Al-1wt%Zn-0.3wt%Mn) and ZEK100 (Mg-1wt%Zn-<1wt%Rare Earths-<1wt%Zr). Initial trials were carried out on a small laboratory scale shear mill. This had bottom and top roll diameters of 229 and 76mm respectively, giving a shear ratio of 3:1[15]. This work lead to the modification of a 48 inch mill at Magnesium Elektron to enable shear rolling. This mill has a bottom roll of diameter 367mm and a top roll of diameter 495mm, giving a shear ratio of 1:1.35. Initial trials on this mill were carried out on 10 inch wide strips to gain a greater understanding of the material behavior at different rolling conditions. The best processing parameters were selected following biaxial testing to assess the formability of the 10 inch wide sheets. Wider sheet was rolled using these parameters. Rolling wider sheet was more complex than the smaller strips; some issues were anticipated while others were not. The focus of this paper is the engineering challenges arising from shear rolling wide magnesium sheet, and the solutions that were implemented to overcome them. Finally, demonstrator parts were formed to demonstrate the viability of shear rolled sheet.

## Experimental Materials and Procedure

<u>Modification of Mill at Magnesium Elektron</u>

A smaller scale industrial mill (known as 8-Mill, see Figure 1) based at MENA was modified to allow shear rolling. 8-Mill has a single drive motor, and thus the conversion to shear rolling required different diameter rolls. The maximum shear ratio that could be carried out on 8-Mill was limited by the design of the mill. The small roll could be no smaller than the chocks in which it is held and the top backup roll could be raised no higher than the top of the screw, limiting the size of the large upper roll. This resulted in a maximum shear ratio of 1:1.35 via the installation of rolls with diameters of 367mm and 495mm. To achieve the higher position of the upper roll, a new spindle was engineered with universal couplings, that replaced the old wobbler joints, to accommodate the steeper angle of operation created in shear mode.

It was known from previous shear rolling that the effect of one roll moving faster than the other caused a curl in the exiting sheet. Therefore, a stripper plate arrangement was designed for the mill to ensure that material exited horizontally. This will be discussed further in the results section.

Figure 1 – 8-Mill

Rolling

Magnesium Elektron supplied commercially rolled sheet for these trials. Both AZ31B and ZEK100 were direct chill cast and rolled from slab down to the feedstock gauge of nominally 2.5mm.

The best rolling conditions for formability were found from the previous work on smaller strips. Lower processing temperatures performed better and this was used for the wider sheet. The conditions chosen were 250°C with 10% reduction. It was intended to carry out the rolling without lubrication. This was to allow maximum friction between the rolls and the material, presumed to cause the maximum amount of shear during rolling. However, as will be discussed later, this was not practical, and lubrication had to be used.

Analysis

Samples for optical microscopy were cut from the rolled sheet and mounted and polished using standard metallographic techniques. An acetic-picral etch was used to reveal the grain structure and imaging was carried out on an Olympus GX51 microscope.

Samples for X-ray diffraction (XRD) analysis were cut from the rolled sheet and polished down to the centreline for texture analysis. A 4-axis diffractometer was used with a cobalt x-ray source, and pole figures were measured for the (0002) reflections.

Elevated temperature biaxial testing was carried out using proprietary equipment at Oak Ridge National Laboratory. A spherical punch of diameter 100mm was used to deform a 138mm diameter disc held in a 103mm inner diameter ring. The material and the ring were heated in a furnace prior to deformation. A speckle pattern was applied to the disks with paint, and with the use of two cameras, a three dimensional strain map was collected during the tests. The final dome height at fracture was used as the measure of formability for the current study. More complex analysis will be shown in further publications.

**Results and Discussion**

Initial Trials

Ten inch wide pieces were rolled on 8-Mill to find suitable conditions for rolling the wider sheet. Samples from a selection of these sheets were tested for biaxial formability (see Figure 2 for an example of a tested sample) and the best performing were selected. Table 1 shows the dome heights of the samples that were studied. It can be seen that rolling at a lower temperature led to greater dome height. Therefore, 250°C was chosen as the rolling temperature for further work.

Table 1 – Biaxial dome heights of 10" wide strips shear rolled on 8-Mill

| Alloy | Temperature (°C) | Reduction (%) | Dome Height (mm) |
|---|---|---|---|
| AZ31B | 250 | 10 | 21.3 |
| | | 50 | 26.7 |
| | 425 | 10 | 22.5 |
| | | 50 | 16.0 |
| ZEK100 | 250 | 10 | 44.3 |
| | | 50 | 43.7 |
| | 425 | 10 | 40.5 |
| | | 50 | 33.8 |

Figure 2 – Example of a biaxial test sample

Wide Sheet Rolling

The arrangement of the stripper plates is shown in Figure 3. It was anticipated that the material would increase in length on the side of the fast roll, and thus curl downwards. This had been observed in previous trials and therefore the stripper plates were designed such that the lower plate was positioned very close to the roll to ensure the material did not wrap around the roll. The top stripper plate was put in place to avoid the sheet buckling upwards. By forcing the sheet though the slot, it was thought that a flatter sheet would be produced. Therefore, the upper plate was not positioned as close to the roll (see Figure 4). An entirely new exit table was designed that incorporated the stripper plates and bracing bars to enable shear rolling, see Figure 3.

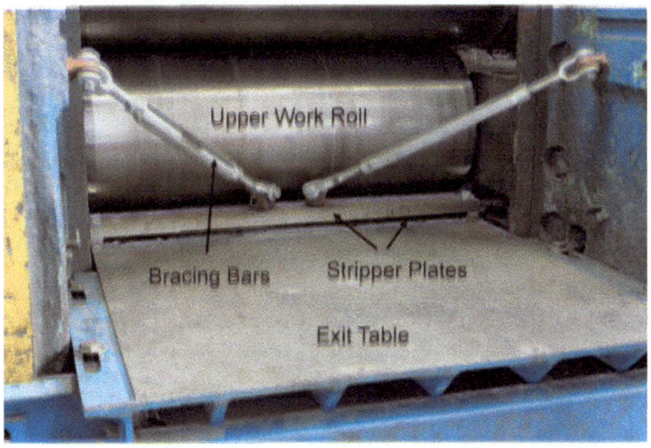


Figure 3 – Exit side of the mill showing the stripper plate arrangement

It was found during rolling that the material did not behave entirely as expected. In many cases, the material curved upwards. This caused problems with the stripper plates and certain sheets came into contact with the plate edges when they exited the mill. Figure 5 shows the end of one sheet that slightly contacted the stripper plate; a small section of the sheet was removed. Figure 6 shows the extreme case of this behaviour when a sheet caught on the stripper plate so badly that it was ripped into 3 pieces. Therefore, it can be seen that the positioning of the stripper plates is critical in a shear rolling mill. This is more complicated for a wide mill as the plates need to be supported across the length. The material itself also seems more prone to catching on the plates as there is a greater chance that some small portion may deviate far enough to touch the edge of the stripper plate.

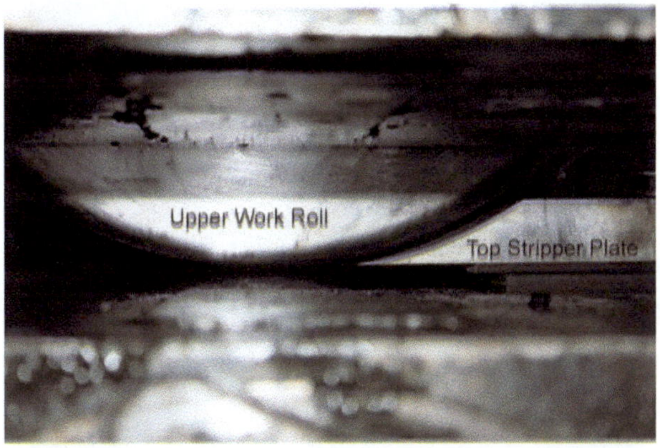


Figure 4 – View from the side of the mill showing the alignment of top roll and stripper plate

The stripper plates needed to be shaped carefully and fitted less than 200μm from the roll surface. This was achieved by placing a sheet of paper between the plate and the roll, pushing the plate in to touch and then removing the paper. Ideally, both plates would be positioned this close to the rolls. However, the upper roll is not stationary during a rolling campaign; not only is the gauge intentionally changed for different rolling passes, but the majority of mill stretch in the system, the elastic stretch of the rolling mill due to separation forces, causes movement of the top work roll. Therefore, a shear mill designed from scratch should have the top stripper plate attached to move in conjunction with the top work roll; this is not generally possible for modifications to existing mills.

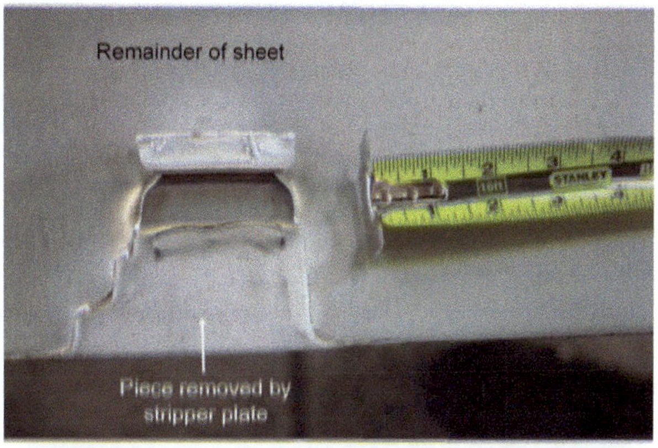


Figure 5 – Photograph of a sheet that contacted the stripper plate (the piece removed was placed back where it belonged to show the damage)

Figure 6 – Photograph of a sheet that caught in the stripper plate and split into three pieces

As well as causing problems with the stripper plates, the curl of the sheet raised questions about the effect of shear rolling. It was expected that the fast roll would stretch the surface of the sheet further than the slow moving roll, leading to a downward curl in the sheet. However, in most cases it was seen that the sheet curled upwards. The exact cause of this is unknown but two theories have been put forward. The first is that the sheet is simply sticking to the upper roll and is being bent after it has exited the roll gap, putting an opposite curve into the sheet. The second is that there is a difference in friction between the two rolls. At the very start of rolling this is unlikely as the rolls were ground to exactly the same finish. However, after several passes are carried out build-up of magnesium and magnesium oxide on the rolls occurs. The amount of build up is likely to be different on the upper and lower rolls due to the speed that each roll is travelling and the contact length. This will change the friction coefficient between the roll and the material, which will alter the forces on the sheet, causing a variation of the total shear.

Figure 7 – Photograph of rolled sheets showing the difference in shape control along the length

The first sheets were rolled with no lubrication. It was thought that this would apply more shear to the material, due to higher friction, and thus create 'better' sheets. However, after a few sheets had been rolled, the material started to 'chatter' very badly through the roll bite, with alternating regions of sticking and slipping, as a result of pickup on the rolls. This produced very uneven surface and was expected to give inconsistent amounts of shear deformation. For this reason, lubrication was used for the remainder of rolling. While there may have been more slipping at the roll surface, and consequently less shear applied to the metal, the overall quality was much better and more consistent. Shape control was also helped by the application of lubricant, in its role as a coolant of the rolls. Figure 7 shows a typical example of the type of problem that was encountered. The leading edge of the sheet is shown in the foreground. It can be seen that there was a clear line between good, flat material at the start and poorly shaped/warped material at the end. The initial good section is approximately one revolution of the rolls, and it is thought that heating from the material is responsible for the change in shape. Rolling hot sheet results in a thermal profile across the rolls, getting hotter nearer the centre line of the mill. The expansion of the rolls at the center will lead to buckling in the sheet if not countered. 8-Mill was not equipped with any shape control mechanics. This issue is more a problem with rolling wide, thin sheet than any specific effect of shear rolling, as longer rolls are more prone to shape control problems, which in thin gauges manifest as buckles and waves in the sheet. However the use of coolant minimized these issues by cooling the rolls during processing.

Microstructure and Texture

An overview of the microstructure and texture is shown here to demonstrate that the wide shear rolling caused the same changes as were seen previously in smaller strips. This evolution will be analyzed in more detail in later publications.

Figure 8 and Figure 9 show the microstructure at the centre line of the AZ31B and ZEK100 sheets respectively. It can be seen that the behavior of the two alloys under the same conditions was quite different. The AZ31B had a smaller grain size than the ZEK100 and many more twins. The grains in the ZEK100 appear to be much more elongated than the approximately equiaxed grains in AZ31B. This indicates that more dynamic recrystallization occurred in the AZ31B. However, the presence of many twins in AZ31B shows that it still contained a significant amount of stored work.

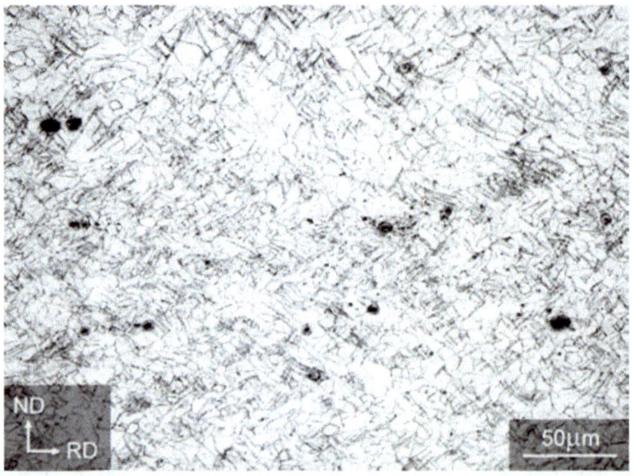


Figure 8 – Microstructure of 36 inch wide shear rolled AZ31B sheet

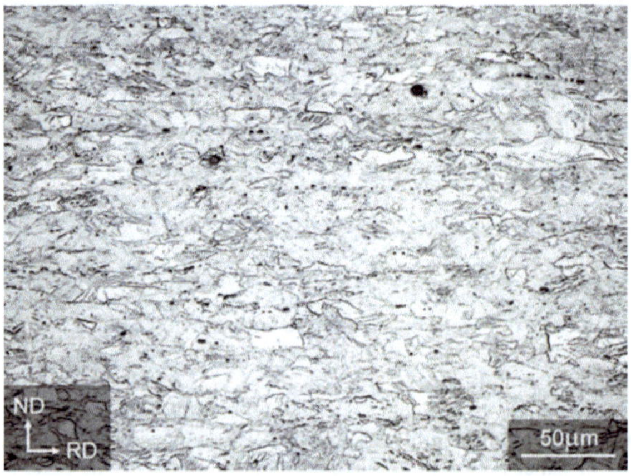


Figure 9 – Microstructure of 36 inch wide shear rolled ZEK100 sheet

The textures of the two alloys are shown in Figure 10 and Figure 11. Again, a significant difference was seen between AZ31B and ZEK100. The distinctly different textures have been seen previously and are responsible for the disparity in mechanical properties. The texture seen in AZ31B was typical, with a split of the basal poles about the normal direction, tilted towards the rolling direction. There is an asymmetry of the split, which may be due to the shear rolling but is too small to be truly significant. The texture in the ZEK100 was also a split basal, as already seen in AZ31B, but with significant spreading in the transverse direction and much weaker intensity.

The microstructure and texture of the wide sheets processed on 8-Mill were comparable to the 10 inch wide sheets. This demonstrated that shear rolling could be scaled up to wider sheet, and will give the same microstructural and textural effects.

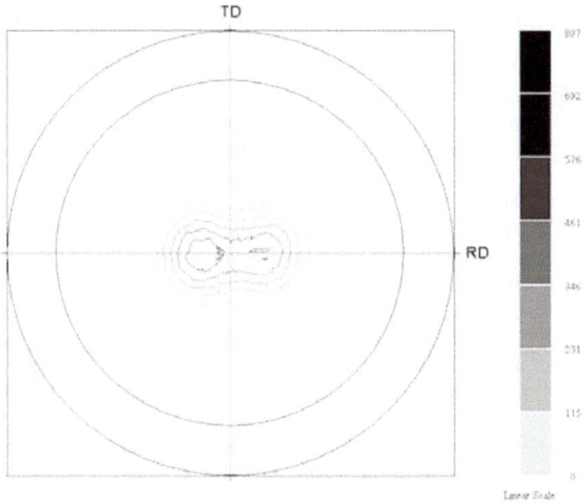


Figure 10 – 0002 pole figure for AZ31B

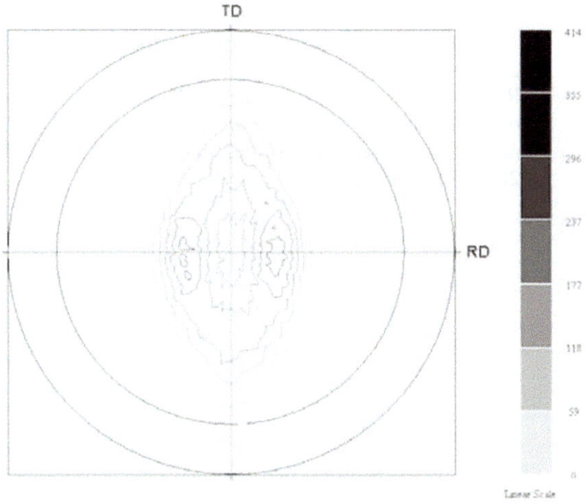


Figure 11 – 0002 pole figure for ZEK100

Formed Parts

Demonstration parts were formed at Superform USA. The part chosen was a section of the firewall for the Panoz Abruzzi. This was chosen as it is a reasonably large part (using a blank approximately 24x28 inches), with a variety of different strains experienced in different areas. An example of a formed part is shown in Figure 12. The parts were formed in a two stage process, consisting of a deep draw followed by pneumatic coining (superplastic forming). Both AZ31B and ZEK100 formed successfully; it was observed that the ZEK100 could be formed faster than the AZ31B, which split at higher forming speeds, see Figure 13.

Figure 12 – Photograph showing a part formed at Superform from shear rolled ZEK100

Figure 13 – Photograph showing the failure of an AZ31B shear rolled sheet formed at a faster rate

## Summary

Wide sheets of AZ31B and ZEK100 have been successfully shear rolled and the following conclusions were drawn.

- Frictional effects during shear rolling are complex and often act opposite to that expected
- Stripper plates both top and bottom are required on the exit side of the mill
- The stripper plates must be positioned closely and accurately with respect to the rolls
- Lubrication was necessary to ensure successful rolling, but may also have decreased the amount of shear experienced by the material
- Microstructure changes were the same in wide sheet as in smaller strips
- Texture evolution during rolling was the same in wide sheet as in smaller strips
- Automotive parts were successfully formed from shear rolled material



## Acknowledgements

This work was carried out in collaboration with Oak Ridge National Laboratory, in part, under sub contract 4000092046 as part of the American Recovery and Reinvestment Act through the Department of Energy. Also, this research through the Oak Ridge National Laboratory's High Temperature Materials Laboratory User Program was sponsored in part by the U. S. Department of Energy, Office of Energy Efficiency and Renewable Energy, Vehicle Technologies Program. Thanks go to those staff members at Oak Ridge National Laboratory that contributed to the research program.

Thanks also go to Superform USA for help with part forming.




## References


1. Agnew, S.R., et al., *Enhanced ductility in strongly textured magnesium produced by equal channel angular processing.* Scripta Materialia, 2004. **50**(3): p. 377-381.
2. Mukai, T., et al., *Ductility enhancement in AZ 31 magnesium alloy by controlling its grain structure.* Scripta Materialia, 2001. **45**(1): p. 89-94.
3. Cheng, Y.Q., et al., *Improvement of Drawability at Room Temperature in AZ31 Magnesium Alloy Sheets Processed by Equal Channel Angular Rolling.* Journal of Materials Engineering and Performance, 2008. **17**(1): p. 15-19.
4. Kim, W.J., et al., *Grain size and texture control of Mg-3Al-1Zn alloy sheet using a combination of equal-channel angular rolling and high-speed-ratio differential speed-rolling processes.* Scripta Materialia, 2009. **60**(10): p. 897-900.
5. Chino, Y., et al., *Enhanced formability at elevated temperature of a cross-rolled magnesium alloy sheet.* Materials Science and Engineering: A, 2006. **441**(1-2): p. 349-356.
6. Watanabe, H., T. Mukai, and K. Ishikawa, *Differential speed rolling of an AZ31 magnesium alloy and the resulting mechanical properties.* Journal of materials science, 2004. **39**(4): p. 1477-1480.
7. Kim, S.H., et al., *Texture and microstructure changes in asymmetrically hot rolled AZ31 magnesium alloy sheets.* Materials Letters, 2005. **59**(29-30): p. 3876-3880.
8. Huang, X., et al., *Microstructure and texture of Mg-Al-Zn alloy processed by differential speed rolling.* Journal of Alloys and Compounds, 2008. **457**(1-2): p. 408-412.
9. Huang, X., et al., *Microstructural and textural evolution of AZ31 magnesium alloy during differential speed rolling.* Journal of Alloys and Compounds, 2009. **479**(1-2): p. 726-731.
10. Watanabe, H., T. Mukai, and K. Ishikawa, *Effect of temperature of differential speed rolling on room temperature mechanical properties and texture in an AZ31 magnesium alloy.* Journal of Materials Processing Technology, 2007. **182**(1-3): p. 644-647.
11. Huang, X., K. Suzuki, and N. Saito, *Textures and stretch formability of Mg-6Al-1Zn magnesium alloy sheets rolled at high temperatures up to 793 K.* Scripta materialia, 2009. **60**(8): p. 651-654.
12. Ji, Y.H., J.J. Park, and W.J. Kim, *Finite element analysis of severe deformation in Mg-3Al-1Zn sheets through differential-speed rolling with a high speed ratio.* Materials Science and Engineering: A, 2007. **454**: p. 570-574.
13. Gong, X., et al., *Microstructure and mechanical properties of twin-roll cast Mg-4.5 Al-1.0 Zn sheets processed by differential speed rolling.* Materials & Design, 2010. **31**(3): p. 1581-1587.
14. Kim, W.J., J.D. Park, and W.Y. Kim, *Effect of differential speed rolling on microstructure and mechanical properties of an AZ91 magnesium alloy.* Journal of Alloys and Compounds, 2008. **460**(1-2): p. 289-293.
15. Randman, D., et al. *The Effect of Rare Earth Elements on the Texture and Formability of Shear Rolled Magnesium Sheet.* in *Magnesium Technology 2011.* 2011. San Diego, CA.

# Magnesium Technology 2012

## Primary Production

### Monday PM

*Session Chairs:*

**Neale Neelameggham**
**(IND LLC, USA)**
**Adam Powell**
**(Metal Oxygen Separation Technologies Inc., USA)**

**Magnesium Technology 2012**
*Edited by: Suveen N. Mathaudhu, Wim H. Sillekens, Neale R. Neelameggham, and Norbert Hort*
*TMS (The Minerals, Metals & Materials Society), 2012*

# CARBOTHERMAL PRODUCTION OF MAGNESIUM: CSIRO'S MAGSONIC™ PROCESS

Leon H. Prentice[1]; Michael W. Nagle[1]; Timothy R. D. Barton[1]; Steven Tassios[1]; Benny T. Kuan[2]; Peter J. Witt[2]; Keri K. Constanti-Carey[1]
[1]CSIRO Process Science and Engineering; Box 312; Clayton South, VIC, 3169, Australia
[2]CSIRO Mathematics Informatics and Statistics; Box 312; Clayton South, VIC, 3169, Australia

Keywords: Carbothermal Magnesium, Carbothermic Reduction, Shock Quenching, Reversion, Laval Nozzle, Supersonic Flow, Pyrophoric Powder, Computational Fluid Dynamics, Scale Up, Pilot Plant, Process Safety

## Abstract

Carbothermal production has been recognized as conceptually the simplest and cleanest route to magnesium metal, but has suffered from technical challenges of development and scale-up. Work by CSIRO has now successfully demonstrated the technology using supersonic quenching of magnesium vapor (the MagSonic™ Process). Key barriers to process development have been overcome: the experimental program has achieved sustained operation, no nozzle blockage, minimal reversion, and safe handling of pyrophoric powders. The laboratory equipment has been operated at industrially relevant magnesium vapor concentrations (>25% Mg) for multiple runs with no blockage. Novel computational fluid dynamics (CFD) modeling of the shock quenching and metal vapor condensation has informed nozzle design and is supported by experimental data. Reversion below 10% has been demonstrated, and magnesium successfully purified (>99.9%) from the collected powder. Safe operating procedures have been developed and demonstrated, minimizing the risk of powder explosion. The MagSonic™ Process is now ready to progress to significantly larger scale and continuous operation.

## Introduction

Magnesium is the lightest of the structural metals and is useful for a wide range of industrial applications [1]. The production of magnesium via carbothermal reduction is not a new concept; indeed, a process using this chemistry was patented as far back as 1932 [2]. The basic chemistry depends on the use of carbon as the reductant to produce magnesium and carbon monoxide from magnesia and carbon:

$$MgO_{(s)} + C_{(s)} \leftrightarrow Mg_{(v)} + CO_{(g)} \quad (1)$$

The well-known primary issue with the carbothermal process is reversion – the reaction above proceeds in the reverse direction very easily; this can be minimized by cooling the gaseous reaction products very rapidly. The largest industrial-scale operation of carbothermal technology was at Permanente, CA [3, 4]. Cooling of the gases was achieved either by heat transfer with surfaces [3] or shock-chilling by dilution with, for example, natural gas [4], nitrogen [5], molten magnesium chloride [6], or even molten magnesium itself [7]. Other versions of the chemistry focused on removing magnesium from the vapor phase by reaction in a solution [8, 9].

The idea of supersonic quenching is also not new. The reaction products (gases) are accelerated to several times the speed of sound (typically $Ma \approx$ 4-5), achieving very rapid gas cooling. Useful reviews of the literature have been presented by the present authors [9-11]. Previous attempts appear to have failed primarily because of blockage of the supersonic Laval nozzle [12] under sustained operation. This paper describes one form of the technology, called MagSonic™, developed by CSIRO (Commonwealth Scientific and Industrial Research Organisation) in Australia.

## Technology Description

Feed Preparation

Magnesite may be commercially mined at relatively high purity and calcined to produce magnesia (MgO) suitable for reaction. Options exist for additional purification of the magnesite or magnesia, which depend on purity and morphology requirements and may include acid leaching, flotation, and other processing. The MagSonic™ process has demonstrated itself remarkably robust, processing magnesia that has been only calcined (not purposely purified). The behavior of impurities in the carbothermal reduction process has been addressed previously [13], reporting that essentially all the impurity species are reduced before the carbothermal magnesium reaction begins. The species may be detected in the cyclone (used to separate the magnesium powder from the gas) by Laser-Induced Breakdown Spectroscopy (LIBS).

Feed preparation is completed by briquetting calcined magnesia with a carbon source. For much of the experimental program, powdered graphite has been used ($d_{50} \approx 12$ μm, >99.5% C, Merck Pty Ltd, Kilsyth, VIC 3137, Australia). Although briquetting graphite without a binder is difficult, briquettes of the size and shape suitable for piloting have been produced without a binder (Figure 1), and have been tested in the laboratory rig.

Figure 1: Magnesia-carbon briquette

### Reaction

The reaction proceeds quickly above 1700 °C, and there is evidence of some reduction/vaporization as low as 1250 °C [13]. In experimental work to date, heating has been achieved by the use of an induction furnace (up to 50 kW), with sustained reaction at temperatures typically around 1800 – 1900 °C [14]. The reaction kinetics have been described by the present authors [14, 15], including achievement of sustained (60+ minutes) operation and confirmation of kinetic modeling for pilot plant and full-scale design.

### Supersonic Quenching

The aerodynamics of supersonic Laval nozzle operation are well known, but the present operation presents a number of novel aspects. These include high temperatures, metal vapor condensation and solidification, and relevant boundary layer interactions. CSIRO has developed a number of novel computational fluid dynamics (CFD) tools that have been implemented and validated, providing significant information for scale-up and further development.

The effect of metal vapor condensation may be illustrated by Figure 2. Condensation adds significant heat to the stream, affecting temperature distributions, expansion shocks, and other aspects of the flow. Early CFD work identifying the condensation effect shown in Figure 2 was reported in [9] with further publication on these topics pending.

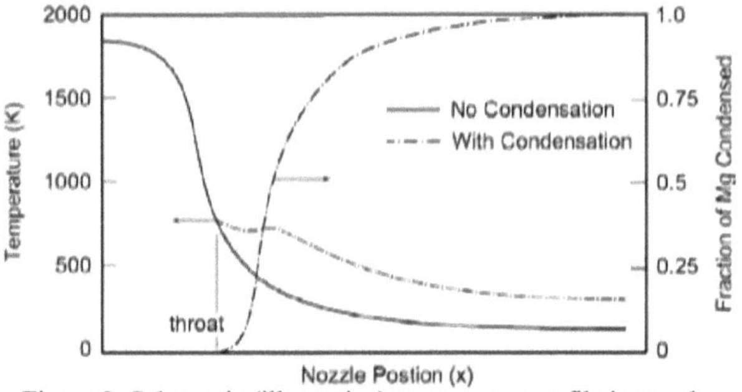


Figure 2: Schematic (illustrative) temperature profile in nozzle calculated with and without metal vapor condensation

### Condensation / Solidification

The effect of supersonic expansion, combined with some heat transfer with the walls of the chamber, result in a stream temperature around 200 – 300 °C. The peak rate of cooling during the process is around $10^6$ $K.s^{-1}$. Following rapid cooling, the magnesium completely solidifies, and the reversion reaction is prevented.

### Separation

Separation may be achieved by a number of methods, the simplest of which is a cyclone. While some reduced efficiency at moderate temperatures may be experienced [16, 17], separation efficiencies above 90 %, often higher than 95 %, have been consistently achieved. The powder may be conveniently recovered, despite its very low bulk density (typically < 0.3 $g.ml^{-1}$), and flows easily.

### Gas Handling

Vacuum systems are typically designed to handle one or two of the relevant process conditions – low pressure, high mass flow, high temperature – but rarely all three at once. As such, the vacuum system is a major user of energy, estimated as 0.46 MWh per tonne Mg at full scale [18]. The vacuum system must maintain the absolute pressure downstream of the nozzle below 50 % (but preferably below 10%) of its upstream value to maintain sonic flow. The lower limit of vacuum is determined primarily by the economics of operation; achieving a pressure below 0.5 kPa is likely to be prohibitively expensive, and unnecessary for optimum operation.

### Purification

The experimental product powder is rendered safe either by controlled surface oxidation ('passivation') or by covering it with oil to prevent contact with air. The product has been purified from both media, either by direct sublimation from the passivated powder, or firstly by distilling off the oil at 400-500 °C at atmospheric pressure and subsequently distilling the magnesium metal at about 980 °C and 0.8 kPa absolute pressure in a graphite-lined steel retort. The resulting magnesium metal is distilled again and yields metal containing low levels of metallic impurities (200 ppm Al, 90 ppm Ca, 80 ppm Fe, 10 ppm Mn and 60 ppm Si).

## Major Technical Challenges

### High Temperature Operation and Control

The control of a kinetically-constrained gas-solid endothermic reaction at very high temperatures is non-trivial, though the total system pressure can be relatively easily maintained by the flow rate of an inert cover gas. For maximum efficiency the magnesium concentration in the vapor should be as high as possible, but any higher than a partial pressure of 50 kPa risks over-pressurizing the reactor. The equilibrium partial pressure of 50 kPa is reached at 1764 °C, but the reactor is likely to operate at 1900 – 2000 °C.

In terms of process control, reactor temperature and cover gas flow rate are controlled in the laboratory batch system. Continuous operation will likely operate at constant temperature, with the reaction rate controlled by the amount of reacting bed, and to a lesser extent by the inert gas flow rate. The authors have presented data on the modeling of the reaction kinetics [14, 15, 19], which are useful for sizing and control.

### Preventing Nozzle Blockage

Blockage of the gas exit/quenching nozzle has been a significant problem for the carbothermal process [12]. Typically, magnesium and/or impurity metals condense on the surface and revert to oxides with high melting points, thereby accreting and blocking the nozzle. The CSIRO research team has overcome these issues and provisionally patented aspects of the method [11]. Despite demonstrated operation without blockage at laboratory scale, safe design includes the possibility of recurrence and must be included in risk assessments.

Pyrophoric Powder and Process Safety

Issues with pyrophoric magnesium powders are well known, and previous industrial operation has resulted in fatalities [20]. Studies have quantified the conditions for spontaneous ignition [21]. National Fire Protection Association standards 69 and 484 [22, 23] are also useful for magnesium powder explosion prevention and process safety. Given that magnesium powder produced by the MagSonic™ process is relatively fine and is certainly spontaneously combustible, and that operation is under vacuum, conventional measures such as deflagration venting, explosion suppression, or dust concentration control are not likely to be practically effective in ensuring safe operation.

Design for scale-up presently focuses on explosion containment, whereby equipment is designed to withstand a magnesium powder explosion without rupturing. Safety methods under development are relatively novel (and are the subject of future publications), but will ensure exceedingly high safety integrity levels are met for practical full-scale operation.

At laboratory scale, residual internally-adhered powder is oil-sprayed for safety. At larger scales, oil spraying is impractical and undesirable. Therefore, the metal powder must be rendered safe from a "dry" state. Options include a) batch passivation by controlled oxidation; b) melting of powder directly, or after briquetting, or by stirring into a liquid pool of magnesium; and c) distillation and condensation. The latter two options lead directly to the purification stage of the process.

Purification

The temperatures required to achieve reaction also means that the other metals in the feed are also reduced and contaminate the product to various extents. The product metal that is collected will require subsequent purification to achieve commercial magnesium metal purity.

Purification of magnesium by distillation is known technology [24, 25], but difficulties with fine powders are also known [3]. At Permanente the fine powder was briquetted with tar prior to vacuum distillation, but it is unlikely that a similar process will be used for MagSonic™ at full scale. The CSIRO team has overcome the difficulty of fine powder purification by dissolution into a molten metal, from which magnesium may be vacuum distilled. The molten metal may a metal with low vapor pressure (such as Sn, Pb, Cu), or indeed magnesium itself. Because of its lower oxygen content than Permanente's powder, incorporation into a melt, and subsequent distillation, are easier to achieve. The CSIRO project team has successfully purified the magnesium produced from the MagSonic™ rig and has data to size pilot plant equipment to purify magnesium.

## Experimental Progress

Progress in the experimental program has been rapid following the initial stage (proof of concept) from 2004 - 2007. The charts below illustrate experimental progress.

The concentration of magnesium in the nozzle is an important factor for efficient economic operation. Dungan [3] reported that the (subsonic) "Permanente magnesium plant circulates to each reduction unit from a common system approximately 13,000 $ft^3.min^{-1}$ of natural gas containing up to 6 per cent carbon monoxide." This implies that the magnesium concentration in the quench section was relatively low. For economic operation of the MagSonic™ process, the concentration target has been 25 % Mg in the nozzle. This corresponds to less than 50 % inert gas, though it may also be recycled as in the Permanente plant.

Figure 3 shows that the 25 % target was consistently met towards the end of this experimental program. Not all experiments aimed for very high concentrations, as other goals of the program included kinetics and control operations and the achievement of 'pseudo-steady-state' control [15].

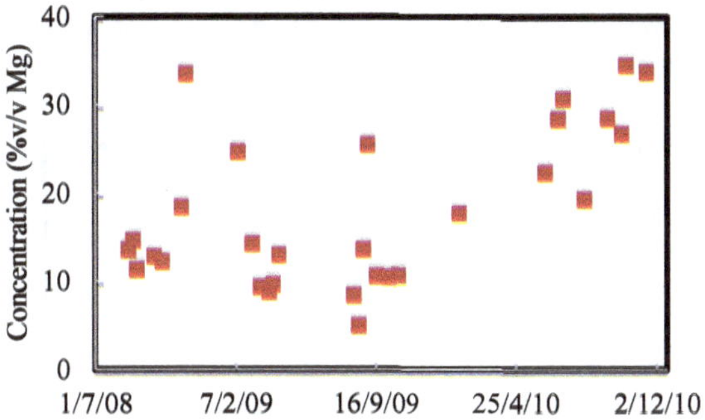

Figure 3: Peak magnesium concentration in nozzle entry gases over main experimental program

The data presented in Figure 3 are higher than reported for similar systems. Odle and McClaine [26] cite experimental work by Mineral Development International (MDI) in Birkerod, Denmark, for which the maximum Mg concentration appears to be around 0.25 %. Hori *et al* [27, 28] worked at reduced reactor pressure, which would also lead to a lower magnesium concentration (per unit volume) than that produced by the MagSonic™ process.

Continuous operation is also imperative. In the laboratory setting the process has been a batch operation, consuming MgO-C pellets. Operation of the reactor, nozzle, vacuum system, and powder recovery process were tested for durations up to around 2.5 hours of operation above 1700 °C (Figure 4). Towards the end of the experimental program, two hours of consistent operation could be reliably achieved, part of which may be considered pseudo-steady-state [15].

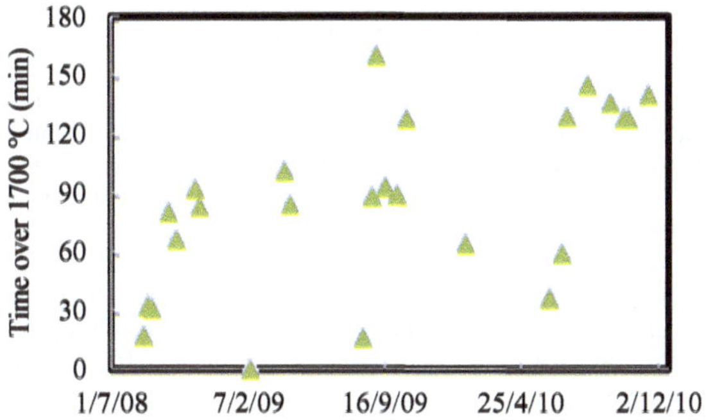

Figure 4: Experiment duration (minutes over 1700 °C) over main experimental program

Finally, another practical measure of continuous operation is the mass of reactants consumed. The laboratory equipment was upgraded in late 2008 to allow up to 2500 g of reactants to be loaded into the reactor. The charge consumption in each experiment is shown in Figure 5. Some failures are included (from, for example, nozzle blockage or operational issues), but since late 2009 consistently good and reliable experiments have been conducted. A 'routine' mass of 2000 g has been used for many of the latter experiments, all of which have gone to effective completion (>99.9 % MgO reacted).

The focus of each experiment was not necessarily maximum charge consumption. As mentioned above, occasionally the purpose is pseudo-steady-state control, or the investigation of alternative cover gases, or modifications to process conditions (e.g. downstream pressure), or simply collection of a batch of powder.

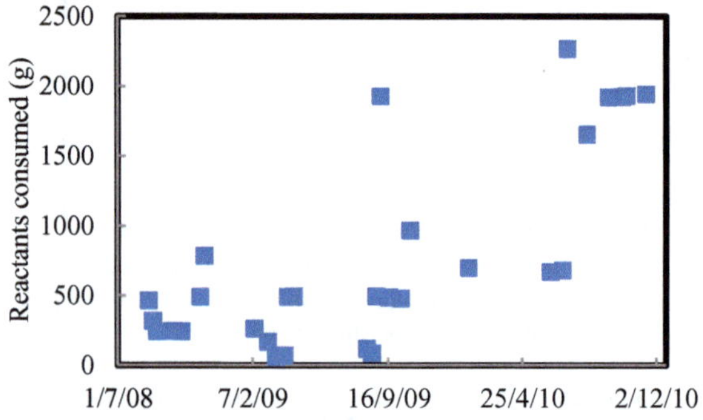


Figure 5: Consumption of reactants in each experiment (g) over main experimental program

The mass of magnesium produced in a batch may be compared with some data available in the literature. Hori *et al* [27, 28] report the consumption of around 1000 g of MgO and carbon, for two experiments listed. Odle and McClaine [26] report that 11 experiments 'processed about 90 to 285 grams of feed'. Engell *et al* [29], whose work may underpin Odle and McClaine's, give examples of consumption of 10-15 g reactants. Larger-scale work was undertaken by Cameron *et al* [30, 31], who reported feeding tens of kg per hour. By contrast, kinetic work by Sano's group [32, 33] operated at the mg scale. Nusheh *et al*'s recent work [34] was also at mg to µg scale. Donaldson and Cordes [12] did not provide data for magnesium experiments.

## Future Development

The development of a novel process for the production of a mass-produced material necessarily involves significant scale-up factors. The MagSonic™ process has proceeded from the gram scale to the kilogram (and from Mg concentrations below one percent to upwards of 25 %), yet significant scale-up factors remain. For example, from Figure 4 and Figure 5 one may deduce a laboratory productivity around 500 g Mg per hour. Full scale may be around 30,000 tonnes Mg per year, which at 80% utilization is around 4.3 tons per hour, or nearly 10,000 times laboratory scale.

The next step in the development of MagSonic™ technology is a pilot plant, for which CSIRO has developed a relatively detailed preliminary design and costing. Such a plant would aim to operate continuously and produce both purified magnesium for industry evaluation and good data for design and operation control. Sustained continuous operation would also validate the environmental claims of MagSonic™ [18], provide data on process robustness, and lead to further scale up and process development.

## Conclusion

CSIRO's MagSonic™ carbothermal magnesium technology has been under development in Australia since mid-2003. Recent advances have demonstrated that sustained operation can be achieved, and the laboratory equipment and experimental program have achieved concentration, reaction time, consumption, recovery, and purification targets. The technology is now ready to progress to the next stage, a continuous Pilot Plant.

## References

1. Abbott, T., *Why Choose Magnesium?* in *4th International Light Metals Technology Conference (LMT 2009)*. 2009. Gold Coast, AUSTRALIA: Trans Tech Publications Ltd.

2. Hansgirg, F., *Production of Metallic Magnesium*. 1932, US1884993.

3. Dungan, T. A., *Production of Magnesium by the Carbothermic Process at Permanente*. in *Transactions of the American Institute of Mining and Metallurgical Engineers*. 1944:(308-314).

4. Byrns, A. C., *Carbothermic Process for Magnesium at Permanente*. Chemical Engineering Progress, 1947. **43**(4): p. 172-173.

5. Dean, K. C., Edlund, V. E., and Lawrence, A. G., *Quenching Carbothermic Magnesium with Nitrogen*. Light Metal Age, 1972. **30**(5-6): p. 21-22.

6. Avery, J. M., *Process for Recovering Magnesium*. 1983, EP75836-A.

7. Avery, J. M., *Method for Producing Magnesium*. 1981, US4290804.

8. Eckert, C. A., Irwin, R. B., and Graves, C. W., *Liquid Metal Solvent Selection: The MgO Reduction Reaction*. Industrial & Engineering Chemistry, Process Design and Development, 1984. **23**(2): p. 210-217.

9. Brooks, G., Trang, S., Witt, P., Khan, M. N. H., and Nagle, M., *The Carbothermic Route to Magnesium*. JOM, 2006. **58**(5): p. 51-55.

10. Brooks, G., Nagle, M., Tassios, S., and Trang, S., *The Physical Chemistry of the Carbothermic Route to Magnesium*. in *Magnesium Technology 2006*. 2006. San Antonio, TX, United States: Minerals, Metals and Materials Society, Warrendale, PA 15086, United States (25-31).

11. Tassios, S., Barton, T. R. D., Constanti-Carey, K. K., Nagle, M. W., and Prentice, L. H., *Manufacture of Metal E.G. Magnesium, Involves Performing Carbothermal Reduction of Metal Oxide, Preventing Reformation of Metal Oxide, and Cooling Stream Using Nozzle Heated with Unit Other Than Gas under Specific Condition*. 2010, WO2010012042-A1.

12. Donaldson, A. and Cordes, R. A., *Rapid Plasma Quenching for the Production of Ultrafine Metal and Ceramic Powders*. JOM, 2005. **57**(4): p. 58-63.

13. Prentice, L., Nagle, M., and Constanti-Carey, K., *Impurities in the Carbothermal Production of Magnesium: To 1500 °C*. in *High Temperature Processing Symposium*. 2009. Swinburne University, Hawthorn, Australia.

14. Prentice, L. and Nagle, M., *Mechanism and Kinetics of Reduction of Magnesium Oxide with Carbon*. in *Magnesium Technology 2009*. 2009. San Francisco, CA: The Minerals, Metals, and Materials Society (35-39).

15. Prentice, L. H., *Psuedo-Steady-State Control of High Temperature Gas-Solid Reaction*. in *Chemeca2011*. 2011. Sydney, Australia: Engineers Australia.

16. Bohnet, M. and Lorenz, T., *Separation Efficiency and Pressure Drop of Cyclones at High Temperatures*, in *Gas Cleaning at High Temperatures*, R. Clift and J.P.K. Seville, Editors. 1993, Blackie Academic and Professional: London, UK. p. 17-31.

17. Bohnet, M., *Influence of the Gas Temperature on the Separation Efficiency of Aerocyclones*. Chemical Engineering and Processing, 1995. **34**(3): p. 151-156.

18. Prentice, L., Wai Poi, N., and Haque, N., *Life Cycle Assessment of Carbothermal Production of Magnesium in Australia*, in *IMA 67th Annual World Magnesium Conference*. 2010, International Magnesium Association: Hong Kong, PRC. p. 77-82.

19. Prentice, L. H., *A Kinetic Model for the Carbothermal Production of Magnesium*. 2009, CSIRO Process Science and Engineering: Clayton, VIC (Internal Report).

20. *Permanente Squeaks Through*, in *Time*. February 8, 1943.

21. Yuasa, S., Kawashima, M., and Sakurai, T., *Spontaneous Ignition of Ultra-Fine Magnesium Powder without an Original Oxide Coat at Room Temperature in $O_2/N_2$ Mixture Streams*. Proceedings of the Combustion Institute, 2009. **32**(2): p. 1929-1936.

22. National Fire Protection Authority, *NFPA 69 - Standard on Explosion Prevention Systems, 2008 Edition*. 2007, NFPA: Quincy, MA.

23. National Fire Protection Authority, *NFPA 484 - Standard for Combustible Metals, 2009 Edition*. 2008, NFPA.

24. Revel, G., Pastol, J.-L., Rouchard, J.-C., and Fromageau, R., *Purification of Magnesium by Vacuum Distillation*. Metallurgical Transactions B, 1978. **9B**(December): p. 665-672.

25. Hideo, T., Noboru, K., Gotou, T., Akiyoshi, K., and Yuuji, K., *Manufacture of Metallic Magnesium*. 1979, JP54130413.

26. Odle, R. R. and McClaine, A. W., *Economic Evaluation of a Nozzle-Based Carbothermal Magnesium Process*. 2007, Metallurgical Viability Inc: Elkton, MD.

27. Hori, F., *Apparatus for Obtaining Mg and Ca through Carbon Reduction*. 1980, US4200264.

28. Hori, F., *Method for Obtaining Mg and Ca through Carbon Reduction*. 1979, US4147534.

29. Engell, J., Frederiksen, J., and Nielsen, K. A., *Method of Producing Metallic Magnesium, Magnesium Oxide, or a Refractory Material*. 1998, US5803947.

30. Warren, G. F. and Cameron, A. M., *Process for Producing Magnesium*. 1985, EP0146986.

31. Cameron, A. M., Lotens, J. P., Ouwehand, C., and Aurich, V. G., *Carbothermic Production of Magnesium*, in *Pyrometallurgy '87*. 1987, The Institution of Mining and Metallurgy: London, UK. p. 195-222.

32. Hong, L., Sohn, H. Y., and Sano, M., *Kinetics of Carbothermic Reduction of Magnesia and Zinc Oxide by Thermogravimetric Analysis Technique*. Scandinavian Journal of Metallurgy, 2003. **32**(3): p. 171-176.

33. Li, R., Wei, P., and Sano, M., *Kinetics and Mechanism of Carbothermic Reduction of Magnesia*. Metallurgical and Materials Transactions B: Process Metallurgy and Materials Processing Science, 2003. **34**(4): p. 433-437.

34. Nusheh, M., Yoozbashizadeh, H., Askari, M., Kuwata, N., Kawamura, J., Kano, J., Saito, F., Kobatake, H., and Fukuyama, H., *Effect of Mechanical Milling on Carbothermic Reduction of Magnesia*. ISIJ International, 2010. **50**(5): p. 668-672.

**Magnesium Technology 2012**
*Edited by: Suveen N. Mathaudhu, Wim H. Sillekens, Neale R. Neelameggham, and Norbert Hort*
*TMS (The Minerals, Metals & Materials Society), 2012*

# MAGSONIC™ CARBOTHERMAL TECHNOLOGY COMPARED WITH THE ELECTROLYTIC AND PIDGEON PROCESSES

Leon H. Prentice[1]; Nawshad Haque[1]
[1]CSIRO Process Science and Engineering; Box 312; Clayton South, VIC, 3169, Australia

Keywords: Carbothermal Magnesium, Carbothermic Reduction, Pidgeon Process, Silicothermic Magnesium, Electrolysis, Technology Comparison, Techno-Economic Evaluation, Environmental Impact, Life Cycle Analysis

## Abstract

A broad technology comparison of carbothermal magnesium production with present technologies has not been previously presented. In this paper a comparative analysis of CSIRO's MagSonic™ process is made with the electrolytic and Pidgeon processes. The comparison covers energy intensity (GJ/tonne Mg), labor intensity (person-hours/tonne Mg), capital intensity (USD/tonne annual Mg installed capacity), and Global Warming Potential (GWP, tonnes $CO_2$-equivalent/tonne Mg). Carbothermal technology is advantageous on all measures except capital intensity (where it is roughly twice the capital cost of a similarly-sized Pidgeon plant). Carbothermal and electrolytic production can have comparatively low environmental impacts, with typical emissions one-sixth those of the Pidgeon process. Despite recent progress, the Pidgeon process depends upon abundant energy and labor combined with few environmental constraints. Pressure is expected to increase on environmental constraints and labor and energy costs over the coming decade. Carbothermal reduction technology appears to be competitive for future production.

## Introduction

Magnesium is the lightest of the structural metals and has significant potential as an important material for future manufactured products, most particularly in the automotive industry for the production of light-weight and more fuel-efficient vehicles. However, difficulties in extracting magnesium from its sources – whether brines, dolomite, magnesite, fly ash, etcetera – have so far limited its production and made it more expensive than other metals, most notably aluminum. Furthermore, present methods of production are generally perceived as excessively capital-intensive or environmentally damaging [1, 2].

A new method of production, based on carbothermal reduction technology with supersonic quenching, has been developed by Australia's Commonwealth Scientific and Industrial Research Organisation (CSIRO). This technology is approaching the point of significant scale-up and commercialization, and promises advantages over existing technologies. Process development over the last 2-3 years has produced the information required for a meaningful comparison with existing technologies, although further proof and refinement of the data will come with increased development.

In this study the MagSonic™ process is compared with the electrolytic and Pidgeon (Silicothermic) processes. Comparisons may be made on the projected energy intensity, labor intensity, capital intensity, and Global Warming Potential. Other technologies, either historical or developmental, are briefly discussed, but no formal comparisons are made. While these are not the only useful indicators of comparison, they can be assessors of the potential new competitor in global production.

## Magnesium Production Technologies

### Electrolytic Production

Historically, the electrolytic process was dominant from the start of commercial production up until the 1990s. Significant production facilities were constructed around the world, the largest of which was Dow's plant at Freeport, TX. Presently, magnesium is produced using the electrolytic process in the United States, Russia, Ukraine, and Israel, with potential start-up in China.

While there are a number of variations of the electrolytic process, all now electrolyze anhydrous magnesium chloride or carnallite. The overall reaction is:

$$MgCl_{2(l)} \rightarrow Mg_{(l)} + Cl_{2(g)} \qquad (1)$$

In the carnallite process, the potassium chloride (KCl) is chemically a spectator, and is recycled. The Dow process electrolyzed the dihydrate $MgCl_2.2H_2O$; the waters of hydration were 'fizzed off' in the electrolysis cell [3].

High purity molten Mg is tapped from the cells and cast into ingots. The details of the electrolytic process have been described in detail elsewhere [3-5].

### Pidgeon (silicothermic) Production

The Pidgeon Process is named after Lloyd Montgomery Pidgeon, who developed the process in the 1940s [6-8]. The process reduces magnesia (in the form of calcined dolomite or 'dolime', $CaO.MgO$) with silicon (in the form of ferrosilicon, $FeSi$). Various species are essentially spectators, but the overall reaction can be described by:

$$2(CaO.MgO)_{(s)} + FeSi_{(s)} \rightarrow 2Mg_{(g)} + 2CaO.SiO_{2(s)} + Fe_{(s)} \qquad (2)$$

The production of magnesium is by batch reaction of briquettes (often with a flux such as $CaF_2$ added), with approximately 40 to 70 kg of magnesium produced from each retort over its 5-10 h cycle. Various fuels may be used to provide heat – the most common fuels are brown coal, natural or coal gas, and electricity from coal or hydroelectric power. High-purity 'crowns' are produced, which are then re-melted to produce Mg ingots.

### Carbothermal (MagSonic™) Production

In carbothermal production, magnesia is reduced directly with carbon in the high-temperature gas-solid reaction:

$$MgO_{(s)} + C_{(s)} \leftrightarrow Mg_{(v)} + CO_{(g)} \qquad (3)$$

The process has been described comprehensively by Brooks *et al* [9, 10]. The reverse arrow in the reaction above indicates that the back-reaction is rapid and spontaneous upon cooling; such reversion may be minimized by very rapid cooling. CSIRO's MagSonic™ process, under development in Australia since 2003, achieves minimal reversion by cooling the reaction products (gases) at up to $1x10^6$ °C.$s^{-1}$ by supersonic acceleration through a Laval nozzle. The magnesium powder is purified and cast into ingots, while the carbon monoxide may be burnt to recover additional energy. The simplicity of direct reaction with carbon is apparent – carbon is plentiful, cheap, and potentially renewable (e.g. biochar) – and the only significant process waste is stoichiometric .

Other Production Methods

Bolzano. The Bolzano process has been modified and expanded and is now operated by Rima in Brazil, producing around 22,000 tonnes Mg per year [11]. It uses Pidgeon chemistry, but in extremely large vertical retorts, producing up to 2,000 kg per ten-hour cycle [11]. While a formal comparison with the Bolzano process is not part of the present work, Rima's claim to be a 'green producer' is supported by Cherubini *et al*'s analysis [12] of an energy intensity (Gross Energy Requirement) around 140 GJ/tonne Mg – half the electrolytic figure and around 60% lower than the Pidgeon process. A low environmental impact is maintained by the use of charcoal and hydroelectric power [11].

Magnétherm. The Magnétherm process – a continuous variation of the Pidgeon chemistry – is no longer in use. A similar process developed by Mintek [13, 14], operating at higher pressure, appears to have not been commercialized. However, Cherubini *et al*'s [12] analysis indicated that the energy intensity of the Magnétherm process was between the Bolzano and Electrolytic processes, at around 233 GJ/tonne Mg.

Solid Oxide Membrane (SOM). The SOM process has been under development by Boston University for some time, and has reported some experimental and scale-up success [15-17]. It is based on electrolysis of MgO in a fluoride-based ionic flux. Das [18] reported in 2008 that the SOM team had achieved up to 100 $g.day^{-1}$ of Mg production. However, more recent work [17] discusses issues in membrane stability, and the future development of the process is uncertain. While Das [18] reported a very attractive commercial parameters – cash operating cost around USD 0.53 per pound, capital intensity around USD 1,345 per tonne installed capacity, and labor intensity only 1.35 person-hours per tonne – the small scale and development uncertainty in the SOM process has excluded it from the present comparison.

## Technology Comparison

Energy Intensity

The most recent data presented on Chinese magnesium production are from the 2011 IMA (International Magnesium Association) meeting, reported by Brown [11]. He reports Chinese Magnesium Association (CMA) data of average energy requirements 'around 5 tce/t' (tonnes of coal equivalents per tonne magnesium) and $CO_2$ emissions of '~12-17' tonnes/tonne Mg. (See below for discussion on emissions.) Coal's calorific value is around 23 MJ/kg [19], hence CMA's energy intensity estimate is around 115 GJ/tonne Mg. Also from 2011, Feng *et al* [2] report that 'the comprehensive energy consumption of 1 [tonne] of primary magnesium... decreases from 360 GJ/t Mg to 297 GJ/t Mg between the year 2005 and 2009.' These figures compare with Du *et al*'s 280 GJ/tonne Mg [19], Cherubini *et al*'s 366 [12], and Ramakrishnan and Koltun's 354.5 [20]. Although detail of CMA's study has not been investigated by the present authors, it appears that the CMA has focused exclusively on the Mg reduction step, and their figure is therefore optimistic. Du *et al* [19], for example, do not take into account *eSi* production, and hence arrive at a low estimate. (Ramakrishnan and Koltun [20] estimate *eSi* production as requiring 113.5 GJ/tonne Mg. They also estimate that the energy intensity of the reduction step alone is 181.4 GJ/tonne Mg, which makes CMA's figure reasonable only for this part of the process.) The true overall figure likely remains around two to three times CMA's figure: Feng *et al* [2] are likely accurate at around 300 GJ/tonne Mg. Given present uncertainty and variations in process details, this figure may be only accurate to within 20%.

The gross energy requirement of the electrolytic process has been described in detail previously [12, 20-22]. Present research into further developing the electrolytic process is limited; developments remain either proprietary or focus on existing process routes [23, 24]. Technology advancement is unlikely to further reduce the estimate of around 270 GJ per tonne Mg. The estimate may be ±10%.

Data presented by the present authors in 2010 [21] disclosed that MagSonic™ carbothermal reduction is expected to require around 56 GJ/tonne Mg. Around 60% of the energy required is for the reaction step, for which an electrically-powered reactor has been estimated at 50% efficiency; the remainder is mostly vacuum systems, calcination, and purification. The data for the reactor are supported by laboratory operation of a 50 kW induction furnace over more than 200 experiments. The most significant advantages of the carbothermal route are the very simple feed preparation (calcination and briquetting), direct reaction, minimal wastes, and continuous operation. Given the present scale of the process, energy intensity estimates may be ±50%.

The energy intensity data are summarized in Table I below.

Table I: Energy Intensity Comparison

| Process | Energy Intensity (GJ per tonne Mg) |
|---|---|
| Pidgeon | 300 (±60) |
| Electrolytic | 270 (±27) |
| MagSonic™ | 56 (±28) |

Global Warming Potential

Foundational work on the life cycle analysis of the electrolytic (and Pidgeon process) was undertaken by Ramakrishnan and Koltun [20, 22, 25]. Their work was updated and expanded by various authors more recently [12, 19, 26]. These data were combined with the present authors' analysis in a comparison with the carbothermal MagSonic™ process in 2010 [21].

Overall, the data showed that the Pidgeon process in China had an environmental impact, measured in terms of Global Warming Potential (GWP), of around 42 tonnes $CO_2$-equivalent per tonne of Magnesium produced [12, 20]. (The data are obviously variable, and significant differences exist for different producers depending on the exact technologies used.) Ramakrishnan and Koltun's [20] estimated range is 37-47. Du *et al* [19] cite a figure of 27 tonnes $CO_2$-eq per tonne Mg, but do not include the impact of coal combustion, which may add between 16 and 40 to this figure, depending on efficiency and type of coal. The CMA's estimate of '~12-17' tonnes $CO_2$ per tonne Mg appears to focus solely on $CO_2$ emissions from the Mg reduction step, for which Ramakrishnan and Koltun [20] estimated 15.9 tonnes $CO_2$ *equivalents* per tonne. Feng *et al*'s data [2] appear more comprehensive, detailing a GWP of around 28 tonnes $CO_2$-eq per tonne Mg for the Pidgeon process in China in 2009. They report that the reduction from 2005 to 2009 is due primarily to improved efficiency, particularly in the 'coal gasification process and exhaust heat recovery in [the] reduction process.'

The average of Feng *et al*'s 2011 [2] data and 'base case' processes from 2005 leads to the overall comparison shown in Table II below. Variation in the Pidgeon process estimate is largely due to process variation, whereas the GWP of the electrolytic and MagSonic™ processes depends primarily on the source of electricity. With hydroelectric power, the MagSonic™ process can have an environmental impact as low as 6 tonnes $CO_2$-eq per tonne Mg.

Table II: Global Warming Impact Comparison

| Process | GWP (tonnes $CO_2$-eq per tonne Mg) |
|---|---|
| Pidgeon | 34 (±8) |
| Electrolytic | 17 (±8) |
| MagSonic™ | 14 (±8) |

Capital and Operating Cost

Data for capital and operating costs are tentative, variable, and highly dependent on location. While this paper reviews data from a number of sources, the final analysis is provided in US dollars for western construction (e.g. in Australia).

Only two Greenfield electrolytic plants have been built – or partially built – to the authors' knowledge, in the last two decades. Capital cost data are not readily available for the Magnola project [27], but data exist for the Australian AMC project, although it collapsed prior to completion. The funding estimate for that project was AUD 1.5bn for a plant to produce 90,000 tonnes Mg per year. Adjusting for inflation and using the long-term exchange rate of 0.8 USD/AUD leads to an installed capital cost around USD (2010) 16,667 per annual tonne Mg. This is similar to figures cited by Das [18]. Given paucity of data and time since estimates were made, an error estimate of 15% may be appropriate.

While the Pidgeon process (as presently operated) would be unfeasible for Australia – see below – it may be worthwhile to estimate its capital cost should such a plant be constructed. The capital cost of the CVM Minerals plant in Malaysia is a useful benchmark: its capital intensity is around USD 2,650 to 3,400 per annual tonne Mg [28]; a value somewhere around USD (2010) 3,250 may be reasonable for Australia. Comparable data on the Pidgeon process in China are difficult to obtain, but are likely lower. At the high end Das [18] also suggests USD (2010) 3,250. (One extremely low estimate is China Magnesium Corporation's Prospectus [29], which cites CNY 3,355 (approximately USD 500) per tonne installed capacity. This figure has not been included in the present analysis, which focuses on western production. Again, paucity of data suggests errors may be around ±20%.

Work by the present authors on the MagSonic™ carbothermal process results in a capital cost of around AUD 210m for a plant producing 30,000 tonnes Mg per year [30]. This equates to USD (2010) 5,600 per annual tonne Mg. The overall breakdown in cost is similar to that of Odle and McClaine [31], where around 25% of the capital cost is reactors, followed by the vacuum system (approximately 17%) and calcining plant (approximately 16%). The initial estimate is only to ±50% confidence.

There have also been literature analyses of the carbothermal process, all of which arrive at lower capital costs. Donaldson and Cordes [32] estimated an installed capital cost of USD (2005) 3,225 per tonne capacity, or USD (2010) 3,900. Odle and McClaine's 2007 study [31] estimated USD 4,500 (based on 90,000 tonnes per year capacity), or USD (2010) 5,040. Das [18] cites Odle an McClaine's data, but does not include contingencies and other costs, arriving at a capital cost of USD 221m for 90,000 tonnes Mg per year, or around USD (2010) 3,000 per tonne installed capacity. These estimates are included as literature values in Table III below.

Table III: Capital Intensity Comparison (western location, 2010)

| Process | Capital Intensity (USD per annual tonne Mg capacity) |
|---|---|
| Pidgeon | 3,250 (±650) |
| Electrolytic | 16,667 (±2,500) |
| MagSonic™ | 5,600 (±2,800) |
| Carbothermal (literature) | 4,000 (±1,000) |

Operating costs are even more location and process-specific, so no detailed comparison is presented here. However, indicative comments and results are discussed below.

Odle and McClaine's 2004 work [33] produced an economic model for the carbothermal process: "the model predicts operating costs between about $0.23 and $0.55 a pound". Adjusting for inflation leads to an estimate of USD (2010) 0.49 (±0.20) per pound. Their paper focused on the structure and rationale of the model, using limited experimental results. Their later study [31] stated a cash cost "as low as $0.30 per pound" (USD (2010) 0.34 per pound). These figures compare with Donaldson and Cordes' estimates [32] of cash cost of production around USD (2010) 0.51 per pound Mg. Excluding capital, Das [18] estimates a cash cost of production via carbothermal reduction around USD (2010) 0.70 per pound.

Developments in MagSonic™ technology also allow for an operating cost estimate, of USD (2010) 1.19 per pound Mg. This figure is obviously significantly higher than those suggested previously [18, 31]. Part of the variation is due to electricity, of which around 15.61 kWh per kg Mg is required [21]. Global electricity prices vary significantly, but assuming USD 0.10 per kWh, the cost of electricity alone is USD 0.71 per pound Mg. It is likely that the cost of energy contributes to the variation in estimates in the literature for the carbothermal operating cost.

The operating cost of the electrolytic process is likewise highly dependent on the cost of electricity, and no meaningful comparisons can be made. A nominal figure somewhere around USD (2010) 0.98 per pound [18] seems reasonable.

In contrast to the carbothermal and electrolytic processes, the Pidgeon process relies primarily on labor (and to a lesser extent ferrosilicon) for its operating cost. Das [18] estimates USD (2010) 0.98 per pound in China, compared with 1.09 in a western location.

Labor Intensity

The labor intensity of the Pidgeon process has been frequently acknowledged [18, 31, 34], but exact figures are rarely found in the literature. Estimates may be made from scale and staffing figures, where available. The 'best case' scenario may be the recent construction of the Perak Magnesium Smelter, operated by CVM Minerals in Malaysia. Their plant is reported to employ 398 people for 15,824 tonnes per year of production [28], which at 1600 person-hours per year per worker equates to approximately 41 person-hours per tonne of production. A 'worst case' scenario may be the 2010 prospectus from Australia's China Magnesium Corporation, which expects labor costs of approximately CNY 956 per tonne [29]. At USD 1.36 per hour [35], this equates to around 105 person-hours per tonne. This latter figure is likely to be more common for the established plants that contribute the bulk of production. An average figure of around 90 person-hours per tonne appears reasonable for the Pidgeon process in China, with an estimated 20% error.

By contrast, CSIRO's work on the MagSonic™ process estimates a labor requirement of 5.8 person-hours per tonne Mg. This is largely due to the simplicity of the process and its likely capability of continuous operation. Given the ongoing development of the process, error is estimated as ±50%. Labor requirements for the electrolytic process are difficult to obtain, but estimates based on the size of the process plant and employment figures for producers suggest around 9-15 person-hours per tonne Mg may be reasonable.

These data are summarized in Table IV below.

Table IV: Labor Intensity Comparison

| Process | Labor Intensity (person-hours per tonne Mg) |
|---|---|
| Pidgeon | 90 (±18) |
| Electrolytic | 12 (±3) |
| MagSonic™ | 6 (±3) |

It is evident from the above analysis that construction of a Pidgeon process plant with its present labor intensity requirements would be prohibitive in countries with medium to high labor costs. In Australia, for example, labor costs may be estimated at around USD (2011) 35 per person-hour [36], rendering the Pidgeon process uneconomic.

## Conclusion

Continued work by CSIRO on the MagSonic™ carbothermal production route for magnesium, combined with data available for competing technologies, has allowed for a comparison to be made. The results highlight the energy and environmental advantages of the carbothermal production of magnesium and indicate that the process is likely to be competitive on a cash operating cost basis. While its capital cost is likely higher than the Pidgeon process, labor intensity is very much lower. The lack of electrolytic construction is explained by its high capital cost, whereas the MagSonic™ process is likely to be competitive for Australian production.

## References


1. Brown, R. E., *Environmental hallenges for the Magnesium Industry*. in *Magnesium Technology 01* . 2011. San Deigo, CA: The Metals, Minerals, and Materials Society (TMS), 7-11.

2. Feng, G., Nie, Z., Wang, Z., Gong, X., and Zuo, T., *A Research on Energy-Saving and Environmental Impacts of Primary Magnesium and Magnesium Alloy Production in hina.* Materials Science Forum, 2011. **685** (Energy, Environment and Biological Materials): 152-160.

3. Lea, D., *Magnesium Extraction Processes Today*. 1982. Detroit, MI, Engl: Int Magnesium Assoc., 28-32.

4. Sivilotti, O., *Developments in Electrolytic ells for Magnesium Production.* Light Metal Age, 1984. **42**(7-8): 16-18.

5. Duhaime, P., Mercille, P., and Pineau, M., *Electrolytic Process Technologies for the Production of Primary Magnesium.* Transactions of the Institution of Mining and Metallurgy Section C-Mineral Processing and Extractive Metallurgy, 2002. **111**: p. C53-C55.

6. Pidgeon, L. M. and Alexander, W. A., *Thermal Production of Magnesium-Pilot-Plant Studies on the Retort errosilicon Process.* Transactions of the American Institute of Mining and Metallurgical Engineers, 1944. **159**: 315-352.

7. Pidgeon, L. M. and King, J. A., *The Vapour Pressure of Magnesium in the Thermal Reduction of Mgo by errosilicon.* Discussions of the Faraday Society, 1948(4): 197-206.

8. Pidgeon, L. M., Matsushima, T., and Thoburn, W., *Metallurgy of Magnesium Production by Thermal Methods.* JOM-Journal of Metals, 1964. **16**(1): 105ff.

9. Brooks, G., Nagle, M., Tassios, S., and Trang, S., *The Physical hemistry of the arbothermic Route to Magnesium*. in *Magnesium Technology 006*. 2006. San Antonio, TX, United States: Minerals, Metals and Materials Society, Warrendale, PA 15086, United States, 25-31.

10. Brooks, G., Trang, S., Witt, P., Khan, M. N. H., and Nagle, M., *The arbothermic Route to Magnesium.* JOM, 2006. **58**(5): 51-55.

11. Brown, R. E., *IMA Meeting in Prague and World Magnesium Update.* Light Metal Age, 2011. **69**(4): 50-52.

12. Cherubini, F., Raugei, M., and Ulgiati, S., *L A of Magnesium Production - Technological verview and Worldwide Estimation of Environmental Burdens.* Resources Conservation and Recycling, 2008. **52**(8-9): 1093-1100.

13. Abdellatif, M., *Mintek Themal Magnesium Process (MTMP): Theoretical and perational Aspects.* in *Southern African Pyrometallurgy 006.* 2006. Johannesburg: South African Institute of Mining and Metallurgy, 329-342.

14. Schoukens, A. F. S., Abdellatif, M., and Freeman, M., *Technological Breakthrough of the Mintek Thermal Magnesium Process.* in *Light Metals onference - Production of Magnesium, Aluminium and Titanium in the 1st entury.* 2005. Pretoria, South Africa.

15. DeLucas, R., Suput, M., Pati, S., and Pal, U. B., *Magnesium Extraction from Magnesium xide Using Solid xide Membrane Process.* 2007: Boston University.

16. Pal, U., *A Lower arbon ootprint Process for Production of Metals from Their xide Sources.* JOM, 2008. **60**(2): 43-47.

17. Gratz, E., Pati, S., Milshtein, A., Powell, A., and Pal, U., *Efficiency and Stability of Solid xide Membrane Electrolyzers for Magnesium Production.* in *Magnesium Technology 01 .* 2011. San Deigo, CA: The Metals, Minerals, and Materials Society (TMS).

18. Das, S., *Primary Magnesium Production osts for Automotive Applications.* JOM Journal of the Minerals, Metals and Materials Society, 2008. **60**(11): 63-69.

19. Du, J. D., Han, W. J., and Peng, Y. H., *Life ycle Greenhouse Gases, Energy and ost Assessment of Automobiles Using Magnesium from hinese Pidgeon Process.* Journal of Cleaner Production, 2010. **18**(2): 112-119.

20. Ramakrishnan, S. and Koltun, P., *Global Warming Impact of the Magnesium Produced in hina Using the Pidgeon Process.* Resources, Conservation and Recycling, 2004. **42**(1): 49-64.

21. Prentice, L., Wai Poi, N., and Haque, N., *Life ycle Assessment of arbothermal Production of Magnesium in Australia,* in *IMA 67th Annual World Magnesium onference.* 2010, International Magnesium Association: Hong Kong, PRC. 77-82.

22. Ramakrishnan, S. and Koltun, P., *A omparison of the Greenhouse Impacts of Magnesium Produced by Electrolytic and Pidgeon Processes.* in *Magnesium Technology 004.* 2004. Charlotte, NC., United States: Minerals, Metals and Materials Society.

23. Long, G., Ma, P., and Chu, M., *Ammoniation Technology for Dehydration of Hydrated Magnesium hloride in hina.* Materials Science Forum, 2005. **488-489** (Magnesium - Science, Technology and Applications): 95-98.

24. Yakovleva, G. A. and Minina, R. G., *Influence of luoride Additives in Magnesium Electrolytes.* Tsvetnye Metally, 2005(12): 69-72.

25. Koltun, P., Tharumarajah, A., and Ramakrishnan, S., *Lifecycle Environmental Impact of Magnesium Automotive omponents.* JOM, 2004. **56**(11): p. 31.

26. Feng, G., Nie, Z., Wang, Z., Gong, X., and Zuo, T., *Life ycle Assessment of Primary Magnesium Production Using the Pidgeon Process in hina.* The International Journal of Life Cycle Assessment, 2009. **14**(5): p. 480-489.

27. Celik, C., Peacey, J., Bishop, G., White, C., Giasson, E., and Avedesian, M., *Magnola - an Innovative Process for Magnesium Production.* in *3rd International Magnesium onference.* 1996. Manchester, England: Inst Materials.

28. CVM Minerals, *Placing and Public ffer.* 2008, Hong Kong: CVM Minerals Ltd. 562.

29. CMC, *Prospectus.* 2010: China Magnesium Corporation Ltd.

30. Prentice, L. H. and Haque, N., *Techno-Economic Evaluation of the arbothermal Production of Magnesium in Australia.* 2010, CSIRO Process Science and Engineering: Clayton, VIC (Internal Report).

31. Odle, R. R. and McClaine, A. W., *Economic Evaluation of a Nozzle-Based arbothermal Magnesium Process.* 2007, Metallurgical Viability Inc: Elkton, MD.

32. Donaldson, A. and Cordes, R. A., *Rapid Plasma Quenching for the Production of Ultrafine Metal and eramic Powders.* JOM, 2005. **57**(4): p. 58-63.

33. Odle, R. R. and McClaine, A. W., *Economic Evaluation of a Nozzle-Based arbothermal Magnesium Process.* 2004, Metallurgical Viability Inc: Newark, DE.

34. Wang, X., Zang, J., and He, J., *Recent Technological Advancement and Developmental Trend of Pidgeon Process in Smelting Magnesium in hina.* 2005. Beijing, China: Trans Tech Publications Ltd.

35. Banister, J. and Cook, G., *hina's Employment and ompensation osts in Manufacturing through 008.* Monthly Labor Review, 2011. **134**(3): 39-52.

36. Australian Bureau of Statistics, *Average Weekly Earnings, Australia, May 011.* 2011 18 August 2011 [cited 2011 14 November]; Available from: http://www.abs.gov.au/AUSSTATS/abs@.nsf/Lookup/6302.0Main+Features1May%202011?OpenDocument.

**Magnesium Technology 2012**
*Edited by: Suveen N. Mathaudhu, Wim H. Sillekens, Neale R. Neelameggham, and Norbert Hort*
***TMS (The Minerals, Metals & Materials Society), 2012***

# SCALING-UP SOLID OXIDE MEMBRANE ELECTROLYSIS TECHNOLOGY FOR MAGNESIUM PRODUCTION

Soobhankar Pati[1], Adam Powell[1], Steve Tucker[1], Steve Derezinski[1]
[1]Metal Oxygen Separation Technologies; 11 Michigan Drive, Natick, MA, 02114, USA

Keywords: Electrolysis, Zirconia, Oxygen, Inert Anode

## Abstract

Metal Oxygen Separation Technologies, Inc. (MOxST) is actively developing Solid Oxide Membrane (SOM) electrolysis technology for production of magnesium directly from its oxide. The vital component of this technology is the oxygen ion-conducting solid zirconia electrolyte separating the molten flux (a mixture of salts and oxide) and the inert anode. The zirconia not only protects the anode from the flux but also prevents anode gas back-reaction, increasing the efficiency. This makes it possible to produce low-cost high-purity magnesium and high-purity oxygen as a by-product with no direct greenhouse gas emissions. In this paper we discuss the design modifications made to address the scaling-up challenges, particularly for producing magnesium in liquid form. The key accomplishment to date is the successful development of a prototype capable of producing few kilograms of magnesium per day. We will also describe the prerequisite properties of an inert anode and suitable materials for the same.

## Introduction

Automakers are keenly looking for appropriate lightweight materials to increase fuel efficiency and decrease greenhouse gas emission. It is estimated that the efficiency of automobiles can be increased ~1% by reducing the automobile mass by 22.5kg (1). Magnesium with a density of 1.74 g/cm$^3$, about 30% and 75 % less than aluminum and steel respectively, is a suitable material for replacing steel and aluminum. Along with its low density, its high stiffness-to-weight ratio makes it an ideal candidate to replace some the structural components of automobiles without compromising safety. It is also potentially the most easily recycled of the proposed light weight vehicle materials, which include polymers and composites. Further benefits of magnesium include lower manufacturing cost and reduced noise, vibration, and harshness. For these reasons, automakers would like to replace 350 kg of steel and aluminum alloy parts per vehicle with 160 kg of magnesium equivalents by the year 2020. This would improve fleet fuel economy by about 1.5 mpg and reduce petroleum demand by over $20 billion per year (2).

Currently, automobiles are made with only few magnesium components, such as bumper support beams, steering support column, and seat/channel guide etc., averaging ~6 kg/car. However, the amount of magnesium used in automobiles is projected to increase with additional applications such as the engine cradle, sub-frame, and body components. As mentioned before, around 160 kg of magnesium is expected to be used in a typical automobile. This will create a huge demand for magnesium. In 2010 approximately 60 million automobiles were produced. If the magnesium content of autos were to increase to 160kg per vehicle, the auto industry will require 9.6 million metric tons. This is almost 10 times the 2010 primary magnesium production. If automakers are to meet these goals, primary magnesium producers will have to scale up their production capacity considerably. This new demand will mandate novel, energy-efficient and disruptive primary magnesium production technologies as the present technologies are very inefficient and not environmentally benign (3).

Solid Oxide Membrane (SOM) electrolysis is such a novel, efficient technology that can produce magnesium at fraction of its present cost (4). The SOM electrolyzer consists of a one-end closed, dense, cubic-phase stabilized zirconia tube immersed in a molten salt bath. It is operated by feeding magnesium oxide into the molten salt bath and applying potential across the cathode, immersed in the molten salt bath and the anode, encased inside the one-end closed zirconia tube. The electricity splits magnesium oxide to produce magnesium metal vapor at the cathode and oxygen gas at the anode. A detail description of the process is given elsewhere (4-5).

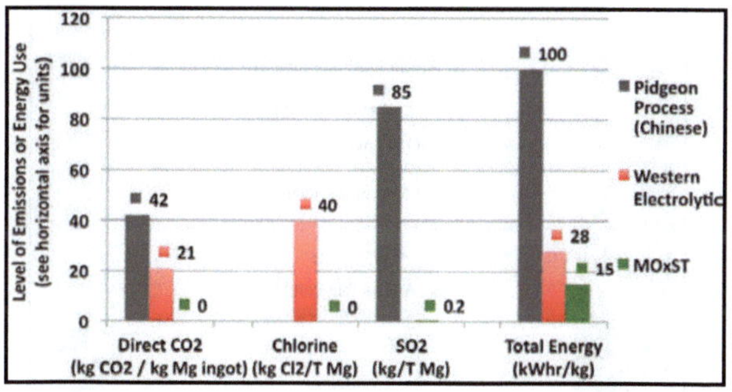


Figure 1. Energy usage and greenhouse gas emission of different primary magnesium production technologies.

Figure 1 shows comparison of the energy usage and greenhouse gas emission of different primary magnesium production technologies (3, 6). The SOM process generates minimal waste compared to other processes. The amount of energy required for per kg is also considerably lower compared to the other processes. It can be operated in modular electrolysis cells, enabling small production facilities at very low capital cost, minimizing startup risks.

In this paper we discuss the recent developments towards scaling up the SOM electrolysis process from producing a few grams per day to a few kilograms per day of magnesium. Nearly all the previous studies report on single-tube SOM electrolyzers. Here we present the results of a scaled-up SOM electrolyzer with three zirconia tubes. We compare results of a three-tube and single-tube

SOM electrolyzer. It is very critical to understand the modularity before scaling up SOM electrolyzer technology. Also, to date SOM electrolysis technology for magnesium production used a reducing environment at the anode. This enabled the operation of the SOM electrolyzer with a metallic current collector at the anode. However, in a commercial system where oxygen is produced at the anode, an oxygen stable current collector is needed for continuous, uninterrupted operation. Towards that, this paper presents some preliminary studies on a suitable design and materials system.

## Design and Experiments

*Set-up*- Figure 2 shows photographs of fabricated welded stainless steel crucible-condenser reaction vessel assemblies for one and three-tube experiments. Except for one experiment where we used 5/16" SS-304 tubes as isolated cathodes, the vessel containing the molten salt served as the cathode for both single-tube and three-tube electrolyzers. The 1" outer diameter (OD) and 7/8" inner diameter (ID) stainless tubes rise from the crucible to the retort top plate surrounding the zirconia SOM tubes and sealing them at the top plate. We used ¼" ID stainless tubes descending from the retort top plate down into the crucible to stir the molten flux electrolyte with argon. The MgO is fed through the 3/8" OD 1/4" ID stainless tube rising from the crucible top to the retort top plate. The argon and magnesium vapor in the crucible was carried to the condenser through the 1/4" OD tubes connecting crucible to the condenser. Since the condenser is at a cooler temperature than the crucible, the magnesium vapor condensed on the wall of the condenser and the argon gas exited through the 1/4" tubes connected to condenser.

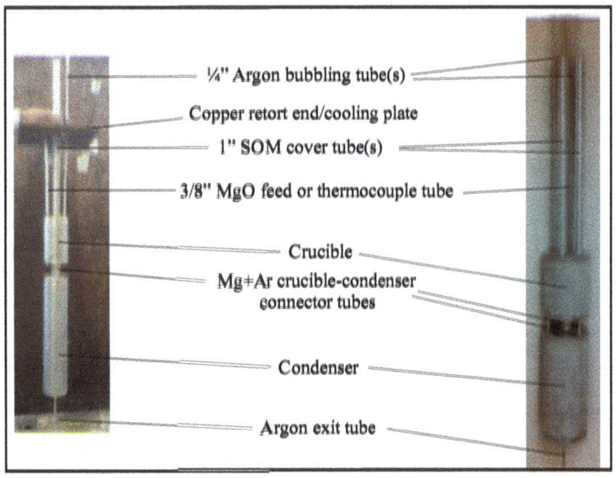


Figure 2. Stainless steel welded reaction vessel for SOM Electrolysis: one-tube vessel on left, three-tube vessel on right.

We used 10 wt% MgO content in the molten salt mixture for all of the electrolysis experiments. The molten salt was composed of either 50wt % $CaF_2$-$MgF_2$ or 50wt % $CaF_2$-$MgF_2$ with a required amount of $YF_3$. We used commercially purchased yttria stabilized zirconia (YSZ) tubes (OD: 0.75", ID: 0.5") for the experiments. The $MgF_2$, $CaF_2$, $YF_3$ and MgO were purchased from Alfa Aesar. We heated the SOM electrolyzer in an inert, low oxygen environment to prevent the weld joint failure from oxidization. We maintained the inert environment by heating the SOM electrolysis cell inside a mullite or a quartz tube and flushed with argon into the tube. The water cooled aluminum flanges sealed with silicone O-rings maintain the inert environment inside the quartz tube.

*Current Collector*- In all of the previous studies liquid copper or tin with a carbon rod current collector or a liquid tin with molybdenum tube current collector were used as anodes (4,5,7). In the first case carbon acted as a reductant while in the second case hydrogen gas was supplied through the molybdenum tube as a reductant. However, this only works if we are not interested in producing oxygen at the anode. When oxygen is generated at the SOM electrolyzer, the anode is expected to behave as the cathode of solid oxide fuel cell (SOFC). Therefore, a perovskite such as lanthanum strontium manganite (LSM), which is a SOFC cathode material, is an obvious choice. However, since the operating temperature of the SOM electrolyzer is around 1150°C, it is expected that the LSM will sinter within a short duration and may delaminate if operated at higher current density.

One way to get around this problem would be to use a liquid metal that is stable in an oxygen environment. Liquid silver is one such promising candidate. Figure 3 (left) shows the assembly with a liquid metal anode and current collector immersed in it. The current collector, as indicated in Figure 3, must provide an electrical connection from the anode to the power supply, while functioning in a pure oxygen environment from room temperature to the operating temperature. Also it should have an extremely low solubility in the liquid anode. Iridium satisfies all of these requirements, but is inordinately expensive.

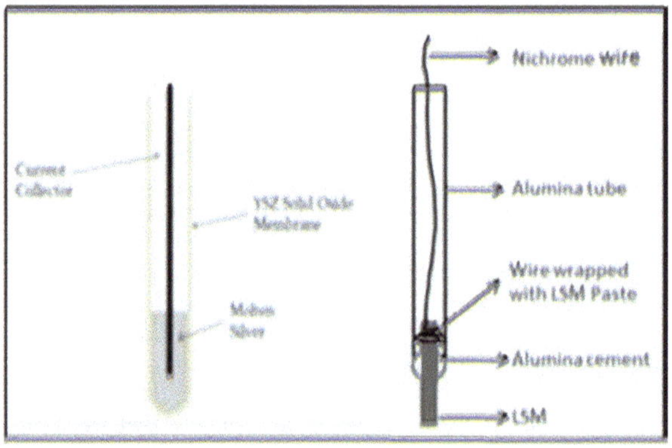


Figure 3. Assembly with liquid metal anode and current collector (left). LSM/ Nichrome current collector (right).

It is also possible to use metals that form a thin protective oxide scale which can conduct electricity. Therefore, we tested nichrome and alumel for some experiments expecting they would form a thin protective oxide layer which might be conductive. Figure 3 (right) shows the design of another current collector used for some experiments. It consists of a ceramic electronically conducting oxide like, LSM immersed in the liquid silver. The LSM is attached to a nichrome wire and the nichrome wire is protected by an alumina tube and alumina cement as shown in Figure 3 (right).

## Results and Discussion

Figure 4 shows a potentiodynamic scan before electrolysis when nichrome current collector was used in the anode of a single-tube electrolyzer. The current density was around 0.7 $A/cm^2$ at 4 V.

The scan shows a reasonably clean sweep with very low current below 1.5 V. The change in slope at 1.5 V is due to dissociation of MgO and formation of metal at the cathode. This is a representative scan for most experiments when a reductant is used at the anode. It is hypothesized that the nichrome oxidizes at the anode to lower dissociation potential of MgO, which would be around 2.3 V when pure oxygen is generated at the anode.

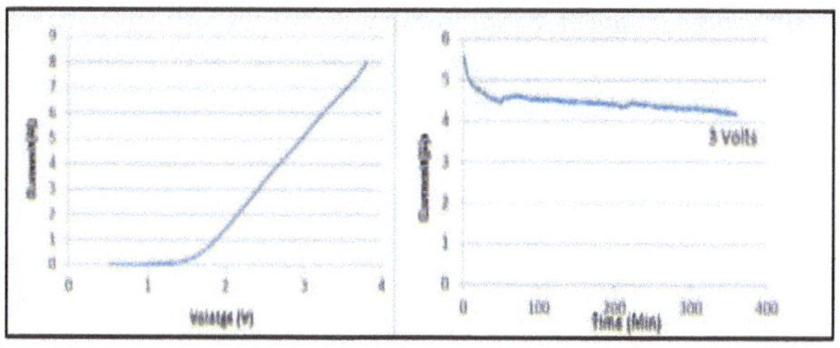


Figure 4. Potentiodynamic scan prior to electrolysis (left), electrolysis at 3 V for 6 hrs (right).

Figure 4 (right) shows the current-time plot during electrolysis. It shows a steady decrease in current with time. This is most likely due to decrease in MgO content in the flux. Figure 5 shows the 304 stainless steel shim which served as a condenser liner, with magnesium deposits. Due to the temperature gradient of the furnace heating zone the top of the shim was at higher temperature than the bottom of the shim. Therefore, the undercooling is greater at the bottom of the shim and the magnesium vapor deposited as fine flaky particles. The metal which condensed at the top of shim collected as liquid magnesium globules (Figure 5), indicating that it is possible to produce liquid magnesium in a SOM electrolysis process. However, proper condenser design is vital to produce a fully-liquid magnesium product.

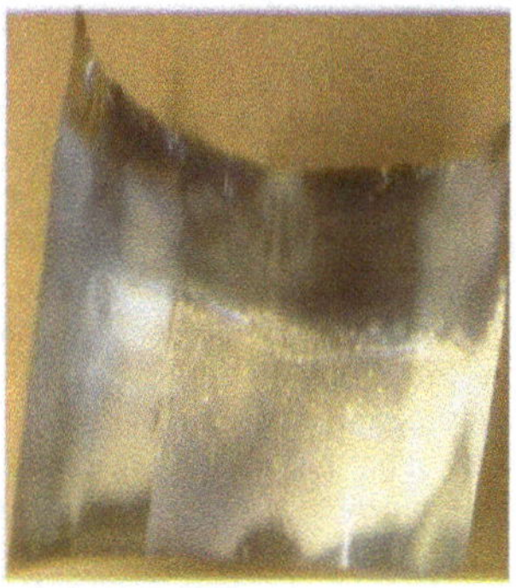

Figure 5. Metal deposit on stainless shim (left.

In order to explore the process efficiency, we used gas fittings to connect the oxygen output to an oxygen sensor and flow meter. Figure 6 shows the current measured, oxygen output of the SOM tube during three brief electrolysis runs experiment, scaled such that they align under Faradaic conditions, i.e. perfect current efficiency. During the first electrolysis run, oxygen output indicates perfect current efficiency after a brief delay, which we believe to be due to oxygen buildup in the liquid silver anode. During the second and third, however, current efficiency is somewhat lower. One possible reason for change in current efficiency may be electronic conduction in the flux. During electrolysis, the magnesium metal product has a tendency to dissolve in the flux, especially when it forms at atmospheric pressure away from argon.

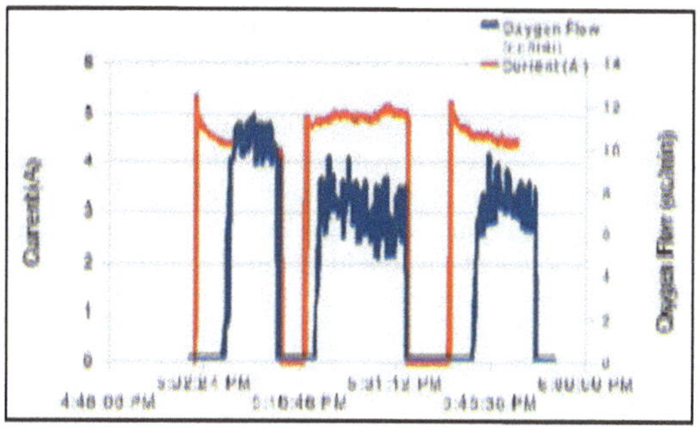

Figure 6. Current and scaled oxygen output during electrolysis.

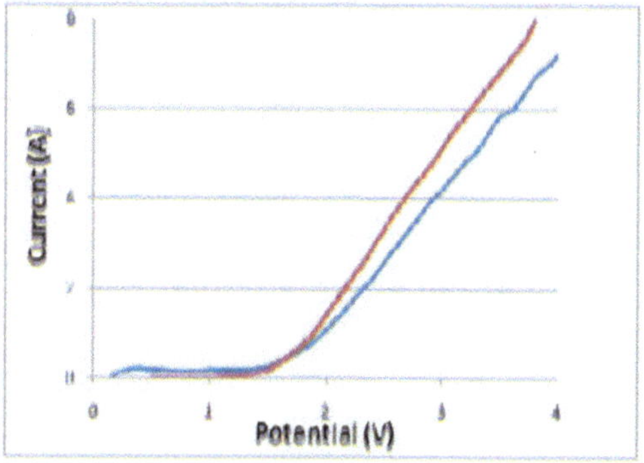


Figure 7. Potentiodynamic scans before (red) and after (blue) electrolysis.

Figure 7 compares potentiodynamic scans before and after electrolysis. The potentiodynamic scan after electrolysis (blue line) shows onset of a small "leakage current" prior to dissociation potential of MgO. This leakage current lowers the current efficiency. Eric et. al., in the abstract titled "Effect of Electronic Current on the SOM Process for Magnesium Production" to be presented in Electrometallurgy 2012 at TMS 2012, give a detail explanation of the origin of the leakage current and how it can be eliminated.

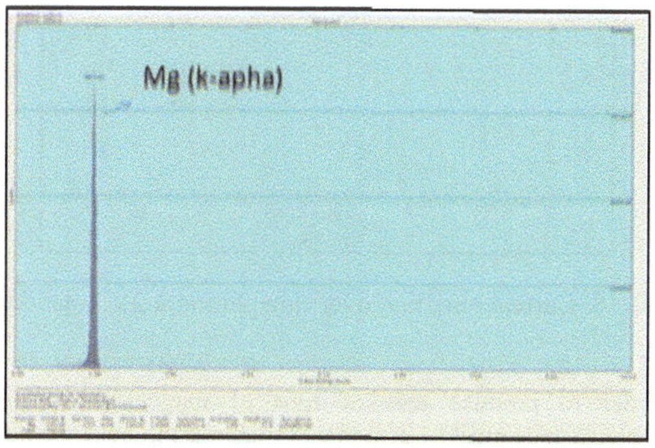


Figure 8. EDAX spectrum of magnesium deposit, showing no detectable impurities.

Figure 8 shows the energy-dispersive X-ray (EDAX) spectrum of the magnesium deposit from this experiment. EDAX can detect elements at 0.1% concentration, so the spectrum and analysis indicate that this magnesium is at least 99.99% pure. This is because this process features in-line distillation, enabling it to make high-purity magnesium from moderate-purity MgO.

Figure 9 shows the typical current-potential characteristics of the three-tube SOM electrolyzer with similar immersion depth as the single-tube experiment. The linear portion of the curve intersects the horizontal axis at about 1.5 V similar to single-tube electrolyzer experiments. It should be noted at any given potential the current is approximately 3 times of the current in a single-tube electrolyzer.

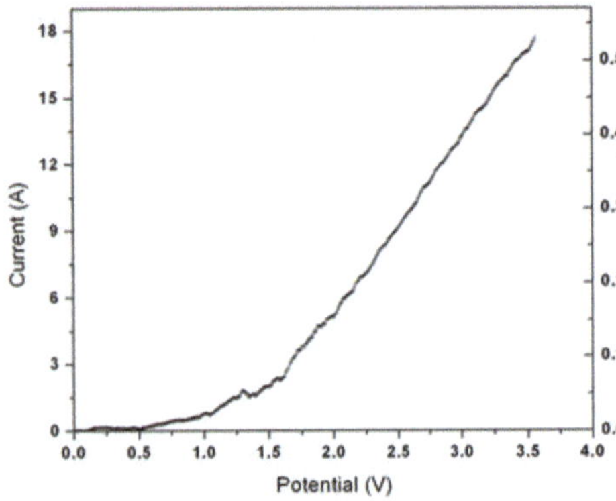


Figure 9. Potentiodynamic scan of the 3-tube apparatus, showing both total current and current density.

After this characterization step, the apparatus ran for one hour at a fixed potential of 3 V to produce magnesium. Figure 10 shows the current during this potentiostatic hold. Current begins at 12A and then decreases to around 9.5A over about 10 minutes as oxide ions are depleted from the zirconia/salt boundary layer. The 9 A current level then slowly falls as the oxide ion concentration in the flux decreases over time.

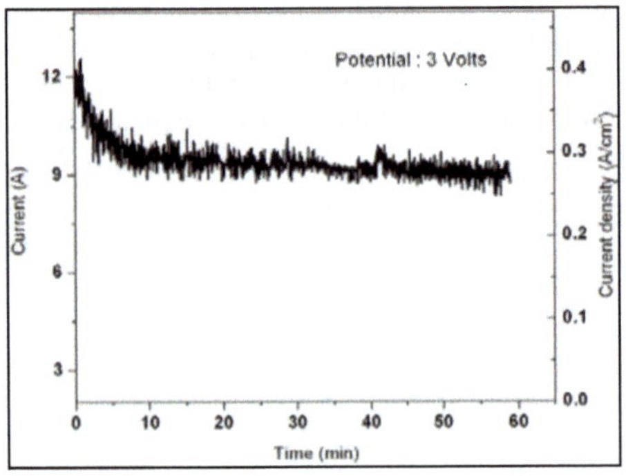


Figure 10. Current variation over time during a 3V potentiostatic hold.

In order to produce tens of tons of magnesium per year, electrolysis cells must be scaled up by increasing both the number of zirconia tubes and the immersion depth of each tube in the molten salt. This will increase the electrode area and total current. We conducted a single-tube deep-immersion experiment. Around 3.5" of the tube was immersed in molten salt during the experiment. Figure 11 shows the comparison of the potentiodynamic scans of a typical shallow-immersion experiment and a single-tube deep-immersion experiment. The potentiodynamic scan of the deep-immersion experiment (blue line) shows a reasonably clean sweep with little current below the dissociation potential of 1.5 V. When compared to a single tube shallow-immersion experiment it can be seen that the current is nearly twice as high for the deep-immersion experiment. This is consistent because electrode area of the deep immersion experiments is approximately twice that of shallow immersion experiments.

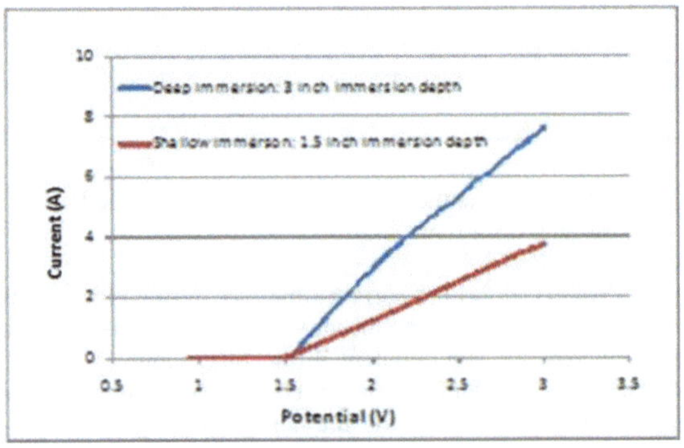


Figure 11. Potentiodynamic scan comparison of deep and shallow immersion experiments; right: deep immersion current.

We built a three-tube SOM electrolyzer which can contain around 8.0 kg of salt and MgO mixture. This electrolyzer is capable of accommodating up to 10" immersion depth, providing large electrode surface area on each tube. The combined electrode area for three tubes is approximately 240 cm$^2$ with an 8" immersion depth. If operated at a current density of 0.5 A/cm$^2$, this SOM electrolyzer would be capable of producing ~1.3 kg/day of magnesium. This SOM electrolyzer was operated at 7 V and 205 A to generate magnesium at a rate 2.2 kg/day. Unfortunately the electrolysis had to be stopped after couple of hours. The long stainless crucible-condenser assembly expands much more than the furnace or its tube during heating from room temperature to 1150-1200°C, and our attempts to accommodate this expansion led to crucible weld failures. The leakage of molten salt and magnesium through the failed weld joints led to breakage of the quartz furnace tube. For this reason we could not collect any magnesium in the condenser. During the electrolysis the anode current collector also failed. The large amounts of oxygen generated at the anode due to the high current led to the oxidation of the nichrome current collector and the oxide scale on the nichrome alloy was not durable.

Based on these results, we fabricated a current collector (described in Figure 3: right) and built a single-tube SOM with a shallow immersion (1" immersion depth) to test durability of the LSM/nichrome current collector. We operated the SOM electrolyzer for more than 80 hours. Figure 12 shows the current during the electrolysis run. The current was low in comparison to previous experiments, due to the use of a high-resistance current collector. The current density was around 0.15 A/cm$^2$ at 5.0 V. The voltage drop was due to high contact resistance at the junction between the 3 cm LSM rod and the thin nichrome wire. Though its resistance is high, this type of current collector has had the best long-duration performance. There was no fluctuation in current for nearly 80 hours.

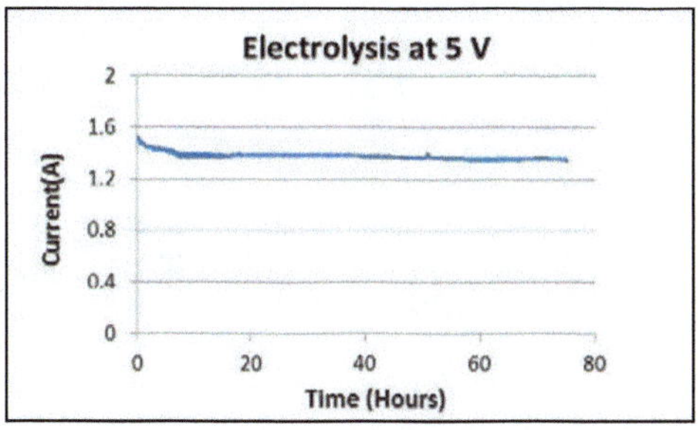


Figure 12. Current-time scan during the electrolysis.

Figure 13: Metal deposit on stainless shim from the 80-hr experiment

Figure 13 shows the metal deposit on the stainless steel shim after the electrolysis. This shows that LSM/nichrome might be good candidate for a current collector if the contact resistance can be lowered. To lower contact resistance we intend fabricate a multi-component current collector current collector shown in Figure 14.

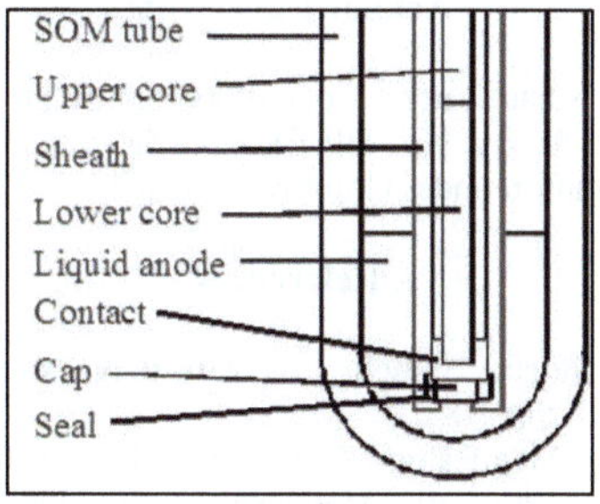


Figure 14. Current collector design which mechanically constrains the cap.

It shows schematically the components of our current collector approach, with material properties as follows:

*Upper core*: very high electrical conductivity, particularly electronic conductivity.

*Lower core*: very high electrical conductivity, melting point above operating temperature, and low solubility in the contact material.

*Contact*: solidus point below operating temperature, and good electrical conductivity in the liquid or semi-solid state at operating temperature.

*Seal*: liquidus point and/or glass transition above operating temperature, minimal solubility in the liquid metal anode, stability in liquid metal anode supersaturated with oxygen, low oxygen diffusivity, ability to provide a hermetic seal between two solids at operating temperature optionally by creep and/or glass flow, but with high enough viscosity or low enough creep rate to not flow out of the cap-sheath gap.

*Sheath*: stability in pure oxygen at operating temperature and between operating and ambient temperatures, low thermal conductivity, and resistance to failure due to temperature gradients or thermal or mechanical shock which would allow oxygen breach.

*Cap*: at operating temperature, minimal solubility in the liquid metal anode, stability in pure oxygen, high electrical conductivity particularly electronic conductivity, and low oxygen diffusivity, implying low oxide ion conductivity. For this application, minimal oxygen diffusivity and minimal porosity are essential, making strontium-doped lanthanum manganite (LSM, La0.8Sr0.2MnO3) a good candidate.

To accommodate the thermal expansion of the electrolyzer and to enable further scaling we have built a three-tube SOM electrolyzer with a new configuration, as shown in Figure 15. In this configuration the salt bath container and the condenser are side-by-side instead of top and bottom. This whole assembly goes in an inert enclosure containing separate heating elements for the salt bath container and the condenser. There are two advantages in this type of configuration. First, the smaller overall dimensions lead to less of a problem with thermal expansion making the weld joints more reliable. Second, there is a better control of the condenser temperature, which is vital for making liquid magnesium.

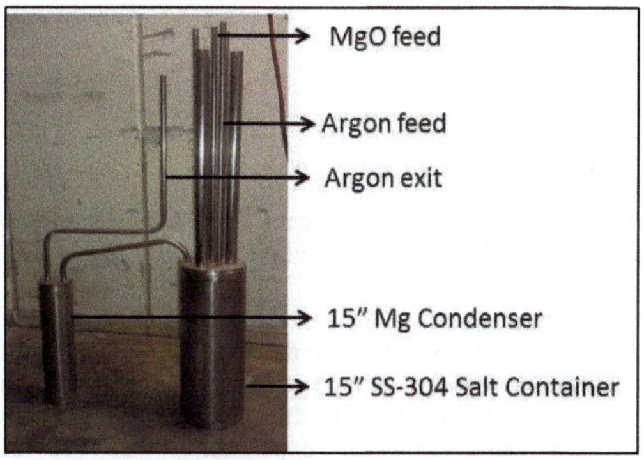


Figure 15: New side-by-side furnace design

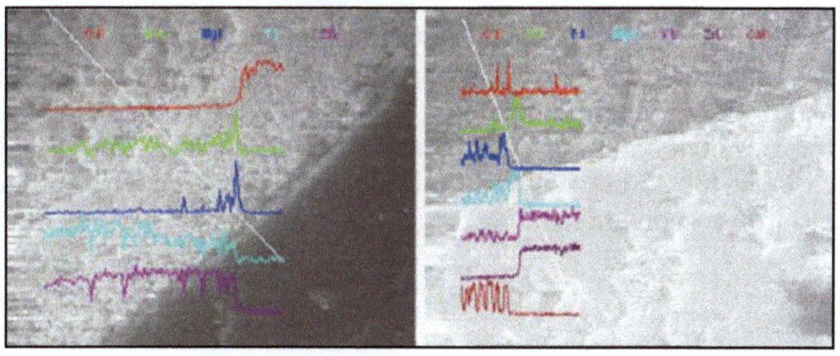

Figure 16. Microstructures for zirconia membranes after electrolysis, with EDAX line scans overlaid. Left: no added YF3, right 3.39 wt% YF3.

After all the experiments we performed micro structural analysis of the zirconia tubes. Figure 16 shows the microstructure of the SOM exposed to molten fluoride salt without $YF_3$ (left) and with $YF_3$ (right) under electrolysis conditions. It clearly shows that YSZ exposed to molten salt that does not contain $YF_3$ develops a yttria-depleted zone near the exposed surface, while YSZ exposed to molten salt with $YF_3$ shows no such degradation. Krishnan (5) observed a similar phenomenon and showed that adding $YF_3$ to the flux inhibits formation of a yttria-depleted layer. Therefore, adding $YF_3$ to the molten salt inhibits yttrium leaching, thus increasing the durability of zirconia tubes.

To further analyze the SOM degradation mechanism of the zirconia tubes, we performed static exposure experiments, which exposed the SOM to molten salt at standard operating temperature with no current. Figure 17 shows a comparison of the yttrium composition at SOM/molten salt interface. It can be clearly seen that the composition profile changes in zirconia in static contact with molten flux are nearly identical to those following an electrolysis experiment. This indicated that the primary mechanism of degradation is not related to the passage of ionic current through zirconia.

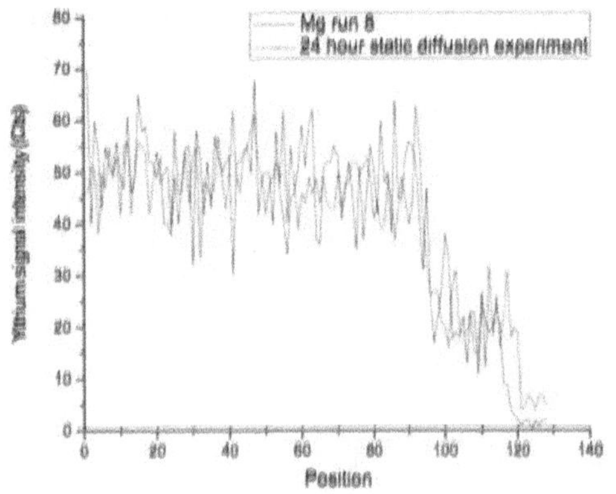


Figure 17: EDAX yttrium line scans from samples immersed in molten salt for 24 h with no added $YF_3$. Red – immersion only. Black – after electrolysis.

Based on these results, we concluded that membrane degradation is primarily due to dissolution of yttria in the flux and the passage of electronic current through the zirconia tubes. Therefore, inexpensive static exposure experiments can provide the information necessary to assess the former corrosion mechanism of the zirconia tubes.

Towards that end, Figure 18 shows the comparison of as-received zirconia tube from the vendor and the zirconia tube after exposure to flux for different durations. The membranes exposed to flux for 100 hours and 500 hours show some degradation, but SOM exposed for 32 hrs shows little to no degradation under static or electrolysis conditions.

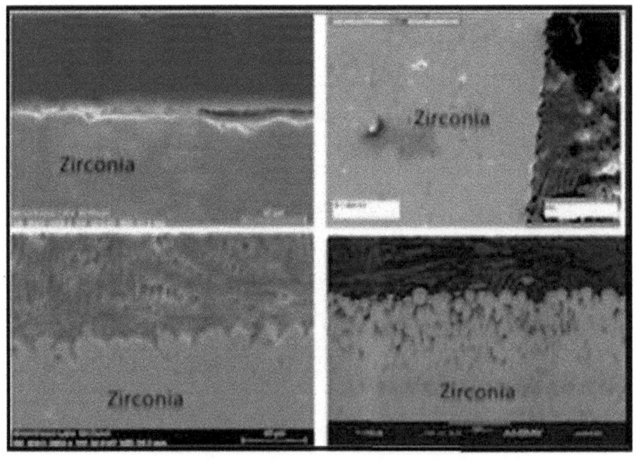


Figure 18. SEM micrograph of the SOM-flux interface: as-received (top-left), 32-hour exposure (top-right), and 100 hour exposure (bottom-left), 500-hour exposure (bottom-right).

## Summary

The SOM electrolysis process produces high-purity magnesium efficiently from available industrial-grade magnesium oxide with zero emissions and with a salable pure oxygen by-product, creating a low-cost magnesium product to facilitate its use in light-weight vehicles. The zirconia solid electrolyte between the molten salt and inert anode protects the anode, blocks oxygen back-reaction with magnesium, and makes the oxygen by- product very pure. The design where the LSM is attached to nichrome can be a durable current collector if the contact resistance is lowered. The temperature of the condenser has to be carefully controlled to generate liquid magnesium. Finally, the zirconia tube durability can be increased by controlling the composition and the microstructure carefully.


## Acknowledgements

This material is based upon work supported by the Department of Energy under Award No. DE-EE0003454, and by the National Science Foundation under Grant No. 1026639.


## References


1. T. M. Pollock, "Weight loss with magnesium" Science 21 May 2010: 986-987.
2. USAMP, "Magnesium Vision 2020" 1st November 2006
3. S Das, "Primary Magnesium Production Costs for Automotive App.," JOM 60(11):63-69, Nov. 2008.
4. A. Krishnan, X.G. Lu and U.B. Pal, "SOM Process for Magnesium Production directly from Magnesium Oxide", Met and Mat Trans, vol. 36B, (2005), p. 463
5. A. Krishnan, "SOM Process for Primary Production of Magnesium from Magnesium Oxide," PhD Thesis, BU, July 2005.
6. G Feng, "Assessing environmental impact of magnesium production using Pidgeon process in China", Trans of Nonferrous Metals, 2007
7. M Suput, R De Lucas, S Pati, G Ye, U Pal and A Powell, "SOM Technology for Environmentally Sound Production of Titanium, Mineral Processing and Extractive Metallurgy 117(2):118-122, June 2008.

**Magnesium Technology 2012**
*Edited by: Suveen N. Mathaudhu, Wim H. Sillekens, Neale R. Neelameggham, and Norbert Hort*
*TMS (The Minerals, Metals & Materials Society), 2012*

# FLUID BED DEHYDRATION OF MAGNESIUM CHLORIDE

K. Adham, C. Lee and K. O'Keefe
Hatch Ltd - 2800 Speakman Drive- Mississauga, Ontario L5K 2R7

Keywords: Magnesium Chloride, Dehydration, Fluid Bed, Dryer

## Abstract

Molten salt electrolysis of $MgCl_2$ is commonly used for the production of magnesium metal. However, the electrolysis feed must be completely dry with minimum oxygen content. Therefore, complete dehydration of the $MgCl_2$ brine or the hydrated prill is a required process, which is very challenging because of the ease of thermal degradation due to hydrolysis of magnesium chloride.

Fluidized bed dryers are often used, under hot air and HCl environments. The key features of three different types of fluid bed technologies, which can be applied to $MgCl_2$ dehydration plants, are described in this paper. In addition, a discussion of chemistry, unit operations and advantages associated with each option, is presented.

The background information is provided based on open literature sources, including papers and patents. Most calculations are performed using commercially available metallurgical software, for the thermodynamic calculations and mass/heat balances.

## Introduction

For electrolytic production of magnesium, the presence of water and other oxygen compounds in the fused-salt bath has a negative impact on both the product quality and the cell efficiency [1].

Oxygen content of the electrolyte is very detrimental to the operation of the electrolysis cells, through anodic attack and/or insoluble-phase formation. Therefore, not only it is required to effect a complete dehydration of the feed to anhydrous magnesium chloride ($MgCl_2$), but the drying process should also be conducted under such conditions that would minimize the extent of magnesium oxidation.

In practice, the oxide content of the electrolyte, prior to its feeding into the electrolysis cells, is regularly monitored (by sampling) and controlled (in the dryers), in order to avoid upset conditions in the electrolysis cells.

The source of magnesium chloride feed to the electrolysis cells can either be an ore that has been leached with HCl, or the evaporation of salty waters that contain $MgCl_2$. Both routes lead to the production of a typical brine solution with approximately 25-35% $MgCl_2$ by weight [2, 3].

The brine can be further evaporated to produce intermediate magnesium chloride hexahydrate ($MgCl_2.6H_2O$) prills, before it is sent to a multi-stage fluid bed dryer to produce magnesium chloride dihydrate ($MgCl_2.2H_2O$) prills [4]. Alternatively, the 25-35% $MgCl_2$ brine can be fed directly into a fluidized bed spray dryer to produce dihydrate prills [5].

Either way, the dihydrate prills are then dried to anhydrous $MgCl_2$ prills, either using a fluid bed dryer with hot HCl gas, or in a molten-salt electrically-heated chlorinator.

This paper discusses the fluidized beds used for the hexahydrate prills air-drying, brine solution spray drying, and dihydrate prills drying with hot HCl gas. The main challenge is demonstrated as the fact that; removing more and more water from magnesium chloride requires the use of higher and higher temperatures, which in turn can result in the pyro-hydrolysis (breakage under temperature and in presence of water) of the $MgCl_2$ to the unwanted hydroxide or even oxide species.

Figure 1 demonstrates that challenge through a simple thermodynamic calculation using the HSC® software. On the y-axis, 100 moles of hydrated magnesium chloride is heated by hot air, resulting in dehydration as well as hydrolysis at higher temperatures. Consequently, only 10 moles of the initial $MgCl_2$ ends up as a fully dried product, with the rest turning into Mg(OH)Cl, at 300°C on the x-axis.

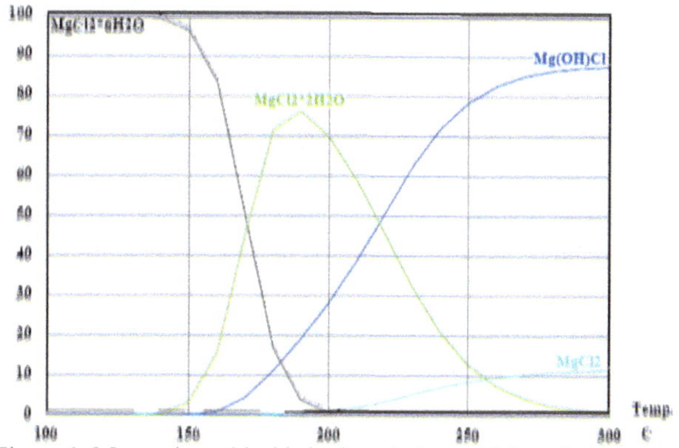


Figure 1. Magnesium chloride hydroxylation, with hot air drying

## Fluidized Hot-Gas Dryers

Most fluidized bed dryers operate in the so called 'bubbling regime'. Bubbling fluidization occurs when the hot gas velocity through the bed of solids is more than the minimum to suspend it, with most of the excess gas moving through the suspended solids in bubble-like pockets [6].

Figure 2 shows a bubbling fluid bed dryer, where solids are agitated and fully contacted with the hot fluidizing gas. Under bubbling conditions, most of the solids remain in a dense and distinct 'bed' phase, with only a fraction of the fines reporting to the off-gas discharge.

As a solid phase reactor, a simple circular bubbling bed is similar to a well mixed single stage unit where, depending on the bed inventory, an average residence time from several minutes up to many hours can be achieved.

However, a rectangular dryer can operate under a plug-flow regime, with wet feed entering at one end and product withdrawn from the other side of the unit.

Alternatively, a similar plug-flow arrangement can be obtained within a circular cross-section unit, through the use of baffles (separating walls) which can direct the solid feed through a labyrinth, from inlet towards the outlet.

Figure 2. Bubbling fluidized bed dryer

**Choice of Fluidization Velocity**

In the design of fluid beds, the minimum fluidizing velocity of the particulate solids must be carefully considered. The minimum fluidization velocity is best measured by simple tests. However, it can be approximated by empirical correlations as well [6].

Bubbling fluidized beds normally operate at 3 to 30 times the minimum fluidization velocity, which increases with the particle size and decreases with temperature.

The fluidizing behavior of particles in hot air versus particles in hot HCl gas is somewhat different. Figure 7 depicts the calculated minimum fluidization velocity curves for hydrated $MgCl_2$ (with 2 and 4 molecules of water) at 120°C and 180°C respectively, and the minimum fluidizing velocity for anhydrous $MgCl_2$ at 180°C and 330°C, for a range of spherical particle diameters expected in the fluid bed dryers.

It is advantageous to run the unit at the highest gas velocity possible, in order to achieve the drying duty within the smallest equipment diameter. However, higher velocities result in more elutriation and dust losses to the off-gas.

Fluidized dryers can operate at high velocities to reduce equipment size. However, as the velocity increases, the fluid bed behavior moves from the bubbling regime into the slugging regime, then the turbulent mode, and eventually the entrainment condition. As the fluid bed behavior approaches the entrainment condition, dust losses to the off-gas increase, and solids residence time in the bed decreases.

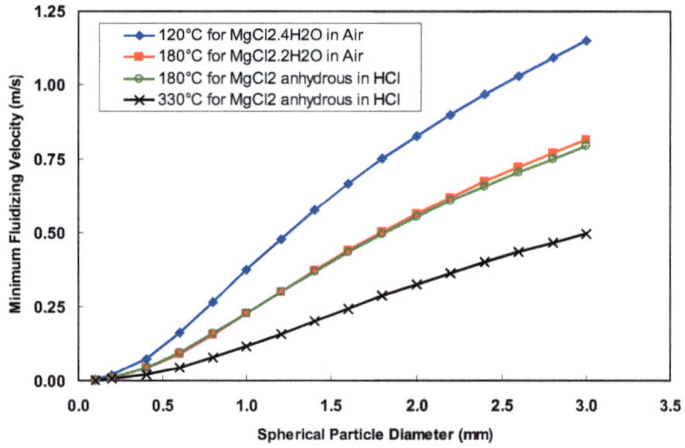


Figure 3. Fluidization velocity and the particulate bed behavior

Dust losses from a fluidized bed are typically related to the terminal velocity of the bed material. At the terminal velocity, a bed particle can be theoretically carried by the exiting hot gas. It is more difficult to measure the actual terminal velocity of specific particles, but given the key physical characteristics, it can be approximated using empirical formulas [6].

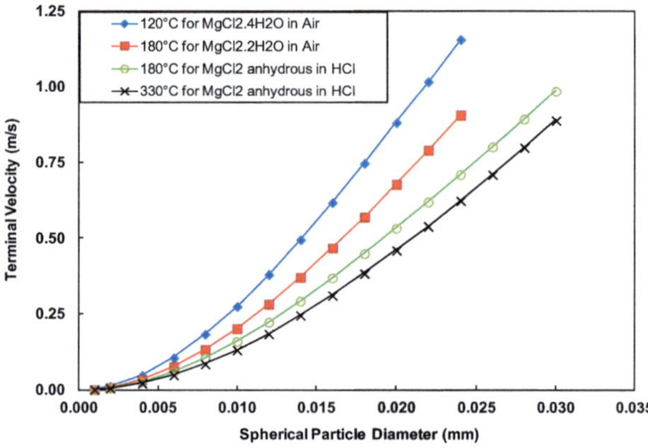


Figure 4. Terminal velocity for single particle carry-over

Knowledge of the minimum fluidizing velocity of prills and terminal velocity for the dust fraction sets the range of velocities at which the fluid bed dryers should be operated.

**Prills Hot Air Drying**

Heated air drying of the magnesium chloride can start with a solid (prill) feed or a solution (brine). Both processes are typically limited to the production of a dihydrate product ($MgCl_2.2H_2O$), as the elevated temperatures required for complete drying can readily form an undesirable oxygenated by-product (Figure 1) [5].

Conventionally, the brine solution is first converted to hexahydrate prills. Prills produced from evaporation and free-fall crystallization of a 30 wt% $MgCl_2$ solution contain a mix of $MgCl_2$ hydrates, but hexahydrate ($MgCl_2.6H_2O$) is predominant [4].

Dehydration of $MgCl_2.6H_2O$ to the dihydrate ($2H_2O$) state can be done using direct contact with hot air or combustion gases, which also serves to fluidize the bed of solids in the dryer [5]. The fluidizing air can be heated through indirect heat transfer or direct flame action.

Direct flame hot gas generation can be achieved in a number of arrangements; however, two commonly utilized systems are the external burner and the in-duct burner. For low and mid-duty applications, the in-duct burners are becoming more popular, due to their simplicity and reasonable cost.

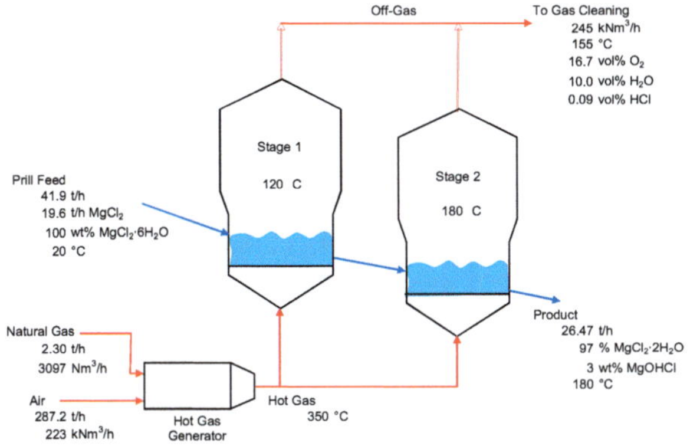


Figure 5. Prill drying for 40,000 t/yr Mg production

The purpose of hot air dryer is to remove moisture from the prills while minimizing oxidation, which produces the undesirable magnesium hydroxychloride. In the presence of water vapor and oxygen, the extent of hydroxychloride production increases rapidly as the moisture content of the prills decreases, likely due to the higher temperatures required to remove the lower hydrates.

The magnesium chloride air-drying is typically carried out in at least two stages, targeting two molecules of hydration-water removal in each stage, so that the drying temperatures can be selected to optimize drying and minimize oxidation. The typical reactions occurring in the dryers are shown below.

$$MgCl_2.6H_2O + 115\ kJ/mole \rightarrow MgCl_2.4H_2O + 2\ H_2O\ (g) \quad (1)$$
$$MgCl_2.4H_2O + 130\ kJ/mole \rightarrow MgCl_2.2H_2O + 2\ H_2O\ (g) \quad (2)$$
$$MgCl_2.2H_2O + 140\ kJ/mole \rightarrow MgOHCl + HCl\ (g) + H_2O\ (g) \quad (3)$$

A quantitative analysis of the energy input to the fluid bed dryer train is presented, by calculating the mass and energy balance equations, using the METSIM® software platform for a two-stage system. Figure 3 shows the drying of 40 kt/a Mg contained in $MgCl_2.6H_2O$ prills. For simplicity, it is assumed that only reaction 1 occurs in the first stage and goes to completion, while reaction 2 only occurs in the second stage and also goes to completion.

In addition, it is assumed that the oxidation reaction 3 only occurs in the second stage and produces 3% MgOHCl by weight. Typically, 2-4% MgOHCl occurs in this stage [7].

The bed temperature of stage 1 is 120°C, which is within the range of temperatures recommended to produce tetrahydrate ($4H_2O$), while the bed temperature for stage 2 is 180°C, which is within the range recommended to produce dihydrate ($2H_2O$) [7].

**Brine Hot Air Drying**

More recently, there has been an attempt for the magnesium chloride brine to be fed directly to a fluid bed (spray) dryer [2]. In this case, the air drying action is imparted on the thin liquid layers that are created on the seed solids (fluidized in the bed), rather than across the whole cross section of every particle.

The drying reaction (for approximately 30 wt% brine) can be approximated as follows:

$$MgCl_2\ (a) + 12\ H_2O\ (a) + 550\ kJ/mole \rightarrow MgCl_2.2H_2O + 10\ H_2O\ (g) \quad (4)$$

Just as in the prills dryer, the hydrolysis reaction 3 can also occur in the spray dryer, producing from 1-5% MgOHCl [10].

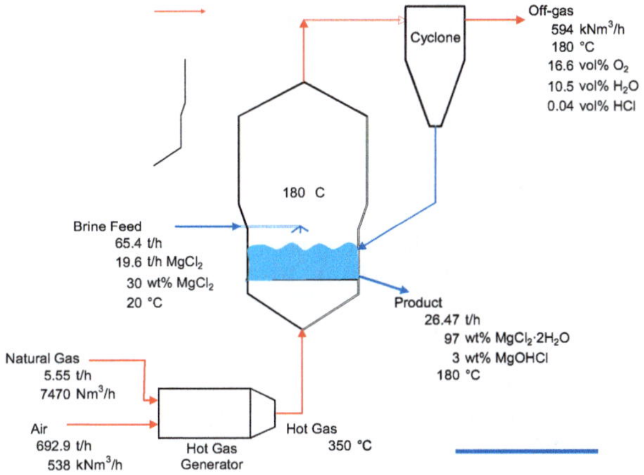


Figure 6. Brine drying for 40,000 t/yr Mg production

The feed brines should be limited to about 30% $MgCl_2$ by weight to avoid crystallization in the spray system. Using a similar mass and energy balance basis as the prills dryer, such a brine containing 40 kt/a Mg is modeled to dry in a spray fluid bed dryer at 180°C, for the production of the dihydrate. The calculated results are presented in Figure 4.

As expected, the natural gas usage increases almost proportionally to the water evaporation rate.

**Prills Hot HCl Drying**

Anhydrous magnesium chloride is normally required for the fused-salt electrolysis cells. As discussed above, drying in a hot air environment at a high temperature can partially oxidize the magnesium into an undesirable byproduct (hydroxychloride).

The key reactions are as follows. Compared to the dehydration reactions, the pyrohydrolysis reaction is slightly exothermic, but normally does not have a significant impact on the overall energy balance.

$MgCl_2.2H_2O$ + 145 kJ/mole ➔ $MgCl_2$ + 2 $H_2O$ (g) (5)

$MgCl_2$ + $H_2O$ (g) ➔ MgOHCl + HCl (g) + 5.5 kJ/mole (6)

The production of hydroxychloride is influenced by the temperature and the gaseous environment.

Figure 1 (calculated using the HSC® software) shows how at the 200-300°C range where the dehydration occurs, the equilibrium state of the $MgCl_2$-$H_2O$-HCl system favors the formation of hydroxychloride, if the solid phase is simply heated in an inert environment, without any extra gaseous HCl addition.

The hydroxychloride formation can be prevented or reversed if excess HCl gas is provided to the dryer as the heating medium.

Figure 5 (also calculated using the HSC® software) shows how at the same 200-300°C range the equilibrium state is shifted towards the formation of anhydrous $MgCl_2$ if a high ratio of HCl to $H_2O$ (in this example 10:1) is maintained in the gaseous phase.

As illustrated in Figure 1 and Figure 5, as the ratio of HCl to $H_2O$ concentration decreases, the amount of hydroxychloride increases.

The thermodynamic example of Figure 5 is conducted at the constant temperature of 300°C, which corresponds to the end point temperature of the Figure 1, where only 10% of magnesium chloride survived pyro-hydrolysis, without the addition of gaseous HCl to the environment.

In case of Figure 5, the starting point is 100 moles of the pyro-hydrolysis product (MgOHCl), which is contacted with an increasing amount of gaseous HCl (up to 300 moles on the x-axis). The result is a chemical shift from the undesirable hydroxychloride to the desired anhydrous $MgCl_2$. Of course at the end of this example case, only 225 moles of HCl gas survive (on the y-axis) as the balance (75 moles) is consumed by $MgCl_2$ generation.

Therefore, the HCl dryer typically requires a large amount of hot HCl gas recycle, as energy source of drying and to create a large HCl over-pressure for the desired chemical equilibrium. However, only a small portion of the HCl reacts and requires a fresh make-up, if the incoming partially-dried magnesium chloride is only slightly pyro-hydrolyzed.

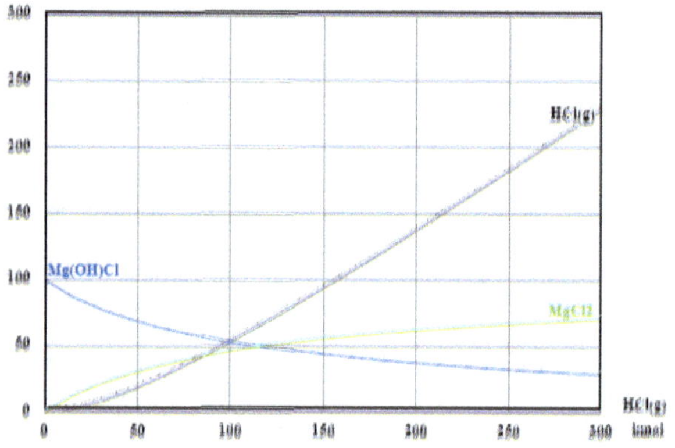


Figure 7. MgOHCl conversion at 300°C with excess HCl

To remedy that problem, the drying of dihydrate to anhydrous can be done inside a dryer fluidized with hot HCl gas. Maintaining a high concentration of HCl and low concentration of moisture drives the pyrohydrolysis reaction in reverse to produce magnesium chloride.

MgOHCl + HCl (g) + 5.5 kJ/mole ➔ $MgCl_2$ + $H_2O$ (g) (7)

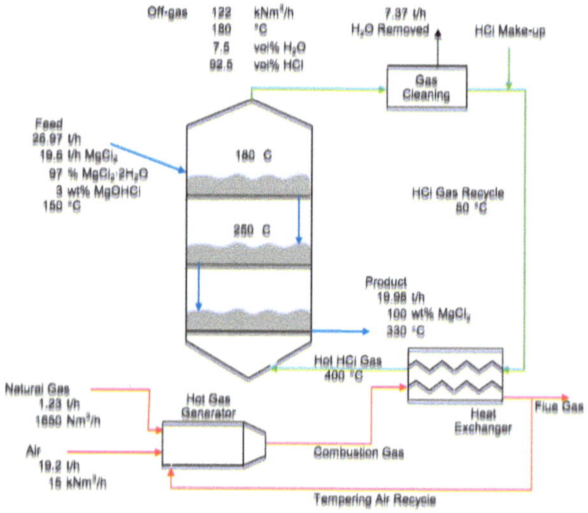


Figure 8. HCl dehydration for 40,000 t/yr Mg production

To maintain the HCl to $H_2O$ ratio throughout the drying process, a multi-stage dryer with dry HCl gas contacting the feed counter-currently can be utilized [4].

A numerical example of the energy input to the HCl fluid bed dryer is provided, for a triple-stage stacked design. Mass and energy calculations results are provided in Figure 6.

## Operating History

Based on the published information, at Norsk Hydro facilities in Bécancour, Canada, a $MgCl_2$ brine was produced from dolomite leaching. At Porsgrunn, Norway, in addition to dolomite leaching, the brine is also obtained through evaporation of $MgCl_2$ laden seawater [3]. The brines are evaporated and sprayed into a tower to produce a mixture of hexahydrate and tetrahydrate prills, which are then dried in an air fluid bed dryer, followed by an HCl fluid bed dryer to produce anhydrous prills with low MgO content ready for the IG Farben electrolysis cells [4].

Similarly, the State Research and Design Titanium Institute reports that air fluidized bed dryers are used to dehydrate carnallite ($KCl.MgCl_2.6H_2O$) feedstocks in Russia, Kazakhstan and Ukraine. It also reports use of HCl injection into a second stage of fluid bed drying as an alternative to electric chlorinators [8].

The Magnola process involved spraying $MgCl_2$ brine produced by leaching serpentine ore, into a fluidized bed dryer to produce granules of mostly $MgCl_2.2H_2O$ [9]. These granules were then sent to their "super-chlorinator", which contacted HCl gas with the granules to reduce MgO content, through melting the granules and removing the remaining moisture. The molten salt was then sent for electrolysis in Alcan-MPC cells [2].

**References**


1. C. B. Wilson, "Modern production process for magnesium", Light Metals 1996, TMS, 1996, pp. 1075-1079.

2. P. Ficara, et al., "Magnola: A novel commercial process for the primary production of magnesium", CIM Bulletin, Vol. 91, April 1998, pp. 75-80

3. N. Hoy-Petersen, "Norsk Hydro's Magnesium Production at Porsgunn, Norway", Light Metal Age, August, 1979, pp. 24 – 25, 40.

4. H. Eklund, et al., "An improved process for the production of magnesium", Magnesium Technology 2002, TMS, 2002, pp. 9-12.

5. O. Boyum, et. al., "Process for the preparation of anhydrous MgCl2 prills", United States Patent, No. 3,742,100, June 26, 1973.

6. D. Kunii and O. Levenspiel, "Fluidization Engineering", Butterworth-Heinemann, Series in Chemical Engineering, 2nd edition, 1991, pp. 137-164.

7. M. Nadler, "Anhydrous Magnesium Chloride Production", United States Patent, No. 3,395,977, August 6, 1968.

8. G. Shekhovtsov, et. al. "Magnesium Electrolytic Production Process", Magnesium Technology 2000, TMS, 2000, pp. 31-34.

9. M. Kreuh, et al., "Dehydration of magnesium chloride prills of Magnola Process in a molten Salt", CMQ, Vol 44, No. 2, 2005, pp. 167-170.

10. B. Langseth, et. al., "Method of producing magnesium chloride granules", United States Patent, No. 6,296 ,790, October 2, 2001.

11. R. E. Brown, "Magnesium industry growth in the 1990 period", Magnesium Technology 2000, TMS, 2000, pp. 3-12.

**Magnesium Technology 2012**
*Edited by: Suveen N. Mathaudhu, Wim H. Sillekens, Neale R. Neelameggham, and Norbert Hort*
***TMS (The Minerals, Metals & Materials Society), 2012***

# DEMONSTRATION OF SOLAR-PUMPED LASER-INDUCED MAGNESIUM PRODUCTION FORM MAGNESIUM OXIDE

Yabe Takashi; Ohkubo Tomomasa; Dinh Thanh Hung; Kuboyama Hiroki; Nakano Junichi; Okamoto Kouta;
Tokyo Institute of Technology, 2-12-1 O-okayama, Meguro-ku, Tokyo 152-8550, Japan

Keywords: Laser-induced magnesium production, solar-pumped laser, fossil-fuel-free energy cycle.

## Abstract

Studies of storing solar energy into chemical energy of magnesium (Mg) through reduction from magnesium oxide (MgO) by solar-pumped laser were conducted. We succeeded in solar-pumped laser-induced Mg production. Laser system consists of a 4 $m^2$ Fresnel lens mounting on a sun tracker platform which focus solar radiation into laser head therefore over 100W (CW) output laser can be irradiated. A single laser beam is focused on a mixture of magnesium oxide and reducing agent silicon. High power density of focused laser leads to high temperature and the reduction reaction resulting in Mg production. The resultant vapor is collected on a copper plate and analyzed in terms of magnesium deposition efficiency. As a result, deposition efficiency of 2.3 mg/kJ was achieved.

## Introduction

More than 80% of the world total primary energy is provided by fossil fuel that otherwise emit greenhouse gases causing global warming. World demand for energy would be more than doubled by year 2050 and more than tripled by the end of century. Current improvements in existing energy networks will not be adequate for this demand in a sustainable way. Finding sufficient supplies of renewable energy for future is one of the challenges that scientist face. There is no doubt that solar energy is an ultimate renewable energy source, however, it is only available in the clear daytime and therefore, needs a reliable storing means as main energy source in future.

Authors proposed to use the chemical potential of magnesium as solar energy reservoir. Magnesium was chosen because it is the eighth most abundant element in the earth's crust (although it cannot be found in its elemental form in nature). It is also the second richest metal element in seawater with concentration of 1.3 g/l. In addition, magnesium is relatively light metal suitable for transportation and has a large energy storage capacity of 43 GJ/$m^3$ compared to 4.3 GJ/$m^3$ of hydrogen stored at 70MPa. Magnesium reacts with steam to generate hydrogen and release 359.8 kJ/mol of thermal energy as follows:

$$Mg_{(s)} + H_2O_{(s)} \rightarrow MgO_{(s)} + H_{2(g)}, \ \Delta H = -359.8 \text{ kJ/mol}.$$

Hydrogen generated further react with oxygen to give steam and additional energy.

$$H_{2(g)} + O_{2(g)} \rightarrow H_2O_{(s)}, \ \Delta H = -241.8 \text{ kJ/mol}.$$

This principle can be used for generation of electricity and power. If MgO can be reduced back to magnesium again, a sustainable energy cycle would be established. For this process, the temperature of at least 4000 K is needed. Laser radiation is favorable for this purpose compared to direct solar radiation because of its high brightness that provides high enough intensity to achieve such temperature. Authors propose to convert solar radiation into laser radiation to realize reduction of Mg by solar power. The principle of laser reducing process resembles the thermal reduction process that is well-developed method known as Pidgeon method [1-4]. The solar-pumped laser using Fresnel lens as solar radiation collector was described as a cost-efficient converter, and feasibility of this proposed cycle as a robust and cost-effective fossil-fuel-free cycle was depicted [1, 5-7]. However, study of laser-induced magnesium reduction was conducted with common laser systems but not direct solar-pumped laser.

In this paper, demonstration of solar-pumped laser-induced magnesium production from MgO is described. A solar-pumped laser system suitable for magnesium energy cycle is proposed and designed here. Authors succeeded in producing magnesium from MgO by direct solar-pumped laser. A single beam 53 W solar-pumped laser is focused on a mixture of magnesium oxide with Si as reducing agent, corresponding to the deposition efficiency is 2.3 mg/kJ.

Fig. 1 Solar-pumped laser setup

Fig. 3 Solar-pumped laser setup a fundamental section of power plant

**Solar-pumped laser system for Mg energy cycle**

Since the natural sunlight does not provide power density sufficient enough for lasing, an additional focusing and guiding devices are usually required. Commonly, the sunlight collecting system of solar-pumped laser consists of two stages. While at the fist-stage, primary concentrator is employed for collecting the sunlight to small spot, the second-stage pumping cavity is used for refocusing the collected solar power into the laser medium for an efficient pumping. A few solar-pumped laser systems have been previously realized with a mirror system [8-10]. Such researches have usually performed experiments using already existing solar facilities designed for other purposes such as solar towers. In order to construct power plants, a bundle of solar-pumped laser beams array scanning over MgO surface would be a section of energy renewal process in Mg cycle, therefore, the concentration devices must be flexible, inexpensive and not bulky.

Solar-pumped laser configuration

Laser output is able to be led to an MgO target using glass fiber and high intensity is realized by focusing techniques. In order dissociation of MgO in equilibrium, a temperature of 4000 K is needed because lower temperature may produce MgO vapor and reduce the deoxidization rate [2]. Such high temperature can be realized by an intensity of $10^5$ W/cm$^2$ over MgO surface, therefore not only laser power but also its beam quality must be required.
Having these facts in mind, a large diameter Fresnel lens was adopted as an optical concentrator device for a new set up of solar pumped laser (Fig.1). Because of low weight, four-segmented Fresnel lens were assembled together to form a complete collector and mounted on a motor-driven bench, which is synchronized with sun trajectory in the sky. This system is one of the prototypes

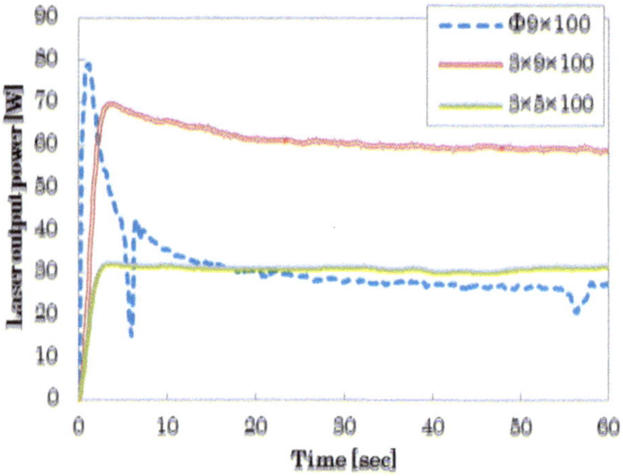


Fig. 2 Behavior of laser output for various laser rod shapes

in a future solar-pumped laser plant (Fig. 2). All equipment including Fresnel lens and laser cavities follow sun trajectory automatically. Once the alignment of the laser cavity is achieved, a stable laser power is obtained while the system tracks the sun. As shown in Fig. 1, Fresnel lens (2×2 m$^2$, f=2 m) focuses the solar radiation toward laser cavities. Since the diameter of sun image at the focal point is limited to 20 mm, we used a conical cavity with input aperture diameter of 80 mm and cone angle of 20 deg (D80-θ20), the shape gave the best in the previous report [6, 7]. By a hybrid pumping scheme combining axial- and side-pumping configurations in which zigzag pumping light come from many directions along, a circular focal spot could be transformed for enhancing solar light absorption to a thin laser rod. Inner wall of pumping cavity is a cone-type mirror made of aluminum, which holds the laser medium along the central axis. One end of laser rod has high-reflection coating at the laser wavelength while the other end has antireflection coating.
Such solar-pumped laser system demonstrated its superiority by high output lasing. The maximum laser power of 80W was achieved with $Nd^{3+}/Cr^{3+}$:YAG rods of 9 mm in diameter and 100 mm in length (Φ9×100 mm) corresponding to total area performance (=laser output/Fresnel lens area) of 20 W/m2 which is 2.8 times larger than other previous works [10].

Feature of laser output

In order to realize a stable Mg production, laser output power must be stabilized. The behavior of laser output power for various laser rod shapes is shown in Fig. 3.
After reached a peak 80 W, laser output power of cylindrical Φ9×100 mm $Nd^{3+}/Cr^{3+}$:YAG rods decrease fast. Since the serious thermal effects along laser rod, stable out power is only about 30 W more than a factor of 2 below the maximum. Furthermore, the absorption is localized at the end of cone-shaped cavity and this leads to the break-up of laser media.
Laser output power is stabilized by using slab-shape laser rods because of their larger cooling surface. The maximum of stable output power of 60 W is achieved with a slab 3×9×100 mm $Nd^{3+}/Cr^{3+}$:YAG rod. However, the thermal effect along laser rod decreasing laser power is still observed here.

Proposal of high performance pumping and cooling cavity

In order to stabilize and improve the laser power, a uniform pumping and cooling effect along laser rod is needed. Authors here propose to use coolant inside a glass cooling blanket as a liquid light guide lens (LLGL) (Fig. 4(b)). The difference of refraction index between air and coolant is used to concentrate and guide pumping light to a thin active material (Fig. 4 (c)). The reflection power at the boundary, which did not penetrate inside the liquid lens, can be recollected by the secondary mirror, so low optical transfer loss is achieved. Furthermore, in liquid light-guide assembled system, lase material and cavity could be cooled separately so that more efficient cooling is realized. The higher flow velocities in cooling blanket leads to turbulent flow, leading to more efficient heat transfer process with a subsequent lower temperature drop.

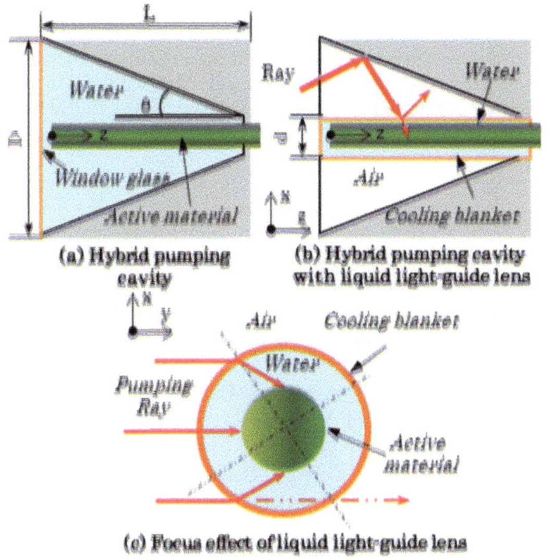


Fig. 5 (a) Cone-shape hybrid pumping cavity and (b), (c) liquid light guide lens configuration

The three-dimensional ray tracing has been performed to characterize the performance of LLGL. It includes:

1. Refraction, reflection, transmission at boundaries such as air-glass, glass-water, inner wall, Fresnel lens surface.
2. Absorption along each passage through various media including Fresnel lens material, water and laser material.

This ray tracing calculation was performed to simulate transmission through Fresnel lens of 2×2 $m^2$ area for all solar spectra considering the 4 mrad divergence angle of natural solar radiation. Absorption coefficient from 280 to 870 nm of $Nd^{3+}/Cr^{3+}$:YAG (Nd: 1.0, Cr: 0.1 at.% ) was measured by Spectrophotometer (Jasco-V-530). Standard solar spectra of one and half air mass (AM1.5G) [11] and absorption coefficient of water in [12] are used as reference data.
Numerical results with slab 3×9×100 mm $Nd^{3+}/Cr^{3+}$:YAG show that the adopt of LLGL in hybrid pumping cavity leads to the increase of the total absorption power in laser material and reduces the strongly localized distribution of absorption power

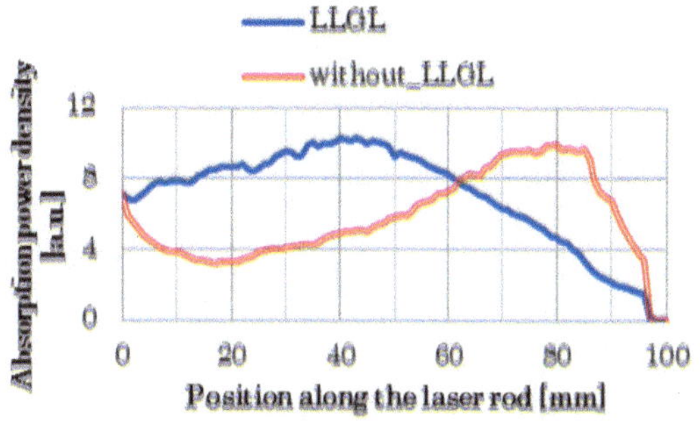


Fig. 4 Calculation of absorption distribution along $Nd^{3+}/Cr^{3+}$:YAG and Nd:YAG laser rod with 14 mm-diameter of LLGL and without LLGL.

density along the rod as in Fig. 5. Because the LLGL enhances the refocusing sunlight by the air-liquid interface, the absorption at the input aperture increases, subsequently absorption peak at around 60-100 mm from the input aperture, that may lead to thermal damage because of high temperature drop on laser rod in the case without LLGL, is moderated. Therefore, high performance of pumping and cooling is achieved and a stable laser output of over 100 W is expected. Implementation of LLGL will be done in near future.

## Experiments of solar-pumped laser-induce Magnesium reduction using reducing agent silicon

Experimental setup

The experimental setup is shown in Fig. 6. Here a chamber was used with the pressure of 7 Pa. The targets were prepared by mixing pure MgO powder and silicon powder, as follows:

1. Mixing magnesium oxide and silicon with molar ratio of 1: 0.5.
2. The mixture is packed into a cylindrical container that has diameter of 35 mm and height of 50 mm.
3. Compressing the target with a static pressure of 20 MPa to minimize the ejection of powder particles during laser irradiation.
4. Annealing the target at 110 °C for 3 hours to remove some of the gases absorbed in the mixtures.

The target was placed in the vacuum chamber. The cw solar-pumped laser of a slab 3×9×100 mm $Nd^{3+}/Cr^{3+}$:YAG rod irradiates vertically the target. Its beam qualities in the thickness direction $M^2_t$ and in the width direction $M^2_w$ are 60 and 193, respectively. Laser is focused by a condensing lens whose focus length is 40 mm. The laser spot side of 0.22×0.5 mm is calculated. The generated magnesium vapor was collected by a horizontally placed copper plate, forming a solid deposit. The mass of deposited material on the copper plate was measured, and the dipped into hydrochloric acid solution (HCl: 1 mol/l) as shown in Fig. 6. The following reactions are expected to take place:

$$Mg_{(s)} + 2HCl_{(aq)} \rightarrow MgCl_{2(aq)} + H_{2(g)}$$

$$MgO_{(s)} + 2HCl_{(aq)} \rightarrow MgCl_{2(aq)} + H_2O_{(g)}$$

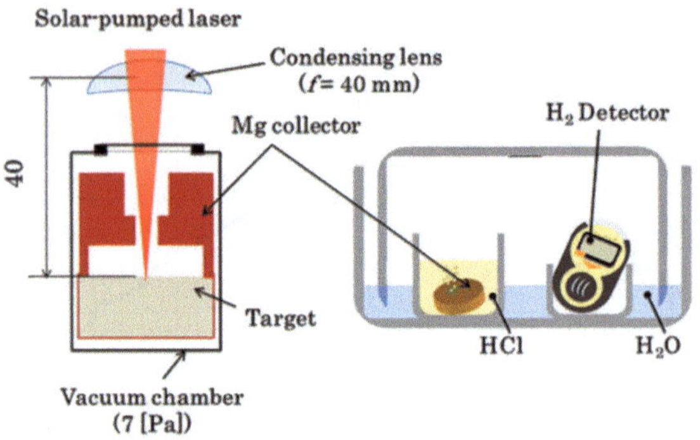


Fig. 6 Schematic of experimental set up and the hydrogen detector system.

Mg only among the deposited materials can produce hydrogen gas. The mount of emitted hydrogen gas was measured by the miniMAX XP hydrogen gas detector manufactured by Lumidor, which detects hydrogen gas in the range of 1-999 ppm (ml/m$^3$). The generated hydrogen gas was accumulated in a water-sealed 1.6-l container. The amount of detected hydrogen was used to calculate magnesium mass, and the deposition efficiency (reduced Mg mass/ input laser energy).

Experiment and result

As results shown in Fig. 7, we have demonstrated successfully solar-pumped laser-induced magnesium reduction from MgO by using reducing agent silicon. At 50 W class of input laser power, the target was kept standstill and the laser duration time *t* was changed. The deposition efficiency of 2.3 mg/kJ was achieved when laser power, calculated laser intensity and laser duration were 53 W, $4.8\times10^4$ W/cm$^2$ and 5sec, respectively, since the laser power was too low and laser spot was too small compared with previous experiment. Such small spot might increase heat loss and so on. The deposition efficiency can be improved by higher input laser intensity. In previous report the deposition efficiency of 12.1 mg/kgJ had been achieved with laser intensity of about $10^5$ W/cm$^2$ and mixing molar ratio of MgO:Si of 1:0.3 [3]. This laser intensity will be satisfied by the high efficient solar-pumped laser system which was proposed in previous section. Furthermore, the other conditions such as laser duration, mixing ratio must be optimized also for an efficient reduction.

## Conclusion

It is important to note that the first time magnesium was produced by direct solar-pumped laser. We here have combined basis studies of laser-induced Mg reduction with solar-pumped laser technology. A solar-pumped laser system suitable for magnesium energy cycle was described. For stabilizing the laser output, slab laser rods was adopted, and high performance pumping and cooling cavity using LLGL was proposed. A single beam 55W solar-pumped laser whose beam qualities in the thickness direction $M^2_t$ and in the width direction $M^2_w$ are 60 and 193, respectively, was focused on a mixture of magnesium oxide with reducing agent silicon. The deposition efficiency of 2.3 mg/kJ was measured. In future study, implementation of solar-pumped laser with LLGL will be done and, focusing techniques of laser array for solar-pumped laser-induced magnesium production will be developed.

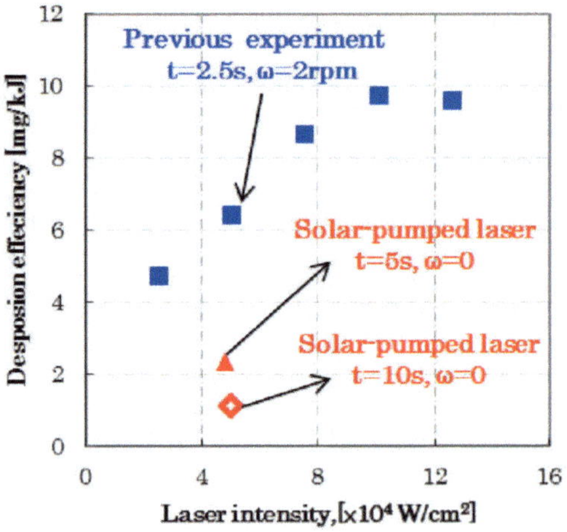


Fig. 7 Solar-pumped laser-induced magnesium reduction in compare with previous result [4]. MgO and Si were mixed with molar ratio of 1: 0.5. *t* and *ω* are the laser duration time and target rotation speed, respectively.

## Reference


1. T. Yabe, S. Uchida, K. Ikuta, K. Yoshida, C. Baasandash, M. S. Mohamed,Y. Sakurai, Y. Ogata, M. Tuji, Y. Mori, Y. Satoh, T. Ohkubo, M. Murahara,A. Ikesue, M. Nakatsuka, T. Saiki, S. Motokoshi, and C. Yamanaka, Appl. Phys. Lett. 89, 261107 (2006).
2. T. Yabe, M. S. Mohamed, S. Uchida, C. Baasandash, Y. Sato, M. Tsuji, and Y. Mori J. Appl. Phys. 101, 123106 (2007)
3. M. S. Mohamed, T. Yabe, C. Baasandash, Y. Sato, Y. Mori, Liao Shi-Hua, H. Sato and S. Uchida, J. Appl. Phys. 104, 113110 (2007).
4. S. H. Liao, T. Yabe, M. S. Mohamed, C. Baasandash, Y. Sato, C. Fukushima, M. Ichikawa, M. Nakatsuka, S. Uchida, and T. Ohkubo, J. Appl. Phys. 109, 013103 (2011).
5. T. Yabe, T. Ohkubo, S. Uchida, K. Yoshida, M. Nakatsuka, T. Funatsu, A.Mabuti, A. Oyama, K. Nakagawa, T. Oishi, K. Daito, B. Bagheri, Y. Nakayama, M. Yoshida, S. Motokoshi, Y. Sato, and C. Baasandash, Appl. Phys. Lett. 90, 261120 (2007).
6. T. Ohkubo, T. Yabe, K. Yoshida, S. Uchida, T. Funatsu, B. Bagheri, T. Oishi, K. Daito, M. Ishioka, Y. Nakayama, N.Yasunaga, K.Kido, Y. Sato, C. Baasandash, K.Kato, K.Yanagitani, and Y.Okamoto, Opt. Lett. 34(2) (2009).
7. T. Yabe, B. Bagheri, T. Ohkubo, S. Uchida, K. Yoshida, T. Funatsu, T. Oishi, K. Daito, M. Ishioka, N.Yasunaga, Y. Sato, C. Baasandash, Y.Okamoto,and K.Yanagitani, J. Appl. Phys. 104, 083104 (2008).
8. C. G. Young, Appl. Opt. 5, 993 (1966).
9. V. Krupkin, Y. Kagan, and A. Yogev, Proc. SPIE 2016, 50 (1993).
10. M. Lando, J. Kagana, B. Linyekina, and V. Dobrusin, Opt. Commun.222, 371 (2003).
11. ASTM International, ASTM G173-03 Reference Spectra Derived from SMARTS v. 2.9.2 (2004).
12. G. M. Hale and M. R. Querry, Appl. Opt. 12(3), 555 (1973).

**Magnesium Technology 2012**
*Edited by: Suveen N. Mathaudhu, Wim H. Sillekens, Neale R. Neelameggham, and Norbert Hort*
***TMS (The Minerals, Metals & Materials Society), 2012***

# MOLTEN SALT ELECTROLYSIS OF $MgCl_2$ IN A CELL WITH RAPID CHLORINE REMOVAL FEATURE

Gökhan Demirci[1], İshak Karakaya[2]
[1]ASELSAN Inc.; Çankırı Yolu 7.km; Akyurt, Ankara, Turkey
[2]METU; Universiteler Mah. Dumlupınar Blv. No:1; Ankara, 06800, TÜRKİYE
Keywords: Magnesium, Molten Salt Electrolysis, Magnesium Chloride

## Abstract

An experimental electrolytic magnesium production cell was designed to remove chlorine gas from the electrolyte rapidly and demonstrate the beneficial effects of reduced chlorine dissolution into the molten salt electrolyte. The back reaction that is the main cause of current losses in electrolytic magnesium production was reduced as a result of effective separation of electrode products and decreased contact time of chlorine gas with the electrolyte. Moreover, smaller inter electrode distances employed and lower chlorine gas present on the anode surface made it possible to work at low cell voltages. Electrolytic cell was tested at different current densities. Energy consumption of 7.0 kWh $kg^{-1}$ Mg that is slightly above the theoretical minimum, 6.2 kWh $kg^{-1}$ Mg, at 0.68 $Acm^{-2}$ anodic current density was achieved for a $MgCl_2$/NaCl/KCl electrolyte.

## Introduction

Magnesium is the lightest structural metal. It is 4.5 times lighter than iron and 1.6 times lighter than aluminum. Magnesium is one of the most abundant elements in the earth. It constitutes about 2% of the earth's crust and it is the third most plentiful element dissolved in seawater with a concentration averaging 0.13% [1]. Despite obvious technical advantages and high abundance, its use is still restricted due to its high production costs that can be attributed to the production methods.

More than half of the magnesium production is carried out by electrolysis. Magnesium production using aqueous electrolytes is not possible because of the highly stable nature of magnesium ions in water. This leaves molten salt electrolysis of $MgCl_2$ as the electrolytic method for magnesium production. On the other hand, pure $MgCl_2$ has a high melting point and magnesium solubility in molten $MgCl_2$ is high enough to lower current efficiencies. Moreover electrical conductivity of the molten $MgCl_2$ is low and its vapor pressure is high. For that reason a chloride salt mixture is used to eliminate these disadvantages.

One of the most important reasons of the high cost is the highly reactive nature of the electrode products, namely magnesium and chlorine. Electrode products react readily when they come into contact and separation of the molten magnesium and chlorine gas has the prime importance to minimize current losses during the electrolysis [2,3,4]. Energy consumption for electrolytic magnesium production is almost double the theoretical value for most efficient industrial cells mainly due to this back reaction.

An energy consumption value of 10-12 kWh $kg^{-1}$ Mg was reported for electrolytic magnesium production, using a high purity $MgCl_2$ feed [5,6,7]. The large difference between the actual and the theoretical, around 6.2 kWh $kg^{-1}$Mg, energy consumptions indicate the possibility of further improvements.

It may be beneficial to have more studies on magnesium production to decrease the gap between the theoretical and the actual energy consumptions. In fact, electro deposition mechanism of magnesium in molten chloride melts has been investigated in considerable detail through the years [8]. However, studies of different electrolysis parameters are scarce for both those representing the industrial production and new designs.

The current losses result from direct contact of the electrode products and back reaction. The investigations of the reaction between dissolved species may help to enhance the cell performance.

Chlorine diffusion was found as the limiting step in recombination reaction between dissolved electrode products in magnesium cell electrolyte [9,10,11]. Thus, it may be possible to reduce back reaction further if the chlorine gas can be removed from the melt in a shorter period of time, considering the fact that the dissolution of the chlorine is proportional to its residence time inside the melt. Current efficiency values are expected to be improved upon decreasing chlorine dissolution into the electrolyte as a result of less interaction of chlorine with the electrolyte.

Therefore, an experimental cell that features rapid chlorine gas removal was constructed. This enabled the removal of chlorine gas with minimized interactions with the floating magnesium metal. Furthermore, it reduced the chlorine dissolution into the electrolyte considerably. Therefore, back reaction was restricted to dissolved magnesium metal and chlorine gas within the melt or fine suspended magnesium particles that might enter the anode compartment due to convection.

## Experimental Details

An externally heated quartz vessel was used to accommodate cell components as in the case of previous studies [9,10]. Electrode leads were extended out of the Teflon lid of the quartz vessel that was placed into a vertical tube furnace. Cell vessel was provided with argon gas flow to create positive pressure inside the cell vessel.

Graphite anode extension rod was placed into a quartz tube which was at the center of the quartz vessel and alumina crucible placed at the bottom. Anode geometry and the quartz tube surrounding the anode made it possible to remove chlorine gas that was produced on the anode surface, from the cell environment, rapidly. The direction of movement of chlorine gas in the cell can be seen by an arrow in anode compartment of Figure 1.

Cathode was designed as circular around the anode and made up of 18 mm wide and 0.2 mm thick electropolished stainless steel ribbons. Windows were opened on stainless steel ribbons by electrochemical machining to increase cathodic current density levels for this setup. One stainless steel rod was spot welded to the cathode to make electrical connection of the cathode. An alumina tube was used to cover cathode lead.

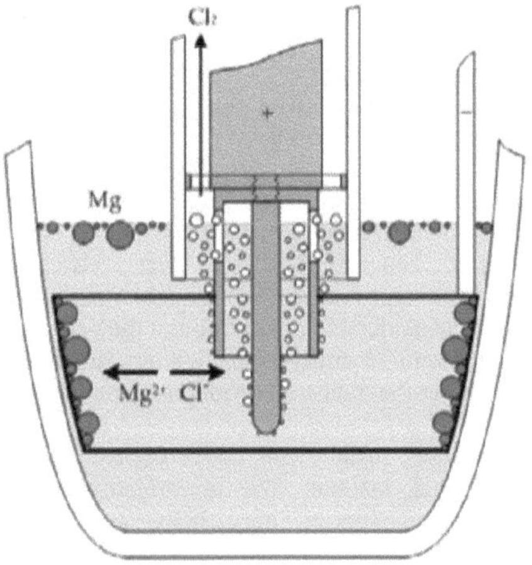


Figure 1 Schematic drawing of the electrolysis cell.

Initial electrolyte composition was; 20% $MgCl_2$, 39.5% KCl, 39.5% NaCl and %1 NaF by weight. $MgCl_2$ obtained from Aldrich had 98% $MgCl_2$ and 1.5% $H_2O$ content. 99.8% NaCl and KCl and reagent grade NaF were from Merck. Electrolysis experiments were carried out at 700°C. Experiments were carried out galvano-statically and cell voltage readings were taken by a portable multimeter and a data acquisition system.

After each experiment, cathode was washed to dissolve adhered salts in the shortest time possible. The cathode was then dried with oil-free compressed air. The cathode weight was recorded and then magnesium deposited to the cathode surface was dissolved in water. A few HCl droplets were added to water to increase the dissolution rate [12]. The amount of magnesium attached to the cathode was determined form the weight difference of the cathode before and after dissolving the magnesium.

After being crushed in a mortar, frozen electrolytes were milled for 24 hours in a rotating glass jar with stainless steel rods inside. Large magnesium particles were collected into a glass cup that was covered with a 50 mesh synthetic fabric. Then the glass cup was flushed with tap water through the fabric to dissolve the salt. Magnesium particles were then dried by applying compressed air to glass cup through the fabric.

Smaller particles, on the other hand, were just subjected to compressed air after milling to prevent their loss in water. The quantities of magnesium attached to the cathode and collected from the frozen electrolyte were added to determine the current efficiency. Pieces of magnesium metal, collected from different parts of cell, are shown in Figure 2.

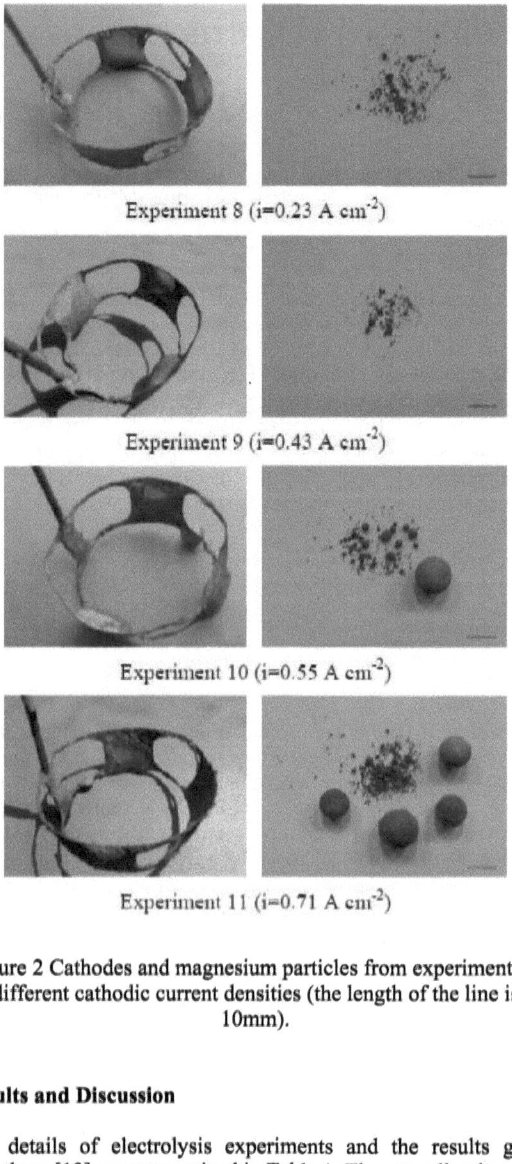


Figure 2 Cathodes and magnesium particles from experiments at different cathodic current densities (the length of the line is 10mm).

**Results and Discussion**

Full details of electrolysis experiments and the results given elsewhere [13] are summarized in Table 1. The net cell voltage vs. time for 13 and 20 mm ACD at indicated anodic current densities are given in Figures 3 and 4 respectively. The net cell voltage was found by subtracting voltage drop along electrical connections from the measured cell voltage. The average net cell voltage is the average of the cell voltage readings after the initial 10 minutes. Anode-Cathode Distance, ACD, is average separation distance between the electrodes before magnesium deposition. The energy consumption, *EC*, per kilogram of magnesium was calculated from;

$$EC = \frac{E_{average\ net} \cdot 100}{0.4534 gA^{-1}h^{-1} \cdot \%CE} \quad (1)$$

where,

$$0.4534gA^{-1}h^{-1} = \frac{3600\frac{second}{hour}\cdot\frac{24.305}{2}\frac{g}{g.equivalent}}{96485\frac{coulomb}{g.equivalent}} \quad (2)$$

**Table 1** Summary of the experimental details and the results obtained.

| Experiment | *1* | *2* | *3* | *4* | *5* | *6* |
|---|---|---|---|---|---|---|
| $j_{anode}$, $Acm^{-2}$ | 0.68 | 0.90 | 1.32 | 1.32 | 1.75 | 2.18 |
| $j_{cathode.}$, $Acm^{-2}$ | 0.24 | 0.31 | 0.43 | 0.44 | 0.58 | 0.77 |
| $V_{avg.\ net}$, V | 3.03 | 3.06 | 3.27 | 3.26 | 3.42 | 3.50 |
| CE, % | 96.0 | 95.1 | 95.1 | 94.4 | 91.1 | 90.3 |
| EC, kWh $kg^{-1}$ | 7.0 | 7.1 | 7.6 | 7.6 | 8.3 | 8.6 |
| ACD, mm | 13 | 13 | 13 | 13 | 13 | 13 |

| Experiment | *7* | *8* | *9* | *10* |
|---|---|---|---|---|
| $j_{anode}$, $Acm^{-2}$ | 0.71 | 1.31 | 1.71 | 2.20 |
| $j_{cathode.}$, $Acm^{-2}$ | 0.23 | 0.43 | 0.55 | 0.71 |
| $V_{avg.\ net}$, V | 3.19 | 3.46 | 3.65 | 3.76 |
| CE, % | 99.1 | 97.7 | 98.5 | 97.0 |
| EC, kWh $kg^{-1}$ | 7.1 | 7.8 | 8.2 | 8.6 |
| ACD, mm | 20 | 20 | 20 | 20 |

Current efficiencies above 90% can easily be achieved for all the tests performed in this study as seen in Table 1, despite the quite short inter-electrode distances employed. The level of current efficiencies obtained is an indicative of good separation of magnesium metal and chlorine gas during the electrolysis. Current efficiency decrease with increased anodic current density is pronounced for smaller ACD experiments.

A rapid initial rise in cell voltage was followed by a decrease in 13 mm ACD, forming a peak as shown in Figure 3. Then the cell voltage remained constant or decreased slightly until the end of the experiments. Both the magnitude of the peak and voltage relaxation rate at the latter portion was pronounced at higher current densities. Similar behavior, but to a less extent, was observed for 20 mm ACD, Figure 4.

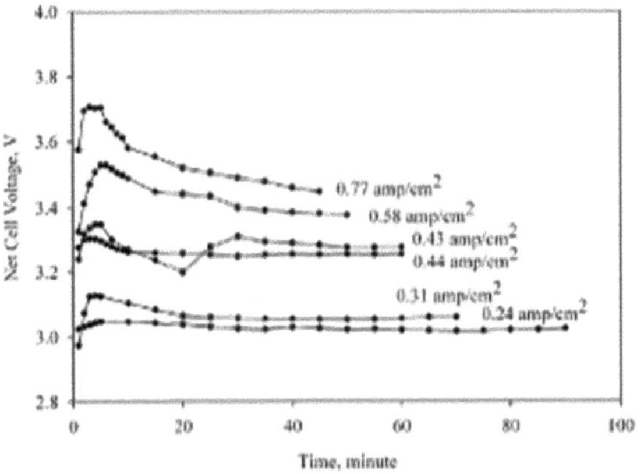


Figure 3 Net cell voltage vs. time at 13 mm ACD.

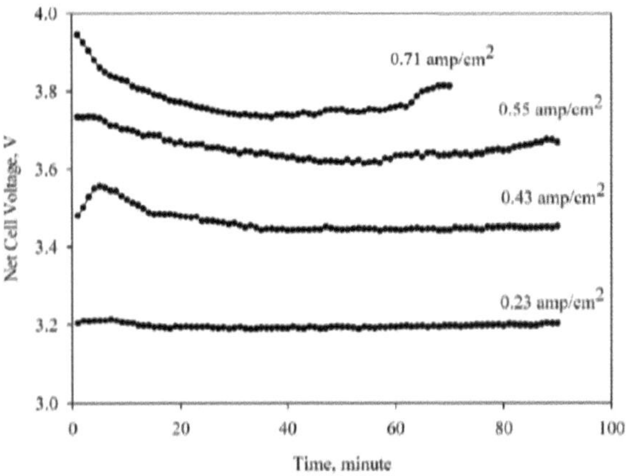


Figure 4 Net cell voltage vs. time at 20 mm ACD.

Characteristic peaks at the beginning of the voltage-time graphs of the experiments are related with the energy required for nucleation. The area under the peaks increased with the increase in current density due to the larger number of nucleation sites.

The final portions of the individual curves were under the control of composition change in general. Voltage relaxation after the initial peaks was due to the decrease in IR drop as a result of the decrease in effective ACD by the magnesium deposition at the cathode surface and increase in electrical conductivity of the melt by the composition change. Cell voltage started to increase at the final portions of the curves due to the dominant effect of decrease in $MgCl_2$ activity when the $MgCl_2$ depleted too much.

The theoretical energy consumption was calculated as 6.2 kWh for present experimental conditions. The computation involves the use of 2.76 V theoretical decomposition potential which was obtained from 2.514 V standard decomposition potential [14] and $MgCl_2$ activity of 0.0025 for the electrolyte [15]. In fact, it is desirable to operate the cell at a voltage higher than theoretical decomposition voltage to supply heat energy required for endothermic $MgCl_2$ decomposition. From the thermo-neutral voltage, which is calculated using the enthalpy change of $MgCl_2$ decomposition reaction, energy consumption of about 7 $kWhkg^{-1}$ was obtained [16]. Furthermore, additional heat must be supplied to compensate the losses from the cell in practice.

As it can be seen in Figure 5, 7.0 kWh $kg^{-1}$ Mg of electrical energy was consumed, at 0.68 $Acm^{-2}$ and 0.24 A $cm^{-2}$ anodic and cathodic current densities respectively for an inter-electrode distance of 13 mm. It is slightly higher than the theoretical value 6.2 kWh for these experimental conditions and well below the lowest energy consumption values industrially practiced, 10.5 to 11.5 kWh $kg^{-1}$ Mg [7].

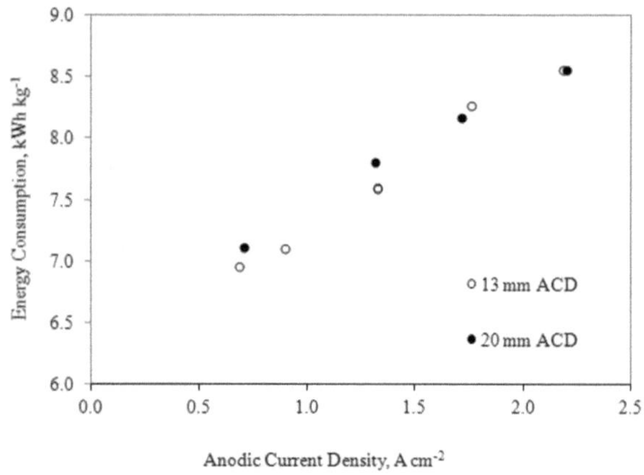


Figure 5 The effect of current density on energy consumption.

There is a small increase in energy consumption with decrease of ACD from 20 mm to 13 mm. This was attributable to the increased turbulence inside the electrolyte with decrease in ACD, since magnesium becomes more susceptible to back reaction when entrapments of chlorine gas or transportation of magnesium occur.

## Conclusions

The cell operates at a low potential due to the low ACD. As a result, 7.0 kWh $kg^{-1}$ Mg was consumed at 0.68 $Acm^{-2}$ anodic current density that is slightly higher than the theoretical value of 6.2 kWh.

The cell demonstrated that, effective separation of electrode products and reduced chlorine diffusion in to the melt due to rapid chlorine removal from the cell made it possible to work with small inter electrode distances and high current efficiencies.

## Acknowledgments

Authors acknowledge the financial support provided by State Planning Organization (DPT) and Scientific and Technological Research Council of Türkiye (TÜBİTAK).

## References

1. Wilson C. B., Claus K. G., Earlam M. R., Hillis J. E., in Kirk-Othmer Encyclopedia of Chemical Technology 4th ed., vol. 14, Wiley, New York, 1995, p. 622.
2. M. Güden, İ. Karakaya, J. Appl. Electrochem. 24 (1994) 791–797.
3. R. Sharma, in: M. Avedesian, R. Guilbault, D. Ksinsik (Eds.), Light Metals 1996, The Metallurgical Society of CIM, Montreal, Canada, 1996, pp.1113–1122.
4. R. Sharma, Electrolytic production process for magnesium and its alloys, US5593566 (1997).
5. H.E. Friedrich, B.L. Mordlike: Magnesium Technology, Springer, Berlin, 2006.
6. R.L. Thayer, R. Neelameggham: JOM, 2001, vol.53, no.8, pp.15-17.
7. P.Ficara, E.Chin, T.Walker, D.Laroche, E.Palumbo, C.Celik: CIM Bulletin, .91, no 1019, April 1998, pp.75-80.
8. A.M. Martinez, B. Borresen, G.M. Haarberg, Y. Castrillejo and R. Tunold: Journal of Electrochemical Society, 151 (7), C508-C513, (2004).
9. G. Demirci, İ. Karakaya: Journal of Alloys and Compounds, 439 (2007) 237–242.
10. G. Demirci, İ. Karakaya: Journal of Alloys and Compounds, 465 (2008) 255-60.
11. G.Demirci, İ. Karakaya: EPD Congress 2011, 597-604 (TMS 2011 Annual Meeting & Exhibition, 2011 San Diego, California).
12. G.Song, A.Atrens, D. St John, X. Wu and J. Nairn: Corrosion Science, 1997, vol. 39 (10-11), pp.1981-2004.
13. G. Demirci, "Electrolytic magnesium production using coaxial electrodes" (Ph.D. Thesis, Middle East Technical University, Ankara, 2006).
14. İ. Karakaya, W.T. Thompson, Canadian Metallurgical Quarterly, 25 (1986) 307-317.
15. E.I. Savinkova, R.P. Lelekova, Zhurnal Prikladnoi Khimii 51 (1978) 1453–1456.
16. İ. Karakaya, "Electrochemical determination of thermodynamic properties of magnesium cell electrolyte: the system $MgCl_2$-NaCl-$CaCl_2$" (Ph.D. thesis, McGill University, Canada, 1985).

**Magnesium Technology 2012**
*Edited by: Suveen N. Mathaudhu, Wim H. Sillekens, Neale R. Neelameggham, and Norbert Hort*
***TMS (The Minerals, Metals & Materials Society), 2012***

# Preparation of aluminum-magnesium alloy from magnesium oxide in $RECl_3$-KCl -$MgCl_2$ electrolyte by molten salts electrolysis method

Shaohua Yang[1],Lin Wu[2], Fengli Yang[1], Mingzhou Li[1], Xianwei Hu[3], Zhaowen Wang[3], Zhongning Shi[3] ,Bingliang Gao[3]
1. School of Materials and Chemistry, Jiangxi University of Science and Technology
2. China Iron&Steel Research Institute Group
3.School of Materials and Metallurgy117#, Northeastern University.

Keywords: Alloy; Preparation; Molten salt; Electrolysis

## Abstract

Aluminum-magnesium alloys were prepared from magnesium oxide by molten salt electrolysis method. 10w%$RECl_3$-63.5w%KCl-23.5w%$MgCl_2$-3w%MgO was taken as electrolyte. The results showed that RE could be attained in aluminum-magnesium alloy, and it was proved that the RE was reduced directly by aluminum. Magnesium in the alloy was produced by electrolysis on cathode. The content of RE in the alloy was about 0.8wt %-1.2wt%, and the content of Mg in the alloy was 1wt%~6wt% with electrolytic times. The highest current efficiency was 81.3% with 0.8A/cm2 current density. The process of electrolysis was controlled together by electrochemical polarization and concentration polarization.

## Introduction

As the lightest of all the commonly used metals, magnesium is very attractive for applications in transportation. It also has other advantageous features, such as good ductility, better damping characteristics than aluminum, and magnesium alloys are readily casting, particularly by high-pressure die casting and exhibit good mechanical properties, they have been promoted as a desirable component for use in many fields, such as aerospace industry, electrical industry, car manufacture and so on. Aluminum-magnesium alloys, in particular, are used wider than others. Aluminum-magnesium alloys containing about 5% magnesium are used in welded structures, such as in ship building and construction field [1~3].

However, commercial applications of magnesium alloys are limited because of poor creep resistance and poor tensile properties at elevated temperatures above 120°C. Rare earths are important alloying elements to magnesium alloys, which can improve casting characteristics, high temperature properties and corrosion resistance without affecting the electrical conductivity of the base alloys. Rare earths have been used in magnesium alloys for many years, whereas the alloys of Mg–Al–RE system have been developed recently. Some investigations have been done on the precipitation, morphology, structure, thermal stability and strengthening mechanisms of inter-metallic phases when RE were added to Mg–Al alloys[4~6].

There are many impossible ways for resolving the contradiction between increasing need of magnesium alloys and high cost of its production, but the best way is to prepare the Al-Mg alloys from MgO by molten salt electrolysis method[7-10]. For many years, many researchers have studied and attained some exciting results on this field[11-19]. Therefore, the method of preparation Al-Mg-RE alloy by molten salt electrolysis is promising way to solve the problems of large productive scale and low productive cost.

This paper focus on preparation of Al-Mg alloys by molten salt electrolysis from magnesium oxide, reduction mechanic was investigated by electrolytic experiments and no-electrolytic experiments.

## Experiment

Experimental apparatus and agents

The purity of rare earth oxide was above 99w%, and its elemental content was shown as table 1:

Table 1 Content of elements in mixed rare earth (w/%)

| Elements | La | Ce | Pr | Nd | Others |
|---|---|---|---|---|---|
| Content | 24.6 | 50.8 | 5.2 | 17.4 | 2 |

$RECl_36H_2O$ was prepared by $RE_2(CO_3)_3$ dissolving in chlorhydric acid, and it was dehydrated at 350°C after amount of ammonium chloride added, $RECl_3$ was prepared by this way. $MgCl_2$ was prepared from $MgCl_2 \cdot 6H_2O$ at 350-400°C in resistance furnace by vacuum dehydration way.

Super low-frequency oscilloscope(SBD6B), Resistance furnace(2kw);Temperature controller(DWK-702); KCl(analytic agent); MgO(analytic agent);$MgF_2$(analytic agent);LiF(analytic agent)

Experimental installing

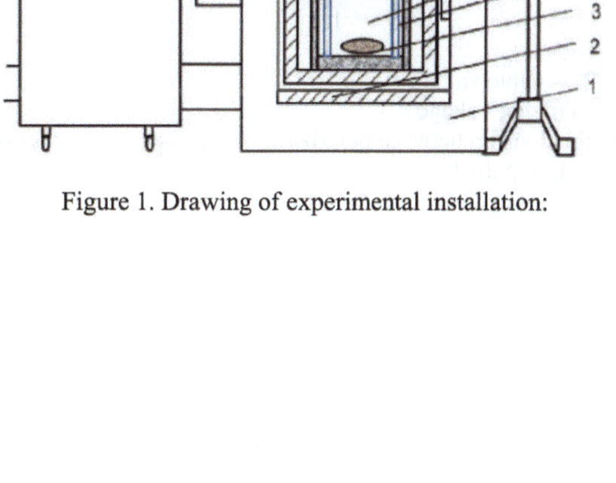


Figure 1. Drawing of experimental installation:

1-Electrical resistance furnace; 2-Refractory brick; 3-SiC board; 4-Corumdum bushing; 5-Electrolyte; 6-Graphile crucible; 7-Tungsten cathode; 8-Iron outer; 9-Furance cover; 10-Supporter; 11-Elevator controller; 12-Tungsten wine;13-Temperature wine; 14-DWK-702 temperature controller.15-Anode.

Process of experiments

All the agents ware dried at 400℃for two hours. According to calculated ratio, the agents were weighed accurately on balance. About 280g 10w% $RECl_3$-23w%$MgCl_2$-67w%KCl electrolyte was fed into graphite cell and melted in resistance furnace. About 30g-40g aluminum was added into corundum tube. After the aluminum was melted into liquid, electrolysis could begin with liquid aluminum as cathode, magnesium oxide as raw material, and graphite as anode. During electrolysis process, amount of magnesium oxide was fed into electrolyte to guarantee stable electrolysis.

## Results and discussion

Reduction of rare earth

10w%$RECl_3$-23w%MgCl-67w%KCl as electrolyte, and liquid aluminum as reduction, experiment was carried through at 720℃ ~780℃temperature, no current was loaded on electrodes from one to three hours. Content of rare earth in alloy was shown in table 2.

Table 2 Contents of Mg and RE in molten Al (w/%)

| | 1 | 2 | 3 | 4 |
|---|---|---|---|---|
| Mg | 0.02 | 0.01 | 0.08 | 0.04 |
| RE | 1.10 | 1.02 | 0.87 | 0.90 |

The results showed that rare earth was reduced directly by aluminum, and its content in alloy was stable from 0.87-1.1w%. This could be matched with demand of practical application, because some researchers had reported that over 1w% of rare earth in alloys had no effect on material properties. Content of magnesium in alloy was very lower than 0.08w%. Therefore, preparation of Al-Mg-RE alloys could be carried out by molten salt electrolysis method, rare earth with stable content in alloy attained by reduction of aluminum, and magnesium was reduced in cell from magnesium oxide. The content of magnesium in alloys was controlled by electrolytic time.

Polarizing voltage and current density

Using 10w% RECl3-23w%MgCl-67w%KCl as the electrolyte, electrolysis was performed at 750℃ temperatures with different current. The back electromotor force (BEMF) was measured by oscilloscope, and polarizing voltage was calculated by following equation.

$E_B=E_0+\varepsilon$ —1

$E_B$--BEMF

$E_0$--Reversible voltage

$\varepsilon$-- Polarizing voltage

The relationship between polarizing voltage and current density is shown in Figure 2.

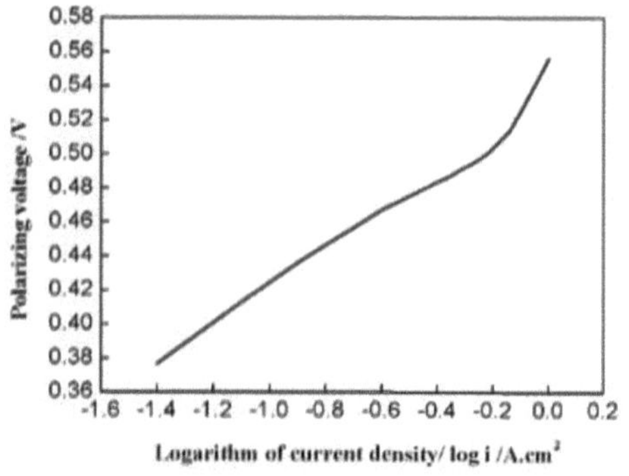


Fig.2 Relationship between polarizing voltage and current density

Process of electrolysis was controlled by electrochemical and concentration polarization together. Under low current density from 0.04-0.6A/ cm$^2$, content of magnesium ion on cathode surface could match with need of electrolysis. Electrolysis could not be affected by depletion of magnesium ion. But, under high current density over 0.6A/ cm$^2$, content of magnesium ion on cathode surface could not match with need of electrolysis. Velocity of magnesium ion diffused in molten salts was not matched with velocity of magnesium ion precipitated on cathode, thus gradient of concentration was created from cathode surface to inner of molten salts. This was main reason of concentration polarization under high current density.

Polarizing voltage and logarithm of current density were fitted with a linear regression line by the method of least squares, and the line equation was:

$\varepsilon_1=0.521+0.099\log I$ --2

$\varepsilon_2=0.555+0.277\log I$ --3

"$\varepsilon_1$" was polarizing voltage with current density under 0.6 A/ cm$^2$, "$\varepsilon_2$" was polarizing voltage with current density over 0.6 A/ cm$^2$, and "I" was current density.

Relationship between current and back EMF, cell voltage

Electrolytic experiment was carried out in different current, and the relationship between current and back electromotor force, cell voltage was shown as Figure 3:

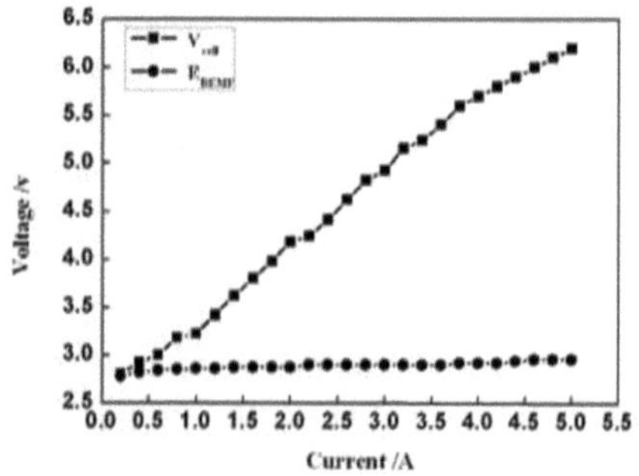


Fig.3 Relationship between current and back EMF, cell voltage

Cell voltage increased with current, but the back electromotor force almost is stable. The value of cell voltage increased from 2.8V to 6.2V, and that of BEMF was from 2.78V to 2.96V. The differences are 3.4V, 0.18V, respectively. The main reason was because the reversible voltage $E_0$ could not change at the same condition, but BEMF could increase with polarizing voltage. Polarizing voltage was influenced with two sides, electrochemical reaction and diffusion. At low current density, the process of electrode was controlled by electrochemical reaction. On the contrary, the process of electrode was controlled by diffusion at high current density. The point was correspondence with Figure 2.

Content of magnesium in alloys

Graphite as anode, aluminum as cathode, the experiment was at 760°C with 0.8A/cm$^2$ current density, the electrolytic time was 30mins,45mins,60mins,75mins and 90mins. The relationship between content of magnesium in alloys and electrolytic time was shown as figure 4

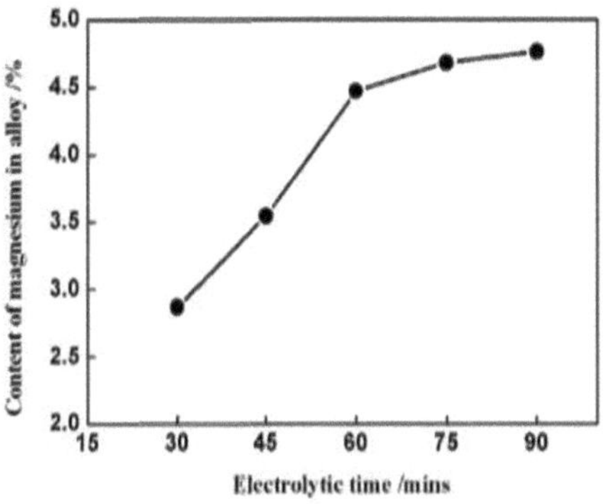


Fig. 4 Relationship between electrolytic time and content of magnesium in alloy

Content of magnesium in alloys increased with electrolytic time, but the increased value was almost at 0.2w%. In another word, content of magnesium in alloys was almost stable after a long electrolytic time. To calculate the current efficiency, the most one was 81.3% when the electrolytic time was about one hour. The phenomena was observed after one hour electrolytic time, bubbles on anode was difficulty in releasing electrolyte surface, and the electrolyte was like viscous. Further study was kept on how to control content of magnesium in alloys with electrolytic time in long operation.

## Conclusions

It is feasible to prepare Al-Mg alloys with 0.8-1.1w% rare earth by molten salt electrolysis from magnesium oxide. The rare earth was reduced directly by aluminum, and magnesium in alloys was attained by decomposed magnesium oxide. The content of rare earth was stable 0.8-1.1w%, and that of magnesium was controlled by electrolytic time. 10%$RECl_3$-23%MgCl-67%KCl as electrolyte, the most current efficiency was in 81.3%. The content of magnesium in alloys increased with electrolytic time, the most was 4.6w%. Process of electrolysis was controlled by electrochemical and concentration polarization together.

## Acknowledgements


The authors wound like to acknowledge the financial supported by "National Natural Science Foundation of China (51164013)", and express great thanks to Pr. Zhaowen Wang for his help in experiments.


## References


1. H.Friedrich, S.Schumann. Research for a "new age of magnesium" in the automotive industry [J].Journal of Materials Procession Technology, 117(2001): 276~281
2. Aghion E. And Bronfin B. Magnesium Alloys Development towards the 21st Century[J]. Materials Science Forum,350~351(2000):19~28
3. Arnek.Dahle, Young C.Lee, Mark D.Nave etal. Development of the as-cast microstructure in magnesium-aluminium[J].Journal of Light Metals,1(2001):61~72
4. Guangling Song, DavidStJohn. The effect of zirconium grain refinement on the corrosion behaviour of magnesium-rare earth alloy MEZ[J]. Journal of Light Metals, 2002 (2):1–16
5. Yizhen Lu, Qudong Wang, Xiaoqin Zeng, Wenjiang Ding, Chunquan Zhai. Effects of rare earths on the microstructure, properties and fracture behavior of Mg–Al alloys[J].Materials Science and Engineering A278 (2000):66–76
6. Friedrich H, Schumann S. Journal of Materials Procession Technology[J], 2001, 117: 276
7. Grobner J, Schmid Fetzer R. Journal of Alloys and Compounds[J], 2001, 320: 296
8. Zeng Xiaoqin, Wang Qudong, Lu Yizhen et al. MaterialsProcessing Technology[J], 2001, 112: 17
9. Ram A. Sharma. An electrolytic process for magnesium and its alloys production [C]. Light Metals, WTMS, 1996: 1113~1122.
10. Ram A. Sharma. A new electrolytic magnesium production process[J]. JOM, 10(1996):39~43
11. Soobhankar Pati, Srikanth Gopalan, Uday B. Pal .A solid oxide membrane electrolyzer for production of hydrogen and syn-gas from steam and hydrocarbon waste in a single step[J] International Journal of Hydrogen Energy, Volume 36, Issue 1, January 2011, Pages 152-159
12. Soobhankar Pati, Marko Suput, Rachel DeLucas, Uday Pal SOLID OXIDE MEMBRANE PROCESS FOR CALCIUM PRODUCTION DIRECTLY FROM ITS OXIDE [C]Light Metals. Warrendale: Minerals, Metals & Materials Soc, 2008:121~126.
13. Rachel DeLucas, Guosheng Ye, Marko Suput, and Uday Pal. Modeling of magnesium extraction from magnesium oxide by SOM process. In Advanced Processing of Metals and Materials Vol. 4: New, Improved and Existing Technologies: Nonferrous Materials Extraction and Processing, pages285–298. TMS, September 2006.
14. Qiu Z X, Zhang M J. Preparation of aluminum master alloy

by electrolysis in molten cryolite[J]. Aluminum, 1990,66 (6): 560~564.

15. Shaohua Yang, Bingliang Gao, Zhaowen Wang, et al. Preparation of aluminum-magnesium alloys starting from magnesium oxide [C]Light Metals. Warrendale: Minerals, Metals & Materials Soc, 2007:283~286.
16. Shaohua Yang, Zhaowen Wang Ying Nie, Jidong Li, Xianwei Hu.Preparation of Al-Mg alloys from MgO through molten salt electrolysis method[A].Light Metals [C]. Warrendale:Minerals, Metals & Materials Soc, 2008:17-20
17. Shaohua Yang, Zhaowen Wang, Ying Nie, Xingliang Zhao.Electrodeposition of Magnesium from BaF2-LiF-MgF2 electrolyte[A].Light Metals [C]. Warrendale : Minerals, Metals & Materials Soc, 2008:13-15
18. Shaohua Yang, Fengli Yang, Guocheng Wang. Effect of KCl on conductivity of BaF2-LiF-MgF2 molten salts[A].Light Metals. Warrendale[C]: Minerals, Metals & Materials Soc, 2010:111-114
19. Fengli Yang, Shaohua Yang, Xianwei Hu, Zhaowen Wang, Zhongning Shi ,Bingliang Gao. Preparation of Al-Mg alloys from MgO in KCl-MgF2-LiF electrolyte by molten salt electrolysis method [A].Light Metals. Warrendale[C] : Minerals, Metals & Materials Soc, 2011:183-186

# EXPERIMENTAL STUDY ON MAGNESIUM EXTRACTED FROM ASCHARITE MINERAL BY ALUMINIUM

Peng Jianping[1], Wu Xiaolei[1], Feng Naixiang[1], Zhou Shigang[1], Di Yuezhong[1]
[1]School of Materials and Metallurgy, Northeastern University, Shenyang, 110819, China

Keywords: ascharite mineral, magnesium, aluminum powder, vacuum thermal reduction

## Abstract

In order to effectively utilize the ascharite mineral, in Liaoning province of China, the paper studied to extract magnesium from ascharite mineral with aluminum powder as reductant by vacuum thermal reduction process. And boron of the raw material was saved in residue that could be used as raw materials to produce non-alkali glass fiber. Environment cannot be polluted during this process, as well as the process can create great economic value. Calcined material and residue were taken XRD analysis. Based on thermodynamic analysis, the Gibbs free energy and critical temperature of reduction ascharite with aluminum were calculated. As a result, the critical temperature is 1302℃ at ordinary pressure, and addition, the critical temperature will reduce farther when the air pressure is reduce in the retort, The recovery rate was effected remarkably by the mass of reductant and the making briquettes pressure.

## Introduction

Currently the ascharite mineral is used to produce borax by carbon dioxide-soda process in China. The boron utilization rate is lower than 75%. At the same time, magnesium resource and the rest boron is remained in the residue, named boron mud, which has never been utilized [1]. In this paper, magnesium was extracted from ascharite mineral by vacuum thermal reduction process, and other components were kept in the residue, which can be used to manufacture non-alkali glass fiber. It avoids pollution of boron mud and effectively utilizes the resource.

Ma Lihong [2] and Xu Dong [3] have studied the vacuum thermal extraction magnesium from ascharite mineral. However, the material they used contained lower amounts of boron and magnesium. As a result, the magnesium's recovery rate was very low and the content of boron in residue was very low too, so that the residue was low value. Wu X L [4] and Peng J P[5] has made researches on vacuum thermal extract magnesium from rich-boron and rich-Mg mineral with silicon and $CaC_2$ as reductant, respectively. While silicon serving as reductant needs adding a lot of CaO, the boron in the residue gotten was diluted. The reduction test was also undesired because the reductive ability of $CaC_2$ was poor.

In this paper, aluminum powder was chosen as reductant because it highly reducing in nature and it needs little CaO in contrast with Si.

## Experiments

### Experiment materials

The chemical compositions of ascharite mineral, from Kuandian, Liaoning province, are shown in table 1.

Table1 Chemical compositions of ascharite mineral

| Chemical composition | MgO | $B_2O_3$ | $SiO_2$ | CaO | $Al_2O_3$ | Others |
|---|---|---|---|---|---|---|
| Content (%.ω) | 44.67 | 26.08 | 6.18 | 3.19 | 1.43 | 18.45 |

Other materials including aluminum powder, $CaCO_3$ and $CaF_2$ are reagents which are produced by Tianjin Kemio Chemical Reagent Co., LTD Production. Particle sizes are less than 74 μm, 100%.

In each test, 80 g ascharite mineral powder was mixed with 151.10 g $CaCO_3$ prior to compaction into several briquettes by 45 MPa. Then these compacted briquettes were fed into a resistance furnace and calcined at 1000℃ for 30 min to eliminate $H_2O$ and $CO_2$. Then the calcined material is milled to -120 μm particle. Figure 1 shows the XRD pattern of the calcined material.

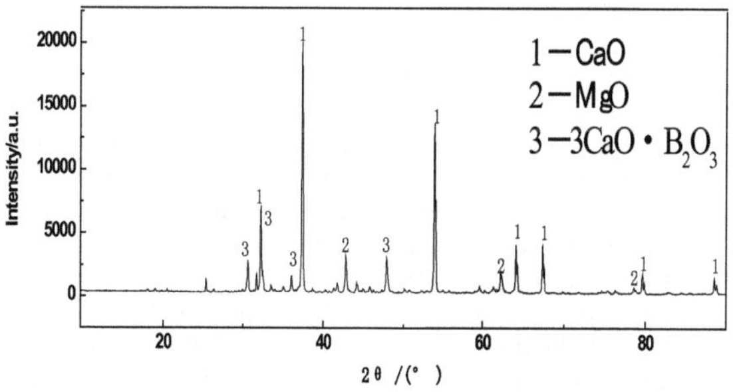


Fig.1 XRD pattern of calcined szaibelyite mineral and $CaCO_3$

According to the Fig 1, the calcined material mainly contains CaO, MgO and $3CaO{\cdot}B_2O_3$.

### Equipment

The equipments in experiments mainly included resistance furnace, temperature detection and control system, vacuum system, reduction equipment and cooling system. The sketch of equipments is shown as Fig 2.

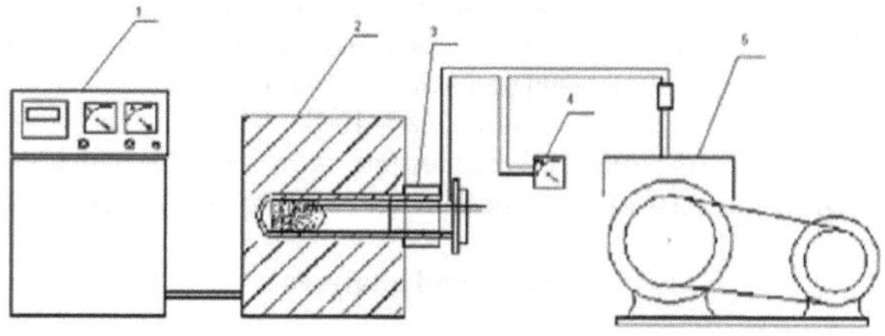

1-Tempareture controlling unit, 2-Reduction furnace, 3-Retort, 4-Vacuum gauge, 5-Vacuum pump
Fig.2 Schematic diagram of experimental apparatus

Experiment Process

The test outline is summarized in the figure below and follows the typical Pidgeon process flow sheet.

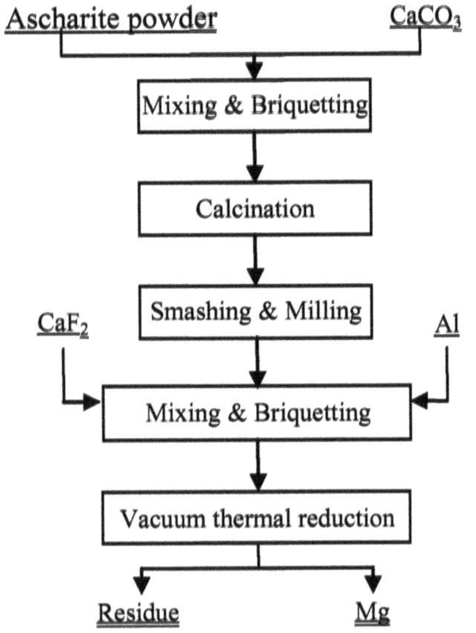


Fig.3 Flow sheet of experiment

The influence of the Al mass and the briquetting pressure prior to reduction on recovery rate was studied in the experiment. In each test, 80 g ascharite mineral powder was mixed with 151.10 g $CaCO_3$. Then the mixed material were compacted into briquettes and undergone calcination. The calcined material was milled and mixed with different mass Al and 5.07 g $CaF_2$, and then the mixed material for reduction tests was compacted into briquettes as 30MPa pressure. Then the influence of the Al mass on recovery rates was researched. At last confirm the charge mass of Al was 18.28 g in each test.

Second part of the experiment, keeping the conditions as before, only changing the briquetting pressure of reduction material, the influence of briquetting pressure on recovery rate was studied.

All experiments were done at 1200℃ for 120 min, and within vacuum system pressure of 4 Pa.

The method of calculating recovery rate

Before reduction, the magnesium collector was weighed as $m_0$. After reduction, the collector was weighed again as $m_1$. So the recovery rate $S_1$ was calculated as following formula,

$$S_1=(m_1-m_0)/m^*\times 100\% \quad (1)$$

where $m^*$ is the mass of magnesium, which can be calculated based on the mass of ascharite mineral contented in the briquettes.

## Results and analysis

The productions and thermodynamic analysis

The experiment produces include magnesium condensed and rich-boron residue. The chemical composition of magnesium condensed is shown in table 2:

Table 2 Chemical analysis result of prepared magnesium compare with ISO Mg-99.8

| | Mg | Fe | Si | Ni | Cu | Al | Mn |
|---|---|---|---|---|---|---|---|
| Experimental value (%) | 99.87 | 0.027 | 0.005 | 0.003 | 0.011 | 0.039 | 0.037 |
| ISO (%) | 99.80 | 0.05 | 0.05 | 0.002 | 0.02 | 0.05 | 0.1 |

It can be see that the produced magnesium in experiments is almost accord with the ISO.

Figure 4 shows the XRD pattern of the reduction residue from the reduction experiment with 18.28 g Al and 30MPa briquetting pressure.

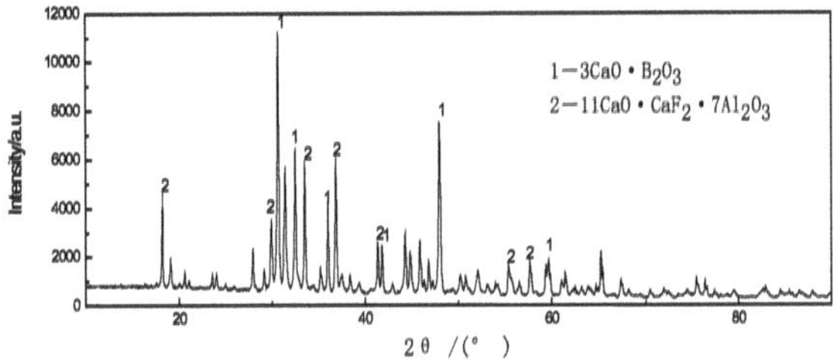


Fig.4 XRD pattern of residue

According to the fig 4, main components of the residue are $3CaO\cdot B_2O_3$ and $11CaO\cdot CaF_2\cdot 7Al_2O_3$. The residue can be used to produce non-alkali glass fiber because it is rich-boron. The chemical compositions of the residue and non-alkali glass fiber are shown in table 3.

Table 3 Chemical composition of experimental residue and non-alkali glass fiber, (%.$\omega$)

| | MgO | $B_2O_3$ | $SiO_2$ | CaO | $Al_2O_3$ | Others |
|---|---|---|---|---|---|---|
| Experimental residue | 3.49 | 14.7 | 3.00 | 56.42 | 21.89 | 0.50 |
| Non-alkali glass fiber | 4.5±0.5 | 8.8±0.5 | 54.1±0.5 | 16.6±0.3 | 14.6±0.4 | |

According to the table 3, the content of $B_2O_3$ in residue is higher than non-alkali glass fiber's demand. The other compositions in residue are also needed in glass fiber.

Fig 4 shows that the MgO is deoxidized and that the $11CaO\cdot CaF_2\cdot 7Al_2O_3$ is made in experiment. The reaction equation of deoxidizing MgO is shown as following:

$$12CaO_{(s)} + 21MgO_{(s)} + 14Al_{(l)} = 21Mg_{(g)} + 12CaO\cdot 7Al_2O_{3(s)} \quad (2)$$

$$\Delta G_2^{\theta} = 3567815-2265.55T \text{ (J/mol)}.$$

At standard atmospheric pressure, the critical reaction temperature of equation (2) is 1302℃. If the system pressure is 10132.5Pa, the Gibbs free energy of equation (2) will be changed as following:

$$\Delta G_2 = \Delta G_2^{\theta} + 21RT\ln(p_{Mg}/p^{\theta}) = 3567815- 2667.57\,T \text{ (J/mol)}$$

As a result, the critical reaction temperature falls to 1064℃. In experiments the system pressure usually was 4 Pa or even low, and the temperature in retort was 1200°C, the conditions in experiments can meet the thermodynamic demands.

The relationship between reductant mass and recovery rate

According to the chemical compositions of ascharite mineral in table 1, there are 35.74g MgO in 80g ascharite mineral. Based on the reaction equation (2), the stoichiometric requirement of Al is 16.08g, but in experiments the Al mass usually is above the stoichiometric requirement. Let the Al mass above 16.08g rate is $\mu$, and the $\mu$ is counted by the following formula:

$$\mu=(M-16.08)/16.08\times100\% \tag{3}$$

where $M$ is the addition mass of Al in test, g. The results are shown in the table 4:

Table 4 The recovery rate obtained by different Al mass experiments

| No. | $\mu$(%) | Al (g) | $CaCO_3$ (g) | Recovery rate (%) |
|---|---|---|---|---|
| 1 | 3.36 | 16.62 | 147.98 | 84.86 |
| 2 | 5.97 | 17.04 | 148.76 | 87.67 |
| 3 | 8.52 | 17.45 | 149.54 | 89.44 |
| 4 | 11.13 | 17.87 | 150.33 | 92.26 |
| 5 | 13.68 | 18.27 | 151.11 | 93.62 |

According to table 4, with the increasing of Al mass, the chance of Al contact with MgO is increased that performs as the material activity is increased and the recovery rate is improved.

Influence of briquetting pressure on recovery rate

Keeping the experimental conditions as before, with 18.28 g Al, changes the briquetting pressure of reduction material. The briquetting pressures of reduction material and the recovery rates are shown in table5.

Table 5 The recovery rate obtained by different briquetting pressure experiments

| No. | Briquetting pressure (MPa) | Recovery rate (%) |
|---|---|---|
| 1 | not being compacted | 82.53 |
| 2 | 15 | 88.86 |
| 3 | 30 | 93.62 |
| 4 | 45 | 90.48 |
| 5 | 60 | 87.51 |

The test No. 1 means that the calcined material powder didn't be compacted and was directly feed in retort for reduction test. According to the theory of Pidgeon process, which is the method currently to extract magnesium, if the material not be compacted, only a little magnesium is gotten.

However in the test No.1 of table 5, Al and ascharite mineral is melt at the experiment temperature, so it still can obtain a high recovery rate because mass and heat transfer in liquid state is easier than in solid state. According to table 5, the best briquetting pressure is about 30 MPa, and under this condition, the highest recovery rate can reach 93.62%. When the briquetting pressure continues growing up the recovery rate will fall.

## Conclusions

There are a lot of $B_2O_3$ in the residue which can be used to produce non-alkali glass fiber.

Based on thermodynamic analysis, with Al as reductant and enough CaO, the critical temperature is 1302°C. If the pressure in retort is 10132.5Pa, the critical temperature will be 1064°C. So the reaction under experiment conditions can progress well.

It is proved by experimental results shows that the recovery rate can be increased by increasing the mass of Al. With 151.10 g $CaCO_3$, 5.07 g $CaF_2$ and 80 g ascharite mineral in each test, the reasonable addition mass of Al is 18.27 g.

The results show that it is not necessary to compact the calcined material powder to achieve high magnesium recovery rate. The reasonable briquetting pressure is 30 MPa, and the recovery rate can reach 93.62%.

## Acknowledgement

The authors wish to thank the financial support provided by the Fundamental Research Funds for the Central Universities (N100302009).

## References

1. Wu X L, Feng N X, Peng J P, et al. Vacuum Thermal Extract Magnesium from Boron Mud [A]. Nyberg E A, Agnew S R, Neelameggham M R, et al. Magnesium Technology Symposium, Magnesium Technology 2009 [C]. San Francisco: TMS, 2009. 61–63.
2. Ma Lihong. Experiment research on thermal reduction magnesium with ascharite minerals [D]. Shenyang: Northeastern University, 2007. 27.
3. Xu Dong, Zhang Xianpeng, Li Jie, et al. Experimental Investigation on Vacuum-Thermal Reduction of Ascharite with Silicon-Aluminum Alloy as Reductant [J]. Journal of Northeastern University (Natural Science), 2008, 29(10): 1455–1458.
4. Wu X L, Feng N X, Peng J P, et al. Experimental Study on Vacuum Thermal Reduction of Ascharite Mineral with Silicon as Reductant [J]. The Chinese Journal of Process Engineering, 2011, 11 (4): 294-298.
5. Peng J P, Wu X L, Zhou S G, et al. Experimental Study on Vacuum Thermal Reduction of Ascharite Mineral with Calcium Carbide as Reductant [A]. Vacuum Technology and Surface Engineering-Proceedings of the 10th Vacuum Metallurgy and Surface Engineering Conference, Vacuum Engineering Conference 2011 and Vacuum Consultancy Workshop 2011 [C]. Beijing: Publishing House of Electronic Industry, 2011. 5

**Magnesium Technology 2012**
*Edited by: Suveen N. Mathaudhu, Wim H. Sillekens, Neale R. Neelameggham, and Norbert Hort*
***TMS (The Minerals, Metals & Materials Society), 2012***

# Electrochemical investigation on chlorine and electrolyte intercalation into graphite anodes during magnesium electrolysis process

Bing Li*, Jingwei Lou, Mengfan Yan, Jianguo Yu
National Reaction Engineering Key Laboratories, National Engineering Research Center for Integrated Utilization of Salt Lake Resource, East China University of Science and Technology, Shanghai, 200237, China

Keywords: Graphite anode; Cyclic voltammetry; Chlorine evolution; Intercalation; Chloride melts

## Abstract

Cyclic voltammetry (CV) was used to investigate chlorine and electrolyte intercalation into three different graphite anodes from NaCl-KCl-$MgCl_2$ melts at 700°C. The three anodes were composed of needle-coke (NC), petroleum-coke (FPC) and common petroleum-coke (CPC), respectively. Chlorine intercalation amount was characterized by the reduction current (reduction electricity quantity) on the reverse scan during the cyclic voltammograms. And the electrolyte intercalation was presented by the increase in oxidation charge between the forward scan and the reverse scan during the CV measurements. The results show that among the three graphite anodes, NC shows the lowest reduction current and nearly no increase in the charges, while about 5-10 times increase in the charge for the PFC and CPC. The above results implied that NC has a better resistance to electrolyte and chlorine intercalation, which were confirmed by the electrolysis experiments results. As a simple and useful method, CV can be used to test graphite anode corrosion resistance to electrolyte and chlorine in chloride system.

## Introduction

Currently, graphite is the only material of choice to be used as the anode in molten chloride melts, such as in Mg, Li, Na, Ca electrolytic processes. Graphite anodes are slowly consumed and have to be replaced after some time use due to oxidation caused by the cell atmosphere and raw material hydration[1-5]. In addition, electrolyte and chlorine can intercalate into graphite anode during the electrolysis, which not only results in graphite disintegration and consumption by the reaction with the evolved chorine, but more importantly in the production of highly toxic C-Cl compounds such as CCl4,olefins and arenes, etc. [6-7].In order to select a long lifetime graphite anode, it is necessary to select kinds of graphite anodes and investigate their corrosion resistance to the chloride melts and chlorine. CV can provide a quick and accurate information on reduction/oxidation currents, which can be used to determine chlorine and electrolyte intercalation currents that is related to the corrosion resistance to the chloride melts and chlorine. In this paper, three graphite materials-NC, PFC, and CPC are selected as the anode and CVs are employed to characterize the chlorine and electrolyte intercalation currents.

Table 1 the tested properties of graphite anodes

| Sample property | NC | FPC | CPC |
|---|---|---|---|
| Bulk density(g/cm3) | 1.705 | 1.575 | 1.811 |
| Open porosity | 14.72% | 17.90% | 8.17% |
| Electrical resistivity (μΩ.m) | 0.74 | 0.96 | 0.85 |
| Graphitization degree | 89.07 | 86.28 | 89.07 |
| Real density | 2.256 | 2.246 | 2.256 |
| Ash | 500ppm | 500ppm | 1000ppm |

## Experimental details

Composition of the electrolyte used in the experiment was composed of NaCl (AR>99.5%), KCl (AR>99.5%), and anhydrous MgCl2 (AR >99.9%). The NaCl and KCl were dried at 200 °C for 6 h and stored in a glove box for use in the experiments. The anhydrous $MgCl_2$ was stored in the glove box and used as received. The three graphite materials –needle-coke (NC), fine petroleum-coke (FPC) and common petroleum-coke (CPC) (all made in China) -were used as the working electrodes, respectively. Prior to use, each electrode was fabricated into a 6 mm. diameter rod, and shielded by an alumina tube with an apparent area of 0.45 $cm^2$ left exposed. A spectroscopic graphite rod was employed as the reference electrode and a W wire as the counter electrode. And each of the three graphite materials was also manufactured to 10mm ×20 mm ×30mm to serve as an anode for electrolysis, respectively, and a carbon steel was used as cathode. The basic properties of the three graphite materials were measured and listed in Table 1. All the graphite electrodes exposed in the melts were polished thoroughly with 1/0 to 5/0 emery papers, and then degreased by acetone and ethanol before immersion into the melts. Both electrolysis experiments and electrochemical measurements were carried out in an alumina crucible housed in a stainless steel vessel under argon atmosphere at 700°C. Electrochemical measurements were performed using Princeton 2273 potentiostats; data acquisition and processing were carried out by a Powersuite software. All the potentials in the paper were referred to the spectroscopic graphite. After electrolysis graphite anode was taken out from the cell and was observed by SEM after the surface electrolyte was cleaned completely.

## Results and discussion

NC material

The typical CVs recorded on NC electrode are shown in Fig.1. A forward scan from 0 to 1.5 V followed by a reverse scan produced Curve 1 of Fig. 1 (a). During the forward scan, there was a small adsorption peak at about 1.0-1.3 V due to chloride ion adsorption on the graphite electrode, it was also observed in the literature [8]. No distinct reduction current appeared during the reverse scan. But in the second scan (Curve 2 of Fig. 1 (a)), an obvious reduction current can be observed during the reverse scan, indicating that the evolved chlorine was reversibly intercalated into the graphite electrode during the forward scan, and then discharged (or deintercalated) in the reverse scan to give rise to the cathodic current. Such a cathodic current was also observed in a NaCl saturated $NaAlCl_4$ melts [7]. Moreover, according to literature [6], $Cl_2$ evolved in the graphite anode can react with the anode to form some C-Cl compounds. In fact, the maximum deintercalated current for the NC was about 6.3 mA/cm$^2$ according to the Curve 2 of Fig.1 (a). $Cl_2$ intercalation into the graphite electrode increases the risk of emitting C-Cl compounds and consuming the graphite anode. The charge consumed during the forward scan from the potential, A, at which chlorine starts to evolve, to 1.6V is the same as that consumed during the reverse scan from 1.6V to the potential, B, at which the oxidization current fell to zero, as shown in Curve 2 of Fig. 1 (a). The results indicate that the electrode area did not change, i.e. nearly no electrolyte and chlorine intercalated into the electrode. While during the reverse scan from the potential B to C, the charge decreased by 0.01 C due to the cathodic current generated.

When further scan to 2.0 V was performed and then reversed, the $Cl_2$ intercalation current increased to 48.1 mA/cm$^2$ (did not give here) and the charge increased by 1.4 C. This is attributed to the increase in working electrode area caused by chloride electrolyte intercalation under $Cl_2$-assistance, similar to the chloroaluminate melts system in which the chlorine intercalation compound has already been determined [7].The consumed charge increased by 3.0 C due to more electrolyte intercalation into the graphite anode under $Cl_2$-assistance when first scanned from the potential A to 4.0 V followed by a reverse scan to the potential B, as shown in Fig.1 (b).

Petroleum-coke (PC) materials

The FPC was similar to the NC in that there was no distinct reduction current for the first scan from 0 to 1.5 V. But for the second scan from 0 to 1.6 V followed by a reverse scan (Fig 2 (a)), the $Cl_2$ deintercalation current density reached 62.5 mA/cm$^2$, far more than that for the NC, which implies more chlorine intercalated into the FPC. The chlorine oxidation current quantity increased by 2.15 C (from about 0.2 C to 2.35 C) when scan from the potential, A, of the onset $Cl_2$ evolution to 1.6V then reversely scan to the potential, B, where the oxidization current fell to zero, as shown in Fig. 2(a), indicating that the working area increased by more than ten times as much as the original one. While the chlorine evolution current quantity increased by 4 C during the reverse scan from 4.0 V (Fig. 2(b)).

The CPC showed a large reduction current even during the first reverse scan from 1.5 to 0 V, reaching a current density of 56.8mA/cm$^2$, indicating a high reactive activity of the graphite material to $Cl_2$ evolution (Fig.2(c)). During the second scan from 0 to 1.6 V and then back to 0 V, the chlorine intercalation current reached 62.5 mA/cm$^2$, and the increase in chlorine evolution current quantity of 3.3 C was the largest among the three graphite anodes, which was caused by the biggest increase in the working electrode area – about 165%. In the anodic scan from the potential, A, where chlorine starts to evolve, to 4V and back to the potential, B, where anodic current fell to zero, the chlorine evolution current quantity increased by 5 C in Figure 2(b), the largest value among the three electrodes. These results demonstrate that the CPC anode is the worst at preventing electrolyte and chlorine intercalation. By comparison of the chlorine evolution potentials shown in the CVs, CPC graphite electrode exhibited a lower chlorine evolution potential than NC and FPC. This may be caused by the fact that it contains more impurities than FPC, and these impurities may act as catalysts for chlorine evolution and for chlorine and electrolyte intercalation. The reaction between the chlorine and the graphite electrode may lead to a depolarization effect on chlorine evolution, so giving a lower chlorine evolution potential than FPC.

**Electrolysis results**

After 6h electrolysis at the current density of 1A/cm$^2$ at 700 °C under argon atmosphere, the NC, FPC and CPC anodes exhibited a different change in the surface morphology, as shown in Figure 3. Obviously, the NC graphite anode is characterized by a less change in the surface after electrolysis compared with the original

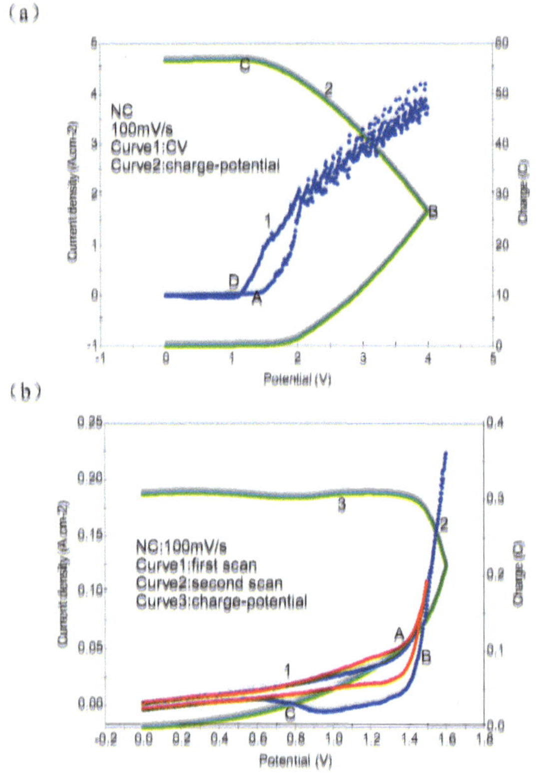


Fig.1 Cyclic voltammograms for NC in NaCl-KCl-$MgCl_2$ melt at 700 °C. WE: C(0.45cm$^2$), RE: spectroscopic graphite, CE: W wire .

morphology before electrolysis, especially no distinctly new cracks occur under the SEM observation scale in Fig.3(a) .While at the same electrolysis conditions, both CPC and FPC graphite anodes exhibited a distinct consumption in the surface and the occurrence of new cracks in the surface in Figure 3 (b) and 3 (c), especially for FPC graphite anodes, long and deep crack was observed, this may related to the relatively worse properties, such as a lower graphitization degree shown in Table 1. Moreover, electrolyte intercalation into CPC graphite anode is remarkable. The higher impurity content of CPC graphite may be the main reason caused by the electrolyte intercalation. The crack appearance in the FPC and CPC is related to chlorine and electrolyte intercalation, but the specific relation remains still unknown. The above results indicate that the NC graphite behaves a better corrosion-resistance to the chloride melts at high temperature than FPC and CPC, in good accordance with the CV analysis. Therefore, CV analysis is a simple and efficient method to compare the graphite anode behaviors in chloride melts and further to be used in selection of a suitable anode in chloride melts.

## Conclusions

CV analysis has determined that the NC anode generated a far smaller chlorine intercalation current, as well as a lower electrolyte intercalation amount than the two PC electrodes, but presented a higher chlorine evolution potential. The CPC exhibited a larger electrolyte intercalation amount, and a lower chlorine evolution potential than that of FPC electrode.

## Acknowledgements

This work was financially supported by 863 projects of Ministry of Science and Technology of China (No.2006AA03Z510) and Key Program of National Natural Science Foundation (50934001) and National Natural Science Foundation (51054004).

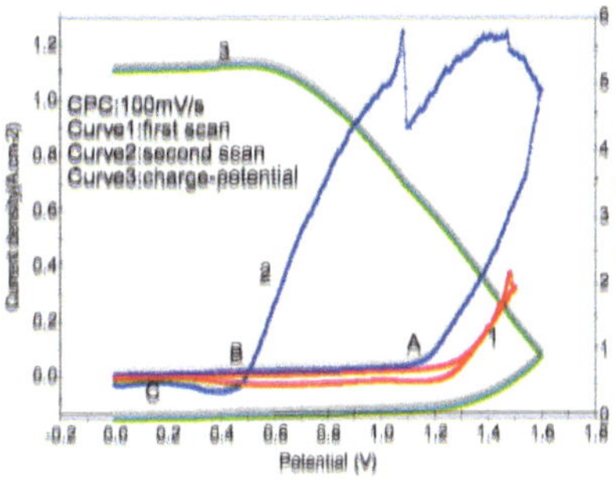


(a)

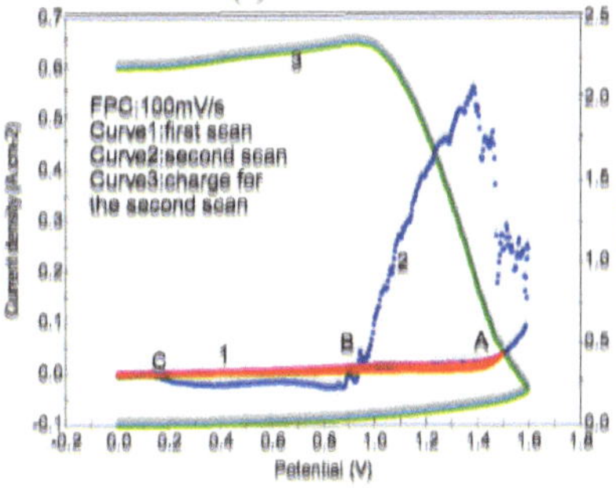


(b)

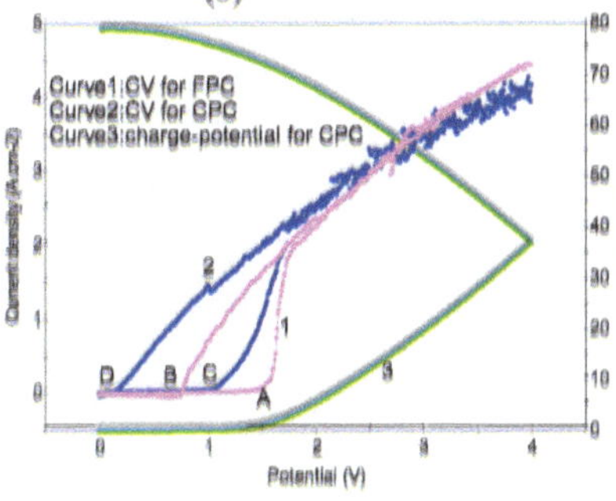


(c)

Fig.2 Cyclic voltammograms recorded on (a) FPC and (b) FPC and CPC (c) CPC in NaCl-KCl-$MgCl_2$ melt at 700 $^0$C. WE: FPC(0.45cm$^2$), CPC(0.45cm$^2$); RE: spectroscopic graphite; CE: W wire .

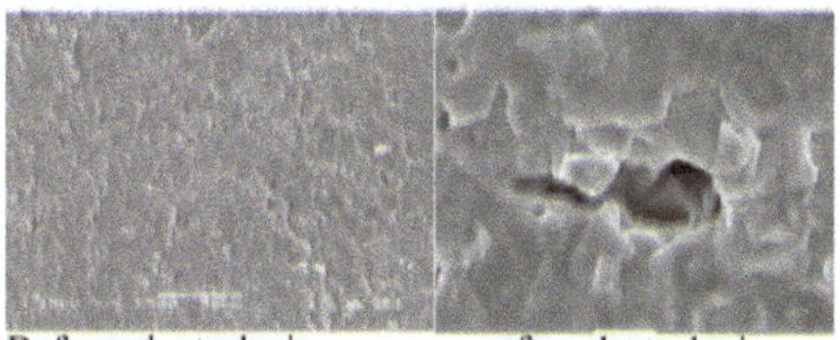

Before electrolysis after electrolysis
(a) NC

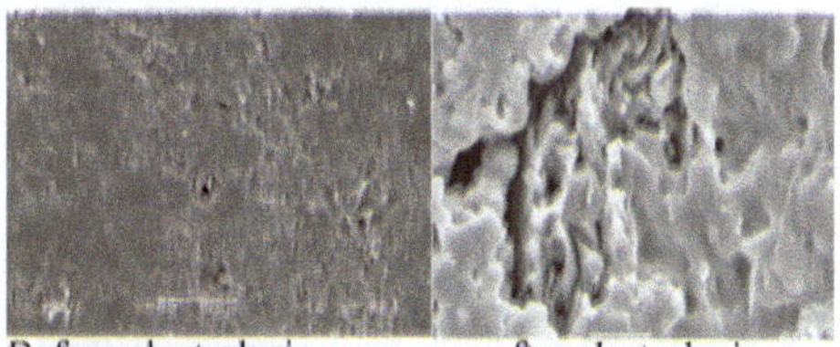

Before electrolysis after electrolysis
(b)FPC

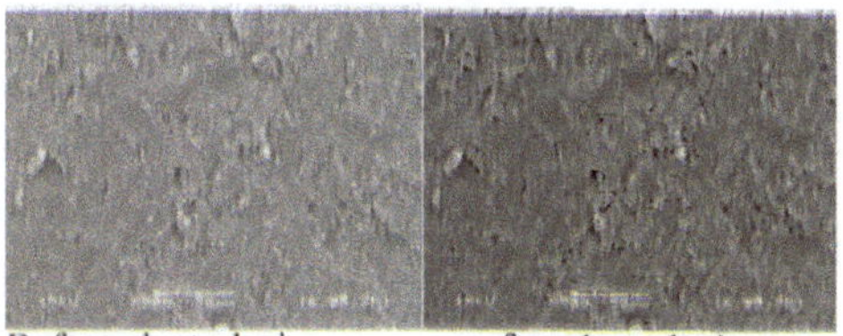

Before electrolysis after electrolysis
(c)CPC

Figure 3 The SEM morphology of NC, FPC and CPC before and after 6h electrolysis at 700 $^0$C

## References

[1]Joshua Rubenstein, Beya Davis Metall. Mater. Trans. B 4 (2007) 193.
[2]R. A. Skar, H. Mediaas, T. Ostvold, Can. Metall. Q. 43 (2004) 381.
[3]D.Laroche, B. R.Davis, K. Watson, P. Ficara, International Symposium on Light Metals as held at the 38th Annual Conference of Metallurgists of CIM; Quebec City, QC, Canada; 2- 26 Aug. 1999, pp. 103-109.
[4]Vindstad, Jens Emil; Mediaas, Heidi; Ostvold, Terje, Acta Chem. Scand. 51 (1997) 1192.
[5]B.Li, J. Lou, C. Zhan, and J. Yu, proceedings of Magnesium Technology 2010 at the 139th Annual Meeting & Exhibition for the Minerals, Metals & Materials Society (TMS),Washington State Convention & Trade Center, Seattle, Washington USA; 14-18Feb., 2010, pp. 3.
[6]H.Wendt,A.Khalil,C.E.Padberg,J.Appl. Electrochem. 21 (1991) 929.
[7]K.S.Mohandas,N.Sanil,M.Noeland P.Rodriguez,J. Appl. Electrochem. 31 (2001)997.
[8]F.Lantelm and H. Alexopoulos, J. Appl. Electrochem. 19 (1989) 649.

Magnesium Technology 2012
*Edited by: Suveen N. Mathaudhu, Wim H. Sillekens, Neale R. Neelameggham, and Norbert Hort*
***TMS (The Minerals, Metals & Materials Society), 2012***

# OPTIMIZATION OF PREPARATION FOR MAGNESIUM OXIDE BY CALCINATION FROM BASIC MAGNESIUM CARBONATE USING RESPONSE SURFACE METHODOLOGY

Bin Zhang[1,2], Jinhui Peng[2,*], Libo Zhang[2], Shaohua Ju[2]
[1]Faculty of Science, Kunming University of Science and Technology, Kunming Yunnan 650093, PR China
[2]Faculty of Metallurgy and Energy Engineering, Kunming University of Science and Technology, Kunming Yunnan 650093, PR China

Keywords: Response surface methodology, Basic magnesium carbonate, Calcination, Magnesium oxide

## Abstract

The conditions of technique to prepare magnesium oxide by calcination from basic magnesium carbonate were optimized using a central composite design (CCD) of response surface methodology (RSM). Two quadratic equation models for decomposition rate and activity of magnesium oxide were built and effects of main factors and their corresponding relationships were obtained. The statistical analysis of the results showed that in the range studied the decomposition rate and activity of magnesium oxide were significantly affected by the calcination temperature and calcination time. The optimized calcination conditions were as follows: the calcination temperature 666.99℃ and the calcination time 2.12 h, respectively. Under these conditions the decomposition rate of basic magnesium carbonate was 92.1971%, and the activity of magnesium oxide was 155.503 mg $I_2$/g MgO. The validity of the model was confirmed experimentally and the results were satisfactory. In addition, the sample was characterized by X-ray Diffraction (XRD).

## Introduction

Magnesium oxide has received a great deal of attention for its applications in the chemical and electronic industries. MgO powder has been utilized as a catalyst and catalyst support [1-4], a fundamental material for chemical heat pump and it is very appropriate for insulation applications due to their low heat capacity and high melting point [5]. The most conventional method of synthesis of MgO is the decomposition of various magnesium salts or magnesium hydroxide.

Generally, the traditional approach "one variable at a time (OVAT)" is used to analyze the thermal decomposition process of basic magnesium carbonate. In OVAT, the effect of each experimental factor is investigated by altering the level of one factor at a time while maintaining the level of the other factors constant. Furthermore, this technique is not only time and work demanding, but completely lacks in representing the effect of interaction between different factors [6]. In order to solve these problems, it is necessary to find a multivariate statistic technique for optimization of preparation processes. Response surface methodology (RSM) might be a useful method to optimize preparation processes. RSM is a collection of mathematical and statistical techniques useful for analyzing the effects of several independent variables. The main advantage of RSM is the reduced number of experimental trials needed to evaluate multiple parameters and their interactions [7]. It can deal with multivariate experimental design strategy, statistical modeling and process optimization [8]. Several previous researchers have proved that RSM was a powerful statistical tool in process optimization, and it has just recently been applied to optimize the process parameters for biosorption of metals [9] or dyes [10] from synthetic solutions. However, as far as known to the authors, there have been very few studies to optimize the preparation of magnesium oxide from basic magnesium carbonate by the RSM approach. The aim of this work was to optimize the thermal decomposition process of basic magnesium carbonate parameters with a consideration of response by applying RSM. A quadratic model was derived to describe the effects of calcination temperature and calcination time on the decomposition rate of basic magnesium carbonate and activity of magnesium oxide. The crystal structure of magnesium oxide under the optimum condition was also analyzed.

## Experimental

### Calcination Experiments

The calcination experiments were carried out at the different calcination temperatures and calcination times of sample. Initially, the Muffle furnace was preheated with a speed of 10 ℃·min$^{-1}$ until the desired temperature was reached. Then the basic magnesium carbonate was weighed and placed inside the ceramic crucible which was located in the center of the conventional Muffle furnace. During the reaction, the temperature was monitored by a temperature controller system, namely the PID (proportional integral derivative). Several cycles of experiment were repeated. For each cycle, a reaction was performed for a fixed duration, once the fixed duration is over, the experiment was stopped immediately. The product was moved out from the Muffle furnace and put into the drier rapidly. They were naturally cooled to the room temperature. The final weight (m) of sample was weighted subsequently, from which the decomposition rate was calculated based on the following equation:

$$Y = \frac{m_0 - m}{m_0 \times w_t} \times 100\% \qquad (1)$$

where M and $M_0$ are final weight and initial weight of sample, respectively; $W_t$ means theory value of weight loss, which is the total weight loss of basic magnesium carbonate during the thermal decomposition; Y is decomposition rate of basic magnesium carbonate.
Iodimetric titration was used to analyze the activity of magnesium oxide.

### Designing Experiment Using Response Surface Methodology

RSM was employed to optimize the calcination conditions in order to obtain a qualified magnesium oxide, and CCD was employed to design the experiments. This method helps to optimize the effective parameters with a minimum number of experiments and also analyze the interaction between the parameters and results [11]. Based on our previous work, the effects of two independent variables, $x_a$ (calcination temperature), and $x_b$ (calcination time), at five level were investigated using central composite design. The rank of values associated with the variables: calcination temperature (℃) was 600 and 700, and calcination time (h) ranged between 1.5 and 2.5 (Table 1). Generally, a second-order polynomial model with main, quadratic and interaction terms can be developed to fit the experimental data obtained from the experimental runs conducted on the basis of CCD. The experimental data obtained from the designed experiment were analyzed by the response surface regression procedure using the following second-order polynomial equation:

$$Y = \beta_0 + \sum_{i=1}^{n} \beta_i \chi_i + \sum_{i=1}^{n} \beta_{ii} \chi_i^2 + \sum_{i<j} \beta_{ij} \chi_i \chi_j \tag{2}$$

where Y is the predicated response, $\beta_0$ is a constant, $\beta_i$ is the ith linear coefficient, $\beta_{ii}$ is the ith quadratic coefficient, $\beta_{ij}$ is ijth interaction coefficient, $x_i$, $x_j$ are the coded values of independent variables, and the terms $x_i x_j$ and $x_i^2$ represent the interaction and quadratic terms, respectively.

Table 1 Independent variables and their levels for central composite rotatable design

| Variables | Symbol | Range and levels | | | | |
|---|---|---|---|---|---|---|
| | | -1.41421 | -1 | 0 | 1 | 1.41421 |
| calcination temperature /°C | a | 579.25 | 600 | 650 | 700 | 720.71 |
| calcination time /h | b | 1.29 | 1.5 | 2 | 2.5 | 2.71 |

XRD Analysis

The X-ray diffraction analysis of the final solid product under optimization conditions were carried out using D/max-2200 Diffractometer (Japan) with Ni-filtered CuKa radiation under air atmospheres. The identification of the completeness of the magnesium oxide was made by comparing the diffraction peaks of each compound in the sample with the ones of the standard magnesium oxide. If the diffraction pattern of the final solid product satisfactorily matched with that of the standard magnesium oxide, it means that the decomposition of basic magnesium carbonate is complete.

## Results and Discussion

Data Analysis and Evaluation of the Model by RSM

The experiments were conducted based on the design matrix under the defined conditions and the responses obtained from the experimental runs are shown in Table 2.

Table 2 shows the total number of 13 experiments as per CCD method. The experimental sequence was randomized in order to minimize the effects of the uncontrolled factors. Five experiments were repeated in order to estimate the experimental error. According to the sequential model, the sum of squares can be obtained, and the models were selected based on the highest order polynomial where the additional terms were significant and the models were not aliased. The responses of decomposition rate and activity of magnesium oxide were considered in studying the effect of process variables. The responses of decomposition rate and activity of magnesium oxide and the independent variables were used to develop two empirical models, which are presented by Eq. (3) and Eq. (4):

Table 2 Experimental design matrix and results

| Run | Calcination variables | | Decomposition rate of basic magnesium carbonate, $Y_1$ / % | Activity of magnesium oxide, $Y_2$ / mg/g MgO |
|---|---|---|---|---|
| | Calcination temperature, a / ℃ | Calcination time, b / h | | |
| 1 | 600(-1) | 1.5(-1) | 85.41 | 136 |
| 2 | 700(+1) | 1.5(-1) | 92.68 | 137 |
| 3 | 600(-1) | 2.5(+1) | 87.20 | 134 |
| 4 | 700(+1) | 2.5(+1) | 93.15 | 135 |
| 5 | 579.29(-1.41421) | 2.0(0) | 83.58 | 124 |
| 6 | 720.71(+1.41421) | 2.0(0) | 93.91 | 127 |
| 7 | 650(0) | 1.29(-1.41421) | 87.41 | 142 |
| 8 | 650(0) | 2.71(+1.41421) | 92.33 | 146 |
| 9 | 650(0) | 2.0(0) | 90.15 | 158 |
| 10 | 650(0) | 2.0(0) | 91.71 | 157 |
| 11 | 650(0) | 2.0(0) | 90.71 | 156 |
| 12 | 650(0) | 2.0(0) | 91.03 | 159 |
| 13 | 650(0) | 2.0(0) | 90.91 | 157 |

$$Y = 90.90 + 3.48\chi_a + 1.15\chi_b - 0.33\chi_a\chi_b - 1.00\chi_a^2 - 0.44\chi_b^2 \tag{3}$$

$$Y = 157.40 + 0.78\chi_a + 0.21\chi_b - 15.76\chi_a^2 - 6.51\chi_b^2 \tag{4}$$

The quality of the two models developed was evaluated based on the correlation coefficient value (Table 3 and Table 4). The $R^2$ value for Eq. (3) was 0.9629 and for Eq. (4) was 0.9902 which indicated that 96.29% variability of the total variation in the decomposition rate was attributed to the experimental variables studied and 99.02% variability of the total variation in the activity of magnesium oxide was attributed to the experimental variables studied. The closer the $R^2$ value to unity, the better the model will be as it will give predicted values which are closer to the actual values for the response. The $R^2$ of 0.9629 for Eq. (3) was considered relatively high, indicating that there was a good agreement between the experimental decomposition rate and the predicted one from this model. The $R^2$ of 0.9902 for Eq. (4) was considered relatively high, indicating that there was a good agreement between the experimental activity of magnesium oxide and the predicted one from this model.

Furthermore, analysis of variance (ANOVA, also a part of RSM) was further carried out to justify the adequacy of the model. The ANOVA for the quadratic model for decomposition rate and activity of magnesium oxide are presented in Table 3 and Table 4. The model's adequacy was tested through the lack of fit F-test, in which the residual error was compared to the pure error. According to the software analysis, "Lack of fit F-values" of 3.34 and 3.44 imply that the lack of fit was not significant relative to the pure error due to noise. The "Model F-values" of 36.31 and 140.99 imply that the two models were significant. Value of "Prob > F" less than 0.05 indicates that the two models terms are significant [12], whereas the values greater than 0.1000 are not significant.

Table 3 Analysis of variance (ANOVA) for response surface quadratic model for the decomposition rate of basic magnesium carbonate

| Source | Sum of squares | $D_f$ | Mean square | F-value | Prob>F |
|---|---|---|---|---|---|
| Model | 115.54 | 5 | 23.11 | 36.31 | <0.0001 |
| Residual | 4.45 | 7 | 0.64 | | |
| Lack of fit | 3.18 | 3 | 1.06 | 3.34 | 0.1374 |
| Pure error | 1.27 | 4 | 0.32 | | |
| Cor total | 119.99 | 12 | | | |

$R^2$=0.9629; $R^2$adj=0.9364; adequate precision=18.155(>4)

Table 4 Analysis of variance (ANOVA) for response surface quadratic model for the activity of magnesium oxide

| Source | Sum of squares | $D_f$ | Mean square | F-value | Prob>F |
|---|---|---|---|---|---|
| Model | 1874.16 | 5 | 374.83 | 140.99 | <0.0001 |
| Residual | 18.61 | 7 | 2.66 | | |
| Lack of fit | 13.41 | 3 | 4.47 | 3.44 | 0.1319 |
| Pure error | 5.20 | 4 | 1.30 | | |
| Cor total | 1892.77 | 12 | | | |

$R^2$=0.9902; $R^2$adj=0.9831; adequate precision=29.455(>4)

Response Surface Analysis

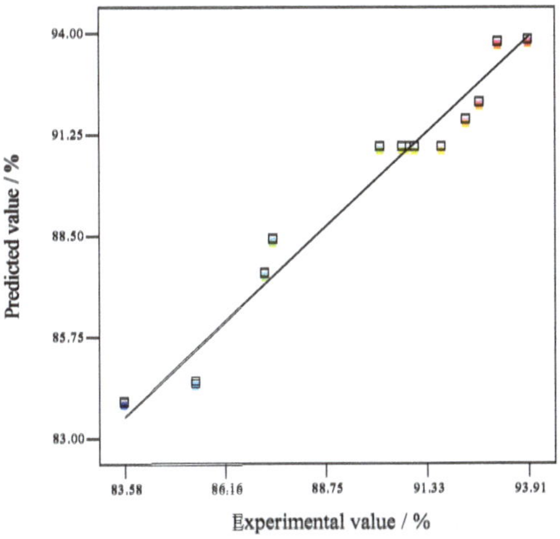

Fig. 1 Comparison of model prediction with the experimental data for decomposition rate of basic magnesium carbonate

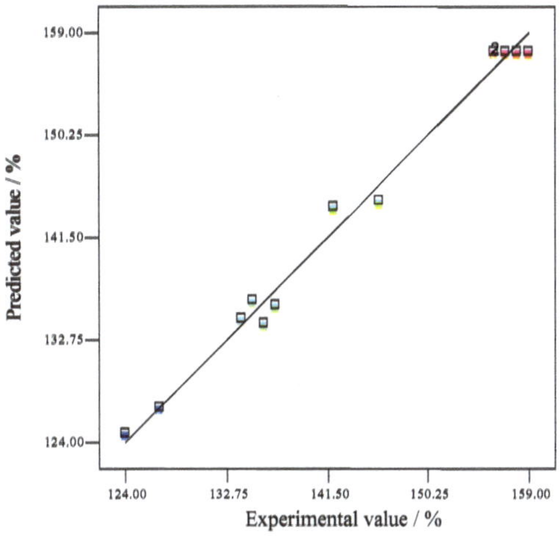

Fig. 2 Comparison of model prediction with the experimental data for activity of magnesium oxide

Fig. 1 and Fig 2 show the predicted values versus the experimental values for decomposition rate and activity of magnesium oxide. Actual response values were measured for a particular run, and the predicted values were evaluated from the model and generated by using the approximating equations. As can be seen, the predicted values obtained were quite close to the experimental values, indicating that the two models developed were reasonable.

The best way to visualize the influence of the independent variables on the response is to draw surface response plots of the model. The three-dimensional response surfaces which were constructed to show the effects of the calcination of basic magnesium carbonate variables on decomposition rate and activity of magnesium oxide using the fitted quadratic polynomial equations obtained from regression analysis was shown in Fig. 3 and Fig. 4.

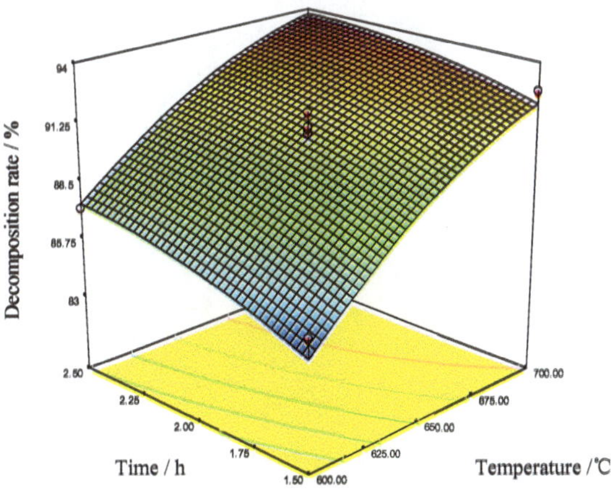

Fig. 3 Effect of calcination temperature and holding time on the decomposition rate of basic magnesium carbonate

Fig. 3 shows the effect of calcination temperature and calcination time on decomposition rate of basic magnesium carbonate. It was observed that the decomposition rate significantly increased with increasing calcination temperature and extending calcination time.

Increasing the calcination temperature up to 700℃, extending calcination time to 2.5 h gave an enhanced effect on the decomposition rate. The maximum predicated value of 93.15% was achieved at this temperature and time. The figure reveals that the effect of the calcination temperature on the field was more significant than calcination time.

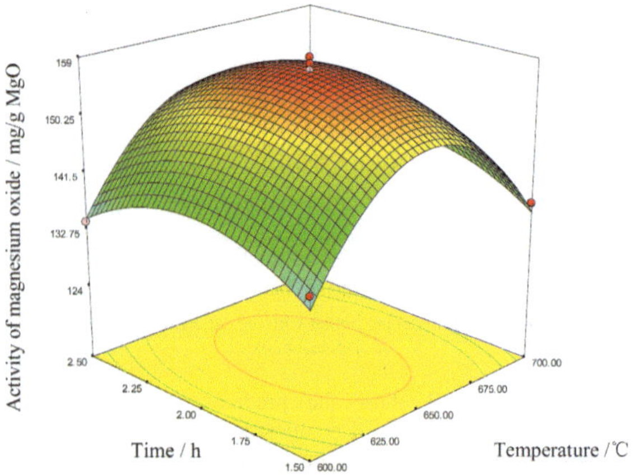


Fig. 4 Effect of calcination temperature and holding time on the activity of magnesium oxide

Fig. 4 shows the effect of calcination temperature and calcination time on activity of magnesium oxide. It was observed that the activity of magnesium oxide significantly increased with increasing calcination temperature and extending calcination time. But, when the calcination temperature was higher than 650℃ and calcination time was longer than 2 h, the activity of magnesium oxide began to decrease with the increasing of calcination temperature and extending calcination time.

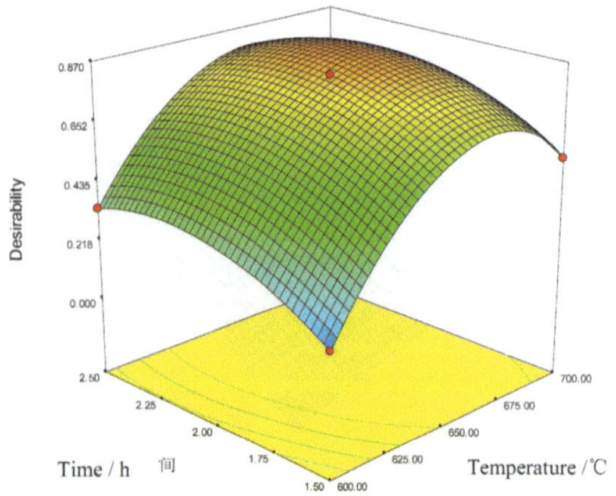


Fig. 5 Effect of calcination temperature and holding time on the comprehensive performance

Fig. 5 shows the effect of calcination temperature and calcination time on decomposition rate and activity of magnesium oxide. It was observed that the optimal condition for decomposition rate and activity of magnesium oxide was at 666.99℃and 2.12 h. The decomposition rate was 92.1971% and the activity of magnesium oxide was 155.503 mg $I_2$/g MgO.

Optimal Conditions and Verification of the Model

Thus, based on the above models, the optimal condition for decomposition rate and activity of magnesium oxide was at 666.99℃, 2.12 h and the decomposition rate was 92.1971%, the activity of magnesium oxide was 155.503 mg $I_2$/g MgO. In order to confirm the optimized conditions, the accuracy of the model was validated with experiments under conditions of optimum. An experiment was carried out with parameters as suggested by the model. The conditions used in the confirmatory experiment were as follows: calcination temperature 667℃ and calcination time 2.1 h, the giving a decomposition rate of 92.78% and activity of magnesium oxide of 155 mg $I_2$/g MgO (Table 5), which concurred with the model prediction. The two models, therefore, were considered to fit the experimental data very well in these experimental conditions. Therefore, the two models are acceptably valid.

Table 5 Optimum calcination conditions with model validation

| Calcination temperature, $_a$/ ℃ | Calcination time, $_b$/ h | Decomposition rate of basic magnesium carbonate, $Y_1$ / % | | Activity of magnesium oxide, $Y_2$ / mg/g MgO | |
|---|---|---|---|---|---|
| | | Predicted | Experimental | Predicted | Experimental |
| 667 | 2.1 | 92.1971 | 92.78 | 155.503 | 155 |

XRD Analysis

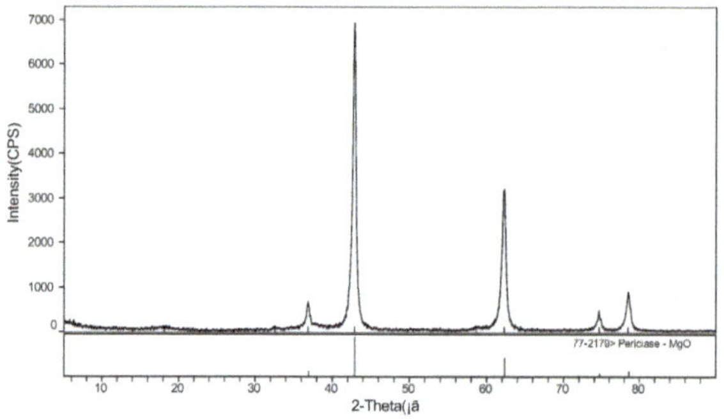


Fig. 6 The XRD pattern for the decomposition final solid products under the optimization conditions

Results of X-ray diffraction studies of the products on optimization conditions are shown in Fig. 6. The results show that magnesium oxide was the most solid product identified in which diffraction patterns satisfactorily matched with that of magnesium oxide and few decomposition products or reaction intermediates were identified in the XRD studies. Furthermore, it indicated that it is feasible to prepare magnesium oxide by calcination from basic magnesium carbonate under optimum conditions.

## Conclusions

This study showed that response surface methodology was a suitable approach to optimize conditions for achieving qualified decomposition rate of basic magnesium carbonate and activity of magnesium oxide. The experimental and predicted values were very close, which reflected the correctness and applicability of RSM. Using RSM to optimize experiments, the optimal condition was found to be at 666.99℃ and 2.12 h, respectively. Under these conditions, the predicted value of decomposition rate of 92.1971%

and activity of magnesium oxide of 155.503 mg $I_2$/g MgO were in good agreement with the actual experimental values (92.78% and 155 mg $I_2$/g MgO). The magnesium oxide prepared under the optimum conditions was characterized by XRD, from which the diffraction pattern satisfactorily matched with that of the standard magnesium oxide.


### Acknowledgements

Financial support for this work from the National Natural Science Foundation Council of China (50734007) and Analysis and Testing Foundation of Kunming University of Science and Technology (2010-284) are gratefully acknowledged.


### References


1 W.J. Climent, A. Corma, S. Iborra, M. Mifsud. MgO nanoparticle-based multifunctional catalysts in the cascade reaction allows the green synthesis of anti-inflammatory agents [J]. J. Catal, 2007, 247:223-230

2 Fereshteh Meshkani, Mehran Rezaei. Effect of process parameters on the synthesis of nanocrystalline magnesium oxide with high surface area and plate-like shape by surfactant assisted precipitation method [J]. Powder Technology, 2010, 199:144-148

3 I.V. Mishakov, A.F. Bedilo, R.M. Richards, V.V. Chesnokov, A.M. Volodin, V.I. Zaikovskii, R.A. Buyanov, K.J. Klabunde. Nanocrystalline MgO as a dehydrohalogenation catalyst [J]. J. Catal. 2002, 206:40-48

4 V.V. Chesnokov, A.F. Bedilo, D.S. Heroux, I.V. Mishakov, K.J. Klabunde, Oxidative dehydrogenation of butane over nanocrystalline MgO, $Al_2O_3$, and $VO_x$/MgO catalysts in the presence of small amounts of iodine [J]. J. Catal., 2003, 218:438-446

5 N. Takahashi. Simple and rapid synthesis of MgO with nano-cube shape by means of a domestic microwave oven [J]. Solid State Sciences, 2007, 9:722-724

6 Bingguo Liu, Jinhui Peng, Libo Zhang, Rundong Wan, Shenghui Guo, Liexing Zhou. Optimization of preparation for $Co_3O_4$ by calcination from cobalt oxalate using response surface methodology [J]. Chemical Engineering Research and Design, 2010, 88(8):971-976

7 Ming-Ju Chen, Kun-Nan Chen, Chin-Wen Lin. Optimization on response surface models for the optimal manufacturing conditions of dairy tofu [J]. Journal of Food Engineering, 2005, 68(4):471-480

8 Jianfeng Fu, Yaqian Zhao, Qiuli Wu. Optimising photoelectrocatalytic oxidation of fulvic acid using response surface methodology [J]. Journal of Hazardous Materials, 2007, 144(1-2):499-505

9 Bala Kiran, Anubha Kaushik, C.P. Kaushik. Response surface methodological approach for optimizing removal of Cr(VI) from aqueous solution using immobilized cyanobacterium [J]. Chemical Engineering Journal, 2007, 126(2-3):147-153

10 Sarayu Mohana, Shalini Shrivastava, Jyoti Divecha, Datta Madamwar. Response surface methodology for optimization of medium for decolorization of textile dye Direct Black 22 by a novel bacterial consortium [J]. Bioresource Technology, 2008, 99(3):562-569

11 R. Azargohar, A.K. Dalai. Production of activated carbon from Luscar char: Experimental and modeling studies [J]. Microporous and Mesoporous Materials, 2005, 85(3):219-225

12 A.L. Ahmad, S.C. Low, S.R. Abd Shukor, A. Ismail. Optimization of membrane performance by thermal-mechanical stretching process using responses surface methodology (RSM) [J]. Separation and Purification Technology, 2009, 66(1):177-186

# Magnesium Technology 2012

## Deformation Mechanisms

**Monday PM**

*Session Chairs:*

**Bin Li**
**(Mississippi State University, USA)**
**Alok Singh**
**(National Institute of Materials Science, Japan)**

**Magnesium Technology 2012**
*Edited by: Suveen N. Mathaudhu, Wim H. Sillekens, Neale R. Neelameggham, and Norbert Hort*
***TMS (The Minerals, Metals & Materials Society), 2012***

# AN ELASTO-PLASTIC MICROMECHANICAL METHOD FOR TWIN DRIVEN PLASTICITY

Pierre Alexandre Juan[1], Laurent Capolungo[1], Stephane Berbenni[2]
[1]Georgia Institute of Technology, UMI-GT CNRS 2958, Metz, 57070, France
[2]LEM3, ile du Saulcy, Metz, 57000, France

Keywords: Twinning, Magnesium, Micromechanics, Plasticity

## Abstract

In this work, a mean-field self-consistent approach based on a generalization of the Tanaka-Mori method is introduced to predict the mechanical response of polycrystalline Mg. The key idea is to homogenize the response of grains containing twin domains such that a direct coupling between the twin and parent domain is introduced. The work is particularly suited to study Mg as it allows for a direct coupling of the mechanical response of twin and parent domains. Such coupling accounts, for the first time, for relative size and shape effects on the development of internal strains in twin domains.

## Introduction

Due to its low symmetry, plasticity in Mg is typically conveyed by the activation of different types of deformation modes: twinning and slip[1]. While the motion of dislocations mediates both mechanisms, their effects on the development of internal strains are different. Indeed, the motion of twinning dislocations engenders substantial lattice rotation.

The role of twinning to plasticity is complex. On the one hand the motion of the twinning dislocations provides for a strain/stress relaxation. On the other hand, the new orientation of the lattice in twin domains can allow for the activation of secondary slip and twin, which can either lead to hardening or softening. The accuracy of the prediction of the activity of secondary slip and twinning is conditional on that of the predictions of the stress state in the twin domains.

Recent high-energy X-ray measurements have revealed that at the onset of twinning, the stress state within the twin and parent domains can be significantly different [2]. Precisely, it was found that the resolved shear stress of the activated tensile twin is of opposite sign in the parent and twin phases. These "backstresses" can be rationalized by the following two considerations. First, at a fine length scale, it is conceivable that, upon penetrating the grain boundary opposite to their emission site, the twinning partial dislocations lead to a distortion of the grain boundary. Such distortion can be the source of long-range (back)stresses. However, adopting a reasoning based on the conservation of Burgers vector to estimate the energy associated with the distortion, it appears unlikely that the dissociation of twinning partial dislocations can lead to such effect.

Second, the lattice rotation associated with twinning and the elastic anisotropy of hexagonal crystals yields an elastic constant mismatch across the twin interface. The mismatch in elastic constants can constrain the twin domain and be the primary source of back-stresses. Essentially, the process is similar to the volume constraint during corrosion- and described with the Pilling Bedworth ratio- or during phase transformation.

While substantial effort is dedicated to the development of full-field methods -both finite element and fast Fourier transform based- to model plasticity, mean-field methods remain widely used. Indeed mean-field Eshelbian micromechanics models allow for a numerically efficient way to model plasticity and texture evolutions in high and low symmetry polycrystals [3, 4]. Currently, the formation of twin domains is treated in two different ways whether coupling between the twin and parent phases is taken into account or not. In the decoupled approach, the mean stress and strain fields within the twin domains are simply obtained by solving the elementary inclusion problem of Eshelby –after linearization of the constitutive response- whereby the twin domain is embedded in a homogeneously equivalent medium. Effects arising from elastic mismatches between the twin and parent domains are thus not accounted for. In the coupled approach, traction continuity conditions between the twin and parent domains are imposed on the averaged fields in each phase. Therefore this coupling method is more appropriate when the twin domains remain relatively thin. In any case, the direct effect of elastic mismatches between the twin and parent phases is not accounted for.

In recent work by the authors [5], a third method was introduced to solve rigorously a double inclusion problem. This topology corresponds to that of a twin domain embedded in a parent phase (see Figure 1). The proposed approach, basically extending the Tanaka-Mori method [6], and was limited to a purely elastic approach and to a double inclusion topology. Nonetheless, it was shown that large effects arising from elastic mismatches are to be expected.

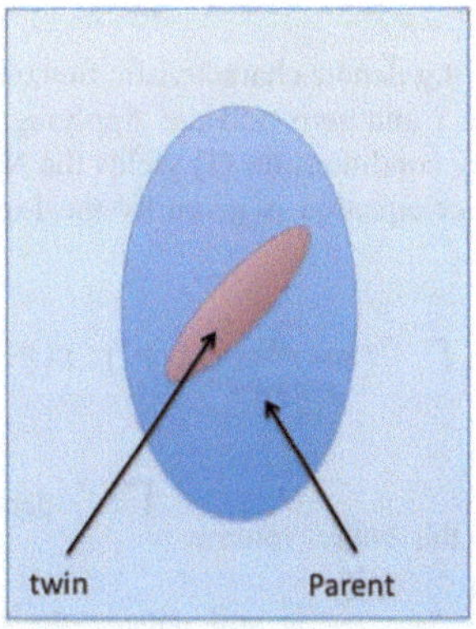


**Figure 1: Schematic of the topological representation of grains containing twins. The red domain denotes the twin volume while the blue domain refers to the grain without twin volume or parent volume.**

In the proposed work an extension of previous work is proposed such as to introduce the generalized Tanaka-Mori method in an elasto-plastic self consistent frameowork (EPSC).

## Generalized elasto-plastic Tanaka Mori method in Elasto plasticity

Let us consider first the topology described in Figure 1 whereby a grain embedded in a medium contains a twin domain –appearing in red-. Denoting $r$ the position vector, in an incremental theory the constitutive response of the material is written as:

$$\dot{\sigma}(r) = L(r)\dot{\varepsilon}(r) \tag{1}$$

Here $\sigma$ and $\varepsilon$ denote the Cauchy stress and small strain tensors, respectively. $L$ denotes the tangent modulus of the material. In what follows, t and g used either as subscripts or superscripts will refer to the twin domain and grain phase –including the twinning domain-. 'gt' is used to denote quantities associated to the parent phase –i.e. grain without the twin domain-.At any point within the medium, the tangent modulus can be written as the sum of a homogensous modulus, corresponding to that of an effective medium and of a spatial variation:

$$L(r) = L^{eff} + \delta L(r) \tag{2}$$

Where,

$$\delta L(r') = (L^g - L^{eff})\delta v_g(r') + (L^t - L^g)\delta v_t(r') \tag{3}$$

Here $\delta v_i$ with i=t,g denote characteristic function taken the value 1 within volume i and zero outside. Applying both equilibrium and compatibility conditions on (1) yields the Navier equation. A solution of Navier equation is given by the Lippman-Schwinger equation:

$$\dot{\varepsilon}(r) = \dot{E} - \int_V \Gamma^{s,\infty}(r - r') : \delta L(r') : \dot{\varepsilon}(r') dV_{r'} \tag{4}$$

here V denotes the entire volume. $\Gamma^{\infty,s}$ denote the modified Green functions As $\delta L$ is null outside the grain, equation 4 reduces to a sum of two integrals:

$$\dot{\varepsilon}(r) = \dot{E} - \int_{V_g} \Gamma^{s,\infty}(r - r') dV_{r'} : (L^g - L^{eff}) : \dot{\varepsilon}^g$$
$$- \int_{V_t} \Gamma^{s,\infty}(r - r') dV_{r'} : (L^t - L^g) : \dot{\varepsilon}^t \tag{5}$$

Here $V_t$ and $V_g$ denote the twin and grain domains –with the twin-, respectively. From (5), one can compute the average strain rate in the twin and parent domains including the twin and in the parent domain without the twin:

$$\dot{\varepsilon}^t = \frac{1}{V_t} \int_{V_t} \dot{\varepsilon}(r) dV_r \tag{6}$$

and

$$\dot{\varepsilon}^g = \frac{1}{V_g} \int_{V_g} \dot{\varepsilon}(r) dV_r \tag{7}$$

and

$$\dot{\varepsilon}^{gt} = \frac{1}{V_g - V_t} \int_{V_g - V_t} \dot{\varepsilon}(r) dV_r$$

In order to proceed with analytical derivations, it is assumed here that the strain rates within the twin and grain (with twin) domains are equal to their average within their respective volumes.

<u>Twin domain</u>

From (5) and (7), the average strain in the twin domain can thus be written as:

$$\dot{\varepsilon}^t = \dot{E} - \frac{1}{V_t} \int_{V_t} \int_{V_g} \Gamma^{s,\infty}(r - r') dV_{r'} dV_r : (L^g - L^{eff}) : \dot{\varepsilon}^g$$
$$- \frac{1}{V_t} \int_{V_t} \int_{V_t} \Gamma^{s,\infty}(r - r') dV_{r'} dV_r : (L^t - L^g) : \dot{\varepsilon}^t \tag{8}$$

If $r \in V_t \subset V_g$, then $\int_{V_g} \Gamma^{s,\infty}(r - r') dV_{r'}$ and $\int_{V_t} \Gamma^{s,\infty}(r - r') dV_{r'}$ are uniform and one obtains the following expression of the average strain rate within the twin domain:

$$\dot{\varepsilon}^t = \dot{E} - P^{V_g} : (L^g - L^{eff}) : \dot{\varepsilon}^g - P^{V_t} : (L^t - L^g) : \dot{\varepsilon}^t \tag{9}$$

With:

$$P^{V_g} = \int_{V_g} \Gamma^{s,\infty}(r-r')dV_{r'} \quad (10)$$

$$P^{V_t} = \int_{V_t} \Gamma^{s,\infty}(r-r')dV_{r'} \quad (11)$$

Grain domain

In the domain associated to the grain (including the twin) (5) reduces to:

$$\dot{\varepsilon}^g = \dot{E} - \frac{1}{V_g}\int_{V_g}\int_{V_g} \Gamma^{s,\infty}(r-r')dV_{r'}dV_r : (L^g - L^{eff}) : \dot{\varepsilon}^g$$
$$-\frac{1}{V_g}\int_{V_g}\int_{V_t} \Gamma^{s,\infty}(r-r')dV_{r'}dV_r : (L^t - L^g) : \dot{\varepsilon}^t \quad (12)$$

However here because $r \in V_g - V_t$,only $\int_{V_g} \Gamma^{s,\infty}(r-r')dV_{r'}$ is uniform such that the second integral needs to be computed by switching the order of the integration signs.

This is done by applying the Tanaka Mori theorem to (12) can be written as:

$$\dot{\varepsilon}^g = \dot{E} - \frac{1}{V_g}\int_{V_g}\int_{V_g} \Gamma^{s,\infty}(r-r')dV_{r'}dV_r : (L^g - L^{eff}) : \dot{\varepsilon}^g$$
$$-\frac{1}{V_g}\int_{V_t}\int_{V_g} \Gamma^{s,\infty}(r-r')dV_{r'}dV_r : (L^t - L^g) : \dot{\varepsilon}^t$$

(13)

Finally the average strain within the grain domain –i.e. containing the twin domain- can be written as:

$$\dot{\varepsilon}^g = \dot{E} - P^{V_g} : (L^g - L^{eff}) : \dot{\varepsilon}^g - \frac{V_t}{V_g} P^{V_{gt}} : (L^t - L^g) : \dot{\varepsilon}^t$$

(14)

With,

$$P^{V_g} = \int_{V_g}\int_{V_g} \Gamma^{s,\infty}(r-r')dV_{r'}dV_r \quad (15)$$

From equations (12) to (13) one can introduce localization and concentrations relations between the strain rate imposed on the boundary and that on each volume. One has:

$$\dot{\varepsilon}^g = A^g : \dot{E}$$
$$\dot{\varepsilon}^t = A^t : \dot{E}$$
$$\dot{\varepsilon}^{gt} = \frac{1}{V_g - V_t}\left[V_g A^g - V_t A^t\right] : \dot{E}$$

(16)

The localization tensors are given by:

$$A^g = \left[\begin{matrix} \frac{V_t}{V_g}(S^g(V_g) - S^{eff}(V_g)) - \\ \left(I + (S^t(V_t) - S^g(V_t))\right) : \left(I + (S^g(V_g) - S^{eff}(V_g))\right) \end{matrix}\right]^{-1} : \left[\frac{V_t}{V_g} I - \left(I + (S^t(V_t) - S^g(V_t))\right)\right]$$

(17)

and

$$A^t = \left(I + (S^t(V_t) - S^g(V_t))\right)^{-1} : \left[\begin{matrix} I - (S^g(V_g) - S^{eff}(V_g)) : \\ \left(\frac{V_t}{V_g}(S^g(V_g) - S^{eff}(V_g)) - \left(I + (S^t(V_t) - S^g(V_t))\right)\right) \\ : \left(I + (S^g(V_g) - S^{eff}(V_g))\right) \end{matrix}\right]^{1} : \left[\frac{V_t}{V_g} I - \left(I + (S^t(V_t) - S^g(V_t))\right)\right] \quad (18)$$

$I$ is the fourth order identity tensor. In (17) and (18), the following tensors were introduced:

$$\mathbf{S}^j(V_i) = \mathbf{P}^{V_i} : \mathbf{C}^j \text{ with j =t,g}$$

(19)

Finally, by identification one obtains an expression of the localization tensor associated with the grain without twin volume:

$$A^{gt} = \frac{1}{V_g - V_t}\left[V_g A^g - V_t A^t\right] \quad (22)$$

**Extension to polycrystals**

In the previous section localization relations were obtained in the case of a double inclusion embedded in an effective medium. These relationships can be used in an elasto-plastic self-consistent method as that derived by Hill and implemented in EPSC [7].

Consider now a different topology whereby it is desired to homogenize the elasto-plastic response of a polycrystal. The effective response of the polycrystal is given by:

$$\dot{\Sigma} = L^{eff} : \dot{E} \quad (23)$$

Here $\dot{\Sigma}$ denotes the macroscopic Cauchy stress rate. Using the same tangent/ incremental approach, the local constitutive response of each crystal is written as:

$$\dot{\sigma}^c = L^c \dot{\varepsilon}^c \tag{24}$$

The superscript *c* refers to a crystal. Note that, with the linearization adopted here the grain modulus is dependent on the grain shape and on its plastic history. In the self-consistent approach it is assumed that each crystal is embedded in an infinite medium with properties and behavior equal to that of the polycrystal.

From this point two cases are to be identified: either a twin domain has formed within a grain or not. Therefore two different types of localization relations have to be used.

Twin free grains

Following similar steps as that of Hill [8], the equivalent inclusion method leads to the following relationship between the

$$(\dot{\sigma}^c - \dot{\Sigma}) = -L^{eff} : \left(S^{-1} - I\right) : \left(\dot{\varepsilon}^c - \dot{E}\right) \tag{25}$$

From (25), the effective stiffness $L^*$ tensor can be defined as:

$$L^* = L^{eff} : \left(S^{-1} - I\right) \tag{26}$$

**S** denotes the elasto-plastic Eshelby tensor. Using the local and global constitutive responses, the following localization relationship can be obtained:

$$\dot{\varepsilon}^c = A^c : \dot{E} \tag{27}$$

where

$$A^c = \left(L^c + L^*\right)^{-1} : \left(L^{eff} + L^*\right). \tag{28}$$

Grains containing a twin domain

In existing approaches whether a grain contains a twin or not, equation (28) is used to quantify the localization tensor of the crystal. In the present approach, it is acknowledged that when a grain contains a twin, the topology of the problem has changed and becomes equivalent to that treated in equations. Therefore, in grains containing twins, it is necessary to separate the twin and parent phases and use:

$$A^c = A^t \text{ for the twin phase} \tag{29}$$

and

$$A^c = A^{gt} \text{ for the parent phase} \tag{30}$$

With this, a direct micromechanical coupling between the different phases is introduced.

Self consistency

Finally, the effective response of the material is obtained by considering macro-homogeneity conditions. Focusing on the strain increment one has:

$$\dot{E} = \left\langle \dot{\varepsilon}^c \right\rangle \tag{30}$$

Here terms in brackets denote volume averages. In addition, if one introduces the localization relations into the macro-homogeneity condition on stress, the effective tangent modulus of the polycrystal can be uniquely determined self-consistently with:

$$L^{eff} = \left\langle L^c A^c \right\rangle \left\langle A^c \right\rangle^{-1} \tag{31}$$

## Conclusion

In this work, a new self-consistent homogenization scheme is introduced in order to accurately predict the development of internal strains during twinning in polycrystals. The model is based on the introduction of new localization functions introducing a direct coupling between parent and twin domains. These are obtained by extending a generalized Tanaka Mori method to an elasto-plastic medium. While this has not been done yet, the model is well suited to be applied to the case of pure polycrystalline Mg. Among others it would be necessary to study the effect of these new localization relations on the predictions of internal stress in twins.

Acknowledgements: L.C and P-A J. would like to acknowledge support from Office of Basic Energy Sciences, Project FWP 06SCPE401, under U.S. DOE Contract No. W-7405-ENG-36.

## References

[1] Partridge, P.G. Metallurgical Reviews 1967; 12.
[2] Aydiner, C.C., Bernier, J.V., Clausen, B., Lienert, U., Tome, C.N., Brown, D.W. Physical Review B 2009; 80.
[3] Lebensohn, R.A.,Tome, C.N. Acta Metallurgica Et Materialia 1993; 41:2611-2624.
[4] Lebensohn, R.A., Tome, C.N., Maudlin, P.J., *An extended self-consistent visco-plastic formulation: application to polycrystal with voids.* 2003, Los Alamos Natl. Lab.
[5] Juan, P.A., Berbenni, S., Capolungo, L. Acta Materialia 2011; in press.
[6] Tanaka, K.,Mori, T. Journal of Elasticity 1972; 2.
[7] Turner, P.A.,Tome, C.N. Acta Metallurgica Et Materialia 1994; 42:4143-4153.
[8] Hill, R. Journal of Mechanics and Physics of Solids 1965; 13:213-222.

**Magnesium Technology 2012**
*Edited by: Suveen N. Mathaudhu, Wim H. Sillekens, Neale R. Neelameggham, and Norbert Hort*
***TMS (The Minerals, Metals & Materials Society), 2012***

# ANOMALOUS TWIN BANDS IN AZ31 MAGNESIUM SHEET BENDING

J.C. Baird [1], B. Li [1], S. Yazdan Parast [1], S.J. Horstemeyer [1], H. El Kadiri [1,2], P. Wang [1], M.F. Horstemeyer [1,2]
[1] Center for Advanced Vehicular Systems, Mississippi State University, USA
[2] Department of Mechanical Engineering, Mississippi State University, USA

Keywords: Twinning; Magnesium; Bending; Plastic deformation.

## Abstract

Three-point bending was performed on an AZ31 magnesium sheet with a grain size ~8.0 μm at room temperature. In-situ electron backscatter diffraction and metallography examinations revealed an anomalous twinning pattern during bending. In the compression zone, the $\{10\bar{1}2\}<10\bar{1}\bar{1}>$ twins appear in an extremely localized fashion, consisting of alternating twin bands. Each band comprises a high density of twins. In between the twin bands, twins are absent. As the bending angle increases, the twin bands grow into the tension zone although the stress disfavors the extension twinning.

## Introduction

In recent years, lightweight Mg and Mg alloys have attracted significant attention for their potentials of improving fuel efficiency of automotive vehicles. A major barrier that bottlenecks extensive engineering applications of Mg alloys is the relatively poor ductility and low formability at room temperature, which inherently stem from the low symmetry hexagonal close-packed (HCP) crystal structure. In sharp contrast to high symmetry structures, e.g., face centered cubic (FCC), plastic deformation in Mg is dominated by mechanical twinning. This twinning dominated deformation gives rise to unique mechanical properties of Mg alloys, such as tension-compression asymmetry and anisotropy in plastic deformation [1].

In practice, forming of Mg alloys, e.g., sheet rolling, bending and drawing, has to be performed at elevated temperatures to prevent fracture. It is desirable to reduce the processing temperature of sheet forming. Hence, understanding microstructural and texture evolution *during* forming instead of post-forming is necessary. The mechanical properties of AZ31 Mg sheet metals have been studied extensively [2,3]. But, little attention was paid to microstructural evolution during forming. The purpose of this work is to examine microstructural evolution during three-point bending at room temperature, using in-situ electron backscatter diffraction (EBSD) and optical microscopy.

## Experimental Methods

The material chosen for the current study was a magnesium alloy AZ31 rolled sheet with a thickness 1.0 mm. The as-received sheet was annealed at 300°C for 2 hours in an Argon atmosphere. Strips with a dimension 60×7×1 mm were cut for subsequent three-point bending tests. One of the through-thickness cross section of the strip was metallographically polished and then electropolished using a Struers electopolisher and the standard C1 Struers electrolyte (160 mg sodium thiocyanate, 800 ml Ethanol, 80 ml Ethylene glycol monobutyl ether, and 20 ml distilled water).

Three-point bending was performed with a specially designed fixture which allows us to conduct in-situ EBSD and metallography. By in-situ we mean that the sample was scanned with EBSD without unloading the sample, in contrast to conventional EBSD scans in which samples are unloaded after deformation. Details of our design will be described elsewhere. With the fixture we designed and fabricated, samples can be deformed continuously and EBSD scans can be carried out to examine the microstructure and texture evolution while the external load remains on the sample. This capability circumvents the problem of detwinning [4-6] after a deformed sample is unloaded, which is common in deformation of Mg and Mg alloys. Detwinning leads to the loss of microstructure and texture information.

The EBSD scans were performed through the thickness of the sheet using a Zeiss Supra 40 Field Emission Gun (FEG) Scanning Electron Microscope (SEM) equipped with an EDAX Hikari Electron Backscatter Diffraction (EBSD) detection system. After data collection, all data sets were rotated 90 degrees about the transverse direction (TD) so that all data analysis could be performed looking at the normal direction (ND). In the present study, all EBSD scans were performed with the load still on the sample. Samples during bending were also examined by metallography with differential interference contrast (DIC) illumination.

## Results and Discussion

Figure 1 displays the inverse pole figure (IPF) which corresponds to the initial texture and microstructure of the rolled sheet. The pole figure is shown as well. The initial texture is typical of rolled materials exhibiting a strong basal (0001) pole, i.e., fiber texture. The grain size of the rolled sheet is about 8 μm.

The three-piont bending of the sheet introduces two specific stress states to the sheet, i.e., a compression zone right under the indenter, and a tension zone right next to the compression zone. In between, a neutral axis separates these two zones and experiences zero strain. These two different stress states provides an ideal scenario for in-situ study twinning-detwinning and other important issues related to materials mechanical responses by EBSD. The compression zone of the sample provides a preferred loading condition for the $\{10\bar{1}2\}<10\bar{1}\bar{1}>$ extension twins which generate an elongation along the <c> axis that is nearly parallel to the normal direction of the sheet (Figure 1). But in the tension zone, the stress state disfavors the $\{10\bar{1}2\}<10\bar{1}\bar{1}>$ twinning.

Figure 2 shows the IPF of a region in the compression zone. As expected, twinning occurred in this zone during bending. But surprisingly, the twins are localized in a band with a band width much larger than the grain size. The band runs roughly 45° from top to bottom. Away from the twin band, no twins are observed at all. Misorientation analysis shows that these twins are indeed $\{10\bar{1}2\}<10\bar{1}\bar{1}>$ extension twins. The compressive stress creates an elongation along the <c> axis of the highly textured microstructure. Notably, the twin variants are aligned in two nearly perpendicular directions. Most of the twin variants run through whole grain, with the twins in the neighboring grains being connected at the grain boundary. The pole figure confirms that the $\{10\bar{1}2\}<10\bar{1}\bar{1}>$ twinning indeed took place, as two new intensity spots appear nearly 90° away from the rod intensity, a definitive signature of the extension twins.

Figure 2 shows only one twin band. Actually multiple twin bands were formed in the compression zone of the sheet, as shown in Figure 3. The EBSD scan was taken near the center line which is in the middle of the image. Similar penetrating pattern can be observed. Multiple twins connect at the grain boundaries of neighboring grains. More interestingly, as the bending angle increases, the twin bands continue to grow across the geometrical center line. Similar pole figure was captured, typical of the $\{10\bar{1}2\}<10\bar{1}\bar{1}>$ twinning. It was found that the farther scans were taken from maximum compression and closer to maximum tension the intensity spot from twinning was reduced.

In addition to the in-situ EBSD scans in local regions of compression and tension, metallography observations were also made while the sample was still loaded. To better reveal the twin bands, differential interference contrast (DIC) illumination technique was used. Figure 4(a) shows one of the observations that macroscopically display the alternating twin bands. The width of the bands narrows toward the neutral axis, and the length of the individual bands decreases as the distance to the indenter increases. The bands roughly point to the same direction, about 45° with the top surface. Interestingly, the tip of the longest (indicated by an arrow) extended across the center line (also shown in Figure 3). As the bending angle increases, the twin bands widen laterally. Meanwhile, more twin bands appear in the direction nearly perpendicular to the original twin bands (indicated by an arrow).

More three-point bending tests were conducted on non-annealed sheet samples and we found that the results of the localized twin bands are repeatable.

## Discussions

To our knowledge, such anomalous, highly localized twin bands in Mg sheet bending were not reported previously. Shear bands formed during sheet forming such as cold/warm rolling of AZ31 Mg sheets were reported by a number of researchers [2,3], but these bands are different from what we observed in the AZ31 sheet bending. Here we only briefly discuss the possible mechanism of the anomalous twin bands.

Despite the relatively fine grain structure of the AZ31 sheet, twinning did occur in a highly localized fashion during three-point bending. Close examinations on the twins in the individual grains show that for grains in the twin bands, one twin variant runs through whole grain, from one side of the grain boundary to the opposite side. At the location where this twin variant stops, another twin variant nucleates and propagates into the neighboring grain, approximately in the same direction. Such nucleation and growth of the twin variants in the neighboring grains cascades until the stress state of the sheet disfavors the $\{10\bar{1}2\}<10\bar{1}\bar{1}>$ twinning. The twin variants conjoin at the grain boundaries and form a continuous band. Such an unusual twinning pattern indicates that the twin variants in the neighboring grains cannot be completely independent of each other, vastly different from dislocation controlled twinning modes where a twin variant must stop at a grain boundary because a twinning dislocation cannot penetrate a grain boundary.

In fact, penetrating twin variants were observed in previous reports where a twin variant can penetrate a grain boundary or another twin variant [7,8]. Hence, the $\{10\bar{1}2\}<10\bar{1}\bar{1}>$ twinning mode must have a growth mechanism that is unique to other twinning modes in HCP metals. In the classical twinning theory [9], a homogeneous shear and atomic shuffling must be involved for low symmetry hcp structures. Atomic shuffling was nicely described by Christian and Mahajan [10], and confirmed by Li and Ma [11] in their molecular dynamics simulations. For the most commonly observed $\{10\bar{1}2\}<10\bar{1}\bar{1}>$ twinning, the magnitude of shuffling is much greater than the magnitude of the theoretical twinning dislocation [11] which equals 0.024 nm for Mg [12]. Such an extremely small Burgers vector indicates that the $\{10\bar{1}2\}<10\bar{1}\bar{1}>$ twinning is dominated by atomic shuffling, not by shear. In other words, a twinning dislocation is not needed for this particular twinning mode. This conclusion is supported by extensive experimental observations in which abnormal properties such as reversible twinning [4-6], non-Schmid effect [13], and large deviation between the actual twin boundaries and the $\{10\bar{1}2\}$ twinning plane [14]. In other twinning modes, a twinning dislocation is required, plus minor atomic shuffling or no shuffling at all [15-16]. The anomalous twin bands observed in this work again support the conclusion that the $\{10\bar{1}2\}<10\bar{1}\bar{1}>$ twinning mode in HCP metals can only be controlled by atomic shuffling, not by twinning dislocations.

## Conclusions

Three-point bending tests were performed on an AZ31 Mg sheet. In-situ EBSD scans and optical microscopy when samples were still loaded were performed. Extremely localized, alternating twin bands were observed. Each twin bands comprise high density $\{10\bar{1}2\}<10\bar{1}\bar{1}>$ twins, whereas in between the twin bands, no twins were present. The results we obtained support the conclusion that the $\{10\bar{1}2\}$ twinning cannot be controlled by twinning dislocations, instead, by atomic shuffling.

## References


[1] J. Bohlen, M. R. Nurnberg, J. W. Senn, D. Letzig, S. R. Agnew, "The texture and anisotropy of magnesium–zinc–rare earth alloy sheets", Acta Mater 55(2007)2101-2112.

[2] Couling SL, Pashak JF, Sturkey L, "Unique deformation and aging characteristics of certain magnesium-based alloys", Trans of the ASM 51: 94-107 (1959).

[3] Barnett MR, Keshavarz Z, Nave MD, "Microstructural Features of Rolled Mg-3Al-1Zn", Metall Mater Trans A 36A: 1697-1704 (2005).

[4] L. Wu, A. Jain, D. W. Brown, g. M. Stoica, S. R. Agnew, B. Clausen, D.E. Fielden, P.K. Liaw, "Twinning–detwinning behavior during the strain-controlled low-cycle fatigue testing of a wrought magnesium alloy ZK60", Acta Mater 56, (2008) 688-695.

[5] A. Jain, S. R. Agnew, "Modeling the temperature dependent effect of twinning on the behavior of magnesium alloy AZ31B sheet", Mater Sci Eng A 462(2007)29-36.

[6] C.H. Caceres, T. Sumitomo, M. Veidt, "Pseudoelastic behaviour of cast magnesium AZ91 alloy under cyclic loading–unloading", Acta Mater 51 (2003) 6211-6218.

[7] B. Li, E. Ma, K.T. Ramesh, "Dislocation configurations in an extruded ZK60 magnesium alloy", Metall Mater Trans A, 39A (2008) 2607-2614.
[8] Q. Ma, H. El Kadiri, A.L. Oppedal, J.C. Baird, B. Li., M.F. Horstemeyer, S.C. Vogel, International Journal of Plasticity, 2011, in press.
[9] B.A. Bilby, A.G. Crocker, "The theory of the crystallography of deformation twinning", Proc Roy Soc. A, 288 (1965) 240-255.
[10] J.W. Christian, S. Mahajan, "Deformation twinning", Prog. Mater Sci 39 (1995) 1-157.
[11] B. Li, E. Ma, "Atomic shuffling dominated mechanism for deformation twinning in magnesium", Phys Rev Lett, 103 (2009) 035503.
[12] N. Thompson, D. J. Millard, "Twin formation in cadmium", Phil Mag 43 (1952) 422.
[13] M.R. Barnett, Z. Keshavarz, A.G. Beer, X. Ma, "Non-Schmid behaviour during secondary twinning in a polycrystalline magnesium alloy", Acta Mater 56 (2008) 5-15.
[14] P.G Partridge, E. Roberts, "The formation and behaviour of incoherent twin boundaries in hexagonal metals", Acta Met 12 (1964) 1105-1210.
[15] B. Li, E. Ma, "Zonal dislocations mediating $\{10\bar{1}1\} < 10\bar{1}\bar{2} >$ twinning in magnesium", Acta Mater 57 (2009) 1734-1743.
[16] B. Li, H. El Kadiri, M.F. Horstemeyer, " Extended zonal dislocations mediating $\{11\bar{2}2\} < 11\bar{2}\bar{3} >$ twinning in titanium", Philo Mag 2011, in press.

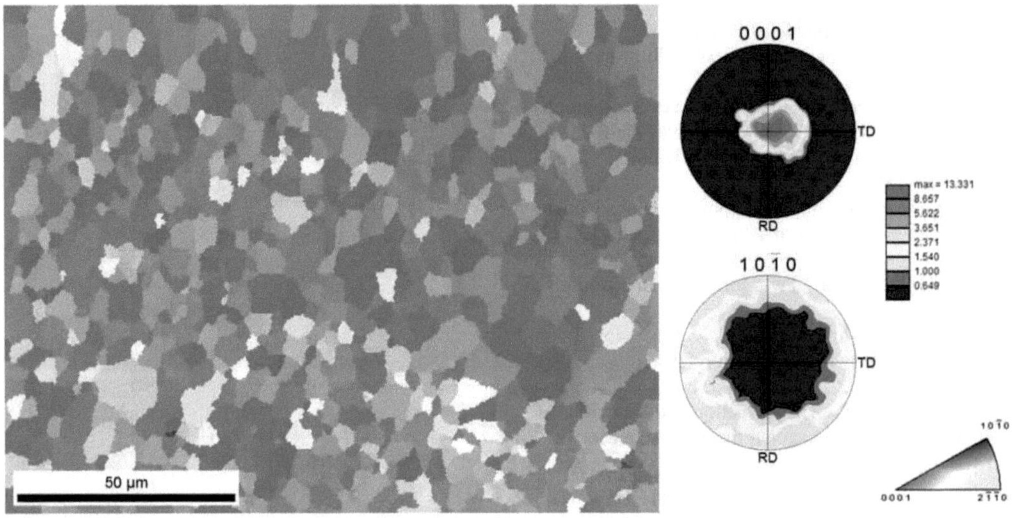


Figure 1. The initial microstructure and texture of the AZ31 rolled sheet. The texture is typical of the (0001) rod texture. The average grain size is about 8 μm.

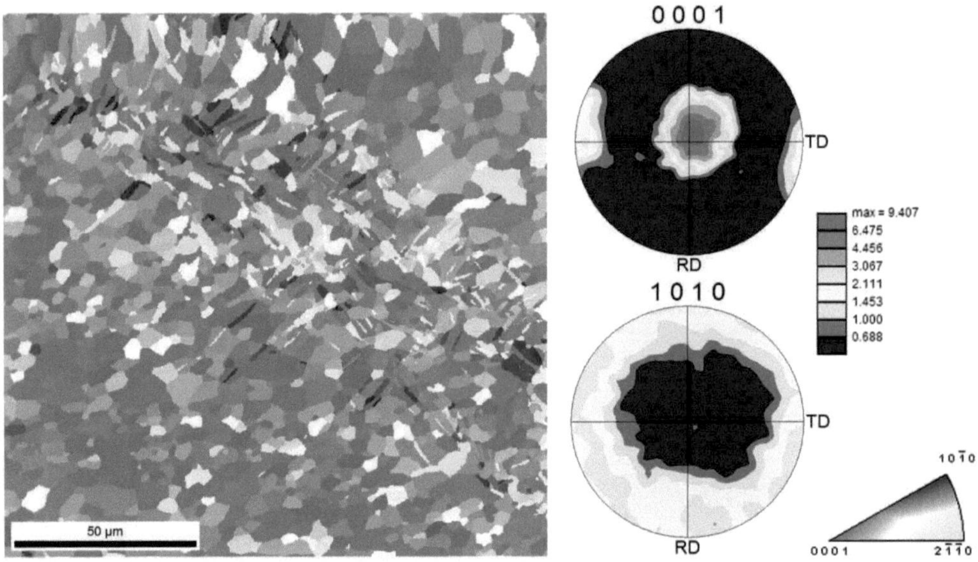


Figure 2. Electron backscatter diffraction (EBSD) shows a high density of twins in the sheet deformed in three-point bending. The twinning is highly localized. Away from the twin band, almost no twins can be observed. It can be seen from the pole figure that the twins are $\{10\bar{1}2\} < 10\bar{1}\bar{1} >$ extension twins. The scan was taken near the top of the cross-section in the compression zone of the specimen.

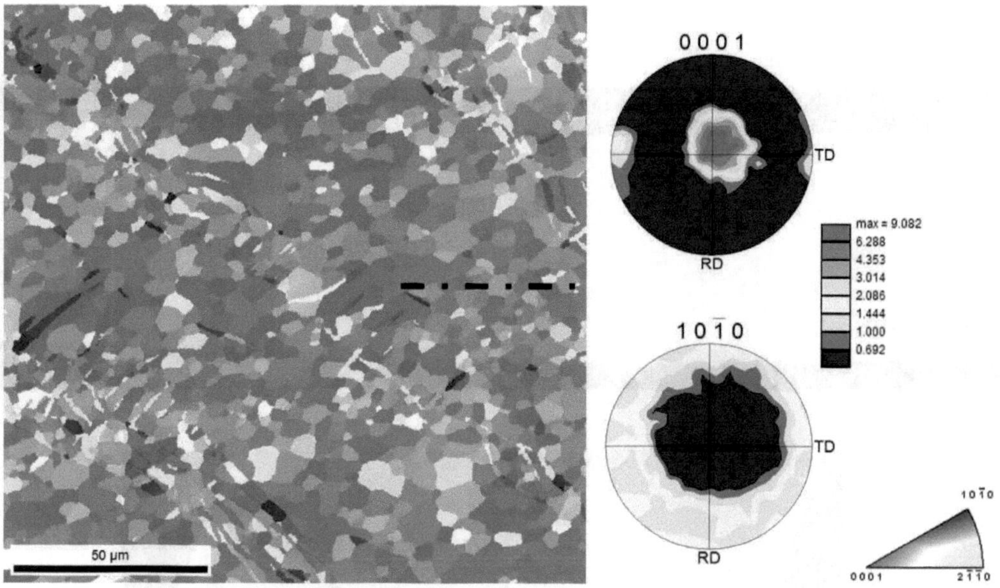


Figure 3. Multiple twin bands were observed near the geometrical center line of the specimen (indicated by a dash-dotted line in the middle). The twin bands cross the geometrical center line.

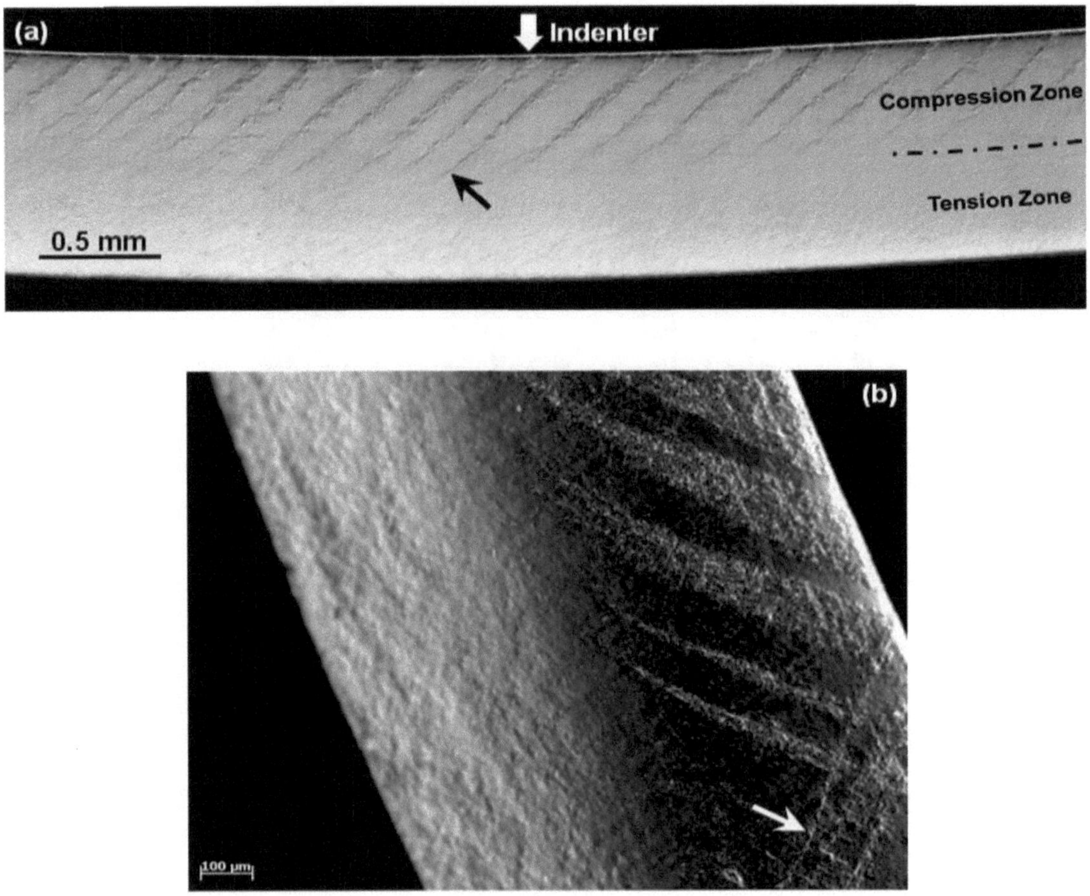


Figure 4. Through-thickness (1.0 mm) optical metallography with the differential interference contrast (DIC) illumination during three-point bending. The load remained on the specimen when the images were taken. (a) Alternating twin bands can be clearly observed. From the indenter toward both ends of the strip, the length and width of the twin bands decreases because the plastic strain decreases. The longest twin band (indicated by an arrow) passes the center line. (b) As the bending angle increases, the twin bands widen by lateral growth. Meanwhile, new bands (indicated by an arrow) that are nearly perpendicular to the original bands are formed.

Magnesium Technology 2012
*Edited by: Suveen N. Mathaudhu, Wim H. Sillekens, Neale R. Neelameggham, and Norbert Hort*
*TMS (The Minerals, Metals & Materials Society), 2012*

# FORMATION OF NANO-SCALE TWINS AND LOW ANGLE GRAIN BOUNDARIES DURING FRACTURE OF A FINE GRAINED MAGNESIUM ALLOYS

Alok Singh[1], Hidetoshi Somekawa[1], Toshiji Mukai[2]

[1]Structural Materials Unit, Research Center for Strategic Materials,
National Institute for Materials Science, Sengen 1-2-1, Tsukuba, Ibaraki 305-0047, Japan
[2]Department of Mechanical Engineering, Kobe University, Kobe, Japan

Keywords: magnesium alloys, fracture toughness, twinning, transmission electron microscopy

## Abstract

Fine grained magnesium alloys show high fracture toughness, associated with void formation. Detailed microstructural evolution during a fracture toughness test of a fine grained Mg-Zn alloy has been studied here by transmission electron microscopy (TEM). Focused ion beam (FIB) technique was used for obtaining samples near crack. Initially, subgrain structures form ahead of the crack tip, after which $\{10\bar{1}1\}$ type twins of width of about 400nm form. They further twin by $\{10\bar{1}2\}$ twinning. Subsequent twinning occurs at a finer scale near the crack, forming configurations of $\{10\bar{1}1\}$-$\{10\bar{1}2\}$ double twins and low angle boundaries. The scale of the twin domains became progressively finer to less than 50nm. In absence of enough dislocations to pile up causing fracture, deformation continues to occur by the twinning.

## Introduction

Fracture toughness in an important criteria in deciding the use of a material for structural application. Fracture toughness of magnesium alloys are known to be low [1]. Formation of $\{10\bar{1}2\}$ type of twins at the beginning of the deformation [2, 3] is a cause of low fracture toughness. These twins become the crack propagation route [4]. However, grain refinement of magnesium alloys can reduce the formation of twins [5]. Consequently, the fracture mechanism changes from brittle to ductile. Ductile fracture is associated with void formation. Thus the fracture toughness of magnesium alloys can be increased by grain refinement [6, 7].

In a single crystal strained in tension parallel to the basal plane, fracture was found to occur primarily by twinning, identified to be $\{10\bar{1}1\}$-$\{10\bar{1}2\}$ double twinning [8]. Reduction of a single crystal along its $c$-axis is accommodated by twinning on $\{10\bar{1}1\}$ and retwinning on $\{10\bar{1}2\}$, followed by a basal shear [9]. Compression perpendicular to unconstrained $c$-axis activates $\{10\bar{1}2\}$ twinning [10].

Ductile fracture and high fracture toughness in magnesium alloys are associated with blunt crack tip of fracture. We reported on a high fracture toughness in a fine grained magnesium alloy and the deformation structures associated with it [11]. From the partly tested and tested to fracture samples of a three point bending specimen, samples for transmission electron microscopy (TEM) were cut out from specific sites by focused ion beam (FIB). Details of these investigations are reported here, in order to understand the fracture mechanism in a fine grained magnesium alloy.

## Experimental Procedure

A Mg-2.4at%Zn alloy was cast and then extruded at 473K. The extruded alloy had grain size in the range of 1-3$\mu$m. Nano-sized precipitates of rounded shape were dispersed inside the grains. Details of initial microstructure and dimpled fracture surface showing ductile character are reported elsewhere [12]. Three-point bending specimens of length 10mm and width 5mm were machined from the extruded bar and a pre-crack inserted by fatigue. TEM specimens were sampled from specific sites by FIB. A specimen was cut ahead of the crack after an interrupted test. Another was selected from under a void ahead of the crack, and a third from under the crack. These specimens were observed by high resolution TEM (HRTEM) using a JEOL 4000EX microscope operated at 400 kV.

## Results

Fig. 1(a) shows a micrograph from the sample taken ahead of the crack tip, in an interrupted fracture toughness test. A large strain-like contrast

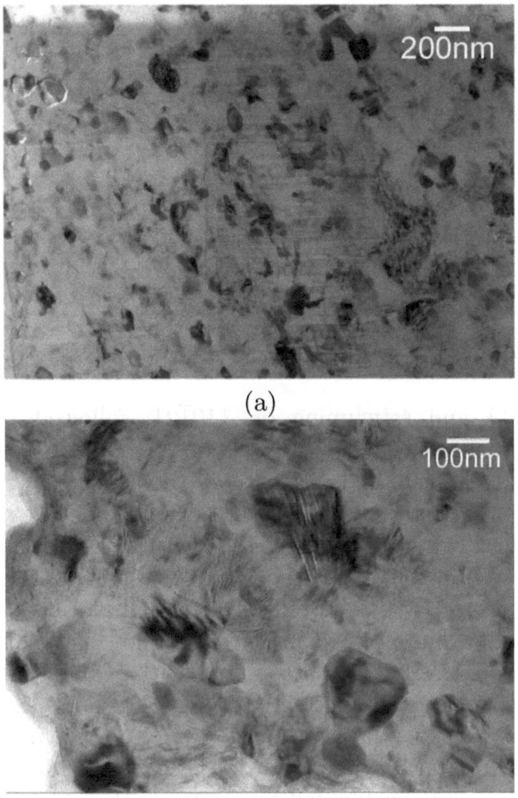


Figure 1: Bright field electron micrographs showing grain and deformation structures (a) ahead of the crack tip and (b) bottom of a void.

is observed, especially in the right of the micrograph. There are dark particles of 100-150nm in size, which are Mg-Zn binary phase believed to be of the monoclinic $Mg_4Zn_7$ phase. Some apparently recrystallized grains of size less then 200nm are also observed.

In the sample taken from the bottom of a void, Fig. 1(b), large strains are also present. These strained regions are more localized than in the sample taken ahead of the crack. Some recrystallized grains are observed too.

Fig. 2 shows the deformation structure just below a crack. A part of the crack surface is visible in the upper right corner. Based on the deformation structures, three regions could be identified in this sample - away from the crack surface is a region marked I, in which parallel twins, two marked by arrowheads, are identified. From diffraction patterns, these twins

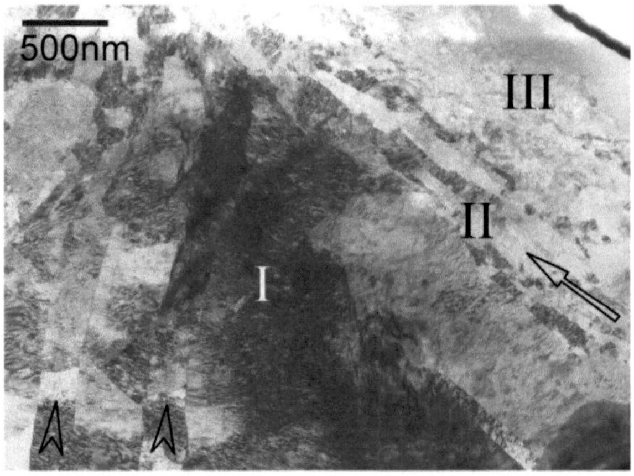


Figure 2: Bright field electron micrographs showing a region just below the crack surface.

were identified to be $\{10\bar{1}1\}$ type twins. The width of these twin bands is similar, of about 400nm. A large strain contrast is evident in this region. In the region marked II, elongated domains parallel (in the direction marked by an arrow) to the crack surface. These domains have sharp boundaries. Closest to the crack surface is the region marked III, in which nano-grains of size about 50nm occur.

The $\{10\bar{1}1\}$ twins parallel to each other observed in Fig. 2 twin further, to make double twins $\{10\bar{1}1\}$-$\{10\bar{1}2\}$. An example is shown in Fig. 3. All domains in this micrograph are in zone axis $\langle 11\bar{2}0\rangle$. These domains are rotated with respect to each other about this zone axis. The direction of the $c$ axis in each domain is marked by an arrow. Diffraction patterns showed that the domains marked 'A' and 'B' are $\{10\bar{1}1\}$ type twins (angle between the $c^*$ axes of about 56°). In between, a third twin marked 'C' is observed. 'C' makes a $\{10\bar{1}2\}$ twin with 'B' (angle between the $c^*$ axes of about 86°). Thus A, B and C make a double twin. This is observed in the composite diffraction of Fig. 3(b). $c^*$ axes of all the three twin domains are marked in this diffraction pattern. There is a coincidence of $\{10\bar{1}1\}$ spots of A and B, confirming a $\{10\bar{1}1\}$ type twin. A $\{10\bar{1}2\}$ of domains B and C coincide, making a $\{10\bar{1}2\}$ type twin.

Towards the crack surface, the twins multiply and become finer. Boundaries between the domains are low angle boundaries (LAB) or twins. Double twin configurations were often observed. The long domains in the region II were similarly oriented with

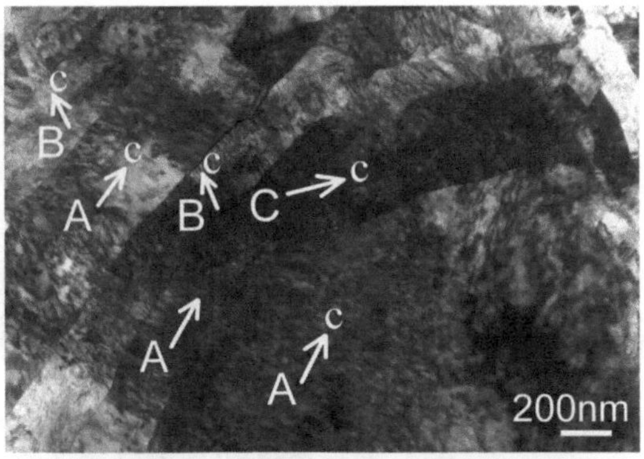


(a)

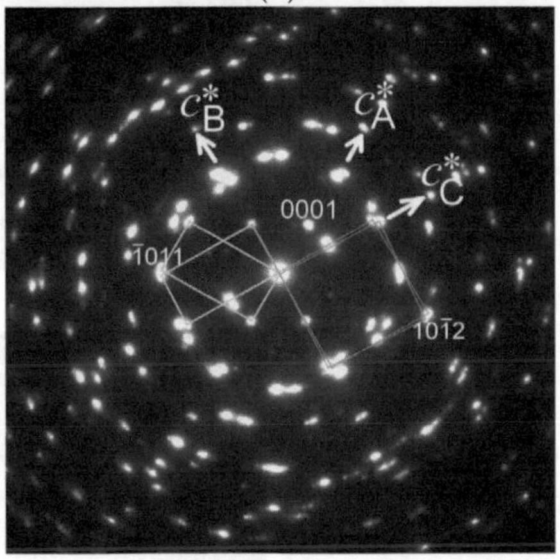


(b)

Figure 3: Bright field electron micrographs showing grain and deformation structures (a) ahead of the crack tip and (b) bottom of a void.

respect to the neighboring domains. This structure became finer in the region III, in which the domain size was about 50 nm in size. In this region, the boundaries between the domains were no longer planar, even if a twin orientation occurred. A domained region is shown in Fig. 4, in which all domains are oriented along one of their $\langle 11\bar{2}0\rangle$ axes. The domains are marked M to S. Strained regions within the domains are often observed. A LAB exists between M and N, on which a regular arrangement of dislocation contrast is visible. In this case, the LAB is formed by 0001 planes. LABs in other orientations were also observed, such as between the domains M and R.

Fig. 5 shows two LABs from the domains in Fig. 4. The LAB between N and M resembles a classic LAB with an array of dislocation contrasts marked by while arrows. This boundary is made up by the basal planes in the two domains. The angle of the

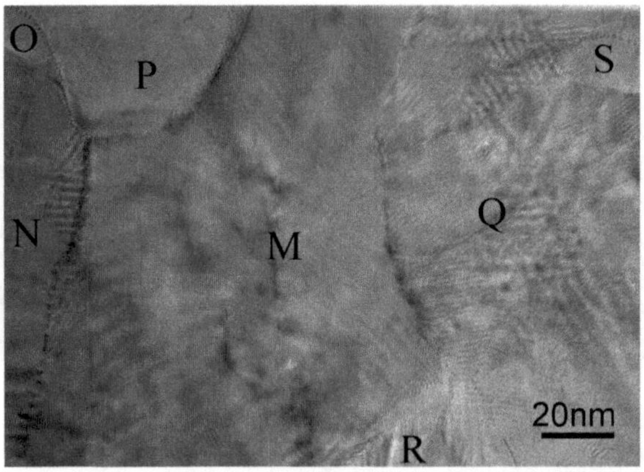


Figure 4: A lattice image showing nano-domains close to the crack surface.

boundary is 6°, which can be precisely measured in the FFT. Another LAB shown in Fig. 5(c,d) is, in some part, perpendicular to the basal planes of M. The basal planes in these two domains M and R make an angle 8°.

Fig. 6(a) shows the interfaces between M, O and P in detail. A Fast Fourier Transform (FFT) of the domains M and O (Fig. 6(b) shows that they form a $\{10\bar{1}1\}$ type twin. The coincident $\{10\bar{1}1\}$ spots from the two domains are marked by circles in (b). The boundary between the two domains is sharp, planar and symmetrically oriented along the common $\{10\bar{1}1\}$ plane, as a twin boundary. A section of the interface between domains M and Q is shown in Fig. 6(c). The FFT in (d) indicates that these two domains form a $\{10\bar{1}1\}$ type twin. The shown section of the interface is roughly planar, but on the (0001) plane of M.

Inside domain Q is embedded another domain S, shown in detail in Fig. 6(e). The accompanying FFT in (f) shows them to be related by $\{10\bar{1}2\}$ type twinning (coincident $\{10\bar{1}2\}$ spots are encircled in (f)). The lower interface is roughly parallel to the common $\{10\bar{1}2\}$ plane; however, the upper part is highly curved.

## Discussion

In a coarse grained magnesium alloy ($> 20\mu$m), the dominant deformation mechanism is twinning, which also become crack propagation route, resulting in a brittle fracture. The dominant deformation

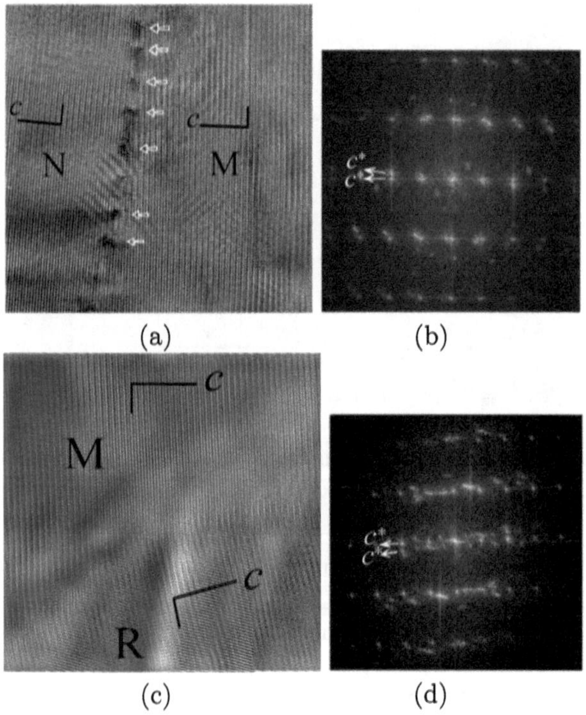


Figure 5: Two examples of low angle boundaries shown in Fig. 4 and their Fast Fourier Transforms. A low angle boundary between domains M and N (a,b) and between domains M and R (c,d).

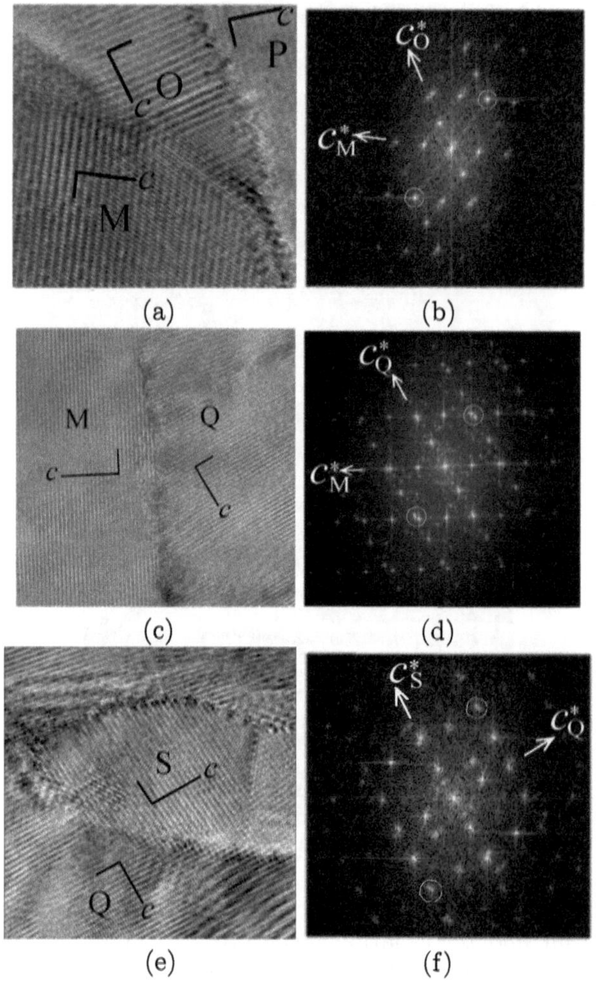


Figure 6: Some twin boundaries from the domains shown in Fig. 4 and their Fast Fourier Transforms. (a,b) $\{10\bar{1}1\}$ type twin between domains M and Q. (c,d) $\{10\bar{1}1\}$ type twin between domains M and O. (e,f) $\{10\bar{1}2\}$ type twin between domains Q and S.

mechanism in a fine grained alloy is dislocation slip. In the beginning of deformation, subgrain structures form ahead of the crack tip, instead of twins. The crack tip blunts because no crack propagation features such as twins exist. This causes high fracture toughness. In addition, voids are nucleated due to stress, by cracking of particles/precipitates or interfaces. Nano-twins are formed due to operation of large stresses around the voids [11].

The slip activity which results in formation of sub-grain like structure, in turn, raises the stress for twin formation. At large stresses required for fracture, formation of $\{10\bar{1}1\}$ type twins occurs. However, it is known that the $\{10\bar{1}1\}$ twins generate substantial stress in the matrix, which are then relieved by formation of $\{10\bar{1}1\}$-$\{10\bar{1}2\}$ double twins.

Our earlier study showed that fracture in a fine grained magnesium alloy occurred by twinning, followed by slipping which caused accumulation of stresses on the twin boundaries [13]. In the present study, however, the domain size are so small that there is not enough slip activity to accumulate stresses on the boundaries. Therefore, deformation progresses by further twinning. To put this in another perspective, continuous twinning is able to relieve the stress necessary for fracture. In this, the role of LABs is crucial. LABs sustain the stresses formed between the various configurations of the twins.

## Conclusions

Fracture mechanism during fracture toughness test has been investigated in a fine-grained Mg-Zn alloy with transmission electron microscopy by sampling specimens from selected sites by focused ion beam.

It was concluded that

1. Ahead of the crack, only subgrains formed, instead of twins. This gives rise to a ductile fracture, instead of a brittle one observed in large grained alloys.

2. At high stresses, $\{10\bar{1}1\}$ type twins formed, with width of about 400nm. These twins further twinned to form $\{10\bar{1}1\}$-$\{10\bar{1}2\}$ double twins.

3. In absence of sufficient slip activity in the subgrains/twins, deformation proceeded by further twinning, leading to smaller domains of 50nm.

4. Thus, before forming a crack, a microstructure consisting of nano-domains related to each other by twinning or low angle boundaries is formed.

## References


[1] H. Somekawa, A. Singh, T. Mukai, "High fracture toughness of extruded Mg-Zn-Y alloy by the synergitic effect of grain refinement and dispersion of quasicrystalline phase", *Scripta Metallurgica* 56 (2007) 1091.

[2] H. Yoshinaga, R. Horiuchi, "", *Mater. Trans. JIM* 4 (1963) 1.

[3] J. Koike, Metall. "Enhanced deformation mechanisms by anisotropic plasticity in polycrystalline Mg alloys at room temperature", *Mater. Trans.* 36A (2005) 1689.

[4] H. Somekawa, A. Singh, T. Mukai, "Fracture mechanism of a coase-grained magnesium alloy during fracture toughness testing", *Phil. Mag. Lett.* 89 (2009) 2.

[5] M. A. Meyers, O. Vohringer, V. A. Lubarda, "The onest of twinning in metals; A constitutive description", *Acta metallurgica* 49 (2001) 4025.

[6] H. Somekawa, T. Mukai, "Effect of grain refinement on fracture toughness in extruded pure magnesium", *Scripta Metallurgica* 53 (2005) 1059.

[7] H. Somekawa, H. S. Kim, A. Singh, T. Mukai, "Effect of precipitate shapes on fracture toughness in extruded Mg-Zn-Zr magnesium alloys", *J. Mater. Res.* 22 (2007) 965.

[8] W.H. Hartt and R.E. Reed-Hill, Trans. TMS-AIME 239 (1967) 1511.

[9] B.C. Wonsiewics and W.A. Backofen, Trans. TMS-AIME 239 (1967) 1422.

[10] E.W. Kelly and W.F. Hosford, Jr., Trans. TMS-AIME 242 (1968) 5.

[11] H. Somekawa, K. Nakajima, A. Singh, T. Mukai, Philo. Mag. Lett. 90 (2010) 831.

[12] H. Somekwa, Y. Osawa, A. Singth, T. Mukai, J. Mater. Res.23 (2008) 1128.

[13] V. Gartnerova, A. Singh, A. Jager and T. Mukai, "Deformation behaviour of ultra-fine-grained Mg-0.3at%Al alloy in compression" (to be communicated)

**Magnesium Technology 2012**
*Edited by: Suveen N. Mathaudhu, Wim H. Sillekens, Neale R. Neelameggham, and Norbert Hort*
***TMS (The Minerals, Metals & Materials Society), 2012***

# TENSILE AND CREEP DEFORMATION MECHANISMS IN ROLLED AZ31

Z. Chen[1], C. Boehlert[1,2,3], I. Gutiérrez-Urrutia[4], J. Llorca[2,3] and M.T. Pérez-Prado[3]

[1]Michigan State University, East Lansing, Michigan 48824-1226, USA
[2]Department of Materials Science, Polytechnic University of Madrid, 28040 Madrid, Spain
[3]IMDEA Materials Institute, Calle Profesor Aranguren, s/n, 28040 Madrid, Spain
[4]Max Planck Institute for Iron Research, Max Planck Strasse, 1, 40237 Düsseldorf, Germany

Keywords: Microstructure, Tensile, Slip, Twinning, Creep, Magnesium, Lightweight Alloys

## Abstract

Tensile experiments were performed on a rolled AZ31 alloy in an SEM at 323K (50°C), 423K (150°C), and 523K (250°C) in order to analyze the deformation mechanisms *in-situ*. Electron backscatter diffraction (EBSD) was performed both before and after deformation. The mechanical anisotropy was considerably reduced with temperature. Extension twinning was observed at 323K (50°C), but disappeared at 423K (150°C), indicating that the CRSS of non-basal systems becomes smaller than that of twinning at T<423K (150°C). From 423K (150°C) to 523K (250°C), a transition occurred in the dominant deformation mode from basal + prismatic <a> to mainly prismatic <a> slip. This is consistent with a decrease of the CRSS of non-basal slip systems with increasing temperature. *In-situ* tensile-creep experiments, performed at approximately the yield stress at 423K (150°C), indicated less slip and more grain boundary cracking occurs under creep deformation as compared to the higher-stress tensile experiments.

## Introduction

The development of lightweight Mg alloys with higher strength requires a detailed knowledge of the deformation mechanisms. Extensive studies have been performed at low strain rates, using conventional *ex-situ* techniques [1-3]. Although widespread values of the critical resolved shear stresses (CRSS) have been reported for the various slip and twinning systems of Mg alloys [4,5], it is generally accepted that the CRSS follows the trend $CRSS_{basal}<CRSS_{\{10\text{-}12\}twinning}<CRSS_{prismatic}<CRSS_{pyramidal}$ at room temperature (RT). The CRSS for basal slip and {10-12} extension twinning are believed to be temperature independent [3]. The CRSS of prismatic and pyramidal systems decrease as the temperature increases, reaching values lower than that of {10-12} extension twinning and, eventually, basal slip [3,6]. Thus, the activity of non-basal slip systems increases with temperature, facilitating intergranular compatibility [1,2,7]. However, this transition temperature is still unclear.

Moreover, the role of grain boundary sliding (GBS) during deformation of Mg alloys at moderate-to-high temperatures is also controversial. Some studies suggested an increasing contribution of GBS at high temperatures, even in materials with conventional grain size [7-9], some also suggested that GBS contributes to deformation even at RT in fine grained Mg alloys [10]. However, these ideas were refuted by some other studies [11].

Highly-textured, rolled AZ31 sheet material shows a significant drop in plastic anisotropy (commonly characterized by the r-value) between 298K (25°C) and 473K (200°C). This behavior was explained in two possible ways. One proposed it to be the result of an increased activity of pyramidal <c+a> slip [2], while the other attributed it to the activation of GBS [7, 12]. Both explanations were supported by their simulation results based on viscoplastic self-consistent modeling.

In this study, *in-situ* deformation characterization was performed during tensile and creep experiments on a rolled AZ31 alloy. Electron backscattered diffraction (EBSD) was performed both before and after the tensile tests. The onset of the different slip and twinning systems as a function of temperature was identified. Further evidence regarding the influence of GBS on the anisotropy was also provided.

## Experimental

Rolled and annealed AZ31 sheet of 3mm thickness, was purchased from Magnesium Elektron (Manchester, UK). The sheet material had a strong basal texture. The bulk chemical composition of the as-rolled material, measured using an Inductively Coupled Plasma Analyzer, was Mg-2.9Al-0.73Zn-0.2Mn(wt.%).

Flat dog-bone tensile specimens with gage dimensions of 3 x 2.5 x 10 mm were electro-discharge machined out of the as-received sheet. The samples were machined with the tensile axis either parallel to (RD) or perpendicular to (TD) the rolling direction of the sheets. *In-situ* tensile tests were performed at a constant displacement rate of 0.004mm/s, equivalent to a strain rate of approximately $10^{-3}s^{-1}$, using a screw-driven tensile stage placed inside a Zeiss (Jena, Germany) EVO MA15 SEM at 323K (50°C), 423K (150°C) and 523K (250°C). The temperature was maintained within 5K of the target using a constant-voltage power supply to a 6mm diameter tungsten-based heater located just below the gage section of the sample. Secondary electron (SE) SEM images were taken before loading and during interruptions at various strains throughout the tensile test. The strain values were estimated from the crosshead displacement of the testing machine taking into account the gage length of the samples. Further details of this apparatus and testing technique can be found elsewhere [13]. After the experiments were terminated (usually after achieving 25% strain) and the samples were unloaded, the width and thickness of the samples were measured and compared to those measured before the experiment. These measurements, performed only on the samples tested in the RD, were used to determine the plastic anisotropy. They were compared to previous results on tensile specimens deformed to approximately 10% strain [2,11,12].

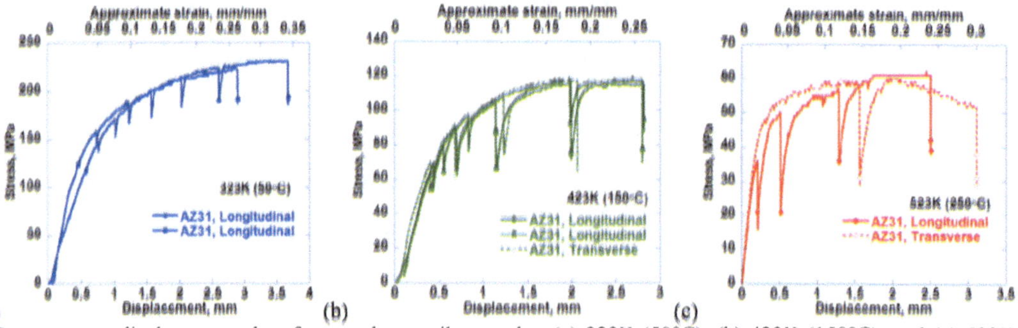

Figure 1. Stress versus displacement plots for samples tensile tested at (a) 323K (50°C), (b) 423K (150°C), and (c) 523K (250°C). The load drops indicate the stress relaxation that occurred when experiment was paused and SEM images were acquired.

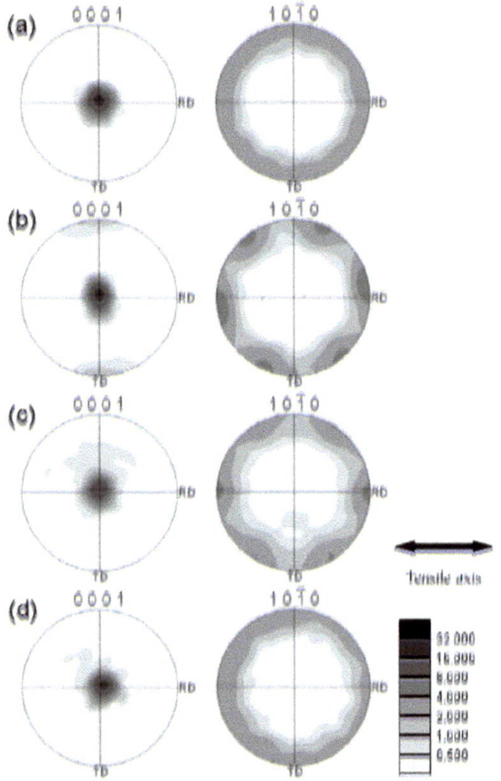

Figure 2. EBSD pole figures along the normal direction (ND) for the AZ31 tensile specimens, (a) un-deformed, the tensile axis is along the RD, (b) tested at 323K (50°C) along RD after ~32.1% strain, (c) tested at 423K (150°C) along RD after ~25.6% strain, and (d) tested at 523K (250°C) along RD after ~24.2% strain. The intensity of the peaks is indicated in the scalebar provided.

The microstructure in a selected area of the specimens (in the center of the gage section) was analyzed before and after each tensile test by EBSD in a 6500 F JEOL field emission gun-scanning electron microscope (SEM) equipped with an EDAX-TSL EBSD system. Sample preparation for EBSD included grinding with 4000 grit SiC paper, mechanical polishing with a 0.5μm silica suspension and final electrochemical polishing for 60s at 20V using the AC2™ commercial electrolyte. The grain size, calculated in the EBSD orientation maps using only grain boundaries with misorientations greater than 15°, was about 13μm. Following a previously described methodology [14], the active slip and twinning systems were identified using the EBSD data and the SEM images acquired during the tensile experiments. Three slip modes and two twinning modes were considered in the deformation analysis: (1) {0001}<-2110> basal <a> slip, (2) {01-10}<2-1-10> prismatic <a> slip, (3) {11-22}<-1-123> second order pyramidal <c+a> slip, (4) {10-12}<-1011> extension twin, and (5) {10-11}<10-1-2> contraction twin.

Two samples, one along RD and one along TD, were creep tested at 423K (150°C) and 75MPa using the same stage described above. SE images were acquired during deformation without interrupting the experiment. EBSD orientation maps were not acquired for the creep-tested specimens, which were not taken to failure but were unloaded after approximately 50 hours of creep.

## Results

### *Tension*

The tensile curves for seven specimens tested at different temperatures are shown in Figure 1. Little difference was observed between the RD and TD tests (see Figures 1b and c). The yield strength (YS) was approximately 110MPa at 323K (50°C), 60MPa at 423K (150°C), and 40MPa at 523K (250°C). These values are comparable to those found in the literature for textured AZ31 [15].

Figure 2 shows the EBSD pole figures along the normal direction (ND), for specimens before testing, and after testing at 323K (50°C), 423K (150°C), and 523K (250°C). The as-rolled AZ31 sheet exhibited a basal fiber texture in which the basal plane was parallel to the rolling plane, typical of rolled and annealed Mg alloys [16]. The basal texture was maintained after deformation at 323K (50°C). The alignment of (10-10) poles with the RD (tensile axis) revealed the occurrence of prismatic slip. A new texture component, (0001) parallel to the TD direction, was developed, which revealed the activation of extension twinning. At 423K (150°C), samples tested along RD and TD developed a similar deformation texture, and it was similar to that observed at 323K (50°C). However, the texture component <0001>//TD was not observed in the sample deformed at 423K (150°C), which indicated that twinning was not active at 423K (150°C). At 523K

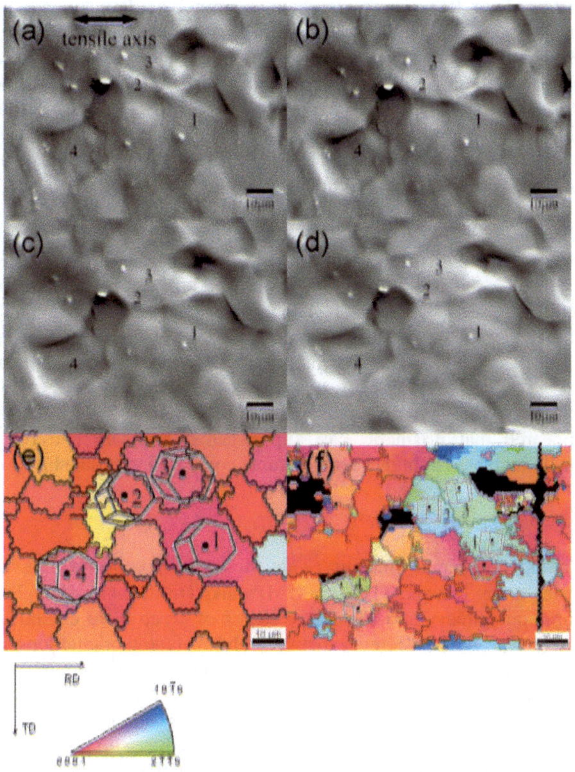


Figure 3. SE SEM images for a sample tested at 323K (50°C) along RD corresponding to approximate strain of: (a) 5.4% strain, (b) 10.9% strain, (c) 21.2% strain, and (d) 32.1% strain, (e) EBSD IPF map in the ND before test, and (f) EBSD IPF map in the ND after test. Twinning was observed in the grains labeled 1-4 after strains of greater than 5.4%.

Table I. Twinning modes observed in Figure 3

| Grain | Orientation relationship between reoriented region and parent grain | Twin type | Schmid Factor |
|---|---|---|---|
| 1 | 89.7°@<11-20> | (-1102)[1-101] | 0.05 |
| 2 | 88.0°@<11-20> | (-1102)[1-101] | 0.05 |
| 3 | 89.7°@<-12-10> | (-1012)[10-11] | 0.08 |
| 4 | 89.6°@<-2110> | (0-112)[01-11] | 0.05 |

(250°C), samples tested along RD and TD also developed a different texture than that formed at lower temperatures (323-423K (50-150°C)). Basically, no texture variation with respect to the initial texture was developed at 523K (250°C), indicating that plastic strain was controlled by different dominant deformation modes than those controlling plasticity at lower temperatures.

At 323K (50°C), twinning was observed during the early stages of deformation (147-171MPa/~2.9-5.4% strain). Slip traces were observed after the sample reached approximately 5% strain, but not in large quantities. Furthermore, some cracks became apparent at the grain boundaries as the deformation increased. Figure 3 shows SEM images of the same region taken at different stages of the tensile test. Deformation twins were found in grains referred as to 1, 2, and 3 at ~5.4% strain, and in grain 4 at ~21.2%

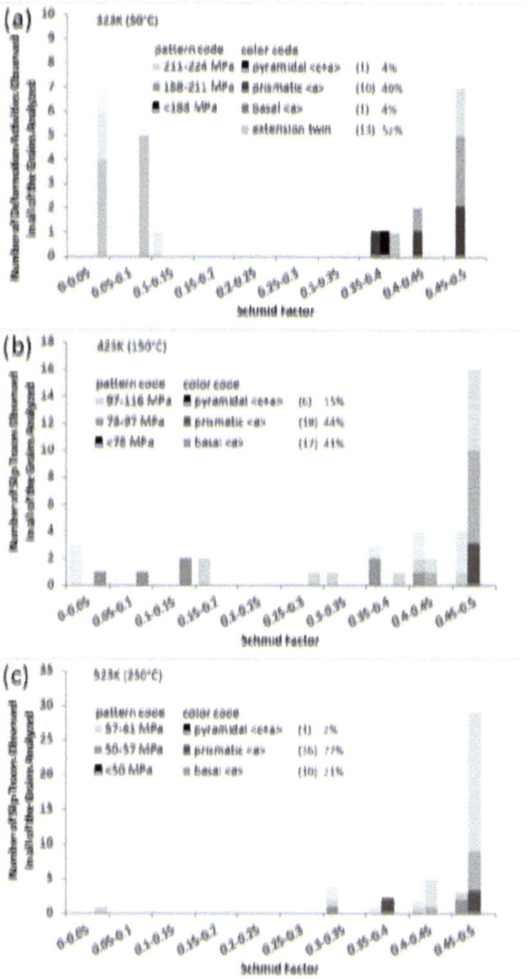


Figure 4. Schmid factor distributions and stress levels at which the activities of each of the deformation modes were observed for samples tensile tested at (a) 323K (50°C), (b) 423K (150°C), and (c) 523K (250°C).

strain. As the deformation proceeded, the twin area fraction increased, and ultimately consumed a large part of the parent grain. The lattice orientations of the grains with twinning are indicated by the unit cells drawn on the inverse pole figure (IPF) map for this region both before and after the tensile test (Figure 3e and 3f). The orientation relationships between the reoriented region and the parent grain, see Table I, suggests the formation of {-1102}<1-101> extension twin. This deformation mode, which ideally reorients the lattice 86.5° about the <11-20> direction and leads to extension along the c-axis, was not favorable in most of the grains considering the basal texture in the material. However, extension twinning under tension along RD in AZ31 sheet has been observed occasionally [15]. The Schmid factors for twinning were small, suggesting that twinning is easily activated at RT compared to the other deformation modes. In the two samples tested at 323K (50°C), 13 grains were observed to reorient due to extension twinning, and no contraction twins were observed. Slip was observed in 12 grains. Figure 4(a) summarizes the Schmid factor distribution and stress levels at which the identified deformation modes were observed. Among the 12 slip traces analyzed, 1 was basal <a> slip, 10 were prismatic <a> slip, and 1 was pyramidal <c+a> slip.

At 423K (150°C), no twinning was observed during the *in situ* test. Slip bands were first observed between approximately 4.0-8.5% strain, and an increase in the density of slip bands can be clearly observed with increasing strain. The appearance of some grain boundary ledges could be distinguished after approximately 1.1% strain. As the deformation proceeded, some grain boundary cracks became apparent. Figure 5 illustrates a region in a sample

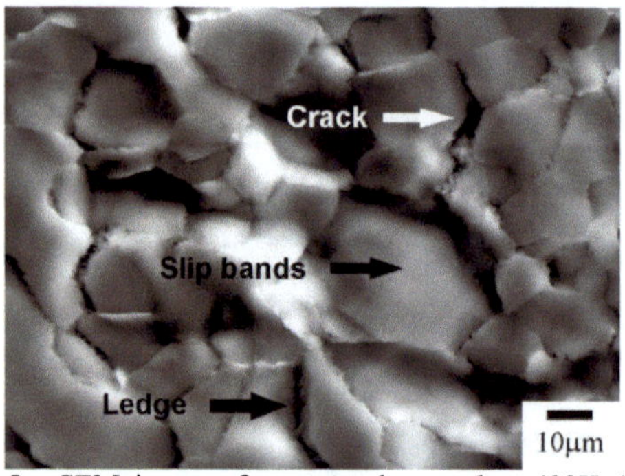


Figure 5. SEM images for a sample tested at 423K (150°C) along RD; at approximately 25% strain.

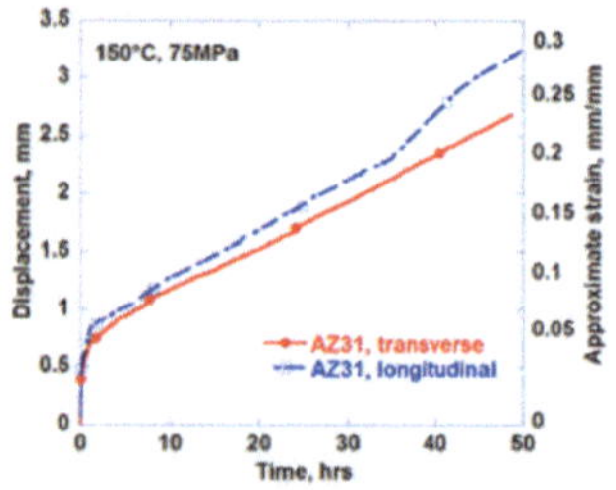


Figure 6. Creep displacement versus time plot for rolled AZ31 at T=423K (150°C) and σ=75MPa.

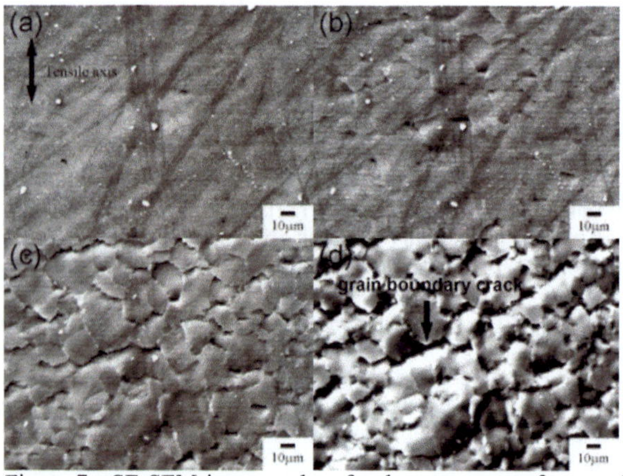


Figure 7. SE SEM images taken for the same area of a sample creep test at T=423K (150°C) along TD. The creep time and approximate strain are (a) 0.1 hrs, 1.4%, (b) 0.4 hrs, 2.9%, (c) 12.0 hrs, 9.4%, and (d) 45.6 hrs, 22.9%. Grain boundary cracking was prevalent.

tested along RD, where slip bands, grain boundary ledges and cracks were indicated. In the three samples tested at 423K (150°C), the active slip systems were determined for 41 different slip trace, among which 17 were basal <a> slip, 18 were prismatic <a> slip, and 6 were pyramidal <c+a> slip, see Figure 4(b). Basal slip and pyramidal <c+a> slip occurred over a wide range of Schmid factors and stresses, while prismatic <a> slip was activated only when a high Schmid factor (>0.4) was present.

In the samples tested at 523K (250°C), the slip traces were more obvious than in the samples tested at 323K (50°C) and 423K (150°C). As the deformation proceeded, some grain boundary relief was also observed. The slip systems were determined for 47 slip traces in the two samples tested at 523K (250°C): 10 were basal <a> slip, 36 were prismatic <a> slip, and 1 was pyramidal <c+a> slip, see Figure 4(c). Most of the active slip systems exhibited a relatively high Schmid factor (>0.30). Thus, it appeared that prismatic <a> slip became dominant and more easily activated at higher temperatures compared to basal slip.

<u>*Creep*</u>

Creep strain versus time curves are shown in Figure 6 for samples tested along RD and TD at σ=75MPa and T=423K (150°C). The common primary and secondary stages of creep were exhibited. There was little difference between the different orientations in terms of the creep strains and strain rates. Grain boundaries served as crack nucleation sites in creep, see Figure 7. No slip traces were observed. It is noted that during the 423K (150°C) tensile test, 3 basal slip traces and 3 prismatic <a> slip traces were observed at stress less than 78MPa, the latter only for Schmid factors between 0.4 and 0.5.

## Discussion

<u>*Twinning and Dislocation Slip*</u>

The EBSD pole figures shown in Figure 2 were consistent with previous studies [2] and revealed that, when the AZ31 sheet alloy was deformed along an in-plane direction at 323K (50°C) and 423K (150°C), prismatic slip was prevalent. This lead to the alignment of the (10-10) poles parallel to the tensile axis. Basal slip should also be favorable in grains which deviated from the perfect basal orientation. However, when deformation took place at higher temperature (523K (250°C)), the initial basal fiber remained, yet the (10-10) poles were not aligned parallel to the tensile axis according to the pole figures (see Figure 2(d)). Thus, the dominant deformation modes at this temperature could not be easily inferred from examination of the pole figures.

The variation with temperature of the CRSS of various slip and twinning systems in Mg alloys has been studied by both crystal plasticity models in polycrystals and experiments in single crystals. The CRSS of basal slip and extension twinning are generally believed to be temperature independent [3]. Using the output of a full constraint Taylor model proposed by Barnett [3], the CRSS of prismatic and pyramidal slip are predicted to become smaller than that of extension twinning at temperatures higher than 506K (233°C) and 473K (200°C), respectively, and they remain higher than that of basal slip until about 733K (460°C). Based on single crystal experiments, Chapuis and Driver [6] reported that the CRSS of pyramidal slip was higher than that of prismatic slip within the temperature range investigated and that neither of them became comparable to that of extension twinning until temperatures as high as 723K (450°C). Earlier studies on the subject [17], carried out also in single crystals in a temperature range from 298-773K (25-500°C), confirmed that the CRSS for pyramidal <c+a> slip is higher than that of prismatic <a> slip, although the latter becomes comparable to that of extension twinning at temperatures higher than 473K (200°C). Thus, large discrepancies still remain between various studies regarding the critical temperatures at which the CRSS of non-basal systems become comparable to that of extension twinning and basal slip. Furthermore, the relative activity of the different deformation mechanisms in the temperatures ranges of interest is not well understood.

Our *in situ* experiments provided new insight on these issues. The relative contributions of the different slip systems at 423K (150°C) were the following: 41% basal, 44% prismatic <a>, and 15% pyramidal <c+a>. The relative contributions of the different slip systems at 523K (250°C) were: 21% basal, 77% prismatic <a>, and 2% pyramidal <c+a>. The present work showed that from 423K (150°C) to 523K (250°C), a transition in the dominant deformation mechanism occurred from basal + prismatic <a> to mainly prismatic <a> slip. Basal slip took place at 423K (150°C) even in grains in which the Schmid factor is very low (as low as 0.05) (see Figure 4(b)). Conversely, 16 out of the 18 prismatic slip traces occurred in grains in which the corresponding Schmid factor was greater than 0.45. This suggests that the CRSS for prismatic slip is still significantly higher than that of basal slip at 423K (150°C). However, it must be already lower than that of twinning, as twinning was absent at 423K (150°C). The increase in the contribution of prismatic slip with temperature, from 44% at 423K (150°C) to 77% at 523K (250°C), suggests that the CRSS for prismatic slip might be on the order of that of basal slip at 523K (250°C). The present observations provided by the *in situ* tests on polycrystals seem to be more in agreement with earlier experimental single crystal studies [17] than with the other works, and close to the transition temperature predicted by crystal plasticity models [3]. These comparisons must be, however, taken with caution, as the CRSS values provided by single crystal experiments do not directly reflect the applied shear stress that is necessary to activate certain deformation systems in polycrystalline materials since most hardening contributions are not proportional to the CRSSs but are additive to them [4]. It is noteworthy that the activity of pyramidal <c+a> slip, which was 4% at 323K (50°C), 15% at 423K (150°C) and 2% at 523K (250°C), remained low in the temperature range analyzed in the current study.

## Grain Boundary Sliding

The role of GBS during deformation of Mg alloys is controversial [2,7-11]. In the current *in situ* observations, no significant differences were revealed between the amount of grain boundary cracking and of jogging of fiducial lines, both potential signs of GBS, between the three different temperatures for tensile test. In turn, as described above, a clear variation between the relative activities of the various slip systems was detected. However, during the creep test at 423K (150°C) and 75MPa, a significant decrease of the slip activity accompanied by an increase of the incidence of grain boundary cracking and jogging of fiducial cracks, was clearly observed. These observations suggest that GBS does indeed contribute to the total strain under these conditions. However, the contribution of GBS to the total applied strain was small even at temperatures as high as 250°C at constant strain rates of $10^{-3}s^{-1}$ for the tensile experiments. This is consistent with the fact that the creep test took place at much lower strain rates (the strain rate corresponding to the secondary stage was approximately $1.1x10^{-6}$ $s^{-1}$), at which it might have been assisted by diffusion, which would take place much more easily.

## Origin of the Decrease in Anisotropy at Higher Temperatures

Figure 8 compares the r-values of rolled AZ31 as a function of temperature, including the results of the current study. Our results showed that there was a considerable drop in the r-value with temperature, and this is in agreement with previous data [2,11,12]. The decrease in anisotropy with increasing temperature has been attributed to an increase of the contribution of GBS [7,12] or to the enhanced operation of pyramidal <c+a> slip [2,11] with increasing temperature. Stanford *et al.* [11] showed that increasing the relative activity of pyramidal <c+a> slip by only 15%, with a commensurate decrease in prismatic slip activity, led to a reduction of the normal anisotropy by a factor of two. Their observations suggested that the reduction in the r-value was likely caused by an increase in non-basal slip activity as they did not observe any significant contribution of GBS. In the current work, the proportion of non-basal slip activity (primarily prismatic <a>) significantly increased with temperature, and there was no significant difference between GBS at each of the temperatures examined. Although our data do not reveal an increase in the activity of pyramidal <c+a> slip, as reported in [2], the present results suggest that the observed decrease in anisotropy might be due to increased non-basal activity, rather than to GBS.

## Creep

As expected, little difference were observed between samples tested in the RD and TD. Grain boundaries served as crack nucleation sites in creep, see Figure 7, and grain boundary cracking may have been accommodating GBS. There are two well described accommodation mechanisms for GBS: diffusion [18] and slip [19]. GBS accommodated by diffusion is typically observed at high temperature, such as those encountered during creep. In this model, the cavity created by the relative movement of one grain past another is filled by diffusing atoms. This mechanism allows for a shape change of the grains during the deformation. The other accommodation mechanism for GBS is slip. When two grains slide past each other, stresses accumulate at the grain boundary, which are released by the intragranular movement of dislocations. Of the two accommodation

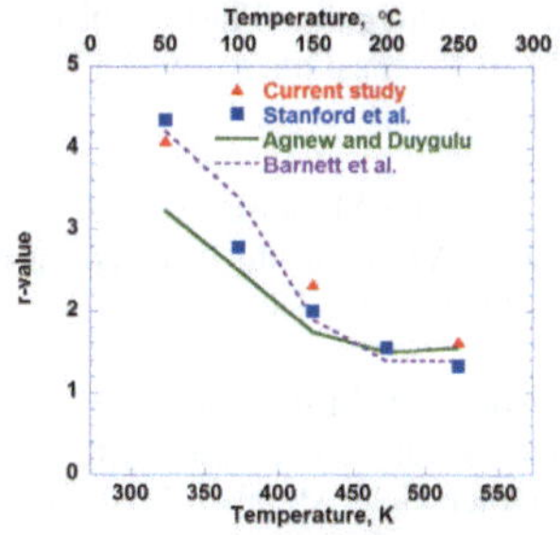


Figure 8. The r-value measured after a tensile strain for the current study as compared with previous studies [2,11,12].

mechanisms, the former is more likely for the conditions evaluated in the present work as grain boundary cracking, rather than slip, was observed [20].

## Conclusions

Rolled AZ31 was *in situ* tensile tested in an SEM at temperatures of 323K (50°C), 423K (150°C), and 523K (250°C), and EBSD orientation maps were acquired both before and after the deformation on the same local microstructural patch. Twinning was observed at 323K (50°C), but disappeared at 423K (150°C). The deformation at high temperature was controlled by a combination of basal and prismatic <a> slip. From 423K (150°C) to 523K (250°C), a transition in the dominant deformation mechanism occurs from basal + prismatic <a> to mainly prismatic <a> slip. A significant drop in the normal anisotropy was exhibited with increasing test temperature. Observation of grain boundary offsets and grain boundary cracking during the tensile deformation confirmed that GBS was present. However, it did not significantly contribute to the overall total strains and appeared to be insensitive to temperature. Therefore the reduction in normal anisotropy with temperature was more likely due to the increased slip activity of the non-basal slip systems rather than to GBS.

In addition, two *in situ* creep tests were performed at $\sigma$=75MPa and T=423K (150°C). Slip traces were not observed, while cracking preferentially occurred at grain boundaries and fiducial markings indicated jogs at grain boundary locations. Thus creep deformation was dominated more by GBS than slip activity, and therefore grain boundaries play a more important role during creep than during tension.

## Acknowledgments

This work was supported by the National Science Foundation Division of Material Research (Grant No. DMR1107117). CJB acknowledges the support from the Spanish Ministry of Education for his sabbatical period in Madrid (SAB2009-0045). The authors would like to thank the vehicle interior manufacturer, Grupo Antolin Ingeniería, S.A., within the framework of the project MAGNO2008-1028-CENIT funded by the Spanish Ministry of Science and Innovation.

## References

1. A. Couret and D. Caillard, "An Insitu Study of Prismatic Glide in Magnesium .2. Microscopic Activation Parameters", *Acta Metallurgica*, 33 (8) (1985), 1455-1462.

2. S.R. Agnew and O. Duygulu, "Plastic Anisotropy and the Role of Non-basal Slip in Magnesium Alloy AZ31B", *nt J Plast*, 21 (6) (2005), 1161-1193.

3. M.R. Barnett, "A Taylor Model Based Description of the Proof Stress of Magnesium AZ31 during Hot Working", *Metall Mater Trans A*, 34A (9) (2003), 1799-1806.

4. W.B. Hutchinson and M.R. Barnett, "Effective Values of Critical Resolved Shear Stress for Slip in Polycrystalline Magnesium and Other HCP Metals", *Scr Mater*, 63 (7) (2010), 737-740.

5. M.R. Barnett *et al.*, "A Semianalytical Sachs Model for the Flow Stress of a Magnesium Alloy", *Metall Mater Trans A*, 37A (7) (2006), 2283-2293.

6. A. Chapuis and J.H. Driver, "Temperature Dependency of Slip and Twinning in Plane Strain Compressed Magnesium Single Crystals", *Acta Mater*, 59 (5) (2011), 1986-1994.

7. B. Hutchinson *et al.*, "Deformation Modes and Anisotropy in Magnesium Alloy AZ31", *nt J Mater Res*, 100 (4) (2009), 556-563.

8. J.A. Del Valle *et al.*, "Deformation Mechanisms Responsible for the High Ductility in a Mg AZ31 Alloy Analyzed by Electron Backscattered Diffraction", *Metall Mater Trans A*, 36A (6) (2005), 1427-1438.

9. J.A. Del Valle and O.A. Ruano, "Separate Contributions of Texture and Grain Size on the Creep Mechanisms in a Fine-Grained Magnesium Alloy", *Acta Mater*, 55 (2) (2007), 455-466.

10. J. Koike *et al.*, "Grain-Boundary Sliding in AZ31 Magnesium Alloys at Room Temperature to 523 K", *Mater Trans*, 44 (4) (2003), 445-451.

11. N. Stanford *et al.*, "Deformation Mechanisms and Plastic Anisotropy in Magnesium Alloy AZ31", *Acta Mater*, 59 (12) (2011), 4866-4874.

12. M.R. Barnett *et al.*, "Role of Grain Boundary Sliding in the Anisotropy of Magnesium Alloys", *Scr Mater*, 61 (3) (2009), 277-280.

13. C.J. Boehlert *et al.*, "In situ Scanning Electron Microscopy Observations of Tensile Deformation in a Boron-Modified Ti-6Al-4V Alloy", *Scr Mater*, 55 (5) (2006), 465-468.

14. C.J. Boehlert *et al.*, "Slip System Characterization of Inconel 718 Using In-Situ Scanning Electron Microscopy", *Advanced Materials & Processes*, 168 (11) (2010), 41-45.

15. A.S. Khan *et al.*, "Mechanical Response and Texture Evolution of AZ31 Alloy at Large Strains for Different Strain Rates and Temperatures", *nt J Plast*, 27 (5) (2011), 688-706.

16. A.D. Rollet and S.I. Wright, "Typical Textures in Metals", *Te ture and Anisotropy*, ed. U.F. Kocks, et al. (Cambridge, Cambridge University Press, 1998), 179-239.

17. T. Obara *et al.*, "[11-22](-1-123) Slip System in Magnesium", *Acta Metallurgica*, 21 (7) (1973), 845-853.

18. M.F. Ashby and R.A. Verrall, "Diffusion-Accommodated Flow and Superplasticity", *Acta Metallurgica*, 21 (2) (1973), 149-163.

19. A.K. Mukherje, "Rate Controlling Mechanism in Superplasticity", *Mater Sci Eng*, 8 (2) (1971), 83-89.

20. C.J. Boehlert *et al.*, " *n situ* Analysis of the Tensile and Tensile-Creep Deformation Mechanisms in Rolled AZ31", *Acta Mater*, in print.

**Magnesium Technology 2012**
*Edited by: Suveen N. Mathaudhu, Wim H. Sillekens, Neale R. Neelameggham, and Norbert Hort*
*TMS (The Minerals, Metals & Materials Society), 2012*

# STRUCTURAL ORIGIN OF REVERSIBLE TWINNING, NON-SCHMID EFFECT, INCOHERENT TWIN BOUNDARIES AND TEXTURE IN HEXAGONAL CLOSE-PACKED METALS

B. Li [1,2], H. El Kadiri [1,3], X.Y. Zhang [4], S.N. Mathaudhu [5], Q. Ma [1]
[1] Center for Advanced Vehicular Systems, Mississippi State University, Starkville, MS 39759, USA
[2] Center for Advanced Metallic and Ceramic Systems, Johns Hopkins University, Baltimore, MD 21218, USA
[3] Department of Mechanical Engineering, Mississippi State University, Starkville, MS 39759, USA
[4] School of Materials Science and Engineering, Chongqing University, Chongqing, China
[5] Weapons and Materials Research Directorate, U.S. Army Research Laboratory, Aberdeen Proving Ground, MD
Keywords: Keyword1, Keyword2, Keyword3, etc.

## Abstract

Notably the most dominant twinning mode in hexagonal close-packed metals, $\{10\bar{1}2\} < 10\bar{1}\bar{1} >$ twinning presents abnormal properties such as reversible twinning and non-Schmid effect. The twin boundaries may significantly deviate from the $\{10\bar{1}2\}$ twinning plane. HCP metals also present a strong propensity to develop texture during processing. Through electron backscatter diffraction and high resolution transmission electron microscopy observations, we show that these properties can be well understood from the perspective of the atomic shuffling that dominates in the twinning.

## Introduction

Due to the insufficient number of easy slip systems in low symmetry hexagonal close-packed (HCP) metals, e.g., magnesium (Mg), titanium (Ti), zirconium (Zr) and cobalt (Co), profuse twinning is activated during plastic deformation to accommodate the strain perpendicular to the basal plane, i.e., along the direction of <c>-[0001]. The ubiquitous $\{10\bar{1}2\} < 10\bar{1}\bar{1} >$ twinning is the most commonly observed and dominant twinning mode in HCP metals. The properties of this particular twinning mode are distinctive in comparison to those of other twinning modes:

(1) *Reversible twinning*. Detwinning [1,[2]] can occur if the applied external load is removed or reversed. Twin boundaries (TBs) may move backward and shrink the twins, giving rise to the pseudoelasticity [3] in HCP metals;
(2) *Non-Schmid effect*. The Schmid Law can be violated. Barnett et al. [4] found that, in the $\{10\bar{1}2\} + \{10\bar{1}1\}$ double twinning, the secondary $\{10\bar{1}2\}$ twinning does not follow the Schmid Law;
(3) *Extremely incoherent TBs*. Partridge and Roberts [5] observed abnormal TB migration in Mg by optical microscopy. They found that, under small stresses from indentation, the existing $\{10\bar{1}2\}$ TBs can migrate and evolve into extremely incoherent structures which considerably deviate from the $\{10\bar{1}2\}$ twinning plane. In other words, the TBs no longer coincide with the $\{10\bar{1}2\}$ twinning plane.

Additionally, Mg and other HCP metals have a strong propensity for developing pronounced texture during mechanical processing. During rolling or extrusion, the basal planes of the crystallites are reoriented roughly parallel to the rolling or extrusion direction [6,7]. Basal texture has also been reported after grain refining by severe plastic deformation (SPD) [8]. Karaman et al. [9] studied the texture of Zr, beryllium (Be), AZ31 Mg alloy and Ti-6Al-4V alloy after each was subjected to one pass of equal channel angular extrusion (ECAE). They found that the basal planes always aligned with the most elongated direction of the sheared grains, irrespective of the *c/a* ratio, grain size and initial texture. Their visco-plasticity-self-consistent (VPSC) modeling showed that basal slip must be the primary slip system in all four materials, contributing to the texture, when twinning was not considered in the modeling. In fact, the $\{10\bar{1}2\} < 10\bar{1}\bar{1} >$ twinning is a major contributor to texture in HCP metals [10] because twinning reorients the parent lattice much more efficiently than dislocation slip.

A question immediately arises: intrinsically, how are these aforementioned properties of HCP metals originated from the perspective of crystal structure? The purpose of this work is to address this important question based upon our experimental observations.

## Experimental Methods

An AM30 Mg alloy (composition) was pre-extruded and then compressed along the extrusion direction (ED) until 2.3% plastic strain. The microstructures of the pre-extruded and the deformed samples were examined with field emission gun scanning electron microscopy (FEG-SEM) equipped with an electron back-scatter diffraction (EBSD) data acquisition and analysis system. The sample was polished at a section normal to radial direction of the extruded billet. Pure, polycrystalline Co samples were also deformed dynamically at room temperature [11], and the microstructure of the tested samples was examined by transmission electron microscopy (TEM) and high resolution TEM (HRTEM).

## Results

Figure 1(a) shows the inverse pole figure mapping the misorientations in the deformed AM30 sample. The elongated grains (in blue) experienced large plastic strain during extrusion, and the elongation paralleled the extrusion direction (ED). The variation in the blue background indicates a relatively high density of dislocations and substructures in these grains. In contrast, the fine, equiaxed grains (in green) are dynamically recrystallized during extrusion with a low density of dislocations. Compression along the ED activated the $\{10\bar{1}2\} < 10\bar{1}\bar{1} >$ twinning. The needles (in red) are deformation twins that cross the elongated grains. With orientation analysis, the traces of the twinning planes (must be one of the $\{10\bar{1}2\}$ planes) can be determined. Figure 1(b) shows a magnified view that highlights the twins. Three intercrossing white lines mark the traces of the $\{10\bar{1}2\}$ planes in the parent lattices. There should be six $\{10\bar{1}2\}$ planes and hence six such traces. Since three of the traces almost coincide with the others, three lines suffice to demonstrate the deviation. Strikingly, the TBs do not coincide with the traces of the $\{10\bar{1}2\}$ plane. Additional scanning in different regions and grains shows similar results, with most TBs deviating from the $\{10\bar{1}2\}$ twinning plane.

Large deviation was also revealed in deformation twins in Co samples. Figure 2 shows a HRTEM micrograph of the $\{10\bar{1}2\}<10\bar{1}\bar{1}>$ TBs and the traces of the twinning planes. Lattice fringes from the basal planes can be seen. In the inset of Figure 2, the diffraction pattern indicates the grains indeed satisfy the $\{10\bar{1}2\}<10\bar{1}\bar{1}>$ twin relationship. The zone axis is parallel to $[1\bar{2}10]$ direction. The TBs, twinning planes, and the basal planes of the twin and the parent are all parallel to the zone axis. In this orientation, there are only two possible twinning planes, and their traces are marked with two broken lines. It can be seen that the actual TBs deviate strikingly far from the $\{10\bar{1}2\}$ twinning plane. In fact, the TB near the top is almost parallel to the basal plane of the parent.

### Analysis and Discussion

The large deviation between the actual TBs and the $\{10\bar{1}2\}$ twinning plane observed in our EBSD and HRTEM examinations indicates that the $\{10\bar{1}2\}<10\bar{1}\bar{1}>$ twinning mode may be mechanistically different from other twinning modes in which twinning dislocations control twin growth [12]. Along with the reversible twinning and the non-Schmid effect, these abnormal properties call into a question whether this twinning mode is truly controlled by twinning dislocations.

The crystallographically-defined Burgers vector of the elementary twinning dislocation [13] on the $\{10\bar{1}2\}$ plane equals

$$b = \frac{(3-\gamma^2)\cdot a}{2\sqrt{3+\gamma^2}}, \tag{1}$$

where $a$ is the lattice parameter (0.321 nm for Mg) and $\gamma$ the /a ratio (1.624 for Mg). We can calculate the magnitude of the Burgers vector for Mg as 0.024 nm, which is less than one tenth of the Burgers vector of the basal dislocations. That the twinning dislocation has such a tiny Burgers vector significantly implies that the twin lattice exists almost in the parent lattice before any twinning shear is applied. In other words, negligible shear is needed and only local atomic shuffling is required to accomplish the twinning.

Figure 3 schematically demonstrates how the $\{10\bar{1}2\}<10\bar{1}\bar{1}>$ twinning is achieved by atomic shuffling rather than twinning dislocation. In fact, the $\{10\bar{1}2\}$ twin can be constructed directly from the parent without involving any well-defined twinning dislocations at the TB. Details of this concept can be found elsewhere [14]. Here we provide a brief description of the theory. In Figure 3, the parent HCP lattice is plotted to show the ...*A  A  A  A* ... stacking sequence of the basal planes perpendicular to the paper plane. If we connect those atoms with distances ~ $a$ (the lattice parameter), immediately we can see that a new hexagonal lattice can be constructed with the new basal planes at a misorientation of 90° with respect to the basal plane of the parent lattice. The two lattices indeed satisfy the $\{10\bar{1}2\}<10\bar{1}\bar{1}>$ twin relationship because the $\{10\bar{1}2\}$ plane (the shaded plane in Figure 3) is common to both lattices. The basal planes of the twin are marked by *A'* and ', parallel to the figure plane. The basal planes of the parent and the twin are perpendicular to each other, obviously satisfying the twin orientation relationship. However, the twin lattice does not have the correct HCP packing, as can be seen both from the constructed prism plane that is flat but should be double-layered, as well as from the constructed basal plane which is double-layered but should be flat. To correct these distortions and packing "errors", atomic shuffling is required: the pink atoms on the new basal (*A'*) plane have to shuffle outward to make the new basal plane flat, while the atoms of the ' plane have to shuffle downward to make the new prism plane double-layered. These shuffles, together with others, are concerted to reach the correct a ratio with magnitudes much smaller than any known dislocation Burgers vectors in the parent lattice; likewise, the shuffles convert the parent basal planes into the twin prism planes and the parent prism planes into the twin basal planes.

It should be noted that the atomic shuffling-dominated twinning mechanism for the $\{10\bar{1}2\}<10\bar{1}\bar{1}>$ mode is *not* at odds with the classical twinning theory [15] in which a homogeneous shear and atomic shuffling must generally be involved at the same time for twinning in HCP metals. As pointed out by Christian and Mahajan [16], for HCP metals, atomic shuffling may be as important as the magnitude of shear in controlling the operative twinning modes. According to our simulations and calculations, the amount of the shear (i.e., the Burgers vector of the elementary twinning dislocation) and the magnitude of the shuffle of each twinning mode, i.e., $\{10\bar{1}1\}<10\bar{1}\bar{2}>$, $\{10\bar{1}2\}<10\bar{1}\bar{1}>$, $\{11\bar{2}1\}<11\bar{2}\bar{6}>$ and $\{11\bar{2}2\}<11\bar{2}\bar{3}>$, drastically vary from one twinning mode to another. The $\{10\bar{1}2\}<10\bar{1}\bar{1}>$ mode is an extreme case in which shear is negligible and shuffle is dominant (as shown in Figure 3); the $\{11\bar{2}1\}<11\bar{2}\bar{6}>$ mode is an extreme case in which shear is dominant and no shuffle is involved; the $\{10\bar{1}1\}<10\bar{1}\bar{2}>$ mode is shear dominant with minor off-plane shuffles, i.e., the atoms have to shuffle slightly off the $\{10\bar{1}1\}$ twinning plane [12]; and the $\{11\bar{2}2\}<11\bar{2}\bar{3}>$ mode is middle-ground where shear and shuffle are nearly equal, and the shuffles are confined in-plane [17]. From this point of view, our shuffling model (Figure 3) is an extreme case that still falls inside the classical framework of twinning theory. Nonetheless, there is no unified dislocation model that is able to describe all these major twinning modes in HCP metals and also account for the diversity of shear and shuffle of the individual twinning modes.

The $\{10\bar{1}2\} < 10\bar{1}\bar{1} >$ twinning reorients the -axis of the parent lattice by about 90° around the $< \bar{1}2\bar{1}0 >$ direction, which is the normal of the plane of shear. If we detach the two lattices in Figure 3, as shown in Figure 4, it is clear that there is a misfit in lengths between the two lattices in the direction of the -axis of the parent lattice. Hence, a misfit strain is created by the $\{10\bar{1}2\} < 10\bar{1}\bar{1} >$ twinning. This misfit strain ($\varepsilon$) along the -axis of the parent equals

$$\varepsilon = \frac{\sqrt{3} - \gamma}{\gamma}. \tag{2}$$

The value of the misfit strain varies as a function of $\gamma$ ($\varepsilon$ = 9.1% for Ti, and 6.7% for Mg). Accordingly, for those HCP metals with a $a$ ratio less than $\sqrt{3}$ (e.g., Mg, Ti, Zr, Co, Be etc.), a positive tensile stain is produced; for those HCP metals with a $a$ ratio greater than $\sqrt{3}$ (e.g., Cd and Zn), a negative compressive strain is produced. Therefore, for Mg alloys, the $\{10\bar{1}2\} < 10\bar{1}\bar{1} >$ twinning efficiently accommodates the elongation if a tensile load is applied parallel to the -axis of the parent, but the shear strain is negligible because this twinning mode is dominated by atomic shuffling. This analysis is essentially the same as Yoo's pioneering work [18] in which the well-known "$\sqrt{3}$ criterion" was established but in a more explicit fashion.

The atomic shuffling-dominated $\{10\bar{1}2\}$ twinning entails very low critical stress for twin growth, which is confirmed by early experimental observations. Partridge and Roberts [5] studied the movement of the $\{10\bar{1}2\} < 10\bar{1}\bar{1} >$ twinning in Mg under small stresses with micro-hardness indenter. Figure 4 immediately reveals that, under an external load, intrinsically HCP metals have a tendency to reorient the basal planes to the direction parallel to the maximum elongation direction of deformed grains, as observed by Karaman et al [9]. As a result, in extrusion and rolling processes, the basal planes will be reoriented parallel to the extrusion or the rolling direction.

Generally, small deviations between the twinning plane and the TB are allowed when twin accommodation is required [5], but such deviations do not exceed a few degrees. Small deviations may also be allowed by the presence of twinning dislocation loops at a TB. The abnormal TB migration observed by Partridge and Roberts [5], and the large deviation between the actual TB and the $\{10\bar{1}2\}$ twinning plane observed in this work, can both be well understood with the shuffling mechanism: because the migration of the TB is dominated by atomic shuffling instead of shear, the actual TB does not have to match the crystallographic twinning plane.

Finally, the structural origin of the reversible $\{10\bar{1}2\} < 10\bar{1}\bar{1} >$ twinning is well explained, as can be readily seen from Figure 3, by shuffling-dominated twinning. The conversions between the parent lattice and the twin lattice are reversible. Upon unloading or removal of the load, the atoms at TBs may shuffle back and move the TBs backward. The driving force is the local strain and the twin/parent interfacial energy.

## Conclusions

Using EBSD and TEM, we measured the traces of $\{10\bar{1}2\}$ twin boundaries in Mg and Co and revealed a large deviation between the twinning plane and the twin boundaries. These results confirm that $\{10\bar{1}2\}$ twinning is dominated by atomic shuffling. This conclusion well explains the structural origin of reversible twinning, non-Schmid effect and texture in HCP metals.

**Acknowledgements**

This work was partly supported at Center for Advanced Metallic and Ceramic Systems, Johns Hopkins University, by ARMAC-RTP, DAAD19-01-2-0003 and W911NF-06-2-0006. B. Li, H. El Kadiri, and Q. Ma gratefully acknowledge support from the Center for Advanced Vehicular Systems, Mississippi State University. Thanks are extended to H.R. Wainwright for her assistance with the preparation of the manuscript.

## References

[1] L. Wu, A. Jain, D. W. Brown, g. M. Stoica, S. R. Agnew, B. Clausen, D.E. Fielden, P.K. Liaw, "Twinning–detwinning behavior during the strain-controlled low-cycle fatigue testing of a wrought magnesium alloy ZK60", Acta Mater 56, (2008) 688-695.

[2] A. Jain, S. R. Agnew, "Modeling the temperature dependent effect of twinning on the behavior of magnesium alloy AZ31B sheet", Mater Sci Eng A 462(2007)29-36.

[3] C.H. Caceres, T. Sumitomo, M. Veidt, "Pseudoelastic behaviour of cast magnesium AZ91 alloy under cyclic loading–unloading", Acta Mater 51 (2003) 6211-6218.

[4] M.R. Barnett, Z. Keshavarz, A.G. Beer, X. Ma, "Non-Schmid behaviour during secondary twinning in a polycrystalline magnesium alloy", Acta Mater 56 (2008) 5-15.

[5] P.G. Partridge, E. Roberts, "The formation and behaviour of incoherent twin boundaries in hexagonal metals", Acta Met 12 (1964) 1105-1210.

[6] J. Bohlen, M. R. Nurnberg, J. W. Senn, D. Letzig, S. R. Agnew, "The texture and anisotropy of magnesium–zinc–rare earth alloy sheets", Acta Mater 55(2007)2101-2112.

[7] S. R. Angew, J. A. Horton, T. M. Lillo, D. W. Brown, "Enhanced ductility in strongly textured magnesium produced by equal channel angular processing", Scripta Mater 50(2004)377-381.

[8] R. Z. Valiev and T. G. Langdon, "Principles of equal-channel angular pressing as a processing tool for grain refinement", Prog. Mater. Sci. 51(2006)881-981.

[9] G. G. Yapici, I. Karaman, "Common trends in texture evolution of ultra-fine-grained hcp materials during equal channel angular extrusion", Mat Sci Eng A, 503 (2009) 78-81.

[10] J. J. Fundenberger, M. J. Phillipe, F. Wagner, C. Esling, "Modelling and prediction of mechanical properties for materials with hexagonal symmetry (zinc, titanium and zirconium alloys)", Acta Mater 45(1997) 4041-4055.

[11] X.Y. Zhang, Y.T. Zhu, Q. Liu, "Deformation twinning in polycrystalline Co during room temperature dynamic plastic deformation", Scripta Mater 63 (2010) 387-390.

[12] B. Li, E. Ma, "Zonal dislocations mediating $\{10\bar{1}1\} < 10\bar{1}\bar{2} >$ twinning in magnesium", Acta Mater 57 (2009) 1734-1743.

[13] N. Thompson, D. J. Millard, "Twin formation in cadmium", Phil Mag 43 (1952) 422.
[14] B. Li, E. Ma, "Atomic shuffling dominated mechanism for deformation twinning in magnesium", Phys Rev Lett, 103 (2009) 035503.
[15] B.A. Bilby, A.G. Crocker, "The theory of the crystallography of deformation twinning", Proc Roy Soc. A, 288 (1965) 240-255.
[16] J.W. Christian, S. Mahajan, "Deformation twinning", Prog. Mater Sci 39 (1995) 1-157.
[17] B. Li, H. El Kadiri, M.F. Horstemeyer, "Extended zonal dislocations mediating $\{11\bar{2}2\} < 11\bar{2}\bar{3} >$ twinning in titanium", Philo Mag 2011, accepted.
[18] M.H. Yoo, "Slip, twinning, and fracture in hexagonal close-packed metals", Met Trans A, 12A, 1981: 409-418.

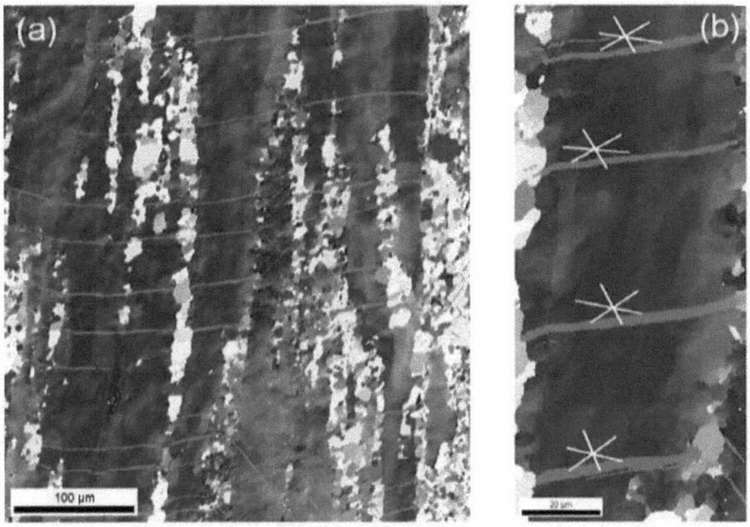


Figure 1. (a) Mapping of misorientations in a pre-extruded AM30 Mg alloy after compression along the extrusion direction (ED, vertical direction). The elongated grains have a relatively high density of dislocations; whereas the fine, equiaxed grains are dynamically recrystallized during extrusion with a low density of dislocations. The needles are deformation twins mostly in the elongated grains. (b) A magnified view of (a) highlights the twins. A group of white lines mark the traces of the $\{10\bar{1}2\}$ planes in the matrix. There are six $\{10\bar{1}2\}$ planes and hence six traces, but three of the traces almost coincide with the others, thus we only plot three markings. As is apparent, the twin boundaries deviate from the traces of the $\{10\bar{1}2\}$ plane.

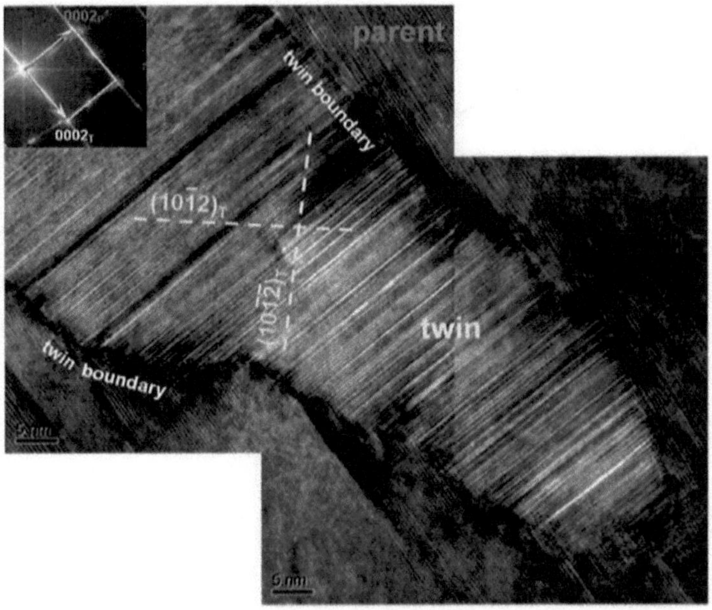


Figure 2. A high Resolution TEM micrograph shows deformation twins in dynamically compressed Co. The diffraction pattern in the inset indicates that the twin and the parent satisfy the $\{10\bar{1}2\} < 10\bar{1}\bar{1} >$ twin relationship (zone axis $[1\bar{2}10]$). In this orientation, the two possible $\{10\bar{1}2\}$ twinning planes are marked with broken lines. We can see the actual twin boundaries deviate strikingly far from the twinning planes (> 45°). The deviation can also be appreciated by noticing that the top twin boundary is nearly parallel to the basal plane of the parent.

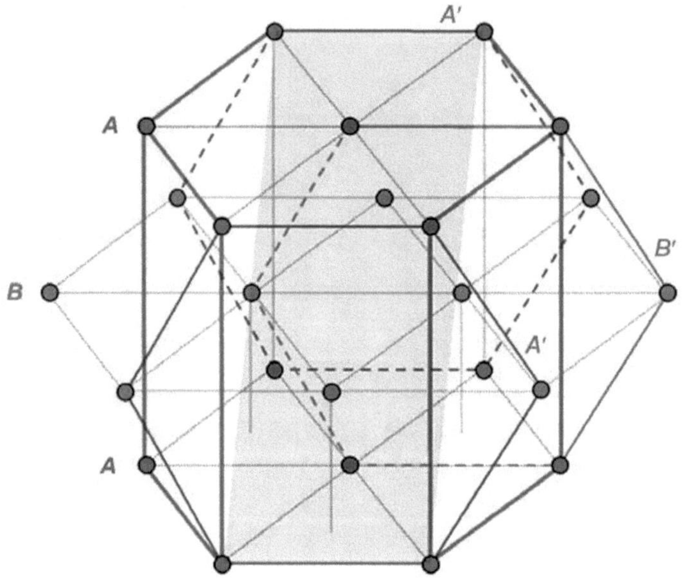


Figure 3. Construction of a new HCP lattice (basal planes are parallel to the paper plane and labeled as *A'B'A'B'*...) from an existing HCP lattice (basal planes are perpendicular to the paper plane and labeled as *ABAB*... ). The two lattices share a common $\{10\bar{1}2\}$ plane (the shaded, inclined plane); hence, they satisfy the twin relationship. Local shuffling is required to correct the distortion in the new HCP lattice such that correct stacking sequence and the *c/a* ratio can be established.

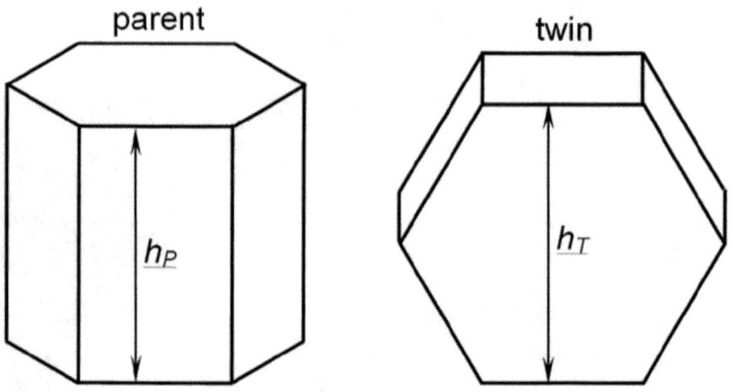


Figure 4. After $\{10\bar{1}2\} < 10\bar{1}\bar{1} >$ twinning, the parent lattice is rotated by 86.3° around $< \bar{1}2\bar{1}0 >$, which produces a misfit strain between the parent and the twin lattices: $\varepsilon = (h_T - h_P)/h_P = (\sqrt{3} - \gamma)/\gamma$ ($\gamma$ is the *c/a* ratio). For Mg ($\gamma$= 1.624), the twinning produces 6.7% tensile strain along the *c*-axis of the parent lattice.

**Magnesium Technology 2012**
*Edited by: Suveen N. Mathaudhu, Wim H. Sillekens, Neale R. Neelameggham, and Norbert Hort*
*TMS (The Minerals, Metals & Materials Society), 2012*

# Length Changes in Extruded Magnesium Alloy Bars Under Large Strain Free-End Torsion

H. Wang[1], P.D. Wu[1], K.W. Neale[2]
[1]Department of Mechanical Engineering, McMaster University
Hamilton, Ontario L8S 4L7, Canada
[2]Faculty of Engineering, University of Sherbrooke
Sherbrooke, Quebec J1K 2R1, Canada

Keywords: Crystal plasticity, Torsion, Texture, Magnesium Alloy

## Abstract

We numerically study the large strain free-end torsion of extruded magnesium alloy bars based on the recently developed Elastic Visco-Plastic Self-Consistent (EVPSC) model, in which both slip and twinning contribute to plastic deformation. It is shown that the predicted second-order length change is very sensitive to the initial texture and texture evolution. Numerical results suggest that the free-end torsion test can provide an effective means for assessing the adequacy of polycrystal plasticity models for magnesium alloys.

## Introduction

The torsion of circular bars in the region of large strain deformations is of considerable practical interest. In particular, a torsion test seems to be ideally suited for the experimental determination of material parameters in the range of large strains. The major advantage for this purpose over tensile tests is that deformations during torsion remain homogeneous in the axial direction until fracture, without showing strain localization in necks etc. A second, slightly more academic interest in large strain torsion relates to attempts to incorporate deformation-induced anisotropy into large strain plasticity models. It has been generally accepted that texture development, which is the prime source of anisotropy in metals, leads to axial strains in the case of specimens with free ends (see e.g. Swift [1]), or leads to axial stresses in specimens with axial constraints or fixed ends (see e.g. Montheillet et al. [2]). The prediction of these phenomena depends strongly on the constitutive model, in particular the description of anisotropic hardening (see e.g. Harren et al. [3], Neale et al. [4], Toth et al. [5], and Wu et al. [6]). Thus, the torsion test can provide an effective means for assessing the adequacy of constitutive models.

Extensive experimental and numerical research has been performed on large strain torsion for polycrystals with the Face Centered Cubic (FCC) and Body Centered Cubic (BCC) crystallographic structures. However, there are rare investigations on the torsion of polycrystalline materials with the Hexagonal Close Packed (HCP) crystallographic structure. Sanchez et al. [7] and Evans et al. [8] studied texture evolution during torsion by simulating simple shear. Barnett [9] experimentally studied the flow stress of AZ31 under torsion. Balasubramanian and Anand [10] numerically studied texture evolution and the mechanical response of titanium under fixed-end torsion of a circular tube by using the finite element method. Beausir et al. [11] experimentally and numerically studied the free-end torsion of AZ71 and pure magnesium. However, all these numerical studies on the large strain torsion of HCP polycrystals excluded twinning. Very recently, Wang et al. [12] studied the large strain fixed-end torsion of HCP polycrystalline solid bars based on the semi-analytical approach proposed by Neale and Shrivastava [13], and the visco-plastic self-consistent (VPSC) model developed by Molinari et al. [14] and Lebensohn and Tomé [15], in which both slip and twinning contribute to plastic deformation. It was shown that the predicted second-order axial force is very sensitive to the initial texture, texture evolution and the constitutive models employed.

In this paper, we numerically study the large strain free-end torsion of magnesium alloy AZ31 extruded bars by using a special purpose finite element designed by Wu and Van der Giessen [16]. All the simulations are based on the large strain elastic viscoplastic self-consistent (EVPSC) model for polycrystalline materials developed by Wang et al. [17], which includes both slip and twinning as plastic deformation mechanisms. Results are presented in terms of predicted torque responses and axial strains. The influences of initial texture and texture evolution on the axial strain are emphasized.

## Constitutive Model

In this section, we briefly recapitulate the constitutive model we adopt throughout this paper, mainly for the purpose of definition and notation. For details we refer to Wang et al. [17].

The elastic constitutive equation for a crystal is:

$$\overset{\nabla *}{\boldsymbol{\sigma}} = \boldsymbol{L} : \boldsymbol{d}^e - \boldsymbol{\sigma} tr(\boldsymbol{d}^e) \tag{1}$$

where $\boldsymbol{L}$ is the fourth order elastic stiffness tensor, $\boldsymbol{d}^e$ is the elastic strain rate tensor and $\overset{\nabla *}{\boldsymbol{\sigma}}$ is the Jaumann rate of the Cauchy stress $\boldsymbol{\sigma}$ based on the lattice spin tensor $\boldsymbol{w}^e$. The single crystal elastic anisotropy is included in $\boldsymbol{L}$ through the crystal elastic constants $C_{ij}$.

Plastic deformation of a crystal is assumed to be due to crystallographic slip and twinning on systems $(\boldsymbol{s}^\alpha, \boldsymbol{n}^\alpha)$. Here, $\boldsymbol{s}^\alpha$ and $\boldsymbol{n}^\alpha$ are the slip/twinning direction and the direction normal to the slip/twinning plane for system $\alpha$, respectively. In the present study we consider slip in the Basal <*a*> ($\{0001\} < 11\bar{2}0 >$), Prismatic <*a*> ($\{10\bar{1}0\} < 11\bar{2}0 >$) and Pyramidal <*c+a*> ($\{\bar{1}\bar{1}22\} < \bar{1}\bar{1}23 >$) slip systems, and twinning in the $\{10\bar{1}2\} < \bar{1}011 >$ tensile twin system (see Fig. 1).

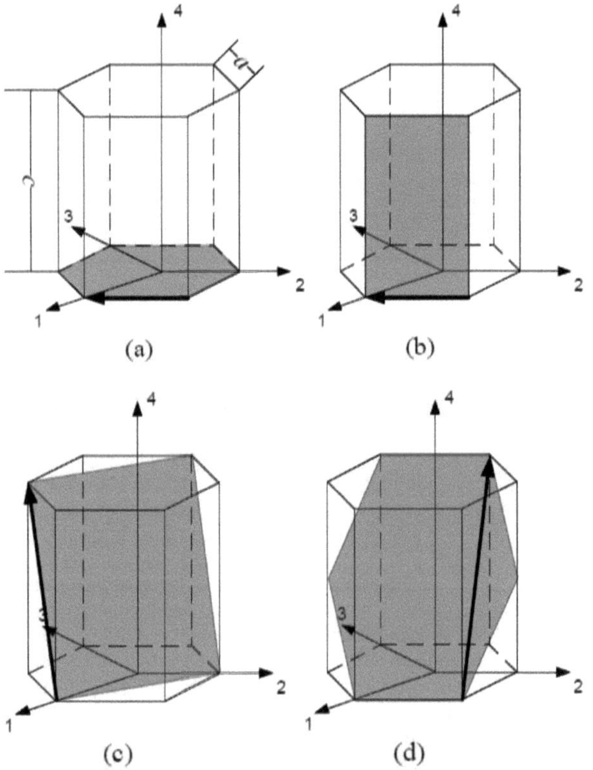


Figure 1. Plastic deformation modes for hexagonal structure: (a) basal <a> slip systems, (b) prismatic <a> slip systems, (c) pyramidal <c+a> slip systems, and (d) tensile twin.

The following equation gives the grain (crystal) level plastic strain rate:

$$d^p = \dot{\gamma}_0 \sum_\alpha P^\alpha \left|\frac{\tau^\alpha}{\tau^\alpha_{cr}}\right|^{\frac{1}{m}-1} \frac{\tau^\alpha}{\tau^\alpha_{cr}} \quad (2)$$

where $\dot{\gamma}_0$ is a reference value for the slip/twinning rate, $m$ is the slip/twinning rate sensitivity, $P^\alpha = (s^\alpha n^\alpha + n^\alpha s^\alpha)/2$ is the Schmid tensor for system $\alpha$, and $\tau^\alpha = \sigma : P^\alpha$ and $\tau^\alpha_{cr}$ are the resolved shear stress and critical resolved shear stress (CRSS) for system $\alpha$, respectively.

The self-consistent approach is applied to obtain the response of a polycrystal comprised of many grains. In a self-consistent approach each grain is treated as an ellipsoidal inclusion embedded in a Homogeneous Effective Medium (HEM), which is an aggregate of all the grains. Interactions between each grain and the HEM are described using the Eshelby inclusion formalism. During each deformation step, the single crystal constitutive rule (which describes the grain-level response) and the self-consistency criteria are solved simultaneously. This ensures that the grain-level stresses and strain rates are consistent with the boundary conditions imposed on the HEM. To date, various self-consistent schemes have been proposed. The EVPSC model for polycrystals recently developed by Wang et al. [17] is a completely general elastic-viscoplastic, fully anisotropic, self-consistent polycrystal model, applicable to large strains and to any crystal symmetry. Very recently, Wang et al. [12, 18-20] evaluated various linearization self-consistent schemes employed in self-consistent modeling by applying them to the large strain behavior of magnesium alloys under different deformation processes, including the fixed-end torsion of solid bars and forming limit diagrams. It was found that the Affine self-consistent scheme gave the best overall performance among the self-consistent approaches examined. Therefore, the EVPSC model with the Affine self-consistent scheme is applied in the present study. In addition, the Predominant Twin Reorientation (PTR) scheme proposed by Tomé et al. [21] is used to model the twinning activities.

## Problem formulation and method of solution

We consider a solid circular bar with radius $R_0$ and length $L_0$ subjected to an angle of twist $\psi$ due to an applied torque $T$. The lateral surface of the bar is stress free, and all properties are assumed to be axisymmetric and homogeneous along the axial direction. The end faces of the bar are constrained to the extent that they remain plane and perpendicular to the axial direction, so that we may assume that any cross-section of the bar remains planar. Although anisotropy will be induced during the deformation process, the behavior remains axisymmetric and the bar remains circular cylindrical with current radius $R$ and current length $L$ (Fig. 2). The bar is assumed to be in the free-end condition, thus allowing the possible development of a uniform axial displacement over the end face.

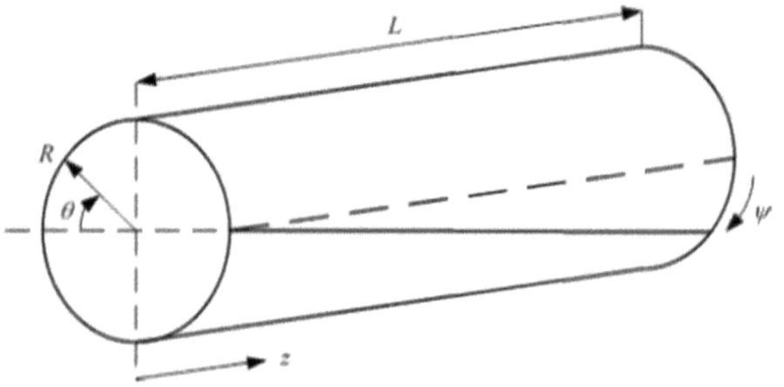


Figure 2. Schematic representation of a solid circular bar under torsion.

A numerical approach is adopted to solve the governing equations using a special purpose finite element developed by Wu and Van der Giessen [16, 22], and used by Van der Giessen et al. [23] and Wu et al. [6]. Each element effectively is a circular cylindrical tube, but computationally it is considered to be one-dimensional along the radial direction, with two nodes. The degrees of freedom of the entire finite element model of the bar consisting of $n$ elements then comprise $n+1$ radial nodal displacements along with the axial displacement $U = L - L_0$ and the angle of twist $\psi$. Furthermore, each element has two material sampling points whose positions coincide with the Gaussian integration points. Each sampling point in the finite element formulation is modeled as a polycrystalline aggregate, and the EVPSC model with the Affine self-consistent scheme is applied locally at each sampling point. Finally, an equilibrium correction procedure is applied to prevent drifting of the solution from the true equilibrium path.

The results of large strain free-end torsion of a circular solid bar are efficiently presented in terms of the following quantities:

$$\Gamma = \frac{R_0}{L_0}\psi, \quad \bar{\tau} = \frac{3T}{2\pi R_0^3}, \quad \varepsilon = \ln\left(\frac{L}{L_0}\right) \quad (3)$$

Here, $\Gamma$ represents the shear strain at the outer surface of the bar, and $\bar{\tau}$ and $\varepsilon$ are the normalized torque and axial strain, respectively.

## Results and discussions

The material considered here is magnesium alloy AZ31 extruded bar, which has been experimentally studied by Agnew et al. [24]. The {0001} pole figure of the initial texture (Fig. 3) shows that the grains tend to have their basal planes oriented parallel (*c*-axis oriented perpendicular) to the extrusion direction (ED) of the bar. For a comparison, an initial random texture is also considered. The reference slip/twinning rate $\dot{\gamma}_0$ and the rate sensitivity $m$ are prescribed to be the same for all slip/twinning systems: $\dot{\gamma}_0 = 0.001 s^{-1}$ and $m = 0.05$, respectively. The room temperature elastic constants of magnesium single crystal are taken from Simmons and Wang [25] to be $C_{11} = 58.0$, $C_{12} = 25.0$, $C_{13} = 20.8$, $C_{33} = 61.2$ and $C_{44} = 16.6$ (units of GPa). Values of the hardening parameters for each mode are estimated by fitting numerical simulations of uniaxial tension and compression along the ED to the corresponding experimental flow curves.

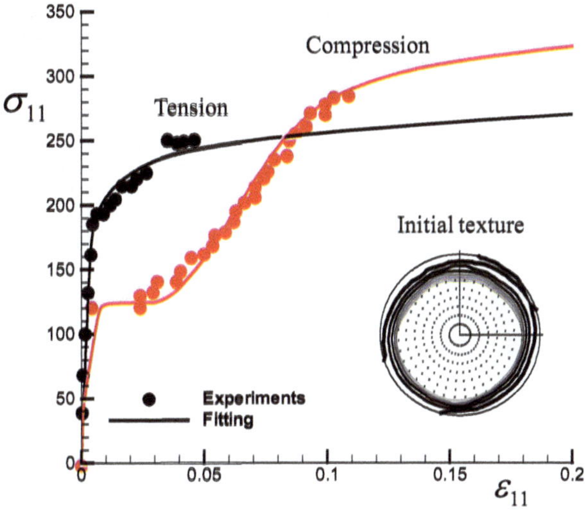


Figure 3. Stress and strain responses of AZ31 bar under uniaxial tension and compression along the ED. The insert is the initial texture represented in terms of the {0001} pole figure. The reflecting surface is normal to the ED.

The uniaxial tension and compression true stress - true strain curves along the ED are presented in Fig. 3. The characteristic S-shape of the compressive flow curve clearly reveals the importance of twinning in compression. The macroscopic yield stresses for uniaxial compression and tension are about 120 MPa and 190 MPa, respectively, showing the strong tension/compression asymmetry associated with twinning. The EVPSC model with the Affine self-consistent scheme fits the experimental curves quite well. The values of these parameters will be used in subsequent simulations.

We proceed by simulating large strain free-end torsion of solid bars with the initial extrusion texture and random texture. Calculations are carried out for solid bars with $R_0 = 1\,\text{mm}$ and $L_0 = 1\,\text{mm}$. For the fixed-end torsion of solid bars, Van der Giessen et al. [23] found, by comparison with the semi-analytical method (Neale and Shrivastava [13]), that sufficiently accurate results were obtained by using only $n=5$ elements. Figure 4 shows stress distributions inside the solid bar with the initial random texture under free-end torsion at $\Gamma = 1.5$. These results are based on two meshes with 5 and 15 elements, respectively. It is found that the two meshes give very similar stress distributions. Although not shown here, the differences in the predicted $\bar{\tau}$ and $\varepsilon$ between 5 and 15 elements are found to be very small. These observations suggest that using a mesh with 5 elements is also sufficient to accurately simulate the free-end torsion of solid bars. However, all the numerical results reported in the present study are based on a mesh with $n = 10$ elements.

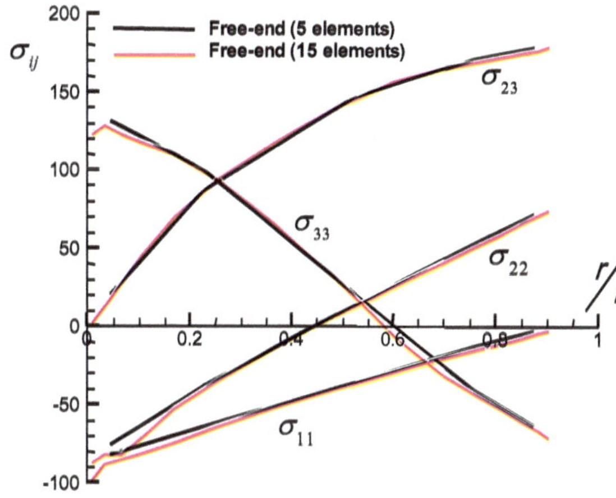


Figure 4. Predicted stress distributions inside the solid bar with the initial random texture under free-end torsion at shear strain $\Gamma = 1.5$ at the outer surface of the bar.

Figure 5 shows the predicted normalized torque $\bar{\tau}$ and axial strain $\varepsilon$ as functions of the shear strain $\Gamma$ for specimens with the initial extrusion texture and random texture. It is found that the predicted response of free-end torsion, especially the axial strain $\varepsilon$, is very sensitive to the initial texture. A noticeable strain softening effect in the $\bar{\tau}$ vs. $\Gamma$ curves is detected at large strains for both initial textures. For the initial extrusion texture, the predicted axial strain exhibits a complicated trend: contracting first and then elongating axially. This seems in a good qualitative agreement with the experimental axial strain $\varepsilon$ vs. $\Gamma$ curve for the magnesium alloy AZ71 extruded bar under free-end torsion reported by Beausir et al. [11]. For the initial random texture, the predicted axial strain is almost zero up to $\Gamma \approx 0.5$ but becomes very large at large strains.

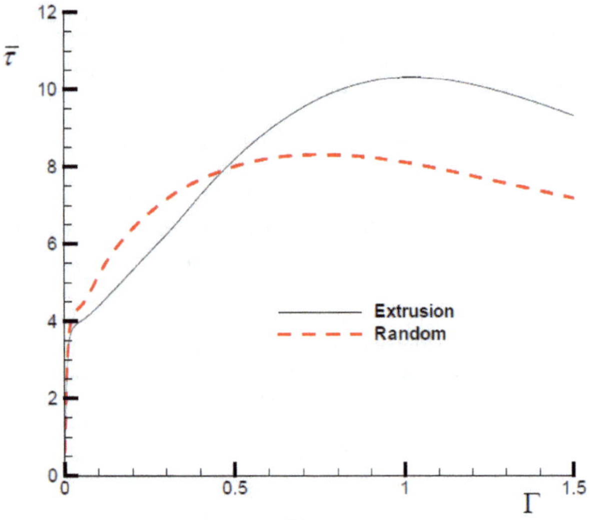


(a)

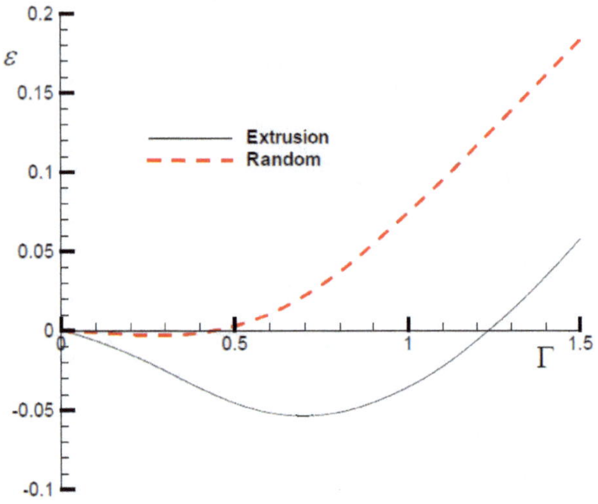


(b)

Figure 5. Predicted normalized torque (a), and elongation (b) for solid bars with the initial extrusion texture and random texture.

Figure 6 shows the predicted deformation textures in terms of $\{0001\}$ and $\{10\bar{1}0\}$ pole figures at $\Gamma = 1.5$ for the initial random and extrusion textures. These pole figures exhibit shear textures analogous to the experimental textures observed in free-end torsion by Beausir et al. [11] for HCP materials. It is also observed that the shear texture from the initial random texture is stronger than the one from the initial extrusion texture.

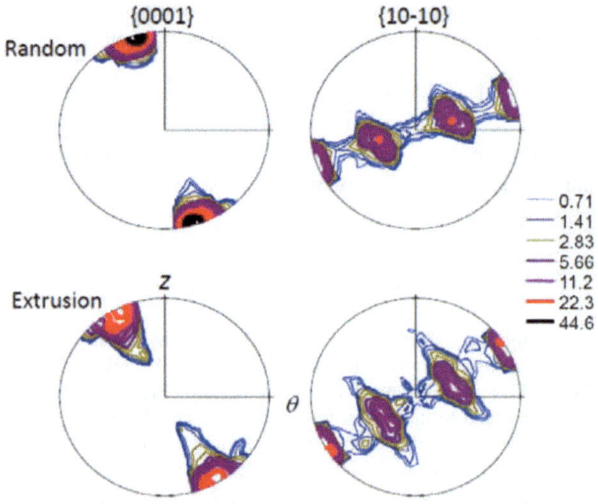


Figure 6. Predicted deformation textures in terms of $\{0001\}$ and $\{10\bar{1}0\}$ pole figures at shear strain $\Gamma = 1.5$ at the outer surface of the bar for the initial random texture (Top) and extrusion texture (Bottom).

The development of the second-order axial strain is believed to mainly result from the initial and deformation induced anisotropy and is thus sensitive to the texture evolution. In this paper, repeating the calculations reported in Fig. 5 but turning off the texture evolution assesses the influence of the texture evolution on free-end torsion. Figures 7 and 8 give the predicted normalized torque and axial strain for the initial extrusion texture and random texture, respectively. In these simulations both texture evolution due to slip and reorientation due to twinning are excluded. It is found that excluding texture evolution dramatically reduces the predicted axial strain. For the initial random texture, the predicted axial strain is nearly zero during the entire deformation process. It is also observed that the softening effect in the $\bar{\tau}$ and $\Gamma$ curves at large strains is due to texture evolution.

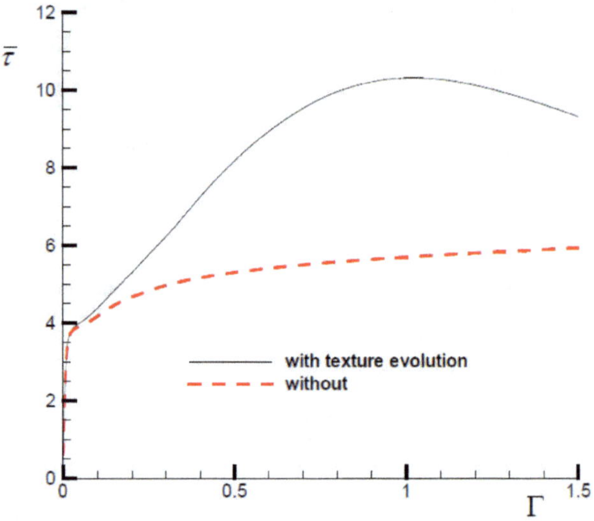


(a)

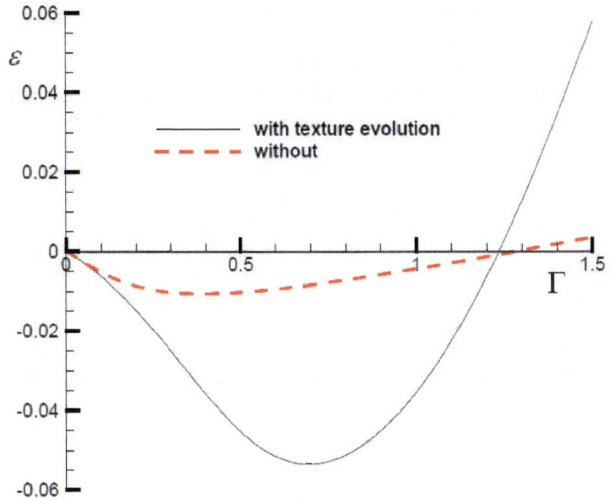


(b)

Figure 7. Predicted effect of texture evolution on normalized torque (a), and elongation (b) for a solid bar with the initial extrusion texture.

## Conclusions

In this paper, the large strain free-end torsion of HCP polycrystals has been studied numerically, based on the EVPSC model with the Affine self-consistent scheme. It has been found that the development of the second-order axial strain is very sensitive to the initial anisotropy due to the initial texture, and the deformation induced anisotropy due to the texture evolution. Numerical results have indicated that excluding texture evolution dramatically reduces the development of the axial strain. For the magnesium alloy AZ31 extruded bar, the predicted axial strain exhibits a complicated trend: contracting first and then elongating axially. This is in a good qualitative agreement with the experimental axial response for the magnesium alloy AZ71 extruded bar under free-end torsion reported by Beausir et al. [11]. Numerical results also suggest that the free-end torsion test can provide an effective means for assessing the adequacy of polycrystal plasticity models for magnesium alloys.

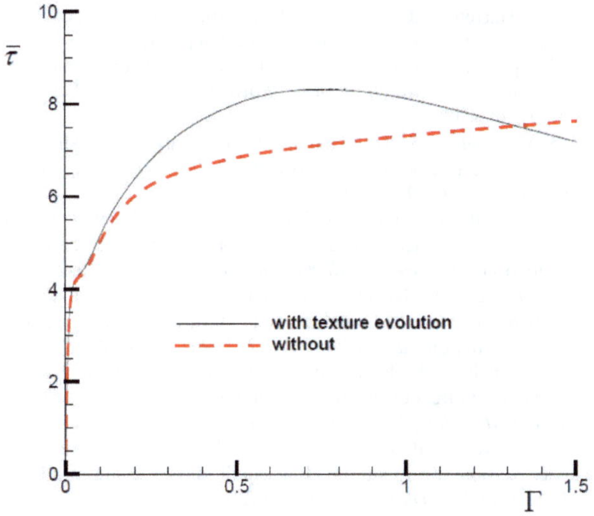


(a)

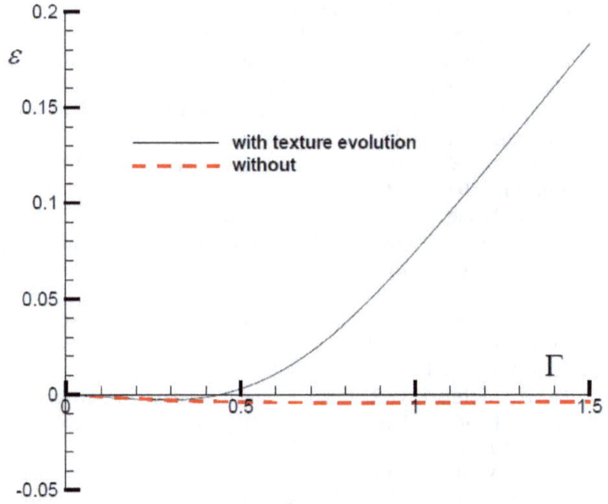


(b)

Figure 8. Predicted effect of texture evolution on normalized torque (a), and elongation (b) for a solid bar with the initial random texture.

## Acknowledgement

This research was supported by Ontario Ministry of Research and Innovation and the Natural Sciences and Engineering Research Council of Canada (NSERC).

## References

1. H.W. Swift, "Length changes in metals under torsional overstrain," *Engineering,* 163 (1947), 253-257.
2. F. Montheillet, M. Cohen, and J.J. Jonas, "Axial stresses and texture development during the torsion testing of Al, Cu and Alpha-Fe," *Acta Metall.*, 32 (1984), 2077-2089.

3. S. Harren, T.C. Lowe, R.J. Asaro, and A. Needleman, "Analysis of large-strain shear in rate-dependent Face-Centred Cubic polycrystals: Correlation of micro- and macromechanics," *Philosophical Transactions of the Royal Society of London*, A328 (1989), 443-500.
4. K.W. Neale, L.S. Toth, and J.J. Jonas, "Large strain shear and torsion of rate-sensitive FCC polycrystals," *International Journal of Plasticity*, 6 (1991), 45-61.
5. L.S. Toth, J.J. Jonas, D. Daniel, and J.A. Bailey, "Texture development and length changes in copper bars subjected to free end torsion," *Texture and Microstructure*, 19 (1992), 245-262.
6. P.D. Wu, K.W. Neale, and E. Van der Giessen, "Simulation of the behaviour of FCC polycrystals during reversed torsion," *International Journal of Plasticity*, 12 (1996), 1199-1219.
7. P. Sanchez, A. Pochettino, T. Chauveau, and B. Bacroix, "Torsion texture development of zirconium alloys," *Journal of Nuclear Materials*, 298 (2001), 329-339.
8. W.J. Evans, J.P. Jones, and M.T. Whittaker, "Texture effects under tension and torsion loading conditions in titanium alloys," *International Journal of Fatigue*, 27 (2005), 1244-1250.
9. M.R. Barnett, "Influence of deformation condition and texture on the high temperature flow stress of magnesium AZ31," *Journal of Light Metals*, 1 (2001), 167-177.
10. S. Balasubramanian, and L. Anand, "Plasticity of initially textured hexagonal polycrystals at high homologous temperatures: application to titanium," *Acta Materialia*, 50 (2002), 133-148.
11. B. Beausir, L.S. Toth, F. Qods, and K.W. Neale, "Texture and mechanical behavior of magnesium during free-end torsion," *Journal of Engineering Materials and Technology*, 131 (2009), article number 011108.
12. H. Wang, Y. Wu, P.D. Wu, and K.W. Neale, "Numerical analysis of large strain simple shear and fixed-end torsion of HCP polycrystals," *CMC-Computers, Materials & Continua*, 19 (2010), 255-284.
13. K.W. Neale, and S.C. Shrivastava, "Finite elastic plastic torsion of a circular bar," *Engineering Fracture Mechanics*, 21 (1985), 747-754.
14. A. Molinari, G.R. Canova, and S. Ahzi, "A self-consistent approach of the large deformation polycrystal viscoplasticity," *Acta Metallurgica*, 35 (1987), 2983-2994.
15. R.A. Lebensohn, and C.N. Tomé, "A self-consistent anisotropic approach for the simulation of plastic-deformation and texture development of polycrystals - Application to Zirconium alloys" *Acta Metallurgica et Materialia*, 41 (1993), 2611-2624.
16. P.D. Wu, and E. Van der Giessen, "Analysis of elastic-plastic torsion of circular bars at large strains," *Archive of Applied Mechanics*, 61 (1991), 89-103.
17. H. Wang, P.D. Wu, C.N. Tomé, and Y. Huang, "A finite strain elastic-viscoplastic self-consistent model for polycrystalline materials," *Journal of the Mechanics and Physics of Solids*, 58 (2010), 594-612.
18. H. Wang, B. Raeisinia, P.D. Wu, S.R. Agnew, and C.N. Tomé, "Evaluation of self-consistent polycrystal plasticity models for magnesium alloy AZ31B sheet," *International Journal of Solids and Structures*, 47 (2010), 2905-2917.
19. H. Wang, P.D. Wu, and K.W. Neale, "On the role of the constitutive model and basal texture on the mechanical behavior of magnesium alloy AZ31B sheet," *Journal of Zhejiang University – Science*, A11 (2010), 744-755.
20. H. Wang, P.D. Wu, K.P. Boyle, and K.W. Neale, "On crystal plasticity formability analysis for magnesium alloy sheets," *International Journal of Solids and Structures*, 48 (2011), 1000-1010.
21. C.N. Tomé, R.A. Lebensohn, and U.F. Kocks, "A model for texture development dominated by deformation twinning - Application to Zirconium alloys," *Acta Metallurgica et Materialia*, 39 (1991), 2667-2680.
22. P.D. Wu, and E. Van der Giessen, "On large strain inelastic torsion of glassy polymers," *International Journal of Mechanical Sciences*, 35 (1993), 935-951.
23. E. Van der Giessen, P.D. Wu, and K.W. Neale, "On the effect of plastic spin on large strain elastic-plastic torsion of solid bars," *International Journal of Plasticity*, 8 (1992), 773-801.
24. S.R. Agnew, D.W. Brown, and C.N. Tomé, "Validating a polycrystal model for the elastoplastic response of magnesium alloy AZ31 using in situ neutron diffraction," *Acta Materialia*, 54 (2006), 4841-4852.
25. G. Simmons, and H. Wang, *Single crystal elastic constants and calculated polycrystal properties* (Cambridge (MA): MIT Press, 1971).

# NANO-INDENTATION STUDIES OF TWINNED MAGNESIUM SINGLE CRYSTALS

Fumiaki Hiura[1], Raja K. Mishra[2], Michael Lukitsch[2], Marek Niewczas[1]
[1]McMaster University (Materials Science and Engineering); 1280 Main St. West; Hamilton, ON L8S4M1, CANADA
[2]General Motors Research & Development Center; 30500 Mound Rd.; Warren, MI 48090, USA

Keywords: Magnesium, Deformation twinning, Nano-indentation

## Abstract

Nano-indentation measurements have been performed on {10-12}<10-11> twin and adjacent matrix regions of deformed magnesium single crystals and the hardness values were analyzed by the Oliver-Pharr method. Although the hardness difference between the twin regions and the adjacent matrix was insignificantly small, the hardness values showed orientation dependence regardless of the twins' size and variants. This observation can be interrupted by texture- softening resulting from the lattice reorientation in the twin regions. In contrast, the experimental evidence for the Basinski hardening mechanism in {10-12}<10-11> twins, which is an increase in strength/hardness as a result of dislocation contributions within twin area, was not obtained from this experiment. This presentation provides framework for the discussion of the hardening/softening effect of {10-12}<10-11> twinning on the plastic flow in single crystalline magnesium and quantitative values for hardening parameters used in the crystal plasticity modeling.

## Introduction

Hexagonal close-packed (HCP) materials such as magnesium deform by slip and twinning. Basal slip system is the dominant deformation mechanism at room temperature in single crystalline magnesium. However, the basal slip system is insufficient to produce arbitrary deformation of a metal and other deformation mechanism is required. Therefore, non-basal slip systems and twinning becomes important modes of plastic deformation. The {10-12}<10-11> twinning is the predominant twinning mode in magnesium, contributing to the ductility and the strength of the metal, because it provides component of the strain along the c-axis of HCP unit cell and the critical resolved shear stress (CRSS) for the {10-12}<10-11> twinning is the second lowest next to the basal slip. So far, three independent hardening mechanisms due to the {10-12}<10-11> twinning have been suggested: (i) dynamic Hall-Petch effect, arising from the effective grain size refinement and the reduction of the dislocation mean free path due to the twin formations, (ii) the Basinski hardening [1], which predicts an increment in strength/hardness developed as a result of dislocation transmutations within twin area, and (iii) texture hardening/softening, resulting from the lattice rotation in the twin regions (see ref. [2,3] as relevant review papers). However, it is still unclear how much the above hardening mechanisms by the deformation twinning contribute to the overall hardening of the material, even if one considers the hardening behavior of magnesium single crystals.

Nano-indentation is a powerful tool for investigating the mechanical properties and deformation mechanism of materials on the submicron scale, and it has been widely used on a wide range of materials, e.g. thin film materials, bulk materials, and biomaterials over the past few years [4]. The nano-indentation measurements on matrix and twin regions of deformed metals may bring new knowledge concerning the hardening mechanism by deformation twinning, especially for the Basinski hardening and texture hardening/softening due to deformation twinning. {10-12}<10-11> twin regions developed in magnesium single crystals have been reported to be relatively large in size, the order of a few microns in thickness. Therefore, the nano-indentation measurements inside the twin regions should provide information about twin strength without the effect of twin boundaries, because the tip radius of the nano-indentation is very small compared to the width of {10-12}<10-11> twins.

In this study, we deal with the evaluation of the strength of the matrix and the {10-12}<10-11> twin regions in deformed magnesium single crystals by using orientation-imaging microscopy (OIM) and nano-indentation measurements.

## Experimental Procedure

Magnesium single crystals of 99.995% purity were used in this experiment. A parent crystal oriented for basal slip was deformed in tension at room temperature, up to a strain of 10%. Smaller secondary samples were cut from the pre-strained parent sample and were subjected to secondary tensile deformation at room temperature. The secondary samples were deformed to a strain of 7% (sample A) and 11% (sample B) and the crystallographic orientation of the samples was such as to induce {10-12}<10-11> twinning during secondary tensile tests. Slip traces and deformation twinning were observed by optical microscopy, and the development of local misorientations and analysis of the type of deformation twins in the secondary samples were characterized by SEM/EBSD measurements. Nano-indentation measurements were carried out on the secondary samples containing matrix and twin regions, with TI 900 Triboindenter$^{TM}$ (Hysitron$^{TM}$) equipped with a diamond cube corner tip with a tip radius of 50nm. During the nano-indentation measurements, the loading force was increased at a rate of 90μN/s to a maximum load of 450μN, held for 2 seconds, and then decreased at an unloading rate of 90μN/s to zero. 30 measurements in 5x6 rectangular arrays geometry separated by 5μm were carried out at one location. The indentation curves that deviate from the behavior of the majority of indentation curves were rejected for the accuracy and reliability of the calibration. The indentation data sets were analyzed by the Oliver-Pharr method [5].

## Experimental Results and Discussion

Indentation curves

Fig. 1 shows typical load-depth curves obtained from the nano-indentation measurements on matrix and twin regions of the sample A. Loading and unloading curves for matrix and twin regions exhibit very similar behavior. Displacement bursts, i.e. "pop-in" behaviors marked as arrowheads in Fig. 1 were observed on the *P-h* characteristics of both matrix and twin regions. The phenomenon has been observed in several bulk and thin-film materials and it arises from the set-in of plastic deformation on the defect-free regions and the associated nucleation and propagation of dislocations when the shear stress is equal to the theoretical strength of the indented materials [4].

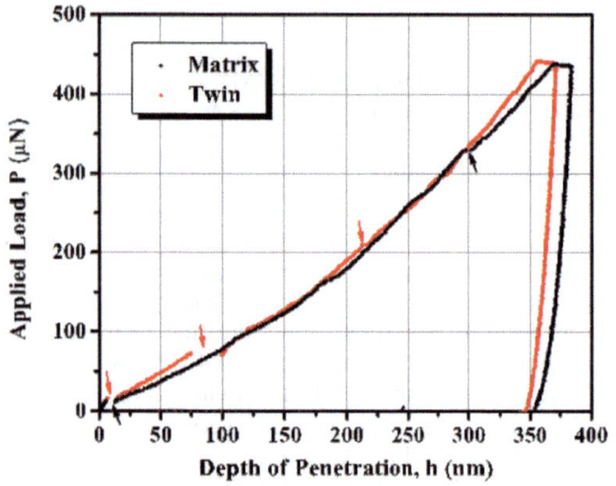


Fig. 1 Typical load-depth curves for matrix and twin regions.

SEM/EBSD analysis and nano-indentation measurements

Fig. 2 shows SEM images of the indentation surface on samples A and B. More complex texture containing various twinning were observed on the surface of Sample B compared to Sample A. Fig. 3 and 4 show inverse pole figure (IPF) maps and 0001 pole figure maps constructed from EBSD data in the location B of Sample A and the locations B and C of Sample B, respectively. SEM/EBSD analysis indicates that all the deformation twins found in the samples were {10-12}<10-11> type.

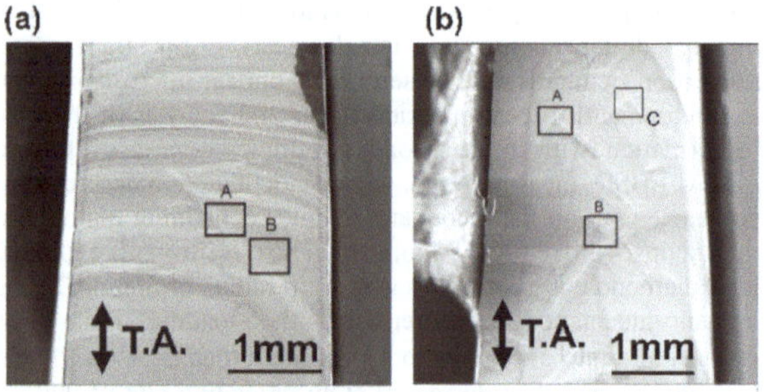


Fig. 2 SEM images of (a) sample A and (b) sample B.

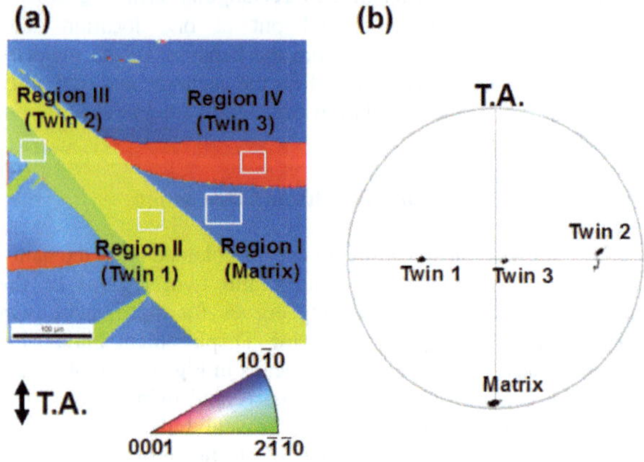


Fig. 3 (a) IPF map and (b) 0001 pole figure on location B of sample A.

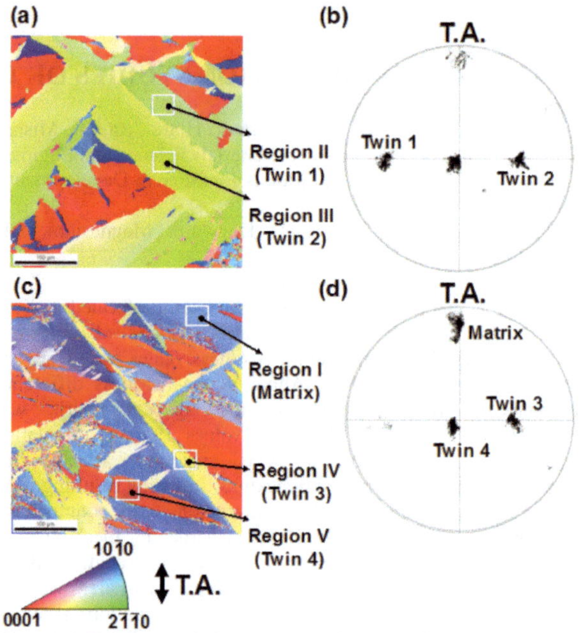


Fig. 4 (a) IPF map and (b) 0001 pole figure on location B; (c) IPF map and (d) 0001 pole figure on location C in sample B.

In the present nano-indentation experiments, the tip radius of 50nm is significantly narrower than the width of the twins, even for higher order twins with very fine size. Therefore, the measurements can be carried out at the area of a homogeneous twin phase far from twin boundaries. The hardness from the Oliver and Pharr method [5] is defined as:

$$H=P_{max}/A \tag{1}$$

,where $P_{max}$ is the maximum indentation force and $A$ is the resultant projected contact area at that load. Nano-indentation experiments were carried out on regions I-IV (sample A) and regions I-V (sample B), which were located at matrix and several {10-12}<10-11> twins. The hardness data for samples A and B are summarized in Table I. The standard deviation and the basal Schmid factor values calculated by the pole figures constructed from the EBSD data are also included in the table. The hardness values clearly depend on the orientation of the region. Deformation twinning causes a lattice rotation of the twinned region. Twinned regions rotated to softer orientation such as twins 1 and 2 in sample A and twins 1, 2 and 3 in sample B exhibited relatively smaller hardness values, while twins having harder orientation such as twin 3 in sample A and twin 4 in sample B showed higher hardness values. Matrix region was originally in a harder orientation and therefore, showed higher hardness values. Thus, the present nano-indentation measurements show that the hardness of the matrix and the twin regions is strongly dependent upon crystallographic orientation and the effect of the structural transformation on the increase in the hardness of twin regions is negligible. Previous nano-indentation studies on polycrystalline α-titanium [6] showed that deformation twins were always harder than the matrix regardless of the twins' size, variants and orientations. The authors argued that the hardness increment in twin regions arises from structural transformation of dislocations of matrix to the twin lattice, which is known as the Basinski hardening [1]. In contrast, the present results did not confirm this strengthening effect of the twin. This may be due to low deformation applied both in primary and secondary tensile tests to induce meaningful dislocation densities responsible for this effect. Also the nano-indentation experiments may not fully probe the three-dimensional distribution of dislocations in the sample tested. Therefore, further studies on the type and distribution of dislocations in matrix and twin regions by TEM are required and are currently in progress.

Table I. Nano-indentation results obtained from matrix and {10-12}<10-11> twins.

| Sample | Locations | Hardness [MPa] | Std. Dev.* | S.F. for Basal slip** |
|---|---|---|---|---|
| A | Matrix | 838 | 40 | 0.06 |
| | Twin1 | 824 | 33 | 0.49 |
| | Twin2 | 747 | 28 | 0.42 |
| | Twin3 | 888 | 23 | 0.07 |
| B | Matrix | 832 | 47 | 0.08 |
| | Twin1 | 774 | 28 | 0.46 |
| | Twin2 | 752 | 31 | 0.49 |
| | Twin3 | 751 | 24 | 0.49 |
| | Twin4 | 798 | 28 | 0.05 |

*Std. Dev.=Standard Deviation, **S.F.=Schmid factor.

## Summary

Nano-indentation measurements on the deformation matrix and several {10-12}<10-11> twin regions have been carried out in deformed magnesium single crystals. The hardness values show orientation dependence, following Schmid law in both matrix and twins and no significant increase of the hardness of twin regions is observed.

## Acknowledgement

The authors would like to thank Mr. Robert Kubic and Ms. Daad Haddad at General Motors R&D center for useful discussion and technical assistance with some experiments: SEM/EBSD analysis and nano-indentation experiments. Finally, this research was supported by funding from Initiative for Automotive Manufacturing Innovation (IAMI) at McMaster University, ON, Canada.

## References

1. Z.S. Basinski et al., "The Transformation of Slip Dislocations during Twinning of Copper-Aluminum Alloy Crystals" *Rev. Metall.*, 94 (1997), 1037-1042.

2. C.H. Caceres, P. Lukac, and A. Blake, "Strain Hardening due to {10-12} Twinning in Pure Magnesium", *hilosophical Magazine*, 88 (2008), 991-1003.

3. H. El Kadiri and A.L. Oppedal, "A crystal plasticity theory for latent hardening by glide twinning through dislocation transmutation and twin accommodation effects", *Journal of the Mechanics and hysics of Solids*, 58 (2010), 613-624.

4. A. Gooldstone et al., "Indentation across Size Scales and Disciplines: Recent Developments in Experimentation and Modeling", *Acta Materialia*, 55 (2007), 4015-4039.

5.W.C. Oliver and G.M. Pharr, "An Improved Technique for Determining Hardness and Elastic Modulus Using Load and Displacement Sensing Indentation Experiments", *Journal of Material Res.*, 7 (1992), 1564-1583.

6. A.A. Salem et al., "Strain Hardening Due to Deformation Twinning in α-titanium: Mechanisms", *Metallurgical and Materials Transactions A*, 37A (2006), 259-268.

# NON-BASAL TEXTURES IN MAGNESIUM ALLOY STRIPS PRODUCED BY EXTRUSION-MACHINING

Dinakar Sagapuram[1], Mert Efe[1], Wilfredo Moscoso[3], Srinivasan Chandrasekar[2], Kevin P. Trumble[1]
Center for Materials Processing and Tribology
[1]School of Materials Engineering, [2]School of Industrial Engineering, Purdue University, West Lafayette, IN, USA
[3]Department of Electro-Mechanical and Electronics Engineering, Pontificia Universidad Catolica Madre y Maestra, Santiago, Dominican Republic

Keywords: Magnesium, Non-basal textures, Secondary slip activity, Hall-Petch effect

## Abstract

A constrained chip formation technique, large strain extrusion machining (LSEM), was used to produce bulk magnesium alloy (AZ31B) strips with fine grain size (2-6 μm) and non-basal textures. These characteristics are known to enhance the final formability. The deformation temperature during extrusion-machining was varied by preheating the workpiece to a range of temperatures (50°C to 375°C). Microstructural refinement and texture evolution were studied as function of deformation temperature. It was possible to refine the grain size down to ~2 μm by restricting the dynamic grain growth at low to moderate deformation temperatures (below 320°C). LSEM was shown to be capable of resulting in non-basal textures at low deformation temperatures (below 220°C) as well as at elevated deformation temperatures (above 420°C). The influence of active deformation mechanisms and dynamic recrystallization on the texture development is also addressed.

## Introduction

As the lightest structural metal, magnesium has attracted much attention due to the significant weight savings it can offer to the discrete product industry. Especially, wrought magnesium in the sheet form is highly desired for the interior automobile body frames. However, wide usage of magnesium sheet is limited by its poor room-temperature formability and high cost. The lack of formability in magnesium and its alloys is a direct result of insufficient number of active independent deformation modes required for the homogeneous deformation. Main deformation modes active at low temperatures (25-200°C) are the basal $\langle a \rangle$ slip and the $\{10\bar{1}2\}$ mechanical twinning [1]. In addition to the limited deformation modes, the plate texture (or basal texture) that develops during conventional sheet rolling also adversely affects the final formability of the sheet. The plate texture is characterized by the $(0002)$ basal poles aligned perpendicular to the plate/sheet surface, and exhibiting fiber symmetry. The near-zero Schmid factor for the basal slip realized during the forming operations, as a result of this plate texture, results in the limited plasticity. Formability is promoted by random texture, or non-basal (or tilted basal) textures, i.e., basal poles tilted at an angle with respect to the sheet normal. However, achieving a random-textured wrought product, especially in hexagonal close packed (HCP) metals, is difficult. Non-basal textures in magnesium achieved by Mukai et al. [2] and Suwas et al. [3] through equal channel angular pressing (ECAP) were previously shown to significantly enhance the ductility/formability. However, ECAP is not suitable for direct sheet making. Besides crystallographic texture, grain size was also known to significantly influence the mechanical properties in Mg. Finer grain size was shown to enhance the strength as well as the ductility [4,5]. Here, large strain extrusion machining (LSEM) is introduced as a deformation processing technique that is capable of producing bulk magnesium strips having non-basal textures, as well as fine grain size (2-6 μm). LSEM offers control over a wide range of shear strains, strain rates, deformation temperatures, etc. In the present study, deformation temperature was varied in order to achieve a variety of textures and microstructures.

## Large Strain Extrusion Machining

LSEM is a single-step deformation technique that is capable of producing metallic strip/sheet forms by employing machining and extrusion at the same time. Fig. 1 shows a schematic of the process. The set-up basically consists of a cylindrical workpiece of width, $w$, rotating on a spindle with a constant surface velocity (also called as deformation speed), $V$. A strip of final thickness, $t_c$, having outcoming velocity $V_c$ is obtained by inserting the LSEM tool assembly into the workpiece at a constant feed, $t$, per revolution. The LSEM tool assembly, consisting of a cutting and a constraining component, results in simultaneous cutting and extrusion. For more detailed description of the process, the reader is suggested to refer Ref. [6].

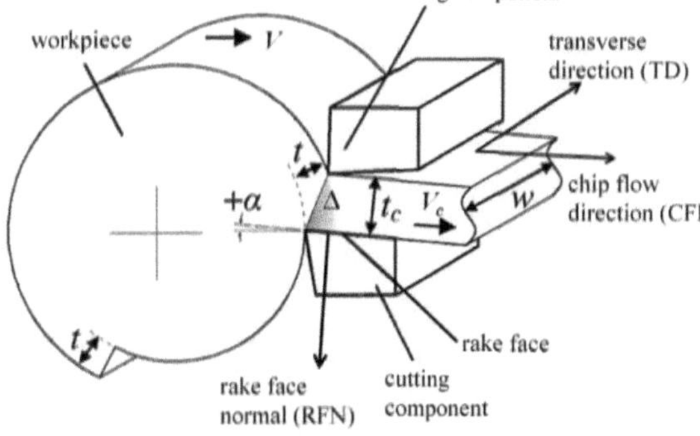


Figure 1. Schematic of LSEM.

The effective strain, $\bar{\varepsilon}$, in LSEM is given by:

$$\bar{\varepsilon} = \frac{1}{\sqrt{3}}\left[\frac{\lambda}{\cos\alpha} + \frac{1}{\lambda\cos\alpha} - 2\tan\alpha\right] \quad (1)$$

where $\lambda = t_c/t$ is the chip compression ratio and $\alpha$ is the rake angle. By controlling $\lambda$ and $\alpha$, very high effective strain (up to 3) can be imposed in a material, which helps in refining the microstructure down to the ultrafine grain size regime [6]. The effective strain rate, $\dot{\bar{\varepsilon}}$, in LSEM is given by:

$$\dot{\bar{\varepsilon}} = \frac{\varepsilon V}{\Delta} \quad (2)$$

where Δ is the thickness of the primary deformation zone (shaded area, Fig. 1), which was experimentally found to be quite narrow (50-150 μm thick). In LSEM, very high strain rates (up to $10^5$ $s^{-1}$) can be obtained by adopting high $\bar{\varepsilon}$ and $V$. These high strain rates then cause high localized heat generation in the deformation zone. The actual deformation temperature, $T_{def}$, is therefore given by:

$$T_{def} = T_o + \Delta T \quad (3)$$

where $T_o$ is the initial workpiece temperature and $\Delta T$ is the temperature rise during deformation due to high strain rate effects. Hydrostatic pressure, $p$, in the deformation zone is given by:

$$\frac{p}{2k} = \frac{1}{2} + 2\tan^{-1}\left(\frac{1}{2\lambda}\right) \quad (4)$$

where $k$ is the shear strength of the material. Hydrostatic pressures as high as 1.7$k$ can be achieved in LSEM. The high heat generation and hydrostatic pressure inherent in the extrusion-machining reduces the need for external heating and aids in smooth extrusion without leading to fracture.

### Experimental Procedure

For extrusion-machining, a disc-shaped workpiece (6.4 mm thick) was machined from a magnesium alloy AZ31B (Mg-3 wt.% Al-1 wt.% Zn) tooling plate, supplied by ThyssenKrupp Materials NA, Inc. It should be noted that the thickness of the workpiece, $w$, corresponds to the final strip width. A constant rake angle, $\alpha$, of 5° was used. The chip compression ratio, $\lambda$, was fixed at ~0.75 by choosing $t$ = 0.25 mm and $t_c$ = 0.19 mm. According to Eq. 1, effective strain, $\bar{\varepsilon}$, was therefore kept constant at 1.1. A constant deformation speed, $V$, of 0.5 m/s was used. Strain rate, $\dot{\bar{\varepsilon}}$, was therefore fixed at 0.55x$10^4$ $s^{-1}$, according to Eq. 2. Deformation temperature, $T_{def}$, was varied by controlling the initial workpiece temperature (or preheating temperature), $T_o$. The temperature rise, $\Delta T$ of 170°C was estimated from the cutting forces measured using a piezoelectric dynamometer. Details of the heat-partition model used for calculating the temperature rise can be found elsewhere [7]. The preheating temperature was found to have a negligible effect on $\Delta T$. Assuming that the heat-partition model is valid even at higher temperatures, $T_{def}$ can be given by: $T_{def}$ (°C) = $T_o$ (°C) + 170°C. A range of preheating temperatures from 50 to 375°C were used. This accordingly corresponds to a variation of $T_{def}$ from 220 to 545°C.

For optical microscopy, chip flow direction (CFD)-rake face normal (RFN) plane (refer to Fig. 1) of the samples was polished according to standard metallographic procedures and etched with a picric acid solution (100 ml ethanol, 10 ml water, 6 g picric acid, and 5 ml acetic acid). Average grain size was estimated using the linear intercept method. Hardness measurements were carried out on the CFD-RFN plane. For crystallographic texture analysis, an X-ray area detector diffraction system (GADDS, Bruker AXS) was used. A Rietveld refinement program, MAUD [8], was used to analyze the diffraction data and construct the pole figures.

### Results

The as-received AZ31B tooling plate exhibited a plate texture (i.e., in-plane basal fiber texture) and an equiaxed microstructure (images not shown here). The grain size was 16 ± 2 μm and the hardness was 58 HV. The effect of $T_o$ on the microstructure development in the LSEM strips is given in Fig. 2. All the strips exhibited equiaxed microstructures. Fig. 3 shows the effect of $T_o$

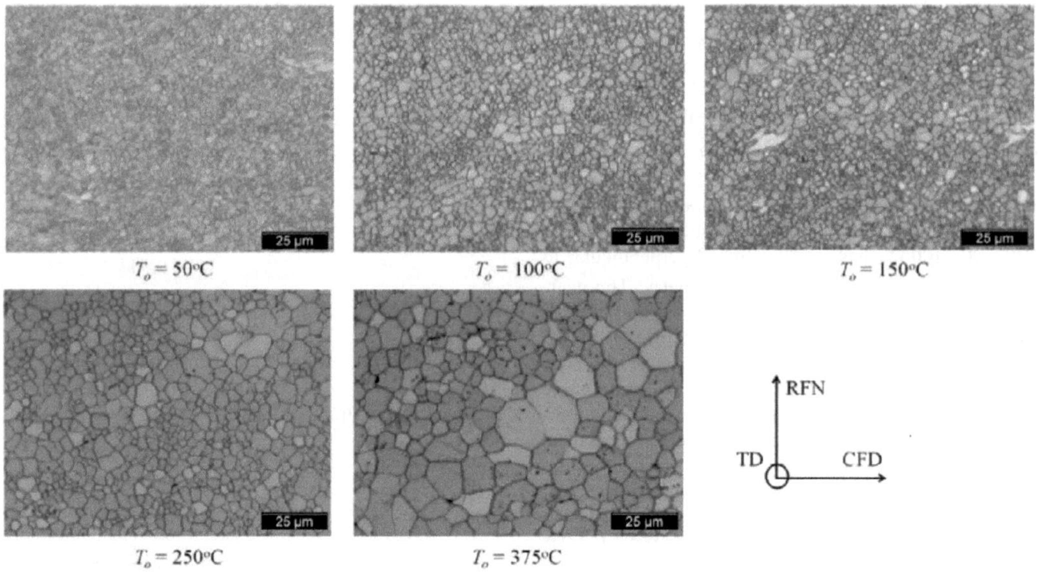

Figure 2. Microstructure development in the LSEM strips as a function of workpiece preheating temperature, $T_o$. (TD: transverse direction).

on the final grain size, $d$. Standard deviations are given by the error bars. It can be seen that $d$ remains almost constant initially and increases rapidly when $T_o$ is greater than 150°C. Increase in $T_o$ from 50 to 375°C lead to a corresponding increase in $d$ from 1.7 to 6.4 μm. It is noteworthy that extrusion-machining is capable of refining the microstructure by 2-10 times in a single step, as opposed to the standard multi-step deformation techniques.

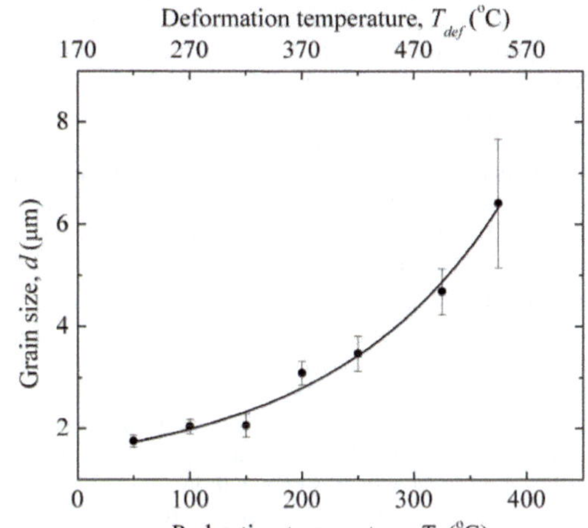


Figure 3. Variation of $d$ as a function of $T_o$.

The Hall-Petch relationship [9] is typically exhibited by metals and alloys when grain size is in the microcrystalline range. It is given by:

$$H = H_o + K_H\, d^{-1/2} \quad (5)$$

where $H_o$ and $K_H$ are called the Hall-Petch constants. In order to examine the Hall-Petch relationship in the AZ31B strips produced

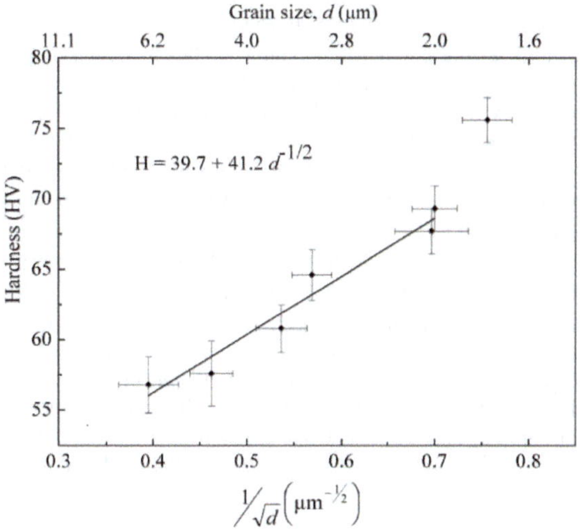


Figure 4. Hall-Petch plot for AZ31B.

by LSEM, hardness is plotted against $d^{-1/2}$ in Fig. 4. The AZ31B strips obeyed the Hall-Petch relationship with $H_o$ = 39.7 and $K_H$ = 41.2.

Evolution of texture with $T_o$ is shown in Fig. 5. The $(0002)$ and $(10\bar{1}0)$ pole figures were measured from the TD-CFD plane (refer to Fig. 1). A mirror symmetry about the CFD axis in the pole figures can be observed. Texture intensity is given in the unit, multiples of a random distribution (mrd). When $T_o$ is low (50°C), extrusion-machining was found to result in two distinct texture components having a similar intensity: a tilted basal texture component, and a basal texture component. The tilted basal texture component is characterized by $(0002)$ basal poles tilted at an angle of ~30° away from RFN toward CFD, while the basal texture component has basal poles aligned parallel to RFN.

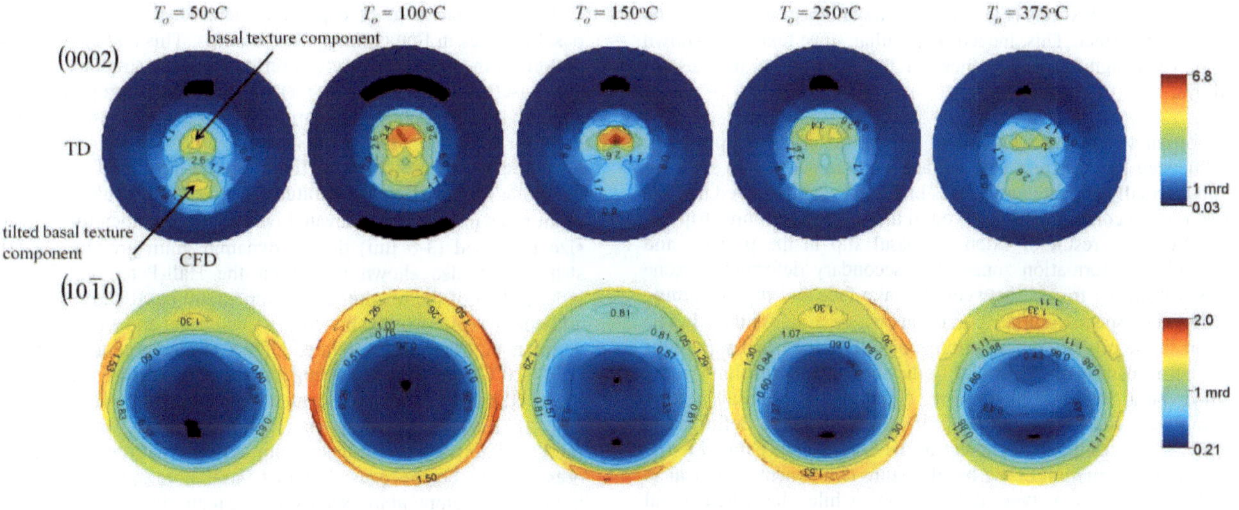


Figure 5. Texture evolution in the LSEM strips with increasing $T_o$.

From the $(10\bar{1}0)$ pole figures, it seems that both the tilted basal and basal texture components exhibit $<0001>$ fiber symmetry. With increasing $T_o$, intensity of the tilted basal texture was found to gradually decrease, while that of the basal texture component increased, and reached a maximum value of 6.8 mrd at $T_o$ = 150°C. Correspondingly, the basal texture component no longer exhibits $<0001>$ fiber symmetry. Instead, $(10\bar{1}0)$ poles are aligned parallel to CFD. Further increase in $T_o$ to 250°C resulted in a decrease in texture intensity and splitting of the basal poles about RFN by ± 10-20° along TD. The corresponding $(10\bar{1}0)$ pole figure indicates that the basal poles have regained the fiber symmetry. With further increase in $T_o$ to 375°C, no qualitative changes in the texture were observed. It is interesting to note that, with increasing $T_o$, the peak intensity value initially increased from 5 mrd at $T_o$ = 50°C, reached a maximum of 6.8 mrd at $T_o$ = 150°C, and then started decreasing, where it finally dropped down to 4 mrd at $T_o$ = 375°C.

### Discussion

The fine equiaxed grain structures (Fig. 2) found in the LSEM strips and the grain size variation with deformation temperature (Fig. 3) can be explained by dynamic recrystallization (DRX) followed by dynamic grain growth. Grain size refinement from 16 μm in the as-received material to 1.7 μm in the strip processed at a low temperature of $T_o$ = 50°C is a result of high shear strain in extrusion-machining and restricted grain growth due to warm deformation temperature ($T_{def}$ = 220°C). Grain size remained almost constant up to $T_o$ = 150°C. However, when $T_o$ is greater than 150°C, grain size started to increase rapidly. This suggests the increased activity of dynamic grain growth at moderate temperatures ($T_{def} > 320°C \approx 0.5T_m$).

The Hall-Petch effect in AZ31B (Fig. 4) confirms the grain size strengthening widely known in metals. The values of the Hall-Petch constants found in this study ($H_o$ = 39.7, $K_H$ = 41.2) are in fair agreement with the literature, where values of $H_o$ = 40-47 and $K_H$ = 42-50 were reported for the extruded and ECAP'ed AZ31 [10]. It should be noted that the LSEM strips, although having quite different textures (Fig. 5) from one another, still exhibit the Hall-Petch effect. This, indeed, is peculiar since texture is known to have a significant influence on the strength in magnesium alloys.

Deformation texture evolution in a material is primarily influenced by the stress state of deformation, active slip systems, and thermally activated processes like DRX. It is believed that the two texture components observed in the low temperature strip ($T_o$ = 50°C) is a result of extensive basal slip in the primary and secondary deformation zones. The secondary deformation zone arises from the friction between the rake face and the outcoming strip [11], especially under conditions where $\lambda$ is less than 1. The shear in the primary deformation zone due to the basal slip aligns the basal planes nearly parallel to the shear plane, resulting in the tilted basal texture component. Similarly, the basal texture component occurs due to the shear in the secondary deformation zone, which results in alignment of the basal planes parallel to the rake face. Formation of the basal texture component is typical in rolling/compression type deformation, while the tilted basal texture component is observed in shear-based deformation as in ECAP [3,12]. The increase in $T_o$ to 100°C lead to the strengthening of basal texture component, accompanied by a decrease of tilted basal texture component intensity. This type of textural change with increasing $T_{def}$ in extrusion-machining was previously found to coincide with the beginning of discontinuous dynamic recrystallization (DDRX) [13]. DDRX is a type of DRX, where new grains nucleate near the original grain boundaries, having orientations different from the orientations of deformed grains. The beginning of DDRX was also found to be accompanied by a sharp hardness drop, which seems to be true even in the present study. In Fig. 4, a sharp hardness drop occurs when grain size is increased from 1.7 μm to 2 μm. Hence, the change in texture with increasing $T_o$ from 50 to 100°C is believed to be due to DDRX. However, the reason why DDRX seems to be strengthening particularly the basal texture component is currently unknown, and needs further investigation. With increasing $T_o$ to 150°C, basal texture is further intensified and displays a maximum peak intensity of 6.8 mrd, presumably due to the increased DDRX as a result of higher $T_{def}$. The reason for the alignment of $(10\bar{1}0)$ poles with CFD for this condition is not clear. Further increase in $T_o$ to 250°C caused splitting of basal poles away from RFN along TD, which is surprising in AZ31B. This kind of basal pole splitting in TD is usually a characteristic of rolling textures of HCP metals like Zr [14], where prismatic $\langle a \rangle$ slip is dominant. ECAP of Mg-4 wt.% Li, which is known to increase prismatic slip activity, was also shown to result in a texture component tilted towards the flow plane normal direction [12]. Since the extrusion-machining geometry can be compared to that of ECAP, it is presumed that increased secondary slip activity (i.e., prismatic $\langle a \rangle$ slip or pyramidal $\langle c + a \rangle$ slip, or both) at elevated temperatures ($T_o$ > 250°C, $T_{def}$ > 420°C) as the reason for the basal pole splitting in TD. The drastic decrease in critical resolved shear stress (CRSS) for secondary slip systems with temperature [15] is also in agreement with this assumption. Further increase in $T_o$ to 375°C lead to no significant changes in texture, except for the decreased peak intensity value. The texture weakening at elevated deformation temperatures is, at present, not clear.

### Conclusions

LSEM was shown to be capable of achieving a variety of non-basal textures in fine grained AZ31B strips. This ability of LSEM to control the texture, as well as refine the microstructure, will be very useful in designing the final mechanical properties like strength and formability. Deformation temperature was found to greatly influence the microstructure and texture evolution. Dynamic grain growth was found to be restricted at low to moderate deformation temperatures (≤ 320°C), resulting in fine grain sizes (~2 μm). At elevated deformation temperatures, grain size increased (3-6 μm) due to dynamic grain growth. AZ31B strips were also shown to exhibit the Hall-Petch relationship independent of the texture. Low temperature basal slip resulted in a texture having two components: a tilted basal and a basal texture component. Thermally activated DDRX at moderate temperatures (270-320°C) was found to strengthen the basal texture and weaken the tilted basal texture. At elevated temperatures (≥ 420°C), basal pole splitting in the strip transverse direction was observed. Increased activity of secondary slip systems with temperature is presumed to be the cause for this basal pole splitting. Further increase in temperature seems to weaken the texture without changing the texture character. Finally, formability enhancement in AZ31B as a result of the LSEM non-basal textures (i.e., textures containing the tilted basal texture component or exhibiting the basal pole splitting) and fine grain size looks promising.

**Acknowledgements**

Support through NSF-CMMI-MPM-0800481 and DOE PNNL-DE-AC06-76RL01830 is gratefully acknowledged.

**References**

1. I.J. Polmear, "Magnesium Alloys and Applications", *Materials Science and Technology*, 10 (1994), 1-16.

2. T. Mukai, M. Yamanoi, H. Watanabe, and K Higashi, "Ductility Enhancement in AZ31 Magnesium Alloy by Controlling its Grain Structure", *Scripta Materialia*, 45 (2001), 89-94.

3. Satyam Suwas, G. Gottstein, and R. Kumar, "Evolution of Crystallographic Texture during Equal Channel Angular Extrusion (ECAE) and its Effects on Secondary Processing of Magnesium", *Materials Science and Engineering A*, 471 (2007), 1-14.

4. Edward F. Emley, *Principles of Magnesium Technology* (Oxford, New York, Pergamon Press, 1966).

5. L. Jin, D. Lin, D. Mao, X. Zeng, and W. Ding, "Mechanical Properties and Microstructure of AZ31 Mg Alloy Processed by Two-step Equal Angular Channel Extrusion", *Materials Letters*, 59 (2005), 2267-2270.

6. W. Moscoso, M.R. Shankar, J.B. Mann, W.D. Compton, S. Chandrasekar, "Bulk Nanostructured Materials by Large Strain Extrusion Machining", *Journal of Materials Research*, 22 (2007), 201-205.

7. J.H. Wiener, "Shear-plane Temperature Distribution in Orthogonal Cutting", *Transactions of the American Society of Mechanical Engineers*, 77 (1955), 1331-1341.

8. L. Lutterotti, M. Bortolotti, G. Ischia, I. Lonardelli, and H.R. Wenk, "Rietveld Texture analysis from Diffraction Images", *Zeitschrift für Kristallographie Supplements*, 26 (2007), 125-130.

9. E.O. Hall, "The Deformation and Ageing of Mild Steel: II Characteristics of Lüders Bands", *Proceedings of the Physical Society B*, 64 (1951), 742-747.

10. W.J. Kim, Y.K. Sa, "Micro-extrusion of ECAP Processed Magnesium Alloy for Production of High Strength Magnesium Micro-gears", *Scripta Materialia*, 54 (2006) 1391-1395.

11. M. Sevier, H.T.Y. Yang, W. Moscoso, and S. Chandrasekar, "Analysis of Severe Plastic Deformation by Large Strain Extrusion Machining", *Metallurgical and Materials Transactions A*, 39A (2008), 2645-2655.

12. S.R. Agnew, P. Mehrotra, T.M. Lillo, G.M. Stoica, and P.K. Liaw, "Texture Evolution of Five Wrought Magnesium Alloys during Route A Equal Channel Angular Extrusion: Experiments and Simulations", *Acta Materialia*, 53 (2005), 3135-3146.

13. D. Sagapuram, M. Efe, W. Moscoso, S. Chandrasekar, and K.P. Trumble, "Deformation Temperature Effects on Microstructure and Texture Evolution in High Strain Rate Extrusion-Machining of Mg-AZ31B", *Proceedings of the 16th International Conference on Texture of Materials*, Bombay, India, Dec, 2011.

14. Y.N. Wang, and J.C. Huang, "Texture Analysis in Hexagonal Materials", *Materials Chemistry and Physics*, 81 (2003), 11-26.

15. S.E. Ion, F.J. Humphreys, and S.H. White, "Dynamic Recrystallization and Development of Microstructure during the High Temperature Deformation of Magnesium", *Acta Metallurgica*, 30 (1982), 1909-1919.

**Magnesium Technology 2012**
*Edited by: Suveen N. Mathaudhu, Wim H. Sillekens, Neale R. Neelameggham, and Norbert Hort*
***TMS (The Minerals, Metals & Materials Society), 2012***

# THE ELASTO-PLASTIC TRANSITION IN MAGNESIUM ALLOYS

K. Vanna Yang, C.H. Cáceres*
ARC Centre of Excellence for Design in Light Metals, Materials Engineering
School of Engineering, The University of Queensland, Brisbane, QL, 4072, Australia
C.N. Tomé
MST Division Los Alamos National Laboratory Los Alamos NM 87545 USA

Keywords: AZ31 alloy; Magnesium; Elasto-plastic transition; Kocks-Mecking analysis; Strain hardening; Twinning; Prism slip

## Abstract

Published neutron diffraction data on internal strains in a textured polycrystal of AZ31 alloy and tensile data for a random polycrystal of Mg are discussed from two complementary points of view, a purely phenomenological and a formal crystallographic one. It is suggested that the transition to full plasticity involves two different fractions of the grains population, at different strains. The role of texture and twinning are discussed. Critical experiments to test some of the hypotheses on how the elasto-plastic transition develops are proposed.

## Introduction

The present paper is the result of an exchange of ideas between the US and Australia based authors concerning the meaning of the elasto-plastic transition in textured AZ31 alloy, and by extension, Mg and its alloys in general. It aims at identifying weak points in the understanding of the phenomenon as well as to advance a few hypotheses. Kocks-Mecking analysis of published data is used as a framework to the discussion.

The onset of gross plastic deformation in randomly oriented Mg and Mg alloys involves a microplasticity stage which may extend to nearly 1 % strain in which basal slip, and sometimes twinning, are active, followed by a linear hardening stage during which general plasticity develops [1, 2]. The linear hardening stage may extend to strains as large as 10 % or even higher, depending on the grain size.

In a recent neutron diffraction study by Agnew *et al.* [3] on textured AZ31 alloy, the linear hardening stage past the first ~2 % and up to nearly 10 % strain was described as a protracted elasto-plastic (e/p) transition, akin to what is observed in composite materials in which the apparent, or macroscopic, strain hardening rate reflects the load-sharing between soft- and hard-oriented grains in the polycrystalline aggregate.

This interpretation of the linear hardening stage is at odds with recent analyses of the strain hardening behaviour of texture-free pure Mg polycrystals [4, 5] which show, based on the Kocks-Mecking method of analysis, that athermal accumulation of forest dislocations should determine the strain hardening rate of polycrystalline Mg in much the same way as in FCC metals at low temperature or with a low value of the stacking fault energy, hence with a strong tendency to planar glide, accounting for the observed linear hardening in the first ~10 % strain.

* Corresponding author, c.caceres@uq.edu.au, Fax:+ 61-7-33467015, Phone: +61-7-33654377.

In the next sections the Kocks-Mecking phenomenological method of analysis [6] is applied to the lattice strain data collected by Agnew *et al.* [3] in tensile specimens of cold rolled alloy AZ31, and subsequently compared with a similar analysis made on the tensile flow curve of a texture-free polycrystal of pure Mg to assess the rate at which the increase of internal stresses in soft- and hard-oriented grains relates to the macroscopic strain hardening rate as the materials are deformed.

## Data for Analysis

The macroscopic flow curve and the internal strain-applied stress curves obtained by the in-situ strain measurements of the alloy are reproduced in Figures 1 and 2. Figure 1 also includes the tensile flow curve of a fine grained (grain size 19 μm [7]) pure Mg polycrystal.

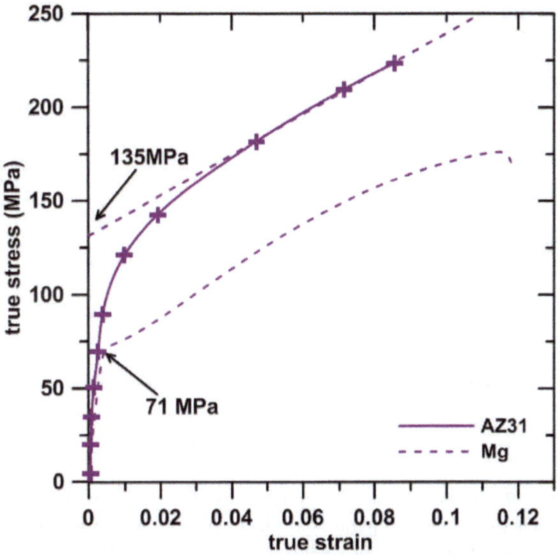


Figure 1 The solid line is the stress-strain curve for in-plane tension of alloy AZ31, reproduced from Figure 1-a of reference [3]. In this and the following figures the symbols (crosses in this case) denote the loci of the in-situ neutron diffraction measurements. The dashed line is the tensile flow curve of a Mg polycrystal reproduced from references [4, 7]. The yield strength of both materials is indicated by arrows (see text).

The flow curves of Figure 3 were obtained by converting the internal (lattice) strains of Figure 2 to internal stresses through Hooke's law (Equation 1). Only the elastic strains parallel to the tensile axis are considered in the analysis.

$$\sigma_l = E(hk.l) * \varepsilon_l \tag{1}$$

where $\sigma_l$ is the internal stress, $E(hk.l)$ is the Young's modulus normal to the relevant crystallographic planes (hk.l), and $\varepsilon_l$ is the lattice strain along the same direction. The values for $E(hk.l)$ listed in Table 1 were calculated using standard formulae and the compliance constants from Table 2, all taken from [8-10]. Equation 1, for simplicity, assumes unconstrained uniaxial loading.

Table 1 Young's modulus in the direction perpendicular to the $(hk.l)$ crystallographic planes, calculated as described in [8]

| plane | (10.0) | (10.1) | (10.3) |
|---|---|---|---|
| $E_{hkl}$ (GPa) | 45.45 | 43.29 | 45.03 |

Table 2 Compliance constants for pure Mg [9, 10]

| Compliance constants ($10^{-11}$ Pa) | | | | crystallographic parameter (nm) | |
|---|---|---|---|---|---|
| $s_{11}$ | $s_{33}$ | $s_{44}$ | $s_{13}$ | $a$ | $c$ |
| 2.20 | 1.97 | 6.10 | 0.50 | 0.318 | 0.5131 |

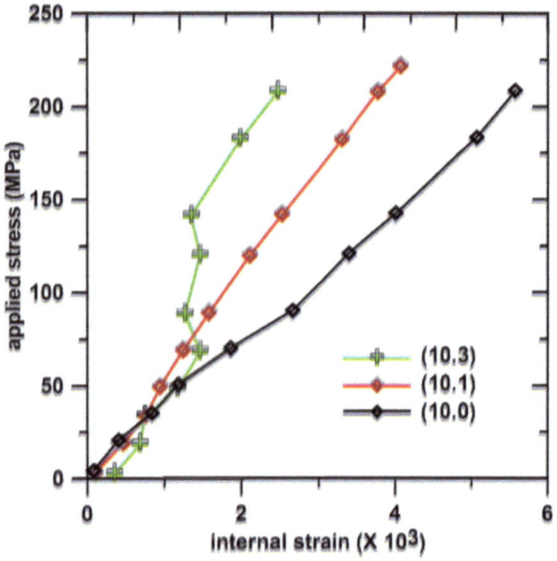


Figure 2 Internal strain and applied stress for grains with normal to their prismatic and pyramidal planes parallel to the stress axis of the AZ31 specimen of Figure 1. In this and remaining figures (hk.l) identifies the crystallographic planes of which the normals are parallel to the stress axis.

In Figure 3-a, the internal stress in all three grain orientations increases almost identically up to ~25 MPa, diverging for larger applied stresses. The (10.3) normal direction parallel to the stress axis shows a relatively short microplasticity regime followed by a plateau, ascribed in the original reference to twinning, and a linear hardening regime at larger strains. The (10.1) and (10.0) normal directions exhibit a longer microplasticity regime and linear hardening past ~2 % strain, without any major perturbation from twinning. The strain hardening rates at large strains are similar, i.e., the stress-strain curves of Figure 3-a are nearly parallel to each other. In Figure 3-b the flow curves are back extrapolated to determine the respective yield strengths $\sigma^*$ for the analysis of the next section. The yield strengths of the macroscopic curves are indicated in Figure 1.

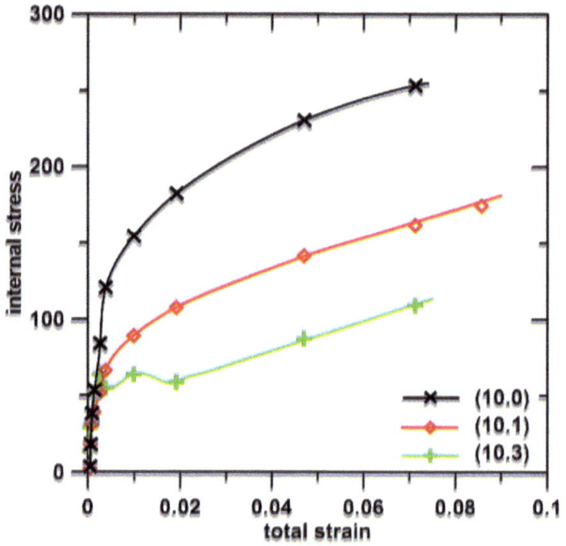


(a)

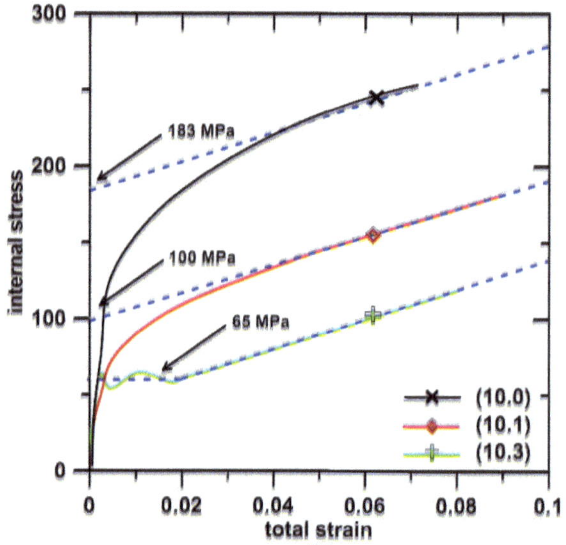


(b)

Figure 3 (a) Internal stress-total strain curves calculated using Equation 1 for the curves of Figure 2; (b) Determination of the offset stresses, $\sigma^*$, by back extrapolation of the large strain flow curves of Figure 3-a.

*Kocks-Mecking analysis* The formalism was applied to the present materials as described in detail in references [4, 11] for pure Mg polycrystals, according to:

$$(\sigma - \sigma^*)\frac{d\sigma}{d\varepsilon} = (\sigma - \sigma^*)\Theta_h = (\sigma - \sigma^*)\frac{E}{C} \tag{2}$$

where $\sigma$ is the flow stress, $\varepsilon$ is the true strain and $\Theta_h = \frac{d\sigma}{d\varepsilon} = E/C$ is the strain hardening rate when hardening is controlled by athermal accumulation of dislocations, $\sigma^*$ is the offset stress determined for the alloy in Figures 1 and 3-b as suggested by Kocks and Mecking [12] by back-extrapolation (dashed lines), and C is a constant whose value depends on the Taylor factor $M$. With $M = 4.5$ [5] and $E = 44$ GPa, $\Theta_h \approx E/32 = 1.4$ GPa, a value closely matched by random polycrystals of Mg [4, 11].

The graphical version of Equation 2, called "$\Theta\sigma$-plot", is shown in Figures 4-a and 4-b for the curves of Figures 1 and 3, respectively; the initial slopes of the respective $\Theta\sigma$ plots are in the range of 1.0 GPa to 2.0 GPa, matching the expected value of 1.4 GPa quite closely.

### Discussion

*Flow behavior* A number of differences and some crucial similarities between the flow curves of Figure 1 are relevant to the present analysis. The onset of the microplasticity stage (i.e., the departure from pure elastic behavior) occurs at fairly low stresses (25-50 MPa) for both materials. For the pure Mg macroscopic flow begins at 71 MPa with an incipient plateau, most likely connected to twinning, and a subsequent linear hardening stage that extends to 8-10 % strain. The alloy, in contrast, exhibits a round shape from about 75 MPa and up to ~2 % strain, suggesting the a gradual development of internal stresses. At large strains the flow curves of both materials become parallel to each other. At about 10% strain the pure Mg flow stress tends to saturate; a similar behavior is hinted by the alloy, although the data available do not go far enough. Further details of the yielding process are discussed below.

*Kocks-Mecking behavior* Under an athermal regime of dislocations accumulation [4, 5], the $\Theta\sigma$-plot of a plastically deforming Mg polycrystal starts with a straight line through the origin of gradient $\Theta_h = 1.4$ GPa. This behavior is shown in Figure 4-a, curve labeled Mg. The initial gradients of the corresponding $\Theta\sigma$-plots for the macroscopic and the internal stresses of the alloy (figures 4-a and b) closely match that of the pure Mg. Therefore it may be asserted that the elastic loading of the hard-oriented grains in the alloy is largely determined by the monotonic accumulation of dislocations in the plastic grains past the first 2 % strain, mimicking what occurs in the pure Mg polycrystal.

An overall picture of how the plasticity evolves in the either one of the materials past the purely elastic regime can be construed as follows, partly following the descriptions given in [4, 5]. Prior to the onset of microyielding, internal (elastic) strain develops linearly in all grains, differing only by the elastic anisotropy, which in Mg is minor. When basal slip is activated in the grains with the most favorable orientation to the stress axis, the microplasticity stage starts. During this stage soft-oriented grains diminish their share of elastic strain and hard-oriented grains compensate by becoming more highly stressed. The overall hardening rate at this stage is therefore determined by the load shedding between the (micro) plastic and elastic grains. The end of the microplasticity stage is determined by the activation of the first available hard slip system, which experiments show is prism slip [13]. This seems to enable sufficient polyslip in the plastically active grains, marking the onset of macroplasticity; the strain hardening rate consequently drops to a value that commensurate with athermal accumulation of dislocations, (1.4 GPa, Figure 4).

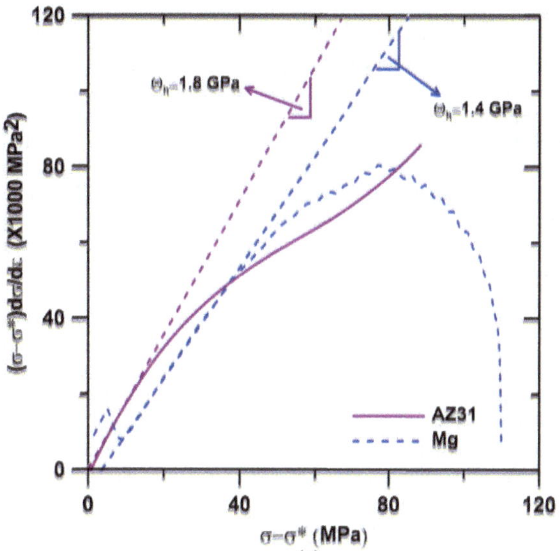

(a)

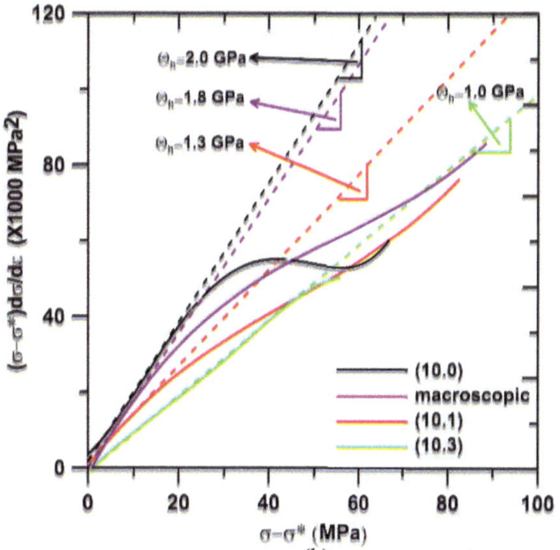

(b)

Figure 4 $\Theta\sigma$-plots for the (a) AZ31 and pure Mg flow curves of Figure 1 and (b) internal stresses curves of Figure 2. The straight dashed lines are lines of best fit to the low stress region of the respective $\Theta\sigma$-plots, with the slope as indicated. The curve labeled 'macroscopic' in 4-b corresponds to the flow curve of alloy AZ31 in Figure 4-a and Figure 1.

The differences in behavior at yield between the alloy and the pure Mg made evident by Figure 1 can be accounted for as follows: in the random polycrystal there are enough soft-oriented grains to ensure that any hard-oriented grains are rapidly wrapped out by fully plastic grains, and the transition from the microplasticity stage to fully plasticity is very short, as shown by Figure 1.

For the alloy, the rounded flow curve in the first 2 % strain is an indication that the fraction of soft-oriented grains is initially not enough to wrap around the hard-oriented ones. As the stress increases, the fraction of plastic grains increases monotonically until they are enough to wrap around the remaining elastic grains. Pat this point, the strain hardening rate is determined by the forest hardening of the plastic grains and the e/p transition is over. Twinning plays an important part in the pure Mg as shown by the incipient plateau in Figure 1 and although the neutron diffraction data indicates limited activity, it seems likely that the combination of slip and twinning is sufficient to mock true polyslip [14] in the alloy as well, and thus avoid the need for 5 independent slip systems in both materials leading to the observed linear hardening.

It is of interest to note in passing that the behaviour of the AZ31 alloy specimen is remarkably consistent with that of polycrystalline CoTi as described by Wollmershauser *et al.* [15]. Due to the plastic anisotropy of the intermetallic, and similarly to Mg, the soft slip system <100>{011} is activated first, creating a steep microplasticity stage, followed by an extended linear hardening stage which extend to strains as large as ~20 %. The transition between the micro and the macro plasticity stages, which replicates the behavior of Mg, is triggered by the activation of the <111>{1-10} hard slip system.

*Yield surface and full plasticity* From a more formal point of view, it may be argued that the "end of the e/p transition" should meet the von Mises-Taylor criterion for full plasticity, which is only possible when the stress "lodges" in a vertex of the yield surface and stops "sliding" along it (at least for a substantial amount of strain) [16]. The concept is of course well established for FCC metals since a vertex provides the 5 independent slip systems required. For the AZ31 alloy specimen the neutron diffraction results showed that no slip systems are active up to ~25 MPa, only basal slip is active between 25 and 75 MPa, at a strain of about 0.5 %, then basal and prism systems are active between 75 and 175 MPa and the onset of pyramidal slips is identified past 175 MPa, at strains above 4~5 %. Since, in principle, basal and prism slips are insufficient to accommodate full plasticity, the internal stress will not reach a vertex until pyramidal slip is fully activated in a significant fraction of crystals, delaying the onset of full plasticity to strains approaching 10 %. This description of course ignores the enabling role of twinning discussed earlier.

Returning to the phenomenological approach, Mecking *et al.*'s suggested that the yield surface of plastically anisotropic materials becomes increasingly isotropic at large strains. The behaviour of the present materials closely conforms to this assertion since, on the one hand, the macroscopic flow curves become parallel to each other past the strain of 2 %; and on the other, at $\Theta_h$ = 1.4 GPa, the strain hardening behaviour of both materials, and in particular that of the alloy's internal stresses, is consistent with athermal forest hardening and independent of the initial texture.

*The role of texture* It is possible to produce a rather simplistic but unified description of the e/p transition for both the alloy and the pure metal considering that there are two e/p transitions, independent of each other, affecting different fractions of the grain population, and occurring at different strains in the flow curve.

The first e/p transition follows the description given in the preceding section. Figure 1 shows that in the Mg specimen this transition takes place at well-defined values of the stress (~71 MPa in Fig. 1) and strain, whereas in the alloy it extends over the first ~2 % strain, leading to the rounded flow curve. In both materials it is followed by a linear hardening stage.

Past ~5 % strain the internal stresses are high enough to activate pyramidal slip in the (still elastic) hardest oriented grains, and true von Mises-Taylor plasticity begins to take place in those grains, gradually propagating to other hard-oriented grains. This in practice constitutes a second e/p transition affecting both the pure Mg and the alloy alike. The second transition develops gradually as it is driven by the athermal forest hardening of the soft-oriented grains. The strain hardening rate decreases slowly during the process, which is completed at ~10 % strain, when the flow stress saturates.

The first transition may be termed "global", whereas the second could be considered "latent" or "alien".

With reference to the curves of Figure 1, the effect of crystallographic texture can now be understood as follows: in the random polycrystal of Mg the proportion of soft-oriented grains is large enough to ensure a rapid and well defined global transition. By contrast, the scarcity of soft-oriented grains leads to an extended "Masing-solid" behavior on the side of the alloy [17], giving raise to the rounded shape of the flow curve. The increasing difference in flow stress level between the alloy and the pure Mg is consistent with internal stresses developing as in dispersion hardened materials [18, 19].

Figure 1 also shows that the linear hardening stage is quite extended in the pure Mg, and this is due to the sharpness of the global transition. For the alloy, texture extends the global transition up to the onset of the latent one, shortening the linear hardening regime. In a sense, it may be asserted that in the presence of a strong texture the global and latent e/p transitions overlap, effectively extending the overall yielding process over a strain of ~ 10 %.

*Limitations to the approach* Despite the generally self-consistent picture produced by the Kocks-Mecking analysis, and largely due its phenomenological nature, the description given above suffers from several shortcomings and limitations, as detailed below.

The use of Equation 1 to create the "internal flow curves" of Figure 3, albeit very convenient, is somewhat misleading as the strain used in the x-axis is coming from a subset of grains which may not be even close neighbors. The "crystallographic stiffness" is therefore not the directional Young's modulus of the single crystal, because none of the contributing crystals in the subset experiences the same elastic stress state. (As a matter of fact, the crystallographic stiffness is different for different textures.) However, since Mg's elastic tensor is close to isotropic, any absolute errors in Figure 3 cannot be too large, and it should be kept in mind that the curves of Figure 3 are only used to determine the internal strain hardening rate, thus the conclusions extracted from Figure 4-b are quite likely realistic .

The strain rate for athermal forest hardening was obtained using a value of $M = 4.5$ for the Taylor factor. Such value was obtained as a relation between strain hardening rates (following Mecking *et al.*'s [20] approach), and as such it represents only a lower bound when it comes to stresses, making the "expected" value for $\Theta_h$ = 1.4 GPa a lower bound to the strain hardening rate by forest hardening.

The above treatment of twining ignores a most important difference between the alloy and the pure metal. It is known that solid solution hardening delays the onset of twinning, and the effect is anisotropic, i.e., it affects the soft and hard slip systems in opposite directions and with different strengths [21, 22]. This difference is made evident in Figure 1 by the significant activity of twinning at low strains in the pure Mg flow curve. In the alloy, only the (10.3) system exhibits limited effects of twinning (Figure 3a).

A further oversimplification relates to the assumption that soft grains physically wrap hard grains, the latter assumed to remain elastic as hard inclusions would in a dispersion hardened material. A similar elasto-plastic behaviour would result from the activation of twinning in a subset of grains and basal slip in others, followed by prismatic slip in still other subset. This alternative conception dispenses with the need to preserve a subset of grains completely elastic, and implies that the strain interval over which some grains remain purely elastic is small.

Likewise, there is no need for the stresses to be on a vertex of the yield surface in every grain for the aggregate to be fully plastic. Simulations show that depending on relative CRSSs and texture 3 or 4 average active systems per grain usually suffice to accommodate deformation (see for instance Figures 8b, 8c and 11b in [10]). In other words, individual grains need not to follow the macro strain tensor, provided their average behavior does. This blurs somewhat the boundary between the global and latent e/p transitions. For instance, the description given above assumes that at 5% strain pyramidal slip gets activated in grains which are still fully elastic, while in practice many of those grains may have basal or prism slip activated already.

*Further experiments and modelling* Experiments that should enable separating the elastic and the plastic components of the strain could involve Bauschinger effect type of tension-compression test, and measuring the x-ray peak broadening. The former would enable quantifying the amount of elastic recovery at any given strain whereas the second would allow assessing the dislocation density evolution. With these data a more detailed EPSC simulation would be possible, incorporating the overall enabling role of twinning, the anisotropy of solid solution hardening and the effect of texture.

## Conclusions

The Kocks-Mecking analysis was applied to the macroscopic and the internal stress-total strain curves of grain populations with different orientations in AZ31 sheet metal with a strong basal texture, and compared with the tensile behavior of a randomly oriented fine grained Mg polycrystal.

Strain hardening rates of 1~2 GPa were obtained for the internal stresses in all three crystallographic directions as well as for the macroscopic flow curve of the alloy. These values match that of the pure Mg polycrystal (1.4 GPa) and appear consistent with athermal accumulation of forest dislocations.

There is a first, or global, elasto-plastic transition involving basal, prism slip and twinning, leading to a linear strain hardening stage due to athermal forest hardening.

The activation of pyramidal slip in the hardest-oriented elastic grains triggers a second, or latent, e/p transition at a strain of ~5 % in both materials. This transition develops slowly with the strain, driven by the forest hardening of the soft-oriented grains. The whole process is completed at a strain of ~10%.

The presence of crystallographic texture in the alloy extends the global transition over a strain of ~2 %, in contrast with the random polycrystal of Mg, for which it occurs at a very well defined point in the flow curve.

## References


[1] A. Akhtar, E. Teghtsoonian, Metall. Trans. A, 2 (1971) 2757-2763.

[2] A. Blake, C.H. Cáceres, in: N.R. Neelameggham, H.I. Kaplan, B.R. Powell (Ed.) *Magnesium Technology 2005*, The Minerals, Metals and Materials Society (TMS), Warrendale, PA, 2005, pp. 403-407.

[3] S.R. Agnew, C.N. Tomé, D.W. Brown, T.M. Holden, S.C. Vogel, Scripta Mater., 48 (2003) 1003-1008.

[4] C.H. Cáceres, A.H. Blake, Mater. Sci. Eng. A, 462 (2007) 193-196, http://dx.doi.org/10.1016/j.msea.2005.12.113.

[5] C.H. Cáceres, P. Lukác, Philos. Mag. A, 88, http://dx.doi.org/10.1080/14786430801968611 (2008) 977-989,

[6] U.F. Kocks, H. Mecking, Prog. Mater. Sci., 48 (2003) 171-273.

[7] C.H. Cáceres, G.E. Mann, J.R. Griffiths, Metall. Mater. Trans. A, 42 (2011) 1950-1959, DOI: 10.1007/s11661-010-0599-2.

[8] J.M. Zhang, Y. Zhang, K.W. Xu, V. Ji, Thin Solid Films, 515 (2007) 7020-7024, 10.1016/j.tsf.2007.01.045.

[9] H.B. Huntington, Sol. St. Physics, 7 (1958) 213-351.

[10] U.F. Kocks, C.N. Tomé, H.-R. Wenk, Texture and Anisotropy, Cambridge University Press, Cambridge, 1998.

[11] C.H. Cáceres, P. Lukác, A. Blake, Philos. Mag. A, 88 (2008) 991-1003, http://dx.doi.org/10.1080/14786430701881211.

[12] S. Sannes, H. Gjestland, H. Westengen, C. Dorum, O. Lohne, in: SAE 2005 World Congress & Exhibition, , , Detroit, MI, USA, 2004, pp. 725-731.

[13] A. Akhtar, E. Teghtsoonian, Materials Transactions (JIM), 9, supplement (1968) 692-697.

[14] U.F. Kocks, D.G. Westlake, Transactions AIME, 239 (1967) 1107-1109.

[15] J.A. Wollmershauser, C.J. Neil, S.R. Agnew, Metall. Mater. Trans. A, 41 (2010) 1217-1229, 10.1007/s11661-009-9990-2.

[16] C.N. Tomé, U.F. Kocks, Acta Metall., 33 (1985) 603-621,

[17] G. Masing, in: (Ed.) *Proc. 2nd Int. Congr. Applied Mechanics*, 1926, pp. 332–335.

[18] L.M. Brown, W.M. Stobbs, Philos. Mag., 23 (1971) 1185-1199.

[19] L.M. Brown, W.M. Stobbs, Philos. Mag., 23 (1971) 1201-1233.

[20] H. Mecking, U.F. Kocks, C. Hartig, Scripta Mater., 35 (1996) 465-471.

[21] D. Nagarajan, C.H. Caceres, J.R. Griffiths, Acta Polonica, In press (2012).

[22] A. Akhtar, E. Teghtsoonian, Acta Metall., 17 (1969) 1351-1356.

# Magnesium Technology 2012

## Casting and Solidification

**Wednesday AM**

*Session Chairs:*

**Norbert Hort**
**(Helmholtz-Zentrum Geesthacht, Germany)**
**Yongho Sohn**
**(University of Central Florida, USA)**

**Magnesium Technology 2012**
*Edited by: Suveen N. Mathaudhu, Wim H. Sillekens, Neale R. Neelameggham, and Norbert Hort*
***TMS (The Minerals, Metals & Materials Society), 2012***

# TWIN ROLL CASTING OF THIN AZ31 MAGNESIUM ALLOY STRIP WITH UNIFORM MICROSTRUCTURE AND CHEMISTRY

Iman Bayandorian, Ian Stone, Yan Huang, Zhongyun Fan

BCAST, Brunel University, Uxbridge, Middlesex, UB8 3PH, UK

Keywords: Twin roll casting, Magnesium alloys, Grain refinement, Melt conditioning

## Abstract

Magnesium alloys produced by the existing twin roll casting (TRC) technique have coarse and non-uniform microstructures and severe centre-line segregation. Consequenty, TRC strip is cast typically no thinner than 5-9 mm, relying on costly subsequent downstream processing to produce thin strip with an improved microstructure. In the research described herein melt conditioning by intensive melt shearing was used prior to TRC to promote heterogeneous nucleation and provide a refined and uniform microstructure without severe macrosegregation. Additionally, a TRC machine with small diameter rolls was used to cast AZ31 strip of less than 2 mm in thickness suitable for direct component manufacturing such as stamping, without the necessity of hot rolling. The effects of process parameters, such as casting speed and melt superheat, on as-cast microstructure were studied. Experimental results show that the melt conditioning process provided considerable reduction in as-cast grain size and elimination of centre-line segregation. The texture and mechanical properties of melt conditioned strips were much improved over conventional TRC strips, offering the potential for higher quality final components.

## Introduction

Wrought Mg alloys in sheet form are of great interest for applications in automotive, electronic packaging and other engineering disciplines, due to their low density, and high specific strength and stiffness. Twin roll casting (TRC), which was originally patented by Sir Henry Bessemer in 1865 [1], can produce magnesium alloy strip directly from the melt [2]. TRC offers a promising alternative route for producing Mg sheet products at high efficiency and low cost, with potentially improved microstructure and enhanced properties due to the high cooling rate compared with conventional ingot casting route [2]. However, the Mg strip produced by the current TRC technology has several problems including the predominance of coarse and non-uniform dendritic columnar grains and severe centre-line segregation, resulting in an increased burden on downstream processing and poor mechanical performance of the final products [2, 3]. Producing thicker strip by TRC followed by intensive downstream rolling operations, alloy development or adding grain refiners can help resolve these problems [4, 5]. As a result, however, the overall cost is increased and the products become less competitive.

The lack of ductility due to the non-uniformity of microstructure and centre-line segregation in TRC Mg strip is compounded by reduced sheet formability resulting from a strong basal texture which develops during lengthy downstream processing steps of annealing and repetitive hot/warm rolling. Significant effort is required to find a means to weaken the basal texture or obtain a random texture in Mg sheet. Non basal or weak textures have been reported recently in Mg alloys with additions of rare-earth elements such as Ce, Nd and Y [6]. However in addition to the high cost of such additions, insoluble particles may impair the mechanical properties and recyclability of the final Mg alloy product.

In contrast to the "chemical approach" to producing a fine grained microstructure by adding grain refiners, it is possible to employ a "physical approach" based on activating naturally occurring heterogeneous nucleant particles in the alloy melt by a physical means, such as intensive melt shearing by the MCAST (Melt Conditioning by Advanced Shearing Technology) process [7]. The MCAST process has been developed to manipulate or condition Mg and Al alloy melts prior to solidification processing, leading to significant grain refinement, considerable reduction of cast defects and substantial improvement of mechanical properties [7-9]. The combination of the MCAST process and a TRC machine (known as MC-TRC) takes full advantage of the TRC process for magnesium strip production whilst obtaining a uniform and refined microstructure, uniform chemistry and significantly improved mechanical properties [3]. The present work builds on the MC-TRC process with the ultimate aim of producing Mg strip sufficiently thin (less than 2mm) and formable for direct stamping of components without recourse to downstream rolling. In particular, this paper investigates the effect of casting parameters on the microstructure and texture development during MC-TRC of AZ31 Mg alloy.

## Experimental Details

Commercial AZ31 alloy (Mg-3.34Al-0.97Zn-0.31Mn, wt.%), supplied by Magnesium Elektron (Manchester, UK) was melted in 10 kg batches in a steel crucible at 670 ºC under a protective atmosphere. The alloy melt was then transferred to a twin screw MCAST unit and subjected to intensive shearing at ~640-650 ºC for 60 s. The screw rotation speed was ~600 rpm, giving a shear rate of ~1633 $s^{-1}$. The conditioned melt was immediately fed, under a protective atmosphere, into either a horizontal TRC machine (HTRC) with 318 mm diameter rolls or a vertical TRC machine (VTRC) with smaller 100 mm diameter rolls. The strip thickness was controlled in the range of 1.5–6 mm and the casting speed was between 1 and 8 m/min. Downstream processing was carried out on some selected as-cast strips by homogenization at 400 ºC, hot rolling (73%) at 400 ºC and followed by annealing at 345 ºC for 2 h. As a result of geometric considerations the smaller rolls in the VTRC process ensure that a narrower roll gap can be used. Thus thinner strip can be cast at higher speed.

Specimens for microstructure and texture characterization were cut from the middle of each strip along the casting or rolling direction and all examinations were carried out on the transverse section. Specimens for optical microscopy were prepared using standard metallographic procedures followed by etching in a solution of 5% $HNO_3$ in ethanol. Optical microscopy was performed on a Carl Zeiss Axioskop MAT 2 microscope equipped with Axiovision image analysis software. Specimens for electron backscatter diffraction (EBSD) were prepared by electropolishing in a solution of 15% nitric acid in ethanol at -30 °C was carried out for 30 s at 12 V. EBSD mapping was carried out on a Zeiss Supera35 FEGSEM equipped with an Oxford Instruments EBSD system. Tensile test samples with a gauge length of 25 mm were machined in accordance with the ASTM-E21 standard, and tested at room temperature and at a strain rate of $6.7 \times 10^{-4}$ $s^{-1}$ using a Lloyd EZ50 test frame.

## Results and Discussion

The Effect of Melt Conditioning

Figure 1 shows typical optical micrographs of the microstructures of as-cast AZ31strip obtained by HTRC, VTRC and MC-VTRC. The microstructures formed without prior melt conditioning treatment (Figure 1a-d) consist of non-uniform, coarse, dendritic columnar grains with large, equiaxed grains in the centre and a

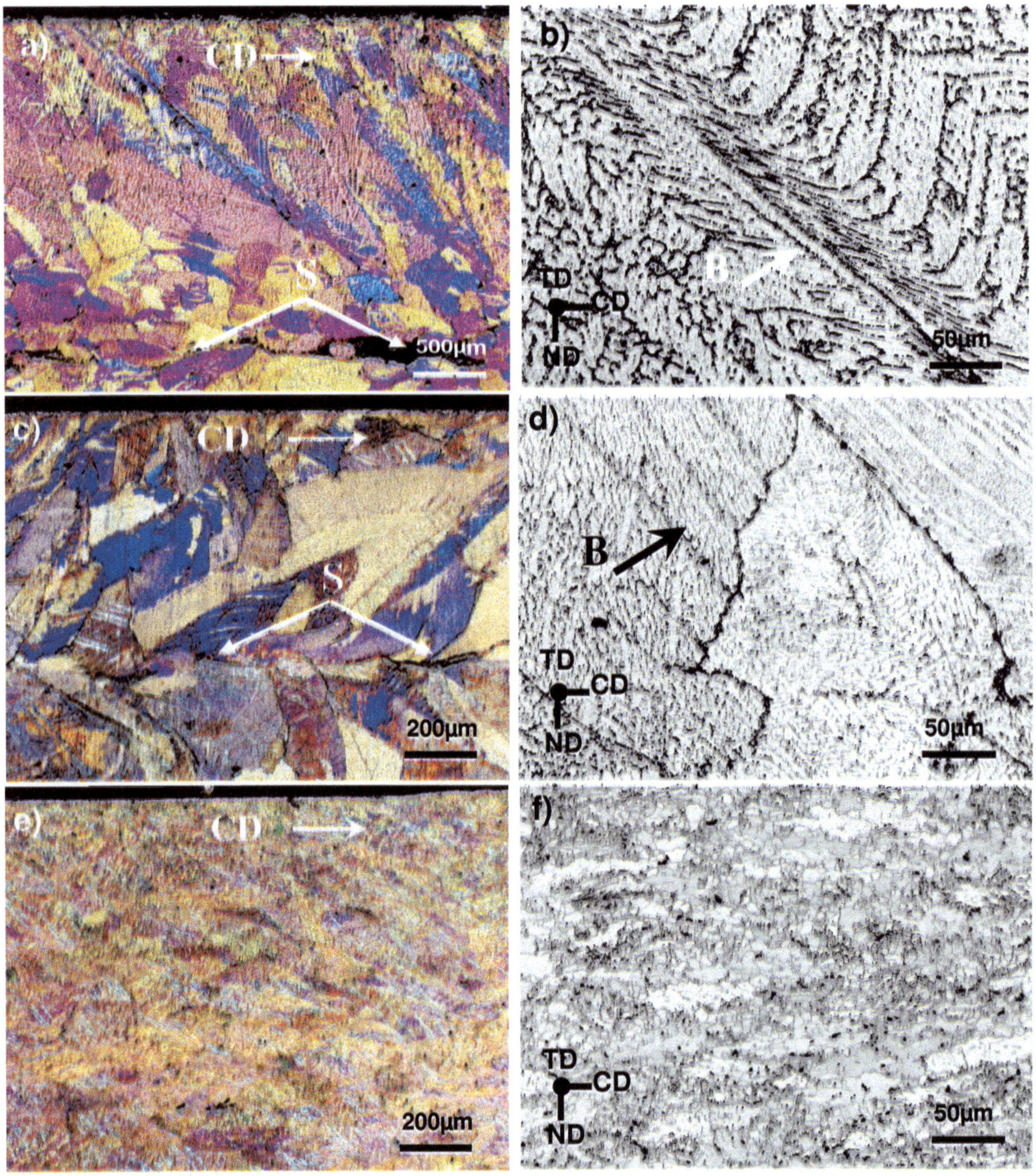


Figure 1. Optical micrographs showing long transverse section as-cast microstructures of AZ31 alloy strip produced by HTRC (a&b), VTRC (c&d) and MC-VTRC (e&f): pouring temperature of 640°C and casting speeds of (a&b) 1.35 m/min (thickness = 6 mm) and (c-f) 5 m/min (thickness = 1.8 mm). Severe chemical segregation (labelled 'S') along the centre-line of the two TRC strips (a-d) is clearly visible.

thin, fine grained chill layer at the surface. The HTRC (Figure 1a&b) and VTRC (Figure 1c&d) strips had grain sizes of ~616 μm and ~380 μm, respectively. In contrast, the MC-VTRC strip had fine ~50 μm sized grains necklaced by very fine recrystallized grains of ~7-9 μm in size (Figure 1e&f). Extensive microstructural examination revealed that the MC-VTRC strips were free from severe centre-line segregation, while macrosegregation (labelled 'S' in Figure 1) was found along most of the centre-line of the HTRC and VTRC strips due to directional crystal growth. These findings are in agreement with previous results published by the authors and other researchers [2, 3]. It has been shown previously that the intensive melt shearing offered by MCAST develops uniform temperature, uniform chemical composition and disperses evenly magnesium oxide films and oxide skins into individual MgO particles of 100-200 nm in diameter that act as potent nucleation sites [7, 8]. Therefore, heterogeneous nucleation during the solidification is enhanced due to the increased number of potential nucleating particles and increased nuclei survival rate, giving rise to the refined and uniform microstructure of the AZ31 strip produced by MC-VTRC. This equiaxed solidification process prohibited the accumulation of solute concentrated liquid in the central region, successfully eliminating the formation of centre-line segregation.

Dynamic recrystallization in Mg alloys accelerates increasingly under greater amounts of plastic deformation and larger numbers of existing grain boundaries (i.e. finer grains). Recent studies show that the increased amount of plastic deformation in the refined microstructure of MC-TRC strip generates higher stored energy in the material compared with TRC strip [10]. This was mainly responsible for the enhanced kinetics of dynamic recrystallization in the MC-VTRC as-cast AZ31 strip (Figure 1e&f). The plastic deformation in the HTRC and VTRC strips mainly occurred at the surface regions. In the coarse columnar dendrites, plastic deformation affects the structure by macroscopic shear banding (labelled B in Figure 1).

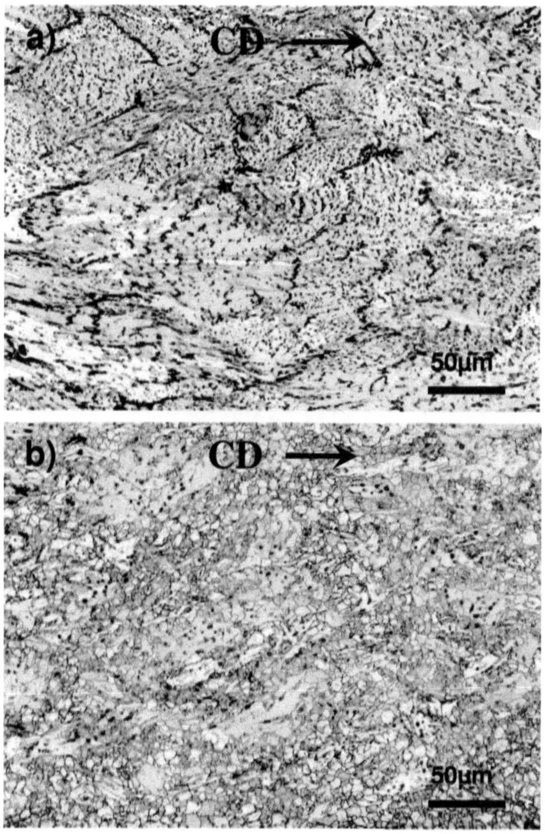


Figure 2. Optical micrographs of as-cast MC-VTRC AZ31 strips obtained at a casting speed of 5 m/min and pouring temperatures of (a) 650 °C and (b) 638 °C.

<u>The Effect of Casting Parameters</u>

So far, the results show that the application of melt conditioning by intensive melt shearing was necessary for obtaining twin roll cast AZ31 strip with a uniform microstructure and chemistry. However, the following results show that an optimised combination of casting parameters is also important for achieving the desirable microstructure.

<u>Melt Superheat</u> Figure 2 provides optical micrographs of MC-VTRC AZ31 strips cast with melt superheats of 18°C and 6°C (melt temperatures of 650 °C and 638 °C) respectively. Figure 2 shows that the 12°C difference in superheat has a significant effect on the microstructural characteristics of MC-VTRC strip. For the strip cast with 18°C superheat an equiaxed dendritic structure dominated, while for the strip cast with 6°C superheat a significant number of new grains formed due to dynamic recrystallization. At lower superheat less cooling is required during casting pushing the liquid/mushy sump towards the entry side of the rolls and so giving rise to a longer region of fully solid material within the rolls and thus a longer zone for plastic deformation. As a result, a greater number of dynamically recrystallized grains are generated for lower superheat. In addition, it is important to point out that intensive melt shearing enable TRC with low superheat. Without melt conditioning, the naturally occurring oxide films causes uneven feeding and possible blockage of the tip of the headbox, while melt conditioning can mitigate such problems by dispersing the oxide films into fine scale particles.

<u>Casting speed</u> Figure 3 illustrates the effect of different casting speeds (2.4-8m/min) on the microstructure of MC-VTRC strips. At 8m/min the microstructure is characterized by equiaxed dendritic grains (Figure 3a). At this very high casting speed solidification was uneven throughout the strip thickness, and some patches of liquid regions remained during the plastic deformation in the nip forming solute rich regions. Such patches manifest themselves in Figure 3a as intermetallic rich regions necklaced by particle stimulated recrystallized grains. At the reduced speeds of 4m/min and 2.4m/min, more extensive deformation features were apparent, such as shear bands, twins (labelled T) and a greater number of new recrystallized grains (labelled C), as shown in Figure 3b&c. With decreasing casting speed, the melt will have more time to release heat through the roll surfaces for a given distance from the tundish and therefore the liquid/mushy sump will be further towards the entry side of the rolls. As a result, the plastic deformation zone is lengthened and more plastic deformation is introduced to the solidifying strip, for a fixed gauge thickness and setback length.

Figure 4 shows the microstructures of MC-VTRC AZ31 strips produced at casting speeds of 5m/min and 8m/min and subsequently homogenized at 400 °C for 90 min. After homogenization, the strip cast at the lower speed developed a uniform and fine grain structure, with an average grain size of approximately 8 μm, whereas the strip cast at the higher speed developed a non-uniform structure of large and irregular

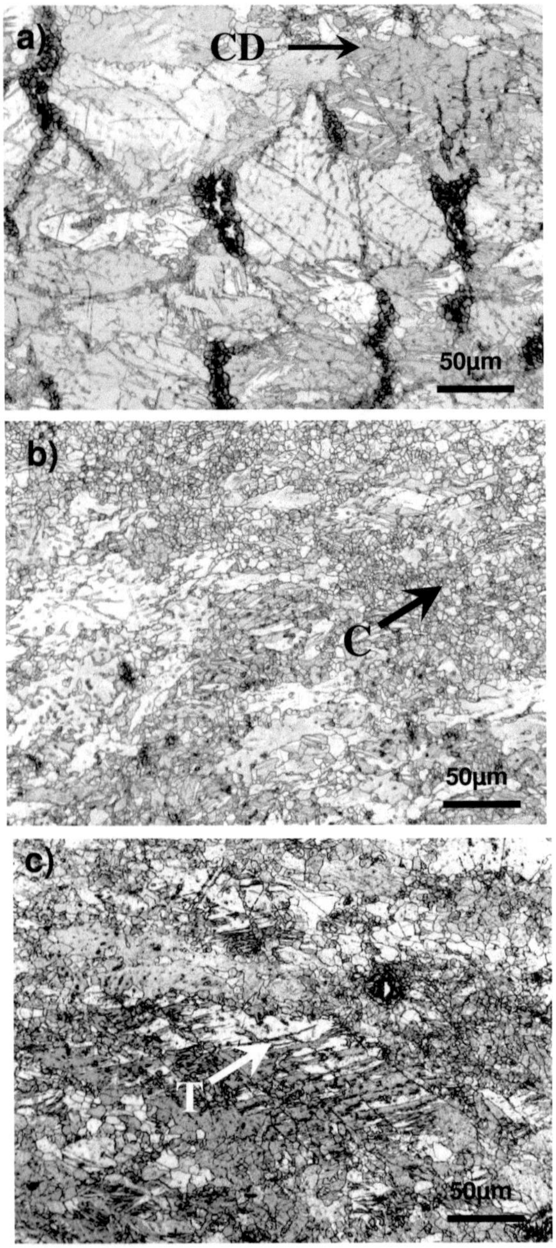


Figure 3. Optical micrographs of as-cast MC-VTRC AZ31 strips, obtained at a pouring temperature of 640 °C and casting speeds of (a) 8 m/min, (b) 4 m/min and (c) 2.4 m/min.

recrystallized and unrecrystallized grains. These results show that the enhancement of recrystallization during subsequent homogenization depends on the amount of energy stored in the material as a result of deformation during the casting stage.

<u>Texture and Mechanical Properties</u>

Figure 5 shows the microstructures and EBSD pole figures of conventional HTRC AZ31 strip (Figure 5a&c) which has undergone a lengthy downstream process of homogenization,

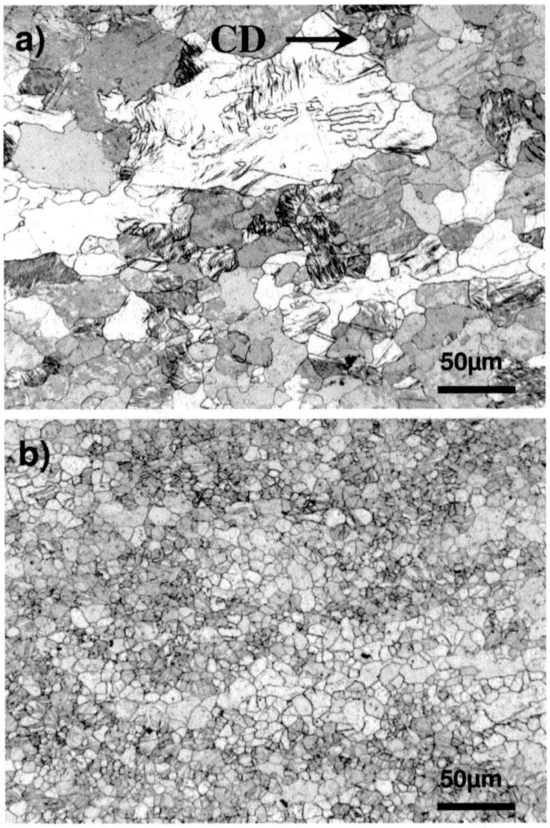


Figure 4. Optical micrographs homogenized (400 °C for 90 min) MC-VTRC AZ31 strips obtained at a pouring temperature of 640 °C and casting speeds of (a) 8 m/min and (b) 5 m/min.

rolling and annealing to provide a strip of 1.7mm thickness, and a homogenized MC-VTRC strip (Figure 5b&d) cast directly to a similar thickness of 1.8mm. The microstructure of homogenized MC-VTRC strip (Figure 5b) comprised uniformly very fine recrystallized grains while the homogenized, rolled and annealed HTRC strip (Figure 5a) had a non-uniform structure of fine recrystallized and coarse un-recrystallized grains. The corresponding EBSD pole figures showed that both strips had similar texture, dominated by the basal component (Figure 5c&d). However, the basal texture of the homogenized-only MC-VTRC strip was significantly weaker in intensity than that of the HTRC strip subjected to homogenization, hot rolling and annealing.

The tensile mechanical properties of the homogenized, hot rolled and annealed HTRC AZ31 strip and the homogenized only MC-VTRC strip are presented in Table I. Compared with the highly processed HTRC strip, the homogenized-only MC-VTRC strip had a similar UTS and lower 0.2% proof stress which makes the MC-VTRC more suitable for further forming processes such as stamping. Moreover, the MC-VTRC strip had a remarkably high tensile elongation in comparison with the HTRC strip which enhances its suitability for forming still further. The enhanced tensile properties of the MC-VTRC strip cast directly to the final gauge compared with the conventional hot rolled HTRC strip are considered to be due to: (i) the more refined and uniform microstructure, (ii) the elimination of severe centre-line

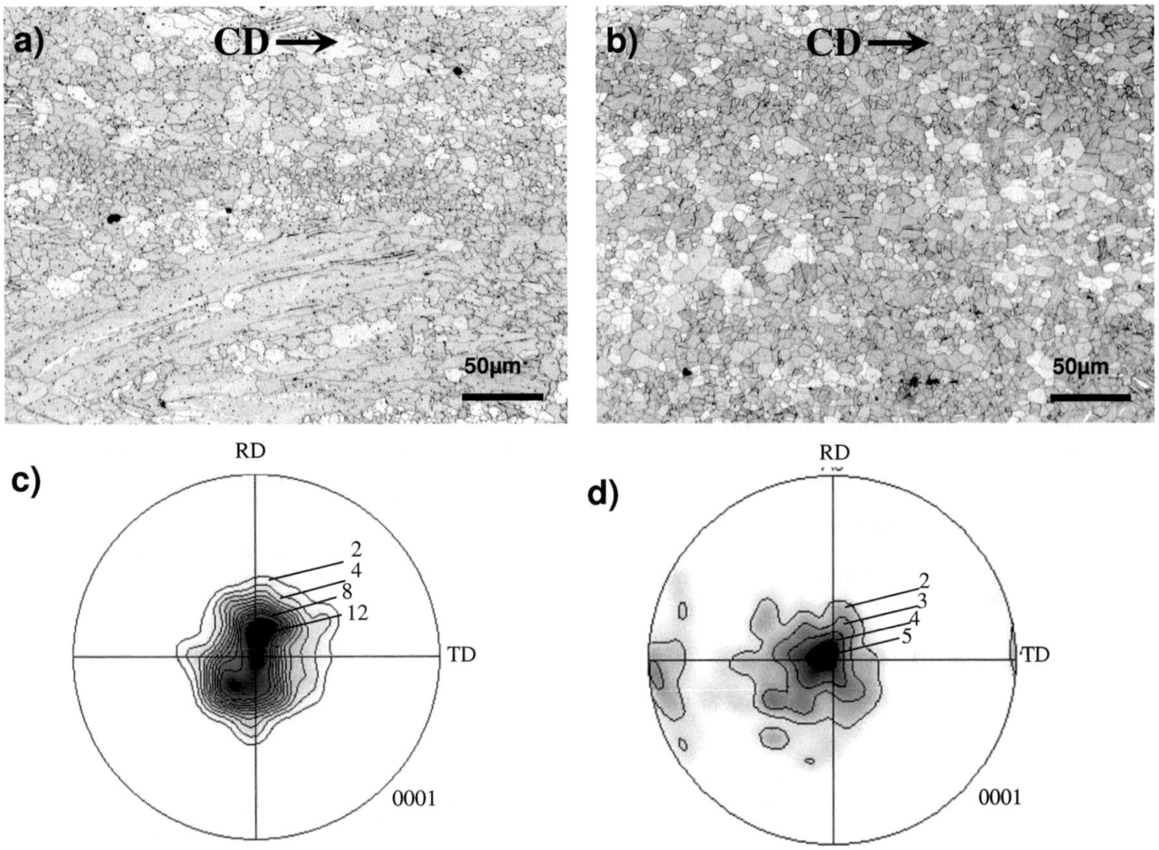


Figure 5. Optical micrographs of (a) conventional HTRC AZ31strip cast at 1.35 m/min (thickness = 6 mm) subsequently homogenized at 400 °C for 1 h, hot rolled at 400 °C to 73% (final thickness = 1.7 mm) and annealed at 345 °C for 2 h, and (b) MC-VTRC strip cast at 2.4 m/min (thickness = 1.8 mm) subsequently homogenized at 400 °C for 1 h; and (0001) EBSD pole figures (c) and (d) showing the textures corresponding to the microstructures in (a) and (b) respectively.

segregation (which remained in the TRC strips even after lengthy downstream processing) and (iii) the weaker basal texture. It has been shown that due to the brittleness and poor strength of the β-phase in AZ31, micro-cracks are likely to nucleate at the α-Mg/β interface and so particularly within the regions of centre-line segregation in conventional TRC strip [9]. Moreover, textures usually have a strong effect on the tensile deformation of wrought Mg alloys. In this work, the homogenized-only MC-VTRC thin strip has weaker basal texture which leads to lower yield stress and enhanced elongation compared with the highly processed TRC strip. The impact of MC-VTRC on the formability of Mg alloy sheet is the subject of ongoing investigation.

Table I. Room temperature tensile properties of the conventional HTRC strip after homogenization at 400 °C for 1 h, hot rolling at 400 °C to 73% and annealing at 345 °C for 2 h, and MC-VTRC strip after homogenization at 400 °C for 90 min.

| **Process** | **Conditions** | **0.2% Proof (MPa)** | **UTS (MPa)** | **Uniform Elongation (%)** | **Elongation-to-failure (%)** |
|---|---|---|---|---|---|
| HTRC | Hom+Hot-rolled+Annealed ($t^*$ = 6 » 1.7mm) | 181.6 | 243.1 | 6.1 | 9.6 |
| MC-VTRC | Hom ($t^*$ = 1.8mm) | 126.4 | 248.1 | 18.5 | 22.7 |

* Strip thickness (mm)

## Conclusions

(1) The application of a vertical twin roll caster with small diameter rolls facilitates the production of very thin Mg alloy strip (less than 2 mm in thickness) at higher casting speeds and lower roll forces.

(2) Melt conditioning by intensive shearing prior to vertical twin roll casting of AZ31 Mg alloy resulted in strips with a fine and uniform microstructure free of centre-line segregation.

(3) It has been demonstrated that an optimised combination of small roll MC-TRC casting parameters and subsequent homogenization heat treatment made it possible to produce near final gauge AZ31 Mg alloy strips with significantly greater ductility than strips of similar gauge produced by conventional TRC and downstream thermomechanical processing.

## Acknowledgement

The support of EPSRC (grant EP/H026177/1) is gratefully acknowledged.

## References

1. H. Bessemer, "Improvement in the manufacture of iron and steel," *US Patent*, 1865, no. 49,053.

2. M. Ferry, *Direct strip casting of metals and alloys* (Woodhead Publishing Limited, Cambridge, 2006), 90, 189-191.

3. I. Bayandorian et al., "Magnesium alloy strip produced by melt conditioned twin roll casting (MC-TRC) process" (paper presented at the TMS 2009 Annual Meeting and Exhibition, Magnesium Technology, San Francisco, CA, USA, 2009), 363-368.

4. S.S. Park et al., "The twin-roll casting of magnesium alloys," *JOM*, 61 (2009), 14-18.

5. H. Chen et al., "Microstructure and mechanical properties of Mg–4.5Al–1.0Zn alloy sheets produced by twin roll casting and sequential warm rolling," *Mater Sci Eng A*, 492 (2008), 317-326.

6. K. Hantzsche et al., "Effect of rare earth additions on microstructure and texture development of magnesium alloy sheets," *Scripta Mater*, 63 (2010), 725-730.

7. H. Men, B. Jiang, and Z. Fan, "Mechanisms of grain refinement by intensive shearing of AZ91 alloy melt," *Acta Mater*, 58 (19) (2010), 6526-6534.

8. Z. Fan et al., "Enhanced heterogeneous nucleation in AZ91D alloy by intensive melt shearing," *Acta Mater*, 57 (16) (2009), 4891-4901.

9. S. Kumar et al., "Influence of intensive melt shearing on the microstructure and mechanical properties of an Al-Mg alloy with high added impurity content," *Metall and Mater Trans A*, 42 (2011), 3141-3149.

10. I. Bayandorian, Y. Huang, and Z. Fan, "Microstructures and mechanical properties of an AZ91 magnesium alloy processed by melt-conditioned twin roll casting and subsequent thermomechanical processing," *J Mater Sci Technol*, in press.

**Magnesium Technology 2012**
*Edited by: Suveen N. Mathaudhu, Wim H. Sillekens, Neale R. Neelameggham, and Norbert Hort*
***TMS (The Minerals, Metals & Materials Society), 2012***

# MATHEMATICAL MODELING OF THE TWIN ROLL CASTING PROCESS FOR AZ31 MAGNESIUM ALLOY - EFFECT OF SET-BACK DISTANCE

Amir Hadadzadeh[1], Mary Wells[1], Elhachmi Essadiqi[2]

1- Mechanical and Mechatronics Engineering Department, University of Waterloo, Waterloo, ON, Canada.
2- CANMET Materials Technology Laboratory, Hamilton, Ontario, Canada

Keywords: Twin Roll Casting, Mathematical Modeling, AZ31 Magnesium Alloy.

## Abstract

A 2-D coupled thermal-fluid-stress model was developed and used to simulate the twin roll casting (TRC) of an AZ31 magnesium alloy using the commercial software package, ALSIM. The model was used to predict the fluid flow, temperature distribution and mechanical behavior of AZ31 magnesium alloy in the roll bite. An important parameter in controlling the TRC process is the set-back distance; the distance between the nozzle entry to the kissing point of the rolls. There are two approaches to increase the set-back: 1) increasing the entry thickness and 2) decreasing the final strip thickness. In this study the effect of set-back distance and casting speed on the thermo-mechanical behavior of the strip during TRC has been studied. The thermo-mechanical behavior of the strip has a significant effect on the final quality as defect formation depends on such behavior.

## Introduction

The interest in magnesium alloys in the automotive industry has been increased in the recent years because of the lightweight properties of these alloys; magnesium is 33% lighter than aluminum and 70% lighter than steel. Currently, most of the magnesium usage in cars is limited to die castings. Additional applications for magnesium alloys sheet is available if reasonable properties and manufacturing costs can be obtained. Conventionally, magnesium alloys used in sheet applications, have been cast using the Direct Chill (DC) casting technology followed by several passes of hot rolling and annealing, where at each pass the reduction is limited to 5-20% with multiple annealing heat treatments in between passes [1]. An alternative process which produces strip from molten metal is twin roll casting (TRC) which incorporates casting and hot rolling in one process and subsequently reduces the costs and energy consumption considerably [2]. This process, known as a near-net-shape manufacturing process, has the capability of producing strips with the thickness of 2-10mm [3, 4]. High cooling rate achieved by this process, in the range of $10^2$-$10^3$°C/s [4], has a significant effect from a metallurgical point of view; increasing the amount of solute in solid solution, enhanced precipitation nucleation within the matrix, a finer distribution of precipitates and a finer grain structure [5].

The need for a quantitative understanding of the process and conditions (e.g. casting speed, set-back distance, strip thickness and casting temperature) that lead to high quality sheet means that it is imperative to develop a knowledge-based process model of the TRC process for magnesium alloys. During the past few years, various models for the TRC process have been developed by considering transport phenomena coupled with solidification effects (Thermal-Fluid models). There were limited attempt to consider mechanical deformation in the model (Thermal-Fluid-Stress models) [6]. In the current study a two dimensional thermal-fluid-stress model has been developed to study the effect of casting speed and set-back distance on the thermal history and stress development during TRC of AZ31 magnesium alloy. A constant heat transfer coefficient (HTC) at the roll/strip interface was chosen based on the validation done by comparing the model results to the measurements done at the Natural Resources Canada government materials laboratory, CANMET located in Hamilton, Ontario.

## Model Development

As shown schematically in Figure 1, the twin roll caster at CANMET consists of two water cooled steel rolls with a diameter of 355mm. The speed of the rolls and exit thickness can vary between 1-6m/min and 2-8mm, respectively. The strip width achievable by this caster is 150-250mm. The following assumptions were considered in developing the model [7]:

I. The process is dominated by transport phenomena in two dimensions. In the third dimension (across the width) there is no significant heat transfer or fluid flow as the simulation is done at the mid-width location,
II. The geometry is symmetrical, hence, only the top half is considered in the model,
III. The fluid flow is laminar due to shape of nozzle, and;
IV. Roll deformation is neglected.

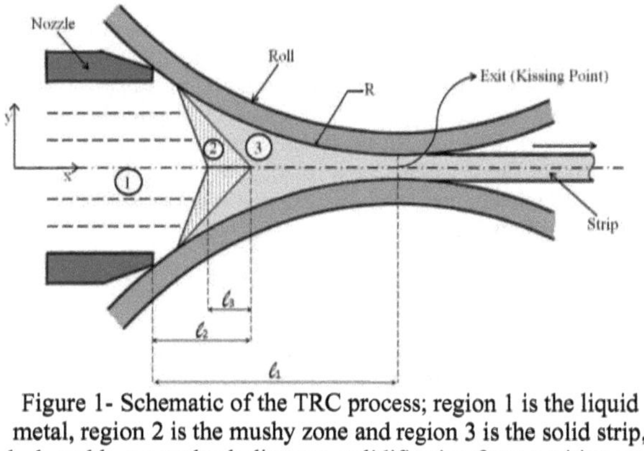


Figure 1- Schematic of the TRC process; region 1 is the liquid metal, region 2 is the mushy zone and region 3 is the solid strip, $l_1$, $l_2$ and $l_3$ are set-back distance, solidification front position and mushy zone depth, respectively [3].

The important physical phenomena which happen during the process and are included in the model are:
I. Heat transfer and fluid flow in the melt sump (liquid metal),
II. Heat transfer, fluid flow and latent heat of fusion release in the mushy zone and deformation in the material once the coherency point is reached,
III. Heat transfer and plastic deformation in the solid phase, and
IV. Heat transfer from the magnesium sheet to the roll surface.

The basis for numerical flow simulation is the conservation laws of mass, momentum and energy [7]. To account for solidification, two important effects of this phenomenon on the fluid flow and heat transfer need to be included; damping fluid flow in the mushy zone and latent heat of fusion. These effects are considered by adding the appropriate source terms to the momentum conservation equation and energy conservation equation, respectively [8].

To model the mechanical deformation occurring during the process, the magnesium alloy is defined as an isotropic elastic-viscoplastic material. The total strain imposed on the material is subdivided into thermal, elastic and viscoplastic components [9] and the constitutive equation shown in Equation 1 models the plastic behavior of material below the coherency temperature [9].

$$\sigma = K(T).(\dot{\varepsilon}_P + \dot{\varepsilon}_{Po})^{m(T)}(\varepsilon_P + \varepsilon_{Po})^{n(T)} \quad (1)$$

where $\sigma$ is the stress in MPa, $\dot{\varepsilon}_P$ the strain rate in $s^{-1}$, $\varepsilon_P$ the strain, $K$ strength coefficient, $n$ the strain hardening exponent and $m$ the strain rate sensitivity exponent. $\dot{\varepsilon}_{Po}$ and $\varepsilon_{Po}$ are small numerical constants needed to ensure that at a plastic strain of zero, the yield stress of the material is correct. For AZ31 a coherency temperature of 595°C (0.9 $f_s$) was assumed.

The commercial FE package, ALSIM, was used to develop the model and solve the above-mentioned equations to simulate the TRC process of AZ31 magnesium alloy. The thermo-physical and constitutive material properties were taken from the literature [7, 10]. Appropriate boundary conditions were applied and the HTC was chosen equal to 11kW/m$^2$°C based on previous work [11]. Table 1 shows the range of casting conditions used in the model. The aim of the present study was first to develop and validate the model for AZ31 TRC and then study the effect of set-back distance and casting speed on the thermal history and stress development in the strip. As illustrated in Figure 1, set-back distance (SB), $l_1$, is the distance between the nozzle entry and kissing point (least gap between rolls). The SB, is influenced by the roll diameter, entry height and exit thickness (reduction), shown by Equation 2.

$$SB = l_1 = \sqrt{R\Delta h - 0.25\Delta h^2} \quad (2)$$

where $R$ is the roll radius in mm and $\Delta h$ is the reduction (the difference between entry and exit thickness) in mm.

There are two approaches to increase the SB; increasing the entry height for a constant final thickness and decreasing the final thickness for a constant entry height. The former approach was used in the current study.

Table 1- Casting conditions used in modeling

| Casting Temperature (°C) | Entry Height (mm) | Exit Thickness (mm) | Set-Back Distance (mm) | Casting Speed (m/min) |
|---|---|---|---|---|
| 677 | 12 | 6 | 32.5 | 1.0, 1.7, 2.0 & 2.5 |
| | 14 | | 37.5 | |
| | 16 | | 41.8 | |

## Results and Discussions

### Thermal History of the Strip

Figure 2 shows the model-predicted thermal history of the strip surface in the roll bite region for a set-back distance of 41.8mm. As the molten material enters the roll bite and is in contact with the roll surface, the temperature drops to the liquidus temperature and solidification starts. Continuing along the arc of contact, more heat extracted from the material as it fully solidifies and then heat is extracted in the solid strip.

As expected, as the casting speed increase the overall heat extraction is reduced as these is less contact time between the strip and the roll. Since the mechanical properties of the material are temperature dependent, it could affect the mechanical behavior of the strip which shown later.

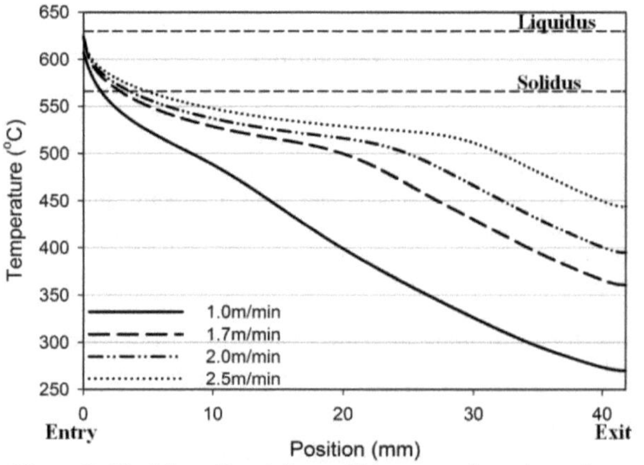

Figure 2- Model-predicted thermal history on the strip surface during TRC for a SB=41.8mm.

Figure 3 illustrates the exit strip temperature for different casting speeds and set-back distances. As expected, the surface temperature increases for higher casting speeds. Moreover, increasing the set-back distance leads to a lower exit strip temperature. This is due to longer arc of contact (strip/roll interface) and more heat transfer from the strip to the roll. The sensitivity of the set-back distance is more pronounced at higher casting speeds; a larger temperature difference is observed by changing the set-back at higher speeds, a maximum of 37°C in the exit temperature for set-back values of 32.5mm and 41.8mm at a casing speed of 1m/min compared to 61°C at a casting speed of 2.5m/min.

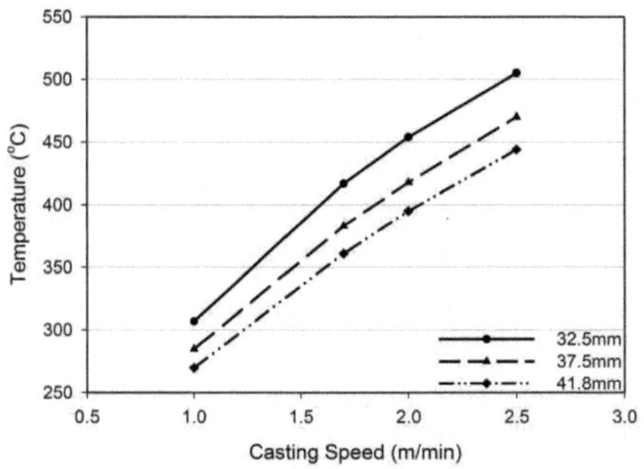


Figure 3- Model predicted effect of casting speed and set-back distance on the strip exit temperature.

Mechanical Behavior of the Strip

By using the model developed in the current study, the mechanical behavior during TRC was examined. Figures 4 shows the evolution in normal stress on the strip surface during TRC (roll pressure) in the roll bite for a set-back distance of 41.8mm and casting speed of 1.7m/min in conjunction with the corresponding contour lines in the strip; showing liquidus and solidus temperatures for this case. It can be seen that the roll pressure remains close to zero until almost all the strip has solidified. The zero level of the pressure is due to the inability of the molten material beneath the solidified shell on the roll surface to develop and sustain the stress. The peak in the normal stress is attributed to the friction at the roll/strip interface and the "relative motion" between the strip and roll as well as the amount of deformation the material experiences; rising prior to the neutral point and dropping afterward.

As shown in Figure 5, increasing the casting speed leads to decreasing the pressure peak. Lower casting speed means there is much more time for heat extraction and solidification; so, there would be more solid material in the roll bite and consequently more plastic deformation occurs and the stress develops to a higher values. As shown in the previous part, for lower casting speed the temperature of the strip will be lower and since the mechanical properties of the material is dependent on temperature, the mechanical behavior of the strip will be affected. The material has higher strength at lower temperature; hence, more stress is needed to deform it which leads to higher pressure at lower casting speeds.

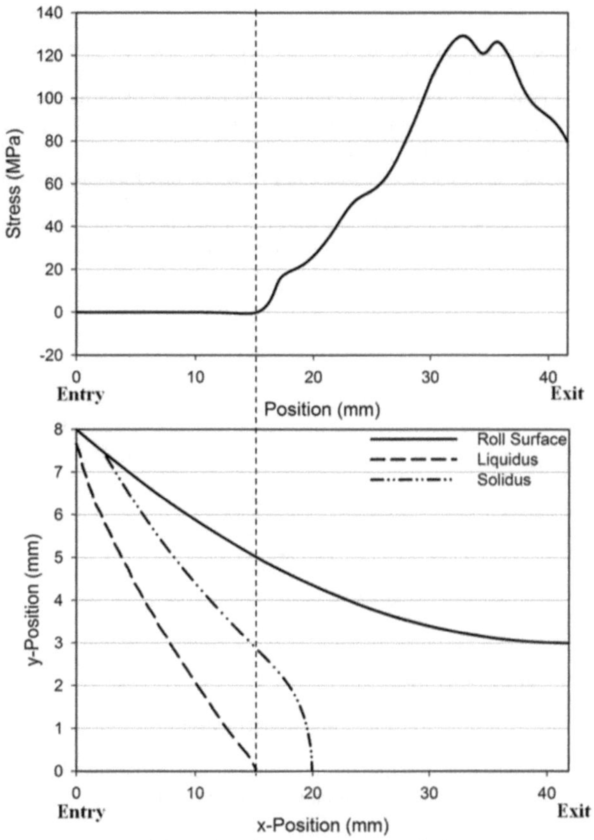


Figure 4- Model-predicted normal stress evolution at the strip surface and contour lines showing liquidus and solidus temperatures during TRC, for SB=41.8mm and casting speed of 1.7m/min.

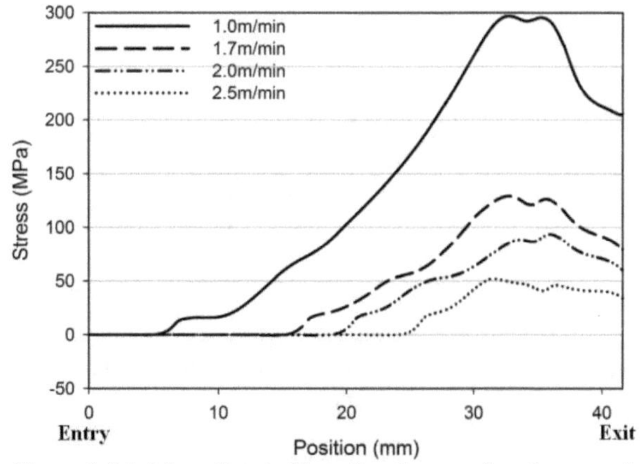


Figure 5- Model-predicted effect of casting speed on the normal stress evolution at the strip surface during TRC, for SB=41.8mm.

Figure 6 shows the effect of set-back on the roll pressure for a casting speed of 1.0m/min on a normalized scale. The procedure of normalizing the x-position (entry to exit) consists of dividing x-position to the value of the set-back distance for each case. Thus, the set-back is scaled to 0 for the entry position to 1 for the exit position, and ranges in this interval. Since for a larger set-back the solidification occurs more quickly in the roll bite and the temperature drops to lower values, hence the peak pressure/stress is greater. The effect of these parameters is similar for roll separating force, as shown in Figure 7. At low casting speeds the process is more sensitive to the set-back distance in terms of roll separating force. At faster casting speeds, less difference in separating force is observed for different set-back distances (794kN difference for set-backs of 32.5 and 41.8mm at 1.0m/min and 125kN at 2.5m/min).

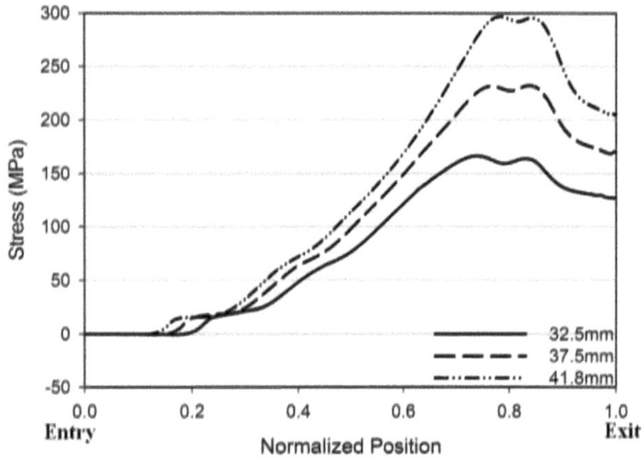


Figure 6- Roll pressure for different set-back distances, the set-back is normalized on the x-axis.

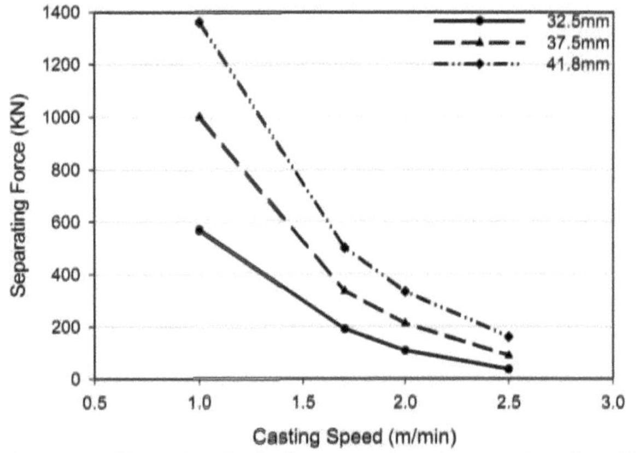


Figure 7- Effect of set-back distance and casting speed on the roll separating force.

**Conclusions**

1- A 2D thermal-fluid-stress model was successfully developed for twin roll casting of AZ31 magnesium alloy and thermal history and mechanical behavior of the strip was studied.

2- The temperature on the strip surface is affected by both set-back distance and casting speed; lower temperature achieved at lower casting speeds and larger set-backs. Exit strip temperature is more sensitive to set-back distance at higher casting speeds.

3- For larger set-backs and at lower casting speeds, the peak pressure at the roll/strip interface is greater.

4- Roll separating force decreases by increasing the casting speed and also casting by smaller set-back distances. The effect of set-back distance is more pronounced at lower speeds.

**References**


[1] D. Liang, C. B. Cowley, *"The Twin-Roll Strip Casting of Magnesium"*, Journal of the Minerals, Metals and Materials Society, 56 (5) (2004), 26-28.

[2] E. E. M. Luiten, K. Blok, *"Stimulating R&D of Industrial Energy-Efficient Technology; The Effect of Government Intervention on the Development of Strip Casting Technology"*, Energy Policy, 31 (2003), 1339-1356.

[3] H. Zhao, P. Li, L. He, *"Coupled Analysis of Temperature and Flow during Twin-Roll Casting of Magnesium Alloy Strip"*, Journal of Materials Processing Technology, 211 (2011), 1197-1202.

[4] H. -S. Di, et al., *"New Processing Technology of Twin Roll Strip Casting of AZ31B Magnesium Strip"*, Materials Science Forum, 488-489 (2005), 615-618.

[5] H. Chen, et al., *"Microstructure and mechanical properties of Mg-4.5Al-1.0Zn alloy sheets produced by twin roll casting and sequential warm rolling"*, Materials Science and Engineering A, 492 (2008), 317-326.

[6] P. Zhang et al., *"Numerical Simulation on the Stress Field of Austenite Stainless Steel duringTtwin-Roll Strip Casting Process"*, Computational Materials Science, 2011, doi:10.1016/j.commatsci.2011.03.039.

[7] A. Hadadzadeh, M. A. Wells, E. Essadiqi, *"Mathematical Modelling of the Twin Roll Casting for Magnesium Alloys - Effect of Heat Transfer Coefficient between the Roll and the Strip"*, Proceedings of 8th International Conference on Magnesium Alloys and their Applications, Weimar, Germany, Oct 26-29, 2009, pp. 138-144.

[8] J. Zeng, et al., *"Numerical Simulation of the Twin-Roll Casting Process of Magnesium Alloy Strip"*, Journal of Materials Processing Technology, 209 (2009), 2321-2328.

[9] M. M'Hamdi, A. Mo, H.G. Fjær, *"TearSim: A Two-Phase Model Addressing Hot Tearing Formation During Aluminium Direct Chill Casting"*, Metallurgical and Materials Transactions A, 37A (2006), 3069-3083.

[10] B. J. Howes, M. A. Wells, R. Bathla, D. M. Maijer, *"Constitutive Behaviour of As-Cast Magnesium Alloy AZ31 Uder Deformation Conditions Relevent to DC Casting"*, Proceedings of the 2nd International Light Metals Technology Conference, [ed.] H. Kaufmann. St. Wolfgang, Austria, 2005.

[11] A. Hadadzadeh, M. A. Wells, E. Essadiqi, *"Mathematical Modeling of the Twin Roll Casting for AZ31 Magnesium Alloy"*, (Paper presented at the 22nd Canadian Materials Science Conference, Waterloo, Ontario, 10 June 2010).

Magnesium Technology 2012
*Edited by: Suveen N. Mathaudhu, Wim H. Sillekens, Neale R. Neelameggham, and Norbert Hort*
*TMS (The Minerals, Metals & Materials Society), 2012*

# INTERMETALLIC PHASE FORMATION AND GROWTH IN THE MG-Y SYSTEM

K. Bermudez, S. Brennan, Y.H. Sohn

University of Central Florida, Advanced Materials Processing and Analysis Center

Department of Mechanical, Materials and Aerospace Engineering, Orlando, FL 32816, USA

**Abstract**

Rare earths have been added to magnesium alloys in order to improve the creep resistance, corrosion resistance and strength. Solid-to-solid diffusion couples were assembled between Mg (99.9%) and Y (99.9%) to investigate the formation and growth of intermetallic phases and interdiffusion in the Mg-Y system. The diffusion anneals were performed at 450, 500 and 550°C for 360, 240 and 120 hours, respectively. The intermetallic layers that developed were the δ-$Mg_2Y$ and ε-$Mg_{24}Y_5$ phases, however the MgY phase did not form. A substantial penetration of Y in Mg was observed, however along with Kirkendall porosity that indicates faster diffusion of Mg than Y in Mg solid solution. The activation energies for parabolic growth in ε-$Mg_{24}Y_5$ and δ-$Mg_2Y$ were calculated to be 84 kJ/mol and 77 kJ/mol, respectively.

Keywords: Magnesium, Yttrium, in ε-$Mg_{24}Y_5$ and δ-$Mg_2Y$

## Introduction

The increasing use of lightweight magnesium (Mg) alloys in aerospace and transportation applications has led the desire for more fundamental materials research in Mg-based systems. Rare earth elements (RE) have been found to improve the mechanical properties of magnesium at both room and high temperatures [1-3]. Yttrium (Y) additions enhance creep resistance and high temperature performance. In this study, solid-to-solid diffusion couples were used to observe the phase layer formation and growth within the binary Mg-Y system. Parabolic growth constants were calculated from measured phase layer thicknesses. The activation energies for growth were determined for the intermediate phase layers that were observed.

## Analytical Framework and Experimental Procedure

The parabolic growth rate of a phase, $k_p$, can be described by:

$$k_p = \frac{Y^2}{t} \tag{1}$$

where Y is the phase layer thickness and t is the annealing time. Typically, this follows an Arrhenius temperature dependence given as:

$$k_p = k_o \exp\left[-\frac{Q_k}{RT}\right] \tag{2}$$

where $k_o$ is the pre-exponential factor and $Q_k$ is the activation energy for growth. R is the universal gas constant (8.314 kJ/mole-K) and T is the absolute temperature (K).

Rods of pure Mg (99.9%) and pure Y (99.9%) were supplied from Alfa Aesar™ and sectioned into disks, approximately 3 mm thick. The diameter of each Mg disk was 7.9 mm and that of Y was 12.7 mm. All disks were metallographically polished down to 1 μm using non-oxidizing lubricants, and contact with water was avoided during the entire preparation stage. The prepared disks were assembled in between two inert alumina spacer disks and clamped in a stainless steel jig as shown in Figure 1.

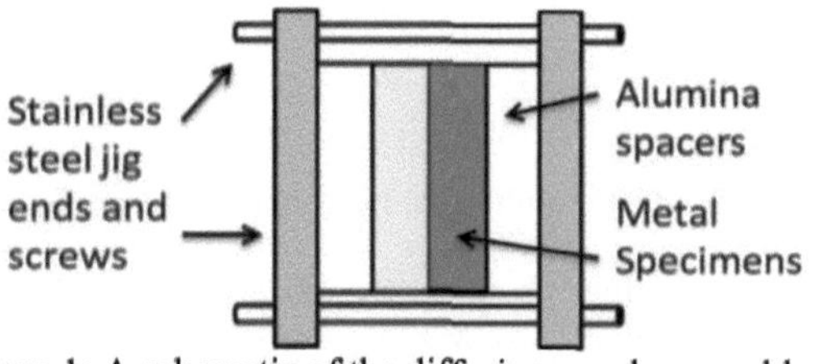


Figure 1: A schematic of the diffusion couple assembly in a stainless steel jig.

The diffusion couple jig assembly was then placed in a quartz capsule that was repeatedly evacuated to approximately $10^{-6}$ Torr and flushed with hydrogen and ultra-high purity argon at least three times, then backfilled with ultra-high purity Ar to a pressure that would be slightly above 1 atm at the respective annealing temperature. The sealed capsule was then placed in a preheated furnace that was monitored with an independent type-K thermocouple and maintained within ±2° of the desired annealing temperature. The annealing times and temperatures were 450, 500, and 550°C for 360, 240 and 120 hours, respectively. After the anneal, the capsule was quickly removed and quenched in water at room temperature. The full jig assembly was mounted in epoxy and then cross-sectioned in a low speed saw with a diamond wafering blade. The cross-sections were then metallographically prepared and examined in an optical microscope to check the quality of the diffusion bond. Field emission scanning electron microscope (Zeiss™ Ultra-55 FE-SEM) was used to further examine the diffusion couple, and identify the phases present using X-ray energy dispersive spectroscopy (XEDS). Concentration profiles are also acquired across the interdiffusion zone by using electron microprobe analysis (EPMA) with a point to point scan using a step size of 5 μm and accelerating voltage of 20 keV. Pure standards of Mg and Y, present at the terminal ends of each diffusion couple were used along with ZAF correction for the determination of concentration profiles.

## Results and Discussion

FE-SEM equipped with XEDS was initially employed to obtain backscatter electron micrographs of the interdiffusion zone, and identify each phase present in the diffusion couple according to the Mg-Y equilibrium binary phase diagram presented in Figure 2.

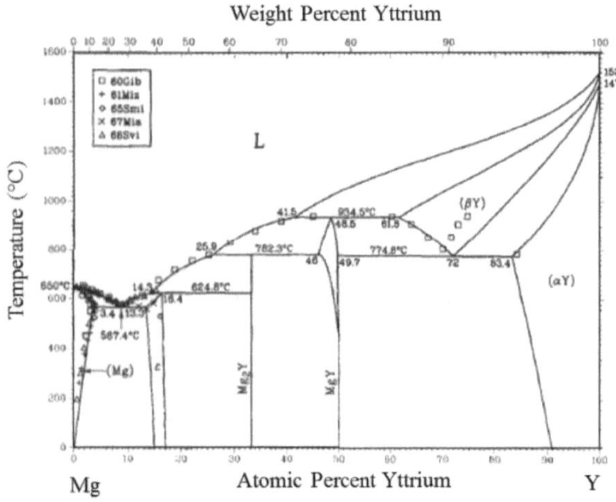


Figure 2: Mg-Y equilibrium binary phase diagram [4].

Two intermetallic phases were observed in all three diffusion couples: the ε-$Mg_{24}Y_5$ and δ-$Mg_2Y$ phases as shown in the backscatter electron micrographs in Figure 3. These same two phases were also identified in the diffusion couple study by Zhao *et* al. [5]. The MgY phase was not observed in any of the diffusion couples examined in this study. In each diffusion couple, there were some cracks across the diffusion zone, most likely attributed to molar volume differences and quenching. Rectangular shaped pores were evident in all three couples in the Mg solid solution. These are Kirkendall pores that indicate the large intrinsic difference in diffusion of Mg: Mg atoms out of the Mg solid solution vs. Y atoms out the δ-phase and into the Mg solid solution.

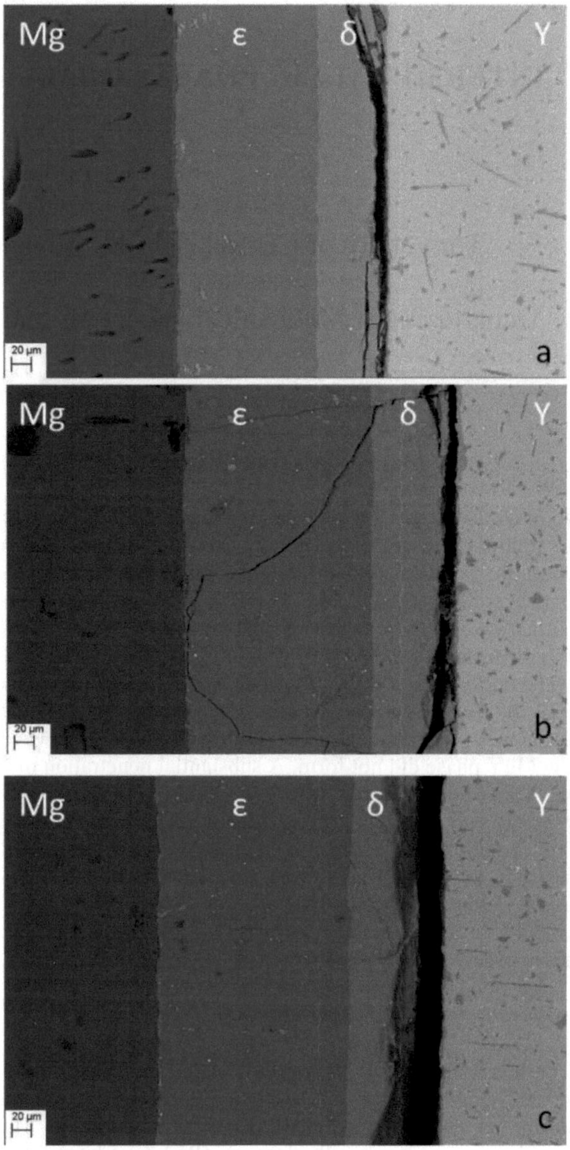


Figure 3: Backscatter electron micrographs from Mg-Y binary diffusion annealed at (a) 450°C for 360 hours, (b) 500°C for 240 hours and (c) 550°C for 120 hours.

EPMA (JEOL™ 733 SuperProbe) was employed to collect concentration profiles of Mg and Y across the cross section of each diffusion couple as presented in Figure 4. There is a significant solubility for Y in Mg in accordance to the equilibrium phase diagram. However, there is little to no solubility for Mg in Y, in contrast to the equilibrium phase diagram. Both intermetallic phases also exhibited some range of solubility, which was expected for the ε-phase, but not so for the $Mg_2Y$ phase. In the equilibrium binary phase diagram, $Mg_2Y$ is a line compound. In the study by Zhao *et* al. [5], solubility in both intermetallic phases was also observed, and an adjustment to the equilibrium phase diagram was proposed. A comparison of the solubility found by Zhao *et* al. [5] and the results of this study by EPMA are reported in Tables I and II, respectively, for ε-$Mg_{24}Y_5$ and δ-$Mg_2Y$. Similar ranges were found for the two intermetallic phases developed in the Mg-Y binary system. The results of this study support the notion of adjusting the equilibrium phase diagram. Relevant coefficient of interdiffusion for Mg, ε-$Mg_{24}Y_5$ and δ-$Mg_2Y$ will be reported elsewhere with due respect for uncertainty in concentration gradients and variation in molar volume.

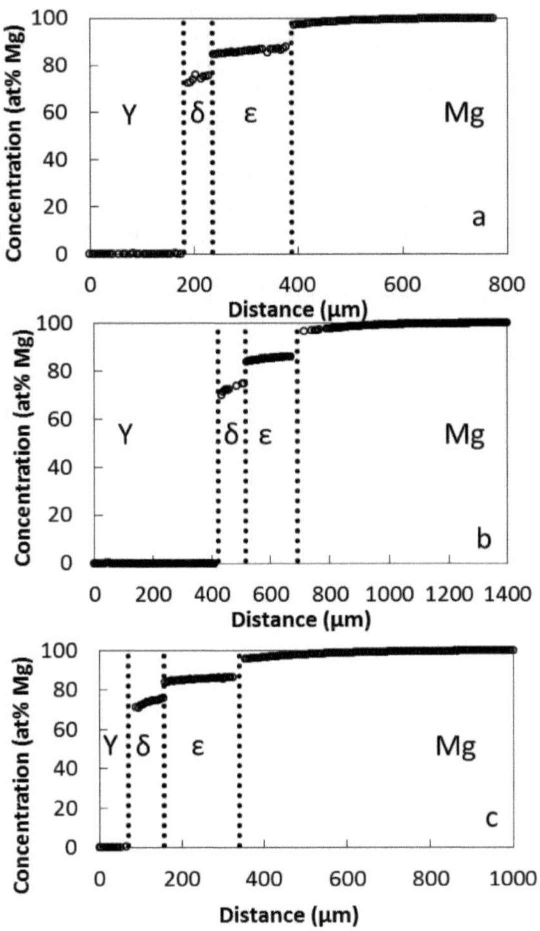


Figure 4: Concentration profiles of Mg and Y determined by EPMA from solid-to-solid diffusion couples annealed at (a) 450°C for 360 hours, (b) 500°C for 240 hours, and (c) 550°C for 120 hours.

Table I. Comparison of the Mg solubility limits in ε-$Mg_{24}Y_5$.

| Temperature (°C) | Equilibrium Compositions of ε-$Mg_{24}Y_5$ phase boundaries (at.% Mg) | | | |
|---|---|---|---|---|
| | This Study | | Zhao *et* al. [5] | |
| | Mg/ε | ε/δ | Mg/ε | ε/δ |
| 400 | N/A | | 88.0 | 84.2 |
| 450 | 88.0 | 84.7 | 88.1 | 84.4 |
| 500 | 86.1 | 84.0 | 88.1 | 84.9 |
| 550 | 86.6 | 84.0 | N/A | |

Table II. Comparison of the Mg solubility limits of δ-$Mg_2Y$.

| Temperature (°C) | Equilibrium Compositions of δ-$Mg_2Y$phase boundaries (at.% Mg) | | | |
|---|---|---|---|---|
| | Current Study | | Zhao et al. [5] | |
| | ε/δ | δ/Y | ε/δ | δ/Y |
| 400 | N/A | | 75.8 | 70.5 |
| 450 | 75.6 | 72.5 | 75.5 | 69.0 |
| 500 | 75.0 | 70.0 | 76.4 | 69.4 |
| 550 | 76.1 | 71.1 | N/A | |

Phase layer thickness measurements were made from backscatter electron micrographs for each diffusion couple, and parabolic growth constants were calculated using Eq. (1). The temperature dependence of the growth constants was determined using Eq. (2), and the activation energies of growth for both the ε-$Mg_{24}Y_5$ and δ-$Mg_2Y$ phases were calculated from the Arrhenius plot presented in Figure 5. The measured layer thicknesses, pre-exponential factors and activation energies for growth are reported in Table III.

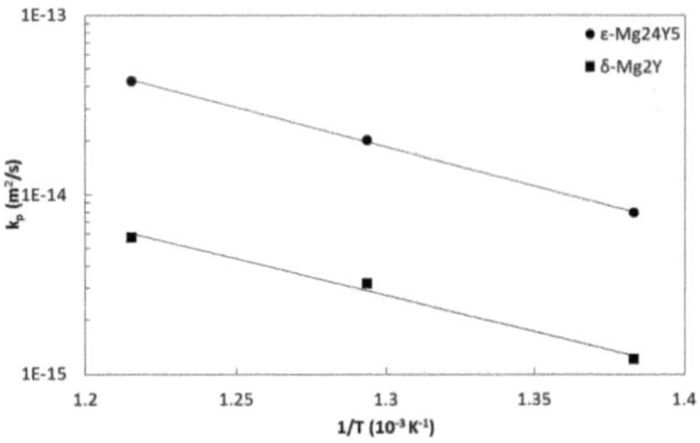


Figure 5: Temperature dependent growth constants for δ-$Mg_2Y$ and ε-$Mg_{24}Y_5$.

The two intermetallic phases have activation energies that are close in magnitude, however, the thickness difference of each phase is apparently influenced by the pre-exponential factor, and suggests that the detailed mechanism of diffusion in δ-$Mg_2Y$ and ε-$Mg_{24}Y_5$ phases may be very different.

## Summary

Solid-to-solid binary diffusion couples between Mg and Y were assembled and annealed for 450, 500 and 550°C for 360, 240, and 120 hours respectively. SEM equipped with XEDS was used to identify the intermetallic phases that formed and grew, namely, the ε-$Mg_{24}Y_5$ and δ-$Mg_2Y$ phases. However, the MgY phase was not observed in any of the diffusion couples. The pre-exponential factor and activation energies for the two intermetallic phases were calculated. The activation energy for ε-$Mg_{24}Y_5$ and δ-$Mg_2Y$ were 84 and 77 kJ/mol, respectively. Despite the similar activation energies, the ε-$Mg_{24}Y_5$ phase was significantly larger than the δ-$Mg_2Y$ phase, owing to a larger pre-exponential factor.

Table III. Activation energy and the pre-exponential factor of the parabolic growth constants for ε-$Mg_{24}Y_5$ and δ-$Mg_2Y$ phases determined based on thickness measurements.

| Parameters | Phase | 550°C 120 h | 500°C 240 h | 450°C 360 h |
|---|---|---|---|---|
| Y (μm) | δ-$Mg_2Y$ | 70.5 (1.7) | 74.4 (2.3) | 56.1 (4.0) |
| | ε-$Mg_{24}Y_5$ | 192.6 (1.3) | 186.7 (2.7) | 143.4 (4.5) |
| $k_o$ ($m^2$/s) | δ-$Mg_2Y$ | $4.89 \times 10^{-10}$ | | |
| | ε-$Mg_{24}Y_5$ | $8.84 \times 10^{-9}$ | | |
| $Q_k$ (kJ/mol) | δ-$Mg_2Y$ | 77.3 | | |
| | ε-$Mg_{24}Y_5$ | 83.6 | | |

Note: values in parenthesis are standard deviation.

**Acknowledgments**

The authors would like to thank the Material Characterization Facility at the University of Central Florida for the use of its equipment and the assistants of its technicians.

**References**

[1] M. Suzuki, H. Sato, K. Maruyama, H. Oikawa, *Mater. Sci. Eng. A* 252 (1998) 248-255.

[2] A. Tharumarajah, P. Koltun, *Journal of Cleaner Production* 15 (2007) 1007-1013.

[3] L. Gao, R. S. Chen, E. H. Han, *J. Alloy Compd.* 472 (2009) 234-240.

[4] H. Okamoto, "Mg-Y (Magnesium-Yttrium),"*Journal of Phase Equilibria*, 13, 1992, 105.

[5] H. Zhao, G. Qin, Y. Ren, W. Pei, D. Chen, Y. Guo, *J. Alloy Compd.* 509 (2011) 627-631

Magnesium Technology 2012
*Edited by: Suveen N. Mathaudhu, Wim H. Sillekens, Neale R. Neelameggham, and Norbert Hort*
*TMS (The Minerals, Metals & Materials Society), 2012*

# MICROSTRUCTURE AND MECHANICAL PROPERTIES OF HIGH PRESSURE DIE CAST AM50 MAGNESIUM ALLOY CONTAINING Ce

Faruk Mert[1,2], Ahmet Özdemir[1], Karl Ulrich Kainer[2] and Norbert Hort[2]
[1] Gazi University, Faculty of Technology, Department of Manufacturing Engineering, TR-06500 Ankara - TURKEY
[2] Helmholtz-Zentrum Geesthacht, Magnesium Innovation Centre, Max-Planck-Straße 1, D-21502 Geesthacht - GERMANY

Keywords: Microstructure, Mechanical properties, AM50, Cerium, High pressure die casting

## Abstract

Over the past decades, the necessity to decrease the weight of automobile parts has created a considerable interest in magnesium alloys. The need to develop magnesium die casting alloys for car components motivates R&D in high temperature magnesium alloy development. In order to improve magnesium alloys with low cost, high strength and with notable elevated temperature properties, the effect of Ce addition (0.5/1.0 wt.-%) on microstructure and mechanical properties of AM50 was studied. Results show that addition of Ce to AM50 resulted in grain refinement and the formation of the secondary phase $Al_{11}Ce_3$. Mechanical properties of investigated alloys at both room and elevated temperature were remarkably increased. AM50 containing 1%wt. Ce showed relatively better refinement and mechanical properties compared to AM50 with 0.5%wt. Ce.

## Introduction

Weight reduction in automotive industry is very attractive in recent years due to potential of fuel efficiencies and lower $CO_2$ emission levels. Magnesium alloys are one of the lightest in engineering materials. High Pressure Die Casting (HPDC) is the most preferred process to produce magnesium parts for automotive as well as for numerous other applications [1, 2]. Some widely used commercial magnesium alloys, such as AZ91, AM50 and AM60, have already been applied to automobile components, such as seat frames, steering wheels, instrument panels, brackets, fans, etc [3]. However, some commercial application of magnesium alloys for important structural components is limited due to poor mechanical properties at elevated temperatures [4-6].

Nonetheless, the demand of magnesium applications in automotive parts has been increasing, so that new magnesium alloys having low cost and high elevated temperature properties are needed to develop. Hence, more studies have been devoted to develop new magnesium alloys mainly based on Mg-Al-RE (RE = rare earth), Mg-Al-Ca/Sr and Mg-Al-Si systems, which have improved elevated temperature performance [7-13]. Among them, Mg-Al-RE system show a significant improvement in heat-resistant Mg-Al based alloys due to formation of thermal stable $Al_{11}RE_3$ precipitates and complete suppression of the $Mg_{17}Al_{12}$ phase [14, 15]. Even though the influences some RE, such as La, Pr and Nd, on the microstructure and mechanical properties of HPDC AM40 alloy was investigated [7-9], studies on effects of single Ce element addition on microstructure and mechanical properties of HPDC AM50 alloy is quite limited. Thus, this study investigates the influence of different amounts of Ce on phase formation and microstructure development as well as its influences on mechanical properties at different temperatures in order to contribute for the development of new Mg-RE alloys.

## Experimental Procedure

Cerium additions of 0.5 wt.% and 1.0 wt.% were made to the base AM50 alloy. Three compositions are referred as AM50, AM50Ce0.5 and AM50Ce1, respectively. The alloys were cast into step-shape plate with four different step thicknesses (2, 6, 10, 14 mm) as seen in Fig. 1. 14 mm section was used for the present studies.

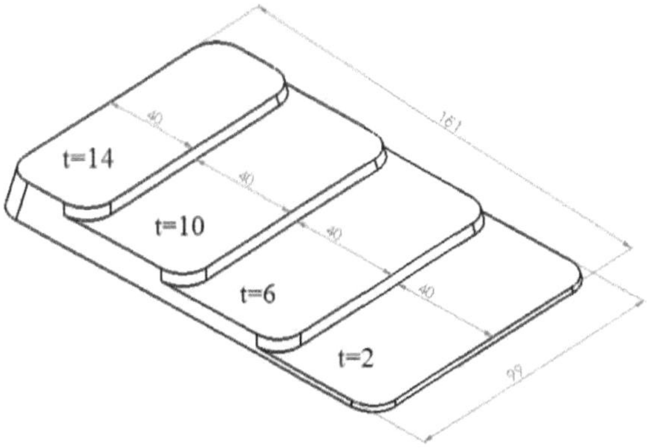


Fig. 1. HPDC step plates with different thickness.

A Frech DAK-450 cold chamber die casting machine was used for casting experiments and 30 shots have been made for each alloy. Die casting conditions and parameters are demonstrated in Table 1.

Table 1. Experimental die casting conditions

| Parameter | Value |
|---|---|
| Melt temperature | 680 °C |
| Casting pressure | 40 MPa |
| Die temperature | 200 °C |
| Shot sleeve speed | 5 m/s |
| Protective gas | Mixture of $N_2$ and 2 vol.% $SF_6$ |
| Machine | Frech DAK 450 (cold chamber) |

Samples (20x10x10 mm) were taken from 14 mm section of the step plates for microstructural observation and hardness tests. These samples were ground with 800, 1200, 2400 grit silicon carbide emery papers. After last grinding step, the specimens were washed with water, rinsed with ethanol and dried. The final polishing was conducted with OPS$^{TM}$ suspension (Struers). The chemical compositions of the alloys were analyzed using a Spark Emission Analyser, Spectrolab9 from SPECTRO (Kleve, Germany). The specimens were chemically etched using a combination of acetic acid, picric acid, ethanol, and purified water to reveal the microstructure. Microstructure was investigated by

Leica DMI 5000 optical microscopy and Zeiss Ultra55 scanning electron microscopy (SEM) equipped with energy dispersive X-ray analysis (EDX). Grain sizes were determined using the line intercept method [16]. The XRD measurements were performed on a Siemens D5000 machine using Cu-K$\alpha_1$ radiation (wavelength $\lambda$=0.15406 nm). The diffraction patterns were measured from $2\Theta$ (15-75°) for each sample. The increment was 0.04° and the step time was 6.4 s. The hardness tests were carried out by Vickers hardness on Emcotest M1C010 automated hardness test machine; the test load, the dwell time and the repetition were 5 kg, 10 s and 10 times. Tension and compression tests were performed at room temperature and elevated temperatures (100 °C, 120 °C, 150 °C) using a Zwick 050 testing machine according to DIN EN 10002 [17] and DIN 50106 [18]. Specimens used in tension tests with a gauge length of 30 mm, a diameter of 6 mm and threaded heads were machined according to ASTM E8M-09 [19]. The compression specimens were cylinders with a height of 16.5 mm and a diameter of 11 mm. Both tension and compression tests were done under a strain rate of $1\times10^{-3}$ $s^{-1}$. In order to obtain homogenous temperature, the samples were soaked at the test temperatures (except RT) for 10 minutes before test run. All tests were repeated at least three times.

## Results and Discussion

Microstructures

The chemical composition of investigated die casting magnesium alloys was analyzed and given in Table 2.

Table 2. Chemical composition of investigated alloys (wt.%).

| Alloy | Mg | Al | Mn | Ce |
|---|---|---|---|---|
| AM50 | Bal. | 4,78 | 0,43 | - |
| AM50Ce0.5 | Bal. | 4,67 | 0,35 | 0,53 |
| AM50Ce1 | Bal. | 4,49 | 0,32 | 1,02 |

The X-ray diffraction (XRD) patterns of investigated alloys are demonstrated in Fig. 2. AM50 alloy is mostly consisted of Mg and $Mg_{17}Al_{12}$ phases [20]. It can be seen in Fig 2, two different phases are detected in the microstructure of AM50 alloy, whereas three different phases are detected in the microstructure of AM50Ce0.5 and AM50Ce1 alloy as a result of Ce addition. When Ce added to AM50 alloy, the diffraction peaks of $Mg_{17}Al_{12}$ phase decrease and go down the detection limit of XRD machine. The main secondary phase in the AM50Ce0.5 and AM50Ce1 alloys is $Al_{11}Ce_3$.

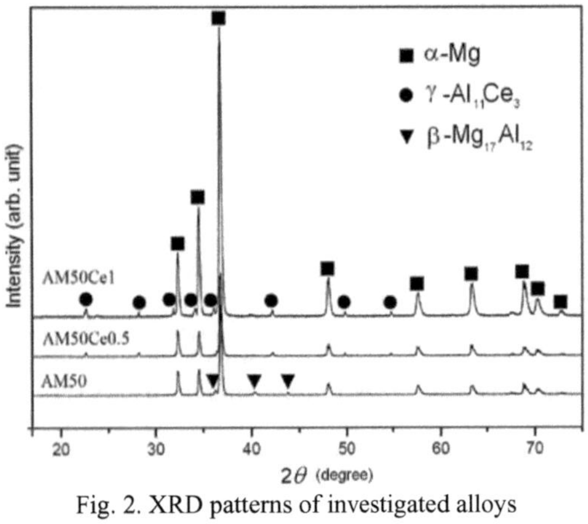


Fig. 2. XRD patterns of investigated alloys

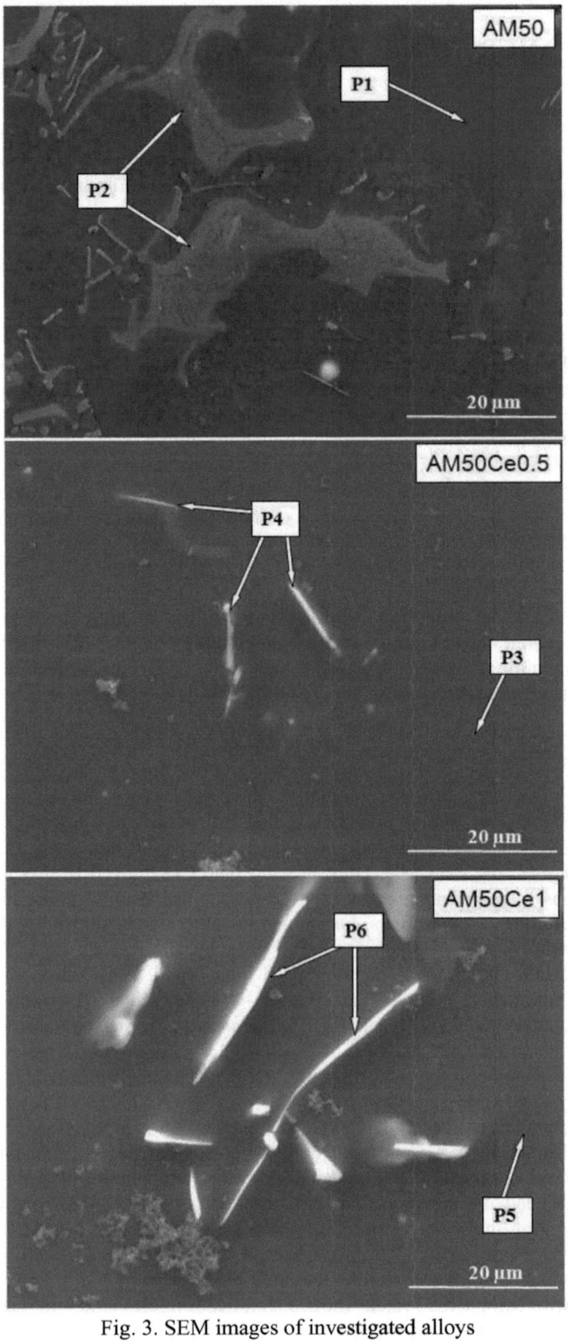


Fig. 3. SEM images of investigated alloys

Fig. 3 shows SEM morphology of investigated alloys. α-Mg and intermetallic constituents appeared in AM50 alloy. EDX study was done on selected points as shown in AM50. The results show that P1 is on α-Mg matrix, P2 is on β-$Mg_{17}Al_{12}$ intermetallic compound. When 0.5wt.% Ce was added to AM50 alloy, as seen in AM50Ce0.5, new γ-$Al_{11}Ce_3$ phase occurred and it appeared as needle-shaped. As Ce amount was increased from 0.5% to 1%, also γ-$Al_{11}Ce_3$ phase amount clearly increased in AM50Ce1 alloy. Similar EDX study was done again in AM50Ce1 and EDX results show that P5 is on α-Mg, and P6 is on γ-$Al_{11}Ce_3$ in AM50Ce1 alloy. Similar situation can be seen in AM50Ce0.5 alloy, but γ-$Al_{11}Ce_3$ phase was less comparing to AM50Ce1 alloy because of Ce amount. As shown in Fig. 3, Ce addition decreases the fraction of β-$Mg_{17}Al_{12}$ phase and facilitates to form γ-$Al_{11}Ce_3$. The overall EDX results of AM50 and with 0.5 and 1wt.% content of Ce are shown Table 3. According to these results, it can be easily said that addition of Ce causes the formation of new γ-$Al_{11}Ce_3$ phase.

Table 3. EDX result of constituent phases of investigated alloys from Fig. 3 (at.%)

| Location | $n_{Mg}/n_{Al}$ | $n_{Al}/n_{Ce}$ |
|---|---|---|
| P1 | 96.54/3.12 | --- |
| P2 | 62.33/27.16 | --- |
| P3 | 96.67/2.98 | --- |
| P4 | --- | 45.18/15.02 |
| P5 | 98.09/1.76 | --- |
| P6 | --- | 53.34/17.11 |

Fig. 4 demonstrates microstructure of studied alloys. It can be seen that the microstructure of AM50 alloy mainly comprises of α-Mg and β-$Mg_{17}Al_{12}$ phases. $Mg_{17}Al_{12}$ phase formed along grain boundaries as network. When Ce is added to AM50, a new phase occurs. Since solid solubility of Ce in Mg-matrix solid solution of AM50 magnesium alloy is low, this phase contains mostly Al and Ce [21]. The Al-Ce phase often appears in between the $Mg_{17}Al_{12}$ phase, possibly providing obstructions to the growth of $Mg_{17}Al_{12}$, breaking up its networked structure. Hence, Ce addition is proved to be beneficial to refining the grain size of AM50 alloy [22]. The grain size decreases with the addition of Ce from 0.5wt.% to 1wt.%, as shown in Table 4.

Table 4. Grain size of investigated alloys.

| Alloy | Grain size [μm] |
|---|---|
| AM50 | 42 (±7) |
| AM50Ce0.5 | 33 (±6) |
| AM50Ce1 | 24 (±4) |

<u>Mechanical properties</u>

The relationship between hardness and Ce addition is demonstrated in Table 5. Hardness of investigated alloys increases clearly with increasing amounts of Ce.

Table 5. Hardness result of the investigated alloys.

| Alloy | Hardness [HV] |
|---|---|
| AM50 | 38 (±6) |
| AM50Ce0.5 | 40 (±9) |
| AM50Ce1 | 45 (±7) |

Fig. 5 shows tensile properties of the investigated alloys tested at room and elevated temperatures. The alloys show the highest tensile yield strength (TYS) and ultimate tensile strength (UTS) at room temperature (RT). Typically, the TYS and UTS decrease, while elongation increases with the increase of the temperature. It can be seen that the addition of Ce can remarkably improve the tensile properties of AM50 alloys at different temperatures.

From Fig. 5a, it is obvious that AM50 shows the lowest TYS, while with addition of Ce, the TYS is remarkably improved both at room and elevated temperatures. The TYS of AM50 is 87 MPa at RT and 69 MPa at 150 °C, whereas for AM50Ce1 the TYS is 98 MPa at RT and 80 MPa at 150 °C.

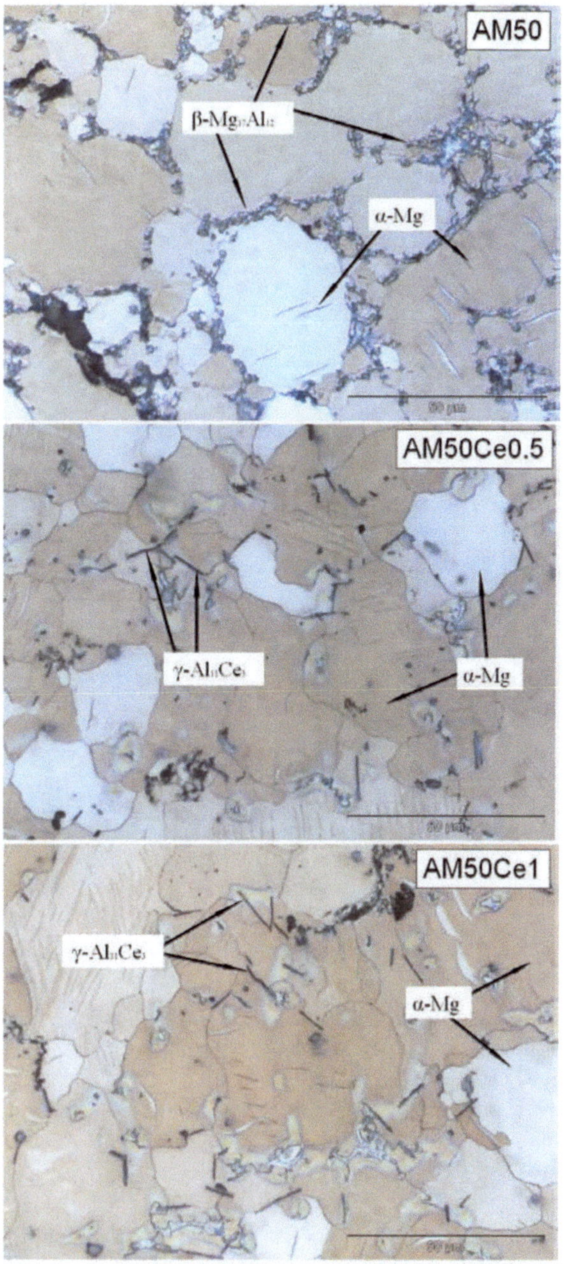


Fig. 4. Microstructures of investigated alloys

The UTS has same trend as TYS with the change of Ce content (Fig. 5b). The UTS of AM50 is 145 MPa at RT and 120 MPa at 150 °C, while for AM50Ce1 the UTS is 168 MPa at RT and 137 MPa at 150 °C. The elongation of the alloys is also improved as the result of Ce addition and at all temperatures the highest value is acquired from AM50Ce1 (Fig. 5c). For example, the elongation of AM50 are 2,80% at RT and 8,31% at 150 °C, whereas for AMCe1 the elongation are 4,82% at RT and 10,30% at 150 °C. The elongation of the investigated alloys increases with the increase of the temperature and reaches a highest value at 150 °C.

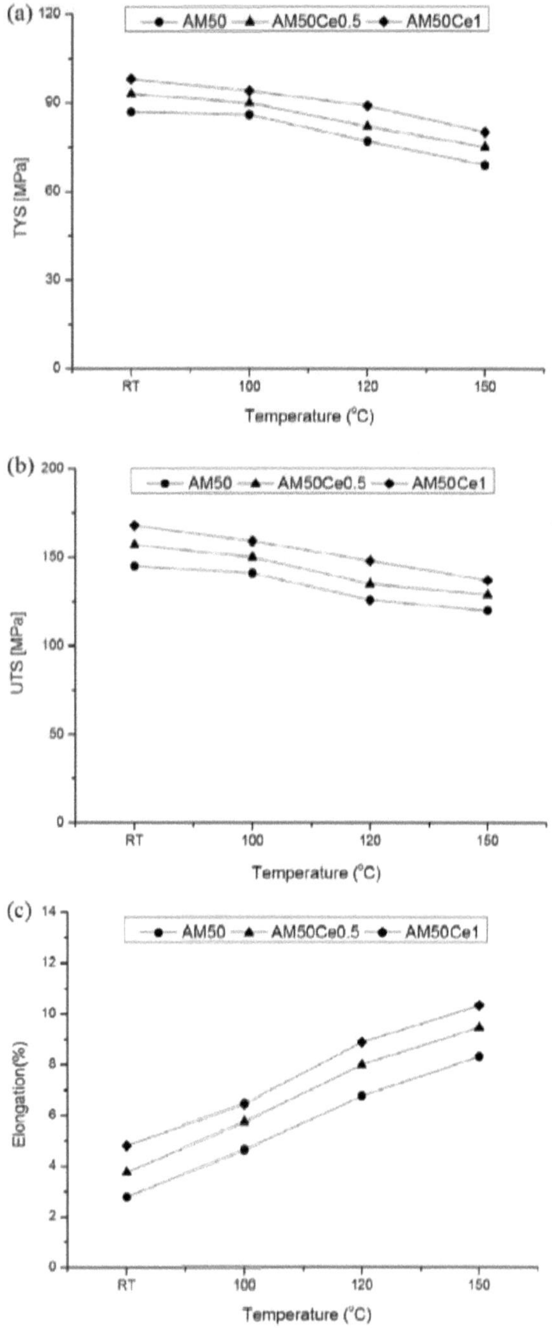


Fig. 5. Tensile properties of investigated alloys at different temperatures: a)TYS, b)UTS, c)Elongation.

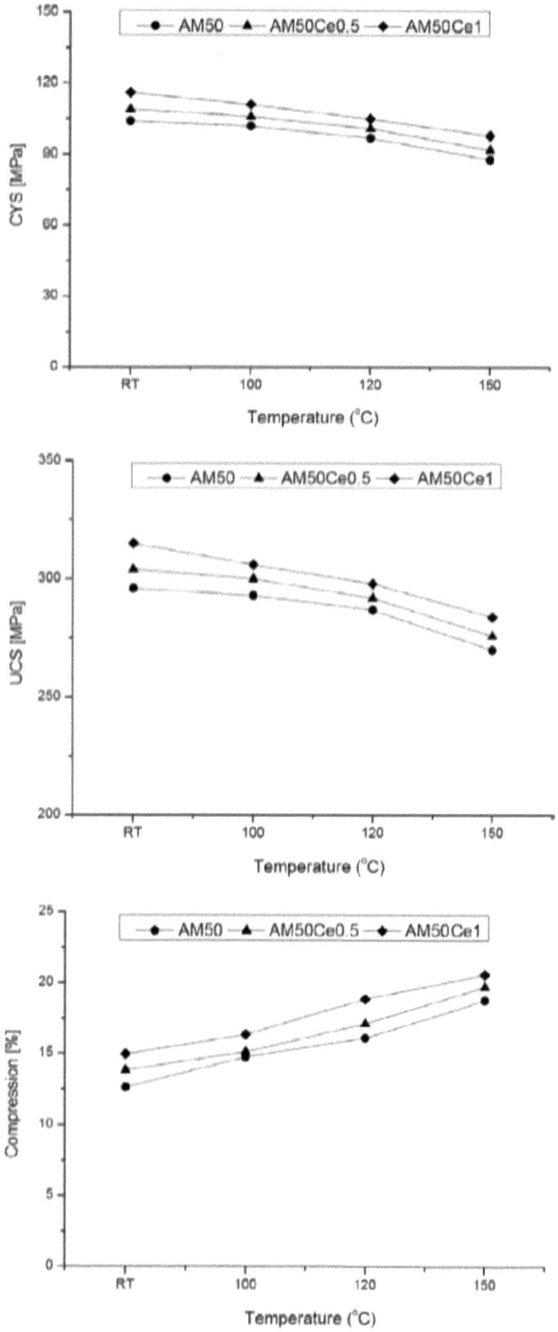


Fig. 6. Compressive properties of investigated alloys at different temperatures: a)CYS, b)UCS, c)Compression.

In previous studies, the effect of RE on the microstructure and mechanical properties of AM series magnesium alloy has been investigated [8, 9, 21-23]. These studies indicate that four RE elements (La, Ce, Pr, Nd) can improve the mechanical properties of AM series magnesium alloys and strengthening effect increases with increasing RE content. The present work is good agreement with previous studies.

Compressive yield strength (CYS) and ultimate compressive strength (UCS) have same trends both at room and elevated temperature as TYS and UTS at different temperatures. From Fig. 6a, it is obvious that AM50 exhibits the lowest CYS, while with addition of Ce, the CYS is remarkably improved both at room and elevated temperatures. The CYS of AM50 is 104 MPa at RT and 88 MPa at 150 °C, whereas for AM50Ce1 the CYS are 116 MPa at RT and 98 MPa at 150 °C. The UCS of the alloys has the general same changing rules with CYS after addition of Ce (Fig. 6b). The UCS of AM50 alloy are 296 MPa at RT and 270 MPa at 150 °C, while for AM50Ce1 alloy the UCS are 315 MPa at RT and 284 MPa at 150 °C, respectively. The compression of the alloys is also developed as the result of Ce addition and at all temperatures; the highest value is obtained from AMCe1 alloy (Fig. 6c). For example, the compression of AM50 are 12,6% at RT and 18,78% at 150 °C, whereas for AMCe1 the compression are 14,96% at RT and 20,6% at 150 °C, respectively. The deformation in compression of the alloys investigated increases with the increase of the temperature and reaches a highest value at 150 °C.

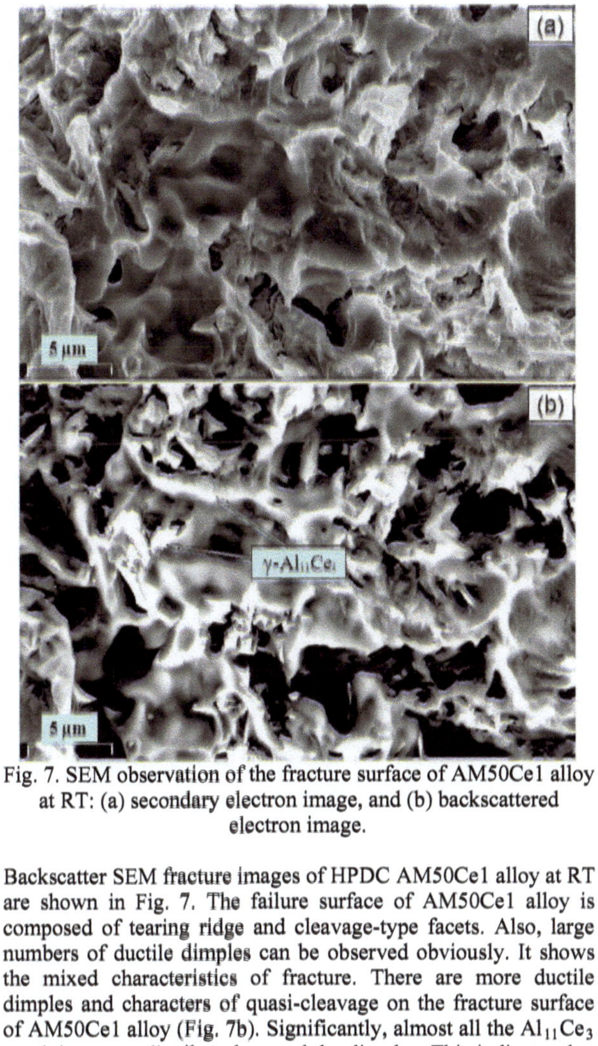


Fig. 7. SEM observation of the fracture surface of AM50Ce1 alloy at RT: (a) secondary electron image, and (b) backscattered electron image.

Backscatter SEM fracture images of HPDC AM50Ce1 alloy at RT are shown in Fig. 7. The failure surface of AM50Ce1 alloy is composed of tearing ridge and cleavage-type facets. Also, large numbers of ductile dimples can be observed obviously. It shows the mixed characteristics of fracture. There are more ductile dimples and characters of quasi-cleavage on the fracture surface of AM50Ce1 alloy (Fig. 7b). Significantly, almost all the $Al_{11}Ce_3$ precipitates are distributed around the dimples. This indicates that before fracture the alloy has undergone certain plastic deformation, and the fine $Al_{11}Ce_3$ particles are not the main crack sources.

With addition of Ce, the grains of α-Mg matrix are refined and eutectic phase β-$Mg_{17}Al_{12}$ is suppressed and substituted by $Al_{11}Ce_3$ with the former being the dominant one. Because obtaining of $Al_{11}Ce_3$ particles when Ce was added to AM50 could refine the grain size significantly. Small grain size can effectively prevent the dislocation slip, so that deformation of fine grain structural alloys needs higher stress. The main strengthening phase in AM50 alloy is β-$Mg_{17}Al_{12}$, which has low melting point (with eutectic temperature of 437 °C) and a poor thermal stability. β-$Mg_{17}Al_{12}$ phase can readily soften and coarsen at the temperature exceeding 125 °C [24]. Addition of Ce to AM50 results in the suppression of β-$Mg_{17}Al_{12}$ phase and the formation of $Al_{11}Ce_3$ phase. $Al_{11}Ce_3$ phase has a high melting point (>1200°C), and then the softening temperature of the alloy containing $Al_{11}Ce_3$ phase can be increased. With the increase of Ce content in the alloys, the amount of $Al_{11}Ce_3$ particles increases significantly, and then the strengthening effect is also improved. Thermally stable $Al_{11}Ce_3$ phase was formed after addition of Ce, which is located at grain boundaries and suppressed the slippage of grain boundary at elevated temperature, so that elevated temperature properties of the alloys are improved.

## Summary

The microstructure and mechanical properties of HPDC AM50 alloys with Ce (0,5% and 1%) have been investigated. The main conclusions could be drawn as follows:
(1) Addition of Ce reduces grain size of AM50 alloy.
(2) The addition of Ce can remarkably improve tensile and compressive properties of AM50 alloys at room and elevated temperatures. This is due to Al-Ce phases on the one hand and grain refinement on the other hand.

## Acknowledgements

The authors want to express their gratitude to G. Wiese, J. Bohlen, G. Meister, W. Punessen and S. Schubert for their technical support. Financial support from Turkey Council of Higher Education (YÖK) scholarship for Faruk's PhD study in Helmholtz-Zentrum Geesthacht is also appreciated.

## References

[1] Külekçi, M.K., "Magnesium and its alloys applications in automotive industry," *International Journal of Advanced Manufacturing Technology*, 39 (2008), 851-865.
[2] Mordike, B.L. and Ebert, T., "Magnesium - Properties - applications – potential," *Materials Science and Engineering A*, 302 (2001), 37-45.
[3] Easton, M., Beer, A., Barnett, M., Davies, C., Dunlop, G., Durandet, Y., Blacket, S., Hilditch, T. Beggs, P., "Magnesium alloy applications in automotive structures," *Journal of the Minerals, Metals and Materials Society*, 60 (2008), 57-62.
[4] Luo A.A., "Recent magnesium alloy development for elevated temperatures applications," *International Material Reviews*, 49 (2004), 13-30.
[5] Powell, B.R., Ouiment, J.L., Allison, J.E., Haynes, J.A., Beals, R.S., Kopka, L., Reid, P.P., "The Magnesium Powertrain Cast Components Project: Part I – Accomplishments of Phase I and the Objectives and Plans for the Magnesium Engine in Phase II," in *Magnesium Technology 2004*, ed. Alan A. Luo, 2004, The Metals, Minerals and Materials Society, Charlotte, North Carolina, p.3-10.
[6] Bronfin, B., Aghion, E., Buch, F., Schumann, S., Katsir, M., "Die casting magnesium alloys for elevated temperatures applications," in *Magnesium Technology 2001*, ed. John N. Hryn,

The Metals, Minerals and Materials Society, 2001, New Orleans. Louisiana, p.127-130.

[7] Zhang, J., Zhang, M., Meng, J., Wu, R., Tang, D., "Microstructures and mechanical properties of heat-resistant high-pressure die-cast Mg-4A1-xLa-0.3Mn (x=1, 2, 4, 6) alloys," *Materials Science and Engineering A*, 527 (2010), 2527-2537.

[8] Zhang, J., Liu, K., Fang, D., Qiu, X., Yu, P., Tang, D., Meng, J., "Microstructures, mechanical properties and corrosion behavior of high-pressure die-cast Mg-4Al-0.4Mn-xPr (x=1, 2, 4, 6) alloys," *Journal of Alloys and Compounds*, 480 (2009), 810-819.

[9] Zhang, J., Wang, J., Qiu, X., Zhang, D., Tian, Z., Niu, X., Tang, D., Meng, J., "Effect of Nd on the microstructure, mechanical properties and corrosion behavior of die-cast Mg-4Al-based alloy," *Journal of Alloys and Compounds*, 464 (2008), 556-564.

[10] Kondori, B., Mahmudi, R., "Effect of Ca additions on the microstructure, thermal stability and mechanical properties of a cast AM60 magnesium alloy," *Materials Science And Engineering A*, 527 (2010), 2014-2021.

[11] TerBush, J.R., Suzuki, A., Saddock, N.D., Jones, J.W., Pollock, T.M., "Dislocation substructures of three die-cast Mg-Al-Ca-based alloys," *Scripta Materialia*, 58 (2008), 914-917.

[12] Czerwinski, F., Zielinska-Lipiec, A., "The microstructure evolution during semisolid molding of a creep-resistant Mg-5Al-2Sr alloy," *Acta Materialia*, 53(2005), 3433-3444.

[13] Srinivasan, A., Swaminathan, J., Pillai, U.T.S., Guguloth, K., Pai, B.C., "Effect of combined addition of Si and Sb on the microstructure and creep properties of AZ91 magnesium alloy," *Materials Science and Engineering A*, 485 (2008), 86-91.

[14] Bakke, P., Westengen, H., "Die casting for high performance-Focus on alloy development," *Advanced Engineering Materials,* 5 (2003), 879-885.

[15] Pettersen, G., Westengen, H., Hoier, R., Lohne, O., "Microstructure of a pressure die cast magnesium-4wt.% aluminium alloy modified with rare earth additions," *Materials Science and Engineering A*, 207 (1996), 115-120.

[16] ASTM E112-10 Standard test methods for determining average grain size, 2004.

[17] DIN EN ISO 10002, Tension Testing of Metallic Materials, Beuth Verlag, Berlin, 2001.

[18] DIN 50106, Compression Testing of Metallic Materials, Beuth Verlag, Berlin, Germany, 1978.

[19] ASTM E8M - 09 Standard Test Methods for Tension Testing of Metallic Materials, 2009.

[20] Materials Science International Team (MSIT), and Gröbner, J., Ran, Q., *Light Metal Ternary Systems: Phase Diagrams, Crystallographic and Thermodynamic Data,* Vol. 11A3, Part 3, Ed: Effenberg, G., Ilyenko, S., 2005, Springer Press, Berlin.

[21] Liao, H.M., Long, S.Y., Guo, C.B., Zhu, Z.B., "Effect of Ce-rich misch metal addition on sequeeze cast microstructure and mechanical properties of AZ81 alloy," *Transactions of Nonferrous Metals Society of China,* 18 (2008), 44-49.

[22] Liu, S.F., Huang, S.Y., Xu, P., "Influence of cerium addition on as-cast microstructure of AZ91 magnesium alloy," *Acta Metallurgica Sinica*, 42 (2006), 443-448.

23] Wang, M., Zhou, H., Wang, L., "Effect of yttrium and cerium addition on microstructure and mechanical properties of AM50 magnesium alloy," *Journal of Rare Earths*, 25 (2007), 233-237.

[24] Lü, Y., Wang, Q., Zeng, X., Ding, W., Zhai, C., Zhu, Y., "Effects of rare earths on the microstructure, properties and fracture behavior of Mg–Al alloys," *Materials Science and Engineering A,* 278 (1999), 66-76.

**Magnesium Technology 2012**
*Edited by: Suveen N. Mathaudhu, Wim H. Sillekens, Neale R. Neelameggham, and Norbert Hort*
***TMS (The Minerals, Metals & Materials Society), 2012***

# MELT CONDITIONED DC (MC-DC) CASTING OF MAGNESIUM ALLOYS

Y B Zuo, B Jiang, Y Zhang, Z Fan
BCAST, Brunel University, Uxbridge, Middlesex, UB8 3PH, UK

Keywords: DC casting, Melt conditioning, Grain refinement, Microstructure, Magnesium alloy.

**Abstract**

A new melt conditioned direct chill (MC-DC) casting process has been developed for producing high quality magnesium alloy billets and slabs. In the MC-DC casting process, intensive melt shearing provided by a high shear device is applied directly to the alloy melt in the sump during DC casting. The high shear device provides intensive melt shearing to disperse potential nucleating particles, creates a macroscopic melt flow to distribute uniformly the dispersed particles, and maintains a uniform temperature and chemical composition throughout the melt in the sump. Experimental results have demonstrated that the MC-DC casting process can produce magnesium alloy billets with significantly refined microstructure and reduced cast defects. In this paper, we introduce the new MC-DC casting process, report the grain refining effect of intensive melt shearing during the MC-DC casting process and discuss the grain refining mechanism.

## Introduction

Magnesium alloys are used in a wide variety of applications including automotive, aerospace, defense, electronics, healthcare and sports equipment because of their low density, high specific strength, high damping capacity, good castability and excellent electromagnetic shielding properties [1, 2]. However, due to their specific crystallographic structure, Mg-alloys have only a limited number of slip systems available, and therefore are difficult to deform at low temperatures. This poor deformability means low productivity and high processing cost, and consequently limited engineering applications. It is well accepted that a fine grain size can result in both improved formability and enhanced mechanical properties [3, 4]. In addition, fine and equiaxed solidification can reduce chemical segregation, hot tearing and formation of other casting defects [5], especially when producing large billets.

DC casting is a major technology for processing wrought magnesium alloys. However, the billets or slabs produced by the conventional DC casting process often have coarse and non-uniform microstructures, severe chemical segregation, porosity and hot tearing, which not only burden the downstream thermo-mechanical processing but also have a negative influence on the mechanical properties of the final products. Grain refinement by inoculation with chemical grain refiners is an effective approach for mitigating such problems [6, 7]. Although grain refinement of magnesium alloys has attracted intensive research [2, 8-10] the search for new and effective grain refiners and methods for grain refinement still continues. Up to date, the research on grain refinement of magnesium alloys is mainly focused on searching for new chemical grain refiners and the application of external physical fields during solidification processing, such as electromagnetic field, electrical pulsing and ultrasonic fields [11]. Recently, we found that intensive melt shearing provided by a twin-screw mechanism has a significant grain refining effect on both aluminium and magnesium alloys [12-15]. Based on a similar principle, a new high shear device has been developed for conditioning liquid metals prior to solidification processing [16], and the high shear device has been integrated into the conventional DC casting process to form a melt conditioned DC casting process for production of high quality Mg-alloy billets or slabs.

This paper introduces the newly developed MC-DC casting process with intensive melt shearing, reports our experimental results on microstructural refinement of MC-DC cast Mg-alloys and discusses the grain refining mechanisms.

## Experimental Procedure

Commercial AZ91D Mg-alloy (Mg-8.67%Al-0.62%Zn-0.21%Mn, all compositions are in wt%), and AZ31 Mg-alloy (Mg-2.92%Al-0.85%Zn-0.36%Mn) were used in the present work. The measured liquidus is 600°C for AZ91D and 630°C for AZ31. The alloys were melted at 680°C in a steel crucible under a protective atmosphere containing a gas mixture of $N_2$ and 0.5 vol% $SF_6$. Mg-alloy billets were DC cast with and without intensive melt shearing. Figure 1 is a schematic illustration of the MC-DC casting process, which is effectively a combination of a conventional DC caster with the high shear device submerged in the sump of the caster [16]. The alloy melt was poured with 50K superheat, and the high shear device was operated at 5000-15000RPM. The same protective gas was supplied to the hot top during the DC casting process. 80mm diameter Mg-alloy billets were produced under the same conditions but with or without the high shear device being switched on.

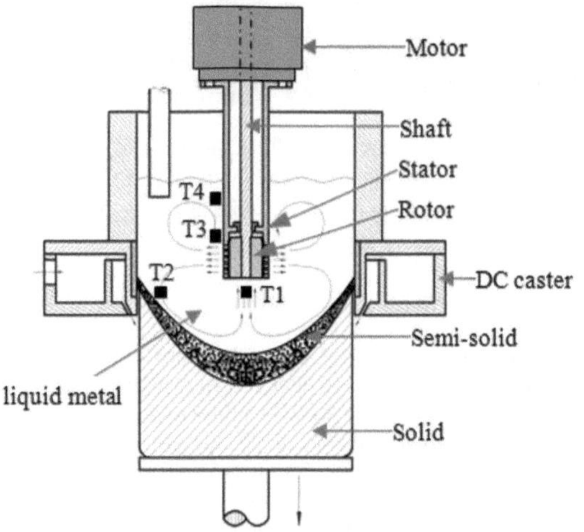


Figure 1. Schematic illustration of the melt conditioned direct chill (MC-DC) casting process, which is effectively a conventional DC caster with a high shear device submerged in the melt sump. The high shear device provides intensive melt shearing to disperse potential nucleating particles and a macroscopic flow pattern to homogenize the melt temperature and the composition in the sump. T1-T4 indicates the positions of thermocouples for temperature measurement.

Samples for both macroscopic and microscopic examinations were cut, prepared and then examined according to standard metallographic procedures. To further understand the effect of melt shearing during DC casting on the microstructure, the temperature field was measured by using thermocouples attached to the high shear device and a data logger during the DC casting process. T1-T4 in Figure 1 schematically indicate the positions of the thermocouples for measuring the melt temperature.

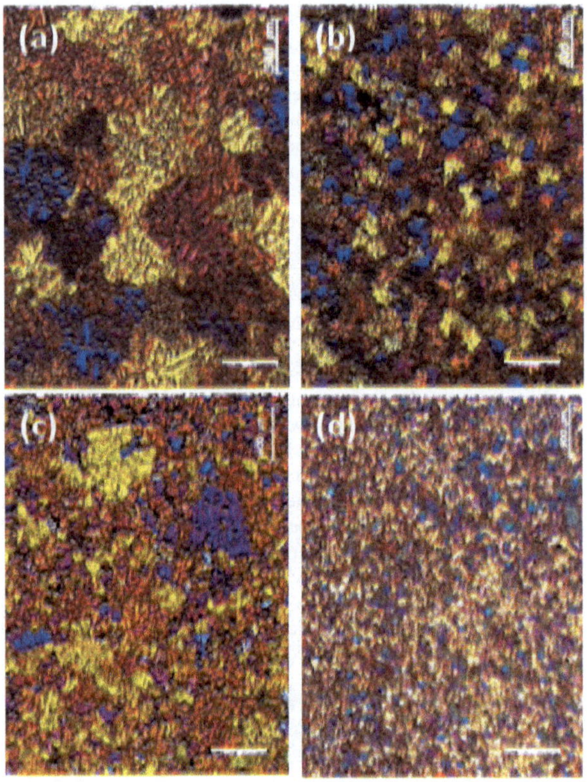


Figure 2. Microstructure of AZ91D magnesium alloy cast in a TP1. (a) 650℃, without shearing; (b) 650℃, with shearing; (c) 620℃, without shearing (d) 620℃, with shearing. The size bars in all the micrographs are 500μm.

## Results

### Microstructure of AZ91D Alloy with Intensive Melt Shearing

To demonstrate the effect of intensive melt shearing on the grain refinement of Mg-alloys, AZ91D alloy melts were sheared at different temperature and cast in TP1 mould to assess the grain size. Figure 2 shows the microstructures of AZ91D alloy cast at 650°C and 620°C in the TP1 mould with and without intensive melt shearing. The TP1 mould was operated under the standard conditions for grain size assessment with a constant cooling rate of around 3.5 K/s for aluminium alloys [17]. With intensive melt shearing, the microstructure becomes more uniform and the grain size is significantly reduced at both casting temperatures. The as-cast grain size was reduced from 550μm to 170μm when cast at 650°C (Figures 2a and 2b), and from 180μm to 90μm when cast at 620°C (Figures 2c and 2d). It is interesting to note from Figure 2 that the grain size of Mg-alloys has a strong dependence on melt superheat, and that such superheat effect is suppressed significantly by intensive melt shearing.

### Microstructure of MC-DC Cast AZ91D Alloy

Figure 3 shows the macroscopic structures of the 80mm diameter AZ91D alloy billets produced by DC casting at 650°C with and without intensive melt shearing. Although both billets have an equiaxed structure, the grain structure is much finer and more uniform in the billet with intensive melt shearing. In addition, chemical analysis using optical emission spectroscopy indicates that there is no macro-segregation along either the longitudinal or the radial directions in both types of billets.

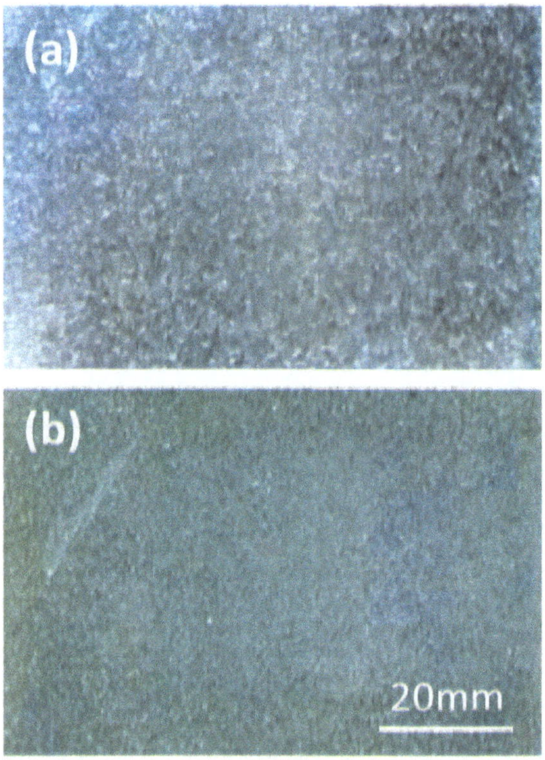


Figure 3. Macrographs of sectioned DC cast AZ91D alloy ingots cast at 650°C with and without shearing.

The microstructures of the AZ91D billets DC cast with and without intensive melt shearing are shown in Figure 4. The conventional DC cast billet (without intensive melt shearing) shows a coarse equiaxed dendritic microstructure with an average grain size of around 500μm near to the ingot surface (Figure 4a) and a coarser equiaxed grain structure with an average grain size of 700μm at the centre of the billet (Figure 4b). However, the AZ91D billet produced by the MC-DC process (with intensive melt shearing) have a consistently fine and equiaxed grain structure throughout the entire billet with an average grain size of 190μm (Figures 4c and 4d). Figure 5 compares the grain size variation between the AZ91D alloy billets produced by DC casting with and without intensive melt shearing. For the conventional DC cast billet, there is a thin layer with finer microstructure with an average grain size of 500μm, and apart form that layer, the microstructhre is dominated by coarse grains with an average grain size of 700μm. However the billet DC cast with intensive melt shearing shows a uniform grain size distribution through the entire cross section, with an average grain size of 190μm.

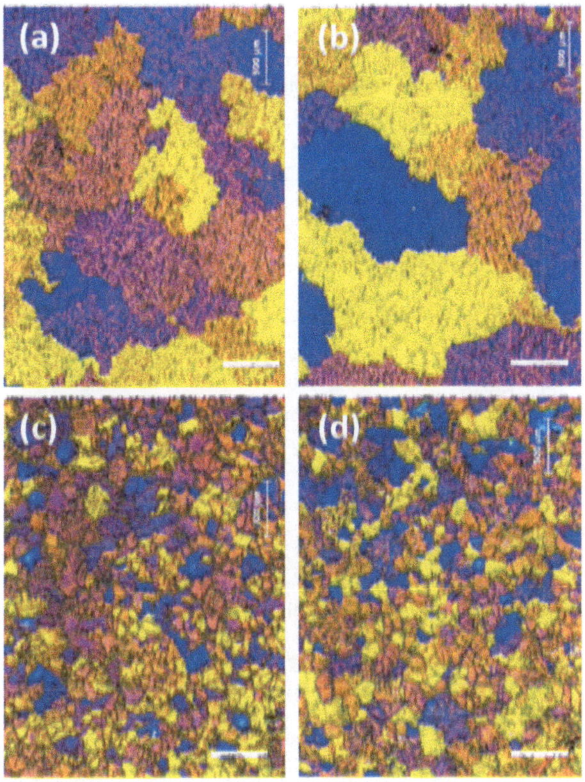


Figure 4. Comparison of the microstructures of DC cast AZ91D alloy billets at 650°C with and without intensive melt shearing. (a) edge without shearing; (b) centre without shearing; (c) edge with shearing; (d) centre with shearing. The size bars in all the micrographs are 500μm.

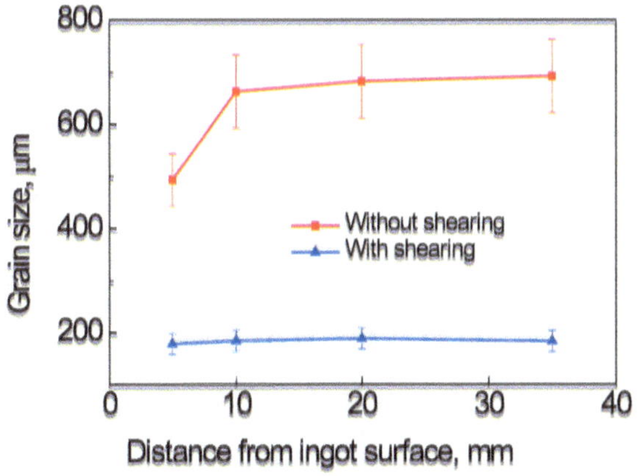


Figure 5. Variation of grain size as a function of the radial distance from the billet surface in AZ91D alloy billets produced by DC casting with and without intensive melt shearing.

In order to further demonstrate the effect of *in situ* intensive melt shearing on the microstructure of DC cast Mg-alloy billets, AZ91D alloy was cast by the newly developed MC-DC casting process by switching on and off the high shear device. The high shear device was submerged in the sump of the DC caster, and was switched off during the first half of the billet casting and switched on during the second half of the billet casting. Figure 6 shows the microstructure of DC cast AZ91D alloy with the transition from the section cast with high shear device off (bottom) to the section with the high shear device on (top). Without intensive melt shearing, the DC cast AZ91D alloy shows a coarse dendritic microstructure, while with intensive melt shearing it turns sharply into a fine and equiaxed microstructure. From the transition up wards, the AZ91D alloy billet maintains the same fine and uniform microstructure.

It is clear from the above results that during the DC casting process, the application of intensive melt shearing has a significant grain refining effect on AZ91D magnesium alloy and can improve the uniformity of the microstructure in the as-cast billets.

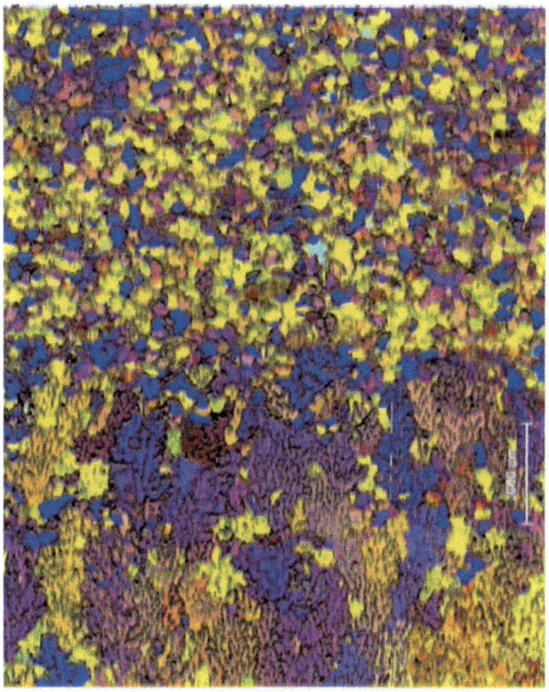

Figure 6. Microstructure of DC cast AZ91D alloy billet at 650°C with (top) and without (bottom) intensive melt shearing, showing the transition from coarse dendritic grain structure without intensive melt shearing (high shear device off) to a fine and equiaxed grain structure with intensive melt shearing (high shear device on).

**Microstructure of DC Cast AZ31 Alloy**

To further confirm the grain refining effect of intensive melt shearing, the AZ31 magnesium alloy ingots were cast at 680°C with *in situ* intensive melt shearing. Figure 7 shows the microstructure of AZ31 Mg-alloy ingot without (Figure 7a) and with (Figure 7b) intensive melt shearing. The microstructure of conventional DC cast AZ31 alloy ingot shows coarse columnar dendrites with a dendrite cell size of a few mm (Figure 7a);

however, with the application of intensive melt shearing the microstructure becomes finer (less than 200μm) and the grain morphology is also changed to equiaxed. This clearly demonstrates the power of intensive melt shear for grain refinement of Al-containing Mg-alloys.

It is well know that grain size is strongly dependent on the concentration of solute elements; the higher the solute concentration, the smaller the grain size due to the increased growth restriction factor with the increase of solute concentration. However, a comparison between Figures 4 and 7 suggests that intensive melt shearing during DC casting may significantly reduce the grain size dependence on solute contents.

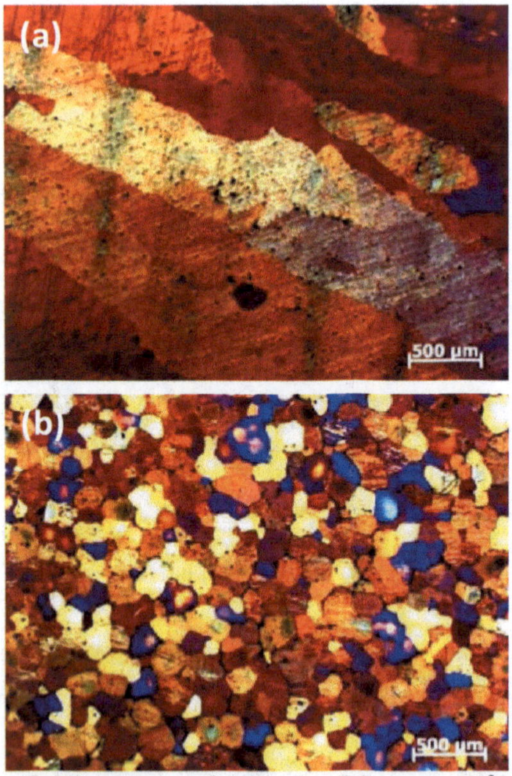


Figure 7. Microstructure of AZ31 alloy DC cast at 680°C (a) without and (b) with *in situ* intensive melt shearing, showing an order of magnitude grain size reduction by application of intensive melt shearing.

**Temperature Uniformity during DC Casting**

To facilitate the understanding of grain refining mechanisms by intensive melt shearing, temperature measurement during DC casting was carried out. The recorded temperatures during DC casting of AZ91D magnesium alloy are shown in Figure 8, which was obtained through the thermocouples attached to the head of the high shear device. The positions of thermocouples (T1-4) are schematically shown in Figure 1. Based on the direct measurement of the sump depth with stainless steel rods, during the casting process, the bottom of the high shear device is about 40 mm above the liquid/solid interface. Without melt shearing, the melt temperature in the sump is expected to be non-uniform; the further the distance from the solidification front, the higher the melt temperature. However, with the start of the high shear device, there is a sharp change of melt temperature distribution in the sump. Intensive melt shearing in the sump made the temperature curves become closer together, resulting in a more or less uniform melt temperature in the sump. The temperature curves also show that after the application of intensive melt shearing, the melt temperature becomes very close to the liquidus and eventually reaches a plateau which is around 7°C lower than the alloy liquidus. This could make an important contribution to the grain refinement of microstructure.

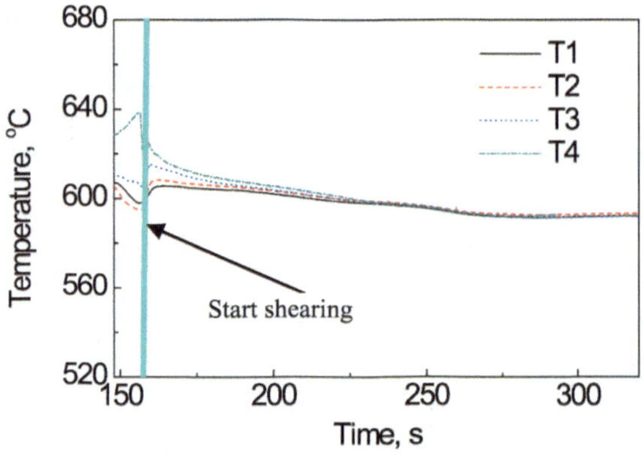


Figure 8. Temperature curves measured during DC casting of AZ91D alloy. T1, centre, 10 mm below the bottom of the high shear unit; T2, edge, 10 mm below the bottom of the high shear unit; T3, near to the centre, 30 mm above the bottom of the high shear unit; T4, near to the centre, 90 mm above the bottom of the high shear unit.

**Discussion**

The experimental results presented in the previous section clearly demonstrate that the MC-DC casting process is a simple and effective technology for production of high quality magnesium alloy billets (or slabs). Compared with the conventional DC casting process, the newly developed MC-DC process has the following anticipated advantages:

- Significant grain refinement
- Uniform and equiaxed microstructure
- Reduced/eliminated macro-segregation
- No large inclusions and oxide films
- Reduced/eliminated hot tearing
- Reduced sensitivity to alloying contents
- Easy to implement within existing DC casting facilities

Further research and development of the MC-DC process is under the way before industrial implementation.

The high shear device is the key element in the MC-DC process to deliver the above mentioned advantages. The high shear device provides *in situ* conditioning of the liquid alloy during the DC casting process. The major function of the high shear device can be summarized as follows:

A high shear dispersive mixing action: The centrifugal force due to the rotation of the rotor generates a negative pressure inside the stator, which in turn sucks the melt below the high shear device into the stator and forces it to escape at high speed through the openings in the wall of the stator. This process subjects the melt to a high shear action in the gap between the rotor and the stator and the in the openings in the stator wall. The shear rate could be as high as $10^5$/s. As a consequence of this intensive melt shearing the clusters of inclusions and oxide films naturally occurring in the melt will be dispersed into individual particles [12], as shown in Figure 9. It is well documented that naturally occurring oxide films are hazardous to casting processes, and more importantly they deteriorate mechanical properties of the final cast products [18, 19]. In the MC-DC process, the intensive melt shearing makes the usually harmful oxide films and other inclusions not only harmless, but also useful for grain refinement.

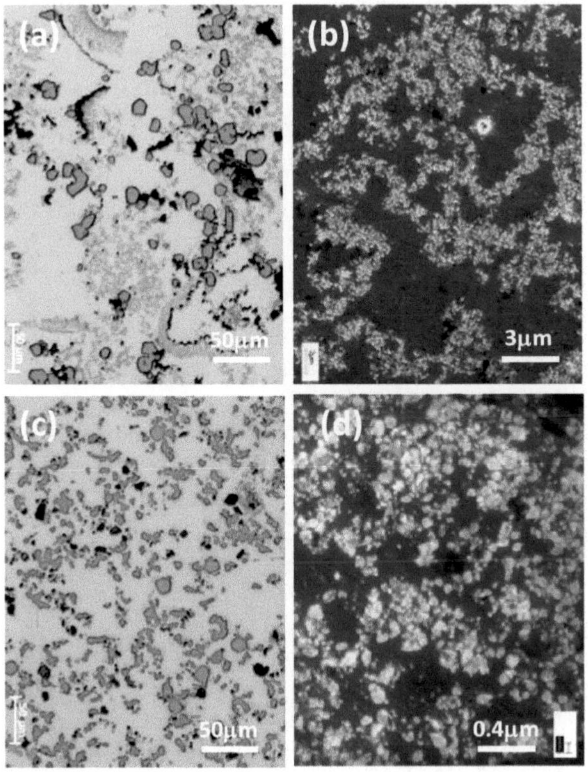


Figure 9. Optical micrographs (a, c) and SEM micrographs (b, d) showing the physical status of oxide in AZ91D alloy. Without intensive melt shearing oxide exists in the melt in the form of oxide films (a, b) consisting of discrete nano-sized MgO particles (b); after intensive melt shearing, oxide films become invisible under optical microscope (c) and are dispersed into more individual particles (d). All the samples here are produced by pressurized melt filtration. The large grey particles in (a) and (c) are Al-Mn intermetallic particles.

A macroscopic flow pattern: The rapid escaping liquid from the openings on the stator will create a macroscopic flow pattern in the sump, as schematically illustrated in Figure 1. This macroscopic flow provides a distributive mixing action in the sump, which homogenizes the temperature and chemical composition of the melt. As a consequence, the melt temperature and chemical composition in the sump will be uniform, at least around the tip of the high shear device (see Figure 8). In addition, this macroscopic melt flow will be strong near the tip of the high shear device, and weak near the surface of the melt. This will prevent the occurrence of gas entrapment from the melt surface.
A reduced solute build-up at the solidification front: Due to the partitioning solute elements in the solid and liquid phases during solidification, there is usually a built-up of solute elements at the advancing solidification front. The forced convection generated by the high shear devise will suck the solute elements at the solidification front and distribute them uniformly in the sump. Such uniform fields of both temperature (slightly below liquidus) and chemical composition will create a condition which promotes heterogeneous nucleation and equiaxed growth.
Enhanced heterogeneous nucleation in the MC-DC casting process is originated from the intensive melt shearing. Firstly, intensive melt shearing generates more potential nucleating particles by dispersing the oxide films (Figure 9). Previous study has confirmed that the potential number of nucleating particles can be increased by three orders of magnitude [20], and that MgO particles in a Mg-alloy melt do nucleate α-Mg [12]. Increased nucleation events lead to grain refinement. Secondly, the uniform temperature field in the sump can increase the survival rate of the existing nuclei by avoiding their dissolution in the hotter melt (temperature greater than the liquidus). This also contributes to finer grains. Thirdly, there is a possibility that intensive melt shearing promotes fragmentation of dendrite arms since the melt temperature in the sump is below the liquidus (Figure 8). This may also contribute partially to the grain refined microstructure in the MC-DC cast billets.

## Summary


A new DC casting process, the melt conditioned direct chill casting (MC-DC) process, has been developed by submerging a high shear device in the sump of a conventional DC caster. The high shear device provides both intensive melt shearing to disperse the naturally occurring oxide films into individual oxide particles as potential nucleation sites and a macroscopic flow pattern to distribute the dispersed particles and to homogenize the melt temperature and chemical composition in the sump. Experimental results have demonstrated that the MC-DC process is capable of producing high quality magnesium billets/slabs with significantly grain-refined microstructure, homogenized chemical composition, reduced/eliminated cast defects. The grain refinement in the MC-DC casting process may be attributed to the increased number of potential nucleating particles by dispersing naturally occurring oxide films, increased nuclei survival rate in the homogenized temperature composition fields and dendrite arm fragmentation due to intensive melt shearing.


## Acknowledgement


The financial support from the EPSRC and the Technology Strategy Board of UK is gratefully acknowledged.


## References


[1] Mordike B L, Ebert T. Magnesium: Properties -- applications -- potential. Mater. Sci. Eng. A 2001;302:37.

[2] Emley E F. Principles of magnesium technology. Oxford: Pergamon Press, 1966.

[3] Zhang S M, Fan Z, Zhen Z. Direct chill rheocasting (DCRC) of AZ31 Mg alloy. Mater. Sci. Technol. 2006;22:1489.

[4] Qiu D, Zhang M X, Taylor J A, Fu H M, Kelly P M. A novel approach to the mechanism for the grain refining effect of

melt superheating of Mg-Al alloys. Acta Mater. 2007;55:1863.

[5] Nadella R, Eskin D G, Katgerman L. Role of grain refining in hot cracking and macrosegregation in direct chill cast AA7075 billetsRole of grain refining in hot cracking and macrosegregation in direct chill cast AA7075 billets. Mater. Sci. Technol. 2007;23:1327.

[6] Eskin D G. Physical metallurgy of direct chill casting of aluminum alloys. Boca Raton CRC Press, 2008.

[7] Murty B S, Kori S A, Chakraborty M. Graun refinement of aluminium and its alloys by heterogeneous nucleation and alloying. Inter. Mater. Rev. 2002;47:3.

[8] David H SJ, Ma Q, Easton M A, Cao P, Hildebrand Z. Grain refinement of magnesium alloys. Metall. Mater. Trans. A 2005;36:1669.

[9] Zhang M X, Kelly P M, Qian M, Taylor JA. Crystallography of grain refinement in Mg-Al based alloys. Acta Mater. 2005;53:3261.

[10] Easton M A, Schiffl A, Yao J-Y, Kaufmann H. Grain refinement of Mg-Al(-Mn) alloys by SiC additions. Scripta Mater. 2006;55:379.

[11] Cui J Z, Zhang Z Q, Le Q C. Direct-chilling casting of Mg alloy under electromagnetic and ultrasonic combined field. Trans. Nonferrous Met. Soc. China 2010;20:s297.

[12] Fan Z, Wang Y, Xia M, Arumuganathar S. Enhanced heterogeneous nucleation in AZ91D alloy by intensive melt shearing. Acta Mater. 2009;57:4891.

[13] Zuo Y B, Li H T, Xia M X, Jiang B, Scamans G M, Fan Z. Refining grain structure and porosity of an aluminium alloy with intensive melt shearing. Scripta Mater. 2011;64:209.

[14] Fan Z, Liu G. Solidification behaviour of AZ91D alloy under intensive forced convection in the RDC process. Acta Mater. 2005;53:4345.

[15] Zuo Y B, Xia M, Liang S, Wang Y, Scamans G, Fan Z. Grain refinement of DC cast AZ91D Mg alloy by intensive melt shearing. Mater. Sci. Technol. 2011;27:101.

[16] Fan Z, Zuo Y B, Jiang B. Apparatus and method for liquid metals treatment. UK patent, Application No. 1015498.7, 2010.

[17] The Aluminium Association. Standard test procedure for aluminium alloy grain refiners (TP-1). Washington DC, USA, 1990.

[18] Mirak A R, Divandari M, Boutorabi S M A, Campbell J. Oxide Film Characteristics of AZ91 Magnesium Alloy in Casting Conditions. Int. J. Cast Met. Res. 2007;20:215.

[19] Campbell J. Entrainment defects. Mater. Sci. Technol. 2006;22:127.

[20] Men H, Jiang B, and Fan Z. Mechanisms of grain refinement by intensive shearing of AZ91 alloy melt. Acta Mater. 2010;58:6526.

# EFFECT OF THE SOLIDIFICATION RATE ON MICROSTRUCTURE OF CAST Mg ALLOYS AT LOW SUPERHEAT

Gregory Poole[1], Nathan Rimkus[2], Aeriel Murphy[1], Paige Boehmcke[1], and Nagy El-Kaddah[1]
[1] Department of Metallurgical Engineering, University of Alabama, Tuscaloosa, Alabama 35487 USA
[2] Los Alamos National Laboratory, Los Alamos, New Mexico 87545 USA

Keywords: Mg alloys, Solidification Rate, Grain Structure, Microsegregation

## Abstract

This paper investigates the effect of cooling rate on the grain size and microstructure of Mg AZ31B alloy cast at a superheat of 8°C using the Magnetic Suspension Melting (MSM) process, which is capable of melting and casting at superheats less than 5°C. In this study, the Mg alloy was unidirectionally solidified in a bottom-chill mold with stainless steel and copper chill blocks. The solidification parameters, namely growth velocity (V) and temperature gradient (G), were determined from numerical simulation of the cooling curves, which was found to be in good agreement with measurements. For the investigated solidification rates, metallographic examination showed globular solidification morphology, and the grain size was inversely proportional to the square root of the cooling rate. Microprobe analysis of the cast ingots also showed that Al segregation occurs primarily at the grain boundaries, and the solidification rate affects the size and distribution of both the secondary α phase and the intermetallic $Mg_{17}Al_{12}$ phase.

## Introduction

In recent years, there has been considerable interest in casting magnesium automotive components as a mean to reduce the overall weight of the cars, thus improving fuel efficiency [1]. To produce quality products with a high degree of reliability, it is critical to minimize the effect of casting defects on mechanical properties of the cast alloy. One such defect is solute segregation in the as-cast alloy, which is known to be detrimental to mechanical properties [2-4]. For typical solidification rates for sand and die castings in the range of 0.1-100°C/s, the grain structure is dendritic, with segregation of the alloying elements in the matrix between the secondary dendrite arms,. Therefore, it is important to reduce the extent of solute segregation in the matrix by reducing the grain size, which generally depends on the solidification rate (V) [5].

Previous studies have shown that increasing the solidification rate reduces both the secondary dendrite arm spacing (SDAS) and the amount of secondary phases between the dendrite arms [6-8]. To achieve the desired mechanical properties of the cast Mg alloys, much effort has been focused on grain refinement via the addition of grain refiners as a means to reduce the SDAS at the same solidification rate [9-11]. This approach, however, does not completely eliminate solute segregation of the alloying elements in cast products.

It is well established that casting at low superheat not only reduces the grain size, but the solidification morphology of the casting from a dendritic to a globular structure, further preventing segregation within the grains. The idea of behind this technique is to suppress the growth of dendrite arms by reducing the temperature gradient (G) at the solidification front.

Previous studies on casting superalloys using the Microcast X technology have shown that a superheat of approximately 10°C produces fine grains around 100 μm [12-14]. Studies on melting and casting Al-Li and Mg alloys at such low superheat using the Magnetic Suspension Melting (MSM) process have shown that the castings are homogenous, oxide free, and exhibit a fine globular grain structure. The average grain size for Al-Li 2090 and 8090 alloys unidirectionally-solidified using a stainless steel bottom chill mold was about 30 μm [15,16], while that for Mg AZ31B alloy was approximately 80 μm [17,18]. These studies have also shown that segregation of secondary and intermetallic phases did not occur within the grain, but rather only at the grain boundaries. This paper examines the effect of solidification rate on the as-cast grain structure and micro/macro segregation of Mg AZ31B alloy at low superheat by varying the thermal properties of the chill block.

## Experimental Technique

Casting of Mg AZ31B at low superheat was carried out using the MSM process, which is an integrated containerless induction melting and casting process specifically developed for reactive metals such as Mg. The experimental system is comprised of two main components: the melting compartment and the casting chamber as illustrated in Figure 1, thus combining melting and casting into a single, seamless operation, where the charge is melted in the melting compartment and pours into the mold located in the casting chamber. A detailed description of the experimental system used in this study is given in Reference [18].

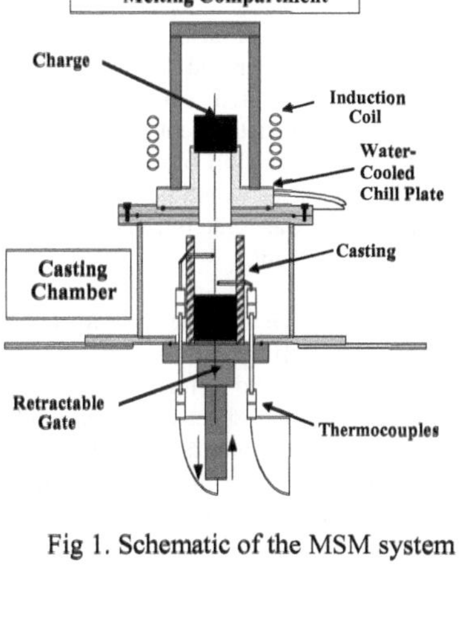


Fig 1. Schematic of the MSM system

All experiments were conducted using an Mg AZ31B alloy with composition Mg-3.143%Al-0.66%Zn-0.377%Mn-0.051%Si. A typical experimental run involved placing short bars 76 mm in diameter and 63 mm high into the melting compartment and flushing the system with argon at a rate of 8.5 L/min . The coil was then energized, adjusting the current as necessary to maintain a constant superheat of 8°C. Upon melting the charge, the metal was poured into a bottom-chilled mold shown in Figure 2. The ceramic mold was a 153 mm silica tube with 76 mm ID and 89 mm OD produced by Zircar Refractory Composites Inc. The stainless steel and copper chill blocks used in this study were 76 mm in diameter and 50 mm high. A battery of thermocouples was used to measure the cooling rates at different heights in the center of the mold, and was attached to an Omega TC-08 data acquisition system.

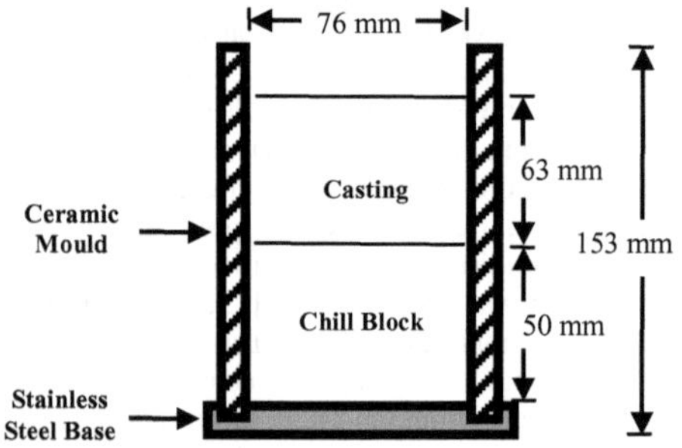


Fig 2. Sketch of the casting mold.

Characterization of the cast microstructures included optical microscopy and microprobe analysis. Optical micrographs were taken using a Nikon Epiphot 200 optical microscope, and EDS mapping of solute distribution was done using a JEOL 8600 FE SEM EPMA. The solidification parameters V and G were determined from numerical simulation of the cooling curves.

## Results and Discussion

In the following, we shall present the results of the thermal analysis of the cooling curves used in the determination of the solidification parameters for MSM cast Mg AZ31B alloy on copper and stainless steel chill blocks at superheat of 8°C, together with the observed grain structure and segregation of Al for these two cases.

Figure 3 shows the computed and measured cooling curves of the cast alloy in the mold for copper and stainless steel chill plates. The lines indicate the computed values while the symbols correspond to experimental measurements. This figure shows that the cooling rate for the copper chill block is about twice that of the stainless steel chill block, as would be expected. It also shows that the solidification model accurately reproduces the measured cooling curves within ±5 percent.

Figure 4 shows an expanded view of the cooling curves within the solidification range. As seen in this figure, the experimental solidification times for copper and stainless steel chill blocks were found to be 24.7s and 36.1s, respectively. Although the model accurately predicts these total solidification times, it reasonably predicts the temperature evolution in the mushy region. This is mainly due to the difference between the actual and modeled evolution of latent heat during solidification.

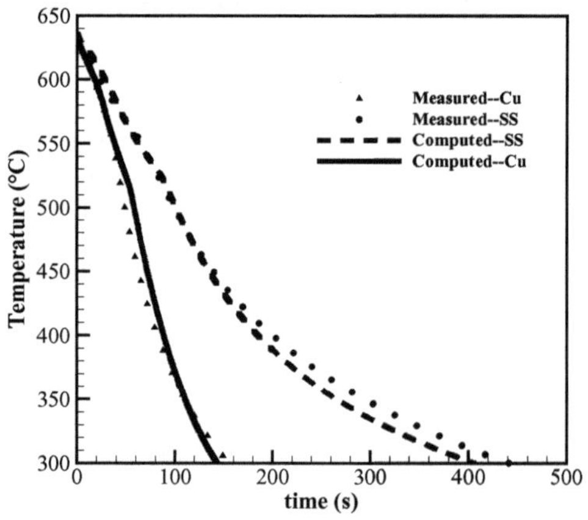


Fig 3. Measured and computed cooling curves for cast Mg AZ31B alloy over copper and stainless steel chill blocks.

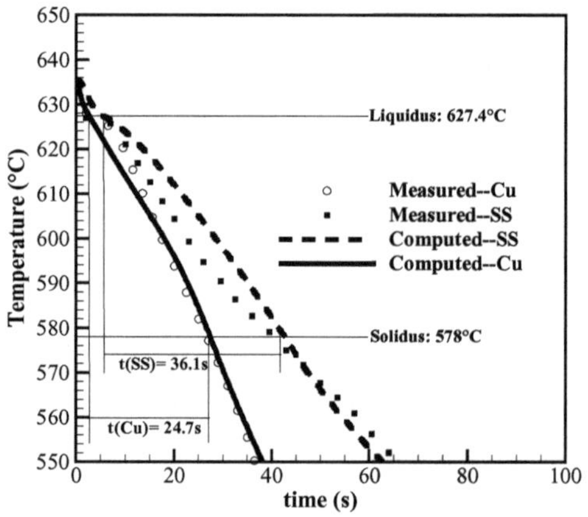


Fig 4. Cooling curves for solidification region.

From these two curves, solidification parameters G and V were evaluated. Figure 5a shows the rate of advance of the solidified layer. As seen in this figure, the growth velocity V is almost constant over the entire height of the casting for both copper and stainless steel chill blocks. The mean growth velocities for the Mg casting on copper and stainless steel chill blocks were 0.75 mm/s and 0.6 mm/s, respectively. In contrast, the value of G at the solidification front depend on the height of the casting, Figure 5b, and is mainly due to the increase of thermal resistance of heat extraction in the chill block during solidification. For solidification on copper chill block, G decreases from 3.5 °C/mm at the bottom of the mold to 2.5 °C/mm at the top of the ingot, with an average value of 3 °C/mm. The corresponding G values for the stainless steel chill block ranged from 3.5 °C/mm to 1.5 °C/mm, with an average value of 2.5 °C/mm. From these values, the average cooling rate, which is used to correlate the grain size with solidification rate [7], was evaluated. For copper chill block, the cooling rate is 2.25 °C/s, while for stainless steel the value was 1.5 °C/s. It is important to note that for castings at high superheats these solidification parameters correspond to equiaxed dendritic solidification morphology [5].

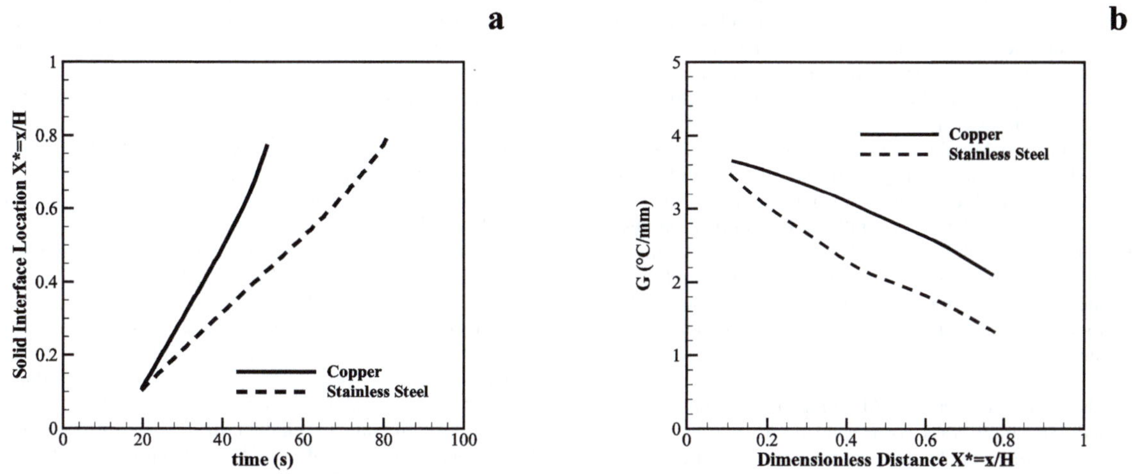


Fig 5. Solidification parameters for copper and stainless steel chill blocks: (a) solidification rate V , (b) Temperature gradient G.

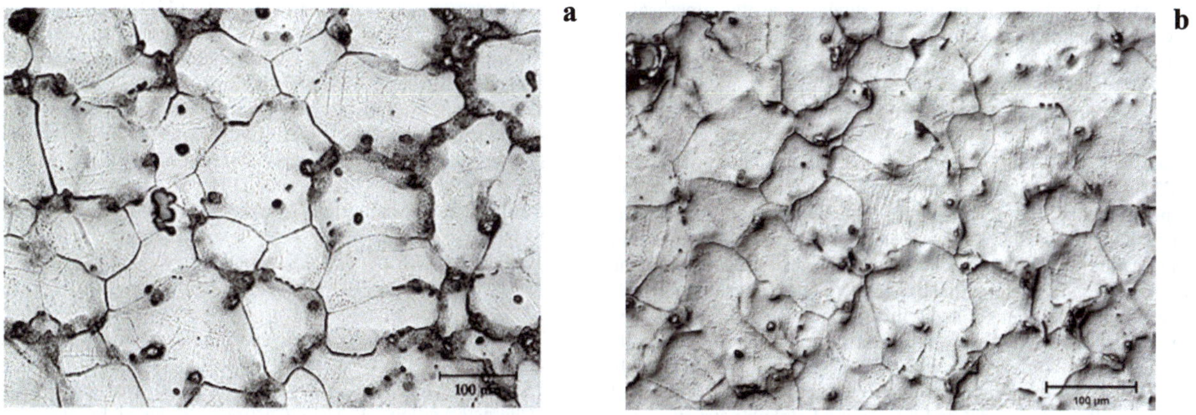


Fig 6. Grain structure of MSM cast Mg AZ31B at 8°C superheat at 200 X for: (a) stainless steel and (b) copper chill blocks

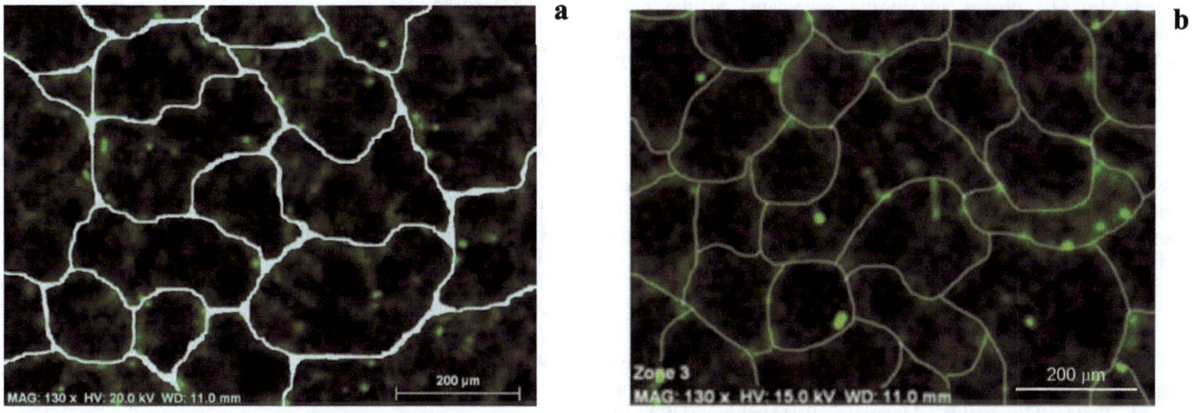


Fig 7. EDS Maps of Al solute distribution in MSM cast Mg AZ31B at 8°C superheat for: (a) stainless steel and (b) copper chill blocks

Figures 6a and 6b show the optical micrographs of the grain structures for cast Mg AZ31B on copper and stainless steel chill blocks at a superheat of 8°C. In these micrographs the brighter areas are the α-Mg phase, while the darker areas at the grain boundaries correspond to the Al-rich secondary-α phase and $Mg_{17}Al_{12}$ intermetallics. This suggests globular solidification morphology for cast Mg alloys at low superheat [17,18]. Further inspection shows that the average grain size for a solidification rate of 0.6 mm/s, which corresponds to stainless steel chill block, is 92 μm, while that of the copper chill block solidification rate of 0.75 mm/s is approximately 75 μm. These results suggest that the grain size is inversely proportional to the square root of the cooling rate. Work is currently in progress to obtain more data to develop a more accurate correlation between the grain size and the cooling rate.

Figure 7 shows the EDS maps of Al segregation for the two cases investigated. The light areas correspond to Al-rich secondary-α phase, while the more bright spots are the $Mg_{17}Al_{12}$ intermetallic. These phases are essentially found at the grain boundaries, which further confirm globular solidification morphology. It is also seen that the solidification rate affects the size and distribution of these phases. Moreover, the observed minor entrapment of secondary-α and $Mg_{17}Al_{12}$ phases in the matrix suggest that the superheat of 8°C corresponds to the transition temperature from a globular to dendritic structure.

## Conclusions

The effect of the rate of cooling on the microstructure of cast Mg AZ31B alloy at low superheat of 8°C was investigated experimentally and theoretically. The experimental study was carried out using the MSM process, which is capable of melting and casting metals at very low superheat. Casting experiments were carried out in a bottom chill copper and stainless steel molds. The solidification rates V and temperate gradients G for these two chill blocks were determined from a heat transfer model of the solidification process, which accurately predicts the experimentally measured cooling curves for these two chill blocks. All castings exhibited fine-grained globular structure with grain size less than 100 μm, and the grain size was found to inversely proportional to the square root of the cooling rate. Microprobe analysis of Al segregation showed that the secondary-α and $Mg_{17}Al_{12}$ phases are mostly segregated at the grain boundaries, thus confirming that casting at low superheat promote the transition from dendritic to globular solidification morphology for Mg alloys.

## Acknowledgements

The authors graciously thank the NSF for funding this project under grant number CMMI-0856320. Also, we would like to thank Rob Holler of the Central Analytical Facility at the University of Alabama for his help in using the Microprobe.

## References

1. "Magnesium Vision 2020: A North American Automotive Strategic Vision for Magnesium," Automotive Metals Division, United State Automotive Materials Partnership (USAMP) (November 2006).

2. S.G. Lee, et al., "Characterization of the effects of process parameters on macrosegregation in a high-pressure die cast magnesium alloy," *Materials Characterization*, 55 (2005), 219-224.

3. D.G.L. Prakash and D. Regener, "Quantitative characterization of Mg17Al12 phase and grain size in HPDC AZ91 magnesium alloy," *Journal of Alloys and Compounds*, 2007, 461, 139-146.

4. S.I. Bakhtiyarov and R.A. Overfelt, "First V-Process casting of magnesium," *Proceedings of Magnesium Technology 2004*, ed. A.A. Luo, (Warrendale, PA: TMS, 2004), 187-192.

5. K. Fisher, *Fundamentals of Solidification* (Aedermannsdorf, Switzerland: Trans Tech Publications, 1986), 88.

6. D.H. Kang, M. Paliwal, E. Essadiqi, and I.H.Jung, "Experimental Studies on the As-Cast Microstructure of Mg-Al Binary Alloys with Various Solidification Rates and Compositions," *Proceedings of Magnesium Technology 2010*, eds. S.R. Agnew et al, (Warrendale, PA: TMS, 2010), 533-536.

7. M. Masoumi and M. Pekguleryuz, "Effect of Cooling Rate on the Microstructure of AZ31 Magnesium Alloy," *AFS Transactions*, 2009, 617-626.

8. Y. He, A. Javaid, E. Essadiqi, and M. Shehata, "Numerical Simulation and Experimental Study of the Solidification of a Wedge-Shaped AZ31 Mg Alloy Casting," *Canadian Metallurgical Quarterly*, 48 (2) (2009), 145-156.

9. Q. Jin, J.P. Eom, S.G. Lim, W.W. Park and B.S. You, "Grain Refining Mechanism of a Carbon Addition Method in a Mg-Al Magnesium Alloy," *Scripta Materialia* 49(2003), 1129-1132.

10. G. Han, X. Liu, and H. Ding, "Grain Refinement of AZ31 Magnesium Alloy by New Al-Ti-C Master Alloys," *Transactions of Nonferrous Metals Society of China*, 19(2009), 1057-1064.

11. P. Cao, M. Quian, and D.H. St. John, "Native Grain Refinement of Magnesium Alloys," *Scripta Materialia*, 53(2005), 841-844.

12. Brinegar, J.R., *et al.*, U.S. patent, No. 4832112 (1989).

13. Y. Ma, et al., *Journal of Materials Processing Technology*, 137(2003), 35-39.

14. L. Liu, et al., *Materials Science and Engineering A* 394(2005), 1-8.

15. D. J. Reynolds, M. Shamsuzzoha, and N. El-Kaddah, "Characterization of Al-Li Castings Produced by the Magnetic Suspension Melting Process," *Proceedings of the 4th Decennial International Conference on Solidification Processing*, eds. J. Beech and H. Jones, (Sheffield, UK: University of Sheffield, 1997), 45-48.

16. C. Mahato, M. Shamsuzzoha, and N. El-Kaddah, "Solidification Morphology and Structure of Cast Al-Li 2090 Alloy at Low Superheats," *Solidification of Aluminum Alloys*, eds. M.G. Chu et al., (Warrendale, PA: TMS, 2004), 321-328.

17. N. Rimkus, L. Vihtelic, and N. El-Kaddah, "Grain and Microstructure of Cast Mg. AZ31B Alloy Produced by the Magnetic Suspension Melting Process," *Proceedings of Light Weight Materials for Vehicles and Components Symposium*, (Warrendale, PA: TMS, 2004), 753-761.

18. N.W. Rimkus, M.L. Weaver, and N. El-Kaddah, "Microstructure and Mechanical Behavior of Cast Mg AZ31B Alloy Produced by the Magnetic Suspension Melting Process," *Proceedings of Magnesium Technology 2011*, eds. W. Silleken et al., (Warrendale, PA: TMS, 2011), 335-343.

**Magnesium Technology 2012**
*Edited by: Suveen N. Mathaudhu, Wim H. Sillekens, Neale R. Neelameggham, and Norbert Hort*
***TMS (The Minerals, Metals & Materials Society), 2012***

# IMPACT AND ENERGY DISSIPATION CHARACTERISTICS OF SQUEEZE AND DIE CAST MAGNESIUM ALLOY AM60

Sante DiCecco, Henry Hu, and William Altenhof

Mechanical, Automotive & Materials Engineering
University of Windsor, 401 Sunset Ave.
Windsor, Ontario, Canada N9B 3P4

Keywords: Squeeze Casting, Magnesium, Fracture properties, Instrumented Charpy Impact testing

## Abstract

High-pressure die cast (HPDC) magnesium alloy AM60 is recognized for its versatility in the manufacturing of weight sensitive components of relatively thin cross section. To further expand practical applications of the alloy, squeeze casting has been proposed to allow for thicker castings. In this study, AM60 alloy specimens of 10mm thickness were squeeze cast using a hydraulic press with an applied pressure of 60 MPa. Fracture energies, following a Charpy Impact Testing protocol, of the squeeze cast specimens were characterized in comparison with the HPDC counterparts using both experimental and numerical techniques. The experimental results show the squeeze cast alloy absorbing approximately 46.2% more energy during impact than its HPDC counterpart. Scanning electron microscopy fractography reveals the favourable quasi-cleavage fracture mode of the squeeze cast alloy AM60, relative to the decohesive rupture fracture mode present in the die cast alloy.

## Introduction

Weight reduction is a driving factor in the transportation sector, where fuel efficiency is paramount in the success of any commercial vehicle design. As such, the development and application of magnesium alloys has seen rapid growth, due to the high specific strength of the alloys, and their excellent die castability [1]. Despite the advances in magnesium alloys, the casting of body-on-frame magnesium alloy components has been restricted to thin-walled applications due to the high porosity and coarse microstructure associated with the die casting of relatively thick components [2]. However, past research has indicated that squeeze casting of magnesium alloys provides a means of casting relatively thick components, while maintaining a fine microstructure with low levels of porosity [2].

Squeeze casting is a process that involves the solidification of molten metal by making use of very large filling gates and imposed high pressures. The high applied pressures keep entrapped gases in solution and squeeze the molten metal from hot spots to incipient shrinkage pores. As a result, the porosity in a squeeze cast component is almost completely eliminated, with an accompanying increase in densification. Furthermore, the elimination of the air gap at the liquid-mould enhances heat transfer, resulting in enhances solidification and cooling rates. Thus, the squeeze casting process promotes improved mechanical, thermal, and microstructural properties, due to grain refinement and porosity reduction within the cast component [2-5].

Previous research has focused on the microstructural behaviour and tensile properties of squeeze cast magnesium alloys as a function of section thickness and casting method. The objective of this study was to determine the impact resistance and fracture behaviour of the alloy for a relatively thick specimen, using both casting methods, and to establish a numerical model which accurately predicts the energy absorbed until failure by each specimen during Charpy Impact Testing. This paper presents the results of instrumented Charpy Impact testing, numerical modeling of the Charpy Impact event, and SEM fractography of the fracture specimens.

## Experimental Procedure

Alloy Preparation

| The Magnesium alloy selected for this study is the conventional magnesium alloy AM60, of which the chemical composition is listed in Table 1. Cylindrical coupons with a diameter of 95mm and section thickness of 10mm were squeeze cast. The squeeze casting processes began with the transfer of the molten alloy (690°C) into the bottom half of a preheated (275°C) die set mounted in a hydraulic | Al (wt.%) | Mn (wt.%) | Zn (wt.%) | Ni (wt.%) |
|---|---|---|---|---|

| | | | | |
|---|---|---|---|---|
| press. The dies were then closed, with the top half lowering into the bottom die. A steadily increasing pressure was applied to the melt by the bottom half of the die by means of a piston until a pressure of 60 MPa was achieved. This pressure was held until the entire casting had solidified. For the purpose of this study, flat rectangular coupons (125mm x 27mm) with a section thickness of 10 mm were cast in a 700 ton cold chamber horizontal high pressure die casting machine.Alloy Symbol | | | | |
| AM60A | 5.89 | 0.335 | 0.0068 | 0.0004 |

Squeeze and die cast castings were cut for density measurement, porosity evaluation, Charpy Impact testing and the subsequent SEM fractography analysis.

Porosity Evaluation

Porosity of the coupons was evaluated via density measurement. Following the measurement of specimen weight in the air and distilled water, the actual density ($D_a$) of each specimen was determined using Archimedes principle based on ASTM standard D3800 [6]:

$$D_a = W_a D_w/(W_a - W_w) \quad [1]$$

where $W_a$ and $W_w$ are the weight of the specimens in air and in water, respectively, and $D_w$ is the density of water. The porosity of each specimen was calculated by the following equation:

$$\%Porosity = [(D_t - D_a)/D_t] \times 100\% \quad [2]$$

where $D_t$ is the theoretical density of the alloy AM60, which is 1.8g/cm$^3$).

Instrumented Charpy Impact Testing

The relative fracture energies of the squeeze and die cast alloy were obtained using a Riehle Impact Testing Machine that has been calibrated and instrumented following ASTM standards E23 [7] and E2298 [8]. Charpy specimens were sectioned from the center of each casting with dimensions of 10mm x 10mm x 55mm and were tested in the unnotched condition. Tests were performed at room temperature to obtain the load/displacement behaviour of the specimens during impact. Absorbed impact energies were subsequently calculated, and the average of six tests was taken for each casting condition.

Fractography Analysis

For the fracture analysis, the fracture surfaces of specimens that had undergone Charpy Impact testing were viewed under low and high magnifications of a JSM-5800LV scanning electron microscope (SEM). Areas of ductile and brittle failure were subsequently noted and correlated with previously calculated porosity measurements.

**Results and Discussion**

Porosity Evaluation

Figure 1 presents the porosity and density measurements taken of the AM60 series alloy for the squeeze and die cast conditions, with a casting thickness of 10mm. As can be seen, the squeeze cast sample has only 0.72% porosity, while the die cast sample has 4.00% porosity. This corresponds to a decrease in porosity of 81.8% relative to the die cast sample. Furthermore, there is a 5.16% increase in density of the squeeze cast sample relative to the die cast sample, due to the reduction in porosity. The porosity reduction and accompanying densification of the squeeze cast AM60 alloy can be attributed to the fact that the applied pressure suppresses gas nucleation, while enabling the melt to penetrate areas of microshrinkage forming in the last solidifying region of the casting.

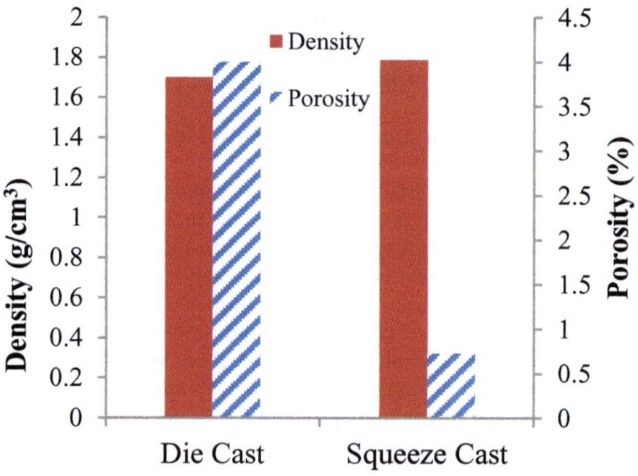

Figure 1: Effect of casting method on density and porosity of AM60.

Instrumented Charpy Impact Testing

Figure 2 shows the absorbed fracture energies of the AM60 alloy after undergoing impact to failure using a Charpy Impact testing machine. From these results, it is apparent that the squeeze cast alloy requires more energy for failure to occur, with the squeeze cast alloy requiring 46.2% more energy relative to the die cast alloy. These results indicate that AM60 cast using the squeeze casting process, for a casting thickness of 10mm, has a greater resistance to high impact loadings relative to the die cast alloy, and is thus more ductile. This finding may be directly attributable to the fact that the squeeze cast specimens had significantly less porosity with respect to the die cast specimens.

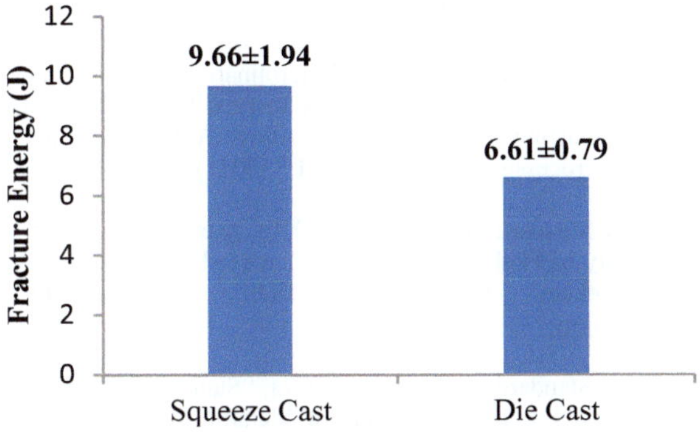


Figure 2: Absorbed energy to failure under Charpy Impact loading of 10mm thickness squeeze and die cast AM60 alloy.

Fractography Analysis

Differences in the fracture behaviour of the die cast and squeeze cast 10mm thickness AM60 alloy are clearly revealed in Figures 3 and 4 by SEM fractography. Figure 3 shows the typical fracture surface that was observed of the squeeze cast alloy. As can be seen from Figure 3(a), the fracture surface contains many flat facets, which are characteristic of the quasi-cleavage fracture mode. The flat facets are accompanied by steps and river markings. The steps are the result of the cleavage propagation interacting with dislocation tangles or twins, while the river markings are attributable to shear connections forming from steps created along grain boundaries with non-zero twist components. Therefore, the formation of river markings is the primary indicator that some plastic deformation has occurred in the squeeze cast alloy prior to failure, which correlates well with the previously discussed absorbed energies during Charpy Impact testing. Also note that the river markings move from right to left, which corresponds to the deformation induced from the Charpy Impact striker. The fractograph with higher magnification, Figure 3(b), better shows the depth of the steps and river markings, which further demonstrate that plastic deformation occurred in the fracture specimen of the squeeze cast alloy.

Figure 4 shows the typical fracture behaviour that was observed after Charpy Impact testing of the 10mm die cast AM60 alloy specimens. Converse to the squeeze cast alloy, the die cast alloy exhibits a fracture mode most consistent with decohesive rupture. This is illustrated in Figure 4(a). There exist a number of sites where intergranular fracture is present. The cause for this mode of fracture may be attributed to the porosity present within the die cast specimen, which causes low cohesion between grains, thus providing an ideal location for crack initiation. Figure 4(b) provides a higher magnification fractograph that is in good agreement with the conclusions drawn from Figure 4(a). However, in the higher magnification fractograph there exists a small amount of cleavage in the top right corner of the image, indicative that some plastic deformation may still occur in regions of lower porosity. Regardless, the results of SEM fractography for both casting methods are in good agreement with the results obtained from Charpy Impact testing.

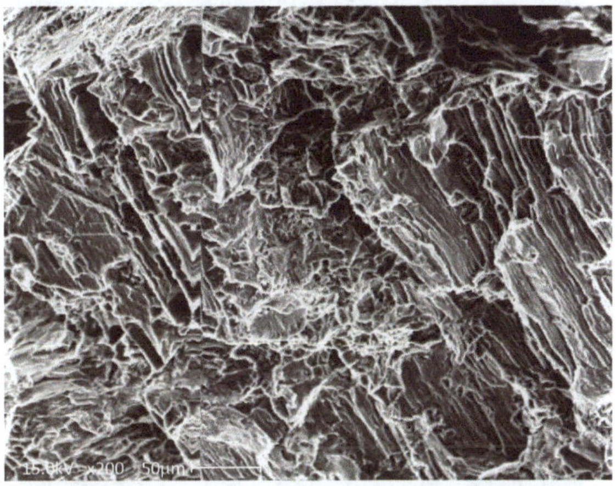

(a)

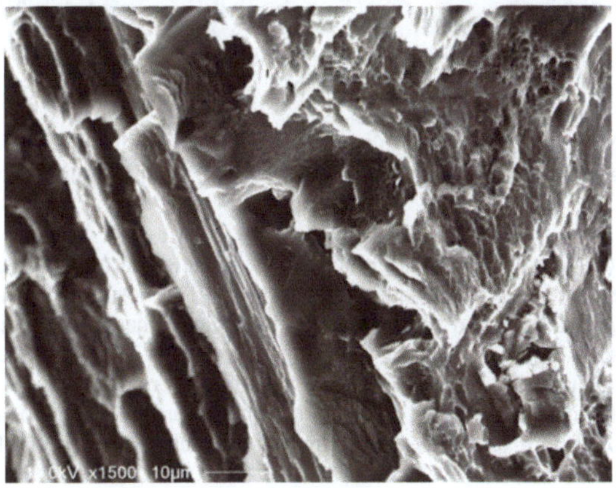

(b)

Figure 3: SEM fractographs of squeeze cast Mg alloy AM60 with a 10mm cast thickness after undergoing Charpy Impact testing, (a) low magnification and (b) high magnification.

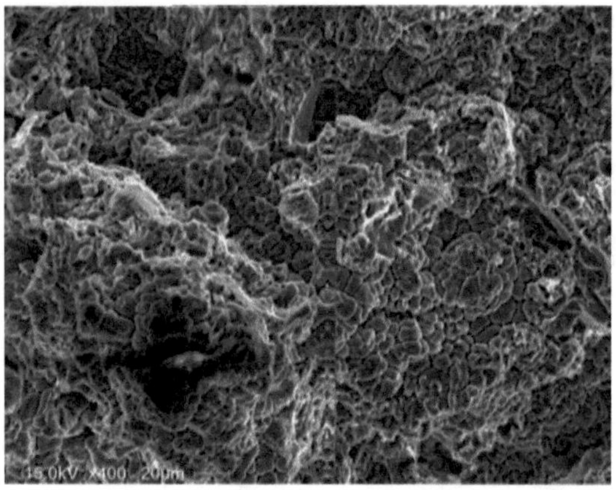


(a)

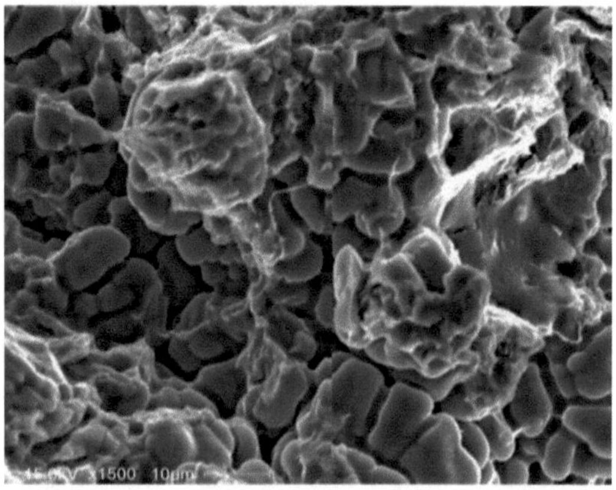

(b)

Figure 3: SEM fractographs of the die cast alloy after Charpy Impact testing (a) low magnification and (b) high magnification.

**Conclusions**

The effect of the casting process was investigated on the fracture mechanics of Magnesium alloy AM60. The results of Charpy impact testing and SEM fractography indicate that for a relatively thick casting, squeeze casting promotes ductile fracture characteristics and an increase in absorbed energy to fracture. The reduction in porosity and accompanying material densification should be responsible for the improved fracture properties of the squeeze cast alloy.

**Acknowledgment**

The authors would like to take this opportunity to thank the Natural Sciences and Engineering Research Council of Canada for supporting this work. One of the co-authors, Sante DiCecco, wishes to acknowledge the government of Ontario and University of Windsor for financial support in the form of an Ontario Graduate Scholarship and a University of Windsor Tuition Scholarship, respectively.

**References**

1. Hu, H., Yu, A., Li, N. & Allison, J.E. "Potential Magnesium Alloys for High Temperature Die Cast Automotive Applications: A Review." Materials and Manufacturing Processes 18.5 (2003): 687-717.

2. Zhou, M., Hu, H., Li, N. & Lo, J. "Microstructure and Tensile Properties of Squeeze Cast Magnesium Alloy AM50." Journal of Materials Engineering and Performance 14.4 (2005): 539-45.

3. Wang, S., Sun, Z. & Hu, H. "Effect of Pressure Levels on Tensile Properties of Squeeze Cast Mg-Al-Sr Alloy." Magnesium Technology (2008): 421-25.

4. Masoumi, M. & Hu, H. "Influence of Applied Pressure on Microstructure and Tensile Properties of Squeeze Cast Magnesium mg-Al-Ca Alloy." Materials Science and Engineering A 528.10-11 (2011): 3589-93.

5. Sun, Z., Hu, H. & Niu, X. "Determination of Heat Transfer Coefficients by Extrapolation and Numerical Inverse Methods in Squeeze Casting of Magnesium Alloy AM60." Journal of Materials Processing Technology 211.8 (2011): 1432-40.

6. ASTM Standard D3800, 1999 (2010), "Standard Test Method for Density of High-Modulus Fibers", ASTM International, West Conshohocken, PA, 2002, DOI: 10.1520/D3800-99, www.astm.org

7. ASTM Standard ASTM E23, 2007a, "Standard Test Methods for Notched Bar Impact Testing of Metallic Materials", ASTM International, West Conshohocken, PA, 2002, DOI: 10.1520/E0023-07AE01, www.astm.org

8. ASTM Standard E2298, 2009, "Standard Test Method for Instrumented Impact Testing of Metallic Materials", ASTM International, West Conshohocken, PA, 2002, DOI: 10.1520/E2298-09, www.astm.org

**Magnesium Technology 2012**
*Edited by: Suveen N. Mathaudhu, Wim H. Sillekens, Neale R. Neelameggham, and Norbert Hort*
***TMS (The Minerals, Metals & Materials Society), 2012***

# SLIDING WEAR BEHAVIOUR OF SQUEEZE CAST MAGNESIUM COMPOSITE AM60-9% $(Al_2O_3)_f$

A. Banerji[1], H. Hu[1], A.T. Alpas[1]

[1]Department of Mechanical, Automotive & Materials Engineering,
University of Windsor, 401 Sunset Avenue,
Windsor, Ontario, Canada N9B 3P4

Keywords: Squeeze cast composites, Sliding wear, Wear loss of alloy and composites, Damage to counterface

## Abstract

Magnesium matrix (AM60) composites reinforced with 9 vol% $Al_2O_3$ fibres were fabricated using $Al_2O_3$fibre preform preparation and squeeze casting techniques. Boundary lubricated sliding wear tests were performed on the composites and the matrix alloy at low loads (1.0-5.0 N) to evaluate their wear characteristics. The Mg matrix alloy underwent scuffing and oxidative type wear leading to high rate of material removal. Under the same conditions, the composites' wear resistance was at least $10^2$ times higher. At 1.0 N load, $Al_2O_3$fibres protected the matrix from damage for up to $1.0 \times 10^6$ sliding cycles. For longer cycles $Al_2O_3$ fibres underwent fracture and exhibited decohesion at the fibre/matrix interface. At higher loads and sliding cycles, the matrix suffered wear damage leading to an increase in wear rate. The counterface, AISI 52100, worn against the base alloy AM60 revealed slight damage in the form of surface scratches and presence of MgO. The counterface experienced higher amount of material removal when it was worn against AM60-9% $(Al_2O_3)_f$ composite.

## Introduction

During the last two decades, automotive engineers have focused on developing and using structural lightweight materials. The motivation for mass reduction is directly related to resistance to vehicle performance and fuel consumption and correspondingly automotive emissions. In this endeavor, Mg alloys and composites have received special interest.

The rapid increase in commercial usage of die cast Mg alloys may be attributed almost entirely to automotive use. Apart from being lighter than aluminum other factors that make use of Mg lucrative are its good castability and machinability. Also, low heat capacity and latent heat of solidification are further advantages to Mg resulting in shorter casting cycles and longer die life [1-3].

Understanding of wear behaviour of cast Mg alloys is critical for their use in engine powertrain applications. Piston cylinder bore assemblies, running under lubricated conditions should result in very little material removal, with the wear rate not exceeding a few nanometers per hour [4]. This feature qualifies engine wear to be classified as ultra-mild wear (UMW), since the rate of wear is few orders lower than mild and severe wear and also the micromechanisms of UMW are different. For the hypereutectic Al-Si alloys (e.g., Al-18.5% Si) with large primary Si particles, the surface damage is limited to the tops of these particles, with the alloy retaining its original microstructure. The UMW in eutectic Al-Si alloys is characterized by sinking-in of the Si particles into the matrix. The Si particles with large aspect ratio are usually fractured. The Al matrix suffers damage until a protective oil-residue layer forms comprising of fractured silicon particles comminuted to nano-size fragments and organic components. Once this layer is formed the matrix is protected from further wear and UMW regime prevails as long as the layer remains intact [4-7].

Reinforcing Mg alloys with discontinuous ceramic fibres and particles helps to improve wear resistance. Alahelisten et al. [8] examined $Al_2O_3$-fibre-reinforced Mg–9Al–1Zn (AZ91) composites in sliding, abrasion and erosion tests. The authors observed that the sliding wear resistance was the highest for an optimum fibre content of 10%. For two body abrasion, abrasion resistance improved continuously with fibre content. Other studies with Mg alloys reinforced with $Al_2O_3$ fibres, SiC and TiC particles have been reported [9-12]. In all these studies dry sliding wear behaviour of Mg composites was studied at different loads and the general observation was that wear resistance of the composite was considerably higher than that of the unreinforced alloy. However, wear properties of Mg composites under lubricated conditions and at low loads, replicating the actual engine operating conditions, is missing in the literature. In this study a Mg alloy, AM60, reinforced with short discontinuous $Al_2O_3$fibres was subjected to boundary lubricated sliding wear under UMW conditions. Understanding the wear mechanisms of Mg matrix composites reinforced with $Al_2O_3$ fibres is expected to facilitate the tribological design of powertrain castings for automotive lightweight strategy.

## Experimental Procedure

Materials

The matrix alloy used for this study was squeeze cast Mg alloy AM60 with a composition (wt%) consisting of Al: 5-6%, Mn: 0.4% etc. Short (20.91±5.9 μm) $Al_2O_3$ fibres (9 vol%) were incorporated in the alloy to form fibre reinforced composites. The composites were prepared by squeeze casting method. $Al_2O_3$ fibres were made into slurry by adding organic and inorganic binders and later dried and sintered to prepare the preform. Both the upper and the lower dies were preheated to 350 °C. The preform and the alloy were heated to 450 °C and 750 °C before pouring it into the lower die. Once the molten magnesium was added to the die containing the $Al_2O_3$ fibres, the dies were closed by raising the lower die and fitting into the upper one. In this way the plunger pushed the molten Mg alloy and the preform into the die cavity under a constant pressure of 90 MPa applied by the plunger to allow liquid Mg alloy to infiltrate into the gaps between the $Al_2O_3$ fibres. This presuure was maintained until the solidification of the casting was complete [12]. The resulting alumina reinforced composite was designated as AM60-9% $(Al_2O_3)_f$. The matrix hardness was 51.0 ± 8.0 HV and the overall hardness of AM60-9% $(Al_2O_3)_f$ was 121.8 ± 11.1 HV. Figure 1 show the setup used to fabricate the composite.

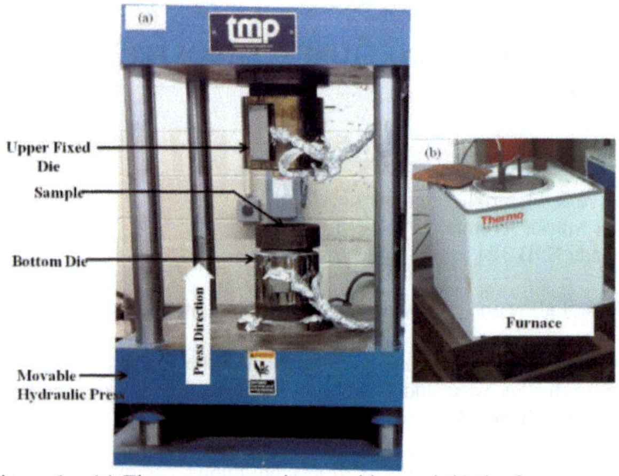


Figure 1 – (a) The squeeze casting machine and (b) the furnace. The dies are closed by raising the lower die against the upper one by the hydraulic press. The magnesium alloy AM60 is melt at 750 °C in the furnace.

Wear Testing

A pin-on-disk type of tribometer was used to perform lubricated sliding wear tests on AM60-9% $(Al_2O_3)_f$ against 6 mm diameter AISI 52100 grade steel balls (848 HV). Sliding wear tests were performed on the composite under 1.0 N, 2.0 N, 5.0 N loads and for sliding cycles between $2.5\times10^4$-$1.0\times10^6$ at a constant sliding velocity of 0.05 m/s. Wear tests were also performed on the matrix alloy AM60 at 1.0 N load for comparison. All wear tests were conducted at room temperature under boundary lubricated conditions using synthetic oil SAE 5W-30. The lubrication regime was evaluated by calculating the ratio ($\lambda$) of minimum lubrication thickness and the root mean square roughness of the surfaces in contact [13] and $\lambda$ was found to be 0.25, 0.24, 0.22 for 1.0, 2.0, 5.0 N loads respectively. The worn surfaces were analyzed using a scanning electron microscope (JEOL 6400) equipped with an energy dispersive spectrometer (EDS) and were quantified using an optical surface profilometer (WYKO NT 1100).

The wear rate of AM60 alloy was calculated by mass difference method from the volume lost vs. sliding distance plot. Wear of composites occurred in the UMW regime where wear losses were less than that could be measured using mass difference method using a balance with high sensitivity ($10^{-5}$ g). Hence, optical profilometry calculations were employed to determine the material removal rates in the composites according to the reference 4. The amount of material removal associated with the groove formations along the wear track were detected in this way. The amount of material removed was obtained from an area A by calculating the cross-sectional area that fell below a reference position with respect to unworn magnesium matrix. The calculated area (Aij) from eight different portions of the wear multiplied by the perimeter of the wear track gives the volumetric wear loss W according to the equation 1 [4]:

$$W = \frac{2}{n}\pi R_w \left[\sum_{i=1}^{n}\sum_{j=1}^{k} A_{ij}\right] \quad (1)$$

where, k is the number of grooves per section and n=24: number of different sections along the wear track from which measurements were made.

Preceding the wear loss calculated as above was the observation that $Al_2O_3$ fibres that were initially protruded above the matrix fractured and buried into the matrix during sliding contact. The difference in elevation of the Mg matrix and the $Al_2O_3$ fibres was quantified by optical profilometry measurements from which histogram showing the distribution (frequency) of Mg matrix and $Al_2O_3$ fibre was plotted. The second Gaussian peak, with higher frequency represents the Mg matrix. This peak serves as the reference point for the other Gaussian peak which represents the $Al_2O_3$ fibre elevation with respect to the matrix.

**Results and Discussion**

Wear of Unreinforced Alloy; AM60

Figure 2 shows the wear losses of AM60 alloy and AM60-9% $(Al_2O_3)_f$ composite plotted as a function of sliding cycles. It is evident that the wear loss from the surface of AM60 alloy was $10^2$ times higher than that of AM60-9% $(Al_2O_3)_f$ despite the fact that the wear losses for the composite reported in this figure were measured at 5.0 N load. Figure 3a shows the SEM micrograph of wear track of the AM60 alloy at 1.0 N for $2.5\times10^4$ sliding cycles and Figure 3b shows the wear track of AM60-9% $(Al_2O_3)_f$ at 2.0 N for $6.0\times10^5$ sliding cycles. The greater extent of wear (width of wear track: 0.73±0.08 mm) for the AM60 alloy at 1.0 N in comparison to the composite (width of wear track: 0.35±0.02 mm) is evident despite being worn at a higher load and for longer sliding cycles.

Secondary scanning electron micrographs of the worn surfaces of the alloy (Figure 4a and b) revealed the prevalent wear features at higher magnification. For low cycles (i.e., for < $5.0\times10^4$ cycles), the damage was in the form of surface plastic deformation. At higher sliding cycles, evidence for oxidative wear was found in addition to the matrix surface deformation. Presence of MgO was detected in the form of wear debris on the wear track and also in the material transferred to the counterface. The material removal rate for the AM60 alloy, $5.67\times10^{-4}$ $mm^3/m$, indicated mild wear throughout the test conditions. According to Fig 2, in the mild wear regime, the volume loss due to sliding wear increased linearly with the sliding distance, indicating that wear progressed under the steady state (i.e. constant wear rate) conditions.

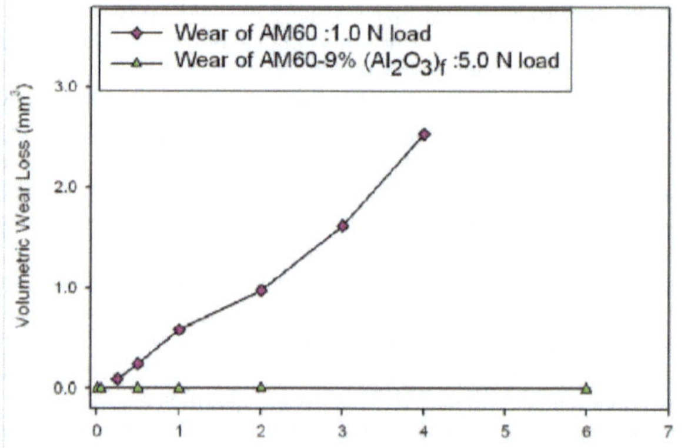


Figure 2- Volumetric wear loss vs. sliding cycles plot for AM60 alloy and AM60-9% $(Al_2O_3)_f$ .The wear loss of the composite is $10^2$ times lower than that of the matrix alloy. Note the wear of the composite was shown at 5.0 N and the matrix at 1.0N load.

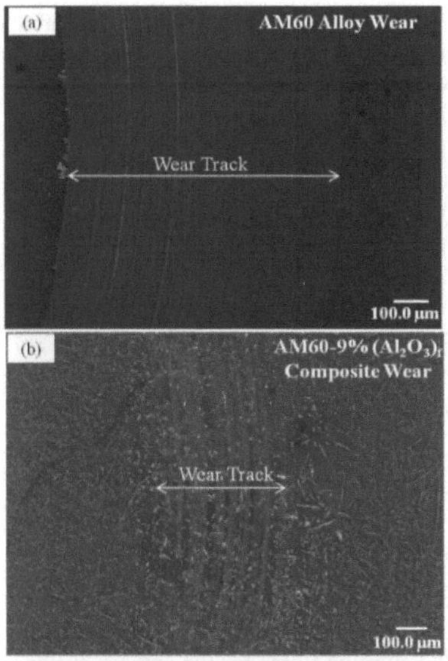


Figure 3- SEM micrographs showing the difference in the width of wear tracks of (a) AM60 alloy at 1.0 N and $2.5\times10^4$ cycles and (b) AM60-9% $(Al_2O_3)_f$ at 2.0 N and $6\times10^5$ cycles indicating the wear resistance of the ceramic reinforced composite is higher than the alloy.

Figure 4- SEM Micrographs showing wear damage of AM60 alloy under 1.0 N load at (a) $2.5\times10^4$ cycles and (b) $4.0\times10^5$ cycles. Plastic deformation and oxidative type wear are detected.

Wear of Composite; AM60-9% $(Al_2O_3)_f$

Wear of AM60-9% $(Al_2O_3)_f$ occurred in the UMW regime. Figure 5 indicates the volumetric wear loss against the sliding cycles at three different loads and helps correlate the material removal rates to the wear features at each load when examined together with the SEM images of contact surfaces (Fig 6). Under 1.0 N load $Al_2O_3$fibres protected the magnesium matrix from damage by counterface for up to the longest sliding cycles ($1\times10^6$) used. Thus, the material removal from the matrix was essentially nil for all sliding cycles (Figure 5). The wear was limited to the tops of the $Al_2O_3$ fibres. Some fibre fracture and decohesion of fibres from the matrix were also observed.

Similar features were observed at 2.0 N load for up to $2\times10^5$ sliding cycles. For $\geq 6\times10^5$ cycles wear damage to the matrix surface was observed in the form of wear grooves covering the entire wear track. Consequently, at this point the wear rate increased to $1.35\times10^{-7}$ $mm^3/m$ as shown in Figure 5.

At 5.0 N load the magnesium matrix damage incurred as early as $1\times10^5$ sliding cycles. The SEM images in Figure 6 show the effects of load on the wear of AM60-9% $(Al_2O_3)_f$: 1.0 N load even for $6.0\times 10^5$ cycles the matrix was unscathed (Figure 6a) whereas at 5.0 N damage to the matrix was observed at $2.0\times10^5$ cycles (Figure 6b). EDS analysis of the worn surfaces of the AM60-9% $(Al_2O_3)_f$ composites showed presence of iron particles which were transferred from the counterface.

Thus at higher loads and higher sliding cycles, the material removal rate was considerably high and correspondingly the wear rate was found to increase cumulatively (Figure 5). However, the highest wear rate estimated at 5.0 N for $2.0\times10^5$ cycles was $1.82\times10^{-6}$ $mm^3/m$ which was of the order of $10^2$ times lower than that of the AM60 alloy.

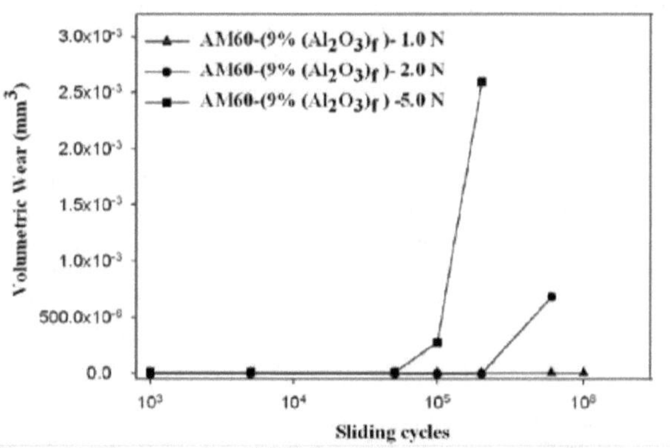


Figure 5- Wear rate for AM60-9% $(Al_2O_3)_f$ composites showing the dependence of wear rate on the applied load and sliding cycles.

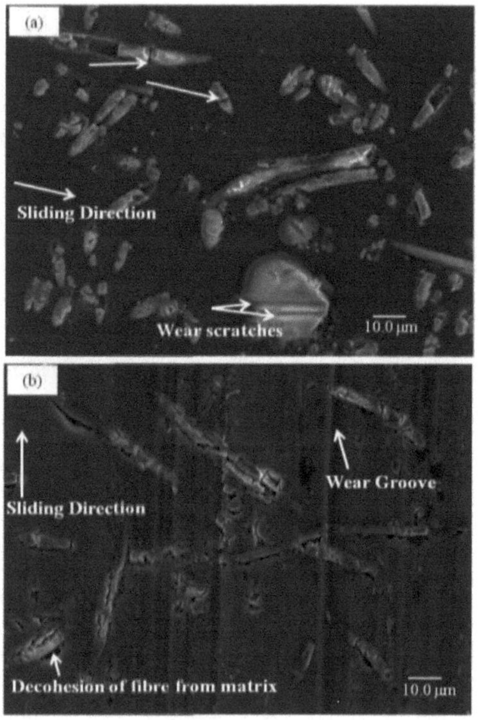


Figure 6-SEM micrographs showing the effect of applied load on the wear of AM60-9% $(Al_2O_3)_f$. (a) At 1.0 N load and $6.0\times10^5$ cycles the matrix is protected by the alumina fibres; (b) at 5.0 N load and $2.0\times10^5$ cycles the matrix was plastically deformed by the counterface.

The higher wear resistance of the AM60-9% $(Al_2O_3)_f$ and its wear rate dependence on load may be related to the increase in the strength of the composite. The addition of discontinuous aluminum oxide short fibres to the AM60 alloy increased the yield strength of the composite from 81 to 125 MPa and tensile strength increased from 171 to 182 MPa [12]. However it is more insightful to consider the wear resistance at microscopic level in the following way: As indicated above the $Al_2O_3$ fibres in the magnesium matrix composite acted as load bearing elements thus protecting the softer matrix from wear by the counterface. Prior to every wear test the sample was finely ground by SiC emery papers to 4000 grit and finally polished by water based diamond suspensions of 3 and 1µm. This process led to an elevation of the $Al_2O_3$fibres by 1.8 ± 0.2 µm, above the magnesium matrix, as determined by optical profilometry (Figure 7). Thus, the AISI 52100 counterface slides against the elevated alumina fibres leaving the softer matrix untouched. As the sliding cycles and applied load was increased the $Al_2O_3$ fibres fractured and the fractured fragments were embedded into the magnesium matrix until the fibres and the matrix came to the same elevation (Figure 7). At this point sliding by the counterface on the soft magnesium matrix caused wear grooves on the wear track. This was reflected in the wear rate plot for the composites (Figure 5). As shown by the change of the slopes of the wear curves in figure 5, the sliding cycles at which the matrix damage initiates becomes shorter with increasing the applied load. However, it should be noted that despite the increase in slope the wear rates remain sufficiently low for all applied sliding cycles so that wear of AM60-9% $(Al_2O_3)_f$ remains in UMW regime. The wear behaviour of AM60-9% $(Al_2O_3)_f$ at higher cycles should be further examined to determine if a transition to higher wear regime would occur.

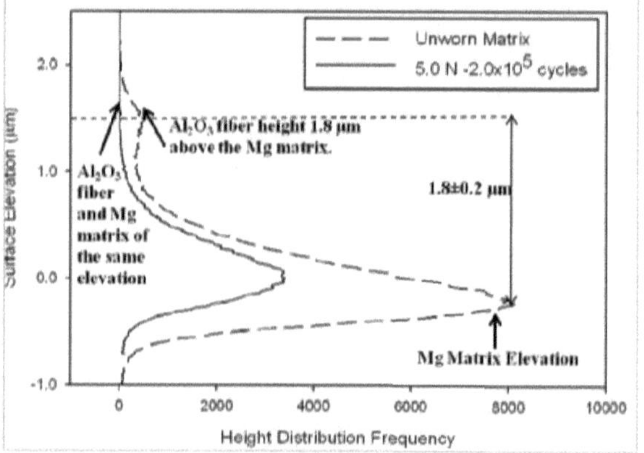


Figure 7- Histogram showing the height differences between the magnesium matrix and the alumina fibres for the unworn surface and also the worn surface after $2\times10^5$ cycles at 5.0 N load. For the unworn surface the average height difference is 1.8 µm whereas for the worn surface the height of the matrix and the fibres are the same after $2\times10^5$ cycles at 5.0 N load.

Wear of Counterface: AISI 52100

When AISI 52100 counterface was worn against the matrix alloy AM60 the wear on the surface of the counterface was limited to the formation of scratch marks (Figure 8a). EDS analysis of the counterface showed presence of transferred magnesium oxide wear debris, which indicated oxidative type wear between the steel ball and the magnesium alloy.

In AM60-9% $(Al_2O_3)_f$ composite-counterface system, initial contact always occurred between the $Al_2O_3$ fibres and the AISI 52100 surface. Since the ceramic fibres were harder (1365 HV) than the steel (848 HV), wear groove formation was detected on AISI 52100 (Figure 8b). EDS analysis of the counterface indicated no material transfer between the composite and the counterface. Presence of Fe particles was detected on the wear track of the composite which probably resulted from the transfer of material from the counterface (Figure 9). Thus $Al_2O_3$ fibres in AM60-9% $(Al_2O_3)_f$ composite proved to be disadvantageous for the wear of the counterface AISI 52100

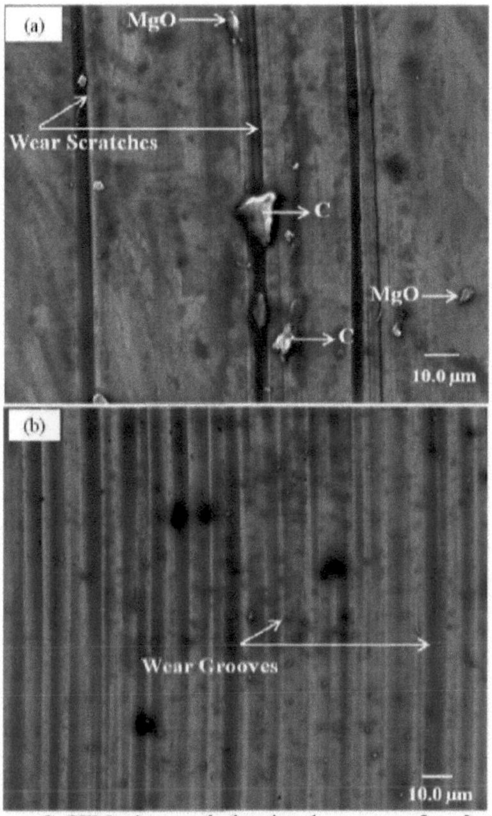


Figure 8- SEM micrograph showing the worn surface features of the counterface due to wear against (a) AM60 alloy at 1.0 N and $2.5\times10^4$ cycles and (b) AM60-9% $(Al_2O_3)_f$ at 1.0 N and $2.0\times10^5$ cycles. Transfer of wear debris (MgO) from the alloy to the counterface was detected.

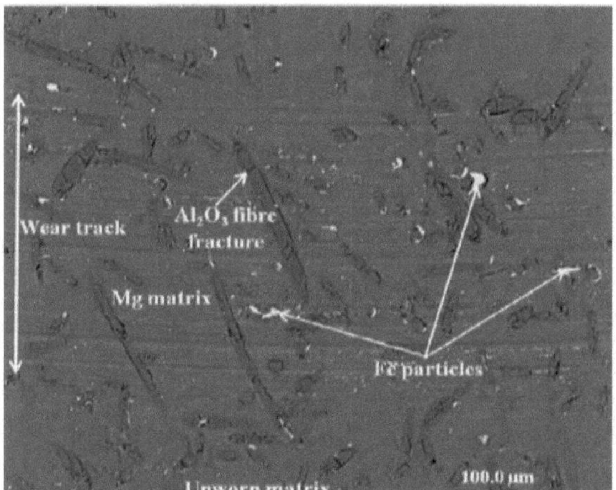


Figure 9- Back-scattered micrograph of worn surface of AM60-9% $(Al_2O_3)_f$ at 5.0 N and $1.0\times10^5$ cycles showing presence of iron particles transferred from the counterface AISI 52100 due to sliding against harder alumina fibres.

**Summary and Conclusions**

Magnesium alloy AM60 incorporating short discontinuous $Al_2O_3$ fibres prepared from a perform were used to fabricate an AM60-9% $(Al_2O_3)_f$ composite by squeeze casting method. The matrix alloy and the composite were subjected to boundary lubricated sliding wear under 1.0-5.0 N load. The main findings of the study are summarized as follows:

1. Wear mechanisms of magnesium alloy, AM60, at 1.0 N consisted of damage by surface plastic deformation and oxidative type wear. Material removal rate for the AM60 alloy, $5.67\times10^{-4}$ $mm^3/m$, indicated occurrence of mild wear throughout test conditions. Formation of wear debris, consisting of MgO particles, was observed on the wear track after $4\times10^5$ sliding cycles. Transfer of magnesium oxide particles from the alloy to the counterface was detected.
2. Wear of AM60-9% $(Al_2O_3)_f$ composite under 1.0-5.0 N loads occurred in ultramild wear regime, with wear rates $< 2.0\times10^{-6}$ $mm^3/m$. With increase of applied load and sliding cycles, there was an increase in rate of Mg matrix removal.
3. The $Al_2O_3$ fibres protected the matrix from damage by the counterface up to a certain sliding distance, which decreased with the increase in the applied load. Fibre fracture and interfacial decohesion led to the embeddeing of the fibres into the matrix and eventually resulting in an increased rate of material removal.
4. Wear of the counterface, AISI 52100, during sliding against the magnesium alloy AM60, was low and manifested in the form of wear scratches. The amount of material removal from the counterface was higher when tested against the composite.

It can be concluded that the AM60-9% $(Al_2O_3)_f$ composites possess enhanced wear resistance over the base alloy AM60. However, the greater amount of wear of the counterface at higher loads necessitates further study with respect to modification of the counterface material to reduce the amount of wear therein.

**References**

1. H. Hu, A. Yu, N. Li and J. Allison, Potential magnesium alloys for high temperature die cast automotive applications: A review, *Materials and Manufacturing Processes*, 18 (2003), 687-717.
2. M. O. Pekguleryuz and A. A. Kaya, Creep resistant magnesium alloys for powertrain applications, *Advanced Engineering Materials*, 5(2003), 866-871.
3. K.M. Asl, A. Masoudi and F. Khomamizadeh, The effect of different rare earth elements content on microstructure, Mechanical and Wear Behavior of Mg-Al-Zn Alloy, *Material Science and EngineeringA*, 527,( 2010), 2027-2035.
4. M. Chen, X. Meng-Burany , T.A. Perry and A.T. Alpas, Micromechanisms and mechanics of ultra-mild wear in Al–Si alloys, *Acta Materialia*, 56 (19), (2008), 5605–5616.
5. M. Chen, T.A. Perry and A.T. Alpas, Ultra-mild wear in eutectic Al–Si alloys, *Wear*, 263(2007),. 552–561.
6. M. Chen and A.T. Alpas, Ultra-mild wear of a hypereutectic Al–18.5 wt.% Si alloy, *Wear*, 265 (1–2) (2008), 186–195.
7. S.K. Dey, T.A. Perry, A.T. Alpas, Micromechanisms of low load wear in an Al-18.5% Si alloy, *Wear*, 267 (2009), 515–524.
8. A. Alahelisten, F. Bergman, M. Olsson, and S. Hogmark, On the wear of aluminium and magnesium metal matrix composites, *Wear*, 165 (1993), 221-226.

9. B. Hu, L. Peng and W. Ding, Dry sliding wear behavior of Saffil fiber-reinforced Mg-10Gd-3Y-0.5Zr magnesium alloy-based composites, *Journal of Composite Materials*, 45(201), 683–693.
10. C.Y.H. Lim, S.C. Lim and M. Gupta, Wear behaviour of SiCp-reinforced magnesium matrix composites, *Wear*, 255(2003), 629–637.
11. S.J. Huang, Y.R Jeng., V. I. Semenov and Y.Z. Dai, Particle Size Effects of Silicon Carbide on Wear Behavior of SiCp-Reinforced Magnesium Matrix Composites, *Tribological Letters*, 42(2011), 79–87.
12. L.F. Franco, E. B. Becerril, J. L. Ruiz, J.G. Gonzalez-Rodriguez, R. Guardian and I.Rosales, Wear performance of TiC as reinforcement of a magnesium alloy matrix composite, *Composites:Part B*, 42(2011), 275-279.
13. Q. Zhang and H. Hu, Advances in Light Weight Materials– Aluminum, Casting Materials and Magnesium Technologies, *SAE 2010 World Congress*, (2010), 107-114.
14. I.M. Hutchings, Tribology: Friction and Wear of Engineering Materials-(Metallurgy & Materials science series), (CRC Press, 621.89, 1992), 77-105.

**Magnesium Technology 2012**
*Edited by: Suveen N. Mathaudhu, Wim H. Sillekens, Neale R. Neelameggham, and Norbert Hort*
***TMS (The Minerals, Metals & Materials Society), 2012***

# SOLIDIFICATION STUDIES OF Mg-Al BINARY ALLOYS

Manas Paliwal[1], Youn-Bae Kang[2], Elhachmi Essadiqi[3] and In-Ho Jung[1]

[1]Department of Mining and Materials Engineering, McGill University, Montreal, QC H3A 2B2, Canada
[2]CANMET-MTL, Ottawa, ON K1A 0G1, Canada
[3]Graduate Institute of Ferrous Technology, POSTECH, Pohang, Republic of Korea

Keywords: Mg-Al binary alloys, Secondary dendrite arm spacing, Cellular-columnar transition

## 1. Introduction

The knowledge of as cast microstructure of Mg alloys is of prime importance to design the downstream thermo mechanical processes like rolling and annealing which then follows with commercial applications of Mg alloys in various industrial sectors [1]. Specifically, the emerging technologies like Twin roll casting that require minimal thermo mechanical processing; it is the solidified microstructure at the casting stage which determines the mechanical properties of the final product. Hence, the as cast microstructural information of Mg alloys becomes imperative but there has been little or no work on the same. In this regard, the present study focuses on the influence of the solidification parameters like thermal gradient (G), solidification velocity (V), cooling rate and solute content on the microstructural features of Mg-Al alloys. Directional solidification experiments and gravity experiments have been performed to explore the relationship between secondary dendrite arm spacing (SDAS) of Mg-Al alloys and the solidification parameters. The transition between cellular-columnar growth modes for Mg-Al alloys has also been investigated in context to the solidification parameters.

## 2. Experimental

Directional solidification experiments have been performed using a modified bridgemann furnace which consists of an induction furnace mounted on top and a middle section which houses the Mg sample resting on water cooled Cu mould. The sample length was fixed at 200 mm and the solidification experiments were carried out in alumina crucibles of 5.5 and 4.5 mm I.D and 300 mm in length. The furnace temperature was set to 710 °C. The experiments are started by placing the sample in the induction furnace column and then the sample is pulled down with the desired solidification velocity for 100 mm and then quenched with He gas. The directional solidification experiments enable to control the solidification parameters like thermal gradient (G) and solidification velocity (V) and the product of these two parameters (G.V) gives the cooling rate. The cooling rates that could be achieved through these experiments for Mg-Al binary alloys were in the range of 0.05 to 2.75 K/sec and the associated G and V are listed in table 1.

Table1: The parameters for directional solidification experiments.

| Thermal gradient (K/mm) | Solidification velocity (mm/sec) | Cooling rate (K/sec) |
|---|---|---|
| 8.5 | 0.00625 | 0.05 |
| 7.5 | 0.05 | 0.38 |
| 7.0 | 0.17 | 1.16 |
| 5.5 | 0.50 | 2.75 |

For intermediate and higher cooling rates, gravity casting experiments were performed. For example, casting in a wedge shaped Cu mould enables to attain different cooling rates along the varying cross section of the tapered mould. Three thermocouples were put at different positions from the tip of the wedge to attain cooling rates of 30, 75 and 110 K/sec. Casting experiments were also done in sand mould and graphite mould to obtain cooling rates of 1 and 20 K/sec respectively. Injection mould experiments were conducted in 4 and 6 mm Cu mould to obtain higher cooling rates but the exact measurement was not possible due to the limitations of data acquisition systems; however the cooling rates could be estimated from the Secondary dendrite arm spacing (SDAS) measurements which is discussed in next section.

## 3. Results and Discussion

### 3.1 Secondary dendrite arm spacing

Secondary dendrite arm spacing (SDAS) is an important length scale in microstructural solidification as it sets the diffusion distance in solid and liquid phase thus affecting the microsegregration and second phase formation in the solidified microstructure. SDAS investigation has been conducted for AZ91 alloy by many researchers [2-8] in the cooling rate range of 0.1 to $10^5$ K/sec. A linear relationship (log-log scale) was proposed by Dube et al. [7] between SDAS and cooling rate for the cooling rate range mentioned above. However, the SDAS measurements from directional solidification experiments by Tensi and Rosch [3] and Pettersen et al. [2] for AZ91 alloy deviate from the proposed linear relationship in the cooling rate range of 0.1 to 1 K/sec. This deviation in SDAS data by directional solidification experiments cannot be ignored as the SDAS was very accurately determined. More over there is no study in literature which investigates the effect of solute content on the SDAS of Mg-Al alloys which could be of technical importance as the solute content is an important parameter for Mg as-cast alloy design.

The microstructural examination is done for directionally solidified Mg-Al alloys and for the gravity cast Mg-Al alloys. The optical micrographs for the directionally solidified Mg-Al (3, 6 and 9 wt. % Al) alloys at 0.375 K/sec is shown in Fig.1. The microstructure shows columnar growth for the alloys with the primary dendrite stem growing against the heat extraction direction. Substantial side branching is seen for all the alloys with microstructure becoming finer as the solute content increases from 3 and 6 to 9 wt. % Al. Secondary dendrite arm spacing (SDAS) is extracted from the optical microstructures using linear intercept method from all the solidified microstructure obtained through directional solidification and gravity casting experiments.

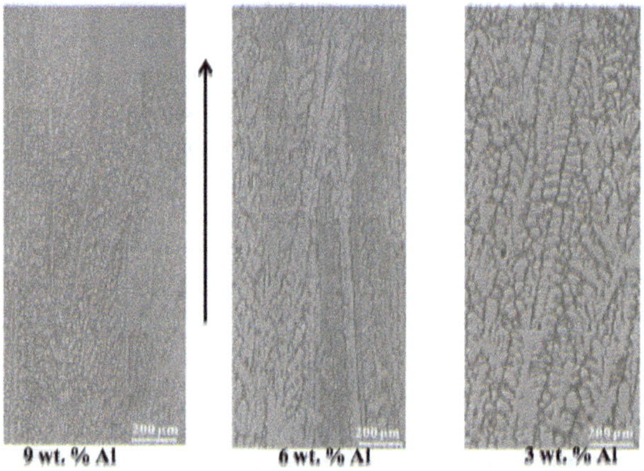


Figure 1: The longitudinal sections of the directionally solidified Mg-Al alloys at the Thermal gradient (G):7.5 K/mm, Solidification velocity (V):0.015 mm/sec and Cooling rate: 0.38 K/sec. Arrow indicated the growth direction.

The variation of SDAS for Mg-Al binary alloys with cooling rate is shown in Fig 2. As seen from Fig. 2, the SDAS for all the Mg-Al binary alloys decreases with cooling rate but the relationship is not necessarily linear in the entire range of cooling rate. Particularly, the SDAS variation at lower cooling rates from 0.375 to 10 K/sec shows a different slope as compared to the SDAS in the higher cooling rate specifically greater than 20 K/sec. In this context, there cannot be a strict linear relationship between SDAS and cooling rate (log-log scale) for Mg-Al binary alloys as deduced for AZ91 alloy by Dube et al. [7].

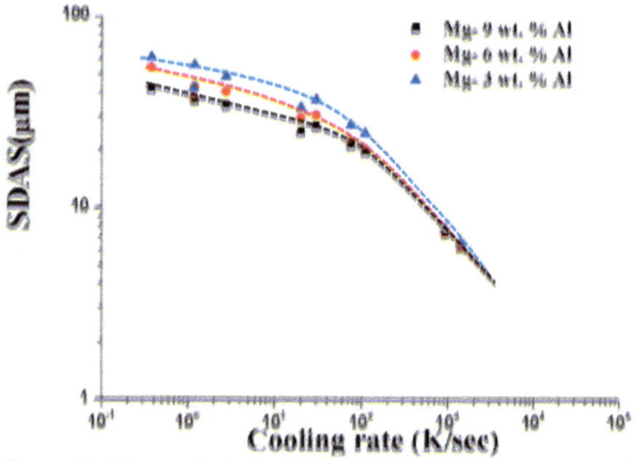


Figure 2: The variation of secondary dendrite arm spacing for Mg-Al alloy with cooling rate and Al concentration.

Interestingly, the SDAS becomes finer with increase in Al concentration and Mg-9 wt. % Al alloy has the lowest SDAS as compared to Mg-3 and 6 wt. % Al alloy at the same cooling rate. SDAS is primarily determined by the growth of the primary dendrite tip and the coarsening of the secondary branches which are already formed [9]. During columnar growth, higher solute content causes more undercooling at the growing dendrite tip which promotes dendrite tip splitting, as the solidification velocity or the cooling rate increases. Higher undercooling can also cause more perturbations to form along the sides of the main primary stem which then coarsen to from secondary branches. Since coarsening is diffusion driven process, higher solute content can also enhance the coarsening process. Thus, the combination of dendrite tip splitting and coarsening process can decrease the SDAS during columnar growth and both these processes are more effective for higher solute content. Even in equiaxed growth, SDAS is dictated by the coarsening of the secondary branches and the growth of secondary branches is similar to the growth of primary dendrite stem in columnar growth. Thus like the columnar growth, SDAS in the equiaxed growth is also expected to decrease as the solute content is increased. As stated earlier, the SDAS is dependent on the dendrite tip splitting of the primary dendrite stem and the coarsening of the secondary branches. One of these effects or the combination of both could be dominant in the lower cooling rate regime than the higher cooling rate which could explain the change in slope in Fig.2. In order to obtain the cooling rates for the injection mould casting experiments a linear relationship (log-log scale) was assumed between the SDAS and cooling rates measured for wedge casting experiments. This relationship gave the cooling rates of 933 and 1412 K/sec for the injection mould casting with 6 and 4mm diameter Cu mould respectively. More experiments are currently in progress for cooling rates greater than 100 K/sec for Mg-Al binary alloys to correctly ascertain the relationship between

SDAS and cooling rate, specifically in the higher cooling rate regime.

### 3.2 Cellular-Columnar transition

Columnar dendrite is the most common solidification form for many commercial alloys. Primary dendrite arm spacing (PDAS) is an important length scale that characterizes the columnar growth. PDAS formed in the solidified microstructure affects the mechanical properties and even the creep performance of cast Mg alloys is affected by PDAS [10]. Under specific conditions of solidification, alloys are expected to exhibit a transition from cellular to columnar growth which has been investigated for Al based alloys as functions of thermal gradient (G) and solidification velocity (V) [11]. However, the effect of solute content on the transition is not well studied. Specifically for Mg-Al binary alloys, there is no study undertaken in this regard.

As presented in table 1, directional solidification experiments were done at low cooling rate of 0.05 K/sec for Mg-Al (3, 6 and 9 wt. % Al) binary alloys and the optical microstructures are presented in Fig. 3. Mg-9 wt. % Al alloy shows columnar growth with secondary branches along the sides of the primary dendrite stem. Mg-6 wt. % Al alloy shows some side branching at the mushy zone near the quenched solid-liquid interface but the microstructure seems cellular far behind the quenched interface. Interestingly, Mg-3wt. % Al alloy looks entirely cellular at 0.05 K/sec with no side branching present at the growing primary dendrite stem.

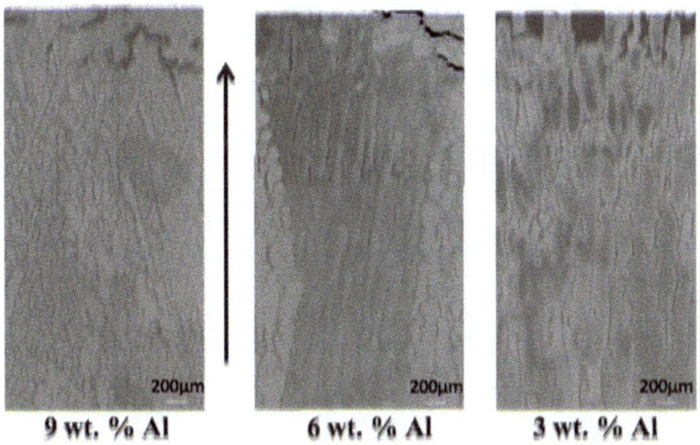


Figure 3: The longitudinal sections of the directionally solidified Mg-Al alloys at the Thermal gradient (G):8.5 K/mm, Solidification velocity (V):0.00625 mm/sec and Cooling rate: 0.05 K/sec. Arrow indicated the growth direction.

Primary dendrite arm or cell spacing could be measured from the microstructures for Mg-Al binary alloys for all the cases of cooling rate given in table 1 and a plot could be drawn showing the variation of PDAS with cooling rate for all the three alloys as shown in Fig. 4. It is noted that PDAS decreases for Mg-9 wt. % Al alloy with increase in cooling rate from 0.05 to 0.375 K/sec, however Mg-6 wt.% Al alloy shows a decrease in spacing at 0.05 K/sec when compared to 0.375 K/sec and the PDAS decreases again as the cooling rate is increased from 0.375 to 2.75 K/sec. Mg-3 wt. % Al alloy shows the similar trend as Mg-6 wt. % Al alloy but the spacing for Mg- 3 wt. % Al is lower when compared to Mg-6 wt.% Al alloys. This indicates that Mg-3wt. % Al will predominantly remain cellular when the cooling rate is less than 0.05 K/ sec; however at this cooling rate Mg-6 wt. % Al is still in the transition mode and will only adopt the cellular mode of growth at cooling rates lower than that required for Mg-3 wt. % Al. In this context, Mg-9 wt. % Al alloy will start showing cellular growth only at very low cooling rates when compared to Mg-3 and 6 wt. % Al alloys. This clearly emphasizes the influence of solute content on the cellular-dendrite transition in Mg-Al alloys.

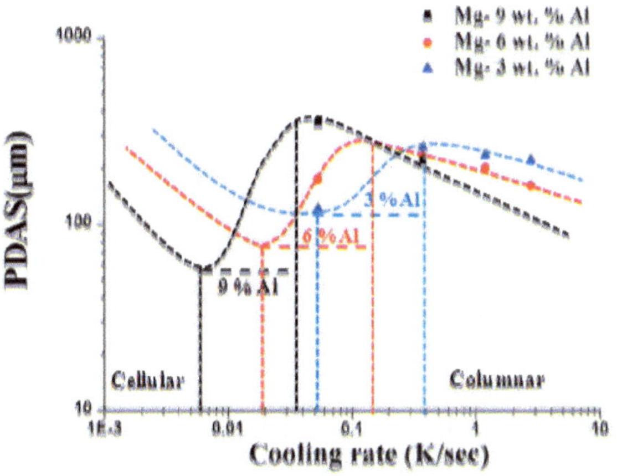


Figure 4: The variation of Primary dendrite arm spacing with cooling rate for Mg-Al binary alloys. The cellular-columnar transition is indicated for the alloys by the dashed lines.

The experimental trends discussed so far clearly emphasises the importance of undercooling at the dendrite tip on the various solidification growth modes. In the cellular mode, the increasing cooling rate decreases the undercooling at the tip of the growing cell which results in more cell flattening (increase in tip radius) and increasing the cooling rate further, splits the cell tip which marks the beginning of columnar growth with substantial side branching. Increasing the cooling rate beyond this point increases the undercooling due to more dendrite tip splitting (decrease in tip radius). Higher solute content will have more undercooling at the tip under both the cellular and columnar growth modes. Hence Mg-9 wt. % Al can achieve the transition to columnar mode at lower cooling rates than Mg-3 and 6 wt. % Al alloys.

## 4. Summary

Based on the directional solidification experiments and gravity casting experiments, following observation were made on the microstructural evolution of Mg-Al (3, 6 and 9 wt. % Al) alloys:

1. Secondary dendrite arm spacing (SDAS) decreases with cooling rate and increasing the Al content in Mg-Al binary alloys.
2. A linear relationship (log-log scale) cannot describe the SDAS variation in cooling rate as previously observed in case of AZ91 alloys and SDAS for the Mg-Al binary alloys showed that the deviation from linearity arises at lower cooling rates.
3. Cellular-columnar transition was studied for Mg-Al binary alloys with the aid of directional solidification experiments and it was observed that the transition is strongly dependent on Al concentration of the alloy.

All the as cast microstructural data of Mg-Al binary alloys are being used to test the solidification model in conjunction with the thermodynamic database.

## 5. Acknowledgement

Financial support from General Motors of Canada Ltd. and the CRD and MagNET research programs of the Natural Sciences and Engineering Research Council of Canada is gratefully acknowledged. We would also like to thank Dr. A. Javaid of CANMET for Mg alloy casting and Dr. D. H. Kang for thoughtful discussions.

## 6. References

1. S.S. Park et al., Scripta Mater. 57 (2007) 793-795.
2. K. Pettersen, O. Lohne and N. Ryum, Metall.Mater. Trans. A. 21 (1990) 221-230.
3. H. M. Tensi and R. Rösch DGM Verlag, Oberursel, Germany, (1992) 283-289.
4. H. Sasaki et al., Proc. 53rd Annual World Magnesium Conference, Virginia, USA (1996) 86-92.
5. W. P. Sequeira et al., TMS, (1997) 169-183.
6. C. Labrecque et al. Can. Met. Quart. 36 (1997) 169-175.
7. D. Dubé et al., Int. J. Cast Metals Res. 11 (1998) 139-144.
8. C. H. Cáceres et al., Mat. Sci. Eng. A 325 (2002) 344-355.
9. W.Kurz and D.J.Fisher Fundamentals of Solidification, Trans Tech Publications (1992).
10. Flemings MC. Solidification processing. New York: McGraw-Hill Book Co. (1974).
11. M.D. Peres, C.A. Siqueira, and A. Garcia, J. of Alloys and Compounds 384 (2004) 168-181.

# Magnesium Technology 2012

## Alloy and Microstructural Design

### Wednesday AM

*Session Chairs:*

Jian-Feng Nie
(Monash University, Australia)

Nack Kim
(POSTECH, South Korea)

**Magnesium Technology 2012**
*Edited by: Suveen N. Mathaudhu, Wim H. Sillekens, Neale R. Neelameggham, and Norbert Hort*
***TMS (The Minerals, Metals & Materials Society), 2012***

# Age hardening behavior of Mg-1.2Sn-1.7Zn alloy containing Al

T. T. Sasaki, T. Ohkubo, K. Hono
National Institute for Materials Science, 1-2-1 Sengen, Tsukuba, Ibaraki, 305-0047, Japan

Keywords: Mg-Sn, Precipitation hardening, Al

### Abstract

We have investigated the age hardening behavior of Mg-1.2at.%Sn (T5) based alloys, which are solution treatable at lower temperature as compared to the previously studied Mg-2.2Sn (T10) based alloys. To enhance the age hardening response with the reduced Sn content, Zn and Al were alloyed with the Mg-1.2at.%Sn alloy. Particularly, Mg-1.7Zn-1.2Sn-2.0Al alloy (TZA543) showed the peak hardness of 79 VHN at 200°C. The peak aged Mg-1.2Sn-1.7Zn-2.0Al alloy consisted of cuboidal $\beta_2'$ precipitates with the size of less than 50 nm, rod shaped precipitates, and lathplate shaped $Mg_2Sn$ phase.

### Introduction

There are strong demands for the development of high strength wrought magnesium alloys applicable to transportation vehicles and aircrafts. Since commercial wrought alloys such as Mg-Zn-Zr (ZK) and Mg-Al-Zn (AZ) show only moderate age hardening responses, not much attention has been paid to the precipitation hardening of wrought alloys. Recent investigations showed yield strength can be readily increased by aging rolled sheets and extruded rods of microalloyed Mg-Zn alloys microalloyed with Ca and Ag **[1]**, demonstrating the potential of heat treatable wrought mangnesium alloys.

Mg-Sn based alloys appear to be promising as heat resistant alloys for both cast and wrought applications because the $Mg_2Sn$ phase has a higher melting temperature than those in Mg-RE (RE: Rare Earth) systems **[2]**. In fact, some of precipitation hardened Mg-Sn based alloys exhibit excellent creep properties **[3, 4]**. For better formability, extrusion and rolling temperature should be selected above the solvus temperature of alloys. However, the solution heat treatment temperature of the previously studied Mg-2.2Sn based alloy (>500 °C) is too high for thermomechanical processing **[5]**. One way to reduce the solution treatment temperature is to decrease the Sn content. However, it also reduces the age hardening responses. Since the typical size of $Mg_2Sn$ precipitates is rather coarse (~1 μm) **[5-8]**, the significant refinement of the precipitates may compensate the degradation of the age hardening response even with the reduced supersaturation. In Mg-Sn-Zn alloys, $MgZn_2$ type phase such as $\beta_1$' phase provides the heterogeneous nucleation site for the $Mg_2Sn$ phase **[8, 9]**, thereby increasing the number density of the $Mg_2Sn$ precipitates. In this work, 1.7at.%Zn was alloyed with Mg-1.2at.%Sn so that the alloy can be solution treated at 450 °C . Since the addition of Al is reported to refine the $Mg_2Sn$ phase in the Mg-2.2Sn-0.5at.%Zn alloy **[8, 10]**, we examined the effect of Al alloying expecting the improvement of the age hardening behavior of the Mg-1.2at.%Sn-1.7at.%Zn alloy (TZ54).

### Experimental Procedure

Alloy ingots with nominal compositions of Mg–1.2Sn-1.7Zn-*x*Al (*x*=0, 0.5, 1 and 2) in at.% were prepared by induction melting using steel crucibles under an Ar atmosphere and by casting into iron moulds. Table 1 shows the compositions of the alloys in both at.% and wt.%.

Table 1: Chemical composition of alloys

| at.% | wt.% |
|---|---|
| Mg-1.2Sn-1.7Zn | Mg-5.5Sn-4.3Zn |
| Mg-1.2Sn-1.7Zn-0.5Al | Mg-5.4Sn-4.3Zn-0.5Al |
| Mg-1.2Sn-1.7Zn-1.0Al | Mg-5.4Sn-4.2Zn-1.0Al |
| Mg-1.2Sn-1.7Zn-2.0Al | Mg-5.4Sn-4.2Zn-2.1Al |

The ingots were subjected to homogenization at 350 °C for 24 hr followed by 450 °C heat treatment for 24 hr in a He-filled Pyrex tube. The small samples that were cut out from the ingots were solution heat treated for 0.5 hr s at 450 °C in the He-filled Pyrex tubes and then water quenched. Single aging was performed on alloys at 200 °C in an oil bath for various aging times. The age hardening responses were measured by the Vickers hardness tester under a load of 500 g. Transmission electron microscope (TEM) specimens were prepared by the twin-jet polishing technique using a solution of 5.3 g LiCl, 11.16 g $Mg(ClO_4)$, 2,500 ml methanol, and 100ml 2-butoxy-ethanol at about −50 ◦C and 90 V. Some of the specimens were finished for surface cleaning by ion milling using a Gatan Precision Ion Polishing System (PIPS). Transmission Electron Microscope (TEM) observation was carried out using FEI TECNAI T20 TEM operating at 200 kV.

### Results and Discussion

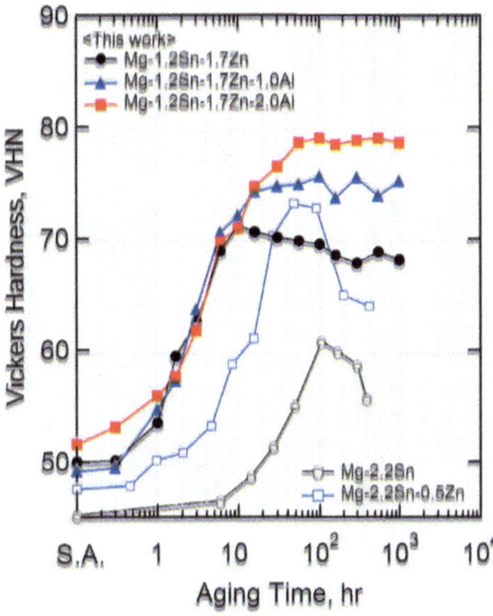


**Figure 1**: Variation in Vickers Hardness as functions of time during aging at 200°C on Mg-1.2Sn-1.7Zn, Mg-1.2Sn-1.7Zn-1.0Al and Mg-1.2Sn-1.7Zn-2.0Al alloys. Included is the data obtained from Mg-2.2Sn and Mg-2.2Sn-0.5Zn alloy **[5]**.

**Figure 1** shows variations in Vickers hardness as functions of time during 200 °C aging for Mg-1.2Sn-1.7Zn (TZ54), Mg-1.2Sn-1.7Zn-1.0Al (TZA541) and Mg-1.2Sn-1.7Zn-2.0Al (TZ542) alloys. The age hardening curves show that the addition of Al substantially increased the peak hardness. The peak hardness increased from 71 to 79 VHN by the addition of 2at.% Al while the time to reach the peak hardness is delayed by the Al addition. Mg-1.2Sn-1.7Zn alloy reached the peak hardness after 10 h of aging at 200 °C, and Mg-1.2Sn-1.7Zn-2.0Al alloy took 100 h to reach the peak hardness. While Harosh et al. reported two hardness peaks appeared in an Mg-5.6Sn-4.4Zn-2.1Al alloy during aging at 225 °C [**11**], the Mg-1.2Sn-1.7Zn-2.0Al alloy showed only one hardness peak. The Mg-1.2Sn-1.7Zn based alloys do not overage as quickly as Mg-2.2Sn and Mg-2.1Sn-0.5Zn alloys [**5**]. The Mg-1.2Sn-1.7Zn alloy maintained the peak hardness of ~ 70 VHN up to 100 h and the Mg-1.2Sn-1.7Zn-2.0Al alloy maintained the hardness level of ~ 80 VHN up to 1000 h during 200 °C aging.

**Figure 2** shows the bright field TEM images obtained from the Mg-1.2Sn-1.7Zn alloy aged at 200 °C. **Figure 2 (a) and (b)** are those after aging at 200°C for 10 h, which corresponds to the peak aged condition, and **Fig. 2 (c) and (d)** are the ones after overaging for 100 h at 200 °C. Note that **Fig. 2 (a) and (c)** are taken from the $[11\bar{2}0]$ zone, and **Fig. 2 (b) and (d)** are from the [0001] zone, respectively. As shown in **Fig. 2 (a) and (b)**, the peak aged Mg-1.2Sn-1.7Zn alloy shows similar microstructure to that seen in the Mg-Zn based alloy subjected to aging [**12, 13**], i.e., rod shaped $\beta_1'$ along the [0001] direction of the magnesium matrix, and the other is $\beta_2'$ plates lying on the (0001) plane of the magnesium matrix. The corresponding SAED pattern in **Fig. 2 (a)** exhibits diffuse streaks along the $[01\bar{1}0]$ directions. This is attributed to the rod shaped $\beta_1'$precipitates along the [0001] direction as well as the crystal structure of the precipitate phase.

The overaging for 100 h resulted in the coarsening of the $\beta_1'$ and $\beta_2'$ phases as shown in **Fig. 2 (c) and (d)**. The $\beta_1'$ phase coarsened to ~ 1 μm in length and 40 nm in diameter, and the size of the $\beta_2'$ phase increased to 200 nm in diameter and 50 nm in thickness. In addition to the coarsened $\beta_1'$ and $\beta_2'$ phases, another kind of precipitates is observed as indicated by an arrow. From **Fig. 2 (c) and (d)**, the precipitates have plate like shape forming on the prismatic plane of the magnesium matrix, $(10\bar{1}0)$. **Figure 3 (a)** shows the microbeam diffraction patterns obtained from one of the plate like precipitates. Note that they are recorded from the $[01\bar{1}0]$ zone. The precipitates are identified as the

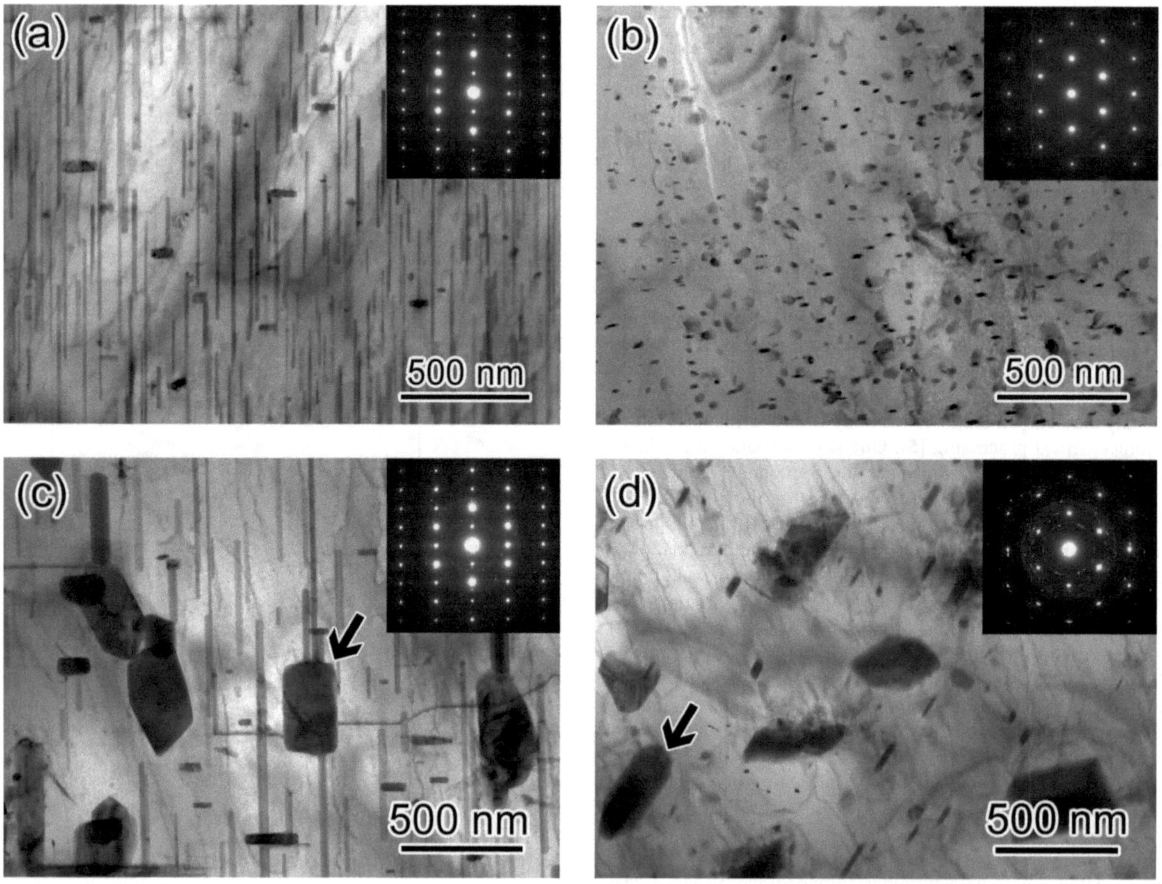


**Figure 2**: Bright field TEM images of Mg-1.2Sn-1.7Zn alloy subjected to 200 °C aging. (a) and (b) are obtained after 200°C/10hr aging, which is consistent with the peak aging condition, and (c) and (d) are are obtained after aging at 200°C for 100hr. (a) and (c) are taken from the zone axis of $[11\bar{2}0]$, and (b) and (d) are taken from the zone axis of [0001].

$Mg_2Sn$ phase as shown in the calculated diffraction pattern, Fig. 3 (b). The orientation relationship (OR) between the plate like $Mg_2Sn$ phase forming on the prismatic and the magnesium matrix is determined to be $(0001)_{Mg} // (110)_{Mg_2Sn}$, $[11\bar{2}0]_{Mg} // [001]_{Mg_2Sn}$, which is the common OR in the Mg-Sn based alloy **[5-8]**. The late precipitation of $Mg_2Sn$ phase after the peak aged condition is considered to compensate the hardness drop due to the coarsening of the $\beta_1'$ and $\beta_2'$ phases.

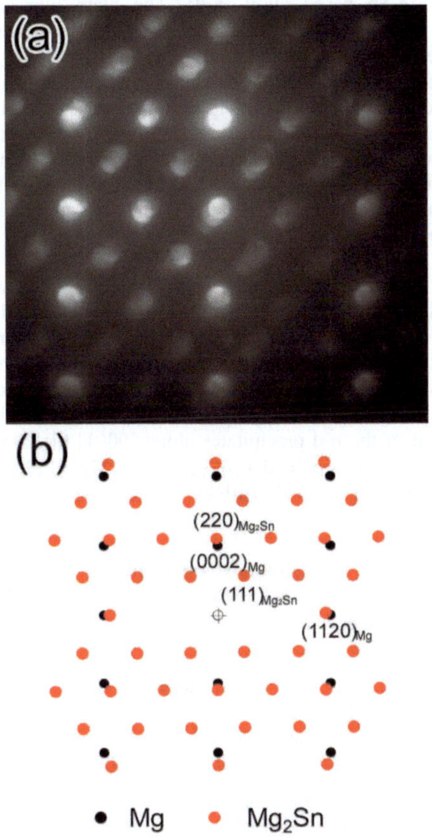


**Figure 3**: (a) microbeam diffraction patterns obtained from a precipitate indicated by arrow in Fig. 2 (a), and (b) Calculated diffraction patterns. Note that these images are taken from the zone axis of $[01\bar{1}0]$.

**Figure 4** shows the bright field TEM images obtained from the Mg-1.2Sn-1.7Zn-2.0Al alloy (TZA542) aged at 200 °C for 100 hr, which corresponds to the peak aged condition. Note that **Fig. 4 (a) and (b)** are taken from the $[11\bar{2}0]$, and [0001] zones, respectively. The peak aged microstructure shows different feature as compared to the peak aged Mg-1.2Sn-1.7Zn alloy. The rod shaped precipitate growing along [0001] direction is refined to ~ 300 nm as indicated by arrowhead 1. In addition, as reported by Oh-ishi et al. **[10]**, the Al addition resulted in the formation of the cuboidal shaped precipitates with the size less than 50 nm as indicated by arrowhead 2. In addition to these precipitates, the lath shaped precipitates forming on the prismatic plane could be also observed as indicated by arrowhead 3. The lath shaped precipitates are the $Mg_2Sn$ phase, and its size was refined to ~200 nm in length and ~75 nm in thickness.

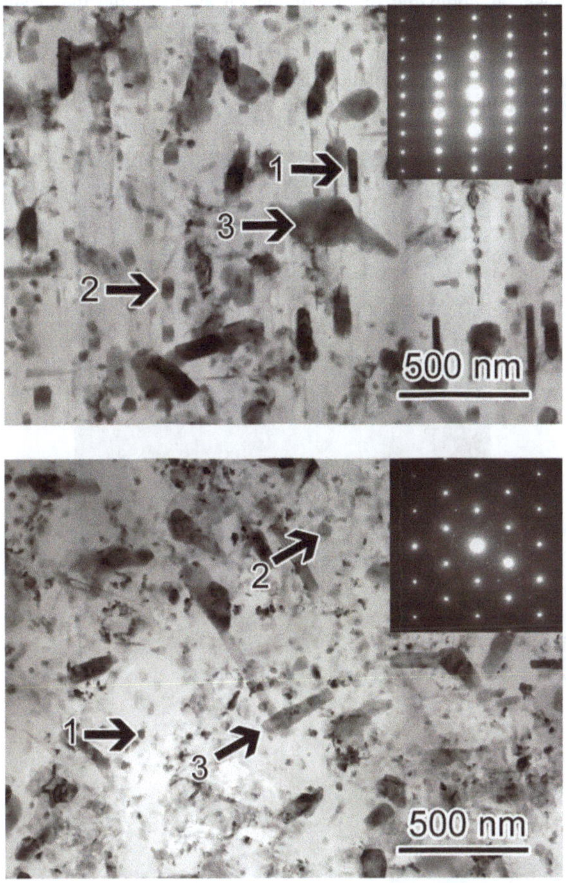


**Figure 4**: Bright field TEM images of Mg-1.2Sn-1.7Zn-2.0Al alloy aged at 200 °C aging for 100 hr, consistent with the peak aging condition. (a) and (b) are taken from the zone axis of $[11\bar{2}0]$, and [0001], respectively.

**Figure 5 (a)** shows the micro-beam diffraction pattern obtained from the cuboidal precipitates observed in the 200 °C / 100 hr aged Mg-1.2Sn-1.7Zn-2.0Al alloy taken from the zone axis of [0001]. As shown in **Fig. 5 (b)**, the structure of the cuboidal precipitate is consistent with that of the $\beta_2$' phase. However, since the structure of the rod precipitates remains unclear, future work will shed light on the clarification of the phase of the rod shaped precipitates.

In this work, we have investigated the age hardening behavior of Mg-1.2Sn-1.7Zn based alloy containing Al, and found that the peak hardness of the Mg-1.2Sn-1.7Zn based alloy could be substantially increased from 71 to 79 VHN by the addition of 2.0at.% of Al. From the microstructure observation of the Mg-2.1Sn-1.7Zn alloy at peak aged and over aged conditions, **Fig. 2**, the peak hardness was attributed to the precipitation of the $\beta_1$'and $\beta_2$' phases, **Fig. 2 (a) and (b)**. Although the hardness is expected to decrease upon the coarsening of the $\beta_1$'and $\beta_2$' phases by the over aging, the Mg-1.2Sn-1.7Zn based alloys showed smaller hardness reduction up to 1000 h. This is because $Mg_2Sn$ phase precipitated after the peak aging, and could compensate the hardness decrease due to the coarsening of the $\beta_1'$ and $\beta_2'$ phases, **Fig. 2 (c) and (d)**. Since the diffusion of the Sn atoms is much slower as compared to that of the Zn atoms in the magnesium matrix, and Zn and Sn atoms behave repulsively due to the positive enthalpy of mixing as reported previously [**5, 14**], we can understand the precipitation of the $\beta_1$'and $\beta_2$' phases prior to the precipitation of the $Mg_2Sn$ phase.

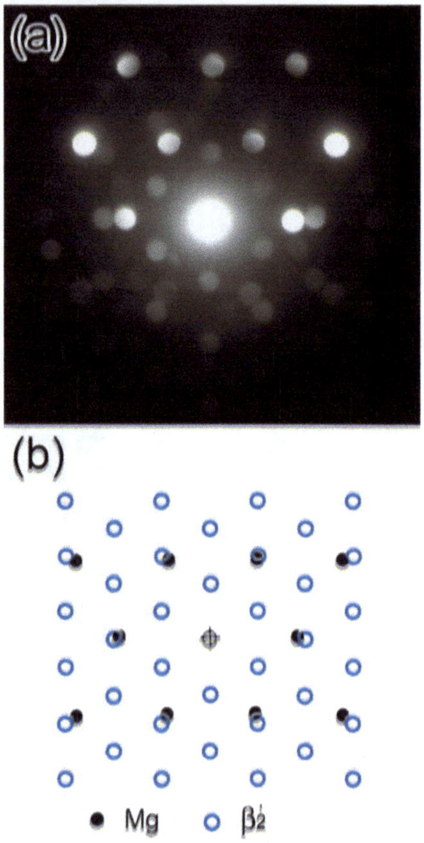


**Figure 5**: (a) microbeam diffraction patterns obtained from one of cuboidal precipitate seen in 200°C/100hr aged Mg-1.2Sn-1.7Zn-2.0Al alloy, and (b) Calculated diffraction patterns. Note that these images are taken from the zone axis of$[10\bar{1}0]$.

In the Mg-1.2Sn-1.7Zn-2.0Al alloy, on the other hand, the $Mg_2Sn$ phase also contributes to the peak hardness in addition to the rod shaped and cuboidal shaped precipitates as seen in **Fig. 4**. The precipitation of the $Mg_2Sn$ resulted in the increased number density of the precipitates in the peak aged condition, resulting in the enhanced peak hardness in the Mg-1.2Sn-1.7Zn-2.0Al alloy. The time to reach peak hardness was shortened in the Mg-2.2Sn-0.5Zn alloy even by the addition of 1.0at.% Al [**8**]. The presence of the $Mg_2Sn$ phase in the peak aged condition is attributed to the change in the kinetics due to the Al.

The enhanced peak hardness in the Mg-1.2Sn-1.7Zn-2.0Al alloy is also attributed to the increase in the number density of the precipitates, and the change in the morphology of the precipitates. The addition of Al refined the precipitate size as shown in the microstructure of the peak aged samples, **Fig. 2 and 4**. The peak aged Mg-1.2Sn-1.7Zn-2.0Al alloy consisted of fine cuboidal $\beta_2$' precipitates with the size of less than 50 nm, rods along [0001], and fine plate shaped $Mg_2Sn$ phase while the Mg-1.2Sn-1.7Zn alloy was mainly strengthened by coarse rod shaped $\beta_1$' phase with over 500 nm in length. In the Mg-2.3Zn-0.5Mn alloy, Oh-ishi et al concluded that plate to cuboidal transition of $\beta_2$' phase is because of the lattice misfit change between the Mg and precipitate phases by Al addition [**10**]. This may also have caused the shape changes of the precipitates in this system.

Finally, the $Mg_2Sn$ phase observed in the Mg-1.2Sn-1.7Zn had different shape from the ones seen in the previously studied Mg-Sn based alloy [**5-9**]. The plate like precipitates forming on the prismatic plane of the magnesium matrix, $(11\bar{2}0)$ could be observed, **Fig. 2 (c) and (d)**, while the $Mg_2Sn$ phase usually forms on the (0001) plane of the magnesium matrix [**5-9**]. Since the precipitation of the $Mg_2Sn$ on the prismatic plane could be seen even in the Al-free Mg-1.2Sn-1.7Zn alloy, the lattice misfit change due to Al addition might not be the sole reason for the precipitation on the prismatic plane. Therefore, future work will also shed light on the detailed formation mechanism of the $Mg_2Sn$ phase.

## Summary

We have investigated the age hardening behavior of the Mg-1.2Sn-1.7Zn-xAl (x=0.5, 1.0 and 2.0). The addition of 2.0at.%Al substantially increased the peak hardness from 71 to 79 VHN while the time to reach peak hardness was delayed. In the Mg-1.2Sn-1.7Zn alloy, the peak hardness is attributed to the precipitation of $\beta_1$'and $\beta_2$' phases. On the other hand, the refinement of the rod precipitates along [0001] direction and the precipitation of fine cuboidal $\beta_2$' phase and $Mg_2Sn$ phase resulted in the increase in the peak hardness in the Mg-1.2Sn-1.7Zn-2.0Al alloy.

## Acknowledgement

The authors gratefully acknowledge Dr. C. L. Mendis for her fruitful discussion.

## References

[1] C.L. Mendis, K.Oh-ishi, Y. Kawamura, T. Honma, S. Kamado, K. Hono, "Precipitation-hardenable Mg-2.4Zn-0.1Zn-0.1Ca-0.16 Zr (at.%) alloys," *Acta Materialia*, 57 (3) (2009) 749-760.

[2] A.A. Nayeb-Hashemi and J.B. Clark, *Phase Diagrams of Binary Magnesium Alloys*, (Materials Park, OH ASM International, 1988).

[3] M.A. Gibson, X. Fang, C.J. Bettles, C.R. Hutchinson, "The effect of precipitate state on the creep resistance of Mg-Sn alloys," *Scripta Materialia*, 63 (8) (2010) 899-902.

[4] T.T. Sasaki, T. Ohkubo, K. Hono, "Heat-treatable Mg-Sn-Zn wrought alloy," *Scripta Materialia*, 61 (1) (2009) 80-83

[5] T.T. Sasaki, K. Oh-ishi, T. Ohkubo, K. Hono, "Enhanced age hardening response by the addition of Zn in Mg-Sn alloys," *Scripta Materialia*, 55 (3) (2006) 251–254.

[6] C.L. Mendis, C.J. Bettles, M.A. Gibson, C.R. Hutchinson, "An enhanced age hardening response in Mg-Sn based alloy containing Zn" *Materials Science and Engineering A*, 435 (2006) 163-171.

[7] C.L. Mendis, C.J. Bettles, M.A. Gibson, S. Grosse, C.R. Hutchinson, "Refinement of precipitate distributions in and age-hardenable Mg-Sn alloy thorough microalloying" *Philosophical Magazine Letters*, 86 (7) (2006) 443-456.

[8] T.T. Sasaki, K. Oh-ishi, T. Ohkubo, K. Hono, "Effect of double aging and microalloying on the age hardening behavior of

a Mg-Sn-Zn alloy," *Materials Science and Engineering A*, (2011) doi: 10.1016/j.msea.2010.05.010.

[9] A. Gorny, A. Katsman, "Precipitation and stress-influenced coarsening in Mg-based Mg-Zn-Sn-Y and Mg-Zn-Sn-Sb alloys" *Journal of Materials Research*, 23 (2008) 1228-1236

[10] K. Oh-ishi, K. Hono, K.S. Shin, "Effect of pre-aging and Al addition on age hardening and microstructure in Mg-6wt.%Zn alloys," *Materials Science and Engineering A*, 496 (2008) 425-433.

[11] S. Harosh, L. Miller, G. Levi, M. Bamberger, "Microstructure and properties of Mg-5.6%Sn-4.4%Zn-2.1%Al alloy," *Journal of Materials Science*, 42 (2007) 9983-9989

[12] J.B. Clark, "Transmission electron microscopy study of age hardening in a Mg-5 wt.%Zn alloy," *Acta Metallurgica*, 13 (12) (1965) 1281.

[13] G. Mima, Y. Tanaka, "The Aging Characteristics of Magnesium-4 wt% Zinc Alloy" *Transactions of Japan Institute of Metals*, 12 (2) (1971) 71-75.

[14] A.B. Eric and G.B. Brook, *Smithells Metals Reference Book, (7th ed.)*, (Oxford, Butterworth-Heinemann, 1992)

# EFFECT OF NANO-ALUMINA AND COPPER MICRON SIZE PARTICULATES ON MICROSTRUCTURE AND MECHANICAL PROPERTIES OF MAGNESIUM ALLOY AZ31

Quy Bau Nguyen, K.S. Tun and M. Gupta
Department of Mechanical Engineering, National University of Singapore
9 Engineering Drive 1, Singapore 117576

Keywords: Magnesium, Nanocomposite, Mechanical behavior, Microstructure, Solidification

**Abstract**

In this paper, magnesium composites are synthesised through the addition of nano-alumina and copper micron size particulates in AZ31 magnesium alloy using the technique of disintegrated melt deposition. The simultaneous addition of Cu and nano-alumina particulates led to an overall improvement in both microstructural characteristics and mechanical response of AZ31. The 0.2% yield strength increased from 180 to 300 MPa (67%), while the ductility increased by almost 24%. The overall tensile properties assessed in terms of work of fracture improved by 66%. An attempt is made to correlate the tensile response of composites with their microstructural characteristics. The results suggest that these alloy composites have significant potential in diverse and wider engineering applications when compared to AZ31 alloy.

**Introduction**

Weight reduction is now a big challenge in civil defence, aerospace, sports and automobile industries. This has been enchanting research scientists around the globe to look for structurally light weight materials. Among the elements in the periodic table of elements, Mg is one such potential candidate which can even bring down the weight of a structure by about 30% and 70% when compared to aluminum and steel structures, respectively]. Magnesium materials exhibit some advantages such as lowest density amongst structural materials, good castability and machinability. However, main disadvantages of current magnesium materials include low ductility, low strength, loss of strength with increase in service temperature and low corrosion resistance [1].

Magnesium alloy AZ31 (Mg–3wt%Al–1wt%Zn) is commonly employed in light weight engineering applications. The literature search in recent years shows that serveral attempts have been made to further improve the properties of this alloy through the addition of different reinforcements and subjecting to different deformational processes. Commonly used reinforcements consist of nano-$Al_2O_3$ particulates [2] and micro-SiC particulates [3]. Among these reinforcements, addition of nano-$Al_2O_3$ particulate reinforcement enhanced the failure strain of AZ31B alloy remarkably [2]. The literature research also reveals that some attempts have been made to alloy magnesium and magnesium alloys with different alloying elements [4]. The results showed that Cu is a potential alloying element to improve strength of magnesium based materials. However, no attempt so far is made to improve overall mechanical properties of AZ31 by the simultaneous incorporation of nano-$Al_2O_3$ and micron Cu using cost effective and industrially viable solidification route.

Accordingly, an attempt is made in the present study to add nano - alumina particulates to improve ductility and copper to enhance microstructural characteristics, hardness and strength of AZ31 alloy. The main objective was to make an attempt to improve the combination of 0.2%YS, UTS and failure strain when compared to AZ31. The choices of these additions, further, are also expected to enhance high temperature properties and oxidation behavior as established before. Disintegrated melt deposition technique is used in the present study to synthesize these materials and all the characterization studies were performed on the extruded samples.

**Materials and Synthesis**

In the present study, AZ31 magnesium alloy ingots with chemical composition of (2.94% Al, 0.87% Zn, 0.57% Mn, 0.0027% Fe, 0.0112% Si, 0.0008% Cu, 0.0005% Ni and balance Mg) were cut into rectangular pieces. To obtain reasonable distribution of particulates into the AZ31 matrix, holes (10mm diameter and 55 mm deep) were drilled into these pieces and required amount of 50 nm alumina particulates and 50 μm average sized copper particulates were subsequently filled in these holes (see Fig. 1). Two volume percentages of Cu (2% and 4%) were chosen for addition into AZ31 magnesium alloy. The amount of 50nm $Al_2O_3$ particulates was kept at 1.5 volume percentage as this composition showed best combination of strength and ductility when compared to other AZ31B-$Al_2O_3$ formulations [2]. Synthesis of AZ31-Cu-$Al_2O_3$ composites was carried out using disintegrated melt deposition (DMD) technique. The deposited preforms were machined to 36 mm diameter and hot extruded to 8 mm diameter rods.

**Results and Discussion**

Fig. 1 shows principally three regions in different colors, matrix in gray, $Mg_{17}Al_{12}$ phase in bright white color and $Mg_2Cu$ phase in light gray color. Microstructural studies conducted on the extruded samples revealed reasonably uniform distribution of the particulates/intermetallics in the AZ31 matrix. This can be attributed to: (i) minimal agglomeration of reinforcements during melting of matrix, (ii) minimal gravity-associated segregation due to judicious selection of stirring parameters, and (iii) the effect of nano-$Al_2O_3$ hard particulates in dispersing second phases of AZ31 [2].

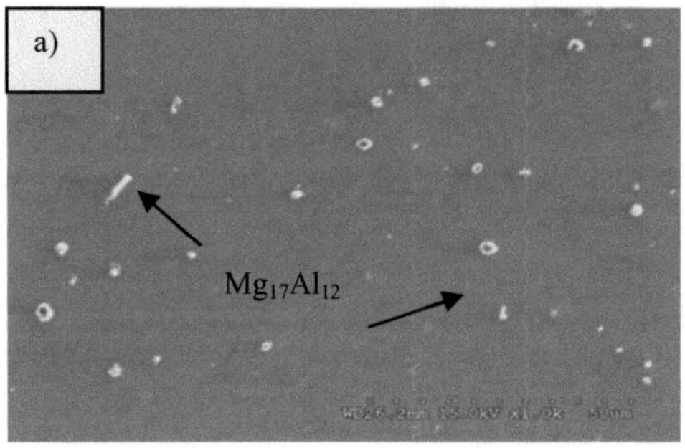

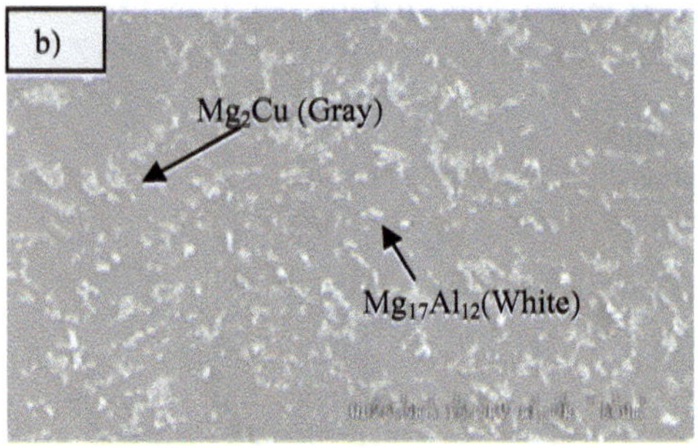


*Fig. 1. Representative FESEM micrographs showing the distribution characteristics of secondary phases in the cases of: a) AZ31 and b)AZ31-2Cu-1.5Al$_2$O$_3$.*

The microstructural characterization reveals the presence of minimal porosity in all samples. Presence of minimal porosity can be attributed to: (i) good compatibility between the AZ31 matrix and particulates/intermetallics, leading to the absence of voids and debonded regions usually associated with ceramic reinforcements, and (ii) judicious selection of experimental parameters during primary and secondary processing. The results of microstructural characterization of monolithic and composite samples also indicated a near defect-free interface formed between secondary phases and the matrix. The interfacial integrity was assessed in terms of interfacial debonding and the presence of microvoids at the interface.

Fig. 2 shows the results of coefficient of thermal expansion measurements obtained from monolithic and composite samples. The results revealed a reduction in CTE of the AZ31 matrix with an increase in amount of Cu and due to presence of nano-Al$_2$O$_3$ particulates. The reduction in CTE can be attributed to much lower CTE values of copper and alumina when compared to AZ31 (17.4 x 10-6K-1 and 7.4 x 10-6K-1 for Cu and Al$_2$O$_3$ respectively) and the ability of the reinforcements to effectively constrain the expansion of the matrix [5]. The lower CTE value of AZ31-Cu and AZ31-Cu-Al$_2$O$_3$ samples may also be attributed to the additional presence of Mg$_2$Cu. The lower CTE values of composite samples suggest that composites are more dimensionally stable as a function of temperature when compared to monolithic AZ31.

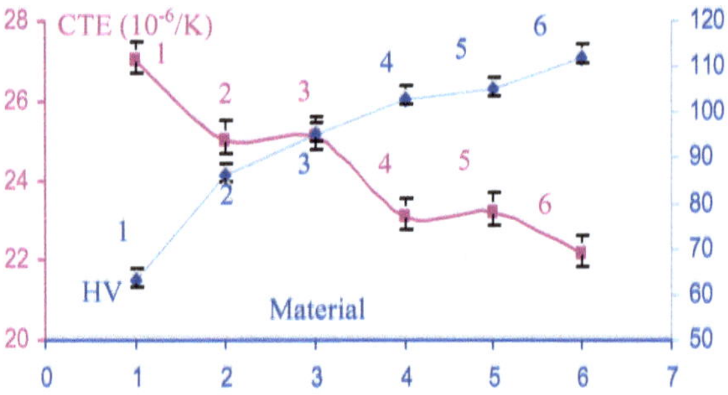


1– AZ31; 2 – (AZ31-1.5Al$_2$O$_3$); 3 – (AZ31-2Cu); 4 – (AZ31-2Cu-1.5Al$_2$O$_3$); 5 – (AZ31-4Cu); 6 – (AZ31-4Cu-1.5Al$_2$O$_3$).

*Fig. 2 Diagram of CTE and microhardness of samples.*

The results of microhardness measurements revealed a significant increase in average microhardness with an increase in amount of nano-alumina and micron copper particulates (see Fig. 2). The increase in hardness of composites with increasing amount of copper and nano-alumina can be attributed primarily to: (i) an increase in the presence of harder intermetallic phases and nano-Al$_2$O$_3$ ceramic particulates and (ii) greater constraints on localized matrix deformation during indentation due to the increased presence of intermetallic phases [2].

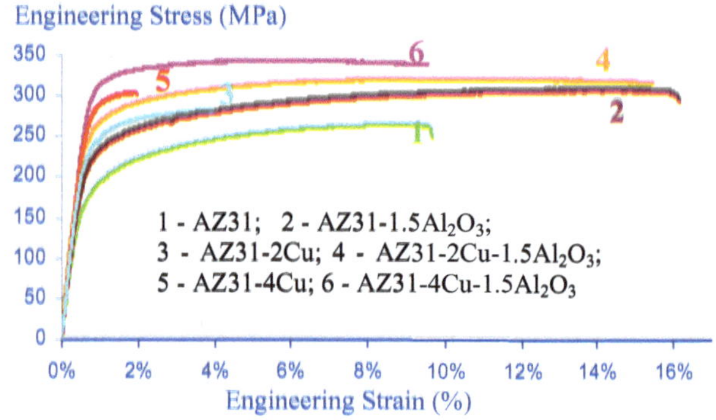


*Fig. 3 Engineering stress- strain curves of AZ31 and its composite samples.*

Fig. 3 shows results of room temperature tensile properties of studied samples. Simultaneous addition of nano-alumina and micron copper particulates led to a significant improvement in 0.2% yield strength and ultimate strength of AZ31 up to 4vol.% Cu. The yield strength increased from 180 MPa to 300 MPa (67% increment), the ultimate strength increased from 265MPa to 350MPa (32% increment). Ductility increased in case of AZ31-2Cu-1.5Al$_2$O$_3$ from 10.9% to 13.5% (~24% increment). However, when only copper was added, strengths went up but the ductility was significantly compromised. The net outcome was an enhanced work of fracture of AZ31-Cu-Al$_2$O$_3$ system from 27.3MJ/m$^3$ and 4.5 MJ/m$^3$ to 45.2 MJ/m$^3$ in the case of AZ31-2Cu-1.5Al$_2$O$_3$ (66% and 904% increments) when compared to AZ31 and AZ31-Cu system, respectively.

The significant increase in 0.2% yield strength and ultimate tensile strength of monolithic AZ31 due to addition of copper and nano-alumina particulates can be primarily attributed to: (i) grain refinement [6]; (ii) The presence of reasonably distributed harder particulates/intermetallics phases [7]; (iii) Orowan strengthening due to a significantly reduced interparticulate spacing of intermetallics phases [8]; (iv) the effective load transfer between matrix and secondary phases; and (v) the formation of internal stresses due to different thermal expansion behavior between particulates/intermetallics and the matrix [9].

Fig. 3 also shows the increase in ductility of AZ31 when nano-alumina or nano-alumina and copper particulates were added. 38% and 24% ductility was increased in the case of AZ31-1.5Al$_2$O$_3$ and AZ31-2Cu-1.5Al$_2$O$_3$ when compared to AZ31 sample, respectively. The increase in ductility can primarily be attributed to: (a) grain refinement; (b) presence and reasonably good distribution of intermetallics; (c) reduction in size and roundness of second phases [2]; and (d) the re-orientation of basal plane (0 0 0 1). Grain refinement particularly benefits hexagonal metals in ductility increment. The increase in ductility of brittle materials such as magnesium can also be attributed to the presence of nano-size particulates such as $Al_2O_3$ and CNT. The nano-particulates have been shown in the previous study to provide sites to open cleavage cracks ahead of advancing crack front and altering the local effective stress from plane strain to plane stress in the neighborhood of crack tip. The reduction in size and roundness of the secondary phase in the microstructure can also be attributed as one of the reason in improvement in failure strain. It has been established before that breakdown of the secondary phase located at grain boundaries and the change in their distribution from a predominantly aggregated type to dispersed type as also observed in the present study can assist in improving ductility [2]. Lastly, the nature of redistribution of basal plane in the extruded directions can also be attributed as a reason for an increase in failure strain.

Further increase in amount of copper (AZ31-4Cu-1.5$Al_2O_3$) led to reduced failure strain but still stayed close to that exhibited by AZ31 and much higher when compared to AZ31-Cu samples (see Fig. 3). The results clearly reveal the inherent characteristic of Cu addition in reducing failure strain [16] and that of alumina particulates at nanolength scale to increase the failure strain of magnesium based materials.

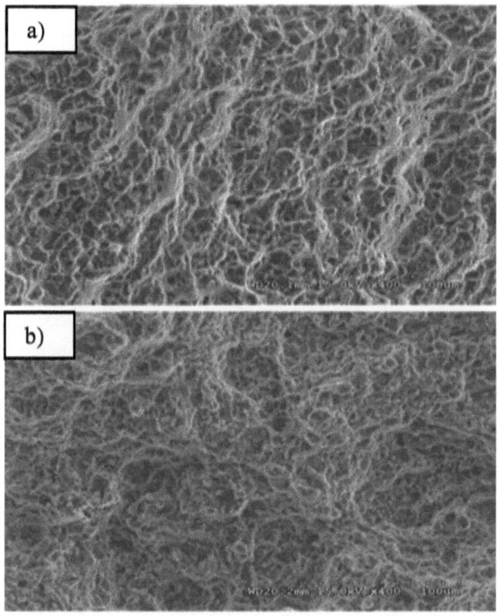


*Fig. 4 Fractographs showing fracture characteristics of samples: a) AZ31 and b) AZ31-2Cu-1.5$Al_2O_3$.*

The tensile fracture surfaces of monolithic and composite samples are shown in Fig. 4. The study of uniaxially deformed fracture surfaces indicated the microstructural effects on fracture characteristics of nanocomposites. The monolithic and AZ31-Cu-$Al_2O_3$ samples showed mixed-mode fracture, presence of microcracks and evidence of plastic deformation [10-11].

In essence, the results of present study clearly indicate the ability of simultaneous addition of copper and nano-$Al_2O_3$ particulates in enhancing the overall ambient temperature mechanical response of AZ31. Considering the ability of copper to improve high temperature strength of magnesiumand that of nano-alumina to enhance the oxidation resistance of AZ31 till 450°C [12], it is expected that AZ31-Cu-$Al_2O_3$ formulations can be strong candidates for a wider range of engineering applications.

## Conclusions

The main conclusions that may be derived from this work are as follows:

1. Addition of Cu particulates leads to formation of $Mg_2Cu$ secondary phase in matrix. Reasonably uniform distribution of secondary phases in the AZ31 matrix is obtained.
2. The increasing presence of copper in AZ31-Cu-$Al_2O_3$ samples leads to a significant improvement in hardness, 0.2%YS and UTS and ductility. The ductility is relatively uncompromised up to 4vol.% Cu.
3. Fracture behavior of AZ31-Cu-$Al_2O_3$ samples shows evidence of mixed-mode fracture with limited presence of microcracks and limited evidence of plastic deformation while AZ31-Cu samples exhibited significant presence of microcracks and the breakage of Cu particulates.

**Acknowledgement:**

Authors would like to acknowledge National University of Singapore and Agency for Science, Technology and Research (ASTAR) Grant 092 137 0015 (WBS# R-265-000-321-305).

**References:**

[1] J. Umeda, M. Kawakami, K. Kondoh, E.L.S Ayman and H. Imai. Materials Chemistry and Physics 123 (2010) 649–657.
[2] Q.B. Nguyen, M. Gupta. Journal of Alloys and Compounds 459 (2008) 244–250.
[3] K.K. Denga, K. Wua, X.J. Wanga, Y.W. Wua, X.S. Hua, M.Y. Zhenga, W.M. Ganb, H.G. Brokmeier. Materials Science and Engineering A 527 (2010) 1630–1635.
[4] Han, L., Hu, H. and Northwood, D.O. Materials Letters 62 (2008) 381–384.
[5] K.S. Tun, M. Gupta. Composites Science and Technology 63 (2007) 2657-2664.
[6] H. Somekawa and T. Mukai. Scripta Materialia, 53 (2005) 541-545.
[7] M. Sumida. Journal of Alloys and Compounds 460 (2008) 619–626.
[8] Q. Han and D.C Dunand. Materials Science and Engineering A 277 (2000) 297–304.
[9] G.E. Dieter, Mechanical Metallurgy, SIMetric Edition, McGraw-Hill Book Company, 1988, pp. 208–212.
[10] B. Dodd, Y. Bai, Ductile Fracture and Ductility, Academic Press Inc. Ltd., London, 1987.
[11] A.S. Ashby, Prog. Mater. Sci., Bruce Chalmers Anniversary Volume, 1981, pp. 1–25.
[12] Q.B. Nguyen, M. Gupta and T.S. Srivatsan Materials Science and Engineering A 500 (2009) 233-237.

**Magnesium Technology 2012**
*Edited by: Suveen N. Mathaudhu, Wim H. Sillekens, Neale R. Neelameggham, and Norbert Hort*
***TMS (The Minerals, Metals & Materials Society), 2012***

# Evaluating the effect of pre-ageing deformation on $\beta'$ precipitate size and distribution in Mg-Zn(-Y) Alloys

Julian M Rosalie[1], Hidetoshi Somekawa[1], Alok Singh[1], Toshiji Mukai[2].

[1]Microstructure Design Group, National Institute for Materials Science (NIMS), Sengen 1-2-1, Tsukuba, Ibaraki, 305-0047, Japan.
[2]Dept. Mechanical Engineering, Kobe University, 1-1 Rokkodai, Nada, Kobe city, 657-8501 Japan.

Keywords: Magnesium alloys, Precipitation

## Abstract

The effect of pre-ageing deformation on the precipitate diameter and length distribution in Mg-Zn(-Y) alloys was examined. Extrusion and pre-ageing deformation were used to introduce dislocations while precluding twin formation. Dislocations provided nucleation sites for rod-like $\beta_1'$ precipitates, resulting in a refinement of the precipitate size distribution. In the binary alloy 5% strain reduced the average precipitate length from approximately 450nm to 60nm, and average diameter from 14 to 9 nm. Substantial reductions in precipitate size were also observed in the Mg-Zn(-Y) alloy. The average interparticle spacing of the rod-like precipitates was measured by Delaunay triangulation. The precipitate distribution was found to be significantly inhomogeneous, with measured interparticle spacings approximately 32% greater than predicted. For 5% pre-ageing deformation the yield strength of the binary alloy approached 95% of the ultimate tensile strength.

## Introduction

Magnesium-zinc based alloys show a particularly strong age-hardening response compared to other commercial magnesium alloys [1]. The precipitate providing the greatest resistance to basal slip in these alloys is termed $\beta_1'$ and takes the form of high aspect-ratio rods with $\langle 0001 \rangle$ habit [1].

Pre-ageing deformation of binary Mg-Zn alloys has been shown to accelerate the ageing process and increase the maximum hardness obtainable [2]. Reports of the precipitation of the $\beta_1'$ precipitate on dislocations indicate that these defects are the primary nucleation sites [1, 3, 4]. While it has been shown that the hardness increment in response to cold-rolling reaches a plateau [2], the extent to which the precipitate size and diameter distribution can be refined by deformation has not been established.

In recent years there has been considerable interest in the judicious use of rare-earth elements and yttrium in magnesium alloys, particularly to improve high temperature properties such as creep resistance. Yttrium is poorly soluble in Mg [5] and generally partitions to Mg-Zn-Y ternary phase precipitates at the grain boundaries [3, 6, 7]. Although precipitation of the $\beta_1'$ phase is not directly influenced by yttrium, the formation of the grain boundaries phases may affect the availability of zinc in the matrix and thus the extent and refinement of the $\beta_1'$ precipitate distribution. Despite the interest in yttrium-bearing Mg-Zn based alloys there has been no yet been any attempt made to quantify how the presence of yttrium impacts on the distribution of the $\beta_1'$ phase.

Nucleation on heterogeneous sites such as dislocations in magnesium alloys may also result in significant localisation of precipitation. It has been shown that that in nickel-based alloys clustering of dispersoids results in a highly inhomogeneous distribution [8] whose interparticle spacing is not well represented by standard stereological formulae. Several methods such as the used of Voronoi tessellation [8] and Delaunay triangulation [9] have been suggested to account for this difficulty. While Delaunay triangulation has been used to calculate more physically representative oxide dispersoid spacings in nickel-based alloys [9] to our knowledge it has not been applied to heterogeneous precipitation in magnesium alloys.

The present work has measured the $\beta_1'$ precipitation length and diameter as a function of tensile pre-ageing deformation and the corresponding effect on the yield strength and ultimate tensile strength of the alloys. The interparticle spacing was measured directly via Delaunay triangulation to determine whether dislocations introduced via pre-ageing deformation lead to an inhomogeneous precipitate distribution.

## Experimental

Billets of binary Mg-3.0at.%Zn and ternary Mg-3.0at.%Zn-0.5at.%Y alloys (hereafter designated 3000Z and 3005ZY, respectively) were prepared by direct-chill casting. The compositions were confirmed via inductively-coupled plasma mass spectroscopy (ICP-MS). Each billet was homogenised under a $CO_2$ atmosphere for 15 h at 300°C and 350°C, respectively for the binary and the ternary alloy. The homogenised billets were then extruded into 12 mm diameter rods at an extrusion ratio of 12:1 and temperatures of 300°C. Tensile samples of gauge length 15 mm and diameter 3 mm were machined from the extruded rods. These were encapsulated in helium and solution treated for 1 h at 300°C for alloy 3000Z and 400°C for alloy 3005ZY, and quenched into water.

Pre-ageing deformation was carried out using an Instron mechanical tester. Tensile deformation was applied parallel to the extrusion axis at a strain rate of $1\times10^{-3}\,s^{-1}$. Samples were deformed to a nominal plastic strain of either 3% or 5% in tension. Ageing was conducted in an oil bath at 150°C with the ageing response measured using Vickers hardness testing. Further tensile tests were then carried out in the Instron mechanical tester, at strain rates of $1 \times 10^{-3}\,s^{-1}$ on samples aged to peak hardness to determine yield strength and ultimate tensile strength.

Samples for TEM analysis were prepared from the aged specimens by first grinding to ~60 − 70 μm then polishing to perforation using a Gatan precision ion polishing system. TEM observations were conducted using a JEOL 4000EX instrument operating at 400 kV. Stereological measurements were carried out on scanned negatives using ImageJ software version 1.44. The cross-sectional areas of individual precipitates were directly determined from the images recorded along [0001] zone axis of the matrix grains by the analysis software (ImageJ). The area of individual precipitates was determined by using image processing software. Precipitate diameters were taken as the diameter for a cylindrical rod of equivalent area. The average centre-centre distance between $\beta_1'$ particles was determined by Delaunay triangulation, using the Delaunay Voronoi plug-in for ImageJ (version 1.44). The mean interparticle spacing was determined by subtracting the mean precipitate diameter from the mean centre-centre distance determined by Delaunay triangulation.

## Results

Pre-ageing deformation acted to significantly refine the diameter of the $\beta_1'$ precipitates from 14 nm to 9 nm in the binary alloy and 19 nm to 11 nm in the ternary alloy. Figure 1 shows the decrease of the average diameter for Mg-Zn and Mg-Zn-Y alloys plotted against the plastic deformation applied before ageing. The error bars indicate the standard deviation the precipitate diameter and are

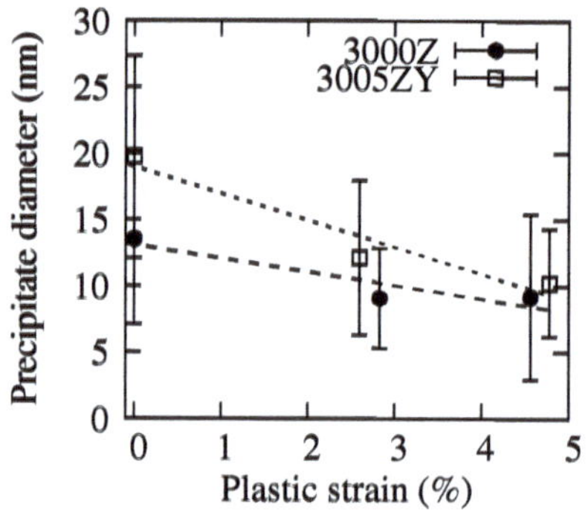


Figure 1: The average $\beta_1'$ precipitate diameter in 3000Z and 3005ZY alloys as a function of pre-ageing strain. The lines are intended as a guide for the eye only. The precipitate diameter in the ternary alloy was greater than that of binary Mg-Zn for approximately equivalent deformation.

not a measure of uncertainty of the average diameter. The lines are intended as a guide for the eye only and to indicate the monotonic decrease in precipitate diameter with increasing pre-ageing strain. The precipitate diameter in the ternary alloy was greater than that of binary Mg-Zn for approximately equivalent deformation, particularly in the non-deformed samples.

The precipitate length was also substantially reduced in samples subjected to pre-ageing deformation. Figure 2 shows the average precipitate length of the $\beta_1'$ rods as a function of the plastic strain applied before ageing. In both alloys the average precipitate length drops substantially after 3% nominal strain, with a much more modest decrease between 3% and 5% nominal strain. Unlike the case of the precipitate diameter, there was little difference between the length values of precipitates in the binary and ternary alloys.

### $\beta_1'$ precipitate size and spacing

The average interparticle spacing, $\lambda$, was calculated using the stereological formula for cylindrical rods normal to the (0001) plane in Mg (Equation 1 [10] ).

$$\lambda = \left(\frac{0.953}{\sqrt{f}} - 1\right) d_1 \tag{1}$$

where $f$ was the volume fraction and $d_1$ was the average precipitate diameter.

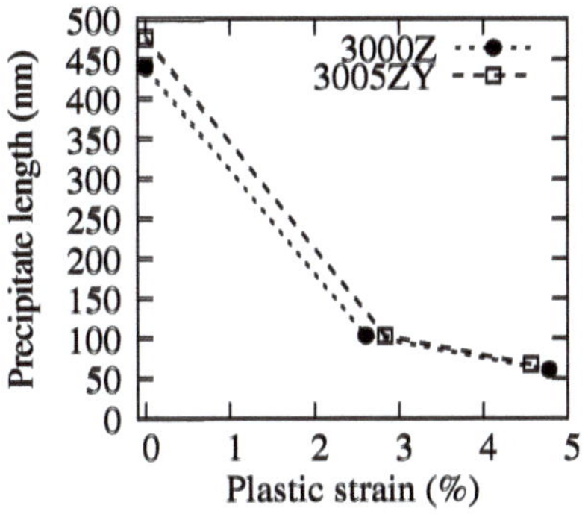


Figure 2: The average $\beta_1'$ precipitate length in 3000Z and 3005ZY alloys as a function of pre-ageing strain. The lines are intended as a guide for the eye only.

This value was compared with average spacings determined by Delaunay triangulation of the images [9, 11]. Figures 3 and 4 present representative images showing $\beta_1'$ precipitates with the triangulation overlaid on the micrograph. Visual inspection is sufficient to show that the precipitates are inhomogeneously distributed, with a tendency to lie along particular lines. In Figure 3(a) it can be clearly seen that such lines correspond to strain contrast indicating the location of dislocations. This resulted in a wide spread of near-neighbour particle spacings, with the particles in a given line being closely spaced, with considerably wider spacings between groups.

The measured and calculated particle spacings are compared in Figure 5. The dashed lines indicate the best fit through the Delaunay triangulation data, while the solid line indicates a homogeneous distribution. The error bars in the measured data indicate the standard error based on the spread in the particle spacing and the uncertainty in the volume fraction. While there is considerable spread in the data, the measured values are consistently greater than for a homogeneous distribution, with the gradient of the dashed line suggesting a discrepancy of approximately 32%. This is true even for the extreme case of the 3005ZY alloy in the non-deformed condition where the measured spacing was 277 nm or an order of magnitude greater than the 5% strained samples.

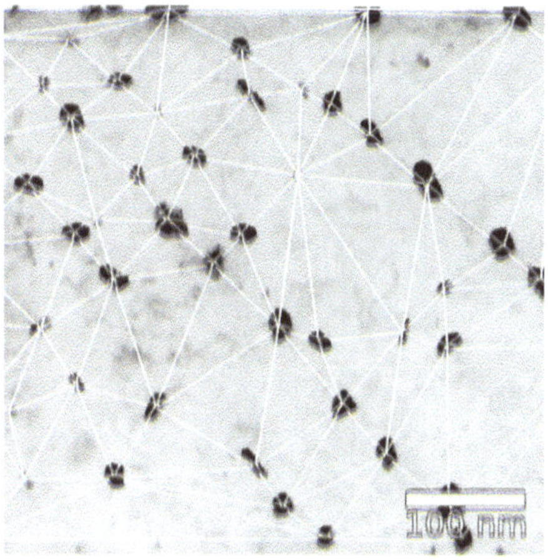


(a) 3000Z, 0% strain

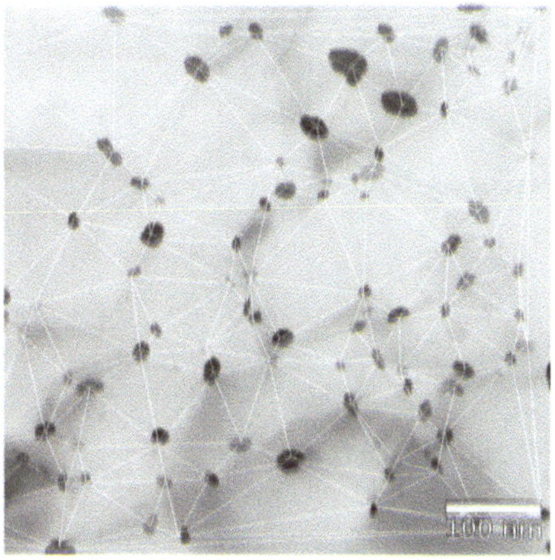

(b) 3000Z, 3% strain

Figure 3: Delaunay triangulations of electron micrographs of $\beta_1'$ precipitates in 3000Z alloys.

## Yield strength & ultimate tensile strength

The initial condition of the two alloys was similar, with yield strengths in the solution treated condition (143 ± 5 and 149±1 MPa, respectively) as shown in Figure 6. However, after isothermal ageing (with or without deformation) the 3000Z alloy displayed greater strength than the 3005ZY alloy for identical treatment conditions. Ageing to optimum hardness resulted in substantial increases in the strength of the alloys, due to precipitation hardening, with the binary 3000Z alloys showing a more significant yield strength increment (273±1 MPa) compared to the ternary alloy (218±3 MPa). The application of pre-ageing deformation resulted in further increases in the yield strength which reached 309±5 for the 3000Z alloy and 308±5 MP for the 3005ZY alloy. The ultimate tensile strength of the 3000Z alloy displayed a gradual increase

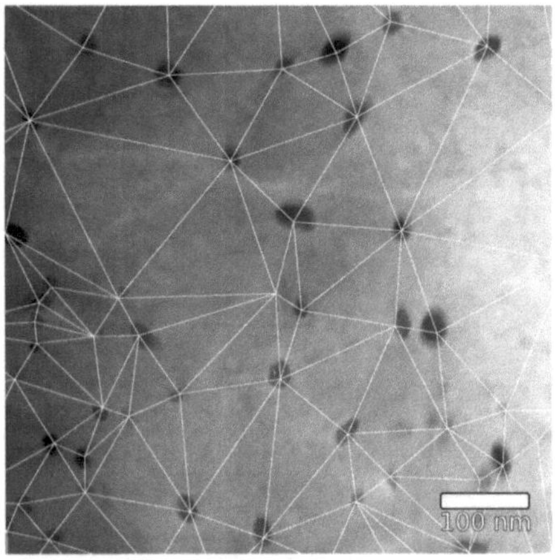


(a) 3005ZY, 3% strain

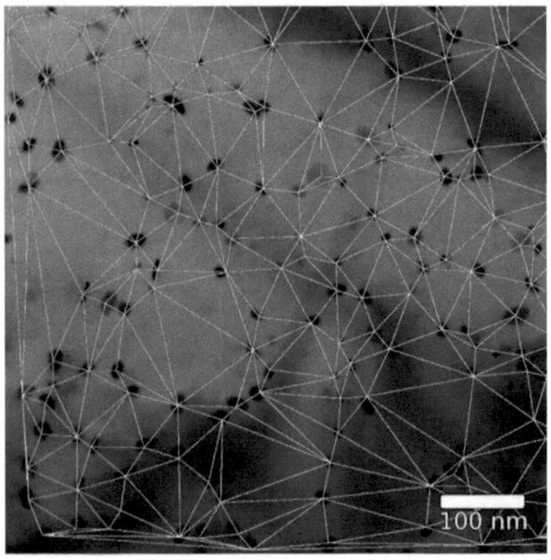


(b) 3005ZY, 5% strain

Figure 4: Delaunay triangulations of electron micrographs showing $\beta_1'$ rod-shaped precipitates in 3005ZY.

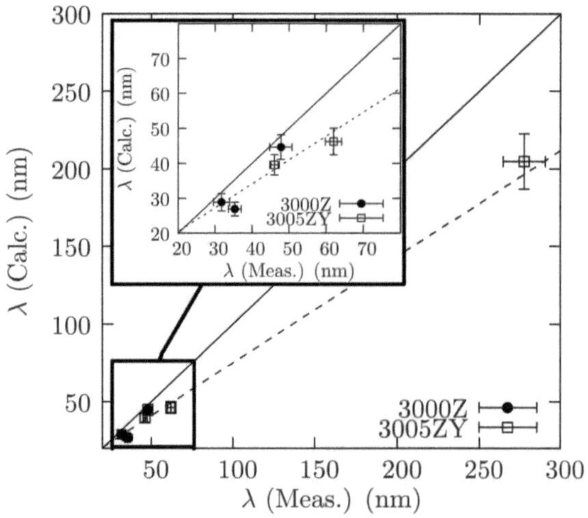


Figure 5: Measured vs calculated interparticle spacings ($\lambda$). Particle spacings measured directly from micrographs using Delaunay triangulation (See Figures 3 & 4). were significantly greater than those calculated assuming a homogeneous precipitate distribution. The dashed lines indicate the best fit through the measured points, while the solid line a homogeneous distribution i.e. equal measured and calculated values.

with increasing pre-ageing deformation (281→322 MPa), however, in terms of true strain of the failure strain of 3000Z was virtually unchanged with and average value of 335MPa and a standard deviation of 2 MPa. The 3005ZY alloy showed a overall similar response, except that the sample aged without deformation showed markedly lower tensile strength. In the 3000Z alloys deformed either 3 or 5% before ageing, the yield strength is approximately 95% of the tensile strength and little further improvement in yield strength through pre-ageing deformation is to be expected.

## Discussion

The application of up to 5% pre-ageing deformation had a significant effect on the size and spacing of the $\beta_1'$ precipitates and a subsequent improvement in the resistance to plastic deformation. The precipitate diameter and length did not appear to show the same behaviour as a function of plastic strain. Precipitate lengths in the binary and ternary alloys showed similar changes as a function of pre-ageing deformation, with a rapid decrease in length for 3% deformation with a slight further decrease after 5% pre-ageing deformation. In addition, for a given heat treatment condition there was $\leq$11% difference in the precipitate length, suggesting similar lengthening behaviour in the two alloys.

Precipitate diameter showed an approximately linear reduction with increasing pre-ageing plastic strain (Fig 1). There were substantially greater differences in the precipitate diameter between binary and ternary alloys, although as the diameter of precipitates in the ternary alloy decreasing more rapidly the values tended to converge. It should be noted that the graph shows only the plastic

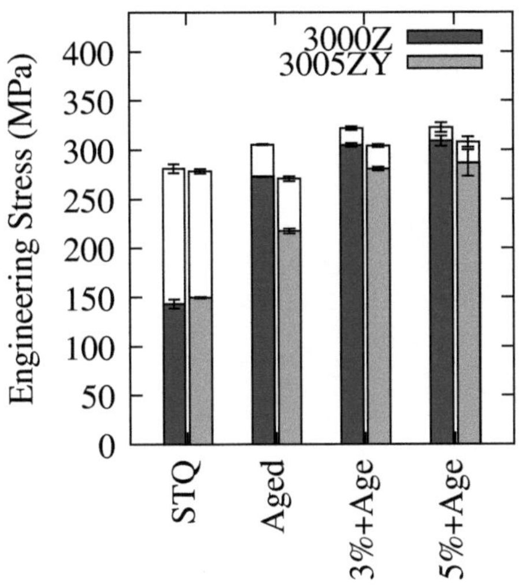


Figure 6: Alloy strength as a function of heat treatment. Filled and empty boxes show the yield and tensile strength of the alloys as a function of heat treatment condition. ("STQ" indicates the solution treated and quenched condition. )

strain (i.e. strain up to the yield point was neglected). Since twinning is negligible, increases in the plastic strain and should be in proportion with the extent of slip and hence (in very rough terms) with the dislocation density.

The difference in the lengthening and thickening response of the precipitates to deformation was thought to reflect competition between precipitate nucleation and precipitate growth. In non-deformed alloys the relatively few precipitate nuclei formed face little competition for solute and are able to grow extensively. Since lengthening of the rods is easier than lateral growth, such precipitates can reach extreme lengths (~450 nm in the case of the Mg-Zn-Y alloy). When heterogeneous sites such as dislocations are introduced, the number of nuclei is dramatically increased. These nuclei will lengthen and thicken – with growth primarily through lengthening – however they will face greater competition from one another and their growth will ultimately be restricted by the available solute.

For alloys in the peak-aged condition, it is generally considered that all available solute will be partitioned to the precipitates and the volume fraction of precipitates should be a function of the overall composition. In this case, then the above reasoning indicates that dispersion strengthening for rod-like precipitates normal to the Mg basal plane may best be improved by multiplying the number of available nuclei and restricting the their lengthening.

The inhomogeneity of the precipitate spacing is expected to have a significant effect on the effectiveness of precipitate strengthening. As shown in Figure 5 the measured interparticle particle spacings averaged roughly 32% more than would be expected for a homogeneous distribution. The Delaunay triangulation technique takes into account the variation in the precipitate spacing in this non-homogeneous system, which produces large non-strengthened regions of the matrix.

It was notable that although increasing deformation substantially refined the precipitate spacing, even for 5% tensile deformation, the measured precipitate spacing exceeded the expected value for a homogeneous system. This suggests that while the introduction of additional dislocations refines the precipitate distribution, it is does not help to overcome the inherent problem with heterogeneous nucleation; namely that precipitation will be concentrated at the defect sites.

Pre-ageing deformation was effective in supplementing the resistance of the alloys to plastic deformation compared to ageing without deformation. The binary alloy displayed greater improvements in yield strength than the ternary alloy, in accordance with the more refined precipitate distribution that was apparent from the the stereological measurements. Using 5% pre-ageing strain it was possible to achieve yield strengths up to 95% (3000Z) or 91% (3005ZY) of the ultimate tensile strength.

The isothermal ageing temperature of 150°C was sufficient for stress relief [12] and it is probable that during the early stages of ageing there is competition between nucleation of $\beta_1'$ particles on dislocations and the loss of dislocations through annealing. This was consistent with the observation that most of the increase in yield strength occurred in the first 3% plastic deformation. The narrow window between yield strength and UTS, along with the diminishing rate of return for deformation suggests that tensile deformation beyond 5% is unlikely to further enhance the yield strength, particularly for Mg-Zn.

## Conclusions

Pre-ageing tensile deformation up to 5% strain had a strong refining effect on the diameter and length of $\beta_1'$ precipitates in Mg-Zn and Mg-Zn-Y alloys. The precipitates were distributed along dislocations in the matrix giving rise to a somewhat inhomogeneous precipitate distribution. Measurements of the interparticle spacing via

Delaunay triangulation were 32% greater than calculated values assuming a homogeneous precipitate distribution. Precipitation in the binary alloy was more responsive to pre-ageing deformation than in the ternary alloy and the binary alloy showed a greater increment in yield stress after pre-ageing deformation. Yield strengths in the binary alloy approached 95% of the ultimate tensile strength, suggesting that more extensive deformation would not greatly improve the resistance to plastic deformation.

## References


[1] J. B. Clark, *Acta Metall.*, **13**:(1965) 1281.

[2] G. Mima, Y. Tanaka, *Trans Japan Inst Met*, **12**(2):(1971) 76.

[3] A. Singh, A. P. Tsai, *Scr. Mat.*, **57**:(2007) 941. doi: 10.1016/j.scriptamat.2007.07.028.

[4] K. Oh-ishi, K. Hono, *et al.*, *Mater Sci Eng*, **A 496**:(2008) 425. doi: 10.1016/j.msea.2008.06.005.

[5] H. Okamoto, *J Phase Equil*, **15**(1):(1994) 129. doi: 10.1007/BF02667700.

[6] L. Y. Wei, G. L. Dunlop, *et al.*, *Metall. Mater. Trans. A*, **26A**(7):(1995) 1705. doi: 10.1007/BF02670757.

[7] A. P. Tsai, A. Niikura, *et al.*, *Philos Mag Lett*, **70**(3):(1994) 169. doi: 10.1080/09500839408240971.

[8] Y. Estrin, S. Arndt, *et al.*, *Acta Mater.*, **47**:(1999) 595. doi: 10.1016/S1359-6454(98)00362-0.

[9] J. Lépinoux, Y. Estrin, *J Physique IV*, **11**(PR4):(2001) 357.

[10] J. F. Nie, *Scr. Mat.*, **48**:(2003) 1009. doi: 10.1016/S1359-6462(02)00497-9.

[11] J. Lépinoux, Y. Estrin, *Acta Mater.*, **48**(17):(2000) 4337. doi: 10.1016/S1359-6454(00)00181-6.

[12] M. M. Avedesian, H. Baker, eds., *Magnesium and magnesium alloys*, 79, ASM specialty handbook, ASM, 1999.

Magnesium Technology 2012
*Edited by: Suveen N. Mathaudhu, Wim H. Sillekens, Neale R. Neelameggham, and Norbert Hort*
***TMS (The Minerals, Metals & Materials Society), 2012***

# EFFECT OF ZINC CONTENT ON THE MICROSTRUCTURE AND MECHANICAL PROPERTIES OF EXTRUDED Mg-Zn-Y-La ALLOYS WITH LPSO PHASE

Jonghyun Kim[1] and Yoshihito Kawamura[2]
[1]Kumamoto Technology and Industrial Foundation; 2080-10 Tabaru, Mashiki-machi, Kamimashiki-gun, Kumamoto, 861-2202, JAPAN
[2]Department of Materials Science, Kumamoto University; 2-39-1, Kurokami, Kumamoto, 860-8555, JAPAN

Keywords: Mg alloy, LPSO, Mechanical properties

## Abstract

$Mg_{98.4}Zn_xY_{1.5}La_{0.1}$ (X=0.25 to 2.0 at.%) alloys were prepared by high-frequency induction melting in an Ar atmosphere. The microstructure and mechanical properties of the extruded alloys were investigated. The microstructure contained very small sized RE intermetallic compound particles within the α-Mg matrix and between the LPSO intergranular phase. Specifically, for extruded $Mg_{98.15}Zn_{0.25}Y_{1.5}La_{0.1}$ alloy, the presence of this RE compound after heat treatment was responsible for the observed improvement in tensile and creep properties, which were similar to those of the $Mg_{96}Zn_2Y_2$ alloy. Excellent tensile yield strength was obtained in the extruded $Mg_{98.15}Zn_{0.25}Y_{1.5}La_{0.1}$ alloys, with a yield strength of 275 MPa at 523K.

## Introduction

Magnesium alloys, as one of the lightest structural metallic materials, have recently attained an increasing interest for use in the weight reduction for the computers, mobile phone and automobile industries because of their low density, high specific strength and stiffness, and good damping property [1, 2]. However, the number of commercially available Mg alloys is still limited, especially for application at room and elevated temperatures. For example, AM- and AZ-series could not be applied to power train parts operating at temperatures higher than 130°C due to their poor elevated temperature mechanical properties [3]. The recent developments of heat resistance Mg alloys are mainly concentrated on the improvement of conventional Mg-Zn and Mg-Al based alloys by alloying with third element, such as Si, rare earth (RE), alkaline earth metals, etc. [4].

Recently, Kawamura *et al.* developed rapidly solidified powder metallurgy (RS P/M) Mg-Zn-Y alloys with superior mechanical properties; a tensile yield strength of 610 MPa and an elongation of 5 % at room temperature was achieved [5]. Extruded $Mg_{97}Zn_1Y_2$ and $Mg_{96}Zn_2Y_2$ alloys also exhibited excellent mechanical properties. The tensile yield strength of $Mg_{96}Zn_2Y_2$ alloy was higher than that of $Mg_{97}Zn_1Y_2$ alloy at room temperature [6,7]. The $Mg_{97}Zn_1Y_2$ alloy consisted of fine α-Mg grains and the long period staking ordered (LPSO) phase and the improvement of mechanical properties in this alloy is considered to originate from the high dispersion of the bent LPSO phase, and the refinement of those LPSO structures and the α-Mg grains. In contrast, the $Mg_{96}Zn_2Y_2$ alloy consisted of 3 phases: fine α-Mg, LPSO phase, and the intermetallic RE containing compound. The volume fraction of intergranular phases (LPSO phase and the intermetallic RE compound) of $Mg_{96}Zn_2Y_2$ alloy was higher than that of the $Mg_{97}Zn_1Y_2$ alloy [7]. Therefore, to improve the mechanical properties of Mg-Zn-Y alloys, phase control is necessary to obtain an appropriate volume fraction of the α-Mg, LPSO, and intermetallic RE compound.

The investigation focused on the influence of Zn contents, which were also effective in strengthening Mg-Zn-Y alloys with ductility at room and elevated temperature and improving creep properties for economical efficiency, on the microstructure and mechanical properties of $Mg_{98.4}Zn_xY_{1.5}La_{0.1}$ alloys.

## Experimental procedure

The alloys, with nominal composition of $Mg_{98.4}Zn_xY_{1.5}La_{0.1}$ (X=0.25 to 2.0 at.%), were prepared by high-frequency induction melting in an Ar atmosphere and then cast into a cylindrical steel mould (31 mm of diameter). Rods of 70 mm height and 29 mm length were cut from the cast ingots and then extruded at 623K with an extrusion ratio of 10 and a ram speed of 2.5 mm/s. Heat treatment was then performed at 473K for 1.5h. The phase structures of as-cast and extruded alloys were investigated through X-ray diffractometry (XRD), optical microscopy (OM), scanning electron microscopy (SEM) and transmission electron microscopy (TEM). The mechanical properties of the investigated alloys were determined using tensile test specimens with a gage section of ϕ 2.5 mm × 15 mm.

## Results and discussion

Figure 1 shows SEM micrographs of the as-cast $Mg_{98.4}Zn_xY_{1.5}La_{0.1}$ (X=0.25 to 1.5 at.%) alloys. As shown in the picture, the secondary phase with fine lamellar contrasts was observed in the grain and also dendrite arm boundaries, and had a sharp and smooth interface. According to the XRD results and EDS analysis, it can be confirmed that the secondary phases are LPSO and/or the Mg-RE compound.

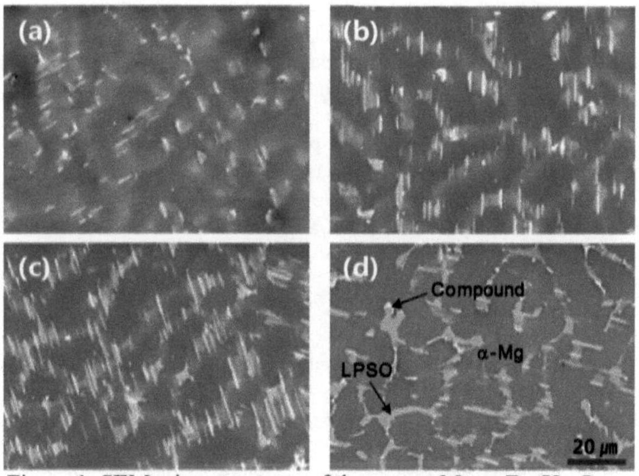


Figure 1. SEM microstructures of the as-cast $Mg_{98.4}Zn_xY_{1.5}La_{0.1}$ alloys: (a) x=0.25 (b) x=0.5 (c) x=1.0 and (d) x=1.5

In the $Mg_{96.9}Zn_{1.5}Y_{1.5}La_{0.1}$ alloys, continuous net-work of LPSO and intermetallic RE compounds form a coarse interdendrite structure. The volume fractions of both secondary phase (LPSO and compound) in $Mg_{98.4}Zn_xY_{1.5}La_{0.1}$ alloys increased with increasing Zn contents.

Fig. 2 shows the tensile properties of the five studied alloy compositions, at (a) room temperature and (b) 523K, after extrusion..,At room temperature, from 0.25 - 1.0 at.% Zn, the yield and ultimate tensile strengths of the alloys at room temperature sharply increased with increasing Zn addition, while the elongation decreased. Above 1.0 at.% Zn, tensile properties of Mg alloys only slightly increased.

The yield strength, ultimate tensile strength and elongation of $Mg_{96.9}Zn_{1.5}Y_{1.5}La_{0.1}$ at room temperature were 345 MPa, 391 MPa and 5.3 %, respectively. However, at 523K, they were 255 MPa, 223 MPa and 28 %, respectively. Therefore, the yield and ultimate tensile strengths are both slightly higher at room temperature than at elevated temperature.

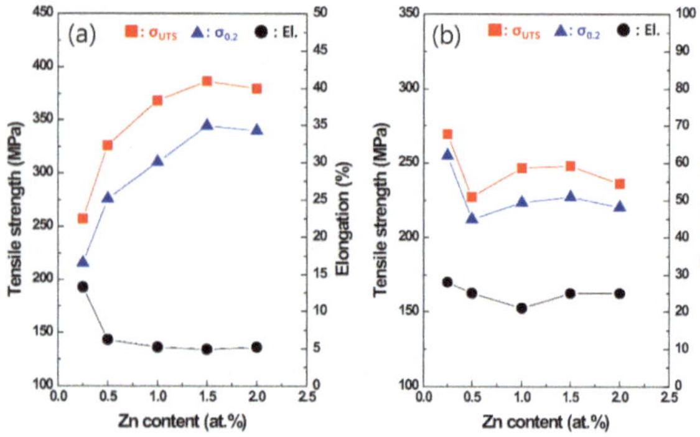


Figure 2. Mechanical properties of $Mg_{96}Zn_2Y_2$ and $Mg_{96}Zn_2Y_{1.9}M_{0.1}$ alloys at room and elevated temperature

As a result the effect of the intermetallic RE compounds on the mechanical properties at elevated temperature needs to be investigated. To do this, the extruded alloys were heat-treated at 473 K for times ranging from 0.5 to 6 h, and then reanalyzed. From micro-hardness test result of $Mg_{98.15}Zn_{0.25}Y_{1.5}La_{0.1}$ alloy after heat treatment, the hardness increased with increasing heat treatment time up to 1.5 h, but changed little afterwards.

Figures 3(a) shows the tensile properties at room temperature of the studied extruded alloys after heat treatment at 473 K for 1.5 h, which shows the same trends observed without heat treatment properties. For $Mg_{96.9}Zn_{1.5}Y_{1.5}La_{0.1}$, the yield strength, ultimate tensile strength and elongation of the $Mg_{96.9}Zn_{1.5}Y_{1.5}La_{0.1}$ extruded alloys after heat treatment at 523 K were 275 MPa, 298 MPa and 23 %, respectively.

The creep tests were carried out at 473 K and at applied stresses between 90 and 210 MPa. The 300 h creep curves for $Mg_{98.15}Zn_{0.25}Y_{1.5}La_{0.1}$ after heat treatment tested at different stresses are shown in Fig. 3. The creep curves at 150 and 210 MPa exhibit well-defined primary and secondary stages for Mg alloys. The creep resistance at 210 MPa was poor. Overall, the creep resistance of this alloy was almost that of $Mg_{96}Zn_2Y_2$ alloy [8].

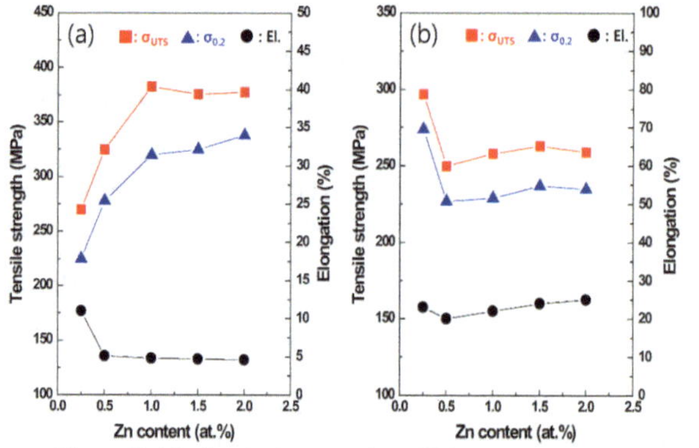


Figure 3. Mechanical properties of $Mg_{96}Zn_2Y_2$ and $Mg_{96}Zn_2Y_{1.9}M_{0.1}$ alloys at room and elevated temperature

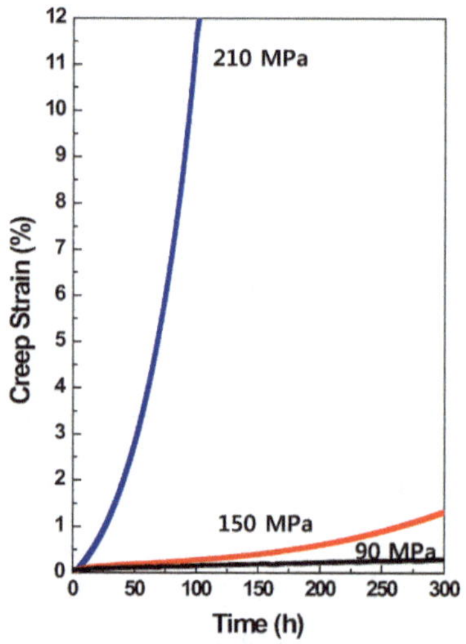


Figure 4. Creep curves of (a) $Mg_{96}Zn_2Y_2$ and (b) $Mg_{96}Zn_2Y_{1.9}M_{0.1}$ alloys at 473 K. the applied stress were 90, 150 and 210 MPa

Figure 5 shows the microstructures of extruded $Mg_{98.15}Zn_{0.25}Y_{1.5}La_{0.1}$ alloy (b) without heat treatment and (a) with heat treatment at 473 K for 1.5 h. The average α-Mg grain size without heat treatment was 1.5 ㎛ and after heat treatment was 5.2 ㎛. The compounds precipitates by annealing at high temperature in the α-Mg matrix.

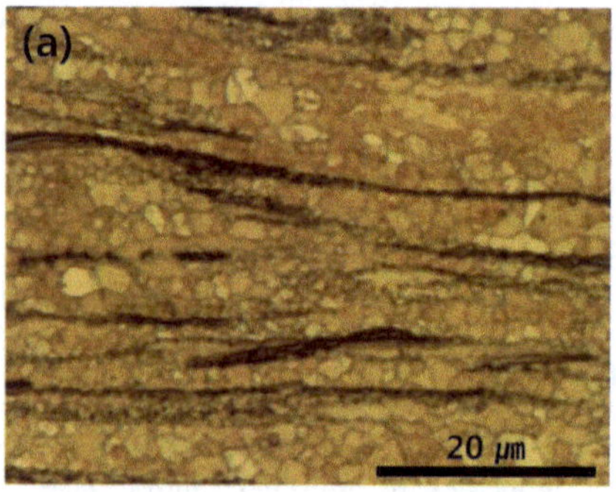

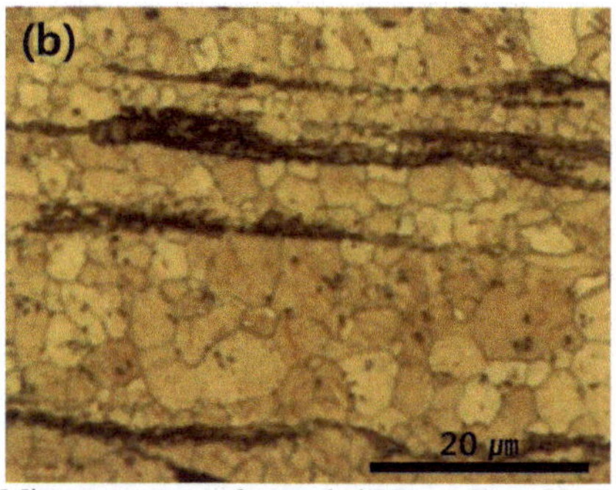


Figure 5. Microstructures of extruded $Mg_{98.15}Zn_{0.25}Y_{1.5}La_{0.1}$ alloy (a) without heat treatment and (b) with heat treatment at 473 K for 1.5 h

Figure 6 (b) shows the microstructures of extruded $Mg_{98.15}Zn_{0.25}Y_{1.5}La_{0.1}$ alloy with subsequent heat treatment at 473 K for 1.5 h. The $Mg_{98.15}Zn_{0.25}Y_{1.5}La_{0.1}$ alloys without heat treatment consist of α-Mg grains, the internetallic RE compounds and the LPSO phase (fig. (a)). The Mg-RE compounds exists between the LPSO phase in the Mg alloys. After heat treatment, the Mg-RE compounds exist within the α-Mg matrix and between the LPSO phases in the $Mg_{98.15}Zn_{0.25}Y_{1.5}La_{0.1}$ alloys. These results indicate that the strengthening of the quaternary $Mg_{98.15}Zn_{0.25}Y_{1.5}La_{0.1}$ alloy system at elevated temperature is dependent on the LPSO phase, fine-grained α-Mg matrix grains and especially, the highly dispersed Mg-RE compounds. Specifically, the amounts of these compounds are very important in this case.

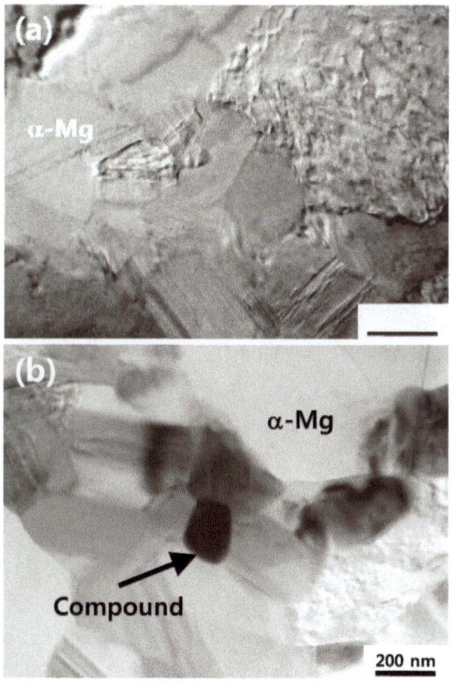


Figure 6. TEM images of extruded $Mg_{98.15}Zn_{0.25}Y_{1.5}La_{0.1}$ alloy (a) without heat treatment and (b) with heat treatment at 473 K for 1.5 h

## Conclusions

(1) Very small sized intermetallic RE compounds precipitates existed in α-Mg matrix in the $Mg_{98.4-x}Zn_xY_{1.5}La_{0.1}$ (X=0.25 to 2.0) alloys.

(2) The precipitation of these compounds phase in extruded $Mg_{98.15}Zn_{0.25}Y_{1.5}La_{0.1}$ alloy during subsequent heat-treatment was responsible for its improved tensile and creep properties.

(3) The tensile strength of extruded $Mg_{98.15}Zn_{0.25}Y_{1.5}La_{0.1}$ alloy at elevated temperature with heat-treatment was higher than that for extruded $Mg_{96}Zn_2Y_2$ alloy

## Acknowledgements


This work has been financially supported by the Development of Key Technology for Next-generation Heat-resistant Magnesium alloys, Kumamoto Prefecture Collaboration of Regional Entities for the advancement of Technology Excellence, Japan Science and Technology Agency.


## References


1. A. A. Luo, "Recent magnesium alloy development for automotive powertrain applications," *Materials Science Forum* , 419-422 (2003) 57-66.
2. Magnesium & Magnesium alloys, ed. M. M. Avedesian & H. Baker: ASM International, Materials Park, OH (1999), 3-6.
3. A. Luo, M.O. Pekguleryuz, "Cast magnesium alloys for elevated temperature applications," *Journal of Materials Science*, 29 (1994) 5259-5271.
4. Y.X. Wang, S.K. Guan, X.Q. Zeng and W.J. Ding "Effects of RE on the microstructure and mechanical properties of Mg-8Zn-4Al magnesium alloy," *Mater. Sci. Eng. A*, **416** (2006) 109-118
5. Y. Kawamura, K. Hayashi and A. Inoue, "Rapidly solidified powder metallurgy $Mg_{97}Zn_1Y_2$ alloys with excellent tensile yield strength above 600 MPa," *Materials Transactions*. 42 (2001) 1171-1174.
6. Y. Kawamura and M. Yamasaki, "Formation and Mechanical Properties of $Mg_{97}Zn_1RE_2$ Alloys with Long Period Stacking Ordered Structure," *Materials Transactions*. 48 (2007) 2986-2992.
7. S. Yoshimoto, M. Yamasaki and Y. Kawamura, "Microstructure and Mechanical Properties of Extruded Mg-Zn-Y Alloys with 14H Long Period Ordered Structure," *Materials Transactions*. 47 (2006) 959-965.
8. Jonghyun Kim and Y Kawamura, "Microstructure and Mechanical properties of Mg-Zn-Y-M (M: Mixed RE) alloys with LPSO phase," *Magnesium Technology* (2011) 229-232.

**Magnesium Technology 2012**
*Edited by: Suveen N. Mathaudhu, Wim H. Sillekens, Neale R. Neelameggham, and Norbert Hort*
***TMS (The Minerals, Metals & Materials Society), 2012***

# EFFECT OF Ca ADDITION ON THE MICROSTRUCTURAL AND MECHANICAL PROPERTIES OF AZ51/1.5$Al_2O_3$ MAGNESIUM NANOCOMPOSITE

Md Ershadul Alam[1], Rowshan Ara Rima[1], Abdelmagid Salem Hamouda[1], Quy Bau Nguyen[2] and Manoj Gupta[2]
[1] Department of Mechanical and Industrial Engineering, College of Engineering, Qatar University, Doha, Qatar 2713
[2] Department of Mechanical Engineering, National University of Singapore, 9 Engineering Drive 1, Singapore 117576

Keywords: Magnesium, AZ51-$Al_2O_3$, Ca, Nanocomposite, Mechanical properties.

## Abstract

In the present study, new AZ51/1.5$Al_2O_3$-1Ca magnesium nanocomposite was successfully synthesized by simultaneously adding 2 wt. % aluminum, 1 wt.% Ca and 1.5 vol.% nano-sized $Al_2O_3$ (50 nm) into AZ31 matrix using an innovative disintegrated melt deposition technique. AZ51/1.5$Al_2O_3$ nanocomposite was developed following the same processing route except adding Ca. All nanocomposite samples were then subsequently hot extruded at 400 $^0$C and characterized. Microstructural characterization studies revealed uniaxial grain size, reasonably uniform distribution of intermetallics and minimal porosity. Results also showed that the Ca addition into AZ51/$Al_2O_3$ nanocomposite helped to reduce the average grain size. Physical properties characterization revealed that addition of Ca reduced the coefficient of thermal expansion when compared to Ca free nanocomposite. The presence of Ca also assisted in improving overall mechanical properties including microhardness, 0.2% yield strength and ultimate tensile strength while the ductility was compromised.

## Introduction

Magnesium based materials, with a density around 1.74 g/cm$^3$, are the lightest engineering materials available in the earth which is about two-thirds of the density of aluminum based (~ 2.70 g/cm$^3$) and one-quarter of that of ferrous based (~ 7.87 g/cm$^3$) materials [1]. Therefore, magnesium based alloys are attracting much attention as light weight materials due to the increasing demand for the reduction of fuel consumption and green house gas ($CO_2$) emission [2-3]. Hence, its use is gaining significance in certain key engineering applications, particularly in automobile, aviation, electronics and consumer industries where weight is one of the most important criteria in the choice of materials [2-7].

Magnesium alloys that contain aluminum (Al) and zinc (Zn) are known as AZ alloys [8-10]. These alloys are easily available, reasonably priced and used in many engineering applications. However, the properties of these alloys can be further improved by applying metal matrix composites (MMCs) technique. This is particularly useful in engineering applications especially when properties of traditional materials like metals, polymers and ceramics are not able to match with expected properties. In recent studies, it has been observed that the addition of nano-sized reinforcements such as ceramic oxides, SiC and carbon nanotubes can lead to a simultaneous increase in strength and ductility of magnesium [11-13]. Among all the reinforcements, it can be observed from the study that the 1.5 vol. % addition of nano-sized alumina particulates in magnesium based matrix showed the best overall combination of microstructural and mechanical properties [11]. Accordingly, AZ51/1.5 vol. % $Al_2O_3$ nanocomposite was developed. Results of literature search also indicate that the addition of Ca assists in grain refinement and enhancing corrosion resistance as well as thermal and mechanical properties of magnesium based alloys and their composites [8-10, 14-15]. Literature search results also indicate that no attempt has been made so far to synthesize and characterize AZ51/$Al_2O_3$-Ca nanocomposite. Accordingly, in the present study, new AZ51/1.5 $Al_2O_3$-1Ca nanocomposite has been synthesized using disintegrated melt deposition technique.

Characterization studies were carried out on these formulations to investigate their physical, microstructural and mechanical properties. These include density and porosity measurements, the coefficient of thermal expansion, grain morphology, x-ray diffraction analysis, optical and scanning electron microscopy, microhardness, tensile tests and fractography studies.

## Experimental Procedures

### Materials

In the present study, AZ31 magnesium alloy ingots (2.94% Al, 0.87% Zn, 0.57% Mn, 0.0027% Fe, 0.0112% Si, 0.0008% Cu, 0.0005% Ni and balance Mg, supplied by Tokyo Magnesium Company Limited, Japan) were used as a matrix material and cut into small pieces so that they can be placed into the graphite crucible easily. The aluminum lumps of 99.5% purity (supplied by Alfa Aesar, USA) and the reinforcement in the form of 50 nm alumina powder of 99.4 % purity (supplied by Baikowski, Japan) were used for developing AZ51/1.5$Al_2O_3$ nanocomposite. Ca granules of 99.6% purity (supplied by Alfa Aesar, USA) were also used with AZ31, Al and nano-sized $Al_2O_3$ to develop AZ51/1.5$Al_2O_3$-1Ca nanocomposite.

### Primary Processing

Synthesis of AZ51/1.5$Al_2O_3$ and AZ51/1.5$Al_2O_3$-1Ca magnesium based nanocomposites was carried out using disintegrated melt deposition (DMD) technique. Synthesis of AZ51/1.5 $Al_2O_3$-1Ca nanocomposite involved heating the AZ31 magnesium alloy pieces with the addition of 2 wt.% Al, 1 wt.% Ca and 1.5 vol. % (~3.3 wt.%) nano-sized $Al_2O_3$ to 750 $^0$C in an inert Ar gas atmosphere in a graphite crucible using a resistance heating furnace. The crucible was equipped with an arrangement for bottom pouring. Upon reaching the superheat temperature, the molten slurry was stirred for 5 min at 450 rev./min using a twin blade (pitch 45$^0$) mild steel impeller to facilitate the incorporation and uniform distribution of reinforcement particulates in the metallic matrix. The impeller was coated with Zirtex 25 (86% $ZrO_2$, 8.8% $Y_2O_3$, 3.6% $SiO_2$, 1.2% $K_2O$ and $Na_2O$, and 0.3% trace inorganic) to avoid iron contamination of the molten metal. The melt was then released through a 10-mm diameter orifice at the base of the crucible. The melt material was disintegrated by two jets of argon gas orientated normal to the melt stream. The argon gas flow rate was maintained at 25 L /min. The disintegrated melt slurry was subsequently deposited onto a metallic substrate. Preform of 40-mm diameter was obtained following the deposition stage. The synthesis of AZ51/1.5$Al_2O_3$ magnesium nanocomposite was carried out using steps similar to those employed for AZ51/1.5$Al_2O_3$-1Ca except that no Ca were added (see Table I).

Table I. Compositions of Synthesized Nanocomposites

| Compositions | Al (Wt. %) | n- $Al_2O_3$ (Vol. %) | Ca (Wt. %) | AZ31 |
|---|---|---|---|---|
| AZ51/$Al_2O_3$ | 2 | 1.5 | - | Balance |
| AZ51/$Al_2O_3$-Ca | 2 | 1.5 | 1 | Balance |

Secondary Processing

Pre-Extrusion: The 40 mm diameter ingots were machined down using a lathe machine to a diameter of 36 mm and cut into billets with heights of approximately 45 mm. They were then lathed in the direction perpendicular to the length of the ingot to ensure both ends are flat and perpendicular to the surface. The billets were then sprayed with colloidal graphite for lubrication purposes.

Extrusion: The billets were first soaked at 400 $^0$C for 60 minutes in a constant temperature furnace before extrusion. Extrusion was performed on a 150 tonne hydraulic press using an extrusion ratio of 20.25:1, producing rods of 8 mm diameter.

Post-Extrusion: After extrusion, the extruded rods were machined to produce tensile specimens. Sections of approximately 4 to 10 mm in height were also cut by a low speed diamond blade and the surface graphite was then cleaned-off to use for various characterization studies.

Density and Porosity Measurements

Density ($\rho$) measurements were performed in accordance with Archimedes' principle on three randomly selected polished samples of AZ51/$Al_2O_3$ and AZ51/$Al_2O_3$-Ca nanocomposites taken from the extruded rods. Distilled water was used as the immersion fluid. The samples were weighed using an A&D ER-182A electronic balance; with an accuracy of ±0.0001 g. Theoretical densities of materials were calculated assuming they are fully-dense. Rule-of-Mixture was used in all calculations. The porosity was calculated by using the theoretical and experimental densities.

Microstructural Characterization

Microstructural characterization studies were conducted on metallographically polished extruded samples to investigate morphological characteristics of grains, reinforcement distribution and interfacial integrity between the matrix and reinforcement. The etching solution (5 ml acetic acid, 6 g picric acid, 10 ml water and 100 ml ethanol) was applied for approximately 10 seconds using a swabbing technique and then washed under running water to reveal the grain boundaries [16-17]. The sample was then analyzed using the Olympus BH2-UMA metallographic optical microscope equipped with Olympus DP-10 microscope digital camera. The grain boundaries were then traced out from the micrographs with the aid of the Adobe Photoshop program. Image analysis using the Scion system was carried out to determine the grain size of the materials.

The presence and distribution of the intermetallic phase was investigated using the JEOL JSM-5600LV Scanning Electron Microscope (SEM) and Hitachi S-4300 FESEM. Polished specimens were observed at 1000x magnification to reveal the intermetallic phase.

X-Ray Diffraction Studies

X-ray diffraction analysis was conducted using the automated Shimadzu LAB-X XRD-6000 x-ray diffractometer. Flat, ground and ultrasonically cleaned specimens of approximately 5 mm in height were exposed to $CuK_\alpha$ radiation ($\lambda$ = 1.54056 Å) with a scanning speed of 2 $^0$/min. The scanning range was 30$^0$ to 80$^0$ for all samples. A plot of intensity against 2θ (θ represents Bragg angle) was obtained, illustrating peaks at different Bragg angles. The Bragg angles corresponding to different peaks were noted; and the values of interplanar spacing (d-spacing) obtained from the computerized output were compared with the standard values from the International Centre for Diffraction Data's Powder Diffraction File (PDF).

Coefficient of Thermal Expansion

The coefficients of thermal expansion (CTE) of all the compositions were determined by measuring the displacement of the samples as a function of temperature in the temperature range of 50 $^0$C- 400 $^0$C using an automated SETARAM 92-16/18 thermo-mechanical analyzer.

Mechanical Characteristics

The mechanical properties carried out on all samples were investigated by conducting tensile and microhardness tests. All experiments were carried out at room temperature.

Microhardness Test: The microhardness tests were conducted on flat and metallographically polished specimens. The test was conducted using a Shimadzu HMV automatic digital microhardness tester with a Vickers indenter (square-based pyramidal-shaped diamond indenter with face angle of 136 $^0$). An indenting load of 25 gf and a dwell time of 15 seconds were used. Testing was performed in accordance with ASTM test standard E384-08. Indentations were performed on the matrix and measurements were recorded in Vickers Hardness (HV).

Tensile Testing: The tensile properties of each sample were determined in accordance with ASTM test standard E8M-08. Round tension test specimens of 5 mm in diameter were machined from the 8 mm extruded rod. An average of five tensile specimens could be obtained from a single rod, and a minimum of four tensile tests were conducted for each sample. Tests were performed on the 810 Material Test System (MTS) with an extensometer of 25 mm gauge length and a crosshead speed of 0.254 mm/min was used. The raw stress-strain data recorded was extracted to be analyzed. Broken test specimens were labeled individually, carefully handled and packed into individual plastic bags to prevent damage of the fractured surface. A Microsoft Excel program was written to analyze the raw data extracted. Then, 0.2% offset yield strength (0.2% YS), ultimate tensile strength (UTS) and failure strain was computed via a series of Microsoft Excel functions.

Fracture Behavior: The fracture surfaces of the broken tensile samples were analyzed to investigate the failure mechanisms that occurred during the tensile tests. Fractography studies were performed on the JEOL JSM-5600LV Scanning Electron Microscope and Hitachi S-4300 FESEM. Images were captured at an accelerating voltage of 15 kV and a working distance of about 16 mm. Macromechanism of these fracture surfaces were also investigated.

## Results

### Macrostructure

The result of macrostructural characterization conducted on the as deposited $AZ51/Al_2O_3$ and $AZ51/Al_2O_3$-Ca nanocomposites did not reveal any presence of macropores, shrinkage, cracks or any other defects. Following extrusion, there was also no evidence of any macrostructural defects.

### Density and Porosity Measurements

The results of the density measurements are shown in Table II. Results show that near dense nanocomposites can be developed using the fabrication methodology adopted in the present study. Results also revealed that all nanocomposite samples exhibited minimal porosity.

Table II. Results of Density and Porosity Measurements

| Compositions | Density ($g/cm^3$) | | Porosity (%) |
|---|---|---|---|
| | Theo. | Expt. | |
| $AZ51/Al_2O_3$ | 1.827 | 1.825 | 0.09 |
| $AZ51/Al_2O_3$-Ca | 1.824 | 1.822 | 0.12 |

### Microstructural Characteristics

Microstructural studies conducted on the extruded specimens showed near equiaxed grain morphology (see Table III and Figure 1). Average grain size of $AZ51/Al_2O_3$ nanocomposite samples decreased with the addition of Ca.

Table III. Results of Grain Morphology

| Compositions | Grain Size (μm) | Aspect Ratio | Roundness[1] |
|---|---|---|---|
| $AZ51/Al_2O_3$ | 3.6 ± 1.5 | 1.7 ± 0.5 | 1.6 ± 0.4 |
| $AZ51/Al_2O_3$-Ca | 1.7 ± 0.5 | 1.6 ± 0.3 | 1.6 ± 0.4 |

*1. Roundness is the shape of the grain expressed by the formula (perimeter)$^2$ / 4π (area)*

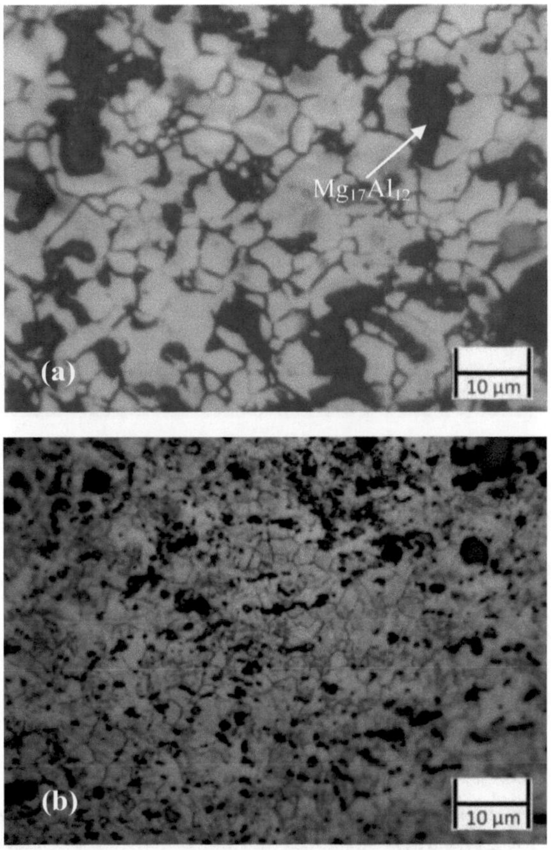


Figure 1. Representative optical micrographs showing the grain morphology (1000x) of: (a) $AZ51/Al_2O_3$ and (b) $AZ51/Al_2O_3$-Ca samples, respectively.

Microstructural characterization studies were also conducted on the extruded and polished samples to observe the shape and distribution of intermetallic phases (see Figure 2). The white color second phase corresponds to $Mg_{17}Al_{12}$ and the light gray color phase corresponds to $(Mg, Al)_2Ca$ phase based on previous study conducted by the same research group (see Figure 2b) [10]. Results showed that amount of $Mg_{17}Al_{12}$ phase decreased noticeably with the addition of Ca into $AZ51/Al_2O_3$ nanocomposite (see Figures 2 (a) and (b)).Figure 2(c) showed the interfacial bonding between the $(Mg, Al)_2Ca$ and the matrix.

### X-Ray Diffraction

X-ray diffraction (XRD) was carried out on all extruded samples. The obtained lattice spacing's (d-spacing) and two-theta (2θ) value were compared with the standard values for magnesium and the Mg-Al-Ca intermetallic phases. Results showed the presence of eutectic $Al_{12}Mg_{17}$ intermetallic phase in the α-Mg phase for the case of $AZ51/Al_2O_3$ samples while both $Mg_{17}Al_{12}$ and $(Mg, Al)_2Ca$ intermetallic phases were found in the α-Mg matrix for the case of $AZ51/Al_2O_3$-Ca sample. However, no nano-sized alumina phase was detected from the XRD analysis.

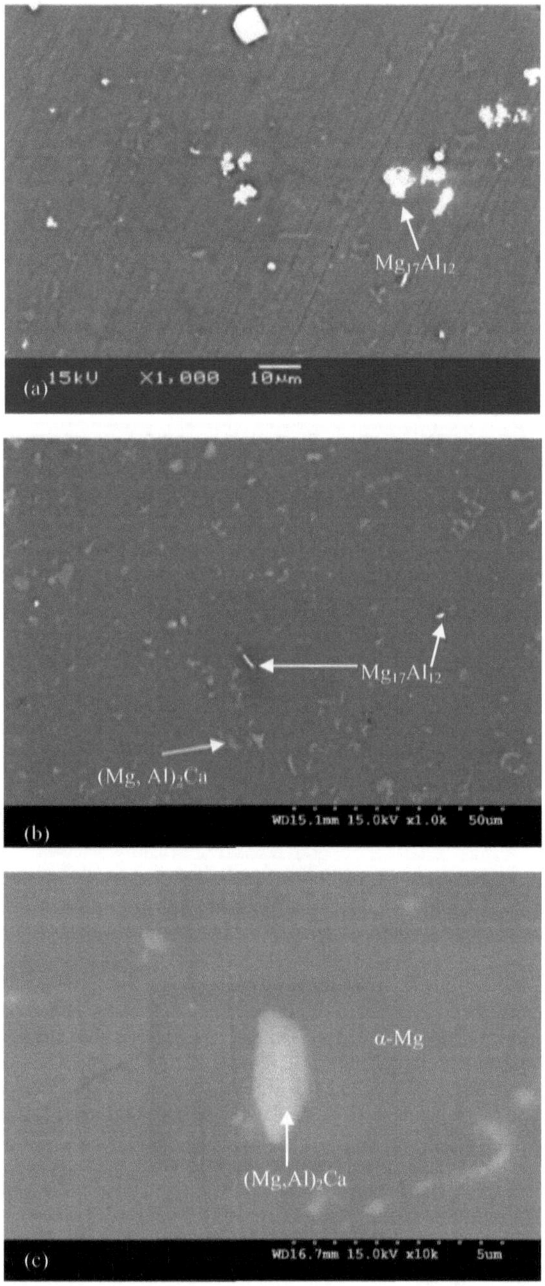


Figure 2. Representative SEM and FESEM micrographs showing the shape and distribution of $Mg_{17}Al_{12}$ and $(Mg, Al)_2Ca$ phases in the case of: (a) $AZ51/Al_2O_3$ and (b) $AZ51/Al_2O_3$-Ca samples, respectively. Figure (c) shows the interfacial bonding between second phase and the matrix.

Coefficient of Thermal Expansion

Table IV shows the results of coefficient of thermal expansion measurements obtained from $AZ51/Al_2O_3$ and $AZ51/Al_2O_3$-Ca samples. The results revealed a reduction in CTE of the $AZ51/Al_2O_3$ nanocomposite with the addition of Ca. This value is much lower than that of AZ31 sample.

Table IV. Results of CTE and Microhardness Tests

| Compositions | CTE (μm/mK) | Microhardness (HV) |
|---|---|---|
| $AZ51/Al_2O_3$ | 26.2 | $110 \pm 3$ |
| $AZ51/Al_2O_3$-Ca | 25.7 | $119 \pm 5$ |
| AZ31[16] | 28.8 | $66 \pm 2$ |
| AZ61A [1] | - | 62 |

Mechanical Characteristics

Microhardness Test: Vickers microhardness tests were conducted on all extruded and polished samples (see Table IV). Results revealed that the addition of Ca increased the microhardness value of $AZ51/Al_2O_3$ nanocomposite. More than 8% microhardness value was increased when 1 wt. % Ca was added into $AZ51/Al_2O_3$ nanocomposite. Moreover, $AZ51/Al_2O_3$-Ca magnesium based nanocomposite exhibited 80% and 92% higher hardness when compared to AZ31 and commercially available AZ61A alloy, respectively (see Table IV).

Tensile Test: Results of ambient temperature tensile test on round tensile specimens revealed that the addition of Ca in $AZ51/Al_2O_3$ nanocomposite increased the 0.2% yield strength (YS) and ultimate tensile strength (UTS) (see Table V). $AZ51/Al_2O_3$-Ca nanocomposite exhibited around 7% and 2% higher YS and UTS, respectively, when compared to $AZ51/Al_2O_3$ samples. Results also showed that the Ca added nanocomposite exhibited 25% and 16% higher 0.2% YS and UTS, respectively, when compared to the AZ31 alloy. However, failure strain (FS) dropped with the addition of Ca.

Table V. Results of Room Temperature Tensile Properties

| Compositions | 0.2 % YS (MPa) | UTS (MPa) | FS (%) |
|---|---|---|---|
| $AZ51/Al_2O_3$ | $211 \pm 4$ | $311 \pm 3$ | $13.4 \pm 1.2$ |
| $AZ51/Al_2O_3$-Ca | $225 \pm 3$ | $317 \pm 4$ | $9.4 \pm 0.5$ |
| AZ31[16] | $180 \pm 3$ | $273 \pm 6$ | $10.6 \pm 1.3$ |

Fracture Behavior: Macroscopic fracture characteristics of the broken tensile samples revealed that all samples exhibited shear-type ductile fracture mode. Microscopic features of fracture surface were also conducted on the tensile fractured surface using SEM/FESEM and dimple-like features were observed in all the cases.

**Discussion**

Macrostructure

Syntheses of $AZ51/Al_2O_3$ and $AZ51/Al_2O_3$-Ca nanocomposites were successfully accomplished by using DMD technique followed by hot extrusion. Observations made during DMD processing revealed: (a) minimal oxidation of melts, (b) undetectable reaction between graphite crucible and melts and (c) absence of blowholes and macropores. The results suggest the appropriateness of processing parameters and methodology used in the present study. These observations are also consistent with the previous findings made on DMD processed magnesium based materials [10-11, 16].

Density and Porosity Measurements

The experimental density obtained by the Archimedes' principle exhibited that the density of newly developed magnesium based $AZ51/Al_2O_3$-Ca nanocomposite slightly decreased with the addition of elemental Ca into $AZ51/Al_2O_3$ (see Table II). This can be attributed to the relatively lower dense Ca (1.55 $g/cm^3$) addition into AZ31 matrix (1.78 $g/cm^3$) [1, 11]. Obtained results also revealed that the experimental values were relatively close to the theoretical values. This indicated that the experimental method used was accurate, and the experimental density of the fabricated materials could be calculated.

Porosity of the materials was calculated using the theoretical and the experimental densities obtained from each sample. The highest porosity obtained among the samples was 0.12% (see Table II), hence indicating that near dense materials was obtained and that the alloying and reinforcing constituents were successfully incorporated into the matrix. Therefore, as demonstrated by prior studies, the fabrication route of DMD followed by hot extrusion is capable of producing new $AZ51/Al_2O_3$-Ca nanocomposite with minimal porosity [11].

Microstructural Characteristics

The results of microstructural characterization revealed the presence of nearly equiaxed grains (see Table III and Figure 1) in Ca-added and Ca-free $AZ51/Al_2O_3$ nanocomposite samples. Second phases predominantly located at grain boundaries were observed in all samples (see Figure 1). Addition of elemental Ca into $AZ51/Al_2O_3$ decreased the average grain size (see Table III and Figure 1). This can be attributed to the pinning of grain boundaries by the higher amount of $(Mg, Al)_2Ca$ second phases resulting in limited grain growth.

SEM and FESEM micrographs of the etched samples revealed the presence and good distribution of the equilibrium intermetallic phases $Mg_{17}Al_{12}$ and $(Mg, Al)_2Ca$ (see Figure 2). This could be attributed to the use of a layered arrangement for the raw materials during the solidification process and effective stirring parameters. The successful disintegration of the melt and the inert atmosphere caused by the argon gas jets were also the contributing factors. In the $AZ51/Al_2O_3$ system, only $Mg_{17}Al_{12}$ intermetallic was found while for the $AZ51/Al_2O_3$-Ca system, both $Mg_{17}Al_{12}$ and $(Mg,Al)_2Ca$ intermetallics were observed (see Figure 2). Amount of $Mg_{17}Al_{12}$ decreased with the addition of Ca. This can be attributed to the formation of $(Mg, Al)_2Ca$ phase where Al reacted with Ca. Figure 2(c) revealed the strong interfacial bonding between matrix and second phase.

X-Ray Diffraction

X-Ray diffraction results confirmed the presence of the intermetallic phase $Mg_{17}Al_{12}$ in $AZ51/Al_2O_3$ nanocomposite while both $Mg_{17}Al_{12}$ and $(Mg, Al)_2Ca$ phases were observed for $AZ51/Al_2O_3$-Ca samples. However, the presence of the nano-sized alumina was not identified. This can be attributed to the limitation of the filtered x-ray to detect phases with less than 2 vol.% of nano-sized particles [16, 18].

Coefficient of Thermal Expansion

The results of CTE measurement showed that the addition of Ca alloying constituent into $AZ51/Al_2O_3$ nanocomposite decreases the average CTE values of the material (see Table IV). The results suggested an appropriate integration of $AZ51/Al_2O_3$composite with elemental Ca leading to low CTE values. Moreover, the lower CTE values of Ca (22.3 μm/mK) than Mg (25.2 μm/mK) also helped to lower the average CTE value of the newly developed compositions [1, 16].

Mechanical Characteristics

Microhardness Test: The experimental results of microhardness measurement are shown in Table IV. Significant increase in microhardness was observed in $AZ51/Al_2O_3$-Ca composite samples when compared to $AZ51/Al_2O_3$ nanocomposite, pure AZ31 and commercially available AZ61A samples (see Table IV). This can be attributed to the lower average grain size of the new compositions when compared to $AZ51/Al_2O_3$ which is associated with larger grain boundary area leading to higher hardness [16, 19]. The increase in hardness can be attributed to: (i) the solid solution hardening, (ii) the reasonably uniform distribution of harder $(Mg, Al)_2Ca$ intermetallic in the matrix and (iii) higher constraint to the localized matrix deformation during indentation due to the higher presence of intermetallic [10, 16].

Tensile Test: Room temperature tensile test results revealed that the addition of elemental Ca increased the strength of $AZ51/Al_2O_3$ composite (see Table V and Figure 3). Tensile results also showed that the $AZ51/Al_2O_3$-Ca nanocomposite exhibited around 25% and 16% higher 0.2% YS and UTS, respectively, when compared to the AZ31 alloy. This can be attributed to the lower average grain size of $AZ51/Al_2O_3$-Ca samples when compared to $AZ51/Al_2O_3$ samples (see Table III and Figure 1). Presence of additional Ca modified the microstructural features, hence increasing strength. This observation is similar with the findings of other researchers obtained from $AZ31B/Al_2O_3$-Ca samples [10].

Failure strain decreased with Ca addition into $AZ51/Al_2O_3$ composite. This can be attributed to the formation and presence of sharp edge intermetallic particles which might act as a stress concentration sites (see Figure 2).

Fracture Behavior:

Macroscopic observations performed on the broken tensile samples revealed shear type fracture characteristics. Microscopic observations showed that dimple like features were predominately present in all compositions, indicating high plastic deformation (see Figure 3). However, presence of microcracks and cleavage step on the fracture surface were observed for new system, hence are responsible for the decrease in failure strain of these composites.

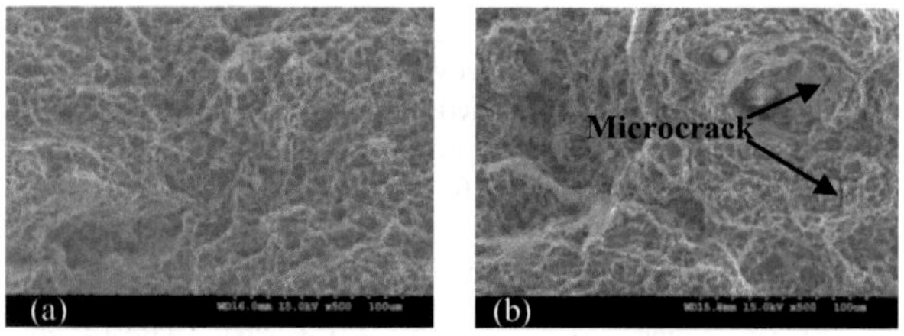


Figure 3. Representative FESEM fractographs showing the dimple like features and microcrack for: (a) $AZ51/Al_2O_3$ and (b) $AZ51/Al_2O_3$-Ca samples, respectively.

### Conclusions

The following conclusions can be made from the present study:

1. $AZ51/Al_2O_3$ and $AZ51/Al_2O_3$-Ca nanocomposites can be successfully synthesized by using the disintegrated melt deposition technique followed by hot extrusion with minimal porosity.

2. The addition of Ca leads to decrease in $Mg_{17}Al_{12}$.secondary phase and form $(Mg,Al)_2Ca$ phase. Ca addition also helps to decrease average grain size. The coefficient of thermal expansion of $AZ51/Al_2O_3$ magnesium nanocomposite also decreases with the addition of Ca.

3. A good combination mechanical properties (in terms of microhardness, 0.2% YS and UTS) was observed for $AZ51/Al_2O_3$-Ca samples when compared to $AZ51/Al_2O_3$ samples.

### Acknowledgments

The authors gratefully acknowledge the support received for this research work ref: NPRP 08-424-2-171 from the Qatar National Research Fund (QNRF), Qatar.

### References

[1] J. R. Davis, "Properties and Selection: Nonferrous Alloys and Special-Purpose Materials", Formerly Tenth Edition, (ASM Handbook, Volume 2, ASM, Metal Park, Ohio, 1993) 480-515, 1099-1100, 1105-1107,1118-1128,1199, 1132-1135.

[2] S. F. Hassan and M. Gupta, "Development of Nano-$Y_2O_3$ Containing Magnesium Nanocomposites Using Solidification Processing", Journal of Alloys and Compounds, 429 (1-2) (2007) 176-183.

[3] R. Fink, in: K. U. Kainer (Ed.), "Die-Casting Magnesium, Magnesium Alloys and Technologies", (Wiley-VCH Verlag GmbH and Co., Germany, 2003) 23-44.

[4] B. L. Mordike and K. U. Kainer, "Magnesium Alloys and Their Applications", (Werkstoff-Informationsgesellschaft, Frankfurt, Germany, 1998).

[5] B. L. Mordike and T. Ebert, "Magnesium: Properties-Applications-Potentials", Mater. Sci. Eng. A, 302 (2001) 137-45.

[6] D. J. Lloyd, "Particle Reinforced Aluminium and Magnesium Matrix Composites", Int. Mater. Rev., 39 (1994) 1-23.

[7] F. Moll and K. U. Kainer, "Magnesium Alloys and Technology", Kainer, K. U. (Ed), (Weinheim: Wiley-VCH, Germany, (2002) 197-217.

[8] G. Wu, Y. Fan, H. Gao, C. Zhai, and Y. P. Zhu,. "The Effect of Ca and Rare Earth Elements on the Microstructure, Mechanical Properties and Corrosion Behavior of AZ91D", Mater. Sci. Eng. A, 408 (2005) 255–263.

[9] B. Jing, S. Yangshan, X. Shan, X. Feng, and Z. Tianbai, "Microstructure and Tensile Creep Behavior of Mg–4Al Based Magnesium Alloys with Alkaline Elements Sr and Ca Additions", Mater. Sci. Eng. A, 419(2006) 181–188.

[10] Q. B. Nguyen and M. Gupta, "Microstructure and Mechanical Characteristics of $AZ31B/Al_2O_3$ Nano-Composite with Addition of Ca", Journal of Composite Materials, 43 (2009) 5-17.

[11] Q. B. Nguyen and M. Gupta, "Increasing Significantly the Failure Strain and Work of Fracture of Solidification Processed AZ31B Using Nano-$Al_2O_3$ Particulates", J. Alloys Comp., 459 (2008) 244-250.

[12] C. S. Goh, J. Wei, L. C. Lee and M. Gupta, "Development of Novel Carbon Nanotube Reinforced Magnesium Nanocomposites Using the Powder Metallurgy Technique", Nanotechnology, 17 (2006) 7-12.

[13] S. F. Hassan and M. Gupta, "Effect of Different Types of Nano-Size Oxide Particulates on Microstructural and Mechanical Properties of Elemental Mg", J. Mater. Sci., 41 (2006) 2229-2236.

[14] D. Wenwen, Z. Yangshan, M. Xuegang, X. Feng, Z. Min, and W. Dengyun, "Microstructure and Mechanical Properties of Mg–Al Based Alloy with Calcium and Rare Earth Additions," Mater. Sci. Eng A, 356 (2003) 1–7.

[15] G. Y. Yuan, Z. L. Liu, Q. D. Wang, and W. J. Ding, "Microstructure Refinement of Mg–Al–Zn–Si Alloys", Materials Letters, 56 (2002) 53–58.

[16] M. E. Alam, S. Han, A. M. S. Hamouda, Q. B. Nguyen and M. Gupta, "Development of New Magnesium Based Alloys and Their Nanocomposites", Journal of Alloys and Compounds, 509 (2011) 8522-8529.

[17] A. Jager, P. Lukac, V. Gartnerova, J. Haloda, and M. Dopita, "Influence of Annealing on the Microstructure of Commercial Mg Alloy AZ31 after Mechanical Forming", Mater. Sci. Eng. A, 432 (2006) 20-25.

[18] B. D. Cullity, "Elements of X-Ray Diffraction", (3rd ed., Prentice Hall, London, 2001).

[19] G. E. Dieter, "Mechanical Metallurgy", ($2^{nd}$ Edition, McGraw-Hill, Inc., USA, 1976) 191-193.

**Magnesium Technology 2012**
*Edited by: Suveen N. Mathaudhu, Wim H. Sillekens, Neale R. Neelameggham, and Norbert Hort*
***TMS (The Minerals, Metals & Materials Society), 2012***

# Effect of Zn Concentration and Grain Size on Prismatic Slip in Mg-Zn Binary Alloys

N. Stanford, M.R. Barnett
Centre of Excellence for Design in Light Metals,
Deakin University, Pigdons Road Geelong, Victoria, Australia 3217

Keywords: magnesium, slip, twinning, deformation, grain size, Hall-Petch

## Abstract

Mg-Zn binary alloys with concentrations between 0 and 2.8wt% Zn have been prepared and processed via hot rolling and annealing to produce specimens with a strong basal texture and a range of grain sizes. These have been deformed in tension, a condition in which the deformation is dominated by prismatic slip. This data has been used to assess the Hall-Petch parameter as a function of Zn concentration for deformation dominated by prismatic slip. Pure magnesium showed non-linear Hall-Petch behaviour at large grain sizes, and this is compared to the values for prismatic slip measured on single crystals. The differences between critical resolved shear stress measurements made through single crystal, polycrystal and mathematical modelling techniques are also discussed.

## Introduction

Due to the close packed hexagonal crystal structure of magnesium, alloys of this material deform by five principle deformation modes: basal <a> slip, prismatic <a> slip, second order <c+a> pyramidal slip, $\{10\bar{1}2\}$ twinning and $\{10\bar{1}1\}$ twinning [1,2]. The critical resolved shear stress (CRSS) values for these deformation modes have been measured in single crystals [3-9], and their values in polycrystals have been estimated by mathematical models such as the VPSC approach eg. [10]. It is noted in a number of publications that these two sets of data do not correlate well [11,12]. The ratio of the CRSS values for basal to prismatic slip, measured in single crystal studies, is in the region of 1:85, while VPSC modelling typically find values around 1:6 for polycrystal alloys [11]. This discrepancy could be either a result of differences in the solute level, or a result of the difference in grain size between the experimentally examined single crystals and the mathematically modeled polycrystals.

It is known that magnesium alloys typically increase in strength with a decrease in grain size, consistent with the Hall-Petch effect, see Fig. 1. However, due to the fact that a number of deformation modes operate in this system, it has not yet been quantified what effect grain size has on the individual deformation modes. In this short paper we restrict our discussion to prismatic slip, but this communication is part of a much larger body of work that also examines basal slip and twinning. The aim of this research is to develop a quantitative measure of the effect of grain size on the stress to activate the different deformation modes, in the case discussed here, prismatic slip. Our measurements are based on the notion that when strongly textured magnesium alloys are deformed in tension, the prismatic slip system is the dominant deformation mode. We therefore deliberately induce strong textures and then deform in tension to activate prismatic slip. These measurements are normalized for texture in order to estimate as closely as possible the stress required to activate the prismatic slip system as a function of grain size.

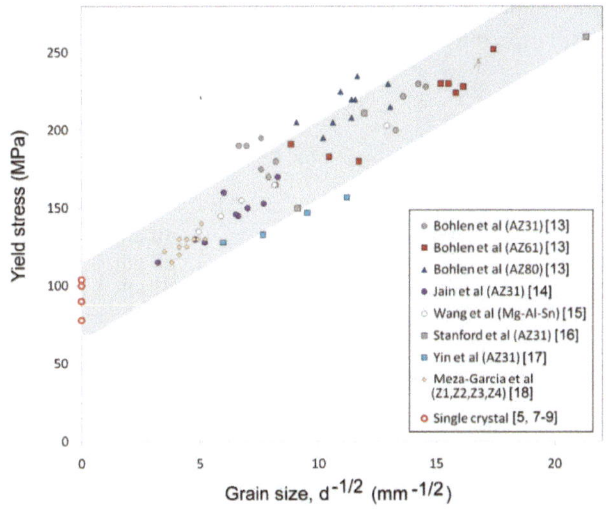


Figure 1 – Hall-Petch plot for a range of alloys from literature. All alloys have been processed by either rolling or extrusion, and are therefore expected to have strong textures. All alloys have been deformed in tension. Single crystal data from crystals orientated for prismatic slip

## Experimental Methodology

This work is based on five alloy compositions (in wt%): Pure Mg, Mg-0.2Zn, Mg-0.5Zn, Mg-1.1Zn, and Mg-2.8Zn. The alloys were produced in 400 g castings that were chill cast into 25 mm thick plate. The alloys were solution treated for 4 hours at 450°C in a flowing argon atmosphere. The alloys were then sectioned and hot rolled at 400°C to 1.25 mm thickness. Tensile specimens with a gage length of 25 mm and width of 3 mm were cut from the rolled plate and then annealed to produce three different grain sizes for each alloy composition. Tensile testing was carried out on an Instron tensile tester equipped with a non-contact video extensometer and a 5 kN load cell. The tests were carried out at an initial strain rate of $10^{-3}$ $s^{-1}$. All specimens were deformed to failure, and were tested in the rolling direction.

Some additional experiments were carried out only on pure magnesium with a larger grain size range. For these experiments the magnesium plate was hot rolled from 10 mm to a thickness of 5 mm and then recrystallisation annealed at various times and temperatures to obtain the desired grain sizes. These larger thickness specimens were used for the largest grain sizes because it has been shown that a minimum number of grains, about 15, are required across the guage thickness to ensure that the yield point is indicative of bulk material behaviour [19,20]. The texture of all specimens was measured using x-ray diffraction with CuK$\alpha$ radiation. Only the basal pole figures were measured.

### Results

Typical examples of the stress strain curves obtained in tension are shown in Fig. 2. There was an increase in yield point with a decrease in grain size, and this was more marked in the higher concentration alloys. For the smaller grain sizes there was a well defined yield point in the more concentrated alloys, Fig. 2b.

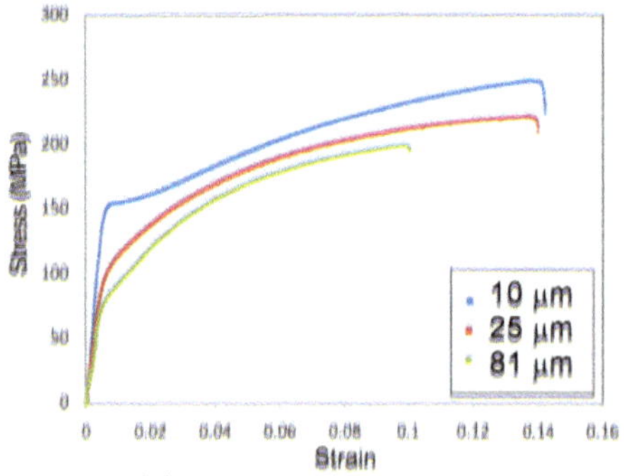


(a) 1.1% Zn, three grain sizes

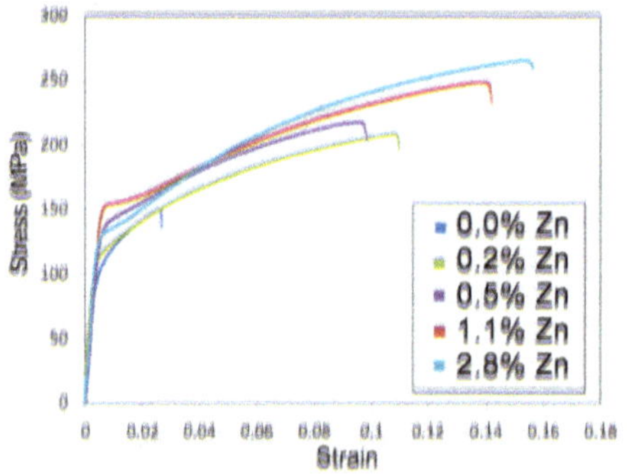


(b) Grain size = 13 (±2) μm

Figure 2 – Tensile curves for selected specimens

The ductility of the pure magnesium specimens was quite low, less than 0.05 in all cases, Fig. 3a. However the ductility was significantly improved at higher Zn concentrations, Figs. 3a and 3b. The work hardening rate was not significantly affected by the Zn concentration, with all the work hardening rate data fitting to a straight line when plotted against $d^{-1/2}$, Fig. 4.

For the pure magnesium, a larger range of grain sizes were tested in order to validate the Hall-Petch behaviour, Fig. 5. Also included on this plot is the stress to activate prismatic slip measured on single crystals [5, 7-9] assuming a Taylor factor of 2. The band of data from Fig. 1 is also included for reference. It is apparent from this plot that there is a plateau in the yield stress for larger grain sizes, and that this plateau corresponds well with single crystal data.

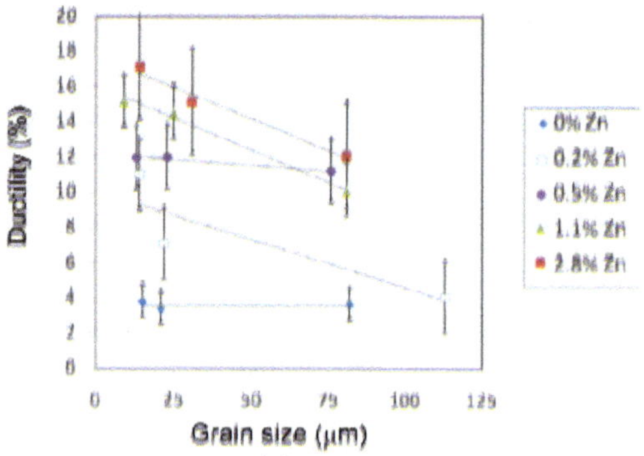


(a)

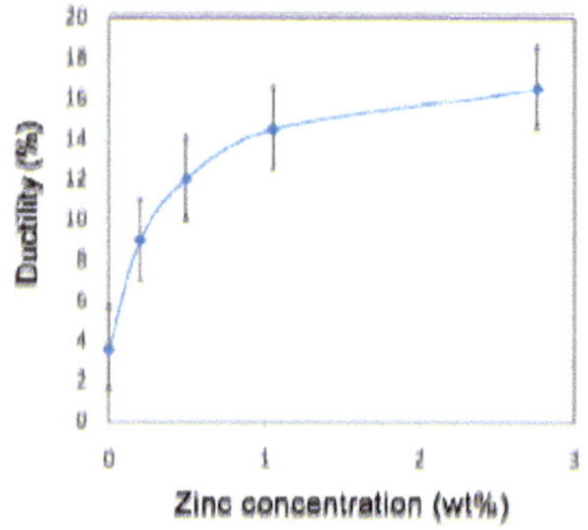


(b)

Figure 3 – (a) Measured ductility for all alloys and grain sizes tested. (b) Effect of Zn concentration on the tensile ductility for a grain size of 20 μm.

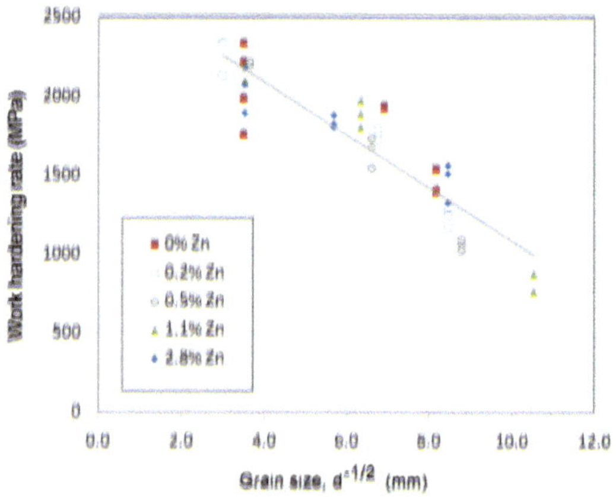


Figure 4 – Work hardening rate at a strain of 0.02 for all alloys and grain sizes tested.

A number of different fits could be applied to the data shown in Fig. 5. For example, this same data set plots well against the inverse of the grain size, $d^{-1}$. Another option would be a linear fit that crosses the y-axis at ~70 MPa. However, taking into consideration that the single crystal data are a minima estimate, we conclude here that the best fit to the available data is a plateau at grain sizes above ≈30 μm. We also draw weight for our argument to plot against $d^{-1/2}$ from the data shown in Fig. 1 where it is clear that the Hall-Petch relationship holds over a range of grain sizes and alloys.

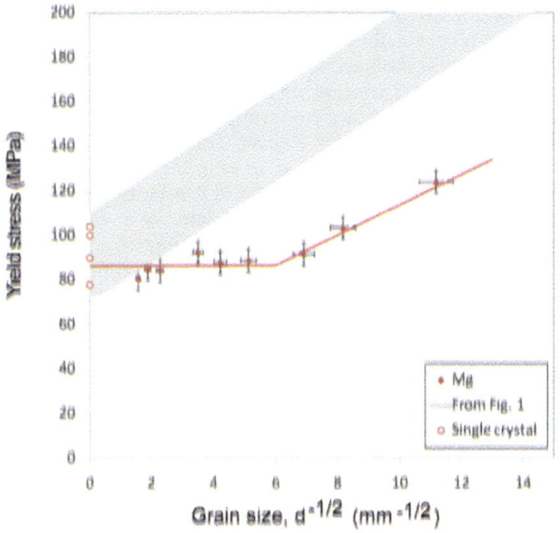


Figure 5 – Hall-Petch plot for pure magnesium. Data zone from Fig. 1 also included.

If we now assume that the strongly textured specimens are deforming predominantly by prismatic slip, we can use the measured textures to estimate the average Schmid factor, m. The average Schmid factors was measured from XRD texture data for each processing condition, and all fell within the range of 0.40 to 0.45. We can now estimate the stress required to activate prismatic slip by multiplying the yield point by the average Schmid factor. The σ.m data for each of the alloys are shown in Fig. 6. It can be seen that the three most concentrated alloys fall essentially on the same line, and that the weakest alloy, 0.2%Zn lies between the pure magnesium and high Zn lines. Extrapolation of the Hall-Petch line for the alloy compositions indicates they would have an intercept ~25 MPa. This value is far below those measured for single crystals. For this reason it is suggested that for the alloy compositions tested here, yielding behaviour at larger grain sizes must plateau, as it does in pure magnesium. We can predict the minimum value of the plateau from measurements made on single crystals by Akhtar *et al.* [5], see Table 1. Using these single crystal values as a guide, the stress to activate the prismatic system has been elucidated for a range of compositions and grain sizes, Fig. 6.

Table 1 – Single crystal values for prismatic slip measured in [5]

| Composition | Measured CRSS [5] | CRSS used here |
|---|---|---|
| Mg | 45 MPa | 38 MPa |
| Mg-0.2Zn | 47.6 MPa | 40 MPa |
| Mg-0.5Zn | 45.5 MPa | 38 MPa |
| Mg-1.1Zn | 41.5 MPa | 35 MPa |
| Mg-2.8Zn | Not measured | 33 MPa |

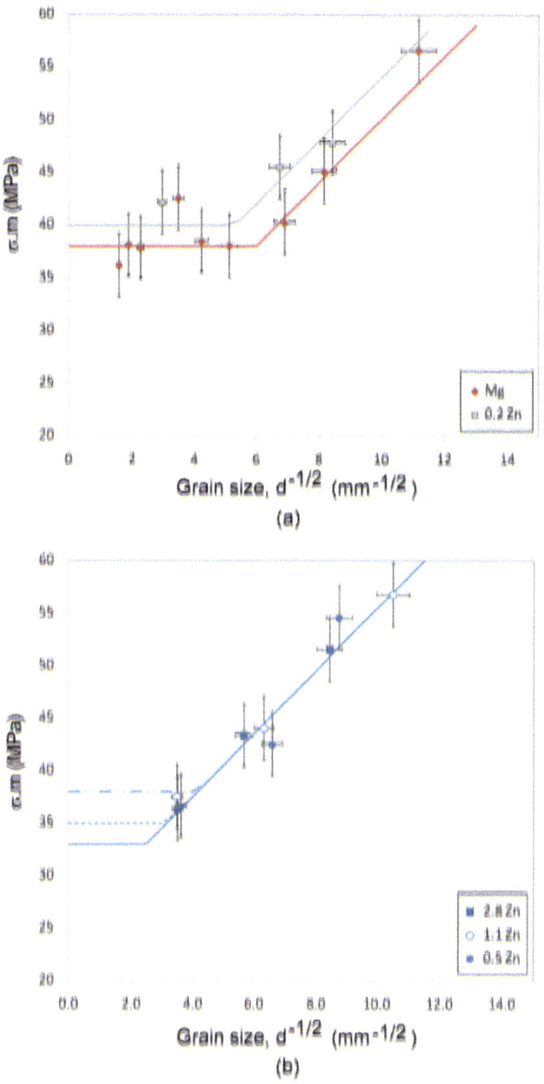


Figure 6 – Stress to activate prismatic slip in strongly textured magnesium plate estimated from the yield point multiplied by the average Schmid factor, m.

## Discussion

Hall-Petch Effect

Let us firstly consider the measured yield stress values for these dilute binary alloys against those measured on more concentrated commercial alloys, Fig. 7. This figure combines the data from Figs. 1 and 5, and also includes some additional data for pure magnesium from the literature. It is firstly apparent that the strongly textured commercial alloys almost all lie within a single linear band, colored grey in Figs. 1 and 7. The different alloys do not shift up or down significantly from one another, see Fig. 1. These data are a combination of rolled and extruded materials, and all will be expected to have strong textures. Thus, they will be deforming predominantly by prismatic slip. This can therefore be considered a Hall-Petch plot of prismatic slip for these materials. It is clear that for the alloys that lie in this grey band, prismatic slip is not strongly affected by the compositional differences between them. It is also apparent that these data all tend towards the single crystal CRSS values at the y-axis, although it is not clear what the trend will be at large grain sizes, > ~100 μm. This lack of sensitivity to composition in commercial type alloys has been noted in previously published literature [11,18]. However, it can be seen from Fig. 7 that pure magnesium is softer than the commercial alloys under conditions that promote prismatic slip. The addition of Zn strengthens the alloy, as shown in Fig. 7, but in the binary alloys tested here the strength exhibited by commercial alloys is not reached. The obvious difference between the alloys plotted in figs 1 and 7 and the Mg-Zn binaries examined here is the presence of large amounts of Al in the alloy systems. The presence of Al in the commercial alloys is likely to account for their higher strength.

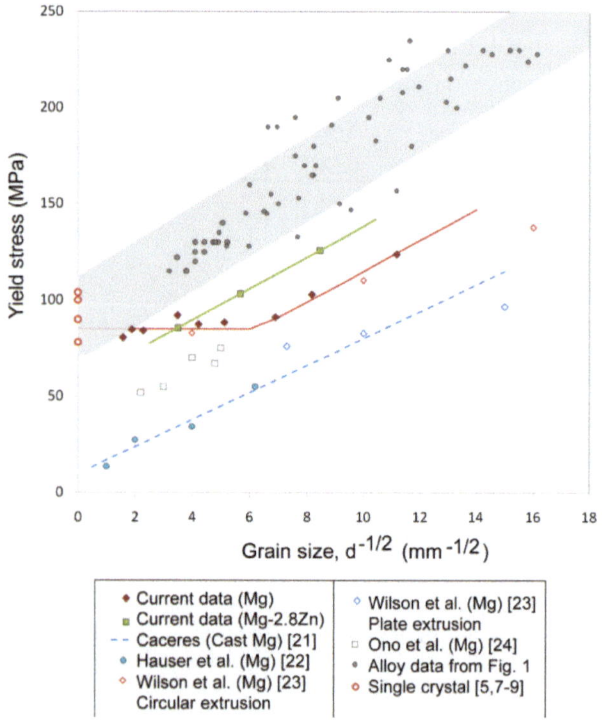


Figure 7 – Hall-Petch plot for pure magnesium taken from literature. Data points from current study also included for pure magnesium and Mg-2.8Zn alloys. Data zone for alloy compositions from Fig. 1 also included.

Also shown in Fig. 7 is previously published Hall-Petch data for pure magnesium. At first glance the data appears to not fit with the measurements made here. However, these discrepancies can be explained by the significant texture differences between the tested samples. The dashed line in Fig. 7 is the Hall-Petch slope measured by Caceres *et al.* on cast magnesium [21]. Being randomly oriented, the deformation of as-cast material will not be dominated by one deformation mode in particular, but will likely be significantly contributed to by both twinning and basal slip, the two softest modes. Therefore the measured Hall-Petch plot for the cast magnesium shows a significantly lower stress compared to the textured magnesium for a given grain size. This measured Hall-Petch slope by Caceres *et al.* fits closely with the data measured by Hauser *et al.* [22] and also that measured by Wilson and Chapman [23] on pure magnesium. We can therefore conclude that the pure magnesium measurements made in those cases differ to the present results due to texture differences. The data published recently by Ono *et al.* [24] sits between these two extremes of randomly textured and strongly textured materials, and it is likely that their samples had a texture that behaved somewhat between these two texture types. This demonstrates that a Hall-Petch plot for magnesium will only be useful in predicting the yield strength for a given grain size if the texture and orientation of testing are consistent with those used to develop the plot.

## Solute Softening

The concept of solute softening, rather than strengthening, was reported by Akhtar and Teghtsoonian [5] who showed that the CRSS for prismatic slip of Mg single crystals decreased with the addition of Zn. Figure 8 shows the effect of Zn content on the stress to activate prismatic slip in the polycrystals examined in this study. This graph was produced by using the data shown in Fig. 6 and plotting it against Zn concentration. This figure shows that although large grain sizes may exhibit solute softening, at grain sizes below ~50 μm the addition of Zn has a strengthening effect in magnesium. It is therefore concluded that for the grain sizes commonly used for wrought product, solute softening is not a phenomena that is likely to occur.

## Single crystal and polycrystal CRSS values

One of the motivating factors for this work was to try and explain why the ratio of the CRSS values for prismatic slip compared to basal slip are so vastly different for single crystals compared to polycrystals. For the data examined here it appears that the plateau in stress required to activate prismatic slip in dilute magnesium alloys could be one contributing factor to this discrepancy. If the basal slip system hardened with increasing grain size, while the prismatic system remained unchanged, then the ratio $Pris_{CRSS}$ / $Basal_{CRSS}$ would be reduced accordingly. However, at this stage there are no quantitative experimental studies on the effect of grain size on the CRSS for basal slip, so a quantitative evaluation of how this ratio changes with grain size cannot at this stage be made.

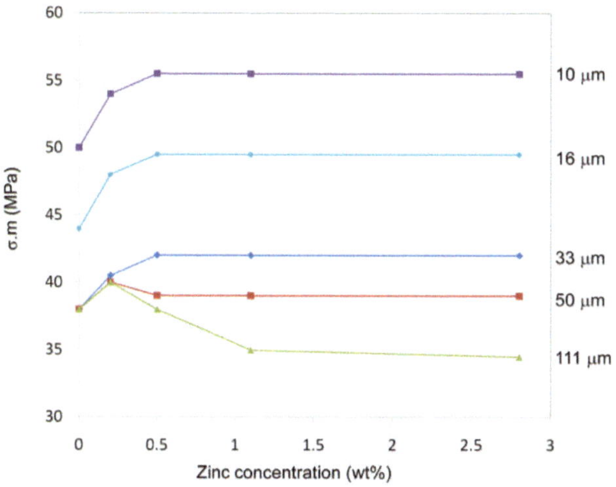


Figure 8 – Data from Fig. 6 re-plotted to show the effect of Zn concentration on the stress to activate prismatic slip for different grain sizes.

## Conclusions

Five alloys in the range of pure Mg to 2.8 wt% Zn have been examined for a range of grain sizes. The strong texture developed during hot rolling ensured that during tensile testing the dominant deformation mode was prismatic slip. This has been used to examine the Hall-Petch effect on the prismatic slip system. It has been found that:

- The ductility of the alloys increased with increasing Zn concentration
- The work hardening rate was not affected by the Zn concentration
- For pure magnesium textured to favour prismatic slip, the Hall-Petch effect does not apply at grain sizes above ~50 μm. Grain sizes larger than this do not show a reduction in yielding stress with increasing grain size, but plateau at a constant yield stress.
- For pure magnesium textured to favour prismatic slip, the addition of Zn produces a solute softening effect for grain sizes above ~50 μm. At grain sizes smaller than ~50μm the addition of Zn produces a solute strengthening effect.
- Collation of a range of data from literature has shown that there is a significant difference in the Hall-Petch plot for strongly textured and weakly textured materials.

### *Acknowledgements*

The work described in this paper was funded by the Centre of Excellence for Design in Light Metals. The authors would also like to thank Mohan Setty, Mahendra Ramajayam, Lynton Leigh and Rob Pow for their assistance with the experimental work.

## References

1. E.F. Emley. *Principles of Magnesium Technology*. Pergamon Press 1966.
2. S.R. Agnew, O. Duygulu. *International Journal of Plasticity* 21 (2005) 1161.
3. E.W Kelley, W.F. Hosford. *Trans. AIME.* 242 (1968) 5.
4. B.C. Wonsiewicz, W.A. Backofen. *Trans. AIME.* 239 (1967) 1422.
5. A. Akhtar, E. Teghtsoonian. *Acta Metallurgica* 17 (1969) 1351.
6. A. Akhtar, E. Teghtsoonian. *Acta Metallurgica* 17 (1969) 1339.
7. R.E. Reed-Hill, W.D. Robertson. *Trans AIME.* 209 (1957) 496.
8. A. Couret, D. Caillard. Acta Metallurgica. 33 (1985) 1447.
9. P. Ward Flynn, J. Mote, J.E. Dorn. *Trans AIME.* 221 (1961) 1148.
10. A. Jain, S.R. Agnew. *Materials Science and Engineering A* 462 (2007) 29.
11. B. Raeisinia, S.R. Agnew, A. Akhtar. *Metallurgical And Materials Transactions A* 42A (2011) 1418.
12. W.B. Hutchinson, M.R. Barnett. *Scripta Materialia* 63 (2010) 737.
13. J. Bohlen *et al. Materials Science and Engineering A* 462 (2007) 302.
14. A. Jain *et al. Materials Science and Engineering A* 486 (2008) 545.
15. H. Wang et al. Materials Science and Engineering A (2011) doi:10.1016/j.msea.2011.07.052
16. N. Stanford, M.R. Barnett. *Journal of Alloys and Compounds* 466 (2008) 182.
17. D.L. Yin *et al. Journal of Alloys and Compounds* 478 (2009) 789.
18. E. Meza-García *et al. Magnesium Technology 2007*, 263.
19. N. Hansen. *Acta Metallurgica* 25 (1977) 863.
20. R.L. Fleischer, W.F. Hosford. *Trans AIME* 221 (1961) 244.
21. C.H. Caceres, G.E. Mann, J.R. Griffiths. *Metallurgical And Materials Transactions A* 42A (2011) 1950.
22. F.E. Hauser, P.R. Langdon, J.E. Dorn. *Journal of Metals* (1956) 589.
23. D.V. Wilson, J.A. Chapman. *Philosophical Magazine* 8 (1963) 1543.
24. N. Ono, R. Nowak, S. Miura. *Materials Letters* 58 (2003) 39.

**Magnesium Technology 2012**
*Edited by: Suveen N. Mathaudhu, Wim H. Sillekens, Neale R. Neelameggham, and Norbert Hort*
***TMS (The Minerals, Metals & Materials Society), 2012***

# Effects of Ca on Microstructure and Mechanical Properties of ZA62 Alloys

YOU Jun-hua[1], ZHANG Wei-gang[1], QIU Ke-qiang[1], REN Ying-lei[1]

[1](School of Materials Science and Engineering, Shenyang University of Technology, No.111, Shenliao West Road, Economic & Technological Development Zone; Shenyang, Liaoning 110870, P.R.CHINA)

Key words: Magnesium alloy, Ca element, Microstructure, Mechanical Properties

**Abstract**

Effects of Ca element on microstructure and mechanical properties of ZA62 alloys were investigated by using zeiss microscope, scanning electron microscope equipped with an energy dispersive spectrometer, X-ray diffractometer, and electronic universal testing machine. The results show that Ca can significantly refine microstructure and cause the precipitation of $Mg_2Ca$ and $Al_2Ca$ phases on grain boundary of ZA62 alloys. The addition of Ca can improve the tensile properties of alloys at room temperature. When the content of Ca is 2%, good properties are obtained, the tensile strength and yield strength of Mg-6Zn-2Al-2Ca alloy reaches 240.79MPa and 82.78MPa respectively.

## 1 Introduction

As one green metal structure material, magnesium alloys will substitute steel, iron, aluminum and plastic due to their low density, high specific strength, high specific stiffness, high modulus of elasticity, good cut processing performance and good EMI performance[1-3]. ZA alloys are promising high temperature creep alloys due to their low cost and good casting performance. However, ZA alloys will enter the hot cracking zone, form serious thermal cracking and shrinkage with increasing the content of Ca element[4]. When aluminum content is lower than 8%, the tensile strength increases while the yield strength decreases with increasing content of Ca. When aluminum content is higher than 8%, the results are contrary. At the same time, the microstructure and mechanical properties of ZA alloys can be affected by the hard and brittle $Mg_2Ca$ phase and $Al_2Ca$ phase.

In this study, taking the change of aluminum / zinc into account, researchers have designed Mg-6Zn-2Al-xCa series alloys, studied the effects of Ca on microstructure and mechanical properties of ZA62 alloys.

## 2 Experimental

The Mg-6Zn-2Al, Mg-6Zn-2Al-1Ca, Mg-6Zn-2Al-2Ca and Mg-6Zn-2Al-3Ca alloys used in this study were prepared by Mg(99.99%), Zn(99.99%), Al (99.99%), and Ca added in the form of Mg-20%Ca (mass fraction) master alloy. The experimental alloys were melted in pit furnace and protected by the gaseous mixture of nitrogen and sulfur hexafluoride. The specimens whose size was 40mm×2mm×4mm were fabricated from the casting for tensile properties test.

The microstructure, fracture pattern and element composition of alloys were examined by Zeiss microscope and SN3400 type scanning electron microscope (SEM) equipped with an energy dispersive spectrometer (EDS). The phase composition of alloys were analyzed by XRD-7000 type X-ray diffractometer (XRD). The tensile properties of alloys at room temperature were tested by WDW-100 type electronic universal testing machine. The tensile strength and yield strength were obtained based on the average value of three tests.

## 3 Results and discussion

### 3.1 Effect of Ca on as-cast alloys microstructure

Figure.1 shows the microstructure of as-cast Mg-6Zn-2Al-xCa series alloys. Seen from the Fig.1(a) to Fig.1(d), grain boundary precipitates transform from discontinuous precipitation into continuous precipitation gradually and grain refines significantly with the increasing of Ca content. The as-cast Mg-6Zn-2Al alloy is mainly composed of α-Mg and MgZn phases according to the XRD analysis in Fig.2. Most of the divorced eutectic MgZn phases exist in the grain boundary and a small amount of MgZn phase exist in the grain interior[5]. Sheet and rod-shaped phases formed in grain boundary except divorced eutectic phase with the increasing of Ca content. The content of Sheet and rod-shaped phases increased with Ca increasing.

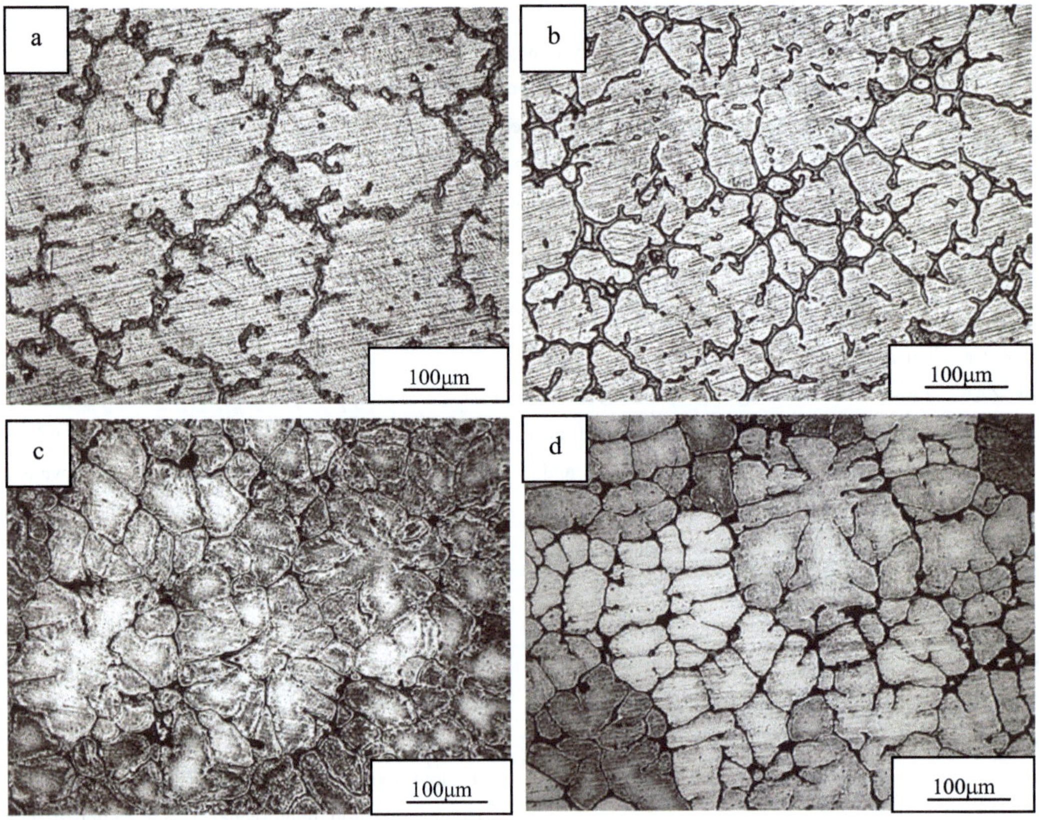


Figure.1 Optical microstructures of as-cast Mg-6Zn-2Al-xCa alloys

a) Mg-6Zn-2Al b) Mg-6Zn-2Al-1Ca c) Mg-6Zn-2Al-2Ca d) Mg-6Zn-2Al-3Ca

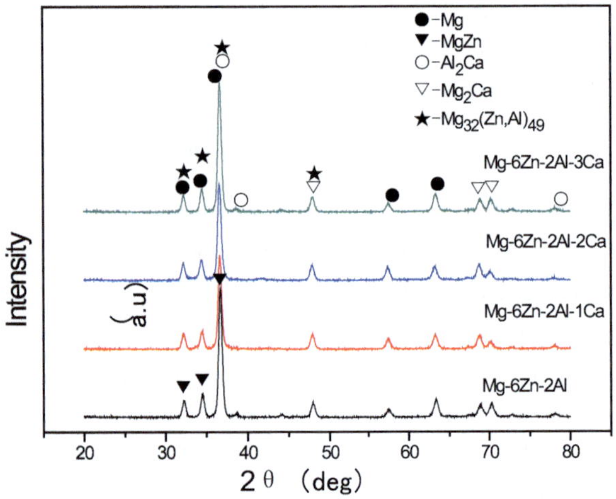


Figure.2 X-ray diffraction patterns of Mg-6Zn-2Al-xCa alloys

Figure.2 shows the phase composition of Mg-6Zn-2Al-xCa series alloys. Seen from the diffraction patterns of Fig.2, Mg-6Zn-2Al alloy is composed of α-Mg phase , MgZn phase and a small amount of $Mg_{32}(Zn, Al)_{49}$ phase. With the addition of Ca element, the MgZn phase disappears, $Mg_2Ca$ phase and $Al_2Ca$ phase precipitates. There all exists α-Mg phase and $Mg_{32}(Zn, Al)_{49}$ phase but no MgZn phase in Mg-6Zn-2Al-1Ca, Mg-6Zn-2Al-2Ca and Mg-6Zn-2Al-3Ca alloys.

The SEM images of the Mg-6Zn-2Al and Mg-6Zn-2Al-2Ca alloys are shown in Figure.3. It can be seen from Fig.3(a) that the intergranular phases are semi-continuous bone-like structure. EDS analysis indicates that the intergranular phases are mainly composed of Mg and Zn elements, at the same, there exist a small amount of Al element. XRD analysis shows that the intergranular phases are MgZn phase containing a small amount of Al. When 2% Ca is added, $Mg_2Ca$ and $Al_2Ca$ emerge in matrix. The intergranular lamellar eutectics phase is composed of $Mg_2Ca$. The lamellar $Al_2Ca$ phase mainly exist in the grain boundary, as is shown in Fig.3(b). The content of $Al_2Ca$ phase and $Mg_2Ca$ phase increases significantly with the increasing of Ca content, and the grain of α-Mg matrix decreases gradually in experimental alloys.

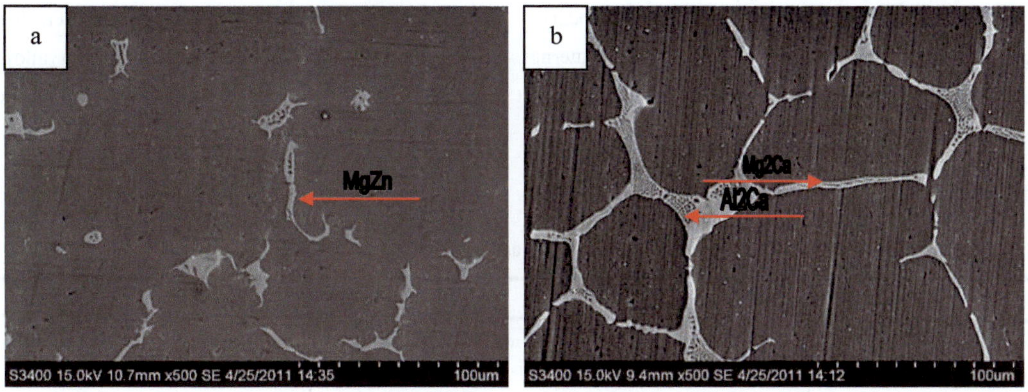


Figure.3 SEM images of ZA62 and ZA62-2Ca alloys

a) Mg-6Zn-2Al b) Mg-6Zn-2Al-2Ca

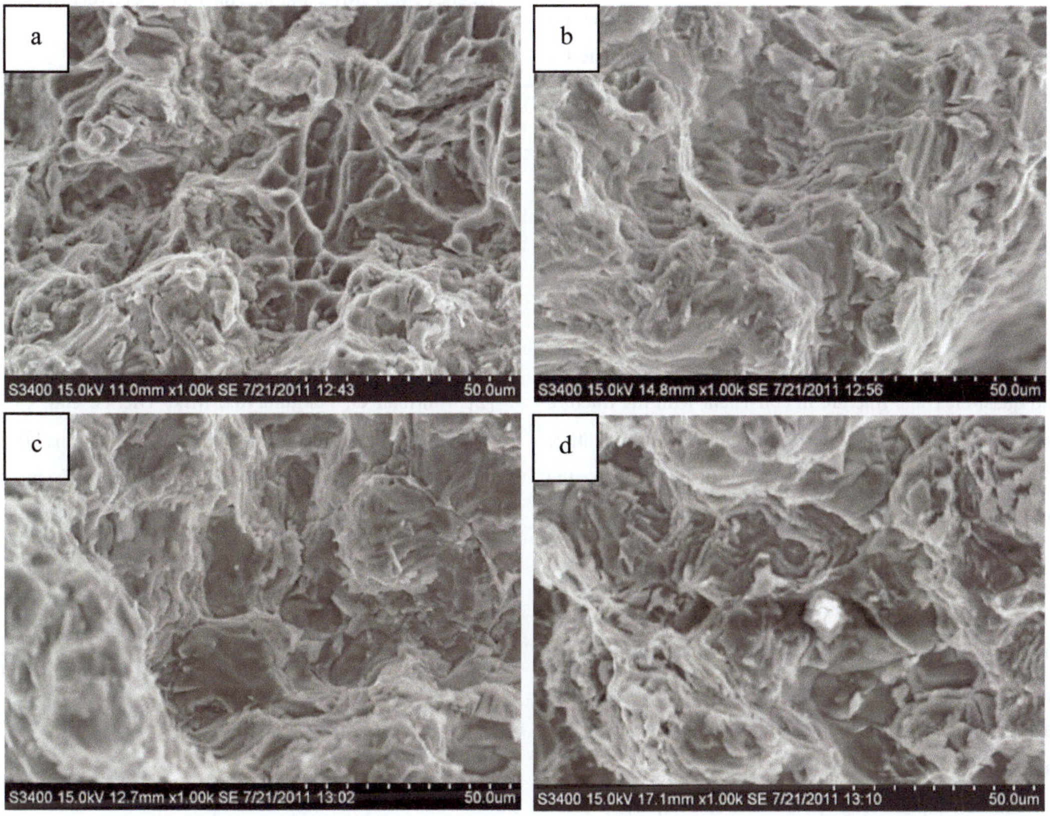


Fig.4 SEM tensile fractographs of as-cast Mg-6Zn-2Al-xCa alloys at room temperature

a) Mg-6Zn-2Al b) Mg-6Zn-2Al-1Ca c) Mg-6Zn-2Al-2Ca d) Mg-6Zn-2Al-3Ca

### 3.2 Effect of Ca on as-cast alloys tensile fractographs

The SEM tensile fractographs of as-cast Mg-6Zn-2Al alloys at room temperature are shown in Figure.4. It can be seen from Fig.4(a) that there exist numerous small size dimple in Mg-6Zn-2Al alloy due to coarse MgZn phases in grain boundary. The fractographs exhibit typical ductile dimple fracture pattern. Seen from Fig.4(b), Fig.4(c) and Fig.4(d), the SEM tensile fractographs of as-cast Mg-6Zn-2Al-xCa (x=1, 2, 3) alloys change greatly due to the increasing of Ca content. Fig.4(b) and Fig.4(c)show that there exist a small amount of large dimples in the process of fracture due to the formation of lamellar and rod-shaped $Mg_2Ca$ and $Al_2Ca$ phases, the tensile fracture of the alloys occurs along inter-granular boundary. Fig.4(d) shows that there no exist dimples in Mg-6Zn-2Al-3Ca alloy, the tensile fracture results from recombination action of quasi-cleavage and cleavage.

### 3.3 Effect of Ca on mechanical properties of alloys

Table 1 Tensile properties of as-cast Mg-6Zn-2Al-xCa alloys

| 合金名称 | $\sigma_b$(MPa) | $\sigma_{0.2}$(MPa) | $\delta(\Delta x)$ |
|---|---|---|---|
| Mg-6Zn-2Al | 199.21 | 68.08 | 2.30 |
| Mg-6Zn-2Al-1Ca | 225.73 | 67.16 | 0.69 |
| Mg-6Zn-2Al-2Ca | 240.79 | 82.78 | 0.59 |
| Mg-6Zn-2Al-3Ca | 213.04 | 0 | 0.28 |

The relationships between mechanical properties and Ca content are shown in table 1. Seen from the table 1, with the Ca content (mass fraction) increasing from 1% to 2%, the tensile strength increases from 225.73MPa to 240.79MPa and yield strength increases from 67.16MPa to 82.78MPa respectively. However, with the further increasing of Ca content, the tensile strength and yield strength of the alloys decrease rapidly. When the Ca content reachs 3%, the tensile strength of the alloy is 213.04MP and the yield strength of the alloy reduces to zero. The reason is that the excessive addition of Ca causes cleavage fracture and quasi-cleavage fracture in consistent with the fracture analysis. It can be seen from the table 1, the elongation percentage of Mg-6Zn-2Al-xCa alloys decrease significantly with the increasing of Ca content.

## 4 Conclusions

(1) The grain size of Mg-6Zn-2Al-xCa magnesium alloys is effectively reduced, the MgZn phase disappears, $Mg_2Ca$ phase and $Al_2Ca$ phase precipitates on the grain boundary as a result of the Ca addition.

(2) The addition of Ca can improve the tensile properties of alloys at room temperature. When the content of Ca is 2%, the tensile strength and yield strength of Mg-6Zn-2Al-2Ca alloy reaches 240.79MPa and 82.78MPa respectively, while the tensile strength and yield strength of Mg-6Zn-2Al-xCa alloys declines due to the precipitation of coarse $Mg_2Ca$ and $Al_2Ca$ phases on grain boundary with the further increasing of Ca content.

## References


[1] SHI Ya-jing, LI Quan-an, REN Wen-liang, etal. Research and development of heat resistant magnesium alloy[J]. Light Alloy Fabrication Technology. 2009. 37(5):1-4

[2] Mihriban O, Pekguleryuz, Mike Dierks, etal. Creep resistant magnesium alloys for power-train applications [J]. Proceedings of Annual World MA 2001 Magnesium conference, 2001, 32(1) 37-45

[3] Munit A Z, Cotler C, Stem A, etal. Mechanical properties and microstructure of gas tungsten are welded magnesium AZ91D plates[J]. Materials Science and Engineering 2001, 302: 68-73

[4] Lou A A. Recent magnesium alloy development for elevated temperature application[J]. International Materials Reviews, 2004, 49(1): 13-30

[5] LIAO Kun, ZHANG Fu-quan, CHEN Ji-hua, etal. Effects of tin on microstructure and mechanical properties of ZA62 alloys[J]. Foundry Technology 2006, 27(4): 378-381

Magnesium Technology 2012
*Edited by: Suveen N. Mathaudhu, Wim H. Sillekens, Neale R. Neelameggham, and Norbert Hort*
*TMS (The Minerals, Metals & Materials Society), 2012*

# MECHANICAL PROPERTIES AND HIGH-TEMPERATURE OXIDATION BEHAVIOR OF Mg-Al-Zn-Ca-Y MAGNESIUM ALLOYS

Young Min Kim[1], Bong Sun You[1], Myeong-Shik Shim[2], Nack J. Kim[2]
[1]Light Metals Group, Korea Institute of Materials Science, 66 Sangnam, Seongsan, Changwon 641-831, Korea
[2]GIFT, Pohang University of Science and Technology, Pohang 790-784, Korea

Keywords: Magnesium alloys, Oxidation, Microstructure, Mechanical properties, Electron probe microanalysis

## Abstract

Ignition resistance and tensile properties of Mg-3Al-1Zn and Mg-6Al-1Zn (wt.%) alloys with combined addition of calcium and yttrium were investigated in the present study. The results of this study clearly show that the combined addition of calcium and yttrium can lead to significant increase in both ignition temperature and tensile properties, comparing to 2 wt.% Ca-added magnesium alloys with lamella-shaped eutectic phases. This is because the reduction of calcium content and the addition of a small amount of yttrium bring about a reduced amount of coarse Ca-containing phases and the formation of duplex protective oxide layers that effectively prevent oxygen penetration into the melt.

## Introduction

Magnesium and its alloys are very active, especially in a molten state, and are rapidly ignited or combusted when the clean surface of the melt comes into contact with oxygen in air. Flux or a cover gas has to be used to prevent ignition of magnesium alloys during melting. However, the above methods may cause many problems such as global warming and increased production costs, so many researchers are paying more attention to finding more essential methods to improve ignition resistance. Obviously, it is simple to prevent the ignition of magnesium alloys by the addition of alloying elements. However, it is difficult to select suitable elements, because only some of alloying elements have been proved to benefit the oxidation resistance of magnesium alloys [1-5].

Calcium is one of the promising elements that can improve ignition resistance. The ignition temperature increases by 250 K after an addition of 3 wt.% Ca-added into pure magnesium [6]. You et al. [7] investigated the oxidation behavior of a Mg-xCa alloy (x=0-3 wt.%), and they confirmed that a dense and compact protective MgO/CaO layer formed at elevated temperatures. However, a large amount of calcium addition generally lowers tensile strength and ductility because coarse Ca-compounds precipitate along the grain boundaries. Unfortunately, most of such studies have concentrated on either ignition resistance or tensile properties of magnesium alloys, but there has been no attempt to improve both them.

In fact, increasing both ignition temperature and tensile properties is difficult to achieve because of their contradictory aspects; that is, a large addition amount of ignition resistance improving alloying element is necessary to make ignition-proof magnesium alloys [2, 6-8], but at the same time it causes the deterioration of tensile properties due to the formation of coarse brittle phases. This indicates that instead of a large amount of a single alloying element, the combined addition of two or more alloying elements may suppress the formation of a large fraction of coarse brittle phase without sacrificing ignition resistance. In the present study, we therefore investigate the effect of the combined addition of calcium and yttrium on ignition resistance and tensile properties of magnesium alloys. That is crucial to make magnesium alloys more reliable materials and thus enlarge their application areas, including in advanced transportation vehicles requiring strict stability in a material selection.

## Experimental procedures

Mg-xAl-yCa (x=3, 6 wt.%, y=1, 2 wt.%) and Mg-xAl-1Ca-0.6Y (x=3, 6 wt.%) alloys were prepared according to the nominal composition by pure magnesium (99.93 wt.%), pure aluminum (99.9 wt.%), pure zinc (99.9 wt.%), pure calcium (99.9 wt.%) and pure yttrium (99.9 wt.%) in an electric furnace using a low-carbon steel crucible heated to 993 K under a protecting gas (10% $SF_6$ and 90% $CO_2$). The melt was poured into a permanent mold preheated to 473 K and then cooled to ambient temperature in air atmosphere. The cast ingots were subsequently heated to 673 K for homogenization treatment and kept at this temperature for 20 hours. The ingots were hot-rolled at 673 K at a reduction rate of 30%/pass with a total reduction of 93%. The rolled sheets with a 1 mm thickness were finally annealed at 523 K for 30 minutes. Ignition tests were carried out by inserting the chipped samples taken from the castings into a tube furnace held at 1273 K. The tests were carried out in air and at least five chipped samples were used. The ignition temperatures were determined as ones associated with both the appearance of flame and a significant temperature increase as shown Fig. 1. Metallographic samples were prepared for microstructure observation according to a standard procedure. The distribution of alloying elements in the surface oxide layer that was formed on the melt was checked using an EPMA (electron probe micro-analyzer). Tensile tests were carried out using an Instron-4206 universal testing machine at a strain rate of 0.001/s at room temperature. Sheet-type tensile specimens with the dimensions of the ASTM E 8M standard were used. All the tensile specimens were taken in parallel to a longitudinal direction of the rolled sheet.

## Results and discussion

Fig. 2 shows the as-cast microstructures of the experimental alloys. In Figs. 2(c) and 2(f), the alloys with the combined addition of 1 wt.% Ca and 0.6 wt.% Y contain a small amount of semi-continuous dihexagonal C36-$(Mg,Al)_2Ca$ Laves phase at interdendritic regions and spherical $Al_2Y$ particles inside grains; however, as shown in Figs. 2(b) and 2(e) only 2 wt.% Ca-added

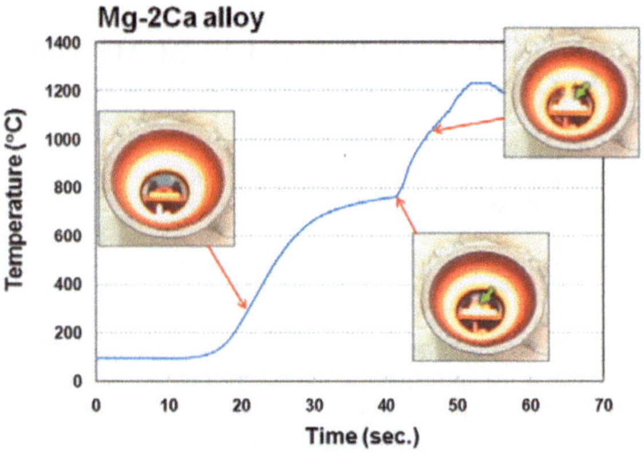

Figure 1. Measurement of ignition temperature. A sudden increase in temperature indicates that the sample begins to be ignited.

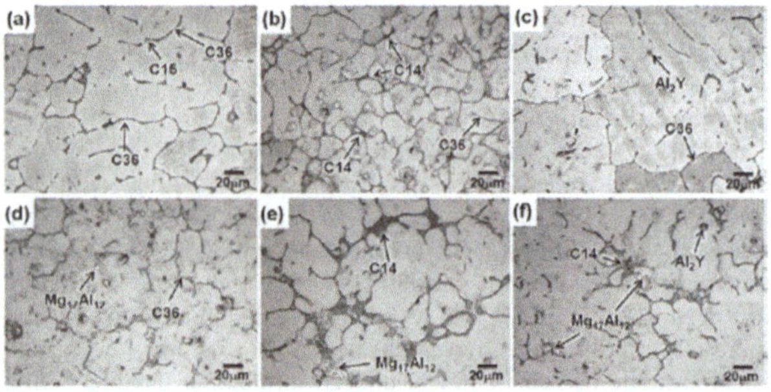
Figure 2. Optical micrographs showing the as-cast microstructures of (a) AZX311, (b) AZX312, (c) AZX311-0.6Y, (d) AZX611, (e) AZX612, and (f) AZX611-0.6Y.

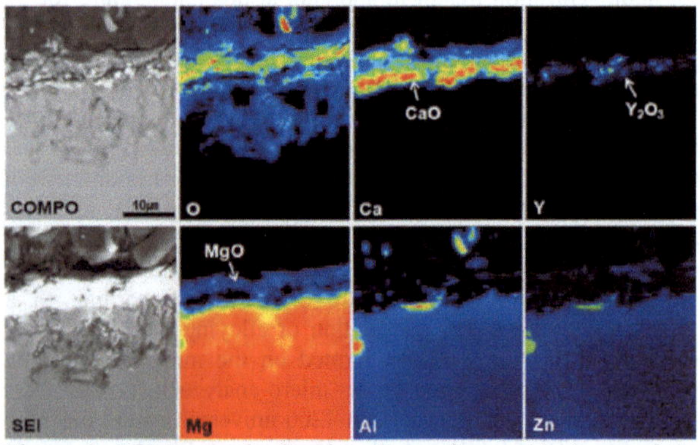

Figure 3. EPMA result showing the surface oxide layer of the AZX611Y alloy held at 950K for 10 min. The white oxide layer in the SEI image is composed of two distinctive layers in the BEI-COMPO image.

alloys contain a large amount of lamella-shaped C14-$Mg_2Ca$ phase [9-10]. According to a thermodynamic calculation, an addition of yttrium does not influence the mass fraction of Ca-containing eutectic phases, namely C14-$Mg_2Ca$, C15-$Al_2Ca$, and C36-$(Mg,Al)_2Ca$, but it is involved only in the formation of $Al_2Y$ particles.

Table 1. Ignition temperatures of the castings and tensile properties of the rolled sheets.

| | $T_{ig}$ [a)] (K) | YS (MPa) | UTS (MPa) | El. (%) | U.El. (%) |
|---|---|---|---|---|---|
| AZ31 | 753 | 176.4 | 274.5 | 25.2 | 17.4 |
| AZX311 | 981 | 191.1 | 276.1 | 24.3 | 16.9 |
| AZX312 | 1020 | 255.2 | 303.6 | 16.5 | 9.7 |
| AZX311-0.6Y | 1041 | 175.8 | 265.2 | 24.7 | 18.4 |
| AZ61 | 794 | 218.7 | 324.0 | 22.0 | 17.2 |
| AZX611 | 976 | 204.4 | 306.2 | 19.7 | 16.0 |
| AZX612 | 1028 | 230.0 | 321.0 | 16.7 | 14.1 |
| AZX611-0.6Y | 1047 | 225.7 | 323.4 | 19.6 | 15.5 |

[a)] Ignition temperature of chip sample

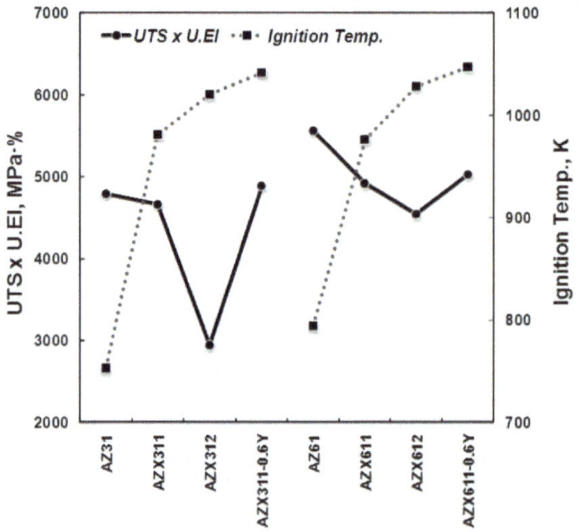

Figure 4. The changes in ignition temperatures and tensile properties with additions of Ca and Y. The combined addition of a small amount of calcium and yttrium improves both ignition resistance and tensile properties of magnesium alloys.

Despite the small amount (no more than 1 wt.%) of calcium and yttrium, the ignition temperatures of the AZX311-0.6Y and AZX611-0.6Y alloys in Table 1 are 1041 K and 1047 K, respectively, and they are approximately 20 K higher than those of the AZX312 and AZX612 alloys and moreover more than 60 K higher than those of 1 wt.% Ca-added alloys. These results of chip ignition tests demonstrate that the alloys with the addition of 1wt.% Ca and 0.6wt.% Y have better ignition resistance than those with the addition of only 2wt.% Ca. The studies in Refs. [6-8] suggest that an individual addition of calcium and yttrium to magnesium alloys should be larger than at least 2 wt.% and 4.3 wt.%, respectively, to obtain practically acceptable ignition-proof performance. Compared to the alloys with an individual addition of calcium or yttrium, our study confirmed that the alloys with a combined addition of those alloying elements exhibited much more superior ignition-proof performance, even by a small addition of no more than 1 wt.%.

Fig. 3 clearly shows that the surface oxide of the AZX611-0.6Y sample held at 950 K for 10 minutes is composed of two distinctive layers. An addition of yttrium to Ca-added magnesium alloys can make a dense and protective oxide layer that contains Ca and Y elements on the surface contacting with a Mg alloy melt in addition to an outer oxide layer that contains Mg, Al, and Ca elements.

The tensile properties of the hot-rolled sheets are shown in Table 1. As calcium content increases from 1 wt.% to 2 wt.%, both yield and tensile strengths increase, but at the same time elongation drops due to a large amount of brittle Ca-containing phases. In particular, the total elongation of the AZX312 alloy decreases approximately 8% compared with the AZX311 alloy. We can infer from Table 1 that such a deterioration of elongation can be restored by reducing the calcium content and adding a small amount of yttrium. In Table 1, the elongations of the AZX311-0.6Y and AZX611-0.6Y alloys are 24.7% and 19.6%, respectively, which are similar to those of the AZX311 and AZX611 alloys. Furthermore, the ignition temperatures of the AZX311-0.6Y and AZX611-0.6Y alloys are higher than those of the AZX312 and AZX612 alloys, respectively. As an increase in elongation commonly gives rise to a decrease in strength, the multiplied values of ultimate tensile strength by uniform elongation are better to represent the overall tensile properties of the samples. In Fig. 4, the values significantly decreased in 2 wt.% Ca-added magnesium alloys, but they were restored to those of 1 wt.% Ca-added magnesium alloys by the combined addition of calcium and yttrium. This is because the reduced amount of brittle and coarse Ca-containing phases is attributed to the improvement in tensile properties of those alloys. On the other hand, compared with 2 wt.% Ca-added magnesium alloys, the large decrease in yield strength of the alloys with the combined addition of calcium and yttrium was caused from only a small amount of fine C15-$Al_2Ca$ precipitates, which can form at grain interiors in alloys with excess calcium content of more than 2 wt.%, and thus contribute to strengthening matrix [10].

## Conclusions

In this study, the combined addition of calcium and yttrium has a very attractive beneficial effect on the ignition resistance and tensile properties of magnesium alloys. The duplex protective oxide layers of CaO-$Y_2O_3$ at the inner part and MgO-$MgAl_2O_4$–CaO at the outer part are attributed to further improvement in ignition resistance, compared to the only Ca-added magnesium alloys. In addition, the reduced amount of brittle and coarse Ca-containing phases in the alloys with the combined addition of a small amount of calcium and yttrium contributes to a large increase in tensile properties, especially elongation. Our approach to improve both ignition resistance and tensile properties provides a strong impact on the development of clean manufacturing technology for reliable magnesium alloys, which in turn will lead to an expansion of their application areas.

## Acknowledgements

This study was financially supported by the R&D Program of Korea Institute of Materials Science.

## References

[1] N.V. Ravi Kumar, J.J. Blandin, M. Suéry, E. Grosjean, Scripta Mater. 49 (2003) 225-230.
[2] T.S. Shih, J.H. Wang, K.Z. Chong, Mater. Chem. Phys. 85 (2004) 302-309.
[3] P.Y. Lin, H. Zhou, W. Li, W.P. Li, N. Sun, R. Yang, Corrosion Sci. 50 (2008) 2669-2675.
[4] X. Zeng, Q. Wang, Y. Lu, Y. Zhu, W. Ding, Y. Zhao, J. Mater. Process. Tech. 112 (2001) 17-23.
[5] Y.M. Kim, C.D. Kim, H.S. Kim, B.S. You, Scripta Mater. 65 (2011) 958-961.
[6] M. Sakamoto, S. Akiyama, K. Ogi, J. Mater. Sci. Lett. 16 (1997) 1048-1050.
[7] B.S. You, W.W. Park, I.S. Chung, Scripta Mater. 42 (2000) 1089-1094.
[8] X.M. Wang, X.Q. Zeng, Y. Zhou, G.S. Wu, S.S. Yao, Y.J. Lai, J. Alloys. Compd. 460 (2008) 368-374.
[9] A. Suzuki, N.D. Saddock, J.W. Jones, T.M. Pollock, Scripta Mater. 51 (2004) 1005-1010.
[10] Y.M. Kim, C.D. Yim, S.S. Park, B.S. You. Met. Mater. Int. 17 (2010) 583-586.

**Magnesium Technology 2012**
*Edited by: Suveen N. Mathaudhu, Wim H. Sillekens, Neale R. Neelameggham, and Norbert Hort*
***TMS (The Minerals, Metals & Materials Society), 2012***

# Effects of Si on microstructure and mechanical properties of Mg-5Sn-2Sr alloy

YOU Jun-hua[1], HAO Shuai[1], QIU Ke-qiang[1], REN Ying-lei[1]

[1](School of Materials Science and Engineering, Shenyang University of Technology, No.111, Shenliao West Road, Economic & Technological Development Zone; Shenyang, Liaoning 110870, P.R.CHINA)

Key words: Magnesium alloy, $Mg_2Si$, Microstructure, Mechanical properties

## Abstract

Phase constitution and microstructure of the Mg-5Sn-xSi-2Sr (x=0, 1, 2) alloys were analyzed by X-ray diffractometer, optical microscope and scanning electron microscope with an energy dispersive spectrometer. The tensile properties of the alloys were tested by electronic universal testing machine, and the fracture surfaces of the alloys were observed with scanning electron microscope. The results show that the microstructure of the Mg-5Sn-2Sr alloy consists of α-Mg, MgSnSr and $Mg_2Sn$ phases. The addition of Si element can promote the formation of $Mg_2Si$ phase on the grain boundary, and the content of $Mg_2Si$ phase increases with increasing the Si element. $Mg_2Si$ is a strengthening phase, which can effectively prevent grain boundary sliding and dislocation motion, and thus, the mechanical properties of the alloys were obviously enhanced. The mechanical properties of Mg-5Sn-2Si-2Sr alloy are better than other alloys, the tensile strength, yield strength, elongation and hardness are 147MPa, 110MPa, 5.0% and 47HB respectively.

## 1 Introduction

Magnesium alloys are the lightest structural alloys and have greatly potential applications in automotive, aerospace and other industries due to the unique combination of low density, high specific strength, superior damping capacity, good castability and excellent machineability[1,2]. However, the elevated temperature properties of magnesium alloys is poor, therefore, improving the elevated temperature properties has become a critical issue for its possible application in heat-resistant components[3,4]. Studies have shown that the Sn and Si elements are beneficial to improve heat resistance of magnesium alloys, which is attribute to the formation of the thermally stable $Mg_2Sn$ and $Mg_2Si$ phases[5]. Si is an effective alloying element for improving the high temperature properties of magnesium alloy since Si combines with Mg and forms the intermetallic compound $Mg_2Si$ which exhibits high melting point, low density, high hardness and low thermal expansion coefficient[6]. Moreover, the intermetallic phase $Mg_2Sn$ in Mg-Sn alloys has much higher melting point (770 ℃) than that of $Mg_{17}Al_{12}$ phase (462 ℃) in Mg-Al alloys. So, Mg-Sn-Si alloy is likely to have superior high-temperature resistance. But the mechanical properties of alloys will reduce due to the formation of the coarse dendrite-shape primary $Mg_2Si$ in Mg-Si alloys with high Si content and the brittle chinese script eutectic $Mg_2Si$ in Mg-Si alloys with low Si content[7]. It is widely known that the refinement of $Mg_2Si$ phase is mainly responsible for improving the mechanical properties of alloys. Recent studies report that the Y addition can modify the morphology of $Mg_2Si$ precipitates[8], but the price of rare earth is expensive, so the RE-containing Mg alloys don't have the market competition. Adding alkaline earth elements calcium and strontium can also refine the $Mg_2Si$ phase and modify the morphology of $Mg_2Si$ crystals[9-11]. The price of alkaline earth elements is cheaper than rare earth, thus, the Mg-Sn-Si alloys containing alkaline earth is one of the future development direction.

In recently years, the Mg-Sn based alloys has been a renewed interest globally in this alloy system as it has potential applications at elevated temperatures, including the Mg-Sn-Ca[3], Mg-Sn-Zn[4], Mg-Sn-Ce[12], Mg-Sn-Nd[13], Mg-Sn-Pb[14] and Mg-Sn-Y[15] alloys. Studies have shown the mechanical properties of Mg-(1-10)%Sn alloys and the mechanical properties of Mg-5%Sn alloy is better than other alloys[16]. Liu[17] also studies the effect of Sr on the microstructure and mechanical properties of as cast Mg-5wt.%Sn alloy at ambient and elevated temperatures, indicating that Sr can improve the thermal stability of Mg-Sn alloys. In this work, Si was added to Mg-5Sn-2Sr alloy for achieving more excellent mechanical property.

## 2 Experimental

The Mg-5Sn-xSi-2Sr (x=0, 1, 2) alloys were prepared using high purity Mg (99.9%), Sn (99.9%), and Mg-30%Si and Mg-30%Sr (mass fraction) master alloys according to the nominal composition in this study. The experimental alloys were melted in a crucible resistance furnace under 0.1%$N_2$+0.15%$SF_6$ (volume fraction) atmosphere. The Mg-Si and Mg-Sr master alloys were added to the melt at 700℃, and the melt was mixed by artifical stirring for 1 min. After the melt was mixed completely, then held at 690℃ for 20 min and poured into a preheated permanent mould with a diameter of 120mm.

The phase constituent of the alloys was analyzed by using XRD-7000 type X-ray diffractometer (XRD). The samples were etched with the 3% (volume fraction) nitric acid alcohol solution, and then the microstructure analysis was carried out using an optical microscope (OM) and S-3400N type scanning electron microscope (SEM) with an energy dispersive spectrometer (EDS). The WDW-100 type electronic universal testing machine was employed to obtain the tensile properties of the alloys, the ultimate tensile strength (UTS), yield strength (YS) and elongation to failure were obtained based on the average of three tests. The tensile specimens shown in Fig.1 were fabricated by linear cutting from the casting. Furthermore, the fracture surfaces of the alloy samples with different compositions were observed by SEM. The hardness of alloys were tested by the brinell hardness tester.

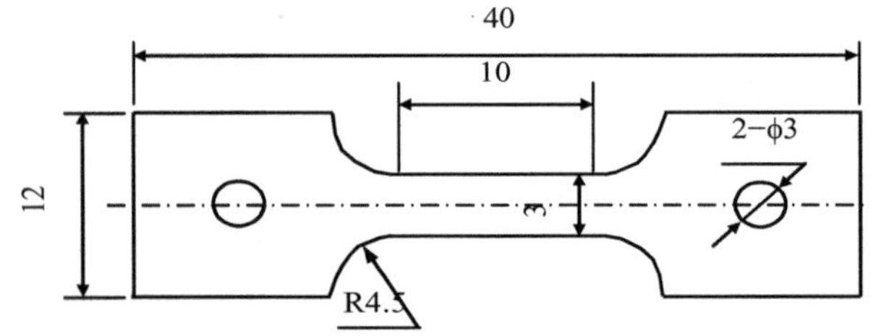


Fig.1 Morphology and size of the tensile tests samples (unit: mm)

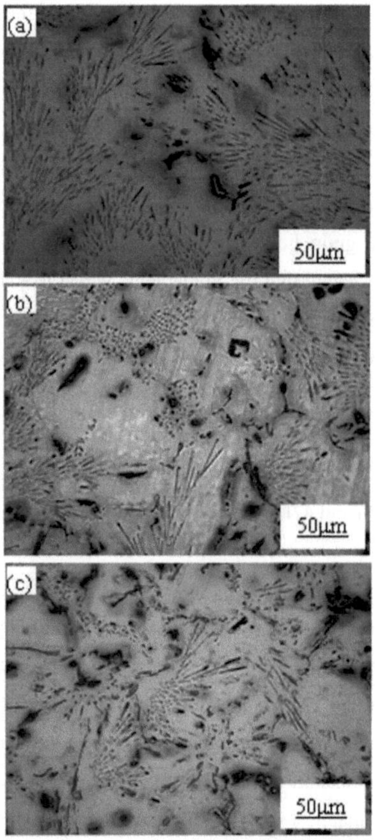


Fig.3 Optical images of the microstructures of the as-cast experimental alloys

(a) Mg-5Sn-0Si-2Sr; (b) Mg-5Sn-1Si-2Sr; (c) Mg-5Sn-2Si-2Sr

## 3 Results and discussions

### 3.1 Microstructure analysis

Fig.2 shows the XRD patterns of the as-cast experimental alloys.

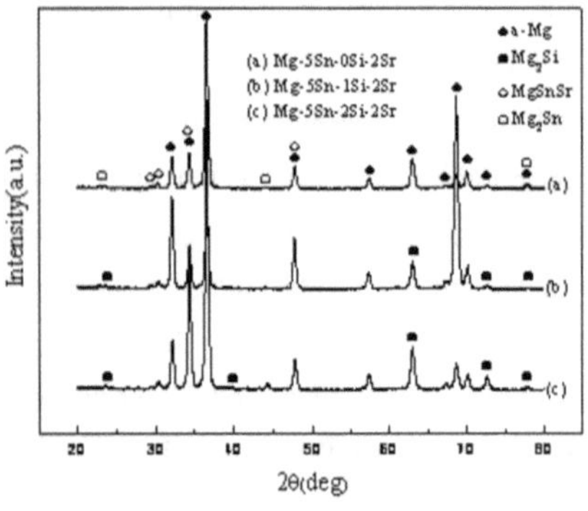


Fig.2 XRD patterns of as-cast experimental alloys

It is observed from Fig.2 (a) that the microstructure of the Mg-5Sn-2Sr consisted mainly of α-Mg, MgSnSr and $Mg_2Sn$ phases. When adding Si to Mg-5Sn-2Sr alloy, new diffraction peaks were observed as labeled "■" in Fig.2 (b) and (c), the new diffraction peaks perfectly matched the characteristic peaks of $Mg_2Si$, and the $Mg_2Si$ phase diffraction peak intensity increased with increasing Si content.

Fig.3 shows the optical microstructure of the investigated alloys in the as-cast condition. Fig.3 indicates that the experimental alloys are composed of α-Mg and secondary solidification phases (black and grey precipitates). The black phase with a blocky morphology was mainly gathered in the grain boundary areas and the volume and size of this phase grew up with the increasing Si additions. The grey phases with needle-like and particle-like morphologies were formed within the grain.

Fig.4 shows the SEM microstructure of the investigated alloys in the as-cast condition, and the EDS results of experimental alloys were shown in Table 1. Fig.4 (a) shows the microstructure of Mg-5Sn-0Si-2Sr alloy. According to XRD and EDS results, the phase with needle-like morphology is MgSnSr, and the blocky phase is $Mg_2Sn$ whose typical morphology is shown in Fig.4 (a). Fig.4 (b) and (c) show the microstructure of Mg-5Sn-xSi-2Sr (x=1, 2) alloys, the particle-like (Fig.4 (b)) and feather-like phases (Fig.4 (c)) within grain are MgSnSr, and the $Mg_2Sn$ and $Mg_2Si$ phases are formed mainly along the grain boundaries. Since a maximum solid solubility of Si into Mg is only 0.003 at.% and Si atoms react with Mg atoms and are precipitated as an intermetallic compound of $Mg_2Si$[8]. In addition, the $Mg_2Sn$ phase can be easily precipitated because of the high solubility limit of Sn in Mg (14.48 mass%) at 560 ℃ and its little solubility at ambient temperature[18].

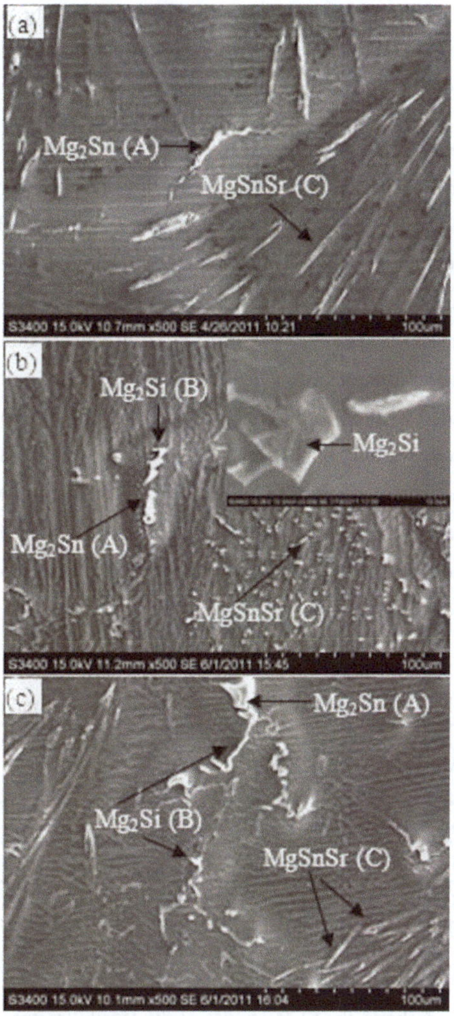


Fig.4 SEM images of the microstructures of the as-cast alloys
(a) Mg-5Sn-0Si-2Sr; (b) Mg-5Sn-1Si-2Sr; (c) Mg-5Sn-2Si-2Sr

Table 1 EDS results of as-cast experimental alloys (molar fraction, %)

| Element | Position in Fig.4(a) | | Position in Fig.4(b) | | | Position in Fig.4(b) | | |
|---|---|---|---|---|---|---|---|---|
| | A | B | A | B | C | A | B | C |
| Mg | 74.05 | 67.28 | 68.35 | 69.39 | 88.25 | 70.80 | 70.26 | 71.46 |
| Sn | 25.95 | 22.62 | 28.05 | 6.49 | 9.24 | 27.97 | 7.99 | 16.51 |
| Si | | | 3.60 | 24.12 | | 1.23 | 21.75 | |
| Sr | | 10.10 | | | 7.51 | | | 12.03 |
| Total | 100 | 100 | 100 | 100 | 100 | 100 | 100 | 100 |

The results show that the microstructure of the Mg-5Sn-2Sr alloy consists of α-Mg, MgSnSr, $Mg_2Sn$ phases. The addition of Si element can promote the formation of $Mg_2Si$ phase on the grain boundary, and the content of $Mg_2Si$ phase increases with increasing the Si element. The chinese script type $Mg_2Si$ particle doesn't observed in the experimental alloys due to the addition of Sr element can refine the $Mg_2Si$ phase[10].

### 3.2 Mechanical properties

The mechanical properties of experimental alloys are shown in Table 2. It can be seen that the ultimate tensile strength (UTS), yield strength (YS) and hardness increases with increasing the Si element. Adding the Si to Mg-5Sn-2Sr alloy can promote the mechanical properties. The $Mg_2Si$ distributed along grain boundaries can effectively prevent grain boundary sliding and dislocation motion, so the tensile strength of alloy was obviously enhanced. The $Mg_2Si$ exhibits a high hardness[8] of $4.5\times10^9 Nm^{-2}$, therefore it can promote the hardness of alloy. From the XRD and SEM results above, the volume fractions of $Mg_2Si$ phase increases with increasing the Si element, therefore the mechanical properties of Mg-5Sn-2Si-2Sr alloy are better than Mg-5Sn-xSi-2Sr (x=0, 1) alloys.

Table 2 Mechanical properties of as-cast experimental alloys

| Alloy | UTS/ MPa | YS/ MPa | Elongation/ % | Hardness/ HB |
|---|---|---|---|---|
| Mg-5Sn-0Si-2Sr | 121 | 82 | 4.3 | 40 |
| Mg-5Sn-1Si-2Sr | 139 | 92 | 4.4 | 43 |
| Mg-5Sn-2Si-2Sr | 147 | 110 | 5.0 | 47 |

Fig.5 shows the SEM images of the tensile fracture surfaces for the experiment alloys. As shown in Fig.5, lots of cleavage planes and steps are present, some river patterns and minute lacerated ridges can also be observed in the local areas of the tensile fracture surfaces. It can be seen that the tensile fracture surfaces of experiment alloys have mixed characteristics of cleavage and quasi-cleavage fractures, which is the characteristic of brittle fracture. The fracture surfaces of Mg-5Sn-0Si-2Sr and Mg-5Sn-1Si-2Sr alloys exhibit relatively large cleavage-type facets (position A in Fig.5 (a) and (b)), while the fracture surfaces of Mg-5Sn-2Si-2Sr alloy are relatively small. Furthermore, some minute lacerated ridges and dimples can be seen from the fracture surfaces of the Mg-5Sn-2Si-2Sr alloy, therefore the Mg-5Sn-2Si-2Sr alloy exhibits better tensile strength and elongation.

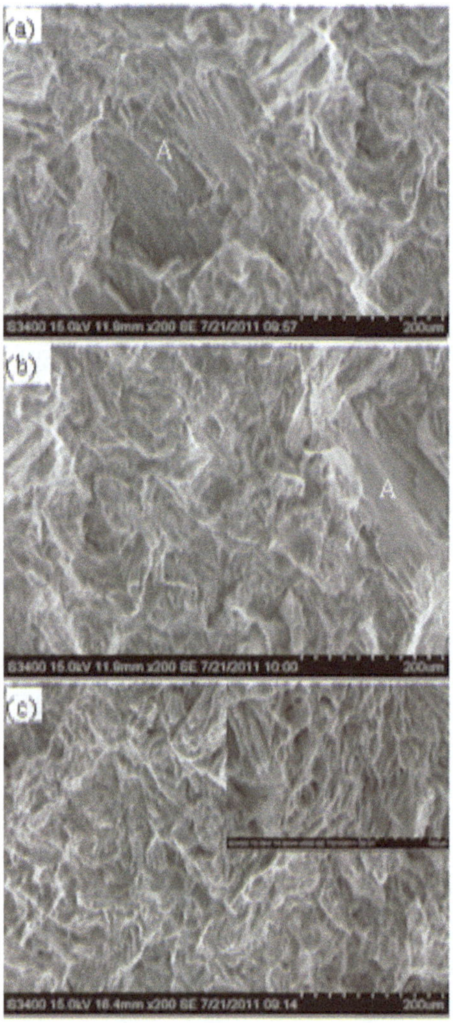


Fig.5 SEM images of tensile fractographs for as-cast alloys at room temperature

(a) Mg-5Sn-0Si-2Sr; (b) Mg-5Sn-1Si-2Sr; (c) Mg-5Sn-2Si-2Sr

## 4 Conclusions

(1) Mg-5Sn-2Sr alloy consists of α-Mg, MgSnSr, $Mg_2Sn$ phases. The addition of Si element can promote the formation of $Mg_2Si$ phase, and the volume fractions of $Mg_2Si$ phase increases with increasing the Si element.

(2) Adding Si to Mg-5Sn-2Sr alloy can promote the mechanical properties because the $Mg_2Si$ can effectively prevent grain boundary sliding and dislocation motion. The mechanical properties of Mg-5Sn-2Si-2Sr alloy are better than other alloys.

**References**

[1] Yang M B, Pan F S. Effects of Y addition on as-cast microstructure and mechanical properties of Mg-3Sn-2Ca (wt.%) magnesium alloy[J]. Materials Science and Engineering A, 2009, 525(1-2): 112-120.

[2] Nayyeri G, Mahmudi R, Salehi F. The microstructure, creep resistance, and high-temperature mechanical properties of Mg-5Sn alloy with Ca and Sb additions, and aging treatment[J]. Materials Science and Engineering A, 2010, 527(21-22): 5353-5359.

[3] Yang M B, Cheng L, Pan F S. Comparison of as-cast microstructure, tensile and creep properties for Mg-3Sn-1Ca and Mg-3Sn-2Ca magnesium alloys[J].Transactions of Nonferrous Metals Society of China, 2010, 20(4): 584-589.

[4] Yang M B, Cheng L, Pan F S. Effects of calcium addition on as-cast microstructure and mechanical properties of Mg-5Zn-5Sn alloy[J]. Transactions of Nonferrous Metals Society of China, 2010, 20(5): 769-775.

[5] Kang D H, Park S S, Kim N J. Development of creep resistant die cast Mg-Sn-Al-Si alloy[J]. Materials Science and Engineering A, 2005, 413-414, 555-560.

[6] Kim J J, Kim D H, Shin K S, et al. Modification of $Mg_2Si$ morphology in squeeze cast Mg-Al-Zn-Si alloys by Ca or P addition[J]. Scripta Materialia, 1999, 41(3): 333-340.

[7] Li X L, Chen Y B, Wang X, et al. Effect of cooling rates on as-cast microstructures of Mg-9Al-xSi (x=1, 3) alloys[J]. Transactions of Nonferrous Metals Society of China, 2010, 20(s2): 393-396.

[8] Jiang Q C, Wang H Y, Wang Y, et al. Modification of $Mg_2Si$ in Mg-Si alloys with yttrium[J]. Materials Science and Engineering A, 2005, 393(1-2): 130-135.

[9] Srinivasan A, Pillai U T S, Swaminathan J, et al. Observations of microstructural r印nement in Mg -Al-Si alloys containing strontium[J]. Journal of Materials Science, 2006, 41(18): 6087-6089.

[10] Nam K Y, Song D H, Lee C W, et al. Modification of $Mg_2Si$ morphology in as-cast Mg-Al-Si alloys with strontium and antimony[J]. Materials Science Forum, 2006, 510-511: 238-24.

[11] Du J, Iwai K, Li W F, et al. Effects of alternating current imposition and alkaline earth elements on modification of primary $Mg_2Si$ crystals in hypereutectic Mg-Si alloy[J]. Transactions of Nonferrous Metals Society of China, 2009, 19(5): 1051-1056.

[12] Yang M B, Ma Y L, Pan F S. Effects of little Ce addition on as-cast microstructure and creep properties of Mg-3Sn-2Ca magnesium alloy[J]. Transactions of Nonferrous Metals Society of China, 2009, 19(5): 1087-1092.

[13] Wang Q, Chen Y G, Xiao S F, et al. Study on microstructure and mechanical properties of as-cast Mg-Sn-Nd alloys [J]. Journal of Rare Earths, 2010, 28(5): 790-793.

[14] Wang Q, Chen Y G, Xiao S F, et al. Microstructures, mechanical properties and compressive creep behaviors of as-cast Mg-5%Sn-(0-1.0)%Pb alloys[J]. Journal of Central South University of Technology, 2011, 18(2): 290-295.

[15] Zhao H D, Qin G W, Ren Y P, et al. Microstructure and tensile properties of as-extruded Mg-Sn-Y alloys [J]. Transactions of Nonferrous Metals Society of China, 2010, 20(S2): 493-497.

[16] Liu H M, Chen Y G, Tang Y B, et al. The microstructure, tensile properties, and creep behavior of as-cast Mg-(1-10)%Sn alloys[J]. Journal of Alloys and Compounds, 2007, 440(1-2): 122-126.

[17] Liu H M, Chen Y G, Zhao H F, et al. Effects of strontium on microstructure and mechanical properties of as-cast Mg-5wt.%Sn alloy[J]. Journal of Alloys and Compounds, 2010, 504(2): 345-350.

[18] Nayyeri G, Mahmudi R. The microstructure and impression creep behavior of cast, Mg-5Sn-xCa alloys[J]. Materials Science and Engineering A, 2010, 527(7-8): 2087-2098.

# Magnesium Technology 2012

## Corrosion and Coating

**Wednesday PM**

*Session Chairs:*

**Guang-Ling Song**
**(General Motors, USA)**
**Michele Manuel**
**(University of Florida, USA)**

**Magnesium Technology 2012**
*Edited by: Suveen N. Mathaudhu, Wim H. Sillekens, Neale R. Neelameggham, and Norbert Hort*
***TMS (The Minerals, Metals & Materials Society), 2012***

# "ELECTROLESS" E-COATING FOR MAGNESIUM ALLOYS

Guang-Ling Song
Chemical Sciences and Materials Systems Lab, GM Global R&D, MC#:480-106-224, 30500 Mound Rd., Warren, MI 48090, USA

Keywords: Mg alloy, corrosion, coating, surface treatment

## Abstract

By utilizing the unique electrochemistry of Mg, a thin organic film can rapidly be deposited on the surface of a Mg alloy by dipping the Mg alloy in a cathodic E-coating bath solution without applying a current or potential. The self-deposited coating is selectively formed on Mg alloy surfaces. Although the "electroless" E-coating pre-film is relatively thin, it can offer sufficient corrosion protection for Mg alloys in a chloride-containing environment. The stability of the film can be significantly improved after curing. The corrosion resistance of the substrate Mg alloy has an important effect on the corrosion protection performance of the coating. The coating is more protective on a corrosion resistant Mg alloy than on a non-corrosion resistant Mg substrate. The coating protection performance is also influenced by the substrate surface condition or pre-treatment process. Wet cleaning + heat-treatment may be a cost-effective surface preparation/treatment for the "electroless" E-coating in industrial applications.

## Introduction

In the automotive industry, the demand for fuel efficient vehicles is currently driving the application of light materials. Magnesium alloys are potential alternative materials to steel and aluminum alloys in vehicles for the purpose of mass reduction. In addition, Mg alloys have many advantages in the automobile application, such as high mechanical strength, no toxicity, great damping capability, excellent casting fluidity, and low heat capacity [1,2]. However, the poor corrosion performance of Mg alloys is currently a challenge to the automotive engineers. Coating is one of the most cost-effective corrosion protection measures. Currently, the auto industry is looking for robust coatings for magnesium alloys in order to significantly improve their resistance to corrosion attack in automotive applications.

Many coating techniques have been proposed or developed for Mg alloys [1, 2]. In the automotive industry, application of an organic coating is a critical process in vehicle production. In the multi-layer auto coating system, the organic coating layer is expected to be the major barrier layer against corrosion attack. E-coating is a well known and the most popular coating on car bodies. It has excellent corrosion resistance and great covering ability on complex metallic components. However, it requires precise electrical control and expensive bath maintenance. If a coating similar to E-coating can be deposited on car bodies without application of a current or voltage, a significant cost reduction will result in the vehicle production, and current/voltage related issues in the traditional E-coating process may be avoided.

GM recently came up with an "electroless" E-coating technique for Mg alloys. It has been theoretically and experimentally verified that such a coating can be formed and can have good corrosion protection performance [3,4].

This paper briefly reviews such an "electroless" E-coating process that can rapidly (in a few seconds) form a thin E-coating on a Mg alloy surface from a cathodic E-coating bath without application of a current or potential. This new coating technique for Mg alloys may be used in many other industries, too.

## "Electroless" E-coating Process

Basically, the "electroless" E-coating is a dipping process. An abraded/ground Mg alloy coupon is dipped into a cathodic E-coat bath solution (for example, a typical bath has a main composition: water 71-82 wt%, epoxy resin 16-26 wt%, titanium dioxide 1.3 wt%) for a short period of time (for example, 5 seconds). The immersion time can be adjusted according to the coating thickness requirement. In order to obtain a stain-free surface, the Mg alloy specimen should be slowly pulled out of the solution, so that the E-coating bath solution can flow from the specimen surface back to the bath. A coating film immediately forms on the Mg alloy surface while the specimen is being dipped/pulled out the bath. The photos in Fig.1 show the "electroless" E-coating process. In Fig.1(a), the dry-abraded AZ91D has a shining surface before dipped into the bath. After pulled out of the bath, its surface becomes dry immediately while the excess solution is flowing back to the bath (see Fig.1 (b)). The dried surface is uniformly duller than its original surface (see Fig.1 (c)).

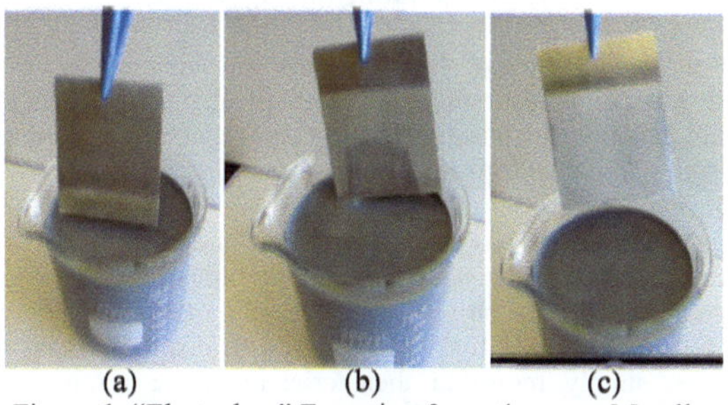

(a) (b) (c)
Figure 1. "Electroless" E-coating formation on a Mg alloy (AZ91D) dipping in a cathodic E-coating bath solution. (a) Before dipping; (b) being pulled out after dipping; (c) about 30 seconds after dipping.

The film rapidly formed on the Mg alloy surface after being dipped is in nature an E-coating layer. It can be cured just like a normal E-coating. For example, curing can be performed in an oven at 171°C for 25 minutes. After curing, the "electroless" E-coating is fully polymerized and has the same basic properties as a commercial E-coating.

## Coating Mechanism

The "electroless" process follows the same coating deposition mechanism as a normal E-coating.

During cathodic E-coating, the following reactions are always involved [5,6,7,8]:

$2H_2O + 2e \rightarrow 2OH^- + H_2\uparrow$ (1)

$R\text{-}NH_3^+ + OH^- \rightarrow R\text{-}NH_2\downarrow + H_2O$ (2)

in which $R\text{-}NH_3^+$ is an aminoepoxy resin dissolved in the E-coating bath. Cathodic reaction (1) generates hydroxyls, resulting in an increase in pH value at the cathode surface. Reaction (2) is an E-coating pre-film deposition or coagulation step, in which the ammonia bearing organic $R\text{-}NH3^+$ reacts with $OH^-$ from reaction (1) and becomes an insoluble organic deposit ($R\text{-}NH_2$) on the cathode surface. After curing, the E-coating pre-film polymerizes and turns into a stable insoluble hydrophobic film. In these steps, the critical species is $OH^-$. As long as a sufficient amount of $OH^-$ is provided, reaction (2) will continue and an E-coating pre-film be formed.

Mg surface always has high alkalinity in an aqueous solution because of its reaction with water:

$$Mg + 2H_2O \rightarrow Mg^{2+} + 2OH^- + H_2\uparrow \quad (3)$$

This is a rapid alkalization reaction. It generates a sufficiently high concentration of $OH^-$ on the Mg surface to trigger reaction (2), resulting in deposition of the E-coating pre-film there. Therefore, no reaction (1) is required, and thus the coating process is electroless.

## Selective Coating for Mg Alloys

The above "electroless" E-coating mechanism predicts that the coating can only occur on Mg surfaces. Steel, galvanic steel or Al cannot generate sufficient $OH^-$. Thus, an "electroless" E-coating pre-film will not be formed on their surfaces.

In experiments, it has been found that the surfaces of steel, galvanic steel and Al coupons are wet for a long time or could not be dried quickly like an AZ91D coupon after being pulled out of the E-coating bath. After being washed in distilled water, their surfaces simply return to their original states; no noticeable coating remains on their surfaces.

Figure 2. The abraded surfaces of steel and AZ91D coupons 10 seconds after being dipped in and pulled out of an E-coating bath.

Figure 2 show a comparison between steel and Mg alloy coupon surfaces after being pulled out from a baker containing a cathodic E-coating bath solution. The steel and Mg alloy (AZ91D) coupons were dipped in a cathodic E-coating bath together. About 10 seconds after they were taken out the bath, half of the dipped AZ91D surface area already became dry, while the whole dipped surface of the steel was completely wet. After 30 seconds, the AZ91D surface was completely dried, but the steel surface was still wet. The drying process is a phenomenon of the "electroless" E-coating deposition. Similar results have also been obtained from comparison between the Mg alloy and an Al alloy or galvanized steel. The quickly drying process selectively occurring only on Mg alloy surfaces after dipping in an cathodic E-coating bath further confirms the "electroless" E-coating deposition mechanism.

## Coating Thickness

According to the proposed "electroless" E-coating mechanism, the coating growth rate should be decreasing with dipping time, and the coating cannot be very thick. This is because that after the first layer of the "electroless" E-coating pre-film is formed on a Mg alloy surface, the diffusion of $OH^-$ through the film to meet the $R\text{-}NH_3^+$ from the bath is limited. The diffusion becomes slower and slower as the film grows.

The pre-film of "electroless" E-coating in theory is porous, just like a normal E-coating. The pores can be the intervals between $R\text{-}NH_2$ molecule trains. They are not straight holes through the pre-film from the substrate/pre-film interface directly to the pre-film/solution interface, but sinuous in the film. During an "electroless" E-coating deposition process, alkalization reaction (3) mainly occurs at the substrate/pre-film (m/f) interface. The alkalization reaction can easily proceed because of the presence of sufficient water in the pores of the pre-film next to this interface. The alkalization reaction can readily make the concentration of $OH^-$ reach the saturated level of $Mg(OH)_2$ at this interface. At the film/bath solution (f/s) interface, the supply of $R\text{-}NH_2$ is much easier than at the m/f interface. $R\text{-}NH_2$ has a relatively large organic molecule which is more difficult to penetrate or transfer than a hydroxide ion in a pre-film. Therefore, pre-film deposition reaction (2) mainly occurs at the f/s interface, and the growth of the pre-film should be controlled by the diffusion of $OH^-$ through the film from m/f interface to the f/s interface.

Based on the above reasonable assumptions, the dependence of the pre-film thickness $L$ on dipping time $t$ can be deduced [3,4]:

$$L = Kt^{1/4} \quad (4)$$

where $K$ is a constant determined by diffusion coefficient of $OH^-$ in the film, film formation reaction constant, porosity and pore sinuosity of the film. This growing kinetics predicts that the film growth rate decreases gradually with dipping time, which is supported by experimental film thickness measurements (see Fig.3). In Fig.3, the experimental power value of dipping time is 0.24, very close to the theoretical value ¼.

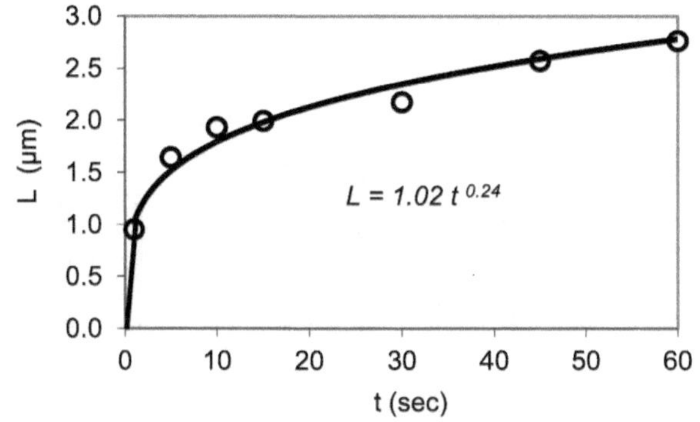


Figure 3. The growth of pre-film of "electroless" E-coating in bath [3,9].

**Coating Microstructure**

The deposited pre-film on Mg is porous. This is an assertion inferred from existing understanding of the commercial E-coating microstructure. It is also an assumption based on which the film growth kinetics (equation (4)) is deduced. In fact, the porosity of the "electroless" E-coating is an experimentally confirmed result. Fig.4 shows the microstructures of cured and uncured "electroless" E-coatings on AZ91D. The pre-film before curing does not appear to be dense. There are numerous tiny micro-pores in the pre-film. After curing, the number of tiny pores reduces, but relatively larger micro-pores form. These micro-pores are defects where corrosion could be initiated. The coating porosity suggests that the film formation and growth models proposed earlier are reasonable.

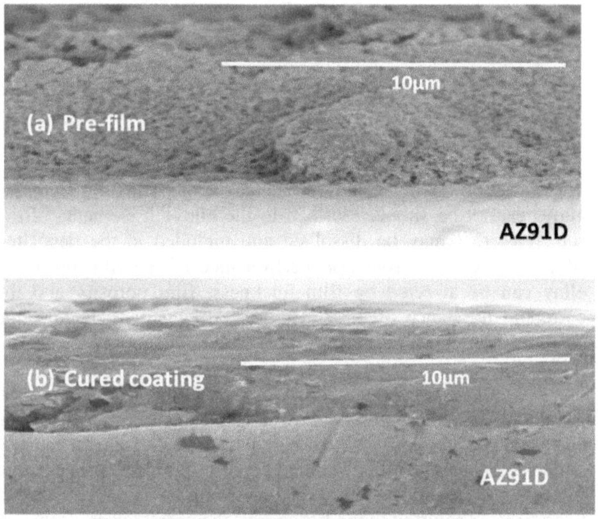


Figure 4. Microstructures of (a) the pre-film of "electroless" E-coating and (b) cured "electroless" E-coating [9]. The pre-film was formed on SiC paper abraded AZ91D dipped in the bath for 1 min. The cured coating was obtained at 171°C for 25 min.

The porosity of a pre-film could result from hydrogen evolution (equation (3)) during the film deposition process. The polymerization of the pre-film during the film curing process is normally accompanied by evaporation of solvent and release of $H_2$. This also leaves porosity in the film. Thus, a cured "electroless" E-coating is also porous.

**Coating Protection Performance**

Since the pre-film consists of insoluble $R\text{-}NH_2$, it should offer a certain level of corrosion protection for its substrate if it uniformly covers the substrate. Even though the film is porous, the presence of the film on a Mg alloy surface can to a great extent retard the ingress of aggressive species and separate the substrate from corrosive environments. Hence, improved corrosion resistance of a Mg alloy by the pre-film can be expected. Curing actually turns the pre-film into a polymer coating, which is very stable material in most natural solution. Thus, curing can dramatically stabilize the coating and further enhance the corrosion resistance of a coated Mg alloy.

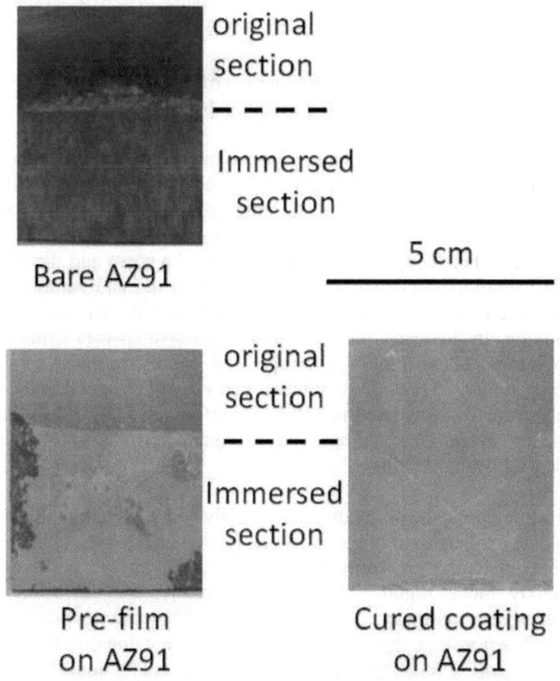


Figure 5. Corrosion damage of bare, pre-film and cured coating coated AZ91D after immersion in 5 wt.% NaCl for 24 hours [9]. The cured coating was obtained at 171°C for 25 min.

Fig.5 presents corrosion morphologies of bare, pre-film and cured coating coated AZ91D after immersion in 5 wt.% NaCl for 24 hours. It shows that the bare AZ91D is badly corroded in the whole immersed surface; the "electroless" E-coating pre-film covered AZ91D has film-damage in some areas of the immersed surface; there is no corrosion damage at all on the AZ91D surface covered with a cured "electroless" E-coating. These experimental results demonstrate that the "electroless" E-coating pre-film can effectively protect a Mg alloy from corrosion attack in 5 wt.% NaCl solution, and curing can greatly further improve the corrosion protection performance of the "electroless" E-coating. It should be noted that both the pre-film and cured "electroless" E-coating have been scratched before corrosion tests. No preferential corrosion damage occurs along these scratches. This is an attractive feature of this coating in applications.

### Corrosion of a Coated Mg Alloy

It is interesting that the porous "electroless" E-coating on a Mg alloy surface does not lead to a severe localized or pitting corrosion damage (see Fig.5). This can be explained by the corrosion mechanism of a coated Mg alloy as illustrated below.

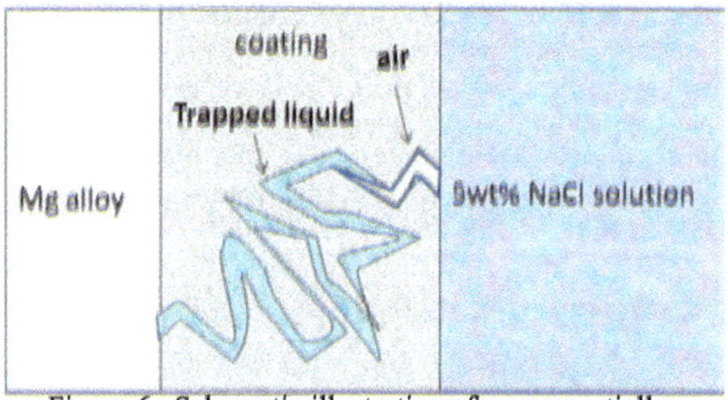


Figure 6. Schematic illustration of a pore partially filled with solution [4].

After an "electroless" E-coating formed on a Mg alloy surface and dried in air, a certain amount of water evaporates from the film pores. When the specimen is immersed into a 5wt% NaCl solution, due to the presence of some trapped air in the pores, the continuity between the liquid remaining in the pores and the bulk NaCl solution will be interrupted. This can result in a hydrophobic interface formed between the coating and the bulk environmental solution. In this case, the ingress of aggressive species from the environmental solution is interrupted by the discontinuity of the pore liquid. This discontinuity caused by trapped air is schematically illustrated in Fig.6. Therefore, an "electroless" E-coating, even though porous, can offer great corrosion protection for a substrate Mg alloy.

In some pores, the pore liquid may finally become continuous and initiate the corrosion of the substrate. Hydrogen bubbles generated from the corroding substrate will be trapped in the pores and quickly block liquid paths again. Therefore, the corrosion is unlikely to continue in the same pore. This may explain the phenomenon that the corrosion damage is not localized on an "electroless" E-coating coated Mg alloy surface. It should also be noted that the formed pre-film or cured coating is an insulating layer. Even after they are broken and the substrate Mg alloy is exposed in some local areas, there is no galvanic effect between the coatings and substrate. In other words, the undamaged coat will not accelerate the corrosion of the substrate in the broken areas. This is another reason why the corrosion damage does not appear to be localized.

### Coating on Various Mg Alloys

Reactions (1) through (3) can occur on any Mg alloy surface. However, the reaction rates may vary from alloy to alloy. It is expected that the "electroless" E-coating pre-films formed on different Mg alloys may be slightly different. Fig. 7 presents such a slight difference in the pre-film thickness ($L_{coat}$) on 8 different Mg alloys.

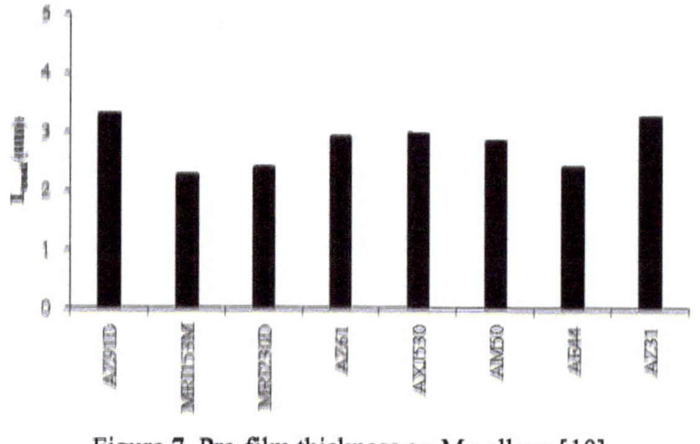


Figure 7. Pre-film thickness on Mg alloys [10].

The pre-film covered Mg alloys show significantly different corrosion performance. The corrosion damage areas of the pre-films on different Mg alloys after immersion in 5 wt.% NaCl solution for 56 hours are shown in Fig.8. It appears that the coated AZ91D is the most corrosion resistant, while the coated AXJ530 the worst in corrosion behavior.

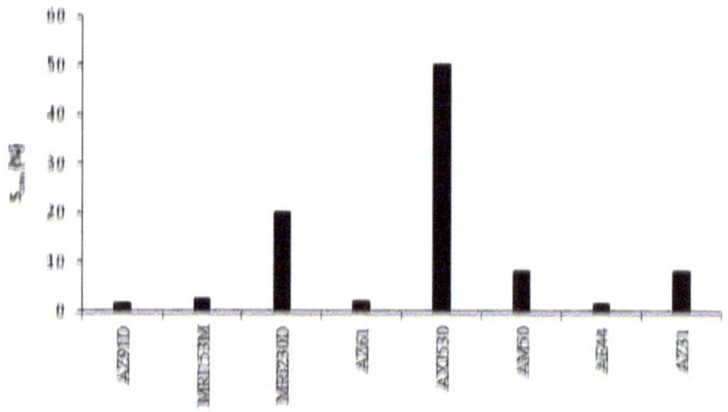


Figure 8. Corrosion damage areas ($S_{corr}$) of pre-film coated Mg alloys after immersion in 5 wt. % NaCl solution for 56 hours [10].

The variation in corrosion resistance of pre-film covered Mg alloys cannot be simply ascribed to the alloying elements of the substrates that may be dissolved and included in the deposited films. In fact, the protection performance of the film on a Mg alloy can be affected by film thickness, film porosity and the corrosion resistance of the substrate. Among these factors, the corrosion resistance of the substrate could be the most critical one.

For example, Fig.9 shows that the corrosion damage areas of the films on the 8 Mg alloys have a good relationship with the corrosion weight loss of the bare substrates. This is actually understandable according to the corrosion model proposed in Fig.6. Since corrosion mainly initiates from the substrate exposed to aggressive solution in the film pores, the corrosion resistance of the substrate can certainly to a great degree influence the corrosion performance of the Mg alloy underneath the coating.

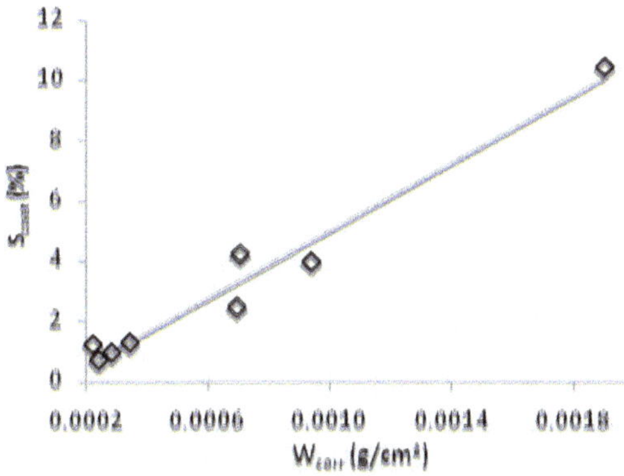


Figure 9. The relationship between the corrosion damage area ($S_{coat}$) of coated Mg alloys after 56 hours of corrosion immersion in 5 wt.% NaCl solution and the weight loss ($W_{corr}$) of corresponding substrate bare Mg alloys after 32 hours of corrosion immersion in the same NaCl solution.

### Surface Preparation Effect

The surface condition can also significantly affect reactions (1) through (3) and thus will influence the coating quality. Different surface pre-treatments have been tried on Mg alloys for the "electroless" E-coating process. For example, it has been found that dry grinding/abrading can produce a good surface for this coating in terms of the coating protection performance. Sandblasting of the substrate is detrimental to the coating quality. Polishing in non-aqueous media does have a beneficial effect on the coating formation.

Among these trial experiments, there is an interesting finding: heat-treatment not only alters the corrosion resistance of the substrate, but also has a beneficial effect on the "electroless" E-coating quality. Fig.10 shows that wet-abraded AZ31 Mg alloy with heated and un-heated surfaces after "electroless" E-coating and curing has very different corrosion damage areas in 5 wt.% NaCl solution.

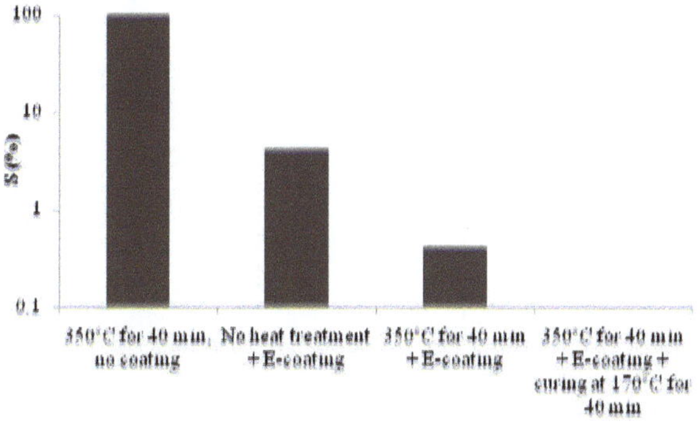


Figure 10. Corrosion area of bare and "electroless" E-coated AZ31 after 24-hour immersion in 5 wt.% NaCl solution. All the AZ31 samples were wet-abraded.

The beneficial effect of heating after abrading on coating protection performance is obvious. As heat-treatment deteriorates the corrosion resistance of AZ31D [11], this beneficial heating effect on the coating cannot be associated with the corrosion resistance variation of the substrate, but probably be ascribed to the influence of Mg oxides on the "electroless" E-coating formation, which is currently under investigation in the lab.

An acid cleaned Mg alloy surface if pre-treated differently can also result in a different coating quality. For example, if simply etched in 10 wt.% H2SO4, AZ31 is not very corrosion resistant (see Fig.11 (a)). Application of a "electroless" E-coating pre-film on the etched and dried surface can significantly improve the corrosion performance (see Fig.11 (b). If the surface after etching is heated before "electroless" E-coating, the corrosion resistance of the coated AZ31 can be further strikingly enhanced (see Fig.11 (c)).

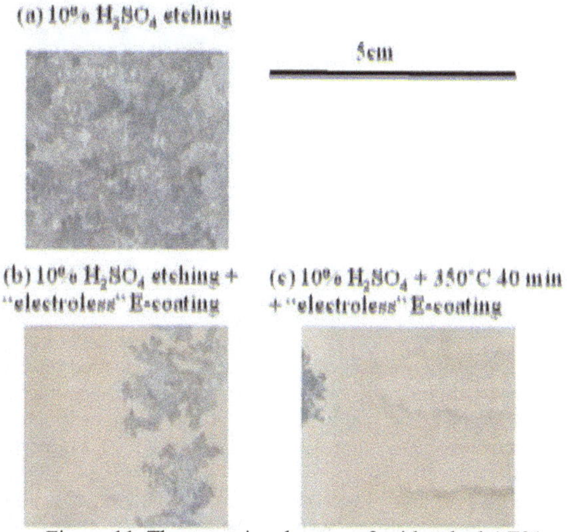


Figure 11. The corrosion damage of acid etched AZ31 samples before and after 24 hour immersion in 5 wt. % NaCl solution.

The results suggest that acid cleaning is also suitable for pre-treating a Mg alloy before "electroless" E-coating. Compared with abrading/polishing, this is a practical process in industries.

Phosphating is a commercial surface treatment technique for many metals. It produces a layer of phosphates on the substrate surface, which has a beneficial contribution to the adhesion of a coating on its top. If a Mg alloy is phophated, the "electroless" E-coating can still be applied on its surface.

It has been found that the coating on such a treated Mg alloy can remarkably enhance the corrosion resistance of the phosphate surface. Fig.12 shows that on a phosphated AZ91D surface, the right half surface has been "electroless" E-coated; after 2 days of immersion in 5 wt.% NaCl solution, the phosphated and "electroless" E-coated area is uncorroded, while the whole phosphated area without "electroless" E-coating has been very badly corroded.

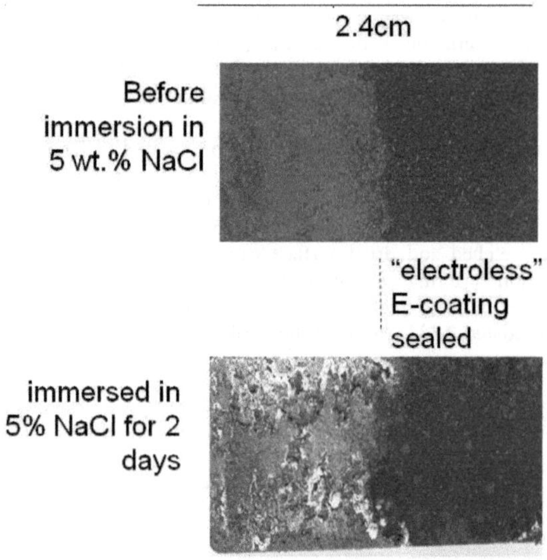


Figure 12. Corrosion morphology of phosphated AZ91D with and without "electroless" E-coating before and after 2 days of immersion in 5 wt.% NaCl solution.

### Summary

1) An E-coating pre-film can be rapidly self-deposited on a Mg alloy by dipping the Mg alloy in a cathodic E-coating bath solution without applying current or voltage. Such an "electroless" E-coating deposition is a result of Mg surface alkalization effect.
2) The "electroless" E-coating is formed only on Mg alloys which have a strong surface alkalization effect. On other metals, such as steel, aluminum and zinc and their alloys, the self-deposited "electroless" E-coating cannot be formed.
3) The "electroless" E-coating is thinner than a conventional E-coating. It contains pores which could be the sites to initiate corrosion.
4) The pre-film formed on a Mg alloy during "electroless" E-coating process provides adequate corrosion protection for Mg alloys in 5 wt% NaCl solution. Curing can lead to polymerization of the pre-film and significantly enhanced protection performance of the "electroless" E-coating. The discontinuity of the pore liquid in the pores of the coating is responsible for the high corrosion resistance of a coated Mg alloy.
5) The protection performance of the coating is closely dependent on the corrosion resistance of the substrate Mg alloy due to the coating porosity nature.
6) The substrate surface state and condition can influence the pre-film formation and thus the coating quality. A cleaned and heated Mg alloy surface is most suitable for "electroless" E-coating.
7) The "electroless" E-coating process can also be applied on a phosphated Mg alloy surface. The coating can significantly improve the corrosion resistance of a phosphated Mg alloy.

### Acknowledgements

The author would like thank Ms. Minghong Liu for her assistance in some experiments.

### References

1. J. Gray, B. Luan, "Protective coatings on magnesium and its alloys—a critical review," *Journal of Alloys and Compounds*, 336 (2002), 88-113
2. C. Blawert, W. Dietzel, E. Ghali, G-L. Song, "Anodizing Treatments for Magnesium Alloys and Their Effect on Corrosion Resistance in Various Environments," *Advanced Engineering Materials*, 8 (2006), 511-533
3. G-L. Song, "Electroless E-Coating: An Innovative Surface Treatment for Magnesium Alloys," *Electrochemical and Solid-State Letters*, 12(10) (2009), D77-D79
4. G-L. Song, "Electroless deposition of a pre-film of electrophoresis coating and its corrosion resistance on a Mg alloy," *Electrochimica Acta*, 55 (2010), 2258–2268
5. F. Beck, "Fundamental aspects of electrodeposition of paint," *Progress in Organic Coatings*, 4 (1976), 1-10
6. N. Vatistas, "Initial Mechanisms of the Electrocoating Process," *Industrial & Engineering Chemistry Research*, 37 (1998), 944-951
7. E. Okada, H. Hosono, T. Nito, et al., "Development of High Throw power E-coat", 2005 SAE World Congres, Detroit, SAE Technical Paper #2005-01-0619 (2005)
8. N. Sato, "Interfacial control between phosphate films and electrodeposition films on coated steels", SAE technical paper #900838, International Congress and Exposition, Detroit, MI (1990)
9. G.-L. Song, "A dipping E-coating for Mg alloys," *Progress in Organic Coating*, 70 (2011), 252-258
10. G.-L.Song, "The Effect of Mg Alloy Substrate on "Electroless" E-Coating Performance," *Corrosion Science*, 53 (2011), 3500-3508
11. G.-L.Song, "Effect of microstructure evolution on corrosion of different crystal surfaces of AZ31 Mg alloy in a chloride containing solution," *Corrosion Science* , 54 (2012), 97-105.

**Magnesium Technology 2012**
*Edited by: Suveen N. Mathaudhu, Wim H. Sillekens, Neale R. Neelameggham, and Norbert Hort*
***TMS (The Minerals, Metals & Materials Society), 2012***

# THE INFLUENCE OF GALVANIC CURRENT ON CERIUM-BASED CONVERSION COATINGS ON MG, AL, AND GALVANIZED STEEL COUPLES

Surender Maddela[1], Matthew J. O'Keefe[1], Yar-Ming Wang[2]
[1]Materials Research Center, Dept. of Materials Science and Engineering, Missouri University of Science and Technology, Rolla, MO 65409, USA
[2]Chemical Science and Materials System Laboratory, GM R&D Center, Warren, MI 48090, USA

Keywords: Mg alloy coating, AZ31 and AZ91D corrosion, bimetallic couples, zero resistance ammeter

## Abstract

The influence of galvanic current on cerium-based conversion coatings (CeCCs) for magnesium (AZ91, AZ31), aluminum (6016), and electro-galvanized steel (EGS) couples has been studied using zero resistance ammeter (ZRA) measurements in prohesion solution. The galvanic current measured between magnesium-aluminum, magnesium-galvanized steel, and aluminum-galvanized steel couples correlated with significant changes in coating morphology and deposition rate. The ZRA galvanic currents (mA) were 0.02 for 6016-EGS, 0.38 for AZ91-EGS, 0.72 for AZ91-6016, 1.08 for AZ31-EGS, and 1.08 for AZ31-6016 couples. The corrosion performance of the coated couples was evaluated by ASTM B117 neutral salt spray testing. Cerium conversion coated couples performed better in salt spray testing compared to uncoated couples. The correlation of galvanic current, cerium deposition, and corrosion performance will be discussed.

## Introduction

Magnesium alloys are being extensively used in automobile industry, aerospace, and electronic applications due to its low density and high strength-to-weight ratio. In an automobile industry, the use of magnesium alloys has doubled in the last decade [1-2]. However, the use of Mg alloys in the automobile industry has been limited due to galvanic corrosion with other structural components [3]. Moreover, the development of a robust paint-line compatible pretreatment process for magnesium intensive vehicles is critical to its expanded use in automotive body and chassis assemblies. Current phosphate electrolyte baths are formulated to treat mixed-metals such as galvanized steel, cold rolled steel, and aluminum in body-in-white (BIW) assemblies. However, because of high solubility of Mg in the phosphate baths magnesium alloys are not compatible with existing phosphate processes used in automotive manufacturing [3-4]. In order to treat a mixed-metal BIW assembly, including magnesium alloy parts, a new conversion coating process needs to be developed.

Chromium-based conversion coatings are one of the most effective ways of preventing corrosion of base metals [5]. However, the carcinogenic nature of chromates restricts the use of chromate baths and extensive effort is being done to find a suitable replacement [6]. Several researchers have reported alternative chromate replacement coatings [7-13]. Among the alternatives, cerium-based conversion coatings (CeCCs) are promising as they offer significant corrosion protection and are environmentally benign [10-15]. Cerium compounds are used in large scale applications such as polishing glass and are not expensive unlike other rare earth materials. The deposition mechanism and corrosion protection of cerium-based conversion coatings on individual alloys have been discussed in numerous publications [10-12]. However, CeCCs on bimetallic couples, such as magnesium-aluminum, magnesium-galvanized steel, and aluminum-galvanized steel, have not been reported. Therefore, the effect of galvanic current on cerium deposition and corrosion protection of cerium based conversion coatings on bimetallic couples will be examined.

## Experimental procedure

Magnesium (AZ91D, AZ31B), aluminum (6016), and electro-galvanized steel (EGS) samples were sectioned into 2.5 x 2.5 cm coupons for galvanic current measurements (zero resistance ammeter) in prohesion solution of composition 0.6 wt.% NaCl and 0.6 wt.% of $(NH_4)_2SO_4$. Similarly, test panels were sectioned into 5 x 10 cm coupons which were then coupled using aluminum rivets for subsequent cerium conversion coating (CeCC). The bimetallic couple schematic diagram is showing in Figure 1. All magnesium alloy samples were first polished using 220, 320, and 400 SiC grit paper cleaned with isopropyl alcohol, and then rinsed with de-ionized water prior to all experiments.

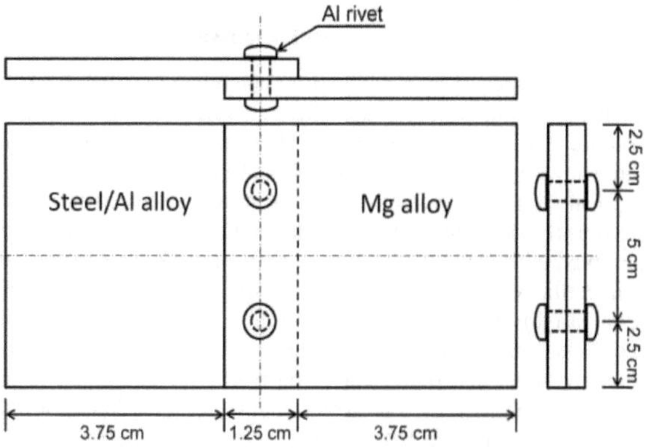


Figure 1. Schematic diagram (not to scale) of bimetallic galvanic couple with aluminum rivets.

The corrosion performance of bimetallic couples with and without cerium conversion coatings was evaluated by ASTM B117 neutral salt spray test. The bimetallic couples were cleaned with 0.5 wt. % tetrafluoroboric acid ($HBF_4$) for 20 s at room temperature, rinsed with de-ionized water, cleaned with 5 wt. % sodium hydroxide (NaOH) for 30 s at room temperature, and them rinsed again with de-ionized water. The cleaned couples were immersed in cerium deposition solution for 2 min at room temperature. The composition of the deposition solution was 205 g of stock solution, 25 g of DI water, 0.8 g of gelatin organic (RDH, Rousselot) and 20 ml of $H_2O_2$ (Fisher Chemical, 30 vol. %) [14]. Cerium coated couples were post-treated in a phosphate solution for 5 min at 85 °C [10].

Cross-sectional structures of the coated panels were prepared by focused ion beam (FIB) milling (Helios NanoLab 600, FEI). The FIB used a gallium ion beam to selectively mill through coatings and into the substrates to reveal the coating interface at selected areas. A platinum layer was deposited on the coating surface to protect the area of interest from damage during milling. The scanning electron microscopy (SEM) and energy dispersive spectroscopy (EDS) capability of the FIB system were then used to image the coating interface and analyze chemical compositions, respectively. All images were taken in secondary electron mode using an acceleration voltage of 5 kV and EDS measurements (Oxford, X-Max Silicon Drift Detector) were carried out at an acceleration voltage of 30 kV.

Zero Resistance Ammeter (ZRA) Technique

Galvanic corrosion current ($I_{gal}$) between two electrodes fixed to a cell was measured by a zero resistance ammeter technique. The potentiostat serves as a zero resistance ammeter and measures the current flowing between the anode (Mg alloy) and cathode (Al, EGS). The galvanic currents between couples in different electrolyte solutions were measured by the ZRA technique as follows. The samples were fixed at two ends of a flat cell (Princeton applied research) by modifying the flange arrangement. The two electrodes (Mg alloys and Al/EGS) were placed in the arrangements so as to expose only 1 $cm^2$ area of each sample to the electrolyte. A saturated calomel electrode (SCE) in saturated KCl served as a reference electrode. The flange for the counter electrode (Pt) was replaced with a flange that contained the working electrode holder. In this manner, samples of 1 $cm^2$ area would be fixed at the ends of the flat cell. Measurements were made as a function of time using the galvanic current routine in Power-Suite software connected to PARSTAT 2273.

## Results and discussion

Optical images of bare and cerium conversion coated AZ91-6016 bimetallic couple before and after salt spray testing is showing in Figure 2, while, AZ91-EGS, AZ31-6016, and AZ31-EGS couples are showing in Figure 3, Figure 4, and Figure 5, respectively. The optical images revealed that cerium conversion coatings uniformly deposited on both aluminum and EGS panels when coupled with magnesium alloys (AZ91 and AZ31). Cerium coated bimetallic couples had better salt spray corrosion performance compared to bare couples. Among the four couples with different combinations, cerium coated AZ91 bimetallic couple combinations had better corrosion resistance compared to AZ31 couple combinations. For AZ91 couple combinations, the AZ91-6016 couple showed better salt spray corrosion performance than the AZ91-EGS couple, as the EGS panel corroded severely compared to the AZ91 alloy. This may be due to the presence of zinc/zinc oxide under the cerium coating. The open circuit potential of zinc is very close to magnesium alloys and therefore the zinc is probably more active corrosion than AZ91 alloy [16]. It is worth noting that there is no trace of iron rust found in the field of the EGS surface, indicating that the zinc/zinc oxide is serving as a sacrificial anode to the steel panel.

Optical images of salt spray tested bimetallic couples indicated significant galvanic corrosion at the galvanic junction compared to the field away from the junction. It can be seen from the images in Figure 2 to 5 that the galvanic current between the two metals has a significant effect on the overall corrosion behavior of the bimetallic couple. The influence of galvanic current on overall corrosion performance of different couples has been discussed in details by Zhang [16]. In particular, Song et. al. have reported that the galvanic current effect on overall corrosion of magnesium AZ91D alloy coupled with alloys of aluminum, steel, and zinc in a sodium chloride solution [17]. They have reported that the galvanic current significantly influences the corrosion of magnesium alloys in the presence of other metals and the galvanic effect is more pronounced at the galvanic junction compared to the field away from the junction. However, CeCCs on bimetallic couple significantly change the overall corrosion performance compared to uncoated couples. The bare couple severely corroded within four days of salt spray testing whereas cerium coated couples were able to last more than a week without any major corrosion damage. The cerium conversion coating significantly enhances the corrosion resistance of the bimetallic couples.

The cross-sections of CeCCs and substrate interface was prepared by FIB milling on individual alloys and bimetallic couples as shown in Figure 6 and Figure 7, respectively. The FIB cross-sections of CeCCs deposited AZ91 and AZ31 alloy panels reveal the interface between the magnesium substrate and cerium coating layer was approximately 500 nm thick (Figure 6 a and b). However, there was little or no cerium conversion layer visible on aluminum and EGS substrates. The EDS analysis detected a cerium signal within a few hundred nano-meters of the interface. The FIB analysis indicated that cerium conversion coatings covered magnesium alloy substrates but that the 6016 and EGS panels were lightly coated, if at all, when deposition was done on individual alloys. On the other hand, FIB cross-sections showed cerium coatings on both the aluminum alloy and EGS panels when coupled with magnesium alloys (Figure 7 a through d). The approximate coating thickness on 6016 and EGS panels was about 300 to 400 nm in both bimetallic combinations. Therefore, the galvanic current significantly influences the cerium deposition mechanism on bimetallic couples. Several researchers have reported the influence of micro-galvanic (multiphase alloy) effect on cerium conversion coating mechanism [5, 10-11]. In particular, Pinc et. al. have discussed the influence of metal-matrix and second phase interactions on spontaneous deposition of cerium conversion coatings [11, 12, 18]. They have reported that the micro-galvanic current assists the spontaneous deposition of cerium conversion coatings on multiphase alloys. The galvanic current between bimetallic couples appears to have a similar effect on cerium deposition mechanism. These results clearly demonstrate that the galvanic current significantly influences the deposition mechanism and/or deposition rate. Therefore, galvanic current has a significant effect on spontaneous deposition of cerium conversion coatings on bimetallic anode/cathode couples.

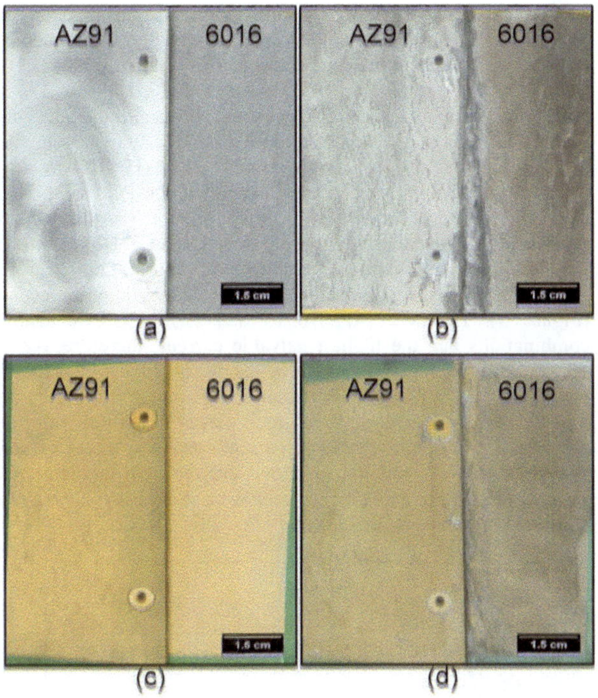


Figure 2. Optical images of AZ91-6016 couple with and without CeCCs, before and after salt spray testing, a) bare couple, b) 4 days salt spray tested bare couple; c) CeCCs deposited couple, and d) 7days salt spray tested CeCCs deposited couple.

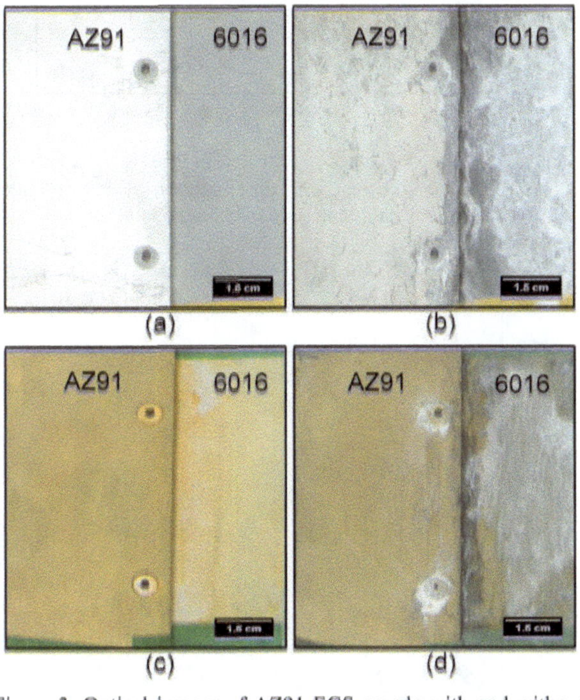


Figure 3. Optical images of AZ91-EGS couple with and without CeCCs, before and after salt spray testing, a) bare couple, b) 4 days salt spray tested bare couple; c) CeCCs deposited couple, and d) 7days salt spray tested CeCCs deposited couple.

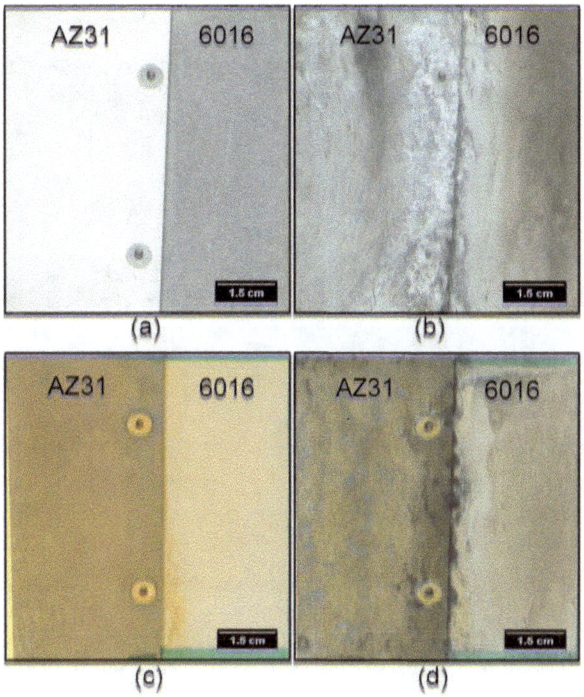


Figure 4. Optical images of AZ31-6016 couple with and without CeCCs, before and after salt spray testing, a) bare couple, b) 4 days salt spray tested bare couple; c) CeCCs deposited couple, and d) 7days salt spray tested CeCCs deposited couple.

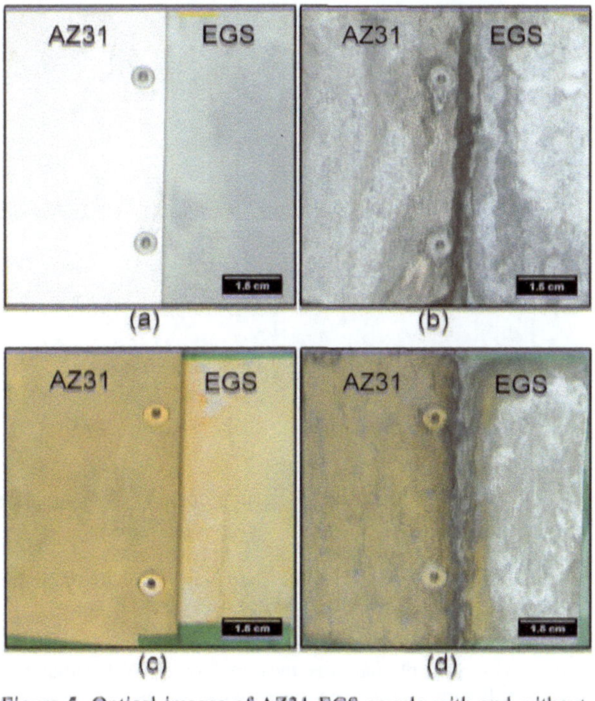


Figure 5. Optical images of AZ31-EGS couple with and without CeCCs, before and after salt spray testing, a) bare couple, b) 4 days salt spray tested bare couple; c) CeCCs deposited couple, and d) 7days salt spray tested CeCCs deposited couple.

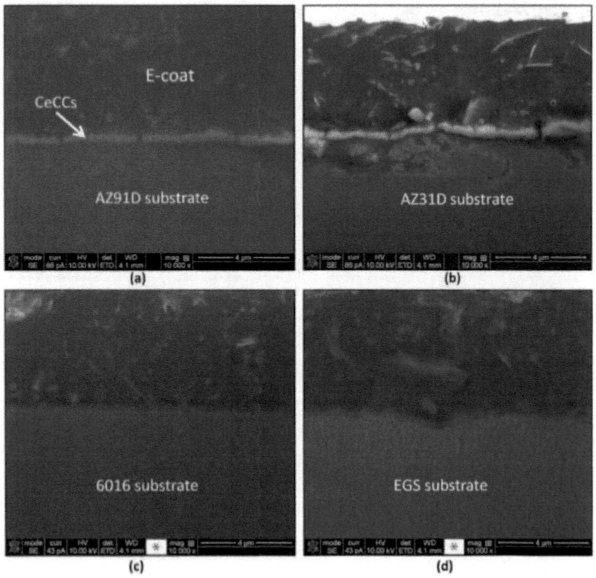


Figure 6. FIB cross-sections of uncoupled CeCCs deposited panels; a) AZ91 alloy, b) AZ31 alloy, c) 6016 alloy, and d) EGS.

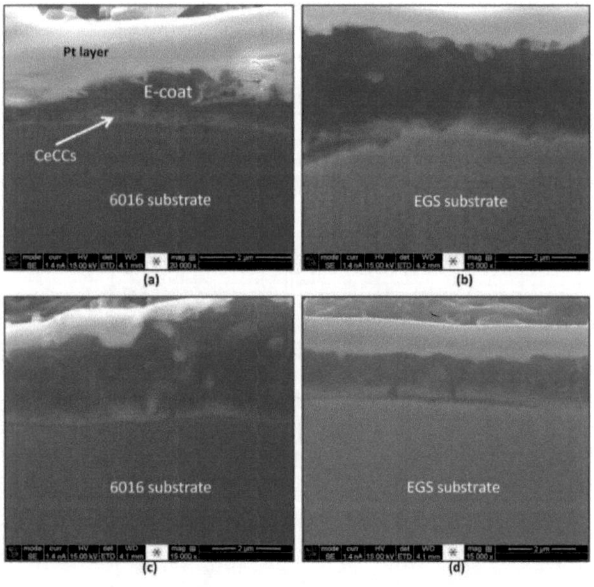


Figure 7. FIB cross-sections of coupled CeCCs deposited aluminum and EGS panels; AZ91 couple a) 6016 alloy b) EGS; and AZ31 couple c) 6016 alloy and d) EGS.

The variation of galvanic current between AZ91-6016 couple in sodium chloride solution was measured by ZRA technique is shown in Figure 8. The experiments were conducted twice with both anodic and cathodic configurations. Similarly, galvanic current was measured between different couples with and without cerium conversion coatings in a sodium chloride solution. The average galvanic current after stabilization with time was derived and is presented for different combinations of bimetallic couples in Figure 9. The ZRA galvanic currents (mA) were 0.02 for 6016-EGS, 0.38 for AZ91-EGS, 0.72 for AZ91-6016, 1.08 for AZ31-EGS, and 1.08 for AZ31-6016 couples. Overall, the galvanic current between AZ31 Mg coupled with 6016 /EGS was higher compared to AZ91 Mg coupled with 6016/EGS. It was also observed that the galvanic current between 6016-EGS couples is low and showed a corresponding low corrosion rate during salt spray testing. The galvanic current between the AZ91-EGS couple was low compared to the other bimetallic couple combinations and hence showed less galvanic corrosion after salt spray testing (Figure 3). However, AZ31-6016 and AZ31-EGS bimetallic combinations had the highest galvanic current; therefore, AZ31 bimetallic couples combinations showed higher corrosion compared to AZ91 couple combinations (Figure 4 and Figure 5). It is shown in Figure 9 that the bare panels in sodium chloride solution showed higher galvanic current compared to the cerium coated panels. As mentioned above galvanic current not only assisted the cerium deposition mechanism, it is also contribute to enhance the corrosion resistance of bimetallic couples.

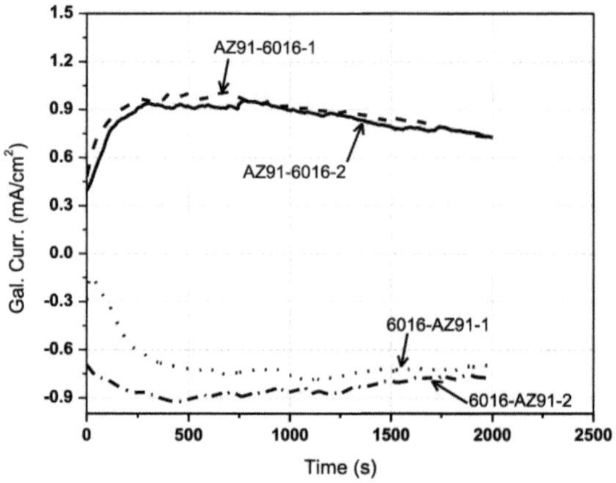


Figure 8. Galvanic current with time between magnesium alloy and aluminum alloy couple in sodium chloride solution.

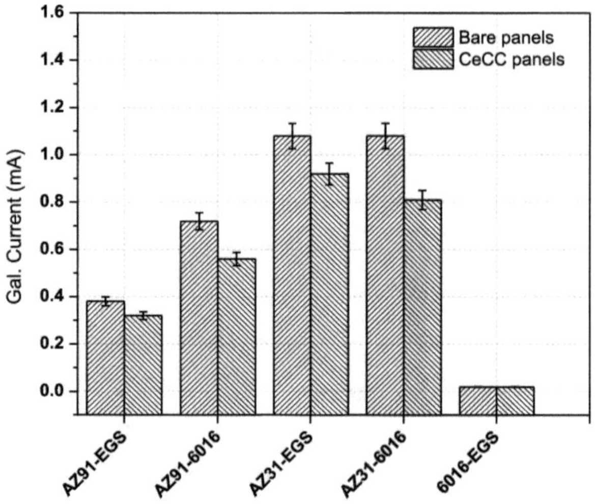


Figure 9. Galvanic current between different couples with and without CeCCs on individual panels in prohesion solution.

## Conclusions

Cerium conversion coatings successfully deposited on bimetallic couples without any contaminating bath. The influence of galvanic current on cerium conversion coatings on magnesium alloys coupled with aluminum and EGS panels has been investigated using ZRA technique in prohesion solution. The ZRA galvanic currents (mA) were 0.02 for 6016-EGS, 0.38 for AZ91-EGS, 0.72 for AZ91-6016, 1.08 for AZ31-EGS, and 1.08 for AZ31-6016 couples. The AZ31-6016/EGS couple combinations had the highest galvanic current compared to the AZ91-6016/EGS couples. Galvanic current significantly influence the conversion coatings on bimetallic couples; moreover, cerium coated bimetallic couples showed better salt spray performance compared to bare couples. Galvanic current significantly changed the coating thickness on coupled Al, and EGS panels compared to uncoupled panels. Cerium coated panels showed less galvanic current compared to the bare couples in prohesion solution.

## Acknowledgement

The authors would like to thank the USAMP team: Adam Stals, Jim Quinn, Kevin Cunningham and Harry Kuo. We also appreciate Elizabeth Kulp, Carlos Castano and Austin Gerlt for assisting in various aspects.

This material is based upon work supported by the Department of Energy National Energy Technology Laboratory under Award Numbers DE-FC26-02OR22910 and DE-EE0003583. This report was prepared as an account of work sponsored by an agency of the United States Government. Neither the United States Government nor any agency thereof, nor any of their employees, makes any warranty, express or implied, or assumes any legal liability or responsibility for the accuracy, completeness, or usefulness of any information, apparatus, product, or process disclosed, or represents that its use would not infringe privately owned rights. Reference herein to any specific commercial product, process, or service by trade name, trademark, manufacturer, or otherwise does not necessarily constitute or imply its endorsement, recommendation, or favoring by the United States Government or any agency thereof. The views and opinions of authors expressed herein do not necessarily state or reflect those of the United States Government or any agency thereof.

## References

1. M. K. Kulekci, "Magnesium and its alloys applications in automotive industry," *International Journal of Advanced Manufacturing Technology,* 39 (2008), 851–865.

2. A. A. Luo, "Magnesium: current and potential automotive applications," *JOM-Journal of Mineral Metals and Material Society,* 54 (2) (2002), 42-48.

3. G. Cole, "Issues that influence magnesium's use in the automotive industry," *Materials Science Forum,* 419-422 (2003), 43-50.

4. G. Gao, and M. Ricketts, "Evaluation of protective coatings on magnesium for phosphate process compatibility and galvanic corrosion prevention," *SAE Technical Paper 010081, SAE International,* Warrendale, PA, 2002.

5. J. E. Gray, and B. Luan, "Protective coatings on magnesium and its alloys- a critical review," *Journal of Alloys and Compounds,* 336 (2002), 88-113.

6. M. Costa, and C.B. Klein, "Toxcity and carcinogenicity of chromium compounds in humans," *Critical Reviews in Toxicology,* 36 (2) (2006), 155-63.

7. A. L. Rudd, C. B. Breslin, and F. Mansfeld, "The corrosion protection afforded by rare earth conversion coatings applied to magnesium," *Corrosion Science,* 42 (2000), 275–288.

8. H. Bhatt, A. Manavbasi, and D. Rosenquist, "Trivalent chromium for enchanced corrosion protection on aluminum surfaces," *Metal Finishing,* 107 (6) (2009), 39-47.

9. E. Mazario, L. Cabrera, and P. Herrasti, "Corrosion inhibition of aluminum by deposition of polypyrrole and silane compounds," *Journal of Advances in Engineering Science,* 1 (2007), 37-44.

10. D. K. Heller, W. G. Fahrenholtz, and M. J. O'Keefe, "The effect of post-treatment time and temperature on cerium-based conversion coatings on Al 2024-T3," *Corrosion Science,* 52(2) (2010), 360-368.

11. W. Pinc, S. Geng, M. J. O'Keefe, W. G. Fahrenholtz, and T. O'Keefe, "Effect of acid and alkaline based surface preparations on spray deposited cerium-based conversion coatings on Al 2024-T3," *Applied Surface Science,* 255 (2009), 4061-4065.

12. W. Pinc, S. Maddela, M. J. O'Keefe, and W. G. Fahrenholtz, "Formation of subsurface crevices in aluminum alloy 2024-T3 during deposition of cerium-based conversion coatings," *Surface and Coating Technology,* 204 (2010), 4095-4100.

13. R. F. Zheng, and C. H. Liang, "Conversion coating treatment for AZ91D magnesium alloys by a permanganate-REMS bath," *Materials and Corrosion,* 58 (2007), 193-197.

14. S. Maddela, M. J. O'Keefe, Y. M. Wang, and H. H. Kuo, "Influence of surface pretreatment on cerium-based conversion coating on Mg AZ91D alloy," *Corrosion,* 66(11) (2010), 115008/1-8.

15. K. Brunelli, M. Dabala, I. Calliari, and M. Magrini, "Effect of HCl treatment on corrosion resistance of cerium-based conversion coatings on magnesium and magnesium alloys," *Corrosion Science,* 47 (2005), 989-1000.

16. Uhlig's Corrosion Handbook, "Galvanic corrosion," in: R. Winston Revie (Eds.), Wiley International Publication, New York, 2nd Edn., (Electrochemical Society Series), 137-165.

17. G. Song, B. Johannesson, S. Hapugoda, and D. St. John, "Galvanic corrosion of magnesium alloy AZ91D in contact with an aluminum alloy, steel and zinc," *Corrosion Science,* 46 (2004), 955-977.

18. W. Pinc, P. Yu, M. J. O'Keefe, and W. G. Fahrenholtz, "Effect of gelatin additions on the corrosion resistance of cerium-based conversion coatings spray deposited on Al 2024-T3," *Surface and Coating Technology,* 203 (2009), 3533-3540.

# EFFECT OF $Sn^{4+}$ ADDITIVES ON THE MICROSTRUCTURE AND CORROSION RESISTANCE OF ANODIC COATING FORMED ON AZ31 MAGNESIUM ALLOY IN ALKALINE SOLUTION

S. A. Salman[1, 3], K. Kuroda[1], N. Saito[2], M. Okido[1]
[1]Graduate School of Engineering, Nagoya University, Furo-cho, Chikusa-ku, Nagoya, 464-8603 Japan
[2]EcoTopia Science Institute, Nagoya University, Furo–cho, Chikusa, Nagoya 464–8603, Japan
[3]Graduate School of Engineering, Al-Azhar University, Nasr City, Cairo 11371, Egypt

Keywords: Mg alloy, Anodizing, Sodium Stannate Trihydrate, Alkaline solution.

## Abstract

Magnesium is the lightest structural metal with high specific strength and good mechanical properties. However, poor corrosion resistance limits its widespread use in many applications. Magnesium is usually treated with chromate conversion coatings. However, due to changing environmental regulations and pollution prevention requirements, a significant push exists to find new, alternative for poisonous $Cr^{6+}$. Therefore, we aim to improve corrosion resistance of anodic coatings on AZ31 alloys using low cost non-chromate electrolyte. Anodizing was carried out in alkaline solutions with tin additives. The effect of tin additives on the coating film was characterized by SEM and XRD. The corrosion resistance was evaluated using anodic and cathodic polarizations and electrochemical impedance spectroscopy (EIS). Corrosion resistance property was improved with tin additives and the best anti-corrosion property was obtained with addition of 0.03 M $Na_2SnO_3.3H_2O$ to anodizing solution.

## Introduction

Magnesium is the 8th most abundant element on the earth [1]. It is the lightest metal among all structural metal materials in the practical use with high strength-to-weight ratio. It has also attractive combinations of properties such as high thermal conductivity, dimensional stability, good electromagnetic shielding characteristics, high damping capacity and good machining characteristics. These desirable properties make it an attractive target and excellent choice for a number of applications especially when weight reducing is required. Unfortunately, magnesium has poor corrosion resistance, especially in chloride and all acids except hydrofluoric or chromic acid.
Surface treatment was performed in order to achieve a good corrosion resistance such as electrochemical plating [2], chemical conversion [3], anodizing [4, 5], physical vapor deposition [6], and electroless nickel plating [7].

Magnesium and its alloys are usually treated with chromate conversion coatings, however, hexavalent chromium compounds have been recognized as carcinogenic. Therefore, developing of non-chromate coatings that do not contain hexavalent chromium attracts much attention today [8]. In our previous works, we investigated the anodizing of Mg alloys in NaOH solution at various applied potentials. The film formed at 3 V has a dark gray color with rough surface and it mainly consisted of magnesium hydroxide. Moreover, the film was thicker than all those films formed at higher potential. On the other hand, the anodic film obtained at 100 V has a smooth surface with a white color. The main phase of the coating film was magnesium oxide. The anodic film obtained at 100 V has the lowest hydrogen evolution rate, and it has the best corrosion resistance [9].

Potassium stannate was used for the conversion treatment of magnesium alloy [10]. Moreover, it was reported that addition of tin to conversion coating solution produces a stable film with high corrosion resistance. The conversion coating consisted of hemispherical particles, which were porous and contained nanograins dispersed within the amorphous matrix. The film mainly consisted of tin, magnesium, oxygen and phosphorus species [11, 12]. In the present research, we will investigate the effect of $Na_2SnO_3.3H_2O$ additive to 1 M NaOH alkaline aqueous solution on the microstructure and anti-corrosion property of the anodic film formed on AZ31 Mg alloy at applied potential 100 V. The anti-corrosion behaviors were evaluated using the anodic polarization curves and the electrochemical impedance spectroscopy (EIS). Furthermore, the surface morphologies and phase structure were detected using SEM and XRD.

## Experimental

### Specimens

Commercially available AZ31 Mg alloys were used as the substrate. The chemical composition of the alloy is listed in Table I. The surface of the alloy was polished up to #2000 emery paper followed by 0.05 μm alumina powders. The specimens were carefully cleaned with water, rinsed with acetone and dried under air. All of the experiment specimens were mounted using polytetrafluoroethylene (PTFE) resin tape, leaving 1 $cm^2$ surface area.

**Table I** Chemical composition of AZ31 Mg alloy (mass%)

| Al | Zn | Mn | Si | Cu | Ni | Fe | Mg |
|---|---|---|---|---|---|---|---|
| 3.0 | 1.0 | 0.43 | 0.01 | <0.01 | 0.001 | <0.003 | Bal. |

### Anodizing

Anodizing was performed at room temperature for 0.5 – 600 s in the aqueous electrolyte of 1 M NaOH using P.R. series power supply (6 A, 500 V, manufactured in Japan) to provide various ranges of constant voltage to the anodizing cell. The applied potential used in this study was adjusted to 100 V. AZ31 magnesium alloy is connected to a positive terminal of a power supply and platinum counter electrode is connected to a negative terminal. Ag/AgCl sat. KCl reference electrode was connected to the cell in order to watch the actual voltage near the surface of the specimen and keep at constant value throughout the anodizing process. After the treatment, the specimens were carefully rinsed using distilled water and dried under air before analysis.

Morphology and Structure of Anodic Film

The morphology and microstructure of the anodic films were observed with a Hitachi S-800 scanning electron microscope (SEM), optical microscopy and digital camera. The phase structure was identified with X-ray diffraction (XRD).

Corrosion Measurements

The potentiodynamic polarization tests were carried out using a Solartron 1285 potentiostat from Solartron Analytical, Farnborough, United Kingdom. The measurements were controlled by Scribner Associates Corrware electrochemical experiment software. The anodic and cathodic polarization curves were measured in 0.1 M NaCl and 0.1 M $Na_2SO_4$ solution at 298 K with a scanning rate of 1 $mVs^{-1}$.

Electrochemical impedance spectroscopy of the immersed samples was measured in 3.5 mass % NaCl solution in order to accelerate the corrosion rate of the anodized specimens. The impedance test was performed at 298 K using a Solarton 1287 electrochemical interface and a Solarton 1260 frequency response analyzer with a frequency range from 100 kHz to 0.01 Hz. The measurements were controlled by Scribner Associates Z plot electrochemical experiment software.

## Results and Discussion

Figure 1 shows the change in the current density as a function of time during the anodizing treatment of AZ31 magnesium alloy in 1 M NaOH with various concentrations of $Na_2SnO_3.3H_2O$ additive.

Addition of 0.01 M $Na_2SnO_3.3H_2O$ to 1 M NaOH solution did not affect the anodizing treatment, and it showed the same behaviours as anodizing in 1 M NaOH without Sn additive.
The current density flowed greatly first, the spark was generated, and the current became small afterwards. Moreover, sparking time was longer when 0.01 M $Na_2SnO_3.3H_2O$ was added. The coating film after treatment has a white color similar to that formed without $Na_2SnO_3.3H_2O$ additive. It has been reported that oxygen evolution accompanies with the intense sparking during magnesium anodizing [13]. Magnesium oxide can be formed according to the following reactions.

$$2Mg + O_2 \rightarrow 2MgO \quad (1)$$

$$Mg(OH)_2 \rightarrow MgO + H_2O \quad (2)$$

Addition of 0.02 M $Na_2SnO_3.3H_2O$ made the current continue to flow from beginning to the end of the anodizing treatment with intensive spark and oxygen evolution. The coating film contains a cream color precipitates on the surface. This precipitate was also increased with the increase in $Na_2SnO_3.3H_2O$ concentration in the anodizing solution.
Figure 2 shows the SEM images of the surface after anodizing treatment. The coating film formed in 1 M NaOH solution has a porous structure with small holes. Addition of 0.01 M $Na_2SnO_3.3H_2O$ did not affect the surface structure of the anodic film. However, some parts of the film were destroyed due to the dielectric breakdown of the anodic films with continuous spark

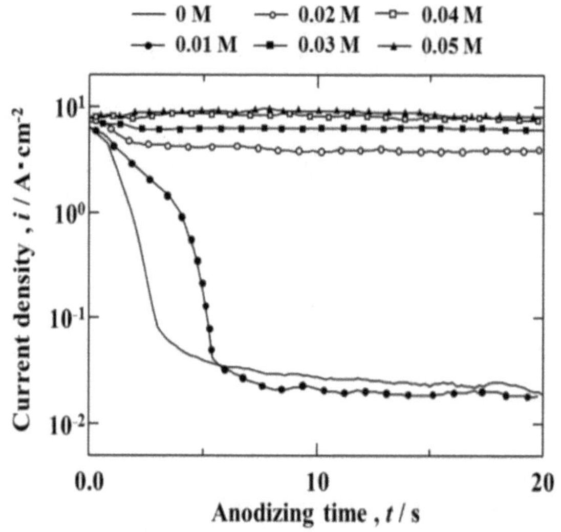


**Figure 1** Change in current density with time during anodizing of AZ31 at 100 V for 20 s in 1 M NaOH with various concentrations of $Na_2SnO_3.3H_2O$ at 298 K.

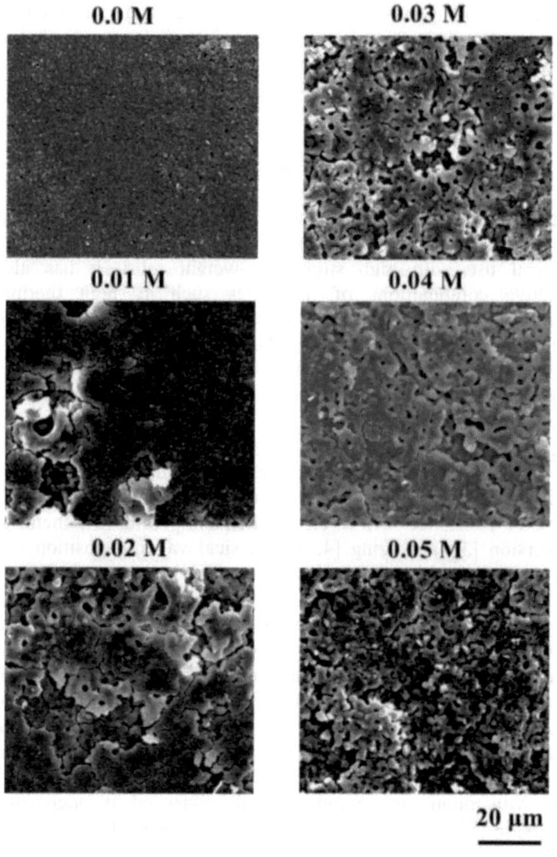


**Figure 2** Surface SEM images of AZ31 anodized at 100 V for 20 s in 1 M NaOH with various concentrations of $Na_2SnO_3.3H_2O$ at 298 K.

for long time. The porosity of the anodic films increased with the increase in $Na_2SnO_3.3H_2O$ concentration to 0.02 M. Addition of 0.03 – 0.04M $Na_2SnO_3.3H_2O$ produced a porous anodic film covers the whole surface and no failure was observed. When 0.05 M $Na_2SnO_3.3H_2O$ was added, some non-adhered powder was observed on the surface due to the increasing of $Na_2SnO_3.3H_2O$ concentration in the anodic solution.

The cross-section of the anodic films was shown in Figure 3. The specimen anodized in 1 M NaOH shows a compact film with thickness of 1.6 µm. Addition of 0.01 $Na_2SnO_3.3H_2O$ leads to a little increase in the film thickness. The anodic film increased to 2.85 µm, 3.55 µm, 5.7 µm, and 4.25 µm with increasing of $Na_2SnO_3.3H_2O$ concentration to 0.02 M, 0.03 M, 0.04 M, and 0.05 M respectively. Moreover, these anodic films show rough surface with observed pores in the cross-section, which suggested that the pores on the surface extend down to the whole films. It is thought that this is an influence where the spark had been always generated throughout the anodizing treatment.

XRD analysis of the anodic films formed in 1 M NaOH solution and 1 M NaOH + 0.01 M $Na_2SnO_3.3H_2O$ solution show two main peaks which have been identified as Mg and MgO as shown in Figure 4. After anodizing in 1 M NaOH + 0.02 M $Na_2SnO_3.3H_2O$ solution, new peaks were detected. These new peaks were identified as $Na_2SnO_3$. The intensity of these peaks increased and the peak widths became narrower with increasing the concentration of $Na_2SnO_3.3H_2O$ from 0.02 to 0.05 M. This indicates that the $Na_2SnO_3$ formed in the anodic films was crystallized into a perfect crystal structure with increasing of $Na_2SnO_3.3H_2O$ concentration in the anodizing solution. $Mg_2SnO_4$ peaks were detected with addition of 0.03 to 0.05 M $Na_2SnO_3.3H_2O$ to the solution. The intensity of these peaks

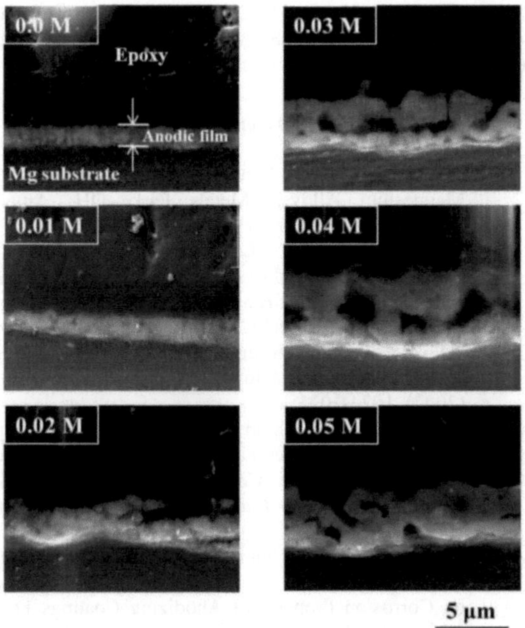


**Figure 3** Cross-section SEM images of AZ31 anodized at 100 V for 20 s in 1 M NaOH with various concentrations of $Na_2SnO_3.3H_2O$ at 298 K.

increases with a nearly linear dependence on increasing of $Na_2SnO_3.3H_2O$ concentration in the solution.

In order to evaluate and compare the anti-corrosion properties of anodized specimens, anodic and cathodic polarization curves were measured in 0.1 M NaCl and 0.1 M $Na_2SO_4$ solution. Generally, the anodic polarization represents the dissolution of magnesium, the cathodic polarization curves are represents the cathodic hydrogen evolution through water reduction. The anodic polarization curves in Figure 5 were revealed clearly that the pitting potentials were moved toward more positive values with

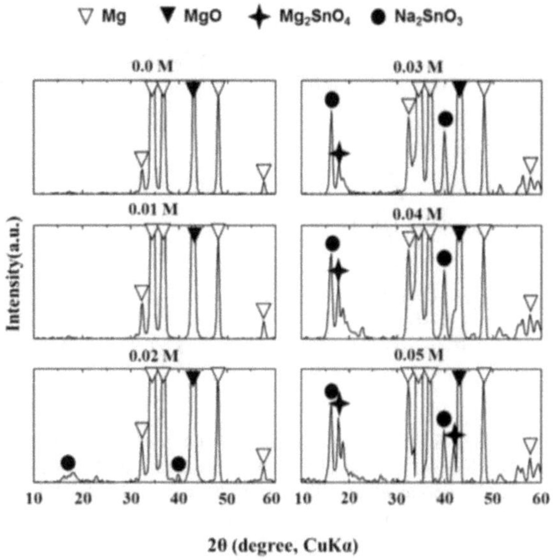


**Figure 4** XRD results of of AZ31 anodized at 100 V for 20 s in 1 M NaOH with various concentrations of $Na_2SnO_3.3H_2O$ at 298 K.

anodizing in 1 M NaOH solution without $Na_2SnO_3.3H_2O$ additives. Therefore, corrosion resistance has improved with the

anodizing treatment in 1 M NaOH with -1.36 V pitting potential. Addition of 0.01 M $Na_2SnO_3.3H_2O$ has moved the pitting potential values to more negative potential of -1.42. With increasing the concentration of $Na_2SnO_3.3H_2O$, the pitting potential moved to more positive values of -1.22, -1.2, -1.2 and -1.19 V at 0.02, 0.03, 0.05, and 0.04 M $Na_2SnO_3.3H_2O$ respectively. The cathodic polarization curves were measured in order to get more information about the corrosion behaviors of the anodic film. The cathodic potential of all anodized specimens has negatively potentials in comparison to non-treated specimens, which indicate the deceasing of hydrogen evolution by the anodic passive films which mainly formed from magnesium oxide and $Mg_2SnO_4$. The hydrogen evolution overpotential of all treated specimens with tin additives was better than those treated in 1 M NaOH solution and non-treated specimens.

Electrochemical impedance spectroscopy measurement (EIS) was performed to ensure the effect of $Na_2SnO_3.3H_2O$ additives on the anti-corrosion property. Corrosion rate determination is associated with the charge transfer resistance ($R_{ct}$) by using EIS technique. The charge transfer resistance is equal to the diameter of the semicircle in the complex plane graph (nyquist diagram).

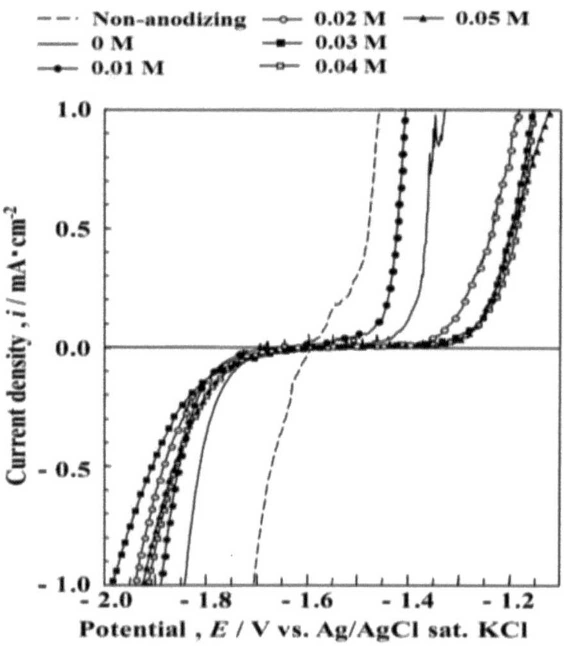


**Figure 5** Polarization curves in 0.1 mass% NaCl + 0.1 M $Na_2SO_4$ solution at 298 K with scan rate of 1 mVs-1 for AZ31 anodized at 100 V for 20 s in 1 M NaOH with various concentrations of $Na_2SnO_3.3H_2O$ at 298 K.

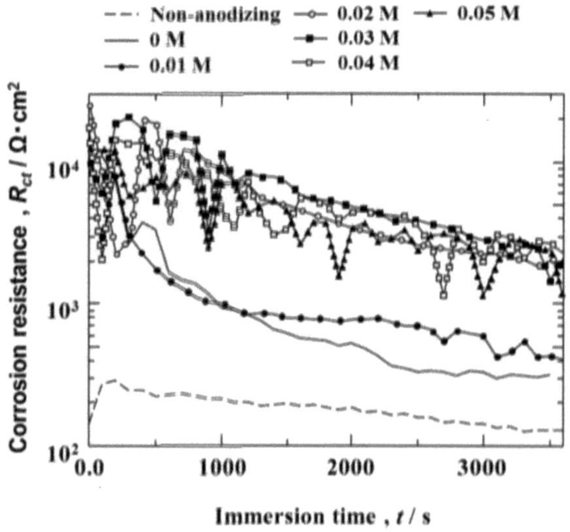


**Figure 6** Corrosion resistance in 3.5 mass% NaCl solution for AZ31 before and after anodizing at 100 V in 1 M NaOH with various concentrations of $Na_2SnO_3.3H_2O$ for 20 at 298 K.

Figure 6 shows the time dependence of the charge transfer resistance, which is corresponding to corrosion rate. The charge transfer resistance increases with anodizing treatment. The non-treated specimen and those anodized at in 1 M NaOH with and without 0.01 M $Na_2SnO_3.3H_2O$ additive have the lowest $R_{ct}$ value. On the other hand, the films formed in 1 M NaOH + (0.02 M-0.05 M) $Na_2SnO_3.3H_2O$ hav higher $R_{ct}$ values. The highest value was observed with addition of 0.03 M $Na_2SnO_3.3H_2O$. This result is in agreement with anodic polarization measurements. The reciprocal values of $R_{ct}$ for the anodic films formed at high $Na_2SnO_3.3H_2O$ concentration, it may be results from shifting of the pH to more alkaline side due to the dissociation of the tin contained films.

## Conclusions

Anodizing of AZ31 Mg alloy was performed at 100 V constant potential in 1 M NaOH solution with various $Na_2SnO_3.3H_2O$ concentrations. The Thickness of the anodic film increased notably with tin additives to alkaline solution. The anodic film increased from 1.6 μm to, 5.7 μm, with increasing of $Na_2SnO_3.3H_2O$ concentration to 0.04 M. Some tin compounds such as $Mg_2SnO_4$ were included in the anodic films. The anti-corrosion property was improved with addition of 0.03 M $Na_2SnO_3.3H_2O$ to anodizing solution due to the increasing of film thickness several orders of magnitude and formation of new tin compounds during anodizing process.

## Acknowledgment


The authors gratefully acknowledge the financial support of the Ministry of the Education, Culture, Sports, Science and Technology, Japan. We wish also to acknowledge Professor Ryoichi Ichino of Ecotopia Science Institute, Nagoya University who gave me a lot of valuable discussion and advices during this research. Special thanks go to Ryuzo Mori and Dai Yamamoto of the Department of Material Science and Engineering, Nagoya University for their assistance during the experiments.


## References


1. H. Baker et-al., eds., Metals Handbook: Magnesium and Magnesium Alloys, (Metals Park, OH: American Society for Metals, 1999), 3-84.
2. J.E. Gray and B. Luan, Protective coatings on magnesium and its alloys – a critical review, *Journal of Alloys and Compounds* 336, 88-113 (2002).
3. S. A. Salman, Ryoichi Ichino, and Masazumi Okido, "Development of Cerium-based Conversion Coating on ZA31 Magnesium Alloy," *Chemistry Letters*, 36 (8) (2007), 1024-1025.
4. S. A. Salman, R. Ichino and M. Okido, "Production of Alumina-Rich Surface Film on AZ31 Magnesium Alloy by Anodizing with Co-Precipitation of Nano-Sized Alumina," *Materials Transactions*, 49 (5) (2008), 1038-1041.
5. S. A. Salman, R. Ichino and M. Okido, "Influence of Calcium Hydroxide and Anodic Solution Temperature on Corrosion Property of Anodizing Coatings Formed on AZ31 Mg Alloys," *Surface Engineering*, 24 (3) (2008), 242-245.
6. B.L. Yu and J.Y. Uan, "Sacrificial Mg film anode for cathodic protection of die cast Mg-9wt. %Al–1wt.%Zn

alloy in NaCl aqueous solution," *Scripta materialia*, 54 (2006), 1253-1257.

7. W.J. Cheong, B.L. Luan and D.W. Shoesmith, "Protective coating on Mg AZ91D alloy-the effect of electroless nickel (EN) bath stabilizers on corrosion behavior of Ni–P deposit," *Corrosion Science*, 49, (2007), 1777-1798.
8. H. Umehara, M. Takaya, & Y. Kojima, "An Investigation of the Structure and Corrosion Resistance of Permanganate Conversion Coatings on AZ91D Magnesium Alloy," *Materials Transactions*, 42 (8) (2001), 1691-1699.
9. S. A. Salman, R. Mori, R. Ichino and M. Okido, "Effect of Anodizing Potential on the Surface Morphology and Corrosion Property of AZ31 Magnesium Alloy," *Materials Transactions*, 51 (6) (2010), 1109- 1113.
10. A. R. Shashikala, R. Umarani, S. M. Mayanna and A. K. Sharma, "Chemical Conversion Coatings on Magnesium Alloys - A Comparative Study," *International Journal of Electrochemcal Science*, (9) (2008), 993-1004.
11. C.S. Lin, H.C. Lin, K.M. Lin, W.C. Lai, "Formation and properties of stannate conversion coatings on AZ61 magnesium alloys," *Corrosion Science*, 48 (2006), 93-109.
12. Lihui Yang, Milin Zhang, Junqing Li, Xiang Yu, Zhongyi Niu, "Stannate conversion coatings on Mg–8Li alloy," *Journal of Alloys and Compounds*, 471 (2009), 197-200.
13. F. A. Bonilla, A. Berkani, Y. Liu, P. Skeldon, G. E. Thompson, H. Habazaki, K. Shimizu, C. John, and K. Stevens, "Formation of Anodic Films on Magnesium Alloys in an Alkaline Phosphate Electrolyte," *Journal of the Electrochemical Society*, 149 (1) (2002), 4-13.

**Magnesium Technology 2012**
*Edited by: Suveen N. Mathaudhu, Wim H. Sillekens, Neale R. Neelameggham, and Norbert Hort*
***TMS (The Minerals, Metals & Materials Society), 2012***

# EFFECT OF THICKNESS ON THE MORPHOLOGY AND CORROSION BEHAVIOR OF CERIUM-BASED CONVERSION COATINGS ON AZ31B MAGNESIUM ALLOY

Carlos E. Castano[1], Surender Maddela[1], Matthew J. O'Keefe[1], Yar-Ming Wang[2]
[1] Materials Research Center, Department of Materials Science and Engineering, Missouri University of Science and Technology; 305 McNutt Hall 1400N Bishop; Rolla, MO, 65409, USA
[2]Chemical Sciences and Materials Systems Laboratory, GM R&D Center; MC: 480-106-224, 30500 Mound Rd; Warren, MI 48090-9055, USA

Keywords: Cerium-based Conversion Coatings, Corrosion of Magnesium Alloys, Transmission Electron Microscopy, Coating Thickness.

## Abstract

Cerium-based conversion coatings (CeCCs) were deposited onto AZ31B magnesium alloy substrates using a spontaneous reaction of $CeCl_3$, $H_2O_2$ and gelatin in a water-based solution. The coating thickness was adjusted by controlling the immersion time in the deposition solution. Prior to deposition, the AZ31B substrates were treated using an acid pickling in nitric acid and then an alkaline cleaning in sodium metasilicate pentahydrate. After deposition, the coated samples were immersed in a phosphate bath that converted cerium oxide/hydroxide into cerium phosphate. Electrochemical impedance spectroscopy, potentiodynamic polarization and neutral salt spray testing studies indicated that ~100 nm thick CeCC had better corrosion performance than ~400 nm coatings. Characterization of the CeCCs by transmission electron microscopy (TEM) revealed a three layer structure with different compositions.

## Introduction

Recently there has been increased interest in using magnesium-based alloys in a number of applications. The main motivation for this interest is that magnesium alloys offer a high potential for use as a lightweight structural material for automotive and aerospace components [1-4]. Comparing the densities of Mg (1.74 g/cm$^3$), Al (2.70 g/cm$^3$) and Fe (7.84 g/cm$^3$), magnesium is 35% lighter than aluminum and 77% lighter than iron [1]. However, the use of magnesium and its alloys in the automotive industry is limited due to poor corrosion performance [1,2]. Magnesium AZ91 alloy is one of the most commonly used alloys in the automotive industry due to its relatively low cost, good yield strength, and acceptable corrosion resistance. However, the amount of Al (~ 9wt %) in AZ91 reduces ductility due to the formation of brittle second phases [5,6]. In contrast, Mg AZ31 alloy is one of the most promising alloys for automobile applications because it offers a good combination of strength and ductility [6]. Nevertheless, the corrosion resistance of AZ31 alloy is still a great concern. Corrosion resistance of magnesium alloys can be increased by the use of conversion coatings that not only provide a barrier between the substrate and its environment but also supply a suitable substrate for paint coatings. Rare earths conversion coatings, principally CeCCs, have demonstrated improvements on the corrosion resistance for magnesium alloys [7-10]. CeCCs are commonly considered to be environmentally friendly corrosion inhibitors on aluminum alloys. Nevertheless, the effectiveness of the CeCCs on the corrosion protection has been highly dependent on surface pretreatment, deposition parameters and post-treatment [2, 7-10]. Current progress at the Missouri University of Science and Technology has developed an adequate surface pretreatment for deposition of CeCCs on AZ91D magnesium alloy consisting of acid then alkaline cleaning that can be applied to other magnesium alloys [8]. In this study, an optimization was carried out to correlate thickness, microstructure, and deposition time of cerium conversion coatings with the corrosion performance of Mg AZ31B alloy.

## Experimental Procedure

Cerium conversion coatings were deposited onto Mg AZ31B alloy coupons of dimensions 100 mm x 50 mm x 2 mm. Prior to deposition, AZ31B alloy panels were mechanically polished using 180 grit abrasive SiC papers, then cleaned with isopropyl, rinsed with deionized water, and finally dried in air at room temperature. The surface of the cleaned samples was pretreated in 1 wt% nitric acid aqueous solution for 30 seconds followed by an alkaline cleaning in 5 wt% of $Na_2SiO_3{\cdot}5H_2O$ aqueous solution for 5 minutes at room temperature. The CeCC aqueous solution consisted of 4 wt% of $CeCl_3{\cdot}7H_2O$ (99.9 %, Alfa Aesar), 6.7 wt% of hydrogen peroxide (Fisher Chemical, 30 vol%), and 0.25 wt% of organic gelatin (RDH, Rousselot) in DI water. Mg AZ31B alloy coupons were immersed in the CeCC solution for 5 s, 60 s, 120 s, and 180 s to produce various layer thicknesses. Finally, coated samples were postreated for 5 min at 85 °C in a 2.5 wt% $NaH_2PO_4$ aqueous solution to convert the coating into $CePO_4$ [11]. Electrochemical measurements were carried out in 1.6 wt% NaCl solution using a flat cell of 1 cm$^2$ of exposed area (model K0235, Princeton Applied Research) with a saturated calomel electrode (SCE) and platinum mesh served as a reference and counter electrodes, respectively. Open circuit potential (OCP) was monitored for 1000 seconds before impedance and potentiodynamic polarization measurements. The cyclic potentiodynamic scans were conducted from -0.3 to +0.8 V and scanned back to -0.25 V with a scan rate of 1 mV/s with respect to OCP using an EG&G potentiostat model 273A (Princeton Applied Research). Corrware software was used to control the analytical equipment and CView and ZView were used for data fitting analysis. The corrosion performance of the coated panels was also evaluated by ASTM B117 neutral salt fog testing. Electrochemical impedance spectroscopy (EIS) was performed using a Schlumberger model SI1255 frequency response analyzer in combination with the above-mentioned potentiostat. Data were collected over a frequency range of 0.01 to 1x10$^5$ Hz with 10 mV AC with five points per decade. The TEM samples were prepared using the Helios NanoLab 600. The dual beam system is equipped with a focused ion beam (FIB) milling system that was used in conjunction with a micromanipulator to mount ~100 nm thick TEM specimens onto Cu grids for subsequent TEM characterization.

## Results and Discussion

A cross-sectional TEM image of a representative microstructure of an as-deposited CeCC on a AZ31B alloy substrate is shown in Figure 1. A three layer coating structure was identified. Initially, a transition layer shows a porous structure and the cerium conversion coating can be divided in two sublayers 1 and 2. From Figure 2 the structure of sublayer 1 revealed a nanocrystalline region of the CeCC and the thickness is approximately constant over the entire sample, whereas the structure of sublayer 2 is practically amorphous and the thickness varies.

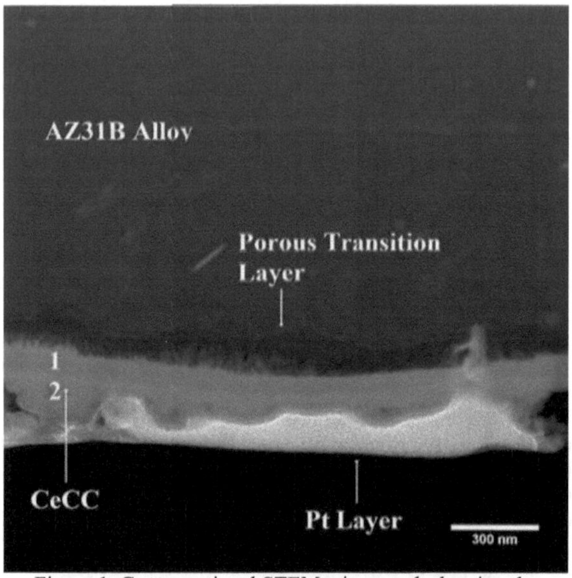


Figure 1. Cross-sectional STEM micrograph showing the microstructure of the 5s immersion sample.

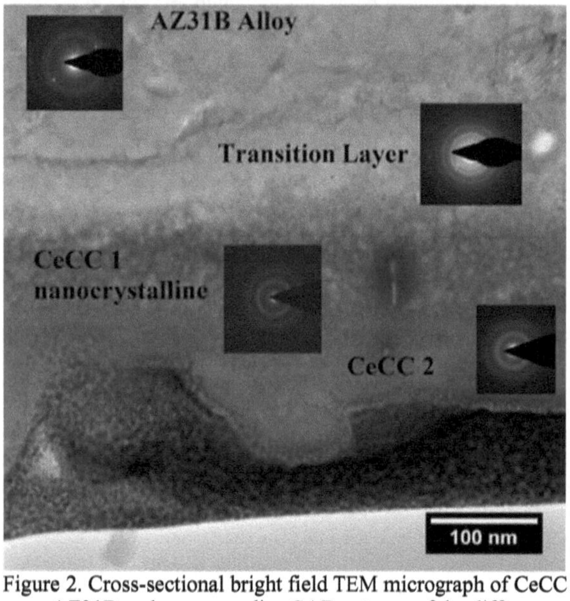


Figure 2. Cross-sectional bright field TEM micrograph of CeCC on AZ31B and corresponding SAD patterns of the different regions.

Chemical analysis by energy dispersive spectroscopy (EDS) revealed that the porous transition layer is composed of Mg, Al, O and C. In addition, EDS on the different sublayers of the CeCC revealed that the conversion coating consisted predominately of Ce, P, and O but the concentration of P throughout the entire thickness of sublayer 2 was higher than in sublayer 1. Based on TEM analysis, the variation of the average thickness of the transition layer, the CeCC coating (sublayer 1 + 2) and the total thickness (transition layer + CeCC coating) with different immersion times are presented in Figure 3. From the analysis, the thickness increased with immersion time except for the 180 s deposition.

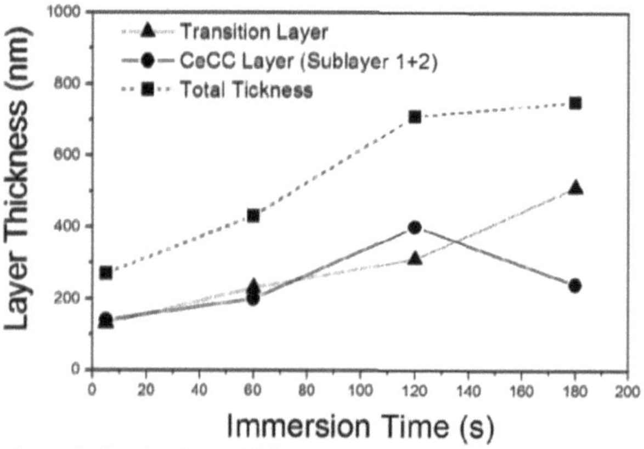


Figure 3. Coating layer thickness as a function of immersion time.

Selective area diffraction patterns from different regions of a representative coated panel are shown in Figure 4.

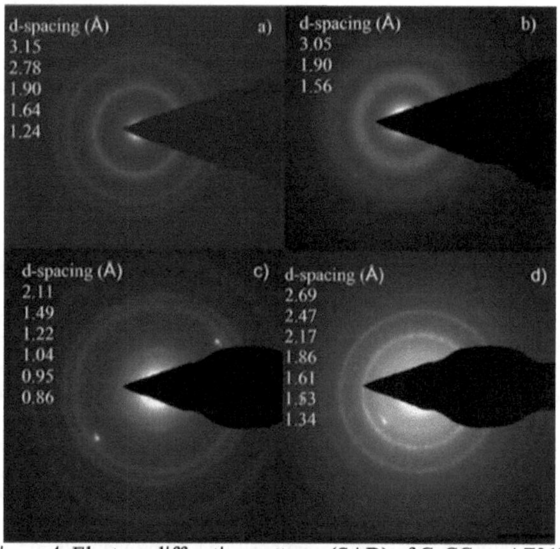


Figure 4. Electron diffraction patterns (SAD) of CeCC on AZ31B from different regions: a) sublayer 1, b) sublayer 2, c) transition layer, and d) AZ31B alloy.

Figures 4 a) and b) show the selected area diffraction patterns (SAD) of the cerium conversion coating. A nanocrystalline structure is suggested since the SAD pattern contained continuous rings. However, a few diffused halos on the pattern showed in

Figure 4 b) are consisted with a structure that is largely amorphous. From Figure 4 a) the rings of the SAD pattern correspond to $CeO_2$ (PDF 00-034-0394).

The electron diffraction pattern collected from the interfacial region would indicated a nanocrystalline structure, and the pattern can be indexed to MgO (PDF 00-045-0946), Figure 4 c.

Figure 4 d) shows the rings and spots of the electron diffraction micrograph, which is identified as hexagonal Mg (PDF 00-035-0821). Summary of measurements performed on the diffraction patterns is presented in Table I.

**TABLE I.** d-spacing (Å) calculated from SAD patterns of the different regions and .

| CeCC 1 | CeCC 2 | *$CeO_2$ PDF 00-034-0394 | Transition Layer | *MgO PDF 00-045-0946 | AZ31B Alloy | *Mg PDF 00-035-0821 |
|---|---|---|---|---|---|---|
| 3.15 | 3.05 | 3.12 | | 2.43 | 2.69 | 2.61 |
| 2.78 | | 2.71 | 2.11 | 2.11 | 2.47 | 2.45 |
| 1.90 | 1.90 | 1.91 | 1.49 | 1.49 | 2.17 | |
| 1.64 | | 1.63 | 1.22 | 1.22 | 1.86 | 1.90 |
| | 1.56 | 1.56 | 1.04 | 1.05 | 1.61 | 1.61 |
| 1.24 | | 1.24 | 0.95 | 0.97 | 1.53 | 1.47 |
| | | | 0.86 | 0.86 | 1.34 | 1.37 |

**Taken from International Centre for Diffraction Data, PDF 4+*

Optical images of the CeCCs obtained at different immersion times after 5 days of salt spray testing are compared to as coated samples in Figure 5. Corroded panels revealed that 5s of immersion in cerium solutions had better corrosion resistance compared to longer immersion times. The panel coated during an 180 s in immersion had numerous pits and salt tails after testing. Although all samples had visible corrosion product after 5 days in salt fog test, the 5s immersion sample exhibited only filiform corrosion while the 60 s, 120 s, and 180 s immersion samples had filiform and pitting corrosion.

Figure 5. Optical images of cerium conversion coatings on AZ31B Mg alloy at different immersion time before and after 5 days of salt spray testing.

Cyclic potentiodynamic curves of bare and coated AZ31B samples at different immersion times are presented in Figure 6. The corrosion parameters calculated from cyclic potentiodynamic polarization curves are presented in Table II.

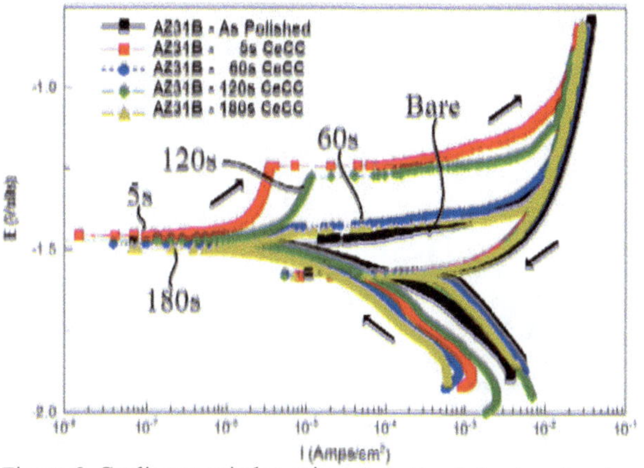


Figure 6. Cyclic potentiodynamic polarization behavior of cerium conversion coatings on AZ31B alloy at different deposition times

The corrosion potential ($E_{corr}$) of AZ31B panels with a CeCC is basically the same value as was obtained for bare samples (as polished). The reverse scans for all measured samples, Figure 6, have a more active potential than the free corrosion potential. This implies a high probability of localized corrosion of the AZ31B alloy even with a conversion coating [7]. However, the corrosion resistance increased for coated samples because the CeCC increases the pitting potential. In particular, an appreciable anodic passivation region (~220 mV) for AZ31B after 5s in immersion is evident in Figure 6. Although a limited passivation region is measured for 60 s and 180 s immersion panels, the pitting potential increased only ~40mV. As a result, it is expected that the specimen will undergo pitting attack once the free corrosion potential is reached. The corrosion current densities ($i_{corr}$) of CeCC samples are significantly reduced compared to the bare sample and the lowest $i_{corr}$ was obtained for the 5s immersion time panel. Calculated values of corrosion rate are directly proportional to the immersion time, which is also directly proportional to the coating thickness.

**TABLE II.** Corrosion parameters derived from the potentiodynamic polarization experiments.

| Sample AZ31B | βa mv/dec | βc mv/dec | $i_{corr}$ μA/cm² | $E_{corr}$ $mV_{SCE}$ | $E_{plt}$ mV | Corr. Rate mm/y |
|---|---|---|---|---|---|---|
| As Polished | 20 | -68 | 5.0 | -1490 | -1490 | 0.11 |
| 5S CeCC | 149 | -57 | 0.8 | -1460 | -1240 | 0.02 |
| 60S CeCC | 75 | -50 | 1.5 | -1480 | -1430 | 0.03 |
| 120S CeCC | 133 | -47 | 1.8 | -1480 | -1270 | 0.04 |
| 180S CeCC | 87 | -79 | 2.2 | -1490 | -1450 | 0.05 |

Impedance spectroscopy results from coated and uncoated samples are shown in Figure 7. The recorded spectra of the bare sample exhibited one capacitive loop and coated samples displayed two capacitive loops. This indicates that coated samples can be modeled by an electrical equivalent circuit that represents a coated alloy with a defect and the bare sample with a simple Randal's model, as shown Figure 8.

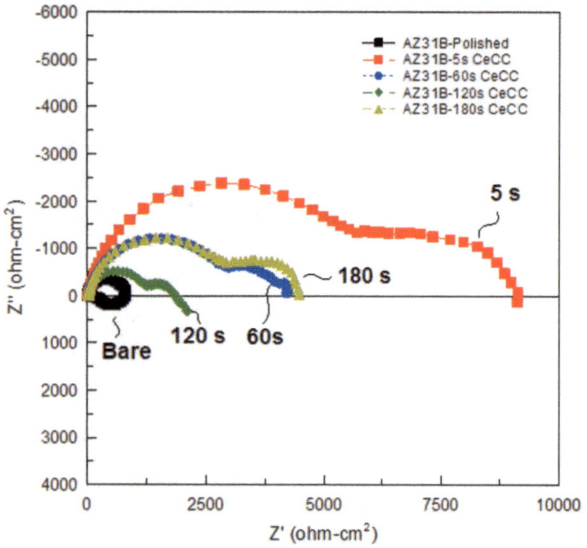


Figure 7. Impedance spectra of cerium conversion coatings on AZ31B alloy at different deposition times.

In figure 8, $R_s$ is the solution resistance, $R_p$ is the coating defect resistance, $R_{ct}$ is the polarization resistance or charge-transfer resistance at the metal interface. The symbols $Q_{ox}$ and $Q_{dl}$ are two constant phase elements used to represent the non-ideal coating capacitance and the double layer capacitance, where $n_{ox}$ and $n_{dl}$ are equal to 1 for an ideal capacitor and less than 1 for non-uniform surfaces. The inductance is labeled L and represents the formation of a surface layer [8].

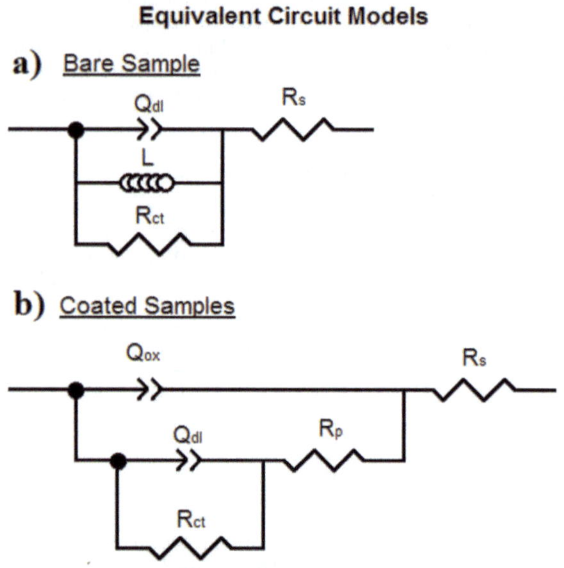


Figure 8. Equivalent circuit models used to fit impedance data: a) simple Randal's model with inductive loop for bare sample and b) two time constant model for cerium coated Mg AZ31B alloy.

The equivalent circuits have been utilized to fit the electrochemical data obtained from bare and cerium-based coated AZ31B panels and the results are summarized in Table III. The 5s CeCC sample had the maximum polarization resistance (5923 Ω cm²) and the highest defect or pore resistance (2921 Ω cm²). Additionally, comparing coated panels the 5s sample has the lowest $Q_{dl}$ value, indicating a lower amount of area that is corroding [9].

**TABLE III.** Corrosion parameters derived from the potentiodynamic polarization experiments.

| Sample AZ31B | As Polished | 5S CeCC | 60S CeCC | 120S CeCC | 120S CeCC |
|---|---|---|---|---|---|
| **Rs** $\Omega\cdot cm^2$ | 20 | 32 | 25 | 29 | 29 |
| **Qox** $\Omega^{-1}\cdot s^n\cdot cm^{-2}$ | | $9.12\times10^{-6}$ | $1.39\times10^{-5}$ | $1.33\times10^{-5}$ | $1.62\times10^{-5}$ |
| $n_{ox}$ | | 0.87 | 0.86 | 0.89 | 0.89 |
| **Qdl** $\Omega^{-1}\cdot s^n\cdot cm^{-2}$ | $1.13\times10^{-5}$ | $5.36\times10^{-4}$ | $1.10\times10^{-3}$ | $1.52\times10^{-3}$ | $9.11\times10^{-4}$ |
| $n_{ox}$ | 0.94 | 0.92 | 0.90 | 0.87 | 0.82 |
| **Rct** $\Omega\cdot cm^2$ | 799 | 2921 | 1159 | 641 | 1626 |
| **Rp** $\Omega\cdot cm^2$ | | 5923 | 3009 | 1232 | 2928 |
| **L** $\Omega^{-1}\cdot\sqrt{s}/cm^2$ | 873 | | | | |

In general the electrochemical results are consistent with the performance observed in salt spray testing. Both electrochemical and salt spray cabinet results indicated that ~100 nm thick cerium-based conversion coatings deposited by 5s immersion provide the best corrosion resistance on Mg AZ31B alloy substrates.

## Conclusions

The effect of immersion time on the morphology, thickness and corrosion behavior of cerium-based conversion coatings on magnesium AZ31B alloy was investigated. The as-deposited coatings system consisted of a three layer structure: (1) a nanocrystalline MgO transition layer in contact with the Mg substrate, (2) a nanocrystalline CeCC layer, and (3) an outer amorphous CeCC layer. The nanocrystalline CeCC layer thickness was a function of immersion time in the deposition solution and ranged from a minimum of ~100 nm (5s immersion) to a maximum of ~400 nm (120s immersion). In general, at all thicknesses the cerium-based conversion coatings improve the corrosion resistance of AZ31B magnesium in chloride media. However, the corrosion resistance of AZ31B magnesium alloy was best for thinner CeCCs. Based on the correlation of deposition time and corrosion behavior, layers of ~100 nm thick obtained by immersion for ~ 5s exhibited a ~5.5X improvement in corrosion resistance vs. bare, as-polished panels. The presence of fewer and smaller cracks in 5s immersion panels, and larger cracks in panels coated at the longer times, provide some explanation for the improved corrosion performance.


## Acknowledgements

The authors would like to extend their thanks to the USAMP team: Adam Stals, Jim Quinn, Kevin Cunningham and Harry Kuo. They also appreciate the technical assistance of Dr. Beth Kulp (FIB/SEM) and Dr. Kai Song (TEM) of the Materials Research Center at the Missouri University of Science and Technology.
This material is based upon work supported by the Department of Energy National Energy Technology Laboratory under Award

Numbers DE-FC26-02OR22910 and DE-EE0003583. This report was prepared as an account of work sponsored by an agency of the United States Government. Neither the United States Government nor any agency thereof, nor any of their employees, makes any warranty, express or implied, or assumes any legal liability or responsibility for the accuracy, completeness, or usefulness of any information, apparatus, product, or process disclosed, or represents that its use would not infringe privately owned rights. Reference herein to any specific commercial product, process, or service by trade name, trademark, manufacturer, or otherwise does not necessarily constitute or imply its endorsement, recommendation, or favoring by the United States Government or any agency thereof. The views and opinions of authors expressed herein do not necessarily state or reflect those of the United States Government or any agency thereof.

## References

1. Friedrich H. E. Mordike B.L. *Magnesium Technology: Metallurgy, Design data, Applications*. Springer – Verlag Berlin Heidelberg (2006).
2. M. K. Kulekci, "Magnesium and its Alloys Applications in Automotive Industry," *The International Journal of Advanced Manufacturing Technology*. 39 (9-10) 851–865 (2008).
3. G. L. Song, A. Atrens, "Corrosion Mechanisms of Magnesium Alloys," *Advanced Engineering Materials*, 1(1) 11-33 (1999).
4. G. L. Song, A. Atrens, "Understanding Magnesium Corrosion: a Framework for Improved Alloy Performance," *Advanced Engineering Materials*, 5(12) 837-858 (2003).
5. E. Ghali, W. Dietzel and K. U. Kainer, "General and Localized Corrosion of Magnesium Alloys: A Critical Review," *Journal of Materials Engineering and Performance,* 13, 7-23 (2004).
6. M. Marya et al., "Microstructural Effects of AZ31 Magnesium Alloy on its Tensile Deformation and Failure Behaviors," *Materials Science and Engineering A*, 418, 341-356 (2006).
7. M. Dabalà, K. Brunelli, E. Napolitani, M. Magrini. "Cerium-based Chemical Conversion Coatings on AZ63 Magnesium Alloy." *Surface and Coatings Technology*, 172 227-232 (2003).
8. S. Maddela, M. J. O'Keefe, Y. -M. Wang, H. -H. Kuo. "Influence of Surface Pretreatment on Coating Morphology and Corrosion Performance of Cerium-Based Conversion Coatings on AZ91 D Alloy." *Corrosion*, 66(11) 115006-1/115006-8 (2010).
9. K. Brunelli, M. Dabalà, I. Calliari, M. Magrini. "Effect of HCl Pre-treatment on Corrosion Resistance of Cerium-based Conversion Coatings on Magnesium and Magnesium alloys." *Corrosion Science*, 47(4) 989-1000 (2005).
10. A. L. Rudd, C. B. Breslin, F, Mansfeld. "The Corrosion Protection Afferded by Rare Earth Conversion Coatings Applied to Magnesium." *Corrosion Science,* 42 275-288 (2000).
11. D. K. Heller, W. G. Fahrenholtz, M. J. O'Keefe, "The Effect of Post-treatment Time and Temperature on Cerium-based Conversion Coatings on Al 2024-T3," *Corrosion Science*, 52(2) 360-368 (2010).

**Magnesium Technology 2012**
*Edited by: Suveen N. Mathaudhu, Wim H. Sillekens, Neale R. Neelameggham, and Norbert Hort*
*TMS (The Minerals, Metals & Materials Society), 2012*

# MECHANICAL AND CORROSION PROPERTIES OF AS-CAST AND EXTRUDED MG10GD ALLOY FOR BIOMEDICAL APPLICATION

Petra Maier[1], Sören Müller[2], Hajo Dieringa[3], Norbert Hort[3]
[1]University of Applied Scienes Stralsund, Zur Schwedenschanze 15, D-18435 Stralsund, Germany
[2]Extrusion Research and Development Center TU Berlin, Gustav-Meyer-Allee 25, D-13355 Berlin, Germany
[3]Helmholtz-Zentrum Geesthacht, Magnesium Innovation Center, Max-Planck-Straße 1, D-21502 Geesthacht, Germany

Keywords: Magnesium Rare-Earth Alloys, Extrusion, Fatigue, Corrosion, Biomedical Application

## Abstract

Due to the good specific strength and the moderate corrosion rate Mg-RE alloys have found growing interest for medical applications as implant material. In this study extruded Mg10Gd has been investigated, once by potentiodynamic method and again under cyclic load. Corrosion exposure is known to reduce the fatigue strength strongly. The data are compared to Mg10Gd as-cast condition and show an increase in fatigue properties and similar corrosion behavior. The form of corrosion and the influence of the temperature during voltammetric tests are discussed. A temperature increase from room to body temperature accelerates the corrosion processes. Due to stress peaks under load pitting corrosion is not preferred. The influence of the microstructure on the corrosion form is discussed. Casted Mg10Gd reveals large dendrites, which change into a globular microstructure during extrusion resulting in improved mechanical properties, mostly the elongation to fracture up to 20 %.

## Introduction

The fact that Magnesium corrodes with time is an advantage in biodegradable material applications. Magnesium has a standard reduction potential of -2.37 Volts and this makes it to be preferentially degraded when in contact with other metals in an aqueous environment [1]. For structural application, this material property is completely abhorred and efforts are made to control the corrosion processes by all means [2,3,4]. However, this is different with biodegradable material for human implantation purposes such as bones or stents [5,6,7,8]. For these applications, permanent retention of implanted materials in the body is considered unnecessary after the healing processes are completed. This leads to the idea of material degradation with time in the body. Different types of materials have been proposed for use in this area taking cognisance of their biocompatibility and corrosion properties. These include Magnesium and its alloys [9,10], pure iron with Manganese [11,12] and polymers especially the so called self expansible poly-l-lactic acid (PLLA) and poly-l-glycolic acid (PLGA) origin [13,14]. Results of some of the past investigations on biodegradable materials for bone or cardio-vascular application are summarised elsewhere [7,15]. Magnesium alloys show very promising results for both bone regeneration and cardio-vascular application purposes, however there are also challenges [9]. Candidate materials should be able to compromise between mechanical properties and material degradation during implantation period. Very fast degradation rate will lead to restenosis or thrombosis in the case of stents for vascular application [7,16] as a result of the early loss of the load bearing capacity of the stents. For bone support and regeneration, that could lead to excessive gas formation, irritation, fracture or complication on the unhealed bones and the surrounding tissues [9]. Since the behavior of the implant materials are not dependent on the materials alone but also on the implant environment, in vitro and in vivo experiments are essential for material screenings and clinical trials respectively [17]. Several factors play a role in determining the final material properties. These include material purity, alloying elements, processing route and post processing operations. Recent work shows that alloying pure Magnesium with Gadolinium improves its mechanical and corrosion properties [18]. However, little is known about the corrosion fatigue of these potential implant alloys. In order to explore this further, forming by extrusion in comparison to the as-cast condition has been adopted to control the microstructure and improve the mechanical properties under supervision of the corrosion behavior. Consequently their cyclic behavior and corrosion fatigue properties were investigated. This work summarises the preliminary results of the microstructure, corrosion response, fatigue and corrosion fatigue behaviors of the extruded alloy in comparison to the as-cast condition. Furthermore the temperature during the corrosion tests has been increased from room to body temperature (37 °C).

## Materials and Experimental

Fatigue and corrosion fatigue tests were carried out on extruded Mg10Gd produced at the Extrusion Research and Development Center TU Berlin. The materials (bars) were directly extruded at an overall temperature of 420 °C, an extrusion speed of 1 mm/s and a deformation degree φ of 1.6 ($\varphi = \ln A_1/A_0$). Before extrusion the material was casted by permanent mould direct chill casting at Helmholtz Centre Geesthacht. Detailed information on the casting process is given in [19]. The microstructure during extrusion has changed dramatically. The obtained optical micrographs from the as-cast permanent mould chill cast show mainly coarse and columnar structures (up to 500 μm), see figure 1. These columnar grains were characterized by dendrite arms according to the preferred orientation during solidification process.

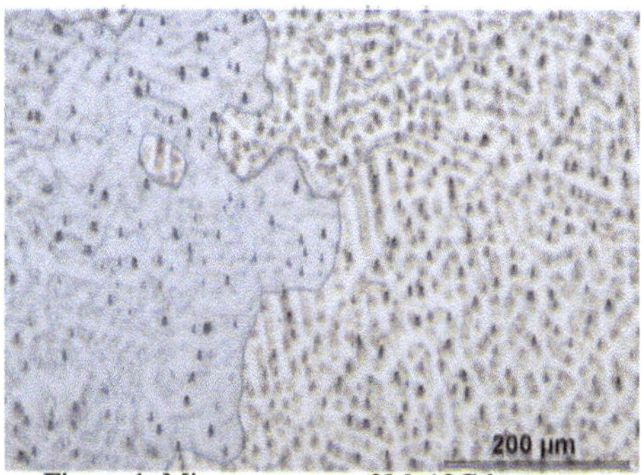

Figure 1. Microstructure of Mg10Gd, as-cast, permanent mould direct chill casting

Figure 2 presents grain refinement of the extruded material with an average grain size of 15 µm. The formations of the black lines, which will consist of aligned second phases, indicate the extrusion direction. The influence of this alignment is part of this study.

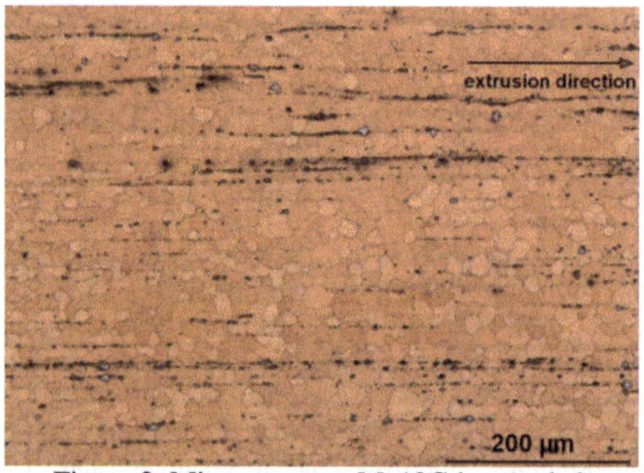


Figure 2. Microstructure Mg10Gd, extruded, extrusion temperature 420 °C, extrusion speed of 1 mm/s, deformation degree φ of 1.6

To evaluate the stress amplitude of the fatigue tests tensile and compression tests were carried out according to EN 10002-1:2001 standard. Tensile stress-strain curves of the extruded material are shown in figure 3 and compression stress-strain curves in figure 4, respectively. The mechanical properties of the extruded material in comparison to the as-cast condition [20] are summarized in table I. A Young's modulus of 42.9 GPa was additionally evaluated from the tensile tests, but not shown in table I. Due to the hexagonal crystal an anisotropic behavior in the elastic moduli (Young's and bulk modulus) occurs, which is a well known fact.

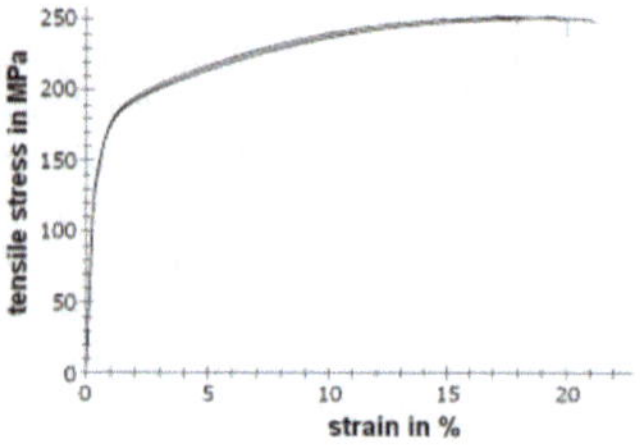


Figure 3. Tensile stress-strain curves of extruded Mg10Gd

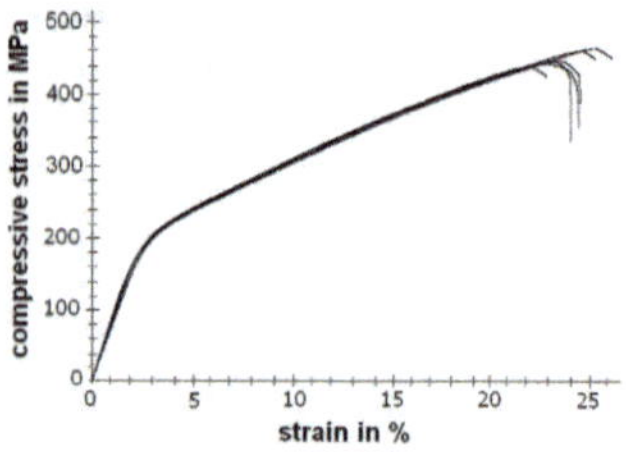


Figure 4. Compression stress-strain curves of extruded Mg10Gd

From table I a strong increase of mechanical properties of the extruded material can be seen. Especially, the elongation to fracture (A) improved from not even 2 % to almost 20 %. The tensile-compressive yield asymmetry in wrought magnesium alloys (different value for tensile yield stress (TYS) and compression yield stress (CYS)) is known for magnesium alloys. The compressive yield stresses in wrought magnesium alloys has been previously found to the 0.6-0.7 times those of the tensile [21]. Such an asymmetric yield behavior, resulting in a poor compressive yield stress, would prohibit the potential application for these wrought Magnesium alloys as implant material being mostly exposed to compression load. However, for the Mg10Gd alloy the ratio of CYS/TYS is above 1 in longitudinal direction, which means that alloying with rare earth elements shows higher compression strength. This behavior might change in other directions. The influence of this asymmetrical plastic deformation behavior on the fatigue properties of wrought magnesium alloys have already been reported in [22, 23], also showing higher compression yield stress values of magnesium alloys alloyed with rare earth elements: cerium, neodymium and lanthanum.

Table I. Mechanical properties of Mg10Gd in different conditions

| | TYS (MPa) | UTS (MPa) | A (%) | CYS (MPa) | ratio CYS/TYS |
|---|---|---|---|---|---|
| as-cast | 60 | 91 | 1.78 | 76 | 1.26 |
| extruded | 148 | 251 | 19.42 | 176 | 1.19 |

Round samples (diameter of 6 mm) were machined with surface roughness of 6.3 µm, according to ISO 1099:2006. Fatigue and corrosion fatigue tests were performed on a closed-loop servo hydraulic testing machine system MTS820. Tests were carried out in stress controlled mode using constant load with a stress ratio of R = -1 and a frequency of 5 Hz. These parameters allow comparison to data given in the literature, even being aware that implants are mostly exposed to compression load, however they are still a dynamic system. Measurements were limited to $2*10^6$ cycles while Ringer-Acetate solution was used as the corrosive media in the corrosion fatigue test. A voltammetric cell (measurement area of 150 mm²) was used to evaluate the free corrosion potential (Open Circuit Potential -OCP-) and passivity. The OCP and the current density - potential curve were measured using an Argenthal reference electrode. The microstructure and fracture surfaces of the investigated alloys were characterized using a Zeiss EVO 40XVP Scanning Electron Microscope (SEM) equipped with an Energy Dispersive X-ray Spectroscopy (EDX) for chemical analysis.

## Fatigue test results

Fatigue tests on as-cast Mg10Gd as evaluated shows fatigue strength (failure probability of 50 percent) of 52 MPa [20]. During fatigue tests, the cyclic deformation behavior of the alloys was monitored by evaluating the movement (cyclic creep) and change in shape (material strengthening / softening) of the hysteresis loop based on the number of cycles. Figure 5 represents the cyclic creep behavior subjected to 50 MPa stress resulting at $1.8*10^6$ cycles to fracture. Due to asymmetrical plastic deformation in the tensile and compressive half cycle, the specimen exhibit pronounced cyclic tensile creep in the second half of the life time.

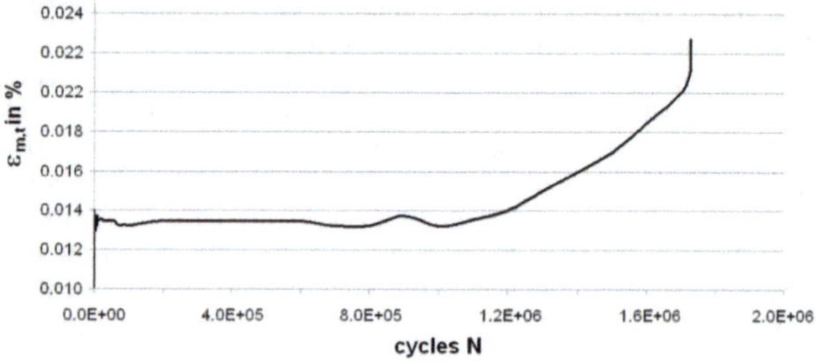


Figure 5. Cyclic creep curves of as-cast Mg10Gd at 50 MPa

The fatigue strength value of the extruded Mg10Gd alloy has not been evaluated yet. However, first fatigue tests have been done to investigate the cyclic creep behavior. According to the higher mechanical properties of the tensile and compression tests of extruded Mg10Gd a stress amplitude of 120 MPa has been chosen. This stress amplitude is higher than the UTS value for the as-cast material. Figure 6 presents the cyclic creep behavior of the extruded material. Extrusion processing has slightly decreased the asymmetric yield behavior (see ratio of CYS/TYS in table I), but the value is still above one. Thus the resistance to plastic deformation is, similar to the as-cast condition, stronger in the second half of the cycle and lead to cyclic tensile creep. However, after $1*10^6$ cycles cyclic creep changes into compressive creep and saturates; hardening effects are expected to appear.

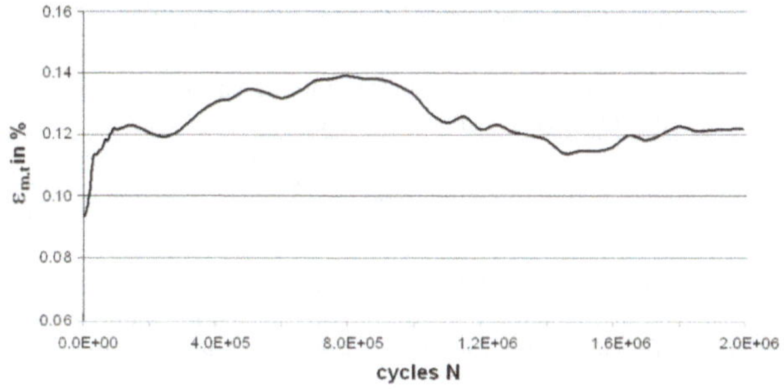


Figure 6. Cyclic creep curve of extruded Mg10Gd at 120 MPa

According to the higher strength and elongation to fracture values of the extruded material and the exposure to a higher stress amplitude the total mean strain ($\varepsilon_{m,t}$) shows higher overall values.

Figure 7 shows the SEM fracture surfaces of the as-cast alloy, loaded at a stress amplitude of 50 MPa, and figure 8 the extruded alloy, loaded at 150 MPa ($2*10^4$ cycles), respectively. The as-cast fracture surface is generally brittle in nature, explained by the low elongation to fracture value due to the coarse grained microstructure (less than 2 %). However, the characteristic fracture surface of fatigue failure has been seen. Due to the brittle behavior more than one crack initiation have been found leading via the area with fatigue striations (growth bands) into the final rupture area, which often shows strong crystal oriented fatigue structures, also seen in other castings [24].

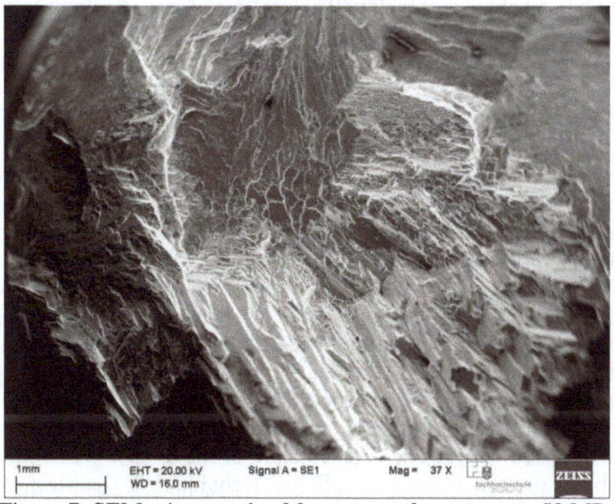


Figure 7. SEM micrograph of fracture surface: as-cast; 50 MPa

The extruded fracture surface indicates a more ductile damage and fracture behavior, which was expected because of the elongation to fracture of almost 20 %. There are no crystal oriented fatigue structures. SEM fractography with higher magnification show spherical dimples, which are characteristic of ductile fracture.

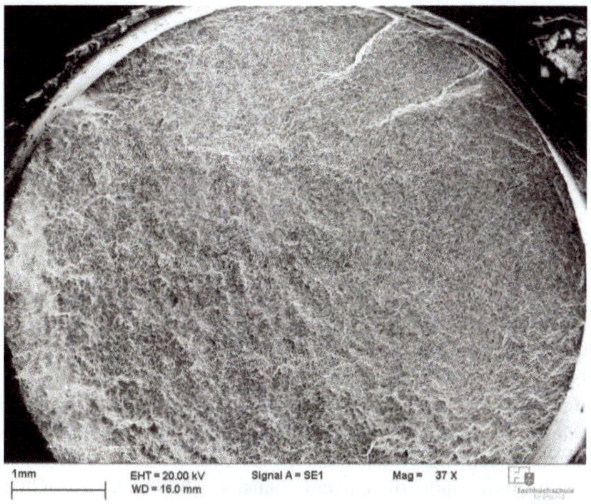


Figure 8. SEM micrograph of fracture surface: extruded, 150 MPa

**Voltammetric measurement results**

The corrosion behavior of the alloys in Ringer-Acetate solution has been evaluated by potentiondynamic measurements using an electrolytic cell. The exposed area of ~ 150 mm² is orientated transversely to the extrusion direction. Figure 9 shows the OCP measurement of the as-cast condition in dependence of the temperature. As-cast Mg10Gd reaches its free potential of -1700 mV at 37 °C already after 150 sec. The OCP value shifts towards noble potential as a result of passivation. However, as-cast Mg10Gd at 20 °C reaches saturation (defined by no change in potential values) after 500 sec, more slowly and the OCP value shows a slight lower free potential of -1690 mV (more noble).

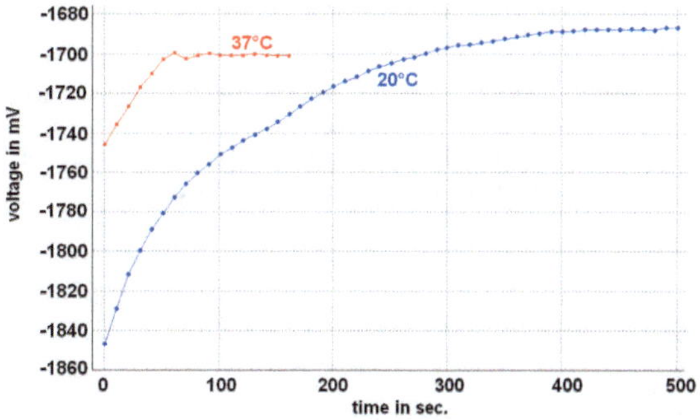


Figure 9. OCP measurements of Mg10Gd at 20°C and 37°C: as-cast

The influence of the grain refinement during extrusion on corrosion of the extruded material is seen in figure 10. The OCP value of the corrosion test at 37 °C is reached after 150 sec, almost unchanged compared to the casting. At 20 °C saturation of the wrought material is reached earlier compared to the casting, after 210 sec, showing more negative free potential.

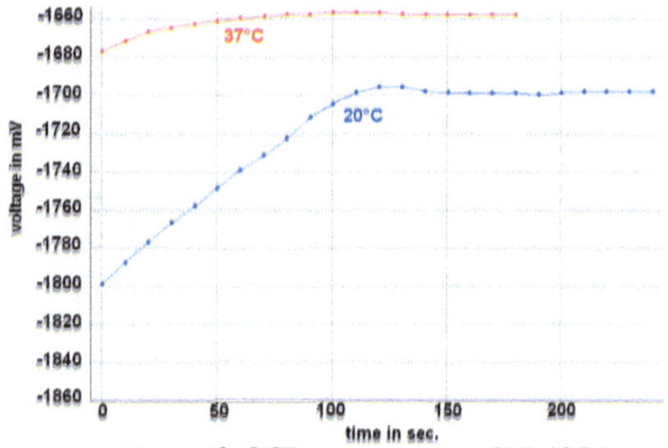

Figure 10. OCP measurements of Mg10Gd at 20°C and 37°C: extruded

The current density - voltage curves in figure 11 (i-E curves) show the difference of corrosion behavior (concentrated on the anodic reaction) in dependence of the temperature of as-cast material, figure 12 of wrought material, respectively. The free potential, passivation and breakthrough values seem not to be influenced strongly. However, increasing the voltage after breakthrough data of current density show an influence of temperature. Increasing the temperature to 37 °C accelerates the corrosion process, for the as-cast as well the wrought alloy. After strong anodic corrosion, passivation starts again. However, the passive layer is not stable by increasing voltage (growth and dissolution can be seen). At high voltage values the influence of the temperature on the corrosion behavior is almost eliminated.

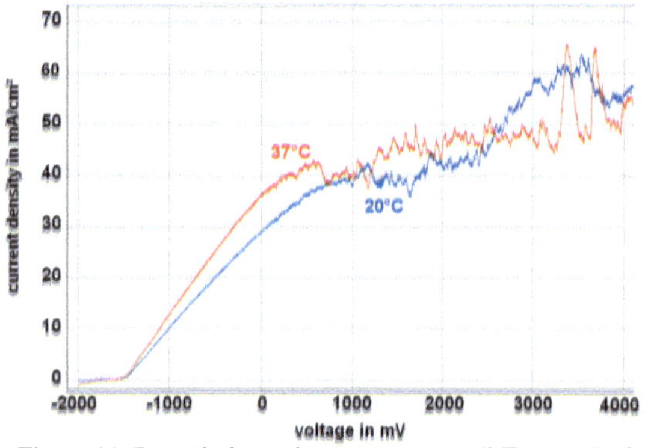

Figure 11. Potentiodynamic measurements (i-E curves) of Mg10Gd at 20 °C and 37 °C: as-cast

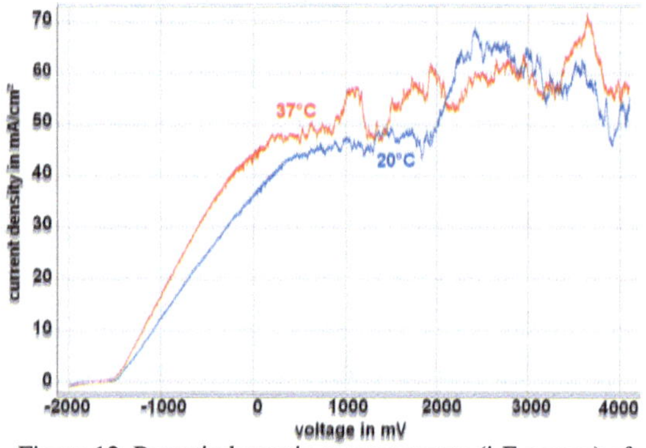

Figure 12. Potentiodynamic measurements (i-E curves) of Mg10Gd at 20 °C and 37 °C: extruded

Microscopic investigations of the corroded material are planned to evaluate the microstructural influence of aligned second phases. The corrosion process is expected to start at areas where high differences in potential occur, but so far the potential value of the precipitate $Mg_5Gd$ is not known yet. However, precipitates are expected to be nobler than the matrix itself.

### Corrosion fatigue results

Corrosion fatigue tests show that the numbers of cycles to fracture are reduced considerably due to Ringer-Acetate solution. Again, extruded Mg10Gd sustained higher cycles to fracture than as-cast Mg10Gd in corrosive medium, obviously depending on the stress amplitude. For stresses of 120 MPa extruded Mg10Gd fractured after 600000 cycles; without the exposure to the electrolyte the samples reached the limited $2*10^6$ cycles, see table II.

Table II. Comparison in fatigue life: without and with exposure to the electrolyte (Ringer-Acetate solution), at 20 °C

| material | cycles, without | cycles, with |
|---|---|---|
| as-cast (50 MPa) | > $1.8*10^6$ | 26,000 |
| extruded (120 MPa) | > $2*10^6$ (limited value) | 600,000 |

Figure 13 (a, b) shows the development of selective corrosion on the surface of the sample in the fatigue corrosion chamber while figure 13 (c) shows the fractured sample with corrosion products.

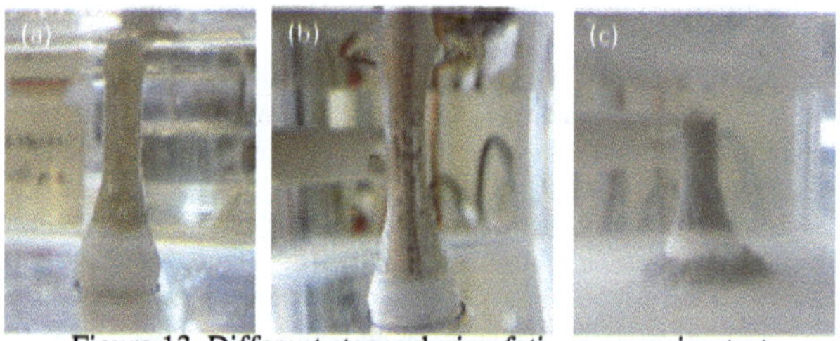

Figure 13. Different stages during fatigue corrosion test
(a) hydrogen evolution (b) selective corrosion (c) fractured sample with corrosion products

To find out if the selective corrosion is developing into corrosion pitting further investigations need to be done. First light micro-structural images of the as-cast and wrought material (cross sections have been looked at) have shown rather homogenous corrosion behavior, which will be preferred for medical implants.

### Discussion

Microstructure

From initial thermodynamic calculations using Pandat software for Mg-Gd binary system [25], HCP and $Mg_5Gd$ precipitates were predicted. The formation of $Mg_5Gd$ is possible at the eutectic temperature of 548 °C with a maximum solubility of 23.49 wt. % [26]. This phase has been observed in previous work and is known to be effective in improving the mechanical property of pure Magnesium [18]. From the two alloys investigated, extruded material shows strongly refined grain size followed by strongly improved mechanical properties. This binary phase will later play a role on the fatigue and corrosion fatigue response of the investigated alloys. The coarse nature of the as-cast grains (>500 µm) of permanent mould chill cast techniques is significantly a weakness to its mechanical properties as this will influence the efficacy of the precipitates potency to opposing dislocation movement. Extrusion at the high deformation degree of 1.6 requires high temperatures (420 °C). The grain refinement as a result of recrystallization improves strength and ductility due to grain size strengthening. However, the uniformly distributed precipitates form lines in extrusion direction. As observed from the EDX analysis results of the casted alloy, it is clearly seen that there are lower amount of rare earth elements dissolved in the solid solution. This phenomenon is known to be associated with oxidation of the rare earth elements during casting. Rare earth elements form oxides that are heavier than the Magnesium melt and in most cases exist as inclusions on the cast material or are removed during casting [18, 27, 28]. During heat exposure of the extrusion process enough energy is given for further diffusion processes between $Mg_5Gd$ precipitates and solid solution. The alignment of second phases will influence the corrosion properties as well as the mechanical properties. Further work will concentrate on microscopy of fractured and corroded samples.

Corrosion

Given that corrosion behavior of Magnesium alloys depends strongly on electrolytes, this work was carried out using electrolyte other than the usual 3 % NaCl solution [1]. Ringer-Acetate solution is close to human blood and is more practical for bio-implantation materials compared to 3 % NaCl solution which is more aggressive. Although the influence of things like protein or serum content in the body may alter the corrosion behavior of the alloy under investigation, Ringer-Acetate solution is however useful for material screening. Figure 13 (a&b) shows selective development of corrosion on the surface of the sample in a Ringer-Acetate solution and (c) represents the fractured sample. Magnesium is usually anodic when in contact with most engineering materials in liquid environment [10]. As an alloy, the α-Mg matrix degrades preferentially due to the cathodic response of many precipitates [29]. The forms and distribution of the precipitates however play a role in determining the nature and dynamics of the corrosion. However, in this work, there is no very strong galvanic effect observed during corrosion apart from the selective attack as observed during the inception of the test as shown in figure 13 (a&b). This could be explained by the nature of the pH-value of the electrolyte which is mostly low (7-7.5) at the inception of the test and soon increases towards alkalinity (10.5) as the duration of the test increases. The increase in the pH-value lowers the rate of corrosion due to formation of $Mg(OH)_2$ which is stable at higher pH-values. Also the temperature has an influence on the corrosion rate; for evaluating OCP the corrosion rate is increased. Also the equilibrium potential is more positive at 37 °C and reaches saturation in shorter time, see figures 9 and 10. Previous work suggested that an accelerated corrosion of Magnesium alloy implant could be expected in vivo as the local pH was estimated at 7.4 or lower due to secondary acidosis resulting from metabolic and resorptive processes after surgery [5]. Another work shows that pH value (> 11.5) will promote a stable protective hydroxide layer on the surface of the Magnesium alloy implant while less than 11.5 will facilitate corrosion of the Magnesium alloy in corrosive media [30]. Zainal Abidin et al. [31] suggest corrosion tests in Hanks solution and to maintain the solution pH constant by $CO_2$ to avoid super-saturation. Furthermore it is mentioned, that increasing temperature increases the effect of microgalvanic corrosion followed by an increase of corrosion rate. Again, the microgalvanic reactions are not clear yet for the Mg10Gd system, but figures 11 and 12 show also increased corrosion rate. After strong corrosion leading into overlaid passivation and breakthrough the influence of temperature and microstructure is eliminated, because the corrosion rate seems to have reached its maximum, especially taking the ph-value increase into account.

Considering the Ringer-Acetate solution used in this work, extruded Mg10Gd shows a slightly more noble response (OCP) due to grain refinement compared to the as-cast material at 37 °C. However, it is planned to correlate each measured data with its local microstructure. As-cast material corrosion response will be influenced by the appearance of grain boundaries and dentritic arm spacing, wrought material has a more homogeneous microstructure (globular, fine grains), but aligned second phases which will promote microgalvanic corrosion, are distributed inhomogeneous. It is likely that the behavioral pattern may change in a different electrolyte as the alloying element reacts differently with respect to various corrosive media. However, bone implant materials are expected to corrode homogeneously while maintaining the necessary mechanical support needed within the healing duration [17].

Fatigue and corrosion fatigue

On the fatigue response of the investigated alloys, extruded Mg10Gd shows an increase in fatigue life over as-cast Mg10Gd in air as well as in Ringer-Acetate solution, even the fatigue strength itself has not been evaluated yet. As seen from the light microscopy images, strong grain refinement after recrystallization is the result of the extrusion process increasing the mechanical properties significantly.

From the fractured surface of as-cast Mg10Gd in figure 7 it is clear that the alloy exhibited a brittle transcrystalline fracture and one sees successive breaking of atomic bonds on the preferred crystallographic planes. The dominant slip systems in magnesium alloys are the basal {0001} <11-20>, prismatic {10-10} <11-20> and pyramidal {10-11} <11-20> systems including pyramidal twinning. At room temperature, plastic deformation occurs predominantly by basal slip and pyramidal twinning [28, 32]. The critical resolved shear stress (CRSS) for slip on {0001} basal planes and the prismatic {10-10} or pyramidal {10-11} planes differ considerably. Previous works [32, 33] show that the (CRSS) of either prismatic or pyramidal slip at 25 °C is several orders of magnitude higher than that of basal slip. This means that the probability of basal slip occurring even at grains unfavourably oriented for basal slip is high at room temperature. To have a complete picture of the fatigue and corrosion fatigue properties (particularly the fracture behavior), experiments should be repeated at 37 °C.

As observed from first microstructure images on as-cast Mg10Gd the grain boundaries alter the direction of crack propagation. Quasi crack blunting for finer grains (due to extrusion) was expected in comparison to coarse or columnar grain structures and could be as seen in this work. When the direction of the crack propagation lies on regions with more grain boundaries, the dislocation motion are delayed giving rise to longer fatigue life. In contrary, when the direction of crack propagation are purely transcrystalline as a result of coarse grains (as-cast condition), opposition to dislocation motion are greatly reduced giving rise to early fatigue failure as witnessed in [20]. The application of stress creates a plastic zone within the sample ahead of the crack tip: due to columnar/ coarse grains, the critical stress to form twins is smaller. Again, boundaries between twins acts as strain build up regions [34]. Since the more driving force at this time is dislocation, crack follows up twin formation and the result for coarse alloys are mainly brittle fracture as observed in the investigated alloys. Reference [35] conducted fracture test on fine and coarse grain Mg-Zn-Al system and show that finer grains resist crack propagation better than coarse Magnesium alloy specimens. This is also true for fatigue tests as observed here for a Mg-Gd system, in as-cast and extruded condition. Finer grains will improve ductility thereby increasing the overall energy needed for crack initiation and propagation in the alloy. As shown in table I, the ultimate tensile strength for permanent mould chill cast Mg10Gd is 91 MPa respectively with ductility less than 2 %. This is very low compared to the standard stainless steel benchmark for stent purposes [7]. Application of grain size reduction due to appropriate manufacturing techniques has been seen to push these values higher (ultimate tensile strength of 251 MPa) and correspondingly improve the room temperature fatigue behavior of extruded Mg-Gd alloy system.

The selected corroded regions act as crack tip during fatigue test and thus strongly reduce the number of cycles to fracture in comparison to the fatigue test in air, in both as-cast and extruded condition. To evaluate the effect of the grain size or rather the number of grain boundaries, dendrite arms in the casting microstructure and the alignment of second phases in the extruded material further microscopic investigations are needed.

The interaction between the Ringer-Acetate solution and alloys lowers the overall fatigue strength of the alloys. However, in the as-cast material a uniform corroded surface was dictated after corrosion fatigue for 18 hours. Uniform corrosion is desirable as that will theoretically ensure the mechanical support of the implants reasonably before eventually degrading compared to strong pitting corrosion. Although most of the known Magnesium alloy precipitates have a more positive electrode potential than α-Mg matrix [29], the net distribution of the eutectic seems uniform on the permanent mould chill cast alloys investigated and as such the coarse segregated precipitates do not lead to strong galvanic situation causing pitting corrosion. For the extruded material no pitting corrosion in fatigue corrosion has been seen yet either, however, longer exposure times (by reducing the stress amplitude) need to be applied to get the overall picture.

From the work of Hort et al. [18, 36], when the Mg-Gd binary phase studied was heat treated to T4 and T6 conditions, its corrosion properties generally improved. For the as-cast Mg10Gd system, the T4 and T6 values were 0.7 mm/y and 0.4 mm/y respectively in 1 % NaCl solution [10, 18, 36]. This improved corrosion resistance was attributed to the nano scaled ß-precipitates which were uniformly distributed within the microstructure. It is well known that appropriate precipitates improve mechanical properties of Magnesium alloys. Heat treatment was not applied in this study, but recrystallization during extrusion also results in uniformly distributed precipitates, even showing some second phase alignment. Again, Peng et al. [37] reported an increase in corrosion resistance after a T4 treatment of GW103K which is a Mg-Gd-Y-Zr alloy in a 5 % NaCl. However, Birbilis and co-workers [38] show that increase in RE elements beyond a certain proportion will be detrimental to the corrosion properties of Magnesium alloys especially in as-cast condition. A combination of adequate alloying elements, Neodymium has already been discovered as an interesting candidate, heat treatment to encourage precipitates formation and uniform distribution will improve both the corrosion and fatigue properties of the Mg10Gd base alloying system.

## Conclusion

Extruded Mg10Gd shows improved mechanical properties compared to the as-cast condition. This is mostly due to the grain refinement. Due to the CYS/TYS ratio above one, cyclic tensile creep occurs during fatigue tests predominantly, when the stress ratio is R=-1.

The improvement of the mechanical properties has no negative influence on the corrosion behavior. Extrusion of Mg10Gd show improved mechanical properties and fatigue strength both in air and in Ringer-Acetate solution.

Extruded Mg10Gd exhibited nobler characteristic (OCP) with respect to as-cast Mg10Gd at 37 °C, which is the environmental temperature for medical application. The improved properties observed in this extruded alloy are connected with the grain refinement. However, extended passivation during potentio-dynamic measurements has not been found. Corrosion rate at body temperature is found to be increased at potentials slightly higher than breakthrough potential. However, looking at the corrosion behavior at higher potentials, there is no difference between cast and wrought material and test temperature. Since the orientation of the exposed area for the potentiodynamic measurements is transversal to the extrusion direction (cross section), further tests on the longitudinal microstructure need to be done. Possible associated differences in behavior might occur.

Extruded Mg10Gd could be a potential implant material with further development; effort is however needed to produce more homogeneous and fine grained feedstock for wrought alloy production. Post casting processes such as heat treatments are sure ways of improving mechanical properties for biodegradable consumption.

Thermo-mechanical forming has shown first improvements. In-vivo experiments are necessary to gain more insight into Mg-Gd alloys. Also the influence of other rare earth elements on the Mg10Gd, for example Neodymium, is of interest. The increase of mechanical properties by alloying Neodymium to Mg10Gd in casting has already been proven in other studies.

## Acknowledgement

The authors appreciate the support of Dipl.-Ing. Ralf Tesch in carrying out fatigue tests and Scanning Electron Microscopy. The assistance of Dipl.-Chem. Hartmut Habeck during corrosion tests, especially potentiodynamic measurements is greatly appreciated.

## References

1. M.M. Avedesian and H. Baker, *Magnesium and Magnesium Alloys* (ASM Specialty Handbook 1999).
2. N. Winzer, A. Atrens, W. Dietzel, V.S. Raja, G. Song and K.U. Kainer, "Characterisation of stress corrosion cracking (SCC) of Mg-Al alloys," *Materials Science and Engineering: A*, 488 (1-2) (2008), 339-351.

3. G. Song, A. Atrens and M. Dargusch, "Influence of microstructure on the corrosion of die cast AZ91," *Corrosion Science*, 41 (2) (1998), 249-273.
4. O. Anopuo, N. Hort, Y. Huang and K.U. Kainer, "The influence of De-icing salts on corrosion of Mg-Alloys," *Magnesium Technology 2007*, ed. R.S. Beals, A.A. Luo, N.R. Neelameggham and M.O. Pekguleryuz, (2007), 439-444.
5. F. Witte, V. Kaese, H. Haferkamp, E. Switzer, A. Meyer-Lindenberg, C.J. Wirth and H. Windhagen, "In vivo corrosion of four magnesium alloys and the associated bone response," *Biomaterials,* 26 (17) (2005), 3557-3563.
6. L. Xu, G. Yu, E. Zhang, F. Pan and K. Yang, "In vivo corrosion behavior of Mg-Mn-Zn alloy for bone implant application," *Journal of Biomedical Research,* 83 (2007), 703-711.
7. H. Hermawan, D. Dubė and D. Mantovani, "Developments in metallic biodegradable stents," *Acta Biomaterialia*, 6 (2010), 1693-1697.
8. M. Peuster, P. Wohlsein, M. Brugmann, M. Ehlerding, K. Seidler and C. Fink, "A novel approach to temporary stenting: degradable cardiovascular stents produced from corrodible metal-results 6-18 months after implantation into New Zealand white rabbits," *Heart*, 86 (5) (2001), 563-569.
9. F. Witte, "The history of biodegradable magnesium implants: A review," *Acta Biomaterialia,* 6 (2010), 1680-1692.
10. A. Atrens, M. Liu and N.I.Z. Abidin, "Corrosion mechanism applicable to biodegradable magnesium implants," *Material Science and Engineering B*, 176 (20) (2011), 1609-1636.
11. H. Hermawan, A. Purnama, D. Dube, J. Couet and D. Mantovani, "Fe–Mn alloys for metallic biodegradable stents: Degradation and cell viability studies," *Acta Biomaterialia*, 6 (2010), 1852-1860.
12. M. Peuster, C. Hesse, T. Schloo, C. Fink, P. Beerbaum and C. Schnakenburg, "Long-term biocompatibility of a corrodible peripheral iron stent in the porcine descending aorta," *Biomaterials*, 27 (28) (2006), 4955-4962.
13. L. Xue, S. Dai and Z. Li, "Biodegradable shape-memory block co-polymers for fast self-expandable stents," *Biomaterials,* 31 (32) (2010), 8132-8140.
14. W.V.D. Giessen, C. Slager, H.V. Beusekom, D.V.I. Schenau, R. Huijts and J. Schuurbiers, "Development of a polymer endovascular prosthesis and its implantation in porcine arteries," *Journal of Interventional Cardiology,* 5 (3) (1992), 175-185.
15. G. Mani, M.D. Feldman, D. Patel and C.M. Agrawal, "Coronary stents: A materials perspective," *Biomaterials*, 28 (2007), 1689-1710.
16. C.J. McMahon, P. Oslizlok and K.P. Walsh, "Early restenosis following biodegradable stent implantation in an aortopulmonary collateral of a patient with pulmonary atresia and hypoplastic pulmonary arteries," *Catheterization and Cardiovascular Interventions*, 69 (2007), 735-738.
17. F. Witte, J. Fischer, J. Nellesen, H. Crostack, V. Kaese, A. Pisch, F. Beckmann and H. Windhagen, "In vitro and in vivo corrosion measurements of magnesium alloys," *Biomaterials,* 27 (2006), 1013-1018.
18. N. Hort, Y. Huang, D. Fechner, M. Störmer, C. Blawert, F. Witte, C. Vogt, H. Drücker, R. Willumeit, K.U. Kainer and F. Feyerabend, "Magnesium alloys as implant materials - Principles of property design for Mg-RE alloys," *Acta Biomaterialia*, 6 (2010), 1714-1725.
19. F.R. Elsayed, N. Hort, M.A. Salgado-Ordorica and K.U. Kainer, "Magnesium Permanent Mold Casting Optimization," *Materials Science Forum*, 690 (2011), 65-68.
20. P. Maier, O. Anopuo, F. Malchau, G. Wienck and N. Hort, "Cyclic deformation of newly developed Magnesium cast alloys in corrosive environment," *Materials Science Forum*, 690 (2011), 495-498.
21. A. Ball and B. Prangnell, "Tensile-compressive yield asymmetries in high strength wrought magnesium alloys," *Scripta Metallurgica of Materialia*, 31 (2) (1994), 111-116.
22. C. Fleck and D. Löhe, "Cyclic deformation behaviour of newly developed microalloyed Mg wrought alloys in corrosive environment," *Magnesium Technology 2007*, ed. R.S. Beals, A.A. Luo, N.R. Neelameggham and M.O. Pekguleryuz, (2007), 275-279.
23. P. Maier and C. Fleck, "Cyclic deformation of newly developed Magnesium wrought alloys microalloyed with Calcium in corrosive environment," (Paper presented at the Eurocorr 2007, The European Corrosion Congress: Progress by Corrosion Control 2007).
24. G. Lange, *Systematische Beurteilung technischer Schadensfälle* (Wiley-VCH Weinheim 2001).
25. Pandat software package (PanMg7).
26. A.A. Nayeb-Hashemi, *Phase diagrams of binary magnesium alloys* (Metal Park, OH: ASM International, 1998).
27. J. Campbell, *Castings* (Oxford: Butterworth Heinemann, 2003).
28. E.F. Emley, *Principles of magnesium technology* (New York: Pergamon Press, 1966).
29. O. Lunder, K. Nisancioglu and R.S. Hanses, "Corrosion of Die Cast Magnesium-Aluminum Alloys," *SAE Technical Paper,* (1993), 930755.
30. M. Pourbaix, *Atlas of Electrochemical equilibria in aqueous solutions* (National Association of Corrosion Engineers, Houston, 1974).
31. N.I. Zainal Abidin, A.D. Atrens, D. Martin and A. Atrens, "Corrosion of high purity Mg, Mg2Zn0.2Mn, ZE41 and AZ91 in Hank's solution at 37 °C," *Corrosion Science,* 53 (2011), 3542-3556
32. P. B. Hirsch and J. S. Lally, "Deformation of magnesium single crystals," *Philosophy Magazine*, 12 (1965), 595-648.
32. P. W. Bakarian and C. H. Matthewson, "Slip and twinning in magnesium single crystals at elevated temperature," *Transactions of the Metallurgical Society of AIME*, 152 (1943), 226-254.
33. B. C. Wonsiewicz and W. A. Backofen, "Plasticity of magnesium crystals," *Transactions of the Metallurgical Society of AIME,* 239 (1967), 1422-1431.
34. H. Somekawa, A. Singh and T. Mukai, "Fracture mechanism of a coarse grained magnesium alloy during fracture toughness testing," *Philosophical Magazine Letters*, 89 (2009), 2-10.
35. H. Somekawa, A. Singh and T. Mukai, "Fracture mechanism and toughness in fine and coarse grained magnesium alloys," *Magnesium Technology 2011*, ed. W.H. Sillekens, S.R. Agnew, N.R. Neelameggham and S.N. Mathaudhu, (2011), 25-28.
36. K.U. Kainer, N. Hort, R. Willumeit, F. Feyerabend and F. Witte, *Processing of Advanced Materials 18,* ed. M. Niinomi, M. Morinaga, M. Nakai, N. Bhatnagar and T.S. Srivatsan, (2009), 975-984.
37. L.M. Peng, J.W. Chang, X.W. Guo, A. Atrens, W.J. Ding and Y.H. Peng, "Influence of heat treatment and microstructure on the corrosion of magnesium alloy Mg-10Gd-3Y-0.4Zr," *Journal of Applied electrochemistry,* 39 (2009), 913-920.
38. N. Birbilis, M.A. Easton, A.D. Sudholz, S.M. Zhu and M.A. Gibson, "On the corrosion of binary magnesium - rare earth alloys," *Corrosion Science*, 51 (2009), 683-689.

# CORROSION BEHAVIOR OF VARIOUS STEELS BY AZ31 MAGNESIUM MELT

Cheuk Kin Tang, Marie-Aline Van Ende, In-Ho Jung,
McGill University; 3610 University Street, Montréal, Québec, H3A 2B2, Canada

Keywords: Corrosion of steel, magnesium melt, oxidation/reduction

## Abstract

During the production of magnesium and its alloys, the melt is in contact with steel at various stages of the process. Steel can be eroded and corroded through chemical reactions with molten magnesium and the protective gases such as $SF_6$ or $SO_2$. The dissolution rate and corrosion mechanism of steels in molten magnesium were investigated by the rotating cylinder method. The corrosion behavior of four steel grades, i.e. high carbon and low carbon steel, stainless steel 316 and 400 were examined for AZ31 melt under $N_2$-1% $SF_6$ gas atmosphere. The corrosion rates were found to be different depending on steel grades. It is found that the corrosion mechanism of steels in magnesium melts involves a two-stage process consisting of oxidation of Fe and its reduction by magnesium melt.

## Introduction

Magnesium and its alloys have become a more and more attractive material primarily owing to its light weight. In order to produce the final Mg products of cast alloys or wrought alloys, all Mg alloys should be once melted in a furnace at about 700 °C. A complete melting and casting equipment consists of a pre-heating and charging equipment, melting and holding furnaces, a protective gas mixing device and a metal feed to the casting machine [1]. The Mg melt in the furnace is always under a protective gas atmosphere to prevent Mg burning. Typical protective gas used in Mg industry is $N_2$-$SF_6$, dry air-$SF_6$ and dry air-$SO_2$ mixture. Since the solubility of Fe in liquid Mg is low (0.04 wt% Fe at 750 °C), steel makes a good candidate for holding and transporting liquid magnesium during its fabrication process. Even most of the tools used for liquid Mg are made of steels. However, Fe is one of the most harmful impurities in Mg alloys as it reduces considerably the corrosion resistance even at low concentration.

Minimizing the corrosion and maintaining the original shape of the steel crucible are important for long service life, easy cleaning and to minimize the contamination of the melt by Fe. Two steel materials have been proven useful for holding the Mg melt. Although expensive, ferritic stainless steel has shown to work best when low wall thicknesses are required for crucible design. However, the most common material used is a bimetal in which a heavy steel plate is joined with an exterior rolled cladding. The Ni-Cr components inherent in stainless steel and the exterior rolled cladding are essential in preventing oxidation at these high temperature conditions [2]. The upper zone of the steel crucible in which the liquid oscillates at horizon is subjected to higher wear than the other zones. However, no comprehensive study on the corrosion of steel near the Mg melt level has been reported. Furthermore, a comparative corrosion study of various steel grades against liquid Mg has not been performed.

In the present study, a laboratory furnace system was designed to investigate the corrosion behavior of steel by molten Mg under a $N_2$-$SF_6$ protective gas atmosphere. The goal of the present study is to provide a preliminary understanding of the corrosion mechanisms of various steels by Mg melt, which will be useful for extending the steel crucible life time and selecting future crucible materials.

## Experimental

### Materials

Four steel grades were investigated using the rotating cylinder experiments: Low carbon steel, high carbon steel, stainless steel 316 and stainless steel 400. These steel types were welded together to form a cylindrical rod exposing each steel types' sides to the liquid. The liquid Mg alloy used for this study is commercial AZ31 magnesium alloy, provided by RIST (Research Institute of Industrial Science and Technology), Korea.

### Rotating Cylinder Experiments

The experimental setup is custom-made for the present experiments and is mainly composed of a resistance furnace containing a steel crucible, a rotation motor with steel specimen rod and a protection gas supply unit. A boron nitride coated steel crucible (diameter 23 cm x height 11.5 cm) is placed inside an open-well type resistance furnace. A removable cover is used to facilitate the sample insertion process as well as provide protection from atmospheric exposure to the Mg liquid. The removable cover has three small holes and one large hole used respectively for the gas inlet, thermocouple, hanging wire and observation glass cover. Temperature of the furnace interior and the liquid melt are measured by means of two thermocouples to control the melt temperature within ± 5 °C. Due to the fact that the process takes place at temperature above the melting point of magnesium (649 °C), the whole liquid system was kept under a protective gas atmosphere during the entire course of the present experiment above 300 °C. The protective gas mixture consists of $N_2$-1% $SF_6$, where the amount of $SF_6$ is higher than the normal industrial standard, and the flow rate of the protective gas was maintained at 200 ml/min, which is enough to protect the Mg melt from burning.

Several pieces of AZ31 ingots amounting to approximately 1.2 kg were placed within the holding crucible to give a liquid pool depth of about 8 cm; after attaining a temperature of 150 °C and evaporating the whole moisture for 30 minutes, the temperature was then raised to 700 °C for melting under protective atmosphere. After melting, the molten metal temperature was stabilized for 1 h. The steel rod was preheated above the melt in the crucible to about 400 °C. The steel rod was immersed up to approximately 2 to 3 cm in the liquid metal. The steel bar was then rotated at a speed of 100 rpm for 5 h. After the rotation experiment, the steel bar was pulled out and suspended in the steel crucible for several minutes to cool down to about 400 °C under the protective gas atmosphere and then pulled out of the crucible to cool it down to room temperature. Fig. 1 illustrates the experimental procedure in the present study.

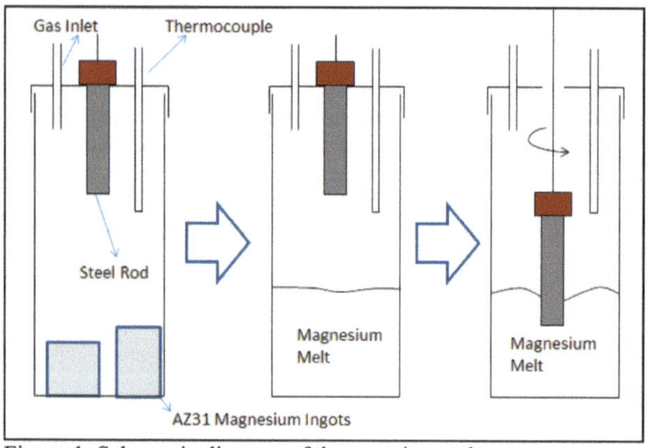

Figure 1. Schematic diagram of the experimental arrangements.

After the experiment, cross section morphologies and chemical compositions of the steel rod were examined using a PHILIPS XL-30 Scanning Electron Microscope (SEM) equipped with an energy dispersive X-ray spectrometer (EDX from EDAX). Standard metallographic techniques were used for specimen preparation. Table I lists the experimental conditions and steel grades used in the present study.

Table I. Experimental conditions for the rotating cylinder experiments and typical composition of the investigated steel grades.

| Experiment | Duration (h) | RPM | % $SF_6$ | Temperature (°C) |
|---|---|---|---|---|
| AZ31 alloy | 5 | 100 | 1 | 700 |

| Steel grade | Chemical composition (wt%) |
|---|---|
| Low carbon | <0.1 C, <0.4 Mn |
| High carbon | 0.6~1 C, 0.3~0.9 Mn |
| Stainless 316 | <0.08 C, 16~18 Cr, 10~14 Ni, 2~3 Mo, <2 Mn |
| Stainless 400 | <0.05 C, <1 Mn, 12~13 Cr, < 1 Si |

## Results and Discussion

Figure 2 shows a sample steel rod after the rotating experiment. Brown crusts were observed to form on both of the carbon steels surfaces. Interestingly, XRD patterns indicate that the brown crust is a mixture of $Fe_2O_3$ (hematite) and $MgFe_2O_4$ (magnesium ferrite). $MgF_2$ and FeS were not detected. On the contrary, the Ni-Cr content inherent to the stainless steel's nature limited excess iron oxide formation and no visible crust formation was observed besides a slight discoloring on the surface. In addition, a white and yellowish powder (XRD pattern indicates the powder to be MgO) was collected from the specimen after experiment just above the melt level. This means that, although the protective $N_2$-1% $SF_6$ gas was used in the experiment, a small amount of oxygen entered into the crucible chamber through the small holes on the steel crucible cover and oxidized the steel bar to form $Fe_2O_3$, $MgFe_2O_4$ and MgO.

The SEM BSE micrographs of the respective steel samples at liquid Mg alloy interface are presented in Figs. 3 to 6. The micrographs were taken at two locations: one above the liquid level (exposed to the protective gas atmosphere) and one below the liquid level (immersed in liquid Mg alloys).

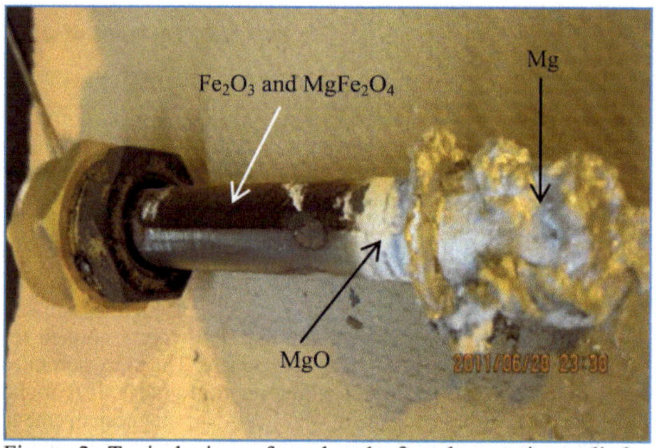

Figure 2. Typical view of steel rod after the rotating cylinder experiment. A thin layer of brown crust developed on top of both high and carbon steel and small amount of liquid Mg is still attached on the bottom of steel.

<u>High Carbon Steel</u>

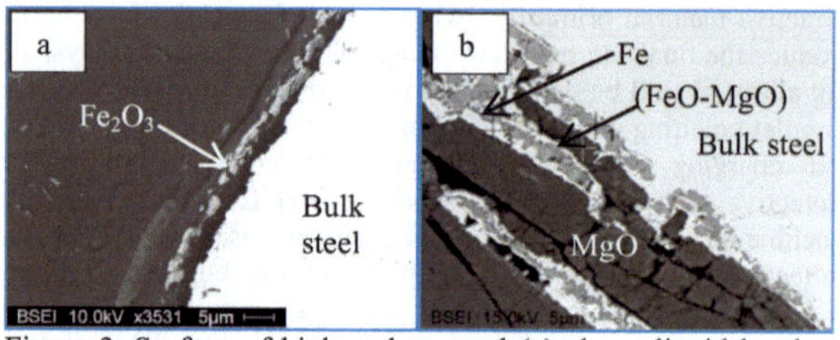

Figure 3. Surface of high carbon steel (a) above liquid level and (b) below liquid level after 5 h rotation cylinder experiment.

Despite being kept under a protective $N_2$-1% $SF_6$ atmosphere, it seems that the high carbon steel was still subjected to a high degree of oxidation throughout the entire experiment from the preheating up to the rotation stage. As can be seen in Fig. 3 (a), EDS analysis showed that a continous $Fe_2O_3$ layer of 2 to 3 μm thickness was formed on the steel surface above the liquid level. It should also be noted that $MgFe_2O_4$ was detected on the brownish surface crust as shown in Fig. 2. This means that $Fe_2O_3$ is formed in the inner side of the oxidation layer and $MgFe_2O_4$ was formed later. Most probably, the $Fe_2O_3$ layer was formed during preheating at about 400 °C before Mg was molten. Then, Mg vapor is generated after the melting of Mg, which can be oxidized on the $Fe_2O_3$ layer to form $MgFe_2O_4$. It should be noted that oxidation of steel can easily occur even under protective gas atmosphere of $N_2$-1% $SF_6$ with very small leakage of atmospheric air, which can happen in real practice of a Mg melting furnace.

Fig. 3 (b) shows the degradation behavior of the pre-oxidized high carbon steel. Interestingly, the $Fe_2O_3$ layer was changed to MgO and Fe. Then, the MgO layer was torn off from the surface of the matrix steel. Occasionally, a (MgO-FeO) mixed oxide layer was also detected between MgO and Fe.

Low Carbon Steel

Similar to the high carbon steel, $Fe_2O_3$ was formed on the steel surface above the liquid level, as can be seen in Fig. 4(a). The layer is less continuous in comparison with the high carbon steel. It also has been reported by Lai [3] that low carbon steel is generally more resistant to corrosion than medium and high carbon steels. The degradation behavior of steel below liquid level by the molten Mg, shown in Fig. 4(b), is similar to that of high carbon steel. However, the amount of MgO formed on the surface was much smaller and therefore less degradation can be expected.

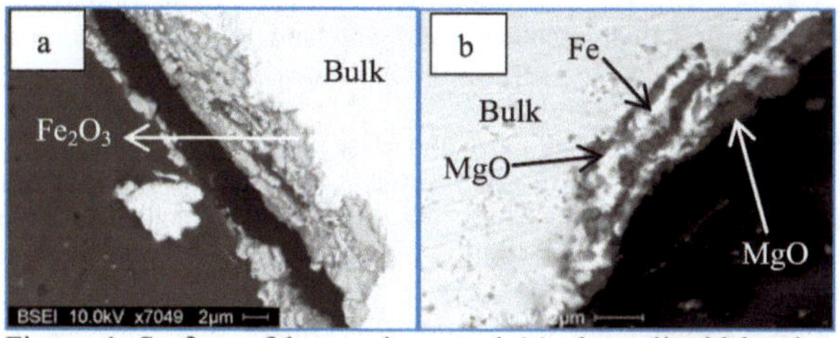


Figure 4. Surface of low carbon steel (a) above liquid level and (b) below liquid level after 5 h rotation cylinder experiment.

Stainless Steel 316

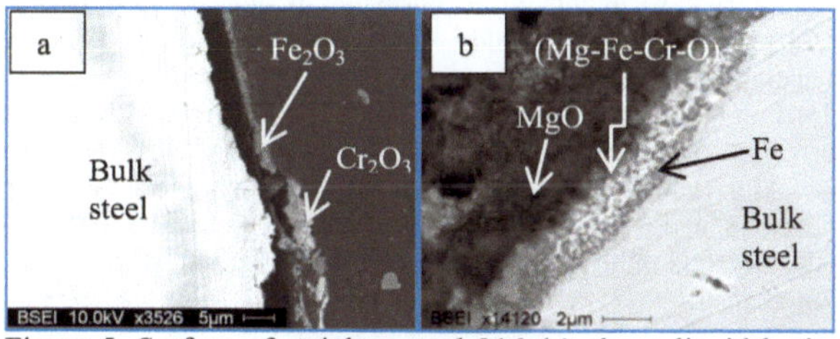


Figure 5. Surface of stainless steel 316 (a) above liquid horizon (b) below liquid horizon after 5 h rotation cylinder experiment.

As shown in Fig. 5(a), well known $Cr_2O_3$ and minor amount of $Fe_2O_3$ oxides were observed at the steel surface. Fig. 5(b) shows a micrograph image of steel sample 316 that was immersed in the Mg melt. The morphology of the steel surface is different from those of the above carbon steels. The external layer was MgO, and a network-shaped metallic Fe is covered by a mixed oxide of Mg-Fe-Cr-O, which seems to be a $MgCr_2O_4$-$FeCr_2O_4$ layer. Although small piece of Fe islands surrounded by MgO oxide was observed around the surface of stainless steel 316, the Mg-Fe-Cr-O layer seems to be strongly attached to the stainless steel matrix by the network-shaped Fe, which reduced the degradation of the stainless steel 316 by molten Mg.

Stainless Steel 400

Similar to stainless steel 316, EDS analysis indicate that $Cr_2O_3$ is formed at the surface of stainless steel 400 exposed to the protective gas, as seen in Fig. 6(a). However, $Cr_2O_3$ is formed inside the steel, while it formed a distinctive layer on the surface of stainless steel 316. That is, internal oxidation occurred for stainless steel 400, while external oxidation occurred for stainless steel 316. Consequently, the morphology of the steel degredation by molten Mg is different from the case of the stainless steel 316. Very complex network-shaped Fe (Fig. 6(b)) is directly connected to the steel matrix, and MgO oxide is formed in between the network. No trace of detached Fe or steel piece was observed near stainless steel 400, which shows the very limited degradation of stainless steel 400 by liquid Mg.

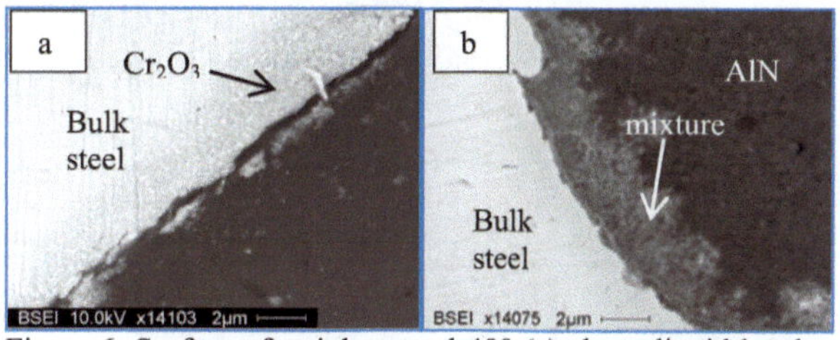


Figure 6. Surface of stainless steel 400 (a) above liquid level and (b) below liquid level after 5 h rotation cylinder experiment.

Corrosion Mechanism: Carbon Steels

Molten metal corrosion of a containment metal is often described by its solubility in the molten metal. This type of corrosion is simply a direct dissolution attack. A containment metal with high solubility to the molten metal would generally exhibit higher corrosion rates. In addition, as the temperature rises, the diffusion rate would also increase. Since the solubility of Fe in Mg at temperature below 700 °C is very limited (400 ppm), a direct dissolution type of corrosion is restricted.

The corrosion mechanism of steel by liquid Mg observed in the present study can be summarized as an "oxidation and reduction" process. Although Fe itself has very limited solubility in liquid Mg, once Fe oxide is formed on the steel surface, it can be reduced by Mg melt. This reaction can be easily understood by the following chemical reactions:

$$x\text{Mg(l)} + \text{Fe}_2\text{O}_3 = x\text{MgO-}(3\text{-}x)\text{FeO} + (x\text{-}1)\text{Fe(s)} \qquad (1)$$
$$x\text{Mg(l)} + y\text{MgO-}(1\text{-}y)\text{FeO} = (x+y)\text{MgO-}(1\text{-}x\text{-}y)\text{FeO} + \text{Fe(s)} \qquad (2)$$

It is well known that MgO is more stable than $Fe_2O_3$. The $\Delta G^0$ of reaction (1) for $x = 3$ is equal to -9.2 $10^5$ J at 700 °C [4-5]. Moreover, MgO and FeO can form a complete solid solution below 1300 °C. Therefore, the reduction of $Fe_2O_3$ by Mg can occur as in reaction (1) with the formation of the MgO-FeO magnesio-wustite solid solution. This can explain the formation of a mixed MgO-FeO layer in low and high carbon steels. Near the liquid Mg interface, the FeO in magnesio-wustite solid solution can be further reduced to lower FeO containing solid solution according to reaction (2) and eventually form pure MgO. The reduced solid Fe is already physically detached from the steel matrix and can be suspended in molten Mg. Part of them can be dissolved in molten Mg but most of them are expected to simply sink to the bottom of the crucible.

This two-stage corrosion mechanism involving the oxidation of the steel interface followed by a reduction by Mg is schematically illustrated in Fig. 7. At the oxidation stage (Fig. 7(a)), the steel surface can be oxidized by air leaked into the protective $N_2$-$SF_6$ gas atmosphere. Then, as this oxidized steel is immersed into molten Mg, the latter can react with $Fe_2O_3$ (Fig. 7(b)). In particular, if there are cracks in the $Fe_2O_3$ layer, liquid Mg can easily penetrate into the solid $Fe_2O_3$ layer and begin to reduce $Fe_2O_3$ to form a MgO-FeO solid solution. With the progress of the reduction, the Fe containing MgO-FeO particles can be detached from the surface of the steel matrix (Fig. 7(c)).

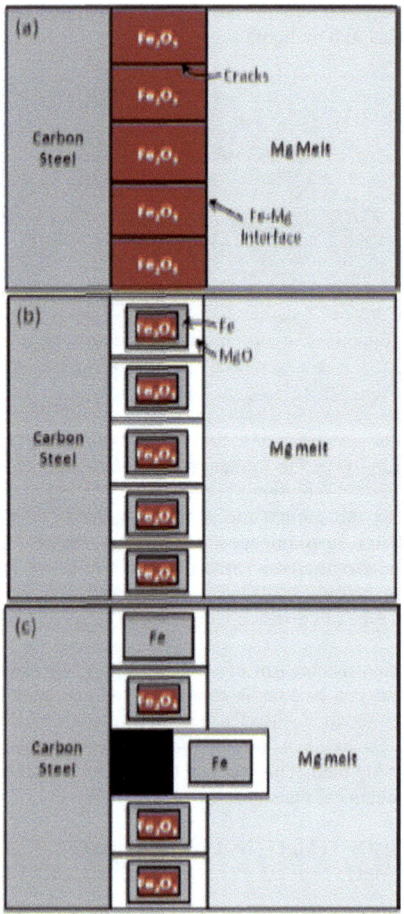


Figure 7. Schematic illustrating the two-step corrosion mechanism of carbon steel by Mg melt: (a) oxidation of the steel surface; (b-c) reduction of $Fe_2O_3$ and MgO-FeO solid solution by Mg melt.

Corrosion Mechanism: Stainless Steels

In the case of stainless steels, their corrosion mechanism is slightly different from that of carbon steel due to the formation of $Cr_2O_3$ oxide. Like Fe, the solubility of Cr in molten Mg is also very limited. Therefore, the reduction reactions can be:

$$3Mg(l) + 2Cr_2O_3 = MgCr_2O_4 + 2Cr(s) + 2MgO \quad (3)$$
$$3Mg(l) + MgCr_2O_4 = 4MgO + 2Cr(s) \quad (4)$$

The $\Delta G^0$ of reactions (3) and (4) are, respectively, -6.7 $10^5$ and -5.7 $10^5$ J at 700 °C [4-5], indicating that the reactions proceed spontaneously towards the reduction of $Cr_2O_3$ and $MgCr_2O_4$ by liquid Mg. The shape of $Cr_2O_3$ determines the reduction behavior. In the case of stainless steel 400, the internal oxidation behavior results in very interesting morphology after reduction as shown in Fig. 6(b).

Corrosion Resistance and Industrial Perspective

Among the four steel grades tested in the present study, the corrosion resistance against liquid Mg can be concluded as (from the more resistant to the less resistant): Stainless steel 400 > stainless steel 316 > low carbon steel > high carbon steel.

In the real operation of a Mg furnace, a steel crucible is typically used. In particular, low carbon steel is typically used in practice. As well-known, the presence of Fe in Mg alloys can cause severe corrosion problems through the formation of galvanic corrosion assemblage with the Mg matrix. In order to reduce the Fe concentration in molten Mg as much as possible, small amount of Al and Mn are added to make stable precipitates, $Al_xMn_yFe_z$. In this way, the soluble Fe concentration in liquid Mg can be reduced from several hundreds wt ppm level to 10~20 wt ppm level. Unfortunately, no alloying elements have been revealed to reduce the Cr or Ni levels in Mg like Fe. As Cr and Ni are enriched in Mg melt and their concentrations are hard to reduce, stainless steels are less common for Mg furnace materials.

However, from a purely corrosion view point, stainless steels 400 or 316 are superior to the carbon steels. In order to reduce the Fe pickup as much as possible, the oxidation of steel should be controlled well. One possibility to reduce Fe and Cr pickup from steel crucible is to replace the crucible of low carbon steel clad by stainless steel 400 only for the liquid Mg line.

**Summary**

Laboratory experiments on the corrosion of steels by liquid AZ31 were performed in this study. Four different steel grades were investigated in the present study: stainless steel 400, stainless steel 316, low carbon steel, and high carbon steel. It is found that the corrosion of the steels occurs by a two-stage process: oxidation of steel and reduction of the oxides by Mg melt. The corrosion resistance against Mg melt was found to be in the order of stainless steel 400 > stainless 316 > low carbon steel > high carbon steel.

**Acknowledgement**

This work was supported by RIST (Research Institute of Industrial Science and Technology), Korea and NSERC (Natural Science and Engineering Research Council of Canada), Canada. The authors also wish to thank Monique Riendeau and Manas Paliwal for assisting in the acquisition of the XRD patterns.

**References**

[1] R. Cibis, A. Kielbus, "Hot chamber Die-casting of Magnesium alloys," *Magnesium: proceedings of the 7th International Conference on Magnesium Alloys and their Applications*, Ed. K.U. Kainer (Weinheim: Wiley-VCH, 2007), 81-86.

[2] H. Dorsam, "Magnesium melting/casting and remelting in foundries," *Magnesium Technology 2000*, Ed. H.I. Kaplan, J.N. Hryn, B.B. Clow (Warrendale, PA: TMS, 2000), 99-105.

[3] G.Y. Lai, *High-temperature corrosion and materials applications* (Materials Park, OH: ASM International, 2007).

[4] FactSage thermochemical software and databases (version 6.2) with FACT pure substance database.

[5] C.W. Bale et al., "FactSage thermochemical software and databases — recent developments," *Calphad*, 33 (2) (2009) 295-311.

Magnesium Technology 2012
*Edited by: Suveen N. Mathaudhu, Wim H. Sillekens, Neale R. Neelameggham, and Norbert Hort*
*TMS (The Minerals, Metals & Materials Society), 2012*

# CORROSION OF ULTRASONIC SPOT WELDED JOINTS OF MAGNESIUM TO STEEL

Tsung-Yu Pan[1], Michael L. Santella[1]
[1]Oak Ridge National Laboratory, One Bethel Valley Road, PO Box 2008, Oak Ridge, TN 37831-6096, USA

Keywords: Magnesium, Steel, Ultrasonic Spot Welding, Galvanic Corrosion, Automotive

## Abstract

Mixed-metal joining, especially between magnesium and steel, is one of the critical technologies in achieving light-weighting vehicle body construction. However, galvanic corrosion between mixed metal joints is inevitable but not well quantified. In this study, 1.6 mm thick Mg AZ31B-H24 was joined to 0.8 mm thick hot-dipped galvanized (HDG) mild steel by ultrasonic spot welding in lap-shear configuration. No specific corrosion protection was applied in order to study worst-case conditions for corrosion behavior. The approach used an automotive cyclic corrosion test - Ford Arizona Proving Ground Equivalent Corrosion Cycle (APGE), which includes cycles of dipping in a salt bath, air drying, then holding in constant humidity environment. Lap-shear strength of the joints decreased linearly with the exposure cycles. All the joints were either taken out of test cycle for mechanical test or they separated within the humidity chamber before 25th cycle. X-ray diffraction analysis confirmed the formation of $Mg(OH)_2$ deposit in the crevice between the AZ31 and steel sheets and on the surface of the AZ31. The deposit grew thicker with cycles with exerting enough force to deform the AZ31 and HDG steel and causing a gradual opening of joints. The corrosion of the AZ31 was localized and non-uniform. The most severe corrosion occurred not at the intersection of AZ31 and the steel but rather 15-20 mm away from the spot welds.

## Introduction

Magnesium (Mg) has great potential for weight reduction as it is the lightest structural metal known (78% lighter than steel, and 36% lighter than aluminum), has a high specific strength and superior damping capacity [1]. Moreover, Mg is one of the most abundant elements in the earth's crust and in seawater, and it can supply industry's needs for the foreseeable future. The ability to significantly increase Mg usage will help the auto industry meet future Federal Corporate Average Fuel Economy (CAFE) targets and reduce exposure to CAFE penalties. United States Automotive Materials Partnership (USAMP) has proposed reducing the weight of an average North American vehicle (1,527 kg) by 132 kg by replacing 286 kg of current aluminum, iron, and steel with Mg [2]. As a result, the weight reduction could benefit in reducing emissions and improving fuel consumption by approximately 5% [3].

Two of the specific areas of concerns for broader use of Mg are reliable joining methods and corrosion. These two concerns are significantly magnified when dealing with Mg components attached to steel structures. The U.S. Department of Energy (DOE) has been sponsoring projects to develop solid-state joining technologies based on friction stir linear and spot welding (FSW/FSSW) and ultrasonic spot welding (USW) for joining Mg alloys to steels. Both processes have demonstrated some potential for meeting this challenge. Some of the results have been published in [4,5,6]. An additional concern for Mg-steel joints is galvanic corrosion because of the direct contact between these metals. The four items needed to initiate galvanic corrosion action are: (1) an anode (e.g. magnesium), (2) direct electrical contact, (3) a cathode (e.g. steel), and (4) the presence of electrolyte (e.g. winter road salt) at the interface between anode and cathode. All these conditions would be met for Mg-steel spot welds in an automotive corrosive environment. It was the objective in this study to evaluate the severity of galvanic corrosion of Mg-steel USW through simple observation of corrosion products and changes of joint strength. It was not the intention to study the details of corrosion mechanisms.

## Experiments

### Mg and Steel Specimens

The materials, machine, and fixture used to make ultrasonic spot welded lap-shear samples were identical to those described in [4,5]. The materials used for these experiments were sheets of 0.8-mm-thick hot-dip-galvanized (HDG) mild steel and 1.6-mm thick AZ31B-H24. Prior to welding the surfaces of the AZ31B sheet were buffed with non-metallic abrasive pads (Scotch-Brite™) to remove surface oxides and produce shiny surfaces. Both metals were cleaned with acetone followed by isopropyl alcohol to remove lubricants and surface debris. The coupons used for welding the lap-shear test specimens had dimensions of 100 mm × 30 mm, and they were overlapped a distance of 75 mm in a fixture. The welds were centered in the overlap region. The only clamping between the coupons was the pressure applied by the welding (sonotrode) tip. Specimens were positioned for welding so that the primary vibration direction of the sonotrode was perpendicular to their long axis.

### Ultrasonic Spot Welding

The machine used for the ultrasonic spot welding was a single-transducer wedge-reed Sonobond CLF 2500. Spot welding was done using a power of 1500 W at impedance setting 6 and welding time of 1.0 sec. The sonotrode tip was made from T1 steel, and it had a flat, rectangular face of 7 mm × 7 mm as shown in [5]. The line pressure to the tip clamping mechanism was adjusted to make the welds under constant nominal pressure of 39 MPa.

### Corrosion Cycling Test

An automotive corrosion cycle test was chosen. Lap-shear coupons were cycled through Ford's "Arizona Proving Ground Equivalent" (APGE) corrosion cycle [7]. The test protocol is:

| *Step* | *Duration* | *Description* |
|---|---|---|
| 1 | 15 min | Immersion in 5% NaCl solution, agitated |
| 2 | 3 hours | Air dry |
| 3 | 20 hr 45 min | Humidity chamber, 50°C 85% RH |

Standard Ford test protocol calls for a complete test to consist of 30 cycles.

In our laboratory, the humidity chamber was a covered glass desiccator with HFP85 potassium chloride saturated aqueous salt solution in a 50°C furnace. The chamber was set up according to the practice defined in ASTM E104-02 [8], which provides ~81% relative humidity at 50°C. No potassium could be detected on exposed samples by energy dispersive spectrometry (EDS) in a scanning electron microscope (SEM).

A total of 30 samples were used for the corrosion cycle test, with an intended plan that one sample would be taken out of test every cycle between 5th and 30th cycles for mechanical lap-shear test. In addition, samples from every 5th cycle were preserved for cross-sectioning and metallographic evaluation.

Lap-shear Mechanical Test

Lap-shear joint strength was measured on an Instron machine at a crosshead speed of 10 mm/min to record the maximum failure load. There was no spacer added to offset the centerline.

X-ray Diffraction (XRD)

Oxidation residue was examined by XRD to study the compounds. Residue from AZ31 samples was ground in a mortar and pestle to a very fine size. Powder x-ray diffraction was performed to determine the composition of the residue. Vaseline was used as the binder and was spread onto a glass slide and the powder was finely dispersed onto this binder. XRD was performed from 15-90°.

## Results and Discussions

Before the 25th cycle, all of the samples were either taken out of the test at specific assigned cycles for mechanical and metallographic analysis, or the joints were completely separated in the humidity chamber.

Joint Strength

The results of the lap-shear testing are presented in Figure 1 where lap-shear strength is plotted against the cycles of exposure. The average strength of non-corroded samples was 4.2 kN [6], shown as the dash baseline in comparing the strength of corroded samples. The joint strength decreases continuously and almost linearly with number of cycles. The last surviving sample was taken out after the 24th cycle and was measured with a failure load of only 0.4 kN. Several samples separated spontaneously in the humidity chamber during the test and were not able to be measured for the strength. Joints up to 17 cycles had what could be considered practically useful strength.

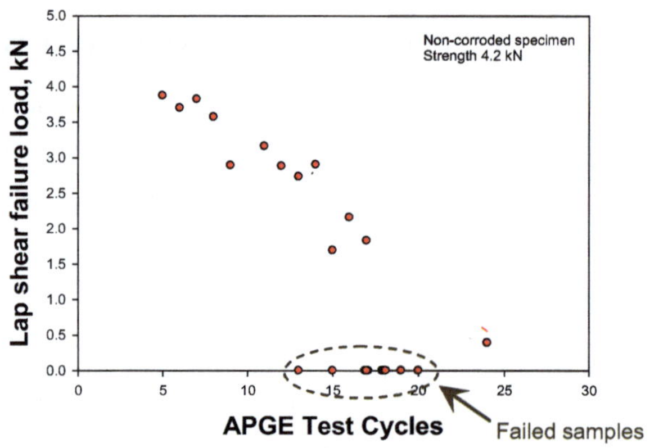


Figure 1. Lap-shear joint strength of USW Mg-steel samples after cyclic corrosion test.

Appearances of the Samples

Deposits grew immediately beginning from the 1st cycle of exposure. Figure 2 shows the appearance of the AZ31 surface where deposits appeared as discrete pits. The pattern of the pits seemed related to the texture of roughness on the surface. Such deposits continued growing in subsequent cycles. After cycle 5, deposits were so dense and thick that the USW joint was hardly visible, as shown in Figure 3. The pits are similar to the pitting corrosion of Mg in aqueous solution with chloride observed by Tunold et al. [9]. They concluded that pitting was initiated by the presence of chloride ions on the metal surface.

Figure 2. Appearance of AZ31 surface after 1 cycle of corrosion test.

Corrosion deposits also grew in the crevices between the AZ31 and steel. Figure 4 is an optical image of the side view of the sample after 5 cycles of corrosion test, clearly showing the deposit grew in-between the AZ31 and steel sheets causing a slight opening of the joint. The deposits in the crevices grew thicker with each successive cycle of exposure. Figure 5 shows a sample after 16 cycles of test. The deposit not only grew thicker, but it also caused significant bending of the steel.

Figure 3. Appearance of AZ31 surface after 5 cycles of corrosion test.

Figure 4. Oxide in the crevice between AZ31 and steel, after 5 cycles of corrosion test. AZ31 was on top.

In addition to the side view of the joint sample, the deposits in the crevices could also be seen after the joints were separated either during the corrosion test or after the mechanical lap-shear test. Figure 6 shows the image of matching (inner) surfaces of the AZ31 and steel after a sample separated after 18 cycles of corrosion test. A large white piece of deposit, about 50 mm in length and across the width of the sheet (30 mm), can be seen adhering to the AZ31. However, from sample to sample, the deposit could be attached to either AZ31 or steel when the joint was separated.

<u>XRD and EDS Analyses</u>

Samples of deposit were analyzed by X-ray diffraction. A representative spectrum is shown in Figure 7. Residues were identified as $Mg(OH)_2$ with major peaks matched exactly as in [10,11]. No sodium (Na) was found in the deposits.

The finding of $Mg(OH)_2$ is in agreement with other published observations [10,12,13,14]. All of those studies were conducted by immersing in aqueous NaCl solutions. It seems, however, that the NaCl immersion, drying, and cyclic exposure to 85% RH humidity produce the same basic corrosion response.

Figure 5. Oxide in the crevice between AZ31 and steel after 16 cycles of corrosion test. AZ31 was on top.

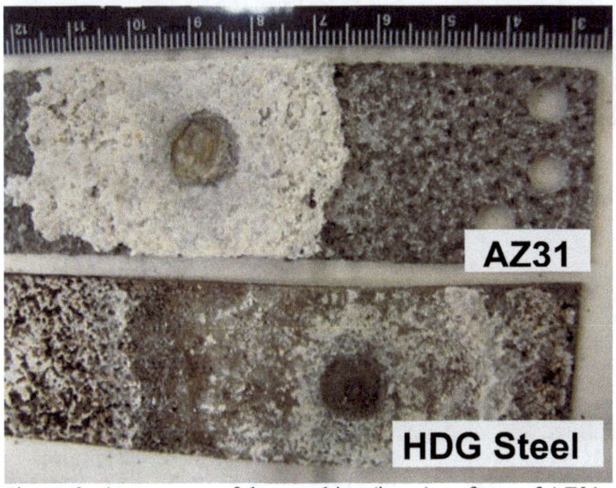


Figure 6. Appearance of the matching (inner) surfaces of AZ31 and HDG steel specimens after 18 cycles of corrosion test.

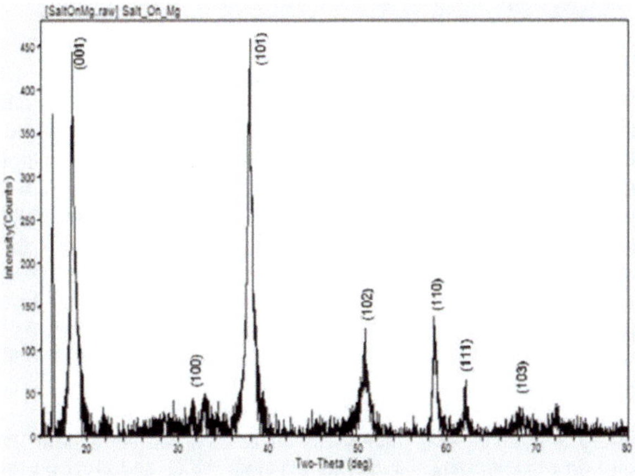


Figure 7. X-ray diffraction analysis result of the deposit.

In Figure 6, the surface of the deposit facing the steel sheet has a layer of ZnO, identified from the EDS analysis, indicating the consumption of the Zn HDG coating on the steel surface. As exposure cycles accumulated the Zn was progressively consumed. That led to exposure of fresh steel surface to the corrosive environment producing brown-color iron oxides.

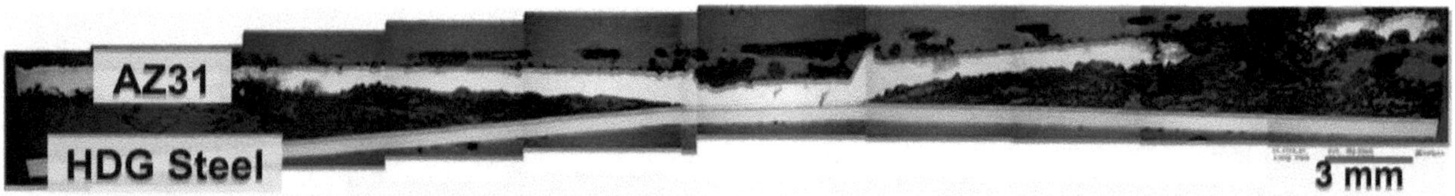


Figure 8. Optical micrograph of cross-section of a sample after 10 cycles of corrosion test. The AZ31 is on top and the HDG steel is at the bottom.

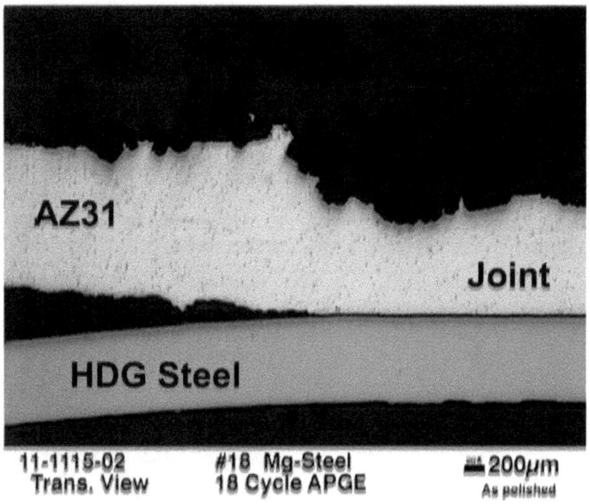


Figure 9. Optical micrograph of cross-sectioned USW after 10 cycles of corrosion test.

Cross-section Micrographs

Samples of every 5th cycle were cross-section, mounted, and polished to examine the depth of corrosion and microstructure of the AZ31. Figure 8 is a composite photo from several micrographs of the sample after 10 exposure cycles. It shows that both AZ31 and steel have been deformed, with steel being deformed more severely than the AZ31. The distance of the gap between Mg and steel was about 5 mm at a location 10 mm away from the joint. Since the original non-corroded samples had very tight fits between AZ31 and steel, the gap, which widened with cycles, was clearly opened by the force from the growth of deposit in the crevice.

Figure 9 is an optical micrograph of an area showing both the joint on the right and the crevice between AZ31 and steel on the left. The corrosion in the AZ31 proceeded by pitting and occurred on both its outer and inner surfaces, as well as into the matrix of Mg. By matching the appearance of the sample before cross-sectioning, the dark circles on top of the AZ31 surface are believed to be the $Mg(OH)_2$ deposit. Similar localized corrosion was shown by Pardo et al [12] for AZ31 immersed in 3.5% NaCl solution. It was said that the $MnAl_2$ inclusions and β-phase ($Mg_{17}Al_{12}$) formed a galvanic couple with the surrounding Mg matrix which caused localized corrosion. The front of corrosion attack was stopped when it reached the lamellar aggregate of the β-phase ($Mg_{17}Al_{12}$). The existence of localized galvanic cells certainly could explain the localized corrosion observed in this study. It may also be influenced by local microstructure, which is affected by the ultrasonic spot welding [4].

The corrosion on the AZ31 sheet was not only localized but also non-uniform across the length of the sample. It can be seen in Figure 8 that the thinning of the Mg sheet was more severe not near the joint but rather at the distance about 15 to 20 mm away from it. This may be related to the "global" galvanic couple between Mg and steel which caused the maximum corrosion to be away from the point of contact. There was no attempt in this study to analyze the detailed mechanism. However, the phenomenon is similar to what was reported in [15] where the corrosion between aluminum and a steel bolt was studied. The maximum corrosion in aluminum was found to be about 1 to 2 mm away from the aluminum-steel contact interface.

Corrosion did not occur at the welded interface between steel and AZ31. Figures 10 (a) and (b) shows the failure surfaces of AZ31 and steel sheet, respectively, following lap-shear testing sample after 14 exposure cycles. A small bump, or a nugget, can be seen on the shiny metal surface. The sample had a failure load of 2.9 kN, about 70% of the non-corroded samples (4.2kN). EDS analysis on the failure surface of this joint showed Mg on both sides of the sheets, indicating the failure occurred within the Mg material. It is also evident from Figure 9 that there was no corrosion at the welded interface.

The size of the fractured Mg interface gets smaller with the number of corrosion cycles. This would be due to the penetration of corrosive medium into the interface when the joint is gradually pushed open from the growth of the deposit in the crevice. The decrease of joint strength was due to the reduced size of the remaining welded metal.

## Conclusions

The ultrasonic spot welds between AZ31B-H24 and HDG steel was evaluated after an automotive cyclic corrosion test. In order to simulate the worst corrosive situation, no protective coating or other means to minimize the galvanic corrosion was used. Joint strength decreased continuously and linearly with the cycles of corrosion from 4.2kN of non-corroded samples to complete spontaneous failure before the 25th exposure cycle. No practical joint strength was preserved after 17 cycles. Deposits of $Mg(OH)_2$, identified by X-ray diffraction, mainly grew in the crevices between the AZ31 and steel sheets, and to a much less degree, on the outer surfaces of the AZ31. Deposits grew thicker with exposure cycles, causing the deformation of the steel and AZ31 sheets.

(a)

(b)

Figure 10. Failure surface of (a) AZ31 and (b) HDG steel after lap-shear test. The sample was gone through 14 cycles of corrosion tests.

Welded joint interfaces between AZ31 and steel were not directly attacked by the corrosion. But the joint area decreased from the consequence of the opening of the joint due to the growth of deposits. Metallographic study showed localized and non-uniform corrosion on Mg, with the most severe corrosion occurring somewhere about 15 to 20 mm away from the joint. Detailed mechanism was not studied but it was believed to be associated with the "global" galvanic cell between Mg and steel.

**Acknowledgements**

This research was sponsored by the U.S. Department of Energy, Assistant Secretary for Energy Efficiency and Renewable Energy, Office of Vehicle Technologies, as part of the Lightweight Materials Program. The authors also appreciated Ms. Katherine Strader for performing the x-ray diffraction analysis, and Dr. Robert C. McCune for his valuable suggestions and discussions for the corrosion test setup and experimental results.

**References**

1. R.S. Beals, C. Tissington, X. Zhang, K. Kainer, J. Petrillo, M. Verbrugge, "Magnesium global development: Outcomes from the TMS 2007 annual meeting", *JOM*, **59**(8) (2007), pp. 39–42.

2. N.N., "Magnesium Vision 2020: A North American Automotive Strategic Vision for Magnesium", United States Automotive Materials Partnership (USAMP) Automotive Metals Division (AMD), released November 2006.

3. E. Wall, R. Sullivan, J. Carpenter, "Automotive Lightweighting Materials FY 2006 Progress Report", http://www1.eere.energy.gov/vehiclesandfuels/pdfs/alm_06/2g_osborne.pdf, October 2007.

4. M.L. Santella, T.J. Franklin, J. Pan, T. Pan, E. Brown, "Ultrasonic Spot Welding of AZ31B to Galvanized Mild Steel", *SAE International Journal of Materials and Manufacturing*, **3**(1) (2010) pp. 652-657

5. T.J. Franklin, J. Pan, M.L. Santella, T. Pan, "Fatigue Behavior of Dissimilar Ultrasonic Spot Welds in Lap-Shear Specimens of Magnesium and Steel Sheets" *SAE Technical Paper* 2011-01-0475, April 2011.

6. S. Jana, Y. Hovanski, G.J. Grant. "Friction Stir Lap Welding of Magnesium Alloy to Steel: A Preliminary Investigation." *Metallurgical and Materials Transactions. A, Physical Metallurgy and Materials Science* **41A**(12) (2010) pp. 3173-3182.

7. K. Lazarz, R.C. McCune, W. Wang, P.K. Mallick, "Effect of Surface Pretreatments on Adhesive Bonding and Corrosion Resistance of AM60B, AZ31-H24, and AM30 Magnesium", *SAE Technical Paper* 2009-01-0037, April 2009.

8. N.N., "Standard Practice for Maintaining Constant Relative Humidity by Means of Aqueous Solutions", ASTM E104 – 02 (2007).

9. R. Tunold, H. Holtan, M. Hägg Berge, A. Lasson, R. Steen-Hansen, "The Corrosion Of Magnesium In Aqueous Solution Containing Chloride Ions", *Corrosion Science*, **17**(4) (1977), pp. 353-365.

10. R. Ambat, "Evaluation of Microstructural effects on corrosion behaviour of AZ91D magnesium alloy", *Corrosion Science* **42** (2000) pp. 1433–1455.

11. B. Xu, H. Deng, Y. Dai, B. Yang," Preparation of Nanoparticles of Magnesium Hydroxide from Bittern", *Transactions of Nonferrous Metals Society of China*, 17 (2007), pp s671-s674.

12. A. Pardo, M.C. Merino, A.E. Coy, R. Arrabal, F. Viejo, E. Matykina, "Corrosion Behaviour of Magnesium/Aluminium Alloys in 3.5 wt.% NaCl", *Corrosion Science* **50** (2008) pp. 823–834.

13. B.J. Schneider, R.C. McCune, M.S. Ricketts, J.A. Hines, "Examination of the Corrosion Behavior of Creep-Resistant Magnesium Alloys in an Aqueous Environment" *SAE Technical Paper* 2007-01-1023, April 2007.

14. G. Song, G. A. Atrens, "Recent Insights into the Mechanism of Magnesium Corrosion and Research Suggestions", *Advanced Engineering Materials*, **9**(3), (2007) pp. 177-183.

15. T. Mizukami, "Simulation of Galvanic Corrosion of Aluminum Materials for Vehicles", *SAE Technical Paper* 2010-01-0724, April 2010.

**Magnesium Technology 2012**
*Edited by: Suveen N. Mathaudhu, Wim H. Sillekens, Neale R. Neelameggham, and Norbert Hort*
***TMS (The Minerals, Metals & Materials Society), 2012***

# EFFECT OF SOME MICROSTRUCTURAL PARAMETERS ON THE CORROSION RESISTANCE OF MAGNESIUM ALLOYS

Yaning Hu[1], Joseph R. Kish[1], Joseph R. McDermid[1], Wenyue Zheng[2]
[1]Centre for Automotive Materials and Corrosion, McMaster University, 1280 Main St. W., Hamilton, ON, L8S 4L7 Canada
[2]CANMET-MTL, 183 Longwood Rd. S., Hamilton, ON, L8P 0A5 Canada

Keywords: Magnesium Alloys, Microstructure, Corrosion

## Abstract

The influence of the β-$Mg_{17}Al_{12}$ phase and, to a lesser degree, the solute content in the α-Mg matrix on the corrosion resistance of Mg alloys was investigated by cyclic potentiodynamic polarization and potentiostatic polarization tests using AZ31B, AM30 and AM60B in contact with a mildly aggressive near-neutral saline solution. Results showed that all three Mg alloys corrode in a partially protective state under open circuit conditions in the test solution. It was also determined that the surface film formed on each exhibits a similar apparent breakdown potential. This indicates that microstructure parameters such as the presence of the β-phase and the solute content of the α-Mg matrix do not strongly influence the factors controlling the breakdown of the surface films formed. It was further determined that the moderately improved protectiveness of the surface film, rather than the distribution of the β-$Mg_{17}Al_{12}$ phase, is responsible for the improved (short-term) corrosion resistance exhibited by AM60B at potentials below the breakdown potential.

## Introduction

Reducing the overall mass of automobiles by increased utilization of Mg alloys is one of the means by which improved fuel efficiency and reduced harmful emissions can be attained. The 2020 strategic goal of the North American-based automotive manufacturers is to substitute 340 lbs. of Mg components for 630 lbs. of current carbon steel and Al alloy parts (290 lbs. weight saved), bringing the total average vehicle Mg content to 350 lbs. [1]. The multi-national MFERD (magnesium front end research and development) project is one major effort towards reaching this goal. The objectives of the project are to develop: (i) key enabling technologies in Mg extrusion, sheet, high-integrity body casting, joining and assembly, and (ii) a knowledge base for increased Mg automotive body applications in corrosion protection, crashworthiness, fatigue, durability and NHV (noise, harshness and vibration) performance [2]. The Mg alloys of choice for the assembly include AZ31 sheet, AM30 extrusions and AM60B die castings.

The corrosion resistance of Mg alloys is a major technological issue preventing widespread application. It is strongly influenced by the alloy composition, microstructure and the formation of a partially-protective surface film. Controlling factors for Mg alloys containing Al, Zn and Mn include the solid solution α-matrix, secondary β-phase ($Mg_{17}Al_{12}$) formed along grain boundaries and intermetallic $Al_xMn_yFe_z$ particles [3-5]. Alloyed Al modifies the composition of the surface film formed on both the α-matrix and β-phase [6]. Recent studies indicate that an increased Al content improves the corrosion resistance in moist air [7], salt spray [8], and aqueous $Cl^-$-containing solutions [9]. The β-phase, which is an effective cathode with a corrosion potential more positive than that of the α-matrix, may accelerate the anodic dissolution of the α-phase by establishing a microgalvanic cell [10-12]. Both the grain size of the α-matrix and the distribution of the β-phase are important factors, as the β-phase can act either as an effective cathode (coarse-grain α-matrix and discontinuous β-phase) or as a corrosion barrier (fine-grain α-matrix and continuous β-phase) [10-12]. The $Al_xMn_yFe_z$ particles, also with a corrosion potential more positive than that of the α-matrix, can act as effective cathodes and accelerate the anodic dissolution of the α-phase, albeit in a more localized manner than that produced by the β-phase [1,11,13].

Standard cyclic salt spray testing is a common method to evaluate the corrosion resistance of Mg alloys exposed to more-realistic, on-vehicle conditions. Comparative corrosion performance data for the three MFERD Mg alloys, using such test methods, is now being reported in the open literature [14]. While these tests are instructive from a relative corrosion performance perspective, they provide little insight into the important, yet subtle differences in electrochemical behavior. Such information is required for a more complete mechanistic understanding of the observed performance. To address this need, in support of the MFERD effort, cyclic potentiodynamic anodic polarization and potentiostatic polarization tests were conducted using the MFERD Mg alloys (AM30, AZ31B and AM60B) in contact with the mildly aggressive near-neutral saline solution utilized in the GMW14872 (Cyclic Corrosion Laboratory Test) Standard.

## Experimental Details

The composition of the three Mg alloys tested is shown in Table I. Significant differences include the Zn content between AZ31B and AM30 and the Al content between AM30 and AM60B. The as-received AZ31B sheet was given a recrystallization heat treatment at 350°C for 6 h. No heat treatments of any kind were performed on the as-received extruded AM30 bar or the die-cast AM60B sheet.

Table I. Composition (wt.%) of Mg Alloys

| Alloy | Al | Zn | Mn | Fe | Si |
|---|---|---|---|---|---|
| AZ31B | 3.1 | 0.96 | 0.43 | 0.002 | 0.02 |
| AM30 | 2.7 | < 0.01 | 0.42 | 0.014 | < 0.01 |
| AM60B | 5.6 | 0.10 | 0.34 | 0.007 | 0.02 |

Figure 1 shows the cross-sectional microstructure normal to the longitudinal direction (rolling direction of AZ31 sheet and extrusion direction of AM30 bar) of the three Mg alloys. The microstructure of both 3 wt.% Al-containing alloys (AZ31B and AM30) revealed an equiaxed-grain α-matrix that contained randomly distributed Al-Mn intermetallic particles (confirmed by SEM-EDS). In contrast, the microstructure of the 6 wt.% Al-containing AM60B revealed a discontinuous network of β-phase precipitates along the α-matrix grain boundaries. Both the extruded AM30 and die-cast AM60B exhibited a distinct surface/skin layer; a coarse-grain skin on a fine-grained interior for the extruded AM30 and a fine-grain skin on a coarse-grained interior for the die-cast AM60B. The spherical pit-like features are an artifact of the mechanical polishing procedure employed.

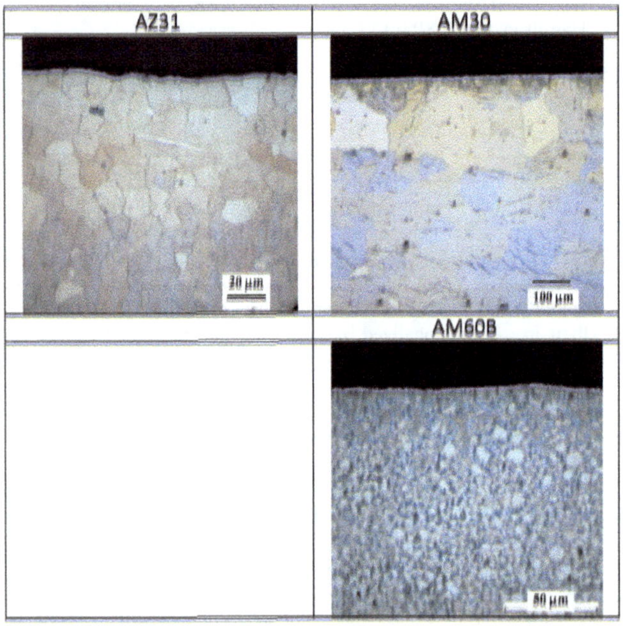


Figure 1. Cross-sectional microstructures normal to the longitudinal direction (rolling direction of AZ31B sheet and extrusion direction of AM30 bar) of the three Mg alloys studied. Note the varying magnifications between the micrographs.

Electrochemical measurements were made in the GMW14872 test solution (9 wt.% NaCl, 0.1 wt.% $CaCl_2$ and 0.25 wt.% $NaHCO_3$) with pH 7.9 at room temperature. No attempt was made to aerate or deaerate the solution. Measurements were conducted using a standard three electrode test cell with a volume of 1000 mL. All working electrodes were prepared by cold mounting a small square sample (1 $cm^2$) in epoxy resin. Electrical contact was obtained by adhesive joining a Cu wire to the back side of the working electrode prior to cold mounting. This Cu lead was inserted into a plastic tube, which was then sealed to the back of the epoxy resin mount using epoxy resin. The majority of working electrodes prepared in this way were ground to a 1200 grit surface finish using SiC paper and absolute ethanol (100%) as a lubricant, then dried in cool air prior to immersion. This mechanically polishing (MP) procedure successfully removed the skin layer in the AM30 and AM60B electrodes. Consequently, additional electrodes were prepared from AM30 and AM60 in the as-received condition (AR), without mechanical polishing, to evaluate the polarization behavior of the skin layer. All electrochemical tests were performed using either an EG&G Model 273 potentiostat or a GAMRY Reference 600 potentiostat with a graphite rod counter electrode. All electrode potentials were measured against a standard calomel (SCE) reference electrode.

All polarization measurements were initiated after the Mg alloy electrode was aged at the corrosion potential for 1h. The cyclic potentiodynamic anodic polarization measurement was initiated at -0.25 V versus the corrosion potential and was reversed at +0.50 V versus the corrosion potential. The scan rate was 1 mV/s. Anodic potentiostatic polarization measurements were made at an applied potential just below the apparent breakdown potential for a hold time of 2 h and at an applied potential just above the apparent breakdown potential for a hold time of 0.5 h. Cathodic potentiostatic polarization measurements were made at an applied potential of -1.8 $V_{SCE}$ for a hold time of 2 h.

## Results

### Potentiodynamic Measurements

Figure 2 shows the cyclic potentiodynamic anodic polarization curves measured on the MP surface of the three Mg alloys. All three curves exhibit a similar general behavior, namely an apparent anodic breakdown potential, above which the anodic current density increases markedly, and an apparent repassivation potential following a hysteresis loop upon reversing the applied potential. The corrosion potential exhibited by AM60B was significantly more positive than that exhibited by either AM30 or AZ31B. In contrast, both the breakdown potential and the repassivation potential exhibited by the three Mg alloys were similar. The breakdown potential for each of the three Mg alloys was more positive than their respective corrosion potentials, but the repassivation potential of AM60B was more negative than its corrosion potential.

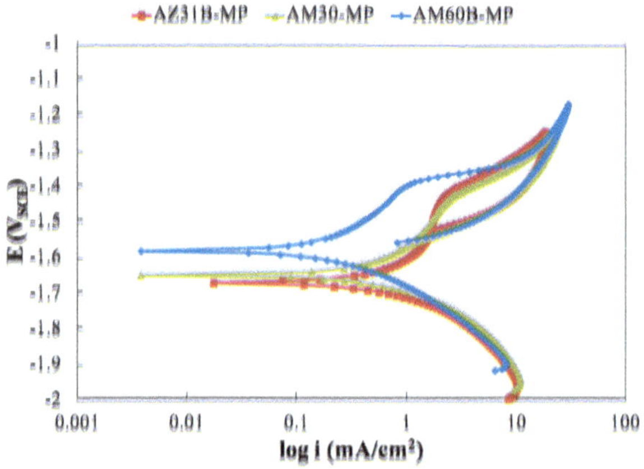


Figure 2. Cyclic potentiodynamic anodic polarization curve for MP surface of AZ31B, AM30 and AM60B after aging for 1 h at the corrosion potential in the GMW14872 test solution at room temperature.

The cyclic potentiodynamic anodic polarization curves measured on the skin and on the interior (polished) surface of extruded AM30 and die cast AM60B are shown in Figure 3. The coarse-grained skin layer of the extruded AM30 exhibited a similar anodic polarization response to that exhibited by the fine-grained interior. In contrast, a marked difference in the response between the fine-grained skin layer and the coarse-grained interior was observed for the die-cast AM60B. The main cause of the marked difference was the more noble corrosion potential exhibited by the fine-grained skin layer. Despite this difference in corrosion potential, both the fine-grained skin layer and the coarse-grained interior of AM60B exhibited a similar breakdown potential and repassivation potential.

Table II lists selected electrochemical kinetic parameters extracted from the curves presented in Figure 2 & 3. The potentials listed support the comparative observations made above regarding the corrosion ($E_{corr}$), breakdown ($E_b$) and repassivation ($E_r$) potentials. The cathodic Tafel slope ($\beta_c$) corresponds to the slope of the linear portion of the cathodic polarization curve, which was assumed to initiate at an applied potential of 50 mV versus the corrosion potential. Both the corrosion current density ($i_{corr}$) and the exchange current density of the hydrogen evolution cathodic reaction ($i_{o,c}$) were computed by extrapolating the linear (Tafel) portion of the cathodic polarization curve back to the respective corrosion potential and the equilibrium half cell potential of the hydrogen evolution reaction (-0.694 $V_{SCE}$ at pH 7.9). The corrosion current density (rate) exhibited by AM60B was an order of magnitude lower than both AZ31B and AM30, which were similar, The corrosion current density of the skin layer and the interior was of the same order of magnitude for both AM30 and AM60B. Moreover, the exchange current density for the cathodic reaction (hydrogen evolution) estimated for all samples was also within one order of magnitude, with the exception of the AM60B-AR sample, which was an order of magnitude lower.

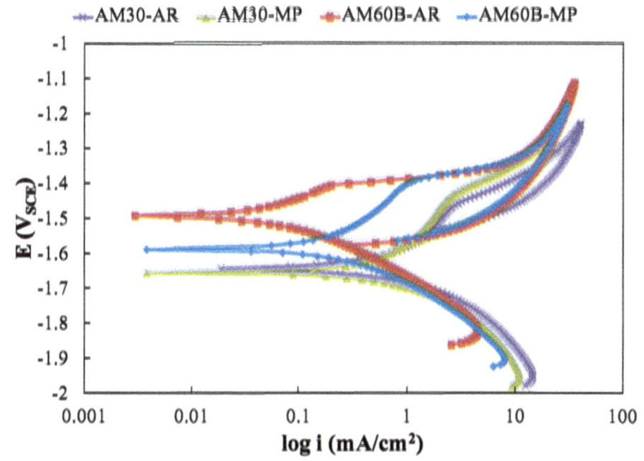


Figure 3. Cyclic potentiodynamic anodic polarization curve for the MP surface and the AR surface of AM30 and AM60B after aging for 1 h at the corrosion potential in the GMW14872 test solution at room temperature.

Table II. Selected Electrochemical Kinetic Parameters

| Alloy | $E_{corr}$ $V_{SCE}$ | $E_b$ $V_{SCE}$ | $E_r$ $V_{SCE}$ | $\beta_c$ (V/dec.) | $i_{corr}$ (A/cm²) | $i_{o,c}$ (A/cm²) |
|---|---|---|---|---|---|---|
| AZ31B-MP | -1.66 | -1.43 | -1.52 | 0.215 | $1.1\times10^{-3}$ | $2.5\times10^{-8}$ |
| AM30-AR | -1.63 | -1.46 | -1.52 | 0.202 | $1.1\times10^{-3}$ | $2.3\times10^{-8}$ |
| AM30-MP | -1.66 | -1.42 | -1.54 | 0.213 | $1.0\times10^{-3}$ | $1.2\times10^{-8}$ |
| AM60B-AR | -1.50 | -1.41 | -1.57 | 0.180 | $1.6\times10^{-4}$ | $5.8\times10^{-9}$ |
| AM60B-MP | -1.56 | -1.40 | -1.58 | 0.218 | $3.7\times10^{-4}$ | $4.1\times10^{-8}$ |

Potentiostatic Measurements

Figure 4 shows the anodic current density transient recorded on the MP surface in response to the anodic potentiostatic polarization at an applied potential just above and just below the breakdown potential for the three Mg alloys. Also included is the anodic current density transient recorded on the AR surface of AM60B at an applied potential just above the breakdown potential. A measurement on the AR surface of AM60B at an applied potential just below the breakdown potential was not attempted since the corrosion potential was 90 mV more negative than the breakdown potential. The AR surface of AM30 was not tested since it exhibited essentially the identical cyclic polarization behavior as observed for the MP surface. All anodic current density transient exhibited a similar behavior, namely an initial rapid increase followed by a much slower decay from the maximum value obtained. The anodic current density was at least one order of magnitude lower at an applied potential just below the breakdown potential than that exhibited at an applied potential just above the breakdown potential for all three Mg alloys. The anodic current density transient measured on the MP surface was lowest for the die-cast AM60B at both applied potentials. The difference in magnitude of the anodic current density exhibited by the AM60B-MP compared to that exhibited by the other two Mg alloys was greater (an order of magnitude lower) for the applied potential below the apparent breakdown potential.

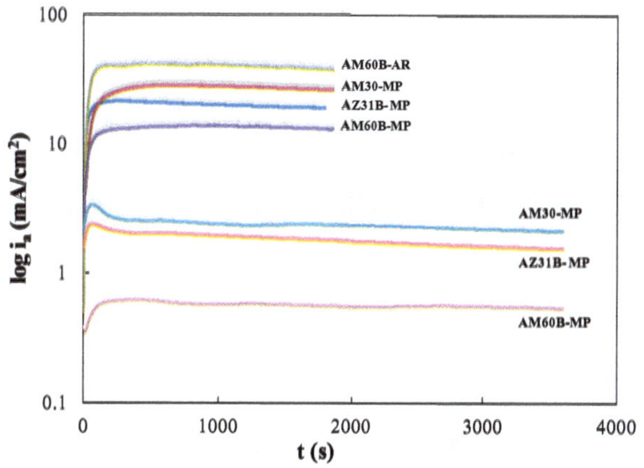


Figure 4. Anodic current transient for the MP surface of the three Mg alloys and the AR surface of AM60B in the GMW14872 test solution at room temperature for an applied potential just above (upper transients) and below (lower transients) the breakdown potential in the GMW14872 test solution at room temperature.

Photographs of the MP surface of AM60B after the potentiostatic anodic polarization experiment are shown in Figure 5. There was little evidence of corrosion observed on the surface exposed at an applied potential just below the breakdown potential. The surface was dull grey and macroscopically smooth in appearance. In contrast, there was significant corrosion observed on the surface exposed at an applied potential just above the breakdown potential. The corrosion mode best resembled extensive pitting corrosion. A similar set of observations was extracted for the MP surface of the two AZ31B and the two AM30 samples exposed to similar conditions.

Light optical microscope images of the MP surface cross-sections for the three Mg alloys after potentiostatic anodic polarization experiments are shown in Figure 6. Superficial corrosion attack was observed on each of the three Mg alloy surfaces held at an applied potential just below the breakdown potential. In contrast, significant corrosion attack was observed on each of three Mg alloy surfaces held at an applied potential just above the breakdown potential. This significant corrosion attack was pit-like in appearance and tended to spread laterally across the surface rather than propagate deep into the Mg alloy sample.

Figure 5. Photograph of the MP surface for AM60B (~1cm$^2$) after potentiostatic anodic polarization at an applied potential just above and just below the breakdown potential ($E_b$) in the GMW14872 test solution at room temperature.

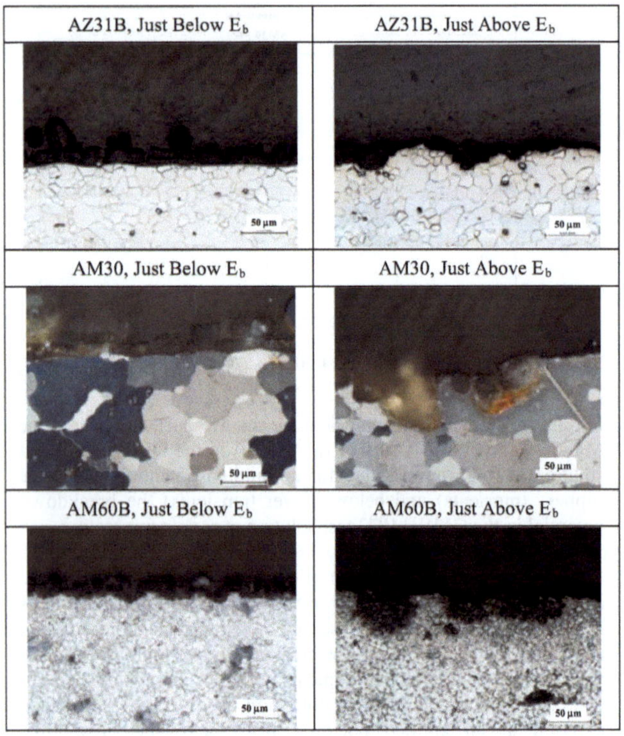


Figure 6. Light optical microscope images of the MP surface, in cross-section, for the three Mg alloys after potentiostatic anodic polarization at an applied potential above and below the breakdown potential ($E_b$) in the GMW14872 test solution at room temperature.

Figure 7 shows the cathodic current density transient recorded on the MP surface of the three Mg alloys in response to the cathodic potentiostatic polarization at an applied potential of -1.8 $V_{SCE}$. Also included is the cathodic current density transient recorded on the AR surface of AM60B. All cathodic current density transients exhibited a similar behavior, namely an initial increase followed by the establishment of a reasonably steady state value. The initial increase was relatively rapid except for AZ31B. Moreover, the steady state current density was established at the maximum value except for the AR surface of AM60B, on which the steady state current density was attained after a decay from the maximum value. Regardless on the initial transient behavior, the steady state current density attained after 1 h exposure was within the same order of magnitude for each case.

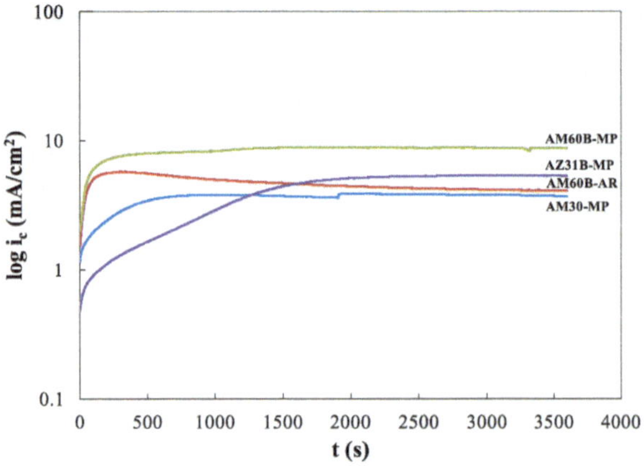


Figure 7. Cathodic current transient for the MP surface of the three Mg alloys and the AR surface of AM60B at an applied potential of -1.8 $V_{SCE}$ in the GMW14872 test solution at room temperature.

## Discussion

Based on the anodic polarization behavior (Figure 2 & 3), all three Mg alloys, regardless of the surface condition exposed, corrode at a potential that is below the breakdown potential after aging for 1 h in the GMW14872 test solution at room temperature. Both the steady state applied anodic current density transients (Figure 4) and the cross-section images (Figure 6) indicate that all three Mg alloys corrode in a partially protected state at these potentials (below the breakdown) for all three Mg alloys. Moreover, the relative differences observed in the anodic current density transients, at an applied potential below the breakdown potential (Figure 4), and in the cathodic current density transients (Figure 7) indicate that the improved relative corrosion resistance of AM60B was controlled primarily by the anodic kinetics. The significance of these findings is discussed in some detail below.

The presence of a breakdown potential in the anodic polarization behavior of AZxx Mg alloys in mildly aggressive aqueous salt ($Cl^-$, $SO_4^{2-}$ and $HCO_3^-$) solutions has been well documented [14-16]. However, a consensus on a micro-mechanisms involved has not yet been attained. The majority of the focus has been placed on delineating the role played by the the microgalvanic interaction between the α matrix and an adjacent effective cathodic phase, such as AlMn intermetallic particles and/or the β phase [4,9,10]. The role by the aggressive anion ($Cl^-$) in the film breakdown process has received little attention in this context. However, the incorporation of the aggressive anion ($Cl^-$) into the surface film at weak points and the subsequent formation of soluble magnesium hydroxychloride complexes has been tenuously proposed to explain the observed behaviour [17,18].

In this study the breakdown potential is essentially the same for all three Mg alloys, regardless of the surface condition exposed. This suggests that a common factor controls the breakdown process. If true, then any factor associated with the β–phase can be ruled out, since it was not present in either AZ31B or AM30. Given that AlMn intermetallic particles are present in all three Mg alloys, any associated factors cannot be directly ruled out. However, at potentials below the breakdown potential, the cathodic (hydrogen evolution) reaction is believed to occur on the relatively thick surface film rather than on the Mg alloy substrate [6,19]. If true, then it seems unlikely that the AlMn intermetallic particles present at the interface between the relatively thick surface film and the Mg alloy substrate could act as effective cathodes causing the surface film to locally break down.

It follows that the controlling factor may indeed be the migration of the aggressive anion ($Cl^-$) into the surface film at weak points. A common breakdown potential would require that the bulk composition of the surface film formed on the three Mg alloys to be largely similar in composition. A recently reported XPS study on the composition of the surface film formed after short exposure to 0.1 N NaCl solutions show this to be the case for AZxx alloys [20]. The surface film is typically comprised of a mixed $Mg(OH)_2$-$MgCO_3$ outer layer on a mixed $Mg(OH)_2$-MgO inner layer. A similar, albeit minor, Al content independent of the concentration in the bulk alloy was observed in the surface film after sputtering.

A comparison of the corrosion rate data (Table II) suggests that there is a meaningful difference between the three Mg alloys, with AM60B exhibiting the highest relative corrosion resistance. It is recognized that a poor manual fit of the cathodic Tafel slope, as done in this study, can change the extrapolated corrosion current density (rate) by a factor between 5 and 10 [21]. Such a maximum error (factor of 10) would discount the observation that there is a meaningful difference in the corrosion rate. However, the order of magnitude improved relative corrosion resistance exhibited by AM60B based on the cathodic Tafel extrapolated corrosion current density (Table II) is consistent with that based on the steady state anodic current density attained at an applied potential below the breakdown potential (Figure 4). This agreement gives a level of confidence that this difference is significant. Thus, the relative ranking of corrosion resistance, at a potential below the breakdown potential, in the GMW14872 solution is given by:

$$\text{AM60B} > \text{AM30} \approx \text{AZ31B}$$

Of the two half-cell processes involved in the corrosion cell, the anodic process was observed to be chiefly responsible for the order of magnitude lower relative corrosion current density (rate) exhibited by AM60B. The cathodic (hydrogen evolution) process is discounted based on the observation that there was no significant difference in either the order of magnitude of the extrapolated exchange current density or the steady state cathodic current density attained an applied potential of -1.8 $V_{SCE}$ exhibited by the three Mg alloys. However, similar cathodic kinetics do indicate that neither the solute (Zn) content of the α-matrix nor the presence of the β-phase has a significant influence at a corrosion potential less than the breakdown potential.

Regarding the anodic process, the root cause of the improved corrosion resistance of AM60B is unclear. However, it is not likely the result of a continuous β-phase barrier layer present at the alloy surface. If true, then differences would be expected in both the breakdown potential and the cathode kinetics, compared to that exhibited by either AZ31B or AM30, based on the reported electrochemical polarization behavior of the β-phase in an aqueous $Cl^-$-containing solution [9]. Based on the essentially identical anodic kinetics exhibited by Zn-containing AZ31B and non Zn-containing AM30, the root cause is also likely not the result of the solute Zn content in the a-matrix

It is noted that this observed improved corrosion resistance (AM60B relative to both AZ31B and AM30) is consistent with a reported study comparing the corrosion resistance between of AZ31 and AZ91 in mildly aggressive aqueous $Cl^-$-containing solutions, in which the corrosion potential was below the breakdown potential [20]. In that study, the improved corrosion resistance was tentatively attributed to MgO as the dominant surface film component. Clearly a detailed compositional analysis of the surface films formed during the exposure at a corrosion potential less than the breakdown potential is required to better understand the root cause of the improved corrosion resistance exhibited by the β-phase containing Mg alloys.

## Conclusions

1. Electrochemical polarization measurements after short exposures were instructive in providing more insight into the important, yet subtle differences in electrochemical behavior exhibited by AZxx and AMxx Mg alloys when exposed at potentials below the breakdown potential.

2. The overall relative ranking of corrosion resistance, at a potential below the breakdown potential, in the GMW14872 solution is given by: AM60B > AM30 ≈ AZ31B. This relative ranking is controlled more so by the anodic kinetics rather than the cathode kinetics. An improved protectiveness of the surface film, rather than the distribution of the β-$Mg_{17}Al_{12}$ phase, appears to be responsible for the improved corrosion resistance of AM60B (regardless of surface condition).

3. All three Mg alloys exhibited no significant differences in breakdown potential. This indicates that microstructure aspects such as the solute content in the α-Mg matrix and the presence of the β-phase do not strongly influence the factors controlling the breakdown of the surface films formed.

## Acknowledgments

This work was financially supported by Natural Resources Canada (CANMET-MTL) and the Natural Sciences and Engineering Research Council of Canada (NSERC). The authors also wish to thank D. Culley of the Department of Materials Science and Engineering at McMaster University for his technical support with metallographic sample preparation. Thanks also to GM for providing samples of the MFERD Mg alloy.

## References


1. G. Cole, "Summary of Magnesium Vision 2020: A North American Automotive Strategic Vision for Magnesium," *Magnesium Technology 2007*, ed. R.S. Beals et al., (Warrendale, PA: TMS, 2007), 35-40.
2. E.A. Nyberg et al., "Magnesium for Future Autos," *Adv. Mater. Processes*, 166 (10) (2008), 35-37.
3. G.L Makar, J. Kruger, "Corrosion of Magnesium," *Int. Mater. Rev.*, 38 (1993), 138-153.
4. G. Song and A. Atrens, "Understanding Magnesium Corrosion: A Framework for Improved Alloy Performance," *Adv. Eng. Mater.*, 5 (2003), 837-858.
5. E. Ghali et al., "General and Localized Corrosion of Magnesium Alloys: A Critical Review," *J. Mater. Eng. Perform.*, 13 (2004), 7-23.
6. G. L. Song et al., "Corrosion Behaviour of AZ21, AZ501 and AZ91 in Sodium Chloride," *Corros. Sci.*, 40 (1998), 1769-1791.
7. M. Jonsson and D. Persson, "The Influence of Microstructure on the Atmospheric Corrosion Behaviour of Magnesium Alloys AZ91D and AM50," *Corros. Sci.*, 52 (2010), 1077-1085.
8. G. Ballerini et al., "About Some Corrosion Mechanisms of AZ91D Magnesium Alloy," *Corros. Sci.*, 47 (2005), 2173-2184.
9. A. Pardo et al., "Corrosion Behaviour of Magnesium/ Aluminum Alloys in 3.5 wt.% NaCl," *Corros. Sci.*, 50 (2008), 823-834.
10. O. Lunder et al., "The Role of $Mg_{17}Al_{12}$ Phase in the Corrosion of Mg Alloy AZ91," *Corrosion*, 45 (1989), 741-748.
11. G. L. Song and A. Atrens, "Corrosion Mechanisms of Magnesium Alloys," *Adv. Eng. Mater.*, 1 (1999), 11-33.
12. G. L. Song, "Recent Progress in the Corrosion and Protection of Magnesium Alloys," *Adv. Eng. Mater.*, 7 (2005), 563-586.
13. O. Lunder et al., "Effect of Mn Additions on the Corrosion Behaviour of Mould-Cast Magnesium ASTM AZ91," *Corrosion*, 43 (1987), 291-295.
14. M-C. Zhao et al., "An Exploratory Study of the Corrosion of Mg Alloys during Interrupted Salt Spray Testing," *Corros. Sci.*, 51 (2009), 1277-1292.
15. L. Wang et al., "Influence of Chloride, Sulfate and Bi-carbonate Anions on the Corrosion Behavior of AZ31 Magnesium Alloy," *J. Alloy Compd.*, 496 (2010) 500–507.
16. L. Wang et al., "Corrosion Behavior of Mg, AZ31, and AZ91 Alloys in Dilute NaCl Solutions," *J. Solid State Electrochem.*, 14 (2010), 1897-1907.
17. R. Tunold et al., "The Corrosion of Magnesium in Aqueous Solution Containing Chloride Ions," *Corros. Sci.* 17 (1977), 353-365.
18. G. Williams and H.N. McMurray, "Localized Corrosion of Magnesium in Chloride-Containing Electrolyte Studied by a Scanning Vibrating Electrode Technique," *J. Electrochem. Soc.*, 155 (7) (2008), C340-C349.
19. G. Song et al., "The Electrochemical Corrosion of Pure Magnesium in 1 N NaCl," *Corros. Sci.*, 39 (1997), 855-875.
20. L. Wang et al., "XPS Study of the Surface Chemistry on AZ31 and AZ91 Magnesium Alloys in Dilute NaCl Solution," *Appl. Surf. Sci.*, 256 (2010) 5807-5812.
21. R.G. Kelly, "Electrochemical Thermodynamics and Kinetics of Relevance to Corrosion," *Electrochemical Techniques in Corrosion Science and Engineering*, ed. R.G. Kelly et al., (New York, NY: Marcel Dekker, Inc. 2003), 45.

**Magnesium Technology 2012**
*Edited by: Suveen N. Mathaudhu, Wim H. Sillekens, Neale R. Neelameggham, and Norbert Hort*
*TMS (The Minerals, Metals & Materials Society), 2012*

# INFLUENCE OF ALUMINUM CONTENT ON CORROSION RESISTANCE OF Mg–Al ALLOYS CONTAINING COPPER AND ZINC

Hiroyuki Kawabata[1], Naohisa Nishino[1], Tsuyoshi Seguchi[2], Yoshikazu Genma[2]
[1]Toyota Central R&D Labs., Inc.; 41-1 Yokomichi, Nagakute, Aichi, 480-1192, Japan
[2]Toyota Motor Corp.; 1 Toyota-Cho, Toyota, Aichi, 471-8572, Japan

Keywords: Magnesium-Aluminum Alloys, Corrosion Resistance, Tolerance Limit, Die Castings

## Abstract

The corrosion resistance of Mg–Al alloys deteriorates significantly with increasing copper content. Consequently, the tolerance limit of copper is commonly fixed at less than 300 ppm in Mg–Al alloys to ensure they have sufficient corrosion resistance. However, it is desirable to use higher copper tolerance limit in actual operations. To realize this, it is very important to prevent the deterioration in the corrosion resistance with increasing copper content. In this study, Mg–Al alloys with different aluminum, zinc, and copper contents were casted by high-pressure die casting and the corrosion resistances of the castings were investigated. The corrosion resistance of Mg–9% Al alloys decreased significantly with increasing copper content. However, adding zinc to copper-rich alloys prevented the reduction in the corrosion resistance due to copper. The corrosion rate of Mg–Al alloys containing 0.5% copper and 3% zinc was minimized at an aluminum content of 9%. The corrosion resistance of Mg–Al alloys containing copper and zinc was highest at a zinc content of over 3% and an aluminum content of about 9%.

## Introduction

Magnesium alloys are used in automobile lighting. In manufacturing processes, biscuits and runners of die castings are put into furnaces and are remelted. Magnesium alloys sometimes become contaminated with impurities during this process. Copper is one of the impurities. The corrosion resistance of Mg–Al alloys deteriorates significantly with increasing copper content, because copper forms copper-based intermetallic compounds [1–2]. Consequently, copper concentrations of less than 300 ppm are commonly used in Mg–Al alloys to ensure they have sufficient corrosion resistance [3]. To enable remelting, this limitation on the copper content needs to be lifted. To realize this, it is very important to prevent the deterioration in the corrosion resistance with increasing copper content.

In this study, the corrosion resistance of Mg–Al alloys containing copper and zinc was investigated. We determine the optimum zinc and aluminum contents for realizing a high corrosion resistance.

## Experimental

Mg–9% Al–(0–6)% Zn–(0–1)% Cu–0.3% Mn alloys were prepared to determine the effect of the copper and zinc contents on the corrosion resistance of Mg–Al alloys. Mg–(5–13)% Al–3% Zn–0.5% Cu–0.3% Mn alloys were then prepared to investigate the influence of the aluminum content on the corrosion resistance of Mg–Al alloys containing copper and zinc. Alloys containing more than 0.7% zinc were casted with commercial AZ91D alloy ingots and grains of pure zinc and copper. Alloys containing less than 0.7% zinc were casted with ingots of pure magnesium and aluminum, Mg–3.3% Mn alloy ingots, and grains of pure zinc and copper. These raw materials were melted at 1023 K and were casted by high-pressure die casting. The cast samples had dimensions of 3×40×150 $mm^3$. The casting temperature was 50 K above the liquidus temperature, the shot pressure was 64 MPa, and the shot velocity was 0.4 m/s. The microstructures of the castings were observed by scanning electron microscopy (SEM) and electron microprobe analysis (EPMA). Samples with dimensions of 2×25×25 $mm^3$ were prepared for corrosion tests by scraping casts. They were polished with #600 waterproof abrasive papers and purified with acetone. For the corrosion test, they were soaked in 5% NaCl aqueous solution. Corroded samples became lighter through the evolution of hydrogen. The corrosion rate was calculated from the weight loss after soaking for 100 h. In addition, alloys were casted at 1023 K by gravity die casting and the cast samples had dimensions of 25×35×300 $mm^3$. The microstructures of the castings were observed by SEM and image analyses of them were performed.

## Results and Discussion

### Influence of copper and zinc contents

Figure 1 shows the relationship between the copper content and the corrosion rate of Mg–9% Al–(0–5)% Zn–(0–1.2)% Cu–0.3% Mn alloys. The corrosion rate increased with increasing copper content irrespective of the zinc content. The rate of increase in the corrosion rate with increasing copper content (i.e., the gradient of the curve) decreased with increasing zinc content for zinc contents between 0 and 2%, whereas it remained almost constant for zinc contents higher than 3%. Thus, the corrosion resistance of Mg–9% Al alloys decreased significantly with increasing copper content. However, adding zinc (preferably more than 3%) to copper-rich alloys prevented the reduction in the corrosion resistance due to copper.

Figure 2 shows the microstructures and EPMA maps of Mg–9% Al–(0.7, 3)% Zn–0.5% Cu–0.3% Mn alloys. Three phases were observed in both alloys. Based on the observed magnesium, aluminum, copper, and zinc distributions, the matrix phase is the α-Mg phase, the gray regions in the secondary-electron (SE) image are β phases ($Mg_{17}Al_{12}$ compounds), and the white regions in the SE image are copper-based intermetallic compounds.

Figure 3 shows the appearance of the surfaces of Mg–9% Al–(0.7, 3)% Zn–0.5% Cu–0.3% Mn alloys after the corrosion test. Both samples appeared porous after the corrosion test. The remains of the alloy containing 3% zinc were thicker than those containing 0.7% zinc. In general, the centers of the α-Mg phases corroded away, whereas the β phases and aluminum-rich regions of the α-Mg phase adjacent to the β phases remained in AZ91D alloys [4–5]. α-Mg phases are anode portions and copper-based intermetallic compounds are cathode portions in Mg–Al alloys containing copper. The centers of α-Mg phases corroded away, whereas the β phase, aluminum-rich zones of α-Mg phases adjacent to the β phase, and copper-based intermetallic compounds remained. Therefore, samples subjected to corrosion test were porous, as shown in Figs. 3(a) and (b).

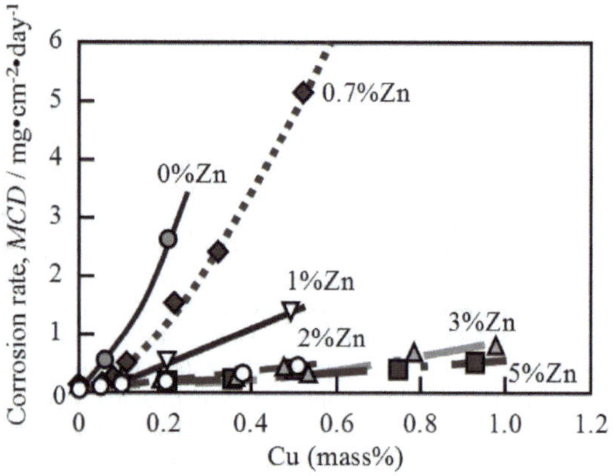


Figure 1. Relationship between copper content and corrosion rate of Mg–9% Al–(0–5)% Zn–(0–1.2)% Cu–0.3% Mn alloys.

Table I shows the chemical compositions of the copper-based intermetallic compounds of Mg–9% Al–(0.7, 3)% Zn–0.5% Cu–0.3% Mn alloys. The zinc content of the intermetallic compounds increased and the copper and aluminum contents decreased when the zinc content was increased from 0.7 to 3%.

For the reasons stated above, the copper-based intermetallic compounds in the alloys containing 0.7% and 3% zinc were cathode portions. However, the copper-based intermetallic compounds in the two alloys had different chemical compositions, which resulted in different cathode reactions. Consequently, the two alloys exhibited different corrosion resistances.

Influence of aluminum content

Figure 4 shows the relationship between the aluminum content and the corrosion rate of Mg–(5–13)% Al–3% Zn–0.5% Cu–0.3% Mn alloys. The corrosion rate of the alloys was minimized when the aluminum content was 9%.

Figure 5 shows SEM-backscattered electron (BSE) images of microstructures of Mg–(5–13)% Al–3% Zn–0.5% Cu–0.3% Mn alloys. α-Mg phases and copper-based intermetallic compounds, which appear white in SEM images, were observed at an aluminum content of 5% (Fig. 5(a)). In addition to α-Mg phases and copper-based intermetallic compounds, β phases, which appear gray in SEM images, were observed at aluminum contents higher than 7% (Figs. 5(b)–(e)). Increasing the aluminum content of the alloys increased the amount of β phases and caused fine dispersion of copper-based intermetallic compounds.

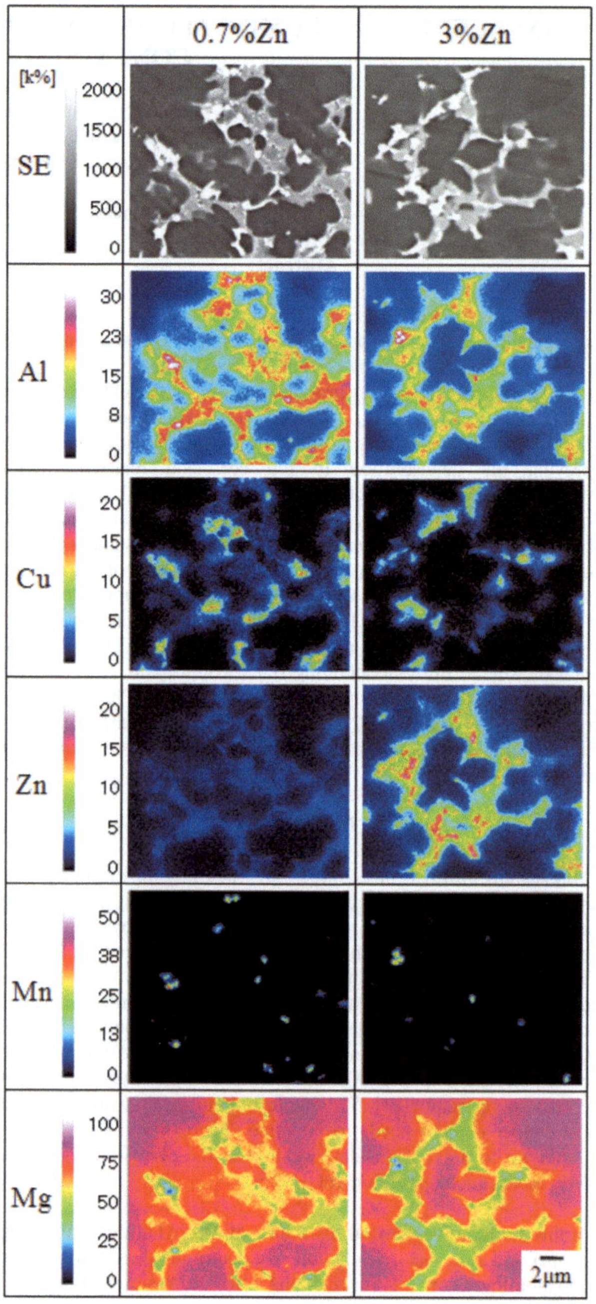


Figure 2. Microstructures and EPMA maps of Mg–9% Al–(0.7, 3)% Zn–0.5% Cu–0.3% Mn alloys.

Figure 6 shows the relationship between the aluminum content and the areas of β phases and copper-based intermetallic compounds. These areas were determined by performing image analysis of the images of the microstructures shown in Fig. 4. The β phase area increases linearly with aluminum content of the alloys. In contrast, the copper-based intermetallic compound area was almost independent of the aluminum content. Aluminum contributes to the formation of β phases and copper-based intermetallic compounds in the alloys. The alloys had the same copper content (0.5%). Consequently, increasing the aluminum content increased only the β phase; the copper-based intermetallic compound areas remained constant.

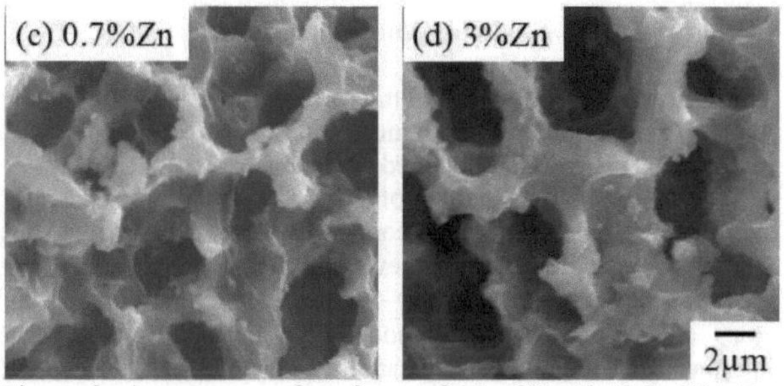

Figure 3. Appearance of surfaces of Mg–9% Al–(0.7, 3)% Zn–0.5% Cu–0.3% Mn alloys after corrosion tests.

Table I. Chemical compositions of copper-based intermetallic compounds of Mg–9% Al–(0.7, 3)% Zn–0.5% Cu–0.3% Mn alloys (mass%)

| Alloy | Al | Zn | Cu | Mg |
|---|---|---|---|---|
| Mg-9%Al-0.7%Zn -0.5%Cu-0.3%Mn | 31.1 | 3.8 | 12.5 | Bal. |
| Mg-9%Al-3%Zn -0.5%Cu-0.3%Mn | 26.1 | 12.5 | 10.6 | Bal. |

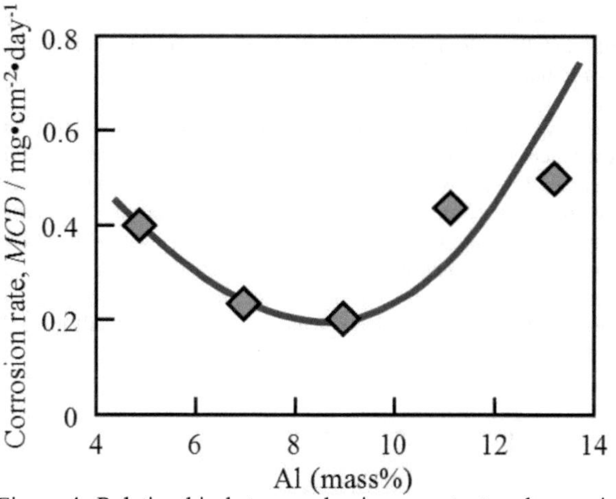

Figure 4. Relationship between aluminum content and corrosion rate of Mg–(5–13)% Al–3% Zn–0.5% Cu–0.3% Mn.

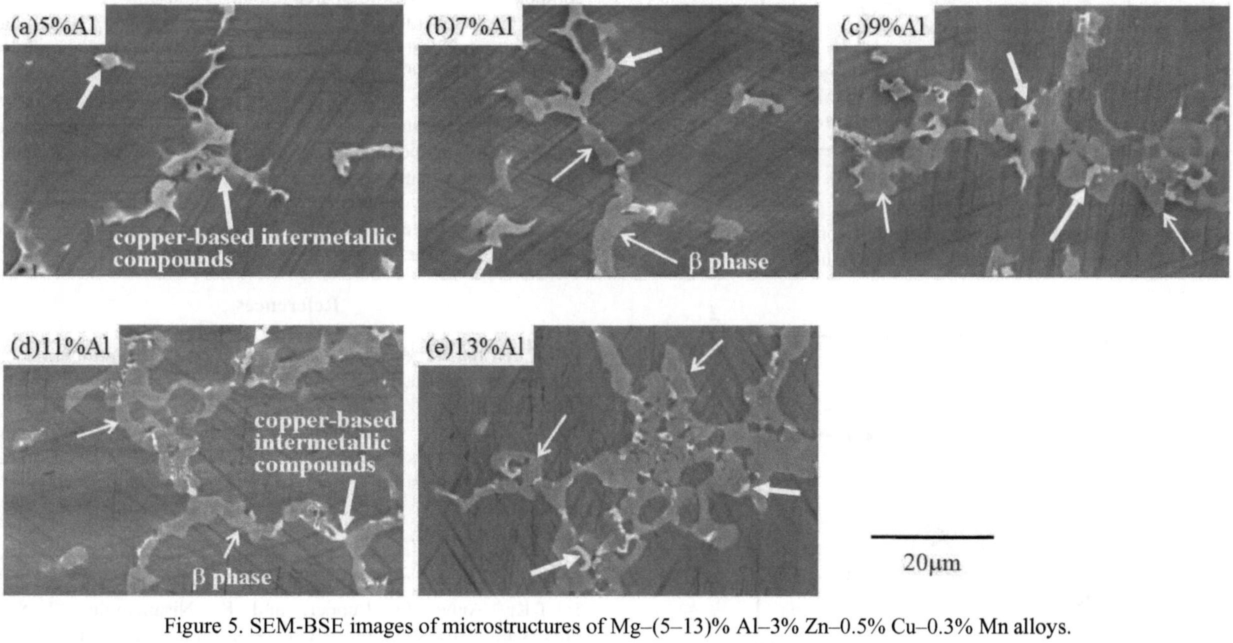

Figure 5. SEM-BSE images of microstructures of Mg–(5–13)% Al–3% Zn–0.5% Cu–0.3% Mn alloys.

Figure 7 shows the relationship between the aluminum content and the number of copper-based intermetallic compounds and their average area per 1 mm2. The number of copper-based intermetallic compounds increased and the average area of the compounds decreased with increasing aluminum content. This indicates that with increasing aluminum content, copper-based intermetallic compounds were finely dispersed while their total area remained constant. During solidification, copper-based intermetallic compounds precipitated in residual liquid phases after the α-Mg phase solidified. With increasing aluminum content, the α-Mg phase grains became finer and residual liquid phases were dispersed more finely. Consequently, copper-based intermetallic compounds were finely dispersed while their total area remained constant.

We now consider the influence of aluminum on the corrosion resistance of the alloys in terms of the above-described structural changes. In Mg–Al alloys containing copper, α-Mg phases form the anode portion and copper-based intermetallic compounds compose the cathode portion. While the centers of α-Mg phases corrode away, β phases and aluminum-rich regions of the α-Mg phase adjacent to a β phase inhibit the progress of corrosion-like barriers. Thus, the corrosion resistances of the alloys will increase with increasing β phase area.

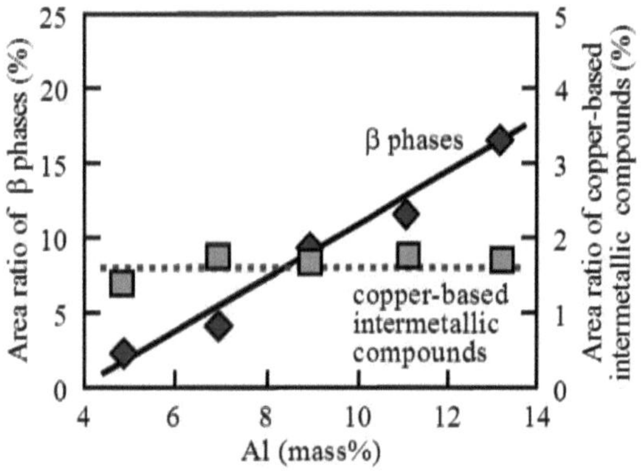


Figure 6. Relationship between aluminum content and areas of β phases and copper-based intermetallic compounds.

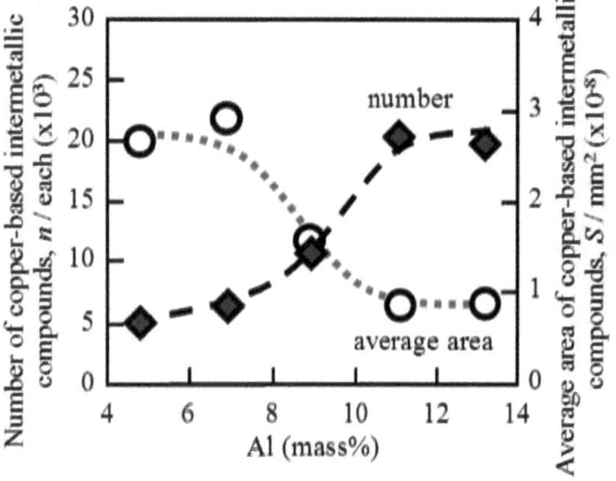


Figure 7. Relationship between aluminum content and the number of copper-based intermetallic compounds and average area.

On the other hand, fine dispersion of copper-based intermetallic compounds, which are cathode portions, increases the corrosion portion and increases the anode–cathode reaction, reducing the corrosion resistance. Thus, increasing the aluminum content in the alloys simultaneously enhances the corrosion resistance by increasing of the amount of β phase and deteriorates the corrosion resistance due to fine dispersion of copper-based intermetallic compounds. At aluminum contents below 9%, the increase in the corrosion resistance due to the increase in the β phase is greater than the reduction in the corrosion resistance due to fine dispersion of copper-based intermetallic compounds. The opposite is true at aluminum contents greater than 9%. Therefore, the corrosion resistance is maximized at an aluminum content of 9%.

## Summary and conclusions

The corrosion resistance and microstructures of Mg–Al alloys with different aluminum, zinc, and copper contents were investigated. The results obtained are summarized as follows:

1. The corrosion resistance of Mg–9% Al alloys decreased significantly with increasing copper content. However, adding zinc to copper-rich alloys prevented the corrosion resistance reduction due to copper. In Mg–Al alloys containing copper and zinc, α-Mg phases were anode portions and copper-based intermetallic compounds were cathode portions. The copper-based intermetallic compounds in alloys with different zinc contents had different chemical compositions and this gave rise to different cathode reactions. Therefore, the alloys had different corrosion resistances.
2. Increasing the aluminum content of the alloys containing 0.5% Cu and 3% Zn resulted in a linear increase in the β phase area and increased the fine dispersion of copper-based intermetallic compounds while their total area remained constant.
3. The corrosion rate of Mg–Al alloys containing 0.5% Cu and 3% Zn was smallest at an aluminum content of 9%. For aluminum contents less than 9%, the improvement in the corrosion resistance due to the increase in the β phase is larger than the reduction in the corrosion resistance due to fine dispersion of copper-based intermetallic compounds. The opposite is true for aluminum contents greater than 9%.
4. In Mg–Al alloys containing copper and zinc, the optimum zinc and aluminum contents for corrosion resistance are more than 3% and about 9%, respectively.

## References


1. J.E. Hillis, "The Effects of Heavy Metal Contamination on Magnesium Corrosion Performance," *SAE Technical paper*, 830523 (1983).
2. C. Blawert et al., "Influence of the Copper Content on Microstructure and Corrosion Resistance of AZ91 Based Secondary Magnesium Alloys," *SAE Technical paper*, 2006-01-0254 (2006).
3. ASTM Standard, B-94.
4. O. Lunder et al., "The Role of $Mg_{17}Al_{12}$ Phase in the Corrosion of Mg Alloy AZ91," *Corrosion*,45(1989),741-748.
5. T.Kr. Aune, O. Lunder, and K. Nisancioglu, "In-Situ Microscopic Investigation of Corrosion on a Magnesium Alloy," *Microstructural Science*, 17(1989),231-241.

# Magnesium Technology 2012

## High Temperature Processing and Properties

### Wednesday PM

*Session Chairs:*

Paul E. Krajewski
(General Motors, USA)
Warren Poole
(University of British Columbia, Canada)

**Magnesium Technology 2012**
*Edited by: Suveen N. Mathaudhu, Wim H. Sillekens, Neale R. Neelameggham, and Norbert Hort*
***TMS (The Minerals, Metals & Materials Society), 2012***

# EFFECT OF ROLLING TEMPERATURE ON THE AZ31B MAGNESIUM ALLOY MICROSTRUCTURE

Litzy L. Choquechambi Catorceno[1], Angelo F. Padillha[1]
([1]PMT-EPUSP Departamento de Engenharia Metalúrgica e de Materiais da Escola Politécnica da Universidade de São Paulo, USP, Brazil); 2463 Av. Prof Mello Moraes; São Paulo, SP, BRAZIL

Keywords: Magnesium alloy, Plastic deformation, AZ31B

## Abstract

The influence of rolling temperature and the effect of strain rate on the microstructure of AZ31B Mg alloy were determined in order to improve its formability. A plate of AZ31 alloy was found to be sensitive to strain rate at high temperature and anisotropy was adversely impacted in cold rolling sheets. Thus, AZ31B has better workability within the temperature range of 200 to 250 °C, due to the grain refinement, caused by recovery and dynamic recrystallization. The effect of rolling temperature was studied on recrystallized sheets (2 mm in thickness) which were deformed by rolling at different temperatures (25, 100, 200 and 250 °C) and the effect of strain rates was evaluated on two different rolling speed (10 and 20 rpm). The microstructural characterization was achieved using several complementary techniques of microstructural analysis, such as optical microscopy, scanning electron microscopy, X-ray analysis by energy dispersive, X-ray diffraction and microhardness.

## Introduction

The AZ31B magnesium alloy is a promising structural light material due to its high specific strength and high specific stiffness. Therefore, it is desired to improve the ductility and workability in order to increase its applications, especially in automotive and aerospace industry [1]. Many researches were carried out to enhance the formability of Mg alloy sheets at high temperature. However, there have been a few investigations on the room temperature formability of Mg alloys. It is known that plastic deformation in metals at low temperatures occurs as a result of two mechanisms: slip and twinning. Twin formation depends on the orientation of the crystals which make up the alloy with respect to the applied stresses [1, 2].

Formability of magnesium alloys greatly increases at high temperatures accompanied by a decrease in flow stress, an increase in strain rate sensitivity, and a decrease in normal anisotropy. The value of the critical resolved shear stress (CRSS) of the non-basal slip systems decreases at high temperature improving the ductility. Furthermore, an increase in the deformed temperature also affects the recrystallization behavior of magnesium alloys. Once a critical strain is exceeded, it is possible that during and after deformation occur both dynamic recrystallization and static recrystallization, respectively. Their mechanisms change the grain morphology, and are function of deformation temperature, strain, strain rate, deformation mode, grain size, and grain orientation [3].

Recently, it was showed that grain refinement is effective for activation of prismatic slip at grain boundary, because grain refinement promotes stress concentration at grain boundary, resulting in activation of prismatic slip. Thus, the deformation mechanisms of Mg alloys depend not only of five independent slips, but also the relaxation mechanisms for the stress concentration at the grain boundary and formation of twins, which is produced at higher strain rates and lower temperatures. Their formation strongly depends on grain size [5]. Therefore, it possible to state that during deformation in a coarse-grained Mg alloy, twinning plays a vital role, not in a fine-grained Mg alloy [10]. Yet, effects of the grain size on roles of twinning in the deformation are not still completely understood.

In the present research, the effect of rolling temperature on microstructural evolution is studied at cold and warm plastic deformation on sheets of AZ31B alloys, using microstructural characterization tools.

## Experimental details

The received material was a plate of 2 mm in thickness of AZ31B magnesium alloy with a nominal chemical composition of 3.34–3.63 wt% Al, 0.45–0.53 wt% Zn, 0.27–0.29 wt% Mn, and balance Mg. It was produced by twin-roll caster by Sinomag Company, China. This sheet was cut into plates with approximate dimensions of 50x10 $mm^2$ for rolling experiments.

Each samples were heated up at rolling temperature (25, 100, 200 and 300 ◦C ) and kept at this temperature for 10 min, then sent to a small sized rolling mill of 108 mm in diameter for rolling at a single pass. The rolling speeds were maintained at 10 rpm, without any lubricant. Such rolling processes were carried out controlling the compressive reduction in thickness near 10-15% for cold rolling (25 and 100 °C), and near 15-20% for warm rolling (200, and 250 °C). For the same temperature range, the rolling speed was increased to 20rpm and the compressive reduction in thickness was increased to 30% for every rolling temperature.

Table 1 – Table showing the temperature and reduced thickness of the sample sheet with one pass.

| 10 rpm | | | 20 rpm |
|---|---|---|---|
| Temperature °C | Reduced thickness (%) | strain rates ($s^{-1}$) | Reduced thickness (%) |
| 25 | 10 | 1.6 | 30 |
| 100 | 15 | 2.3 | 30 |
| 200 | 15 | 2.3 | 30 |
| 250 | 20 | 2.8 | 30 |

a) 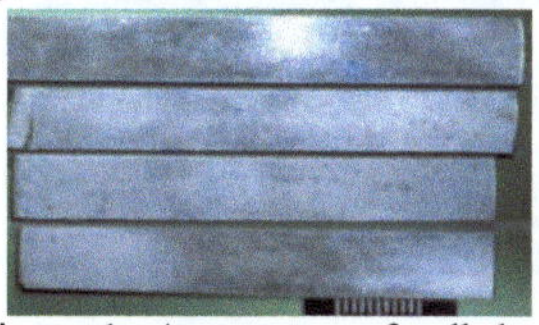 b) 

Figure 1. Appearance of rolled samples at different temperature and rolling speeds: a) 10 rpm and b) 20 rpm for 5.7 ($s^{-1}$) strain rates.

The appearance of the thinnest sheets after rolling is showed on Figure. 1 and the summary of the reduced thickness, strain rates and variation of rolling speeds are on Table 1. In the figure 1b is evident that the deformation strain is accommodated partly by large amounts of shear bands, which are very common features for AZ31 deformed by rolling.

For metallographic analysis, the as-rolled samples were cut into a square of 10mm×10mm in dimensions. The observed positions were on rolling sections parallel to rolling direction. All samples were ground with SiC papers and etching the samples surface with two mixed solution of nitric acid in diethylene glycol and acetic-picral to reveal the grain structure. Metallurgical structure was observed by using an Olympus PME-3 optical microscopy. The crystal structure and network parameters were determined using X-ray diffraction (XRD) with a Cu Kα radiation (Rigaku). The precipitates were analyzed using scanning electron microscopy (SEM, Philips EM400T) equipped with an energy dispersive X-ray (EDX) spectrometer. Vickers hardness was measured with 0.3 kg load and dwell time of 10 s. The mean linear intercept method was used for grain size analysis.

## Results and discussion

### Microstructure

The microstructure of the as-received AZ31B-alloy had a mean grain diameter of 13 μm (as seen in Fig. 2b) and Vickers hardness 49 approximately.

a) b)

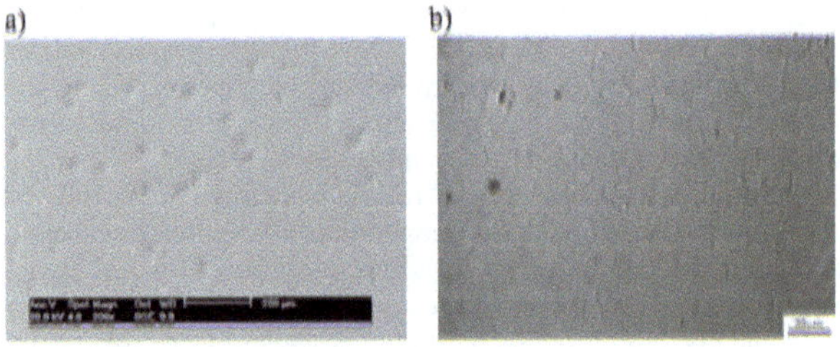

Figure 2. Images (a. MEV; b. OM) the as-received magnesium alloy normal to rolling direction.

The surface of magnesium alloy plate AZ31B as received material presents micro-pores of different sizes and white eutectic precipitates (Figure 2) that according of literature [7]; they are possibly precipitates of $Mg_{17}Al_{12}$, $Al_6Mn$, $Al_4Mn$ e AlMn.

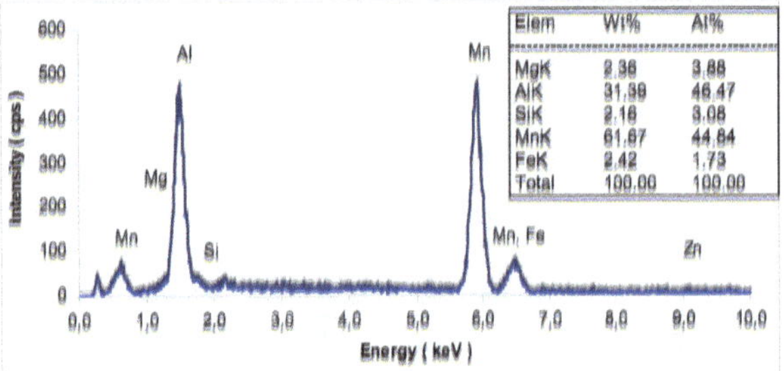


| Elem | Wt% | At% |
|---|---|---|
| MgK | 2,36 | 3,88 |
| AlK | 31,39 | 46,47 |
| SiK | 2,16 | 3,08 |
| MnK | 61,67 | 44,84 |
| FeK | 2,42 | 1,73 |
| Total | 100,00 | 100,00 |

Figure 3. EDX precipitates spectrum at surface as-received material of AZ31B-alloy.

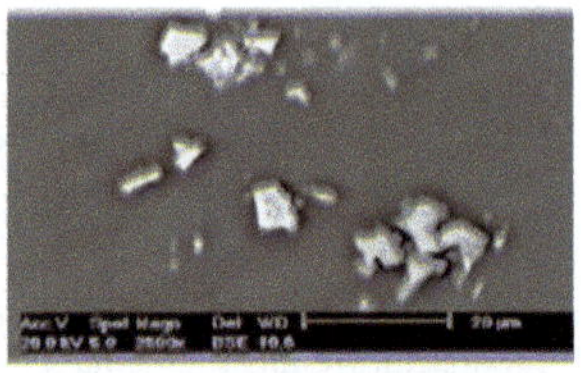

| Elem | Wt% | At% |
|---|---|---|
| MgK | 90,76 | 93,15 |
| AlK | 6,26 | 5,78 |
| CeL | 0,68 | 0,12 |
| MnK | 0,56 | 0,25 |
| FeK | 0,35 | 0,16 |
| ZnK | 1,39 | 0,53 |
| Total | 100,00 | 100,00 |

Figure 4. Precipitates image (MEV) and composition of Mg matrix at surface as-received material.

Big precipitates were observed on micrographs (Figure 4.) and Al and Mn were detected by SEM-EDX. After rolling at up 200 °C temperature, eutectic phases were partly dissolved, resulting in shape changes of eutectic precipitates from big to small particles by increasing strain and temperature (Figures 5). The precipitates were remarkably refined to less than 3 μm (Fig. 5b) and redistributed on the α-Mg matrix, as the rolled temperature increased to 250 °C.

a) b)

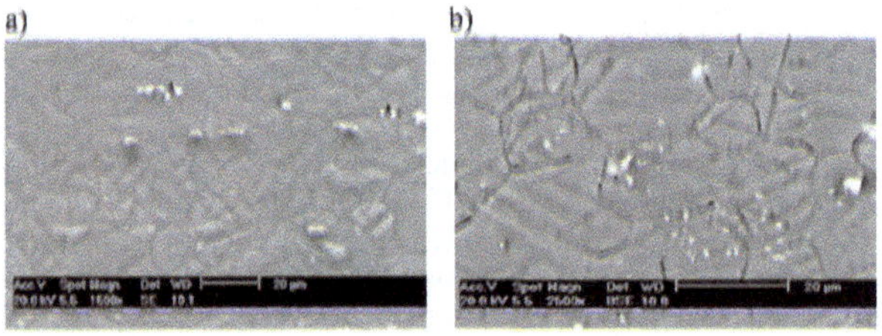

Figure 5. Redistribution of eutectic precipitates of samples rolled at 10 rpm a.) Sample rolled at 25°C and b.) Sample rolled at 250°C

The EDX spectrum from matrix alloy showed that samples rolled at 250 °C contain much more Al and Mn than the as-received material. It confirms that there were lots of eutectic precipitates by dissolution. Otherwise, the atomic percentages of Al and Mn for warm-rolled sample were found to be less than those for cold-rolled one. This fact implies that the precipitates were gradually broken by the large deformation during cold rolling and partly dissolved during the pre-heating and during warm rolling, resulting in redistribution of precipitates.

a) b) c) d)

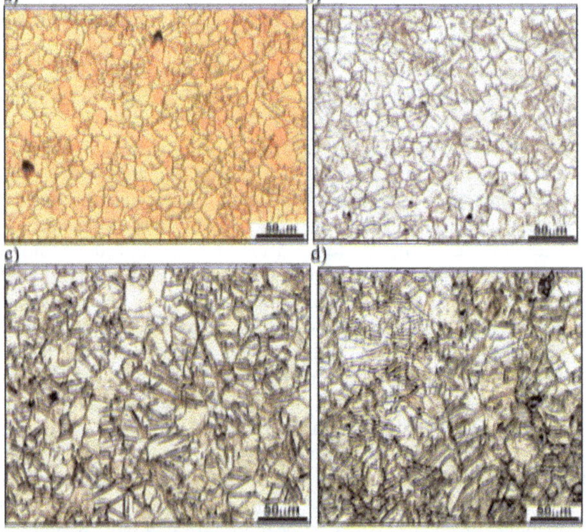

Figure 6. Optical micrographs of single pass rolled AZ31B sheets at 10rpm rolling speed and a) 25 °C, b) 100 °C, c) 200 °C e) 250 °C.

After rolling at 10 rpm, the microstructure of samples was characterized by the mixture of small grains, large amounts of shear bands and twin boundaries (Figure. 6). The grain size of cold rolled sample (25 °C and 10 % reduced thickness) decreased from 13 μm to about 10 μm, inducing inhibition of mechanical twinning. Increasing reduction to 15% and temperatures up to 100 °C, plastic deformation of magnesium produced extensive formation of twinning and dense dislocation pile-ups within grains, especially, on bigger grains where twins were more prominent as was found by Barnett's research. Whereas at temperature up to 200 °C, the microstructures of 15 % and 20% reduced thickness (figure 6c. and 6d.) revealed bimodal distributions of grain sizes with smaller grains about 10 μm and coarser ones about 20 μm. The appearance of these refined grains was mainly induced by recrystallization during the pre-heating and warm rolling. So the static recrystallization may occur during the preheating process for 10 min before the rolling pass.

In order to determine the effect of strain rate, the rolling speed and the reduced thickness were increased to 20 rpm and 30%, respectively. At room temperature (25°C ) rolling to 30% reduction in thickness causes cracks, breaking into many small pieces; hence, 10% reduced thickness represents the upper limit of maximum deformation at room temperature.

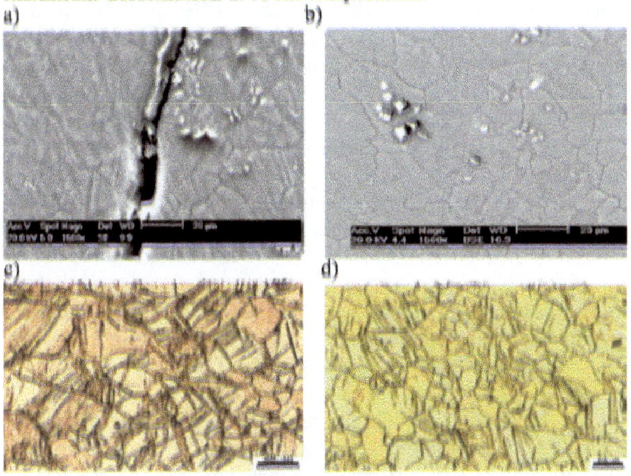


Figure 7. Micrographs of the samples rolled at 20rpm, 30% reduced thickness and a) 25 °C, b) 100 °C, c) 200 °C c) 250 °C.

Comparing samples rolled at 25 °C and 100°C, they had low grain size variations, showing almost a normal distribution of grain sizes about 15 μm (figure 7a and 7b). Increasing temperature up to 200°C, the resultant microstructure became a bimodal mix of finer recrystallized grains (8 μm) and coarser deformed grains (20μm).

## Hardness

Table 4. Microhardness values of the as-rolled samples at different temperature and roll speed.

| | 10 rpm | | 20 rpm and 30% of reduced thickness |
|---|---|---|---|
| T °C | Reduced thickness (%) | Vickers hardness ($kg_f/mm^2$) | Vickers hardness ($kg_f/mm^2$) |
| 25 | 10 | 73±1.1 | 76±1.5 |
| 100 | 15 | 72±1.8 | 81±1.2 |
| 200 | 15 | 73±1.3 | 84±1.5 |
| 250 | 20 | 70±1.1 | 81±1.7 |

The AZ31B alloy as-received material exhibited a relative low hardness of 49±1.8 HV. After rolling at low roll speed (10 rpm), hardening considerably increased to 73HV, and reached the highest value of 84HV at 200 °C with 20rpm roll speed. It means that AZ31B alloy exhibits a strong hardening behavior by plastic deformation and by age precipitation at temperature up to 100 °C because of redistribution of precipitates. Such as samples with heavy-reduction (30% ) at high rolling speed showed 71% increase in hardness (100 and 200 °C). Although at 250 °C, there was a little decrease in hardness due to the nucleation of new strain free grains which removed the strain hardening effects and also likely to be contributed in part from recovery prior to recrystallization.

## Normal anisotropy

For small strains and thickness samples, a parameter that can characterize the formability of rolled sheets is the normal anisotropy, which would be determined by invoking the volume (V) constancy, $DV/V=\varepsilon_w+\varepsilon_t+\varepsilon_l=0$, where w, t, and l denote sample width, thickness, and length, respectively, and i and f denote initial and final values [9].

$$r=\frac{\ln(w_f/w_i)}{-\ln(l_f/l_i)-\ln(w_f/w_i)} \quad (1)$$

Table 3. The planar anisotropy r-value obtained in samples rolled at different rolling speed and temperature.

| | 10 rpm | | 20 rpm strain rate: $5.7s^{-1}$ |
|---|---|---|---|
| T (°C) | strain rate ($s^{-1}$) | R | R |
| 25 | 1.5 | 5.4 | - |
| 100 | 2.3 | 4.3 | 5.5 |
| 200 | 2.3 | 3.9 | 4.2 |
| 250 | 2.8 | 3.7 | 3.4 |

The magnesium alloy AZ31 showed a significant drop in the normal anisotropy between room temperature and 250 °C (Table 3). This is likely the result of an increase in pyramidal slip activity at higher test temperatures as indicated Koike's researches [4]. This anisotropy provides indirect evidence that there is a change in deformation mechanisms and is attributed to well develop [0002] basal plane, which is a typical texture of warm or cold rolled Mg alloy sheets [10]. Nevertheless, the recovery and dynamic recrystallization processes also decrease r-values at high temperatures on hcp metals, improving the ductility. An increment of roll speed from 10 to 20rpm, improved the strain rate sensitivity from $1.6s^{-1}$ to $5.7s^{-1}$ and induced better formability at temperature up to100°C.

## X-ray diffraction

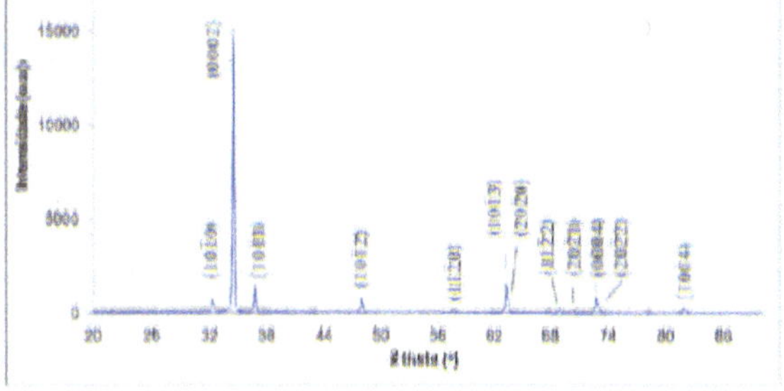


Figure. 8. X-ray diffraction patterns of the as-received AZ31B-alloy.

Figure 8 shows X-ray diffraction patterns of the as-received AZ31B-alloy. All diffraction peaks coincide with those of α-Mg (01-089-4894), suggesting that the AZ31 Mg alloy consists mainly of α-Mg basic phase. The strongest diffraction peak for the AZ31B alloy belongs to (0 0 0 2) and network parameters were calculated as a and c parameters, where a = 0,3198nm; c= 0,5170nm, and c/a = 1,6166.

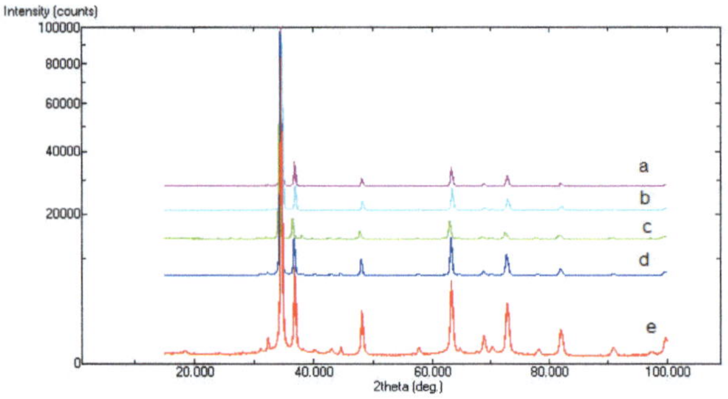

Figure 9. XRD diffraction patterns of AZ31B Mg alloy for rolled samples.

Table 5. Rolling condition and designed name of XRD diffraction patterns.

| Temperature °C | Reduced thickness (%) | X-ray diffraction pattern | Rolling speed (rpm) |
|---|---|---|---|
| 25 | 10 | d | 10 |
| 100 | 15 | c | 10 |
| 200 | 15 | b | 10 |
| 250 | 20 | a | 10 |
| 250 | 30 | e | 20 |

The figure 9 shows the X-ray diffraction patterns of rolled samples with different thickness for AZ31 Mg alloys. The varying of intensity showed a transformation of precipitates into magnesium matrix. Comparing X-ray diffraction peaks of the specimens rolled at 10rpm, for 10% and 20% reduction (curve d and a). At high reduction the intensity of peaks decreases resulting on broad peaks diffraction peaks due to mainly temperature increment from 25 to 250 °C. As reported by Peng-Xi Li that a decrease in intensity results on broad peaks.

At 20 rpm rolling speed (curve e), the X-ray diffraction pattern presented broad and sharp peaks due to the precipitation of crystalline phases. The intensity of reflection was increased with increasing speed of the rolling. The microstructures of these alloys showed that the amount of precipitates increases at high rolled temperature as was obtained by X-ray diffraction analyses. The volume fraction of β phase grows during the preheating process before rolling, as shown by the growth of its associated Bragg peaks. The lattice constants change as aluminum exits the matrix to form precipitates, as shown by the Bragg peak's movement to lower 2θ values. However, the X-ray diffraction of warm rolled samples needs more complement data to determine completely the effect of rolling temperature.

## Summary


AZ31B Magnesium alloy sheets with about 2 mm in thickness rolled by one-pass with different thickness at cold and warm temperature showed large deformation results with high rolling speed.

The grain refinement was achieved at cold rolling with small reduction in thickness (10%) and low strain rate, twinning was predominantly formed in coarse grains, not in fine grains and with; whereas at temperature up 200°C fine grains were formed by discontinuous recrystallization during the preheating and warm-rolling process. The high speed rolling is very effective for grain refinement.

The AZ31B alloy had a large hardening response on cold and warm rolling temperature resulting of plastic deformation, and precipitation of a transition phase β-($Mg_{17}Al_{12}$), consequently offering a combination of good strength and ductility.

AZ31 alloy was found to be sensitive to strain rate in the high temperature, besides the increase in the flow stress, appears as a decrease in material formability. The anisotropy was adversely impacted in cold rolling sheets and was characterized by high anisotropy at room temperature. Thus, the formability of AZ31B alloy increases with temperature when plastic deformation is performed at low rolling speed, while an increasing of rolling speed leads a decrease in formability at high temperature.


## Acknowledgements


The authors would like to thank the " Departamento de Engenharia Metalúrgica e de Materiais da Escola Politécnica da Universidade de São Paulo" and The National Council for Scientific and Technological Development (Cnpq) for financial support.


## References


[1] B. L. Mordike, T. Ebert, "Magnesium properties-applications-potential," *Materials Science and Engineering A302*, (2001), 37–45.

[2] M. R. Barnett, Z. Keshavarz, A.G. Beer, D. Atwel, "Influence of grain size on the compressive deformation of wrought Mg-3Al-1Zn," *Acta Materialia*, 52, (2004), 5093- 5103.

[3] S. E. Ion, F. J. Humphreys, S. H. White, "Dynamic recrystallization and the development of microstructure during the high temperature deformation of magnesium," *Acta Metallurgica, 30*, (1982), 1909-1919.

[4] J. Koike, T. Kobayashi, T. Mukai, H. Watanabe, M. Suzuki, K. Maruyama, K. Higashi, "The activity of non-basal slip systems and dynamic recovery at room temperature in fine-grained AZ31B magnesium alloys," *Acta Materialia 51*, (2003), 2055–2065.

[5] D. Letzig, J. Swiostek, J. Bohlen, P. A. Beaven and K. U. Kainer, "Wrought magnesium alloys for structural applications," *Materials Science and Technology, Vol. 24,* (2008), 991-996.

[6] Zhao Hong-liang, Guan Shao-kang, Zheng Fei-yan, Li Qing-kui, Wang Li-guo, "Microstructure and properties of AZ31 magnesium alloy with rapid solidification," *Trans. Nonferrous Met. Soc. China 15,* (2005), 144–148.

[7] Soon Chan Park, Jong Dae Lim, Dan Eliezer, Kwang Seon Shin, "Microstructure and mechanical properties of Mg–Al–Ag Alloys," *Materials Science Forum Vols. 419-422*, (2003), 159-164.

[8] Yasumasa Chino, Kensuke Sassa, Akira Kamiya, Mamoru Mabuchi, "Enhanced formability at elevated temperature of a cross-rolled magnesium alloy sheet," *Materials Science and Engineering A 441*, (2006), 349–356.

[9] S.R. Agnew and O. Duygulu, "Plastic anisotropy and the role of non-basal slip in magnesium alloy AZ31B," *International Journal of Plasticity 21*, (2005), 1161-1193.

[10] Matthew R. Barnett, Alireza Ghaderi, Iichat Sabirov, Bevis Hutchinson, "Role of grain boundary sliding in the anisotropy of magnesium alloys," *Scripta Materialia 61*, (2009), 277–280.

[11] Peng-Xi Li, Bo-Ping Zhang, Jing-Yi Chen, Yi-Fan Tu, "Microstructure and Mechanical Properties of AZ31B Magnesium Alloy Sheets Prepared by Extruding and Rolling," *Advanced Materials Research Vols. 11-12*, (2006), 653-656.

# Hot Formability Curves for Four AZ31B Magnesium Alloy Sheets Obtained by the Pneumatic Stretching Test

Fadi Abu-Farha[1], Ravi Verma[2], Louis G. Hector, Jr.[2]
[1]Department of Automotive Engineering, Clemson University; 4 Research Drive; Greenville, SC 29607, USA
[2]GM R&D, General Motors; 30500 Mound Road; Warren, MI 48090, USA

Keywords: Formability, Composite FLDs, Mg AZ31B, Pneumatic Stretching Test, Orientation Effects

## Abstract

The formability of four Mg AZ31B sheets produced by either direct chill or twin roll continuous casting, and having different initial grain sizes, was investigated at 400 °C and $5 \times 10^{-3}$ $s^{-1}$ using the pneumatic stretching test. Blanks were pneumatically bulged through four elliptical die inserts, with aspect ratios ranging between 1.0 and 0.4, producing ellipsoidal domes with different biaxial strain combinations. Testing was carried out in two ways: with the major strains being aligned either along or across the rolling direction of the material. Strain combinations were measured in the deformed specimens, and the forming limit curves were constructed for each of the four sheets in two orientations. The results show great impact of sheet orientation on material formability limits. Additionally, the results reveal significant differences between the four sheets; those differences were correlated back to disparities in grain structure and material inhomogeneities.

## Introduction

Increasing demand for more fuel efficient vehicles is driving mass reduction in new vehicle designs. This is not surprising since an excess of 50% of total fuel consumption is found to be mass dependent; a fact that is stimulating interest in lightweight alloys. The low density of magnesium alloys, in particular, makes them ~35% lighter than aluminum alloys and ~78% lighter than steels. In spite of the significant weight-savings promised by magnesium alloys, their use in automotive components is predominantly found in die castings. Poor formability of wrought magnesium alloys, hampered in part by limited room temperature ductility owing to strong crystallographic texture and the large anisotropy between basal and prismatic slip, has precluded broader use of these materials. Basal slip and twinning of the HCP crystal structure of magnesium are the only active deformation mechanisms at room temperature [1]; thus, elevated temperature forming processes are needed to activate the five independent slip systems required by the von-Mises criterion. In addition, other diffusion-assisted mechanisms, such as grain boundary sliding and dislocation climb, may become relevant at elevated temperatures [2]. The most common wrought magnesium sheet alloy is the AZ31B, the elevated temperature forming of which has been directed towards vehicle closure panels [3]. A considerable increase in the ductility of this alloy occurs above 200-225 °C [4-6]; therefore, practical use of the alloy is generally realized through either warm stamping (~200-300 °C) [7] or hot pneumatic forming (~300-450 °C) [3].

Critical to any sheet metal forming operation are forming limit curves (FLCs); empirical relationships between major and minor biaxial strains that describe the limits of sheet formability [8-9]. Accurate FLCs are essential for finite element (FE) simulations of sheet forming, since they establish the boundaries between successful forming and failure. Mechanical stretching tests, according to the Marciniak [10] and Nakazima [11] methods are the most common procedures for measuring sheet formability limits, and they are standardized in ASTM E2218 [8] and ISO 12004 [12]. While these formability tests have been extensively applied in the literature, several issues affect their applicability and accuracy. Frictional effects are difficult to eliminate, even under well-lubricated conditions. The Marciniak test, in particular, is designed to have minimal frictional effects through the use of a carrier blank that isolates the deforming specimen from the punch; yet several examples can be found in the literature where material failure takes place away from the center of the deformed specimen [13]. On the other hand, available equipment for mechanical stretching tests is generally limited to ~300-350 °C [13-15]; this precludes their use in studying the hot formability of lightweight Al and Mg alloys. Moreover, these sheet alloys are often formed pneumatically at high temperatures (400-550 °C) via superplastic forming (SPF) or its lower temperature variant quick plastic forming (QPF); therefore, mechanical stretching tests do not resemble the actual loading conditions the material undergoes during pneumatic forming operations. Finally, it is important to construct the FLCs at near-constant strain rates, since high strain rate sensitivity is expected at warm and high temperatures; the latter is not easy to achieve in mechanical stretching tests since they are mainly carried out at constant punch speeds.

The "pneumatic stretching test" resolves most of the issues encountered in mechanical stretching tests, although with some limitation on the extent to which major/minor strain space can be readily sampled. It is based on the free bulging of sheet specimens through a series of elliptical die inserts, thus producing material deformation at different biaxial strain ratios. The latter can be evaluated by circle grid analysis (CGA) or digital image correlation (DIC) to construct the material's FLCs. The test can be conducted at near-constant strain rates, at very high temperatures, and without the significant frictional effects associated with mechanical stretching tests. In previous efforts, the test was validated by investigating the formability of the AA5083 at selected hot forming conditions [16]. The test was also used to construct the first set of "composite FLDs" which show the effects of sheet orientation on material formability limits for both AA5083 and Mg AZ31B sheets [17]. In this study, the pneumatic stretching test is employed in a detailed investigation on the hot formability (at 400 °C and $5 \times 10^{-3}$ $s^{-1}$) of four Mg AZ31B sheets. The sheets, some of which have been the subject of earlier studies [18-19], have the same composition; however, they were produced through either twin roll continuous (TRC) casting or direct chill (DC) casting and underwent distinct processing routes, resulting in differences in their nominal grain sizes. A composite FLD was generated for each of the four sheets; disparities in their formability limits were related to differences in their initial microstructures and material inhomogeneities.

## Experiments

### Materials and Specimen Preparation

The four Mg AZ31B sheet materials, which we arbitrarily label as (A, M, N and O), have similar compositions, as detailed in Table I. They were received in the O-temper condition, with a nominal thickness of ~1mm. Sheet (M) was produced by DC casting, while the other three sheets were produced by TRC casting. Standard microstructural examination was performed on specimens taken from each sheet; the resulting optical micrographs are shown in Figure 1. Analysis of the latter revealed that sheets (A, N and O) (all of which were produced by TRC casting) have similar average grain size ($d_{avg}$ = ~5 μm). Sheet (M), on the other hand, was found to have an average grain size of ~10 μm; nearly twice as large as that of the other three sheets.

Table I. Composition, manufacturing process and average initial grain size of the four Mg AZ31B sheets investigated in this study.

| Sheet | Composition (%) | | | | Process | $d_{avg}$ |
|---|---|---|---|---|---|---|
| | Al | Zn | Mn | Mg | (Casting) | (μm) |
| A | 2.8 | 0.81 | 0.30 | Balance | TRC | 5.3 |
| M | 2.6 | 0.71 | 0.32 | Balance | DC | 10.0 |
| N | 3.0 | 0.74 | 0.35 | Balance | TRC | 4.9 |
| O | 3.0 | 0.74 | 0.32 | Balance | TRC | 4.8 |

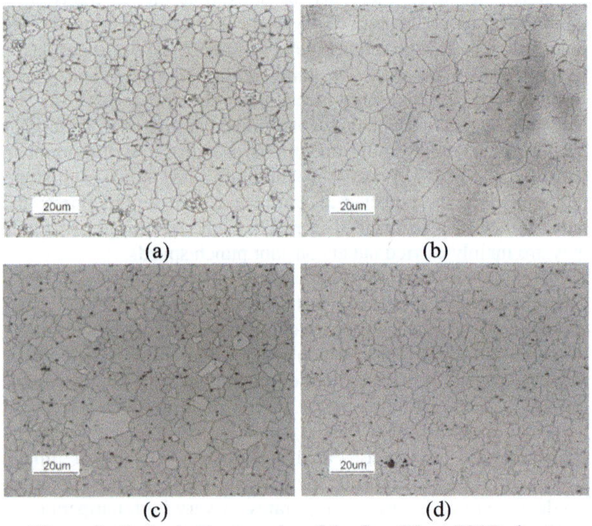


Figure 1. Optical micrographs of the four Mg AZ31B sheets investigated in this study: (a) A (b) M (c) N (d) O.

In preparation for testing, the sheets were first cut into ~100mmx100mm (4"x4") square blanks using a hydraulic shear. Each blank was lightly polished using 400 grit sanding paper, maintaining polishing consistency with the rolling direction of the sheet. Lectroetch LNC-3 solution was then used to electrochemically etch a 2.54mm (0.1") circular grid pattern on the surface of each blank. Particular attention was given to aligning the grid with the sheet's rolling direction; this is crucial for correlating the major and minor strains in the deformed specimen with the longitudinal and transverse directions of the sheet. This is also important for accurately establishing the relationship between a specific sheet orientation and the obtained FLCs, as will be detailed in later sections.

### Testing Setup

The core of the setup is the forming die assembly, which is detailed in Figure 2. A die body provides the base on which one of multiple die inserts can rest. A blank is secured between the die insert and the die cover, and a 15-25 kN clamping force seals the chamber on top of the blank in preparation for testing (bulging). Four elliptical die inserts were used in this work to generate different biaxial strain ratios in the deformed specimens. The inserts share the same major axis of 63.5mm (2.5"); the minor axis varies progressively to produce four die aspect ratios (k=1.0, 0.8, 0.6 and 0.4). All inserts were prepared with a die entry radius of 4.76mm (0.1875"), in order to facilitate sheet deformation and minimize the possibility of premature failure along the die insert's perimeter. The forming die assembly is secured to a universal load frame, and enclosed within a custom-built heating chamber (furnace) capable of achieving temperatures in excess of 700 °C. Further details about the pneumatic stretching test setup can be found in earlier publications [16-17]

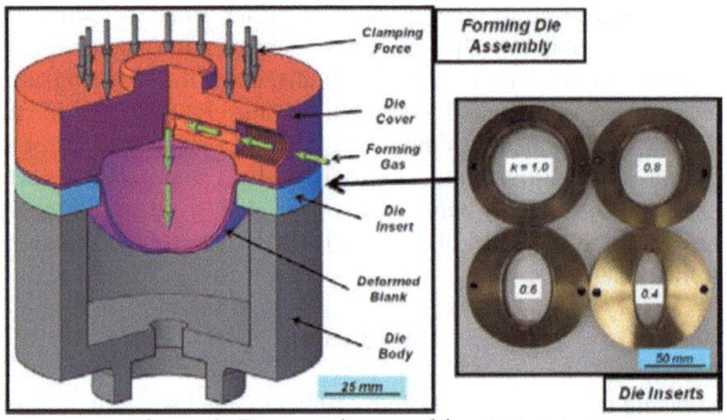


Figure 2. Pneumatic stretching test setup.

### Material Deformation Rate

Control of sheet deformation rate to near-constant levels is essential in this test, and it is accomplished by varying the pressure of the forming gas accordingly. The latter is typically achieved through finite element simulations; yet due to the simplicity of the die inserts, an analytical approach was followed here. The closed-form solution for free bulging into elliptical die inserts, derived by Banabic et al. 2001 [20], describes the necessary gas pressure-time (P-t) profiles as follows:

$$P(t) = \frac{2 s_0 \bar{\sigma}}{b_0 a_0} \frac{\left(1+\alpha k^2\right)}{\rho} e^{\frac{-3\dot{\bar{\varepsilon}} t}{2\rho}} \left( e^{\frac{2-\alpha}{2\rho}\dot{\bar{\varepsilon}} t} - 1 \right) \quad (1)$$

$$k = \frac{b_0}{a_0}, \quad \alpha = \frac{1}{2}\left(1 + e^{1-\frac{1}{k}}\right), \quad \rho = \sqrt{1-\alpha+\alpha^2} \quad (2)$$

where $P$ is the forming pressure, $t$ is the forming time, $\bar{\sigma}$ is the effective stress, $\dot{\bar{\varepsilon}}$ is the effective strain rate, $s_0$ is the initial sheet thickness, $a_0$ and $b_0$ are the major and minor half axes of the elliptical die insert, respectively. Using the above equations, the P-t profiles required for forming the four Mg AZ31B sheets, at a selected temperature of 400 °C and effective strain rate of $5 \times 10^{-3}$ $s^{-1}$, were obtained. The required flow stresses were obtained using the Sellars-Tegart hyperbolic sine equation [21]. Details on the fit to the hyperbolic sine equation for various Mg AZ31B sheets covered in this study can be found in Dryer et al. 2009 [18].

Testing Procedure

The entire forming die assembly is first heated to 400 °C and allowed to equilibrate for a minimum of 60 minutes. A blank is then inserted, and 5 minutes are allowed to guarantee thermal equilibrium. Towards the end of the heating period, the forming die assembly is closed, and the clamping force is applied in preparation for testing. A test is then begun by the introduction of the forming gas (Argon in this case) following the P-t profile for the selected material and the corresponding die insert. A test is ultimately terminated when sheet rupture is detected; the latter is sensed by a sudden drop in the applied pressure. The deformed blank is then retrieved and another testing cycle is started over. At least two tests were performed for each material/die insert to assure repeatability of the results.

Finally, to investigate the effects of sheet orientation on material formability limits, forming experiments were carried out in two orientations: (1) In the first, sheet specimens were positioned in the forming die assembly such that the rolling direction was aligned with the minor axis of the elliptical die insert (which is the direction of major strains); domes formed under this condition and the corresponding results are denoted by "0° orientation". (2) In the second, sheet specimens were positioned such that the rolling direction was transverse to the minor axis of the elliptical die insert; domes formed under this condition and the corresponding results are denoted by "90° orientation".

**Results and Discussion**

After pneumatic testing and deforming the sheet specimens into ellipsoidal domes, circle grid analysis (CGA) was performed on each dome to measure the planar strains associated with the deformed grid elements (deformed circles). The collected strain data points were distinguished by deformation zone (safe, marginal and failure); this is important for delineating the material FLCs, and was accomplished by following the ASTM E2218 guidelines [9]. The concept of a "composite FLD" was recently introduced as a comprehensive tool (formability map) for quantifying orientation effects on the formability limits of the material [17]. The latter is constructed here for sheet (A) in Figure 3. Note that (for each of the two orientations) the lower "green" curve was drawn such that most of the safe data points fall below it; similarly, the upper "red" curve was drawn such that most of the failure data points lie above it.. This plot clearly shows that the material exhibits greater formability limits in the 0° orientation over those in the 90° orientation. Although the curves meet in the balanced biaxial region, this composite FLD indicates that ~20% increase in the major strains can be achieved in the 0° over the 90° orientation, as the strain state moves toward the plane-strain condition. Should this composite FLD be plotted in terms of the engineering strains, a difference of more than 30% between both orientations is observed.

Similar detailed analysis was carried out on the remaining three AZ31B sheets, and the results are summarized in Figure 4. While the general trends noted for sheet (A) were similarly observed in the other sheets, disparities were found to exist among the four sheets. Close examination of Figure 4a (0° orientation) and Figure 4b (90° orientation) shows that sheet (A) exhibits the highest formability limits of all four magnesium alloy sheets in both orientations. Sheets (M) and (O) follow with nearly equal formability limits, although sheet (O) exhibits higher limits in the 90° orientation. Finally, sheet (N) is found to exhibit the lowest formability limits of all four sheets, regardless of the biaxial strain ratio and/or sheet orientation.

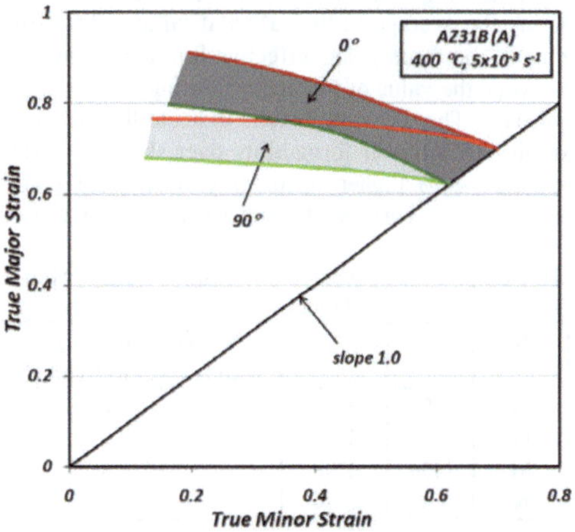

Figure 3. A "Composite FLD" for the Mg AZ31B sheet (A) showing its FLCs in both the 0° and 90° orientations.

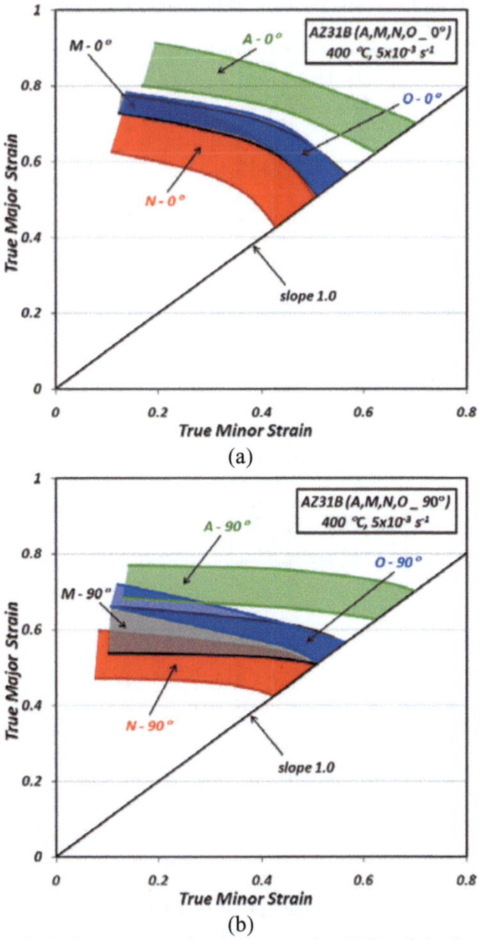

Figure 4. A direct comparison between the FLCs of the four Mg AZ31B sheets in the (a) 0° orientation (b) 90° orientation.

An example of a quantitative comparison between the formability limits of the four Mg AZ31B sheets is provided in Table II, for the balanced biaxial strain region (along the line with slope = 1.0). The major limiting strain ($\varepsilon_{major}$) is extracted for each sheet by considering the average of the safe and failure FLCs (median of each zone in the figure). The effective limiting strain ($\varepsilon_{effective}$) is equal to twice the value of the major limiting strain ($\varepsilon_{major}$), based on von-Mises. The extracted results indicate that sheets (M) and (O) exhibit ~15% higher formability over sheet (N), while sheet (A) exhibits ~40% higher formability over sheet (N). Similar comparisons can be performed at other biaxial strain ratios.

Table II. Comparison between the limiting strains for the four Mg AZ31B sheets at k = 1.0 (according to the FLCs in Figure 4).

| Sheet | $\varepsilon_{major}$ | $\varepsilon_{effective}$ | % change in $\varepsilon_{effective}$ with respect to sheet (N) |
|---|---|---|---|
| A | 0.66 | 1.32 | 40.4% |
| O | 0.54 | 1.08 | 14.9% |
| M | 0.54 | 1.08 | 14.9% |
| N | 0.47 | 0.94 | ---- |

The impact of sheet orientation on material formability limits can be also observed through examination of the maximum dome height prior to failure; the latter is plotted in Figure 5 below.

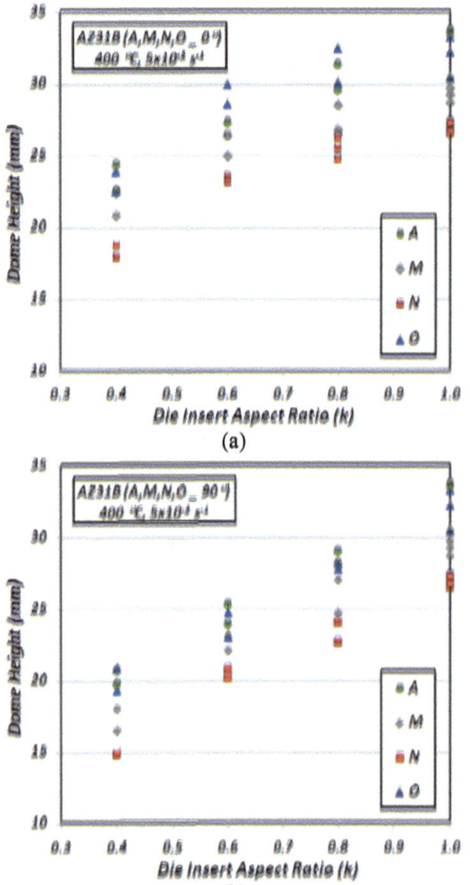


(a)

(b)

Figure 5. A direct comparison between the four Mg AZ31B sheets in terms of the maximum dome height: (a) 0° (b) 90° orientation.

The trend observed here is found to be consistent with the formability limit results in Figure 4, except in the case of sheet (O). Note that sheet (O) achieves greater dome height values than those achieved by sheet (M). In fact, the dome heights achieved with sheet (O) exceed those recorded for sheet (A) in the 0° orientation (Figure 5a). Since sheet (A) clearly achieves higher limiting strains than sheet (O) (according to Figure 4), the higher dome heights observed with sheet (O) must be the results of more uniform deformation (thickness distribution) in the latter compared to the former. To verify this hypothesis, equivalent domes formed from each of the two sheets at k = 1.0 were considered. Images that show cross-sectional profiles in Figure 6 indicate that the dome formed out of sheet (O) is closer to a semi-circular contour than the dome formed out of sheet (A). Moreover, by taking thickness measurements across each of the two domes, the profile corresponding to sheet (O) was found to exhibit more uniformity (i.e. less thinning) relative to that of sheet (A), as demonstrated by the lower part of the figure.

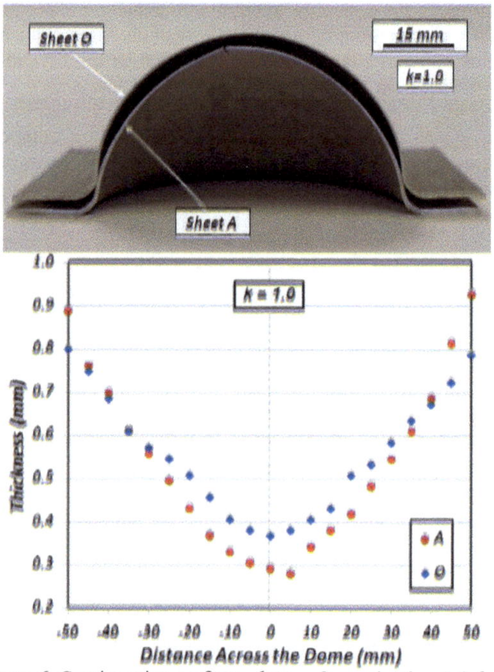


Figure 6. Section views of two domes formed at k = 1.0 from sheets (A) and (O), and the corresponding thickness distributions.

In summary, the results presented so far suggest that sheet (A) is the most formable of the four Mg AZ31 sheets, while sheet (O) exhibits more uniform deformation. Since the chemical composition and initial grain structure of these two materials are similar, the discrepancy in their behaviors can be attributed to the differences in sheet rolling practices of the different suppliers. The formability of sheet (M), which is the only material prepared by DC casting, falls behind that of sheets (A) and (O); this is expected, given its generally larger initial grain size (see Figure 1 and Table I). As for sheet (N), having a fine-grained structure (which was shown to be similar to those of sheets (A) and (O)) does not justify its significantly-lower formability limits, especially in comparison with sheet (M), which has a larger initial grain size. Consequently, further examination was carried out to explain this discrepancy.

Qualitative examination of domes formed from the four Mg AZ31B sheets showed that sheet (N) exhibits remarkably less uniform deformation, accompanied by significant surface damage that's evident over a large portion of the formed dome. An example illustrating the observations in regard to sheet (N) is given in Figure 7, which shows a dome formed at k = 0.8 in the 90° orientation. Note the excessive surface damage and fracture over multiple locations across the surface of the dome. Clearly, rupture does not occur near the dome pole as expected from the pneumatic stretching test. It is speculated that this could be a result of internal material defects that promote premature sheet rupture. A common defect that can lead to premature sheet failure under elevated temperature forming conditions, such as the ones employed in this study, is the clustering of second phase particles [22-23]. A TRC cast sheet with its limited rolling reduction is more prone to this kind of defect than a DC cast sheet. The multiple failure locations seen in the domes formed with sheet N is suggestive of this kind of defect.

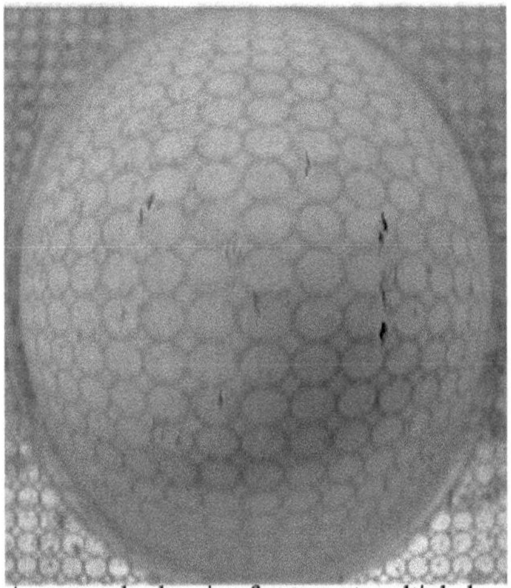

Figure 7. An example showing fracture at multiple locations in a domes formed out of sheet (N) at k = 0.8 in the 90° orientation.

The fracture surface of domes formed out of the four sheets (at k = 1.0) were closely examined by scanning-electron microscopy (SEM); samples of the obtained images are shown in Figure 8. Sheet (A) shows a regular ductile fracture, whereas sheet (N) exhibits material fracturing in multiple layers parallel to the sheet plane. Sheet (O) exhibits some tendency to fracture in layers like sheet (N), whereas sheet (M) is closer in fracture to sheet (A). The aforementioned layering of fractured surface is rather unusual. Therefore, analysis of a fragment between the fracture layers was performed using energy dispersive X-ray spectroscopy (EDS); the analysis revealed metal oxide particles. Furthermore, the optical micrographs in Figure 9 show second phase particle distribution in each of the four sheets, in the as-received condition. Sheets (A) and (M) display a uniform distribution of fine particles, which are mostly submicron in size. In addition to fine particles, sheet (N) exhibits agglomeration of the particles in the form of long (100-200 μm) stringers, parallel to the sheet rolling direction. A similar observation for Mg AZ31 sheet with a 2.0 mm initial thickness was made in the work of Carter et al. 2008 [23], although the morphology and chemistry of the stringers were not investigated in any detail. Sheet (O) shows some tendency for the second phase particles to align along the rolling direction, but nothing close to the level seen in sheet (N).

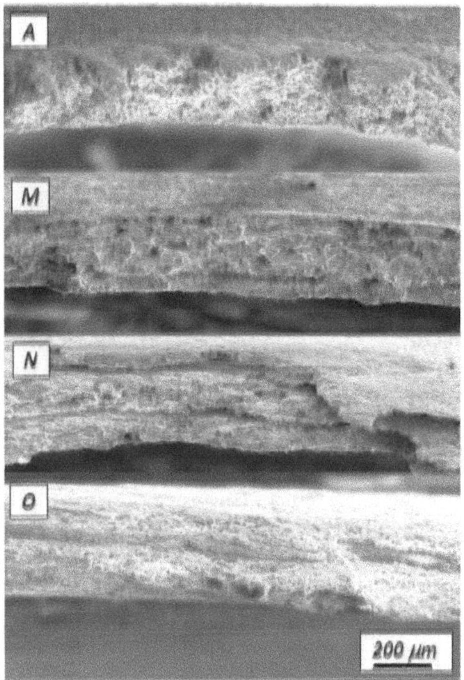


Figure 8. SEM images of fracture surfaces taken from domes formed out of the four Mg AZ31B sheets at (k = 1.0).

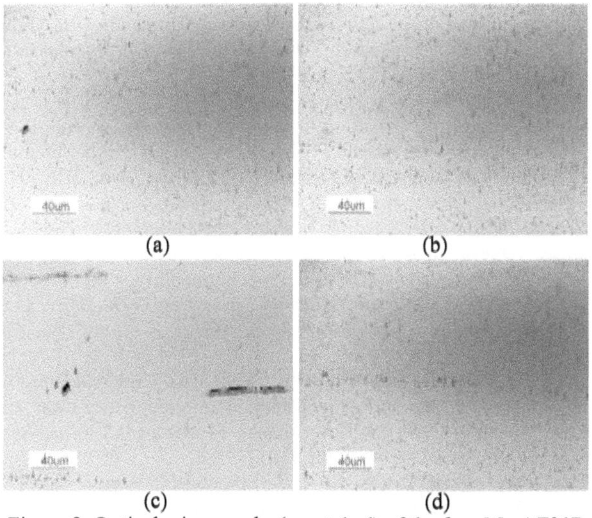


Figure 9. Optical micrographs (un-etched) of the four Mg AZ31B sheets showing second phase particles: (a) A (b) M (c) N (d) O.

EDS analysis of a stringer and a large particle in sheet (N) revealed the second phase to be essentially Mg oxide. It appears that sheet (N) has a significant amount of magnesium oxide that is not uniformly distributed in the microstructure. During elevated temperature forming, the close proximity of these oxide particles, present in the form of long stringers, makes it easier for the cavities formed at the oxide/matrix interface to coalesce and result in premature sheet failure. Failure occurring at multiple locations in sheet (N) corresponds to multiple stringers in the material; unlike the case with the (A, M and O) sheets, in which failure occurs at the dome pole strictly due to geometrical reasons.

**Summary**

A detailed investigation on the formability behaviors of four Mg AZ31 sheets (labeled A, M, N and O) was carried out under hot forming conditions (400 °C and $5x10^{-3}$ $s^{-1}$) using the pneumatic stretching test. For each sheet, friction-independent FLCs were constructed in two orientations; along and across the material's rolling direction. The 0° orientation showed greater formability limits than the 90° orientation for all four sheets. Comparison between the results obtained for the four sheets revealed that sheets (A) and (O), both produced by TRC casting and have a fine grain structure, outperformed the other two sheets. Sheet (M), the only one produced by DC casting, fell behind sheets (A) and (O), mainly due to its larger initial grain size. The formability limits of sheet (N) were found to be inferior to those of the other three sheets, despite its small initial grain size (~5 μm; similar to sheets (A) and (O)). In addition, excessive surface damage and fracture at multiple locations were observed in all the domes formed out of that particular sheet. This was shown to be the result of internal material defects in the form of long oxide stringers, which lead to premature failure and hence reduced formability limits.

**References**

1. J.A. Yasi, L.G. Hector, and D.R. Trinkle, "First-Principles Data for Solid-Solution Strengthening of Magnesium: From Geometry and Chemistry to Properties," *Acta Materialia*, 58 (2010) 5704-5713.

2. H. Li, E. Hsu, J. Szpunar, R. Verma, and J.T. Carter, "Determination of Active Slip/Twinning Modes in Mg Alloy Near Room Temperature," *Journal of Materials Engineering and Performance*, 16 (2006) 321-326.

3. J.T. Carter, V. Savic, L.G. Hector, A.R. Melo, and P.E. Krajewski, "Structural Evaluation of an Experimental Aluminum/Magnesium Decklid," SAE Paper No. 2011-01-0075.

4. E. Doege, and K. Dröder, "Sheet Metal Forming of Magnesium Wrought Alloys - Formability and Process Technology," *Journal Materials Processing Technology*, 115 (2001) 14-19.

5. S. Agnew, and O. Duygulu, "A Mechanistic Understanding of the Formability of Magnesium: Examining the Role of Temperature on the Deformation Mechanisms," *Materials Science Forum*, 419-422 (2003) 177-188.

6. F. Abu-Farha, and M. Khraisheh, "Analysis of Superplastic Deformation of AZ31 Magnesium Alloy," *Journal of Advanced Engineering Materials*, 9 (2007) 777-783.

7. F.K. Chen, and C.K. Chang, "Warm Stamping of Cell-Phone Cases with AZ31 Magnesium-Alloy Sheets," *Advanced Materials Research*, 154-155 (2010) 1826-1829.

8. S.P. Keeler, and W.A. Backofen, "Plastic Instability and Fracture in Sheets Stretched Over Rigid Punches," *ASM Transactions*, 56 (1963) 25-48.

9. ASTM E2218 – 2002, "Standard Test Method for Determining Forming Limit Curves".

10. Z. Marciniak, and K. Kuczynski, "Limit Strains in The Processes of Stretch-Forming Sheet Metal," *International Journal of Mechanical Sciences*, 9 (1967) 609-620.

11. K. Nakazima, T. Kikuma, and K. Asuka, "Study on the Formability of Steel Sheet," (Yawata technical report no. 264, 1971, 678-680).

12. "Metallic Materials - Guidelines for the Determination of Forming Limit Diagrams," (ISO 12004 1997).

13. E. Hsu, J. Carsley, and R. Verma, "Development of Forming Limit Diagrams of Aluminum and Magnesium Sheet Alloys at Elevated Temperatures," *Journal of Materials Engineering and Performance*, 17 (2008) 288-296.

14. T. Naka, G. Torikai, R. Hino, and F. Yoshida, "The Effects of Temperature and Forming Speed on the Forming Limit Diagram for Type 5083 Aluminum-Magnesium Alloy Sheet," *Journal of Materials Processing Technology*, 113 (2001) 648-653.

15. D. Li, and A. Ghosh, "Biaxial Warm Forming Behaviour of Aluminum Sheet Alloys," *Journal of Materials Processing Technology*, 145 (2001) 281-293.

16. F. Abu-Farha, L.G. Hector, and P.E. Krajewski, "Forming Limit Curves for the AA5083 Alloy under Quick Plastic Forming Conditions," SAE Paper No. 2011-01-0235.

17. F. Abu-Farha, and L.G. Hector, "Sheet Orientation Effects on the Hot Formability Limits of Lightweight Alloys," *ASME Journal of Manufacturing Science and Engineering*, In Press.

18. C. Dreyer, F. Polesak, T. Shultz, and S. Agnew, "Blind Study of the Effect of Processing History on the Constitutive Behaviour of Alloy AZ31B," *Proceedings of the 138$^{th}$ TMS Annual Meeting and Exhibition*, San Fransisco, CA (2009) 491-496.

19. R. Verma, J.T. Carter, and P.E. Krajewski, "High Temperature Deformation Behavior of Three Rolled Sheets of Magnesium Alloy AZ31," *Proceedings of the 138$^{th}$ TMS Annual Meeting and Exhibition*, San Francisco, CA (2009) 399-404.

20. D. Banabic, T. Balan, and D. Comsa, "Closed Form Solution for Bulging Through Elliptical Dies," *Journal of Materials Processing Technology*, 115 (2001) 83-86.

21. C.M. Sellars, and W.J. Tegart, "Relationship between Strength and Structure in Deformation at Elevated Temperature," *Mem. Sci. Rev. Metall.*, 63 (1967) 731-745.

22. L.T. Park, S.H. Myunga, D.H. Shin, and C.S. Lee, "Size and Distribution of Particles and Voids Pre-Existing in Equal Channel Angular Pressed 5083 Al Alloy: Their Effect on Cavitation during Low-Temperature Superplastic Deformation," *Materials Sci. & Eng. A*, 371 (2004) 178-186.

23. J.T. Carter, P.E. Krajewski, and R. Verma, "Hot Blow Forming of AZ31 Mg Sheet: Formability Assessment and Application Development," *JOM*, 60 (2008) 77-81.

Magnesium Technology 2012
Edited by: Suveen N. Mathaudhu, Wim H. Sillekens, Neale R. Neelameggham, and Norbert Hort
TMS (The Minerals, Metals & Materials Society), 2012

# TEXTURE EVOLUTION DURING HOT DEFORMATION PROCESSING OF Mg-3Sn-2Ca-0.4Al ALLOY

C. Dharmendra[1], K.P. Rao[1], Y.V.R.K. Prasad[2], N. Hort[3], K.U. Kainer[3]

[1]Department of Mechanical and Biomedical Engineering
City University of Hong Kong, 83 Tat Chee Avenue Kowloon, Hong Kong SAR, China
[2]processingmaps.com (formerly at City University of Hong Kong, Hong Kong, China)
[3]Helmoltz-Zentrum Geesthacht, Max-Planck Str. 1, Geesthacht 21502, Germany

Keywords: Mg-Sn-Ca-Al alloy, Hot Compression, Processing Maps, EBSD, Texture

## Abstract

An experimental investigation of texture evolution during high temperature compression of Mg-3Sn-2Ca (TX32) alloy containing 0.4%Al using electron back scatter diffraction (EBSD) technique is reported. Isothermal uniaxial compression tests were performed in the temperature and strain rate ranges 300-500 °C and 0.0003-10 $s^{-1}$ to examine the influence of processing conditions on the dynamic recrystallization (DRX) behavior and texture evolution. The onset of DRX during compression at low temperatures (300 and 350 °C) and low strain rates (0.0003 and 0.001 $s^{-1}$) gave rise to a fine, partially recrystallized and necklaced grain microstructure, with the basal poles located at 15-30° from the compressive direction although they were split. Specimens deformed at temperatures higher than 450 °C resulted in a fully recrystallized microstructure and an almost random crystallographic texture. It is clear from Schmid factor analysis that the contribution of pyramidal slip system is significant for deformation at temperatures above 450 °C.

## Introduction

Magnesium alloy products, because of their low density, high specific strength and favorable recycling capability, offer great potential for automotive and aircraft industries to replace steel and other high density materials [1]. However, a major obstacle to wider use of magnesium alloys is their poor ductility. The deformation of magnesium alloys is extremely orientation dependent due to their hexagonal close-packed (hcp) crystal structure and the availability of limited number of slip systems at room temperature. Basal slip and the tension twinning are the most readily available slip systems at room temperature [2]. In the hcp metals, the only plane with high atomic density is basal plane (0001). The plastic deformation of poly crystals requires five independent slip systems. For extensive workability in magnesium alloys, slip on (0001) basal, $(10\bar{1}0)$ prismatic and $(11\bar{2}2)$ second order pyramidal planes is essential [3]. Studies on single crystal deformation [2, 4] showed that the critical resolved shear stresses (CRSS) required for activating non-basal slip modes are considerably higher than for basal slip system. Increasing deformation temperatures decreases CRSS and thus, non-basal slip modes also activate and become more important as temperature raises. For the operation of a slip system, there are two conditions, i.e. energy condition and the geometric condition. The energy condition can be described by CRSS while the geometric condition can be described by Schmid factor. The geometric condition for a slip system to operate, under an uniaxial compression condition be expressed by Schmid's law [5]

$$T_c = \pm \sigma \cos \varphi \cos \lambda \quad (1)$$

where $T_c$ is the critical resolved shear stress, $\sigma$ is the applied stress, $\varphi$ is the angle between the slip direction and compressive direction, $\lambda$ is the angle between the normal direction of the slip plane and the compressive direction. The Schmid factor is defined as $m = \cos \varphi \cos \lambda$. For a given slip system with a CRSS $T_c$, the required yield stress is determined by the Schmid factor '$m$' or the crystallographic orientation of the grain, with small Schmid factor being unfavorable for slip to operate.

Magnesium develops strong textures during processing and affects the ductility of Mg-based alloys [6]. In practice, ductility can be improved by working at elevated temperatures [7], finer grain sizes and solute additions [8]. All of these can be rationalized in terms of non-basal slip activation [9]. As-cast and wrought materials both differ in grain size and texture, factors which are known for a strong influence on deformation behavior of magnesium alloys even at high temperatures.

Along with ductility, good corrosion and creep resistance properties are also significant to employ the magnesium alloys as structural materials. To improve these properties, newer Mg-Sn-Ca (TX series) alloys are being developed and the studies of some Mg-Sn-Ca alloys indicate reasonable corrosion and good creep resistance, particularly Mg-3Sn-2Ca (TX32) alloy has exhibited good creep resistance [10-12]. Aluminum addition improves room temperature mechanical properties of Mg by solid solution strengthening due to large atomic size difference (16%) and also precipitation of ($\beta$-$Mg_{17}Al_{12}$) intermetallic phase [13]. However, 0.4 wt% Al to the Mg-3Sn-2Ca alloy does not cause the formation of any new phases. Keeping that in view, 0.4 wt% Al was added to TX32 alloy. Alloying additions can alter the deformation mechanisms, which in turn influence the texture. The aim of the present work is to study the hot compressive deformation behavior of Mg-3Sn-2Ca-0.4Al alloy in as-cast condition using the processing map technique and to evaluate the development of microstructures and micro-textures in the deformed specimens using SEM based EBSD technique. The *processing-map* technique has been extensively used to understand the workability of many materials. The technique is developed on the basis of dynamic materials modelling (DMM) and the detailed description of the model has been given by Prasad et al. [14, 15].

## Experimental

Mg – 3 wt.% Sn – 2 wt.% Ca – 0.4 wt.% Al (TX32 – 0.4Al) alloy was prepared using 99.99% pure Mg, 99.96% pure Sn, 98.5% pure Ca and 99.9% Al. The molten alloy at about 720 °C was kept under a protective cover of Ar+3% $SF_6$ gas before casting in a pre-heated permanent mold to obtain cylindrical billets of 100 mm diameter and 350 mm length. Cylindrical specimens of 10 mm diameter and 15 mm height were machined from the as-cast billet for compression testing. A 1 mm diameter hole was machined at mid height of the specimen to reach its centre for inserting a thermocouple to measure the specimen temperature as well as the adiabatic temperature rise during deformation.

The data for developing processing maps were obtained from isothermal uniaxial compression tests conducted at constant true strain rates in the range 0.0003 – 10 $s^{-1}$ and temperature range 300 – 500 °C using a computer controlled servo-hydraulic testing machine. Details of the test set-up and procedure are described in an earlier publication [16]. Constant true strain rates during the tests were achieved using an exponential decay of actuator speed in the servo hydraulic machine. Graphite powder mixed with grease was used as the lubricant between the specimen and the compression platens in all the experiments. The specimens were deformed up to a true strain of about 1.0 and then quenched in water. The load - stroke data were converted into true stress - true strain curves using standard equations. The flow stress values were corrected for the adiabatic temperature rise at different temperatures and strain rates and this correction was less important at lower strain rates and higher temperatures. The deformed specimens were sectioned in the center parallel to the compression axis and the cut surface was mounted, polished and etched for metallographic examination. Selected samples were prepared for EBSD by grinding with 320 grit SiC paper, oil based diamond polishing 9, 3, and 1 μm and then vibratory polished followed by cleaning with methanol and ultrasonic cleaner.

The texture of the selected area of the deformed samples was examined using a JEOL 5600 SEM equipped with a NordlysF EBSD detector, and HKL Channel 5 software was used for data collection. EBSD was performed at 20KV, 18-20 mm working distance, a tilt angle of 70° and a scan step size of 2 μm. The portion of the target area that was indexed varied between 75% and 85%. Output texture data is shown as pole figures, in which the horizontal axis of the figure is the compression direction.

## Results and Discussion

Processing Map

The processing map of Mg-3Sn-2Ca-0.4Al alloy obtained at a strain of 0.5 is shown in Figure 1 which exhibits two domains of dynamic recrystallization (DRX) in the temperature and strain rate ranges given as follows: (1) 300–350 °C and 0.0003–0.001 $s^{-1}$ with a peak efficiency of 43% occurring at 300 °C and 0.0003$s^{-1}$, and (2) 400–500 °C and 0.005–0.6 $s^{-1}$ with a peak efficiency of 39% occurring at 450 °C and 0.03 $s^{-1}$. The higher strain rate diagonal region starting from 300 °C / 0.001 $s^{-1}$ and ending at 470 °C/10 $s^{-1}$ represents flow instability.

The strain rate sensitivity ($m$) and stress exponent ($n$) values obtained at the temperature of 500 °C were 0.22 and 4.54 respectively, and the apparent activation energy was 175 KJ/mole in the lower strain rate regime and 195 KJ/mole in the higher strain rate regime [16,17]. It is expected that decrease in flow stress with increase in temperature causes the change in rate controlling process as well as the dominant slip systems.

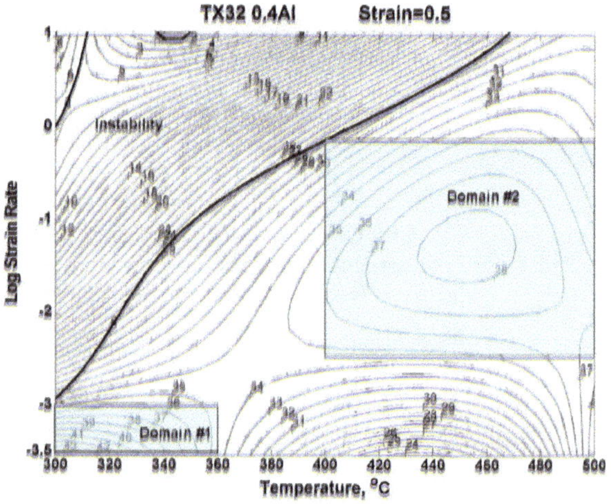


Figure 1. Processing map for Mg–3Sn–2Ca–0.4Al alloy. The numbers against the contours represent efficiency of power dissipation in percent.

Microstructures

The optical microstructures recorded on the specimens deformed in the DRX domain #1 at 300 °C / 0.0003 $s^{-1}$ and 350 °C / 0.001 $s^{-1}$ are shown in Figure 2(a) and (b).

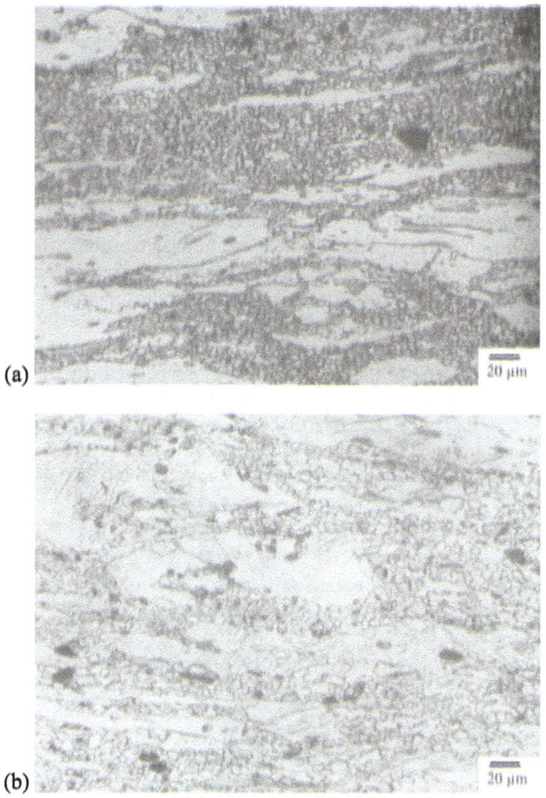


Figure 2. Microstructures of Mg–3Sn–2Ca–0.4Al alloy deformed at (a) 300 °C / 0.0003 $s^{-1}$ and (b) 350 °C / 0.001 $s^{-1}$ (domain #1). Compression axis is vertical.

Both these microstructures indicate that the initial as-cast microstructure is partially converted into fine grained necklaced type of microstructure through a mechanism involving DRX. The optical microstructures of the specimens deformed in the DRX domain #2 at 450 °C / 0.01 $s^{-1}$ and 500 °C / 0.1 $s^{-1}$ are shown in Figure 3(a) and 3(b) respectively. The microstructures were influenced by the deformation conditions. In both the cases, initial as-cast microstructure has transformed into wrought microstructure and all the grains are completely recrystallized. Grain size increased with increase in deformation temperature.

(a)

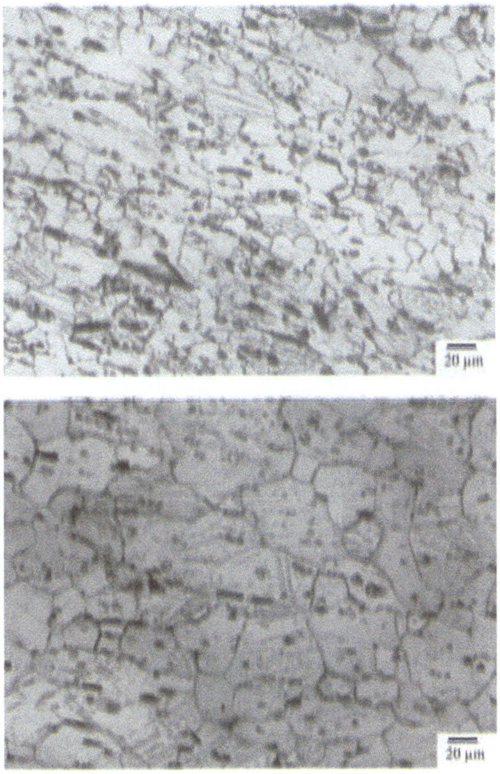


(b)

Figure 3. Microstructures of Mg–3Sn–2Ca–0.4Al alloy deformed at (a) 450 °C/0.01 $s^{-1}$ and (b) 500 °C/0.1 $s^{-1}$ (domain #2). Compression axis is vertical.

Texture Evolution

The important slip systems that operate in magnesium materials are: (1) Basal slip (0001) $\langle 11\bar{2}0\rangle$, (2) Prismatic slip $(10\bar{1}0)$ $\langle 11\bar{2}0\rangle$, (3) first order pyramidal slip $(10\bar{1}1)$ $\langle 11\bar{2}0\rangle$ and $(10\bar{1}2)$ $\langle 11\bar{2}0\rangle$, (4) second order pyramidal slip $(11\bar{2}2)$ $\langle 11\bar{2}\bar{3}\rangle$. They get activated when the critical resolved shear stress (CRSS) exceeds relevant threshold levels. During the room temperature deformation of magnesium, basal slip plays the dominant role as it has the lowest CRSS. Prismatic slip becomes extensive at temperatures higher than 225 °C and pyramidal slip system contributes significantly at temperatures higher than 350 °C. As for CRSS, the second easiest slip system is prismatic slip system for Mg alloys. Even the combination of basal and prismatic slip systems is not sufficient to cause large plastic deformation. Among the other slip systems reported for Mg, the second order pyramidal slip system also gets active.

Figure 4 shows pole figures for the Mg-3Sn-2Ca-0.4Al in the as-cast condition and the alloy exhibited near random crystallographic texture.

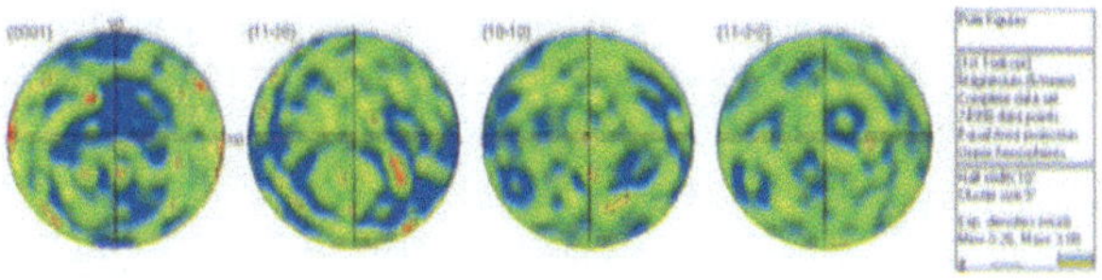

Figure 4. Pole figure corresponding to Mg-3Sn-2Ca-0.4Al alloy in as-cast condition.

Figure 5 shows pole figures for the specimens compressed at 300 °C / 0.0003 $s^{-1}$ and 350 °C / 0.001 $s^{-1}$. If the original grains in the compressed specimens have their basal planes oriented perpendicular to the compression direction, crystallographic orientation in that position is unfavorable for basal slip. However, fine DRX grains have their basal planes inclined to the compressive direction that favors basal slip. From the pole figure of (0001) in Figure 5(a), it can be noticed that basal poles are spread around 15-30° to the compression direction which makes the crystallographic orientation in the position somewhat favorable for basal slip. The point to note here is the textures of the compressed specimens showed in Figure 5 are the combination of two types of crystallographic orientations with respect to the basal planes of DRX (necklaced) grains and the matrix grains.

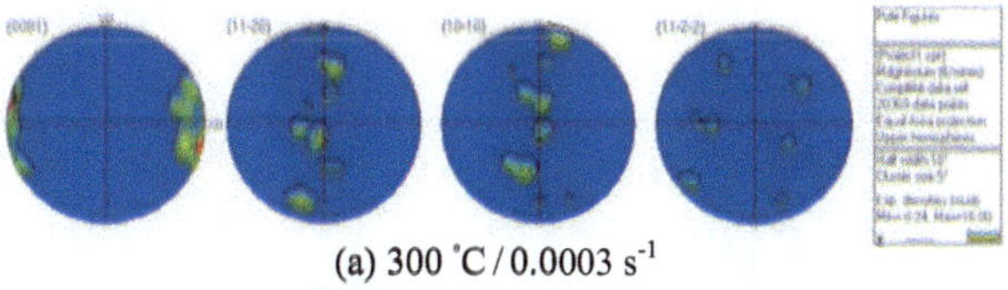

(a) 300 °C / 0.0003 $s^{-1}$

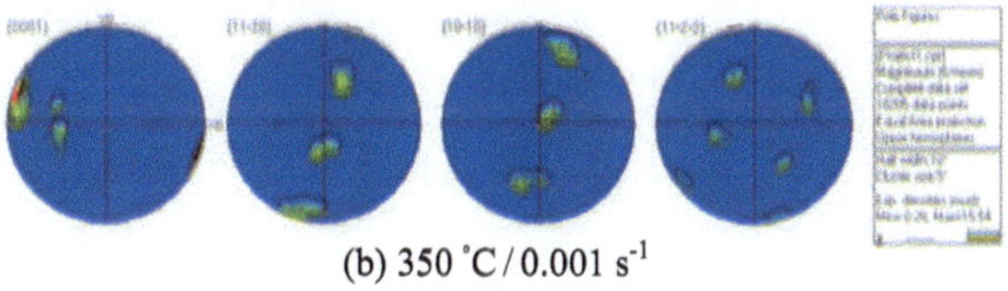

(b) 350 °C / 0.001 $s^{-1}$

Figure 5. Pole figures corresponding to (a) 300 °C / 0.0003 $s^{-1}$, (b) 350 °C / 0.001 $s^{-1}$ (domain #1 and X-axis is compression axis).

Maximum frequencies or activation of (0001) $\langle 11\bar{2}0\rangle$ basal slip for the specimens compressed at 300 °C/0.0003 $s^{-1}$ and 350 °C / 0.001 $s^{-1}$ are found to be at Schmid factor values of 0.41 and 0.48 respectively (shown in Figure. 6(a)). Intensity of texture is almost same in both the conditions. From the pole figures of $(10\bar{1}0)$, it can be observed that prismatic slip planes are mostly aligned perpendicular to compressive direction and also rotated by about 10-15° towards compressive direction. As shown in Figure 6(b), high frequency of activation for the prismatic plane through Schmid analysis is found at only 0.02 and 0.04 for the samples deformed at 300 °C / 0.0003 $s^{-1}$ and 350 °C / 0.001 $s^{-1}$ respectively. These low Schmid factor values for prismatic slip indicate that this slip system is not favorably oriented in the used configuration. Even if the stress level is high enough to initiate the prismatic slip, the resolved shear stress in prismatic planes is practically zero due to new grain orientation. At this point, strain along the c-axis cannot be accommodated during the deformation. The results indicate that by increasing the deformation temperature from 300 ˚C to 350 ˚C, basal slip occurs extensively which reorients the basal poles parallel to the compression axis pushing the prismatic poles to become normal to the compression direction. Thus, prismatic slip becomes unfavorable for further deformation.

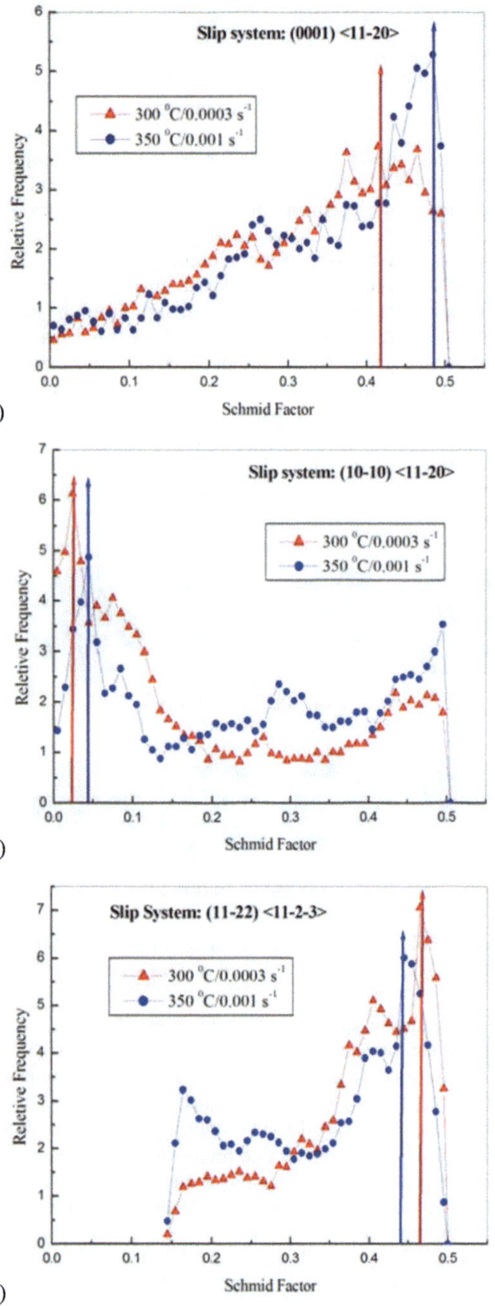


(a)

(b)

(c)

Figure 6. Schmid factor distributions of the grains in the samples deformed at 300 ˚C / 0.0003 $s^{-1}$ and 350 ˚C / 0.001 $s^{-1}$ for slip systems (a) basal (b) prism and (c) pyramid slip.

When the flow stress reaches CRSS for pyramidal <c+a> slip, dynamic recrystallization occurs, reducing the density of dislocation tangles and restoring the ductility to the material. To compensate the imposed deformation, <c+a> pyramidal slip is needed and in fact, $2^{nd}$ order pyramidal system is favorably oriented for the chosen deformation geometry with high Schmid factors of 0.46 and 0.44 for conditions 300 ˚C / 0.0003 $s^{-1}$ and 350 ˚C / 0.001 $s^{-1}$ respectively and it is evidenced from Figure 6(c).

The two DRX domains exhibited in the map are separated by 40 ˚C and it can be viewed as change-over region. Pole figure of the sample deformed at 400 ˚C / 0.1 $s^{-1}$ is shown in Figure 7. For this condition, texture sharpness was decreased compared to the deformed conditions from domain#1 and this can be attributed to due to the difference in operating slip systems. Prismatic slip also contributed during the deformation of compressed specimen for the said condition. However, with further deformation or with increase in strain, other slip modes take over the deformation causing a shift in the maximum pole density of texture towards <0001> pole (i.e. the c-axis in the grains rotated towards the compressive direction).

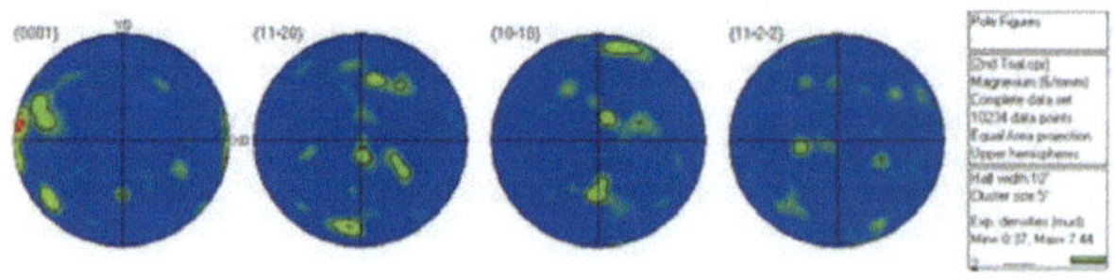

Figure 7. Pole figures corresponding to the deformed condition 400 ˚C / 0.1 $s^{-1}$

Texture Randomization (Domain #2)

The pole figure for the specimen deformed at 450 ˚C / 0.1 $s^{-1}$ is shown in Figure 8. The maximum in the band is laid parallel to compression direction although the planes are split and rotated. (10$\bar{1}$0) prismatic slip planes are also split and randomized. The geometrical condition for the slip systems is changing with deformation conditions. Similar to the analysis for specimens in domain #1, an analysis on slip systems considering Schmid factors has also been considered for the domain #2 samples. Figure 9 shows the EBSD orientation images for the sample deformed at 450 ˚C and 0.1 $s^{-1}$ with Schmid factor considerations for basal slip system (0001) <11$\bar{2}$0> (Figure 9(a)) and $2^{nd}$ order pyramidal slip system (11$\bar{2}$2) <11$\bar{2}$ $\bar{3}$> (Figure 9(b)). It can be seen that high to moderate (red and green colors or dark and light regions marked as A and B respectively) values of Schmid factors dominate the entire images, implying favorable conditions for basal and pyramidal slip systems to operate under this deformation condition.

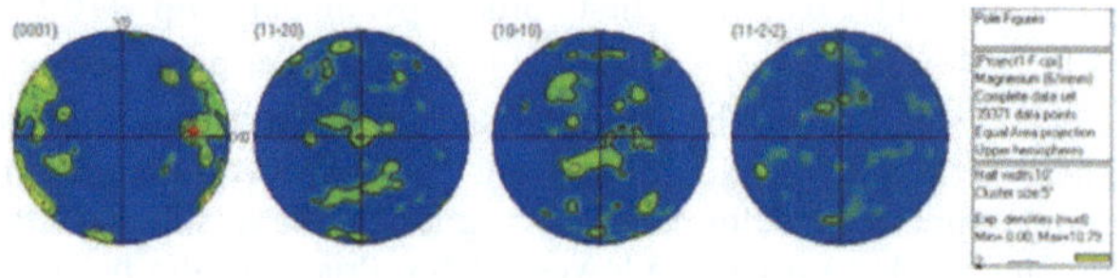

Figure 8. Pole figures corresponding to the deformed condition 450 ˚C / 0.1 $s^{-1}$

The pole figures for the specimens deformed at 500 ˚C / 0.01 $s^{-1}$ and 500 ˚C / 0.1 $s^{-1}$ are shown in Figure 10. The obtained texture evolution demonstrated that with increase in temperature and strain rate progressively texture was randomized. This can be attributed to the dominance of slip on (11 $\bar{2}$ 2) pyramidal planes and to the cross-slip associated with the softening process. This is because of the high stacking fault energy (SFE) on pyramidal slip systems (173 mJ/m$^2$) [18] compared to that on the basal slip systems (60-78 mJ/m$^2$) [19].

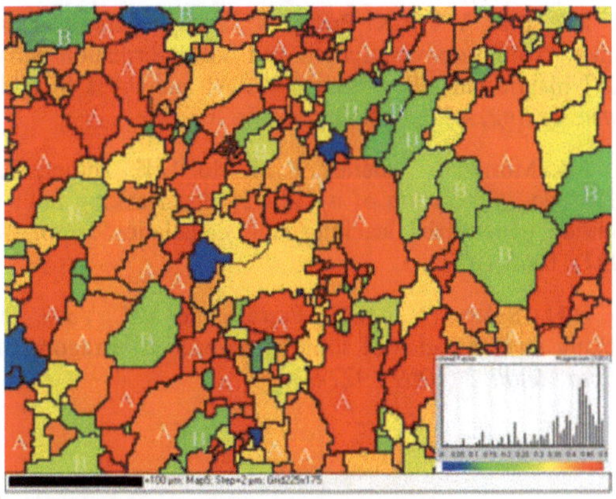
(a)

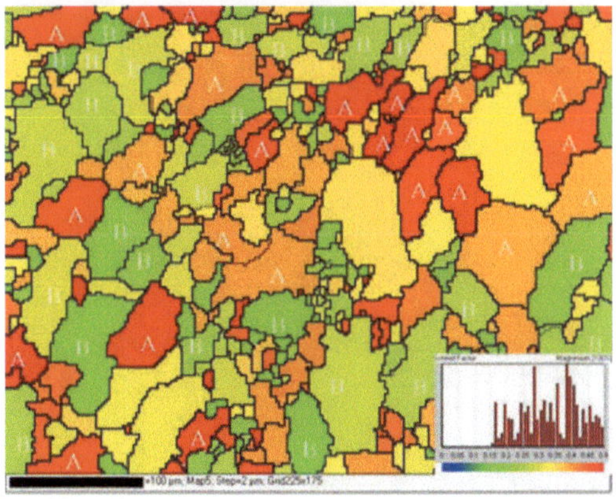
(b)

Figure 9. EBSD Schmid factor distribution maps for the sample deformed to 450 °C/0.1 s$^{-1}$ (a) for (0001) <11 $\bar{2}$ 0> slip system (b) for (11 $\bar{2}$ 2) <11 $\bar{2}$ $\bar{3}$ > slip system. The red and green colors or dark and light regions marked as A and B represent high to moderate Schmid factors.

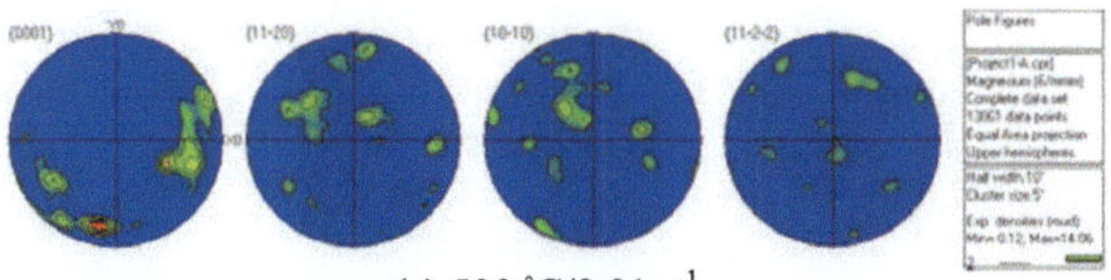
(a) 500 °C/0.01 s$^{-1}$

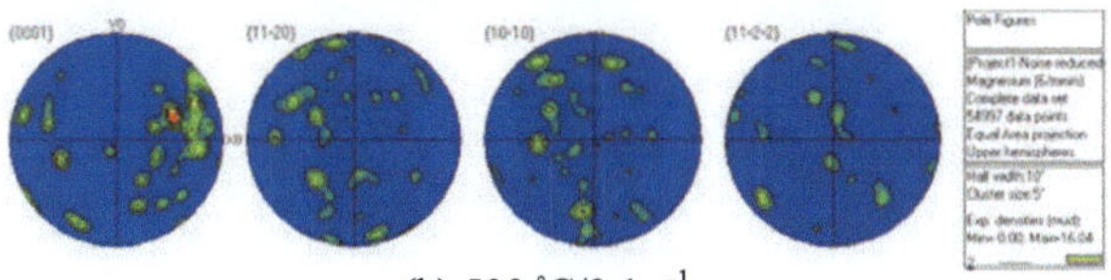
(b) 500 °C/0.1 s$^{-1}$

Figure 10. Pole figures corresponding to conditions (a) 500 °C / 0.01 s$^{-1}$ and (b) 500 °C / 0.1 s$^{-1}$

## Summary and Conclusions

Isothermal uniaxial compression tests were performed on as-cast Mg-3Sn-2Ca-0.4Al alloy to study the effect of deformation conditions on microstructure and texture evolution. The main results are summarized as follows:

(1) Occurrence of the dynamic recrystallization was confirmed by microstructure observation and texture evolution.
(2) The strain rate and temperature have significant influence on the microstructure and texture evolution during uniaxial compression of Mg-3Sn-Ca-0.4Al alloy.
(3) Most basal planes are aligned 15-30° to the compression direction at relatively low temperature (300 °C, 350 °C) and low strain rate (0.0003 s$^{-1}$, 0.001 s$^{-1}$). Texture is randomized at higher temperature (500 °C) and higher strain rate (0.1 s$^{-1}$).
(4) All the three slip systems <a> basal, <a> prismatic, <c + a> pyramidal are necessary for the accommodation of deformation of Mg-3Sn-2Ca-0.4Al alloy. <c + a> pyramidal slip is of crucial importance for achieving high workability.
(5) The optimized condition for hot working of Mg-3Sn-2Ca-0.4Al alloy is 500 °C/0.1 s$^{-1}$.


## Acknowledgement

This work was supported by a grant (Project #115108) from the Research Grants Council of the Hong Kong Special Administrative Region, China.


## References


1. A.A. Luo, "Recent magnesium alloy development for elevated temperature applications", *International Materials Reviews*, 49 (2004), 13-30.

2. B.C. Wonsiewicz and W.A. Backofen, "Plasticity of Magnesium Crystals", *Trans Metall Soc AMIE*, 239 (1967), 1422-1431.

3. C.S. Roberts, *Magnesium and its alloys* (John Wiley and Sons, New York, NY, 1960), 180.

4. R.E. Reed-Hill and W.D. Robertson, "The Crystallographic Characteristics of Fracture in Magnesium Single Crystals", *Acta Metall*, 5 (1957) 728-737.

5. W. Hosford and R.M. Caddle, *Metal forming: Mechanics and Metallurgy*, (Prentice-Hall, 1993), 43-44.

6. W. Hosford, *The Mechanics of Crystals and Textured Polycrystals* (New York: Oxford University Press, USA, 1993), 122-124.

7. E. Schmid and W. Boas, *Plasticity of Crystals: With special reference to Metals* (London: Chapman and Hall, 1935), 140-141.

8. J.A. Chapman and D.V. Wilson, "The room-temperature ductility of fine-grain magnesium", *J Inst Met*, 91 (1962), 39-40.

9. J. Koike, "The activity of non-basal slip systems and dynamic recovery at room temperature in fine-grained AZ31B magnesium alloys", *Acta Mater*, 51 (7) (2003), 2055-2065.

10. T. Abu Leil et al., "Corrosion Behavior and Microstructure of a Broad Range of Mg-Sn-X Alloys," *Magnesium Technology 2006*, eds. A.A. Luo, N.R. Neelameggham and R.S. Beals (Warrendale, PA: The Minerals, Metals and Materials Society, 2006), 281-286.

11. T. Abu Leil et al., "Effect of Heat Treatment on the Microstructure and Creep Behavior of Mg-Sn-Ca Alloys," *Mater Sci Forum*, 546-549 (2007), 69-72.

12. T. Abu Leil et al., Development and Charecterization of a series of Mg-Sn-Ca Alloys, in: *Magnesium Technology in global age*, eds. M.O. Oekguleryuz and L.W.F. Mackenzie, COM 2006, Montreal, Canada, 2006, 739-749.

13. K.M. Asl, A. Tari, and F. Khomamizadeh, "The Effect of Different Content of Al, RE and Si Element on the Microstructure, Mechanical and Creep Properties of Mg-Al Alloys," *Mater Sci Eng*, A523 (2009), 1-6.

14. Y.V.R.K. Prasad et al., "Modeling of dynamic material behavior in hot deformation: Forging of Ti-6242", *Metall Trans*, 15A (1984), 1883-1892.

15. Y.V.R.K. Prasad and T. Seshacharyulu, "Recent advances in the science of mechanical processing", *Indian J Technol*, 28 (1990), 435-451.

16. K.P.Rao, Y.V.R.K. Prasad, N. Hort, K.U. Kainer, "Effect of aluminum addition on the strengthening and high temperature deformation bahavior of Mg-3Sn-2Ca alloy", *Magnesium Technology 2010*, eds. S.R. Agnew, N.R. Neelameggham, E.A. Nyberg and W.H. Sillekens (The Minerals, Metals and Materials Society, 2010), 201-205.

17. K.P. Rao, Y.V.R.K. Prasad, C. Dharmendra, N. Hort, K.U. Kainer, "Compressive strength and hot deformation behavior of TX32 magnesium alloy with 0.4% Al and 0.4% Si additions", *Mater Sci Eng*, A528 (2011), 6964-6970.

18. J.R. Morris, J. Scharaff, K.M. Ho. D.E. Turner, Y.Y. Ye, M.H. Yoo, "Prediction of a {11-22} hcp stacking fault using a modified generalized stacking-fault calculation", *Phil Mag*, 76 (1997), 1065-1077.

19. D.H. Sastry, Y.V.R.K. Prasad and K.I. Vasu, "On the stacking fault energies of some close-packed hexagonal metals", *Scripta Metall*, 3 (1969), 927-930.

**Magnesium Technology 2012**
*Edited by: Suveen N. Mathaudhu, Wim H. Sillekens, Neale R. Neelameggham, and Norbert Hort*
***TMS (The Minerals, Metals & Materials Society), 2012***

# THE EFFECTS OF STRAIN AND STRESS STATE IN HOT FORMING OF MG AZ31 SHEET

Paul A. Sherek[1], Alexander J. Carpenter[1], Louis G. Hector, Jr.[2], Paul E. Krajewski[2], Jon T. Carter[2], Joshua Lasceski[2], and Eric M. Taleff[1]
[1]The University of Texas at Austin; Department of Mechanical Engineering; 1 University Station C2200; Austin, TX 78712-0292, USA
[2]General Motors Corp.; Research and Development; MC 480-106-212; 30500 Mound Rd.; Warren, MI 48090-9055, USA

Keywords: AZ31, bulge forming, material constitutive model

## Abstract

Wrought magnesium alloys, such as AZ31 sheet, are of considerable interest for light-weighting of vehicle structural components. The poor room-temperature ductility of AZ31 sheet has been a hindrance to forming the complex part shapes necessary for practical applications. However, the outstanding formability of AZ31 sheet at elevated temperature provides an opportunity to overcome that problem. Complex demonstration components have already been produced at 450°C using gas-pressure forming. Accurate simulations of such hot, gas-pressure forming will be required for the design and optimization exercises necessary if this technology is to be implemented commercially. We report on experiments and simulations used to construct the accurate material constitutive models necessary for finite-element-method simulations. In particular, the effects of strain and stress state on plastic deformation of AZ31 sheet at 450°C are considered in material constitutive model development. Material models are validated against data from simple forming experiments.

## Introduction

In the automotive and aircraft industries, weight reduction is an effective method of increasing both the fuel efficiency and performance of a vehicle [1-3]. For this reason, lightweight structural materials, such as magnesium alloys, are often preferred over heavier materials. Magnesium is attractive because of its low density, compared to steel or aluminum [4-6]. Because of its low density, a magnesium component can be designed with less weight than a corresponding steel or aluminum component with the same bending strength and stiffness [7]. However, magnesium alloys typically possess poor ductility at room temperature [8]. To obtain sufficient ductility for forming complex parts, high-temperature forming operations, such as superplastic forming (SPF) and quick plastic forming (QPF), can be used [9, 10]. In both SPF and QPF, a sheet metal blank is clamped between two dies and gas-pressure is applied to form the sheet into a die cavity. Such technologies have been used to produce demonstration automotive panels from Magnesium alloy AZ31 sheet [11].

Finite-element-method (FEM) simulations can assist in determining the processing parameters (gas-pressure, temperature, etc.) necessary for gas-pressure forming of complex components. For FEM simulations, the deformation response of a workpiece material is described mathematically through a constitutive model that relates strain to stress. Thus, FEM simulation results are only accurate if an accurate constitutive model is used. Constitutive models describing material deformation behaviors at elevated temperatures are most commonly developed using data from tensile tests [12, 13]. However, data from gas-pressure, bulge-forming experiments at elevated temperature have also been used to inform constitutive models [14-16].

The high-temperature deformation behavior of magnesium sheet materials has most frequently been described in the literature with single-term material constitutive models [12-16], which are designed to predict forming behavior when only one deformation mechanism is active, such as grain-boundary-sliding (GBS) creep. A previous study determined that two-term material constitutive models can be more accurate than single-term models when deformation is controlled by two active deformation mechanisms in an aluminum alloy, which is commonly the case in SPF and QPF operations [17]. A two-term material constitutive model was developed using strain-rate-change (SRC) tensile tests of fine-grained AZ31 at temperatures from 300 to 500°C and strain rates from $10^{-4}$ to $3\times10^{-2}$ $s^{-1}$ [18]. The constitutive equation used is,

$$\dot{\varepsilon} = A_1\left(\frac{\sigma}{E}\right)^{n1} \exp\left(\frac{-Q_1}{RT}\right) + A_2\left(\frac{\sigma}{E}\right)^{n2} \exp\left(\frac{-Q_2}{RT}\right), \quad (1)$$

where $\dot{\varepsilon}$ is the true-strain rate, $A$ is a material constant, $n$ is the stress exponent, $\sigma$ is the true flow stress, $E$ is the Young's modulus, and $Q$ is the activation energy for creep. The subscripts 1 and 2 denote GBS and dislocation-climb creep terms respectively, which are the deformation mechanisms potentially active over the range of conditions investigated. A constitutive model in the form of Equation (1) was derived from tensile data [18]. While it provided accurate predictions of behavior under uniaxial tension, it was far less accurate in predicting deformation under biaxial tension, as determined by comparing predictions to data from gas-pressure bulge tests. A second model was created to improve the agreement of FEM simulation predictions with data from bulge forming experiments. This second model, the "1.3 $\sigma$" model, reduces the effective stress in the dislocation-climb creep term of Equation (1) by dividing the flow stress in the second term by 1.3. Simulation predictions using the 1.3 $\sigma$ model reasonably predict pole heights from gas-pressure bulge tests at 450°C for a range of pressures. The 1.3 $\sigma$ model, however, does not account for significant hardening in the AZ31 material at slow strain rates [18].

The purpose of the present study is to construct new two-term strain-dependent material constitutive models from AZ31 tensile data at 450°C. These new constitutive material models are used in FEM simulations, the predictions of which are compared to the results of simulations using the older models at 450°C and to data from both tensile tests and gas-pressure bulge forming experiments.

## Material Constitutive Model Creation

### Tensile Tests

Material constitutive models were created using data from tensile tests of AZ31 sheet at 450°C. The sheet composition is provided in Table I. Dog-bone shaped tensile coupons with a gauge length of 25 mm, gauge width of 6 mm, and a shoulder radius of 3 mm were machined from sheet with an as-received thickness of 2 mm. The tensile axis of coupons was oriented at 0°, 45°, or 90° relative to the rolling direction. Each coupon was pulled in tension to a true strain of 0.6 at 450°C and a constant true-strain rate in the range of $10^{-4}$ to $10^{-1}$ $s^{-1}$. Testing followed the procedures described in ASTM E 2448-06 [19]. Load and cross-head extension were measured as functions of time during each test and were used to calculate true stress and true strain. The stress-strain data were corrected for machine stiffness by enforcing the elastic modulus of pure Magnesium at 450°C, 32 GPa [20].

Table I. Composition of the AZ31 sheet material in wt. pct.

| Al | Zn | Mn | Fe | Cu | Ni |
|---|---|---|---|---|---|
| 3.1 | 1.0 | 0.42 | 0.006 | 0.003 | <0.003 |
| **Si** | **Ca** | **Be** | **Sr** | **Ce** | **Mg** |
| <0.1 | <0.01 | <0.005 | <0.005 | <0.01 | Bal. |

Analysis of Tensile Data

At strain rates from $10^{-4}$ to $3\times10^{-3}$ $s^{-1}$, flow stress increases with true strain. The Voce strain-hardening law

$$\sigma = \sigma_o + a\left(1 - \exp\left(-c\varepsilon\right)\right) \tag{2}$$

was fit to these data, where $\sigma$ is flow stress, $\sigma_o$ is initial yield strength, $a$ and $c$ are fitting parameters, and $\varepsilon$ is true strain. Flow stress is approximately constant with strain for strain rates from $10^{-2}$ to $10^{-1}$ $s^{-1}$, and only a mean flow stress, $\sigma_m$, was calculated for these conditions. The Voce law parameters and average flow stresses calculated during fitting are presented in Table II. The values in Table II indicate that specimen orientation has little effect on flow stress for a given strain rate. Thus, subsequent analyses assumed flow stress to be the average of all three orientations.

Table II. Parameters for the Voce law and the mean flow stress

| Strain rate, $s^{-1}$ | Orientation, degrees | $\sigma_o$, MPa | $a$, MPa | $c$ | $\sigma_m$, MPa |
|---|---|---|---|---|---|
| $10^{-4}$ | 0 | 1.73 | 8.02 | 1.96 | - |
| | 45 | 1.77 | 6.85 | 2.15 | - |
| | 90 | 1.81 | 6.68 | 2.42 | - |
| $3\times10^{-4}$ | 0 | 4.01 | 6.65 | 3.23 | - |
| | 45 | 3.89 | 6.72 | 3.40 | - |
| | 90 | 4.18 | 6.36 | 3.83 | - |
| $10^{-3}$ | 0 | 8.78 | 4.72 | 4.41 | - |
| | 45 | 8.79 | 4.73 | 4.56 | - |
| | 90 | 8.90 | 4.63 | 4.83 | - |
| $3\times10^{-3}$ | 0 | 14.2 | 2.05 | 6.34 | - |
| | 45 | 14.2 | 2.21 | 5.95 | - |
| | 90 | 14.7 | 1.72 | 7.11 | - |
| $10^{-2}$ | 0 | 19.8 | - | - | 20.0 |
| | 45 | 20.0 | - | - | 20.3 |
| | 90 | 20.4 | - | - | 20.6 |
| $3\times10^{-2}$ | 0 | 24.0 | - | - | 24.6 |
| | 45 | 24.2 | - | - | 24.7 |
| | 90 | 24.7 | - | - | 25.1 |
| $10^{-1}$ | 0 | 28.7 | - | - | 30.1 |
| | 45 | 29.0 | - | - | 30.4 |
| | 90 | 29.2 | - | - | 30.5 |

The fitted parameters in Table II were used to calculate flow stress for each test strain rate at strains from 0 to 0.6, in increments of 0.1. These calculated data are shown in Figure 1. At each strain, the following phenomenological equation for creep deformation was fit to the data,

$$\dot{\varepsilon} = A_1\sigma_1^{n1} + A_2\sigma_2^{n2} \tag{3}$$

where $\dot{\varepsilon}$ is true-strain rate, $A$ and $n$ are fitting parameters, and $\sigma$ is true stress. This equation represents the effect of two independent creep deformation mechanisms. Mechanism 1 (first term) governs deformation at the slowest rates, while mechanism 2 (second term) governs deformation at the fastest rates. These are the GBS and dislocation-climb creep mechanisms, respectively, previously discussed.

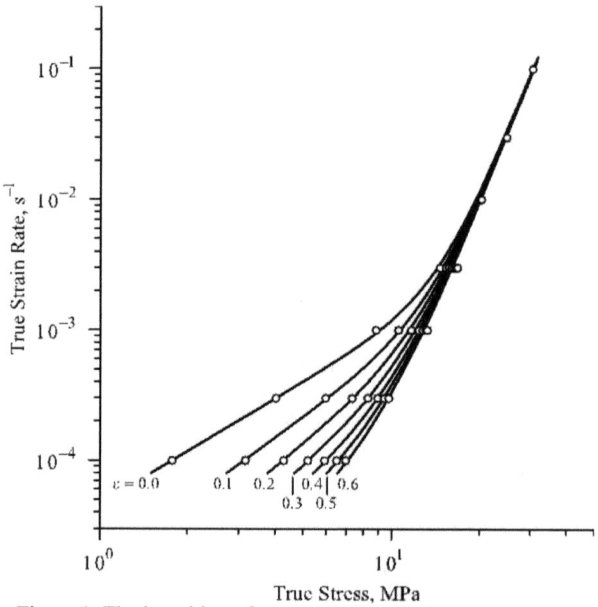


Figure 1. The logarithm of true-strain rate is plotted against the logarithm of true stress for AZ31 tensile data at strains from 0 to 0.6. All specimens were tested at 450°C. Curves represent fits of Equation (3) to the calculated data.

The fits in Figure 1 produced values of $A_2$ and $n_2$ which are constant with strain

$$A_2 = 5.90\times10^{-10}\ s^{-1}\ MPa^{-n2} \tag{4}$$

$$n2 = 5.54 \tag{5}$$

This value of $n2$ suggests that dislocation-climb creep is the active mechanism at the fastest strain rates [22]. However, $A_1$ and $n1$ were observed to vary significantly with strain. The fitted values of $n1$ vary between 1.333 and 2, and are described reasonably by,

$$n1(\varepsilon) = 1.333 + 0.667\tanh\left(4.63\varepsilon\right) \tag{6}$$

Assuming these values for $n1$, the following equation was fit to $A_1$, with units of $s^{-1}$ $MPa^{-n1}$, as a function of true strain,

$$A_1(\varepsilon) = \exp\left(-9.98 - 12.59\varepsilon + 17.85\varepsilon^2 - 11.10\varepsilon^3\right) \quad (7)$$

The range of values observed for $n1$ suggests that GBS creep is the active mechanism at the slowest strain rates [22]. The material model described by Equations (3) through (7) is designated the strain-dependent tensile data (SDTD) material model.

After comparison with data from bulge forming experiments, the effective flow stress in the dislocation-climb creep term, the second term of Equation (2), was reduced. This was accomplished by dividing the flow stress in this term only by 1.3, as was previously done for the 1.3 $\sigma$ model [18]. This produced a new value of $A_2$,

$$A_2 = 1.38 \times 10^{-10}\ \text{s}^{-1}\ \text{MPa}^{-n2} \quad (8)$$

The model described using Equation (3) with Equations (5) through (8) is designated the SDTDM1 (SDTD modification #1) model.

## Model Validation Procedure

Gas-Pressure Bulge Forming Experiments

Gas-pressure bulge forming experiments were performed using a custom apparatus described in Ref. [23]. For these experiments, specimen blanks machined from AZ31 sheet were formed at 450°C using gas pressure. The blanks are disks with a diameter of 90.2 mm and the as-received sheet thickness of 2 mm. A schematic of the experiment is shown in Figure 2. During each experiment, an AZ31 blank was placed between two die halves. The lower die half has a cylindrical cavity with an inner diameter of $D_{cavity}$ = 55.4 mm, a sealing bead diameter of $D_{bead}$ = 64.8 mm, and a die-entry radius of 1.5 mm, see Figure 2. Forming began after the blank reached 450°C, as measured by a thermocouple placed adjacent to the blank. Normal preheat times to reach and stabilize at the test temperature varied from 20 to 60 minutes. In selected short-preheating experiments, the blank was preheated for only 5 minutes prior to forming. Upon the completion of preheating, gas-pressure was applied to the top of the blank, causing it to form downward into a bulge shape similar to an inverted dome. A balanced biaxial stress state occurs at the dome peak during forming. The pressures applied ranged from 500 to 1240 kPa. Dome pole displacement was measured *in situ* during deformation using a digital micrometer with a remote-contact gauge assembly.

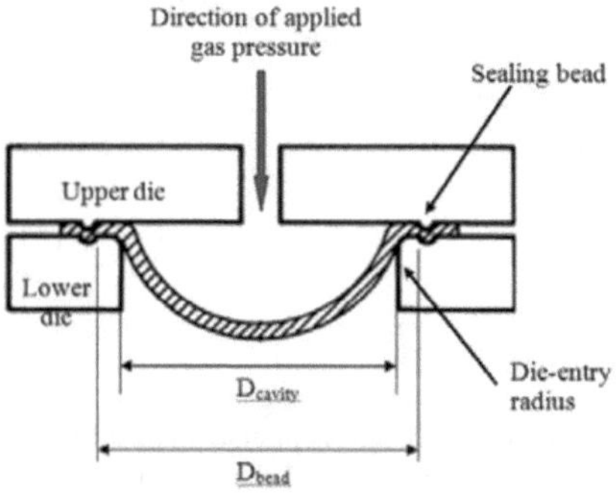


Figure 2. The bulge forming experiment is shown [23].

Finite Element Method Simulations

Tensile test experiments of AZ31 sheet material were simulated using the FEM software package Abaqus™ 6.8 [24]. The gauge section of a tensile coupon 25 mm long, 6 mm wide, and 2 mm thick was meshed using quadrilateral elements. One end of the mesh was held fixed in the tensile direction, while the other end was displaced along the tensile direction to reproduce the constant true-strain rate of the tensile test. Tensile simulations were conducted using the SDTD and SDTDM1 material models.

Bulge forming tests of AZ31 sheet were simulated using Abaqus™ 6.8. The AZ31 blank mesh consisted of a circle with diameter 64.8 mm, which is the diameter of the sealing bead used in experiments. The die mesh had the same diameter and entry radius used in forming experiments. Both the blank and die were modeled using quadrilateral membrane elements. The mesh sensitivity of a similar geometry was previously investigated, and the used here mesh was shown to be sufficient [17]. Bulge simulations used the 1.3 $\sigma$, SDTD, and SDTDM1 material models.

## Results

Uniaxial Tension

The stress-strain curves from tensile simulations are presented with experimental data from tensile tests in Figures 3(a) and (b) for true-strain rates of $10^{-4}$ and $10^{-1}$ $s^{-1}$, respectively. At the slowest rate of $10^{-4}$ $s^{-1}$, there is substantial strain hardening in uniaxial tension. The SDTD model accurately predicts these stress-strain data. The SDTDM1 model agrees well with experimental data up to a strain of approximately 0.4, but it predicts a slightly higher flow stress beyond this strain. The only difference between the SDTD and SDTDM1 models is the dislocation-climb creep term. This further confirms that the governing creep mechanism at this rate is primarily grain-boundary sliding. At $10^{-1}$ $s^{-1}$, almost no strain hardening is observed in the tensile data, and deformation occurs almost entirely by dislocation-climb creep at this fast rate. The SDTD model accurately predicts tensile data at this strain rate, while the SDTDM1 model predicts much higher flow stresses than observed experimentally. This relative accuracy between the two models will later be contrasted against predictions for bulge tests. Predictions and test data at intermediate strain rates reveal a gradual transitions between the behaviors exhibited in Figures 3(a) and (b).

Biaxial Bulge Forming

Figures 4(a) through (c) show bulge pole height as a function of time from gas-pressure bulge forming experiments (normal preheat) for gas pressures of 550, 830, and 1100 kPa, respectively. For each pressure, experimental data are compared with FEM simulation results using the 1.3 $\sigma$, SDTD, and SDTDM1 models. Note that results from two experiments, each represented by a separate curve, are shown for each test condition. At 550 kPa, both the SDTD and SDTDM1 models provide similar predictions of bulge height. The two models predict slightly larger bulge heights than observed experimentally. The 1.3 $\sigma$ model, which contains no strain-dependent term, predicts significantly faster forming than observed experimentally. These results confirm that strain hardening must be accounted for to accurately predict forming at slow forming rates. At 830 kPa, the SDTD model predicts faster forming than the 1.3 $\sigma$ model, and both models predict faster forming than observed experimentally. The SDTDM1 model provides the most accurate predictions of bulge heights at this pressure. At 1100 kPa, the accuracy of the 1.3 $\sigma$ model increases to provide predictions similar to the SDTDM1 model, while the SDTD model continues to predict much larger dome heights than observed experimentally. These results suggest that the reduction in effective flow stress in *only* the dislocation-climb creep term, i.e., the 1.3 $\sigma$ correction factor, is necessary to accurately predict deformation under a biaxial stress state.

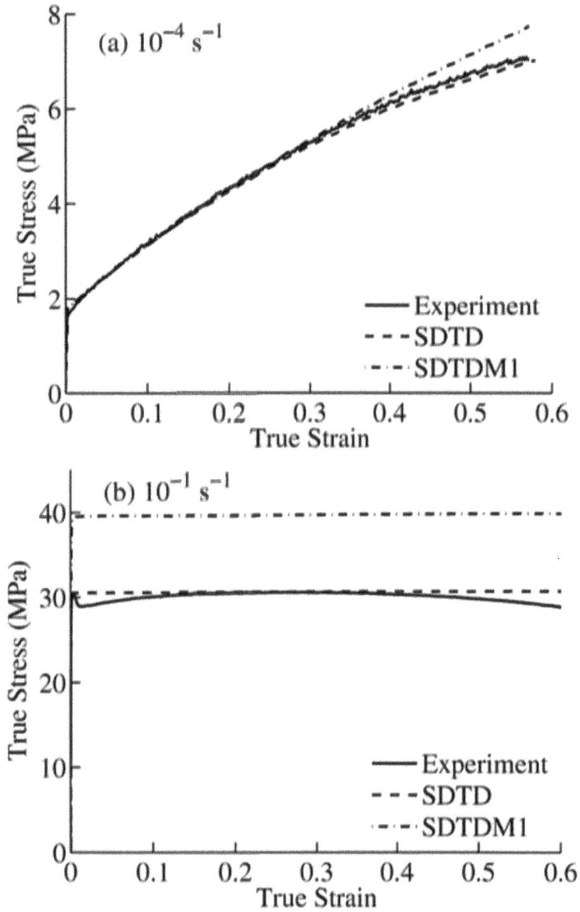


Figure 3. Stress-strain curves are shown for tensile data and simulations at 450°C and strain rates of (a) $10^{-4}$ and (b) $10^{-1}$ s$^{-1}$.

Short-preheat bulge forming experimental data are compared with normal-preheat data and SDTDM1 simulation results in Figures 5(a) through (c) for pressures of 550, 830, and 1100 kPa, respectively. The short-preheat experiments were designed to determine if there is any effect of preheating time on forming, such as might occur if static grain growth during preheating slowed GBS creep. Note that results from two long-preheat experiments, each represented by a separate curve, are shown for each test condition. At 550 kPa, the material forms more rapidly during the short preheat experiment than during the normal-preheat experiment, though forming is still slower than predicted by the SDTDM1 model. At 830 kPa, the SDTDM1 model results more closely match the short preheat data than the normal preheat data. At 1100 kPa, there is little difference between short-preheat and normal-preheat data. These results indicate that the material hardens slightly during preheating, but this effect slows only deformation attributable to GBS creep.

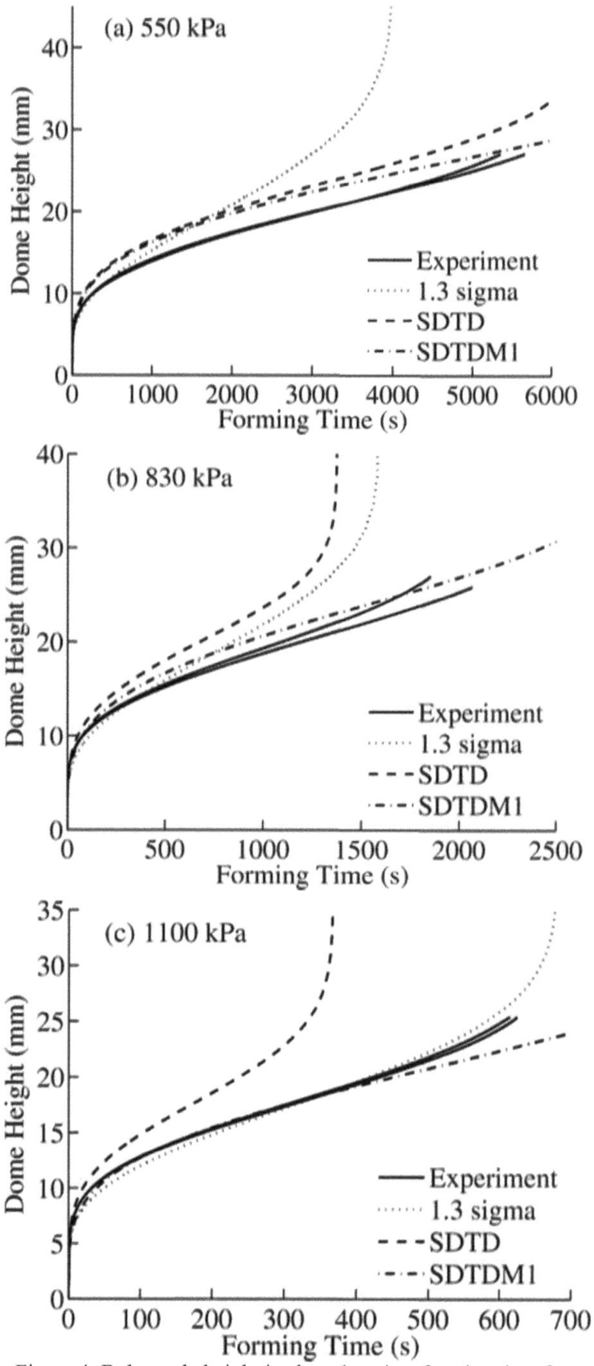


Figure 4. Bulge pole height is plotted against forming time for bulge forming experiments and simulations at 450°C and gas-pressures of (a) 550, (b) 830, and (c) 1100 kPa.

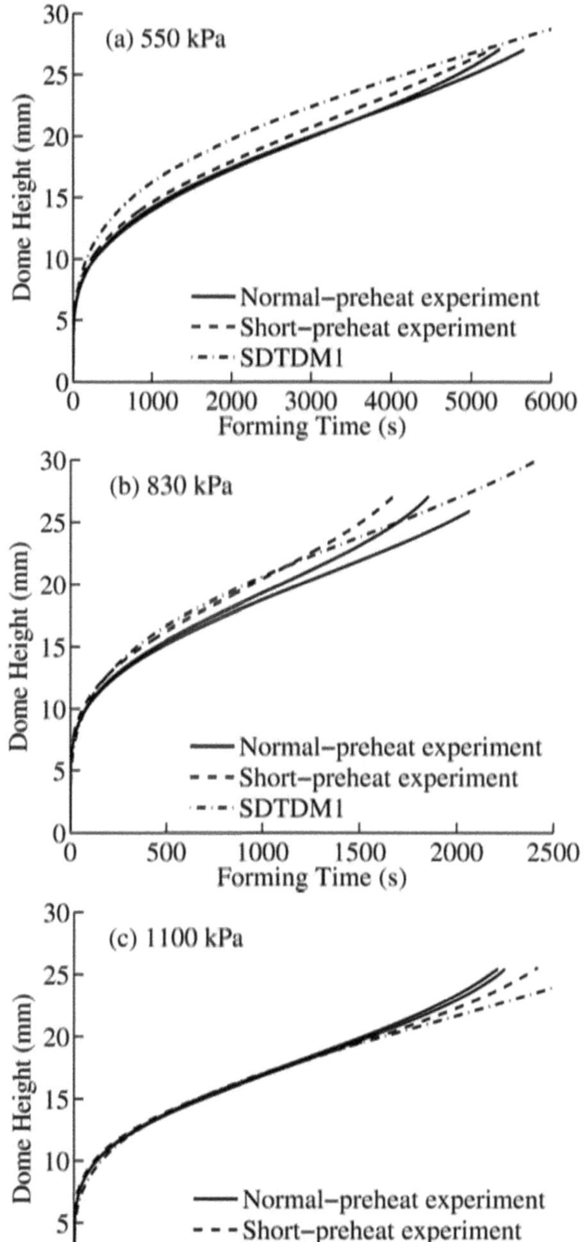


Figure 5. Bulge pole height is plotted against forming time for SDTDM1 simulations and experiments using both a normal preheat and a short preheat at 450°C and gas-pressures of (a) 550, (b) 830, and (c) 1100 kPa. Normal preheat times generally varied between 20 and 60 minutes, while short preheating was 5 minutes.

**Discussion**

Tensile test data at the slowest strain rate of $10^{-4}$, see Figure 3, exhibit significant strain hardening. At the fastest strain rate of $10^{-1}$ $s^{-1}$, flow stress is approximately constant with strain. These results indicate that strain hardening only occurs during GBS creep, which controls deformation at slow strain rates. At fast strain rates, dislocation-climb creep dominates and no strain hardening is observed. A two-term material constitutive model with a strain-dependent GBS creep term can account for this behavior. For bulge forming at 550 kPa, the lowest pressure examined, simulations based on either strain-dependent material models, SDTD or SDTDM1, are in better agreement with bulge data than simulations using the 1.3 $\sigma$ model, which contains no strain dependence. Accounting for strain hardening is required to accurately predict AZ31 forming behavior at the slow strain rates for which GBS creep dominates. However, the results of bulge forming experiments and simulations at 830 and 1100 kPa show that the SDTD model is less accurate than the 1.3 $\sigma$ and SDTDM1 models at these pressures. The effective stress for dislocation-climb creep flow stress is reduced in both the 1.3 $\sigma$ and SDTDM1 models, but not the SDTD model. This indicates that a reduction in effective flow stress for dislocation-climb creep, through the 1.3 correction factor, is required to accurately predict biaxial forming of AZ31 at the fast strain rates for which dislocation-climb creep dominates.

Strain hardening at slow strain rates may be explained by dynamic grain growth. At constant temperature, GBS creep is dependent on grain size in the following manner,

$$\dot{\varepsilon} = \frac{A}{d^p}\sigma^n \tag{9}$$

where $\dot{\varepsilon}$ is strain rate, $A$ is a material constant, $d$ is grain size, $p$ is the grain-size exponent, σ is stress, and $n$ is the stress exponent [22]. As grain size increases, flow stress must also increase to maintain a constant true-strain rate. Thus, the strain-hardening term in the SDTD and SDTDM1 models might be interpreted as a method to account for grain growth during deformation. Similarly, static grain growth may explain the differences between experimental data from normal-preheat and short-preheat bulge tests, shown in Figure 5. As preheat time prior to deformation increases, grain size is expected to increase from static grain growth. In this way, the normal-preheat specimens will have a larger initial grain size than the short-preheat specimens at the onset of deformation, which manifests as a slower forming rate when GBS creep controls deformation. This behavior is observed at forming pressures of 550 and 830 kPa, which is consistent with GBS creep dominating. There is no significant difference in forming rates between normal-preheat and short-preheat specimens at the highest pressure, 1100 kPa. This is consistent with deformation by dislocation-climb creep at the highest pressure. Grain size has no effect on dislocation-climb creep [22] at the fast strain rates associated with the highest test pressure.

The SDTD model consistently provides better predictions of deformation under uniaxial tension than does the SDTDM1 model, see Figure 3. However, the SDTDM1 model consistently provides better predictions of deformation under biaxial tension than does the SDTD model, see Figure 4. This result indicates a fundamentally important effect of stress state on dislocation-climb creep for the AZ31 sheet material at 450°C. The rate of dislocation-climb creep is slower under balanced-biaxial tension than under uniaxial tension, and this difference requires an artificial reduction in the effective stress for dislocation-climb creep by the 1.3 factor used in the SDTDM1 model. The reason for this effect of stress state on dislocation-climb creep is not clear. In a previous study of hot deformation in aluminum alloy AA5083, an effect of stress state was observed for GBS creep only, such that increasing hydrostatic stress increased the creep rate [17]. However, no effect of stress state is observed for GBS creep in the AZ31 material. Thus, the case of AZ31 is unique. One possible source for the effect of stress state on dislocation-climb creep in AZ31 is plastic anisotropy arising from crystallographic texture. The crystal structure of magnesium is HCP, which produces a limited number of slip systems, even at elevated temperatures. Indeed, it is the activation of additional slip systems with increasing temperature that causes significant ductility increases for AZ31 as temperature increases. A strong initial crystallographic texture, in conjunction with a limited number of potential slip systems, could easily lead to plastic anisotropy in dislocation-climb creep. Such plastic anisotropy could readily produce differences between deformation under uniaxial and biaxial tension. The possibility of normal plastic anisotropy, such as might arise out of the basal textures common in rolled magnesium sheet materials, would be worthy of further investigation; normal anisotropy is still consistent with the isotropic in-plane flow stresses reported in Table II. Plastic anisotropy would be expected to affect dislocation-climb creep, which proceeds by the usual slip processes, but not GBS creep. GBS creep involves relative sliding and rotation of grains, which makes it less susceptible to plastic anisotropy and can also lead to randomization of crystallographic texture [25, 26].

## Conclusions

Two strain-dependent material constitutive models, the SDTD and SDTDM1 models, were developed to account for the effects of strain and stress state on the deformation of AZ31 sheet at 450°C. AZ31 sheet material was tested under uniaxial tension at constant true-strain rates from $10^{-4}$ to $10^{-1}$ $s^{-1}$, and the SDTD material model was constructed from these tensile data. Comparisons between FEM simulations and 450°C gas-pressure bulge test data were used to construct the SDTDM1 model, which accounts for the effect of stress state on dislocation-climb creep. At slow strain rates, significant strain hardening occurs during tensile deformation, which is accounted for by a strain-dependent GBS creep term. This strain-dependent term is required to accurately predict biaxial bulge forming behavior at low forming pressures, for which GBS creep dominates deformation. The observed strain hardening during GBS creep is likely a product of dynamic grain growth. Static grain growth qualitatively explains the observed decrease in forming rate with increasing preheat time. Experimental data indicate a difference in the flow stress for dislocation-climb creep between uniaxial and biaxial tension. This effect of biaxial tension is accounted for in the SDTDM1 model by reducing the effective stress in only the dislocation-climb creep term through division of stress by a 1.3 factor. It is proposed that this effect of stress state is the result of plastic anisotropy resulting from crystallographic texture, but this proposition has not yet been explored.

## References


1. L.H. Pomeroy: *Automotive Engineering*, 1922, vol. XI, no. 6, pp. 508-19.
2. G.S. Cole, A.M. Sherman: *Mater. Char.*, 1995, vol. 35, pp. 3-9.
3. J.J. Lee, S.P. Lukachko, I.A. Waitz, A. Schafer: Annual Review of Energy and the Environment, 2001, vol. 26, pp. 167-200.
4. D. Eliezer, E. Aghion, F.H. Froes: *Adv. Perform Mater.*, 1998, vol. 5, pp. 201-12.
5. M.M. Avedesian, H. Baker: *Magnesium and Magnesium Alloys*, ASM International, Materials Park, OH, 1999, pg. iv.
6. B.L. Mordike, T. Ebert: *Mater. Sci. Eng. A*, 2001, vol. 302, pp. 37-45.
7. A.A. Luo: *JOM*, 2002, vol. 54, no. 2, pp. 42-48.
8. M.W. Toaz, E.J. Ripling: *J. Met.*, 1956, vol. 8, pp. 936-46.
9. J.G. Schroth: in Advances in Superplasticity and Superplastic Forming, E.M. Taleff, P.A. Friedman, P.E. Krajewski, R.S. Mishra, and J.G. Schroth, eds., TMS, Warrendale, PA, 2004, pp. 9-20.
10. P.E. Krajewski, J.G. Schroth: *Mat. Sci. Forum*, 2007, vol. 551-552, pp. 3-12.
11. R. Verma, L.G. Hector, Jr., P.E. Krajewski, E.M. Taleff: *JOM*, 2009, vol. 61, no. 8, pp. 29-37.
12. M.K. Khraisheh, F.K. Abu-Farha, M.A. Nazzal, K.J. Weinmann: *Annals of CIRP*, 2006, vol. 55, pp. 233.
13. F.K. Abu-Farha, M.K. Khraisheh: *Adv. Eng. Mater.*, 2007, vol. 9, pp. 777-83.
14. G. Palumbo, D. Sorgente, L. Tricarico, S.H. Zhang, W.T. Zheng, L.X. Zhou, L.M. Ren: *Mat. Sci. Forum*, 2007, vol. 551-552, pp. 317-22.
15. G. Giuliano, S. Franchitti: *Int. J. Mach. Tool Manu.*, 2008, vol. 48, pp. 1519-1522.
16. A.-W. El-Morsy, K. Manabe, H. Nishimura: *Mater. Trans.*, 2002, vol. 43, pp. 2443-48.
17. E.M. Taleff, L.G. Hector, Jr., J.R. Bradley, R. Verma, P.E. Krajewski: *Acta Mater.*, 2009, vol. 57, pp. 2812-22.
18. E.M. Taleff, L.G. Hector, Jr., R. Verma, P.E. Krajewski, J.-K. Chang: *J. Mater. Eng. Perform.*, 2010, vol. 19, pp. 488-94.
19. ASTM E 2448-06. *Standard Test Method for Determining the Superplastic Properties of Metallic Sheet Materials*. ASTM: West Conshohocken, PA. (2006).
20. W. Köster: *Z. Metallkd.*, 1948, vol. 39, pp. 1-9.
21. ASTM E 112-96. *Standard Test Methods for Determining Average Grain Size*. ASTM: West Conshohocken, PA. (1996).
22. O.D. Sherby and P.M. Burke: *Progr. Mater. Sci.*, 1968, vol. 13, pp. 325-90.
23. R.V. Moller: "Design and Fabrication of an Instrument to Test the Mechanical Behavior of Aluminum Alloy Sheets During High-Temperature Gas-Pressure Blow Forming," Thesis, Mechanical Engineering, The University of Texas at Austin, 2008.
24. Dassault Systèmes Simulia Corp., Abaqus (Version 6.8), Providence, RI.
25. T.E. Langdon : *J. Mater. Sci.*, 2009, vol. 44, pp. 5998-6010.
26. M.T. Pèrez-Prado, G. González-Doncel, O.A. Ruano and T.R. McNelley: *Acta Mater.*, 2001, vol. 49, pp. 2259-2268.

**Magnesium Technology 2012**
*Edited by: Suveen N. Mathaudhu, Wim H. Sillekens, Neale R. Neelameggham, and Norbert Hort*
***TMS (The Minerals, Metals & Materials Society), 2012***

# EFFECT OF STRAIN RATE ON DYNAMIC RECRYSTALLIZATION IN A MAGNESIUM ALLOY UNDER COMPRESSION AT HIGH TEMPERATURE

Q. Ma[1], B. Li[1], A.L. Oppedal[1], W. Whittington[1], S.J. Horstemeyer[1], E.B. Marin[1], H. El. Kadiri[1,2], P.T. Wang[1], M.F. Horstemeyer[1,2]
[1]. Center for Advanced Vehicular Systems, Mississippi State University, Starkville, MS 39759, USA
[2]. Department of Mechanical Engineering, Mississippi State University, Starkville, MS 39762, USA

Keywords: Magnesium; Texture; Twinning; Non-basal slip; Dynamic Recrystallization.

## Abstract

Interrupted uniaxial compression tests were performed on an extruded Mg-Al-Mn magnesium alloy (AM30) at 450°C and various strain rates of 0.001 $s^{-1}$, 0.1 $s^{-1}$, 0.5 $s^{-1}$ and 0.8 $s^{-1}$. Texture and microstructure evolution were examined using electron back scattered diffraction (EBSD) and X-ray diffraction (XRD) techniques. Twinning was found to be ubiquitous at high temperature but provided that a highly enough strain rate was applied. The results show that the deformation microstructure evolves with strong dependence on the strain rate as twinning triggers and non-basal slips become harder. Schmid's factor and slip traces analyses elucidated the driving forces controlling dynamic recrystallization mechanisms and emphasized on the role played by extension twinning and non-basal slips.

## Introduction

Dynamic recrystallization (DRX) is of particular importance for Mg and its alloys because DRX alters the microstructure of Mg during thermomechanical processing at temperatures normally above Mg's recovery and recrystallization point. DRX in Mg has been investigated in substantial experimental studies where it was correlated with deformation modes which are temperature dependent [1-4]. Sitdikov and Kaibyshev [1,2] investigated DRX of Mg in compression over a wide range of temperatures (20-500°C) at a constant strain rate $3\times10^{-3}$ $s^{-1}$. They found that between 20°C and 350°C, twinning plays an important role in which crystallites result from mutual intersection of primary twins and/or secondary twins serve as nuclei. Between 250°C and 500°C, continuous dynamic recrystallization (CDRX) dominates via nucleation by subgrain rotation. At 450°C, grain boundary bulging also occurs via parent grain boundary migration in coarse grains. In addition, the pre-existing annealing twin boundaries can also evolve to large angle boundaries and then migrate by bulging mechanism [1,2]. Ion et al. [3] studied the DRX of Mg-0.8Al% alloy under compression between 150°C and 370°C at a strain rate $10^{-5}$ $s^{-1}$. They assumed a "rotation" mechanism of DRX, which explained well the phenomenon of "necklace" grains in Mg. Galiyev et al. [4] suggested other DRX mechanisms based on experimental results of Mg compression between 150°C and 450°C. At low temperatures (below 200°C), DRX is associated with operation of twinning, basal slip and <c+a> slip; between 200°C and 250°C, CDRX dominates via cross-slip; at high temperatures (between 300°C and 450°C), both bulging and subgrain growth play a crucial role in DRX. These mechanisms were revealed in deformation at a constant strain rate in their study.

It is well known that deformation mechanisms in crystalline metals strongly depend upon strain rate [5]. Hence, strain rate should presumably have a vital influence on DRX because activation of twinning, for example, strongly depends upon strain rate. On the other hand, non-basal slip modes, i.e. prismatic and pyramidal slip, become more active at high temperatures [6,7]. Thus, change in strain rate can lead to mechanistic complication in DRX in Mg alloys. In thermomechanical processing of Mg alloys, the strain rate can significantly vary depending upon the geometry of the work piece and tooling fixture and the speed (e.g. ram speed in extrusion). Even in the same work piece that is undergoing thermomechanical processing, the strain rate can be inhomogeneous and different from one location to another. Therefore, it is necessary to investigate how change in strain rate influences DRX. The objective of this work is to study DRX mechanisms in an extruded commercial AM30 Mg alloy under uniaxial compression at 450°C at various strain rates: 0.001 $s^{-1}$, 0.1 $s^{-1}$, 0.5 $s^{-1}$ and 0.8 $s^{-1}$, based on electron back scattered diffraction (EBSD) analysis.

## Experimental

An extruded commercial AM30 magnesium alloy (mass: Al 2.54%, Mn 0.40%, Mg balance) was selected as the experimental material. Right cylindrical samples 10 mm in diameter were cut by electrical-discharge machining (EDM), with the axial direction parallel to the extrusion direction (ED). Before uniaxial compression, the sample to be tested was heated to 450°C in the furnace that houses the sample stage of the INSTRON 5869 machine, and the temperature was maintained constant. Then the samples were compressed inside the furnace at different strain rates: 0.001 $s^{-1}$, 0.1 $s^{-1}$, 0.5 $s^{-1}$ and 0.8 $s^{-1}$. The compression test of each sample was interrupted during plastic flow, and the sample was immediately quenched in water to preserve the microstructure. Then the micro-textures of the samples were examined using EBSD.

The true stress-strain curves of the AM30 at 450°C at different strain rates are presented in Figure 1. The yield stress and the flow stress increase with increasing strain rate. At strain rates 0.001 $s^{-1}$ and 0.1 $s^{-1}$, the strain hardening is minimal. The stress-strain curve of 0.5 $s^{-1}$ shows barely recognizable hardening, followed by a slight softening. When the strain rate increases to 0.8 $s^{-1}$, the stress-strain behavior exhibits an apparent sigmoidal shape of appreciable hardening, with a peak stress followed by a pronounced softening. The stress-strain curves at 0.001 $s^{-1}$ and 0.1 $s^{-1}$ present a parabolic shape in which dislocation slip dominates the plastic flow.

The microstructure and microtexture evolution during compression in ED at various strain rates is shown in Figure 2. The inverse pole figure (IPF) maps are shown as well. Figure 2a shows the initial micro-texture and microstructure of the untested sample. We can see that the initial microstructure consists of a mixture of coarse grains and dynamically recrystallized fine grains created during extrusion. In the pole figure, the basal planes are roughly parallel to the ED (or $\langle 10\bar{1}0\rangle \parallel$ ED), which is typical of extruded Mg alloys. Figure 2b shows homogenous, equiaxed grains at 0.001 $s^{-1}$ and $\varepsilon = -0.15$, indicating a full recrystallization, i.e., nucleation and ripening in comparison to the initial state (Figure 2a). Figure 2c presents the microstructure at 0.1 $s^{-1}$ and $\varepsilon = -0.15$ in which lenticular twins in red start emerging. Twinning is confirmed by the ~90° rotation in {0001} pole figure which is caused by the presence of $\{10\bar{1}2\}\langle10\bar{1}\bar{1}\rangle$ twins. When the strain rate increases to 0.5 $s^{-1}$, as shown in Figure 2d, profuse, large $\{10\bar{1}2\}\langle10\bar{1}\bar{1}\rangle$ twins can be seen in a coarse, non-recrystallized grain, in addition to the small twins in the fine grains. A strong basal texture can be observed in the pole figure. In the mean time, dynamically recrystallized, "necklace" grains nucleate and grow, surrounding the parent grains, then some invading into the matrix along twin boundaries. Figures 2e and 2f show the microstructure at 0.8 $s^{-1}$, $\varepsilon = -0.04$ and $-0.15$, respectively. The $\{10\bar{1}2\}\langle10\bar{1}\bar{1}\rangle$ twinning dominates the deformation and leads to an enhanced basal texture. At $\varepsilon = -0.04$, high density twins can be observed. Meanwhile, fine "necklace" grains surrounding parent grains nucleate and grow. At $\varepsilon = -0.15$, the twins grow significantly, coalesce and consume most of the grains. The dynamically recrystallized necklace grains are sandwiched between the impinged twins. New grains may also nucleate inside the twins and grow, eventually segmenting the large twins into irregular islands. No $\{10\bar{1}1\}\langle10\bar{1}2\rangle$ contraction twin or $\{10\bar{1}1\}-\{10\bar{1}2\}$ double twins were found in this study.

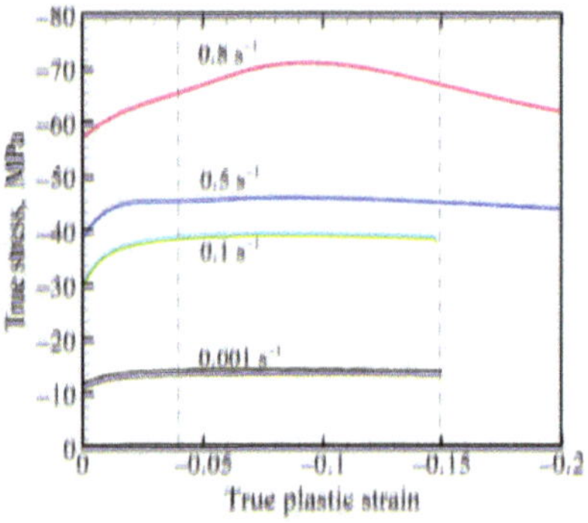


Figure 1: The uniaxial compression true stress true strain curves of AM30 at 450°C and strain rates: 0.001 $s^{-1}$, 0.1 $s^{-1}$, 0.5 $s^{-1}$ and 0.8 $s^{-1}$.

Macrotexture of the AM30 samples was also measured on a Rigaku SmartLab X-Ray Diffractometer (XRD). The six in-completed pole figures, $\{10\bar{1}0\}$, {0002}, $\{10\bar{1}1\}$, $\{10\bar{1}2\}$, $\{11\bar{2}0\}$ and $\{10\bar{1}3\}$ were measured by X-ray, and background and defocusing effect was eliminated. The Orientation Distribution Function (ODF) was calculated based on the measured pole figures. The recalculated pole figures were calculated based on the ODF. The macrotexture of AM30 samples are presented in Figure 3. The macrotexture measured by XRD definitely prove the presence of $\{10\bar{1}2\}$ twinning (Figures 3). Twinning induced a strong basal texture at $\varepsilon = -0.04$ and at higher strain rates of 0.5 $s^{-1}$ and 0.8 $s^{-1}$ (Figure 3b,3c)

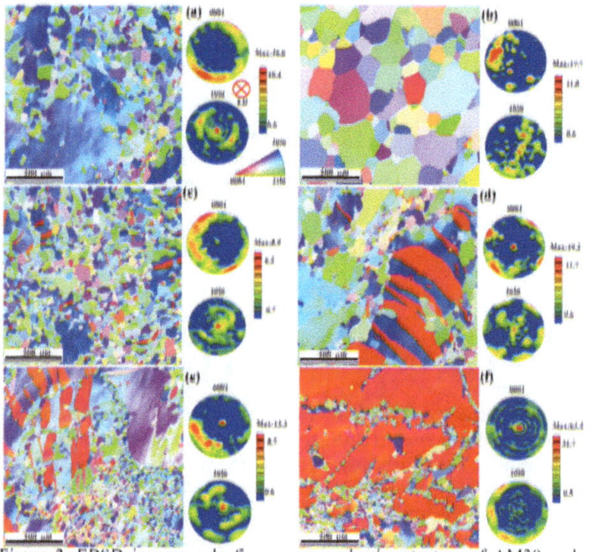

Figure 2. EBSD inverse pole figure map and micro-texture of AM30 under uniaxial compression at 450°C at (a) initial status; (b) 0.001 $s^{-1}$, $\varepsilon = -0.15$, grains become coarse and equiaxed; (c) 0.1 $s^{-1}$, $\varepsilon = -0.15$, visible thin $\{10\bar{1}2\}\langle10\bar{1}\bar{1}\rangle$ twins appear; (d) 0.5 $s^{-1}$, $\varepsilon = -0.04$, thick twins locate in parent grain and "necklace" grains appear and some of them grow along twin interfaces; (e) 0.8 $s^{-1}$, $\varepsilon = -0.04$, twinning propagates in both parent grains and recrystallized grains, and (f) 0.8 $s^{-1}$, $\varepsilon = -0.15$, twinning sweeps almost whole parent grains and results in a very strong basal texture, and DRX evolves inside twinned parent grain. The pole figures are calculated based on the corresponding EBSD maps. The inverse pole figure represents the compression direction, namely the extrusion direction (ED).

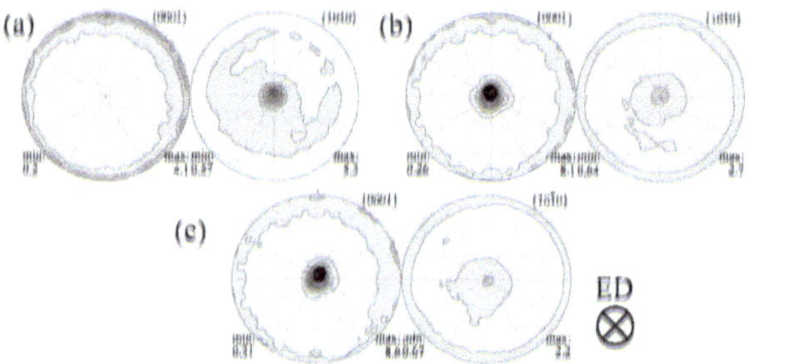


Figure 3. XRD macrotexture of AM30 uniaxial compression along ED at 450°C and at (a) initial state; (b) $\dot{\varepsilon} = 0.5$ $s^{-1}$, $\varepsilon = -0.04$, a strong basal texture appears which induced by $\{10\bar{1}2\}$ twinning; (c) $\dot{\varepsilon} = 0.8$ $s^{-1}$, $\varepsilon = -0.04$, a strong basal texture resulted by $\{10\bar{1}2\}$ twinning. Intensity levels: 1,2,3,4,5,6,7,8.

## Discussion

At the quasi-static strain rate 0.001 $s^{-1}$, the stress-strain curve of 0.001 $s^{-1}$ (Figure 1) shows no hardening. This is because at this slow strain rate, the overall dislocation density is balanced between dislocation accumulation by proceeding deformation and dislocation annihilation by recovery, DRX and subsequent grain growth. Galiyev et al. [4] suggested that DRX is accomplished by subgrain rotation or new grain nucleation and growth inside the parent grains. The main active slip modes should be prismatic and pyramidal slip because (1) the critical resolved shear stress (CRSS) of the prismatic and pyramidal slip is close to that of basal slip at 450°C and low strain rate [6]; (2) the Schmid factor (SF) of prismatic and pyramidal slip is greater than 0.4 near

$\langle 10\bar{1}0\rangle$, in contrast to nearly zero for basal slip (see calculations in Figures 4c to 4e). The initial texture of AM30 (Figures 3a, 4a), measured by X-ray diffraction (XRD), definitely favors $\{10\bar{1}2\}\langle 10\bar{1}\bar{1}\rangle$ twinning because this twinning mode has a maximum SF of 0.5 around $\langle 10\bar{1}0\rangle$ (Figure 4b). However, twinning is suppressed by slip because the flow stress does not reach the CRSS of twinning at strain rate 0.001 $s^{-1}$.

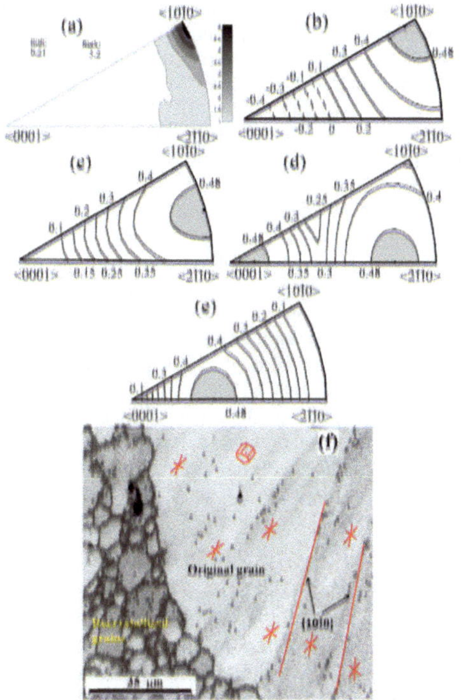

Figure 4. (a) Inverse pole figure of the extrusion direction (ED) for initial AM30 at 450°C by XRD. The Schmid factor distribution in the inverse pole figure compression by the deformation mode of (b) $\{10\bar{1}2\}\langle 10\bar{1}\bar{1}\rangle$ twinning; (c) prismatic $\langle a\rangle$: $\{10\bar{1}0\}\langle 11\bar{2}0\rangle$ slip; (d) second order pyramidal $\langle c+a\rangle$: $\{11\bar{2}2\}\langle 11\bar{2}\bar{3}\rangle$ slip; and (e) basal $\langle a\rangle$: $\{0001\}\langle 11\bar{2}0\rangle$ slip. (f) Prismatic $\langle a\rangle$: $\{10\bar{1}0\}\langle 11\bar{2}0\rangle$ slip traces in EBSD image quality map at 0.8 $s^{-1}$, $\varepsilon$=-0.04. The slip lines prove abundant prismatic $\langle a\rangle$ slips in the parent grain. The red lines and stars in Figure 3f are the $\{10\bar{1}0\}$ plane traces, the dark lines represent large angle grain boundaries (LAGB) >15°.

Barnett et al. [8,9] studied flow stress of AZ31 Mg alloy deformation at 300°C and 350°C at various strain rates. With increasing strain rate, the plateau of the flow curve shifts to a marked sigmoidal flow curve, which is attributed to the possible twinning. Ishikawa et al. [10] and Myshlyaev et al. [11] observed twins in Mg at 400°C and 450°C first at high strain rate ($10^3$ $s^{-1}$) by optical microscopy (OM), then at low strain rate of 0.01 $s^{-1}$ by transmission electron microscopy (TEM). In our previous study [12], profuse $\{10\bar{1}2\}\langle 10\bar{1}\bar{1}\rangle$ twinning was observed in an AZ61 Mg alloy extruded at 450°C, which played an important role in DRX. There are others' results that show absence of deformation twinning in Mg at temperatures greater than ~400°C and strain rates less than 0.01 $s^{-1}$ [1-4,13]. In fact, twinning is athermal in hcp metals [7]. The CRSS of twinning should be insensitive to temperature, while twin nucleation and growth propagation mainly depend upon local stress state that is rate dependent.

When strain rate increases to 0.1 $s^{-1}$, new grains mainly nucleate at the interior of the parent grains (Figure 2c). Similar to the quasi-static case (0.001 $s^{-1}$), pyramidal and prismatic slip are activated mainly in the parent grains. The grain structure is much finer compared with strain rate 0.001 $s^{-1}$, owing to growth kinetics (Figure 2b, 2c). Sporadic $\{10\bar{1}2\}\langle 10\bar{1}\bar{1}\rangle$ twins can be seen inside some of the parent grains (Figure 2c), indicating the flow stress may be just above the CRSS of twinning.

At relatively high strain rates, 0.5 $s^{-1}$ and 0.8 $s^{-1}$, prismatic slip should still be active in the parent grains because it has high SF values around $\langle 10\bar{1}0\rangle$ (Figure 4c). Figure 4f shows slip trace analysis by EBSD at strain rate 0.8 $s^{-1}$ and strain level $\varepsilon = -0.04$. It is clear that prismatic slip is indeed the main active slip mode in the parent grain. The stress concentration near parent grain boundaries may trigger the pyramidal slip to accommodate strain because the prismatic slip has only two independent systems. The pyramidal slip alone can provide five independent slip systems that fulfill the von Mises criteria to accommodate polycrystalline plastic strain. Thus, parent grain boundaries become favorable nucleation sites for DRX due to the activation of the non-basal slip modes.

The formation of the "necklace" grains was attributed to the "rotation recrystallization" induced by the pyramidal slip in Mg, as suggested by Ion et al. [3]. Some grains grow along twin boundaries due to the possible pyramidal slip at twin/matrix interfaces. Yoo [14] assumed that the twin boundaries, in addition to grain boundaries, are another important source of pyramidal slip. High density pyramidal dislocations at twin boundaries were observed in TEM by Agnew et al. [15] and Morozumi et al. [16]. Accordingly, the dynamically recrystallized grains grow at the expense of the twins, giving rise to the softening by DRX as seen in the stress-strain responses at strain rates 0.5 $s^{-1}$ and 0.8 $s^{-1}$. In Figure 2f, some new grains are nucleated inside the largely twinned grains. This can be understood from the fact that the orientation of the twins favors pyramidal slip because the SF is high (Figure 4d). As such, pyramidal slip is activated in the twins (Figure 2f) and facilitates DRX inside the twins. Note that some grains are nucleated along the prismatic slip traces (Figure 4f), indicating that the prismatic slip may also be favorable for nucleation due to the high plastic energy and short-circuit diffusion paths in the dislocation pipe-lines.

Moreover, twinning almost saturates at the strain level $\varepsilon = -0.15$ at strain rate 0.8 $s^{-1}$ (Figure 2f). Accordingly, a distinct softening occurs following the peak flow stress (Figure 1). It is the competition between hardening by twinning versus softening by DRX that results in an apparent peak stress followed by a softening, as shown in the stress-strain curve at strain rate 0.8 $s^{-1}$.

Finally, according to observation and analysis, the effect of strain rate on DRX of the AM30 compressed at 450°C is schematized in Figure 5. At low strain rate, nucleation and grain growth in the parent grain interior dominates the DRX. Due to the low strain rate, the recrystallized grain size ripens, as shown in Figure 4a. At intermediate strain rate, DRX proceeds primarily at the grain boundaries, most likely via the activation of non-basal slip; meanwhile, twinning propagates inside the parent grains. Some new grains may grow along twin boundaries (Figure 5b). At high strain rates, twinning dominates the plastic deformation, with fine grains nucleate along parent grain boundaries; subsequently DRX evolves quickly in the twinned parent grain (Figure 5c). As our experiments clearly demonstrate, the size of the dynamically recrystallized grains, while at the same strain level, decreases as the strain rate increases.

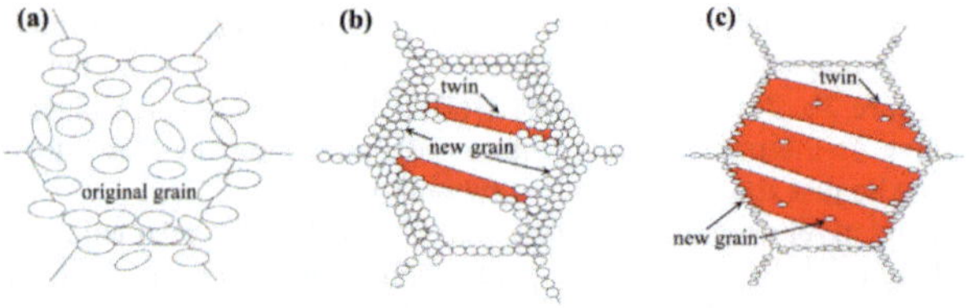


Figure 5. Schematic illustration of dynamic recrystallization in the AM30 alloy compression at 450°C and various strain rates: (a) Low strain rate, (b) intermediate strain rate, and (c) high strain rate.

## Conclusions

In summary, we investigated the effect of strain rate on DRX at 450°C in an extruded AM30 alloy. Compression tests were performed at strain rates 0.001 $s^{-1}$, 0.1 $s^{-1}$, 0.5 $s^{-1}$ and 0.8 $s^{-1}$. Texture evolutions in quenched samples were examined by EBSD and XRD. The stress-strain responses strongly depended on the strain rate. At quasi-static strain rate 0.001 $s^{-1}$, no strain hardening was observed and equiaxed, ripened grains were formed. DRX is accomplished by subgrain rotation or new grain nucleation and growth inside the parent grains [4]. At relatively low strain rate, 0.1 $s^{-1}$, a finer grain structure was obtained and sporadic deformation twins were present; again, no strain hardening was observed. The mechanism of DRX is nucleation and growth of new grains inside the interior of the parent grains. At intermediate strain rate, 0.5 $s^{-1}$, strain hardening was barely recognizable, followed by a slight softening; an increase in twin volume fraction was observed. At high strain rate, 0.8 $s^{-1}$, the stress-strain response presents a sigmoidal behavior, with a peak flow stress followed by a distinct softening. At high strain rates, the dominant mechanism of DRX is the activation of non-basal slip modes at the grain boundaries, and fine, necklace grains are nucleated surrounding the parent grains. In the mean time, the growth of the dynamically recrystallized grains along the twin boundaries consumes the twins, leading to irregular morphology of the twins as the strain is increased.

## Acknowledgements

The authors are grateful to the financial support from the Department of Energy, Contract No. DE-FC-26-06NT42755, and the Center for Advanced Vehicular Systems (CAVS) at Mississippi State University.

## References

1. O. Sitdikov and R. Kaibyshev, "Dynamic recrystallization in pure magnesium", *Materials Transactions*, 42 (2001), 1928-1937.
2. R.O. Kaibyshev and O. Sh. Sitdikov, "On the role of twinning in dynamic recrystallization", *The Physics of Metals and Metallography*, 89 (2000), 70-77.
3. S.E. Ion, F.J. Humphreys and S.H. White, "Dynamic recrystallization and the development of microstructure during the high temperature deformation of magnesium", *Acta Metallurgica*, 30 (1982), 1909-1919.
4. A. Galiyev, R. Kaibyshev and G. Gottstein, "Correlation of plastic deformation and dynamic recrystallization in magnesium alloy ZK60", *Acta Materialia*, 49 (2001), 1199-1207.
5. M.A. Meyers, O. Vöhringer and V.A. Lubarda, "The onset of twinning in metals: a constitutive description", *Acta Materialia*, 49 (2001), 4025-4039.
6. A. Chapuis and J.H. Driver, "Temperature dependency of slip and twinning in plane strain compressed magnesium single crystals", *Acta Materialia*, 59 (2011), 1986-1994.
7. A. Jain and S.R. Agnew, "Modeling the temperature dependent effect of twinning on the behavior of magnesium alloy AZ31B sheet", *Materials Science and Engineering*, A 462 (2007), 29-36.
8. M.R. Barnett, "Influence of deformation conditions and texture on the high temperature flow stress of magnesium AZ31", *Journal of Light Metals*, 1 (2001), 167-177.
9. A.G. Beer and M.R. Barnett, "Microstructural development during hot working of Mg-3Al-1Zn", *Metallurgical and Materials Transactions A*, 38A (2007), 1856-1867.
10. K. Ishikawa, H. Watanabe and T. Mukai, "High strain rate deformation behavior of an AZ91 magnesium alloy at elevated temperatures", *Materials Letters*, 59 (2005), 1511-1515.
11. M.M. Myshlyaev, et al., "Twinning, dynamic recovery and recrystallization in hot worked Mg-Al-Zn alloy", *Materials Science and Engineering*, A 337 (2002), 121-133.
12. Q. Ma, et al., "Twinning-induced dynamic recrystallization in a magnesium alloy extruded at 450 °C", *Scripta Materialia*, 65 (2011), 823-826.
13. T. Al Samman and G. Gottstein, "Dynamic recrystallization during high temperature deformation of magnesium", *Materials Science and Engineering*, A490 (2008), 411-420.
14. M.H. Yoo, "Slip, twinning, and fracture in hexagonal close-packed metals", *Metallurgical Transactions*, A12 (1981), 409-418.
15. S.R. Agnew, J.A. Horto and M.H. Yoo, "Transmission electron microscopy investigation of <c+a> dislocations in Mg and α-solid solution Mg-Li alloys", *Metallurgical and Materials Transactions*, 33A (2002), 851-858.
16. S. Morozumi, M. Kikuchi and H. Yoshinaga, "Electron microscope observation in and around {1-102} twins in magnesium", *Transactions of the Japan Institute of Metals*, 17 (1976), 158-164.

**Magnesium Technology 2012**
*Edited by: Suveen N. Mathaudhu, Wim H. Sillekens, Neale R. Neelameggham, and Norbert Hort*
***TMS (The Minerals, Metals & Materials Society), 2012***

# Effect of strain rate on the kinetics of hot deformation of AZ31 with different initial texture

M.Sanjari[1], A. Nabavi[2], S.A.Farzadfar[1], , In-Ho Jung[1], S.Yue[1], E.Essadiqi[3]
[1]Department of Materials Engineering, McGill University, Montreal, Canada
[2]Department of Mechanical Engineering, McGill University, Montreal, Canada
[3] CANMET – Materials Technology Laboratory, Ottawa, Canada

Keywords: Magnesium alloy, Texture evolution, Dynamic Recrystallization

## Abstract

In this work, the effects of strain rate and initial texture on the flow behaviour and microstructure evolution on AZ31 Mg alloy were studied by compression tests. Cast plates were, homogenized and hot rolled and then compression tests were performed on samples with longitudinal axes either parallel to the rolling direction (RD) or the normal direction (ND). Compression tests were performed to various strains and samples were quenched to investigate the effect of dynamic recrystallization on the texture and microstructural evolution. Results show that for the samples machined in both rolling and normal direction, the rate of texture evolution is increased by increasing strain rate. The deformation mechanism was changed by increasing the strain rate for ND samples and at strain rate of 1 s-1 from slip dominated flow to twin dominated flow. By using EBSD, the deformation mechanism and twinning types were investigated and double and tension twins were detected.

## 1. Introduction

Of all the industrially significant metals and alloys, magnesium alloys are known as lightweight high strength material and therefore are very attractive in applications such as automotive, railway and aerospace industries [1, 2]. One of the major obstacles for using magnesium alloys is their poor formability because of their hexagonal close-packed (hcp) crystal structure and therefore insufficient number of slip systems [3]. The main deformation mechanisms for Mg-alloys are slip and twinning. At room temperature the principal slip system is basal. At higher temperatures, several non-basal slip systems such as prismatic and pyramidal are activated due to temperature dependency of critical resolved shear stress of these slip systems [1, 4]. In the case of twinning, there are two commonly observed twin planes, $\{10\bar{1}2\}$ and $\{10\bar{1}1\}$ , that accommodate extension and contraction along the c-axis, respectively [5, 6]. These slip systems usually generate the basal texture during deformation with the basal poles aligning with the compression axis in compression test and the sheet normal direction during rolling [7, 8]. Current research is focused on generating a fine grained uniform microstructure and desirable (i.e. non-basal) texture using conventional production processes combined with alloying design. To optimize the parameters for hot deformation processing, it is mandatory to understand the effects of temperature and strain rate and also initial texture and microstructure on flow behaviour and microstructure evolution. The effect of initial texture on the plastic deformation of magnesium alloys at different temperature has been studied [9-11]. However, the effect of strain rate has rarely been examined systematically. Therefore the aim of this work is to study the effect of crystallographic texture and strain rate on the subsequent texture and microstructure development during hot deformation of AZ31. To do this, the kinetics of hot deformation are studied using compression test on the samples were machined from as-cast and rolled samples in two normal and rolling direction.

## 2. Experimental procedure

This study was carried out on an AZ31B alloy with the following chemical composition (%wt): 3% Al, 0.9% Zn, 0.67% Mn and Mg (balance). The as-cast material was homogenized at 450 °C for 4 hr, which was found to give a consistent hot deformation behavior. This resulted in a coarse grain size of about 200 μm measured by image analysis. To investigate the effect of strain rate on the microstructure and texture evolution, two series of tests were performed on the material in the temperature range from 300 oC to 450 °C. In order to apply low strain rates (0.01 and 1s-1), a computer controlled servo-hydraulic materials testing system (MTS) with a 100 kN capacity was used. Specimens were deformed to strains of about 0.3 and 0.7 in order to observe the effect of strain on the microstructure. For higher strain rates, a cam plastometer was used, which is a reliable method for determining dynamic behavior at strain rates of the order of 10 to 100s-1. Cylindrical samples, 9 mm in height and 6 mm in diameter, were used for lower compression tests. Because the anvil motion is fixed in the cam plastometer, in order to apply the strains of 0.3 and 0.7, two different sizes, 12 mm height with 9 mm diameter, and 21 mm height with 14 mm diameter were used. The deformed samples were sectioned through parallel to the longitudinal direction in order to examine microstructure and the texture. The samples were mounted and ground to 1200 grit and then polished with alcohol based 3 and 1 micron diamond suspensions for microstructural examination by optical microscopy. The samples were finally etched with an acetic-picral solution (10 mL acetic acid + 4.2 g picric acid + 10 mL water + 70 mL ethanol (99.5 pct)).

In texture analysis, a Siemens D-500 X-ray diffractometer equipped with a texture Goniometer using Co radiation was used. The samples for X-ray diffraction examination were prepared by grinding with SiC papers up to 1200 grit, and then polished with diamond suspension up to the grades of 3 and 1 micron. The data were analyzed to calculate orientation distribution function and to recalculate the pole figures.

## 3. Results and discussion

### 3.1 As-cast samples

#### 3.1.1 Effect of strain rate on the flow behavior

Figure 1 shows typical flow curves at 375°C and at true strain rates of 0.01-100 s-1. Generally, with increasing temperature or decreasing strain rate the flow curves are shifted to lower stress levels. In all curves, a peak stress was observed, although this was not as pronounced at 100 s-1. The peak stress indicates a dynamic softening. After the peak, the stress decreases gradually, Figure 2 shows the effect of strain rate on the peak strain at constant temperatures. As can be seen, by increasing the strain rate the peak strain increases. This can be explained by the fact of time and temperature dependency of dynamic recrystallization. At lower strain rate there is more time available to move boundaries and the higher temperature increases the rate of boundary motion [12].

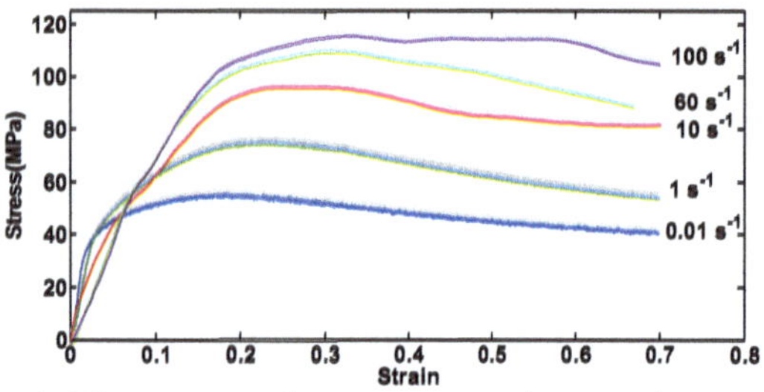


Fig. 1 The stress-strain curve at various strain rate and temperature of 375 °C.

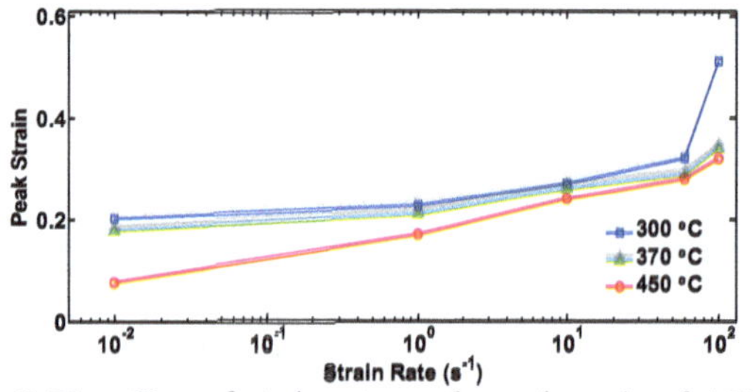


Fig. 2 The effect of strain rate on the peak strain of AZ31 at different temperature and strain rates

### 3.1.2 Microstructural characterization of as-cast material

The microstructural evolution was studied by examining the specimens obtained after deformation according to two criteria: (i) samples deformed up to a fixed strain but at different strain rates and temperatures, and (ii) specimens obtained after deformation at constant strain rate and temperatures at two different strains. Figure 5 shows optical micrographs of samples deformed at temperature 375 °C and strain rates of 0.01, 1, 10, 60 and 100 $s^{-1}$. At the lower strain rates of 0.01 and 1 $s^{-1}$ the necklace structure can be seen in the grain boundaries of initial grains. The mechanism of DRX in these deformation conditions has been analysed by McQueen and Knopleva [13]: Basal slip in favourably oriented grains takes place and twinning also occurs to reorient grains not suited for slip. When the strain reaches a sufficiently high level, DRX can occur at locations with high misorientations, which have been generated by accommodation of dislocations. Due to several slip systems activated near the grain boundary, there are more dislocations and therefore new fine grains form a necklace along grain boundaries. These regions appear to continually undergo recrystallization, possibly because deformation is easier compared to the grain core.
By increasing the strain rate, more twins were observed and this can be proved by the shifting of the shape of curves from convex to concave indicating that deformation mode changes from slip-dominated flow to twin-dominated flow [14]. The volume fraction of DRX grains is higher than at lower strain rates. There is limited time for the movement of dislocations at the high strain rate deformation, and therefore twins are formed to help deformation even at the higher temperatures [10]. In general, when twins exist in the deformed microstructure, DRX takes place at both grain and twin boundaries. However, the influence of twin boundaries increases when the alloys are deformed at higher strain rates simply because the volume fraction of twins is increasing. Grosvenor et al [15] observed twinning induced DRX in cast AZ31. In AZ31 at 250°C and low strain rate, twinning induced DRX was not been observed; however, at 400°C and at high strain rate, twins can be a suitable place for DRX nucleation [16]. Three steps were suggested for twinning induced DRX: creation of the twinning boundary, transformation to a high angle boundary and migration of this boundary [16]. Yin et al [17] suggested that distortion energy accumulated by twinning and the activation of dislocations plays an important role for the nucleation of DRX. Note that because of the initial large grain size in the AZ31B used in this work, grain boundary sliding would be very difficult.
As Fig. 3e revealed, the recrystallization volume fraction rises with the increase of strain rate. Beer et al. [18] suggested that the kinetics of recrystallization is promoted with increasing the strain rate and decreasing the temperature due to higher stored energy at larger values of Z. Since this statement is valid for metadynamic recrystallization [18, 19], it may explain why there is increasing volume fraction of recrystallised microstructure at higher strain rates.

### 3.1.3 Texture evolution

Fig. 4 shows the (0002) pole figures of the samples deformed under different strain rates at a temperature of 375 °C and strain of 0.3. The deformation texture after compression is similar to the strong basal texture observed in common Mg alloys , i.e. the c-axis of in most of the grains aligns with the direction of compressive strain [20]. However, the results show that the rate of texture evolution is increased with strain rate and the maximum intensity is increased from 2.4 to 3.2 for the samples deformed at 0.01 and 100 $s^{-1}$, respectively.

## 3.2 Effect of initial texture on deformation behavior (rolled material)

### 3.2.1 Flow curves

Selected flow curves obtained during the hot deformation of the as-cast and rolled material in two directions are shown in figure. 4. As can be seen from this figure, in both strain rates the stress level is much higher for the as-cast material compared to either of the rolled samples. Comparing the rolled materials, the effect of initial texture on the flow stress is significant pre-peak strain, but is non existent after the peak strain has been reached. The pre-peak differences are much more pronounced at the higher strain rate of 1$s^{-1}$.

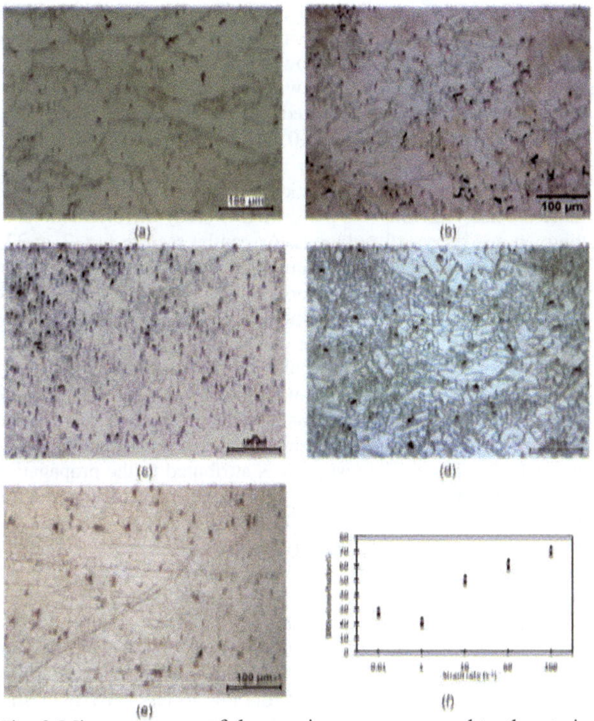


Fig. 3 Microstructures of the specimens compressed to the strain of 0.3 at 375 °C and strain rate of (a) 0.01 $s^{-1}$, (b) 1 $s^{-1}$, (c) 10 $s^{-1}$, (d) 60 $s^{-1}$, and (e) 100 $s^{-1}$ (f) DRX volume fraction at different strain rates

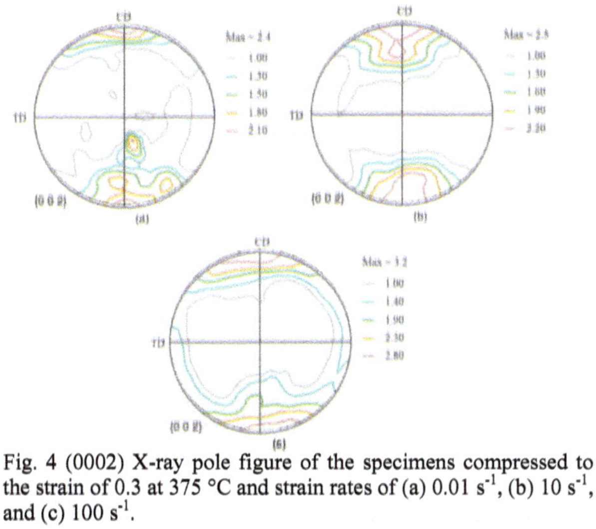


Fig. 4 (0002) X-ray pole figure of the specimens compressed to the strain of 0.3 at 375 °C and strain rates of (a) 0.01 $s^{-1}$, (b) 10 $s^{-1}$, and (c) 100 $s^{-1}$.

### 3.2.2 Slip systems

To find reasons for the above observations, the [0002] pole figure representing the texture of the samples before deformation for ND and RD samples was calculated from the ODF and shown in Fig. 6. As can be seen the basal poles are parallel and perpendicular to the compression axis for the ND and RD samples, respectively. Based on the initial configuration of the basal poles, the potential slip systems in the samples during compression test are discussed below.

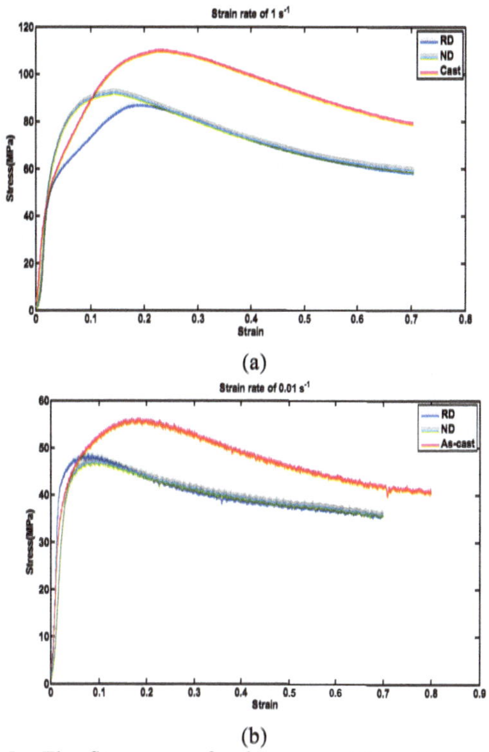


Fig. 5. The flow curves for the as-cast, RD and ND samples deformed at 375°C and strain rate of (a)1 $s^{-1}$ and (b) 0.01 $s^{-1}$

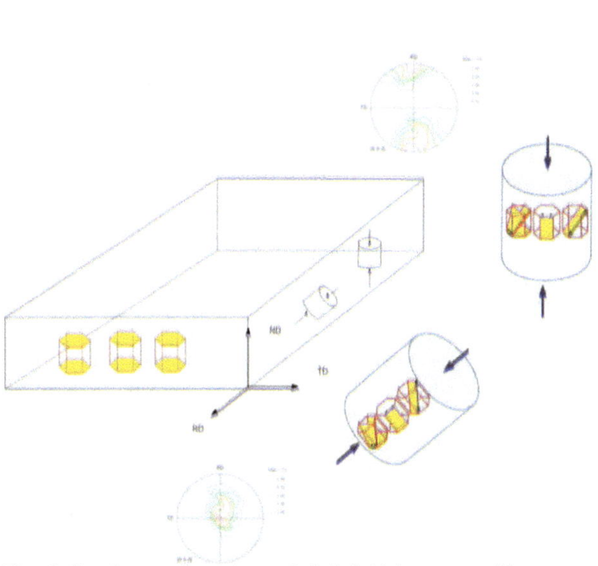

Fig. 6. Specimen orientation and their initial texture with respect to the rolled sheet geometry.

#### 3.2.2 .1 ND samples:

In this sample the basal planes are almost perpendicular to the compression axis and therefore basal slip is very much reduced. The prismatic slip system is also unfavourable since $\{10\bar{1}0\}$ planes are parallel to the compression axis. Although the first-order pyramidal planes are favourably oriented, the slip direction $\langle 11\bar{2}0\rangle$ is almost normal to the direction of applied stress therefore their activation is difficult. In the case of second-order pyramidal, the slip direction is also favourable for slip; however, the relatively high value of the CRSS for this slip would prevent slip at room temperature, but at elevated temperature this system may be activitatedof. In summary, in the ND samples the orientation of the basal planes is favourable for second-order pyramidal slip, and has all other slip systems having much smaller contributions to slip.

#### 3.2.2 .2 RD samples

In RD samples, as can be seen in figure 6, the {0002} planes are oriented parallel to the compression axis and therefore the activity of the basal slip system reduces considerably. The prismatic planes can be oriented parallel, perpendicular or at about 30° respect to the compression axis. The latter orientation can be activated to a limit amount. If the basal planes are considered to be completely parallel to the compression axis, the first-order pyramidal slip planes, $\{10\bar{1}1\}$, will be oriented at about 28° or 46° with respect to the compression axis. The other first-order pyramidal plane, $\{11\bar{2}0\}$, is oriented at 46° or 60° respect to the compression axis which is favourable for slip. It is important to note that the slip direction $\langle 11\bar{2}0\rangle$ can be oriented in 90° from the compression axis and therefore cannot contribute to the slip. The second-order pyramidal slip planes $\{11\bar{2}2\}$ can also be activated during the deformation. In summary, from the initial orientation point of view, for the samples compressed parallel to the rolling direction, basal slip is not a favourable slip system, limited prismatic slip can be activated if the grains have the appropriate direction. The first and second order pyramidal slip systems are highly favoured at the temperature (375 °C) that deformation was performed and cross-slip can also occur on the pyramidal slip systems.

### 3.2.3 Twinning

As mentioned in the introduction, due to lower value of critical shear stress of the basal slip system, plastic deformation of Mg alloys appears to occur entirely by the basal slip modes at lower temperature. However basal slip provides only two independent slip systems. Therefore, it is expected that twinning, in particular $\{11\bar{1}2\}$ type, is increased by reducing the temperature. Lee et. al reported that increasing the strain rate has the same effect of reducing the temperature on increasing the number of twins [11]. The formation of $\{11\bar{1}2\}$ twins in magnesium lattices decreases the length of the crystal in a direction parallel to the basal plane and reorients the basal plane 86° into orientations which resist slip, with the c-axis more or less parallel to the loading axis. Therefore at least a part of rapid rise in the flow curves is due to twinning and an almost a linear part can be distinguished just prior to the peak. The twinning signature in the flow curves can also be investigated by plotting $d^2\sigma/d\varepsilon^2$ versus strain [5]. The present of a local maximum at strain lower than peak strain can be considered as sign of twinning in the deformed samples. As can be seen in Fig. 7 at temperature of 370 °C by increasing the strain rate from 0.01 to 1 $s^{-1}$ a local maximum can be detected in the curve. In the flow curve that exhibits the twinning signature, the strain to the peak flow stress is extended out to the higher value. By increasing the strain rate at constant temperature, more twins were observed and this can be proved by the shifting shape of curves from convex to concave indicating that deformation mode changes from slip-dominated flow to twin-dominated flow [14].

As can be seen in Figure 8, twins can also be detected in the microstructure. It was reported that due to the preferential activation of twining on the $\{10\bar{1}2\}$ $[\bar{1}011]$ in RD samples, this type of twinning plays a major role during the deformation [4]. Therefore, to study the deformation mechanism and twinning type more precisely, EBSD was performed to study the orientations of the twin and recrystallized areas. As can be seen in Fig. 7, at 375 °C and a strain rate of $1s^{-1}$, twinning can easily be detected in the microstructure. From the misorientation angle distributions, the samples show local maxima in the range of 30-40°, 55-60° and 85-90°. The maximum in the range of 30-40° with a dominant <-12-10> rotation axis can be attributed to the presence of {10-11}-{10-12} double twins. The second peak, 55-60°, is related to the clustering of rotation axes near <01-10>. These misorientation relationships occur between twins that form on either the (10-12) and (01-12) planes. The last peak is attributed to the propagation of {1-210} twinning [5, 21]. The most common misorientation angle at this temperature, 85-90°, indicates the presence of the extension twins. Twining activity increases with increasing strain rates and decreasing temperature [21]. Therefore, more twins can be detected in samples deformed at strain rate of 1 $s^{-1}$. This is because individual twins frequently form with high effective interface velocities and they can apparently develop much more rapidly than a slip band [11]. Therefore, at higher strain rates, twinning plays a more important role due to less activity of slip. In ND samples, because most of the grains are oriented in a way that their c-axes are parallel to the compression axis, contraction twins can form and the basal planes are reoriented by 56° around <1021 > axis. This change facilitates glide on the basal planes and generates zones of flow localization. These twins can undergo extension twinning in their interiors that is referred to double twinning.

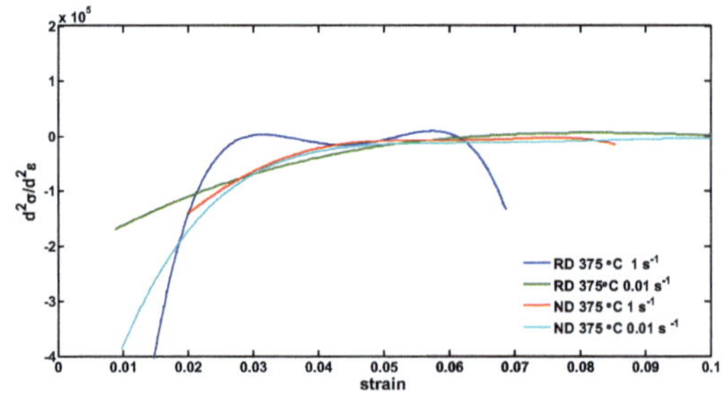


Fig. 7. A plot of change in the work hardening ($d^2\sigma/d\varepsilon^2$) against strain fo RD and ND samples deformed at 375 °C.

### 3.2.4 Texture evolution

Fig .8 shows the (0002) pole figures of the RD samples deformed under different strain rates and strains. The initial basal pole distribution is perpendicular to the compression axis and as deformation is performed the basal poles tilt to the compression axis. The results show that the rate of texture evolution is increased by increasing strain rate. At a strain of 1 $s^{-1}$ a basal pole

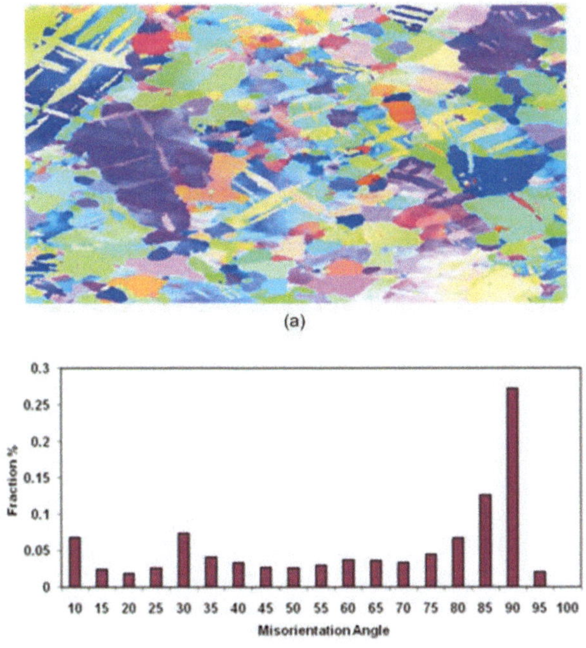


Fig. 7. (a) Orientation map and (b) misorientaion angle distribution of the samples deformed at strain rate of 10 $s^{-1}$ and temperature of 375 °C and strain of 0.1.

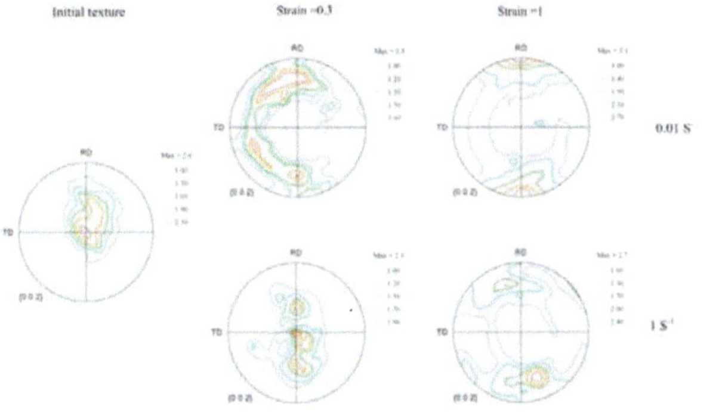

Fig 8. Texture evolution for RD samples deformed at strain rate of 1 $s^{-1}$and different strains and temperature of 375 °C

figure typical for compression testing was achieved after strain of 1; however, for strain rate of 0.01 $s^{-1}$ the pole figure still affected by the initial texture. As mentioned in the previous section by increasing the strain rate, the fraction of extension twinning increases in the deformed RD samples and basal planes are reoriented by 86° around a <1021 > axis; therefore the basal texture forms. It was observed that the recrystallization of Mg alloys is not accompanied with an obvious change of crystallographic texture [12]. Because of a higher level boundary energy grain boundaries and twins can both be the nucleation sites for DRX. DRX on twins is more important than the DRX on grain boundaries for the microstructure evolution during higher strain rate deformation because there are more twins. On the other hand, the c-axis of the twin DRX grain is parallel to the compression stress. Such a grain orientation cannot deform easily and is not favorable for basal slip because the resolved shear stress is nearly zero. Therefore, these grains are less hardened compared with the grains with other orientations and relatively stable and insensitive to DRX. However, new grains with other orientations may easily be consumed by repeated DRX. With deformation progress, more and more grains with their c-axes parallel to compression stress would survive. Therefore, the basal texture is progressively strengthened.

**Conclusion**

The hot compression flow stress and microstructure evolution of AZ31b alloys at different strain rates and different initial texture were investigated. The following conclusions can be drawn from the analysis:

(1) At higher strain rates ($\geq 10\ s^{-1}$), twins form, and since DRX occurs at twins, this explains why DRX is more extensive at higher strain rates. This also results in a more homogeneous microstructure compare.

**(2)** The EBSD results show that the type of twins activated in RD samples are double twins and extension twining.

**(3)** Correlations of the orientation of slip systems and the axis of compression revealed that the hot deformation of the RD specimens involves activation of first- and second-order pyramidal slip while that of ND specimens occurs by second order pyramidal slip.

**(4)** The results show that the rate of texture evolution in RD samples is increased by increasing strain rate because of increasing the fraction of extension twining.


**Acknowledge**

The authors would like to thank Prof. In-Ho Jung for his kind support. Thanks are also due to the Natural Resources Canada's CANMET materials technology laboratory for careful technical support. This study was supported by the NSERC Magnesium Strategic Research Network (MagNET) and MEDA Fellowship from Faculty of Engineering of McGill University.



[1] H.B. M. Avedesian, ASM specialty handbook. Magnesium and magnesium alloys. , ASM International, Materials Park, Ohio, 2000.

[2] T. Mukai, Materials Science and Technology, 16 (2000) 1314-1319.

[3] S.R. Agnew, M.H. Yoo, C.N. Tomé, Acta Materialia, 49 (2001) 4277-4289.

[4] S. Spigarelli, M.E. Mehtedi, M. Cabibbo, E. Evangelista, J. Kaneko, A. Jäger, V. Gartnerova, Materials Science and Engineering: A, 462 (2007) 197-201.

[5] M.R. Barnett, Materials Science and Engineering: A, 464 (2007) 1-7.

[6] J.W. Christian, S. Mahajan, Progress in Materials Science, 39 (1995) 1-157.

[7] Q. Guo, H.G. Yan, H. Zhang, Z.H. Chen, Z.F. Wang, Materials Science and Technology, 21 (2005) 1349-1354.

[8] Q. Jin, S.-Y. Shim, S.-G. Lim, Scripta Materialia, 55 (2006) 843-846.

[9] M.T.S. E. Essadiqi, A. Javaid, C. Galvani, K. Spencer, S. Yue, and R. Verna, in: International Conference on Light Metals Technology, Saint-Sauveur, Quebec, Canada, 2007, pp. 177-182.

[10] K. Ishikawa, H. Watanabe, T. Mukai, Materials Letters, 59 (2005) 1511-1515.
[11] B.H. Lee, W. Bang, S. Ahn, C.S. Lee, Metallurgical and Materials Transactions A: Physical Metallurgy and Materials Science, 39 A (2008) 1426-1434.
[12] A.G. Beer, M.R. Barnett, Materials Science and Engineering: A, 423 (2006) 292-299.
[13] H.J. McQueen, E.V. Konopleva, in, 2001, pp. 227-234.
[14] M.R. Barnett, Z. Keshavarz, A.G. Beer, D. Atwell, Acta Materialia, 52 (2004) 5093-5103.
[15] A. Grosvenor, C.H.J. Davies, in, 2003, pp. 4567-4572.
[16] S.E. Ion, F.J. Humphreys, S.H. White, Acta Metallurgica, 30 (1982) 1909-1919.
[17] D.L. Yin, J.T. Wang, J.Q. Liu, X. Zhao, Journal of Alloys and Compounds, 478 (2009) 789-795.
[18] A.G. Beer, M.R. Barnett, Materials Science and Engineering A, 485 (2008) 318-324.
[19] A. Mwembela, E.B. Konopleva, H.J. McQueen, Scripta Materialia, 37 (1997) 1789-1795.
[20] T. Al-Samman, G. Gottstein, Materials Science and Engineering A, 490 (2008) 411-420.
[21] K. Ishikawa, H. Watanabe, T. Mukai, Journal of Materials Science, 40 (2005) 1577-1582.

**Magnesium Technology 2012**
*Edited by: Suveen N. Mathaudhu, Wim H. Sillekens, Neale R. Neelameggham, and Norbert Hort*
*TMS (The Minerals, Metals & Materials Society), 2012*

# PRECIPITATION BEHAVIOUR OF MICRO-ALLOYED MG-AL-CA ALLOYS DURING HEAT TREATMENT AND HOT COMPRESSION

Jing Su, Shirin Kaboli, Abu Syed Humaun Kabir, Phuong Vo, In-Ho Jung and Stephen Yue

Department of Mining and Materials Engineering, McGill University, Montreal, Canada

Key words: Mg-Al-Ca, precipitation, heat treatment and hot compression

## Abstract

Based on thermodynamic calculations, two micro-alloyed Mg-Al-Ca alloys, Mg-0.3Al-0.2Ca and Mg-0.1Al-0.5Ca, were selected in terms of the equilibrium precipitation temperatures of $Al_2Ca$ and $Mg_2Ca$, respectively. The basic idea is to form precipitates during hot compression to examine their effect on hot deformation. Both alloys, cast by copper mould, were solution treated at 500°C for 8 hours to dissolve eutectic precipitates which formed in the as-cast microstructure, and then isothermally heat treated at 350°C for different times. SEM analysis of the heat treated alloys generally agreed with thermodynamic calculations. Hot compression tests were also conducted on solution treated alloys at 350°C with a strain rate of $0.01s^{-1}$ and different strains (10%, 30%, 60%, and 90%). The precipitation behaviour and microstructural evolution which was characterized by optical microscopy and SEM with BSE and EDS detectors during isothermal heat treatment and hot compression were compared.

## Introduction

Recently Mg alloys have drawn great attention as structural materials for automotive application due to their low density, high strength to weight ratio and high stiffness to weight ratio compared with steel and aluminium. However, due to the poor formability of Mg alloy sheets, the development of new alloys is required for future application [1,2]. It is known that the addition of calcium to magnesium is effective in grain refinement during casting [3-7] as well as improve oxidation resistance, increase ignition temperature and improve creep resistance [7,8]. Furthermore, calcium is cheaper and has lower density ($1.55g/cm^3$) than rare-earth elements (Neodymium, yttrium, thorium etc.) which are now being considered to weaken texture [10-13]. Aluminium is regarded as a strengthening element due to forming solid solution with magnesium and a significant difference of the atomic diameter with magnesium [14]. As well, Mg–Al–Ca system alloys are also considered due to their good castability [9].

Previous studies have reported that Mg-Al-Ca alloys have high strength at elevated temperature due to formation of thermally stable intermetallic compounds $Al_2Ca$, $Mg_2Ca$ or $(Mg,Al)_2Ca$ [8,15,16]. In comparison with the $Mg_{17}Al_{12}$ phase, $Mg_2Ca$ and $Al_2Ca$ phases have higher melting points (460°C for $Mg_{17}Al_{12}$, 715°C for $Mg_2Ca$ and 1079°C for $Al_2Ca$) and thus, are relatively stable at elevated temperatures [17]. An increase in the Ca content results in a reduction of the $Mg_{17}Al_{12}$ phase as well as the formation of thermally stable secondary phases containing Ca [5]. However, these precipitates in magnesium alloys tend to be coarse and brittle [18]. Therefore, workability is rather poor due to the tendency of de-bonding from the magnesium matrix during plastic deformation. Both decreasing the grain size and refining the coarse compounds into small precipitates can improve the formability during plastic working [8].

In addition, precipitates can be used to modify the texture and control grain size during thermal-mechanical processing. Particles may either promote or retard recrystallization depending on their size, spacing and fraction [19]. It has been shown in some prior investigations of magnesium alloys that the change in texture was a result of the operation of PSN in systems containing particles larger than 1μm [12].

In this work, two Mg-Al-Ca alloys were chosen on the basis of forming high temperature precipitates during hot deformation. The effect of precipitates on hardening during isothermal heat treatment and dynamic precipitation during hot deformation were investigated.

## Experimental Procedures

### Materials and Casting Conditions

Two Mg-Al-Ca alloys were prepared using 99.9% pure Mg, 99.9% pure Al and Mg-30% Ca master alloy. The pure magnesium was melted in a graphite crucible using a high-frequency induction furnace at 690°C. Pure Al and Mg–30wt.% Ca master alloy were then added to the melt at 700°C and held for 15min to dissolve. The molten metal was cast under an $SF_6$ and $CO_2$ gas mixture atmosphere from 730°C into a preheated cylindrical copper mould to obtain billets of 15mm in diameter and 145mm in length.

### Heat Treatment and Hardness Test

The as-cast specimens were homogenized at 500°C for 8 hours and quenched in water to dissolve all eutectic precipitates and obtain a supersaturated solid solution. They were then isothermally heat-treated at 350°C for periods ranging from 10min to 32 hours. Hardness tests were conducted by using a Clark CM-100AT micro-hardness tester and a load of 50gf with a minimum of ten measurements for each specimen.

Compression Test

Compression specimens were machined from the solution heat-treated bars into cylinders with a length of 11.4mm and a diameter of 7.6mm (a ratio of 1.5). Hot compression tests were conducted using a 100kN MTS servo-hydraulic testing machine at 350°C with a strain rate of $0.01s^{-1}$ and strains from 10% to 90%. The specimens were heated to the deformation temperature and held for 10min prior to compression in order to ensure a uniform temperature along the specimen. During the test, the specimen was protected under an argon gas atmosphere to avoid oxidation while a combination of boron nitride powder on mica was used to minimize friction between the anvil and specimen. The specimens were quenched within 2 seconds after compression to freeze the deformed microstructure.

Microstructure Characterization

Samples were ground with SiC papers (from No.400, No.600, No.800 to No.1200), and then polished with 3μm and 1μm diamond suspension, and 0.05μm alumina powder. Picric–acetic solution (6g picric acid, 15ml acetic acid, 10ml $H_2O$ and 100ml ethanol) and 2% Nital solution were used to etch samples. Optical microscopy (Nikon EPIPHOT 200) and SEM (Philips XL30 and Hitachi SU8000) with backscattered electron detector (BSE) was used to characterize the microstructure while energy dispersive spectroscopy (EDS) was used to identify composition of the second phase.

**Results and Discussion**

Thermodynamic Calculations and Alloys Design

The design purpose of this study is to form thermally stable precipitates either during hot deformation or annealing after deformation, which may influence recrystallization behaviour and texture evolution depending on the size and distribution of precipitates. In this study, 350°C was chosen to be the deformation temperature for Mg-Al-Ca alloys, as the expected industrial hot deformation temperature range for Mg alloys is between 300°C to 400°C. The criterion for selecting alloy composition was the presence of a second phase ($Al_2Ca$ or $Mg_2Ca$) at 350°C without $Mg_{17}Al_{12}$ phase. According to the ternary phase diagram of Mg-Al-Ca alloy at the Mg corner at 350°C (Fig.1) calculated from FactSage software with FTlite database, additions of less than 0.8% Al do not form $Mg_{17}Al_{12}$ phase. The composition of the alloys in the present study was determined based on the calculated Mg-Al-Ca equilibrium phase diagram at 350°C (Fig.2). The selected alloys of Mg-0.3Al-0.2Ca (alloy 1) and Mg-0.1Al-0.5Ca (alloy 2) feature a Mg matrix with single second phases of $Al_2Ca$ (laves 15) and $Mg_2Ca$ (laves 14), respectively.

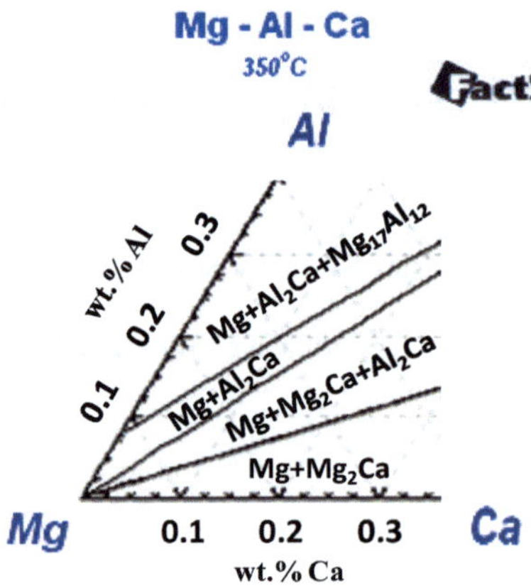


Fig.1 Ternary phase diagram of Mg-Al-Ca alloy at Mg corner at 350°C calculated from Factsage software with FTlite database

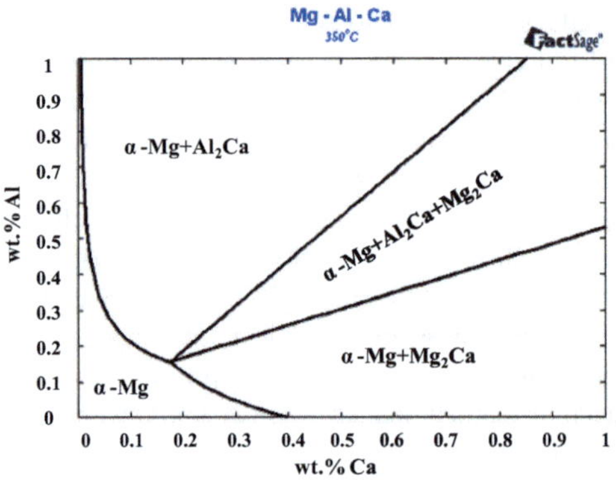


Fig.2 Isothermal phase diagram of Mg-Al-Ca alloy at 350°C calculated from FactSage software with FTlite database

As-cast Microstructure

The optical microstructures of as-cast alloy 1 and 2 are shown in Figure 3. As shown at low magnification (Fig.3 (a),(b)), the grain size of alloy 1 was slightly larger than that of alloy 2. The finer microstructure with increasing Ca level is consistent with previous studies [3-7]. The amount and size of precipitates in alloy 2 was larger than that in alloy 1, as shown at higher magnification in Fig.3(c),(d) The precipitates in alloy 2 were irregular in shape and tended to locate along grain boundaries and grain interiors, while precipitates were almost round in alloy 1 and preferentially distributed inside grains.

With closer analysis by SEM (Fig. 4), the precipitates in both alloys were observed to have a eutectic structure with Mg, Al and Ca peaks shown in the EDS spectrum. In the study of Suzuki et al. [20], the eutectic region was examined by TEM and reported to consist of a-Mg and a ternary compound of $(Mg,Al)_2Ca$ with dihexagonal C36 structure. Besides, there was a small amount of inclusions randomly dispersed in the as-cast microstructure of both alloys. These inclusions were faceted and around 500nm in size with impurity elements of Fe identified from EDS analysis.

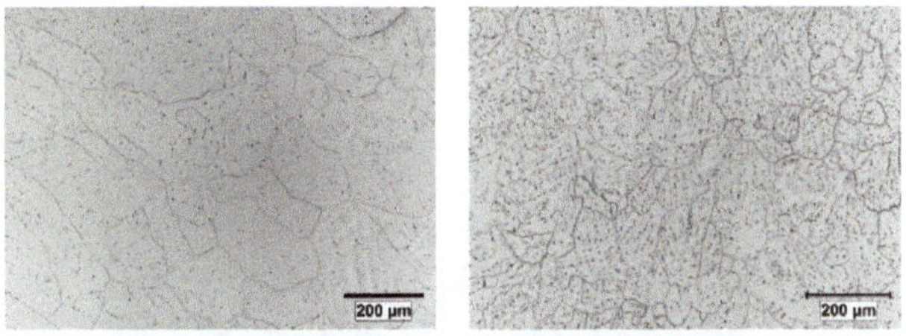


(a)Alloy 1 with a Mag. of 100 (b)Alloy 2 with a Mag. of 100

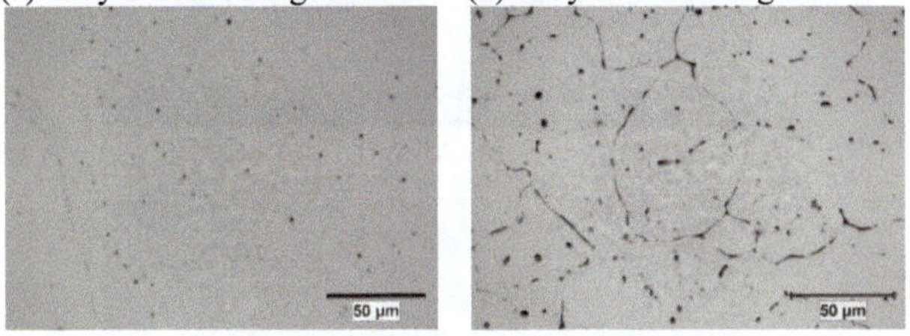


(c)Alloy 1 with a Mag. of 500 (d)Alloy 2 with a Mag. of 500

Fig.3 Optical microstructure of as-cast alloy 1 and alloy 2 with magnifications of 100 and 500

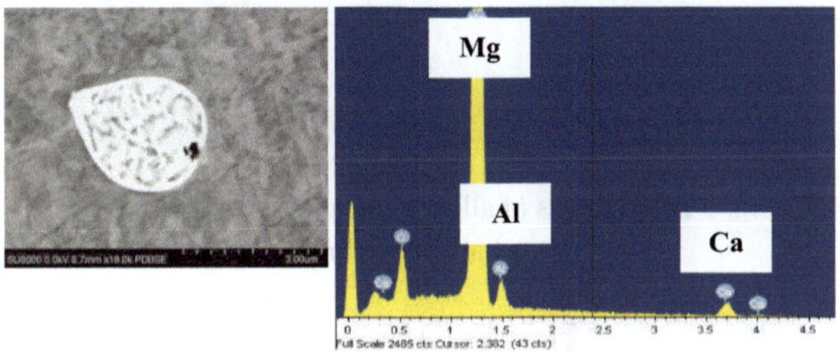


(a) (b)

Fig.4 BSE image (a) and EDS spectrum of eutectic precipitate

Microstructure of Solution Heat Treated Alloys

From FactSage calculations (Figs. 5(a) and (b)), the solution heat treatment temperature ranges are above 400°C for alloy 1 and 430°C for alloy 2. It was found that after heat treatment at 500°C for 8h, eutectic precipitates formed in the as-cast samples were dissolved into the matrix. However, faceted inclusions consisting of Fe still remained, as shown in Fig. 6. The hardness values of alloy 1 and alloy 2 after solution heat treatment were approximately 40HV and 46HV, respectively, which was similar to the hardness of as-cast samples. The grain size of alloy 1 and 2 after solution heat treatment are approximately 190um and 200um, respectively.

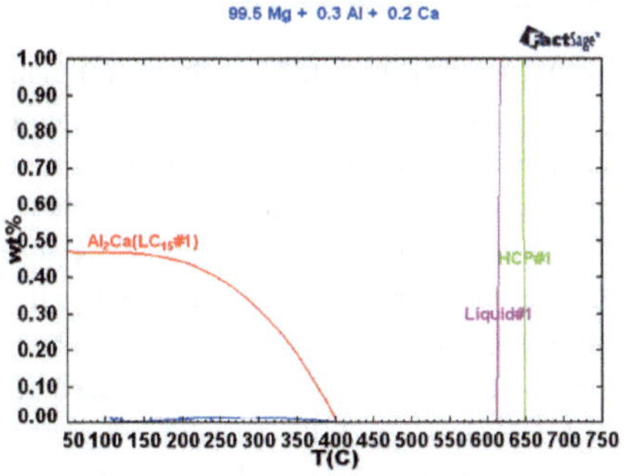


(a) Alloy 1

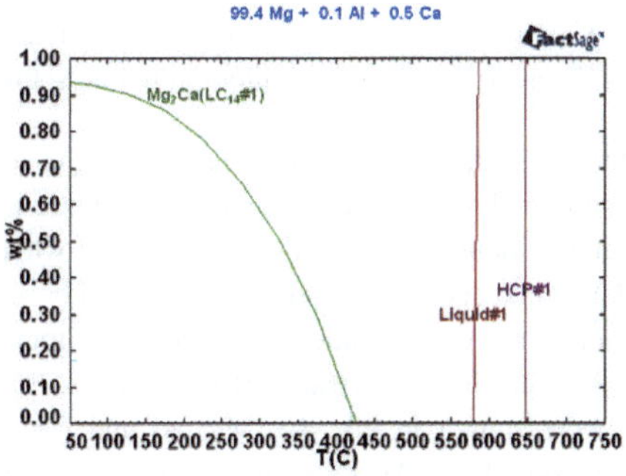


(b) Alloy 2

Fig. 5 Equilibrium phase distribution of alloy 1 (Mg-0.3Al-0.2Ca) (a) and alloy 2 (Mg-0.1Al-0.5Ca) (b)

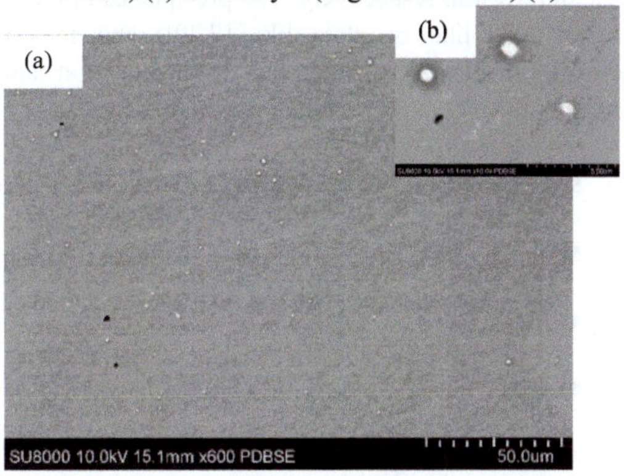


Fig.6 BSE images of solution heat treated alloy 2 with (a) a Mag. of 600 and (b) a Mag. of 10000

Microstructure and Hardness During Isothermal Heat Treatment

In Fig.7, the hardness curves of alloy 1 and alloy 2, which were isothermally heat-treated at 350°C from 10 min to 32 hours after solution heat treatment, were shown. In general, the hardness of alloy 2 is higher than that of alloy 1 for all times of the 350°C heat treatment. This results from larger number of precipitates and slightly smaller grain size of alloy 2 than that of alloy 1. Concerning of the solution hardening effect of aluminium on these two alloys, alloy 1 was added larger amount of aluminium than that of alloy2, which may compensate for the hardness of alloy 1 a little bit.

The variation of hardness value of alloy 1 and alloy 2 were both very small along all time range. The precipitates had already formed at 10min heat treatment and the amount of precipitates increased with time (Figs.11 (a) and (b)), but there is no obvious increase of hardness value (only 1 to 2 HV). There are two possible reasons. First, the volume fraction of precipitates in both alloys was too small, even at longer times. Second, this is because of the morphology of precipitates or the crystallographic relationship between precipitates and matrix of precipitates, since precipitates prefer to grow along the basal plane of magnesium alloys, which is ineffective to inhibit the basal slip and hardening the alloys [18, 20]. The slight decrease of hardness in alloy 2 was mainly due to coarsening of precipitates, as seen from Fig. 8 (a) and (b), as there is no apparent grain growth. The hardness of both alloys from 8 hours to 32 hours was more or less constant. This suggests that the grain boundary precipitates inhibit grain coarsening.

From the FactSage simulation (Figs.5 (a) and (b)) in the equilibrium condition, the amount of second phase in alloy 2 is approximately double that of alloy 1 (about 0.4 wt.% vs 0.2wt.% at 350°C). This level of difference is supported by BSE images (for example, alloy 1 and alloy 2 heated for 4 hours shown in Fig. 9). As well, EDS spectrums of precipitates in alloy 1 and alloy 2 were shown in Fig. 10. Mg, Al and Ca peaks were shown in alloy 1, while only Mg and Ca peaks appeared in alloy 2. According to the EDS spectrum (Fig.10) and the FactSage simulation, the precipitates in alloy 1 and alloy 2 were $Al_2Ca$ and $Mg_2Ca$, respectively. The precipitates in both alloys are oval (disk-like or plate like [7,20]) and the size of precipitates in alloy 1 is smaller than that in alloy 2 at same heat treatment time.

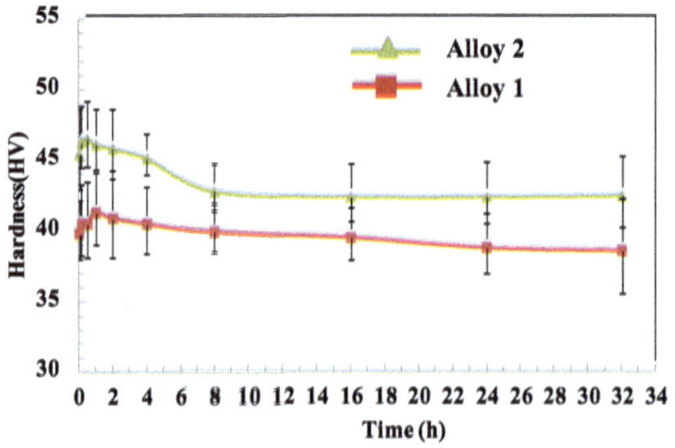


Fig.7 The hardness curves of alloy 1 and alloy 2 at 350°C from 10 min to 32 hours

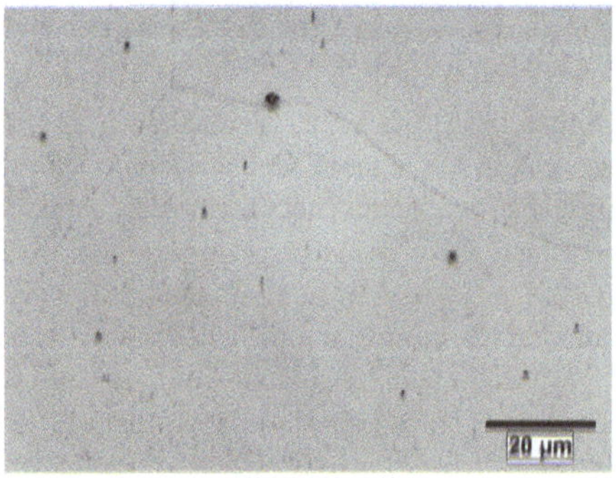


(a)Alloy 2 for 4hrs

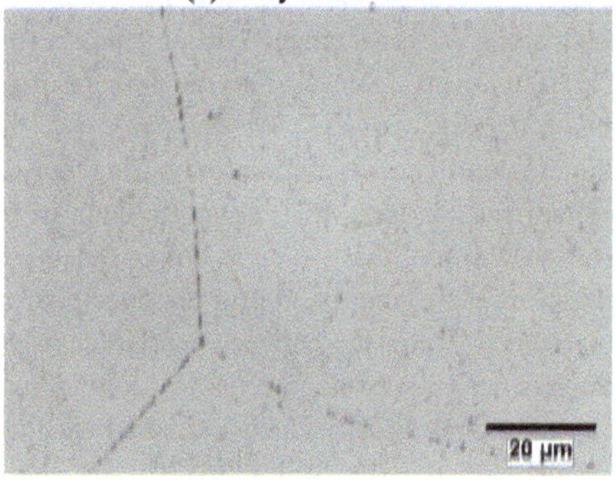


(b)Alloy 2 for 8hrs

Fig. 8 Optical microstructure of alloy 2 heated at 350°C for 4 hours(a) and 8 hours(b) with a Mag. of 1000

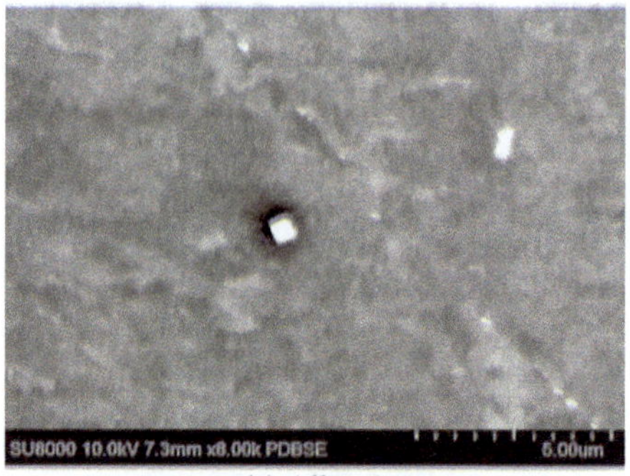


(a) Alloy 1

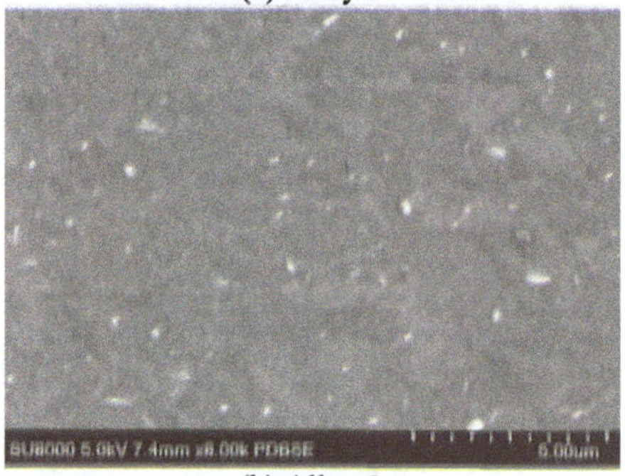

(b) Alloy 2

Fig. 9 BSE images of alloy 1 and alloy 2 isothermally heat treated at 350°C for 4 hours

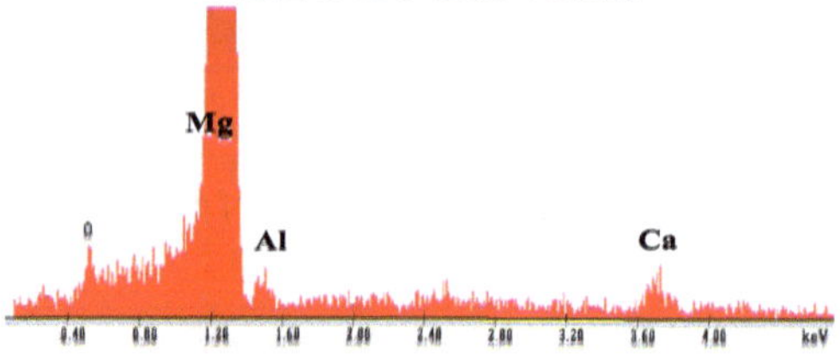


(a)Alloy 1

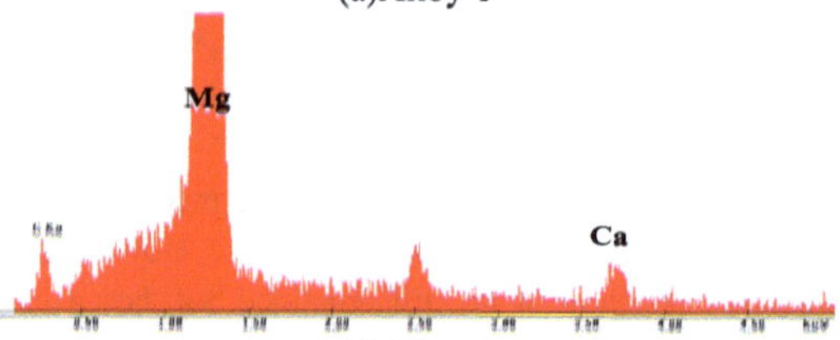


(b)Alloy 2

Fig.10 Spectrum of precipiates of alloy1 (a) and alloy 2 (b) during heat treatment at 350°C by EDS

Microstructure evolution during hot compression

Specimens at 10min heat-treated condition were characterized, since they have to be kept at deformation temperature for 10min prior to compression to achieve uniform temperature. After a 10min heat treatment at 350°C in alloy 1, a small number of precipitates formed along grain boundaries, as shown in Fig. 11(a). After deformation at 350°C and a strain rate of $0.01s^{-1}$, the amount of precipitates did not significantly changed at 10% and 30% strain (Fig.11(c)) and slightly increased at 60% to 90% strain (Fig.11(e)). In alloy 2, the variation in precipitates level was small at low strains (Fig.11(d)) compared with the heat treated sample (Fig.11(b)) and displayed a clear increase in the 60% and 90% strained samples (Fig.11(f)). Therefore, strain induced precipitation occurred in both alloys. However, alloy 2 has more obvious enhancement of precipitation with increase of strain. For 10mins heat treatment without deformation and with deformation at 350°C, more precipitates formed in alloy 2 compared with alloy 1. The sizes of precipitates forming in alloy 1 and alloy 2 are both around 200nm and there is no apparent coarsening of precipitates with increasing of strains. From the recrystallization point of view, alloy 2 is expected to have more effect on dynamic or static recrystallization behaviour [19], since more precipitates form in alloy 2 during hot compression. The relationship of dynamic recrystallization and precipitation behaviour will be done in the future work.

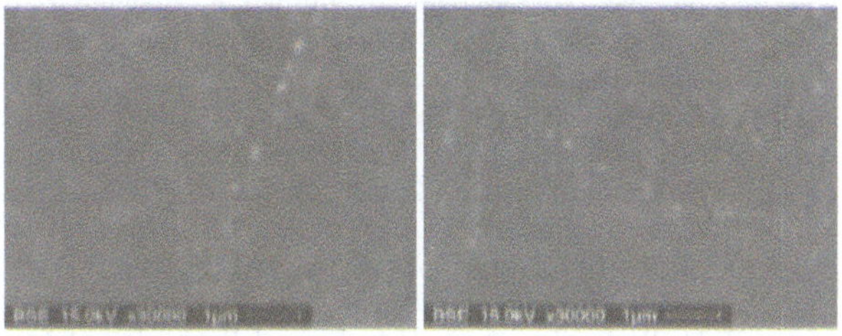

(a)A1 10min heat treatment (b) A2 10 min heat treatment

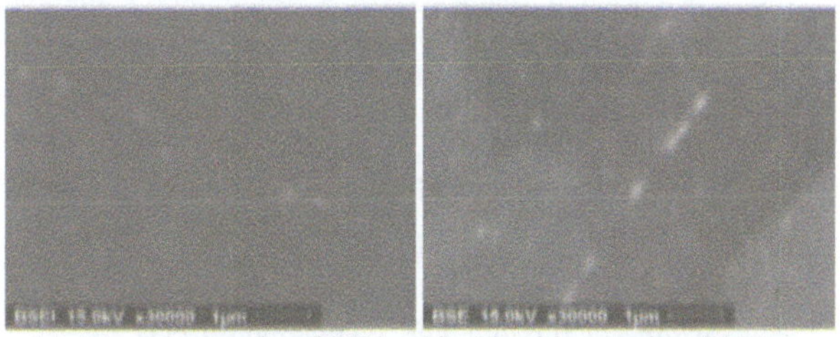

(c) A1 with a strain of 30% (d) A2 with a strain of 30%

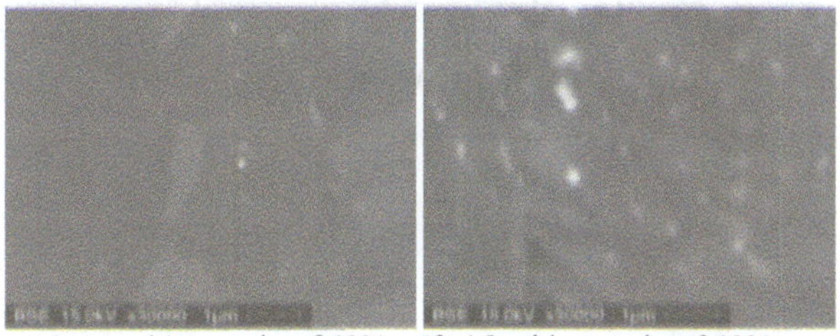

(e) A1 with a strain of 60% (f) A2 with a strain of 60%

Fig.11 BSE image of alloy 1 and alloy 2 heat treated for 10mins(a)(b) and compressed at 350°C with a strain rate of $0.01 s^{-1}$ and strains from 30%(c)(d) and 60%(e)(f)

The flow curves of alloy 1 and alloy 2 were shown in Fig. 12. Below 20% deformation, alloy 2 had higher strength than alloy 1. The top stress values for alloy 1 and alloy 2 were both around 52MPa at a strain of 40%. And the softening behaviour of two alloys was similar at larger stains (60% to 90%). The hardness of both alloys went up with increasing strains, as shown in Fig. 13, which is partly due to precipitation induced by strains. As similar as isothermal heat treatment, the hardness of alloy 2 was higher than that of alloy 1 at all strains. In addition, hardening effect on compressed samples was more efficient compared with that on heat-treated ones. The possible reasons include work hardening effect as well as a large amount of small size and dispersed precipitates in deformed specimens.

During hot compression, the amount of precipitates increases with increasing strains. This is mainly related to more nucleation of precipitates, since more nucleation sites, such as dislocations, deformation bands or sub-grain boundaries, caused by deformation. Compared with the dynamic precipitates during deformation, the volume fraction of precipitates increases with time during isothermal heat treatment, which is primarily because of growth of precipitates.

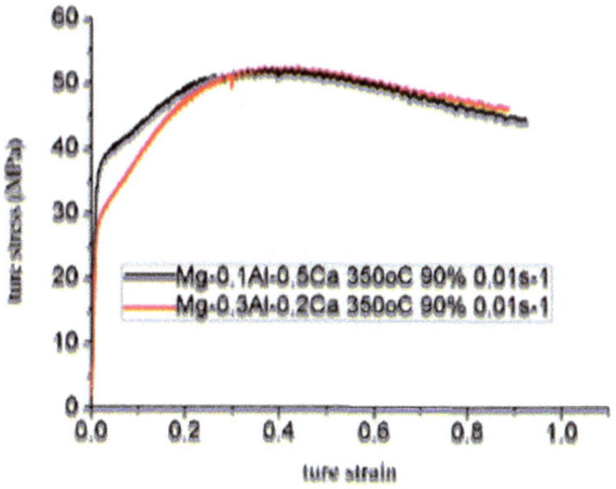


Fig. 12 Flow curves of alloy 1 and alloy 2 compressed at 350°C with a strain rate $0.01 s^{-1}$ and a strain of 90%

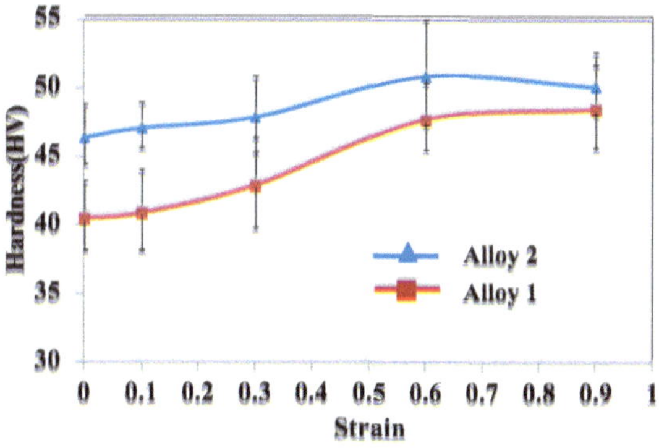


Fig.13 The hardness curves of alloy 1 and alloy 2 compressed at 350°C with a strain rate of $0.01 s^{-1}$ and stains of 10%, 30%, 60% and 90%

## Summary

(1) In the as-cast microstructure, a-Mg and eutectic $(Mg,Al)_2Ca$ precipitates formed in both alloys, which can be dissolved into matrix at 500°C for 8 hours heat treatment.

(2) During isothermally heat treatment at 350°C, $Al_2Ca$ phase formed in alloy 1, while $Mg_2Ca$ formed in alloy 2. Larger amount of precipitates formed in alloy 2 than that in alloy 1, which resulted in higher hardness value of alloy 2. And the hardness of both alloys didn't apparently drop even for 32 hours due to the precipitates pinning on the grain boundaries.

(3) During hot compression at 350°C with a strain rate of $0.01s^{-1}$, strain induced precipitation more obviously occurred in alloy 2 compared with alloy 1. And the hardness of both compressed alloys increased with strains.

(4) Increase of volume fraction of precipitates during heat treatment is due to precipitates growth, while during hot compression it relies on enhancing nucleation of precipitates.

**Acknowledgement**

This work was support by General Motor and NSERC, Canada.

**Reference**

1. B.L. Mordike, and T. Ebert, "Magnesium: Properties-Applications-Potential," *Materials Science and Engineering A,* 302 (1) (2001), 37-45.
2. R. VanFleteren. "Magnesium for Automotive Applications," *Advanced Materials and Processes,* 149 (5) (1996), 33-34.
3. L. Han, H. Hu, and D.O. Northwood, "Effect of Ca Additions on Microstructure and Microhardness of an as-Cast Mg-5.0 wt.% Al Alloy," *Materials Letters* 62(3) (2008), 381-384.
4. B. Kondori, and R. Mahmudi, "Effect of Ca Additions on the Microstructure, Thermal Stability and Mechanical Properties of a Cast Am60 Magnesium Alloy," *Materials Science and Engineering A,* 527 (7-8) (2010): 2014-2021.
5. P. Li, B. Tang, and E.G. Kandalova, "Microstructure and Properties of AZ91D Alloy with Ca Additions," *Materials Letters* 59 (6) (2005), 671-675.
6. L. Shang, et al., "Effect of Microalloying with Calcium on the Microstructure and Hot Working Behavior of AZ31 Magnesium Alloy," *TMS 2007.*
7. J.F. Nie, B.C. Muddle, "Precipitation Hardening of Mg-Ca(-Zn) Alloys, " *Scripta Materialia,* 37 (1997), 1475-1481.
8. W.J. Kim, et al., "Microstructure and Superplasticity of Mg–Al–Ca Electromagnetic Casting Alloys after Hot Extrusion," *Materials Science and Engineering A,* 494 (2008), 391-396.
9. S. W. Xu, et al, "High Temperature Tensile Properties of as-Cast Mg-Al-Ca Alloys." *Materials Science and Engineering A,* 509 (1-2) (2009), 105-110.
10. J.W.Senn, and S.R.Agnew, "Texture Randomization of Magnesium Alloys Containning Rare Earth Elements, " *TMS* 2008.
11. T. Laser, et al., "The Influence of Calcium and Rare Earth Metals on the Microstructure and Mechanical Properties of Mg-3al-1zn During Extrusion, " *TMS* 2008.
12. J.D. Robson, D.T. Henry, and B. Davis, "Particle Effects on Recrystallization in Magnesium–Manganese Alloys: Particle-Stimulated Nucleation," *Acta Materialia,* 57 (2009), 2739-2747.
13. N. Stanford, and M. Barnett, "Effect of Composition on the Texture and Deformation Behaviour of Wrought Mg Alloys," *Scripta Materialia,* 58 (2008), 179-182.
14. K.P. Rao, et al., "Effect of Aluminum Addition on the Strengthening and High Temperature Deformation Behavior of Mg-3Sn-2Ca Alloy, " *TMS 2010.*
15. L.L. Rokhlin, and N.I. Nikitina, "Effect of Calcium on the Properties of the Mg-Al System," *Matal Science and Heat treatment,* 45 (2003), 171-173.
16. H. Watanabe, et al., "Mechanical Properties of Mg–Al–Ca Alloy Processed by Hot Extrusion," *Materials Science and Engineering A,* 454-455 (2007), 384-388.
17. L.L. Rokhlin, N.I. Nikitina, and V. A. Volchenkova. "Magnesium-Rich Mg–$Al_2$ca Alloys," *Russian Metallurgy (Metally,)* 2 (2006), 185-188.
18. B. Langelier, and S. Esmaeili, "The Effect of Zn Additions on Precipitation Hardening of Mg-Ca Alloys," *TMS 2010.*
19. F.J. Humphreys, *Recrystallization and Related Annealing Phenomena / F.J. Humphreys and M. Hatherly,* (Oxford, U.K. : Pergamon, 1995).
20. A. Suzuki, et al., "Precipitation Strengthening of a Mg-Al-Ca–Based AXJ530 Die-Cast Alloy," *METALLURGICAL AND MATERIALS TRANSACTIONS A,* 39A ( 2008 ), 696-702.

**Magnesium Technology 2012**
*Edited by: Suveen N. Mathaudhu, Wim H. Sillekens, Neale R. Neelameggham, and Norbert Hort*
*TMS (The Minerals, Metals & Materials Society), 2012*

# DIFFUSION COUPLE INVESTIGATION OF THE Mg-Zn SYSTEM

Sarah Brennan[1], Katrina Bermudez[1], Nagraj Kulkarni[2], Yongho Sohn[1]

[1]University of Central Florida, Orlando, FL, USA
[2]Oak Ridge National Laboratory, Oak Ridge, TN, USA

**Keywords: Magnesium, Zinc, Diffusion, $MgZn_2$, $Mg_2Zn_3$**

## Abstract

Growing use and development of lightweight Mg alloys has been the catalyst for more fundamental research in Mg based material systems to be completed. Zinc is one of the most common alloying elements in Mg alloys. Phase layer growth and interdiffusion in the binary Mg-Zn system was investigated utilizing solid-to-solid diffusion couples. Anneals were carried out at 295°, 315° and 325°C for 384, 168 and 120 hours, respectively. The diffusion microstructures that developed were examined by optical and scanning electron microscopy (SEM). Concentration profiles were determined using X-ray energy dispersive spectroscopy (XEDS) and electron microprobe analysis (EPMA). The phases observed were the Mg solid solution, $Mg_2Zn_{11}$, $MgZn_2$ and $Mg_2Zn_3$ in all three couples as well as the high temperature, $Mg_{51}Zn_{20}$ phase in the 325°C couple. The $MgZn_2$ phase was observed to grow the thickest layer, followed by the $Mg_2Zn_3$ and the $Mg_2Zn_{11}$ phases. Parabolic growth constants were determined for each phase. Activation energies for the growth of the intermetallic phases were calculated as 105 kJ/mol for the $Mg_2Zn_3$ phase and 207 kJ/mol for the $MgZn_2$ phase.

## Introduction

The drive for more lightweight performance materials has led to increasing interest in magnesium alloys for use in applications in the transportation and electronic industries due to their unique properties, especially their light weight [1-3]. Diffusion plays an important role in many material properties and phenomena such as solidification, precipitation, creep, corrosion, oxidation, etc. Reliable diffusion parameters aid in the design, optimization and understanding of degradation of materials, and are crucial to material development. The scarcity of diffusion data in Mg based systems is the motivation behind this study.

Zn is a common alloying element in commercial Mg alloys, typically added for solid solution and precipitation strengthening. However, little data is available regarding the phase layer growth and interdiffusion behavior in the Mg-Zn binary system. An impurity diffusion study of the radioisotope $^{65}Zn$ in polycrystalline Mg was reported by Lal in 1967 [4]. Cermak and Stloukal also reported the diffusion coefficient as a function of Al concentration of $^{65}Zn$ in Mg-*x*Al solid solutions (*x*=1.77, 3.9 and 9 wt.% Al) and in the Mg alloy AZ91 as well as the diffusion of $^{65}Zn$ in polycrystalline Mg via serial section and the residual activity method [5].

Intermetallic phase formation and growth in the Mg-Zn binary system was studied using the solid diffusion couple technique at 295, 315, and 325°C for 384, 168 and 120 hours, respectively. Three diffusion couples were successfully bonded and examined via optical and scanning electron microscopies. In all three couples, the $MgZn_2$, $Mg_2Zn_3$ and $Mg_2Zn_{11}$ phases were identified as well as the $Mg_{51}Zn_{20}$ phase in the 325°C couple in accordance with the phase diagram shown in Figure 1. Parabolic growth constants were calculated for each phase and activation energies for growth were determined for the $MgZn_2$ and $Mg_2Zn_3$ phases.

## Experimental Procedure

Rods of pure, polycrystalline Mg (99.9%) and pure Zn (99.9%) from Alfa Aesar™ were sectioned into disks, 10 mm in diameter and 2 mm in thickness. These disk specimens were

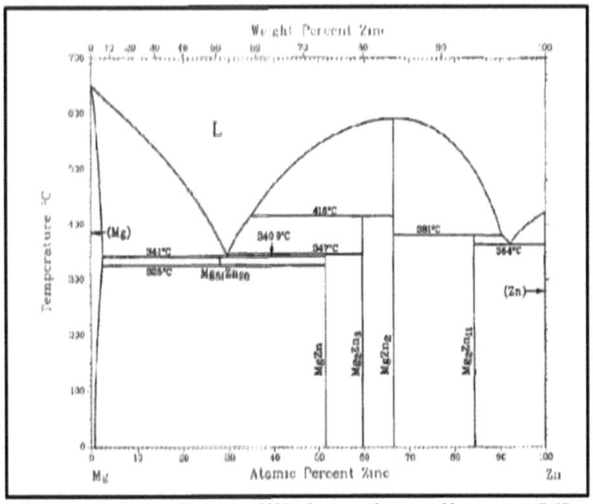

Figure 1: Mg-Zn equilibrium phase diagram [6].

then metallographically polished down to 1 μm using a non-oxidizing lubricant at each grinding and polishing step. Any contact with water was avoided during the preparation process due to the high reactivity of Mg. Mg vs. Zn diffusion couples were assembled as illustrated in Figure 2, with the polished faces of the disks touching, sandwiched between inert, Alumina ($Al_2O_3$) spacers and clamped in a stainless steel jig. The entire jig assembly was then placed in a quartz capsule, which is repeatedly evacuated to ~$10^{-6}$ torr and flushed alternately with hydrogen and ultra-high purity argon. The capsule is then backfilled with ultra-high purity argon to a pressure slightly above 1 atm at the respective annealing temperature.

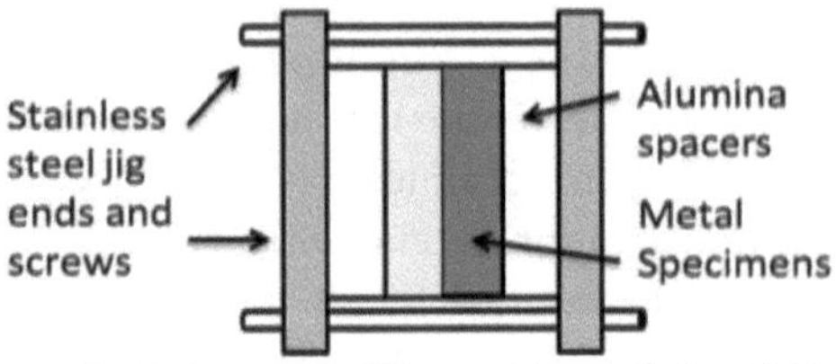

Figure 2: Schematic illustration of the diffusion couple jig assembly.

A Paragon Bluebird[TM] furnace was preheated to the annealing temperature, and the capsule was placed inside. The temperature was monitored with an independent type-K thermocouple and stabilized within ± 2° of the desired temperature. Three diffusion couples were assembled and annealed at 295, 315, and 325°C for 384, 168 and 120 hours, respectively. After the diffusion anneal, the quartz capsule was then quickly removed and quenched in cold water. The whole jig assembly was mounted in epoxy, cross-sectioned, and metallographically prepared for examination. OM was used to check the bond quality and field emission-SEM (Zeiss Ultra55™) equipped with XEDS (X-ray energy dispersive spectroscopy) was utilized to identify the intermetallic phases present along with the equilibrium phase diagram presented in Figure 1. Electron micrographs of the resulting diffusion microstructures were used to measure phase thicknesses via image analysis. Parabolic growth constants, $k_p$, were determined for each phase as described by the equation [7]:

$$k_p = \frac{Y^2}{t} \tag{1}$$

where Y is the layer thickness measured, and t is the annealing time. Activation energy for growth of a phase, $Q_k$, was determined using the relation:

$$k_p = k_o \exp\left[-\frac{Q_k}{RT}\right] \tag{2}$$

where $k_o$ is the pre-exponential factor, R is the ideal gas constant (8.314 J/mol-K) and T is the absolute annealing temperature.

## Results

<u>Diffusion Microstructures</u>

Figure 3 shows electron micrographs of the resulting diffusion microstructures for each temperature studied. In all three diffusion couples the $MgZn_2$, $Mg_2Zn_3$, and $Mg_2Zn_{11}$ phases were identified, and the $Mg_{51}Zn_{20}$ phase was found in the 325°C diffusion couple. The MgZn phase was not observed in any of the diffusion couples studied. The $Mg_2Zn_{11}$ phase is extremely brittle, resulting in the pullout of the phase in the thin layer seen in Figure 3a in the 295°C couple. A small amount of the native oxide layers on both the Mg and Zn disks were present on the surfaces of both disks and served as markers in these experiments. Evidence of this can be seen clearly in the 315 and 325°C couples, where the original interface, or marker plane, is identified in the $Mg_2Zn_{11}$ phase.

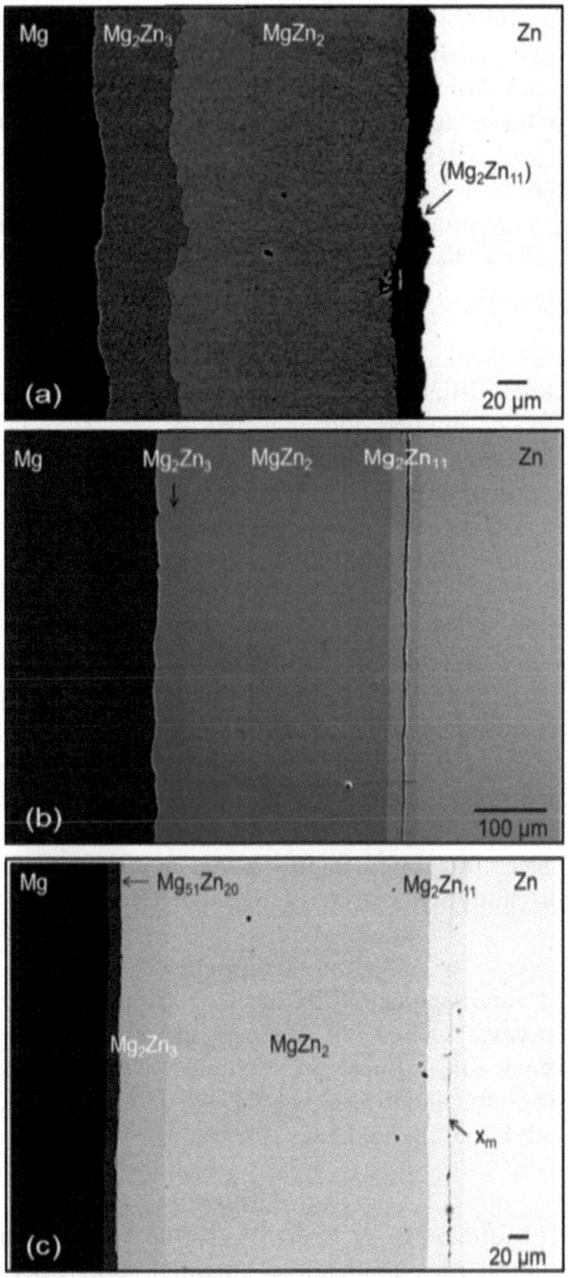


Figure 3: Representative electron micrographs of microstructure from Mg vs. Zn diffusion couple annealed at (a) 295°C for 384 hours, (b) 315°C for 168 hours, and (c) 325°C for 120 hours. $X_m$ is the marker plane. (Note: $Mg_2Zn_{11}$ in (a) is mostly pulled out from the diffusion couple).

<u>Growth Constants and Activation Energies</u>

A minimum of 15 random-location measurements was made to determine the thickness of each phase present in each diffusion couple using image analysis. Parabolic growth constants were calculated using Eq. [1], and are reported along with the thickness measurements in Table I. Figure 4 displays the temperature-dependence of the growth constants for the $Mg_2Zn_3$ and $MgZn_2$ phases. $MgZn_2$ grew thicker, as evident in the above micrographs and the thickness measurements in Table I, and has the correspondingly higher parabolic growth constants. The phase diagram in Figure 1 shows that the $MgZn_2$ phase has the highest melting point of the intermetallic phases present in this binary system and is therefore more thermodynamically stable.

Table I: Intermetallic phase layer thicknesses, Y, and parabolic growth constants, $k_p$.

| | $Mg_2Zn_3$ | | $MgZn_2$ | |
|---|---|---|---|---|
| Couple | Y (µm) [SD] | $k_p$ ($m^2/s$) | Y (µm) [SD] | $k_p$ ($m^2/s$) |
| 295°C 384 hr. | 45.0 [4.1] | $7.3 \times 10^{-16}$ | 157.0 [3.4] | $9.0 \times 10^{-15}$ |
| 315°C 168 hr. | 37.3 [1.4] | $1.1 \times 10^{-15}$ | 244.6 [1.6] | $4.9 \times 10^{-14}$ |
| 325°C 120 hr. | 45.7 [1.1] | $2.4 \times 10^{-15}$ | 258.0 [1.7] | $7.7 \times 10^{-14}$ |
| | | | | |
| | $Mg_2Zn_{11}$ | | $Mg_{51}Zn_{20}$ | |
| 295°C 384 hr. | - | - | - | - |
| 315°C 168 hr. | 38.0 [1.0] | $1.2 \times 10^{-15}$ | - | - |
| 325°C 120 hr. | 36.2 [1.1] | $1.5 \times 10^{-15}$ | 11.6 [0.8] | $1.6 \times 10^{-16}$ |

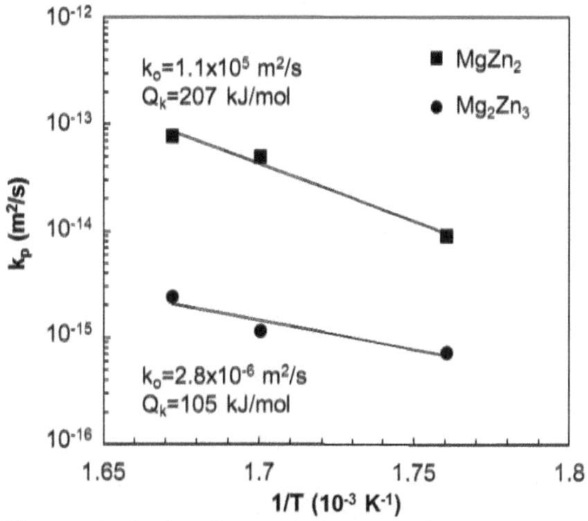


Figure 4: Arrhenius temperature-dependence of the parabolic growth constants for $MgZn_2$ and $Mg_2Zn_3$ intermetallic phases.

## Discussion

In all three of the diffusion couples studied, the MgZn phase was not observed and the $MgZn_2$ Laves phase was observed to grow the thickest layer. These two observations may be explained by some fundamental guidelines of growth and diffusion theory. In general, as shown by a number of previous studies [8-11], an intermetallic phase layer will grow thicker if:

- The phase has a wider homogeneity range,
- The phase has a higher diffusion coefficient than its surrounding phases,
- The phase has a higher heat of formation (or higher melting temperature),
- The crystal structures of the adjoining phases are similar.

Both the $Mg_2Zn_3$ and $MgZn_2$ phases are stoichiometric line compounds according to the equilibrium phase diagram given in Figure 1, therefore the first guideline is not applicable. Also, the crystal structures of each phase in this binary system are dissimilar. The $MgZn_2$ phase has a higher melting temperature and thus is possibly more thermodynamically favorable to form. Its larger thickness and strict following of the parabolic growth law suggests the growth of the phase is diffusion controlled, and therefore has a larger diffusion coefficient than its surrounding phases. The MgZn, on the other hand, not having grown to an observable level, should have a significantly lower diffusion coefficient than its surrounding phases.

The location of the marker plane, $x_m$, as denoted in Figure 3, may give insight into the intrinsic diffusion behavior of the elemental species in this binary system. The marker plane is located in the $Mg_2Zn_{11}$ phase, near the Zn side of the diffusion couple. This suggests Zn is the faster diffusing species in the system. It is somewhat typically observed that the species with the lower melting temperature diffuses faster. Zn melts at ~420°C and Mg melts at 650°C, concluding this observation holds true for this system. Zn is well known to have a high vapor pressure as well.

Zn additions to Mg alloys are mainly added for their solid solution strengthening and precipitation strengthening. The main precipitate in these age hardenable alloys is $MgZn_2$. The growth rates for this phase reported in this study may be used to determine the fractional amount of precipitates and time required to achieve the desired hardening response in these alloys systems.

## Summary

Three diffusion couples of pure Mg vs. pure Zn were assembled and annealed at 295, 315, and 325°C for 384, 168, and 120 hours, respectively. Intermetallic phase formation and growth for the binary system were observed. The $MgZn_2$, $Mg_2Zn_3$, and $Mg_2Zn_{11}$ phases were identified in all three couples as well as the $Mg_{51}Zn_{20}$ phase in the 325°C couple. Parabolic growth constants were determined for each phase at each temperature with $MgZn_2$ having the highest growth rates, followed by the $Mg_2Zn_3$ phase and the $Mg_2Zn_{11}$ phase. The activation energy for growth was calculated as 105 kJ/mole for the $Mg_2Zn_3$ phase and 207 kJ/mole for the $MgZn_2$ phase. Integrated interdiffusion coefficients for this study will be reported elsewhere.


## Acknowledgments

Research sponsored by the U. S. Department of Energy, Office of Energy Efficiency and Renewable Energy, Vehicle Technologies Program, under contract DE-AC05-00OR22725 with UT-Battelle, LLC.


## References


[1] Mordike, B., & Ebert, T. (2001). Magnesium Properties-applications-potential. *Mater. Sci. Eng. ,A* Vol. 302, 37-45.

[2] Luo, A. (2002). Magnesium: Current and potential automotive applications. *JOM,* Vol. 54 No. 2, 42-48.

[3] Kulekci, M. (2008). Magnesium and its alloys applications in automotive industry. *Int. J. Adv. Manuf. Tech.,* Vol. 39, 851-65.

[4] Lal, K. (1967). *CEA Report R* 3136, 54.

[5] Cermak, J. S. (2006). Diffusion of 65Zn in Mg and in Mg-xAl solid solutions. *Physica Status Solidi a,* Vol. 203, No. 10, 2386-2392.

[6] J.B. Clark, L. Z. (1988). Mg-Zn. In E. H. Baker, *ASM Metals Handbook Volume 3:*

*Alloy Phase Diagrams.* Metals Park, OH: ASM International.

[7] Philibert, J. (1991). *Atom Movements Diffusion and Mass Transport in Solids.* France: Les Editions de Phusique.

[8] Kidson, G. (1961). Some Aspects of the Growth of Diffusion Layers in Binary Systems. *J. Nucl. Mater.,* Vol. 3 No.1, 21-29.

[9] Gibbs, G. (1966). Diffusion Layer Growth in a Binary System. *J. Nucl. Mater.,* Vol. 20 No. 1, 303-306.

[10] Pretorius, R., Marais, T., & Theron, C. (1993). Thin film compound phase formation sequence: An effective heat of formation model. *Mater. Sci. Rep.,* Vol. 10, 1-83.

[11] Kajihara, M. (2004). Analysis of kinetice of reactive diffusion in a hypothetical binary system. *Acta Metall.,* Vol. 52, 1193-1200.

# Biaxial deformation behavior of AZ31 magnesium alloy at high temperatures

Daisuke Yamashita[1], Masafumi Noda[2], Kunio Funami[2]
[1]Major in Mechanical Science and Engineering, Chiba Institute of Technology,
2-17-1 Tsudanuma, Narashino, Chiba, 275-0016, Japan
[2] Department of Mechanical Science and Engineering, Chiba Institute of Technology,
2-17-1 Tsudanuma, Narashino, Chiba, 275-0016, Japan

Keywords: Magnesium alloys, Biaxial deformation, High temperature, Fine grain, Tensile twin

## Abstract

The evaluation of plastic deformation processing of materials has been based mainly on uniaxial tensile tests. However, for the analysis of multiaxial stress conditions in sheet-metal plastic forming, more complex tests, such as the controlled biaxial test with cruciform specimens, are required. In this report, we examine the relationship between the biaxial deformation behavior and microstructure of AZ31 magnesium alloy for initial strain rates of $3.0 \times 10^{-3}$ $s^{-1}$ and $2.8 \times 10^{-4}$ $s^{-1}$ at 573 K and 623 K using cruciform specimens. The results show that the flow stress increases with decreasing sheet thickness, and the occurrence of tensile twins is confirmed during biaxial tensile deformation at an elongation of 30 % in a cross-section of a specimen under 623 K for a strain rate of $2.8 \times 10^{-4}$ $s^{-1}$. This deformation behavior and evolution of the microstructure is unique in biaxial tensile tests, and is not recognized in uniaxial tensile tests with the same elongation. Since this controls the width direction of the specimen during deformation, we can only guess that the plate thickness direction can be deformed freely and without restraint. In other words, for biaxial tensile test conditions, the transition point of the microstructure under the influence of the constraint of grain boundary sliding is at 623 K for a strain rate of $2.8 \times 10^{-4}$ $s^{-1}$.

## Introduction

Magnesium alloys are the lightest metals in practical use with strong specific strengths. In recent years, these alloys have begun to find application in auto parts to address environmental issues and to achieve weight reduction of the vehicle [1]. However, because of the crystal structure of these alloys, only the basal plane of the slip system at room temperature, the material is known as a difficult work forming. Therefore, for the improvement of the strength and plastic workability of magnesium alloys, the structural refinement, texture control by thermomechanical treatment, and high-temperature deformation behavior of many slip systems are being actively studied [2-5]. However, evaluation of the strength, ductility, and plastic processes at room temperature and high temperatures was carried out under uniaxial stress. Considering the actual manufacturing process, the strength, ductility, and structural changes under biaxial stress should be examined [6-9]. Such analysis methods allow the plastic stress–strain components to be separated. Therefore, it is possible to create biaxial stress and separate the stress–strain components; thus, a biaxial tension test is required on cross-shaped specimens.

In this report, biaxial tensile tests were conducted at a high temperature using the magnesium alloy AZ31 by rolling processing and structural refinement MAF. We investigated the relationship between the mechanical properties and structural changes, and compared the results with those of plastic processing and the evaluation of uniaxial deformation.

## Experimental Procedures

### Material and structure refinement forming

The materials used were commercially extruded AZ31 Magnesium alloys. The chemical composition was as follows: Al 3.02, Zn 0.92, Mn 0.38, Fe 0.0035, Si 0.023, Mg bal (mass%). A cubic block of length 80 mm was cut out from the extruded AZ31 Mg alloy and subjected to MAF(Multiaxial Alternative Forging) processing at a high temperature. Figure 1 shows the schematic illustration of the working process for MAF [10]. In addition, this block was cut parallel to the extrusion direction to give a sheet with a thickness of 2 mm, and rolling was performed.
Subsequently, recrystallization annealing was carried out at 473 K for 1.2 ks to make the fine-grained structure.

In forging, the pressed sides of the cube were extruded sequentially from the surface. The forging temperatures were 573, 543, and 513 K, the press speed was 1 mm/min rolling speed, and the reduction ratio was 30%. The material was cooled with water after forging in each pass. The rolling conditions were a temperature of 493 K and holding for 0.6ks, and two - pass rolling was performed; the final sheet thickness was 1 mm. Herein, this material is referred to as fabricated material processing material processing rolling after MAF.

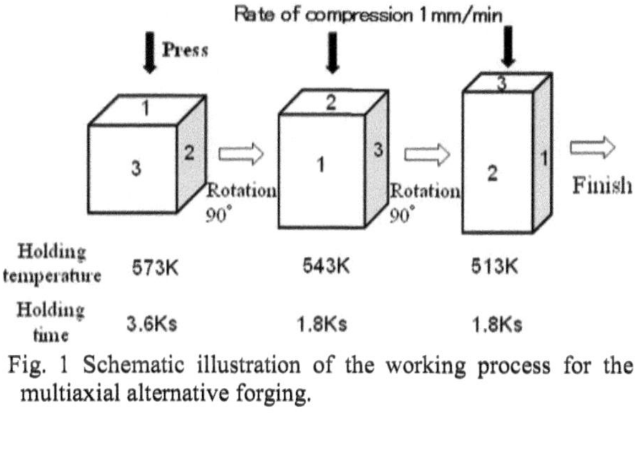

Fig. 1 Schematic illustration of the working process for the multiaxial alternative forging.

### Experiment method

The uniaxial tensile specimen consisted of a parallel portion with a width b of 6 mm, gauge length l of 6 mm, and thickness t of 1 mm. Figure 2 show the cruciform specimen geometry. The center of the cruciform shaped specimen represents, has been pocketing at 0.25 mm increments from each φ6 mm back. Thus, the observed thickness at the center of the test section is 0.5 mm, and the others are 1 mm.

The test temperature was 523 – 623 K, the strain rates were $2.8 \times 10^{-4}\ s^{-1}$ and $3.0 \times 10^{-3}\ s^{-1}$, and holding tests were performed under each set of test conditions 10 min after the set temperature was reached. The sample was heat - treated at 493 K for 1.2 ks prior to testing.

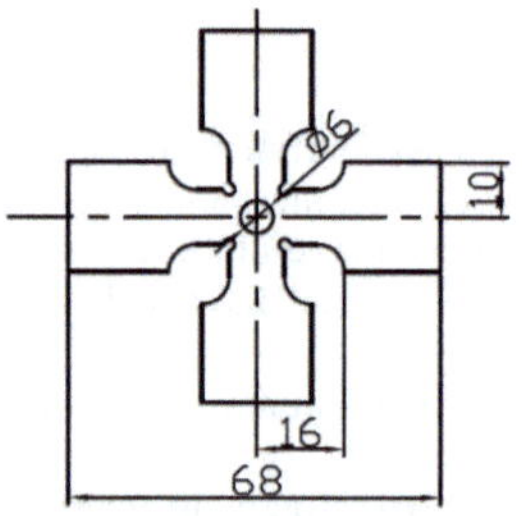


Fig.2 Shape and dimension of cruciform specimen.

## Results and Discussions

### Effect of pretreatment on structure

Figure 3 shows optical micrographs of the structure of the extruded AZ31 Mg alloy and the fabricated material (rolling after MAF processing). As shown in Figure 3, the extrusion material had an average grain size of 16 μm containing coarse grains, whereas in the rolled material, the structure showed a relatively uniform average particle size of 4 μm. This is due to the hot rolling and MAF processing, which causes dynamic recrystallization. Table 1 shows the mechanical properties of these materials at room temperature. For the extruded material, the proof stress showed a remarkable difference in the tension and compression sides, indicating a large anisotropy. In contrast, for the fabricated material, the difference in the proof stress according to direction was small, and the strength and growth improved.

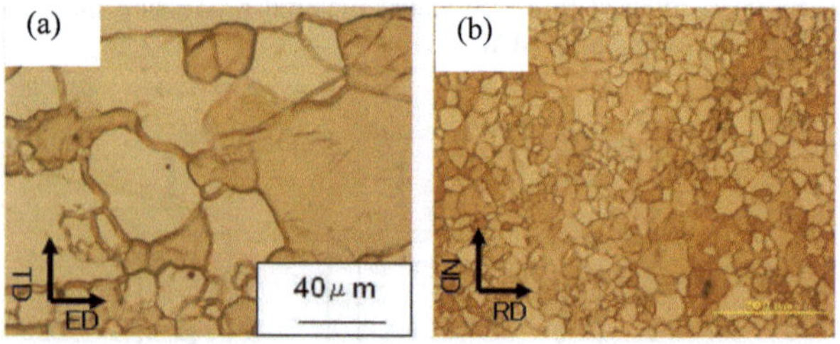


Fig. 3 Optical microstructure extruded material (a) and processing material by forging and heat-treatment (b) of AZ31 magnesium alloy.

Table 1 Mechanical properties of used materials at room temperature.

Extruded material

| Direction | Proof stress/MPa 0° | 45° | 90° | Tensile strength/MPa 0° | 45° | 90° | Nominal strain(%) 0° | 45° | 90° |
|---|---|---|---|---|---|---|---|---|---|
| | 180 | 80 | 60 | 255 | 277 | 229 | 24 | 31 | 24 |

MAF+Rolling material

| Direction | Proof stress/MPa 0° | 45° | 90° | Tensile strength/MPa 0° | 45° | 90° | Nominal strain(%) 0° | 45° | 90° |
|---|---|---|---|---|---|---|---|---|---|
| | 210 | 180 | 180 | 290 | 270 | 280 | 28 | 34 | 23 |

### Tensile properties at high temperatures

The nominal stress – strain diagram and the equivalent stress–strain diagram obtained from the biaxial and uniaxial tensile tests are shown in Figure 4.

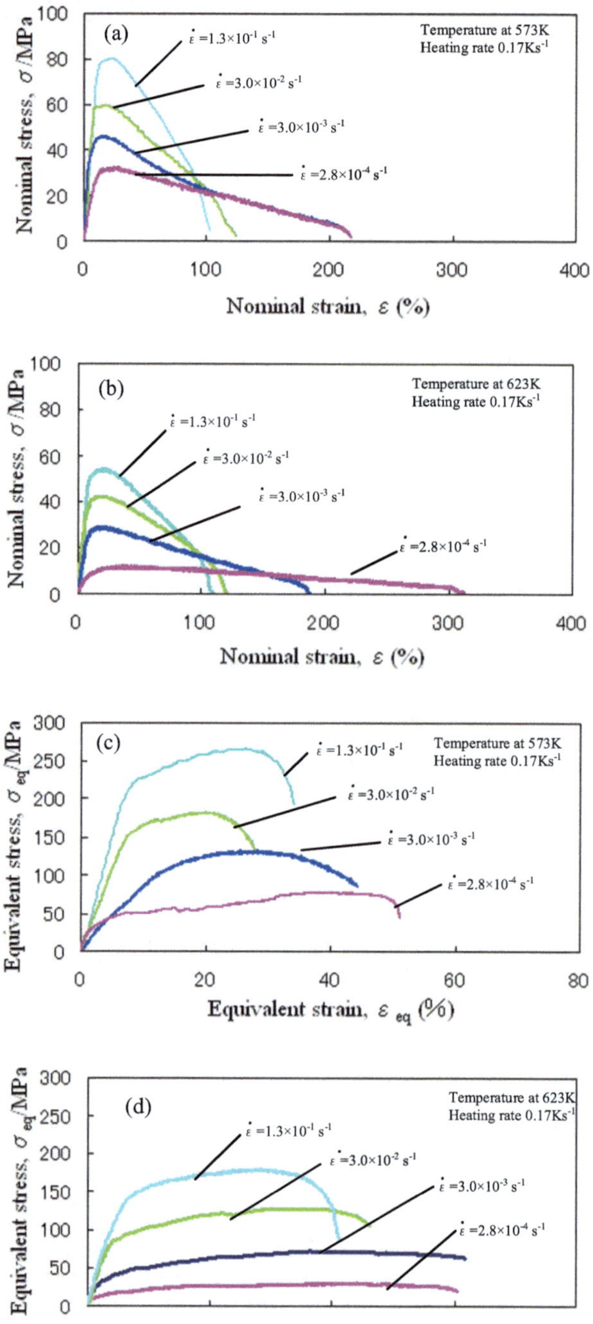


Fig. 4 Nominal stress - strain curves of uniaxial deformation at 573K (a) and 623K (b). And biaxial deformation at 573K (c) 623K (d).

A comparison of uniaxial and biaxial deformation shows that, in uniaxial deformation, after reaching the maximum stress, the stress decreases, and the fracture extends by more than 60%. In particular, at 623 K, the strain of $2.8 \times 10^{-4}\ s^{-1}$ has increased by more than 200%. In the biaxial deformation, it can be seen that the flow stress is higher than in uniaxial deformation. In addition, the biaxial deformation is lower compared to the elongation. Figure 5 shows the relationship between the maximum stress and the strain rate. The strain-rate sensitivity of the biaxial deformation indices tends to have a higher value than for the uniaxial deformation.

For a temperature of 623 K and a strain rate of $2.8 \times 10^{-4}\ s^{-1}$ , in uniaxial deformation, superplastic deformation was observed. However, despite the fact that the biaxial deformation had a value of 0.3, a large elongation was not obtained. This uniaxial deformation proceeds by a reduction in the width direction and thickness. In biaxial tensile deformation, the deformation is in the direction of the tensile stress, and although it acts from two directions it will decrease in the thickness direction only. Because of the poor plastic flow, the low elongation results in a higher flow stress.

The relationship between the strain rate and n value is shown in Figure 6. The n value in uniaxial deformation at a slow strain rate tends to be lower. At 623 K, with a strain rate of $2.8 \times 10^{-4}\ s^{-1}$ , it takes a higher value. Under these conditions, superplastic deformation occurs. Sliding occurs because of the large grain boundary, almost without having to act on the grain deformation, which indicates that the hardening values do not increase. In biaxial deformation, a strain rate of $3.0 \times 10^{-3}\ s^{-1}$ has a peak value of n.

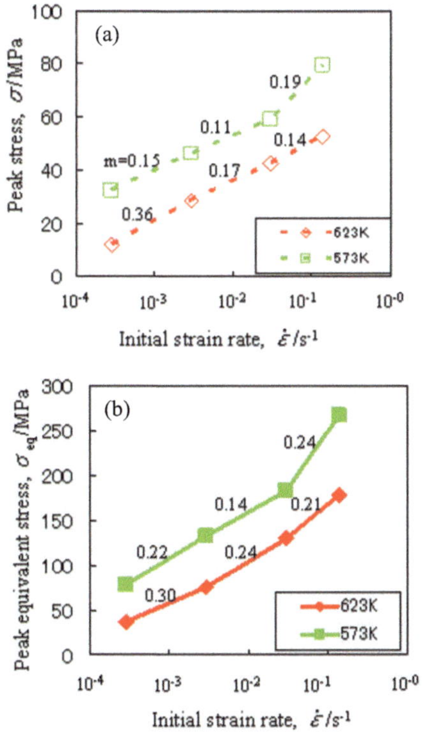


Fig. 5 Relation of peak stress and initial strain rate under uniaxial deformation (a) and biaxial deformation (b).

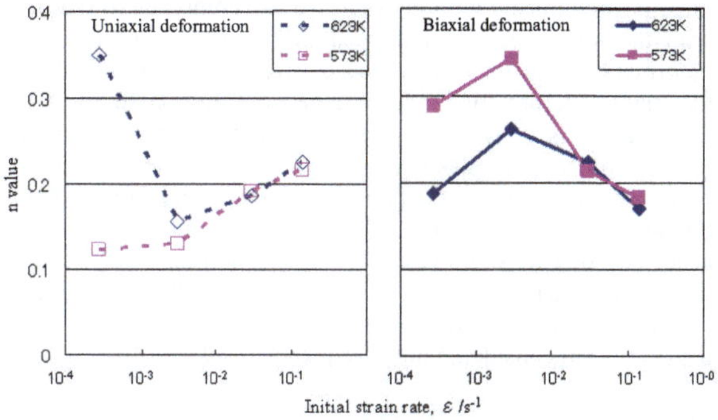


Fig. 6 Relationship between n value and initial strain rate under uniaxial deformation and biaxial deformation.

**Structure after uniaxial and biaxial deformation**

Figure 7 shows the SEM microstructure of the plate surface structure for an applied strain of 30% on a single axis. Uniaxial deformation is understood that a cavity is formed with biaxial deformation.

A comparison of the structures at a strain rate of $3.0 \times 10^{-3}\ s^{-1}$ and those at strain rate of $2.8 \times 10^{-4}\ s^{-1}$ reveals that a larger cavity is formed at a slower strain rate. In addition, the cavity grows in the direction perpendicular to the tensile direction. In biaxial deformation, cavity growth can be observed in many directions.

Deformation under biaxial tensile stress is applied in two directions subject to the grain - boundary sliding resistance caused by grain-boundary segregation.

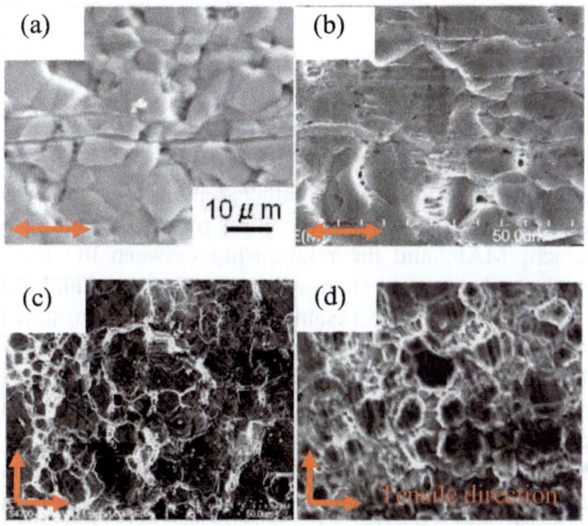


Fig. 7 SEM microstructure at 623K under uniaxial deformation for initial strain rate $3.0 \times 10^{-3}\ s^{-1}$ (a) and $2.8 \times 10^{-4}\ s^{-1}$ (b). Biaxial deformation $3.0 \times 10^{-3}\ s^{-1}$ (c) and $2.8 \times 10^{-4}\ s^{-1}$ (d).

The IPF maps from the EBSD analysis are shown in Figure 8. Figure 8 is an IPF map in cross-section from the results at 30% strain. Under uniaxial deformation, the twinning formation is not seen for any conditions. However, under biaxial deformation at a test temperature of 623 K, and strain rates of $3.0 \times 10^{-3}\ s^{-1}$ and $2.8 \times 10^{-4}\ s^{-1}$ , the formation of {10-12} twins occurred. Under uniaxial deformation, since the unconstrained plane tensile deformation can be reduced in the width and thickness directions, there is good plastic liquidity deformation compared to that found with the lower grain-boundary sliding resistance of the biaxial deformation, and the occurrence of twins is not observed. With the biaxial deformation, the surface contribution to the tensile deformation is constrained only by the reduction in thickness. Because of the poor plastic flow, it is believed that the formation of deformation twins in the grains increases with increasing grain-boundary sliding resistance. The conditions under which twins occur are: 623 K, strain rate $3.0 \times 10^{-3}$ $s^{-1}$and $2.8 \times 10^{-4}$ $s^{-1}$ . The incidence of twins is 27% at $3.0 \times 10^{-3}$ $s^{-1}$ and 30% at $2.8 \times 10^{-4}$ $s^{-1}$ . At this strain rate, the average grain size is larger. Therefore, the results are considered high-frequency results.

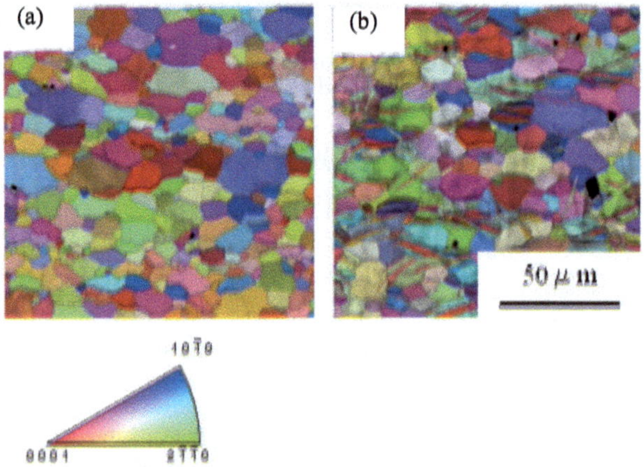


Fig. 8 IPF map of AZ31 magnesium at 623K. Under uniaxial deformation (a) and biaxial deformation (b) on initial strain rate 2.8 $\times 10^{-4}$ $s^{-1}$ .

## Conclusion

Biaxial tensile tests at were conducted at high temperatures using a magnesium alloy AZ31 by rolling processing and structural refinement MAF, and the relationship between the mechanical properties and structural changes were investigated and compared with those of plastic processing and uniaxial deformation. The following conclusions were obtained:

In biaxial deformation, for surface constraints, plastic flow is deteriorated only by the decrease in the thickness. The flow stress is high, and the growth is low.

In biaxial deformation, grain-boundary separation occurs much more in the direction of cavity growth, and slip deformation is constrained at the grain boundary.

In biaxial deformation, the formation of tensile twinning occurs only at 623 K for strain rates $3.0 \times 10^{-3}$ $s^{-1}$ and $2.8 \times 10^{-4}$ $s^{-1}$ .

## Acknowledgment

This part of examination of biaxial tensile test went by the furtherance for from Research Aid Fund 2011, The Japan Society for Technology of Plasticity.

## References

[1]S. Saikawa. JILM,Vol.60 No11, (2010), 571-577,(In Japanese).

[2]M. Kohzu., K. Kii, Y. Nsgsts, H. Nishio, K. Higashi, H. Inoue, Mate. Trans Vol.51, No.4,(2010),749-755.

[3]D.L.Yin, K.F.Zhang, G.F.Wang, W.B.Han, Mate. Let, 59, (2005), 1714-1718.

[4]X. Yang, H. Miura, T. Sakai, JILM, Vol.52, No.7, (2007). 318-323.

[5]J. Xing, X. Yang, H. Miura, T. Sakai, JILM, Vol.56, No.12, (2006), 716-720.

[6]Y. Tozawa, JILM, Vol.51, No.10, (2001), 492-497, (In Japanese).

[7]M. Noda, and K. Funami, Mater. Sci. Forum, 503-504 (2006), 475-480.

[8]E. Yukutake, J. Kaneko and M. Sugamata, JSTP, 44(2003), 82-86.

[9]F. K. Abu-Farha and M.K. Khraisheh, J. Eng. Mater. Tech, 127, (2005), 159-164.

[10] M. Noda, M. Hirohashi and K.Funami, Mater. Trans, 44, (2003), 2288-2297.

# Magnesium Technology 2012

## Processing-Microstructure-Property Relationships I

### Thursday AM

*Session Chairs:*

Hidetoshi Somekawa
(National Institute of Materials Science, Japan)
Kyu Cho
(U.S. Army Research Laboratory, USA)

**Magnesium Technology 2012**
*Edited by: Suveen N. Mathaudhu, Wim H. Sillekens, Neale R. Neelameggham, and Norbert Hort*
***TMS (The Minerals, Metals & Materials Society), 2012***

# MICROSTRUCTURE MODELING OF MAGNESIUM ALLOYS FOR ENGINEERING PROPERTY PREDICTION

Erin Barker[1], Dongsheng Li[1], Xin Sun[1], Mohammad Khaleel[1]
[1]Pacific Northwest National Laboratory, 902 Battelle Blvd, Richland, WA, 99352, USA
Keywords: Mg alloys, microstructure, porosity, casting

## Abstract

Magnesium alloys have found increasing application in the transportation industry due to their low weight and high strength. However, wider application is hindered by limited ductility. Microstructural features, such as porosity, brittle eutectics, and grain size, can significantly influence the macroscopic response of a component. These features can vary widely throughout a component. Our approach to studying the microstructures influence on bulk properties begins with measuring microstructural features in different regions of a component. These measurements are used to create statistically equivalent, 3D synthetic samples of the microstructure. The synthetic microstructures are meshed using finite elements and used to simulate the response and investigate the influence of specific features. We will demonstrate how the digital microstructure samples are generated, how variations in microstructural features influence the bulk properties, and how this methodology can be used to predict component performance and optimize processing.

## Introduction

Magnesium alloys are finding increasing applications in lightweight vehicles because Mg and its alloys are the lightest metallic structural materials. High strength and low density make Mg alloys prospective alternatives to aluminum alloys. However, Mg's limited ductility presents a critical hurdle to more wide spread use. It is well established that microstructural features such as amount and distribution of porosity, grain size, and brittle eutectic phase and precipitates can significantly influence the ductile of Mg alloys. Unfortunately, these features vary spatially throughout a casting and from specific alloy to alloys. The specific casting process used and the processing parameters can also influence the final microstructure.

The characteristic microstructure of high-pressure die-cast (HPDC) AM series alloys consists of α-Mg dendrite cells and a divorce eutectic of α-Mg and $\beta$-$Mg_{17}Al_{12}$ [1,2]. Intermetallic Mn-rich particles are also observed. Some of the α-Mg dendrite cells found in the center of a casting prematurely solidified in the show sleeve and carried into the die. This results in a bimodal distribution of dendrite cells sizes between those formed in the sleeve and in the die [2].

The microstructure closer to the die walls consists of a very finer dendrite cell size, approximately 5-7 μm, with little to no porosity observed. This "skin" region results from higher cooling rates and exhibits higher strength and ductility compared to the interior [3,4]. The dendrite cells in the interior are 10s of μm in diameter depending on processing parameters with externally solidified cells being the largest.

A third region has been observed in some castings. A porosity segregation zone (PSZ) is sometimes observed between the skin and interior regions consisting of one or more bands of high porosity that follow the contours of the die [1,2,5]. The PSZ has a negative effect on the strength and ductility of the casting and is thought to be linked to processing parameters. Dahle et al. have developed a model for prediction when and where the PSZ forms based on casting parameters [6].

The Al content of Mg alloys leads to changes in the microstructure morphology. Dahle et al. [5] shows that low Al content alloys, i.e. 1% by weight, have globular, equiaxed dendrite cells. Increasing the Al content to 5%, results in pools of eutectic beginning to form between the dendrite cells. By 9\% Al, the dendrite structure is fully formed with eutectic between the dendrite arms. Even the slight increase in Al from AM50 to AM60 results in an increased volume fraction of the β-phase.

Porosity is also observed in HPDC Mg alloys. The bulk porosity level is often reported and used to model ductility and fracture. However, Weiler et al. [7] and Lee et al. [8,9] have suggested that the areal fraction of pores along the fracture surface and largest pore volume mater more than the bulk value. Weiler et al. utilize a critical local strain model [10-12] with the areal fraction of porosity to model plastic instability resulting from internal necking due to the reduced cross section area. This was used to calculate the fracture strain. This method requires testing samples to calibrate the material parameters used to describe the bulk response of the Mg alloy as well as x-ray imaging to determine the porosity distribution and largest pore volume.

Chawla et al. [13] have also explored how to include microstructural features in computational models. Serial section of 2070 aluminum composite containing SiC particles was conducted. A 3D geometric representation and finite element mesh was created from the series of images. The resulting model included the irregular shapes of the SiC particles. Commonly used simplifications for the SiC particles are a single cube or sphere in a continuum matrix. The complex geometric representation was compared to a single cube and a single sphere. The single sphere was shown to under predict the stress-strain response while the cube slightly over predicted the hardening response. The complex geometry matched the experimental response the closest and demonstrated the internal stress concentrations due to the sharp features of the SiC particles. This demonstrates the importance of actually representing the microstructure to capture the response of the material. However, having a single geometric representation is limiting and using serial sectioning to generating each new model is expensive and time consuming.

The various features of the AM series microstructure can be generally categorized into two types: intrinsic and extrinsic. Intrinsic factors include features intrinsic to the specific Mg alloy such as phase composition, grain size, morphology, volume fraction, and mechanical properties of the α-Mg matrix and eutectic β phase. Extrinsic factors come from the external processes applied to the alloy such as casting and heat treatment processes and include: porosity, segregation, incomplete fill, hot tear, and cold shut. Combing a particular alloy and casting process result in a specific set of intrinsic and extrinsic features and in some cases flaws.

Computational analyses have been carried out to model the effect of extrinsic features such as porosity on the bulk ductility. However, previous work has focused on approximating the presence of pores and not on how the distribution of location and size influence the bulk response. Experimental observations have demonstrated that the amount of porosity as well as size and distribution vary throughout a casting and from casting to casting. The following work will outline a computational framework for samples containing random distributions of porosity for finite element analysis. First, the process for generating samples that are statistical equivalent to experimental observations is described. Next, the material models being utilized are described. Finally, results from preliminary finite element simulations are presented.

### Generating Statistically Representative Samples

The exact microstructure of each casting is unique. Variation exists not only from sample to sample but throughout a single casting. As a result, reconstructing a microstructure from a single image from a single sample or even a handful of images from a single or multiple samples is extremely limiting. To combat this limitation a framework is created to generate unique representations that are statically equivalent to the target microstructure.

The current focus is how porosity affects the overall ductility of a sample. Weiler et al. [8] characterized the porosity of a series of samples cut from different regions of an AM60B high-pressure die-casting. Data collected includes number of pores, average pore volume, largest pore, volumetric porosity, density of pores, and average distance to the nearest pores. This information is used to generate a series of random samples containing pores that demonstrate the same parameters.

Random samples of porosity are generated using the volumetric porosity and average pore volume as input parameters. Individual pores are randomly placed inside a bounding box and the porosity is calculated. This is repeated until the required porosity is reached. Pores are represented by ellipsoids. An ellipsoid is described by 2 equatorial radii a and b and a polar radius c. If a=b=c, the ellipsoid simplifies to a sphere. For simplicity, the equatorial radii are defined to be equal, meaning a=b. This leaves a and c to define each pore. The average radii are calculated to result in the average pore volume. For each pore added to the sample each radius is allowed to vary +/- 50% of the average radius value, resulting in various sized pores. If a pore intersects the bounding box or an existing pore, the pore is rejected. Once a sample is generated the porosity, average pore volume, and number of pores is reported and can be compared to the experimental observations. Figure 1 shows two samples generated using the same porosity of 5% and external dimensions of 3mm x 6mm x 25mm. Sample 1 contains 20 pores with an average pore volume of 0.18 mm3 while Sample 2 contains 48 pores with an average pore volume of 0.075 $mm^3$.

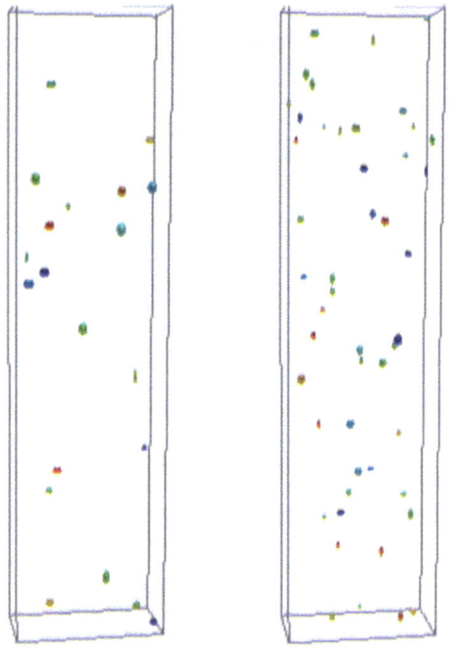

(a) (b)

Figure 1: Randomly generated samples with 5% porosity. (a) Sample 1 contains 20 pores. (b) Samples 2 contains 48 pores.

The geometries generated can then be automatically meshed for finite element analysis using tetrahedral elements. The result meshes conform to the pore surfaces. The interior of the pores is not meshed leaving voids.

### Material Modeling

For the current analyses porosity is represented explicitly in the geometry while the details of the rest of the Mg microstructure are smeared out. The goal is to understand the influence of porosity parameters on the bulk ductility. Computational simulation provides the ability to isolate individual features for study. Future analyses will increase in complexity as additional features of the microstructure are added to the geometry.

$$\sigma = \sigma_{yld} + K\varepsilon^n_p \quad (1)$$

The Mg matrix is modeled using elasto-plasticity with isotropic hardening. The plastic flow stress follows the exponential form of Eqn. 1.

$\sigma_{yld}$ represents the initial yield stress, $\varepsilon_p$ is the plastic strain, K is the hardening coefficient, and n is the strain-hardening exponent. Standard values for Young's modulus, Poisson's ratio, and yield stress of 45 GPa, 0.25, and 125 MPa, respectively, are taken from available literature. The values of K and n reported by Shan et al. [4] of 467 MPa and 0.22 are also used.

### Modeling Results

Using the material parameters in the previous section the samples shown in Figure 1 were subjected to simple tension was applied to each sample. The pores in sample 1 are primarily clustered in the upper portion of the sample while the pores in sample 2 are more uniformly distributed throughout the sample. As seen in Figure 2, the clustering results in a clear stress concentration region where failure will occur while the more uniform distribution lacks a clear single path for failure.

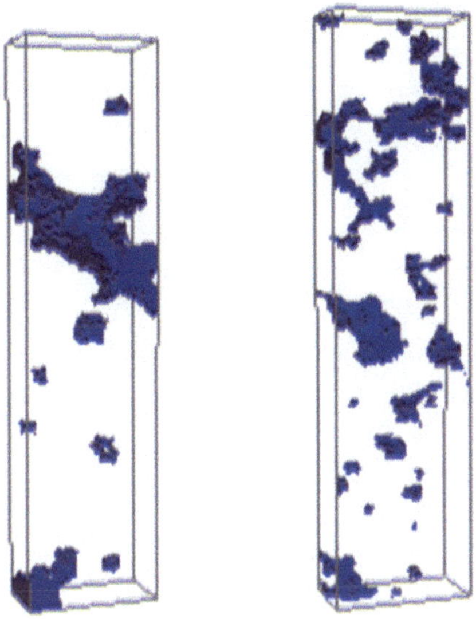

(a) (b)

Figure 2: Location of failure stresses in samples containing 5% porosity.

## Summary

Mg alloys have many intrinsic and extrinsic microstructural features that contribute to the bulk response of a casting. How these features effect the ductility is keep to improving the performance for lightweight automotive applications. Augmenting traditional experimental testing with computational simulation has been useful for isolating individual features and physical responses. Here we have outlined the process for generating a suite of unique digital samples for computational analysis focusing on specific microstructural features. This allows the variation seen in different regions of a casting and between castings to be captured. The effect of variation from sample to sample can be studied to determine the impact on the range of resulting property values.

## Acknowledgments

Pacific Northwest National Laboratory is operated by Battelle Memorial Institute for the US Department of Energy under Contract No. DE-AC05-76RL01830. This work was funded by the Department of Energy Office of Energy Efficiency and Renewable Energy.

## References

1. J. Song, S.M. Xiong, M. Li, and J. Allison, "The Correlation Between Microstructure and Mechanical Properties of High-Pressure Die-Cast AM50 Alloy," *Journal of Alloys and Compounds*, 477 (2009), 863-869.

2. G. Chadha, J.E. Allison and J.W. Jones, "The Role of Microstructure on Ductility of Die-Cast AM50 and AM60 Magnesium Alloys," *Metallurgical and Materials Transactions A*, 38A(2) (2007), 286-297.

3. Z. Shan and A.M. Gokhale, "Utility of Micro-Indentation Technique for Characterization of the Constitutive Behavior of Skin and Interior Microstructures of Die-Cast Magnesium Alloys," *Materials Science and Engineering*, A361 (2003), 267-274.

4. H.Hu, M.Zhou, Z. Sun, and N. Li, "Tensile Behaviour and Fracture Characteristics of Die Cast Magnesium Alloy AM50," *Journal of Materials Processing Technology*, 201 (2008), 362-368.

5. A.K. Dahle, Y.C. Lee, M.D. Nave, P.L. Schaffer and D.H. St. John, "Development of the As-Cast Microstructure in Magnesium-Aluminium Alloys," *Journal of Light Metals*, 1 (2001), 61-72.

6. A.K. Dahle and D.H. St. John, "Rheological Behaviour of the Mushy Zone and its Effect on the Formation of Casting Defects During Solidification," *Acta Materialia*, 47(1) (1999), 31-41.

7. J.P. Weiler, J.T. Wood, R.J. Klassen, E. Maire, R. Berkmortel and G. Wang, "Relationship Between Internal Porosity and Fracture Strength of Die-Cast Magnesium AM60B Alloy," *Materials Science and Engineering*, A395, 2005, 315-322.

8. S.G. Lee, G.R. Patel, A.M. Gokhale, A. Sreeranganathan and M.F. Horstemeyer, Quantitative Fractographic "Analysis of Variability in the Tensile Ductility of High-Pressure Die-Cast AE44 Mg-alloy," *Materials Science and Engineering*, A427 (2006), 255-262.

9. S.G. Lee, G.R. Patel, A.M. Gokhale, A. Sreeranganathan and M.F. Horstemeyer, "Variability in the Tensile Ductility of High-Pressure Die-Cast AM50 Mg-alloy," *Scripta Materialia*, 53 (2005), 851-856.

10. C.H. Cáceres, "On the Effect of Macroporosity on the Tensile Properties of the Al-7%Si-0.4% Mg Casting Alloy," *Scripta Materialia et Materialia*, 32(11) (1995), 1851-4856.

11. C.H. Cáceres and B.I. Selling, "Casting Defects and the Tensile Properties of an Al-Si-Mg Alloy," *Materials Science and Engineering*, A220 (1996), 109-116.

12. A.K. Ghosh, "Tensile Instability and Necking in Materials with Strain Hardening and Strain-Rate Hardening," *Acta Materialia*, 25, (1977), 1413-1424.

13. N. Chawla, V.V Ganesh and B. Wunsch, "Three-Dimensional (3D) Microstructure Visualization and Finite Element Modeling of the Mechanical Behavior of SiC Particle Reinforced Aluminum Composites," *Scripta Materialia*, 52 (2004), 161-165.

**Magnesium Technology 2012**
*Edited by: Suveen N. Mathaudhu, Wim H. Sillekens, Neale R. Neelameggham, and Norbert Hort*
***TMS (The Minerals, Metals & Materials Society), 2012***

# MICROSTRUCTURE MODIFICATION AND DEFORMATION BEHAVIOR OF FINE GRAINED AZ61L SHEET PRODUCED BY THIXOMOLDING® AND THERMOMECHANICAL PROCESSING

T.D. Berman[1], W. Donlon[1], V.M. Miller[3], R. Decker[2], J. Huang[2], T.M. Pollock[3], J.W. Jones[1]

[1]Materials Science and Engineering, University of Michigan, 2300 Hayward St; Ann Arbor, MI, 48109, USA
[2]nanoMAG, LLC., 13753 Otterson Court; Livonia, MI, 48150, USA
[3]Materials Department, University of California, Santa Barbara; Santa Barbara, CA, 93106-5050, USA

Keywords: AZ61L, Microstructures, Texture, Mechanical Properties, Thixomolding, Thermomechanical Processing

## Abstract

X-ray diffraction and hardness measurements are used to study recrystallization in fine-grained AZ61L sheet produced by warm-rolling of Thixomolded® material. The as-rolled sheet is partially dynamically-recrystallized, with a strong basal texture and a sub-micron grain size. Significant increases in ductility with moderate reductions in tensile strength were produced by annealing at temperatures greater than 250 °C. A weakening in basal texture was observed in samples annealed at over 250 °C. Static recrystallization was determined to be responsible for the reduction in texture and associated increase in elongation.

## Introduction

Lightweighting requires engineering high strength magnesium sheet with good formability; the limited formability is the most significant hindrance to the implementation of Mg-sheet. A strong basal texture exists in Mg-alloy sheet due to the preponderance of deformation by slip along the close-packed basal planes [1]. This basal sheet texture is unfavorable for through-thickness deformation. Weakly textured or non-textured sheet is desirable. In addition to texture modification, grain size refinement is also an essential mechanism to improve both the strength and ductility of Mg-alloys [2, 3].

The fine grain structure of Thixomolded Mg-alloys is advantageous for thermomechanical processing to finer grain sizes. Sheet produced by Thixomoldeding and Thermomechanical Processing (TTMP) has show strength and ductility in excess of 300 MPa and 6% respectively [4]. Post-processing annealing of TTMP material has produced sheet with lower strength ($\sim$ 230 MPa) but ductilities near 20% [4]. The strong basal texture imparted by rolling has been alleviated in post-processing annealing [4].

## Experimental

Preheated Thixomolded plates, with dimensions of 200 mm x 200 mm x 3 mm, were warm-rolled with a roll temperature near 200 °C to a thickness reduction of approximately 50%. Table 1 gives the composition of the AZ61L plates. Dogbone tensile specimens with a gauge length of 31.75 mm and cross section of 7.94 mm, were machined from the sheets with the tensile axes parallel to the rolling direction. Room temperature tensile tests were performed with a displacement rate of 0.71 mm/min. An extensonometer was used to measure tensile elongation. Vickers hardness measurements were taken with a Clark Microhardness tester with a dwell time of 15 s and a 200 g load.

All microscopy and texture measurements examine the polished surface of the sheet. SEM and XRD samples were prepared by polishing to 1 $\mu$m diamond paste and then etching for 10 s at room temperature in a 1:9 solution of o-phosphoric acid in ethanol. TEM specimens were prepared by dimpling and ion milling using a Gatan Precision Ion Polishing System (PIPS). TEM was conducted with a Phillips CM12 AEM system. Electron Backscatter Diffraction (EBSD) specimens were prepared by polishing to 1 $\mu$m diamond paste followed by ion polishing in a Gatan PIPS. EBSD examination was conducted with a Philips XL30 FEG equipped with a TexSEM Laboratories OIM system. A step size of 0.25 $\mu$m was used for EBSD measurements. A Rigaku rotating anode XRD system was used to make XRD pole figure measurements.

**Table 1:** Composition of received AZ61L in wt %

| Al | Zn | Mn | Si | Fe | Mg |
|---|---|---|---|---|---|
| 6.5 | 0.46 | 0.14 | 0.01 | 0.003 | bal. |

**Table 2:** Tensile properties of annealed sheet

| Sample | YS (MPa) | UTS (MPa) | El (%) |
|---|---|---|---|
| as-Rolled | 311.2 ± 15.5 | 369.6 ± 13.3 | 8.3 ± 2.6 |
| TTMP + 250 °C 3 min Anneal | 331.5 ± 6.4 | 372.3 ± 5.4 | 5.5 ± 2.4 |
| TTMP + 250 °C 10 min Anneal | 313.0 ± 17.2 | 365.3 ± 8.0 | 5.6 ± 3.4 |
| TTMP + 250 °C 20 min Anneal | 321.7 ± 3.1 | 361.3 ± 2.3 | 7.2 ± 1.4 |
| TTMP + 275 °C 15 min Anneal | 231.7 ± 1.7 | 306.6 ± 4.4 | 13.8 ± 4.0 |
| TTMP + 300 °C 3 min Anneal | 236.7 ± 4.1 | 315.3 ± 1.3 | 15.5 ± 2.1 |
| TTMP + 300 °C 5 min Anneal | 266.2 ± 5.6 | 336.0 ± 4.4 | 14.1 ± 3.1 |
| TTMP + 300 °C 10 min Anneal | 223.6 ± 1.2 | 308.5 ± 4.8 | 16.6 ± 3.0 |
| TTMP + 300 °C 20 min Anneal | 216.5 ± 5.7 | 307.6 ± 1.4 | 22.5 ± 1.5 |
| TTMP + 300 °C 30 min Anneal | 203.7 ± 6.5 | 306.8 ± 1.8 | 19.7 ± 4.3 |

## Results

### Mechanical Properties

Vickers microhardness measurements (Figures 1 and 2) illustrate that softening occurs when annealing above 250 °C. The as-rolled sheet has a Vickers microhardness measurement of 98 ± 3 HV. Coupons subjected to thermal exposures at 130, 170, and 210 °C exhibit a small drop in hardness following five minutes of annealing. After a few hours at these low temperatures, the hardness stabilizes at approximately 94 HV.

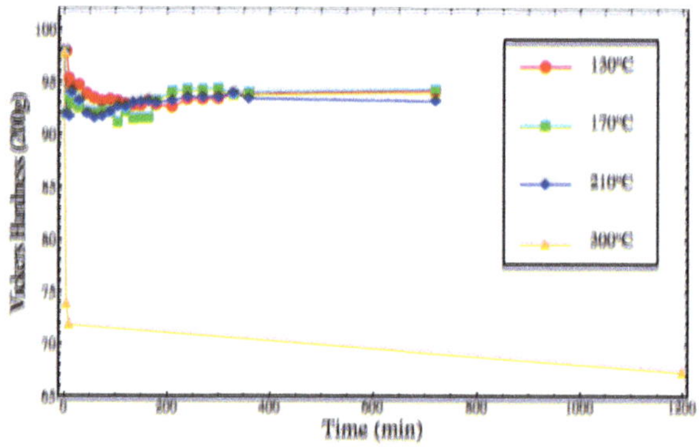


**Figure 1:** Isothermal microhardness measurements of annealed sheet. When going from 210 °C to 300 °C a significant difference in behavior is seen. To improve readability, error bars are not shown; the maximum error is ± 3.1 HV.

A different behavior is seen at annealing temperatures above 250 °C . Ten minute annealing treatments reveal a precipitous decay between the maximum and minimum hardness. The drop in hardness after annealing for 5 minutes at 300 °C is significant, especially when compared to the lower temperature anneals. After annealing at 300 °C for 20 h, the longest annealing treatment considered in this study, the hardness has dropped 30 points to 68 ± 3 HV.

A similar temperature effect is seen in the strength measurements (Table 2). The tensile properties of the samples annealed at 250 °C for 3, 10, or 20 minutes show little change from the as-rolled condition. Samples subjected to annealing above this temperature lose about 90 MPa in yield strength. Elongation of the samples annealed above 250 °C doubles.

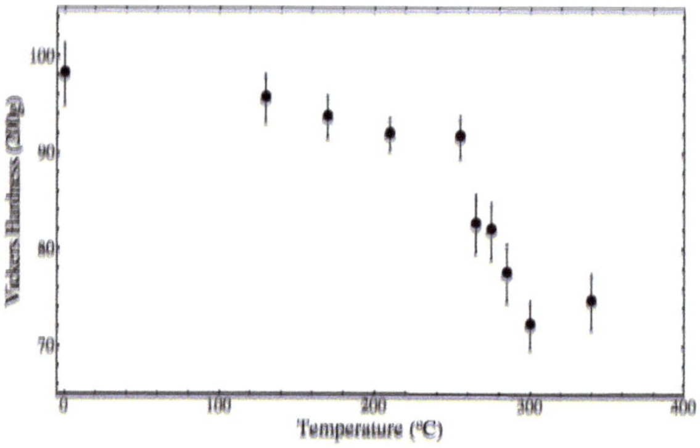


**Figure 2:** Microhardness measurements of sheet annealed for 10 minutes.

### Microstructure

The Thixomolded microstructure consists of $\alpha$-Mg grains with an average diameter of 4 $\mu$m surrounded by a divorced eutectic of $\beta$-$Mg_{17}Al_{12}$ (Figure 3). The $\beta$ particles are heterogeneously distributed throughout the plate, with a volume fraction of approximately 3%. The as-molded material is untextured.

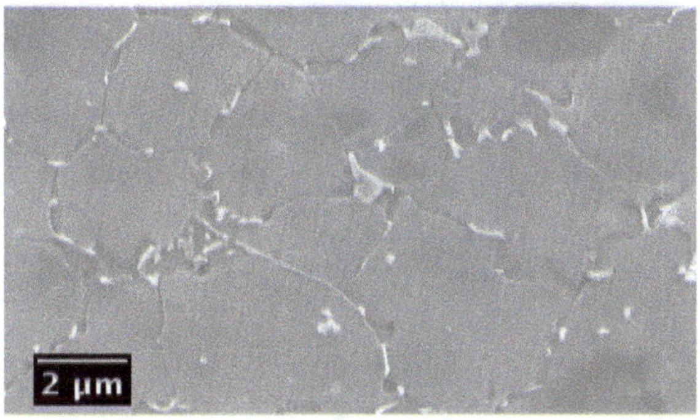


**Figure 3:** SEM image of Thixomolded AZ61L microstructure. The bright particles are the $\beta$ phase.

Partial dynamic recrystallization occurs during the rolling process, as shown in the inverse pole figure map displayed in Figure 4. The fraction recrystallized, roughly estimated by the fraction of points indexed as Mg with a confidence index greater than 0.5%, is 17%. The recrystallized regions have a sub-micron grain size (Figure 5). The $\beta$ particles in the as-rolled sample are smaller than those observed in the as-molded alloy. Coherent lathe $\beta$ precipitates are seen in some regions, these precipitates have a Burgers orientation relationship, $[1\bar{2}10]_{Mg}//[1\bar{1}1]_{\beta}$ (Figure 6). The overall spatial distribution of $\beta$ particles remains largely unchanged from that seen in the as-molded condition.

Crystallographic Texture

The macroscopic as-rolled sheet texture, as measured by XRD, shows a double-peaked basal texture with a maximum intensity of 3 multiples of random distribution (MRD) (Figure 7). The distribution spread is wider along the transverse direction. This double peak is likely due increased activity of the non-basal ⟨c+a⟩ slip mode [5, 6].

XRD basal pole figures of annealed samples illustrate weakening of the basal texture at annealing temperatures above 250 °C. The texture remains unchanged after annealing at 210 °C for 12 hours. In contrast, a large change in texture can be seen after 5 minutes at 300 °C. Further weakening of the basal texture occurs after annealing for 35 minutes at 300 °C. The double-peak is no longer present in this annealing condition.

## Discussion

The hardness and tensile measurements indicate that static recrystallization is occurring during the annealing process. The kinetics of the process seem to match well

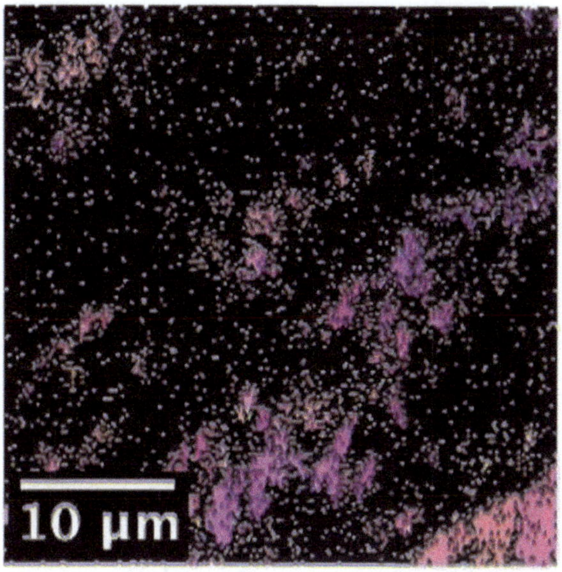


**Figure 4:** Inverse pole figure map of as-rolled sheet. Most of the sheet is not recrystallized, and thus not indexable by the EBSD technique.

with those of the Johnson-Mehl-Avrami-Kolmogorov (JMAK) theory. The fraction recrystallized can be approximated as the fraction softened [7, 8], defined as:

$$X = \frac{H_{AR} - H}{H_{AR} - H_{min}} \quad (1)$$

Here $H_{AR}$ is the hardness of the as-rolled sheet and $H_{min}$ is the hardness of completely recrystallized material. Here $H_{min}$ is approximated as 72 HV, the average hardness of the four highest temperature annealing treatments. This data was then fit with the JMAK model expressed below [8, 9].

$$X = 1 - exp\left(-0.693\left(t/t_{0.5}\right)^n\right) \quad (2)$$

where $t_{0.5}$ is the time to 50% recrystallization, which is calculated using Equation 3 [8].

$$t_{0.5} = BZ^r exp\left(Q_{rex}/RT\right) \quad (3)$$

Z is the Zener-Holloman parameter:

$$Z = \dot{\epsilon} exp(Q_d/RT_d) \quad (4)$$

R is the ideal gas constant (8.314 $JK^{-1}mol^{-1}$). The activation energy for self-diffusion, 135 $kJmol^{-1}$, is used to approximate, $Q_d$, the activation energy for deformation. $T_d$, the temperature of deformation, was approximately 300 °C. The expressions for $t_{0.5}$ and Z were substituted into Equation 2 and fit to our experimental

**Figure 5:** Microstructure of as-rolled sheet. This particular region exhibits a sub-micron dynamically recrystallized grain size.

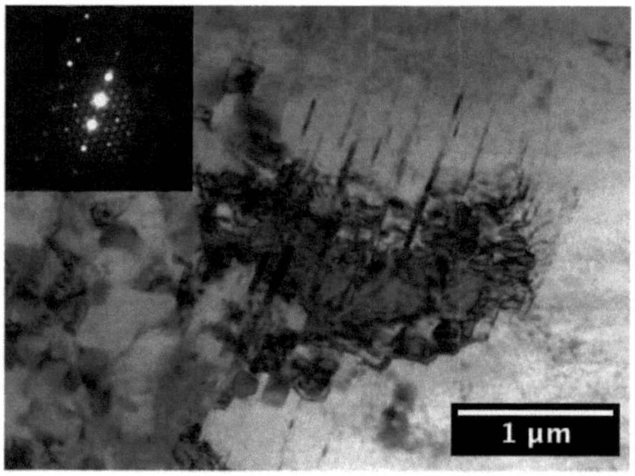


**Figure 6:** Coherent lathe-shaped $\beta$ precipitates in the as-rolled material. The selected area diffraction pattern indicates a Burgers orientation relationship.

**Table 3:** Parameters used for JMAK model

| Parameter | Value |
|---|---|
| n | 0.94 |
| $Q_{rex}$ | 137 kJ$mol^{-1}$ |
| B | 2.68 |
| r | -0.87 |

data for the fraction recrystallized (Figure 8). A non-linear regression algorithm was used to solve for n (the Avrami exponent), $Q_{rex}$ (the activation energy for recrystallization), and the fitting constants r and B. The best fit parameters are presented in Table 3.

An Avrami exponent of 0.94 is slightly lower than what has been calculated in deformed AZ-series alloys in other studies; reported Avrami exponents range between 1.18 and 1.38 [8, 10, 11]. As the Avrami exponent models the slope during the recrystallization process, more hardness measurements near 250 °C could indicate that a steeper slope, and thus slightly higher value of n is more appropriate. Regardless, we observe an Avrami exponent are lower than the value of n=4 which is expected if the assumptions of the model for continuous nucleation and spherical growth are satisfied [12]. This low exponent may be due to non-random nucleation in the deformed material, for instance an exponent of 1 has been found when modeling nucleation of random sites on grain boundaries [13]. Clusters of $\beta$ particles along grain boundaries may be preferential nucleation sites.

Particle stimulated nucleation (PSN) would also help to explain the texture weakening indicated in the XRD pole figures. It is generally believed that static recrystallization does not appreciably alter the deformation texture [14, 15]. Limited twinning has been observed in this fine grained material, so that mechanism in texture reduction is unlikely. Dense clusters of $\beta$ particles, produced in the warm-rolling process may be responsible for nucleating new, non-basal oriented grains. Future work will explore this further.

The reduction of basal texture leads to an improvement in the tensile elongation along the rolling direction. As annealing at 300 °C progresses, and the basal texture weakens, the elongation increases (Table 2). After 3 minutes, at which point our model predicts the material is ~60% recrystallized, the elongation increases from 8% to 16%. At 20 minutes the elongation reaches 23%. No further improvement of elongation is seen by increasing the duration of annealing. This is expected if the improvement is indeed due to texture evolution during recrystallization, as our model predicts that recrystallization will be complete by 20 minutes at 300 °C.

## Conclusions

TTMP AZ61 sheet is only partially dynamically recrystallized, with a sub-micron grain size. Static recrystallization occurs during the annealing of the as-rolled sheet. A simple empirical model, based on JMAK theory, can be used to describe the recrystallization process. A low Avrami exponent of 0.94 was fit using this model, indicating non-random nucleation. XRD pole figure measurements reveal a weakening of the as-rolled near-basal texture during the static recrystalliza-

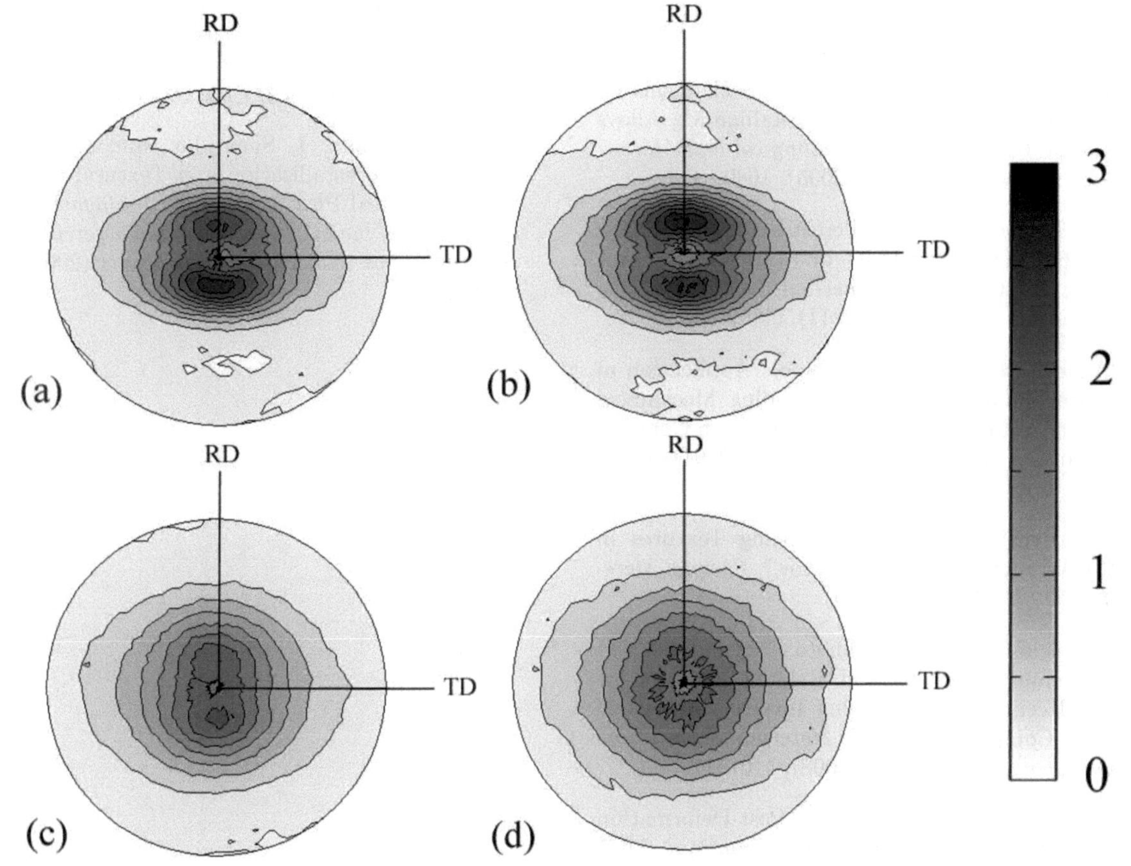


**Figure 7:** XRD (0002) pole figures of (a) as-rolled sheet, (b) sheet annealed for 12 hours at 210 °C, (c) sheet annealed for 5 minutes at 300 °C and (d) sheet annealed for 20 hours at 300 °C.

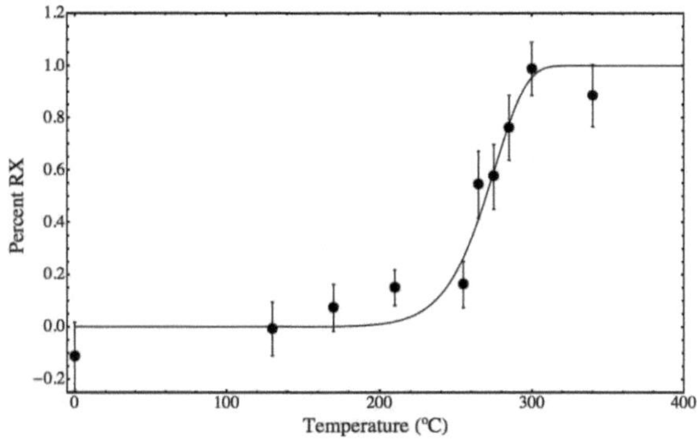


**Figure 8:** JMAK model for fraction recrystallized fit to fraction softening data. Isochronal curve for 10 minute anneal.

tion process. Ductility in the sheet improves as the texture is weakened. An understanding for the mechanism behind this texture randomization is needed to allow for better forming of Mg-alloy sheet.

### Acknowledgements

This material is based upon work supported by the National Science Foundation under Grant No. 0847198. Many thanks to Garret Huff and Eric Hung for their assistance on this project.

### References

1. Y. Wang and J. Huang, "Texture Analysis in Hexagonal Materials," *Materials Chemistry and Physics*, 81 (1) (2003), 11–26.

2. M. Barnett, D. L. Atwell, and A. Beer, "Grain Size in Mg Alloys : Recrystallization and Mechanical

Consequences," *Materials Science Forum*, 558-559 (2007), 433–440.

3. J. Koike, "Enhanced Deformation Mechanisms by Anisotropic Plasticity in Polycrystalline Mg Alloys at Room Temperature," *Metallurgical and Materials Transactions A*, 36 (7) (2005), 1689–1696.

4. T. D. Berman et al., "Microstructure Evolution in AZ61L During TTMP and Subsequent Annealing Treatments," *Magnesium Technology*, eds. W. Sillekens et al. (Wiley, 2011), 599 – 603.

5. S. Agnew, M. Yoo, and C. Tome, "Application of Texture Simulation to Understanding Mechanical Behavior of Mg and Solid Solution Alloys Containing Li or Y," *Acta Materialia*, 49 (20) (2001), 4277–4289.

6. A. Styczynski et al., "Cold Rolling Textures in AZ31 Wrought Magnesium Alloy," *Scripta Materialia*, 50 (7) (2004), 943–947.

7. A. Yanagida and J. Yanagimoto, "Formularization of Softening Fractions and Related Kinetics for Static Recrystallization Using Inverse Analysis of Double Compression Test," *Materials Science and Engineering: A*, 487 (1-2) (2008), 510–517.

8. A. Beer and M. Barnett, "The Post-Deformation Recrystallization Behaviour of Magnesium Alloy Mg 3Al 1Zn," *Scripta Materialia*, 61 (12) (2009), 1097–1100.

9. J. Li et al., "The Effect of Prestrain and Subsequent Annealing on the Mechanical Behavior of AA5182-O," *Materials Science and Engineering: A*, 528 (10-11) (2011), 3905–3914.

10. A. Beer and M. Barnett, "Microstructure Evolution in Hot Worked and Annealed Magnesium Alloy AZ31," *Materials Science and Engineering A*, 485 (2008), 318–324.

11. H. Chao et al., "Static Recrystallization Kinetics of a Heavily Cold Drawn AZ31 Magnesium Alloy Under Annealing Treatment," *Materials Characterization*, 62 (3) (2011), 312–320.

12. M. C. Weinberg, D. P. Birnie III, and V. Shneidman, "Crystallization Kinetics and the JMAK Equation," *Journal of Non-Crystalline Solids*, 219 (1997), 89–99.

13. J. Cahn, "The Kinetics of Grain Boundary Nucleated Reactions," *Acta Metallurgica*, 4 (12) (1956), 449–459.

14. X. Yang, H. Miura, and T. Sakai, "Recrystallization Behaviour of Fine-Grained Magnesium Alloy After Hot Deformation," *Transactions of Nonferrous Metals Society of China*, 31 (2007), 1139–1142.

15. R. K. Nadella, I. Samajdar, and G. Gottstein, "Static Recrystallisation and Textural Changes in Warm Rolled Pure Magnesium," *Magnesium: Proceedings of the 6th International Conference Magnesium Alloys and Their Applications* (2005), 1052–1057.

# Development of High Strength and Toughness Magnesium Alloy by Grain Boundary Control

Hidetoshi SOMEKAWA[1)], Alok SINGH[1)] Tadanobu INOUE[1)] and Toshiji MUKAI[2)]

1) National Institute for Materials Science, 1-2-1 Sengen, Tsukuba, Ibaraki 305-0047 Japan
2) Kobe University, 1-1 Rokkodai, Nada-ku, Kobe 657-8501 Japan

Keywords: Magnesium alloy; Fracture toughness; Strength; Caliber roll; Microstructural control

## Abstract

The mechanical properties, such as strength and fracture toughness, were investigated using caliber rolled Mg-6wt.%Al-1wt.%Zn (AZ61) alloy, which is the material consists of a constrained plane other than the rolling direction plane and is compressed non-simultaneously from two directions. The initial microstructural observations showed that the caliber rolled AZ61 alloy had a high fraction of low-angle grain boundaries and an average grain size of 2.2 μm. In addition, particles with an average size of 100 nm existed in the matrix. This alloy showed a yield strength of 423 MPa and a fracture toughness of 34.1 MPam$^{1/2}$. A combination of grain refinement, formation of low-angle grain boundaries and dispersion of fine particles is one of the effective microstructural controls to produce the magnesium alloys with fracture toughness similar to the conventional high strength aluminum alloys.

## Introduction

Magnesium alloys are the lightest among all the structural alloys in use, but their fracture toughness, which is one of the mechanical properties for the judgment of reliability and safety requirement, has generally been reported to be lower than that in aluminum alloys [1,2]. The lower fracture toughness in magnesium results from the deformation twins, which form at the very beginning of plastic deformation [3]. Since the accumulated plastic strain by the twins is smaller than that of the grain boundary migration by the dislocation slip, the boundaries around twins define the crack propagation route [4]. On the contrary, the deformation twins are reduced or prevented by refining the structure [5], and thus, the fracture toughness tends to increase with grain refinement [6].

Some reports recently pointed out that the formation of a low-angle grain (or sub-grain) boundary structure is effective for the enhancement of ductility in metallic materials [7-9]. Although the grain boundary generally prevents the dislocation movement, it provides the nucleation site for the micro-void and/or -crack formation in the plastic deformation. On the other hand, the amount of dislocation pile-up at the low-angle grain boundaries is lower than that at the high-angle grain boundaries, leading to an advantage towards ductility. Our earlier study has reported the effect of low-angle grain boundaries on the mechanical properties and fracture in magnesium alloy [10]. The previous results showed that the fine-grained Mg-3wt.%Al-1wt.%Zn (AZ31) alloy with a high fraction of low-angle grain boundaries enhanced the fracture toughness, and these boundaries did not become the origin of the micro-void formation.

Particle dispersion is also recognized as an influential microstructural factor because of being the void nucleation site [11,12]. Since finer particles are more difficult to become the void nucleation site compared to coarser particles, control of the particle size, i.e., refinement of particle size, is one of the useful methods for the improvement of fracture toughness. Thus, the combination of grain refinement, high fraction of low-angle grain boundaries and a dispersion of particles has the potential to produce a magnesium alloy with high strength and fracture toughness; however, there are no reports. Therefore, the initial microstructures and mechanical properties were investigated by using a Mg-6wt.%Al-1wt.%Zn alloy, which was produced by caliber rolling, in this study.

## Experimental Procedure

A commercially extruded rod of Mg-6wt.%Al-1wt.%Zn (AZ61) alloy with a diameter of 42 mm was used in the present study. The billet was kept at a temperature of 473 K for 30 minutes in a furnace, and then subjected to plastic working by the caliber roll at room temperature. The billet was passed through 14 times and twice for the final groove to control the cross-sectional shape of the bar, and then was quenched in water. After rotating the rolled billet 90° clockwise around the rolling axis, the billet is subjected to rolling in the same way as the former pass. The detailed experimental procedure is reported elsewhere [13]. The caliber rolled AZ31 alloy was also prepared by the same procedures as AZ61 alloy [10].

The initial microstructure of the rolled alloy was observed by the electron back-scattered diffraction (EBSD) and transmission electron microscopy (TEM). The observed area in the initial sample was the RD-TD plane. The tensile test was performed at an initial strain rate of $1 \times 10^{-3}$ s$^{-1}$ at room temperature. Tensile specimen with a gauge length of 15 mm and a gauge diameter of 3 mm was machined from the rolled bar to make the tensile axis parallel to the rolling direction.

Fracture toughness was measured using the *J*-integral method according to the ASTM E1890 [14]. The specimen that was used for investigating the fracture toughness consisted of a three point bending sample with a width of 10 mm, a thickness of 5 mm and a length of 44 mm. The specimen was machined directly from the rolled bar, and a V-notch was made normal to the rolling direction. The fracture surface was also observed by scanning electron microscopy (SEM) after the fracture toughness test in the ASTM E399 [15].

## Results and Discussion

Typical initial microstructure by EBSD observation in caliber rolled alloys is shown in Fig. 1: (a) inverse pole figure image and (b) distribution of misorientation angle. The average grain size, which consisted of the high-angle grain boundary (θ ≥ 15°), was obtained to be 2.2 ± 0.5 μm by EBSD measurement. Figure 1(b) shows that the fraction of grain structures with the misorientation angles of 5° ≤ θ < 15°, i.e., low-angle grain boundary, was obtained to be 0.25 in the present alloy. This figure also includes the result of the AZ31 alloy, which was produced by the same process [10]. Both the alloys show a similar distribution of misorientation angle, and thus, they are found to have a similar fraction of low-angle grain boundaries. The bright field image of the AZ61 alloy is shown in Fig. 2. The grain size was ~ 3 μm, which is the same as the EBSD observation. The existence in the spherical shaped particles with an average size of 100 nm is observed in the matrix. The particle was identified as the β-phase ($Mg_{17}Al_{12}$) by EDS analysis.

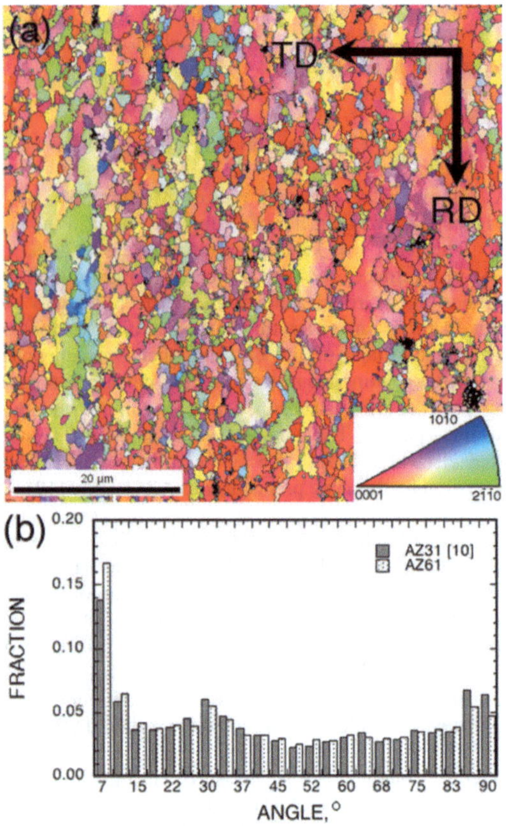


Fig. 1: Typical initial microstructures by EBSD observation in the caliber rolled alloys: (a) inverse pole figure and (b) distribution of misorientation angle. The black lines in Fig. (a) are the high-angle grain boundary (θ ≥ 15°), and Fig. (b) includes the previously reported data of the caliber rolled AZ31 alloy [10].

The mechanical properties of the AZ61 alloy are summarized in Table 1. Since the fracture toughness in the as-received commercially AZ61 alloy is around 23 $MPam^{1/2}$, the fracture toughness is improved by the caliber rolling. This table includes the data for the caliber rolled AZ31 alloy [10]. The present caliber rolled AZ61 alloy also shows higher strength and fracture toughness balance compared to those in the AZ31 alloy. In addition, the fracture toughness of the conventional aluminum alloys, i.e., 7xxxAl alloys, are reported to be ~ 35 $MPam^{1/2}$ [16], which is found to be similar to the present AZ61 alloy.

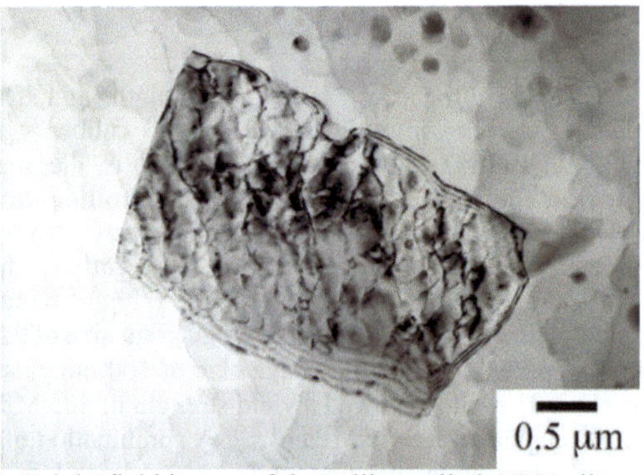


Fig. 2: Bright field image of the caliber rolled AZ61 alloy.

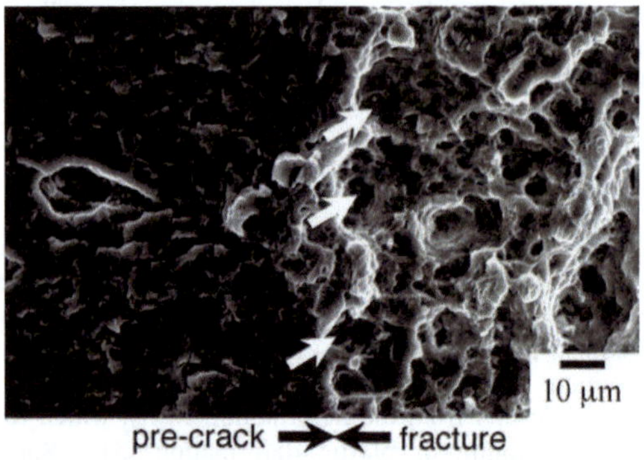


Fig. 3: Typical fracture surface by SEM observation. White arrows mark the dimples.

Hereafter, the reason for the higher fracture toughness in the present alloy is discussed by a comparison between the caliber rolled AZ31 and AZ61 alloys, which have high fraction of low-angle grain boundaries. The grain size is one of the most influential microstructural factors for the fracture toughness; however, the AZ31 and AZ61 alloys have similar sizes of 2 ~ 3 μm. The fracture toughness is reported to be proportional to one-third power of the concentration of solid solute atoms: $\Delta K_{IC} \propto C^{1/3}$ [17]. When the additional concentration of solid solute atom, i.e., aluminum, is 3 wt.%, the improvement for fracture toughness, $\Delta K_{IC}$, is estimated to be ~ 2 $MPam^{1/2}$. However, this estimated value of $\Delta K_{IC}$ is smaller than that in the increment of 5 $MPam^{1/2}$ in the experimental result. Thus, the increment in the fracture toughness in the AZ61 alloy is due to not only the solid solution strengthening but also the other microstructural factors.

Typical fracture surface by SEM observation is shown in Fig. 3. This figure shows that there are many dimples, which are typical of a ductile fracture pattern, marked by white arrows.

The dimple nucleation originates from the coarse particle, and the fracture occurs at the interface between the matrix and the particle [18] or in the particle-itself [19]. A small amount of the dimple nucleation is observed in this figure due to no existence of coarse particles. Thus, the dimple nucleation site is assumed to the presence of particle at the grain boundaries. On the contrary, the fracture toughness increases with the dispersion of fine particle, since they act as obstacles for the dislocation movements around them. Our previous report shows creation of a dislocation net-work around fine particles by the deformed microstructural observations, and predicts the increment ratio of fracture toughness based on Orowan strengthening mechanism [20]. The increment ratio for fracture toughness is estimated to be ~ 10 %, when spherical shaped particles with an average size of 100 nm are dispersed into the matrix. The AZ61 alloy contains particles, as shown in Fig.2, but there are no particles in the caliber rolled AZ31 alloy [10]. Thus, the AZ61 alloy is estimated to show 10 % higher fracture toughness, which corresponds to ~ 3 $MPam^{1/2}$, than that in the AZ31 alloy; the sum of the slid solution strengthening and the dispersion of spherical shaped fine particle is in agreement with the experimental result of 5 $MPam^{1/2}$. Therefore, the superior mechanical properties in the present alloy are related to not only the high fraction of low-angle grain boundaries but also the dispersion of the fine precipitate particles into the fine grains.

## Summary

1) The initial microstrucutal observations showed that the caliber rolled AZ61 alloy had an average grain size of 2.2 μm and high fraction of low-angle grain boundaries. The β-phase particles with an average size of 100 nm were dispersed into the matrix.
2) The caliber rolled AZ61 alloy showed a balance of high strength and fracture toughness, which was the similar to the conventional high strength aluminum alloys.

## Acknowledgement

This work was partly supported by JSPS Grant-in-Aid for Young Scientists (B) Grant No.21760564 and No.23760675.

## References

[1] H. Somekawa, A. Singh, T. Mukai, "High fracture toughness of extruded Mg-Zn-Y alloy by the synergitic effect of grain refinement and dispersion of quasicrystalline phase", *Scripta Mater.* 56 (2007) 1091.
[2] K. Purazang, P. Abachi P, K. U. Kainer, "Investigation of the mechanical behavior of magnesium composites", *Composite.* 25 (1994) 296.
[3] J. Koike, "Enhanced deformation mechanisms by anisotropic plasticity in polycrystalline Mg alloys at room temperature", *Metall. Mater. Trans.* 36A (2005) 1689.
[4] H. Somekawa, A. Singh, T. Mukai, "Fracture mechanism of a coase-grained magnesium alloy during fracture toughness testing", *Philo. Mag. Lett.* 89 (2009) 2.
[5] M. A. Meyers, O. Vohringer, V. A. Lubarda, "The onest of twinning in metals; A constitutive description", *Acta Mater.* 49 (2001) 4025.
[6] H. Somekawa, T. Mukai, "Effect of grain refinement on fracture toughness in extruded pure magnesium", *Scripta Mater.* 53 (2005) 1059.
[7] J-Q. Su, M. Demura, T. Hirano, "Grain boundary fracture strength in Ni3Al bicrystals", *Philo. Mag.* A82 (2002) 1541.
[8] S. Hanada, T. Watanabe, O. Izumi, "Deformation behavior of recrystallized Ni3Al", *J. Mater. Sci.* 21 (1986) 203.
[9] H. Lin, D. P. Pope, "The influence of grain boundary geometry on intergranular crack-propagation in Ni3Al", *Acta Metall. Mater.* 41 (1993) 553.
[10] H. Somekawa, A. Singh, T. Inoue, T. Mukai, "Enhancing fracture toughness of magnesium alloy by formation of low angle grain boundary structure", *Adv. Eng. Mater.* 12 (2010) 837.
[11] R. O. Ritchie, A. W. Thompson, "On macroscopic and microscopic analyses for crack initiation and crack growth toughness in ductile alloys", *Metall. Mater. Trans.* 16A (1985) 233.
[12] J. F. Knott, "Micromechanisms of fibrous crack extension in engineering alloys", *Mater. Sci.* 14 (1980) 327.
[13] T. Inoue, F. Yin, Y. Kimura, "Crystallographic texture of warm caliber rolled low carbon steel", *Mater. Trans.* 48 (2007) 2028.
[14] ASTM E1890-99a, Standard Test Method for Measurement of Fracture Toughness, American Society for Testing and Materials, (West Conshohocken, PA).
[15] ASTM E399, Standard Test Method for Plane-Strain Fracture Toughness of Metallic Materials, American Society for Testing and Materials, (West Conshohocken, PA).
[16] Aluminum and Aluminum alloys, ASM Specialty Handbook, ASM International, Materials Park, OH, (1993).
[17] H. Somekawa, Y. Osawa, T. Mukai, "Effect of solid-solution strengthening on fracture toughness in extruded Mg-Zn alloys", *Scripta Mater.* 55 (2006) 593.
[18] K. Tanaka, T. Mori, T. Nakamura, "Cavity formation at interface of a spherical inclusion in a plastically deformed matrix", *Philo. Mag.* A21 (1970) 267.
[19] S. H. Goods, L. M. Brown, "Nucleation of cavities by plastic deformation", *Acta Metall.* 27 (1979) 1.
[20] H. Somekawa, A. Singh, T. Mukai, "Effect of precipitate shapes on fracture toughness in extruded Mg-Zn-Zr magnesium alloys", *J. Mater. Res.* 22 (2007) 965.

Table 1: List of the mechanical properties in the caliber rolled AZ31 [10] and AZ61 alloys.

| material | d, μm | $\sigma_{ys}$, MPa | $\sigma_{UTS}$, MPa | δ, % | $K_{IC}$, $MPam^{1/2}$ | |
|---|---|---|---|---|---|---|
| AZ61 | 2.2 | 423 | 438 | 9.4 | 34.1 | this study |
| AZ31 | 2.5 | 369 | 385 | 10.9 | 29.4 | [10] |

where $d$ is the grain size, $\sigma_{ys}$ is the yield strength, $\sigma_{UTS}$ is the ultimate tensile strength, $\delta$ is the elongation-to-failure and $K_{IC}$ is the plane-strain fracture toughness.

# EFFECTS OF DIRECT EXTRUSION PROCESS ON MICROSTRUCTURE, TEXTURE EVOLUTION AND YIELD STRENGTH OF MAGNESIUM ALLOY AZ31

Shiyao Huang[1], Mei Li[1], John E. Allison[2], Shaorui Zhang[3], Dayong Li[3], Yinghong Peng[3]

[1]Research and Advanced Engineering Laboratory, Ford Motor Company, Dearborn, MI 48121, USA
[2] Department of Material Science and Engineering, University of Michigan, Ann Arbor, MI 48109, USA
[3]School of Mechanical Engineering, Shanghai Jiao Tong University, Shanghai 200240, P.R. China

Keywords: Magnesium alloy AZ31, Extrusion, Microstructure, Texture, Yield strength

## Abstract

Direct extrusions of commercial casting AZ31 alloy were carried out at elevated temperatures with different extrusion velocities. Microstructure and texture distribution of extruded rods were investigated with optical microscopy (OM) and electron backscattered diffraction (EBSD). Tensile tests were conducted at room temperature using samples from both casting billets and extruded rods. The experimental yield strength can not be solely described by average grain size. In this paper, the grain size and orientation in the extruded samples were characterized by EBSD, and Hall-Petch equation was applied to each individual grain with the input from EBSD results (individual grain size and orientation). The yield strength of tensile sample (polycrystalline aggregate) and individual grain was related by Taylor assumption. The predicted yield strength showed the same trend as experiment results.

## Introduction

The development of magnesium alloys for the automotive and aerospace industry has attracted significant attention due to its light weight and consequent potential to reduce both fuel consumption and green house effect, in addition to their excellent damping capacity and recyclability. Major commercial magnesium alloys are produced either as cast products (e.g. die, sand and mold castings) or wrought products (e.g. extrusion, stamping, forging and rolling). Wrought Mg alloys are of special interest for some applications due to the improved properties compared with casting products, such as fracture toughness, yield strength, fatigue strength, ductility, and corrosion behavior [1-5]. Although extrusion process of Mg alloy AZ31 has been studied extensively, most researches are concentrated on equal channel angular extrusion (ECAE) [6,7], hydrostatic extrusion [8,9], cyclic extrusion [10], and indirect extrusion [11,12]. Detailed studies on microstructure development and texture evolution during direct extrusion process of AZ31 alloy are still rare.

The effects of grain size and texture distribution of magnesium alloy on yield strength have been studied widely [13-17]. The relationship between yield strength $\sigma$ and grain size $d$ could be empirically described by the Hall-Petch equation (1) [18,19],

$$\sigma = \sigma_0 + kd^{-1/2} \qquad (1)$$

where $\sigma_0$ represents the friction stress for dislocation movement, and $k$ is the constant representing the grain boundary as an obstacle for dislocations to slip across the grain boundaries [17]. By plotting yield strength $\sigma$ against average grain size $d$, $\sigma_0$ and $k$ can be determined by parameter fitting. However, this fitting method does not take into account of texture explicitly, and thus the effect of texture on yield strength is not clearly considered. Furthermore, the deformation of casting AZ31 alloy results in very inhomogeneous grain size distribution and textured grain orientation, which makes it necessary to consider the contribution of the grain size and its orientation of each grain to yield strength. Fromm [20] introduced grain size and orientation distribution function (GSODF) into crystal plasticity model to predict large yield strength anisotropy of rolled α-titanium. But this method requires experimental data from multi-section orientation imaging microscopy and the calculation of GSODF is complex.

In the present study, effects of hot extrusion process on microstructure, texture distribution and yield strength of AZ31 alloy were studied. Forward extrusions of AZ31 were carried out at 9 deformation conditions. Microstructure and texture distribution of extruded rods were investigated with optical microscopy (OM) and electron backscattered diffraction (EBSD). Then tensile tests were conducted at room temperature using samples from both casting billets and extruded rods. Hall-Petch equation was applied to each individual grain with the input of EBSD results (individual grain size and orientation), the yield strength of tensile sample (polycrystalline aggregate) and individual grain was related by Taylor assumption.

## Experimental Procedures

Commercial direct chill (DC) casting feedstock of AZ31 alloy was machined into billets (Φ100mm) for extrusion trials. The alloy compositions are listed in Table I. Direct extrusion tests were carried out at 9 deformation conditions with combination of different temperatures (643K, 673K and 703K) and extrusion speeds (0.8 m/min, 1.4 m/min and 2.2 m/min). The extruded rods (Φ40mm) were cooled down in air after extrusion.

Specimens for optical microscopy (OM) were mounted in epoxy resin, grounded with 1200 grit SiC paper, mechanically polished using 6, 3 and 1 μm diamond paste and 0.3, 0.05 μm alumina solution. Specimens were etched for 5s in acetic picral (10ml acetic, 10ml Distilled Water, 70 ml picral).

Table I The Compositions of AZ31 (wt%)

| Alloy | Mg | Al | Be | Cu | Fe | Mn | Pb | Si | Zn |
|---|---|---|---|---|---|---|---|---|---|
| AZ31 | Bal. | 2.66 | <0.0005 | <0.005 | 0.01 | 0.36 | <0.01 | 0.01 | 0.84 |

Specimens for electron backscattered diffraction (EBSD) were ground with 2000 grit SiC paper, followed by mechanically polishing with 3μm diamond paste. Specimens were prepared by electro-polishing in AC2 electrolyte for the final stage. The microstructure was imaged using a LEO 1450-VP SEM. Grain orientations were examined and analyzed using Zeiss Suppa 55VP SEM equipped with HKL Channel 5 system.
In order to measure the mechanical properties improvement after extrusion process, tensile tests were performed on an Instron tensile testing machine at room temperature. Three samples were tested for each extrusion condition. Dimensions of tensile test sample were shown in Figure 1.
All specimens for OM, EBSD and tensile testing were cut from center of the extruded rods, parallel to the extrusion direction.

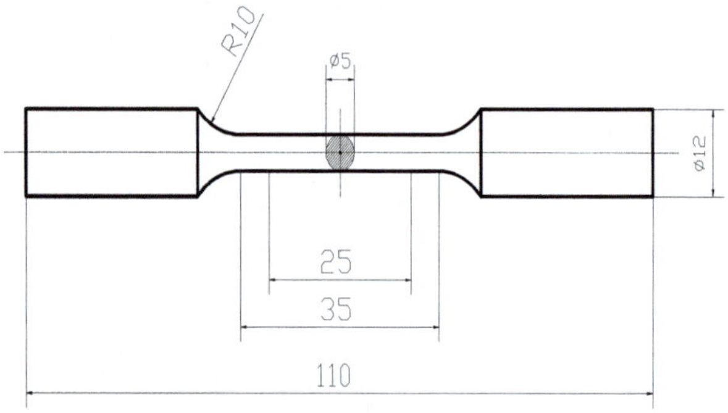


Figure 1. Dimensions of tensile test sample (in millimeters)

**Results and Discussion**

Microstructure Development

Figure 2 shows the microstructure of AZ31 casting billets used in this study. It consists of large grain with mean grain size of about 200μm. Figure 3 shows the microstructures of the AZ31 rods extruded at different conditions. The extruded microstructures are very inhomogeneous with averaged grain size decreasing dramatically compared to casting billet. Large grains in cast billets have been elongated along extrusion direction. Large amount of small grains formed at the junctions of grain boundaries. Fine DRX grains also forms "necklace" structure along grain boundaries.

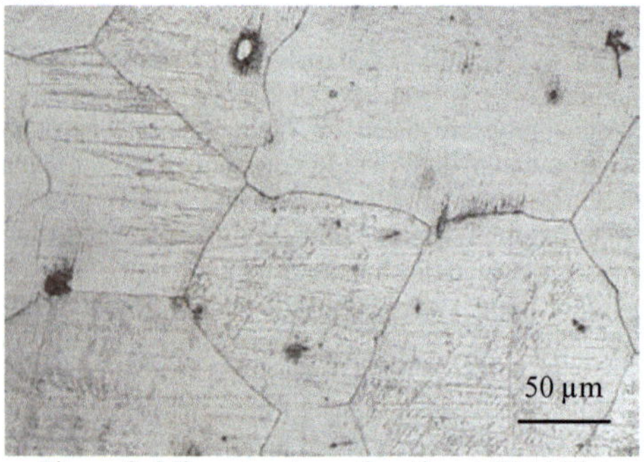


Figure 2. Microstructure of casting AZ31 billet

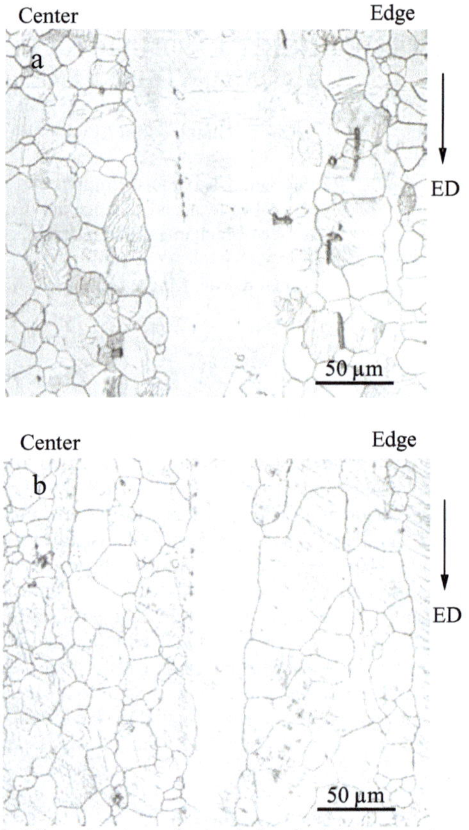


Figure 3. Microstructure of extruded rods. (a) Temperature=643K, velocity=1.4m/min. (b) Temperature=673K, velocity=1.4m/min.

Grain boundaries act as obstacles to the movement of dislocations. Thus, dislocations accumulate and subsequently generate local stress concentrations on the grain boundaries, resulting in the generation of serrated boundaries. The development of serrated grain boundaries leads to the nucleation of DRX by bulging [21]. The serrated grain boundaries are frequently observed in extruded rods, as shown in Figure 3. Figure 4 illustrates two examples of nucleation by bulging. In Figure 4(a) and Figure 4(b), the bulging area of grain A is almost cut off by grain B and grain C, the potential nucleation area is indicated by $A^*$.

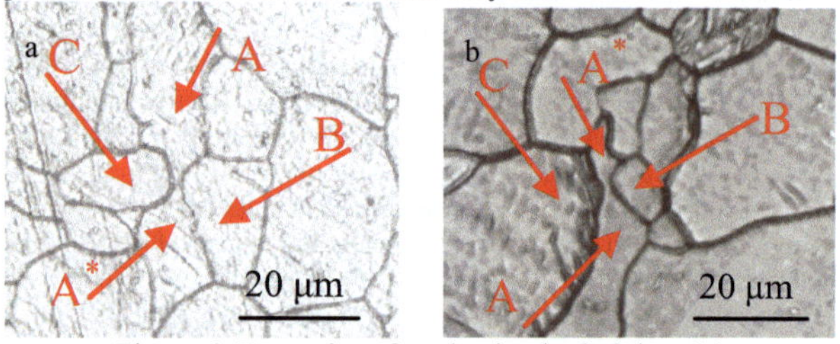


Figure 4. Examples of nucleation by bulging. (a) Temperature=673K, velocity=1.4m/min. (b) Temperature=673K, velocity=2.2m/ min.

During compression test of the as-cast AZ31 alloy, Beer [22] reported a process where a progressive increase in misorientation from the center to the boundary of grain and the conversion of low angle boundaries into the high angle ones. In the current study of direct extrusion process, the development of high angle grain boundary is also observed. As indicated by white arrow in Figure 5, the serrated grain boundaries in the elongated grain can gradually increase in misorientation, inducing the sub-boundaries behind bulges. Further rotation can lead to higher angle grain boundary and may induce the development of DRX grains.

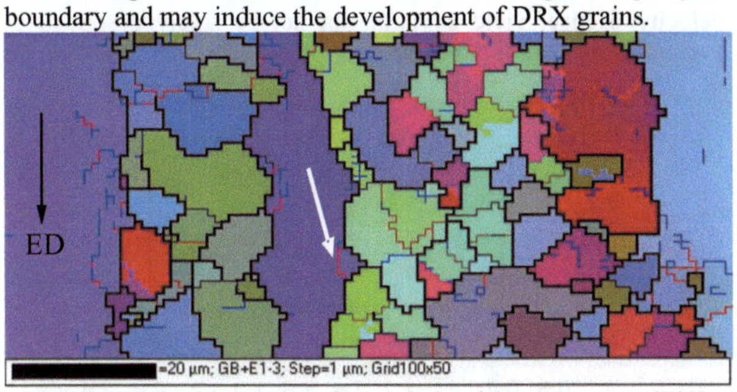


Figure 5. Orientation map of extruded rod (temperature=673K, velocity=0.8m/min). (Thick black line >15° misorientation, thin black line >10° misorientation, red thin line >5° misorientation, blue thin line>3° misorientation.)

Once the "necklace" structure forms along pre-existing grain boundary, dynamic recrystallization grains in AZ31 alloy cease to grow up and consume original grains. This has been observed by other researchers [21]. Upon further deformation, the fraction of recrystallized material will accommodate a number of strains. The continuous strain-induced process is responsible for the increase in misorientation near the grain boundary of DRX grains (indicated by white arrows in Figure 6). This process in recrystallization grains will make the small grains finer.

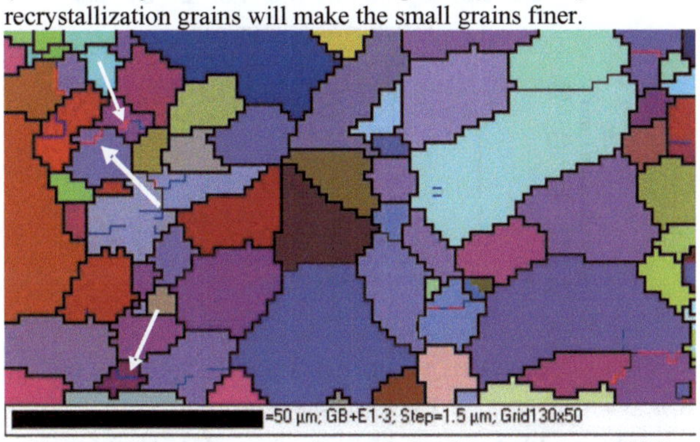


Figure 6. Orientation map of extruded rod (temperature=703K, velocity=1.4m/min). (Thick black line >15° misorientation, thin black line >10° misorientation, red thin line >5° misorientation, blue thin line>3° misorientation.)

Previous research showed that twin boundaries could act as sites for the nucleation of DRX grains [23,24]. Twinning related DRX is also observed in present study. As indicated by white arrow in Figure 7, twining with lens morphology has been subdivided into two grains. This can be attributed to the development of low angle boundaries that transformed to high angle boundaries caused by continuous straining.

Figure 7. Orientation map of extruded rod (temperature=643K, velocity=0.8m/min). (Thick black line >15° misorientation, thin black line >10° misorientation, red thin line >5° misorientation, blue thin line>3° misorientation.)

Formation of Basal Texture

The texture evolution for the 9 different extrusion processes has also been studied by EBSD. Figure 8 presents the orientation distribution function (ODF) sections of casting billet, which do not show an obvious distribution pattern. As shown in Figure 9, there are strong (0002) fiber textures in the extruded AZ31 rods. This is in accord with the conclusion that fiber texture is usually formed for rods during uniaxial deformation of hexagonal structured metals [24]. However, texture component <0002> is not exactly parallel to the extrusion direction, the deviation is roughly 10°-20°.

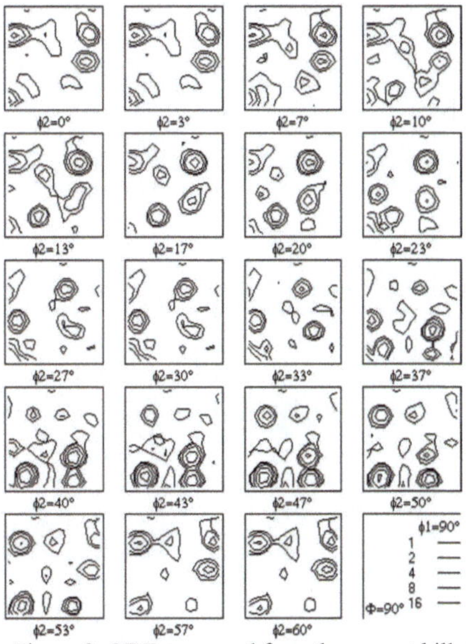


Figure 8. ODF measured from the as-cast billet.

Previous research [15] shows that $\{10\bar{1}2\}$ tension twining is favored at the early stage of deformation when the casting grain size is large, intensive twinning will reorient the initial grain orientation to a basal orientation. Once basal texture is formed, grains act as if they initially have a basal orientation during subsequent deformation.

Figure 3 shows that necklace DRX can gradually take place at the grain boundary. The DRX grains produced by buldging have orientations similar to those of the parent grains and serve to strengthen the basal texture, while nucleation brought by the development of high angle grain boundary does not have such orientation correlation which tends to lead to random texture [25]. Furthermore, twinning and non-basal slip are in a competitive relation [26,27]. With the proceeding of deformation, twinning is suppressed and non-basal slip is activated because of the grain refinement caused by DRX. Non-basal slip also tends to reorient the basal texture towards the radial direction.

a

b

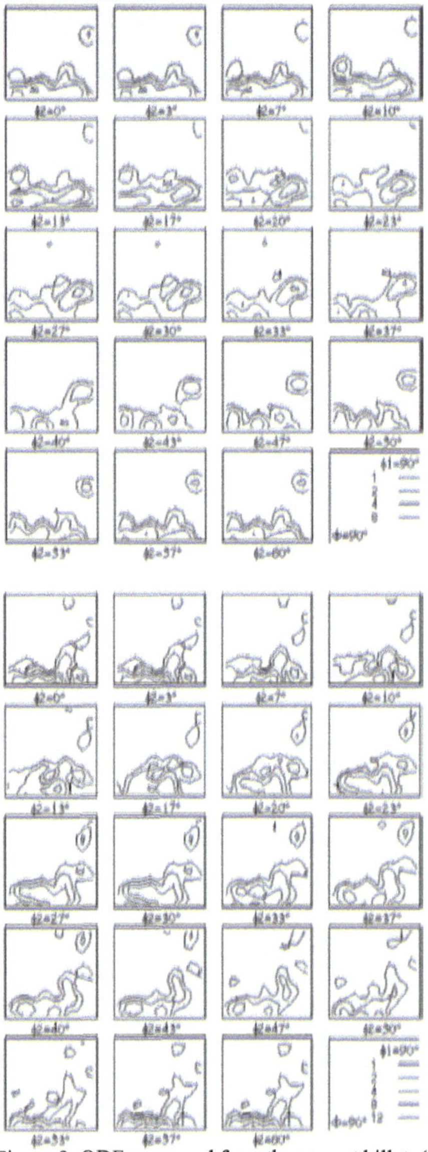

Figure 9. ODF measured from the as-cast billet. (a) Temperature=673K, velocity=0.8m/min. (b) Temperature =703K, velocity=2.2m/min.

Yield Strength Improvement

Grain refinement is proved to be effective in improving yield strength. As presented in Figure 10, average grain sizes of extruded rods decline dramatically compared with grain size of casting billet (about 200μm). Each value comes from the average of about 500 grains. The yield strength of casting billet is 83 Mpa.

As shown in Figure 11, yield strength was greatly improved (up to 200 to 210 MPa) by hot extrusion process. However, under the same extrusion temperature, yield strength of extruded rods decreases with increasing extrusion velocity, while average grain sizes of extruded rods are almost stable with increasing extrusion velocity. Hence, yield strength can not be solely described by average grain size.

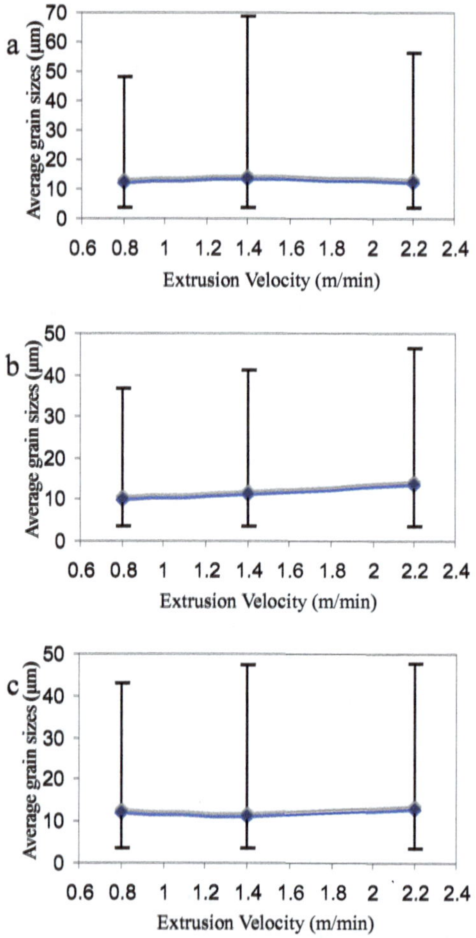


Figure 10. Average grain sizes of extruded rods (Measured by EBSD, transverse direction). (a) Temperature=643K. (b) Temperature=673K. (c) Temperature=703K.

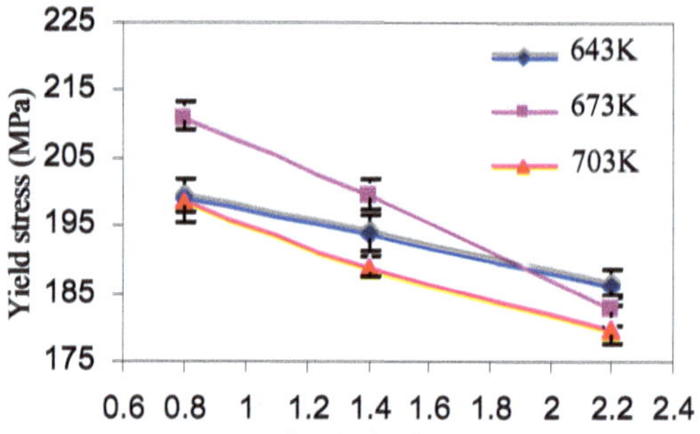


Figure 11. Yield strengths of extruded rods

In the current study, the influence of grain size and its orientation of each grain on yield strength is considered by applying Hall-Petch equation to each individual grain with the input of EBSD results. The procedure is summarized as follows.

The friction stress $\sigma$ in Hall-Petch equation is thought to be related to the deformation system with lowest yield strength [28] which depends on two factors: the critical resolved shear stress (CRSS) of the deformation system and grain orientation. For AZ31 alloy, the potential operative deformation systems are basal <a>, prismatic <a>, pyramidal <c+a> slip systems and tensile twinning systems. The yield strength $\sigma$ of deformation system $\alpha$ along the load direction is represented as $\sigma^{\alpha}$:

$$\sigma_0^{\alpha} = \frac{\tau^{\alpha}}{P^{\alpha}} \qquad (2)$$

where $\tau^{\alpha}$ represents the CRSS of deformation system $\alpha$, and $P^{\alpha}$ is Schmid factor, which is defined as

$$P^{\alpha} = \cos\theta \cdot \cos\varphi \qquad (3)$$

where $\theta$ is the angle between the slip/twinning direction and the stress axis, and $\varphi$ is the angle between the slip/twinning plane normal and the stress axis. The yield strength of individual grain $\sigma_c$ is defined as

$$\sigma_c = \min(\sigma_0^{\alpha}) + kd_c^{-1/2} \qquad (4)$$

where $\min(\sigma^{\alpha})$ represents lowest yield strength among all deformation systems. For grains with different orientations and grain sizes, the yield strength of individual grain $\sigma_c$ can be different. By employing Taylor assumption, the yield strength of polycrystal $\sigma_M$ is calculated as the average of $\sigma_c$:

$$\sigma_M = \langle \sigma_c \rangle \qquad (5)$$

In this formulation, several parameters are required to calculated $\sigma_c$, including the critical resolved shear stress $\tau^{\alpha}$, Schmid factor $P^{\alpha}$, slope $k$ and individual grain size $d_c$. The individual grain size $d_c$ and Schmid factor $P^{\alpha}$ can be obtained with the input of EBSD results (grain size and Eulerian angles). The remaining unknown parameters are CRSS $\tau^{\alpha}$ and slope $k$, which can be obtained by parameters fitting. However, the CRSSs differ significantly for different deformation systems in AZ31 alloy, suggesting four different CRSS values are necessary. Combining with slope $k$, five parameters can lead to a difficulty in determining values by parameters fitting. To simplify the fitting procedure, CRSS ratio is set to basal <a>:prismatic <a>:pyramidal <c+a>: tensile twinning = 1:8.8:11.1:5.2 according to the recent research on AZ31 alloy [29,30]. Here, the CRSS values are normalized by the CRSS of basal <a> slip.

Based on the above fitting procedure, slop k has a value of 12 $Mpa \cdot mm^{-1/2}$, and CRSS of basal <a> slip is 7.5 Mpa. The CRSSs for other deformation systems can be calculated by its ratio to basal slip. Figure 12 plots the yield strength obtained from this fitting method. Under the same extrusion temperature, the calculated yield strength decreases with the increase of extrusion velocity. This is similar to the experiment result. However, the value predicted by the fitting method is not exactly the same as experiment result (The largest deviation is 8.8Mpa, where the extrusion condition is 673k, 0.8m/min). This is possibly caused by the insufficient amount of individual grain data since each fitted value comes from 500 grains. A better result can be obtained if a larger number of grain size and orientation is available.

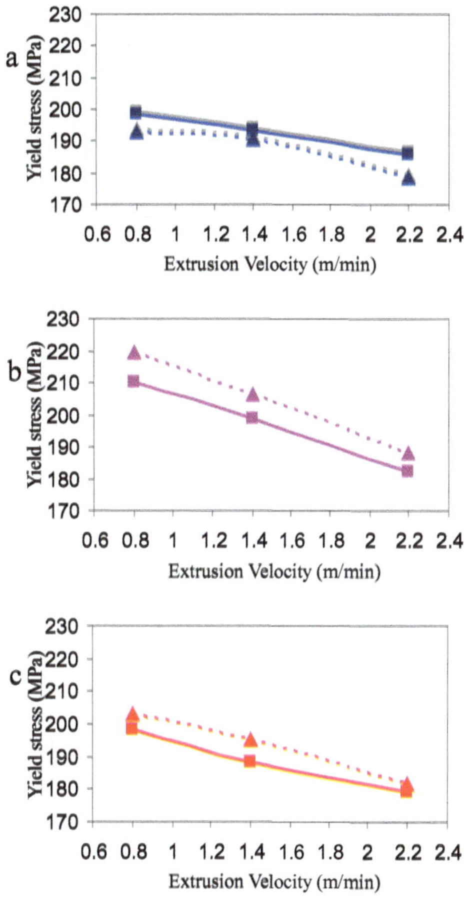


Figure 12. Yield strengths of extruded rods (Experiment Value: thick line and rectangular symbol, fitted value: dashed line and triangle symbol). (a) Temperature=643K. (b) Temperature=673K. (c) Temperature=703K.

## Summary

In this paper, the effects of hot extrusion process on microstructure development, texture evolution and tensile properties of AZ31 alloy were studied. The following conclusions were drawn:

1. The extruded microstructures are very inhomogeneous with average grain size decreasing dramatically which can be attributed to DRX process. DRX grains can be generated in pre-existing grain boundaries, DRX grain boundaries and twinning boundaries.
2. The extruded AZ31 rods have basal planes almost parallel to the extrusion direction within a certain degree. The texture evolution during extrusion process is influenced by twinning, slip and DRX collectively.
3. Yield strength is greatly improved by hot extrusion process. However, under the same extrusion temperature, yield strength of extruded rods decreases with increasing extrusion velocity, while average grain sizes of extruded rods are almost stable with increasing extrusion velocity. Hence, yield strength can not be solely described by average grain size. To get a better prediction of yield strength, EBSD data (individual grain size and

orientation) for individual grains with Taylor assumption (with a large data set) is helpful.

### Acknowledgement

The authors are thankful to the assistance of Dr.J.Boileau on Optical Microscopy work. The authors are thankful to the work of Dr. M.Li on EBSD.
The authors from Shanghai Jiaotong University acknowledge the support from the National Natural Science Foundation of China (No. 51175335), Ministry of Education of China (No.311017), Shanghai Science & Technology Projects (No.10QH1401400, 10520705000, 10JC1407300), and University Research Program of Ford Motor Company. The authors also appreciate the supported of the Research Fund of State Key Laboratory of China (No MSV-MS-2010-05).

### References

1. H. Somekawa and T. Mukai, "Effect of texture on fracture toughness in extruded AZ31 magnesium alloy," *Scripta Materialia*, 53(2005), 541-545.
2. W.J. Kim et al., "Texture development and its effect on mechanical properties of an AZ61 Mg alloy fabricated by equal channel angular pressing," *Acta Materialia*, 51(2003), 3293-3307.
3. Y. Uematsu et al., "Effect of extrusion conditions on grain refinement and fatigue behaviour in magnesium alloys," *Materials Science and Engineering A*, 434(2006), 131-140.
4. V.J. del, F. Carreno, and O.A. Ruano, "Influence of texture and grain size on work hardening and ductility in magnesium-based alloys processed by ECAP and rolling," *Acta Materialia*, 54(2006), 4247-4259.
5. G.B. Hamu, D. Eliezer, and L. Wagner, "The relation between severe plastic deformation microstructure and corrosion behavior of AZ31 magnesium alloy," *Journal of Alloys and Compounds*, 468(2009), 222-229.
6. G.G. Yapici and I. Karaman, "Common trends in texture evolution of ultra-fine-grained hcp materials during equal channel angular extrusion," *Materials Science and Engineering A*, 503(2009), 78-81.
7. L. Jin et al., "Microstructure evolution of AZ31 Mg alloy during equal channel angular extrusion," *Materials Science and Engineering A*, 423(2006), 247-252.
8. J. Swiostek et al., "Hydrostatic extrusion of commercial magnesium alloys at 100 C and its influence on grain refinement and mechanical properties," *Materials Science and Engineering A*, 424(2006), 223-229.
9. J. Bohlen et al., "Microstructure and texture development during hydrostatic extrusion of magnesium alloy AZ31," *Scripta Materialia*, 53(2005), 259-264.
10. Y.J. Chen et al., "Microstructure evolution in magnesium alloy AZ31 during cyclic extrusion compression," *Journal of Alloys and Compounds*, 462(2008), 192-200.
11. K. Mueller and S. Mueller, "Severe plastic deformation of the magnesium alloy AZ31," *Journal of Materials Processing Technology*, 187-188(2007), 775-779.
12. S.S. Park, B.S. You, and D.J. Yoon, "Effect of the extrusion conditions on the texture and mechanical properties of indirect-extruded Mg-3Al-1Zn alloy," *Journal of Materials Processing Technology*, 209(2009), 5940-5943.
13. M. Shahzad and L. Wagner, "Influence of extrusion parameters on microstructure and texture developments, and their effects on mechanical properties of the magnesium alloy AZ80," *Materials Science and Engineering A*, 506(2009), 141-147.
14. V.J. del, F. Carreno, and O.A. Ruano, "Influence of texture and grain size on work hardening and ductility in magnesium-based alloys processed by ECAP and rolling," *Acta Materialia*, 54(2006), 4247-4259.
15. Y.N. Wang and J.C. Huang, "The role of twinning and untwinning in yielding behavior in hot-extruded Mg-Al-Zn alloy," *Acta Materialia*, 55(2007), 897-905.
16. S. Graff, W. Brocks, and D. Steglich, "Yielding of magnesium: From single crystal to polycrystalline aggregates," *International Journal of Plasticity*, 23(2007), 1957-1978.
17. L.L. Chang et al., "Grain size and texture effect on compression behavior of hot-extruded Mg-3Al-1Zn alloys at room temperature," *Materials Characterization*, 60(2009), 991-994.
18. N.J. Petch, "The cleavage strength of crystals," *Journal of the Iron and Steel Institute*, 174(1953), 25-28.
19. E.O. Hall, "The Deformation and Ageing of Mild Steel: III Discussion of Results," *Proceedings of the Physical Society. Section B*, 64(1951), 747-753.
20. B.S. Fromm et al., "Grain size and orientation distributions: Application to yielding of -titanium," *Acta Materialia*, 57(2009), 2339-2348.
21. S.M. Fatemi-Varzaneh, A. Zarei-Hanzaki, and H. Beladi, "Dynamic recrystallization in AZ31 magnesium alloy," *Materials Science and Engineering A*, 456(2007), 52-57.
22. A.G. Beer and M.R. Barnett, "Microstructural development during hot working of Mg-3Al-1Zn," *Metallurgical and Materials Transactions A: Physical Metallurgy and Materials Science*, 38(2007), 1856-1867.
23. A.G. Beer and M.R. Barnett, "The influence of twinning on the hot working flow stress and microstructural evolution of magnesium alloy AZ31," *Materials Science Forum*, 488-489(2005), 611-614.
24. I.L. Dillamore and W.T. Roberts, "Preferred orientation in wrought and annealed metals," *Metallurgical Reviews*, 10(1965), 271-380.
25. D. Ponge and G. Gottstein, "Necklace formation during dynamic recrystallization: Mechanisms and impact on flow behavior," *Acta Materialia*, 46(1997), 69-80.
26. M.R. Barnett et al., "Influence of grain size on the compressive deformation of wrought Mg-3Al-1Zn," *Acta Materialia*, 52(2004), 5093-5103.
27. Y. Chino et al., "Mechanical anisotropy due to twinning in an extruded AZ31 Mg alloy," *Materials Science and Engineering A*, 485(2008), 311-317.
28. R.W. Armstrong, "Theory of the tensile ductile-brittle behavior of poly-crystalline h.c.p. materials, with application to beryllium," *Acta Metallurgica*, 16(1968), 347-355.
29. B. Raeisinia, S.R. Agnew, and A. Akhtar, "Incorporation of solid solution alloying effects into polycrystal modeling of Mg alloys," *Metallurgical and Materials Transactions A: Physical Metallurgy and Materials Science*, 42(2011), 1418-1430.
30. H. Wang et al., "Evaluation of self-consistent polycrystal plasticity models for magnesium alloy AZ31B sheet," *International Journal of Solids and Structures*, 47(2010), 2905-2917.

**Magnesium Technology 2012**
*Edited by: Suveen N. Mathaudhu, Wim H. Sillekens, Neale R. Neelameggham, and Norbert Hort*
***TMS (The Minerals, Metals & Materials Society), 2012***

# COMPARISON OF TENSILE PROPERTIES AND CRYSTALLOGRAPHIC TEXTURES OF THREE MAGNESIUM ALLOY SHEETS

Junying Min[1], Ying Cao[1], Jon T. Carter[2], Ravi Verma[2]

[1]Tongji University, Mechanical Engineering, 1239 Siping Road, Shanghai, China
[2]General Motors R&D, 30500 Mound Road, Warren, MI, 48090, USA

Keywords: AZ31, ZEK100, magnesium, sheet, texture, tensile properties.

## Abstract

The most common commercially available rolled magnesium sheet alloy is AZ31B (typ. 3% Al, 1% Zn, 0.4% Mn, balance Mg). One of the often-cited shortcomings of this sheet is its limited formability at room temperature, which is attributed in part to a strong crystallographic texture in which the basal planes of the hexagonal unit cell are parallel to the plane of the sheet. Attempts have been made to avoid this rolling-induced texture by changing either (a) the alloy composition or (b) the rolling process. Specifically, sheet has been made using the conventional rolling practice, but changing the alloy to ZEK100 (typ. 1% Zn, 0.2 % Nd, 0.2% Zr, balance Mg), or by keeping the AZ31B composition but rolling at a much higher temperature. In this report, both types of sheet are evaluated and compared with conventionally rolled AZ31B sheet. Both show reduced texture and attractive tensile properties, and therefore both are expected to show greater room-temperature formability than conventionally rolled AZ31B.

## Introduction

Magnesium is the lightest (lowest density) of the common structural metals and therefore is potentially attractive for use in automotive applications to achieve mass reduction and the associated improvements in fuel efficiency and handling. Currently, automotive usage of magnesium is primarily in the form of die castings, and includes steering wheels, transmission housings, sunroof frames [1], and inner panels for doors and liftgates [2,3]. Numerous potential automotive applications of wrought magnesium alloys, including seat frames and IP beams, have been prototyped [4-7], but production implementation of sheet has been limited to a very low volume center console cover [8]. The issues which preclude greater implementation of the wrought alloys include the high cost of sheet, limited formability at room temperature, difficulty joining to other materials, and corrosion. Potential solutions to all of these issues are being studied by various groups around the world.

Mg sheet is formable at hot temperatures [9,10] and warm temperatures [11,12]. Ultimately, however, sheet formers would prefer to stamp magnesium alloy sheet at room temperature, similar to what is done with steel. This would reduce the costs associated with heaters, hot tools, and high-temperature-capacity forming lubricants, and might enable beneficial strain hardening of the part being formed. Such stamping might be done using blanks which benefit from having either a different chemical composition of the alloy, or which may have been thermomechanically processed differently so as to enhance formability.

There are many papers in the literature about the common commercial AZ31 wrought magnesium alloy, and several will be discussed here. Beer and Barnett [13] investigated the effects of temperature, strain rate, and annealing temperature after hot deformation, on the microstructure of AZ31. Both the deformation and annealing conditions influence the time required for stable recrystallization. Higher temperature and strain rate are good for obtaining high rate of recrystallization and coarse grains. Perez-Prado and Ruano [14] found that different recrystallizations occurred in the mid layer and outer surface of an as-extruded thick Mg sheet AZ31B during annealing, and that the $<11\bar{2}0>$ prismatic component fiber texture predominates throughout the thickness. During the annealing of AZ31B extruded rod at 400°C for 30 minutes, Yi et al. [15] found that the texture changed from $<10\bar{1}0>$ into the $<11\bar{2}0>$ component.

In the aspect of formability, Takuda et al. [16] gave a flow stress model expressed in a simple form with the work hardening exponent, the strain rate sensitivity exponent and the strength coefficient, and it has a good fit in the investigated strain rate and temperature ranges. Ulacia et al. [17] proposed a new constitutive model based on modifying Johnson-Cook and accounting for the variation of strain hardening with strain rate over a wide range of strain rate and temperatures. Kim et al. [18] investigated the formability of AZ31B at elevated temperatures, and developed two failure criteria based on the effective strain or damage parameters. Both were formulated as a function of Zener–Holloman parameter, and deep drawing tests of AZ31B sheet at elevated temperatures were performed to validate the failure criteria.

Wu et al tensile tested 0.5mm AZ31B-H24 rolled sheet at 250 °C and 370 °C, and found that bars pulled at 45° to the RD exhibited lower flow stress and greater elongation than those pulled at either 0° or 90° [19]. Fatemi-Varzaneh [20] also studied the mechanical properties of AZ31B after rolling between 300 and 450 °C. They found that a low rolling temperature and a high rolling strain led to increased strength and ductility in subsequent room temperature tensile testing. They attributed those increases to the uniform and fine (5 micron) grain size which forms by dynamic recrystallization under those rolling conditions. The normal plastic anisotropy (r value) of AZ31B was studied by Agnew and Duygulu [21]. At room temperature, the r value for rolled sheet in the H24 temper increases from ~1 in the RD to ~3 in the TD. It decreases with increasing temperature up to ~200 °C, especially in the transverse direction.

Magnesium alloys with rare earth (RE) additions have attracted many scholars in recent years because such additions weaken the texture and increase the formability. Liu et al. [22] studied the mechanical properties, bore expanding performance, and limiting drawing ratio of ZE10 sheet in a temperature range from room temperature to 300°C. They showed that the ZE10 has better formability at 150-300 °C than at lower temperatures.

Zhang et al. [23] investigated the effects of Er additions on the properties of a semi-continuously cast Mg-1.5Zn-0.6Zr alloy. They found those additions increased the elongation and resulted in a refined grain size and higher yield strength. A yield point occurred when the Er content was greater than 2%, but the mechanism responsible for the yield point was not identified.

Stanford et al. [24] found the addition of small amounts of Gd decreased the grain size, and due to the Gd additions, the solute strengthening of the prismatic slip system was above 100 MPa. The Gd addition also reduced the texture and this resulted in greater ductility. The upper and lower yield point were observed when the Gd content was greater than 2.75%. They suggested that it was due to the Gd solute locking dislocations during the recrystallization annealing process which was done before tensile testing.

Mishra et al. found that the addition of 0.2% Ce to Mg reduced the tensile yield strength and greatly increased the ductility of extruded rounds. They attributed these changes in tensile properties to Ce-induced changes in the recrystallization [25]. Bohlen et al. have studied the texture and anisotropy of several as-hot-rolled Mg–Zn–RE alloy sheets [26]. They reported that the overall texture strength and the basal pole intensity aligned with the sheet normal direction was lower for RE-containing alloys than for conventional alloys. The anisotropy of the yield and flow strengths was reversed and the planar anisotropy was reduced to ~1 in comparison to conventional alloy, which was related to the fact that the dominant texture components in the Mg–Zn–RE alloys placed more grains in favorable orientations for basal slip and tensile twinning. A very recent study on sheet texture modification in Mg-based alloys by Gd, Nd, Ce, La and mischmetal alloying was performed by Al-Samman and Li [27]. They found that the different rare earths have different solid solubilities, and give rise to distinct microstructures, average grain sizes, and second phase formations. The Mg-based sheets with La, Nd, Gd, and mischmetal additions have less anisotropy of UTS and ductility than the one with Ce. The r-values of all the Mg-Zn-RE sheets are around 1.0 except the Mg-Zn-Ce sheet whose r values ranged from 1.2 in the RD to 1.7 in the TD (compared to 2.2 – 4.0 for AZ31).

Another approach to making a more formable magnesium alloy sheet is to change the rolling process rather than the alloy composition. For example, Hitachi Metals, has developed a rolling process in which a commercial magnesium alloy sheet is rolled at about 100 °C higher temperature than that in the conventional rolling process (lower than 400 °C) [28]. The resulting sheet is reported to exhibit a weak crystal orientation, resulting in excellent room-temperature formability which is comparable to those of aluminum alloy sheets.

Other non-conventional rolling processes have been applied to, or proposed for application to, magnesium sheet alloys on a lab scale, and show some attractive benefits to either the production of sheet or the properties of the resulting sheets, or both. These processes include high-speed heavy rolling [29], asymmetric shear rolling and asymmetric cross rolling [30].

The purpose of this work is compare the tensile properties and crystallographic texture of conventionally-rolled AZ31B magnesium alloy sheet with two alternative magnesium alloy sheet products.

**Experimental Method**

Rolled sheets were obtained from two manufacturers in different thicknesses and tempers. Specifically, the conventionally rolled (CR) AZ31B sheets were 1.6 mm thick, and in the H24 temper ("strain hardened and partially annealed"). The CR ZEK100 sheets were 1.5 mm thick, and in the F temper ("as-fabricated"). The specially rolled (SPR) AZ31B sheets were 0.7 mm thick, and in an unspecified temper.

Tensile bars were waterjet cut at 0°, 45°, and 90° to the rolling direction (RD). The dimensions of the reduced section are 50 mm x 6 mm x sheet thickness. Annealing was done for 15 min in an electric furnace which was filled with argon gas to protect the tensile bars from oxidization. The AZ31B bars were annealed at 350 °C, and the ZEK100 bars were annealed at 500°C. These temperatures were chosen based on unpublished annealing studies in which we observed that AZ31B-H24 rolled sheet softened much at 350 °C, but that the ZEK100-F rolled sheet required 450-500 °C to soften. Rockwell F hardness was measured on the sheet surface with a 1/16" diameter indenter and 60 kg major load. Because the SPR AZ31B sheet is relatively thin and soft, its hardness was measured on the Rockwell superficial HR15T scale with 1/16" diameter indenter and 15 kg load, then converted to Rockwell F. Five measurements were repeated for each datum.

Crystallographic texture of the sheets was studied using an x-ray diffraction (XRD) technique which is often used to study powders, but was applied to intact sheets. Specifically, the x-ray diffraction data was collected with a D8 ADVANCE-DaVinci system using Cu K-alpha radiation (40 kV and 40 mA). The scan was made using Bragg Brentano mode, which is also known as 'coupled theta/two-theta' mode. The relative peak intensities were measured and compared to a standard Mg powder (random texture) provided by the International Crystallography Diffraction Data (ICDD) database, PDF 00-035-0821. This technique samples only those crystallographic planes which are parallel to the surface of the sheet.

The line intensities of sheet samples were normalized by those of the random sample to give a texture intensity factor TIF, with the help of the following equation. (The "o" subscript indicates the powder standard PDF 00-035-0821.)

$$TIF = (I_{(hkl)} / \sum I_{(hkl)}) / (I_{o(hkl)} / \sum I_{o(hkl)}) \quad (1)$$

An Instron tensile machine and an extensometer with a gauge length of 25 mm were employed. The crosshead speed was set at 10 mm/min for all tests. When performing tests for each type of sheet material, three bars in each orientation were pulled to failure, and one bar in each orientation was pulled to 10% elongation to determine r value.

Samples for metallography were carefully cut, cold mounted, ground, and polished in order to avoid introducing twins or other artifacts. They were etched with fresh mixtures of 4.2 g picric acid, 10 ml acetic acid, 70 ml ethanol, and 10 ml distilled water in order to reveal the grains and grain boundaries .

## Results and Discussion

Compositions

Elemental compositions of the sheet materials are as follows: CR AZ31B contains 3.0 %Al, 1.1 %Zn and 0.48 %Mn. SPR AZ31B contains 2.9 %Al, 1.1 %Zn and 0.44 %Mn. CR ZEK100 contains 1.3 %Zn, 0.2 %Nd, 0.25 %Zr and 0.01 %Mn. Al and Zn are added for strength, and Zn and Mn are added for corrosion resistance. Nd is for reduced crystallographic texture, and Zr is for grain refinement.

Hardness

The CR AZ31B was received in H24 temper, and therefore softened from 69 to 51 HRF as expected during 350 °C annealing to the O temper. The CR ZEK100 was received in F temper, and softened from 55 to 44 HRF during 500 °C annealing. In contrast, the SPR AZ31B was received in an unspecified temper, and did not soften (61 vs. 60 HRF) during 350 °C annealing.

Compared to the CR AZ31B, the lower hardnesses of the CR ZEK100 in both as-received and tempered conditions are likely due to the lower alloy content, lower amount of residual cold work in as-received sheets, and larger grain sizes (as will be discussed later in this report). The hardness of SPR AZ31B is less than that of the CR AZ31B-H24, likely due to a lower level of residual cold work, and is greater than that of CR AZ31B-O, likely because of a finer grain size.

Tensile Data

These data are summarized in Figures 1-3, and are discussed below.

Effect of Alloy and Rolling Process. Generally speaking, the CR AZ31B sheet showed the greatest strength, the ZEK100 sheet showed lowest strength and the greatest ductility, and the SPR AZ31B showed intermediate strength and lowest ductility. The CR AZ31B exhibits r values (normal plastic strain anisotropy) in the range 1.2 – 4.0, compared to 0.5 – 1.2 for the other materials, as shown in Figure 4. R value is often called "the resistance to sheet thinning", and therefore in some discussions is casually equated with good formability, at least for certain types of metals and certain types of forming processes. Whether or not the high values of r measured for CR AZ31B translate into improved formability remains to be determined.

Effect of Temper. Annealing of the CR AZ31B sheet reduced YS and UTS, but increased % elongation. This trend was also shown by the ZEK100 sheet. However, for the SPR AZ31B sheet, annealing only slightly reduced YS, didn't change UTS, and increased % elongation. (The magnitudes of all of the annealing-induced changes vary with orientation of the bar's tensile axis relative to sheet RD.) Both n value and k value of CR AZ31B and CR ZEK100 increase obviously after annealing but only slightly for SPR AZ31B. That's attributed to the as-received CR AZ31B and CR ZEK100 having initial work hardening introduced by the rolling process, but the as-received SPR AZ31B having almost none. For CR AZ31B and CR ZEK100, since the annealing process removes the initial work hardening, the O-tempered materials have greater hardening behavior.

As regards r values, as shown in Figure 4, annealing increased the already-high r values of CR AZ31B, but had little effect on those of the other two materials.

Effect of Orientation. The YS, UTS, and % elongation for CR AZ31-H24 are all somewhat greater at both 45° and 90° than at 0°. The same pattern holds true for that sheet after annealing. In contrast, both before and after annealing, the CR ZEK100 bars lose much strength and gain much ductility as the orientation is changed from 0° to 45° and 90°. The effect of orientation on SPR AZ31B is rather small both before and after annealing. Tensile properties of this sheet are generally lower at 45° than at 0° and 90°, but only very slightly.

The cause of in-plane anisotropy of tensile properties, especially YS, is likely due to details of the crystallographic textures and, therefore, the active deformation mechanisms. For example, even though the hcp basal plane poles are typically well aligned approximately normal the surface of rolled magnesium sheet, they often are reported to be tilted +/- a few degrees in the RD. Such tilting will facilitate deformation by both the basal slip mechanism and the extension twinning mechanism when pulling is along the RD [26].

Orientation also has a strong effect on the hardening parameters, n & k, of the CR ZEK materials (Figure 2) and on the normal plastic anisotropy, r, of the CR AZ31B (Figure 4). In both cases, the values increase as the tensile bar orientation changes from 0° to 90°. Both are attributed to crystallographic textures, and particularly to the slight tilting of basal plane poles away from the sheet normals [26].

Texture

TIF data are shown in Figure 5. If the orientations of the crystallographic axes of the grains in the sheets were distributed randomly in space, the TIFs would all have a value of one. Instead, for example, the TIF for the (002) basal planes in the CR AZ31B-H24 midplane sample is over five, indicating a strong tendency for these planes to be parallel to the plane of the sheet in this sample. Correspondingly, the TIFs of most of the other planes are much less than one. This type of texture is commonly reported for both rolled and extruded magnesium alloys. This texture is slightly stronger at the sheet midplane compared to the surface, and slightly stronger for H24 temper compared to O temper.

The TIF data for CR ZEK100 and SPR AZ31B show that these materials are less textured than the CR AZ31B, as expected. Specifically, the maximum TIFs are approximately two rather than five. The through-thickness variation of texture appears to be greater for the CR ZEK100 sheet than for the (thinner) SPR AZ31B sheet. The effect of annealing on TIFs is greater for the CR ZEK100 than for the SPR AZ31B.

It is interesting to note that whereas the literature, as discussed in [26], indicates that the rolling textures are similar for many wrought Mg alloys, but the recrystallization textures differ. Figure 5 shows the rolling textures for CR AZ31B-H24 and CR ZEK100-F to be much different. (The annealed textures are even more different.) This may be due to the former having more microstructural evidence of "cold work" than the latter.

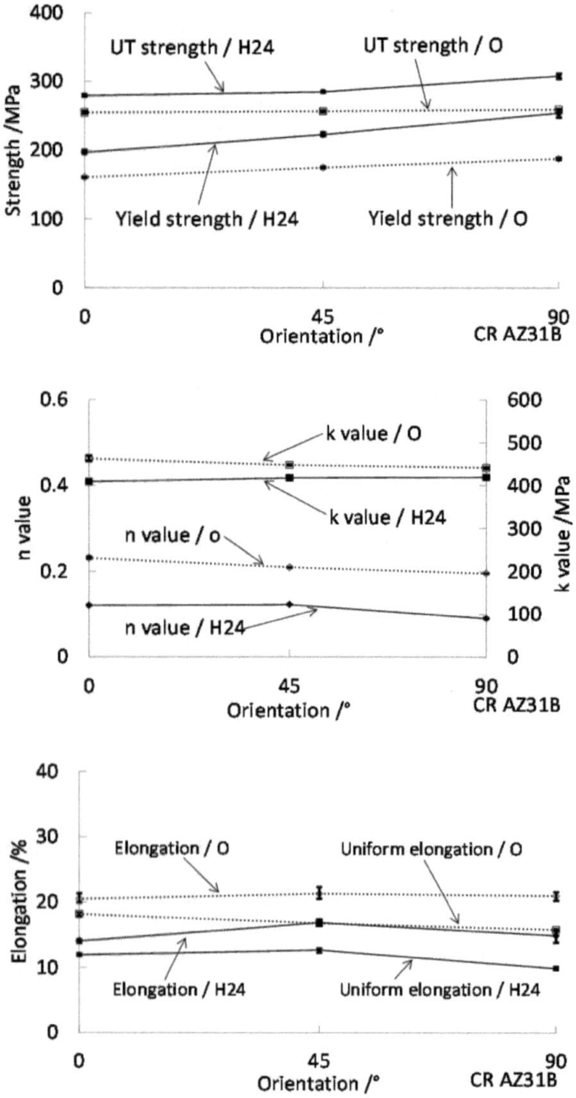


Figure 1. Orientation dependence of tensile properties of conventionally rolled AZ31B sheet in two tempers.

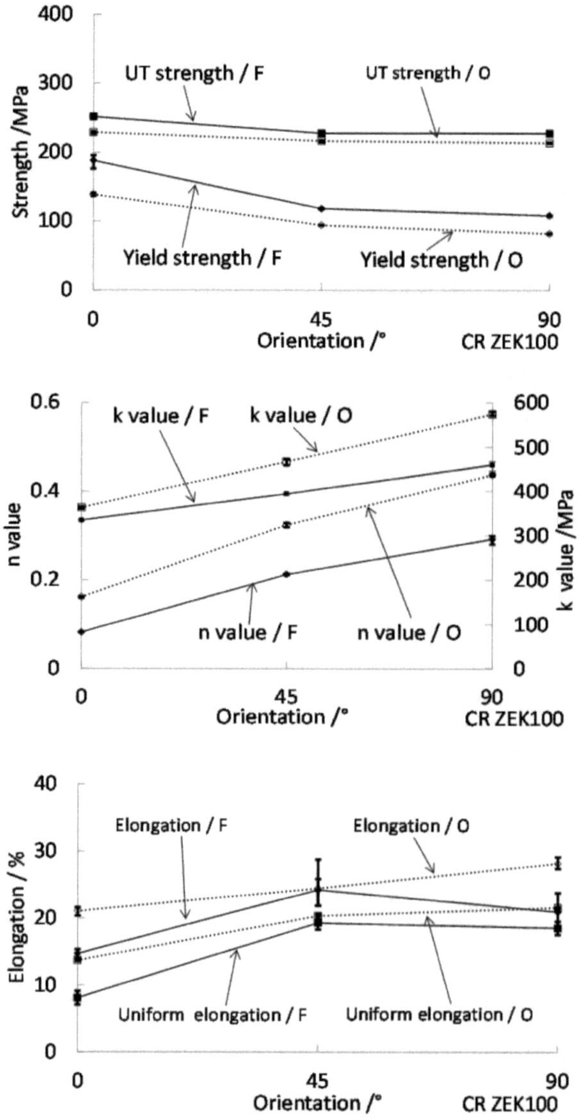


Figure 2. Orientation dependence of tensile properties of conventionally rolled ZEK100 sheet in two tempers.

Microstructures

Microstructures of the three sheet materials in each of two conditions are compared in Figure 6. There were plenty of twins in the as-received CR AZ31B and CR ZEK100, but much less in the as-received SPR AZ31B. During annealing heat treatment most of twinning was eliminated, and the grain sizes, measured in the rolling direction, increased from 4.2 μm to 8 μm for CR AZ31B, and from 15.3 to 19.3 μm for CR ZEK100. See Figure 7. However, the grain size of SPR AZ31B remained nearly constant at 5.6 μm during the annealing process. That is attributed to the work hardening in the as-received CR AZ31B and ZEK100, as it supplies more energy for nucleation and recrystallization than the SPR AZ31B in the annealing process.

## Conclusions

Compared to conventionally rolled (CR) AZ31B magnesium alloy sheet:

1. Both a CR ZEK100 alloy sheet and a specially rolled (SPR) AZ31B alloy sheet exhibit significantly reduced crystallographic texture. Annealing the CR ZEK100 sheet further reduces the already weak basal texture, at least at the midplane of the sheet.
2. The CR ZEK100 is significantly weaker, more ductile, and shows more in-plane anisotropy of tensile properties. The pattern of in-plane anisotropy of strength is opposite that in CR AZ31B.
3. The SPR AZ31B is slightly weaker, less ductile, and shows less in-plane anisotropy of tensile properties.

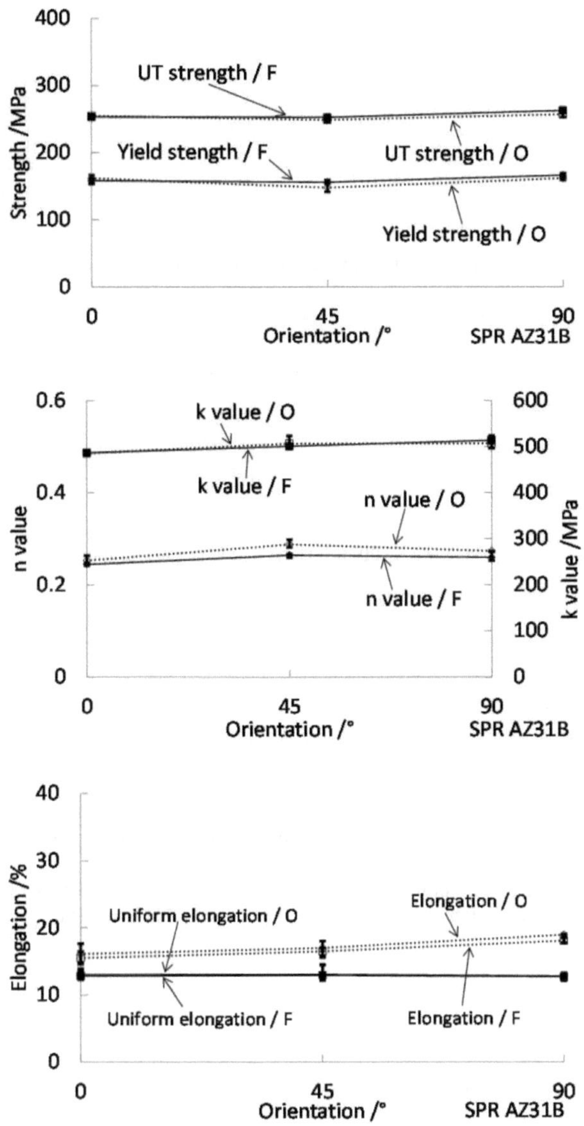


Figure 3. Orientation dependence of tensile properties of special rolled AZ31 sheet in two tempers

Also:

4. Annealing reduces the strength of both conventionally rolled alloys, but has no effect on the strength of the specially rolled alloy.
5. Annealing increases the ductility of all three alloys.

## References


1. A.A. Luo, , JOM, February 2002, pp. 42-48.
2. P.J. Blanchard, G.T. Bretz, S. Subramanian, J.E. deVries, A. Syvret, A. MacDonald, P. Jolley, SAE Paper No. 2005-01-0338.
3. http://media.ford.com/article_display.cfm?article_id=30231.
4. J.T. Carter, A.R. Melo, V. Savic, L.G. Hector, Jr., P.E. Krajewski, SAE Paper No. 2011-01-0075.
5. S. Logan, A. Kizyma, C. Patterson, S. Rama, SAE Paper No. 2006-01-0523.
6.www.superlightcar.com/public/docs/SLC_presentation_TRA08_VW_March08.pdf
7. A.A. Luo, E.A. Nyberg, K. Sadayappan, W. Shi, in Magnesium Technology 2008, eds. M.O. Pekguleryuz, N.R. Neelameggham, R.S. Beals, and E.A. Nyberg, TMS, Warrendale, PA, 2008, pp. 3-10.
8. http://en.wikipedia.org/wiki/Porsche_Carrera_GT.
9. R. Verma and J.T. Carter, SAE Paper No. 2006-01-0525.
10. Automotive Material Data Sheet, AZ31B-O, Issue 1, 2006, Superform Aluminium.
11. P. Krajewski, P. Friedman, K. Oikarinen, D. Cedar, Paper presented at MS&T'07, September 16-20, 2007, Detroit.
12. Stolfig Group, patents DE 101 56 034.6, DE 102 47 129.0, CN 152 76 83 A, Nr. DE 302 08 803, DE 202 02 591.8
13. A.G. Beer, M.R. Barnett, Materials Science and Engineering A, Vol. 485 (2008) 318-324.
14. M.T. Pérez-Prado, O.A. Ruano, Scripta Materialia, Vol. 46, 2002, pp. 149-155.
15. S. Yi, H.G. Brokmeier, D. Letzig, Journal of Alloys and Compounds, Vol. 506, 2010, pp. 364-371.
16. H. Takuda, T. Morishita, T. Kinoshita, N. Shirakawa, Journal of Materials Processing Technology, Vol.164-165, 2005, pp. 1258-1262.
17. I. Ulacia, C.P. Salisbury, I. Hurtado, M.J. Worswick, Journal of Materials Processing Technology, Vol. 211, 2011, pp. 830-839.
18. W.J. Kim, H.K. Kim, W.Y. Kim, S.W. Han, Materials Science and Engineering A. Vol. 488, 2008, pp. 468-474.
19. H.Y. Wu, P.H. Sun, F.Z. Lin, Materials Science and Engineering A. Vol. 528, 2011, pp. 2522-2531.
20. S.M. Fatemi-Varzaneh, A. Zarei-Hanzaki, M. Haghshenas, Journal of Alloys and Compounds, Vol. 475, 2009, pp. 126-130.
21. S.R. Agnew, O. Duygulu, International Journal of Plasticity, Vol. 21, 2005, pp. 1161-1193.
22. Y. Liu, Y. Li, W. Li, Journal of Rare Earths, Vol. 25, 2007, pp. 480-484.
23. J. Zhang, X. Zhang, W. Li, F. Pan, Z. Guo, Scripta Materialia, Vol. 63, 2010, pp. 367-370.
24. N. Stanford, D. Atwell, M.R. Barnett, Acta Materialia, Vol. 58, 2010, pp. 6773-6783.
25. R.K. Mishra, A.K. Gupta, P.R. Rao, A.K. Sachdev, A.M. Kumar, A.A. Luo, in Magnesium Technology 2008, 269-274. Warrendale, PA: The Minerals, Metals & Materials Society, 2008.
26. J. Bohlen, M.R. Nurnberg, J.W. Senn, D. Letzig, S.R. Agnew, Acta Materialia, Vol. 55, 2007, pp. 2101-2112.
27. T. Al-Samman, X. Li, Materials Science and Engineering A, Vol. 528, 2011, pp. 3809-3822.
28. "Development of a New Rolling Process for Commercial Magnesium Alloy Sheets with High Room-temperature Formability", (Translation of AIST press release of January 26, 2010) http://www.aist.go.jp/aist_e/latest_research/2010/20100217/20100217.html.
29. H. Utsunomiya, T. Sakai, S. Minamiguchi, and H. Koh, Magnesium Technology 2006, edited by A.A. Luo, N.R. Neelameggham, and R.S. Beals, TMS, Warrendale, PA, 2006, pp. 201-204.
30. B. Beausir, S. Biswas, D.I. Kim, L.S. Toth, S. Suwas, Acta Materialia, Vol. 57, 2009, pp. 5061-5077.

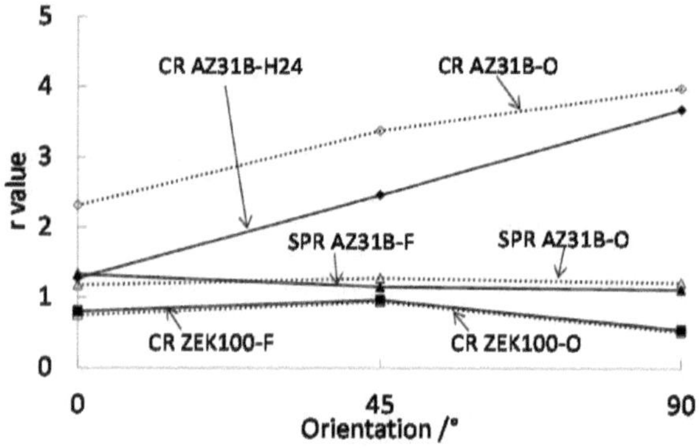

Figure 4. Orientation dependence of r values at 10% strain.

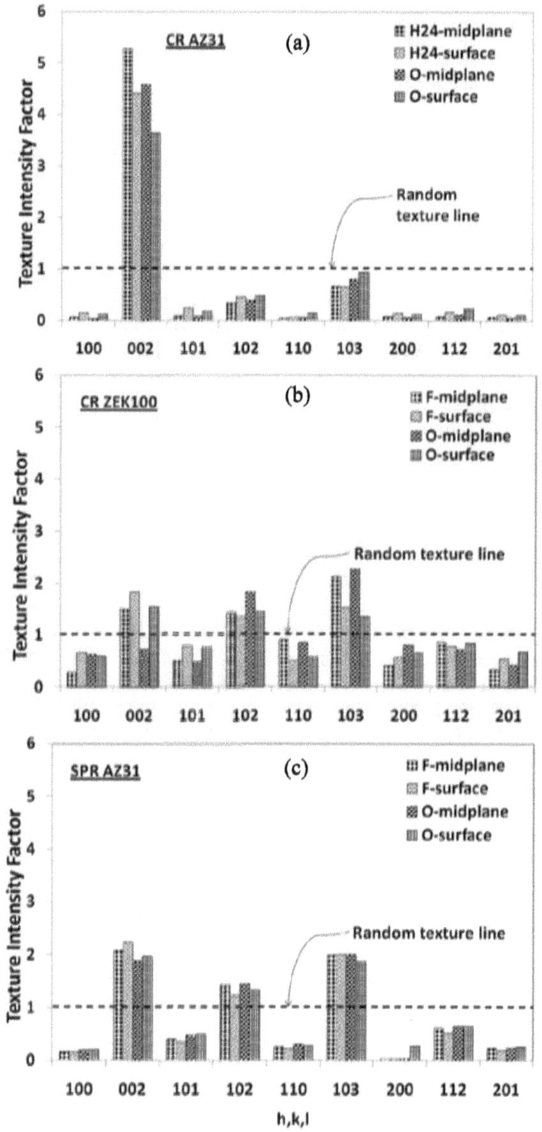

Figure 5. Charts showing texture intensity factors at sheet surface and sheet midplane for three alloys in two tempers each.

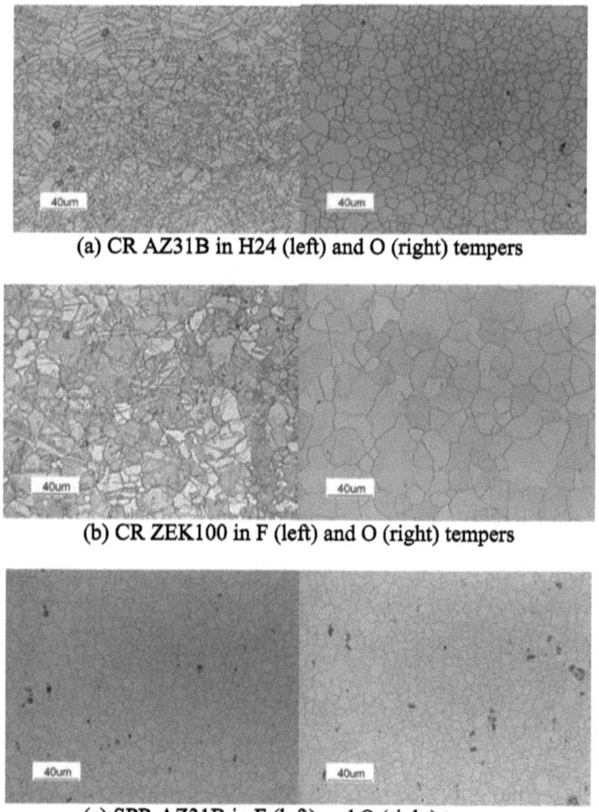
(a) CR AZ31B in H24 (left) and O (right) tempers

(b) CR ZEK100 in F (left) and O (right) tempers

(c) SPR AZ31B in F (left) and O (right) tempers

Figure 6. Photomicrographs of the three magnesium sheets before and after annealing. RD is horizontal. TD is vertical.

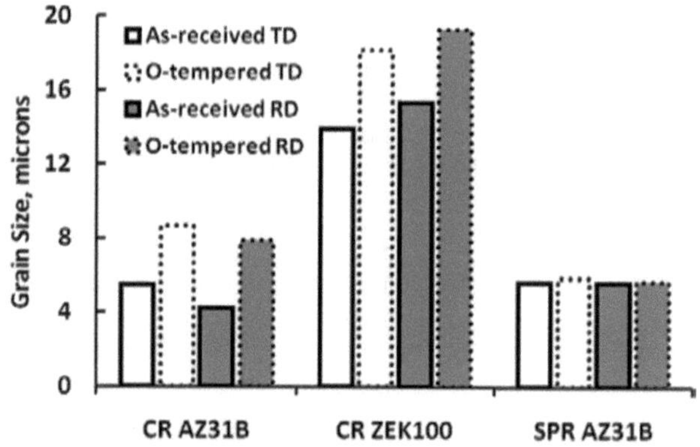

Figure 7. Mean intercept length of grains of three magnesium alloys in two tempers measured RD and TD.

**Magnesium Technology 2012**
*Edited by: Suveen N. Mathaudhu, Wim H. Sillekens, Neale R. Neelameggham, and Norbert Hort*
*TMS (The Minerals, Metals & Materials Society), 2012*

# Strain Hardening of ZK60 Magnesium Alloys

[1]Jae-Hyung Cho*, Suk-Bong Kang* and Sang-Ho Han*

**Korea Institute of Materials Science, 797 Changwondaero, Swongsan-gu, Changwon, Gyeongnam 641-831, Republic of Korea*

Keyword: ZK60; magnesium alloys; compression; texture and microstructure; EBSD

## ABSTRACT

ZK60(Mg-Zn-Zr) alloys possessed precipitate hardening behavior during aging, and the frequency and size of rod and disc shaped precipitates changed with aging. Strain hardening and texturing of ZK60 alloys were investigated during uniaxial warm compression. Uniaxial warm compression tests were carried out at various deformation rates and temperatures. Heat treatments of solid solution (T4) and aging (T6) affected flow stress behaviors. Aging heat treatment samples (T6) had higher yield stresses and lower strain hardening rates than solution heat-treated samples (T4). Both hardening by deformation and softening by dynamic recrystallization (DRX) at elevated temperatures were observed. Most of precipitates formed during aging contained Zn and Zr elements. Linear strain hardening behavior was examined using a visco-plastic self consistent (VPSC) model.

## INTRODUCTION

The Mg-Zn systems are typical of magnesium alloys with precipitation hardening, and microhardness and strength changed with aging [1, 2, 3, 4, 5]. Age hardening effect was also found in Mg-Ca-Zn [6] and Mg-Sn-Zn alloys [7]. Enhanced room-temperature tensile and creep properties of extruded Mg-Sn-Zn were reported because of $Mg_2Sn$ and $MgZn_2$. The ZK60 alloys are designed to add small amount of Zr elements to Mg-Zn systems and better mechanical properties than the usual Mg-Zn alloys were obtained.

In order to better understand deformation mechanism and texture evolution, a viscoplastic self-consistent (VPSC) formulation of polycrystalline aggregates has been widely used for hexagonal materials including Mg alloys [8, 9, 10, 11, 12, 13]. During deformation of the Mg alloys, basal slip systems usually operate, and other non-basal slip systems rarely operate at room temperature. This is because critical resolved shear stresses of non-basal slip systems are much higher than those of basal slip system at room temperature.

Here, we studied the variation of mechanical properties of the ZK60 alloys with aging heat treatments. Two kinds of specimens of solid solution heat treatment (T4), and followed by artificial aging (T6) were prepared to investigate aging effect on texture, microstructure and mechanical properties during uniaxial warm compression. Uniaxial compression tests were carried out at various temperatures and deformation rates. Evolution of texture and twinning was investigated using electron backscatter diffraction (EBSD). A visco-plastic self consistent (VPSC) model was used to predict flow curves, and texture evolution.

## EXPERIMENTAL PROCEDURE

The ZK60 magnesium alloys used in this study had a chemical composition of 5.47Zn-0.58Zr-Mg (in wt%). The alloys were originally fabricated by conventional direct chill casting (DC), and then solution heat-treated at 673 K (400 °C) for 15 hours (T4), and followed by artificial aging (T6) at 448 K (175 °C) for 12 hours. Fig. 1 shows hardness variation of ZK60 alloys during aging. Increase and decrease in hardness were related to variations of precipitate density with aging. Compression samples with artificial aging (T6) were made to have a hardness peak based on hardness variation in [4]. The hardness values after solution heat treatment (T4) were approximately 66 HV. It increased up to 74 HV after 12 hours aging at 448 K (175 °C). The microhardness values were measured using a load of 100 gf. The hardness dropped after 100 hours due to overaging.

Uniaxial compression tests were carried out using

[1] jhcho@kims.re.kr

**TABLE 1.** Chemical composition of ZK60 alloys.

| | Al | Zn | Mn | Si | Fe | Cu | Ni | Ca | Zr | Mg |
|---|---|---|---|---|---|---|---|---|---|---|
| ZK60 | 0.01 | 5.47 | 0.009 | 0.022 | 0.003 | 0.003 | 0.007 | 0.005 | 0.58 | Bal. |

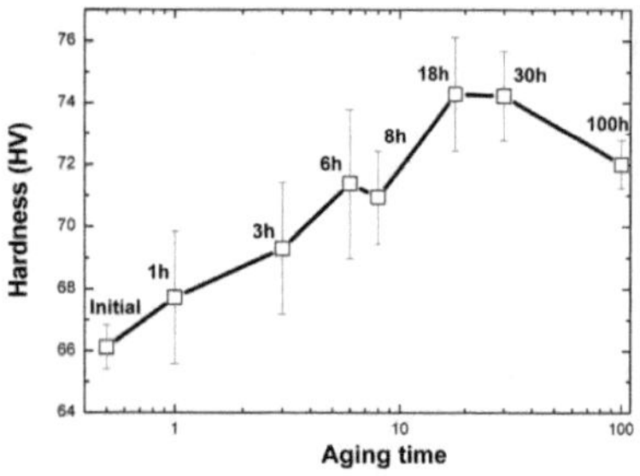


**FIGURE 1.** Hardness variation of ZK60 alloys with aging at 175°C [4]

Thermecmastor-Z (Fuji Electronic Industrial Co.). Cylindrical compression samples were 12 mm in length and 8 mm in diameter. Deformation temperatures for the compression were 473 K (200 °C), 523 K (250 °C) and 573 K (300 °C), and strain rate of 0.32/s.

Texture and microstructure were characterized using electron backscatter diffraction (EBSD), transmission electron microscopy (TEM), and neutron diffraction. Samples observed by OM were prepared using mechanical polishing and subsequent chemical etching with a solution of picric acid (3 g), acetic acid (20 ml), distilled water (20 ml), and ethanol (50 ml). Microstructure and microtexture analyses were made using an automated HR-EBSD (JEOL7001F) with HKL Channel5 and a generalized EBSD data analysis code, REDS [14]. EBSD samples were mechanically polished and then, electropolished using a solution of butyl cellosolve (50 ml), ethanol (10 ml) and perchloric acid (5 ml) at a voltage of 10 V and temperature of 253 K (−20 °C) – 258 K (−15 °C). More detailed examination of microstructure and precipitates were made using a JEM-2100F TEM operating at 200 kV. TEM samples were prepared by mechanical polishing down to about 100$\mu$m thick and then, electropolished using chemicals of methanol (60 ml), glycerin (30 ml) and nitric acid (10 ml). Temperature and voltage of the elctropolishing were 273-278K and 20-25V, respectively. TEM and SEM sample preparations were finished using an ion milling to prevent oxidation or other surface contamination. Quantitative analysis was carried out using energy dispersive spectroscopy (EDS).

## POLYCRYSTAL MODEL

Polycrystal plasticity models usually assume that the loading and displacement conditions at the boundary of the polycrystal are uniform, and the volume average of stress over all grains coincide with the overall stress at the boundary [10, 15, 16]. Taylor assumption of uniform strain throughout the polycrystal confirms strain compatibility and has been favored for modeling cubic materials which have multiple slip in the grains and mild anisotropy. Low symmetry materials, such as hcp are characterized by a variety of active deformation modes, twinning activity and severe directional anisotropy in the single crystals. These crystals cannot accommodate certain deformation components because of lack of the necessary deformation systems or because of high resolved shear stress (RSS). Self-consistent approach successfully modeled heterogeneous deformation of the materials with hcp symmetry.

The viscoplastic constitutive behavior in a single crystal is described by means of the non-linear rate-sensitivity equation,

$$\dot{\varepsilon}_{ij} = \sum_s m^s_{ij}\dot{\gamma}^s = \dot{\gamma}_0 \sum_s m^s_{ij} \left(\frac{m^s_{kl}\sigma_{kl}}{\tau^s_0}\right)^n \tag{1}$$

where, $\tau^s$ and $m^s_{ij}$ are the threshold stress and the symmetric part of Schmid tensor with slip and twinning system(s). $\dot{\gamma}_0$ and n is the normalization factor and rate-sensitivity exponent. $\dot{\gamma}_0$ and m are the normalization factor and strain-rate sensitivity exponent. It was assumed that the rate sensitivity depended on temperature and increased linearly from 0.05 at 273 K to 0.1 at 873 K.

The threshold stress, $\hat{\tau}^s$, describes the resistance to activation of slip and twinning systems, and increases with deformation. The evolution of the threshold stress is characterized by a Voce hardening model,

$$\hat{\tau}^s = \tau^s_0 + (\tau^s_1 + \theta^s_1\Gamma)(1 - \exp\left(-\Gamma\left|\frac{\theta^s_0}{\tau^s_1}\right|\right)) \tag{2}$$

where $\Gamma = \sum_s \triangle\gamma^s$ is the accumulated shear in the grain. The $\tau_0, \theta_0, \theta_1$ and $(\tau_0 + \tau_1)$ are initial critical resolved shear stress (CRSS), initial hardening rate, asymptotic hardening rate and the back-extrapolated CRSS, respectively.

A visco-plastic self consistent(VPSC) polycrystal model with a predominant twinning scheme was used to predict flow curves, twinning and texture evolution during uniaxial compression. And thus, both crystallographic slip and twinning were considered. One

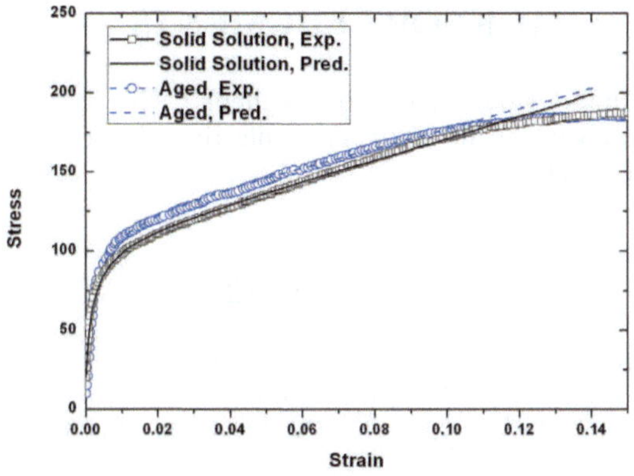


**FIGURE 2.** Comparison of experiments and predictions of flow curves. Warm compression tests were carried out at a strain rate of 0.32/s at 523 K (250 °C)

thousand single crystals with a Gaussian distribution (full width at half maximum, FWHM=15°) were generated from the orientation distribution function (ODF) computed from EBSD data, and were used for verification of VPSC modeling. Flow curves and twinning volume fractions were used to determine strain hardening parameters. Compression tests under a strain rate of 0.32/s at 523K were carried out to obtain flow curves. Experimental flow curves are compared with predicted flow curves in Fig. 2. Yielding followed by linear strain hardening was well predicted.

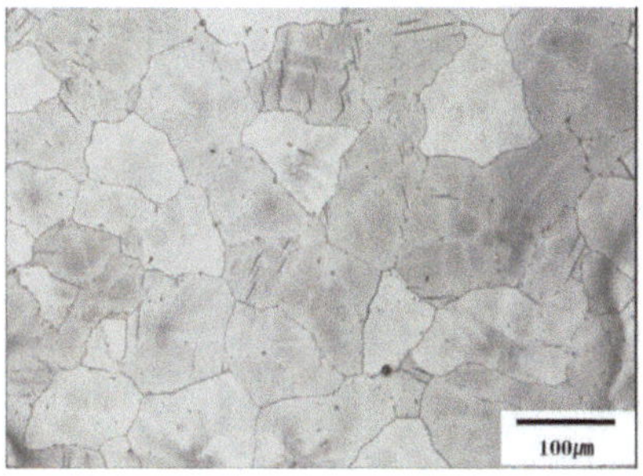


**FIGURE 3.** Optical micrographs of the initial ZK60 alloy

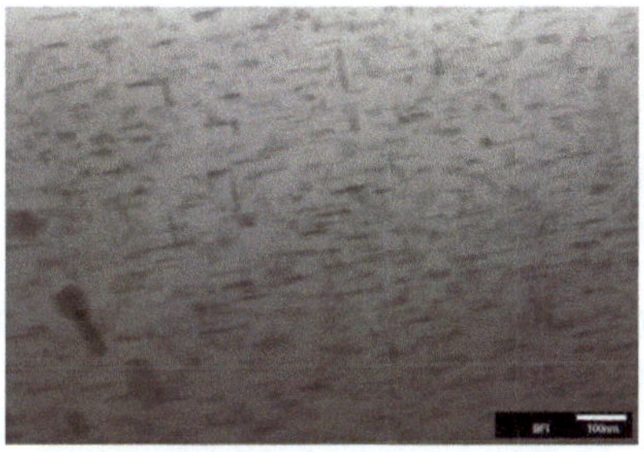

**FIGURE 4.** TEM micrographs of ZK60 alloys (the aged sample)

**FIGURE 5.** TEM micrographs of ZK60 alloys (the twinned area)

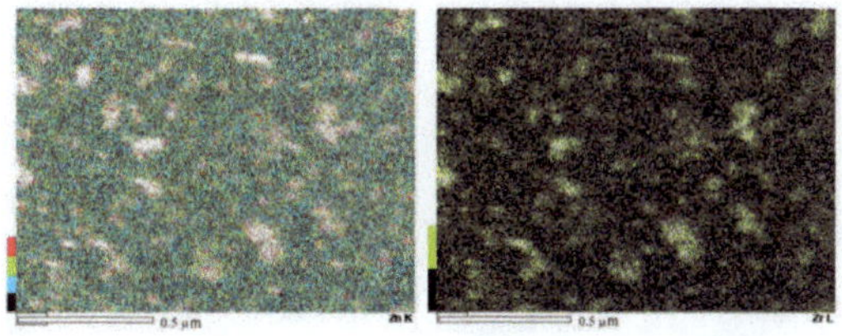

**FIGURE 6.** EDS of precipitates of ZK60 alloys

## RESULTS

Optical and TEM micrographs of as-cast, aged samples were presented in Figs. 3, 4. Most of the grain sizes of IC samples are more than $100\mu m$. Precipitates after aging of 12 hours at 448K are shown in Fig. 4. The aged samples were compressed with a reduction in area (RA) of 45% at 523K, and a dark field image (DFI) with a twin and its SAD (selected area diffraction) in the inset were displayed Fig. 5. Many precipitates were located inside the twinned region. Frequently, the precipitates were found along the boundary between twinned and untwinned region [17]. It seemed that twin propagation was negligibly affected by the nano-sized precipitates. EDS of the particles were presented in Fig. 6.

Stress-strain curves during uniaxial compression at various temperature are shown in Fig. 7. Flow curves at elevated temperatures displayed elastic, plastic strain hardening and softening behaviors. The softening after peak stress came from dynamic recrystallization during hot working process. Plastic strain hardening usually displays nonlinear flow stress curves with strain. Frequently,

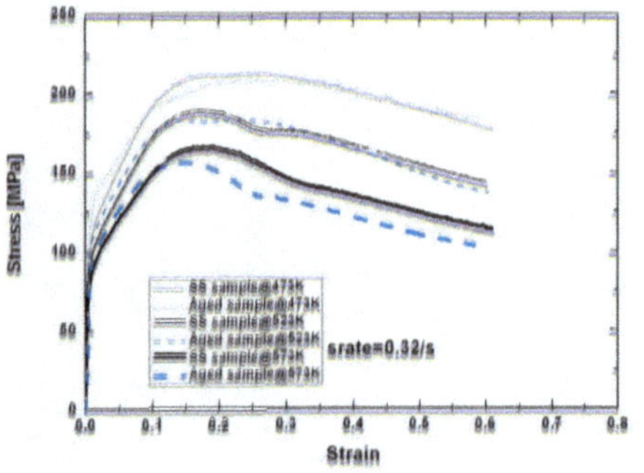


**FIGURE 7.** Flow curves at various temperature and strain rates, $\dot{\varepsilon}$, of 0.32/s

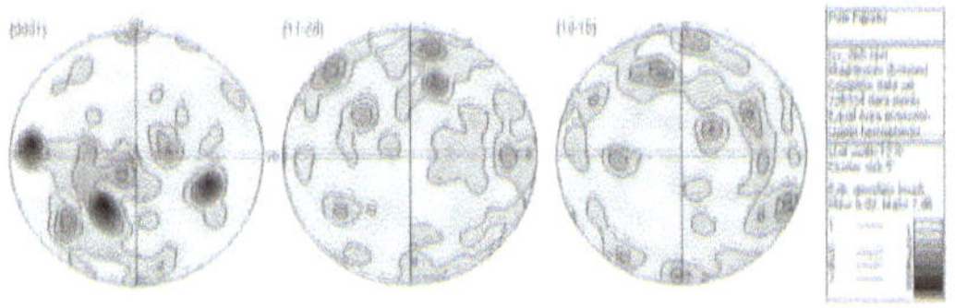

**FIGURE 8.** Pole figure measured from EBSD (T4 sample, $\varepsilon$=2.5%)

some mixture of slip and twinning even makes a more complicated distribution of sigmoidal curves in magnesium alloys [18, 19, 20].

The flow stress of ZK60 increased with strain rate. Negative temperature dependence of the flow stress was also found at all strain rates. There were several features to consider for a better understanding of aging effect on flow curves, i.e., yielding, strain hardening and softening behaviors.

Texturing of ZK60 alloys with a strain of $\varepsilon$=2.5% was compared in Figs. 8, and 9. At a strain of $\varepsilon$=2.5%, more basal intensity was observed in the aged sample than in the solution heat treated sample. At a further deformation, neutron diffraction was used and four complete pole figures were measured for texture analysis. More detailed discussion on aging effect on texturing and mechanical properties will be further presented based on VPSC modeling, EBSD and neutron diffraction.

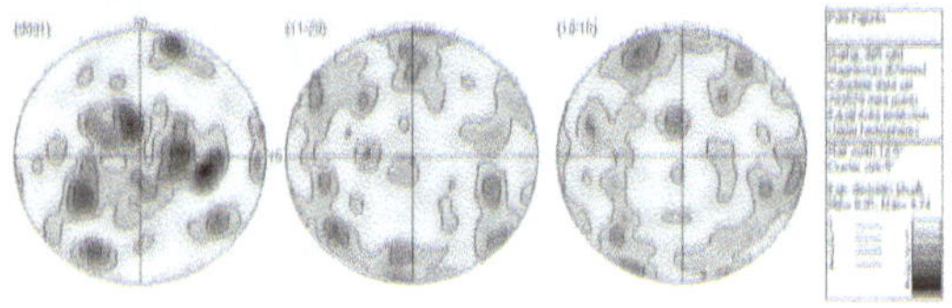

**FIGURE 9.** Pole figure measured from EBSD (T6 sample, $\varepsilon$=2.5%)

## ACKNOWLEDGMENTS

The authors would like to partly thank C. N. Tome and R.A. Lebensohn for their Visco-plastic Self Consistent (VPSC) code.

## REFERENCES

1. Wei, L. Y., Dunlop, G. L., and Westengen, H., *Metall. Trans. A*, **26A**, 1947–1955 (1995).
2. Wei, L. Y., Dunlop, G. L., and Westengen, H., *Metall. Trans. A*, **26A**, 1705–1716 (1995).
3. Gao, X., and Nie, J. F., *Scripta Materialia*, **56**, 645–648 (2007).
4. Cho, J. H., Jin, Y. M., Kim, H. W., and Kang, S. B., *Materials Science Forum*, **558-559**, 159–164 (2007).
5. Chen, H., Kang, S. B., Yua, H., Cho, J. H., Kim, H. W., and Mina, G., *Journal of Alloys and Compounds*, **476**, 324–328 (2009).
6. Oh, J. C., Ohkubo, T., Mukai, T., and Hono, K., *Scripta Materialia*, **53**, 675–679 (2005).
7. Sasaki, T. T., Ju, J. D., Honoa, K., and Shin, K. S., *Scripta Materialia*, **61**, 80–83 (2009).
8. Tome, C. N., Lebensohn, R. A., and Kocks, U. F., *Acta mater.*, **39**, 2667–2680 (1991).
9. Lebensohn, R. A., and Tome, C. N., *Acta mater.*, **41**, 2611–2624 (1993).
10. Tome, C. N., *Modeling Simul. Mater. Sci. Eng.*, **7**, 723–738 (1999).
11. Agnew, S. R., Tome, C. N., Brown, D. W., Holden, T., and Vogel, S., *Scr. Mater.*, **48**, 1003–1008 (2003).
12. Agnew, S. R., and Duygulu, O., *Int. J. Plast.*, **21**, 1161–1193 (2005).
13. Choi, S. H., Shin, E. J., and Seong, B. S., *Acta mater.*, **55**, 4181–4192 (2007).
14. Cho, J. H., Rollett, A. D., and Oh, K. H., *Metallurgical and materials transactions A*, **36A**, 3427–3438 (2005).
15. Hill, R., *J. Mech. Phys. Solids*, **15**, 79–95 (1967).
16. Kocks, U., Tome, C., and Wenk, H.-R., *Texture and Anisotropy*, Cambridge University press, The Edinburgh Building, Cambridge CB2 2RU, UK, 2000, ISBN 0-521-46516-8.
17. Cho, J. H., Chen, H. M., Choi, S.-H., Kim, H.-W., and Kang, S.-B., *Metallurgical and Materials Transactions A*, **41**, 2575–2583 (2010).
18. Choi, S. H., Kim, J. K., Kim, B. J., and Park, Y. B., *Materials Science and Engineering A*, **488A**, 458–467 (2008).
19. Barnett, M. R., Davies, C. H. J., and Ma, X., *Scripta Materialia*, **52**, 627–632 (2005).
20. Nave, M. D., and Barnett, M. R., *Scripta Materialia*, **51**, 881–885 (2004).

Magnesium Technology 2012
*Edited by: Suveen N. Mathaudhu, Wim H. Sillekens, Neale R. Neelameggham, and Norbert Hort*
*TMS (The Minerals, Metals & Materials Society), 2012*

# STRAIN-RATE EFFECTS OF SAND-CAST AND DIE-CAST MAGNESIUM ALLOYS UNDER COMPRESSIVE LOADING

J.P. Weiler[1], J.T. Wood[1]

[1] Department of Mechanical and Materials Engineering, University of Western Ontario, London, Canada, N6A 5B9

Keywords: Compressive impact testing; Strain-rate effects; Cast magnesium alloys; Constitutive modeling

## Abstract

The strain-rate effects of cast magnesium alloys were investigated with uniaxial compression and compressive impact testing. The compressive material response of specimens cut from sand cast AZ91, AE44, and AM60, and high-pressure die-cast AM60 was determined for strain-rates ranging from quasi-static levels to typical rates experienced during crash situations. Several different constitutive material models (Johnson-Cook, Cowper-Symonds, etc.) were used in an attempt to characterize the experimental results. These material models are typically available in commercial finite-element packages and can be used to model the resulting material response of die-cast automotive components produced with these alloys to more complex loading conditions. The resulting deformed microstructures and fracture surfaces of each alloy at different strain-rates were also analyzed.

## Introduction

Magnesium alloys are increasingly being employed in the automotive industry in a variety of different applications due to their excellent specific strength, ease of castability and potential for reductions in fuel consumption. Automotive components varying from instrument panel beams, steering wheel armatures and seat frames (AM60), to cam covers, clutch housings and steering wheels (AZ91), to high-temperature applications such as engine cradles (AE44) have been manufactured with cast magnesium alloys [1]. With increasing use of, and diversity of components, the response of these alloys to mechanical and dynamic loading must be fully understood. With an understanding of the material behaviour to these loading conditions, comprehensive material models can be developed in numerical simulation packages to characterize the structural behaviour of magnesium alloys. The material behaviour at increased strain-rates, or the crashworthiness of these alloys is of importance to the automotive industry to optimize the use of cast magnesium alloys in the automobile, and facilitate progress in achieving these material models. The crashworthiness of automotive materials can be characterized at a laboratory-scale by performing compression or impact testing at speeds in the range of 1-3 m/s. These tests are capable of deforming specimens at strain rates greater than $100s^{-1}$, approximating the conditions of automobile crashes; a 100 mm component compressed in a car crash at 50km/h realizes a strain-rate of approximately $140s^{-1}$.

In this work, the strain-rate effects of specimens from sand cast AE44, AZ91, and AM60, and high-pressure die-cast AM60 plates were investigated. Uniaxial compression testing and compressive impact testing were performed to achieve strain-rates ranging from $0.0002s^{-1}$ to $490s^{-1}$. Constitutive material models available in structural numerical packages were compared with the results. The resulting microstructures from these quasi-static and dynamic tests were compared with undeformed microstructures of each alloy.

## Strain-rate sensitivity in Mg alloys

Several researchers have investigated the strain-rate effects of cast magnesium alloys during both tensile and compressive loading at varying strain-rates [2-7]. These researchers find that both alloy content and grain size influence the strain-rate sensitivity of magnesium alloys. Increases in the aluminum content [2,4,8] and grain size [6] decreases the strain-rate sensitivity of magnesium-based alloys for strain-rates ranging from quasi-static levels to $10s^{-1}$. In this range of strain-rates, the strain-rate sensitivity of AZ91 is considered nearly negligible [2,4], while the strain-rate sensitivity of magnesium alloys with 2% aluminum content – AS21 and AM20 – is considered significant [4-5]. Conversely, small additions of the rare-earth element gadolinium to pure magnesium results in no affect to the tensile strain-rate sensitivity of pure magnesium at strains up to $10s^{-1}$, and the elongation values of the binary alloy are unaffected by increases in the strain-rate [7]. At a range of greater strain-rates, $15s^{-1}$ to $130s^{-1}$, the ultimate tensile strength of AM50, AM60 and AZ91 magnesium alloys increases, indicating a positive strain-rate sensitivity of each of these alloys [3]. At a strain-rate of approximately $1700s^{-1}$, the compressive ultimate stress values of magnesium alloys AM20, AM50 and AM60 increase, while the elongation values decrease compared with measured values at a strain-rate of $1s^{-1}$ [5]. The strain-rate sensitivity of AM20 was found to be the greatest; indicating that even at these large strain-rates, the strain-rate sensitivity of magnesium alloys is influenced by the aluminum content [5].

## Dynamic constitutive material models

Numerous constitutive material models characterizing strain-rate effects are available in literature and available for in numerical packages, such as LS-DYNA [8]. LS-DYNA implements several of these in the available library of material models including the Johnson-Cook model [9], the Cowper-Symonds model [10], and the Zerilli-Armstrong model [11-12], among others. The Johnson-Cook constitutive material model is based upon torsion and tension tests of fifteen different metals (copper, steel, etc.) at different strain rates and temperatures. This material model defines the flow stress, σ, of a material as [9]:

$$\sigma = \left(A + B\varepsilon_p^{\,n}\right)\left(1 + C\ln\frac{\dot{\varepsilon}}{\dot{\varepsilon}_o}\right)\left(1 - T^{*m}\right) \quad (1)$$

where: $\varepsilon_p$ is the plastic strain, $\dot{\varepsilon}$ is the strain rate, $\dot{\varepsilon}_o$ is the reference strain rate (usually $1s^{-1}$), $T^*$ is the homologous temperature, and the parameters *A, B, C, m* and *n* are experimentally determined constants. The quasi-static flow curve of a material can be characterized using *A* as the yield strength, and $B\varepsilon_p^{\,n}$ as the plastic portion of the flow curve. Aune et al. [3] determine Johnson-Cook parameters for die-cast AZ91D, AM50A and AM60B magnesium alloys. They report values of *C*, the strain-rate sensitivity parameter, of 0.019 for AZ91D and AM50A, and 0.023 for AM60B. Altehnof and Ames [13] use the material parameters reported by Aune et al. in numerical simulations of the impact of an AM50 steering wheel armature. They find that the strain-rate effect is a minimal contribution to the rate of deformation of the material at the strain-rates investigated.

The Cowper-Symonds constitutive material model is based upon bending impact tests of cantilever beams with a mass on one end. This material model defines the flow stress of a material due to increased strain-rate, as [10]:

$$\sigma = \sigma_o + \left(\frac{\dot{\varepsilon}}{D}\right)^{1/p} \qquad (2)$$

where: $\sigma_o$ is the quasi-static reference stress, and *D* and *p* are experimental determined material constants. The material constant *D* is the strain-rate at which the flow stress is twice the quasi-static reference stress. Song, Beggs, and Easton [5] propose the material parameters *D* and *p* for several die-cast magnesium alloys including AM60. They report that $D = 3300s^{-1}$ and $p = 1.1$. These researchers also report, however, that the Cowper-Symonds constitutive model does not characterize magnesium alloys well at strain-rates greater than $1000s^{-1}$, as it does not account for the strain-rate sensitivity of the work-hardening behaviour of these alloys.

The Zerilli-Armstrong material model [11] is a 'dislocation-mechanics' based constitutive equation building upon the analysis and data presented by Johnson and Cook [9]. Zerilli and Armstrong give equations for the flow stress of face-centered cubic (FCC) and body-centered cubic (BCC) metals in Ref. [11]. They report in Ref. [12] that since the flow curves of hexagonal close-packed (HCP) metals, such as magnesium alloys, are somewhere between FCC and BCC metals, the dislocation-mechanics based constitutive equation for HCP metals is characterized by a combination of interactions predominant in BCC and FCC metals. However, as of the writing of this paper, the material model library available in LS-DYNA [8] only includes constitutive equations for the flow stress for BCC and FCC metals. The Zerilli-Armstrong constitutive material model defines the flow stress of a BCC material, for example, as [11]:

$$\sigma = c_1 + c_2 \exp\left(-c_3 T + c_4 T \ln \dot{\varepsilon}\right) + c_5 \varepsilon^n \qquad (3)$$

where: $c_1$, $c_2$, $c_3$, $c_4$ and $c_5$ are experimentally determined constants, *T* is the ambient temperature, and $c_1 + c_5\varepsilon^n$ characterizes the quasi-static flow curve (in this work, we use similar values of $c_3$ and $c_4$ as those reported in [11]).

## Experimental Procedures

Materials. The sand casting was performed at CANMET-MTL, in Ottawa, Canada with three different commercially available magnesium alloys: AM60, AZ91 and AE44. The sand casting experiments resulted in a step-casting measuring 330 mm in length, 100 mm in width, and thicknesses of increasing steps, ranging from 4 mm to 40 mm at the feeding riser (Figure 1). The 4 mm step of these castings was used in other studies [14-15]. For compressive and impact testing, rectangular specimens were cut from the nominally 8 mm thick step measuring 5 mm x 5 mm in cross-sectional area (Figure 1). This region of the casting solidified at a rate of approximately 0.5 °C/s, producing grain sizes in the range of 30-50 μm [15].

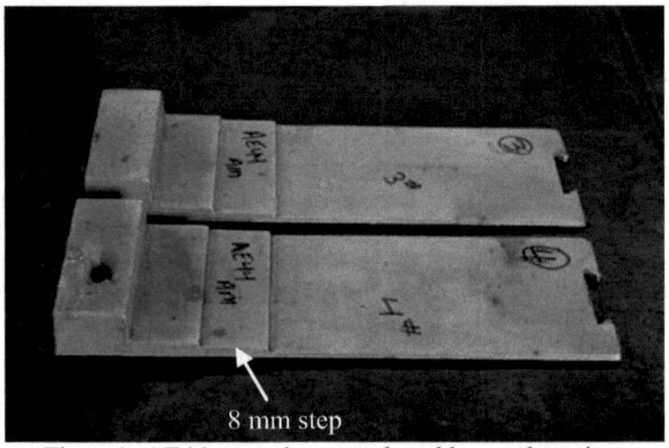


Figure1: AE44 step-plates produced by sand-casting. Specimens were cut from the 8 mm thick step.

The high-pressure die-cast AM60 (HPDC-AM60) test plates were cast at Meridian Lightweight Technologies, in Strathroy, Canada. Rectangular specimens measuring 5 mm x 5 mm in cross-sectional area were cut from the nominally 3 mm casting thickness of the test plates. The chemical composition of each of the alloys investigated is given in Table 1.

Quasi-static and dynamic testing. Uniaxial compression testing was completed on an Instron 8804 load frame with two parallel steel plates until fracture. The crosshead speed was used to determine the initial strain-rate of each test. The tests were completed over three or four magnitudes of strain-rate ranging from $0.0002s^{-1}$ to $0.2s^{-1}$. The strain was calculated from the crosshead displacement corrected for machine compliance. The absorbed energy was calculated using the central difference method on each compressive stress-strain curve.

Compressive impact testing was completed on an Instron Dynatup 9250HV instrumented impact drop tower with a maximum dynamic force of 44 kN. During each test, the time, deflection, velocity, change in kinetic energy, and force of the impactor were recorded. To obtain the widest range of strain-rates, impact tests were performed: 1) at the minimum velocity (approximately 0.8 m/s) with sufficient mass to ensure fracture of the specimen, and 2) with the minimum mass possible at sufficient velocity to ensure fracture. Additional tests were completed at velocities between the velocities of 1) and 2). At least six compression and impact tests were performed at each strain-rate to realize statistical significance.

Table 1: Chemical composition, in weight percentages, of AM60, AZ91 and AE44 magnesium alloys (the Re component in AE44 is a combination of the rare earth elements Ce, La, Nd, and Pr).

| Alloy | Mg | Al | Zn | Mn | Si | Cu | Fe | Ni | Be | Re |
|---|---|---|---|---|---|---|---|---|---|---|
| AM60 | 93.6 | 6.1 | ----- | 0.32 | 0.005 | 0.001 | 0.001 | 0.001 | 0.001 | ----- |
| AZ91 | 90.3 | 8.7 | 0.77 | 0.23 | 0.006 | 0.002 | 0.002 | 0.002 | 0.001 | ----- |
| AE44 | 92.4 | 3.7 | 0.01 | 0.20 | 0.004 | 0.001 | 0.000 | 0.001 | 0.001 | 3.8 |

Specimen Analysis and Metallography. During impact testing, an AOS Technologies Motionscope M3 high-speed video camera was used to capture the deformation and fracture process of several of the specimens of each alloy at different strain-rates. High-speed video of the impact test was captured at rates of 1000, 2000, 4000 and 8000 frames per second for several of the magnesium alloy specimens. Each frame captured during an impact test was correlated to the corresponding stage of mechanical deformation.

Following testing, the microstructures and fracture surfaces of the typical specimens from each strain-rate and alloy were analyzed and compared with the microstructure of undeformed specimens. Specimens were cold-mounted in an epoxy mount, and successively ground with 180, 400, and 1200 silicon carbide grit paper. The specimens were then finely polished in a slurry of first 0.3μm, and then 0.05μm, alumina. The grain-structure of each specimen was revealed with an etchant consisting of 75 mL of distilled water, 25 mL of glycol and 1 mL of nitric acid.

## Experimental Results

Uniaxial compression and impact testing was performed on the sand-cast alloys (AM60, AE44 and AZ91) with the 8mm length of the specimen parallel to the direction of loading. The testing was performed on the high-pressure die-cast AM60 alloy with the 3mm thickness in both parallel and perpendicular orientations to the loading direction, to investigate the effects of the skin region on the mechanical properties and to obtain a wide range of strain-rates using the impact drop tower.

Compression Testing. Uniaxial compression testing was performed at low strain-rates ($0.0001 s^{-1}$ to $0.01 s^{-1}$) to establish the quasi-static compressive response of these alloys, and at strain rates of approximately $0.1 s^{-1}$ to investigate effects of small increases in the strain-rate. Table 2 shows the average yield strength (for HPDC-AM60 specimens; the yield strength of the sand cast alloys is approximately 50 MPa [14], and was not easily determined during impact testing due to the initial inertial load peak), ultimate compressive strength, fracture strain, and absorbed energy for each of the magnesium alloy specimens at the strain-rates investigated. The fracture strain and absorbed energy, for the purpose of this study, were defined as the strain or energy value at the maximum load or point of crack initiation. The fracture of the specimen subsequently occurred with growth of the crack(s).

Table 2 shows that there is no significant increase in the ultimate compressive strength or fracture strain (and as a result, absorbed energy) over the 3 or 4 magnitudes of strain-rates investigated for each alloy, although there is a small increase in ultimate compressive strength for AE44 at a strain-rate of $0.1 s^{-1}$. The table also shows that there is an increase in yield strength, a minimal decrease in fracture strain, and essentially no change in ultimate compressive strength for the HPDC-AM60 specimens with the skin region parallel to the loading direction. The differences are possibly due to the influence of the skin region under direct axial compression. Although not shown, the absorbed energy was also determined for each specimen at final fracture of the specimen. These results indicated that the fracture process geometry of the specimen contributes to this measure of absorbed energy. HPDC-AM60 specimens with a 3mm initial height (perpendicular) resulted in two shear cracks meeting in a *'V'*, while HPDC-AM60 specimens with a 5mm initial height (parallel) resulted in one shear crack. The sum of these crack lengths was directly proportional to the difference in absorbed energy between the two specimen geometries. As a result, we focus our attention on the initiation of fracture for this study. The results shown in Table 2 were used to determine the typical quasi-static response to compare with impact testing results at increased strain-rates ($>100 s^{-1}$).

Impact Testing. The signal recorded by the impact testing equipment contains oscillations in the load versus time output curves due to mechanical resonance and an initial load spike due to inertia loading. Using filtering techniques, the mechanical properties of the specimen under impact loading can be calculated from the load signal [16]. Further, the resulting images from the high-speed video captured of the impact event were used to validate the filtering process. Figure 2 shows the stress-strain curve of an unfiltered signal and the resulting filtered signal of a HPDC-AM60 magnesium alloy specimen under a strain rate of $165 s^{-1}$. The figure also shows high-speed video images corresponding to the mechanical response, recorded at 4000 frames per second.

Figure 2 shows the initial load peak due to inertia and the mechanical resonance resulting in several oscillations. The filtered signal smooths these oscillations and ignores the initial interial load peak. Figure 2 also shows images captured from high-speed video characterizing the deformation and fracture of the HPDC-AM60 specimen. The time stamp of these images corresponds with the load signal captured by the drop tower equipment. The crack visible in Figure 2-iii corresponds to the decreasing load values. Figure 2 also shows an increase in the ultimate compressive stress, fracture strain (and absorbed energy) when compared with the quasi-static mechanical propreties given in Table 2 (HPDC-AM60 parallel specimens).

Table 2: Average mechanical properties from uniaxial compression testing of the magnesium alloys.

| Alloy | Strain-rate ($s^{-1}$) | Yield Strength (MPa) | Ultimate Compressive Strength (MPa) | Fracture Strain | Absorbed Energy ($MJ/m^3$) |
|---|---|---|---|---|---|
| AM60 | 0.0002 | | 290.6 | 0.150 | 30.4 |
| | 0.002 | | 287.7 | 0.145 | 29.4 |
| | 0.02 | | 288.3 | 0.153 | 30.8 |
| | 0.1 | | 293.8 | 0.139 | 28.0 |
| AZ91 | 0.0002 | | 293.8 | 0.135 | 26.6 |
| | 0.002 | | 298.6 | 0.135 | 27.7 |
| | 0.02 | | 302.9 | 0.146 | 30.6 |
| | 0.1 | | 293.1 | 0.132 | 25.6 |
| AE44 | 0.0002 | | 239.0 | 0.135 | 21.5 |
| | 0.002 | | 239.5 | 0.122 | 18.9 |
| | 0.1 | | 248.8 | 0.121 | 20.1 |
| HPDC-AM60 perpendicular | 0.002 | 141.2 | 372.3 | 0.149 | 37.8 |
| | 0.02 | 142.0 | 377.1 | 0.152 | 38.8 |
| | 0.2 | 144.4 | 372.8 | 0.142 | 36.5 |
| HPDC-AM60 parallel | 0.001 | 148.8 | 370.7 | 0.130 | 33.0 |
| | 0.01 | 154.7 | 375.5 | 0.135 | 35.8 |
| | 0.1 | 153.6 | 376.5 | 0.134 | 35.4 |

Microstructure and Deformation. The microstructure and deformation of the magnesium alloy specimens were investigated using high-speed video of the impact testing and metallography of specimens of typical results for each strain-rate to compare with the undeformed microstructures. High-speed video analysis shows that a small amount of specimen barreling occurs while the specimen is under load (Figure 2-ii), and compression and subsequent fracture occur due to the initiation and growth of shear cracks (Figure 2-iii). The high-speed video analysis also shows evidence of two shear cracks meeting in a *'V'* for the HPDC-AM60 specimens with an initial height of 3mm. Metallographic techniques on the magnesium alloys reveal that at increasing strain-rates, a larger number of small shear cracks appear in the specimen. At quasi-static rates, typically, the specimen contains only the crack that fractures the specimen. The microstructure of quasi-static specimens also reveals minimal shear banding at the fracture surfaces. However, at increasing strain-rates of approximately $250s^{-1}$, there is evidence of several shear bands in the deformed microstructures of several of the specimens, predominantly near the shear cracks (Figure 3).

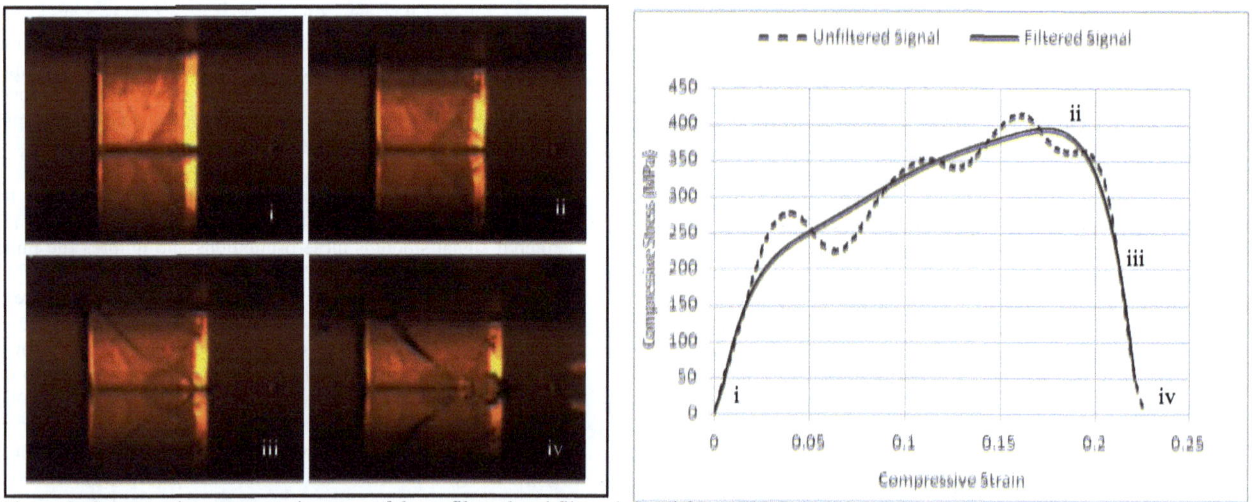


Figure 2: Compressive stress-strain curve of the unfiltered and filtered signal for a HPDC-AM60 specimen under an impact load at a strain rate of $165s^{-1}$, and corresponding images captured from high-speed video showing the deformation and fracture of the specimen.

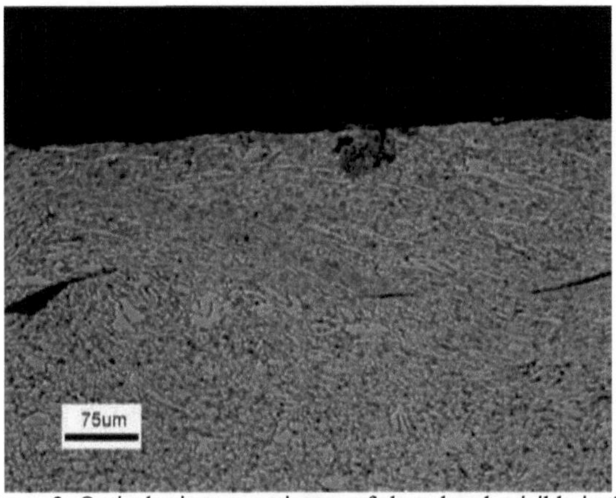

Figure 3: Optical microscope image of shear bands visible in the deformed microstructure of a HPDC-AM60 specimen following impact testing at a strain-rate of 300s$^{-1}$ (X200).

## Discussion

The impact testing results indicate greater values of the ultimate compressive strength at increased strain-rates compared with the quasi-static values in Table 2 (i.e. Figure 2). We attempt to characterize this increase of these magnesium alloys with the constitutive models given in Equations (1) – (3) found in typical numerical simulation packages, such as LS-DYNA. The flow stress characterized by each of these constitutive models at increased strain-rates can be normalized to the quasi-static reference stress either by division (Johnson-Cook and Cowper-Symonds), or subtraction (Zerilli-Armstrong). As a result, we can investigate the *increase* in ultimate compressive strength from the quasi-static values given in Table 2 due to strain-rate effects. Figure 4 shows the normalized compressive strength (impact compressive strength /quasi-static compressive strength) plotted as a function of strain-rate for each specimen (the HPDC-AM60 data for specimens with the skin region perpendicular and parallel to the loading direction are plotted as one data set). Figure 4 shows that at strain-rates of approximately 100s$^{-1}$, the data indicates that the compressive strength increases by approximately 2-3% for each alloy. However, the experimental scatter in the ultimate compressive strengths shown in Figure 4 precludes any definitive evaluation of the data at this strain-rate.

Figure 4 also shows that at strain-rates up to 200s$^{-1}$, there is no significant increase in the compressive strength of sand-cast AE44 specimens. Although not shown, there was also no significant increase in fracture strain or absorbed energy at these strain-rates for this alloy. The lack of strain-rate sensitivity for this rare-earth magnesium alloy may be due to the rare-earth alloying elements, as reported in Ref. [7], or the larger grain sizes found in sand castings, as reported in Ref. [6], or a combination of both factors. Further investigation into the relationships between the strain-rate sensitivity of this alloy and the alloying content and grain size is ongoing.

Figure 4 shows that for the sand cast AZ91 alloy, at strain-rates up to 300s$^{-1}$, there is an increase of approximately 4%, while for the sand cast AM60 alloy, at strain-rates up to 250s$^{-1}$, there is an increase of approximately 3% in the ultimate compressive strengths compared with the quasi-static values. Similar increases were observed for the absorbed energies of these alloys. The figure further shows that the increases in compressive strength for the HPDC-AM60 alloy are greater in magnitude than the sand cast AM60 alloy at similar strain-rates, and increase to approximately 14% at a strain-rate of 490s$^{-1}$. Absorbed energy values at these strain rates are also greater. The discrepancy between the HPDC-AM60 and sand cast AM60 sensitivity to strain-rate is possibly due to the differences in grain sizes between these two casting methods [6]. However, further work with a wider range of grain sizes is required to determine any definitive relationship between the strain-rate sensitivity and grain size of cast AM60.

The constitutive models presented in Equations (1) – (3) were used with strain-rate sensitivity parameters of AM60 from literature – a Johnson-Cook parameter of $C = 0.023$ from Aune et al. [3], and Cowper-Symonds parameters of $D = 3300\text{s}^{-1}$ and $p = 1.1$ from Song, Beggs, and Easton [5] – in an attempt to characterize the effects of strain-rate on the compressive strength of this magnesium alloy. We restrict our analysis to AM60 due to the number and range of strain-rates at which we were able to perform tests. The parameters for the Zerilli-Armstrong constitutive model are typically determined from a suite of tests performed at different strain-rates and temperatures [11-12], however, as of the writing of this manuscript, no Zerilli-Armstrong parameters for magnesium alloys have been reported. For the purpose of this study, we empirically fit the curve using parameters of similar magnitude to those found in Ref. [11] to our experimental data ($c_3 = 0.005$ K$^{-1}$, $c_4 = 0.0008$ K$^{-1}$, and a $c_2$ value of 50 MPa, to obtain results comparable in magnitude to our experimental data). Figure 5 shows the data from Figure 4 for the AM60 magnesium alloy plotted with the Johnson-Cook, Cowper-Symonds and Zerilli-Armstrong constitutive material models normalized to the quasi-static reference stress in each equation. The figure shows that the Johnson-Cook and Zerilli-Armstrong models overestimate the compressive strength increase of AM60 at strain-rates up to approximately 500s$^{-1}$. The Cowper-Symonds model accurately characterizes the effects of the strain-rate on the compressive strength of AM60, and is determined to best fit the experimental data. While the Cowper-Symonds equation is limited to strain-rates less than 1000s$^{-1}$ [5], as it lacks a parameter to account for the increase of work-hardening due to strain-rate effects, it is sufficient for the range of strain-rates investigated here (100s$^{-1}$-500s$^{-1}$), and typically experienced in automotive crashes.

## Conclusions

This work investigates the compressive strain-rate effects on the mechanical properties of sand-cast AE44, AZ91 and AM60 and die-cast AM60 magnesium alloys using uniaxial compression and impact testing. The results from this work indicate that the lower strain-rate sensitivity of the sand cast alloys may result from the larger grain sizes. Further, there was no indication of any strain-rate effects for the rare-earth magnesium alloy, AE44. It was determined that the Cowper-Symonds constitutive material model best characterizes the strain-rates effects of the HPDC-AM60 alloy at strain-rates between 150s$^{-1}$ and 500s$^{-1}$. At these strain-rates, a greater number of shear cracks and shear banding develop leading to an increase in the absorbed fracture energy of the material.

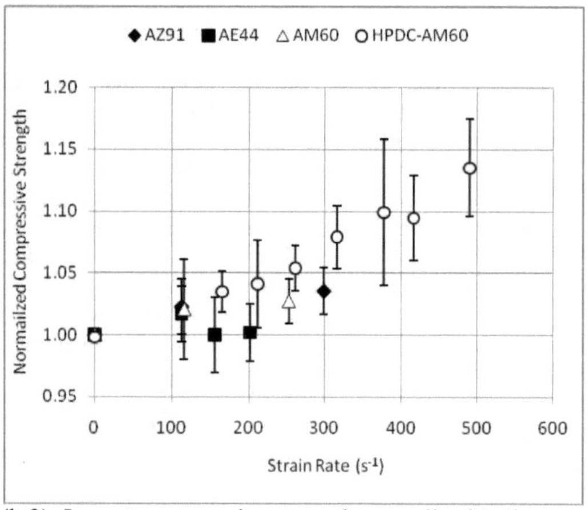


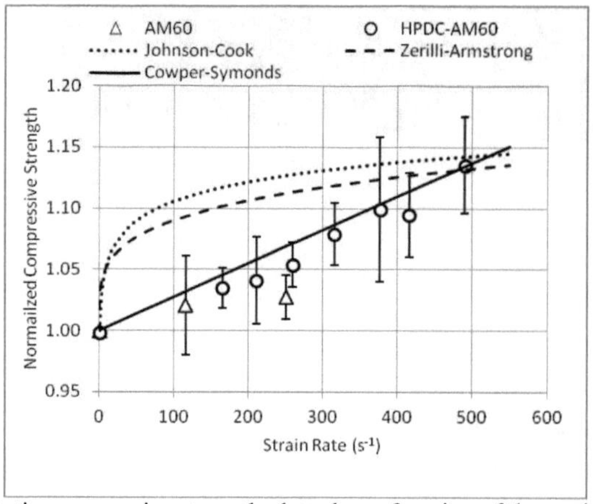


Figure 4 (left): Impact compressive strength normalized to the quasi-static compressive strength plotted as a function of the strain rate for each sand cast alloy: AE44, AZ91 and AM60, and for HPDC-AM60.
Figure 5 (right): Constitutive material models plotted to characterize the compressive strengths in Figure 4 of sand cast and HPDC-AM60.


## Acknowledgements

The authors of this work would like to gratefully acknowledge the significant contributions of Mr. Junming Li in performing the casting experiments. We are grateful as well to Meridian Lightweight Technologies Inc. for access to the materials, equipment and facilities that made this work possible and the Resource for the Innovation of Engineering Materials (RIEM) program funded by CANMET (Canada Centre for Mineral and Energy Technology), a Natural Resources Canada laboratory. This research is funded by Meridian Lightweight Technologies Inc. and the AUTO21 Network of Centres of Excellence, an automotive research and development program focusing on issues relating to the automobile in the 21st century. AUTO21, a member of the Networks of Centres of Excellence of Canada program is funded by the Natural Sciences and Engineering Research Council (NSERC), the Social Science and Humanities Research Council (SSHRC) and multiple industry and government partners.


## References


1. T. Rzychoń, and A. Kiełbus, "The influence of wall thickness on the microstructure of HPDC AE44 alloy," *Archives Mater. Sci. Eng.*, 28 (8) (2007), 471-474.
2. C.A. Newland, and M.T. Murray, "Strain rate dependent behaviour of Magnesium-based alloys," *Proceedings of the First Australasian Congress on Applied Mechanics: ACAM-96*, ed. R.H. Grzebieta (Barton, ACT, Australia: Institution of Engineers, Australia, 1996), 73-76.
3. T.K. Aune et al., "Behavior of die cast Magnesium alloys subject to rapid deformation," *SAE 2000 World Congress*, (2000) TP#: 2000-01-1116.
4. T. Abbott, M. Easton, and R. Schmidt, "Magnesium for crashworthy components," *Magnesium Technology 2003*, ed. H.I. Kaplan (Warrendale, PA: The Minerals, Metals and Materials Society, 2003), 227-230.
5. W.Q. Song, P. Beggs, and M. Easton, "Compressive strain-rate sensitivity of magnesium-aluminum die casting alloys," *Mater. Design*, 30 (2009), 642-648.
6. H.J. Choi et al., "Deformation behavior of magnesium in the grain size spectrum from nano- to micrometer," *Mater. Sci. Eng. A*, A527 (2010), 1565-1570.
7. N. Stanford et al., "Effect of Al and Gd solutes on the strain rate sensitivity of magnesium alloys," *Metall. Mater. Trans. A*, 41A (2010) 734-743.
8. *LS-DYNA Keyword User's Manual*, Version 971, (Livermore, CA: LSTC, 2007).
9. G.R. Johnson, and W.H. Cook, "A constitutive model and data for metals subjected to large strains, high strain rates and high temperatures," *Proc. 7th Inter. Symposium Ballistics* (ADPA, 1983), 541.
10. G.R. Cowper, and P.S. Symonds, "Strain hardening and strain-rate effects in the impact loading of cantilever beams," *Brown University Division of Applied Mathematics Report 28* (1957).
11. F.J. Zerilli, and R.W. Armstrong, "Dislocation-mechanics-based constitutive relations for material dynamics calculations," *J. Appl. Phys.*, 61 (5) (1987), 1816-1825.
12. F.J. Zerilli, and R.W. Armstrong, "Constitutive equation for HCP metals and high strength alloy steels," *ASME AD-Vol. 48*, (New York, NY: American Society of Mechanical Engineers, 1995), 121-126.
13. W. Altenhof, and W. Ames, "Strain rate effects for aluminum and magnesium in finite element simulations of steering wheel armature impact tests," *Fatigue Frac. Eng. Mater. Struct.*, 25 (2002), 1149-1156.
14. J.P. Weiler, J.T. Wood, and I. Basu, "Process-structure-property relationships for cast magnesium alloys," *Accepted to THERMEC 2011*.
15. J.T. Wood, J.P. Weiler, and I. Basu, "Study of solidification of magnesium alloys during high pressure die casting," *Accepted to THERMEC 2011*.
16. N.D. Alexopoulos, and A. Stylianos, "Impact mechanical behaviour of Al-7Si-Mg (A357) cast aluminum alloy. The effect of artificial aging," *Mater. Sci. Eng. A*, A528 (2011), 6303-6312.

**Magnesium Technology 2012**
*Edited by: Suveen N. Mathaudhu, Wim H. Sillekens, Neale R. Neelameggham, and Norbert Hort*
***TMS (The Minerals, Metals & Materials Society), 2012***

# MECHANICAL PROPERTIES OF MG ALLOYS AMX602 AND AZXE7111 UNDER QUASI-STATIC AND DYNAMIC LOADING

J. Shen[1, 2], K. Kondoh[3], T.L. Jones[4], S.N. Mathaudhu[4], L.J. Kecskes[4] and Q. Wei[1]

[1] Department of Mechanical Engineering, University of North Carolina at Charlotte, Charlotte, NC 28223-0001, USA
[2] School of Aeronautics, Northwestern Polytechnical University, Xi'an 710072, China
[3] Joining and Welding Research Institute, Osaka University, 11-1 Mihogaoka, Ibaragi, Osaka 567-0047, Japan
[4] US Army Research Laboratory, 4600 Deer Creek Loop, MD 21005-5069, USA

Key words: Magnesium alloys, Mechanical properties, Quasi-static loading, Dynamic loading

## Abstract

Mg and its alloys have drawn great interest in the materials community due to their high specific strength and potential applications in industry. In this work, we present some experimental results on two Mg-alloys, AMX602 and AZXE7111. We have evaluated the dependence of the mechanical properties of these alloys on the extrusion temperatures under both quasi-static (strain rate ~$1\times10^{-3}$ $s^{-1}$) and dynamic (strain rate ~$4\times10^{3}$ $s^{-1}$) uniaxial compressive loadings. We have observed that the quasi-static yield strength of AMX602 exhibits a slight dependence on the extrusion temperature, whereas the ultimate strength and the deformation-to-failure do not show such dependence. On the other hand, the dependence of the mechanical properties of AZXE7111 is much more complicated. We have also found that the deformation-to-failure of both alloys increases at increased strain rate. For comparison, we have tested two conventional Mg-alloys, WE43 and AZ91C under very similar conditions. We have found that AMX602 and AZXE7111 show significantly improved mechanical properties compared to WE43 and AZ91C under either quasi-static or dynamic loading. Such knowledge is particularly important for applications involving impact loading.

## Introduction

Magnesium is one of the lightest metals of practical interest in many application areas [1, 2]. Magnesium-based alloys have been used, and are showing prominently strong potential as candidates in applications such as automotive, railway, electronic packaging and especially aerospace in light of their extraordinarily low specific weight. In the past few decades, alloy design and fabrication has been an arena of intensive attention in the materials science and engineering community. Among the drawbacks associated with Mg-alloys are their limited tensile ductility (due to the lack of five independent slip systems in single crystal Mg), their relatively poor corrosion resistance, and their anisotropy in mechanical properties. The limited tensile ductility which entails undesirable workability of Mg-alloys has particularly become one of the stumbling blocks for their wide application.

In view of this, in the past dozens of years, a great amount of efforts have been exercised to improve the mechanical properties of pure Mg and Mg-alloys, primarily through the various ways of refining the grain size and modifying the texture [3-5]. For instance, a number of effective approaches have been used in obtaining ultrafine-grain (UFG, grain size d within 100 nm < d ≤ 1000 nm) or nanocrystalline (NC, d<100 nm) microstructures. Techniques involved in achieving such microstructures include electro-deposition [6], powder metallurgy [7], and severe plastic deformation (SPD) [8]. In recent years, SPD methods have been widely used in producing UFG/NC microstructures in Mg-alloys as well as in other metals and alloys such as copper, nickel, steels, tungsten, iron, niobium and tantalum [9-11]. Equal-channel angular extrusion (ECAE) has attracted special interest in this respect, which was also employed to modify the texture of Mg and its alloys [3, 5, 12].

In the past, a great deal of research efforts as evidenced by the large volume of literature has been focused in part on modifying the microstructures of traditional AZ-series Mg-alloys by SPD processes [13-15]. This is based on the observation that although as-cast Mg/Mg-alloys exhibit brittle behavior at ambient temperature, the ductility can be improved by subjecting the ingots to SPD at moderate temperatures. Mukai and co-workers [16] explained that the enhanced ductility is because of the redistribution of basal planes (0001) within the alloy during the SPD process. However, the increase in ductility is at the cost of the yield stress. By using X-ray diffraction (XRD), Feng et al. [17] confirmed that the strong texture initially observed in the as-extruded Mg alloy can be mitigated by ECAE processing. Agnew and co-workers [5] reported a strong orientation effect in eight-path ECAE processed AZ31B and argued that the orientation and its distribution are much more important than any other effects such as the Hall-Petch relationship in determining the yield strengths of the Mg alloys with hexagonal close packed (HCP) structure. Kim et al. [13] also observed ductility improvement and some loss in the yield strength of AZ61 after ECAE. They explained that the change in texture has contributed to the overall change in mechanical behavior.

Besides grain refinement and texture modification, a few researchers have also paid attention to alternative approaches in order to improve the mechanical properties of Mg-alloys, such as Mg-based metal matrix composites [18, 19]. Among such efforts, we have recently developed some new Mg-alloys based on certain concepts of alloy design and fabrication. In particular, in the present work, we will describe some mechanical testing results on two newly developed Mg-alloys, AMX602 and AZXE7111. Fine powders of these alloys were produced by the spinning water atomization process (SWAP), followed by compaction and extrusion at different temperatures. Details of alloy design and production have been described elsewhere [20-22]. Elsayed et al. [23], reported that the newly developed Mg-alloys show great improvement in quasi-static tensile properties against conventional cast ones that are extruded in the same way as the SWAP ones. They have also investigated the microstructure of these newly developed alloys including grain size, crystal orientations and second-phase particles [20-23]. In the present work, we have investigated the mechanical properties of these two Mg-alloys under uniaxial compressive quasi-static and dynamic loading. These alloys were prepared by the spinning water atomization process (SWAP), in which gas atomization is combined with water atomization to produce very high solidification rates (of about $10^6$ K/s), resulting in fine microstructures and super-saturation of alloying elements, and then followed by cold compaction at room temperature with a pressure of 600 MPa. They were finally extruded at temperatures of 573~673 K at an extrusion ratio of about 37 [22]. For the purpose of comparison, we have also examined the traditional Mg-alloys WE43 and AZ91C.

## Experimental

### Materials and Mechanical Testing Samples

In the present study, Mg-alloys of AMX602 (Mg-6%Al-0.5%Mn-2%Ca, all in weight percentage) and AZXE7111 (Mg-7%Al-1%Zn-1%Ca-1%La, all in weight percentage) have been investigated. Both alloys were fabricated at the Joining and Welding Research Institute, Osaka University, Japan. Briefly, spinning water atomization (SWA) technique was used in producing the powders of the alloys with desired chemical composition. The powders were then collected, compacted and consolidated by extrusion at various temperatures. The details of the alloy production have been provided by Kondoh and co-workers elsewhere [20]. Three different rods of each alloy (from which the samples were machined) were extruded at three different temperatures. Specifically, for AMX602 the extrusion temperatures are 350 °C, 300 °C and 250 °C, and for AZXE7111 the extrusion temperatures are 350 °C, 250 °C and 200 °C. Samples were cut from these rods, both in the flow (extrusion) direction and in the transverse direction. Quasi-static compression samples have a rectangular geometry with square loading faces and a height/width ratio of 2:1, following the ASTM standard for such experiments. High-strain rate uniaxial compression samples have cuboid shape, which is also commonly used and well characterized in the community [24, 25]. For comparison, as-cast Mg-alloy WE43 and as-extruded Mg-alloy AZ91C, supplied by commercial venders, were also tested under both quasi-static and dynamic conditions. More specifically, the dimensions of the specimens for uniaxial quasi-static compressive tests is 2.5 mm × 2.5 mm × 5.0 mm (5.0 mm is the gauge length), while the geometry for uniaxial dynamic compressive tests is 2.5 mm × 2.5 mm × 2.0 mm (2.0 mm is the gauge length).

### Quasi-static and High Strain-Rate Mechanical Testing Experiments

The loading faces of all specimens were polished with diamond paper down to 3 μm. The mechanical testing at quasi-static strain rate (strain rate $\sim 1\times10^{-3}$ $s^{-1}$) was performed by using an MTS hydraulic universal testing machine with self-alignment features. Lubricant (grease) was applied between the compression platens and the loading faces of the specimens to reduce or eliminate frictional effect. The prescribed desirable strain rate was achieved by controlling the velocity of the loading actuator (the speed of the cross-head).

We used a Split Hopkinson Pressure Bar (SHPB, or Kolsky bar) system to perform high strain rate (strain rate $\sim 1\times10^{3}$ $s^{-1}$) uniaxial compression. Details of such a system, its working principle, data processing and caveats can be found in several handbook articles [24, 26] and the monographs [25, 27]. Briefly, in this experiment, the specimen that has the required dimension is sandwiched between two co-axial elastic bars of the same diameter made of very high-strength maraging steel (usually called the input/out, or incident/transmitter bars). During the experiment, a projectile (or bullet) made of the same steel with the same diameter is launched by the pressurized gas released from a gas tank and strikes one end of the input bar. The impact generates a stress wave which travels down the input bar and reaches the input bar/specimen interface. Part of the stress wave is reflected back into the input bar, and part of it is transferred into the specimen and then to the output bar. Strain gages are attached to the surfaces of the input/output bars, the signals of which are recorded by a multiple channel high-speed oscilloscope. The data can then be processed to result in the high strain-rate stress-strain curve of the specimen. In this study, due to the limited dimensions of the specimens, we use a desk-top Kolsky bar (DTKB) to evaluate the high strain-rate stress-strain behavior of the specimens. The diameter of the various bars is 4.975 mm. We used a pulse shaper made of a thin copper disk at the impact end of input bar to modify the input stress wave. The interfaces between the specimen and the input/output bars were also lubricated to mitigate frictional effects. At least three samples were tested for each condition to get an indication of reproducibility.

For the dynamic experiments, the data were all collected as a function of time. They were then converted into true stress-strain curves. The incident pulse signal $\varepsilon_i(t)$, reflected pulse signal $\varepsilon_r(t)$, and transmitted pulse signal $\varepsilon_t(t)$ were processed to calculate the strain and stress data based on one-dimensional wave theory [28].

## Results and Discussion

Figure 1 shows the mechanical properties (strength) under uniaxial compression in dependence on the extrusion temperature for AMX602 (a) and AZXE7111 (b). In these plots, "Flow" designates that the loading direction is in the flow or extrusion direction of the material, while "Trans" means that the loading is in the transverse direction of the material. QS and DY indicate quasi-static and dynamic loading, respectively. YS and US are for yield strength and ultimate (maximum) stress, respectively. The stress quantity is the true stress.

We can see that when the extrusion temperature increases, the strength (both yield and ultimate) of both AMX602 and AZXE7111 decreases. This trend holds true at both quasi-static strain rate ($1\times10^{-3}$ $s^{-1}$) and high strain rate ($4\times10^{3}$ $s^{-1}$). We can also notice that, when tested in the flow direction, the specimen generally exhibits slightly higher strength levels than when tested in the transverse direction. It points to some minor anisotropy of the mechanical properties in both Mg-alloys. However, compared to conventional extruded Mg-alloys, the anisotropy is much alleviated. Another point of interest based on the data points in Figure 1 is that, for the same alloy under the same loading condition (either quasi-static or dynamic) the ultimate strength is much higher than the yield strength. This suggests that both alloys have strong strain hardening under uniaxial compression.

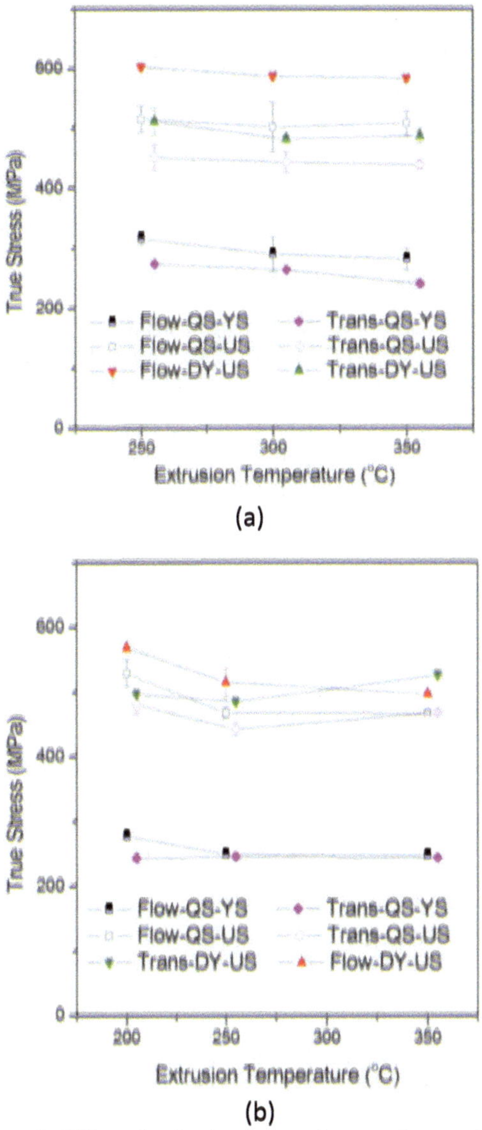


Figure 1. Effect of extrusion temperature on the quasi-static and dynamic compressive strengths of AMX602 (a) and AZXE7111 (b) Mg-alloys. The legends within the plots have the same meaning: "Flow" means that the specimen was loaded in the flow (or extrusion) direction; "Trans" means that the specimen was loaded in the transverse direction; QS and DY indicate quasi-static and dynamic loading, respectively; YS and US suggest the data are for yield strength and ultimate strength, respectively.

In Figure 2 we have plotted the deformation to failure or the total compressive strain as a function of extrusion temperature for AMX602 (a) and AZXE7111 (b). The strain values are the true strain. We should point out that most of the specimens did fail upon either quasi-static loading or dynamic loading. Therefore, in the case where the specimen did not fail, the data point corresponds to the total (prescribed) strain. Figure 2 provides some information associated with the ductility or malleability of the two Mg-alloys. Figure 2 (a) suggests that the ductility of AMX602 alloy does not depend very much on the direction of loading; that is, it is more or less plastically isotropic regardless of the extrusion temperature. Also clear from this plot is that the extrusion temperature only has a marginal effect on the deformation to failure of AMX602. This is nearly true also for AZXE7111.

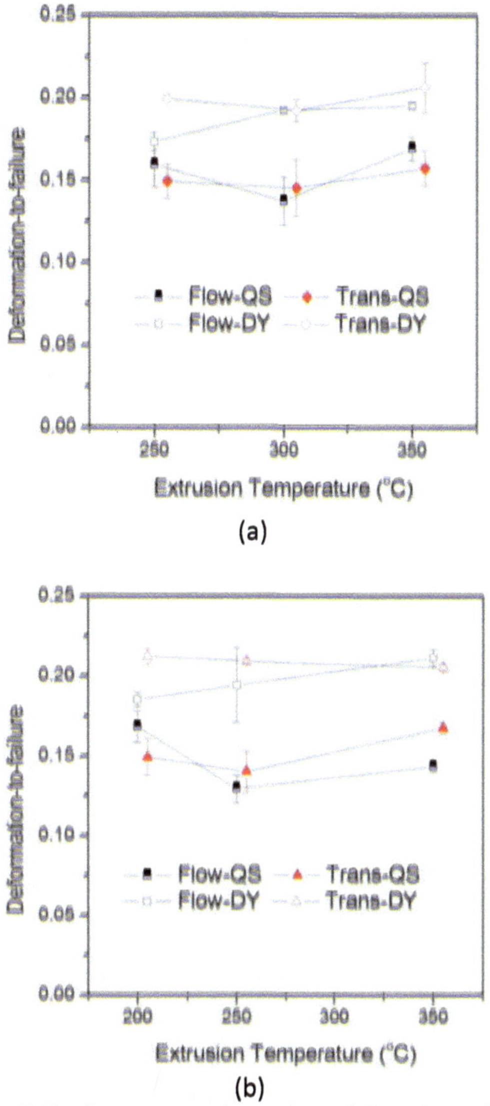


Figure 2. Total true compressive strain, or deformation-to-failure of AMX602 (a) and AZXE7111 (b) as a function of extrusion temperature. The meaning of the legend in the plots is the same as that of Figure 1.

Such observations indicate that during warm extrusion, the change in extrusion temperature only has some marginal effect on the microstructural evolution of the alloy. Or we may postulate that dynamic recrystallization dominates during extrusion and grain growth might have been limited due to lack of time. In other words, the chemical composition design of the alloy has effectively resulted in an extrusion temperature insensitive Mg-alloy. The same conclusion applies to AZXE7111 as indicated by Figure 2 (b).

Our experimental results show that either in quasi-static or dynamic loading, AZXE7111 is less anisotropic than AMX602 in terms of the differences of yield stress and ultimate compressive stress between flow direction and transverse direction. This observation is in line with the results reported by Elsayed et al. [22]. Elsayed and co-workers tried to explain the results on the basis of the more random distribution of crystal orientations induced by the effect of rare-earth addition in AZXE7111 alloy.

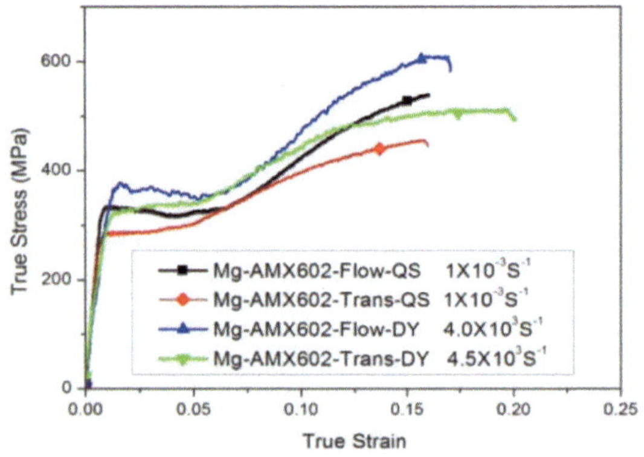


Figure 3. True stress vs. true strain curves of AMX602 at low and high strain rates.

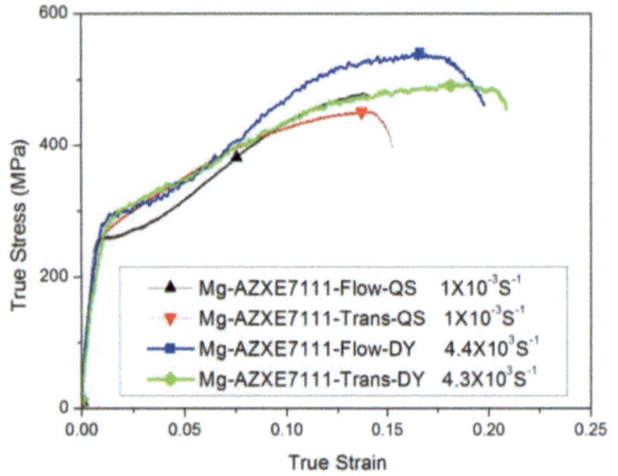


Figure 4. True stress vs. true strain curves of AZXE7111 at low and high strain rates.

As we have noticed, large differences between yield stress and ultimate stress are observed for both AMX602 and AZXE7111, which indicates that strong strain hardening has taken place in the materials in the process of mechanical loading. Actually, alloy AMX602 shows extensive nearly perfect plasticity immediately after yielding, lasting for a large range of plastic strain, and then followed by significant strain hardening. The micro-mechanism of such behavior of plasticity could be initial easy basal slip followed by more difficult prismatic, pyramidal slip, or even mechanical twinning. All these mechanisms of plastic deformation have been reported in Mg and its alloys. Particularly the sharp increase in strain hardening is associated with the operation of mechanically harder slip systems or twinning processes. As shown in Figure 3 and Figure 4, the nearly perfect plasticity was not found in AZXE7111, but rather almost steady strain hardening after yielding until failure or discontinuation of the loading process. This is in sharp contrast to the behavior of alloy AMX602. This difference may be explained by the presence of the rare earth La atoms and the larger amount of other alloying elements in AZXE7111 than in AMX602, which elements and the corresponding compound particles have been well recognized by Elsayed et al. [21]. Additionally, preliminary transmission electron microscopy (TEM) and Energy Dispersive Spectroscopy (EDS) (not shown in this article) have asserted this presumption. As a result, more second-phase particles such as Al-Ca and Al-La are present in the alloy that renders basal slip more difficult. The alloy starts to have plastic deformation by directly appealing to mechanically more difficult plastic deformation mechanisms such as prismatic, pyramidal slip or even deformation twinning.

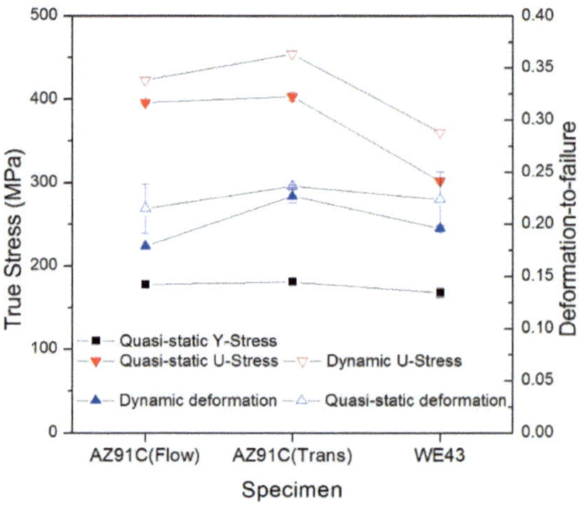


Figure 5. Total compressive strain or deformation-to-failure of AZ91C and WE43. Y and U indicate yielding and ultimate, respectively.

For comparison, conventional Mg alloys AZ91C and WE43 were also tested in the same conditions as those for the newly developed alloys. Figure 5 shows the strength (the yield stress and ultimate stress) and deformation-to-failure of these two conventional Mg alloys. Both flow and transverse directions of AZ91C exhibit no more than 200 MPa of yield stress and around 400 MPa of ultimate stress. In contrast, the newly developed alloys exhibit yield stresses and ultimate stresses around 300 MPa and 500 MPa, respectively. Meanwhile, WE43 is not even comparable with AZ91C as shown in the results, not to mention the new alloys. Though the deformation-to-failure of the conventional alloys shows a little bit higher values than those of the new alloys, the overall performance of AMX602 and AZXE7111 still outplays the conventional alloys.

## Conclusions

We have examined the quasi-static and dynamic uniaxial mechanical properties of two new Mg-alloys AMX602 and AZXE7111. The latter has La as alloying element. Both alloys were produced by spinning water atomization followed by compaction and extrusion at various temperatures. The quasi-static and high strain-rate mechanical testing (uni-axial compression) were performed on these two alloys in both the extrusion (flow) and transverse directions. The results show that both alloys exhibit marginal anisotropic behavior, both in terms of mechanical strength and plasticity under compression. The strength in the extrusion direction is slightly higher than in the transverse direction. Difference in compressive plasticity between the two directions is yet quite small. Both alloys show significant strain hardening. AMX602 exhibits extensive nearly perfect plasticity after yielding followed by strain hardening, while AZXE7111 displays strong strain hardening immediately after yielding. We attribute this difference to the presence of La in AZXE7111 which facilitates precipitation hardening.

## Acknowledgements

This work is supported by the Army Research Laboratory under Contract No. W911QX-08-0073. J.H. Shen would like to thank the Chinese Ministry of Education for the support during his tenure at UNC-Charlotte, and for the support from China Scholarship Council. We thank W. Yin and Z. Pan for assistance with some of the experiments.

## References

[1] M.M. Avedesian and H. Baker, Magnesium and Magnesium Alloys. Materials Park, OH 44073: ASM International, 1999.

[2] B.L. Mordike and T. Ebert, "Magnesium: Properties-applications-potential," *Materials Science and Engineering A*, vol. 302, pp. 37-45, 2001.

[3] A. Yamashita et al., "Improving the mechanical properties of magnesium and a magnesium alloy through severe plastic deformation," *Materials Science and Engineering A*, vol. 300, pp. 142-147, 2001.

[4] M. Furukawa et al., "Influence of magnesium on grain refinement and ductility in a dilute Al-Sc alloy," *Acta Materialia*, vol. 49, pp. 3829–3838, 2001.

[5] S. Agnew, "Enhanced ductility in strongly textured magnesium produced by equal channel angular processing," *Scripta Materialia*, vol. 50, pp. 377-381, 2004.

[6] H. Gleiter, "Nanostructured materials basic concepts and microstructure," *Acta Materialia*, vol. 48, pp. 1-29, 2000.

[7] C. Suryanarayana, "Nanocrystalline materials," *International Materials Review*, vol. 40, pp. 41-63, 1995.

[8] R.Z. Valiev et al., "Bulk nanostructured materials from severe plastic deformation," *Progress in Materials Science*, vol. 45, pp. 103-189, 2000.

[9] R.Z. Valiev et al., "Bulk nanostructured materials from severe plastic deformation," *Progress in Materials Science*, vol. 45, pp. 103-189, 2000.

[10] R.Z. Valiev et al., "Producing bulk ultrafine-grained materials by severe plastic deformation," *JOM*, vol. 58, pp. 33-39, 2006.

[11] A.P. Zhilyaev and T.G. Langdon, "Using high-pressure torsion for metal processing: fundamentals and applications," *Progress in Materials Science*, vol. 53, pp. 893-979, 2008.

[12] K. Xia et al., "Equal channel angular pressing of magnesium alloy AZ31," *Materials Science and Engineering A*, vol. 410-411, pp. 324-327, 2005.

[13] W.J. Kim et al., "Mechanical properties and microstructures of an AZ61 Mg Alloy produced by equal channel angular pressing," *Scripta Materialia*, vol. 47, pp. 39–44, 2002.

[14] H. Kim and W. Kim, "Microstructural instability and strength of an AZ31 Mg alloy after severe plastic deformation," *Materials Science and Engineering A*, vol. 385, pp. 300-308, 2004.

[15] K. Ishikawa et al., "High strain rate deformation behavior of an AZ91 magnesium alloy at elevated temperatures," *Materials Letters*, vol. 59, pp. 1511-1515, May 2005.

[16] T. Mukai et al., "Ductility enhancement in AZ31 magnesium alloy by controlling its grain structure," *Scripta Materialia*, vol. 45, pp. 89-94, 2001.

[17] X.-M. Feng and T.-T. Ai, "Microstructure evolution and mechanical behavior of AZ31 Mg alloy processed by equal-channel angular pressing," *Transactions of Nonferrous Metals Society of China*, vol. 19, pp. 293-298, 2009.

[18] L. Li et al., "Improvement of microstructure and mechanical properties of AZ91/SiC composite by mechanical alloying," Journal of Materials Science, vol. 35, pp. 5553-5561, Nov 2000.

[19] J. Corrochano et al., "The effect of ball milling on the microstructure of powder metallurgy aluminium matrix composites reinforced with $MoSi_2$ intermetallic particles," *Composites Part A: Applied Science and Manufacturing*, vol. 42, pp. 1093-1099, 2011.

[20] K. Kondoh et al., "Microstructures and mechanical responses of powder metallurgy non-combustive magnesium extruded alloy by rapid solidification process in mass production," *Materials & Design*, vol. 31, pp. 1540-1546, 2010.

[21] E. Ayman et al., "Application of rapid solidification powder metallurgy to the fabrication of high-strength, high-ductility Mg-Al-Zn-Ca-La alloy through hot extrusion," *Acta Materialia*, vol. 59, pp. 273-282, 2011.

[22] A. Elsayed et al., "The texture and anisotropy of hot extruded magnesium alloys fabricated via rapid solidification powder metallurgy," *Materials & Design*, vol. 32, pp. 4590-4597, Sep 2011.

[23] A. Elsayed et al., "Microstructure and mechanical properties of hot extruded Mg-Al-Mn-Ca alloy produced by rapid solidification powder metallurgy," *Materials & Design*, vol. 31, pp. 2444-2453, May 2010.

[24] S. Nemat-Nasser, "High strain rate testing," in ASM Handbook. vol. 8, H. Kuhn and D. Medlin, Eds., Materials Park, Ohio: ASM, 2007, p. 425.

[25] W.N. Chen and B. Song, Split Hopkinson (Kolsky) Bar. New York: Springer, 2011.

[26] P.S. Follansbee, "High strain rate compression testing," in ASM Metals Handbook. vol. 8, ed: American Society of Metals, 1985, p. 190.

[27] M.A. Meyers, Dynamic Behavior of Materials. New York: John Wiley & Sons, Inc., 1994.

[28] H. Kolsky, "An investigation of the mechanical properties of materials at very high rates of loading," *Proceedings of the Royal Society*, vol. B62, pp. 676-700, 1949.

# Magnesium Technology 2012

## Advanced Processing and Joining

### Thursday AM

*Session Chairs:*

Tyrone Jones
(U.S. Army Research Laboratory, USA)

Brian Jordan
(University of Alabama, USA)

**Magnesium Technology 2012**
*Edited by: Suveen N. Mathaudhu, Wim H. Sillekens, Neale R. Neelameggham, and Norbert Hort*
***TMS (The Minerals, Metals & Materials Society), 2012***

# THE EFFECT OF FRICTION STIR PROCESSING ON MICROSTRUCTURE AND TENSILE BEHAVIOR OF THIXOMOLDED AZ91 MAGNESIUM ALLOY

B. Mansoor[1], R.F. Decker[2], S. Kulkarni[2], S.E. LeBeau[2], M.K. Khraisheh[1]
[1]Masdar Institute of Science and Technology, Abu Dhabi, UAE
[2]Thixomat Incorporated, Ann Arbor, Michigan, USA

Keywords: Thixomolding, Mg, Ultrafine grain, Friction stir processing, Severe plastic deformation, Aging

## Abstract

Friction Stir Processing (FSP) to partial sheet thickness can be utilized to engineer unique microstructures in metallic alloys. These composite microstructures consist of three distinct layers associated with stirred, transition and core microstructural regions. The stirred region is of particular interest where severe plastic deformation imparted by the rotating and translating FSP tool under frictional heat leads to grain refinement down to ~ 1 μm grain size. In this work, partial depth penetration into thixomolded AZ91 Mg plate from the top and bottom surfaces by friction stir processing is explored. Furthermore, low temperature aging treatments are applied to the processed material. The present results with AZ91 Mg show that FSP processed material exhibits higher strength (> 300 MPa), and improvement in ductility (> 7 % tensile elongation). It is found that in addition to Hall-Petch strengthening produced by ~ 1 μm grain size in the stirred region, the enhanced strength levels and ductility are strongly influenced by dispersoids of the intermetallic precipitates found in this alloy.

## Introduction

Magnesium (Mg) and Mg-based alloys have drawn significant interest as attractive lightweight structural materials with high specific mechanical properties. However, magnesium alloys exhibit moderate strength and limited ductility at room temperature due to their hexagonal close-packed (HCP) crystal structure [1-3]. Presently, most of the magnesium based structural parts are produced by casting processes and despite alloy additions; they have lower strength and formability as compared to their wrought alloy counterparts. Therefore, development of several different high strength wrought Mg alloys with enhanced ductility is imperative for fabricating a variety of parts and products to increase the consumption of Mg. Earlier studies on several metallic alloys have demonstrated that grain refinement and control of precipitates, attained by thermo-mechanical processing can significantly enhance strength and ductility [4-7]. Severe plastic deformation is an effective method for grain refinement of bulk metallic materials. Some noteworthy severe deformation processes include: Equal Channel Angular Extrusion ECAE [8], High Pressure Torsion HPT [9], Accumulated Roll Bonding [10] and Friction Stir Processing [11]. Ultra-fine grained microstructure with < 1 μm grain size can be produced by such methods.

AZ91D is a commercial Mg alloy which is typically available in die-cast or thixomolded condition. Thixomolding is a semisolid injection molding process that was derived from semisolid metal casting techniques developed in the 1970s [12]. The details of the thixomolding process have been reported extensively in the past [13-14]. In addition to near-net shape parts, thixomolding routinely produces low porosity Mg plates. Microstructure of thixomolded alloys typically consists of 3-15 vol.% $\alpha_p$-Mg globules (~ 30-100 μm) in a matrix of homogenous and fine, secondary $\alpha_s$-Mg grains (4-5 μm). Furthermore, the $\alpha_s$-Mg grains are surrounded by a eutectic aggregate (1-2 μm) that consists of $\alpha_e$-Mg and an intermetallic compound β ($Mg_{17}Al_{12}$) [15]. Commercially available, wrought Mg alloys in hot-rolled or extruded condition are characterized by coarse and highly bimodal microstructures [15]. In comparison, thixomolded alloys with fine and relatively homogeneous microstructure are therefore, suitable candidates for secondary thermomechanical processing to enhance mechanical properties by further grain refinement and homogenization. In this work, severe plastic deformation was applied to thixomolded AZ91 Mg by friction stir processing to homogenize and refine as-molded microstructure. Friction-stir processing is an innovative, emerging technology that is based on friction stir welding (FSW) and many recent studies have been reported on FSW/P of magnesium alloys [16-19]. FSP involves plunging a rapidly rotating tool with a profiled probe and larger

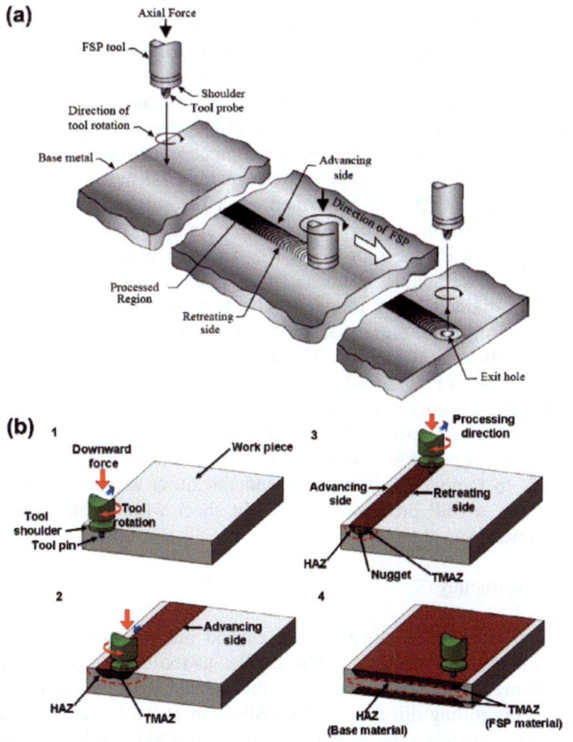


Figure 1. (a) Schematic of FSW/P process, adapted from the *Welding Handbook*, Vol. 3, 7th Ed. [20]. (b) Schematic shows several stages of overlapping FSP. Also, stirred and HAZ (base material) have been identified.

diameter shoulder beneath material surface and then traversing it across (Figure 1 (b)). Frictional heating and intense plastic deformation causes plasticized material (constrained by the shoulder) to flow around the tool and consolidate thereby improving mechanical properties.

The aim of the present study is to reveal the microstructural evolution during partial depth FSP of AZ91 Mg alloy and study the effect of fine grain size and aging treatments on room temperature mechanical properties. The effect of texture on mechanical properties is not investigated at present and will be the subject of a later study.

## Experimental

Thixomolded AZ91D Mg alloy was provided by Thixomat Inc. in the form of 3.25 mm thick plate. The nominal composition is 9.0 wt. % Al, 0.75 wt. % Zn, 0.3 wt. % Mn, Fe and Ni < 0.001 wt. % and Mg as balance. In order to achieve grain refinement in thixomolded AZ91 Mg alloy, severe plastic deformation was applied by FSP.

### Processing

Friction stir processing to partial depth was performed on top and bottom faces of thixomolded AZ91 Mg plate by inserting a rotating FSP tool into the metal to homogenize and refine as-molded microstructure. A schematic illustration of the FSP process and FSP tool is shown in Figure 1 (a). Important stages of overlapping, multi-pass FSP are shown in Figure 1 (b); stirred zone and Heat-Affected Zone (HAZ) are also identified. Overlapping steps were performed such that the retreating side from the previous pass was overlapped by the advancing side of the subsequent pass. Therefore, the processed region is a collection of advancing sides of multiple overlapping passes. In this work a 100 mm x 100 mm work-piece was processed by FSP. A total of 25 overlapping FSP passes with a 1.5 mm overlap were carried out. FSP was performed on a Doosan milling machine using a friction stir tool machined from H13 tool steel, heat-treated to a hardness of 52-55 Rc. The tool used in the present study had a shoulder diameter of 25.4 mm and the tool pin is 3 mm in dia and 1 mm in length. Dwell periods of 2-3 mins were applied prior to the initial FSP step to reach sufficiently high temperature (300-350 $^{0}$C). The dwell periods were progressively decreased for the subsequent steps. The FSP processing parameters were empirically determined. The FSP was performed at a rotational speed of 2100 rpm and a linear speed of 101.4 mm/min. A post FSP milling step was performed on the processed sheet to remove the roughness and thickness variations. In this manner, a FSP processed AZ91 Mg sheet with 3 mm uniform thickness was obtained.

### Microstructure

Microstructure examination was carried out on as-molded, FSP processed and aged Mg alloys. The as-molded samples were sectioned in a plane containing the normal direction (ND) and molding/rolling direction (RD). FSP samples were sectioned in a plane containing plate normal direction (ND) and transverse direction (TD). The metallographic samples were mechanically polished by the standard methods and etched by acetic-picral solution (4.2 g picric acid, 70 ml ethanol, 10 ml acetic acid and 10 ml distilled water). A Quanta FEG 250 SEM operating at 20-30 KeV was employed to observe the very fine microstructures. Grain sizes were measured from the micrographs by an area-based approximation. In order to determine the optimum conditions for aging; several low temperature aging treatment were performed at 110°C, 150°C and 180°C for 1, 5, 10 and 24 hours. The aging treatments were carried out in air; inside a Nabertherm box electric furnace.

### Mechanical Properties

Room-temperature tensile tests were performed on samples with 12.5 mm long × 3 mm wide gauge sections in a computer-controlled, 5982 Instron machine at a constant crosshead speed of 0.5 mm/min. Vickers hardness was measured at an indentation load of 300g for 30 seconds loading time.

## Results and Discussions

### Microstructure of Thixomolded AZ91 Mg

The typical microstructural features of as-molded AZ91 Mg alloy are shown in Figure 2. The microstructure can be described as equiaxed primary particles of (un-melted) $\alpha_p$-Mg, dispersed in a continuous matrix of solidified, fine, secondary eutectic $\alpha_s$-Mg.

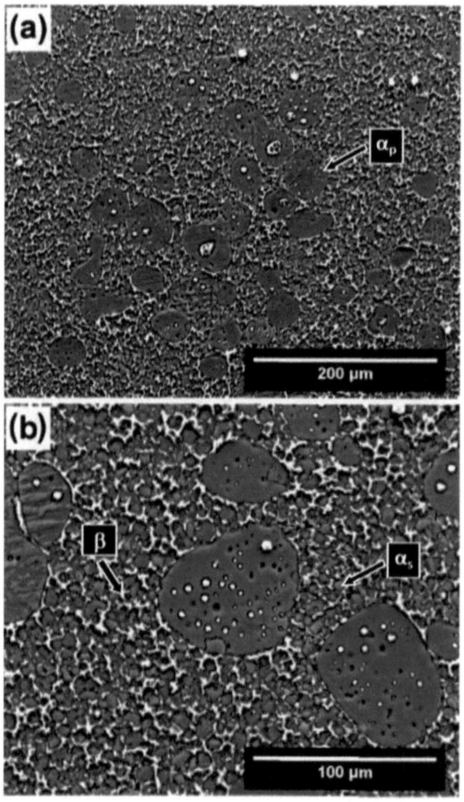


Figure 2. Typical microstructure of as-molded AZ91 Mg alloy. SEM image in (a) shows the overall structure and different morphologies of primary solid phase $\alpha_p$-Mg particles. (b) shows a matrix of equiaxed secondary $\alpha_s$-Mg grains with the eutectic (β) intermetallic network distributed in the spaces around the secondary $\alpha_s$-Mg grains (indicated by the arrow).

The secondary α-Mg grains are surrounded by a fine eutectic β phase network. This corresponds well with the microstructure of a rapidly cooled liquid (~100 °C/s) as no long dendrites (typical of casting microstructure) are visible. Several distinct morphologies of the primary solid particles; spheroidized during thixomolding are found in the micrograph as shown in Figure 2(a), e.g. one is essentially un-modified, solid while another one with entrapped liquid particles is also visible. Figure 2(b), shows morphologically equiaxed, fine secondary $\alpha_s$-Mg grains; delineated by the network of eutectic component i.e. an intermetallic compound β ($Mg_{17}Al_{12}$). The average grain size of $\alpha_s$-Mg grains is ~ 5 μm while the eutectic intermetallic film, distributed between the spaces of $\alpha_s$-Mg grains has a thickness of ~1 μm. In this work, no evidence of coupled eutectic growth was found. it was in fact, difficult to distinguish (both morphologically and chemically) between the fine $\alpha_s$-Mg grains and the eutectic $\alpha_e$-Mg phase [21].

Microstructural Evolution During FSP Processing

A low magnification overview of partial depth FSP processed AZ91 Mg is shown in Figure 3. Since the FSP tool pin partially penetrates the depth of work-piece during processing, sheet thickness can be divided in to three distinct and easily identifiable zones.

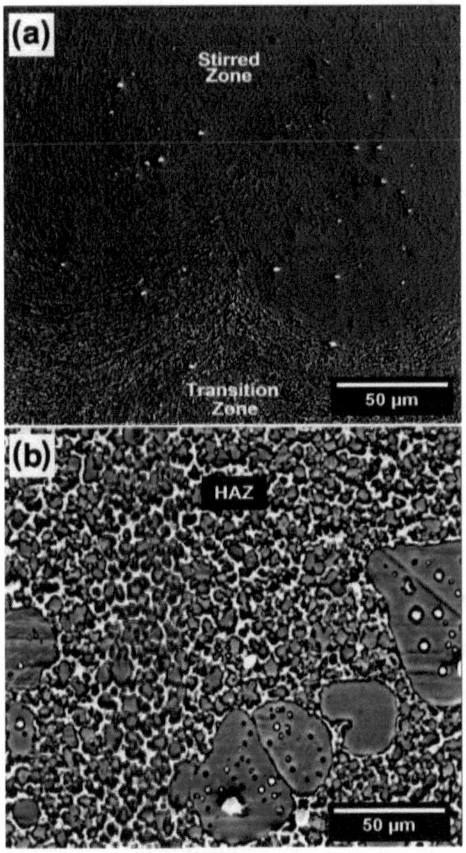


Figure 3. The microstructure AZ91 Mg alloy processed by overlapping multi-step partial depth friction stir process. A unique layered microstructure comprising of (a) stirred zone, transition zone and (b) heat affected zone is created.

The stirred zone, HAZ and transition zone have been identified in Figure 3. The HAZ is located in the middle and it retains the grain structure similar to the as-molded condition. FSP process has no apparent effect on microstructure of HAZ. It has both the as-molded microstructure and the spherical coarse particles (~ 50 μm in diameter). These are the spherical, coarse primary solid particles of nearly pure Mg. However, the continued exposure to elevated temperature leads to solutioninzing and a reduction in primary solid particle size and distribution as compared to as-molded condition.

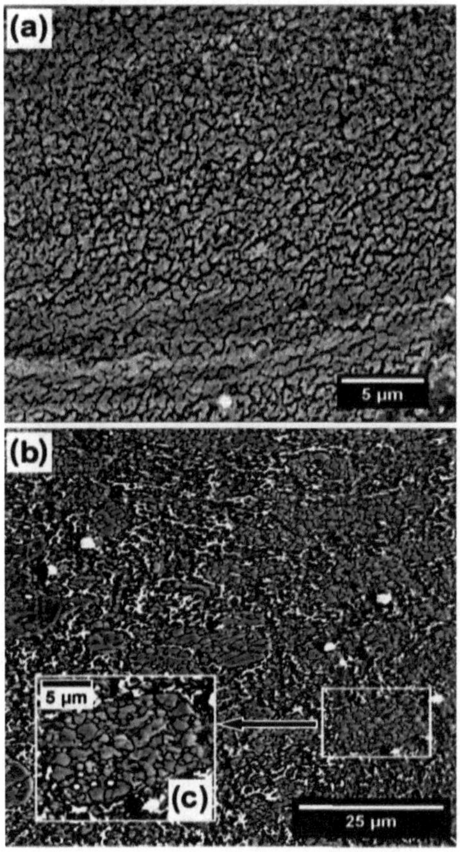


Figure 4. SEM image of FSP processed AZ91 Mg alloy. (a) Shows UFG microstructure with 0.85 μm average grain size in stirred zone, (b) transition zone (near root of tool pin) between FSP processed heat affected zone, (c) in the inset shows the subdivision of a primary solid particle in to several small grains.

The depth of stirred zone corresponds to the FSP pin length ~ 1 mm (not shown here). The general features of the stirred zone microstructure are consistent with earlier reported FSW/P findings for Mg alloys [16-17]. The microstructures for different zones of FSP processed AZ91 Mg are shown in Figure 4. Intense plastic deformation and frictional heating by FSP result in the generation of a ~ 1 μm ultrafine microstructure throughout the stirred zone, as shown in Figure 4(a). These micrographs show that by over-lapping pass FSP, the as-molded microstructure of the thixomolded material can be replaced with a fine and relatively homogeneous α-Mg microstructure. The stirred zone makes up for about 60 vol. % of the processed material. FSP results in complete dispersion and fragmentation of primary solid grains and secondary phases in the stirred zone. No defects, such as cracks and/or porosity, are observed in the stirred zone. According to Al-Mg binary phase diagram, the eutectic structure should completely dissolve into the α-Mg matrix at temperatures > 370 °C. Additionally, the α-Mg matrix should also dissolve when the temperature during FSP exceeds the solidus temperature of AZ91 (~500 °C). This should result in an as-molded microstructure after subsequent cooling. As shown in Figure 3 and Figure 4, as-molded microstructure is not observed in the stirred zones. It is known that this zone is heated to temperatures between 370-500 °C during FSP [21]. Exposure to this temperature range by frictional heating results in a supersaturated solid solution of the α-Mg phase. Also, intense plastic strain at rapid rate is simultaneously introduced into the material. This results in consolidation, mixing and shearing of the intermetallic network during FSP. Thus frictional heating and intense plastic strain at rapid strain rates may cause dynamic recrystallization in FSW/P of Mg alloys [22]. Therefore, the stirred zone experiences some dynamic recrystallization in addition to the dissolution of the eutectic structure. The heating cycles from the subsequent overlapping FSP step may result in recovery and or grain growth in the stirred zone belonging to the previous step. Therefore, a careful selection of dwelling times and processing parameters is absolutely crucial in obtaining a uniform microstructure in the processed regions of the plate. In Figure 4(b), the transition zone is seen below or near the root of the FSP tool pin and is present between the recrystallized stirred zone and the HAZ. It shows a bimodal grain structure with the coarse island shaped primary solid grains, reminiscent of the as-molded microstructure, sub-divided in to several small units. The higher magnification image of the selected area in inset (Figure 4 (c)), shows a large primary α-Mg grain, completely consumed by the smaller grains. Necklace structure, typical of dynamic recrystallization and recovery processes is also observed. The morphology and distribution of eutectic β ($Mg_{17}Al_{12}$) network is significantly altered. It has been fragmented into dispersoids that intermittently exists in the form of a much finer film. Some twinning is also observed in the sub-divided, primary α-Mg grains. This unique combination of layered structure and with UFG grain structure in stirred zone and two phase grain structure in the HAZ is likely to affect strength and ductility. A TEM analysis to understand the size, distribution and orientation of second phase particles and precipitates after FSP and aging will also be the subject of a later study.

Mechanical Property Improvement

The variation in Vicker's hardness measured along the sheet thickness for the FSP and aged conditions is presented in Figure 5(a). The regions corresponding to the stirred zone exhibit the highest hardness while the un-processed heat affected core region retains the hardness of the base material. Over-all there is a slight, through the thickness variation in hardness and it corresponds well with the corresponding grain sizes and phase distributions. Conventionally, AZ91 alloy processed by gravity or low-pressure casting techniques is subjected to a T6 heat treatment. This treatment involves two separate steps, solution treatment at (430 °C for 16–24 hrs) and subsequent aging at 168 °C for 16 hrs or alternatively 216 °C for 6 hrs [23]. In this work, peak hardness (stirred zone) is achieved by aging alone, without a solutioninzing treatment because FSP results in a super-saturated solid solution of the α-Mg phase. Also, aging is done at a much lower temperature of 150°C as compared to the relatively higher aging temperature typically used for this alloy. Figure 5(b). Values of Yield Strength (YS), Ultimate Tensile Strength (UTS), percent elongation and fracture-strain are summarized in Table I. In FSP condition, the material exhibits highest strength; accompanied with some improvement in tensile ductility. However, after suitable low temperature aging treatment, further improvement in strength (>300 MPa) with a combination of good ductility (>7 %) is observed. The UTS of the FSP processed material is approximately 1.5 times that of the as-molded condition and the UTS of the aged material is approximately 1.4 times that of the as-molded condition. On the other hand, the tensile elongation of the as-molded and FSP condition are similar but the aged alloy shows a significant improvement in ductility. This could be explained by the fact that the aging results in a lower dislocation density compared to the FSP condition. This can also explain the lower strength exhibited by the aged alloy as compared to the FSP material. The stress strain curves also show different strain hardening behavior for the three conditions. The FSP condition exhibits the highest strain hardening and the greatest UTS while the as- molded condition exhibits the lowest strain hardening and the lowest UTS.

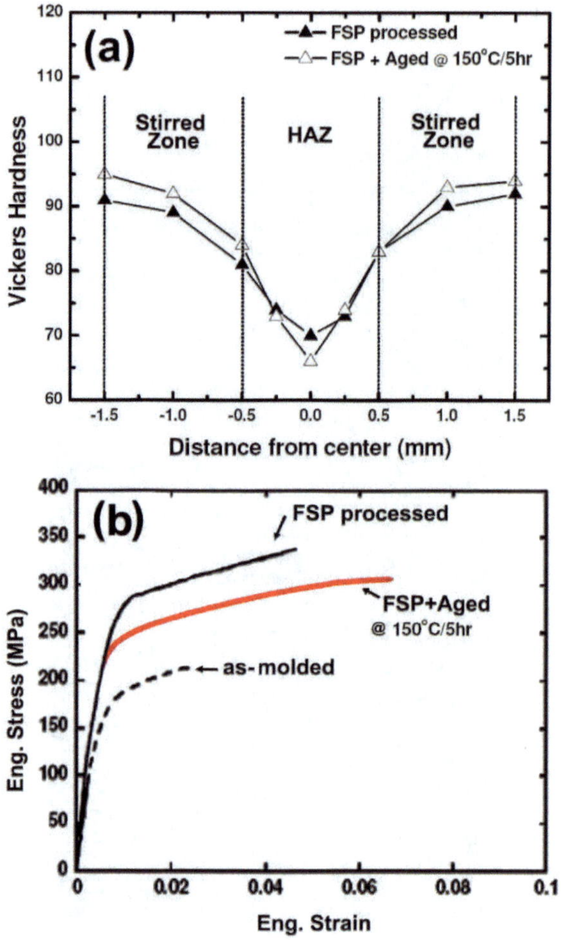


Figure 5. (a) The variation in Vickers hardness along the sheet thickness, (b) Room temperature tensile stress–strain curves for (a) as-molded, FSP processed and aged AZ91 Mg alloy.

Table I Mechanical Properties of thixomolded AZ91 Mg at different processing stages.

| Condition | Yield Strength (MPa) | Ultimate Tensile Strength (MPa) | Elongation, % | Fracture Strain |
|---|---|---|---|---|
| **Thixomolded** | 180 | 215 | 2.6 | 0.038 |
| **FSP** | 285 | 338 | 4.8 | 0.082 |
| **FSP + Aged** | 225 | 302 | 7.6 | 0.091 |

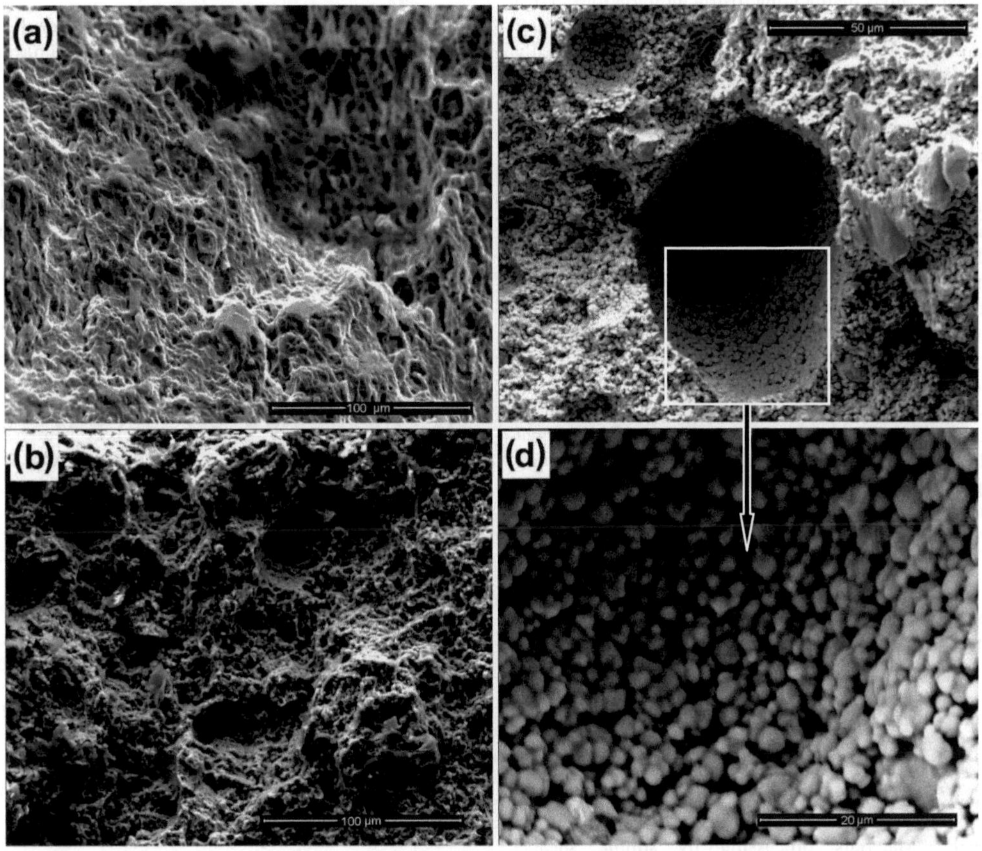


Figure 6. (a) SEM images show the fracture surface after tensile test for (a) as-molded condition (b) FSP processed and aged condition. The SEM images in (c) and (d) show the pull-out of a primary solid grain on the fracture surface observed in (b) at higher magnifications. This region of the sample corresponds to the un-processed HAZ present in the middle of the AZ91 plate.

It is well known that strain incompatibility between the (bcc) β-intermetallic phase and the (hcp) α-Mg phase leads to brittle interfaces in the as-molded condition [24]. This causes low ductility in the as-molded condition. The FSP processed material with UFG in stirred zones and bimodal microstructure in HAZ exhibits high strength. Improvement in ductility here is determined by reduction in the bcc/hcp brittle interfaces. This happens by complementing effects resulting from morphological changes and dissolution of eutectic films at α-Mg grain boundaries and the incorporation of eutectic rich material into the stirred zone. However, since the FSP material contains residual stresses and possibly some cold work, the over-all ductility is still low (< 5%). Further improvement in ductility is noticed after low temperature (150 °C) aging treatment. Aging/annealing causes recovery (relaxation of residual stresses by dislocation rearrangement), possibly some degree of recrystallization (and precipitation of fine nano-sized particles in the stirred zones [15]. It is nevertheless interesting that the 5 hrs aging treatment at a temperature much lower than recrystallization temperature of Mg (~200 °C) causes evolution of equiaxed microstructure in stirred zone (not shown here). These processes, in addition to Hall-Petch strengthening are most likely responsible for the improved mechanical behavior [25-27]. Figure 6 shows the fracture surfaces of tensile tested samples for as-molded and FSP processed and aged condition. The SEM images in (c) and (d) show several small recrystallized grains located in and around a crater possibly left by the pull-out of a primary solid particle on the fracture surface of the aged material. This region of the tensile tested sample corresponds to the heat affected core zone present in the middle of the AZ91 plate. Some precipitates are also visible in the

SEM images of the fracture surfaces. Ductile dimples were evident on the aged material (Figure 6(b)) but the fracture surface for the as-molded condition (Figure 6(a)) shows sharp and brittle characteristics.

## Conclusions

Severe plastic deformation is imparted to a thixomolded AZ91 Mg plate by friction stir process. This creates a unique, layered microstructure with three distinct microstructural regions: (i) a stirred zone with homogeneous, ultrafine (~1 μm) microstructure; (ii) a transition zone with bimodal microstructure and partially dissolved, fragmented β network; and (iii) a HAZ, which largely remains unchanged but shows partial dissolution of $\alpha_p$-Mg globules along with morphological changes in β network. After FSP processing, significant improvement in strength is observed but with some improvement in ductility. The super-saturated α-Mg phase stirred region by FSP eliminates the need for a high temperature solutioninzing treatment. Also, the aging treatment can be carried out at a much lower temperature of 150 °C without causing grain growth. The FSP processed and aged material exhibits superior mechanical properties with a good combination of strength and ductility as compared to the as-molded condition.

## Acknowledgements

This research was performed with support from the Government of Abu Dhabi, UAE. The authors acknowledge this financial support with gratitude. The authors also acknowledge the support from Thixomat, Inc. for providing the Magnesium materials.

## References

1. B.L. Mordike and T. Ebert, "Magnesium properties-applications-potential," *Materials Science & Engineering A, 302 (2001)*, 37-45.
2. J. Goken, J. Bohlen, N. Hort, D. Letzig and K.U. Kainer, "New development in magnesium technology for light weight structures in transportation industries," *Material Science Forum, 426(2003)*, 153-160.
3. E.F. Emley, *Principles of Magnesium Technology* (Oxford, UK: Pergamon Press Ltd., 1966), 1013.
4. V.M. Segal, "Materials processing by simple shear," *Materials Science & Engineering A*, 197(1995), 157-164.
5. R.Z. Valiev, A.V. Korznikov and R.R. Mulyukov, "Structure and properties of ultrafine-grained materials produced by severe plastic deformation," *Materials Science & Engineering A*, 168 (1993), 141-148.
6. Y. Saito, N. Tsuji, H. Utsunomiya, T. Sakai and R.G. Hong, "Ultra-fine grained bulk aluminum produced by accumulative roll-bonding (ARB) process," *Scripta Materialia*, 39(1998), 1221-1227.
7. L. Sturkey and J.B. Clark, "Mechanism of age-hardening in magnesium-zinc alloys," *Institute of Metals - Journal*, 88(1959), 177-181.
8. M. Mabuchi, K. Ameyama, H. Iwasaki and K. Higashi, "Low temperature superplasticity of AZ91 magnesium alloy with non-equilibrium grain boundaries," *Acta Materialia*, 47(1999), 2047-2057.
9. A. Galiyev, R. Kaibyshev and G. Gottstein, "Correlation of plastic deformation and dynamic recrystallization in magnesium alloy ZK60," *Acta Materialia*, 49(2001), 1199-1207.
10. M.T. Pérez-Prado, J.A. del Valle and O.A. Ruano, "Grain refinement of Mg-Al-Zn alloys via accumulative roll bonding," *Scripta Materialia*, 51(2004), 1093-1097.
11. P.B. Berbona, W.H. Bingela, R.S. Mishra, C.C. Bamptonc, M.W. Mahoney, "Friction stir processing: a tool to homogenize nanocomposite aluminum alloys," *Scripta Materialia*, 44(2001), 61-66.
12. J. Campbell, "Rheocasting and Thixocasting - a review of progress to-date" *Foundry Trade Journal*, 138(1975), 291-295.
13. L. Pasternak, R.D. Carnahan, R.F. Decker, R. Kilbert, "Semi-solid production processing of magnesium alloys by Thixomolding," *Proceedings of the 2nd International Conference on Processing of Semi Solid Alloys and Composites*, 1993, 159-169.
14. R.F. Decker, R.D. Carnahan, E. Babij, J. Mihelich, G.Spalding and L.Thompson, "Magnesium semi-solid metal forming," *Advanced Materials & Processes*, 149(1996), 41-42.
15. B. Mansoor, "Microstructural evolution and mechanical properties of high strength Mg alloys fabricated by deformation processing" (Ph.D. thesis, Univ. of Michigan, Ann Arbor, 2010), 39-72.
16. S.H.C. Park, Y.S. Sato and H. Kokawa, "Microstructural evolution and its effect on Hall-Petch relationship in friction stir welding of thixomolded Mg alloy AZ91D," *Journal of Material Science*, 38 (2003), 4379–4383.
17. J.A. Esparza, W.C. Davis and L.E. Murr, "Microstructure-property studies in friction-stir welded, Thixomolded magnesium alloy AM60," *Journal of Material Science*, 38 (2003), 941–952.
18. Y.S. Sato, S.H.C. Park, A. Matsunaga, A. Honda and H. Kokawa, "Novel production for highly formable Mg alloy plate," *Journal of Material Science*, 40 (2005), 637–642.
19. M. Santella, A. Frederick, C. Degen and T.Y. Pan, "The use of friction-stir technology to modify the surfaces of AM60B magnesium die castings," *JOM*, 58(2006), 56–61.
20. Welding Processes, Chapter 7 (212-258), AWS *Welding Handbook*, Vol. 3, Part 2, 7th Ed. (Miami, FL, USA: American Welding Society, 2007).
21. B.M. Darras, M.A. Omar and M.K. Khraisheh, "Experimental thermal analysis of friction stir processing," *Material Science Forum*, 539(2007), 3801-3806.
22. S.H.C. Park, S.Y. Sato and H. Kokawa, "Microstructural evolution and its effect on Hall-Petch relationship in friction stir welding of thixomolded Mg alloy AZ91D," *Journal of Material Science*, 38(2003)4379-4383.
23. A. Stevenson, Heat Treating of Magnesium Alloys (899-906), ASM *Handbook* Vol. 4, 9th Ed., (Materials Park, OH, USA: ASM International, 1990).
24. Y.Z. Lu, Q.D. Wang, W.J. Ding, X.Q. Zeng and Y.P. Zhu, "Fracture behavior of AZ91 magnesium alloy," *Materials Letters*, 44(2000), 265-268
25. Y. Wang, G. Liu, Z. Fan, "Microstructural evolution of rheo-diecast AZ91D magnesium alloy during heat treatment," *Acta Materialia*, 54 (2006), 689–699.
26. E.O. Hall, "The deformation and ageing of mild steel III. Discussion of results," *Proceedings of the Physical Society. Section B*, 64 (1951), 747-753.
27. N.J. Petch, "Cleavage strength of polycrystals," *Iron and Steel Institute- Journal*, 174 (1953), 25-28.

**Magnesium Technology 2012**
*Edited by: Suveen N. Mathaudhu, Wim H. Sillekens, Neale R. Neelameggham, and Norbert Hort*
***TMS (The Minerals, Metals & Materials Society), 2012***

# EFFECT OF WELD STRUCTURE ON FATIGUE LIFE OF FRICTION STIR SPOT WELDING IN MAGNESIUM AZ31 ALLOY

H.M. Rao[1], J.B. Jordon[1]
[1]The University of Alabama, Dept. of Mechanical Engineering, Tuscaloosa, AL, 35487, USA

Keywords: Friction stir spot welding, Fatigue, Magnesium alloy

## Abstract

In this paper the fatigue behavior in friction stir spot welded coupons of magnesium AZ31 alloy manufactured under different welding conditions are investigated. Two sets of lap-shear coupons were welded based on varying the plunge depth and tool geometry. Metallographic analysis of the untested lap-welds revealed differences in microstructural and geometrical features. Results from the load controlled cyclic tests showed that one set of welds exhibited better fatigue performance compared to the other set. Optical fractography of the failed fatigue coupons revealed that fatigue cracks initiated at the weld interface in both sets of coupons. However, the fracture mode showed variability between the two sets of coupons. As such, the main conclusion of this study is that the effective top sheet thickness, which is largely determined by the shoulder plunge depth, plays a significant role in the fatigue behavior of the friction stir spot welds in magnesium alloys.

## Introduction

Friction stir spot welding (FSSW), which is similar to friction stir seam welding, is a thermo-mechanical process for spot lap-joining of sheet metals [1,2]. The rotating tool comprises a probe pin that is plunged into the two sheets of metal to be joined. The probe pin typically penetrates the upper sheet completely and then passes into the bottom sheet to various depths depending on process conditions. Here, the downward force and rotational speed generate localized heat as the pin interacts with the upper sheet. The heat generated by friction softens the two sheets and the softened metal adjacent to the tool deforms plastically and a solid state bond is made between the surface of upper and bottom sheet [3]. Since the dwell time is relatively short, 2 to 5 seconds, the tool rotation speed and plunging motion determine the heat generation [4], joint formation and weld mechanical properties [5]. In general, FSSW is free of many of the defect formation issues commonly associated with fusion welding. This is due to the assumption that, the temperature attained during friction stir welding is less than the melting point of base material. In addition, FSSW is more energy efficient and clean when compared to other types of spot welding. Some of the other advantages include no need of cooling agent, filler materials, or post weld treatment.

As industry continues the push for more fuel-efficient designs, magnesium alloys are receiving increased interest. Thus, FSSW is a favorable technique for spot welding lightweight metals, such as magnesium alloys. Some recent work on FSSW of magnesium AZ31 alloy has focused on process-property-performance relationships [2,5-7]. One of the key findings was related to the size and shape of the interfacial hook. During FSSW, trapped oxide films present between the overlapping sheets are oftentimes displaced in an upward direction toward the top sheet into a "hook-like" shape. This hooking is largely due to the plastic flow of the material resulting from the downward plunge of the pin into the bottom sheet. In fact, higher joint static strengths were found when the hook region was curved more outwards from tool axis and the distance from the hook extremity to the top surface of the weld was the greatest [7]. A similar observation was also found in Al alloys [8]. While fundamental relationships governing the static strength of FSSW in magnesium alloys have been developed, only a few studies exist regarding fatigue [9-10]. In lap-shear fatigue testing, fatigue cracks were found to initiate from this interfacial hook [10]. For cyclic loads approaching the static strength of the joint, the dominant cracks initiated from the interfacial hook and final failure occurred by nugget pullout. However, for lower cyclic loads, fatigue cracks grew from the interfacial hook and then propagated through the top sheet. In this prior study, the height of the interfacial hook was thought to be influential in determining the final number of cycles to failure in magnesium AZ31 alloys.

Beyond the work of Mallick and Agarwal [9], and Jordon et al. [10], the fundamental understanding of fatigue mechanisms in FSSW lap-joints is primarily restricted to aluminum alloys. Lin et al. [11-12] studied fatigue failures in friction spot welded aluminum 6111-T4 sheets, processed using different geometric tools. Under different loading conditions, the failure mode in fatigue was found to vary. In quasi-static loading, it was observed that fatigue cracks initiated from the upper sheet near the stir zone inside the weld. Under cyclic loading conditions, three main types of crack initiations were observed: the first type was similar to quasi-static loading, the second type initiated from stir zone into the bottom sheet and the third initiated from the bend in the upper sheet. In addition, Lin et al. [11] compared the failure modes of FSSW made using different tool geometries. Spot welds created using a flat tool initiated cracks from several locations compared to a concave tool. The amount of material deformation during the welding process due to the shape of the tooling was identified as the likely reason for the difference in failure modes of the flat and concave tooling. The fatigue cracks were observed to initiate near the possible original hook tip in the stir zone and propagated along the circumference of the nugget, then through the sheet thickness and finally along the width of the specimen [3].

To best of authors' knowledge, the effect of weld structure due to variation in welding conditions on fatigue has not been established for magnesium alloys. In this paper, fatigue performance and failure mechanisms of the FSSW magnesium AZ31 alloy coupons manufactured under different process conditions has been studied. Furthermore, we attempt to establish cause and effect relationships for fatigue in FSSW with respect to microstructural and geometrical features.

## Material and Experiments

Magnesium AZ31 alloy sheets of 2.0 mm thickness were chosen for the present study. Two sets of FSSW coupons were welded in an overlap configuration. For process #1, the individual sheet dimensions were: length of 100 mm, width of 35 mm, and a welded overlap area of 35 X 35 mm. For process #2, the individual sheet dimensions were: length of 100 mm, width of 38 mm and a welded overlap area of 38 X 38 mm. For process #1, the FSSW tool was made from standard tool steel (H13) material, having a shoulder with a diameter 12 mm, pin length of 3.2 mm and left hand threads (M5) as shown in Fig. 1a. For process #2, the FSSW tool is identical to the tool used in process #1; however the tool had a triangular pin as shown in Fig. 1b. Both tools had a 10-degree concave shoulder as shown in Fig. 1c. Table I lists the welding process conditions used to join the two sets of lap-shear coupons employed in this study. Figure 2 shows representative samples from both process #1 and #2.

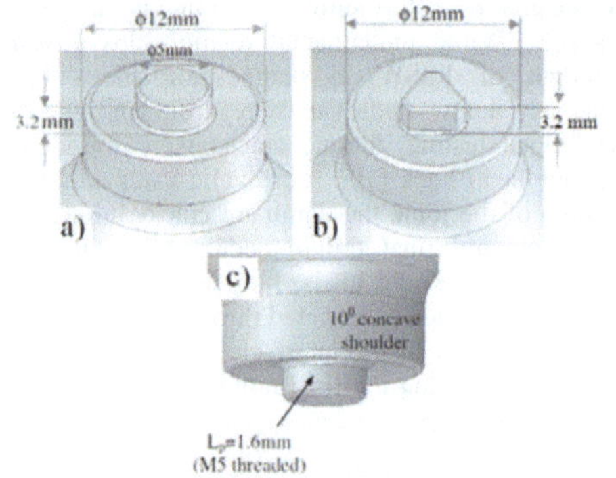


Figure 1. Schematic of the FSSW tool geometry employed for (a) process #1 and (b) process #2 [8]. The tool for process #1 had a cylindrical pin with a diameter of 5 mm and M5 threads. The tool for process #2 had a triangular pin. Both tools had a (c) concave shoulder of 10 degrees and a diameter of 12 mm with a pin length of 1.6 mm.

Figure 2. Representative lap-shear coupons produced from process #1 and process #2.

Table I: Friction stir spot welding conditions.

| Welding Parameters | Process #1 | Process #2 |
|---|---|---|
| Tool rotation (RPM) | 1000 | 750 |
| Tool Plunge Speed (mm/min) | 20 | 20 |
| Shoulder Plunge Depth (mm) | 0.5 | 0.1 |
| Dwell Time (sec) | 2.5 | 2.5 |

For the fatigue tests, the lap-shear coupons were cyclically tested in an MTS 810 servo-hydraulic load frame under load control with a sinusoidal waveform at three different load ratios (0.1, 0.3, and 0.7). The fatigue tests were tested at frequencies that ranged from 5–30 Hz. While most of the tests were conducted at 10 Hz, higher frequencies were employed for the low load amplitudes to expedite failure. Steel shims were employed in the test frame to prevent the addition of bending forces. The grip-to-grip distance employed for each coupon was 110 mm. For weld characterization, samples were sectioned through the center of the weld nugget coupon and parallel to the loading direction. Samples were cold mounted in epoxy and mechanically ground and polished. Using a Keyence VHX-1000 digital optical microscope, key features of the weld were quantified including size and shape of the interfacial hook, the bond width, and distance from the shoulder of the nugget to the root of interfacial hook. In order to reveal the microstructure of the FSSW coupons, the samples were etched using a 4.2 g picric acid, 10 ml acetic acid, 10 ml H2O, and 70 ml ethanol [13]. Typical optical micrographs of the stir zone, thermomechanically affected zone (TMAZ), and the heat-affected zone (HAZ) were obtained. Representative failed coupons from process #1 and process #2 were also observed under the Keyence microscope to study the crack propagation and failure modes under different loading conditions. In addition, fracture surfaces of failed fatigue coupons were examined under a Jeol 7000 scanning electron microscope with the intent to estimate the fatigue crack growth rate.

## Results

Microstructure

Representative microstructure images of coupons from process #1 and process #2 are shown in Fig 3. Overall, the grains were similar in size in both processes. In particular the average grain size in the stir zone, which comprised recrystallized grains, was approximately 9 μm in both processes. Elongated grains were observed in the TMAZ and partially recrystallized grains were observed in the HAZ.

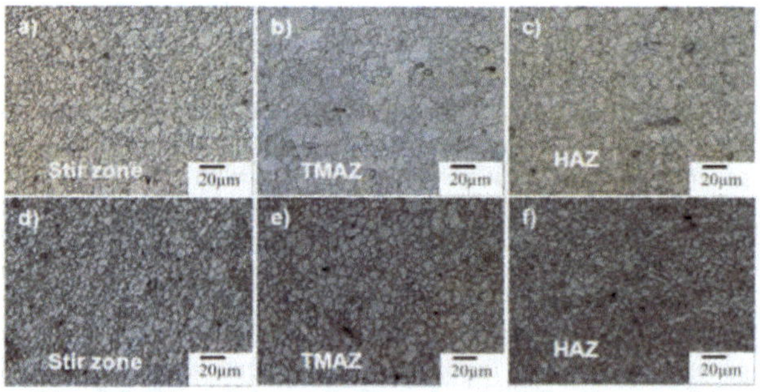


Figure 3. Microstructure features showing: (a) stir zone, (b) TMAZ, (c) and HAZ in coupons from process #1, and (d) stir zone (e), TMAZ, (f) and HAZ in coupons from the process #2.

Interfacial Hooking and Geometrical Features

Figure 4 shows magnified views of the interfacial hook in process #1 (4a) and process #2 (4b). For clarification purposes, we label the two peaks of the interfacial hook as primary and secondary as noted in Fig. 4. In both processes, primary and secondary hooks were observed. Moreover, process #2 produced a smoother interface and more well defined hooking. In contrast, the interfacial hook exhibited a more erratic path in process #1 and the distinction between the primary and secondary hook was not as clear as in process #2. Lastly, the height of the hooks varied greatly between both processes, where in process #2 the hook heights were approximately three times as large as compared to process #1. This large difference in hook height between the two processes is likely due the difference in tool plunge depth. However, additional experiments are needed to confirm this.

In addition, the sheet thicknesses in the weld nugget were measured, where for process #2, the top sheet thickness in the weld nugget was almost identical to the nominal thickness of the magnesium alloy sheets (2 mm). However, process #1 had an effective sheet thickness in the weld nugget of 1.3 mm. As such, the link between this effective sheet thickness is directly related to the shoulder plunge depth during welding. In process #2, the shoulder plunge depth was 0.1 mm, compared to 0.5 mm for process #1. As such, the deeper the shoulder of the tool is plunged, the smaller the thickness of the sheet in the weld zone. Lastly, the bond widths of the welds were similar in size for both processes.

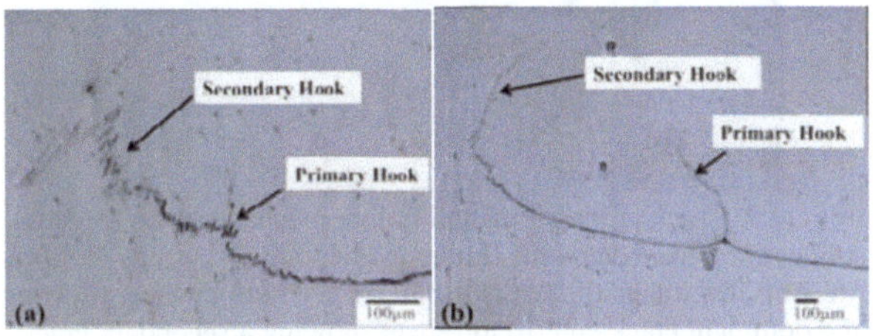

Figure 4. Magnified views of the primary and secondary hooks formed in (a) process #1 and (b) process #2.

Cyclic Testing

Figure 5 shows the results of the fatigue tests for both processes. In general, the fatigue lifetimes associated with final separation were greater for process #2 when compared to process #1. The results also show that the FSSW lap-joint did not exhibit a load ratio effect commonly found in uniaxial base material experiments [cf. 14].

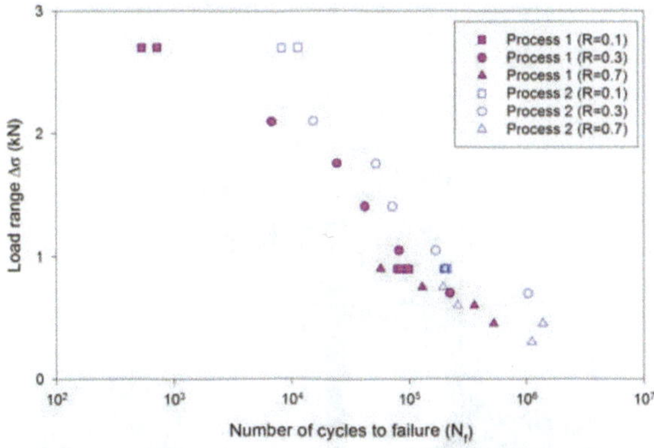

Figure 5. Experimental results of the load range versus the number cycles to failure for complete separation for friction stir spot weld lap-shear coupons tested at three load ratios (R=0.1, 0.3, 0.7) .

Fractography

Figure 6 shows an overview of representative fractured coupons from both process #1 and #2. Several macroscale observations were made regarding the failure modes. For the coupons from the process #1 and #2 tested at a maximum cyclic load of 2 kN or greater, fracture of the weld occurred by failure of the weld nugget. Below 2 kN, the joints failed by fatigue crack propagation through the sheet thickness. For process #1, the dominant crack grew from the primary hook towards the top surface of the top sheet and then propagated along the outer circumference of the nugget. Halfway around the nugget circumference, the crack grew along the width of the top sheet and failed by top sheet pullout. This failure mode was similar to failure modes observed in aluminum FSSW [3]. As shown in Fig. 7a, the nugget was seen intact in the remaining top and bottom sheet. Figure 7b shows the location of where the primary crack initiated on the outer surface of the weld. It is evident from Fig. 7c that the dominant crack propagated from the faying surface. Higher magnification of the interfacial hook region showed that the primary hook propagated towards the surface of the top sheet. Also evident was that the dominant crack initiated and propagated from the primary hook as shown in Fig. 7d. No evidence of a secondary crack was found elsewhere as shown in the magnified view of Fig 7e, which is located at the opposite side of the weld (180 degrees around the circumference).

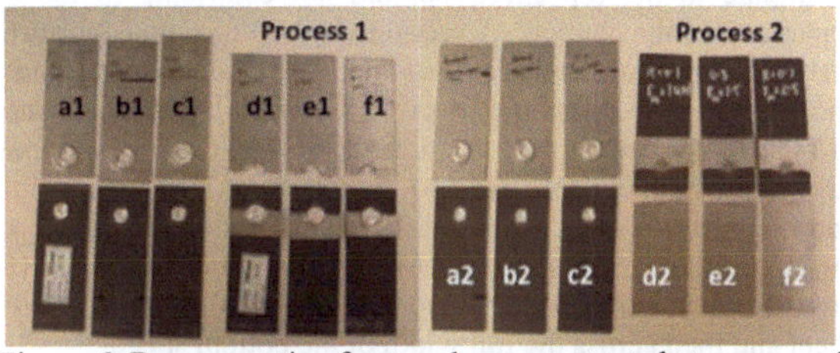

Figure 6. Representative fractured coupons tested at a range of maximum cyclic loads of 1-3 kN at R=0.1, 0.3, and 0.7 (Refer to Table II for a summary of failure modes).

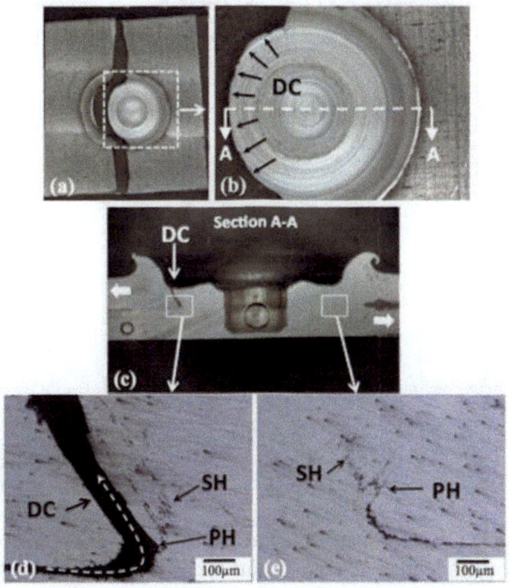

Figure 7. Crack propagation and failure mechanisms for process #1 coupon loaded at a maximum cyclic load of 1 kN at R=0.3. (a) An overview of the surface of the top sheet and (b) a magnified view of the location of surface crack initiation. (c) A section view of the weld nugget shows that the dominant crack (DC) propagated from the weld interface, and (d) and (e) show higher magnification images of the interfacial primary (PH) and the secondary (SH) hooks. Bold arrows in (c) indicate direction of applied loading.

At higher loads in process #1, differences in the failure mode were observed. Coupons tested at and above a maximum cyclic load of 2.5 kN failed by nugget pullout. The dominant crack grew from the primary hook towards the top sheet along the outer circumference of the nugget and failed by nugget pullout as shown in Fig. 8a and 8b. The cross-section view of the weld nugget revealed cracks propagated from the interfacial hooks as shown in Fig. 8c. As shown in Fig. 8d, the dominant crack grew from the primary hook (PH) to the top sheet and secondary cracks (SC) were observed to propagate partially into the bottom sheet as shown in Figure 8e. However, these secondary cracks were not the dominant cracks and did not contribute to final failure. These cracks either propagated at a significantly lower rate, or as the dominant crack became larger it shielded the secondary crack such that they arrested.

Coupons from process #2 tested below a maximum cyclic load of 2.5 kN failed by bottom sheet failure as shown in Fig. 9a. As shown in Fig. 9b, secondary non-dominant cracks were observed to initiate to the top sheet, but did not contribute to the final separation of the coupon. A section view of the nugget confirmed that the dominant crack propagated into the bottom sheet as shown in Fig. 9c. Magnified views of the interfacial hooks shown in Fig. 9d confirmed that secondary cracks (SC) on the opposite side of the nugget propagated to the top sheet from both primary and secondary hooks. However, these cracks did not contribute to the final failure. As shown in Fig. 9d, the dominant crack (DC) grew from the root of the secondary hook towards the bottom sheet and then across the full width of the bottom sheet.

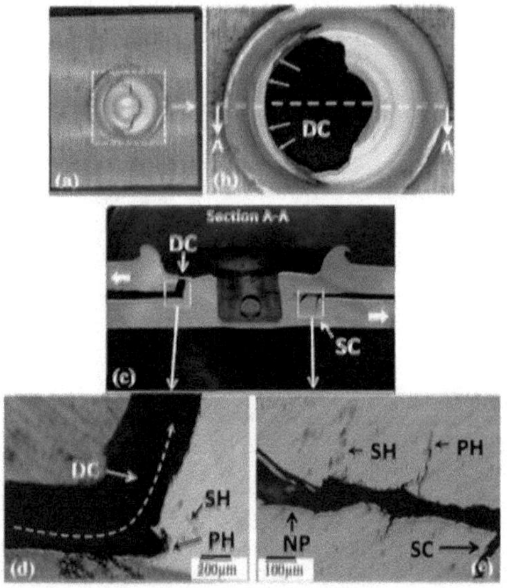


Figure 8. Crack propagation and failure mechanisms for process #1 coupon loaded at a maximum cyclic load of 3 kN at R=0.3. An overview image of the (a) surface of the top sheet and (b) a magnified view of the location of surface crack initiation. (c) A section view of the weld nugget shows that the dominant crack (DC) propagated from the weld interface, and (d) and (e) show higher magnification images of the interfacial primary (PH) and the secondary (SH) hooks. Bold arrows in (c) indicate direction of applied loading.

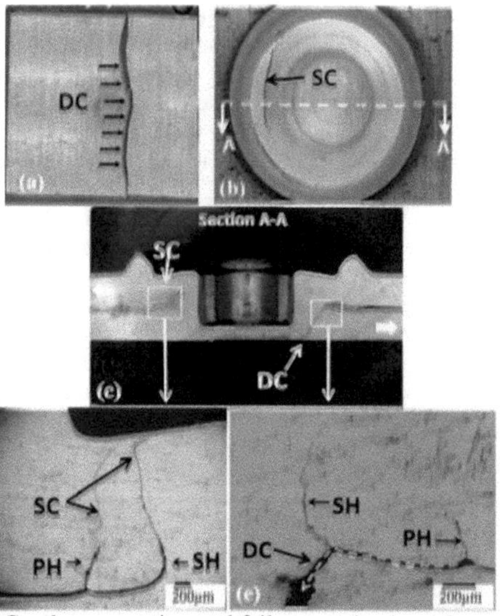


Figure 9. Crack propagation and failure mechanisms for process #2 coupon loaded at a maximum cyclic load of 1 kN at R=0.3. An overview image of the (a) surface of the bottom sheet showing the dominant crack and (b) top sheet showing a secondary crack. (c) A section view of the weld nugget shows that the dominant crack (DC) propagated downward into the bottom sheet and (d) and (e) show higher magnification images of the interfacial primary (PH) and the secondary (SH) hooks. Bold arrows in (c) indicate direction of applied loading.

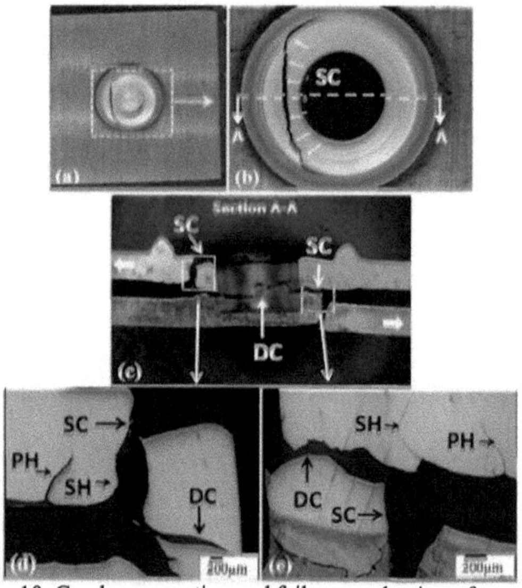


Figure 10. Crack propagation and failure mechanisms for process #2 coupon loaded at a maximum cyclic load of 3 kN at R=0.3. An overview image of the (a) surface of the top sheet and (b) a magnified view of the location of surface crack initiation of the secondary crack (SC). (c) A section view of the weld nugget shows that the dominant crack (DC) was primarily a mode II crack resulting in shear failure, and (d) and (e) show higher magnification images of the interfacial primary (PH) and the secondary (SH) hooking regions. Bold arrows in (c) indicate direction of applied loading.

Coupons from process #2 tested at a maximum cyclic load of 2.5-3 kN failed by interfacial shear failure. Figure 10a shows the top surface of the process #2 coupon tested at 3 kN at R=0.3. It is important to note that the top of the weld nugget remained intact and that a secondary crack was observed to penetrate the top surface of the nugget as shown in Fig 10b. However, this was not the dominant crack. The section view of the nugget region shows evidence of multisite cracking; however the dominant crack grew parallel to the loading direction in mode II as shown in Fig. 10c. Figure 10d and 10e show magnified views of the hooking regions, which indicate that secondary cracks (SC) grew through the primary and secondary hooks. Also shown in Fig. 10e is that secondary cracks were observed to propagate into the bottom sheet. Table II summarizes the fracture modes for the FSSW Mg AZ31 coupons subjected to cyclic loading for both process #1 and #2 based on the optical fractography conducted in this study.

Table II: Summary of failure modes in coupons of process #1 and process #2 (ID correlates with specimens shown in Fig. 6).

| ID | Load Ratio (R) | Max Load (kN) | Mode of Failure |
|---|---|---|---|
| **Process #1** | | | |
| a1 | 0.1 | 3 | Dominant crack grew from primary hook into the top sheet, then halfway around the outer circumference of the nugget, finally failing by nugget pullout. |
| b1 | 0.3 | 2.5 | |
| c1 | 0.7 | 3 | |
| d1 | 0.1 | 1 | Dominant crack grew from primary hook into the top sheet, halfway around the outer nugget circumference and then along the width of the top sheet. |
| e1 | 0.3 | 1.5 | |
| f1 | 0.7 | 1 | |
| **Process #2** | | | |
| a2 | 0.1 | 3 | Dominant crack grew across the weld nugget and parallel to loading direction in mode II propagation. |
| b2 | 0.3 | 3 | |
| c2 | 0.7 | 3 | |
| d2 | 0.1 | 1 | The Dominant crack grew from root of the secondary hook into the bottom sheet, propagated around the outer circumference of the nugget region and then along the width of the bottom sheet. |
| e2 | 0.3 | 1.5 | |
| f2 | 0.7 | 2.5 | |

SEM Analysis

In order to confirm crack failure modes and to estimate crack growth rates in both sets of coupons, representative fracture surfaces of the dominant cracks were imaged using an SEM under high magnification with the intent to measure striation spacing. As shown in Figure 11a, the striation spacing was in the range of 1-1.6 μm for the process #1 coupons testing at a maximum cyclic load of 1 kN. At higher loads, the striation spacing was observed to be in the 3.3-3.5 μm range, as shown in Fig. 11b. In coupons from process #2 tested at a maximum cyclic load of 1 kN, the average striation spacing was around 0.80 μm and in coupons tested at 3kN, the striation spacing was approximately 0.60 μm. We note that the difference between the 1 kN and 3 kN for process #2 is small and this observed difference could also be due to scatter. We also note that the striations were less distinct in process #2 and as such, were considerably more difficult to find compared to process #1.

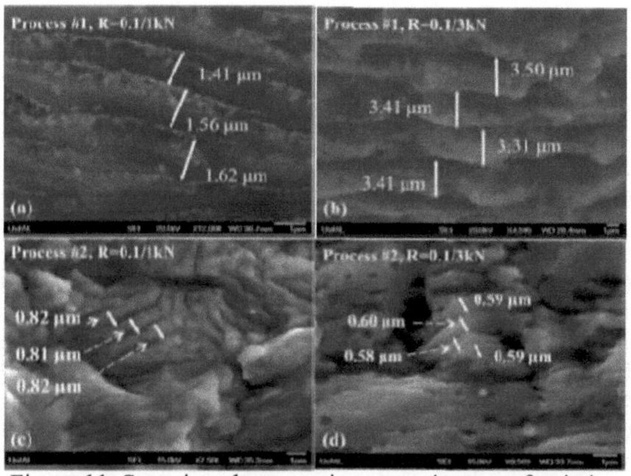

Figure 11. Scanning electron microscope images of striation spacing for: (a) process #1 coupons tested at maximum cyclic load of 1 kN, (b) process #1 coupons tested at maximum cyclic load of 3 kN, (c) process #2 coupons tested at maximum cyclic load of 1 kN, and (d) process #2 coupons tested at maximum cyclic load of 3 kN. Each of these coupons were tested at R=0.3.

**Discussion**

Analysis of the fractography results presented in the previous sections give an understanding of the underlining mechanisms of fatigue damage in FSSW lap-shear joints of various welding parameters. First, it is worth noting that Yin et al. [7] found that the monotonic strength of FSSW Mg AZ31 coupons increased when the bonded width is large, the distance from the hook extremity to the top surface of the weld was the greatest, and if the hook region is curved outward from the axis of rotating tool. In this study, the main difference between the two processes is related to the effective sheet thickness and the height of the hook, where the distance from the hook extremity was greatest in process #2. As such, it appears that Yin et al. [7] conclusions may be also true for fatigue performance. However, in lieu of establishing an empirical relationship between geometrical and microstructural features to fatigue performance as a way to explain the differences in two set of welds, we focus on the mechanics of fatigue crack damage. As such, the concepts of linear elastic fracture mechanics provide the basis for this discussion. If the spot weld is analyzed in terms of a stress intensity factor, we would expect process #2 to exhibit better fatigue resistance since the effective top sheet thickness was considerably larger in process #2 (2 mm) when compared to process #1 (1.2 mm). In the lap-joint configuration, the effective top sheet thickness largely determines the global stress field, which in turn affects the driving force behind crack initiation and crack growth. When we consider process #1, the smaller effective top sheet thickness would likely produce a higher stress intensity factor which would drive the dominant crack to propagate through interfacial hook. As shown in Fig. 7-8, the experimental results do indeed show that the dominant crack initiated and grew through the interfacial hooks for process #1 over the full range of low cyclic loads testing in this study. Likewise, in process #2, it is the effective top sheet thickness that again determines the facture mode. The stress intensity factor would likely be lower in process #2 due to the significantly thicker top sheet thickness. This lower stress intensity would reduce the driving force for dominant mode I crack growth through the top sheet and allow for an alternant location of crack initiation and crack growth. This was confirmed experimentally, where at the higher maximum cyclic loads, the dominant crack was observed to be a predominantly mode II type crack, which resulted in better fatigue lifetimes. At lower maximum cyclic loads, the initial cycles likely do not fully open the interfacial hook, which would further repress the stress intensity factor and allow for the dominant crack to propagate through the bottom sheet as was observed experimentally and shown in Fig. 10.

Lastly, we note that the SEM images of the fracture surface suggest that the cracks propagated at a higher rate in process #1 compared to process #2. This supports the hypothesis that the effective top sheet thickness is the dominant factor controlling fatigue performance in FSSW. In fact, based on the analysis of the experimental results presented here, one can conclude that the better fatigue performance of process #2 is largely determined by the tooling and welding conditions which, in turn, affected the structural integrity of the weld.

## Conclusions

In this study, the influence of microstructural and geometrical features on fatigue lifetimes are evaluated on two different sets of friction stir spot welded lap-shear coupons. Several differences in the initial state of the weld structure including size and shape of the interfacial hooks and the effective top sheet thickness were observed between the two processes. Fractography analysis conducted in this study suggested that the effective top sheet thickness largely determined the failure mode, which in turn influenced the final number of cycles to failure. While the height of the interfacial hook was greater in the process with better fatigue performance, it was the larger effective top sheet thickness that promoted crack propagation modes more favorable to greater fatigue resistance. As such, the shoulder plunge depth during the friction stir welding process may likely be the dominant factor in producing durable spot welds using this novel welding technique.


## Acknowledgements

The authors would like to recognize Mark Horstemeyer, Jim Quinn, Joy Forsmark, Xuming Su, Alan Luo, Robert McCune, AK Khosrovaneh, and Lin Zhang for their encouragement of this study. The authors would also like to thank Harsha Badarinarayan for providing the FSSW coupons. In addition, the authors would like to thank Jenna Grantham for conducting the fatigue tests. This material is based upon work supported by the Department of Energy National Energy Technology Laboratory under Award Numbers DE-FC26-02OR22910 and DE-EE0003583. This work was prepared as an account of work sponsored by an agency of the United States Government. Neither the United States Government nor any agency thereof, nor any of their employees, makes any warranty, express or implied, or assumes any legal liability or responsibility for the accuracy, completeness, or usefulness of any information, apparatus, product, or process disclosed, or represents that its use would not infringe privately owned rights. Reference herein to any specific commercial product, process, or service by trade name, trademark, manufacturer, or otherwise does not necessarily constitute or imply its endorsement, recommendation, or favoring by the United States Government or any agency thereof. The views and opinions of authors expressed herein do not necessarily state or reflect those of the United States Government or any agency thereof.


## References


1. S.M. Chowdhurya, D.L. Chena, S.D. Bholea, and X. Caob, "Tensile Properties of a Friction Stir Welded Magnesium Alloy: Effect of Pin Tool Thread Orientation and Weld Pitch," *Materials Science and Engineering A*, 527 (2010), 6064–6075.

2. Y.H. Yin, N. Sunb, T.H. North, and S.S. Hua, "Hook Formation and Mechanical Properties in AZ31 Friction Stir Spot Welds," *Journal of Materials Processing Technology*, 210 (2010), 2062–2070.

3. D.A. Wang, and C.H. Chen, "Fatigue Lives of Friction Stir Spot Welds in Aluminum 6061-T6 Sheets," *Journal of Materials Processing Technology*, 209 (2009), 367-375.

4. A. Gerlich, P. Su, and T.H. North, "Tool Penetration During Friction Stir Spot Welding of Al and Mg Alloys," *Journal of Materials Science*, 40 (2005), 6473–6481.

5. Y.H. Yin, A. Ikuta, and T.H. North, "Microstructural Features and Mechanical Properties of AM60 and AZ31 Friction Stir Spot Welds," *Materials and Design*, 31 (2010), 4764–4776.

6. N. Sun, Y.H. Yin, A.P. Gerlich, and T.H. North, "Tool Design and Stir Zone Grain Size in AZ31 Friction Stir Spot Welds," *Science and Technology of Welding and Joining*, 14 (2009), 747-752.

7. Y.H. Yin, N. Sun, T.H. North, and S.S. Hu, "Influence of Tool Design on Mechanical Properties of AZ31 Friction Stir Spot Welds," *Science and Technology of Welding and Joining*, 15 (2010), 81-86.

8. H. Badarinarayan, Y. Shi, X. Li, X, and K. Okamoto, "Effect of Tool Geometry on Hook Formation and Static Strength of Friction Stir Spot Welded Aluminum 5754-O Sheets," *International Journal of Machine Tools and Manufacture*, 49 (2009), 814–823.

9. P.K. Mallick, and L. Agarwal, "Fatigue of Spot Friction Welded Joints of Mg-Mg, Al-Al and Al-Mg Alloys," (Paper presented at Society of Automotive Engineers, Warrendale, PA, 2009) 2009-01-0024.

10. J.B. Jordon, M.F. Horstemeyer, S.R. Daniewicz, H. Badarinarayan, and J. Grantham, "Fatigue Characterization and Modeling of Friction Stir Spot Welds in Magnesium AZ31 Alloy," *Journal of Engineering Materials and Technology*, 132 (2010), 041008-1-10.

11. P.C. Lin, J. Pan, and T. Pan, "Failure Modes and Fatigue Life Estimations of Spot Friction Welds in Lap-Shear Specimens of Aluminum 6111-T4 Sheets. Part 1: Welds Made by a Flat Tool," *International Journal of Fatigue*, 30 (2008), 74-89.

12. P.C. Lin, J. Pan, and T. Pan, "Failure Modes and Fatigue Life Estimations of Spot Friction Welds in Lap-Shear Specimens of Aluminum 6111-T4 Sheets. Part 2: Welds Made by a Flat Tool," *International Journal of Fatigue*, 30 (2008), 90–105.

13. N. Afrin, D.N. Chen, X. Cao, and M. Jahazi, "Microstructure and Tensile Properties of Friction Stir Welded AZ31B Magnesium Alloy," *Materials Science and Engineering A*, 427 (2008), 179–186.

14. S. Suresh, *Fatigue of Materials* (United Kingdom: Cambridge University Press, 1998), 204.

Magnesium Technology 2012
*Edited by: Suveen N. Mathaudhu, Wim H. Sillekens, Neale R. Neelameggham, and Norbert Hort*
*TMS (The Minerals, Metals & Materials Society), 2012*

# EFFECT OF CORROSION ON THE TENSILE PROPERTIES OF FRICTION-STIR WELDED AZ31B SHEET

Jennifer M. Thuss, Joseph R. Kish, Joseph R. McDermid
Centre for Automotive Materials and Corrosion, McMaster University, Hamilton, ON L8S 4L7

Keywords: Friction stir welding; Magnesium alloys; Residual stress; Scanning electrochemical measurements

## Abstract

To facilitate the use of magnesium and its alloys within automotive structures, it is necessary to characterize their possible mechanical property degradation in typical application environments. This work examines the effect of exposure to NaCl-based corrosive environments on the mechanical properties of friction stir welded (FSW) AZ31B magnesium alloy sheet. A complete microstructural, electrochemical, residual stress and mechanical property characterization of the as-received FSW panels was performed. Samples were subsequently exposed to 0.01 M and 0.1 M NaCl solutions at the corrosion potential for 24 h and any changes in mechanical properties, as a function of exposure, to the corrosive environment monitored. Friction stir welded panels exhibited large decreases in ductility compared to the base AZ31B material. Exposure to 0.1 M NaCl for 24 hours resulted in a degradation in mechanical properties for the base material, as well as the friction stir welded panel, whereas no significant changes were found for samples exposed to 0.01 M NaCl for 24 hours.

## Introduction

Magnesium alloys are promising candidates for use as structural components in the automotive and aerospace industries due to their relatively high specific strength. To facilitate the use of magnesium alloys within automotive structures, it is necessary to characterize the possible mechanical property degradation in typical application environments [1]. These components can be joined using welding techniques such as metal inert gas welding (MIG), tungsten inert gas welding (TIG) and friction stir welding (FSW), and so it is also critical to assess the combined effects of welding and exposure to corrosive environments on possible mechanical property degradation.

Friction stir welding is a relatively new solid state joining technique that creates a weld bead with limited to no porosity within the weld or heat affected zone (HAZ) [3]. This technique was originally designed for use with aluminum alloys, however magnesium alloys have been successfully welded using this technique [4]. One of the many advantages to using FSW is that the weld zone, created below the melting point of the material, contains fine equiaxed grains [3]. This is due to the rapid dynamic recrystallization that occurs during the welding process. FSW process parameters such as tool rotation speed, tool geometry and tool transverse speed are critical to the evolution of weld microstructures and properties [3].

The objective of the present study is to examine the effect of FSW on the mechanical properties of AZ31B sheet, both in the unexposed condition and after exposure to 0.01 M and 0.1 M NaCl solutions at room temperature.

## Experimental Procedures

### Material

The experimental materials comprised 2 mm thick wrought AZ31B magnesium alloy sheet. The sheets were hot rolled and partially annealed to the H24 condition. The composition of the AZ13B sheet was analyzed using standard wet chemistry techniques and Inductively Coupled Plasma Optical Emission Spectroscopy (ICP-OES), the results of which are shown in Table I.

Table I: ICP-OES Analysis of Material Composition

| Element (in wt%) | | | | | | |
|---|---|---|---|---|---|---|
| Al | Zn | Mn | Fe | Si | Ni | Ca |
| 2.938 | 1.103 | 0.448 | 0.005 | 0.005 | 0.001 | 0.001 |

Prior to fabricating the FSW joints, a steel brush was used to remove any surface oxide present on the AZ31B sheet. A final cleaning step using ethanol wiping was conducted prior to welding. A H13 steel tool with a 9.5 mm shoulder and 3.175 mm left-handed pin was used to weld all sample panels. Welding was completed under position control with a tilt angle of 0.5°. All friction stir welds were fabricated using a rotation speed of 1000 rpm and a transverse speed of 5 mm/s.

### Microstructure Characterization

Weld sections were cross-sectioned and cold mounted in epoxy resin. Samples were then polished to 4000 grit SiC paper. The samples were then electrolytically polished using a solution of 3 parts phosphoric acid (85%) and 5 parts ethanol (100%) maintained at 2°C and stirred vigorously. The sample was subjected to a 2 V potential for 120 s, followed by 1.5 V for 120 s. The sample was then chemically etched for 3 s in an acetic picral solution containing 2.1 g picric acid, 5 mL acetic acid and 80 mL ethanol (95%). Optical microscopy of the weld microstructure was conducted using a Nikon Eclipse LV100 microscope.

### Tensile Tests

Sub-size tensile specimens conforming to ASTM B557 [5] were machined perpendicular to the weld direction for each panel. Specimens were tested in a variety of environmental exposure conditions in order to characterize possible mechanical property degradation. The environmental conditions used were; unexposed and 24 h pre-exposure to room temperature 0.1 M NaCl and 0.01 M NaCl solutions at the corrosion potential. All tensile specimens were tested to fracture at a strain rate of $5.5 \times 10^{-5}\ s^{-1}$ using a standard tensile frame equipped with a 10 kN load cell. Digital image correlation techniques were utilized to calculate strain as a function of time. Tensile sample fracture surfaces were examined using a Philips 515 scanning electron microscope at an acceleration voltage of 15 keV.

Scanning Electrochemical Measurements

The electrochemical behavior of the weld cross-section was analyzed using a M370 scanning electrochemical workstation and PG580R Potentiostat. Potentiodynamic polarization experiments were completed on a number of micro-sections of the weld region, HAZ and base material at a scan rate of 5 mV/s in a 0.1 M NaCl solution at room temperature. These experiments were conducted using a stationary PTFE droplet cell with a 1 mm diameter probe tip. This M370 scanning droplet system (SDS) module is fit with a platinum counter electrode and an Ag/AgCl reference electrode.

Residual Stress Analysis

A 2D X-ray diffractometer, using Cu $K_{\alpha 1}$ radiation, was used to collect residual stress information from six points within the weld cross section. To conduct this analysis a 20 mm section was cut from the welded panel. The data was then analyzed using GADDS software. It is important to note that cutting the specimen might have allowed for relaxation of the residual stresses present. The trends in tensile versus compressive residual stress were the focus of this experiment, however.

## Results and Discussion

Microstructure

Optical images of the microstructure of the stir zone, apparent HAZ and base material are shown in Figure 1 a) – c), respectively.

a)
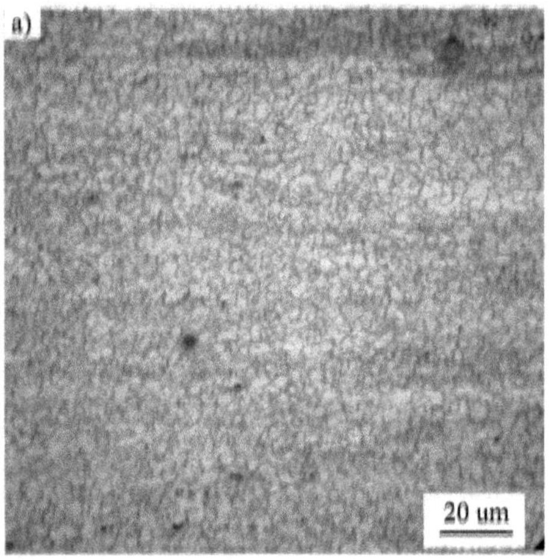


b)
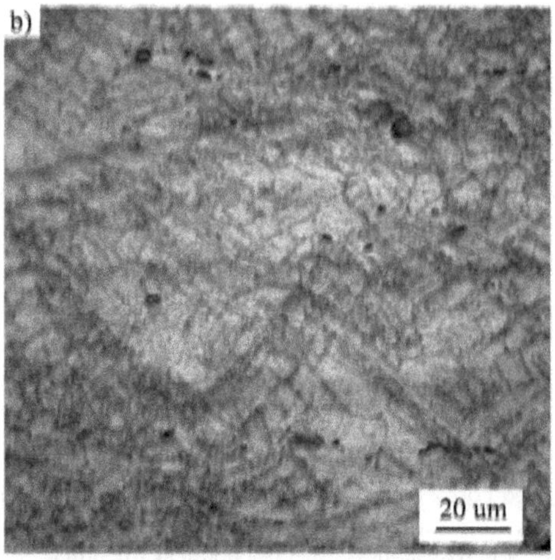


c)
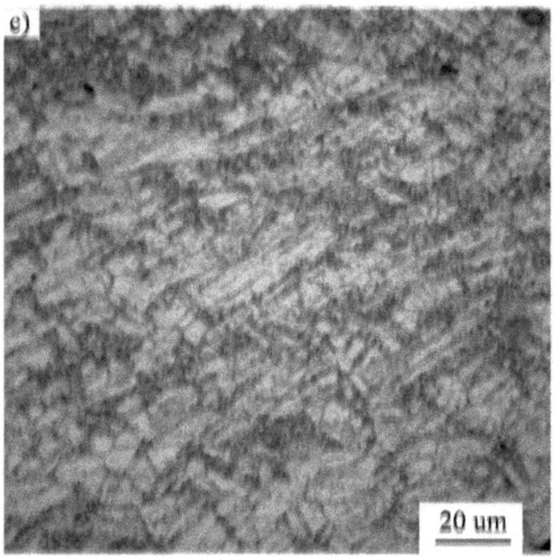


Figure 1: Microstructure of a) stir zone (x = 0 μm) b) apparent HAZ (x = 6500 μm) and c) base material (x = 20,000 μm)

It is clear that the microstructure within the stir zone consisted of fine recrystallized grains due to the friction stir welding process. Both the HAZ and base material exhibited a larger grain size, which contained twins [6]. The presence of twins within the HAZ is believed to be caused by the alloy manufacture as there was no significant difference between HAZ and base material microstructures. This lack of significant HAZ microstructure difference may have arisen from the relatively low heat input for these welding parameters.

Mechanical Properties

Results of the tensile tests for both the base and FSW AZ31B are shown in the form of true stress-strain curves in Figure 2. A summary of the mechanical properties for the base and FSW AZ31B as a function of exposure condition is shown in Table II.

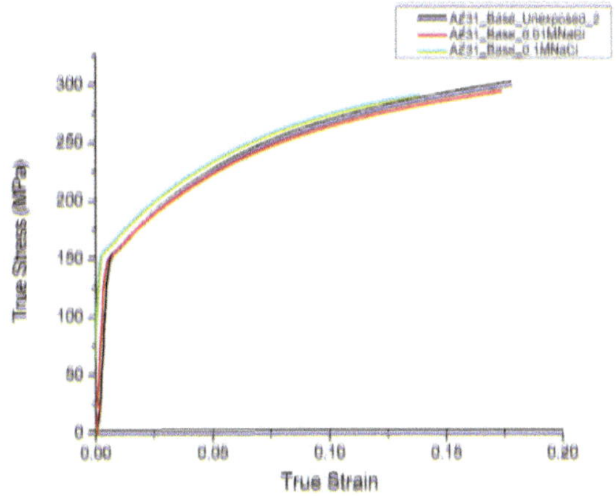


a) Base AZ31B

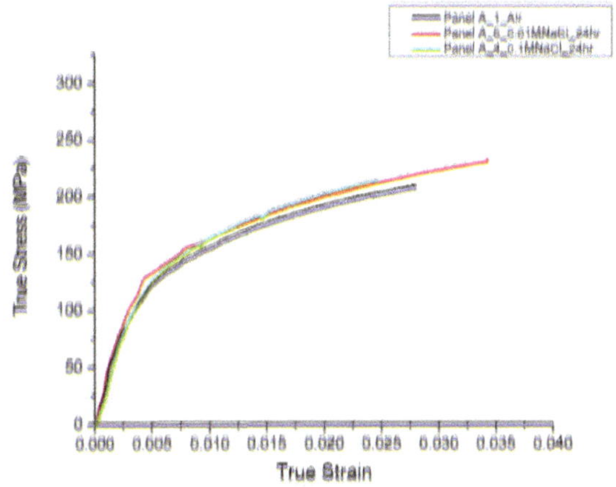


b) Friction stir welded AZ31B

Figure 2: True stress-true strain curves of base and FSW AZ31B in various environments. Note the change in scale for the strain axis between a) and b)

The exposure of base AZ31B to 0.01 M NaCl solution for 24 h at the corrosion potential had no significant effect on the ultimate tensile strength (UTS) or uniform elongation of the material versus the unexposed condition (Figure 2a, Table II). However, exposure to 0.1 M NaCl had a negative impact on the uniform elongation where a decrease in uniform elongation to 14% versus 18% for the base material was observed (Figure 2a, Table II). No significant change in UTS versus the base material was observed for the 0.1 M NaCl exposure. Song et al. found a decrease in ductility and UTS for AZ31 tested in both 0.1 M NaCl and 0.01 M NaCl at a strain rate of $1 \times 10^{-6}$ $s^{-1}$. Larger decreases in ductility, up to 34%, are thought to be an effect of the in-situ corrosion occurring and slower strain rate [2].

All FSW tensile specimens failed at the interface between the stir zone and HAZ on the retreating side of the weld. A large decrease in uniform elongation was observed for the FSW specimens as compared to the base material (Figure 2a, b, Table II). The friction stir welding process did not have as large of an effect on the UTS; a decrease of approximately 20 MPa was observed (Table II). Similar to the findings for the base material, exposure to 0.01 M NaCl did not have a significant effect on the tensile properties of the FSW material (Table II).

Table II: Summary of mechanical properties for as-received and welded AZ31B, as a function of exposure condition

| Specimen | Uniform Elongation (%) | UTS (MPa) |
|---|---|---|
| Base – Unexposed (Average) | 17.8 | 242 |
| Base – 0.01 M NaCl for 24 h | 17.4 | 248 |
| Base – 0.1 M NaCl for 24 h | 13.8 | 254 |
| FSW – Unexposed (Average) | 3.0 | 230 |
| FSW – 0.01 M NaCl for 24 h | 3.0 | 226 |
| FSW – 0.1 M NaCl for 24 h | 2.5 | 217 |

SEM Fractography

Tensile specimen fracture surfaces were analyzed using a scanning electron microscope. The base material was found to exhibit ductile fracture in the unexposed state (Figure 3a) and exhibited some brittle fracture features in the 0.1 M NaCl exposure condition (Figure 3b). This behavior was expected since a decrease in uniform elongation was observed (Figure 2a, Table II). The FSW specimens displayed brittle and cleavage facets in all exposure conditions (Figure 3c, d), as would be expected from the significant decrease in uniform elongation observed versus the base material (Figure 2b, Table II).

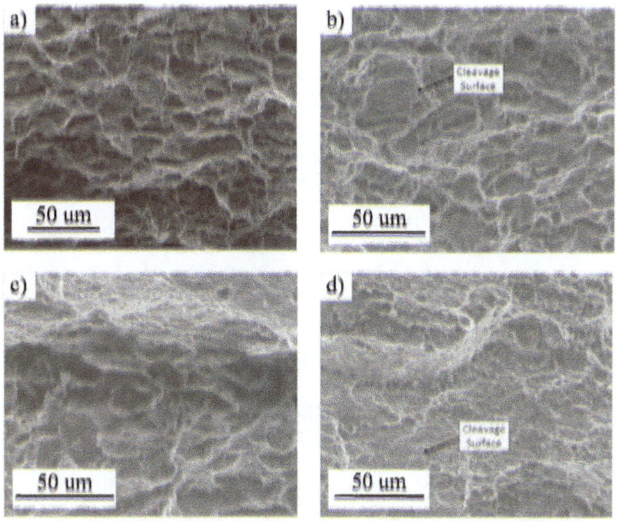


Figure 3: SEM Imaging of a) unexposed base material fracture surface b) base material fracture surface tested after exposure to 0.01 M NaCl solution at the corrosion potential for 24 h c) unexposed FSW material fracture surface and d) FSW material fracture surface tested after exposure to 0.1 M NaCl solution at the corrosion potential for 24 h.

## Scanning Electrochemical Measurements

Potentiodynamic polarization scans were completed for 5 micro-sections on a cross-sectioned sample of welded AZ31B. Electrochemical behavior as a function of distance from the weld centerline (x = 0 μm) were recorded and are shown in Figure 4.

Fluctuations within the potentiodynamic polarization scans during the scanning droplet experiment were observed, due to the existence of a short circuit within the electrochemical cell. The cause of this short circuit and effective fluctuation is not confirmed but could be due to air entering the system or by the generation of gas bubbles during corrosion processes [7]. These short instabilities, however, still allow corrosion analyses to be completed. For clarity, the data affected by these fluctuations within the system was not presented.

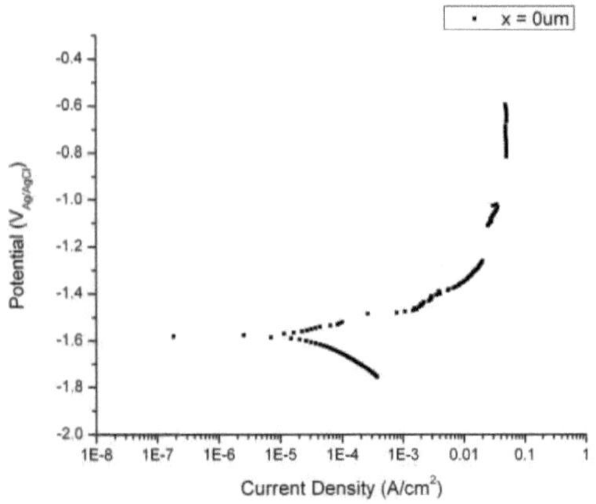


a)

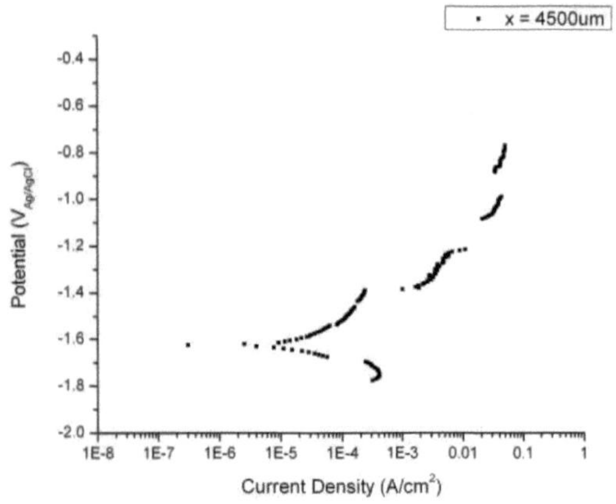


b)

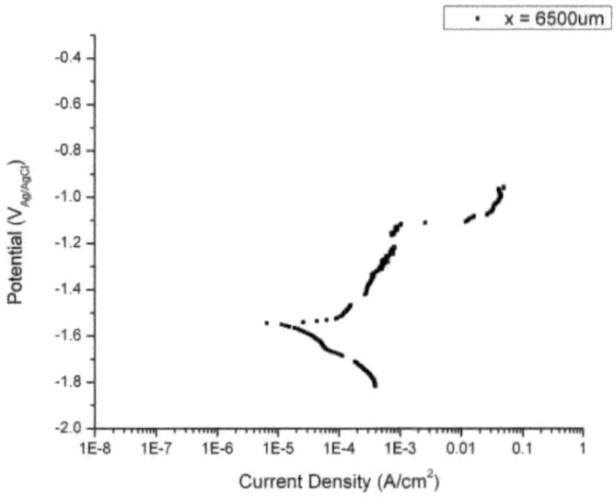


c)

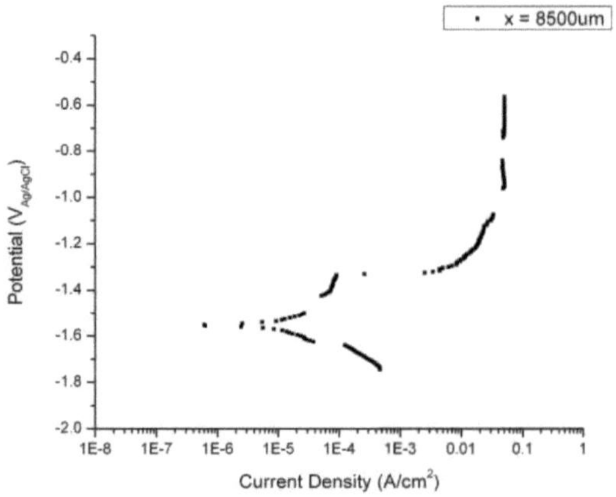


d)

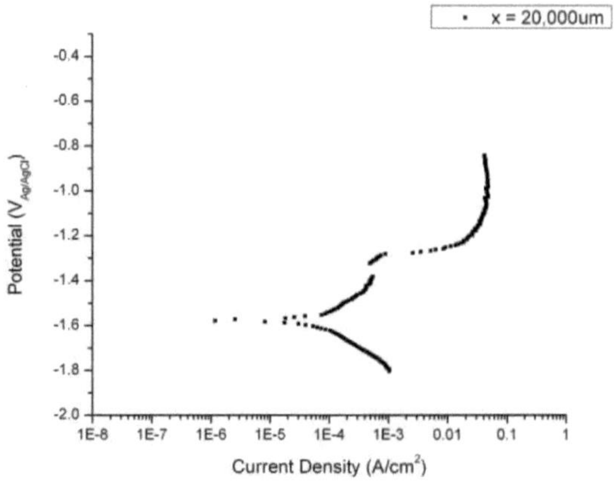


e)

Figure 4: Potentiodynamic behavior of a) stir zone (x = 0μm) b) stir zone edge (x = 4500μm) c) outside edge of stir zone (x = 6500μm) d) HAZ (x = 8500μm) and e) base material (x = 20,000μm).

The data collected from the potentiodynamic scans is summarized in Table III. From this data, it can be seen that there were small variances in corrosion potential and corrosion current density as a function of distance from the weld centerline. All breakdown potentials are more noble than their respective corrosion potentials and all mechanical tests were completed at the corrosion potential. The cathodic exchange current density calculations show that there were no significant differences in the cathodic reaction kinetics across the weld section.

Hydrogen induced cracking and pitting corrosion are two types of corrosion mechanisms that effect mechanical properties. The cathodic hydrogen evolution reaction can cause hydrogen embrittlement which is proposed to aid in the cracking mechanism [8]. Mechanical properties can also be affected by the presence of localized pits which act as stress concentrators [8]. The understanding that there were no significant differences in cathodic kinetics as a function of distance correlates well to the observed mechanical properties. If there was an observed localized cathodic site it would be expected to coincide with the location of fracture due to the effect of hydrogen embrittlement. This is not the case since all tensile specimens failed in the same region on the retreating side of the weld zone. The location of fracture also would (be expected to) have changed if there was an effect of pitting corrosion. Again this is not the case since all tensile specimens were exposed at the corrosion potential, which was less noble than all observed breakdown potentials.

Table III: Electrochemical behavior summary of FSW welded AZ31B.

| Position (μm) | $E_{CORR}$ (V) | $E_{BK}$(V) | $i_{CORR}$ (A/cm$^2$) | $i_{o,c}$ (A/cm$^2$) |
|---|---|---|---|---|
| 0 | -1.58 | -1.49 | $5.0\times10^{-5}$ | $3.6\times10^{-10}$ |
| 4500 | -1.63 | -1.39 | $2.0\times10^{-5}$ | $4.2\times10^{-10}$ |
| 6500 | -1.55 | -1.11 | $1.9\times10^{-5}$ | $1.2\times10^{-11}$ |
| 8500 | -1.55 | -1.33 | $2.6\times10^{-5}$ | $1.7\times10^{-10}$ |
| 20,000 | -1.58 | -1.28 | $6.5\times10^{-5}$ | $3.0\times10^{-10}$ |

Residual Stress Analysis

X-ray analysis was completed on a plane view section of FSW material; a macrograph of this sample is shown in Figure 5.

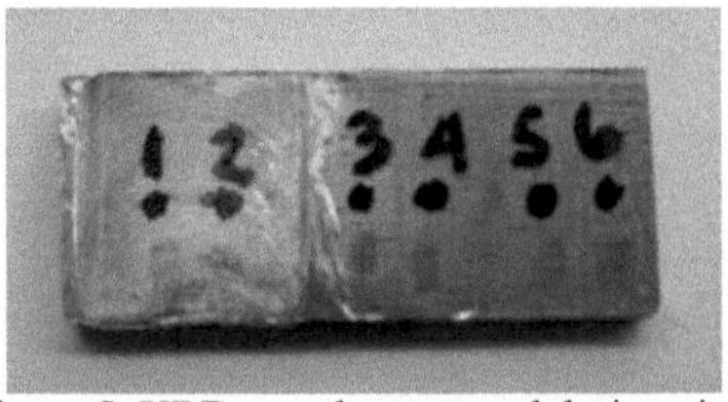

Figure 5: XRD sample setup and designation

The maximum principle stresses calculated from this experiment are shown in Figure 6. From this analysis it was found that the stir zone contained large tensile residual stresses, ranging from 200 to 150 MPa, moving towards small compressive residual stresses within the HAZ and base material. These residual stresses are present due to the FSW process [9]. Again it is important to take note of the trend that exists for the residual stresses found within the weld versus the magnitude of the stresses calculated.

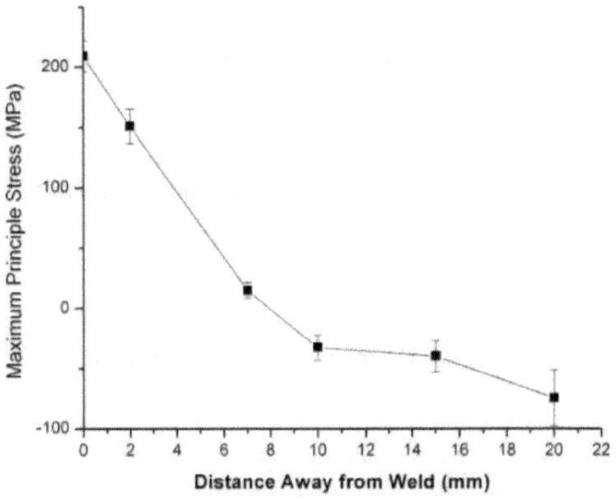

Figure 6: Residual Stress Analysis using 2D X-Ray Diffraction

The large gradient in residual stresses at the interface between the stir zone and HAZ correlates to the tensile specimen fracture that occurs on the edge of the stir zone. This trend is comparable to that found in an aluminum alloy friction stir weld, researched by Prime et al. The 7050-2024 FSW system contained tensile residual stresses within the stir zone, ranging from 30 to 40 MPa, and compressive residual stresses within the base material [9]. The difference in magnitude is assumed to be due to the material system as well as the welding parameters.

## Summary

1) The microstructure within the stir zone consisted of fine, re-crystallized grains. Larger grains containing twins were found within the HAZ as well as the base material.
2) A slight decrease in tensile strength and large decrease in ductility were observed for FSW panels, as compared to the base AZ31B material, for all environmental exposure conditions tested.
3) Exposure to 0.01 M NaCl solution for 24 h at room temperature did not have a significant effect on the mechanical properties (UTS, elongation at fracture, etc) for both the base material and the welded panels, whereas degradation in mechanical properties was observed in both materials exposed to 0.1 M NaCl for 24 h at the corrosion potential.
4) No significant relative differences in cathodic and anodic corrosion behaviour were found across the weld section.
5) A transition between tensile residual stress and compressive residual stress exists from the stir zone out to the base material. The location of this transition zone coincides with the fracture location in the FSW joints.

## Acknowledgements


The authors would like to thank General Motors for the donation of experimental materials and the National Research Council of Canada for fabrication of the friction stir welded panels. The financial support of the Natural Sciences and Engineering Research Council of Canada (NSERC) is also acknowledged.

**References**

1. G. Song, and A. Atrens, "Understanding Magnesium Corrosion, A Framework for Improved Alloy Performance," *Advanced Engineering Materials*, 5 (2003), 837-858.

2. R.G. Song, C. Blawert, W. Dietzel, and A. Atrens, "A study on stress corrosion cracking and hydrogen embrittlement of AZ31 magnesium alloy," *Materials Science and Engineering A*, 399 (2005), 308-317.

3. R.S. Mishra and Z.Y. Ma, "Friction stir welding and processing," *Materials Science and Engineering R: Reports*, 50 (2005)

4. R. Johnson, "Friction stir welding of magnesium alloys," *Magnesium Alloys 2003, Pts 1 and 2*, 419-4 (2003), 365-370

5. ASTM Standard B557, 2010, "Standard Test Methods for Tension Testing Wrought and Cast Aluminum- and Magnesium-Alloy Products," ASTM International, West Conshohocken, PA, 2010, DOI: 10.1520/B0557-10, www.astm.org.

6. M.B. Kannan, W. Dietzel, R. Zeng, R. Zettler, and J.F. dos Santos, "A study of the SCC susceptibility of friction stir welded AZ31 Mg sheet," *Materials Science and Engineering A-Structural Materials Properties Microstructure and Processing*, 460 (2007), 243-250.

7. K. Fushimi, S. Yamamoto, R. Ozaki, and H Habazaki, "Cross-section corrosion-potential profiles of aluminum-alloy brazing sheets observed by the flowing electrolyte scanning-droplet-cell technique," *Electrochimica Acta*, 53 (2008), 2529-2537.

8. N. Winzer, A. Atrens, G. Song, E. Ghali, W. Dietzel, K.U. Kainer, "A critical review of the Stress Corrosion Cracking (SCC) of magnesium alloys," *Advanced Engineering Materials*, 7 (2005), 659-693.

9. M.B. Prime, T. Gnaupel-Herold, J.A. Baumann, R.J. Lederich, D.M. Bowden, and R.J. Sebring, "Residual stress measurements in a thick, dissimilar aluminum alloy friction stir weld," *Acta Materialia*, 54 (2006), 4013-4021.

**Magnesium Technology 2012**
*Edited by: Suveen N. Mathaudhu, Wim H. Sillekens, Neale R. Neelameggham, and Norbert Hort*
***TMS (The Minerals, Metals & Materials Society), 2012***

# ON THE EFFECT OF $Ti_2AlC$ ON THE FORMATION OF THERMALLY STABLE Mg NANO GRAINS

Babak Anasori and Michel W. Barsoum
Department of Materials Science and Engineering, Drexel University, Philadelphia, PA 19104, USA

Keywords: Nanocrystalline Mg matrix composite, Liquid melt infiltration, Powder metallurgy.

## Abstract

When Mg or Mg-alloys are reinforced with $Ti_2AlC$ – using a simple pressureless melt infiltration method – the result is nanocrystalline, nc, Mg-matrix composites, with outstanding mechanical properties. As an added bonus, the nc Mg-matrix is extraordinarily thermally stable. When AZ61 is used to infiltrate the $Ti_2AlC$ preforms, ultimate tensile stresses of 800 MPa are achieved. The reasons that lead to the formation of the nc-Mg are as of yet not understood. In this study, the different composites' microstructures are investigated by X-ray diffraction, scanning and transmission electron microscopy. A correlation was found between the presence of oxygen and the nc Mg grains. Nanometer sized Mg grains were also found in between some of the $Ti_2AlC$ layers.

## Introduction

Mg alloys have attracted a great deal of interest over the past decades due to their high specific strength, which renders them good candidates for applications in the aerospace and automotive industries [1, 2]. One method that can be used to achieve better properties is to fabricate materials wherein the grain size is at the nanometer scale [3]. The challenge for this approach, however, is that the grains tend to coarsen, even at room temperature [4]. Another approach to enhance the mechanical properties is to fabricate composites [5-7]. Herein, we show that by combining both approaches exceptional properties can be achieved.

In 2009, we reported on the fabrication of novel, near net-shape nanocrystalline-Mg, nc-Mg, composites, reinforced with $Ti_2AlC$ [8]. The latter is a member of a large family of ternary carbides and nitrides, referred to as the *MAX phases*, with unique combinations of metal and ceramic properties. Like metals, they are electrically and thermally conductive, most readily machinable [9, 10], not susceptible to thermal shock, plastic at high temperatures, and exceptionally damage tolerant. Like ceramics, some of them are elastically rigid (Young's mod. > 300 GPa), lightweight (≈ 4 g/cm$^3$) and maintain their strengths to high temperatures [11]. The ternaries $Ti_3SiC_2$ and $Ti_2AlC$ are creep, fatigue and oxidation resistant [12,13]. The Ti-based MAX phases are also more conductive, both electrically and thermally, than Ti metal at all temperatures [14, 15].

Because of their layered structure, the MAX phases' major deformation mechanism is basal slip, which in turn results in the formation of kink bands. Due to this deformation mechanism, they have excellent low and high-cycle fatigue and damping properties [16]. The latter increase with the applied stress squared [17]. The unusual damping properties have been explained by the formation and annihilation of incipient kink bands, IKBs. IKBs are parallel, co-axial, basal dislocation loops. More details on IKB-based deformation mechanisms can be found elsewhere [17, 18].

The near net shape nc-Mg composites were made by pressureless infiltration of Mg into a 50 vol.% porous $Ti_2AlC$ preform in a vacuum furnace at 750°C for 1 h. The fabrication details can be found elsewhere [8, 18, 19]. The resulting composites, with no additional steps, possessed outstanding properties. The best results were reported for a Mg-50 vol.% $Ti_2AlC$, henceforth referred to as 50Mg-211. According to X-ray diffraction, XRD, transmission electron microscopy, TEM, and differential scanning calorimetery, DSC, results, the Mg grain size was found to be ≈ 35±15 nm. The density of the composite was 2.9 g/cm$^3$. The ultimate compressive strength, UCS, was 700 ± 10 MPa [8, 20]. Although both starting components had tensile strengths lower than 300 MPa, the composite's ultimate tensile strength, UTS, was 345 ± 40 MPa. This enhancement in UTS was ascribed to the small grain size of the Mg. In contrast to most nc-Mg microstructures, the nano grains were reported to be so exceptionally thermally stable that heating three times to 700°C – i.e. even melting the nc-Mg – did not lead to grain growth [20]. When the same infiltration method was used with $Ti_3SiC_2$ as the reinforcement, the UCSs were not as high. The reason for this state of affairs was again related to Mg grain size, which was not at the nano scale in the Mg/$Ti_3SiC_2$ composites [8].

More recently, we reported on 80 vol.% nc-Mg composites, reinforced with 20 vol.% $Ti_2AlC$. In that case, the Mg average grain size was 90±15 nm. The density was 2 g/cm$^3$ [18]. The UCS was 350±10 MPa. Although this composite was comprised of 80 vol.% Mg, the nano grains were still thermally stable up to 700°C.

The aim of this study is to explore: i) the reasons for the formation of the nc-Mg, and, ii) what keeps the grains from coarsening. XRD, scanning electron microscopy, SEM, and TEM results are discussed below.

## Experimental Details

Both a 50Mg-211 and a Mg-50 vol.% $Ti_3SiC_2$ composite, henceforth referred to as 50Mg-312, were fabricated. The processing method is detailed elsewhere [8, 18, 19]. XRD was carried out on a diffractometer (Model 500D, Siemens, Karlsruhe, Germany) and the patterns were collected using step scans of 0.01° in the range of 10°–80° 2θ and a step time of 2 s. Scans were made with Cu Kα radiation (40 kV and 30 mA). The polished surfaces were observed in a SEM (Zeiss Supra 50VP, Germany) and elemental maps were taken using the energy-dispersive spectroscope (EDS) (Oxford Inca X-Sight, Oxfordshire, UK) in the SEM.

TEM samples were prepared first by cutting ~300 μm thick slices from the bulk samples with a low speed diamond saw. Small 3 mm diameter disks were then made using an ultrasonic disk cutter (Model 170, Fischione, Export, PA). Both sides were polished with a dimpling grinder (Model 200, Fischione, Export, PA) using 3 μm, 1 μm and 0.1 μm diamond pastes successively. The disk thickness after dimpling was about 25 μm. Final perforation was made with an ion mill (Model 1010 Fischione, Export, PA) at 5 kV. TEM characterization was performed using a field emission TEM (JEOL JEM-2010F, Akishima, Japan) operating at 200 kV. Images were collected with a multi-scan CCD digital camera. EDS analysis in the TEM was carried out with an attached EDAX ultra-thin window X-ray energy dispersive spectrometer.

## Results and Discussion

As reported in our earlier studies on these composites [8, 18], Mg peaks from the XRD patterns of all samples were compared with those of pure Mg powder ($d_{av} \approx 150$ μm). Figure 1 compares the main XRD peak of Mg in 50Mg-211, 50Mg-312 with that of pure Mg. Comparing the full width at half maximum, FWHM, of the Mg peaks revealed that peak broadening occurred only for the 50Mg-211 composite. Applying the Scherrer formula [21], the Mg grain size is estimated to be 35±15 nm, in agreement with previous work [8, 18]. The FWHMs of the 50Mg-312 composite are comparable to those of the initial Mg powder (Fig. 1).

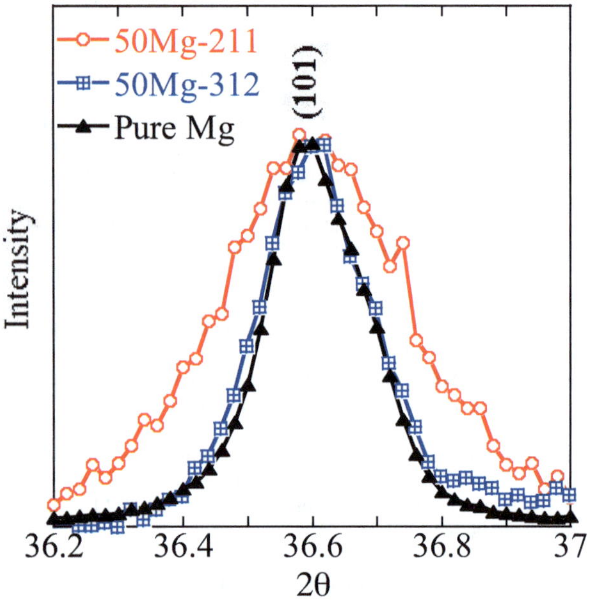


Figure 1. XRD diffractograms of the main Mg peak in the 50Mg-211, 50Mg-312 composites and pure Mg.

SEM micrographs of 50Mg-211 polished surfaces revealed that Mg infiltrated all the $Ti_2AlC$ particles and penetrated into some of the smallest cracks and fissures in the original $Ti_2AlC$ grains. EDS across different $Ti_2AlC$ grains (Fig. 2) showed the presence of Mg even in cracks/delaminations of the order of a few nm (Fig. 2b). This implies that the molten Mg wets the $Ti_2AlC$ grains exceptionally well and infiltrates all the preform. It is worth noting that the as-received $Ti_2AlC$ powder particles are more micro-cracked and delaminated than the $Ti_3SiC_2$ powder particles.

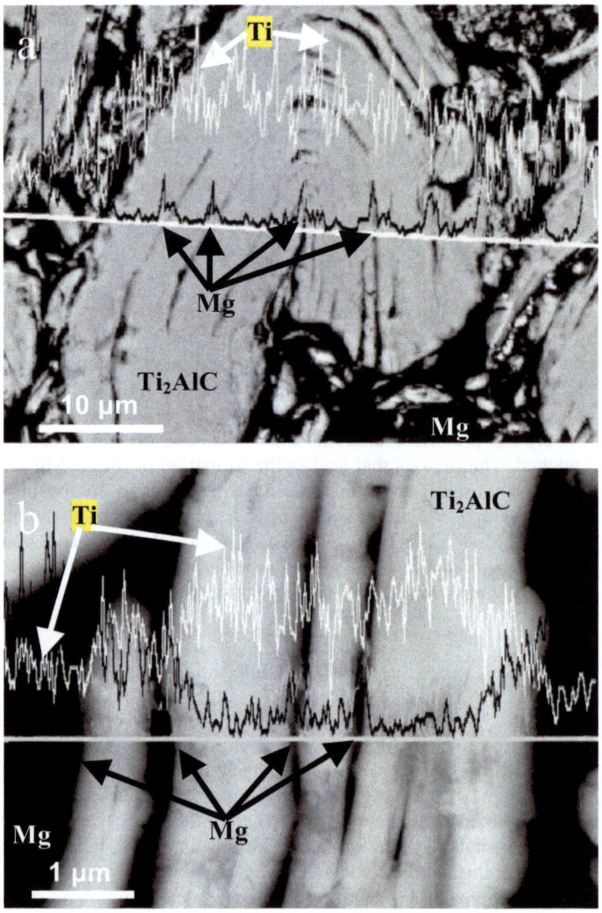


Figure 2. a) SEM micrograph of polished 50Mg-211 surface superimposed with elemental line scans (solid line) showing the presence of Mg in between delaminations in individual $Ti_2AlC$ grains, b) same as a, but at higher magnification. In these maps, black represents Mg; white Ti.

The presence of Mg in between delaminations, cracks and fissures in individual $Ti_2AlC$ grains is also confirmed when fractured surfaces are examined (Fig. 3a). Here two morphologies were observed. Some surfaces (see arrow in Fig. 3a) were not decorated with Mg, but were simply Al or Ti-terminated surfaces typical to those present when bulk $Ti_2AlC$ samples are fractured. Other surfaces, on the other hand, were covered in small (<100 nm) dimples that EDS identified as Mg (see Fig. 3b and bottom inset). The Mg was also found stretched between two $Ti_2AlC$ surfaces (top inset in Fig. 3b). The presence of Mg in such narrow regions is important because it could at least partially explain the broadening observed in the XRD patterns. More importantly, these thin Mg layers would also explain the exceptional mechanical properties obtained in some of these composites, since such thin metallic layers would be difficult to deform. For example, when AZ61 50 vol.% $Ti_2AlC$ composites were fabricated, their UTSs ranged from 600 to 800 MPa.

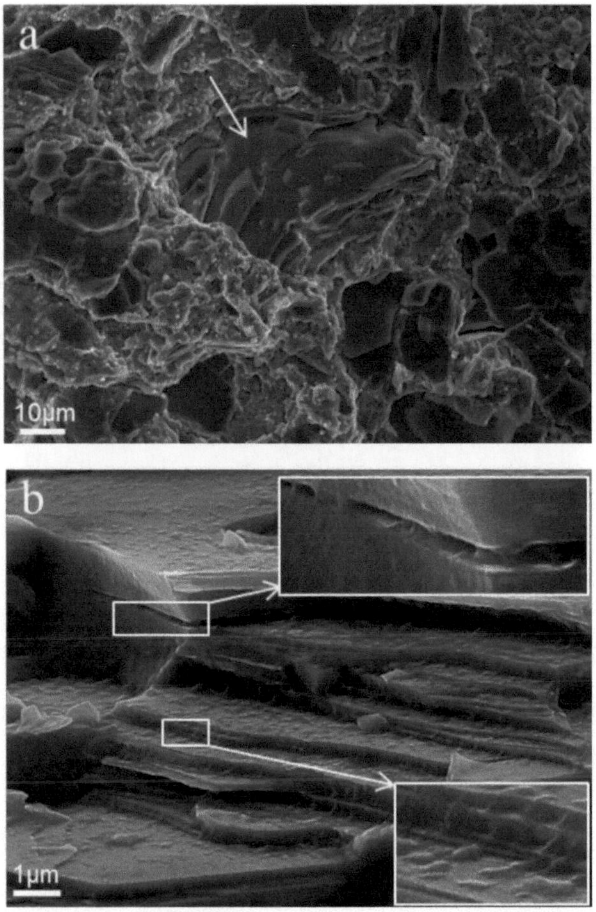


Figure 3. SEM fractographs of 50Mg-211 showing, a) a fracture surface mostly covered with Mg, which also surrounds the $Ti_2AlC$ particles, and b) fractured $Ti_2AlC$ particle covered with a thin layer of Mg. Insets show higher magnification micrographs of select regions on the image. Arrow in (a) point to a $Ti_2AlC$ fractured surface that is not decorated with Mg.

When the same sample was imaged in a TEM, thin Mg layers (not shown) were observed in between the $Ti_2AlC$ layers confirming the SEM observations. In addition, relatively large regions of Mg were observed (Fig. 4a). At higher magnification (Fig. 4b), it is evident that the Mg is present at the nano scale. Selected area diffraction, SAD, of this region (not shown) confirmed that these grains are at the nanometer scale. EDS revealed the presence of oxygen in this region. Its presence can be explained by postulating that thin MgO layers somehow separate, or exist between, the Mg nano grains. This is in agreement with what was reported earlier [8, 18, 20]. The conclusion is bolstered by the facts that, i) MgO peaks are observed in the XRD diffractograms and, ii) the peaks were quite broad [8]. Using the Scherrer formula the size of the MgO grains was estimated to be ~ 3 nm [8]. The thermal stability of nano Mg grains can be explained by the presence of these MgO layers in between the Mg nano grains, as well as the presence of nano Mg layers in between the $Ti_2AlC$ layers. These comments notwithstanding, more work – some of it ongoing – is needed to unambiguously show that thin MgO layers are present in between the Mg grains.

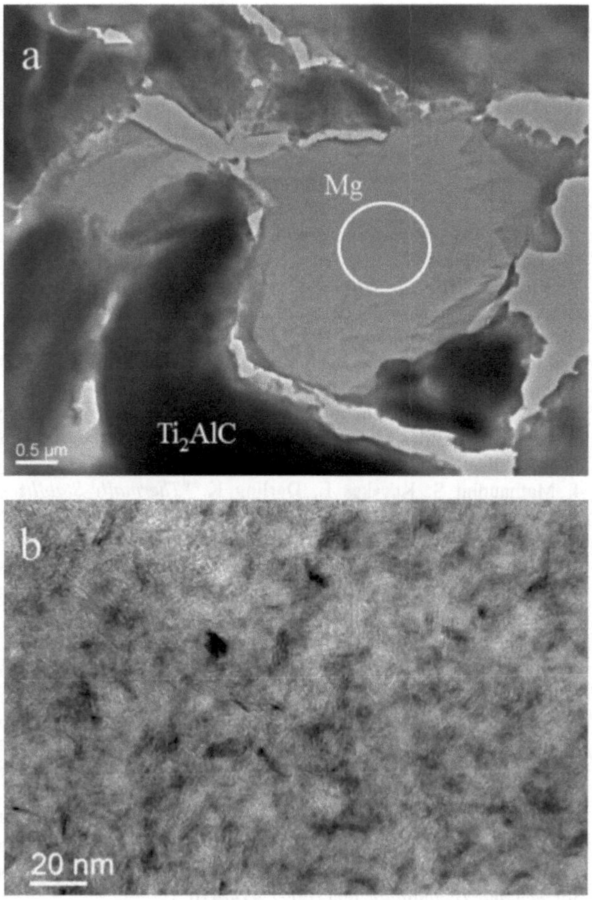


Figure 4. TEM micrographs of 50Mg-211 composite, a) showing large Mg region next to a $Ti_2AlC$ grains, b) higher magnification of circled Mg region in (a) showing Mg nano grains.

It is important to note that not all the Mg regions are comprised of nanograins. In some regions, relatively large Mg grains have been found. The oxygen content in these regions was less than in those where the grains were at the nanoscale. Understanding what role these large grains play on the mechanical properties is ongoing as well.

## Summary and Concluding Remarks

Using a simple pressureless melt infiltration method, near net shape nc-Mg composites reinforced with $Ti_2AlC$, that exhibit good mechanical properties were fabricated. Mg peak broadening in the XRD shows the presence of nano-Mg grains in the matrix. The latter was also confirmed by TEM images and SAD patterns. The reason for the formation of the nano-Mg grains is still unclear. However, given that oxygen is detected in all the nano-Mg regions, it is reasonable to assume that oxygen plays an important role. In addition, the wetting of Mg in these composites is excellent, which leads to Mg penetration into the thinnest of cracks and delaminations in the $Ti_2AlC$ grains and formed nc-Mg between them. The presence of MgO in between the nano-Mg grains and the confinement of Mg in between the $Ti_2AlC$ layers in some regions may explain the thermal stability of Mg nano grains.

**Acknowledgements**

This work was supported by the Army Research Office (No. W911NF-07-1-0628). Use of the equipment of the Centralized Research Facility (Drexel University) is acknowledged.

**References**

1. Pollock TM. "Weight Loss with Magnesium Alloys," *Science*, 328 (2010), 986-987.
2. Kulekci MK. "Magnesium and its alloys applications in automotive industry," *International Journal of Advanced Manufacturing Technology*, 39 (2008), 851-865.
3. Inoue A, Kawamura Y, Matsushita M, Hayashi K. "High strength nanocrystalline Mg-based alloys". ISMANAM 2001: International Symposium on Metastable, Mechanically Alloyed and Nanocrystalline Materials, vol. 386-388. Switzerland: Trans Tech Publications, (2002). p.509-517.
4. Mathaudhu S, Kecskes L, Darling K, "*Thermally-Stabilized Nanocrystalline Mg-Alloys*", in *Magnesium Technology 2011* (John Wiley & Sons, Inc., 2011), pp. 453-456.
5. Contreras A, Lopez VH, Bedolla E. "Mg/TiC composites manufactured by pressureless melt infiltration," *Scripta Materialia*, 51 (2004), 249-253.
6. Ferkel H, Mordike BL. "Magnesium strengthened by SiC nanoparticles," *Materials Science and Engineering: A*, 298 (2001), 193-199.
7. Zheng MY, Zhang WC, Wu K, Yao CK. "The deformation and fracture behavior of SiCw/AZ91 magnesium matrix composite during in-situ TEM straining," *Journal of Materials Science*, 38 (2003), 2647-2654.
8. Amini S, Ni C, Barsoum MW. "Processing, microstructural characterization and mechanical properties of a $Ti_2AlC$/nanocrystalline Mg-matrix composite," *Composites Science and Technology*, 69 (2009), 414-420.
9. Barsoum MW, Brodkin D, ElRaghy T. "Layered machinable ceramics for high temperature applications," *Scripta Materialia*, 36 (1997), 535-541.
10. Barsoum MW, ElRaghy T. "Synthesis and characterization of a remarkable ceramic: $Ti_3SiC_2$," *Journal of the American Ceramic Society*, 79 (1996), 1953-1956.
11. Barsoum MW, Radovic M. "Elastic and Mechanical Properties of the MAX Phases," *Annual Review of Materials Research*, 41 (2011), 195-227.
12. Amini S, McGhie AR, Barsoum MW. "Isothermal Oxidation of $Ti_2SC$ in Air," *Journal of the Electrochemical Society*, 156 (2009), P101-P106.
13. Low IM, Lee SK, Lawn BR, Barsoum MW. "Contact damage accumulation in $Ti_3SiC_2$," *Journal of the American Ceramic Society*, 81 (1998), 225-228.
14. Barsoum MW, Flemings MC, Kramer EJ, Mahajan S, Veyssiere P. "Physical properties of the MAX phases," *In: Encyclopedia of materials science and technology*, (2006).
15. Barsoum MW, Yoo HI, Polushina IK, Rud VY, Rud YV, El-Raghy T. "Electrical conductivity, thermopower, and hall effect of $Ti_3AlC_2$, $Ti_4AlN_3$, and $Ti_3SiC_2$," *Physical Review B*, 62 (2000), 10194-10198.
16. Kontsos A, Loutas T, Kostopoulos V, Hazeli K, Anasori B, Barsoum MW. "Nanocrystalline Mg–MAX composites: Mechanical behavior characterization via acoustic emission monitoring," *Acta Materialia*, 59 (2011), 5716-5727.
17. Barsoum MW, Basu S, "*Kinking Nonlinear Elastic Solids*", in *Encyclopedia of Materials Science and Technology*, edited by K. H. J. Buschow RWC, M. C. Flemings, B. Ilschner, E. J. Kramer, S. Mahajan and P. Veyssiere (Elsevier, Oxford, 2010).
18. Anasori B, Amini S, Presser V, Barsoum MW, "*Nanocrystalline Mg-Matrix Composites with Ultrahigh Damping Properties*", in *Magnesium Technology 2011* (John Wiley & Sons, Inc., 2011), pp. 463-468.
19. Amini S, Barsoum MW. "On the effect of texture on the mechanical and damping properties of nanocrystalline Mg-matrix composites reinforced with MAX phases," *Materials Science and Engineering A*, 527 (2010), 3707-3718.
20. Amini S, Cordoba JM, Daemen L, McGhie AR, Ni CY, Hultman L, Oden M, Barsoum MW. "On the Stability of Mg Nanograins to Coarsening after Repeated Melting," *Nano Letters*, 9 (2009), 3082-3086.
21. Cullity B, "*Elements of X-ray diffraction*". (Addison-Wesley, London, UK, 1978).

# EXPERIMENTAL INVESTIGATIONS INTO THE DEFORMATION BEHAVIOR OF THIXO-MOLDED Mg AZ61L SHEET ALLOY

Muammer Koç[1,3], Ömer Necati Cora[2,3], Ryan Snell[3], Ray Decker[4], Jack Huang[4]
[1] İstanbul Şehir University, İstanbul, Turkey
[2] Department of Mechanical Engineering, Karadeniz Technical University, Trabzon, Turkey
[3] NSF I/UCR Center for Precision Forming, Virginia Commonwealth University, Richmond, VA, USA
[4] Thixomat Inc., Ann Arbor, MI, USA

Keywords: Thixomolding, magnesium alloys, AZ61L, warm hydroforming, uniaxial test, bulge test

## Abstract

Deformation behavior and formability limits of thixo-molded Mg AZ61L alloy and regular Mg AZ61 alloy sheets (with similar chemical compositions except for Mn content) were experimentally investigated at elevated temperature. Tensile (uniaxial), hydraulic bulge (biaxial) tests as well as closed-die hydroforming tests were conducted to understand the material behavior at temperatures ranging from 25 °C to 300 °C and strain rates at 0.0013, 0.013 and 0.13 $s^{-1}$. It was found that flow stress and the maximum plastic strain increased with increasing temperature and decreasing strain rate. Closed-die warm hydroforming tests were also performed to determine the process window for the sheet alloy. Die cavity filling ratios and thinning of the sheet blanks were measured with non-contact optical photogrammetry. Results indicated that lower strain rates and higher temperatures increase formability, particularly above temperatures of 200 °C.

## Introduction

Due in large part to the current state of the global economy along with concerns over global warming and going green, there has been an increasing quest to make vehicles more efficient. Utilization of lightweight materials such as aluminum and magnesium in automotive components is accepted as part of solution attempts to reduce fuel consumption and emissions. Depending on the choice of material, it has been known that the use of lightweight materials can lower the weight of vehicles 15-75% [1]. Mg, for example, is one the most promising materials in weight-saving efforts due to its low density of 1.74 g/cm$^3$ [2]. It is also reported that a 10% reduction in automotive body weight in turn can increase the fuel efficiency by as much as 8% [3]. However, the majority of the lightweight parts in the present vehicles are die castings, for mainly underbody structures and casings. Another realm of automotive components that can utilize lightweight materials to reduce weight is thin-gauge body and structural parts (sheet metal-based) that can be potentially fabricated using magnesium alloys to attain higher levels of fuel savings and emission reductions [4].

Despite the obvious advantages of lightweight materials, they have considerably lower formability at room temperature when compared to mild steel alloys. The formability of aluminum alloys at cold conditions is reported to be limited by its high alloy percentage [5] while the low formability of magnesium alloys at room temperature is known to be mainly caused by the limited number of slip systems in their hexagonal closed-packed (HCP) lattice structure and low tendency to twinning at ambient conditions [6]. To alleviate this problem, magnesium sheets should be formed at elevated temperatures below their recrystallization ranges, where the thermal activation of additional slip systems is realized [7–11]. Previous studies indicated that temperature levels of up to 350 °C for Mg AZ31 [12] and 200-250 °C for AZ31B and AZ61B [10] would increase the elongation. Furthermore, higher elongation could be obtained when low strain rate is used as these materials have intrinsically high strain-rate sensitivity, especially at elevated temperatures (i.e., strain-rate hardening, "m", increases 15 folds from room temperature to 200 °C for magnesium sheet) [7]. The strain-rate effect was shown to be less significant if the forming is carried out at lower temperatures [6]. On the other hand, lower strain rates lead to slow production rates. Thus, it is vital to find the optimum balance between production rate and strain-rate control. This optimization will require a complete understanding of the response of these materials at different temperature and strain-rate levels for accurate analysis and design of the forming processes.

The process of thixomolding takes advantage of the principle that magnesium, aluminum and zinc alloys become semi-solid between the liquidus and the solidus temperatures. Mechanical shearing of the semi-solid metal generates a thixotropic structure that allows these materials to be molded utilizing a process similar to plastic injection molding while eliminating the environmental impacts of die casting. Due to its similarity to injection molding of polymers, thixomolding is capable of creating parts with thin walls, superior surface quality, and be formed to virtually net shape. In addition, thixomolding allows for faster cycle rate, longer die life, and increased worker safety due to the lower temperature of the material. The resulting thixomolded magnesium microstructure characterized as primary Mg rich particles surrounded by rapidly quenched liquid (fully or partly amorphous) is believed to delay crack propagation during fracture, resulting in improved ductility [13]. The AZ61L alloy used in this study was manufactured by means of the Thixomolded-Thermomechanical Processed (TTMP) technique in which sub-micron grain size is obtained. This process also results in nanometer sized dispersoid generation in β phase in the alloy.

The main focus of this study was to understand the formability of thixo-processed Mg sheet material under uniaxial and bi-axial loading conditions at elevated temperatures. It was also intended to compare the mechanical properties of nano-structured Mg AZ61L alloy with the regular AZ61 alloy. One similar alloy, AZ31, was studied by the authors earlier, as well [14]. The following subsections cover the analyses of uniaxial (tensile), biaxial (hydraulic bulge), and closed-die hydroforming tests at different temperature and strain-rate conditions.

### Material Characterization at Elevated Temperatures

Tensile Tests

Nanostructured thixomolded magnesium AZ61L alloy sheets (170x170 mm) with an initial thickness of 1.56 mm were provided by Thixomat (Thixomat Inc., Ann Arbor, MI). Tensile test coupons were prepared from the sheets according to the ASTM standard E8-04. To ensure a proper examination of anisotropy of the alloy, tensile coupons were prepared both parallel to the rolling direction and transverse to the rolling direction. A minimum number of three specimens was used to ensure the repeatability of the results at each testing condition.

Four different temperature levels (25, 100, 200 and 300 °C) and two different initial strain rates (0.0013 and 0.013 $s^{-1}$) were selected as test conditions. Figure 1 shows stress-strain curves obtained at each condition set for nano-structured and regular alloys. As expected, the flow stress values generally decrease with increasing temperature and/or decreasing strain rate. On the other hand, the effect of the strain rate on both flow stress and maximum strain becomes more distinct with increasing temperature. In addition to these effects, the direction in which load was applied had a strong bearing on total strain at failure. For higher temperatures (200 and 300 °C), consistently higher elongations were achieved for AZ61L alloy samples that were loaded parallel to the rolling direction (Figure 1a vs. 1b). When stress and strain values are compared for the alloys tested (Figure 1a versus 1c), it was noticed that both yield and tensile strengths did not differ significantly; however, AZ61L alloy exhibited considerably higher elongation especially at high temperatures (200 and 300 °C). However, at temperatures below 200 °C the samples showed less ductility. This is likely the result of the limited slip planes at temperatures below the critical temperature at which additional slip planes are activated.

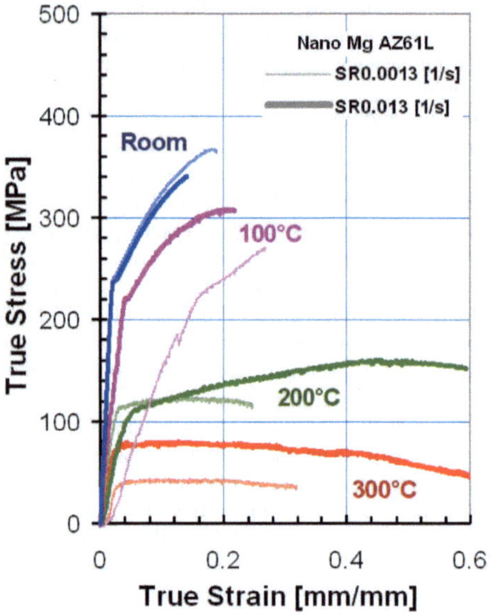


a)

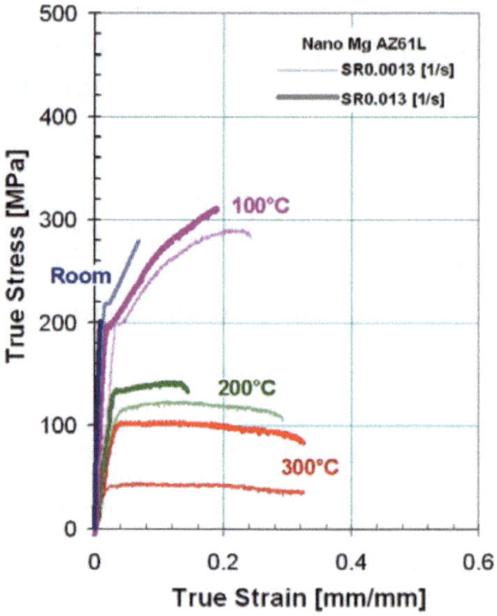


b)

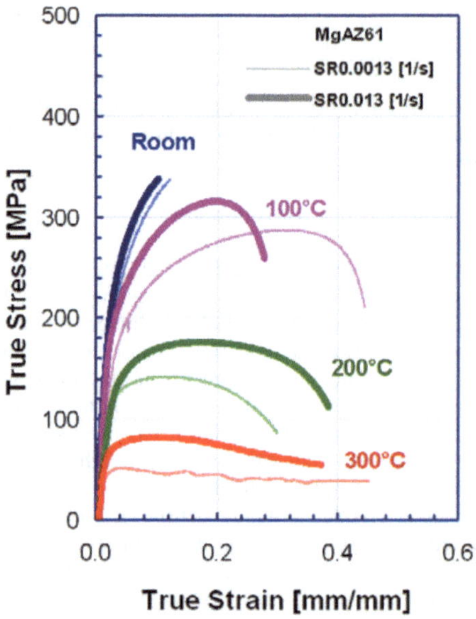


c)

Figure 1. Material flow curves obtained from tensile tests for (a) Nano-structured AZ61L (in rolling direction), (b) (in transverse direction), (c) regular Mg AZ61 (in rolling direction) alloys

The trend that lower strain rates result in increased elongation matches previous research [12]. This is more prevalent in the samples strained parallel to rolling direction than the samples strained in transverse direction. The maximum stresses on the samples before failure are consistent with the results of previous research along with the strain softening behavior at elevated temperatures for both strain orientations.

Hydraulic Bulge Tests

As sheet metal forming operations mostly involve a biaxial state of stress, the hydraulic bulge test is known to provide a more reliable stress-strain curve. Detailed comparison of the uniaxial and the biaxial test can be found in an earlier publication by the authors [15]. The hydraulic bulge test setup and its feedback control loop are presented in Figure 2. The test setup is composed of four major sub-systems: (1) a hydraulic system: pump and pressure transducer, (2) dies and inserts: upper and lower die with a bulge diameter of 100 mm and die corner radius of 6.5 mm; clamping and sealing mechanism between the two die halves using copper gaskets and 1,000 kN blank holder force provided by the Instron Satec 5596 machine, (3) heating system: cartridge and band heaters, temperature controller, and thermocouples to measure temperatures of the both dies.

The experimental setup is capable of conducting bulge tests at elevated temperatures approaching 400 °C. The heating of each die half was performed independently using separate sets of cartridge/band heaters and controlled with thermocouples attached to each die half. With PID control loop, the temperature deviation during the test was below 5 °C with respect to the set value. In addition, with a feedback control loop the strain rate can also be controlled based on the instantaneous dome height measurements.

Three bulge tests were performed for each case where temperature and strain rate were the main variables. During each bulge test, the internal pressure (*P*) and dome height (*h*) were continuously measured and recorded using a pressure transducer and either LVDT or laser sensor. The recorded pressure and dome height data were synchronized with the time step during the test. A schematic diagram of the bulge test setup, and the notation of parameters used in bulge test analysis are shown in Figure 3.

Both nano-structured AZ61L and regular AZ61 alloy sheet blanks were subjected to bulge tests at three strain rates (0.0013, 0.013 and 0.13 $s^{-1}$) and at four temperatures (25, 100, 200 and 300C). Under most conditions it would be expected that the location of rupture of the sample would be at the apex of the "dome". This is due to the fact that this location experiences the highest levels of straining and therefore would have the thinnest cross-section at time of rupture. On the contrary, Figure 4 illustrates that the location of rupture varies based on the temperature and strain rate used in the tests. Samples fractured in the die corner radius, also known as "shear fracture" typically at lower temperatures and/or higher strain rates. The fact that dome apex ruptures occurred at higher temperatures (at and above 200 °C) and lower strain rates indicates that both strain rate and temperature are highly important factors along with the $R_c/t_o$ (die radius corner to initial sheet thickness ratio) and blank holder force (BHF) values.

Based on the observations on the fractured samples, some of the bulged samples did not achieve certain dome height value so that the spherical shape assumption for bulged shape was not valid. Similar observations were made for AZ31 [14] and some other aluminum alloys by the authors earlier. It is assumed that the problem is intrinsic to certain magnesium and aluminum alloys as authors did not experience such premature bulge height values for several other alloys and steels. Possible reasons and mechanisms for premature failure of the sheets are under investigation and therefore, stress-strain curves based on the hydraulic bulge test were not included here. Instead, the relations between strain rate, temperature and maximum dome height achieved are established based on the bulge test results and presented in Figure 5.

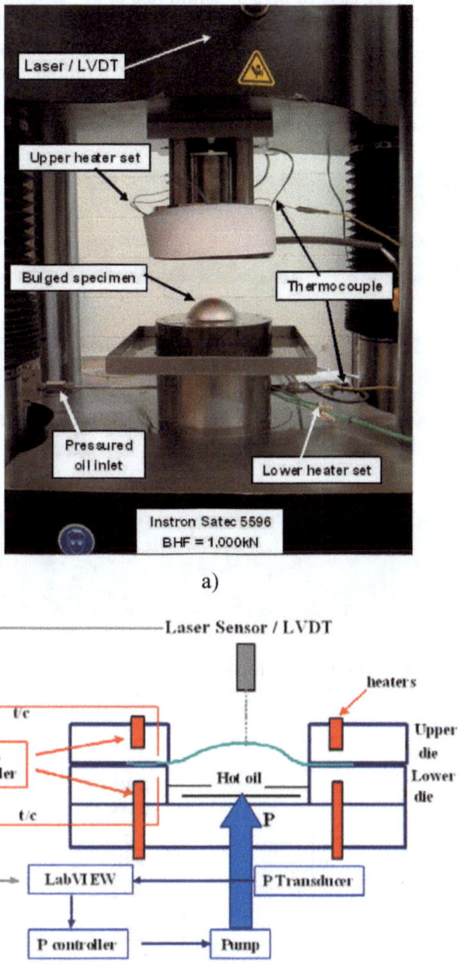


a) b)

Figure 2. (a) Main components of the hydraulic bulge test setup, (b) feedback control loop.

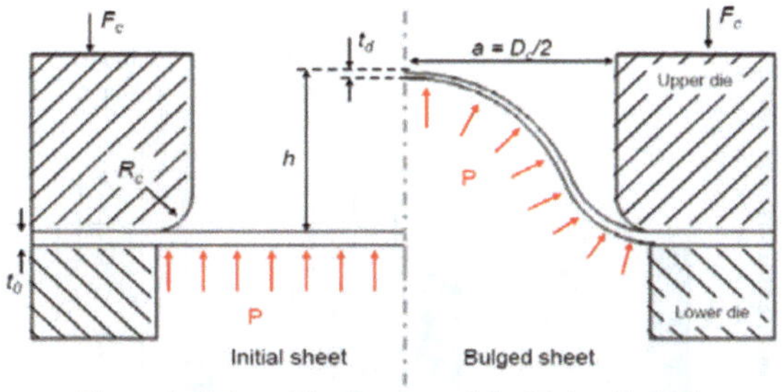


Figure 3. Schematic diagram of the hydraulic bulge test

In general, higher dome height, and therefore improved formability, was obtained when forming at higher temperatures and/or lower strain rates. As expected, higher pressure values were required to form the sheets at lower temperatures as compared to higher temperatures, a notable exception was observed for the samples at 25 °C, though. The noticeably lower pressure in the case of 0.13 $s^{-1}$ strain rate at 25 °C is likely due to the fact that the samples ruptured early in bulging and therefore never had an opportunity to build as much pressure. Strain rate, in general, had an increased effect on required pressure at temperatures above 200 °C, with higher pressure levels being required at higher strain rates.

a)

b)

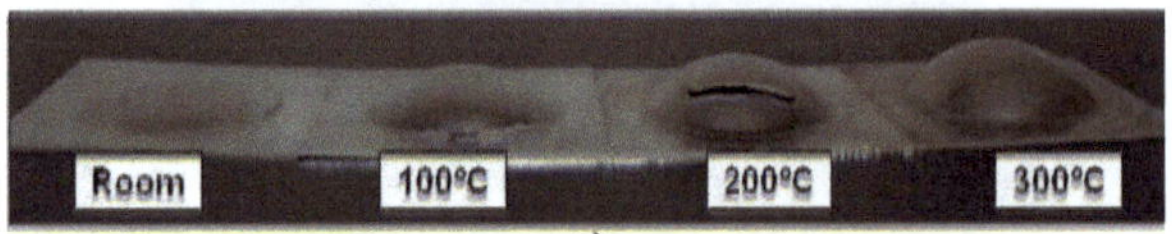


c)

Figure 4. Bulged specimens at strain rate of (a) 0.0013 $s^{-1}$, (b) 0.013 $s^{-1}$ and (c) 0.13 $s^{-1}$

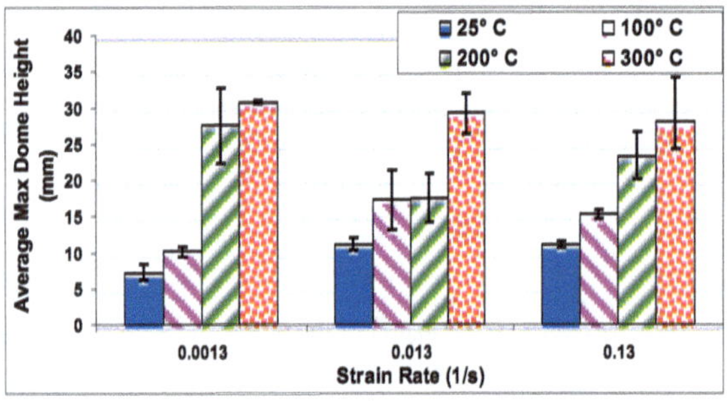


a)

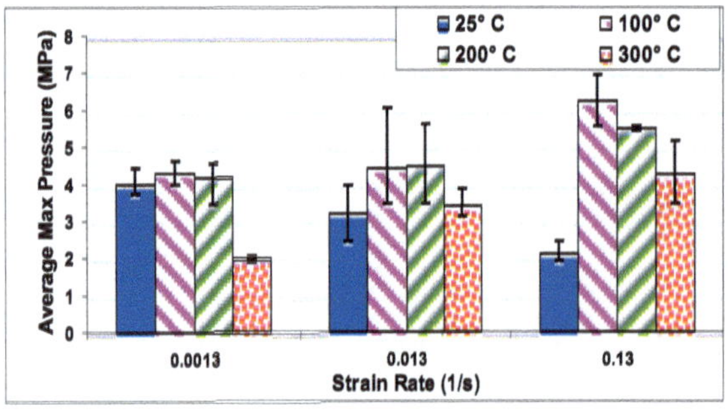


b)

Figure 5. Bulge test results for (a) maximum dome height, and (b) maximum pressure at different test conditions for hydraulic bulge test

As seen in Figure 5a relatively low dome height values obtained at room temperature and 100 °C imply that material flow curves at these temperature levels are well below 0.1 strain value. Contrarily, elongations for the sheets increased significantly at the 200 and 300 °C temperature levels. Examination of the bulge stress and strain graphs reveals the significant effects of temperature and strain-rate conditions. That is, the flow stress decreases and elongation increases with higher temperatures and/or lower strain rate. In addition, the effect of strain rate was observed to be more pronounced at the elevated temperature levels. These conclusions match with the previous studies on warm tensile tests available in the literature [6,12].

Figure 6 shows the optical microscopy analyses performed on the cross-section of bulged samples. Microanalyses showed that the zone under the blank holder (where no bulging effect is experienced) has different surface patterns when compared to that for deformed region.

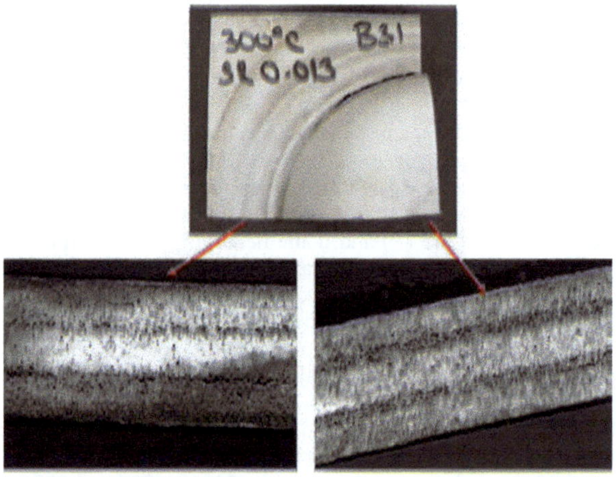

Figure 6. Optical microscopy for bulged samples from different locations at cross-section (test conditions: 300 °C, strain rate: 0.013 $s^{-1}$)

**Closed-die Warm Hydroforming Experiments**

In order to assess the forming limits of the thixo-processed Mg alloy under warm hydroforming conditions, a set of closed-die hydroforming experiments was also conducted at different temperatures and maximum pressure values by using a non-axisymmetric die geometry as shown in Figure 7. Previous studies by Doege et al. showed that only a few seconds of flat contact between the blank and the die are sufficient for heating to temperatures of approximately 200 °C due to the high coefficient of thermal conduction and the low heat capacity of magnesium alloys [6]. In this study, the holding time was chosen to be approximately 10 minutes to assure a uniform temperature distribution (i.e., isothermal condition) on the die set and the pressurizing oil in the lower die cavity. The hydroforming pressure rate was set to a constant value of 0.22 MPa/s eliminating variability. Once the pressure reached the set values, the experiment was stopped and the hydroformed part was taken out of the die for thickness and profile measurements using micrometer and 3D non-contact photogrammetry system ARAMIS (GOM mbH, Braunschweig, Germany). Note that the rolling direction is along the long side of the formed shape. With this system, both 2D and 3D profiles were obtained, revealing the precise geometries of hydroformed parts as seen in Figure 7 for thinning and cavity filling analyses.

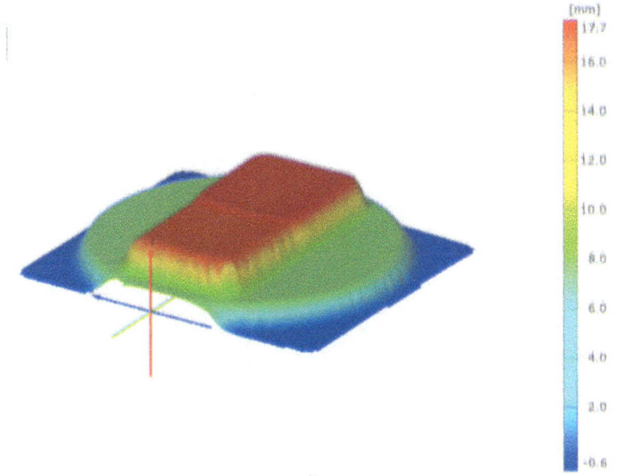


a)

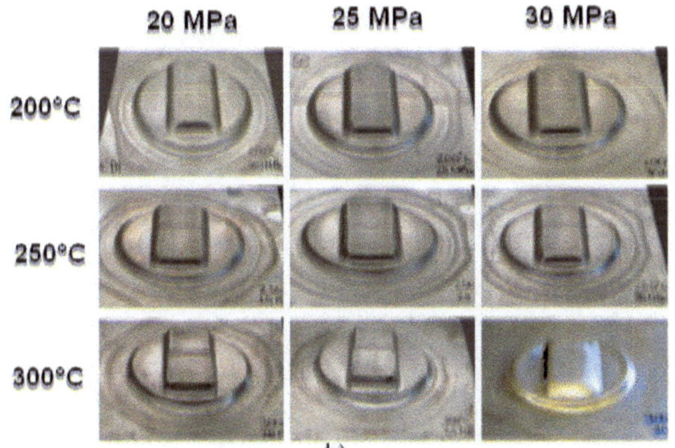


b)

Figure 7. (a) Optical photogrammetry of closed-die hydroformed samples for cavity filling ratio calculation, (b) Closed-die formed samples at each condition set

First, the effects of the temperature and pressure levels were investigated. The closed-die hydroforming tests were conducted at 200, 250 and 300 °C with the pressure ranging between 20 and 30 MPa. This set of conditions was selected based on results of the tensile and bulge tests mentioned previously. The 2D profiles were then compared to the profile of the forming die to evaluate the percentage filling (cavity fill ratio). The cavity fill ratio is defined as the area under the samples profile geometry divided by the area under the die profile to create an unbiased evaluation of the total forming. Although its reason was not understood well, the cavity filling ratio tends to decrease with the increasing pressure level at 200 and 250 °C levels while there is an increase-increase trend pattern at 300 °C. Finally, at 300 °C forming temperature, it was even possible to form a quite small feature (groove) with cross-sectional dimensions of approximately 4 mm in width and 0.3 mm in height.

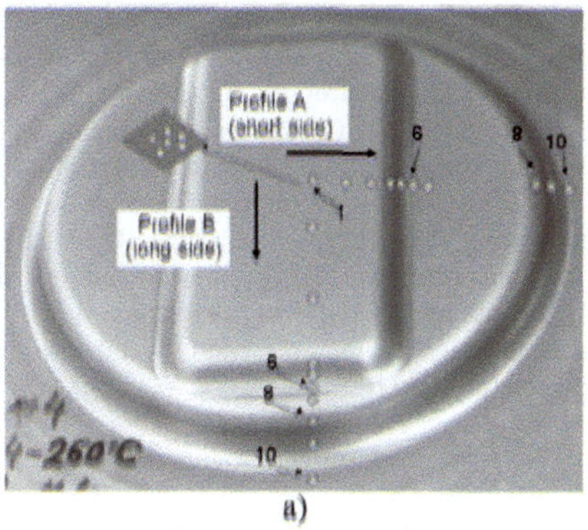


a)

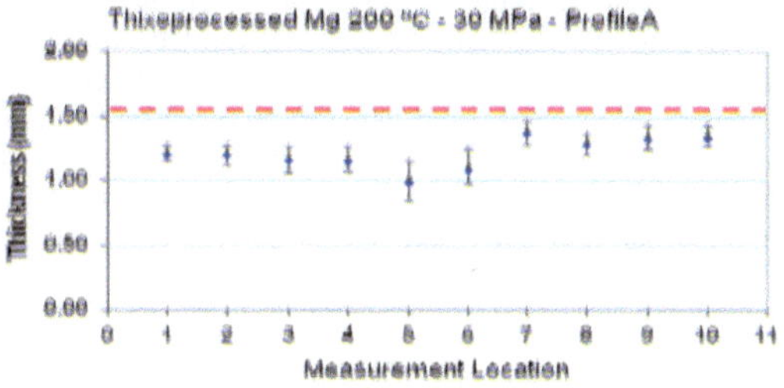


b)

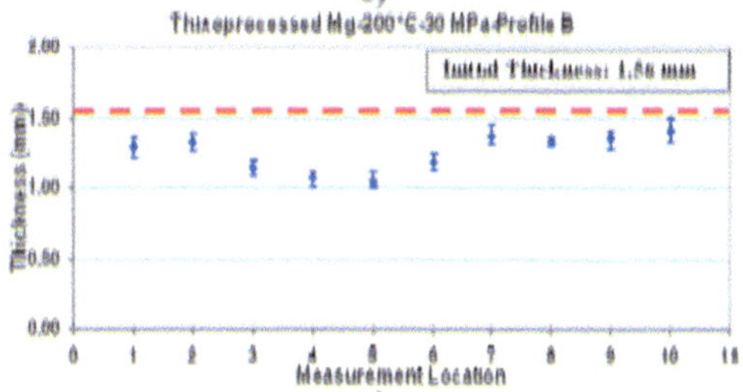


c)

Figure 8. (a) Thickness measurements locations on the formed parts, (b) thickness variation along the transverse direction, (c) along the rolling direction for the closed-die hydroformed nano-structured Mg alloy sheet

In order to measure the thickness of the formed parts, the specimens were marked at ten locations in each direction, as indicated in Figure 8a. A measurement setup including a stationary pin and dial indicator with a spherical tip was used to increase the accessibility and accuracy of the measurement, especially in the corner regions. To minimize the bias based on instrument operator and other random errors, three operators made a minimum number of three independent measurements at each location along the sample profiles. As expected, the thickness values in both profile A (short side) and profile B (long side) decreased with an increase in forming pressure at the same temperature, and the maximum thinning location for each profile is at location #6 and #5, which are the most stretched zones of the blanks in each direction.

## Conclusion

Uniaxial tensile and bulge tests were performed to investigate the formability of nano-structured Mg AZ61L and regular AZ61 alloys. It was shown that the nanostructed alloy differs from the regular version significantly in terms of elongation at higher temperatures (200, and 300°C). Strain softening behavior was also observed at elevated temperature levels. Strain rate did not affect both ultimate tensile strength and maximum elongations at room temperature and 100 °C, whereas a higher tensile strength was observed for the remaining temperature levels at the higher strain rate. Bulge test results, on the other hand, were not useful in terms of generating flow curves due to insufficient dome height value for bulged samples and premature failure during the bulging. It is assumed that the deformation characteristics (combined stress states in bulge test) and crystal structures played an important role in these results.


### Acknowledgments

The authors are thankful to the National Science Foundation (NSF) for the partial support on this project through NSF ENG/CMMI Grant 0703912 and NSF IIP IUCRC Grant 0638588. Thixomolded Mg sheet blanks were kindly provided by Thixomat Inc. Authors also greatly acknowledge the experimental work on AZ61 alloy by former group member Dr. Sasawat Mahabunpachai, and Dr. Sean Agnew of University of Virginia for sharing his laboratory capabilities.


### References


[1] W.F. Powers, "Automotive Materials in the 21st Century", *Advanced Materials & Processes*, 157-5, (2000), 38-41.

[2] J.A. Carpenter, "The FreedomCAR challenge and steel", American Iron and Steel Institute, Great Designs in Steel Seminar, February 24, 2004, Livonia, MI.

[3] R.A. Schultz, "Aluminum for light vehicles – An objective look at the next ten to twenty years" (Ducker Research), 14th International Aluminum Conference, September 15, 1999, Montreal, Canada.

[4] M. Easton, "Mg Alloy Applications and Automotive Structures", *Journal of Materials*, 60, (2008), 57-62.

[5] S. Novotny, and M. Geiger, "Process design for hydroforming of lightweight metal sheets at elevated temperatures", *Journal of Materials Processing Technology*, 138, (2004), 594-599.

[6] E. Doege, and K. Dröder, "Sheet metal forming of magnesium wrought alloys – formability and process technology", *Journal of Materials Processing Technology*, 115, (2001), 14-19.

[7] S.R. Agnew, and O. Duygulu, "A mechanistic understanding of the formability of magnesium: Examining the role of temperature on the deformation mechanisms", *Materials Science Forum*, 419-422, (2003), 177-188.

[8] M.R. Barnett, "Twinning and the ductility of magnesium alloys Part I: "Tension" twins", *Materials Science and Engineering A*, 464, (2007), 1-7.

[9] H. Takuda et al., "Modeling of formula for flow stress of a magnesium alloy AZ31 sheet at elevated temperatures", *Journal of Materials Processing Technology,* 164-165, (2005), 1258-1262.

[10] R. Neugebauer et al., "Sheet metal forming at elevated temperatures", *Annals of the CIRP*, 55-2, (2006), 793-816.

[11] S. Kaya et al., "Determination of the flow stress of magnesium AZ31-O sheet at elevated temperatures using the hydraulic bulge test", *International Journal of Machine Tools and Manufacture*, 48, (2008), 550-557.

[12] K. Siegert, S. Jäger, and M. Vulcan, "Pneumatic bulging of magnesium AZ31 sheet metals at elevated temperatures", *CIRP Annals-Manufacturing Technology*, 52-1, (2001), 241-244.

[13] J. M. Kim et al., "Microstructure and mechanical properties of thixocast Mg-Cu-Y alloy", *Scripta Materialia*, 49-7, (2003), 687-691.

[14] S. Mahabunpachai and M. Koç, "Material behavior and formability of magnesium AZ31 sheet alloy under warm hydroforming conditions", *Materials Forum*, 34, (2010), 105-109.

[15] M. Koç, E. Billur, and Ö.N. Cora, "An experimental study on the comparative assessment of hydraulic bulge test analysis methods", *Materials and Design*, 32, (2011), 272-281.

**Magnesium Technology 2012**
*Edited by: Suveen N. Mathaudhu, Wim H. Sillekens, Neale R. Neelameggham, and Norbert Hort*
***TMS (The Minerals, Metals & Materials Society), 2012***

# EFFECTS OF HIGH-TEMPERATURE SHOT PEENING ON SURFACE CHARACTERISTICS AND FATIGUE PROPERTIES OF FORGED AZ31 MAGNESIUM ALLOYS

Yuki Ichihara[1], Masafumi Noda[2], Kunio Funami[2]
[1]Graduate School of Engineering, Chiba Institute of Technology,
2-17-1 Tsudanuma, Narashinoshi, Chiba 275-0016, Japan
[2]Department of Mechanical Science and Engineering, Chiba Institute of Technology,
2-17-1 Tsudanuma, Narashino, Chiba, 275-0016, Japan

Keywords: Magnesium, Shot peening, Grain refinement, Twin, Fatigue properties.

## Abstract

In this present study, the effects of high-temperature shot peening (HSP) on the surface characteristics and fatigue properties of forged AZ31 magnesium alloys were investigated. The HSP process was performed with a low-carbon steel ball under a shot pressure of 0.2 MPa and time of 6 s at a temperature of 523 K. Gradable microstructures were formed in the near-surface region, about 150 μm from the surface edge. These were composed of an ultra-fine-grained region, a residual working strain region, and a twin region. The HSP-processed material showed increased fatigue strength compared with the non-HSP-processed material. From the measurement of the crack-propagation rate, it was found that the improvement in fatigue properties led to a delay in the crack-propagation rate in the ultra-fine-grained region and the residual working strain region.

## Introduction

The magnesium alloy used here is one of the most light-weight metals in practical use. Since it has many excellent properties such as its specific strength, damping capacity, and machinability, it has started to be utilized for transport machine components in automobile and railroad vehicles. However, until now, its practical use was limited to the parts on which only a comparatively light load would act, such as in electric devices [1-3]. In recent years, the application of plate materials (which have excellent mechanical properties compared with casting materials, in which defects such as blowholes are a concern) to structural members has been considered.

However, the magnesium alloy has strong anisotropy in the crystal itself, showing unique mechanical properties in the texture formed by plastic working processes such as rolling and extrusion [4]. Therefore, in terms of fatigue properties, different modes of action of the applied stress have been reported to give rise to particular behaviors [5,6]. Moreover, it has also been reported that the initiation/propagation characteristics of a crack depend on the crystal grain size or orientation [7-10]. Therefore, full examination of the relationship between the structure and the fatigue characteristics is necessary for the application of Mg alloy expansile materials in structural members.

Traditionally, shot peening (SP) has been used to improve the fatigue characteristics of steel. In recent years, it has been reported for aluminum and titanium alloys that nanoscale fine-grained structures can be formed on the surface through shot peening [11,12]. However, if SP is performed on a magnesium alloy at room temperature, cracks occur easily on the SP surface because the deformation ability is low. To address this problem, we performed SP at various processing temperatures, and found that a graded-structure material with a fine-grained structure layer on the material surface can be produced without crack initiation [13,14].

In this present study, we investigate the relationship between the surface and fatigue characteristics of AZ31 magnesium alloys with surface gradable structures, which were produced by the high-temperature SP process.

## Materials and Experimental Procedure

The materials used were obtained from a commercial AZ31 Mg alloy extrusion material forged with multiaxial forging processing (forging temperature: 1-pass: 543 K, 2-pass: 513 K, 3-pass: 473 K; reduction ratio: 30%; compression speed: 2 mm/min). Recrystallization heat treatment at 523 K with 20-min holding was performed for structure homogenization. Herein, this material is referred to as the Non-SP material. Through this refinement process, the average grain size became 18 to 12 μm, and the random crystalline structure and anisotropy of the proof stress on the tensile side and compression side caused by the texture formed by extrusion was no longer observed. The inverse pole figure (IPF) map of the Non-SP material is shown in Figure 1. The high-temperature SP process used air-type sandblast, and it was performed with a shot pressure of 0.2 MPa and a time of 6 s with a low-carbon steel ball. SP processing was performed at 523 K with 20-min holding in order to homogenize the structure. This material, after the high-temperature SP process, is referred to herein as the HSP material. Tensile tests were carried out at room temperature with a tension speed of 0.5 mm/min. The tensile specimen was cut down by electrical-discharge machining to give a parallel portion (11 mm in length, 5 mm in width, and 1 mm in thickness). SP was carried out on only two faces of the board.

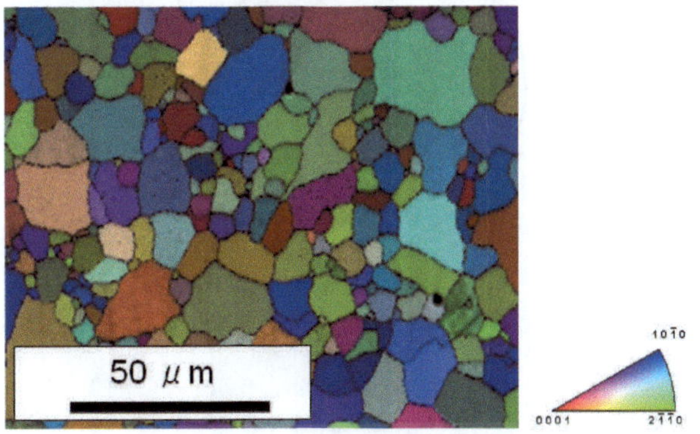


Figure 1 IPF map of AZ31 magnesium alloy structure processed by forging and heat treatment (Non-SP)

Using a servo-type fatigue machine with a capacity of 9.8 kN, we performed fatigue testing at room temperature with a test frequency of 10 Hz and a stress ratio, $R$, of −1.0. To determine the fatigue strength, we carried out high-cycle fatigue testing; the specimen used was a round bar specimen of 4 mm in diameter and with a parallel portion 12 mm in length. The crack-observation specimen was a form with a U-shaped notch (6 mm in diameter) attached to a plate (55 mm in height, 12 mm in width, and 5 mm in thickness), with a stress-concentration factor of 2.6. Lathe processing or electrical-discharge machining from the forging material, such that the length direction and the direction of extrusion became equal, produced two kinds of specimen shape. The surfaces of the specimens were machine-polished with emery papers #180 to #2000, and then buff-finished. SP processing was carried out over the parallel portion of the specimen for fatigue testing. For the crack-observation specimen, SP processing was performed only on the side of the U-shaped notch in order to facilitate the observation of the crack. In addition, the replica method was used for crack observation with optical microscopy and scanning electron microscopy (SEM).

## Results and Discussion

### Surface characteristics of the high-temperature shot-peening materials

Figure 2 shows IPF maps obtained by electron backscatter diffraction (EBSD) of the surface layers of the HSP material (b), and the relation between the distance from the surface and the residual stress (a). It was found from Figures 2 (a) and (b) that the graded-structure zone of about 200 μm on the surface was divided into three zones according to the HSP process: a recrystallized zone, a severe residual strain zone, and a twin zone. It seems that the formation of this graded structure is related to the strain distribution of the SP process and the cooling rate. Moreover, the diameter of a crystal grain in the recrystallization layer was about 1 μm, as determined from the transmission electron microscopy (TEM) images shown in Figure 3 (a). This was further confirmed from the pole figure of the HSP material, and the results for the Non-SP material in Figures 3 (b) and (c) show that the degree of random orientation in the HSP process is stronger than in the static recrystallization process.

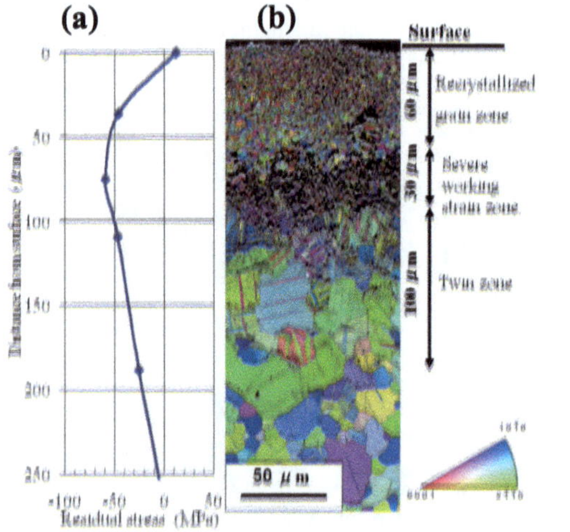


Figure 2 Relation between distance from surface and residual stress of HSP material (a), and IPF map of surface structure of HSP material at 523 K (b)

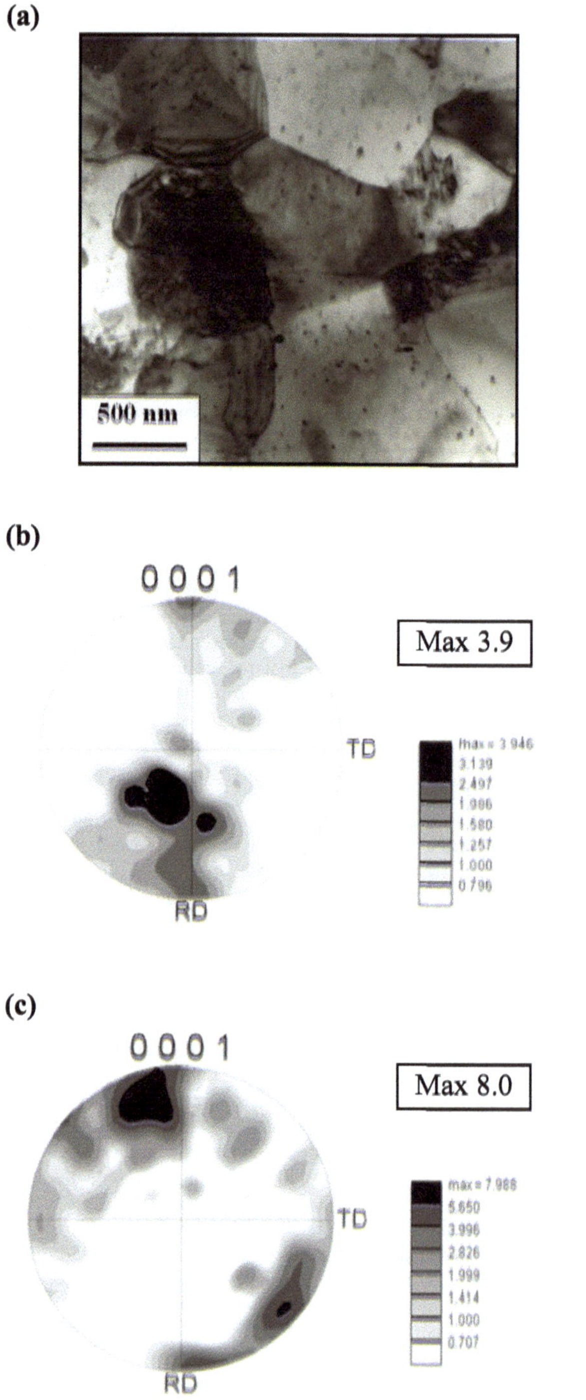


Figure 3 TEM micrographs and pole figure of HSP and Non-SP materials. TEM microstructure of recrystallized grain zone of HSP material (a), pole figure of recrystallized grain zone of HSP material (b), and pole figure of non-SP material (c)

**Mechanical characteristics of HSP material**

The tensile and fatigue test results are shown in Table 1. From the tensile test results, it is found that there is an improvement in strength and elongation with HSP processing. This is considered to arise from the effect of grain refinement and the random orientation seen in Figure 3. Furthermore, the fatigue-testing results showed a 5 MPa improvement in the fatigue limit with HSP processing. SEM micrographs of the fracture surfaces of the HSP and Non-SP materials fractured by a stress amplitude of 120 MPa are shown in Figure 4. In the fracture surface of the HSP material, a ratchet mark is seen, indicated by the arrow in Figure 4 (a). On the other hand, for the Non-SP materials shown in Figure 4 (b), such a trace is not observed. Therefore, from the fracture form of the HSP material, it is thought that two or more cracks are initiated and propagated, resulting in the fracture of the materials.

On the basis of detailed SEM observations, it was confirmed that crack initiation was promoted by the stress concentration in the peening dimples formed by the SP process. Therefore, we suppose that the crack-propagation process is a factor in the improvement of the fatigue characteristics of HSP materials.

Table 1 Mechanical characteristics of non-SP and HSP materials

| | Tensile test | | | Fatigue test |
|---|---|---|---|---|
| | Tensile strength (MPa) | Proof stress (MPa) | Elongation (%) | Fatigue limit ($5\times10^6$) |
| Non-SP | 248 | 128 | 22 | 90 |
| HSP | 258 | 135 | 27 | 95 |

**Fatigue crack-propagation properties of HSP materials**

Figure 5 shows the relation between the crack length and the number of cycles for the HSP and Non-SP materials. This figure shows that the HSP material has a short lifetime to crack initiation compared with the Non-SP material. As described above, it is considered that the stress concentration originates in a peening dimple. Moreover, a delay in the fatigue-crack growth rate is seen in a zone with a surface crack length of about 70 μm. This is in agreement with the length of the recrystallization layer shown in Figure 2. Moreover, SEM micrographs of fatigue crack in the HSP materials (stress amplitude: 120 MPa; number of cycles: 3200) are shown in Figure 6. This figure shows that a crack in the recrystallization layer propagates in a zig-zag form. It has been reported that crack propagation of the polycrystalline AZ31 magnesium alloy progresses remarkably along a grain boundary, and that the fatigue-crack growth rate changes by a factor of about five with the crystal direction [15]. Therefore, the delay in the fatigue-crack growth rate is considered to be caused by the grain boundary increasing on the crack route (which is, in turn, due to the grain refinement by the HSP process) and the random grain orientation . Next, the IPF maps of the microstructures of the twin layer and fatigue crack are shown in Figure 7. This figure shows that the crack does not necessarily propagate along the boundary between the twin and base materials. It has been reported that, although double twin formation leads to crack initiation, however the {10-12} twins do not contribute directly to fatigue failure [9]. Subsequently, direction analysis was carried out in the twin layer of Figure 2. All the twins in this twin layer were {10-12} twins. From this fact, it seems that there was no delay or participation in the fatigue-crack growth rate within the twin layer. Therefore, as shown in Figure 5, the lifetime to crack initiation decreases with the HSP process, but it is thought that, through the delay in crack propagation in the recrystallization layer, this leads to an improvement in fatigue strength. Moreover, further improvement in fatigue strength is expected with suppression of the formation of peening dimples.

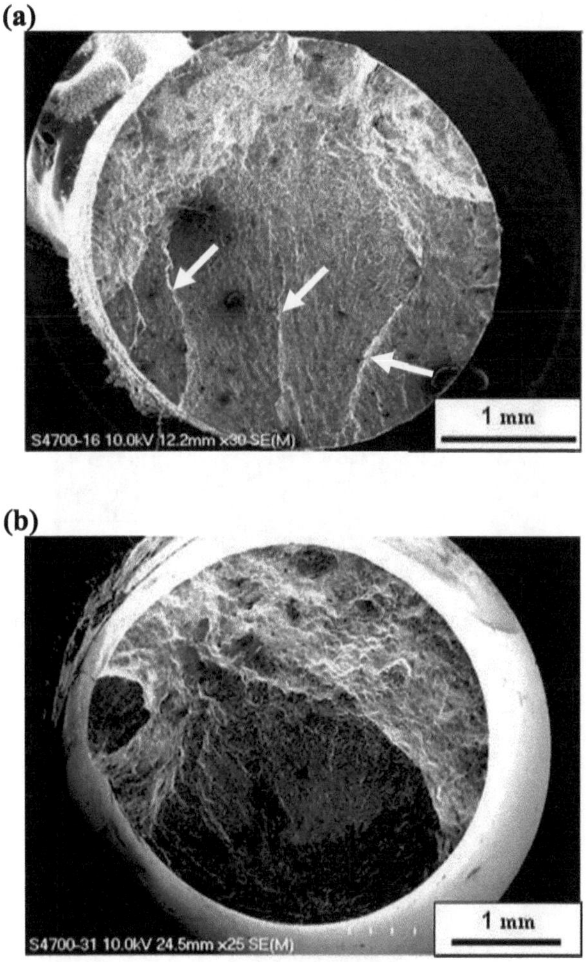


Figure 4 SEM micrographs of fracture surface. (a) HSP material (stress amplitude is 120 MPa); (b) Non-SP material (stress amplitude is 120 MPa)

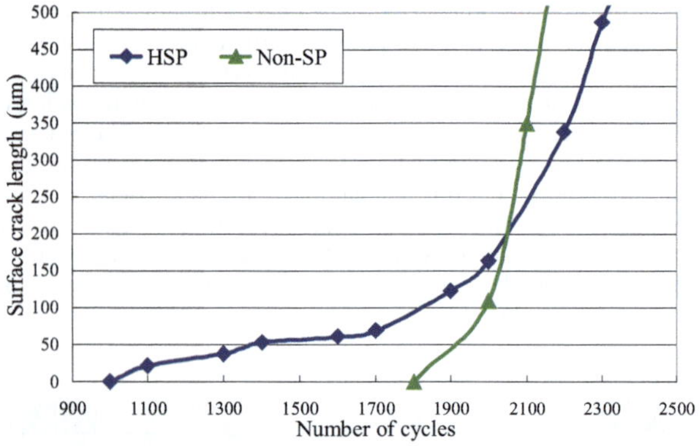


Figure 5 Relation between surface crack length and number of cycles

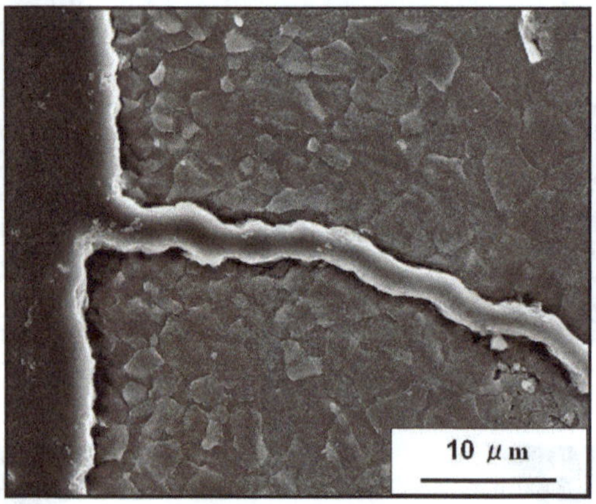


Figure 6 SEM micrographs of the fatigue crack of the HSP material

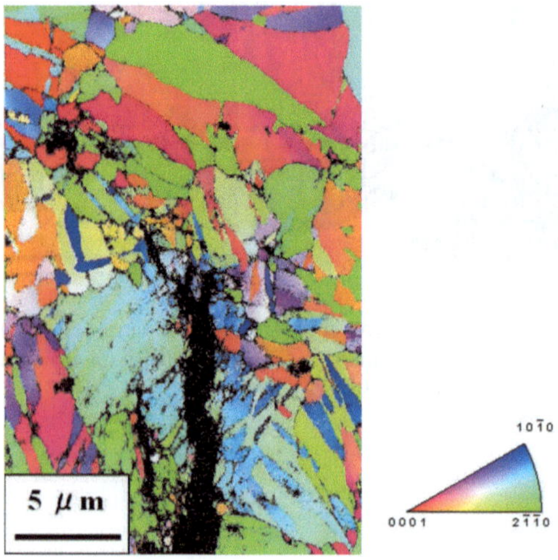


Figure 7 IPF map of the microstructure near the fatigue-crack tip in the twin zone

## Conclusion

In this present study, the effects of high-temperature shot peening (HSP) on the surface characteristics and fatigue properties of forged AZ31 magnesium alloys have been investigated. It has been shown that a graded microstructure consisting of three layers (a recrystallization layer of about 1-μm grain size, a severe working strain layer, and a twin layer) can be formed by the HSP process. Moreover, because of the grain refinement and random orientation in the recrystallization layer, there is an improvement in the tensile strength, elongation, and fatigue strength.

## Acknowledgment

In carrying out the EBSD analysis in this research, the authors received helpful comments and suggestions from everyone at TSL Solutions K.K.

## References

[1] K. ITAKURA, Journal of Japan Institute of Light Metals, Vol. 60, No. 6 (2010), pp. 305–311.
[2] S. SAIKAWA, Journal of Japan Institute of Light Metals, Vol. 60, No. 11 (2010), pp. 571–577.
[3] M. SAKAMOTO, H. UENO, Synthesiology, Vol. 2 No. 2 (2009), pp. 127–136.
[4] J. KANEKO, M. SUGAMATA, Journal of Japan Institute of Light Metals, Vol. 54, No. 11 (2004), pp. 484–492.
[5] K. SHIOZAWA, Journal of the Society of Materials Science, Japan, Vol. 58, No. 12 (2009), pp.982–989.
[6] S. MORITA, Journal of Japan Institute of Light Metals, Vol. 59, No. 10 (2009), pp. 548–554.
[7] M. KAMAKURA, Journal of the Society of Materials Science, Japan, Vol. 54, No. 3 (2005), pp. 245–250.
[8] S. ANDO, Journal of the Society of Materials Science, Japan, Vol. 70, No. 8 (2006), pp. 634–637.
[9] J. KOIKE et al., Scripta Materialia, Vol. 63 (2010), pp. 747–750.
[10] S. MORITA, N. OHNO, F. TAMAI, Y. KAWAKAMI, Materials Transactions, Vol. 51, No. 9 (2010), pp. 1543–1546.
[11] N. NAKAMURA, Journal of Japan Institute of Light Metals, Vol. 61, No. 4 (2011), pp. 155–159.
[12] M. UMEMOTO, Journal of the Society of Materials Science, Japan, Vol. 70, No. 6 (2006), pp. 473–477.
[13] K. FUNAMI, Program of The Conference of Japan Institute of Light Metals, Vol. 120 (2011), pp. 387–388.
[14] F. CZAWINSKI, Magnesium Alloy-Design, Processing and Properties. INTECH (2011), pp. 245–264.
[15] Y.UEMATSU, Journal of the Japan Society of Mechanical Engineers, Vol. 76, No. 763 (2010), pp. 311–316.

**Magnesium Technology 2012**
*Edited by: Suveen N. Mathaudhu, Wim H. Sillekens, Neale R. Neelameggham, and Norbert Hort*
***TMS (The Minerals, Metals & Materials Society), 2012***

# SOLID SOLUTION HARDENING EFFECT OF ALUMINUM ON THE CREEP DEFORMATON OF AZ91 MAGNESIUM ALLOY

Farhoud Kabirian[1], Reza Mahmudi[2]

1) Department of Mechanical Engineering, University of Maryland, Baltimore County, USA
2) School of Metallurgical and Materials Engineering, University of Tehran, Tehran, Iran

Keywords: AZ91, Creep Resistance, Solid Solution Hardeing

**Abstract**

It has been previously shown that separate additions of 2% rare earth elements (RE) and 0.6% Zr to the AZ91 base alloy improve creep properties. However, in this investigation using impression creep method, it is shown that simultaneous addition of these elements not only cannot improve creep properties but also leads to its deterioration. According to the creep deformation mechanism of the base alloy and microstructural evidences, it is believed that this drop in creep resistance mainly stems from depletion of grain interiors and the areas adjacent to grain boundaries from aluminum solute atoms. Reduction of aluminum content as solute atoms is due to the great affinity of RE and Zr atoms for aluminum. This leads to the weakening of Al solute atoms role as obstacles against dislocation movement, even though the volume fraction of the thermally stable particles tends to be high.

**Introduction**

In recent years, concurrent with the rise of energy issues the transportation industries have paid a great attention to employ light weight materials. Mg-Al-based alloys are most promising for more development. These alloys possess a reasonable combination of enhanced mechanical properties [1–3], good corrosion resistance [4,5], and excellent castability [6]. Despite these advantages, the main problem involved in using Mg-Al alloys is the drastic drop in their high temperature mechanical properties and creep resistance with increasing temperature to above 400 K [7]. Generally, the low creep resistance of Mg-Al alloys, with aluminum contents greater than 3 wt.%, is attributed to the formation of β-Mg17Al12 precipitates at grain boundaries and inter-dendritic regions [1–3]. Due to their low melting point, these precipitates are thermally unstable and thus they can accelerate grain boundary diffusion, resulting in considerable instability of the microstructure in regions adjacent to grain boundaries [3,4].

The AZ91 alloy containing 9 wt.% Al and 1 wt.% Zn is the most commonly used cast magnesium alloy for low-temperature applications. The high percentage of Al in this alloy makes it more susceptible to microstructural instability. Most of the attempts for improving the high-temperature structural stability of this alloy have concentrated on suppressing discontinuous precipitation of β-Mg17Al12 and creating more thermally stable intermetallics in the magnesium matrix. Due to the high affinity of aluminum atoms for elements such as Ca, Ti, Zr, and rare earth (RE), it is possible to reduce the amount of the $Mg_{17}Al_{12}$ by consuming the aluminum content of the alloy to form Al-containing intermetallic compounds with relatively high melting points [8].

Kabirian and Mahmudi [9] have reported that AZ91-2RE, owing to the fine distribution of Al11RE3 intermetallic with branch type morphology, shows the best creep properties among RE containing AZ91 alloys with 1, 2 and 3 wt% RE. In another work, they showed that AZ91-0.6Zr advantaging from favorable combination of solid solution hardening and fine distribution of the thermally stable Al2Zr and Al3Zr2, particles has the superior creep resistance compared to the AZ91-0.2Zr and AZ91-1.0Zr alloys [10,11]. Based on these findings, the aim of the present work is to investigate the influence of simultaneous additions of 0.6 wt% Zr and 2 wt% RE on the microstructure of the AZ91 alloy and its creep resistance via impression creep testing.

**Experimental Procedure**

The materials used were AZ91 alloy with an actual chemical composition of Mg-8.9 wt.% Al-0.78 wt. %Zn-0.32 wt.% Mn, and the same material with separate and simultaneous additions of 2.0 wt.% RE and 0.6 wt.% Zr. Melting was carried out in an electrical furnace under the Foseco MAGREX 36 covering flux to protect molten magnesium from oxidation. The cast bars were cut into 4-mm slices and tested in an impression tester. The details of the testing arrangement is explained elsewhere [12] and will only be briefly described here. An MTS universal tensile testing machine equipped with a three-zone split furnace was used to perform constant-load impression tests. Impression tests were run at two different temperatures of 425 and 455 K at different stress levels. The cast samples were studied by scanning electron microscopy (SEM) to examine the as-cast microstructure as well as to identify the phases present in the microstructure. X-ray diffraction (XRD) analysis was carried out on selected specimens to identify the existing phases.

**Results and Discussion**

Figures 1a through 1d show the SEM micrographs of the tested alloys. It can be observed that the microstructure of AZ91 is basically composed of the network-shape eutectic β-$Mg_{17}Al_{12}$ distributed in the magnesium matrix, the volume fraction of which decreases in the other alloys containing Zr and/or RE elements. This can be attributed to the fact that RE elements and Zr consume some of the aluminum to form Al-RE and Al-Zr intermetallic compounds, respectively, leaving less aluminum to combine with the Mg matrix to form β-$Mg_{17}Al_{12}$. Some fine Al-Zr

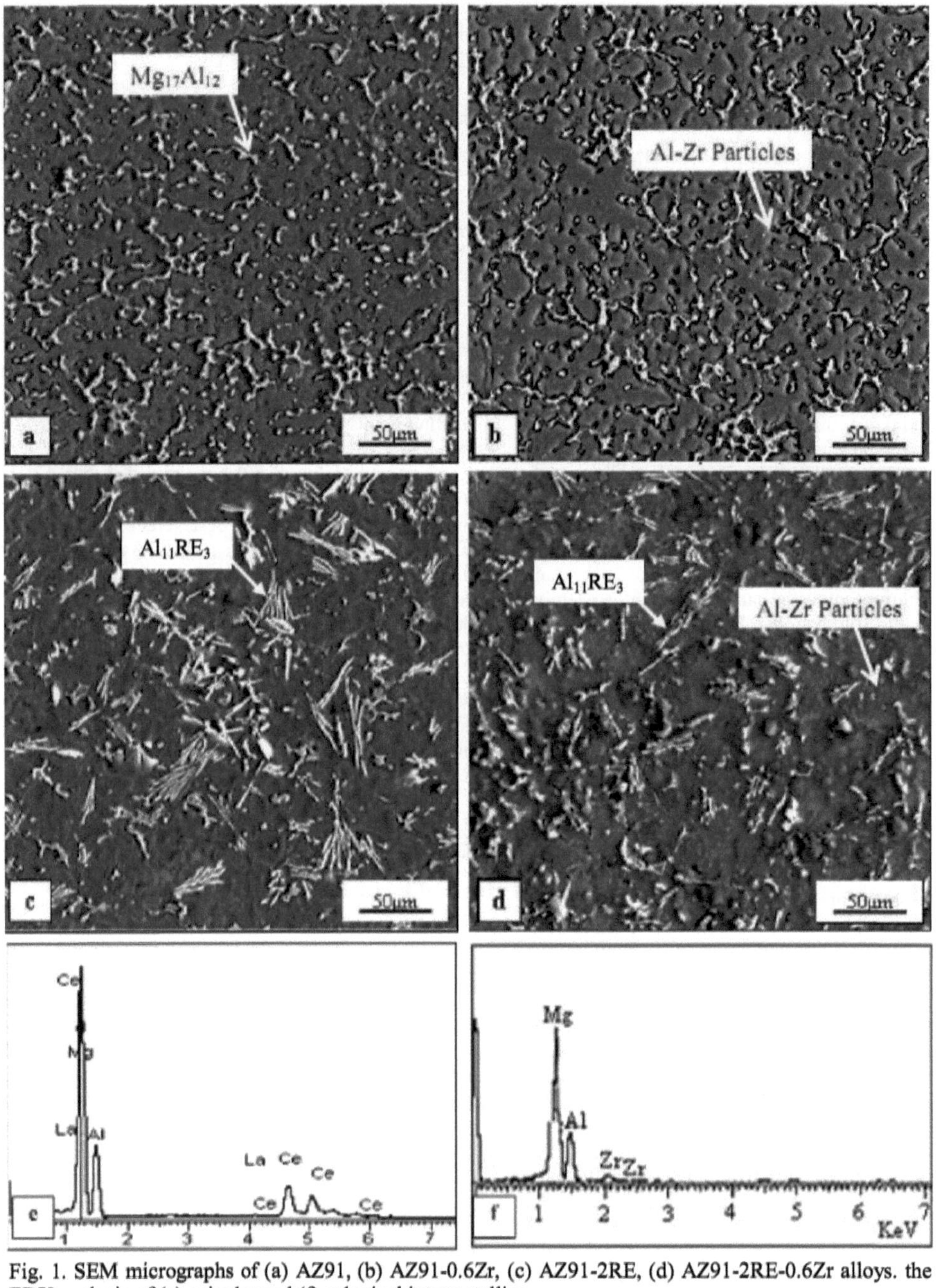


Fig. 1. SEM micrographs of (a) AZ91, (b) AZ91-0.6Zr, (c) AZ91-2RE, (d) AZ91-2RE-0.6Zr alloys. the EDX analysis of (e) acicular and (f) spherical intermetallics.

spherical particles with an average size of about 2 μm have appeared in the microstructure of the AZ91-0.6Zr alloy (Fig. 1b), while needle-shape $Al_{11}RE_3$ particles are dispersed throughout the microstructure of AZ91-2RE, as shown in Fig. 1c. It is worth mentioning that both types of particles have also been detected in the microstructure of AZ91-2RE-0.6Zr in Fig. 1d. The qualitative analysis of acicular and spherical particles determined by the EDS analysis, are shown in Figs. 1e and 1f, respectively. It is evident that La, Ce, and Al are the main constituents of acicular particles (Fig. 1e), while Al and Zr atoms constitute spherical particles (Fig. 1f). To identify the phases present in the tested alloys, XRD spectra were obtained and are shown in Fig. 2. As can be seen, α Mg and β-$Mg_{17}Al_{12}$ are the only phases in the AZ91 alloy, whereas in the AZ91-2RE-0.6Zr alloy, additional peaks are evident. These new peaks correspond to the Al11RE3, Al2Zr, and $Al_3Zr_2$ intermetallics.

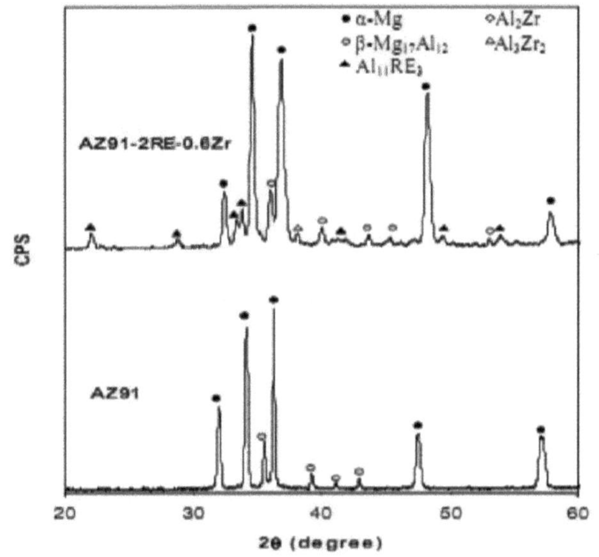


Figure 2. XRD patterns of the tested materials.

Typical impression creep data, presented as the variation of punch penetration depth with dwell time under different stress levels for all materials tested at 425 K, are shown in Fig. 3. As these curves show, increasing stress at a constant temperature results in higher penetration rates. It can further be seen that after a rather short primary creep stage almost all of the curves show a relatively long steady-state region where depth increases linearly with time. Since the impression creep test is compressive in nature, necking and fracture of the specimen do not occur and hence, it is obviously not possible to record a third stage of the curve, as opposed to that which normally happens in an ordinary creep test [13].

Figs. 4a and 4b show comparisons of impression creep curves under the punching stresses of 500 MPa at 425 K and 450 MPa at 455 K, respectively. It can be observed that the introduction of RE and Zr into AZ91 generally improves its creep resistance. The lowest level and slope in the steady state region belong to the AZ91-2RE alloy, the highest to the AZ91 alloy, with the remaining alloys lying between. In other

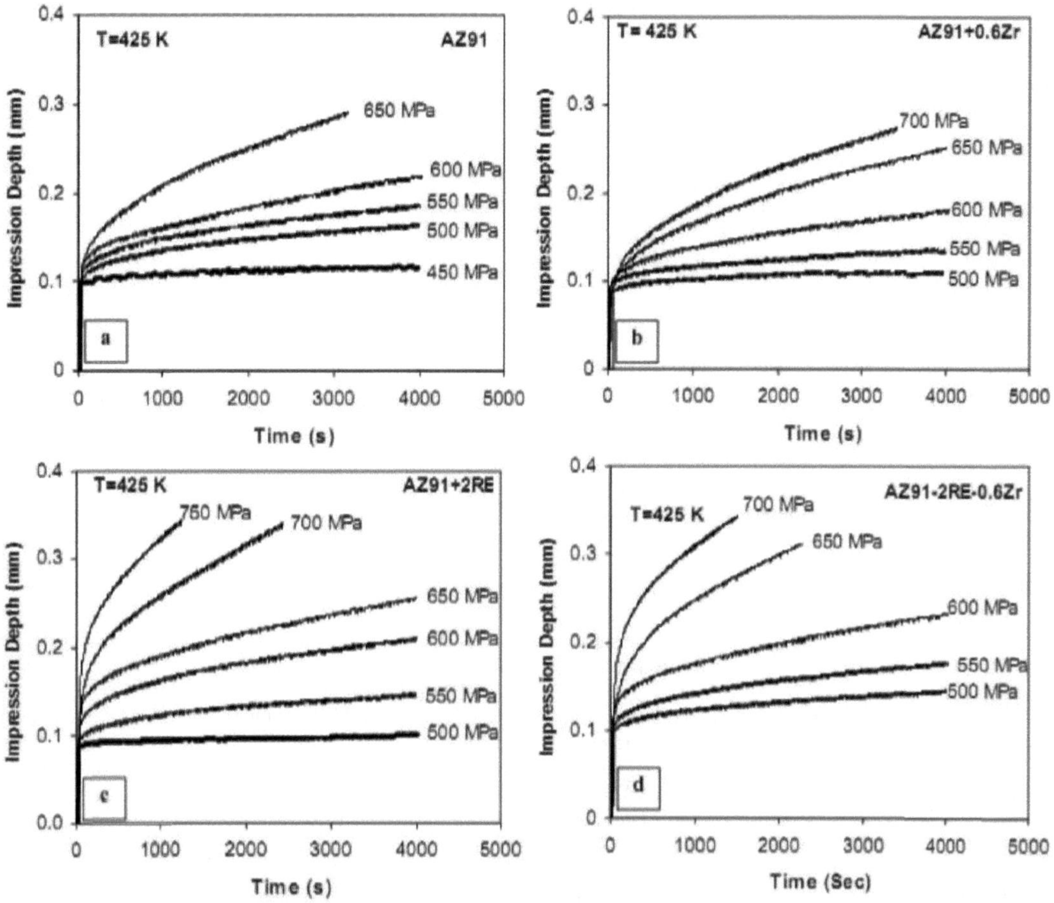


Figure 3. Typical creep curves of the investigated alloys at 425 K under different stress levels: (a) AZ91, (b) AZ91-0.6Zr, (c) AZ91-2RE, and (d) AZ91-2RE-0.6Zr alloy.

words, 2%RE addition has the strongest effect on the creep resistance, followed by 0.6%Zr addition. Simultaneous addition of RE and Zr, however, has the least effect on the enhancement of creep resistance. As shown in Fig. 4, the creep performance of AZ91-2RE-0.6Zr is inferior to that of AZ91-2RE and AZ91-0.6Zr, even though AZ91-2RE-0.6Zr contains both Al-Zr and Al-RE intermetallics. This may emphasize the effect of solid solution hardening as one of the most important strengthening mechanism in AZ91 with high percentage of aluminum content. Since testing temperatures in this investigation are not high enough to accelerate diffusion process through the matrix, it can be concluded that solute aluminum atoms are still playing an important role in the hardening of the material. Formation of higher volumes of Al-containing intermetallics in AZ91-2RE-0.6Zr, compared to both AZ91-2RE and AZ91-0.6Zr, leaves less Al solute atoms through the microstructure of the former alloy. In other words, drop of creepresistance due to the loss of solid solution hardening cannot be compensated by formation of larger volumes of the thermally stable intermetallics through the microstructure of AZ91-2RE-0.6Zr.

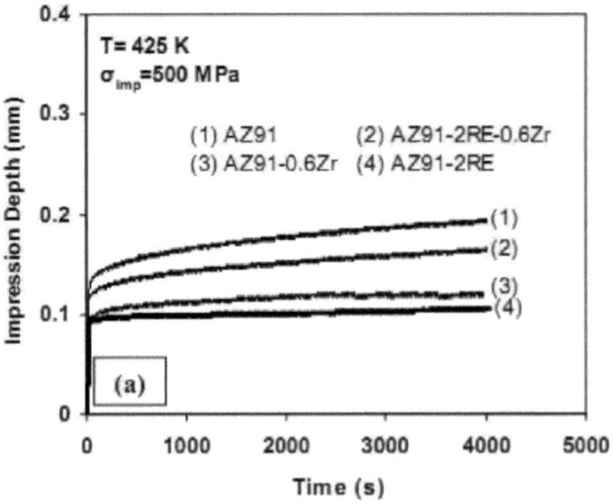


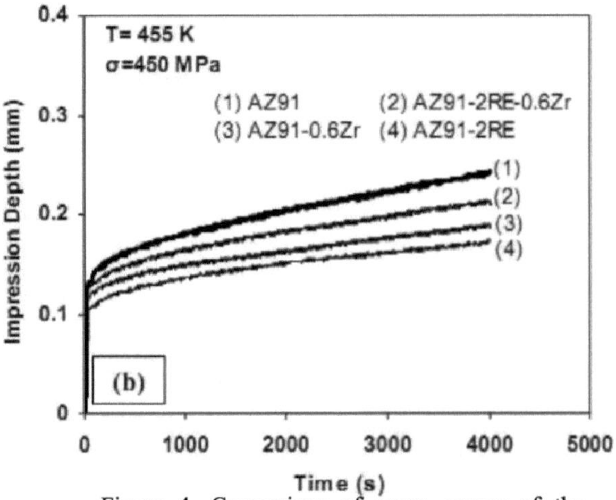


Figure 4. Comparison of creep curves of the investigated alloys at; (a) T = 425 K, $\sigma_{imp}$ = 500 MPa, and (b) T = 455 K, $\sigma_{imp}$ =450 MPa.

**Conclusions**

Following previous works in which separate addition of Zr and RE to AZ91 caused significant improvement of creep properties, it was expected simultaneous presence of these elements results in greater enhancement. However, the newly developed AZ91-2RE-0.6Zr alloy turned out to have poorer creep properties than the AZ91-2RE and AZ91-0.6Zr alloys. This is interesting when considering that both of the latter alloys had previously shown the highest levels of creep resistance among different RE-containing and Zr-containing AZ91 alloys by taking advantage of simultaneous solid solution hardening and particle hardening. Despite the higher volume fraction of the thermally stable particles distributed in the microstructure of AZ91-2RE-0.6Zr, it seems that its creep performance suffers from lack of enough solid solution hardening.

**References**

1. I.J. Polmear: Light Alloys, 2nd ed., Chapman and Hall, Inc., New York, NY, 1989, 170.

2. A.A. Luo and M.O. Pekguleryuz, "Cast Magnesium Alloys for Elevated Temperature Applications", Journal of Materials Science, 29 (1994), 5259-71.

3. A.A. Luo, "Recent Magnesium Alloy Development for Elevated Temperature Application", International Materials Review, 49 (2004), 13-30.

4. G.L. Song and A. Atrens, "Corrosion Mechanisms of Magnesium Alloys", Advanced Engineering Materials, 1 (1999), 11-33.

5. G.L. Song and A. Atrens, "Understanding Magnesium Corrosion. A Framework for Improved Alloy Performance", Advanced Engineering Materials, 5 (2003) 837-858.

6. E. Aghion, B. Bronfin, F. Von Buch, S. Shumann and H. Friedrich, "Newly Developed Magnesium Alloys for Powertrain Applications", JOM, 55 (2003), 30-33.

7. B. Kondori and R. Mahmudi, "Impression Creep Characteristics of a Cast Mg Alloy", Metallurgical and Materials Transactions A, 40 (2009) 2007-2015.

8. B. Kondori and R. Mahmudi, "Effect of Ca Additions on the Microstructure, Thermal Stability and Mechanical Properties of a Cast AM60 Magnesium Alloy", Materials Science and Engineering A, 527 (2010) 2014-2021.

9. F. Kabirian and R. Mahmudi, "Effects of Rare Earth Element Additions on the Impression Creep Behavior of AZ91 Magnesium Alloy", Metallurgical and Materials Transactions A, 40 (2009), 2190-2201.

10. F. Kabirian and R. Mahmudi, "Effects of Zr Additions on the Microstructure and Impression Creep Behavior of AZ91 Magnesium Alloy", Metallurgical and Materials Transactions A, 41 (2010), 3488-3498.

11. F. Kabirian and R. Mahmudi, "Effects of Zirconium Additions on the Microstructure of As-Cast and Aged AZ91 Magnesium Alloy", Advanced Engineering Materials, 11 (2009), 189-193.

# Magnesium Technology 2012

## Processing-Microstructure-Property Relationships II

### Thursday PM

*Session Chairs:*

**Alan Luo**
**(General Motors, USA)**
**Fabrizio D'Errico**
**(Politecnico di Milano, Italy)**

**Magnesium Technology 2012**
*Edited by: Suveen N. Mathaudhu, Wim H. Sillekens, Neale R. Neelameggham, and Norbert Hort*
***TMS (The Minerals, Metals & Materials Society), 2012***

# Enhancement of Strength and Ductility of $Mg_{96}Zn_2Y_2$ Rolled Sheet by Controlling Structure and Plastic Deformation

Masafumi Noda[1], Yoshihito Kawamura[2], Hiroshi Sakurai[3], Kunio Funami[1]
[1] Department of Mechanical Science and Engineering, Chiba Institute of Technology, 2-17-1 Tsudanuma, Narashino, Chiba, 275-0016, Japan
[2] Department of Materials Science, Kumamoto University, 2-39-1, Kurokami, Kumamoto, 860-8555, Japan
[3] Materials Engineering Department, Component Engineering Development Division, Nissan Motor Co., Ltd. 560-2, Okatsukoku, Atsugi, Kanagawa, 243-0192, Japan

Keywords: magnesium alloy, rolling, long period stacking order phase, high strength, ductility, microstructure

## Abstract

Mg-Zn-Y alloys are well known to possess greatly enhanced strength during plastic deformation because of the presence of kink bands in the LPSO phase and refinement of the grains of the alpha Mg phase. On the other hand, Mg–rare earth (RE) and Mg–Zn–RE alloys with a long period stacking order (LPSO) phase show a high tensile yield strength when subjected to an extrusion process but it is not known whether the LPSO and alpha Mg phases develop during plastic deformation. We examined the effect of the finely dispersed LPSO phase and the alpha Mg phase on the development of high strength in sheets of $Mg_{96}Zn_2Y_2$ subjected to a few passes of rolling. The mechanical properties and thermal stability of the alloy were also investigated. The tensile yield strength of rolled sheets of $Mg_{96}Zn_2Y_2$ was 360 MPa and its elongation was 5% when the material was subjected to thermomechanically controlled processing at 673 K with a four-pass rolling schedule. However, the tensile yield strength decreased and the elongation increased at annealing temperature of 623 K or above, because of the presence of grain growth in the alpha Mg phase and the restoration of kink bands in the LPSO phase.

## Introduction

Lightweight Mg alloys with excellent shock-absorption properties are being actively adopted for electronic information devices and automotive parts [1,2]. For use in such structural applications, Mg alloys need to have adequate ductility, thermal stability, and strength. However, Mg alloys often exhibit low ductility and low tensile yield strength at room temperature and above as a result of a scarcity of slip systems in their hexagonal close-packed structure [3]. Among the known effective ways for improving the ductility and tensile yield strength of Mg alloys are grain refinement [4] and control of the texture [5], because these promote prismatic slips and facilitate the creation of large plastic deformations [6].

Recently, Mg–Zn–Y alloy has been found to have superior mechanical properties to those of other Mg alloys [7]. Mg–Zn–Y alloy has a long period stacking order (LPSO) phase as a secondary phase in the main alpha Mg phase [7–9]. In general, Mg alloys with LPSO phases are known to have greatly enhanced mechanical properties, whereas their ductility can be maintained only by extrusion and/or plastic deformation of the cast metal. It has been suggested that kink deformations in the LPSO phase and microstructural refinement of the alpha Mg phase occur during extrusion deformation. The tensile yield strength, microstructure, fatigue properties, and thermal stability of extruded Mg–Zn–Y alloy have been reported [7–10]. On the other hand, many of the plastic deformation processes of LPSO-type Mg alloy are the result of the extrusion process, and there have been few attempts to examine the high strength developed by the rolling process and the working process. In addition, it is generally known that rolling of Mg alloys containing added rare earth (RE) element requires many working cycles in comparison with commercial Mg alloys [11–12].

In this present study, we examined the control of the microstructure of the LPSO phase and the alpha Mg phase of extruded $Mg_{96}Zn_2Y_2$ alloy by means of a short annealing treatment, and we studied the strength, elongation, and annealing properties of $Mg_{96}Zn_2Y_2$ material that had been subjected to the optimal working and rolling processes to produce rolled sheets of high strength and good ductility.

## Experimental Procedures

In the present study, we used extruded $Mg_{96}Zn_2Y_2$ (atomic%) alloy in the form of the as-received material. The alloy was annealed at 673 K in an electric furnace for 3.6 ks to decrease the anisotropy resulting from increased elongation and the specimens were subsequently cooled in water. The alpha-Mg phase showed static recrystallization, grain growth. Kink bands in the LPSO phase were restored by the annealing treatment. Specimens 5 mm thick, 20 mm wide, and 50 mm long were machined from the annealed and extruded materials. The direction of rolling was perpendicular to the direction of extrusion. The rolling process was performed at a roll temperature of 493 K and a roll speed of 0.17 m $s^{-1}$. The specimen was subsequently heated at 673 K for 0.6 ks in an electric furnace and finally the rolled specimens were quenched in water.

Tensile specimens with a gauge section 2.5 mm in width, 1 mm in thickness, and 12 mm in length were machined from the rolled sheet and various annealed sheets in the direction parallel to the rolling direction. Tensile tests were carried out at an initial strain rate of $5 \times 10^{-4}$ $s^{-1}$. Annealing treatment was carried out at various temperatures between 473 and 773 K in an electric furnace for 3.6 ks, and the specimens were subsequently cooled in water.

The microstructures of the as-rolled specimen and the annealed specimens were observed by optical microscopy (OM), scanning electron microscopy (SEM), transmission electron microscopy (TEM), and electron back-scattered diffraction (EBSD). EBSD was carried out in a 200 × 200 μm region with a measurement step of 0.3 μm. In our study, the cross section of the microstructure was observed from the rolling direction. Figure 1 shows SEM micrographs of an as-extruded sample of $Mg_{96}Zn_2Y_2$ alloy and a sample annealed at 673 K for 3.6 ks. The alloy contained an LPSO phase, an alpha Mg phase, and $Mg_3Zn_3Y_2$ compounds. The bright areas in both figures are believed to correspond the LPSO phase. The LPSO phase in the annealed materials appears to be more finely dispersed than that in the extruded alloy. LPSO phase does change from a plate type to a block type on heat treatment, and the plate-type LPSO phase was formed in the grains and grain boundaries.

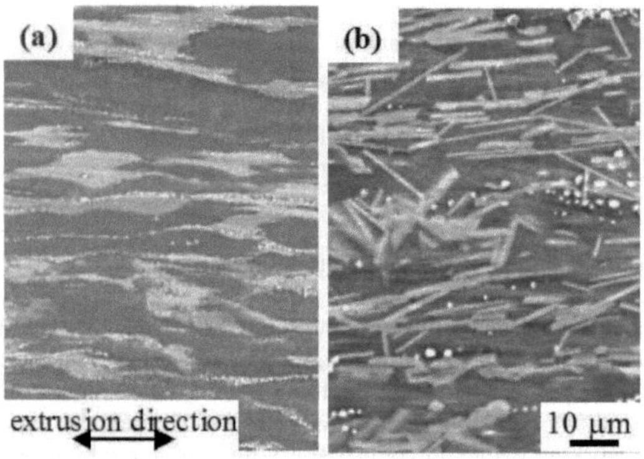


Figure 1. SEM micrographs of (a) the as-extruded $Mg_{96}Zn_2Y_2$ alloy and (b) Mg96Zn2Y2 alloy annealed at 773 K for 3.6 ks.

## Results and Discussions

**Anisotropy of the $Mg_{96}Zn_2Y_2$ rolled sheet**

It is well known that plasticity in LPSO-type Mg–Zn–Y alloy can be generated by processing. The LPSO phase, which is the secondary phase, acts as a fiber reinforcement and is anisotropic. It has been reported [13] that in the as-extruded material, the 0.2% proof stresses in the direction of extrusion shows a difference of about 30% from that in the direction perpendicular to the extrusion direction. Figure 2 shows the results of our investigations of the anisotropy of the tensile properties of the alloy; here, the angle 0° is parallel to the rolling direction. The maximum value of the anisotropy of the 0.2% proof stresses in the rolled material was about 14%, which is less than half that of the extruded material.

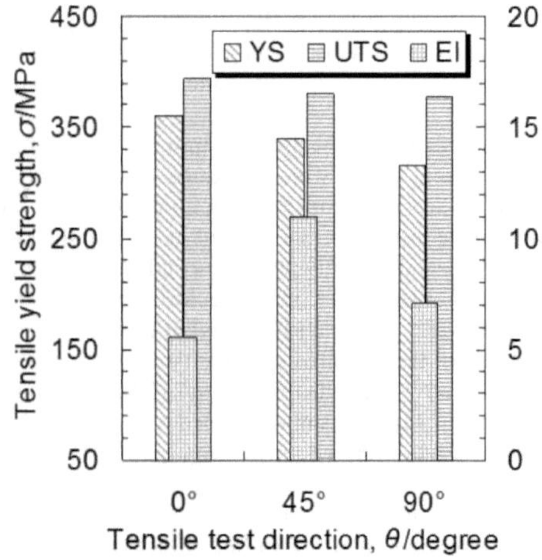


Figure 2. Relationship between mechanical properties and the tensile direction for a $Mg_{96}Zn_2Y_2$ rolled sheet. The tensile test direction of 0° was parallel to the direction of rolling. YS = yield strength; UTS = ultimate tensile strength; El =elongation.

The decrease in anisotropy may have resulted from dispersal of the LPSO phase in the grain boundary of the alpha Mg phase through the heat treatment and rolling processes. The dispersal of the LPSO phase by heat treatment, as shown in Fig. 1, is effective in reducing the anisotropy, and even low-deformation processing results in a high strength of over 360 MPa. The tensile strength was almost constant and was independent of the test direction. However, elongation showed a maximum at a tensile direction of 45°. An important factor in controlling the tensile yield strength is the fine dispersion of the LPSO phase before plastic deformation so that its anisotropy is low; this can be achieved by appropriate heat treatment and processing of the LPSO-type Mg alloy.

High strength and elongation of $Mg_{96}Zn_2Y_2$ rolled sheet

Nominal stress–strain curves of as-rolled and annealed $Mg_{96}Zn_2Y_2$ alloys are shown in Fig. 3. It is well known that for extruded materials, the 0.2% proof stress falls to 200 MPa whereas the elongation increases to 20% on annealing at 673 K for 3.6 ks. In our present study, a 1-mm-thick rolled sheet was produced by a four-pass rolling process. This four-pass schedule is the minimum number of passes for rolling LPSO-type Mg alloy, and the strain ratio introduced by the rolling processing is 1.8, which is less than that produced by the extrusion process. However, the 0.2% proof stress and elongation of $Mg_{96}Zn_2Y_2$ rolled sheet material were 360 MPa and 5.5% respectively, and these values did not show any marked decrease on annealing at 573 K (Fig. 3). In other words, the Mg alloy regains its high strength even if its LPSO phase is not subject to severe plastic deformation. However, the 0.2% proof stress of rolled sheet materials was lower than that of extruded alloys by about 40 MPa, because the grain size of the alpha Mg phase of the rolled sheet materials was 5 μm, which is 3.6 times larger than that of the extruded alloy.

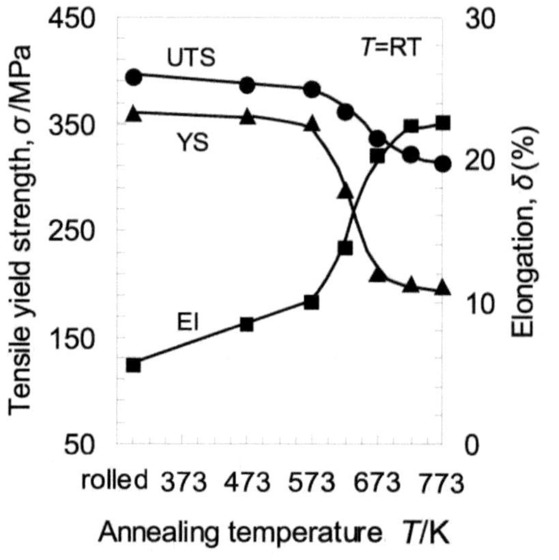


Figure 3. Relationship between the tensile yield strength and annealing temperature for a rolled sheet of $Mg_{96}Zn_2Y_2$ alloy. Tensile tests were performed at room temperature with the tensile direction parallel to the rolling direction.

When annealing was performed for 3.6 ks at 573 K, above which temperature the strength of the specimen decreased, the 0.2% proof stress was 351 MPa, the tensile strength was 381 MPa, and the elongation was 10%. It has been reported [6] that annealing of conventional Mg alloys at 523 K for 3.6 ks disrupts the crystalline orientation and causes a reduction in strength. On the other hand, the alloy that we investigated exhibited small reductions in the 0.2% proof stress and tensile strength of 10 and 12 MPa, respectively, at annealing temperature of up to 573 K, whereas the elongation was markedly improved from 5.5 to 10%. The ductility of the alloy can therefore be improved while maintaining a high strength, despite the fact that it is a Mg alloy that has a low ductility at room temperature.

**Microstructure of $Mg_{96}Zn_2Y_2$ as rolled sheet**

Optical micrograph and inverse pole figure (IPF) maps are show in Fig. 4 and in Fig. 4(b) indicated the orientation key of Mg alloy. In Figures 4(b) and (c), the darkest regions correspond to the LPSO phase. The LPSO phase has a large curvature in the rolling direction and shows considerable deformation. Furthermore, the formation of kink bands can be recognized in the LPSO phase. Normally, the LPSO phase does not show a large curvature in extruded materials, and we suggest that the large deformation of the LPSO phase observed in this case is responsible for the development of high strength in the rolling process. From Fig. 4(b), we estimated that the grain size of the alpha-Mg phase in the extruded material was 1.4 µm, and that in the rolled material was 5.0 µm. In addition, from IPF map and orientation key were shows the microstructure of rolled sheet were random and the proportion of recrystallized region of the alpha Mg phase was more than 70% in both the extruded and the rolled alloy. These materials therefore consisted of an alpha Mg phase (containing recrystallized and nonrecrystallized regions), an LPSO phase (25% by volume), and $Mg_3Zn_3Y_2$ compounds.

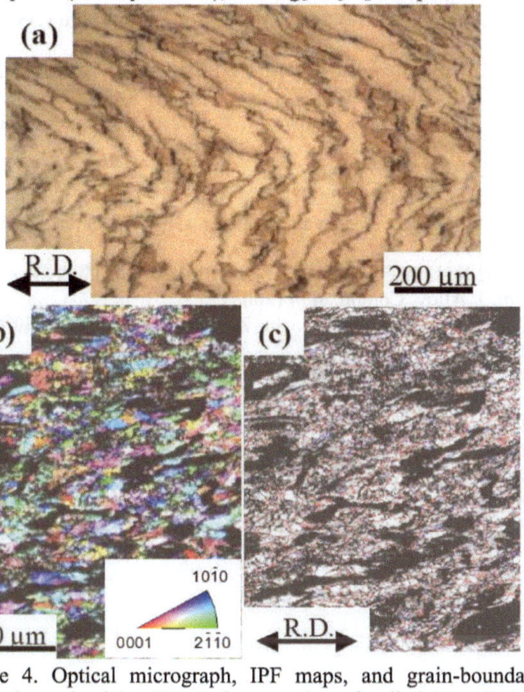


Figure 4. Optical micrograph, IPF maps, and grain-boundary maps determined by EBSD for samples of rolled $Mg_{96}Zn_2Y_2$ alloy. The black region in (b) and (c) correspond to the LPSO phase.

In other words, the introduction of deformation of the LPSO phase is important in producing a high-strength rolled sheet by the rolling process. From Fig. 4(c), the frequencies of low-angle and high angle-grain boundaries in the $Mg_{96}Zn_2Y_2$ rolled sheet were determined to be 19.3% and 80.7%, respectively. The rolling process was able to developed high frequency of high angle grain boundary and a random crystal orientation in $Mg_{96}Zn_2Y_2$ alloy.

TEM micrographs of the LPSO phases in an as-rolled sheet of $Mg_{96}Zn_2Y_2$ alloy are shown in Fig. 5. The LPSO phase contained both an LPSO structure and a 2H–Mg structure. In the rolled sheet, the LPSO phase was continuously deformed, but boundaries were observed in some regions, suggested that the LPSO phase recovered, at least partially, from the kink deformation introduced during rolling process. Although it is difficult to compare results of a previous study [14] directly with those of the present study, for FCC metals, cellular microstructures at the boundary between the deformation band and the deformation matrix acts as recovery nuclei during subsequent reheating. We assume that the kinks acted as recovery nuclei in the Mg–Zn–Y alloy and led to the formation of a static structure when edge dislocations accumulated as a result of reheating of the kink deformation that formed within the LPSO phase. The continuous deformation of kink bands introduced in the LPSO phase explains why the LPSO phase, which is a hard phase, showed large deformations in the rolling process. It is clear from Fig. 5 that the LPSO phase with continuous kink deformations was formed after the rolling process and showed a large curvature. The presence of microvoids at the interface between the LPSO phase and the alpha Mg phase was not confirmed. The present study clarified for the first time that continuous deformation of kink bands is necessary to permit the LPSO phase to show large deformations and that recovery of kink bands during reheating is important.

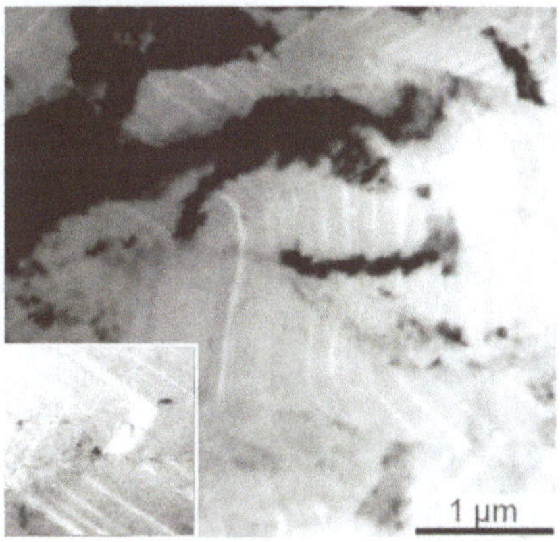


Figure 5. TEM micrographs of the as-rolled sheet of $Mg_{96}Zn_2Y_2$ alloy. The inset figure shows a high-magnification image of a kink deformation.

**Grain growth in annealed $Mg_{96}Zn_2Y_2$ materials**

Optical micrographs of samples of rolled alloy annealed at various temperatures are shown in Fig. 6. In Figs. 6(a)–6(c), the darker regions correspond to the LPSO phase and the brightest regions are alpha Mg phases. The grains of alpha Mg phase grew as the annealing temperature was increased, but up to an annealing temperature of 623 K, both the LPSO and alpha Mg phase showed little difference compared with those before annealing [Fig. 6(a)]. The alpha Mg phase grains became coarser as the annealing temperature was increased, but this occurred very slowly because the grain growth was controlled by the LPSO phase. The structure of the LPSO phase of extruded $Mg_{96}Zn_2Y_2$ alloy changes from a block type to a plate type as the annealing temperature is increased [10], whereas the structure of the LPSO phase in rolled $Mg_{96}Zn_2Y_2$ alloy did not change, suggesting that structure of the LPSO in rolled $Mg_{96}Zn_2Y_2$ alloy is thermally stable. Figure 6 also shows how the slight changes in the mechanical properties and microstructure of the rolled $Mg_{96}Zn_2Y_2$ alloy on annealing at 573 K might have occurred. In the tensile tests (Fig. 3), annealing at 623 K caused marked decreases in the 0.2% proof stress and tensile strength of about 50 MPa and 30 MPa, respectively. From Fig. 3 and Fig. 6, it is apparent that the static recrystallization temperature of the alpha Mg phase of this alloy is 623 K or higher. When the alloy was annealed at above 673 K, its 0.2% proof stress and tensile strength fell to about 200 MPa and 330 MPa, respectively, whereas the elongation was improved to more than 20%. Su et al. [15] reported that AZ31B Mg alloy, processed by equal channel angular pressing and annealed at 773 K for 3.6 ks, showed a 0.2% proof stress of 50 MPa, a mean alpha Mg grain size of as much as 20 μm, and an improved elongation of 15%. The alloy investigated in our study showed even better mechanical properties.

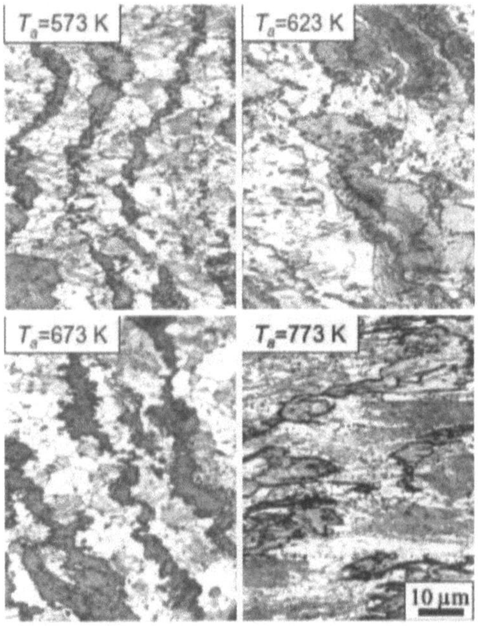


Figure 6. Optical micrographs of samples of rolled $Mg_{96}Zn_2Y_2$ alloys annealed for 3.6 ks at (a) 573 K, (b) 623 K, (c) 673 K, and (d) 773 K.

## Conclusion

We investigated the effects of annealing at various temperatures for 3.6 ks on the mechanical properties and microstructure of rolled samples of Mg96Zn2Y2 alloy. The anisotropy of the rolled sheet alloy was low compared with that of the extruded alloy when rolling was performed at 673 K after annealing treatment. The important factors in the development of Mg96Zn2Y2 alloy with a high strength and a good elongation are as follows: 1. Fine dispersion of the LPSO phase in the grain boundary of the alpha Mg phase; 2. Continuous kink deformation in the LPSO phase; and 3. Large curvature in the LPSO phase after processing by rolling. The presence of microvoids at the interface between the LPSO phase and the alpha Mg phase could not be confirmed. The high tensile yield strength was retained at an annealing temperature of 573 K, with no change in the LPSO phase and slow growth in the grain size of alpha Mg phase, as well as an increasingly randomized texture.

## Acknowledgement


The authors are grateful for the financial supports from the Kumamoto Prefecture Collaboration of Regional Entities supported this work for the Advancement of Technological Excellence, Japan Science and Technology Agency. The examination of rolling process went by the furtherance from Grant in Aid for Young Scientists for (B), Japan Society for the Promotion of Science.


## References


[1] Kinner. W.K., Mater. Eng., **90** (1979), 49-51.

[2] H. Friedrich, S. Schumann, J Mater. Process. Technol. **117** (2001), 276-281.

[3] G.S. Rao, Y.V.R.K. Prasad, Metall. Trans., **13A** (1982) 2219-2226.

[4] S. Mueller, K. Mueller, H. Tao, W. Reimers, Int. J. Mater. Res. **97** (2006) 1384-1391.

[5] Agnew S.R, Horton J.A, Lillo T.M, Brown D.W, Scr. Mater., **50** (2004), 377-381.

[6] N. Stanford, M. Barnett, Mater. Sci. Eng., A **496** (2008) 399-408.

[7] Y. Kawamura, K. Hayashi, A. Inoue, T. Masumoto, Mater. Trans., **42** (2001) 1172-1176.

[8] S. Yoshimoto, M. Yamasaki, Y. Kawamura, Mater. Trans., **47** (2006) 959-965.

[9] X. Zeng, Y. Zhang, C. Lu, W. Ding, Y. Wang, Y. Zhu, J. Alloy. Compd., **395** (2005) 213-219.

[10] M. Noda, T. Mayama, Y. Kawamura, Mater. Trans., **50** (2009) 2526-2531.

[11] H.K. Lim, S.W. Sohn, D.H. Kim, J.Y. Lee, W.T. Kim, D.H. Kim, J. Alloy. Compd., **454** (2008) 515-522.

[12] B. Wang, Y. Liu, A. Su, Z. Cao, Mater. Trans., **50** (2009) 261-265.

[13] T. Mayama, M. Noda, R. Chiba, M. Kuroda, Int. J. Plasticity, doi: 10.1016/j.ijplas.2011.02.007.

[14] K. Higashida, J. Takamura, N. Narita, Mater. Sci. Eng., **81** (1986) 239-258.

[15] C.W. Su, L. Lu, M.O. Lai, Mater. Sci. Technol., **23** (2007) 290-296.

**Magnesium Technology 2012**
*Edited by: Suveen N. Mathaudhu, Wim H. Sillekens, Neale R. Neelameggham, and Norbert Hort*
*TMS (The Minerals, Metals & Materials Society), 2012*

# Microstructural Characteristics of High Rate Plastic Deformation in Elektron™ WE43 Magnesium Alloy

Joseph Hamilton[1], Sarah T. Brennan[1], Yongho Sohn[1], Bruce Davis[2], Rick DeLorme[2], Kyu Cho[3]
[1]University of Central Florida; 12443 Research Parkway, Suite 304; Orlando, FL 32826, USA
[2]Magnesium Elektron North America;1001 College St.; P.O. Box 258,; Madison, IL 62060, USA
[3]US Army Research Laboratory; 2800 Powder Mill Rd.; Adelphi, MD 20783, USA

Keywords: Magnesium, High-Strain Rate, Severe Plastic Deformation, Shear Localization

## Abstract

High strain rate deformation of WE43 magnesium alloy was carried out by high velocity impacts, and the characteristics and mechanisms of microstructural damage were examined. Six samples were subjected to a variety of high velocity impact loadings that resulted in both partial and full damage. Optical, scanning and transmission electron microscopy analyses were performed in order to identify regions of shear localization. These regions were used to map, both quantitatively and qualitatively, the effects of deformation on the microstructure. Shear localization was observed in every sample, and its depth was measured. Evidence of shear localization was observed to a greater extent in samples with partial damage while fracturing was observed more frequently in samples with full damage.

## Introduction

Over the past two decades, there has been an increasing interest in the commercial use of magnesium alloys in various applications where a high strength-to-weight ratio is important, such as in aerospace, automotive, and defense industries [1]. Not only is magnesium an extremely lightweight metal, weighing 30% less than aluminum, and 70% less than steel, it also has the highest strength-to-weight ratio among any of the commonly used non-ferrous and ferrous metallic materials [2].

As magnesium is used more often in structural components, there is an increasing demand for a better understanding of the microstructural evolution during deformation. Due to the limited number of independent slip systems in a hexagonal close packed (HCP) crystal structure, basal slip and twinning are the prevalent mechanism of deformation at lower temperatures, such as 200°C and below, with low strain rates ranging between 1.8 x $10^{-5}$ to 1.8 x $10^{-3}$ $s^{-1}$ [3]. Wei *et* al., showed that at elevated temperatures ranging from 350 to 425°C, and strain rates ranging from $10^{-3}$ to $1^{-1}$ $s^{-1}$, magnesium alloys exhibited excellent superplasticity, due to activation of non-basal slip systems, and grain boundary sliding (GBS) deformation mechanism under these conditions [4]. It has been noted that as the temperature is increased above 200°C the critical resolved shear stress for the non-basal slip systems is reduced, and the activation of these additional mechanisms results in an increase in ductility and a decrease of yield and flow stress [5]. However, at high strain rates, such as greater than $10^3$ $s^{-1}$, the primary mechanism of deformation in metals has been shown to be shear localization [6] [7].

The shear localization seen in high strain rate deformation is characterized by adiabatic shear bands (ASB). These regions of shear deformation are the result of adiabatic shear, shear instabilities, and shear localization. The main mechanisms of deformation within the ASBs are dynamic recrystallization (DRX), and GBS. It is through these mechanisms that the microstructure accommodates the high kinetic energy seen in severe plastic deformation occurring at high strain rates, such as ~$10^3$-$10^6$ $s^{-1}$ [8].

The post-mortem characterization of a deformed microstructure is important to the understanding of the microstructure during deformation, and until a technique is developed that allows real-time diagnostic of the microstructure as it evolves, it will remain as the primary means of investigation of the phenomenon that occur inside shear localization. The purpose of this study was to analyze, through a variety of microscopy techniques, the microstructure of WE43 magnesium alloy samples post-deformation.

## Experimental Methods

A post-mortem analysis was performed on six 80mm thick WE43 samples (3.7-4.3% Yttrium, 2.4-4.4% Rare Earths, 0.4% Zirconium, balance Magnesium), that were produced in the T5 condition, thermally quenched after high-temperature rolling, and then artificially aged. The samples were then subjected to high velocity impacts that resulted in varying levels of severe plastic deformation (SPD). The samples were assigned the nomenclature of A, B, and C based on increasing level of damage. The samples were also labeled as Full, or Partial damaged samples based on the extent of SPD. Examples of the cross sectioned regions are shown in Figures 1 and 2.

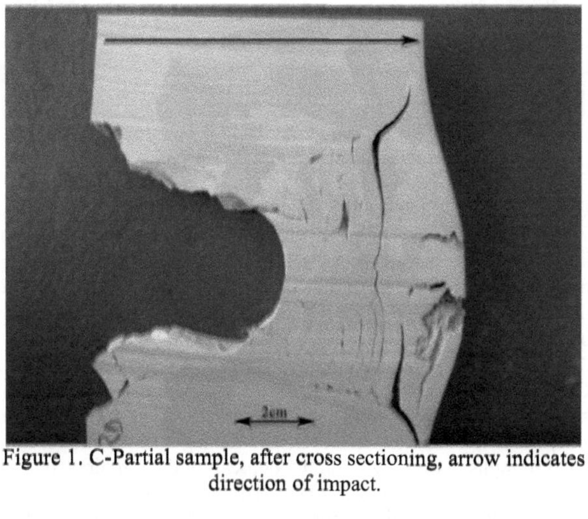


Figure 1. C-Partial sample, after cross sectioning, arrow indicates direction of impact.

In order to preserve the microstructure, preparation methods were chosen that would induce the least amount of damage. Therefore, waterjet cutting was used to cross-section the deformed samples, and a diamond wafer saw was used to perform further sectioning at low-speed. A map is shown in Figure 3 of the sectioning performed on the sample A-Partial, and this is representative of the method used to section all of the samples. The samples were then mounted in epoxy discs, and mechanically polished to 0.05-μm using alumina. Then the samples were etched by submersion in a acetic-picral etchant (100mL ethanol, 6g picric, 10ml water, 5ml acetic).

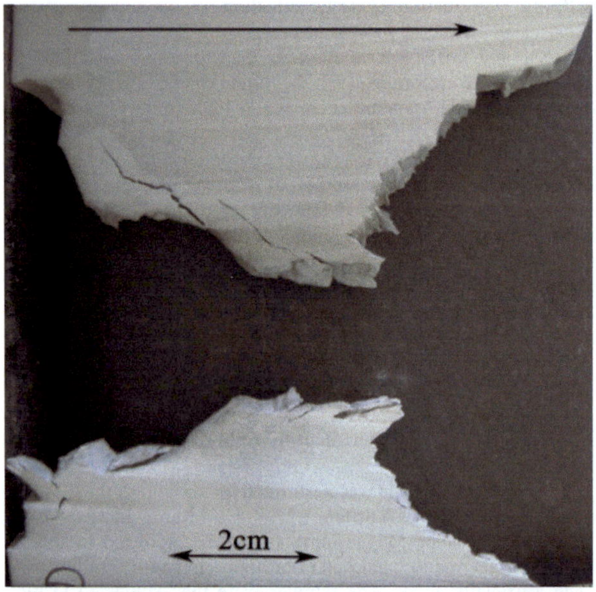


Figure 2. C-Full sample after cross sectioning, arrow indicates direction of impact.

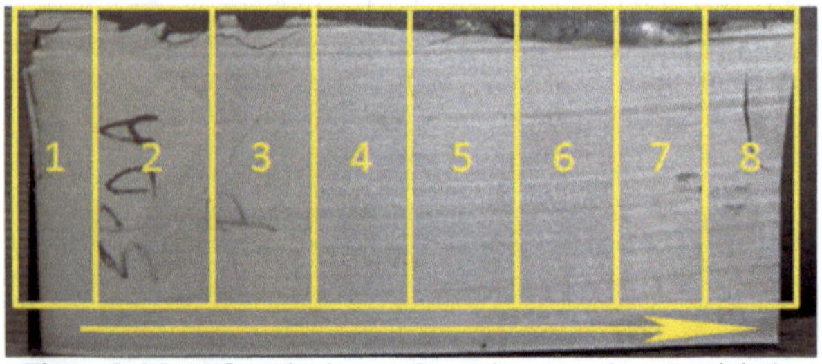


Figure 3. Map of sectioning performed on sample A-Partial, the arrow indicates direction of impact.

In order to characterize different aspects of the deformation a variety of approaches were utilized. Optical microscopy (OM), scanning electron microscopy (SEM), and transmission electron microscopy (TEM) were performed on both deformed and undeformed samples in order to identify and characterize regions of interest.

OM was used for both quantitative and qualitative investigation of shear bands, regions of dynamic recrystallization (DRX), grain size, shear localization, fractures, and damage on the microscale. An Olympus LEXT OLS3000 confocal optical microscope in conjunction with Olympus LEXT software was used for OM imaging. IQ Materials software was used on composite OM images in order to quantify the extent of shear localization. Measurements were made for the depth of the shear localization that extended from the edge of deformation. The regions of deformed and undeformed material were qualitatively distinguished by the features observed with OM. The regions of shear deformation exhibited elongated grains, with shear bands and DRX. The undeformed regions were composed of equiaxed grains and were similar in microstructure to the bulk material. Measurements were taken normal to the edge, across the deformed microstructure, and stopped once an undeformed region was reached, examples of these types of measurements can be seen in Figure 4. The measurements for each sample were averaged, and then the averages for all of the samples were plotted to show the progression of shear localization through the cross-sectioned sample.

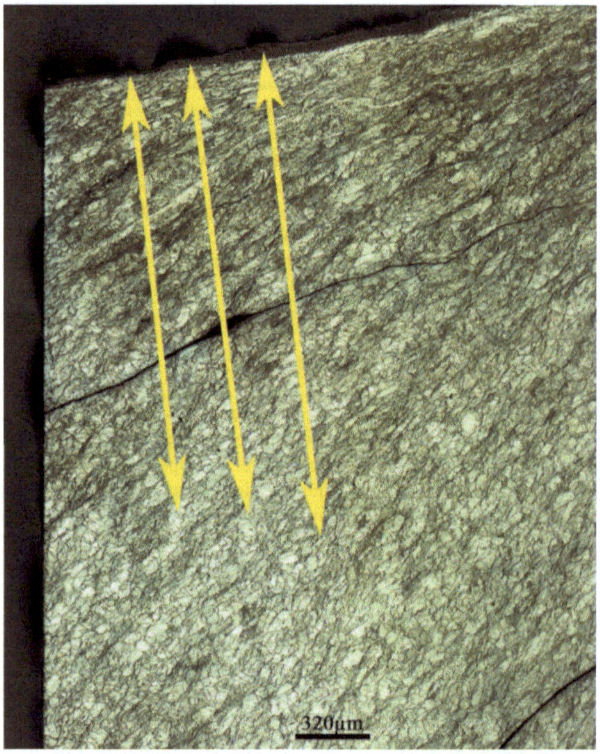


Figure 4. Composite OM image, arrows are example of measurements made of region of shear localization.

SEM was used to further investigate regions of interest at the microscale. A Zeiss Ultra-55 field emission SEM equipped with x-ray energy dispersive spectroscopy (XEDS) was used. TEM was used to investigate the grain size, and formation of precipitates at the nanoscale in regions of DRX found in shear bands, as well as suspected twinning seen throughout the samples. A FEI/Tecnai F30 300keV TEM was utilized for this.

## Results and Discussion

### Microstructural Characteristics

A section of bulk material was removed and analyzed for comparison to damaged regions. The bulk microstructure (i.e. undamaged) is shown in Figure 5. It was assumed that any twinning, or other characteristics of deformation seen in the bulk microstructure, was a result of the processing methods used in manufacturing the samples. Although a significant amount of twinning was observed near the surface of the penetration channel, this paper focuses on the shear localization and the formation of shear bands. The presence of profuse twinning was examined by TEM, and will be reported in future publications.

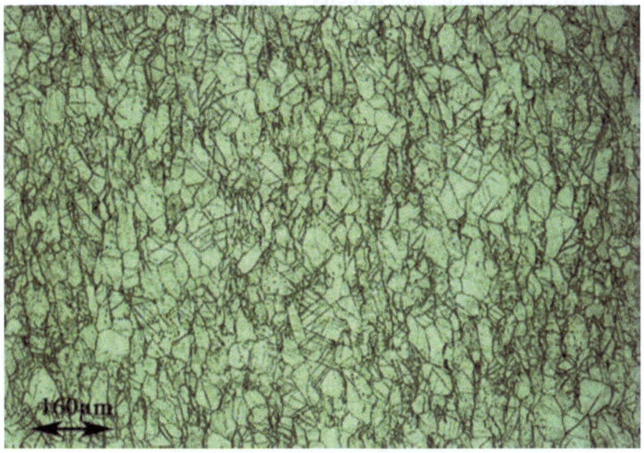


Figure 5. OM of bulk microstructure.

The deformed regions of the samples exhibited a large amount of shear localization, characterized by shear bands and DRX. Shear localization was observed in regions of SPD, and was seen to vary throughout the samples. However, the level of shear localization was highly concentrated in certain areas, and in particular was seen to a greater extent in the Partial-group of samples. These regions were examined further using TEM, and quantified using IQ Materials software.

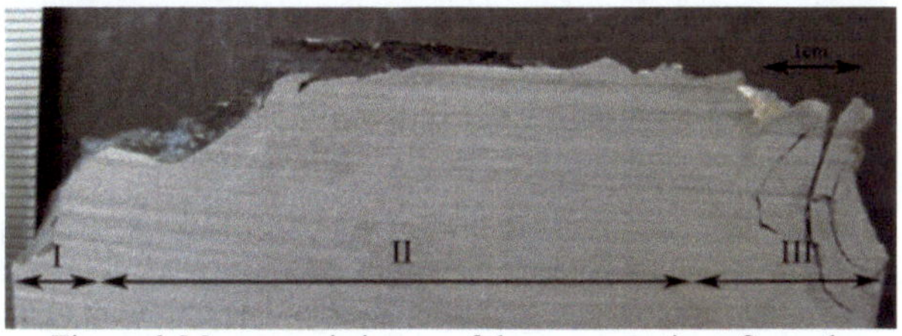


Figure 6. Macroscopic image of the cross-section of sample B-Full, showing microstructural deformation regions I, II, and III

In general, throughout all of the samples, there were three separate regions of distinctive deformation characteristics. Figure 6 shows these regions on the B-Full sample. Region I, the initial area of impact, was characterized by a rough fracture surface along the edge, and a microstructure that was similar to that of the bulk microstructure. An OM of region I is shown in Figure 7. The microstructure of region II showed extensive shear localization, with ASB, DRX, and a smooth edge along the channel as presented in Figure 8. Similar to Region I, region III was composed of a microstructure very similar to that of the bulk microstructure, however, there is a large amount of shear fracturing along the edge. Figure 9 shows the microstructure of region III. The Partial-damage samples did not have a region III, but instead were composed of regions I and II, except in the case of C-Partial which exhibited a large amount of bulging petal fractures.

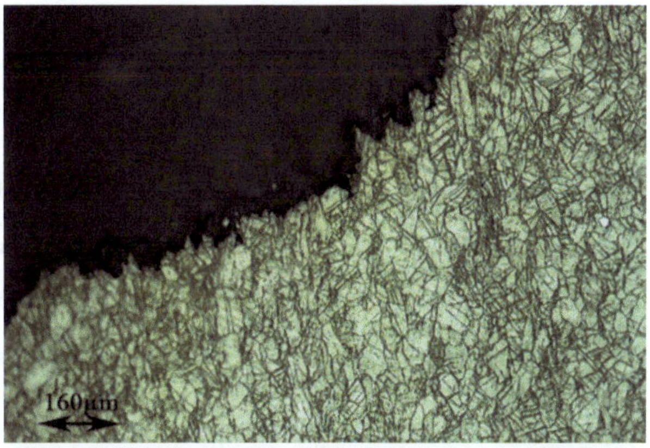


Figure 7. OM image showing typical microstructure of region I.

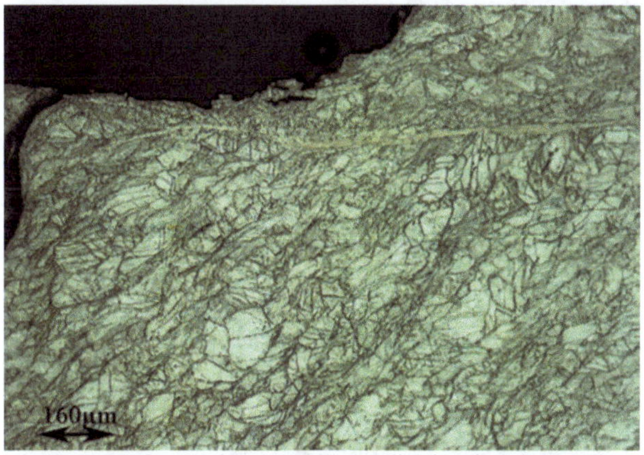


Figure 8. OM image showing shear localization that characterized region II.

Figure 9. OM image of region III, similar fracture edge and microstructure to that of Region I.

The major variation in these regions, as seen when comparing between each of the sample sets, was the overall size of the regions. Of all of the sample sets, those in sample set A showed the smallest regions I and III, while sample set C showed the largest. As regions I and III were composed of large quantities of shear fracture surfaces, it is likely the higher level of energy transferred during the impacts for sample set C were the largest, thus resulting in a greater amount of shear fracturing instead of localized shear deformation.

Shear Localization Depth

Shear localization was present in every deformed cross-sectioned sample. In order to appreciate the extent of shear localization, composite OM images were formed, and the region of shear deformation was measured using IQ Materials software. Each sectioned sample (as shown in Figure 3 for sample A-Partial), of both the Partial and Full samples of each sample set was analyzed. The measurements for each sample were averaged, and the plots are presented in Figures 10, 11, and 12.

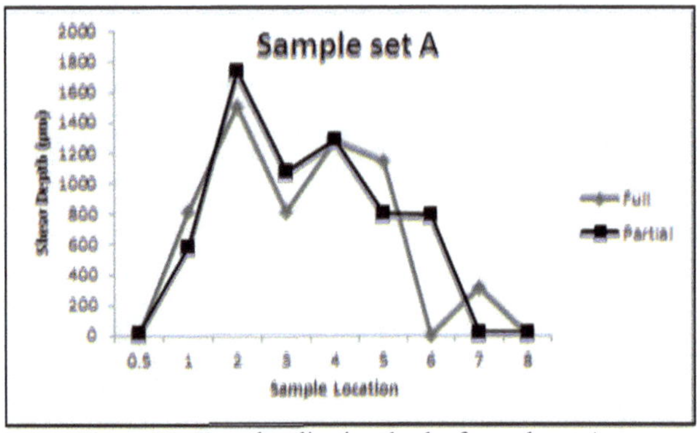


Figure 10. Shear localization depth of sample set A.

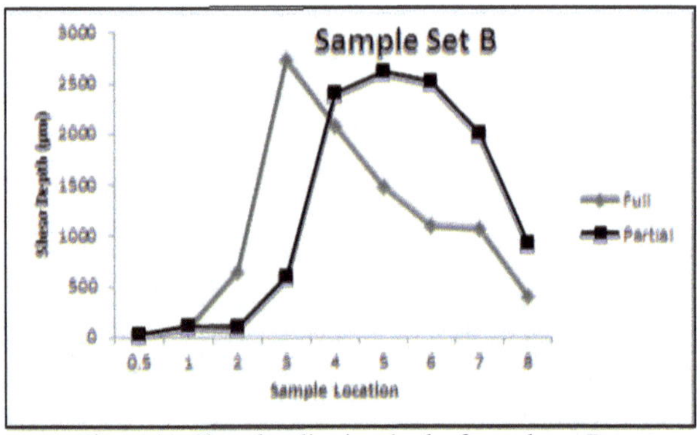


Figure 11. Shear localization depth of sample set B.

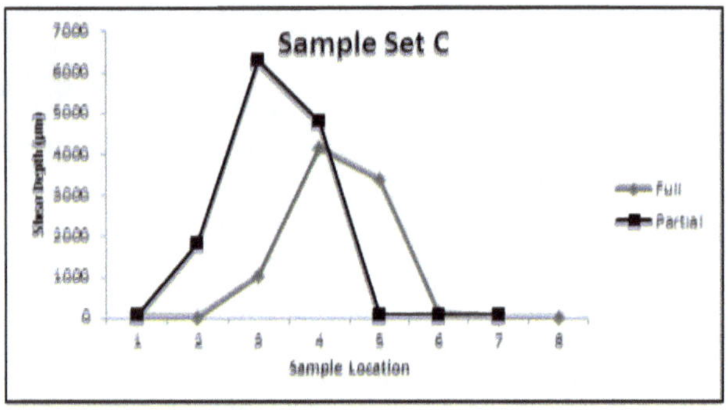


Figure 12. Shear localization depth of sample set C.

Sample set A exhibited a similar quantity of shear localization in both the Full and Partial samples. Furthermore, the overall magnitude of shear depth was to a much lesser degree than either the B or C samples which were both subjected to a much greater level of plastic deformation. The lower level of shear present in sample set A was a result of the lower level of kinetic energy transferred from the high velocity impact.

In sample set B, there was a marked difference in the level of shear localization present in the Partial sample, when compared to the Full sample. Overall, the Partial sample exhibited a greater amount of shear localization, as can be seen in locations 5-8 of Figure 12. In the Full sample, in the regions of lower shear localization, there was a high level of shear fracturing observed.

The greatest magnitude of shear localization was measured in sample set C, as was expected due to the gross amount of severe plastic deformation observed. Of note, the sample locations 5, 6, and 7 of the Partial sample were positioned in such a way, in relation to the damaged edge, to make it impractical to measure shear depth. However, qualitatively these samples had a low amount of shear localization, and there was an extensive amount of fracturing observed, as can readily be seen on a macroscopic level on the right side of the sample in Figure 13.

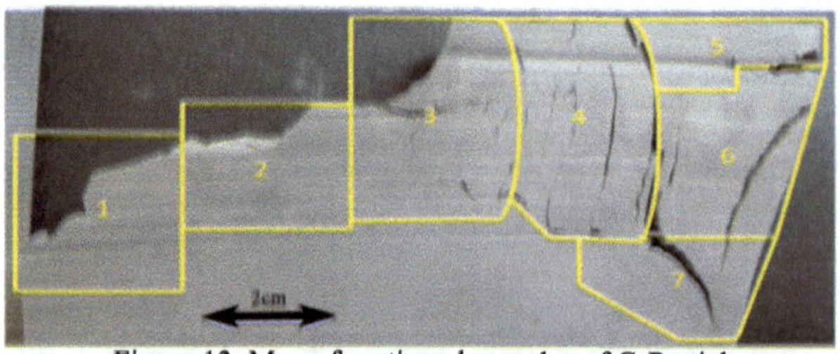


Figure 13. Map of sectioned samples of C-Partial

Overall, shear localization was seen to the greatest extent in the Partial sample group. The shear localization is a result of the microstructure of the material accommodating the kinetic energy from the high velocity impact. When the kinetic energy exceeded a certain amount, full penetration of the sample would occur, and this was characterized by a much higher level of fracturing in the microstructure. Below that level, the material would accommodate the kinetic energy not by fracturing, but instead by shear localization.

A limitation of this study is that any region of the sample that was completely fractured away was no longer attached to the sample, and as a result, not available for post-mortem analysis. It is possible that shear localization was present in these portions of the sample, but no analysis was able to be performed.

Adiabatic Shear Bands

Adiabatic shear bands were a ubiquitous characteristic of regions of shear localization. The shear bands were composed of areas of DRX, which is a result of adiabatic temperature rise [8]. Both regions of DRX, and the transition zone between the DRX and neighboring microstructure, were examined by TEM. The shear band that was analyzed is shown in Figure 14.

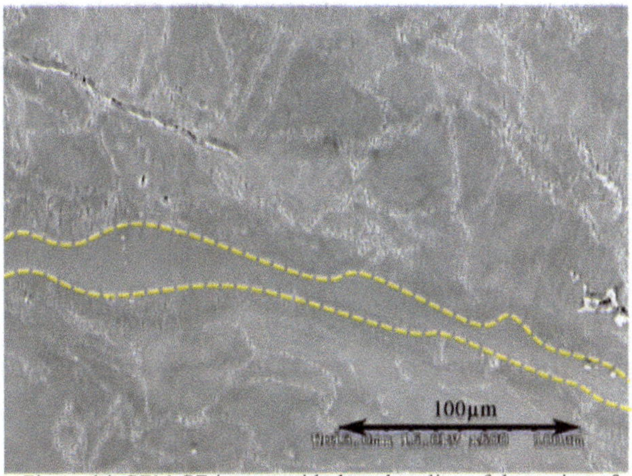


Figure 14. SEM-SE image, with dotted outline of the region of DRX within shear band, that was analyzed by TEM.

Figure 15. Diagram of shear band, and neighboring microstructure.

The shear band was composed of two distinct regions as schematically illustrated in Figure 15. The first region of DRX was composed of small equiaxed recrystallized grains (<500nm) that contained a very limited amount of precipitates, while the second region was a transition zone (TZ) that contained both the larger grains of the bulk matrix, and the smaller grains similar in size to the DRX region. The TEM of the DRX region is shown in Figure 16, while that of the TZ region is shown in Figure 17.

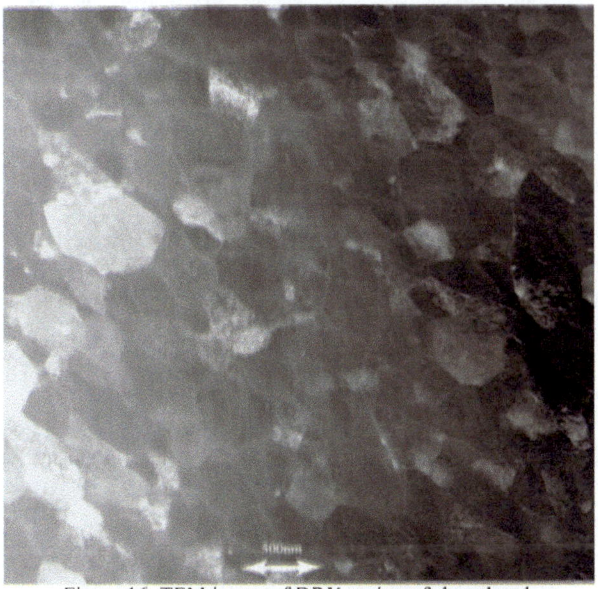

Figure 16. TEM image of DRX region of shear band.

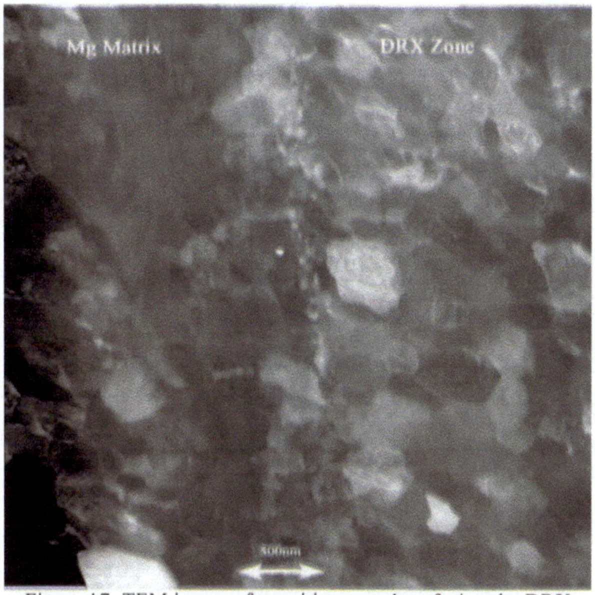


Figure 17. TEM image of transition zone interfacing the DRX region and neighboring microstructure.

The neighboring matrix was similar to that of the bulk material shown in Figure 18, containing large grains with smaller precipitates, and smaller grain regions with large secondary phase precipitates. Indexing performed on the diffraction pattern from a large secondary phase precipitate showed that it was β phase: $Mg_{14}Nd_2Y$ (fcc, a = 2.223nm). Figure 19 shows the diffraction pattern.

Figure 18. TEM image of bulk microstructure, the large precipitate that was indexed is marked.

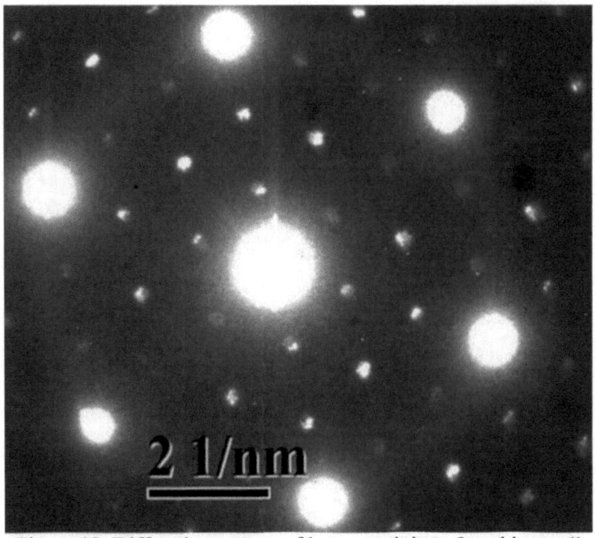


Figure 19. Diffraction pattern of large precipitate found in small grain region of bulk material, indexing to be β phase.

The smaller precipitates located in the coarse grains were analyzed by XEDS as shown in Figure 20, and the composition was quantified as $Mg_{74.8}Y_{17.2}Nd_{8.0}$ (at%). This suggests that these percipitates are most likely β" phase: $Mg_3(Y_{0.85}Nd_{0.15})$.

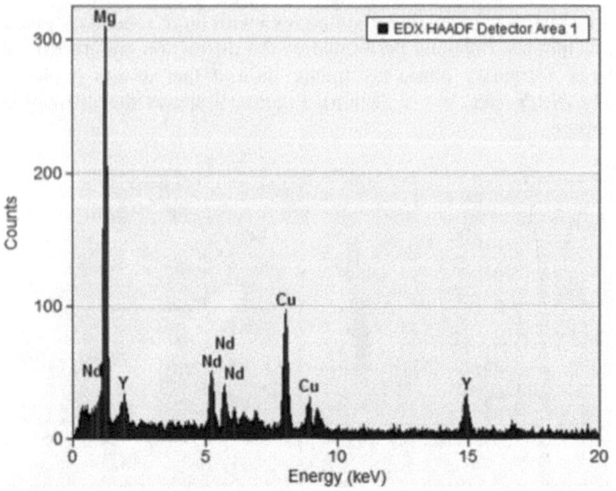


Figure 20. XEDS results of smaller precipitates contained in coarse grains of bulk microstructure.

## Conclusions

After a set of six WE43 magnesium alloy samples were subjected to different high-velocity impacts, which resulted in high-strain rate severe plastic deformation, the microstructures of the deformed samples were analyzed and compared to the bulk material. Shear localization, characterized by bands of dynamic recrystallization, was seen throughout the damaged regions. However, samples that were subjected to impacts that resulted in only partial damage were seen to have a larger quantity of shear localization, compared to those samples that experienced impacts that resulted in full damage. Furthermore, the samples with full damage had a lower occurrence of shear localization in the regions of high shear fracture. This indicates that the material would accommodate the lower level of kinetic energy, which was insufficient to result in shear fracture, by shear localization in the microstructure.

The regions of shear localization, when analyzed by TEM showed distinct variation from the bulk material microstructure. The bulk material was composed of regions with coarse grains containing small elongated precipitates of the β" phase, and smaller grain regions with large β phase precipitates. Microstructure within the ASB was distinctively different. The shear bands were composed of small (<500nm) grains, with a very limited amount of precipitates. In addition, there was a transition zone between the two regions that showed a mixture of both the larger grains seen in the bulk material, and the smaller grains seen in the region of DRX.

## Acknowledgements

Research was sponsored by US Army Research Laboratory and was accomplished under Cooperative Agreement W911NF-07-2-0073. The views, opinions, and conclusions made in this document are those of the authors and should not be interpreted as representing the official policies, either expressed or implied, of Army Research Laboratory or the US Government. The US Government is authorized to reproduce and distribute reprints for Government purposes notwithstanding any copyright notation herein.

## Works Cited

[1] B. Mordike T. Ebert, "Magnesium Properties - applications - potential," *Materials Science and Engineering A*, vol. 302, no. 1, pp. 37-45, 2001.

[2] *Magnesium Die Casting Handbook.* New York, USA: NADACA, 1998.

[3] M. Janecek, P. Dobron, F. Chmelik, V. Supik, F. Hollander R. Kral, "Mechanisms of plastic deformatoin in AZ31 magnesium alloy investigated by acoustic emission and transmission electron microscopy," *Materials Science and Engineering A*, vol. 462, no. 1-2, pp. 311-315, 2007.

[4] Y.H. Wei, Y.P. Zhu, H.T. Zhou, W.J. Ding, Y. Chino, M. Mabuchi Q.D. Wang, "Superplasticity and grain boundary sliding in rolled AZ91 magnesium alloy at high strain rates," *Materials Science and Engineering A*, vol. 360, no. 1-2, pp. 107-115, 2003.

[5] S. Agnew, Ö. Duygulu, "Plastic anisotropy and the role of non-basal slip in magnesium alloy AZ31B," *International Journal of Plasticity*, vol. 21, no. 6, pp. 1161-1193, 2005.

[6] M. Meyers, "Shear localization in dynamic deformation of materials: microstructural evolution and self-organization," *Materials Science and Engineering A*, vol. 317, no. 1-2, pp. 204-225, 2001.

[7] C.W. Tan, L. Wang, Z.Y. Chen, F. Wang, H. Cai, H.L. Ma S. Xu, "Deformation behavior of AZ31 magnesium alloy at different strain rates and temperatures," *Trans. Nonferrous Met. Soc. China*, vol. 17, pp. S347–S352, 2007.

[8] L. Murr, S. Pappu, C. Kennedy E. Trillo, "Adiabatic shear bands and examples of their role in severe plastic deformation," *Journal of Materials Science*, vol. 37, no. 16, pp. 3337–3360, 2002.

**Magnesium Technology 2012**
*Edited by: Suveen N. Mathaudhu, Wim H. Sillekens, Neale R. Neelameggham, and Norbert Hort*
*TMS (The Minerals, Metals & Materials Society), 2012*

# MICROSTRUCTURE AND MECHANIAL PROPERTIES OF AS-EXTRUDED Mg-Sn-Al-Zn ALLOYS

Sung Hyuk Park, Young Min Kim, Chang Dong Yim, Ha-Sik Kim, Bong Sun You
Light Metals Group, Korea Institute of Materials Science, Changwon 641-831, South Korea

Keywords: Magnesium alloy, Indirect extrusion, Precipitation, Yield asymmetry

## Abstract

Mg-(9-x)Sn-xAl-1Zn (x=1, 2, 3 and 4 wt.%) alloys were subjected to indirect extrusion, and the microstructure and mechanical properties of these as-extruded Mg-Sn-Al-Zn (TAZ) alloys were investigated. The TAZ alloys contain only fine $Mg_2Sn$ second-phase particles in the α -Mg matrix, and the amount of precipitates increase with increasing Sn contents. In addition, the TAZ 811 alloy shows finer grain structure than the TAZ 541 alloy due to a larger amount of $Mg_2Sn$ particles, which act as nucleation sites for recrystallization and/or prevention of grain growth by the pinning effect. The textures of the alloys show a typical basal pole orientation and an unusual <11-20> component in which one axis of the HCP unit cell is parallel to the extrusion axis. Tensile yield strength increases with Sn content, while ultimate tensile strength is almost identical due to the increased strain hardenability with Al content. The tension-compression yield asymmetry decreases with reducing Sn content, which is mainly due to the decrease in the amount of particles promoting twin nucleation.

## Introduction

One of the major obstacles to overcome in Mg extrusions is their low extrusion speed, which is directly related to cost and energy efficiency. Normally, the extrusion speed of commercial high-strength Mg alloys, such as AZ80 and ZK60, is restricted to 0.5 to 2.5 m/min, which is well below the speeds attainable with Al alloys [1]. This is mainly due to an increase in the susceptibility to surface tearing during extrusion with increases in alloying elements, such as Al and Zn, which results from the incipient melting of second-phase particles with low melting temperatures. In regard to high-speed extrusion, it is believed that Mg–Sn-based alloys have great potential due to their higher incipient melting temperatures than conventional AZ and ZK series Mg alloys [2,3]. In addition, indirect extrusion facilitates extrusion at a higher speed than conventional direct extrusion since there is no friction between the billet and the walls of the container and therefore no friction heat can occur. In the present study, experimental Mg-(9-x)Sn-xAl-1Zn (x=1, 2, 3 and 4 wt.%) alloys were subjected to indirect extrusion, and the microstructure and mechanical properties of these as-extruded alloys were comparatively investigated.

## Material and Experimental Procedures

The composition of the Mg-(9-x)Sn-xAl-1Zn (TAZ 541, 631, 721, 811) alloys used in this study are listed in Table 1. To prepare the billets for extrusion, the alloys were melted under an inert atmosphere containing a mixture of $CO_2$ and $SF_6$ and were then stabilized at 700°C and poured into a steel mould pre-heated to 200°C. After casting, the billets were homogenized under the heat treatment conditions shown in Table 1 and then water-quenched to induce a supersaturated solid solution. The billets were 80 mm in diameter and 200 mm long. Prior to indirect extrusion, the billets were pre-heated in a resistance furnace set at 250°C for 1 h. The extrusions were conducted at an initial billet temperature of 250°C and a ram speed of 1.3 mm/s (equal to the rod speed of about 2 m/min). This temperature is somewhat lower than those usually used in the conventional direct extrusion of Mg alloys. A rod profile was extruded with a fixed extrusion ratio of 25. Microstructural examinations were conducted on the midsections parallel to the extrusion direction (ED). The texture was examined using X-ray diffraction on the cross-sectional area of the extrudates. The longitudinal tensile and compressive properties of the as-extruded condition were measured at room temperature at an initial strain rate of $1.0\times10^{-3}$ $s^{-1}$ using round tensile specimens with a gauge diameter of 6 mm and a gauge length of 25 mm as well as cylindrical compressive specimens that were 8 mm in diameter and 12 mm in high.

Table 1. Chemical compositions (wt.%) and homogenization conditions of Mg-(9-x)Sn-xAl-1Zn alloys.

| Alloy | Sn | Al | Zn | Mg | Temp. (°C) | Time (hour) |
|---|---|---|---|---|---|---|
| TAZ541 | 5.01 | 3.96 | 0.96 | Bal. | 460 | 8 |
| TAZ631 | 5.95 | 3.01 | 0.97 | Bal. | 480 | 6 |
| TAZ721 | 7.02 | 1.96 | 0.95 | Bal. | 500 | 4 |
| TAZ811 | 7.95 | 0.95 | 0.95 | Bal. | 500 | 4 |

## Results and Discussion

### Second-phase particle

Various second-phase particles, such as $Mg_2Sn$, $Mg_{17}Al_{12}$, and $MgZn_2$ can be formed during extrusion by dynamic precipitation of alloying elements Sn, Al, and Zn, which are in a solid solution state before extrusion. To predict the type and amount of second-phase particles formed in extrudates, the phase diagram of the Mg-(9-x)Sn-xAl-1Zn system was calculated using the PANDAT 7 software based on the CHAPHAD with the thermodynamic database. Although the initial billet temperature and extrusion conducting temperature were 250°C, the exit temperature, which is considered to be closer to the actual temperature of alloys experienced during the extrusion process, was approximately 290°C. According to the calculated phase diagram shown in Fig. 1a, the microstructure of the as-extruded TAZ alloys used in this study consists of α -Mg and $Mg_2Sn$ phases with no other second-phase particles irrespective of the amount of alloying element (Sn or Al), and the calculated amount of $Mg_2Sn$ phase increases from

2.6 to 4.6 wt.% as the Sn content increases (Fig. 1b). These results coincide with the X-ray diffraction pattern results of the as-extruded TAZ alloys, which show only two types of peaks of α-Mg and $Mg_2Sn$ even in TAZ541 alloys with a relatively large amount of 4 wt.% Al.

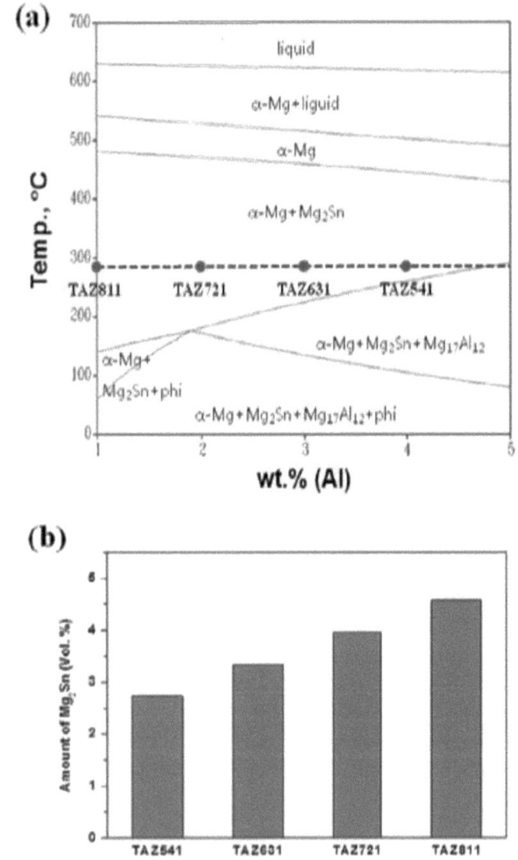


Figure 1. (a) Stable phase diagram and (b) amount of $Mg_2Sn$ phase of TAZ alloys extruded at 290°C calculated with PANDAT software.

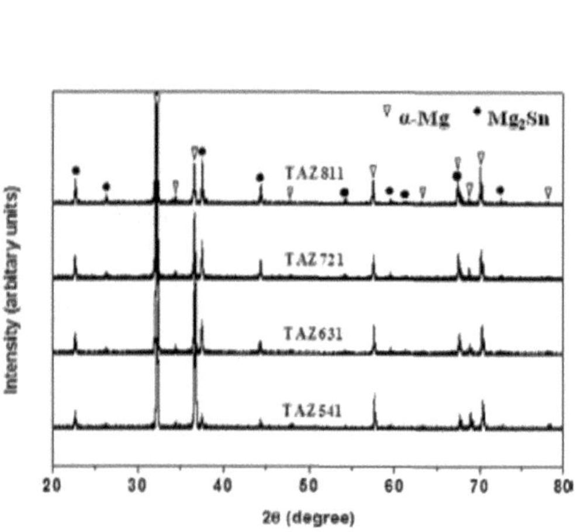


Figure 2. X-ray diffraction patterns of TAZ alloys

Microstructure

The microstructures of the as-extruded TAZ alloys are shown in Fig. 3. Coarse grains with an average size of ~300 ㎛ prior to extrusion were changed into fine grains as a consequence of dynamic recrystallization (DRX) during extrusion. The grain size of the as-extruded TAZ alloys slightly decreases with increasing Sn content. This is due to the volume fraction of fine $Mg_2Sn$ particles that precipitate out during extrusion and act as nucleation sites for recrystallization and/or prevention of grain growth by the pinning effect. Therefore, the TAZ 811 alloy with the largest amount of precipitates shows finer grain structure (~2.2 ㎛) than the TAZ 541 alloy (~5.8 ㎛) with a relatively small amount of $Mg_2Sn$ particles.

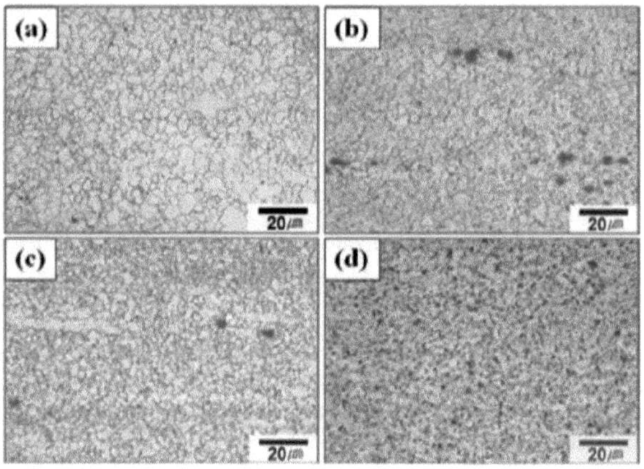


Figure 3. Optical micrographs of as-extruded (a) TAZ541, (b) TAZ631, (c) TAZ721, and (d) TAZ811 alloys

Fig. 4 shows inverse pole figures of the as-extruded TAZ alloy specimens. Most of the extruded Mg alloys have a typical extrusion texture, where the <0002> basal poles and <10-10> prismatic poles are mainly oriented perpendicular and parallel to the ED, respectively [4,5]. Although the textures of as-extruded TAZ alloys show a typical basal pole orientation, they have a strong <11-20> component in which one axis of the HCP unit cell is parallel to the extrusion axis. Although <10-10> texture component preferentially aligned with the extrusion axis is very dominant in extruded Mg alloys, it is possible to develop other texture components by controlling the chemical composition and deformation temperature. For example, AZ31 Mg alloy indirectly extruded at 250°C shows ring type texture in which prismatic poles are randomly oriented without preferred orientation [6]. On the other hand, a double texture component with <10-10>+<11-20> is developed in Mg-Mn alloy extruded at 300°C [7], and some RE additions generate a <11-21> peak [7,8]. Therefore, <11-20> texture of TAZ alloys is attributed to the addition of Sn, which may affect the activation of slip systems or temperature during extrusion. As the Schmid factor values of slip systems and/or twin variants activated during the deformation depend on not only basal pole orientation but also prismatic pole orientation, these prismatic texture components play an important role in deformation behavior [9-11]. Therefore, further research on the DRX mechanism of TAZ alloys is required to understand the relationship between the chemical composition and texture developed during extrusion.

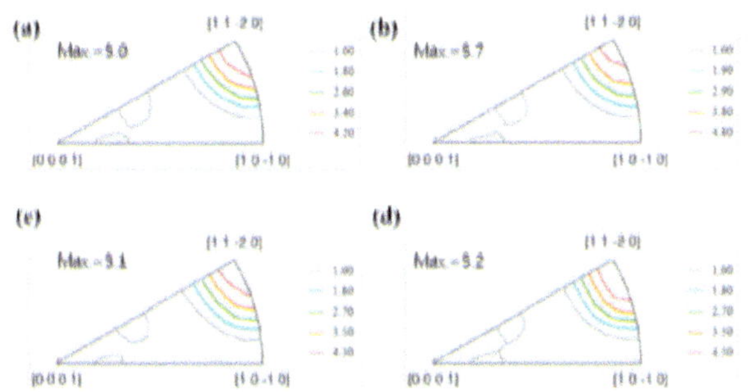


Figure 4. Inverse pole figures of (a) TAZ541, (b) TAZ631, (c) TAZ721, and (d) TAZ811 alloys with respect to the extrusion direction.

Mechanical properties

The results of the tensile tests are summarized in Table 2. As the grain size decreases and the amount of precipitates increases with increasing Sn content, the TAZ alloys with high Sn content show higher yield strength due to the combined effects of grain boundary hardening and precipitation hardening. While strain hardenability increases with Al content (Table 2), so that the TAZ alloys with low Sn content exhibit nearly identical tensile strength with those having high Sn content in spite of their low yield strength.

Fig. 5 shows the tensile and compressive yield stresses and tension-compression yield asymmetry, CYS/TYS. Because a decrease in grain size enhances the activation stress of slip and twinning, both tensile and compressive yield strengths increase with Sn content. However, precipitate particles promote {10-12} twin nucleation as the Orowan stress on twinning dislocations leads to an increase in stress around particles [12]. Accordingly, although an increase of particles causes an increase in tensile yield strength by precipitation hardening, compressive yield strength decreases with an increasing amount of particles due to the reduction of twinning stress, which is activation stress driving twin nucleation. For this reason, the increment of tensile yield strength with Sn content is slightly larger than that of compressive yield strength, resulting in a decrease of yield asymmetry with decreasing Sn content as shown in Fig. 5.

Table 2. Tensile properties of as-extruded TAZ alloys.

| Alloy | YS (MPa) | UTS (MPa) | EL (%) | n |
|---|---|---|---|---|
| TAZ541 | 217 | 340 | 19.6 | 0.24 |
| TAZ631 | 238 | 332 | 18.5 | 0.20 |
| TAZ721 | 261 | 342 | 16.6 | 0.17 |
| TAZ811 | 265 | 333 | 17.7 | 0.15 |

* YS, UTS, EL and n represent the 0.2 % offset yield strength, ultimate tensile strength, elongation, and work hardening exponent, respectively.

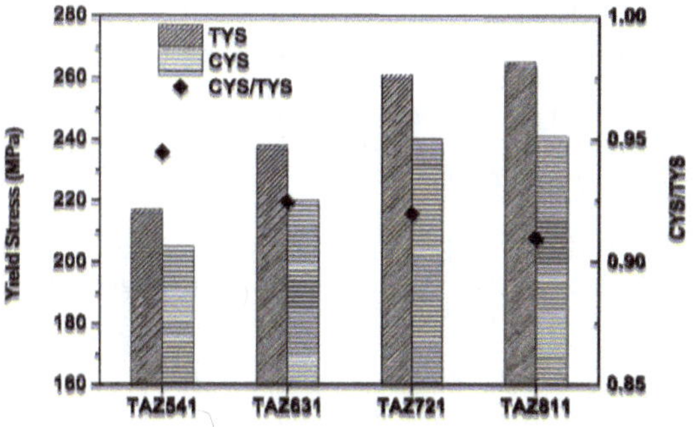


Figure 5. Tensile and compressive yield stresses and CYS/TYS

## Conclusions

Mg-(9-x)Sn-xAl-1Zn (x=1, 2, 3 and 4 wt.%) alloys were subjected to indirect extrusion, and the microstructure and mechanical properties of these as-extruded alloys were investigated. The TAZ 811 alloy shows smaller grain size than the TAZ 541 alloy due to a larger amount of $Mg_2Sn$ particles, which act as nucleation sites for recrystallization and/or prevention of grain growth by the pinning effect. Tensile yield strength increases with Sn content due to the combined effects of grain boundary hardening and precipitation hardening. The yield asymmetry between tension and compression reduces with decreasing Sn content, which is mainly due to the decrease in the amount of particles promoting twin nucleation.

## References


1. C.J. Bettles and M.A. GibsonY, "Current wrought magnesium alloys: strengths and weaknesses" *JOM*, 57 (2005) 46-49.

2. D.H. Kang et al., "Development of creep resistant die cast Mg-Sn-Al-Si alloy" *Mater. Sci. Eng. A*, 413-414 (2005), 555-560.

3. W.L. Cheng et al., "Strength and ductility of novel Mg-8Sn-1Al-1Zn alloys extruded at different speeds," *Materials Letters*, 65 (2011) 1525-1527.

4. S. Yi, "Investigation on the deformation behavior and the texture evolution in magnesium wrought alloy AZ31", *Fakultät für Natur- und Materialwissenschaften*, Technischen Universtität Clausthal (2005).

5. J. Bohlen et al., "Microstructure and texture development during hydrostatic extrusion of magnesium alloy AZ31", *Scripta Mater.* 53 (2005) 259-264.

6. S. Mueller et al., "Microstructure and mechanical properties of the extruded Mg-alloys AZ31, AZ61, AZ80", *Int. J. Mater. Res.* 97 (2006), 1384-1391.

7. J. Bohlen et al., "Influence of the alloying additions on the microstructure development of extruded Mg-Mn alloys", *Magnesium Technology* (2009) 225-230.

8. N. Stanford and M.R. Barnett, "The origin of "rare earth" texture development in extruded Mg-based alloys and its effect on tensile ductility", *Mater. Sci. Eng. A* 496 (2008) 399-408.

9. S.H. Park et al., "Activation mode dependent {10-12} twinning characteristics in a polycrystalline magnesium alloy", *Scripta Mater.* 62 (2010) 202-205.

10. S.G. Hong et al., "Role of {10-12} twinning characteristics in the deformation behavior of a polycrystalline magnesium alloy", *Acta Mater.* 58 (2010) 5873-5885.

11. S.G. Hong et al., "Strain path dependence of {10-12} twinning activity in a polycrystalline magnesium alloy", *Scripta Mater.* 64 (2011) 145-148.

12. J.D. Robson et al., "Effect of particles in promoting twin nucleation in a Mg-5 wt.% Zn alloy", *Scrtipa Mater.* 63 (2010) 823-826.

**Magnesium Technology 2012**
*Edited by: Suveen N. Mathaudhu, Wim H. Sillekens, Neale R. Neelameggham, and Norbert Hort*
***TMS (The Minerals, Metals & Materials Society), 2012***

# TENSILE PROPERTIES OF THREE PREFORM-ANNEALED MAGNESIUM ALLOY SHEETS

Junying Min[1], Jon T. Carter[2], Ravi Verma[2]

[1]Tongji University, Mechanical Engineering, 1239 Siping Road, Shanghai, China
[2]General Motors R&D, 30500 Mound Road, Warren, MI, 48090, USA

Keywords: AZ31, ZEK100, magnesium, sheet, preform, anneal, tensile properties.

## Abstract

Magnesium alloy sheet metal is potentially attractive for use in automotive structural applications due to its high strength-to-weight ratio. However, application has been hindered by the low room-temperature formability of typical sheet alloys. One approach to effectively increase formability is to change the forming process from one which involves a single stamping hit to one which utilizes two hits plus an intemediate anneal (i.e., "preform anneal process"). The purpose of the intermediate anneal is to restore some of the softness and ductility which were reduced by deformation during the first hit.

In this report, the preform annealing behavior of three rolled magnesium alloy sheets was studied using uniaxial tensile tests. The sheets studied were: conventionally rolled (CR) AZ31B, CR ZEK100, and specially rolled (SPR) AZ31B. The preform annealing process was found to increase the total elongation of all three sheets compared to the elongation in the annealed O-temper. The CR ZEK100 with a thickness of 1.5 mm showed more attractive tensile properties than the 1.6 mm CR AZ31B. Although the SPR AZ31B has a thickness of only 0.7 mm, it still has elongation comparable to the 1.6 mm CR AZ31B.

## Introduction

Magnesium is lightest (lowest density) of the common structural metals and therefore is potentially attractive for use in structural automotive applications to achieve mass reduction and the associated improvements in fuel efficiency and handling. Currently, automotive usage of magnesium is primarily in the form of die castings, and includes steering wheels, transmission housings, sunroof frames [1], and inner panels for doors and liftgates [2,3]. Numerous potential automotive applications of wrought magnesium alloys have been prototyped [4-7], but production implementation has been extremely limited [8]. The issues which preclude greater implementation of the wrought alloys include the high cost of sheet, limited formability at room temperature, difficulty joining to other materials, and corrosion. Potential solutions to all of these issues are being studied by various groups around the world.

Mg sheet is formable at hot temperatures [9,10] and warm temperatures [11,12]. Ultimately, however, sheet formers would prefer to stamp magnesium alloy sheet at room temperature, similar to what is done with steel. This would reduce the costs associated with heaters, hot tools, and high-temperature-capacity forming lubricants, and might enable beneficial strain hardening of the part being formed. Such stamping might be done using blanks which benefit from having either a different chemical composition of the alloy, or which may have been thermomechanically processed differently so as to enhance formability, or by changing the forming process. All three of these options are described briefly below.

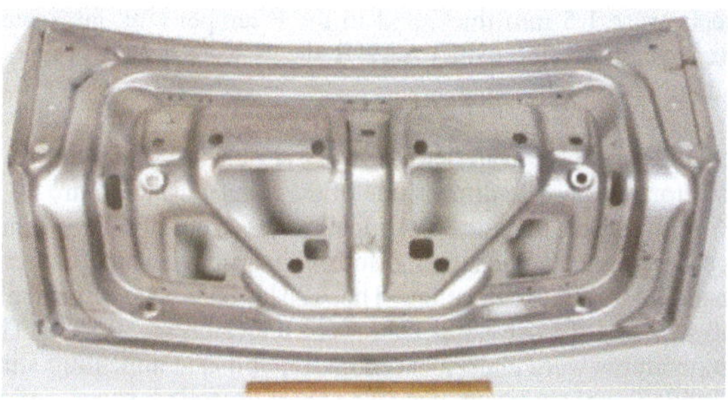

Figure 1. Photograph of experimental decklid inner panel which was formed by hot gas blow forming process, then trimmed/punched.

Wrought magnesium alloys containing rare earth (RE) additions such as Gd, Ce, Nd, and Er, have been found to exhibit greater ductility than their RE-free counterparts. This increase is typically attributed to the the weaker crystallographic texture present in the rolled or extruded forms of RE-bearing alloys upon which forming tests are conducted, especially after recrystallization annealing [13-18].

Alternatively, the texture and/or ductility of rolled magnesium alloy sheet may be reduced by altering the rolling process. For example, beneficial effects have been reported for increasing the rolling temperature [19], increasing the rolling speed [20], and increasing the asymmetry of rolling [21].

Finally, the forming process itself may be altered to effectively increase the formability of metal sheets. Specifically, rather than completely forming a panel with a single press stroke and getting splits, the panel may be partially formed in a first press stroke, annealed to restore ductility to strained areas, cooled back to near room temperature, then formed to completion without splitting in a second press stroke [22]. This process is called "preform annealing". In this process it is beneficial to introduce a large amount of strain during the first press stroke, but less than that at which strain localization, necking, or fracture begin. Components with complex contours, which cannot be manufactured by the conventional stamping process, can be formed successfully by the preform annealing process. Both an aluminum rear door inner panel [23] and an aluminum lift gate inner panel [24] were perfectly produced by using preform annealing process. The latter used localized induction heating in order to reduce cycle time and conserve energy. Some of the more scientific fundamentals of the preform annealing process have been studied both via simulation [25] and experiments [26].

The purpose of the work reported herein was to compare the tensile properties and microstructure of conventionally-rolled AZ31B magnesium alloy sheet with two alternative magnesium alloy sheet products, in both the annealed condition and the preform annealed condition.

### Experimental Method

Rolled sheets were obtained from two manufacturers in different thicknesses and tempers. Specifically, the conventionally rolled (CR) AZ31B sheets were 1.6 mm thick, and in the H24 temper ("strain hardened and partially annealed"). The CR ZEK100 sheets were 1.5 mm thick, and in the F temper ("as-fabricated"). The specially rolled (SPR) AZ31B sheets were 0.7 mm thick, and in an unspecified temper.

Tensile bars were waterjet cut at 0°, 45°, and 90° to the rolling direction (RD). The dimensions of the reduced section are 50 mm x 6 mm x sheet thickness. Annealing was done for 15 min in an electric furnace which was filled with argon gas to protect the tensile bars from oxidization. The AZ31B bars were annealed at 350 °C, and the ZEK100 bars were annealed at 500°C. These temperatures were chosen based on unpublished annealing studies in which we observed that AZ31B-H24 rolled sheet softened much at 350 °C, but that the ZEK100-F rolled sheet required 450-500 °C to soften. These treatments were applied to all of the bars regardless of in-coming temper, and are considered to result in O-temper.

Rockwell F hardness (HRF) was measured on the sheet surface with a 1/16" diameter indenter and 60 kg major load. Because the SPR AZ31B sheet is relatively thin and soft, its hardness was measured on the HR15T superficial scale, then converted to Rockwell F. Five measurements were repeated for each datum.

An Instron tensile machine and an extensometer with a gauge length of 25mm were employed. The crosshead speed was set at 10 mm/min for all tests. When performing tests for each type of sheet material and bar orientation relative to the sheet rolling direction (RD), three O-tempered bars were pulled to fracture. From these tests, the average engineering strain at which maximum load occurred was determined for each orientation. This is referred to as the uniform strain (US). Next, for the preform anneal study, four O-tempered bars in each orientation were pulled to 90% of US, re-annealed, then returned to the tensile machine and pulled to fracture. Finally, in order to determine the normal plastic anisotropy ratio (r value), additional bars were marked with ink at five locations within the gauge length, then pulled to 10%, 15% and 20% strain, removed from the tensile machine, then measured with vernier calipers to determine width and thickness at the five axial locations.

For microstructure observation, samples were cold mounted, ground, polished, and etched. The picric etchant which revealed the grains and grain boundaries was mixed fresh with 4.2 gram picric acid, 10 ml acetic acid, 70 ml ethanol, and 10 ml distilled water. Samples were immersed in the etchant for about 1 second, then immediately washed with soap water and distilled water, rinsed with ethanol, and dried with warm flowing air. Metallographic observations were made using a Leica microscope. The grain sizes were measured in different directions using several micrographs from horizontal and vertical sections and a computer aided linear intercept measurement.

### Results and Discussion

Compositions

Elemental compositions of the sheet materials are as follows: CR AZ31B contains 3.0 %Al, 1.1 %Zn and 0.48 %Mn. SPR AZ31B contains 2.9 %Al, 1.1 %Zn and 0.44 %Mn. CR ZEK100 contains 1.3 %Zn, 0.2 %Nd, 0.25 %Zr and 0.01 %Mn. Al and Zn are added for strength, and Zn and Mn are added for corrosion resistance. Nd is for reduced crystallographic texture, and Zr is for grain refinement.

Hardness

Hardness data for the three Mg alloy sheets are shown in Figure 2. For all three materials the pattern is similar. First, the hardness in the PA condition is the same as that in the O-temper if the amount of pre-deformation (between anneals) is zero. This is shown by the black and white bars. If, however, deformations of 90% of US are introduced between the two anneals, then the hardness measured after the second anneal is less than that measured after the first. Compare the black and dashed bars. If there is no second anneal, the hardness increases due to work hardening, as shown by the black and gray bars. Since both the strength and grain size influence the hardness, the hardness results will be discussed more in later sections.

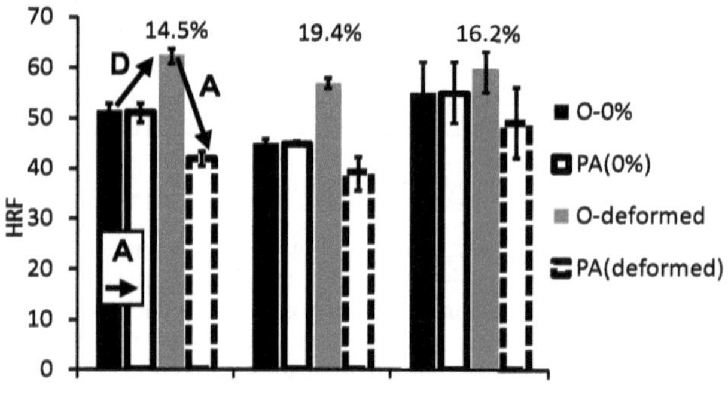


Figure 2. Chart showing hardness data, Rockwell F scale, of CR AZ31B, CR ZEK100, and SPR AZ31B sheets in different conditions. O-0% means O-tempered material with no deformation; PA(0%) means the deformation before 2[nd] anneal is 0%, namely, just annealed two times; O-deformed means O-tempered material with deformation and the deformation levels are marked above the column; PA(deformed) means the material has deformation before 2[nd] anneal and the deformation levels are the same as marked above the column. The arrows indicate the changes due to either "A" annealing processes or "D" deformation processes. Samples were pulled in the 90° orientation

Tensile Data

These data are summarized in Figures 3-6, and discussed below.

Effect of alloy and processing: The two AZ31 sheets have strengths which are comparable, and both are stronger than the ZEK100. This is attributed to the higher alloy content and smaller grain sizes in the AZ31 sheets. In contrast, the ductility of the

SPR AZ31 is less than that of the CR AZ31, and both of those are less than the ductility of the ZEK100. The ZEK100 shows the greatest in-plane anisotropy of n, k, YS, and %el, but CR AZ31B shows the greatest in-plane anisotropy of r value, especially in O temper.

Effect of temper: The preform annealing process can be studied in its entirety, or broken into component parts. Because the ultimate goal is to enable the forming of more complex sheet metal parts at room temperature, we'll first look at the overall effect of the process on ductility. Figures 3c, 4c, and 5c show the uniform elongation and elongation-at-fracture for the three alloys, in both the O temper and the PA condition (i.e., annealed, then pulled to 90% of US, then annealed). In addition, they show a "total elongation", which includes both the prestrain (90% of US) and the elongation-at-fracture measured in the PA condition. Large values of total elongation would suggest that the material might be more easily formed in a multi-strike operation. The levels of total elongation achieved are on the order of 35% for CR AZ31B, 44% for CR ZEK100, and 30% for SPR AZ31B. These are all significantly greater than the measured elongations-at-fracture for the O-temper materials of approximately 23%, 25%, and 18%, respectively. Therefore, it appears that the preform anneal process would be beneficial in increasing the effective formability of these sheets.

The preform annealing process can also be analyzed by comparing the tensile test results of each material in both the O temper and the PA condition. For example, as shown in Figures 3a, 4a, and 5a, the preform annealing process very slightly reduces the UTS of all three materials. It also has little effect on YS, except in the case of the 90° bars of ZEK100, for which preform annealing reduces the YS from 125 MPa in O temper to 80 MPa in the PA condition.

The effect of preform annealing on n value and k value (from the strain-hardening equation $\sigma=k\varepsilon^n$) vary greatly with alloy and orientation, as shown in Figures 3b, 4b, and 5b. For the CR AZ31B sheet there is no effect. For SPR AZ31B, there are decreases of approximately 20% for n and 10% for k. For CR ZEK100 the decreases in n and k vary from 0-55% and 0-35%, depending on orientation.

The stated goal of the preform anneal is to "restore the ductility" of the strained material in the partially formed part. As shown in Figures 3c, 4c, and 5c, the effectiveness of the anneals varied with alloy and orientation. They were about 75% effective for CR AZ31B, 70-100% effective for ZEK100, and 65-85% effective for SPR AZ31B.

Effect of orientation: The effects on tensile properties of bar orientation relative to the sheet rolling direction are shown in Figures 3-6. There are some small effects, such as the increase of YS with orientation angle for CR AZ31B-O shown in Figure 3a, and some large ones, such as the effect on n value of CR ZEK100-O, shown in Figure 4b. The former is typically attributed to the typical crystallographic texture of rolled magnesium sheet, in which the hexagonal unit cell basal planes are largely parallel to the sheet surface, but tilted a few degrees toward the RD. This tilting facilitates basal slip and extension twinning when pulling is in the RD. The root cause of the strong orientation dependence of n and k, shown in Figure 4b, is not immediately clear. It is fascinating that preform annealing largely eliminates the dependence on orientation. The CR ZEK100-O also shows in-plane anisotropy of elongation (Figure 4c), which again is largely eliminated by preform annealing. In contrast to this, the orientation dependence of elongation of SPR AZ31B-O gets increased slightly by the preform annealing.

Considering the r values (the ratios of width true strain to thickness true strain measured at 10% elongation) shown in Figure 6, it is clear that the CR AZ31B shows greater orientation dependence than do the other alloys, especially in the O temper. Preform annealing significantly reduces this anisotropy, and reduces the r values at 45° and 90° to approximately 3. R values for the SPR AZ31B are insensitive to orientation, and increase from 1.3 to 1.7 during preform annealing. R values for the ZEK100-O sheet are about 1 at 0° and 45°, and drop to about 0.7 at 90°, opposite the trend seen in CR AZ31B. Those values increase slightly during preform annealing.

Generally speaking, the three alloy sheets exhibit different values of various tensile properties, different in-plane anisotropies of those values, and different responses to the preform annealing process. Judging from the relatively high values of elongation (29-44%) achieved when this process is applied to uniaxial tensile testing, it may be anticipated that these sheets would form well if the process were to be applied to a stamping operation.

Microstructures of the three alloys in four conditions are show in in Figure 7. (Each micrograph also includes grain size data which will be discussed below.) The top row of micrographs shows the as-annealed structures, which have nearly twin-free grains which are equiaxed. The second row corresponds to samples which received a second annealing treatment, with no intermediate deformation, and shows no great change from the once-annealed microstructures. The samples shown in the third row were annealed once, then pulled to 90% of the uniform strain measured for the annealed sheets. These show twins which are attributed to the tensile deformation, and which are most evident in the ZEK100 material, perhaps due to its large grain size. Subsequent annealing of the pulled samples leads to the microstructures shown in the fourth row. Here the twins are largely gone in the two AZ31 sheets, but less so in the ZEK100 sheet. Grain sizes changed during the post-deformation anneal, as discussed below.

Approximate grain size measurements, in microns, are shown numerically in the bottom right corner of each micrograph in Figure 7. The first number is the intercept length in the sheet normal direction, and the second is that in the sheet transverse direction. These bars were pulled in the transverse direction. The intercept length of grains of CR ZEK100 in O-tempered condition is the largest, about 18um, while the O-tempered SPR AZ31B has the finest grains with the size of ~5.5um. The second annealing (with no intermediate deformation) has little influence on the grain size of CR ZEK100 and SPR AZ31B, but for the CR AZ31B, the grain size increases from ~8um to ~10um. In the case of preform annealing shown in the fourth row of micrographs, it has totally different effects on the grain size of the three materials. For both AZ31B Mg sheets, the grain size increases during the second annealing process, compared to the pre-deformed grain size. This is attributed to recrystallization, which is influenced by the stored energy of preforming. The grain size of the SPR AZ31B increases from ~5.5um in O temper to 6.5um in the preform annealed state. In contrast, the grain size of the CR AZ31B doubles from 9 to 19um ! Things are different for CR

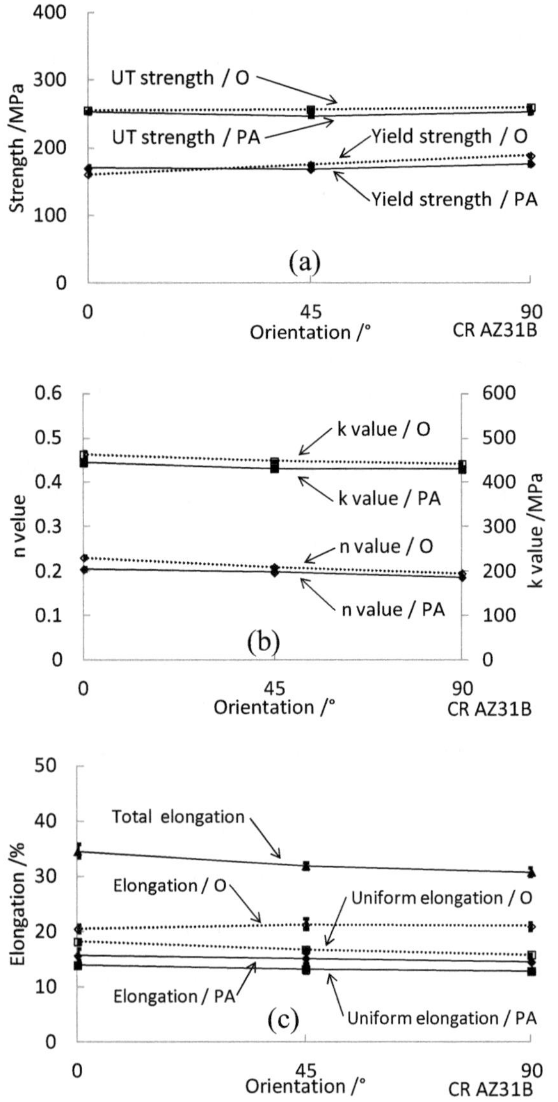


Figure 3. Effect of orientation on tensile properties of CR AZ31B rolled sheet in O-tempered and PA conditions.

ZEK100. Those grains are 18um and equiaxed after first anneal, elongated by pulling, then 17um after the preform anneal. We hypothesize that the apparent stability of grain size of the ZEK100 is due to either the Zr addition or the relatively large initial grain size. The cause of the large difference in change of grain sizes between the two AZ31 materials is not currently known.

## Conclusions

1. The preform anneal process increases the overall ductility (%el at fracture in uniaxial tension) of three rolled magnesium sheets by 30-70%. The increase varies with alloy and orientation within the sheet. The greatest ductility observed was 44% elongation, and that was with the CR ZEK100 sheet pulled at 45° or 90°.

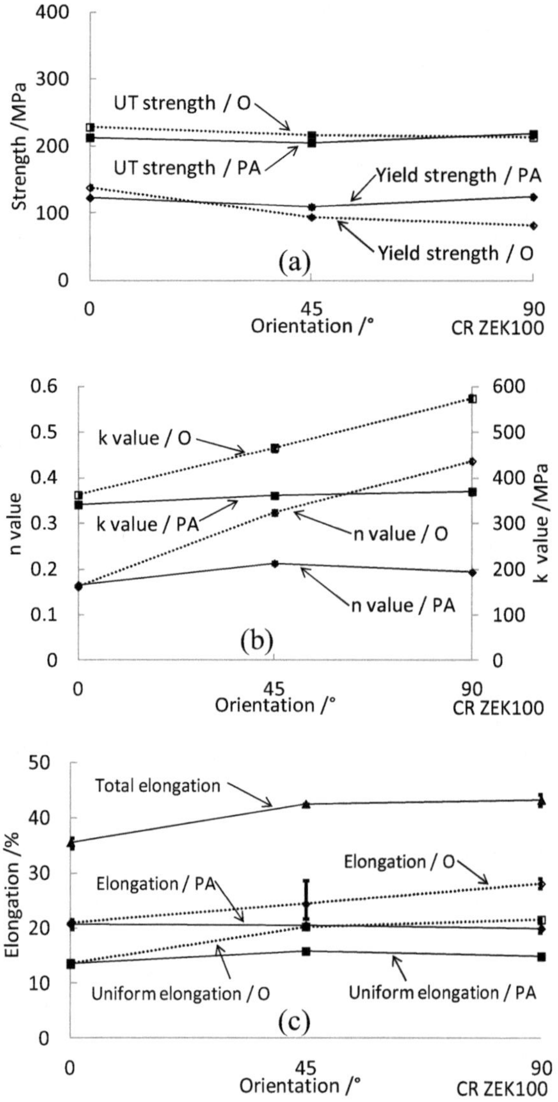


Figure 4. Effect of orientation on tensile properties of CR ZEK100 rolled sheet in O-tempered and PA conditions.

2. Generally, the preform annealing process reduces the hardness and has a little influence on the strength compared with the material in O-tempered condition

3. It reduces the in-plane anisotropy of certain tensile properties, such as the r value of CR AZ31B; and the n, k, YS, and % elong. of CR ZEK100.

4. It does not fully restore the ductility to that of the O-temper material, but typically restores at least 70% of that.

5. The effect of preform annealing on grain size also differs from material to material. The grain size of CR ZEK100 dropped slightly from 18 to 17 microns, that of SPR AZ31B increased slightly from 5.5 to 6.5 microns, and that of CR AZ31B doubled from 9 to 19 microns.

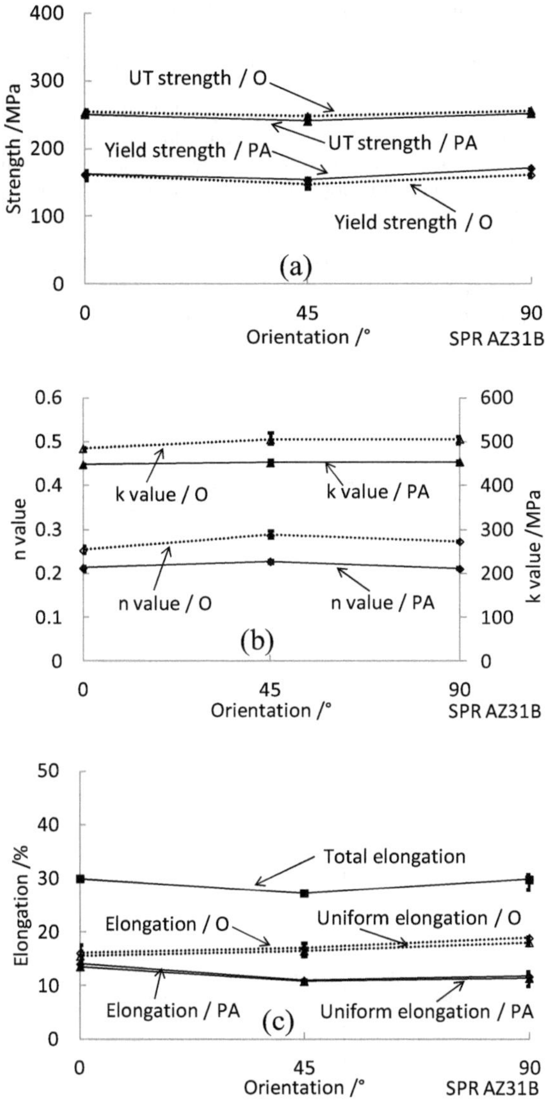


Figure 5. Effect of orientation on tensile properties of SPR AZ31B rolled sheet in O-tempered and PA conditions.

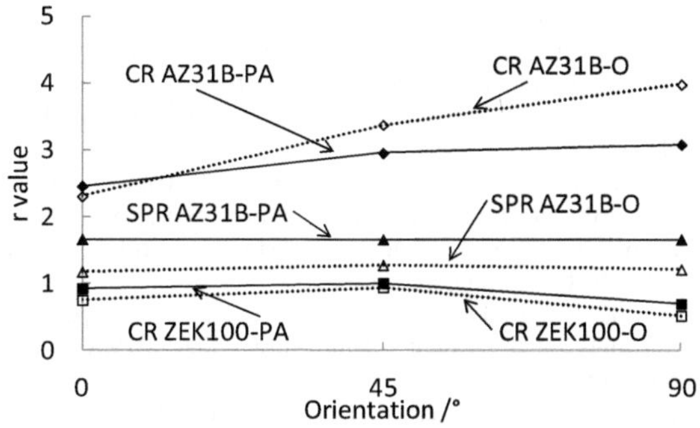


Figure 6. Comparison of r value in each orientation of three rolled magnesium alloy sheets in O-tempered and PA conditions.

### References


1. A.A. Luo, "Magnesium: Current and Potential Automotive Applications", JOM, February 2002, pp. 42-48.
2. P.J. Blanchard, G.T. Bretz, S. Subramanian, J.E. deVries, A. Syvret, A. MacDonald, P. Jolley, "The Application of Magnesium Die Casting to Vehicle Closures", SAE Paper No. 2005-01-0338
3. http://media.ford.com/article_display.cfm?article_id=30231.
4. J.T. Carter, A.R. Melo, V. Savic, L.G. Hector, Jr., P.E. Krajewski, "Structural Evaluation of an Experimental Aluminum/Magnesium Decklid", SAE Paper No. 2011-01-0075.
5. S. Logan, A. Kizyma, C. Patterson, S. Rama, "Lightweight Magnesium Intensive Body Structure", SAE Paper No. 2006-01-0523.
6. www.superlightcar.com/public/docs/SLC_presentation_TRA08_VW_March08.pdf
7. A.A. Luo, E.A. Nyberg, K. Sadayappan, W. Shi, "Magnesium Front End Research and Development: A Canada-China-USA Collaboration", in Magnesium Technology 2008, eds. M.O. Pekguleryuz, N.R. Neelameggham, R.S. Beals, and E.A. Nyberg, TMS, Warrendale, PA, 2008, pp. 3-10.
8. http://en.wikipedia.org/wiki/Porsche_Carrera_GT.
9. R. Verma and J.T. Carter, "Quick Plastic Forming of a Decklid Inner Panel with Commercial AZ31 Mg Sheet", SAE Paper No. 2006-01-0525.
10. Automotive Material Data Sheet, AZ31B-O, Issue 1, 2006, Superform Aluminium.
11. Paul Krajewski, Peter Friedman, Kenneth Oikarinen, Dennis Cedar, "USAMP Project AMD307: Pan Forming of Aluminum and Magnesium", Paper presented at MS&T'07, September 16-20, 2007, Detroit, Michigan.
12. Stolfig Group, patents DE 101 56 034.6, DE 102 47 129.0, CN 152 76 83 A, Nr. DE 302 08 803, DE 202 02 591.8
13. Y. Liu, Y. Li, W. Li, "Stamping Formability of ZE10 Magnesium Alloy Sheets Journal of Rare Earths, Vol. 25, 2007, pp. 480-484.
14. J. Zhang, X. Zhang, W. Li, F. Pan, Z. Guo, "Partition of Er among the constituent phases and the yield phenomenon in a semi-continuously cast Mg-Zn-Zr alloy". Scripta Materialia, Vol. 63, 2010, pp. 367-370.
15. N. Stanford, D. Atwell, M.R. Barnett, "The effect of Gd on the recrystallisation texture and deformation behaviour of magnesium-based alloys". Acta Materialia, Vol. 58, 2010, pp. 6773-6783.
16. Raja K. Mishra, Anil K. Gupta, P. Rama Rao, Anil K. Sachdev, Arun M. Kumar, Alan A. Luo, "Influence of cerium on texture and ductility of magnesium alloy extrusions." In Magnesium Technology 2008, 269-274. Warrendale, PA: The Minerals, Metals & Materials Society, 2008.
17. J. Bohlen, M.R. Nurnberg, J.W. Senn, D. Letzig, S.R. Agnew, "The texture and anisotropy of magnesium–zinc–rare earth alloy sheets". Acta Materialia, Vol. 55, 2007, pp. 2101-2112.
18. T. Al-Samman, X. Li, "Sheet texture modification in magnesium-based alloys by selective rare earth alloying". Materials Science and Engineering A, Vol. 528, 2011, pp. 3809-3822.
19. "Development of a New Rolling Process for Commercial Magnesium Alloy Sheets with High Room-temperature Formability", (Translation of AIST press release of January 26, 2010), http://www.aist.go.jp/aist_e/latest_research/2010/20100217/20100217.html.

20. H. Utsunomiya, T. Sakai, S. Minamiguchi, and H. Koh, "High-speed heavy rolling of magnesium alloy sheets", Magnesium Technology 2006, edited by A.A. Luo, N.R. Neelameggham, and R.S. Beals, TMS, Warrendale, PA, 2006, pp. 201-204.

21. B. Beausir, S. Biswas, D.I. Kim, L.S. Toth, S. Suwas, "Analysis of microstructure and texture evolution in pure magnesium during symmetric and asymmetric rolling", Acta Materialia, Vol. 57, 2009, pp. 5061-5077.

22. Paul E. Krajewski, Method for production of stamped sheet metal panels, US Patent Application 2005/0199032 A1, September 15, 2005.

23. Theresa M. Lee, Susan E. Hartfield-Wünsch, Siguang Xu. Demonstration of the Preform Anneal Process to Form a One-Piece Aluminum Door Inner Panel. SAE Technical Paper 2006-01-0987.

24. John E. Carsley, Susan Hartfield-Wunsch, Theresa M. Lee, A Demonstration of Local Heat Treatment for the Preform Annealing Process, SAE Technical Paper 2011-01-0538.

25. J.J. Li, Y. Zhou, S.J. Hu, L.E. Izquierdo, P.E. Krajewski, and T.M. Lee, Optimization of localized annealing for preform anneal forming of aluminum alloys, Transactions of the North American Manufacturing Research Institution of SME, 2010, 411-418.

26. Jingjing Li, Sooho Kim, Theresa M. Lee, Paul E. Krajewski, Hui Wang, S. Jack Hu. The effect of prestrain and subsequent annealing on the mechanical behavior of AA5182-O. Materials Science and Engineering A 528 (2011) 3905–3914.

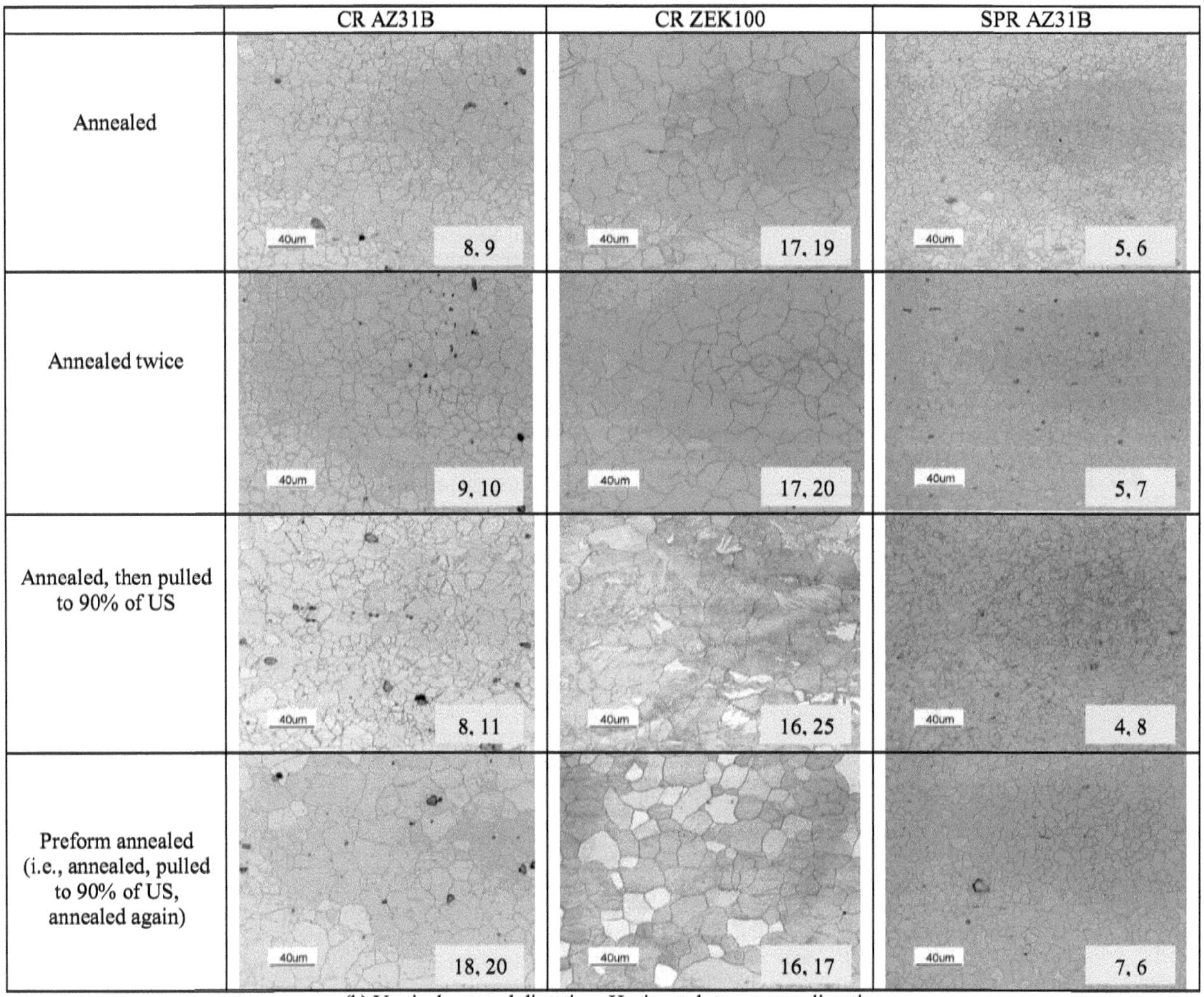


(b) Vertical: normal direction; Horizontal: transverse direction. .

Figure 7. Photomicrographs of tensile bars of three magnesium alloy sheets in four different stages of processing. In each image, the sheet ND is vertical and the sheet TD is horizontal. Approximate grain sizes, in microns, are shown for measurements made in the ND and TD.

**Magnesium Technology 2012**
*Edited by: Suveen N. Mathaudhu, Wim H. Sillekens, Neale R. Neelameggham, and Norbert Hort*
***TMS (The Minerals, Metals & Materials Society), 2012***

# The Role of Intermetallics on Creep Behaviour of Extruded Magnesium Alloys

Michelle Fletcher[1], Lukas Bichler[1], Dimitry Sediako[2]
[1]UBC Okanagan; 1137 Alumni Way; Kelowna, British Columbia, V1V 1V7, Canada
[2]NRC-CNBC; Chalk River Laboratories; Chalk River, Ontario, K0J 1J0, Canada

Keywords: Magnesium, Neutron Diffraction, Creep, Microstructure

## Abstract

This research examined in-situ creep behavior of three extruded high performance magnesium alloys (AE42, AJ32 and ZE10). Neutron diffraction was used to measure compressive creep behavior at 175°C in the extrusion direction. The AE42 and AJ32 alloys exhibited higher creep strains than the ZE10 alloy. The highest strain was recorded for AE42 (2.4%), while ZE10 exhibited greatest creep resistance (0.2% strain). Microstructure analysis has shown that the distribution and composition of secondary phases was critical for creep resistance. The aluminum containing alloys had acicular and globular intermetallics, whereas the ZE10 alloy contained fine and irregular intermetallics dispersed along grain boundaries, effectively contributing to pinning of grains under high temperature loads. Significant grain re-crystallization was also observed in the aluminum containing alloys, but was absent in ZE10.

## Introduction

Ever increasing demand for energy efficient, yet high performance vehicles in the automotive and aerospace industries has resulted in a high demand for light weight structural components. Magnesium (Mg) alloys are the lightest structural materials available with a high specific strength. Further, Mg alloys are easy to machine and cast, with a good potential for recycling [1], establishing themselves as good candidate materials for diverse industrial applications. However, Mg alloys do not have the required creep resistance for automotive applications with operating temperatures above 125°C [2]. This restriction greatly limits their use for automotive components, such as engine block parts, where the greatest impact on fuel efficiency and performance could be achieved [3]. The poor creep strength has been traditionally related to grain boundary movement and plastic deformation which leads to inter-granular failure [4].

### Effect of Alloying Elements on Creep Performance

The addition of aluminum (Al) has been known to increase castability and room temperature strength; however, this strength drastically diminishes above 120°C [3]. The main contributing factor to this drop in strength has been related to the presence of the metastable $Mg_{17}Al_{12}$ intermetallic compound (β phase) along the grain boundaries. This eutectic compound has a low incipient melting temperature, thus enabling relative sliding of adjacent grains during creep [5]. The $Mg_{17}Al_{12}$ precipitates form in a lamellar structure along grain boundaries as well as within grains through a solid state phase transformation [6].

The addition of rare earth elements (REs) such as lanthanum, praseodymium, cerium, and neodymium are known to increase creep resistance of Al-containing Mg alloys. Their beneficial effect has been attributed to the formation of thermally stable intermetallic compounds of diverse morphologies along the grain boundaries. These compounds tend to take the composition of $(Mg, Al)_xRE_y$ due to the low solubility of REs in the Mg matrix [7]. The formation of $Al_xRE_y$ decreases the Al concentration within the alloy matrix, thereby decreasing the amount of available Al for the formation of $Mg_{17}Al_{12}$ [6]. This results in an improvement of creep resistance.

The addition of strontium (Sr) to Al-containing Mg alloys has a similar effect on the creep resistance as adding REs. The Sr addition causes the formation of thermally stable $Mg_xAl_ySr_z$ intermetallic compounds along the grain boundaries [8]. This inhibits Al saturation within the matrix and subsequent $Mg_{17}Al_{12}$ formation. One benefit of adding Sr over REs is the decreased relative cost associated with the Sr addition.

Zinc (Zn) addition to Mg alloys increases room temperature strength, forms intermetallic particles along grain boundaries and raises the eutectic temperature. The increased eutectic temperature raises the alloy operating temperature [9]. However, Zn addition also increases the alloy's susceptibility to microporosity and embrittlement [5], thus negatively affecting alloy performance.

The addition of REs to Zn-containing Mg alloys increases creep resistance through the formation of thermally stable intermetallic compounds and decreases the susceptibility to microporosity and alloy embrittlement [6]. In these alloys the intermetallics tend to take the composition of $(Mg, Zn)_xRE_y$ and form along the grain boundaries, effectively pinning grains at elevated temperatures as high as 420°C [1, 9].

In this research, the morphology and distribution of intermetallic compounds in the AE42, AJ32 and ZE10 alloys was investigated and related to the compressive creep performance of the alloys.

## Experimental Setup

The alloys analyzed in this research were produced by Timminco Corp., and their nominal compositions are provided in Table I. The three alloys studied (AE42, AJ32 and ZE10) were processed via hot extrusion. No processing defects, such as cracking or solute segregation were observed. Material samples with a torus geometry were produced, as seen in Figure 1a, with an outer and inner diameter of 25 mm and 7 mm, respectively. The microstructure was studied by sectioning the sample into radial and cross-section directions, as can be seen in Figure 1b. Of particular interest was the cross-section surface, since grain elongation in the cross-section direction coincided with the extrusion direction.

Table I. Alloy compositions.

| | AE42 | AJ32 | ZE10 |
|---|---|---|---|
| Al | 3.5-4.5 | 3-3.5 | ≤0.05 |
| Zn | ≤0.2 | ≤0.3 | 2.5-3.1 |
| Mn | 0.2-0.5 | 0.2-0.5 | 0.2-0.5 |
| Fe | ≤0.01 | ≤0.01 | ≤0.01 |
| Ni | ≤0.005 | ≤0.005 | ≤0.005 |
| Cu | ≤0.05 | ≤0.05 | ≤0.05 |
| Si | ≤0.05 | ≤0.05 | ≤0.05 |
| RE | 1.5-2.5 | | 2.5-3.5 |
| Sr | | 1.5-2.5 | |
| Zr | | | 0.45-0.7 |

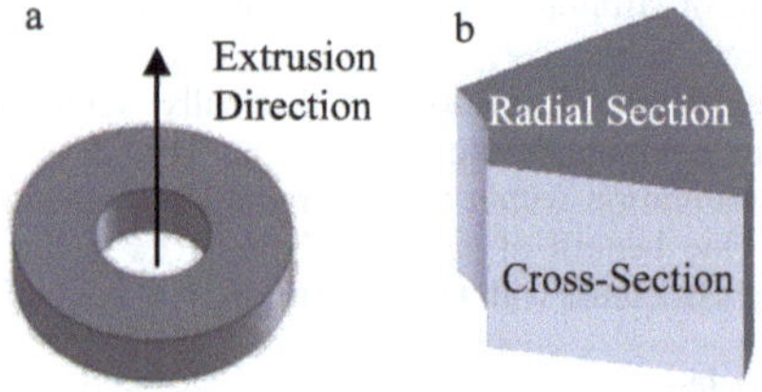


Figure 1. a) Extruded torus and b) sample surfaces.

Neutron Diffraction: Strain Measurement

Compressive creep strain measurement was performed at the Canadian Neutron Beam Center (CNBC) in Chalk River, ON, Canada. A detailed description of the testing procedure was published earlier [10]. Elastic strain ($\varepsilon$) of the atomic lattice was measured using the peak shift method. This method calculates strain using the change in lattice spacing, as seen in Equation 1:

$$\varepsilon = {(d_0 - d)}/{d_0} \quad (1)$$

Where $d_0$ and $d$ are the lattice spacing before and after creep, respectively. The lattice spacing was calculated using Bragg's law (Equation 2):

$$n\lambda = 2dsin(\theta) \quad (2)$$

Where $\lambda$ is the neutron wavelength (1.792 Å), n=1 for first order diffraction, and $\theta$ is the neutron scattering angle measured during the neutron diffraction experiment.

The strain in the $10\bar{1}0$ and $2\bar{1}\bar{1}0$ planes were measured parallel to the applied load in the extrusion direction during the compressive creep test.

The experimental setup can be seen in Figure 2, where the sample was held at 175°C and a compressive load of 50MPa was applied. An extensometer was also attached to the sample to measure the total sample deformation (plastic and elastic combined) over the test period of 24 hours.

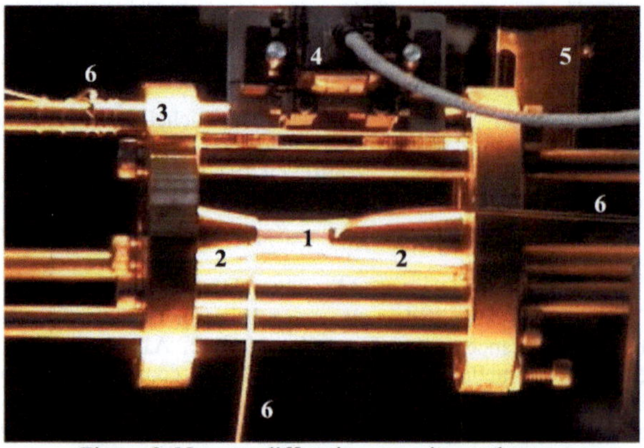


Figure 2. Neutron diffraction experimental setup 1) sample, 2) clamps, 3) loader, 4) extensometer, 5) neutron incidence slit, and 6) thermocouples.

Microscopic Evaluation

As-extruded samples were sectioned, mounted, polished, and etched using standard procedures for Mg samples. Optical microscopy was performed using a Zeiss AxioObserver A1m microscope with differential interference contrast (DIC) capability. Image analysis was then performed using Buehler Omnimet software. SEM analysis was also performed, using a JEOL-6410 SEM with XEDS capability.

## Results and Discussion

Extensometer Results

The total compressive creep deformation (measured using an extensometer) of the alloys under constant load and temperature during testing is plotted in Figure 3. From this figure it can be seen that AE42 shows the highest compressive creep strain of 2.4% after 24 hours of testing. The second highest deformation was measured for AJ32 at 0.6%, followed by ZE10 at 0.2%. The AE42 alloy also had the highest deformation rate during the creep experiment, as measured by the slope of the deformation curve. In contrast, the ZE10 alloy exhibited a high degree of stability throughout the creep experiment.

Figure 3 was also used to estimate the onset of secondary creep deformation by analyzing the creep rate. The results are provided in Table II. The relatively long duration of the primary creep phase in AE42 and ZE10 suggests that work hardening processes were proceeding at much slower rate than in AJ32, which had the shortest primary creep interval. However, the mechanisms of work hardening in the AE42 and ZE10 alloys were possibly unique, due to the size, shape and distribution of second phases, as will be discussed in the following sections of the paper. Further, the magnitude of strain rate difference between the AE42 and ZE10 alloys also suggests that the creep mechanism itself was unique for individual alloys. Since both alloys had a similar grain size, the role of solutes on the diffusion of vacancies or solute in the alloys was likely of relevance. In the AE42 alloy, Al is the primary solute with a nearly 100% solubility in Mg; In contrast, the solubility of REs or Zn in Mg is an order of magnitude lower. Thus, it is possible that the AE42 alloy experienced significant grain boundary diffusion (Coble) creep in addition to dislocation creep, which is controlled by the movement of dislocations. In the case of AJ32 and ZE10, it is believed that dislocation creep was dominant, since creep deformation was mainly accommodated by plastic deformation, as will be discussed in the following section.

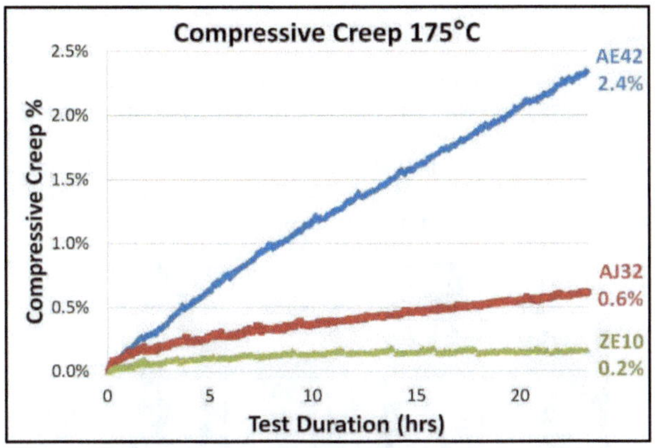


Figure 3. Total sample deformation at 50MPa and 175°C measured with extensometer.

Table II. Onset of secondary creep.

| Alloy | Onset of Secondary Creep (hrs) |
|---|---|
| AE42 | 7.0 |
| AJ32 | 3.5 |
| ZE10 | 6.0 |

Neutron Diffraction Results

Neutron diffraction was used to measure the compressive strain at several stages during the creep test:

1. Sample under no external load while the temperature increased from 25°C to 175°C.
2. Upon application of external load (50MPa), at 175°C.
3. During the creep test (50MPa at 175°C for 24 hours).
4. Load release and sample cooling to 25°C.

These stages are indicated in Figure 4, which shows the compressive creep strain results for the $(10\bar{1}0)$ crystallographic plane. Similar trend was observed for the $(2\bar{1}\bar{1}0)$ planes represented in Figure 5.

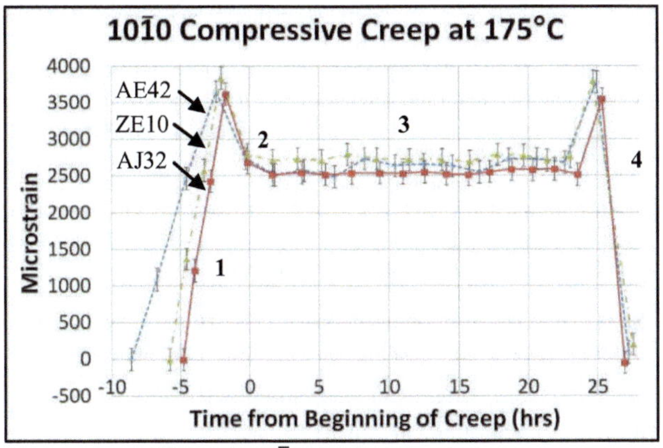


Figure 4. $10\bar{1}0$ elastic creep strain.

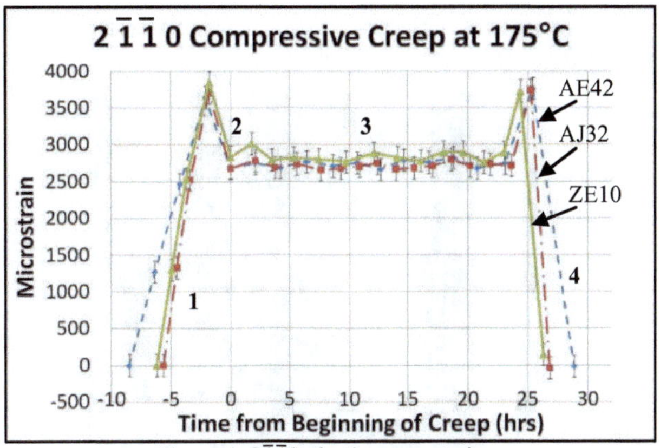


Figure 5. $2\bar{1}\bar{1}0$ elastic creep strain.

As can be seen in Figure 4 and Figure 5, the elastic strains of the three alloys follow similar trends throughout the test. The strain increased under pure thermal loading from 25°C to 175°C. Once the external load was applied a decrease in the strain was recorded as the material compressed elastically. During constant temperature and load conditions, the elastic strains for AJ32 and ZE10 remained nearly constant and within the margin of error as outlined by the error bars. This constant strain value indicates that the total creep deformation observed in Figure 3, was likely the result of plastic deformation occurring along the grain boundaries, again suggesting that dislocation creep dominated in these alloys.

The AE42 alloy exhibited a slight increase in strain after approximately 7 hours of creep. This change in strain corresponds to the onset of secondary creep in AE42, as seen in Table II. It is likely that the secondary creep deformation relieved some compressive strain within the alloy matrix, which allowed the lattice spacing to expand. The majority of material deformation, however, was plastic deformation along the grain boundaries, as the change in strain in Figure 4 does not account for the large creep deformation seen in Figure 3.

Once the 50MPa load was released, the strain in all three alloys increased back to the values seen before external loading was applied. When the material temperature decreased to 25°C the strain returned to zero. This indicates that there was no residual strain present in the material after creep testing and any deformation which was recorded (in Figure 3) was the result of plastic deformation.

Metallographic Analysis

The as-extruded cross-section microstructure of the three alloys is shown in Figure 6. The extrusion direction is aligned vertically. The figures show that in all three alloys, significant grain texture evolved with grain elongation in the extrusion direction.

Figure 6 shows that re-crystallization has occurred in both Al-containing alloys after extrusion. In general, re-crystallization is accompanied by a reduction of strength and a simultaneous increase of ductility. These two factors would have a significant impact on the subsequent creep performance of the Al-containing alloys. In the case of the ZE10 alloy, no re-crystallization (or sub-grains) was observed, suggesting that this alloy has retained work-hardened microstructure (imposed during the extrusion process) with a concomitant retention of extensive dislocation network in the material. Such dislocation network would likely contribute to the superior creep resistance of the ZE10 alloy.

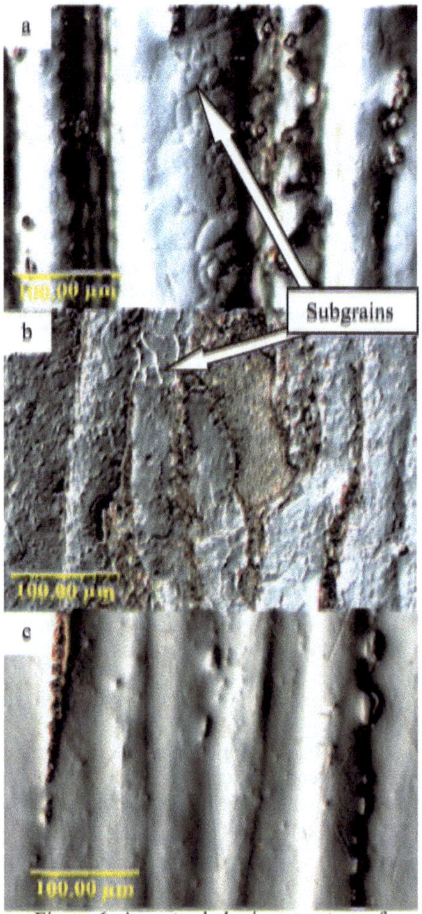


Figure 6. As-extruded microstructure of a) AE42, b) AJ32 and c) ZE10 at 500x.

The average alloy grain widths are provided in Table III. It can be seen that AE42 had ~ 18% smaller grain size than the remaining alloys. A small grain size is known to be detrimental for creep-strength of an alloy, as more grain boundary area where sliding can occur is formed.

Table III. Average grain width.

| Alloy | Grain Widths (µm) |
|---|---|
| AE42 | 12.9 ± 0.3 |
| AJ32 | 16.5 ± 0.3 |
| ZE10 | 15.6 ± 0.3 |

The intermetallic compound $Mg_{17}Al_{12}$ was found in both AE42 and AJ32, as can be seen in Figure 7. In AE42, $Mg_{17}Al_{12}$ formed primarily around polygonal $Al_4RE$ particles, as can be seen in Figure 7a. These polygonal RE-rich particles could have undergone thermal degradation to $Al_2RE$ and $Mg_{17}Al_{12}$, resulting in the appearance of the β phase along the grain boundaries, which would negatively impact the grain boundary strength and contribute to sliding during creep.

AJ32 contained a significantly lower amount of $Mg_{17}Al_{12}$ phase, which was observed in the vicinity of the $Mg_xAl_ySr_z$ particles, as can be seen in Figure 7b. Again, it is expected that diffusion of Al from the $Mg_xAl_ySr_z$ particles enabled formation of the β-phase, however the high thermal stability of the Sr-rich intermetallic was seen to retain the Al within the compound.

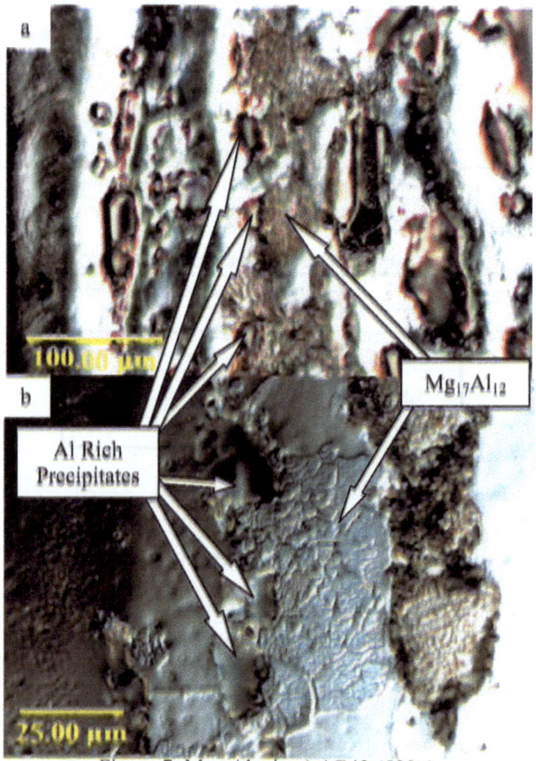


Figure 7. $Mg_{17}Al_{12}$ in a) AE42 (500x) and b) AJ32 (1500x).

SEM imaging was used to investigate the morphology of intermetallic compounds in the three alloys. The intermetallics in the AE42 (Figure 8a) were acicular, globular and polygonal in shape. The two main intermetallics identified by XEDS technique were the $Al_2RE$ and $Al_4RE$ (also reported as $Al_{11}RE_3$). Both $Al_2RE$ and $Al_4RE$ were seen to agglomerate along the grain boundaries. The polygonal particles were likely ineffective in grain pinning. However, acicular $Al_4RE$, seen in Figure 9, were seen to connect adjacent grains. Thus, it is believed that these intermetallics would effectively pin the grain boundaries.

SEM imaging also showed that AJ32 (Figure 8b) contains intermetallics of two distinct shapes and sizes; large, globular, polygonal intermetallics as well as fine, jagged and irregularly shaped intermetallics. These two different intermetallics take on different compositions of $Mg_xAl_ySr_z$ and tend to group together with the large particles usually surrounded by the fine and irregular intermetallics. The fine intermetallics have a large surface area, providing more resistance to grain boundary sliding. The irregular shapes of these fine particulates also contribute to a higher ability to pin grains. The high thermal stability of these intermetallics and the fact that they are not surrounded by $Mg_{17}Al_{12}$, results in an increase in creep resistance compared to AE42. Currently it is not clear if the extrusion process was responsible for fracturing the fine particles.

The precipitates found throughout ZE10 (Figure 8c) consist of irregular shaped particles (~10μm) along grain boundaries as well as extremely fine intermetallic particulates (~1μm) along grain boundaries and throughout the matrix. These intermetallics along with the absence of $Mg_{17}Al_{12}$ likely resulted in a high resistance to dislocation climb or glide during plastic deformation of the material, thereby improving the creep resistance of the alloy.

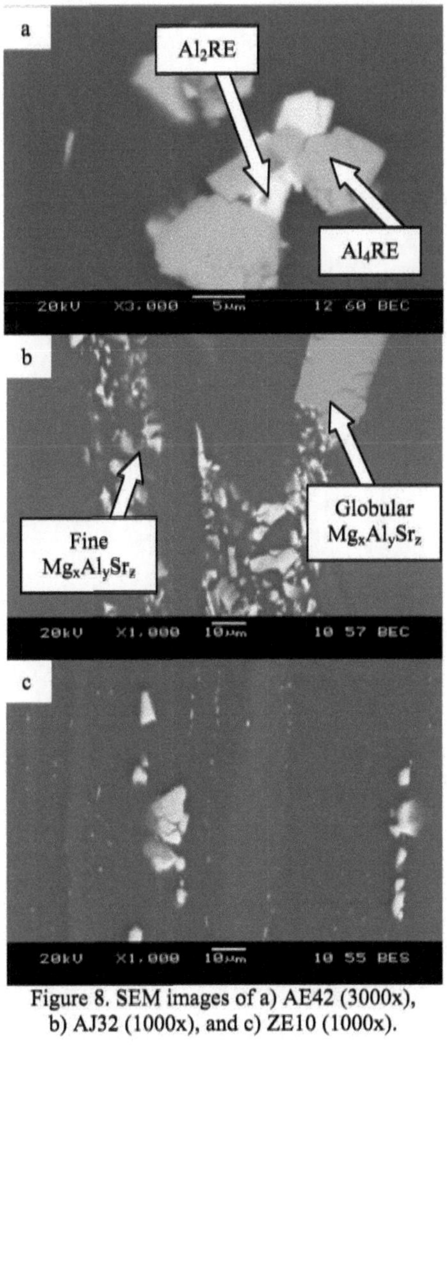


Figure 8. SEM images of a) AE42 (3000x), b) AJ32 (1000x), and c) ZE10 (1000x).

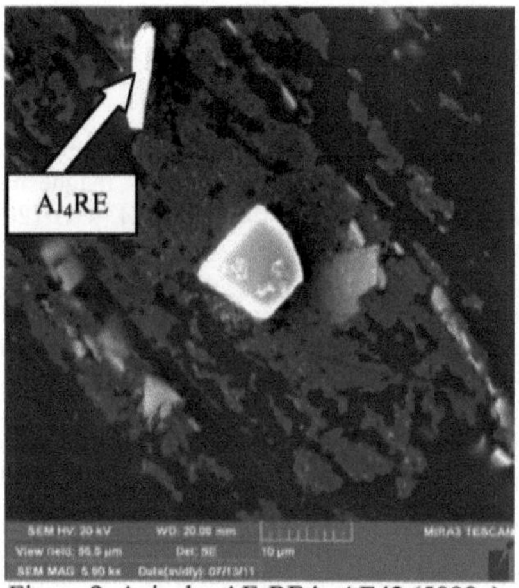


Figure 9. Acicular $AE_4RE$ in AE42 (5000x).

## Conclusions

1. AE42 showed the highest amount of plastic deformation during creep followed by AJ32 and ZE10.

2. Creep deformation primarily took the form of plastic deformation along grain boundaries, indicating the grain boundaries as being the weakest point for creep resistance.

3. No residual strain was present in the alloys post-creep testing.

4. Both AE42 and AJ32 exhibited re-crystallization in their as-extruded microstructure.

5. Significant amounts of $Mg_{17}Al_{12}$ was found surrounding $Al_2RE$ and $Al_4RE$ particles and along grain boundaries throughout AE42, therefore suggesting that the grain-pinning effect of the $Al_xRE_y$ intermetallics was counter-weighted by the presence of β phase during high temperature creep.

6. AJ32 and AE42 contained acicular and globular intermetallics between grains whereas ZE10 contained fine and irregular intermetallics with a higher grain pinning efficiency.

## References

1. Michael Avedesian, and Hugh Baker, *ASM Specialty Handbook; Magnesium and Magnesium Alloys* (Materials Park, OH: ASM International, 1999).

2. Q. Guo et al., "Elevated temperature compression behavior of Mg-Al-Zn alloys," *Materials Science and Technology*, 22 (6) (2006), 725-729.

3. Edward F. Emley, *Principles of Magnesium Technology* (Long Island City, NY: Pergamon Press, 1966).

4. J. Yan et al., "Creep deformation mechanism of magnesium-based alloys," *Journal of Material Science*, 43 (2008), 6952-6959.

5. H. Deming et al., "Indentation creep behavior of AE42 and Ca-containing AE41 alloys," *Materials Letters*, 61 (2007), 1015-1019.

6. B.R. Powell et al., "Microstructure and creep behavior in AE42 Magnesium die-casting alloy," *Journal of Metals*, 54 (8) (2002), 34-38.

7. B. Nami et al., "Effect of Ca and rare earch elements on impression creep properties of AZ91 magnesium alloy," *Metallurgical and Materials Transactions A*, 41A (2010), 1973-1982.

8. A.A. Luo, "Recent magnesium alloy development for elevated temperature applications," *International Materials Reviews*, 49 (1) (2004), 13-30.

9. T. Rypaev et al., "Microstructure of superplastic QE22 and EZ33 magnesium alloys," *Materials Letters*, 62 (2008), 4041-4043.

10. D. Sediako and M. Gharghouri, "Neutron Diffraction Measurements of Residual Stress in Creep Resistant Mg Alloys," *Magnesium Technology 2008*, TMS, 2008, 407-409.

**Magnesium Technology 2012**
*Edited by: Suveen N. Mathaudhu, Wim H. Sillekens, Neale R. Neelameggham, and Norbert Hort*
***TMS (The Minerals, Metals & Materials Society), 2012***

# HIGH PERFORMANCE MG-SYSTEM ALLOYS FOR WEIGHT SAVING APPLICATIONS: FIRST YEAR RESULTS FROM THE GREEN METALLURGY EU PROJECT

Fabrizio D'Errico[1], Gerardo Garces Plaza[2], Markus Hofer[3], Shae K. Kim[4]
[1]Politecnico di Milano, Department of Mechanical Engineering, Via La Masa 34, 20156 Milan, Italy
[2]Centro Nacional de Investigaciones Metalúrgicas, Department of Physical Metallurgy, Avda. Gregorio del Amo, 8, Madrid 28040
[3]Buhler AG, New Technologies AMT Advanced Materials Division, Gupfenstrasse 5, Uzwil 9240, Switzerland
[4]Casting Research Center, Korea Institute of Industrial Technology, 7-47 Songdo-dong, Yeonsu-gu, Incheon 406-840, Korea

Keywords: Magnesium alloys, ultrafine microstructure, sustainability, $CO_2$ reduction

## Abstract

The GREEN METALLURGY Project, a LIFE+ project co-financed by the EU Commission, has just concluded its first year. The Project seeks to set manufacturing processes at a pre-industrial scale for nanostructured-based high-performance Mg-Zn(Y) magnesium alloys. The Project's goal is the reduction of specific energy consumed and the overall carbon-footprint produced in the cradle-to-exit gate phases. Preliminary results addressed potentialities of the upstream manufacturing process pathway. Two Mg-Zn(Y) system alloys with rapid solidifying powders have been produced and directly extruded for 100% densification. Examination of the mechanical properties showed that such materials exhibit strength and elongation comparable to several high performing aluminum alloys; 390 MPa and 440 MPa for the average UTS for two different system alloys, and 10% and 15% elongations for two system alloys. These results, together with the low-environmental impact targeted, make these novel Mg alloys competitive as lightweight high-performance materials for automotive components.

## Introduction

The European Project "Green Metallurgy", a 2 ML€ project co-financed by the European Community, is one of the successful proposals submitted to the EU LIFE+ 2009 Programme. It was presented at the 2011 TMS Annual Exhibition, Magnesium Technology 2011 Symposium. The goal of the 3-year project, which officially started in September 2010, is to overcome the primary historic barriers that currently prevent magnesium from being widely applied as an ultra-lightweight material for the transport sector. The objective of the Green Metallurgy project is to demonstrate that it is possible to manufacture structural elements that are complex in shape using high performance magnesium alloys. Thanks to the Cenim team, nanostructured magnesium based alloys have been developed with drastically increased mechanical properties suitable for scaled-up processes. These lighter structural materials can substitute high performance aluminum alloys generally used in the transport sector and also substitute some low-carbon steel-made products, with greater weight saving possibilities. An additional objective is to set up and use a process route that targets a general reduction of energy intensity, and move the material from the laboratory directly to actual industrial use. A general reduction of total specific energy consumed and the overall carbon-footprint produced in cradle-to-exit gate phase is also targeted.

With the possibility to develop a fully integrated process lay-out, namely single or double process step prototype machinery, and demonstrate the possibility to industrialize an innovative single-step manufacturing process (or quasi double-step process for sheet parts) with a substantially reduced carbon footprint. With the additional capability to demonstrate a completely new nano-Powder Warm Compaction process (nano-PWC) route for Mg based products, based on a combination of warm extrusion and superplastic press-forging, the project could ultimately achieve:

- the avoidance of a secondary melting step,
- reduction of more than 25% of the process energetic input,
- elimination of any green-house gasses,
- facilitation of recycling operations,

which would lead to a compact and lean integrated product cycle that drastically reduces the number of process step and eliminates logistical displacement.

The key features of the new generation of magnesium alloys are the contemporary reduction of energy intensity per kg of finished part produced, the elimination of green-house gasses in the secondary production process steps (downstream steps after primary raw magnesium extraction and master alloy generation), dramatic increases in the specific strength (YS and UTS vs density) ratio while maintaining safe values in elongation and strength. The positive impact from the increased usage of such featured materials has already been evaluated. It is feasible to reduce the $CO_2$ emission approximately 15-20% over the total lifespan of such lighter weight vehicles as these multi-materials mature, especially in the design phase and could potentially reduce the weight of an average vehicle with an internal combustion engine by approximately 180 kg [1-4].

## Perspective of Green Metallurgy process

Once past the primary cast step necessary for production during the magnesium extraction and master alloy generation, various process steps using different equipment often in different plants, are needed for conventional manufacturing routes to produce products needed.

For example, in Figure 1, to use process Route 1 to fabricate a magnesium cast-wheel it is necessary to start with ingots (raw materials), which are then melted at a temperature up to 600°C in a furnace using cover gases. Melted magnesium is then cast inside a die, leaving excess cast materials as recyclable scrap. Furthermore, depending on the final use of the component, further machine work and finishing operations may be required. Alternatively, to increase the mechanical properties that are too low for cast structure, a secondary forging step, perhaps using another machine or another plant, may be necessary. More complex is the route 2 shown in figure 1, but it is needed when higher mechanical property material is required. In fact a hot formed semi-finished product can achieve good mechanical properties up to 230 MPa Yield Strength and 11% elongation. Some semi-finished products currently available require long and complex machining operations.

Compared to the above mentioned routes, the goal of the Green Metallurgy process is to realize semi-finished products, bar and plate forms, starting with raw materials in the form of rapidly solidified powders[1] with very refined microstructure by employing a single-step[2] process. Furthermore, the already tested [5, 6] superplastic behavior of such semi finished products would allowed to net-shape complex component by deploying forging in superplastic regimes.

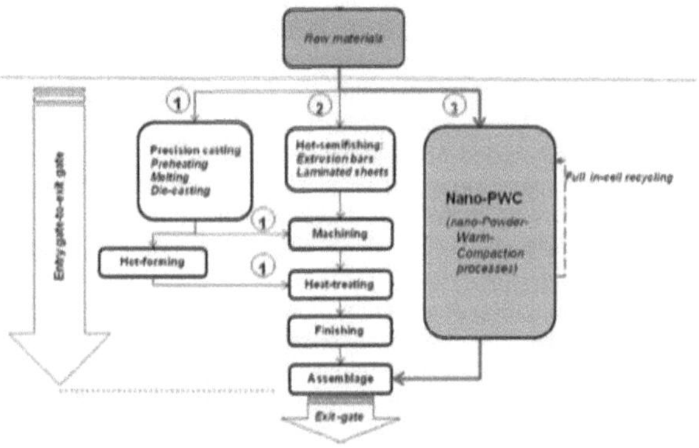


Figure 1. Schematic routes for common mass-part manufacturing: 1) casting into die and post-hot forging; 2) machining from semi-finished products; 3) GREEN METALLURGY process. The thick line identifies the Project manufacturing green-path for magnesium mass-part production.

With this premise, the basic Green Metallurgy process stages to be implemented into the final lay-out are described below :

1. raw material fabrication: a) atomized powders by rapid solidifications, namely the Basic Alloy System MgZn(Y) alloy based; b) machined chips of Eco-Magnesium® for the Alternative Alloy System to be tested.
2. Fabrication of Precursor Material (PMT);
3. Direct extrusion of PMT into semi-finished bars (SSB).
4. Secondary shaping operations (rolling for plates and forging for complex net-shape parts).

A fundamental innovation provided by the project is related to the environmental impact of Mg based alloys process manufacturing, compared to classic approaches. In the second year of the project, a Life Cycle Analysis is planned to measure the environmental impact of the best solution process lay-out resulting from the current preliminary actions.

[1] Further test campaign on machined chips of Eco-Magnesium® alloys is already planned to be set in parallel with test campaign on Basic Alloy System made of atomized powders of MgY (Zn) alloys.

[2] Single or two-step lay-out process will actually define when optimized lab-scale viable layouts shall be analyzed in order to be scale-up to pre-industrial process lay-out considering economical and technological constraints.

**Executive summary of project actions performed**

Based on previous lab-scale experiences developed in the laboratory, the early process-layout is established using the following steps:

a) Preparation of precursor materials (PMT) made of cold compaction of atomized powders produced by conventional casting of master alloys, Basic Alloy System materials, BAS; BAS are produced basing on two chemical compositions (see table I) developed at CENIM laboratories. In figure 2 are details of the micro-structure of rapid solidified powders, while figure 3 shows the PMT result after the cold compaction of BAS powders.

Table I. Nominal chemical composition finely designed at Cenim laboratories.

| Element | BAS1 (MgZn6Y1) | BAS2 (MgZn8Y1.5) |
|---|---|---|
| Zn | 5,5-6,5 wt% | 7,5-8,5 wt% |
| Y | 0,8-1,2 wt% | 1,3-1,7 wt% |
| Fe content | <50ppm | <50ppm |
| Si content | <0,1wt% | <0,1wt% |
| Cu content | <0,03Wt% | <0,03Wt% |

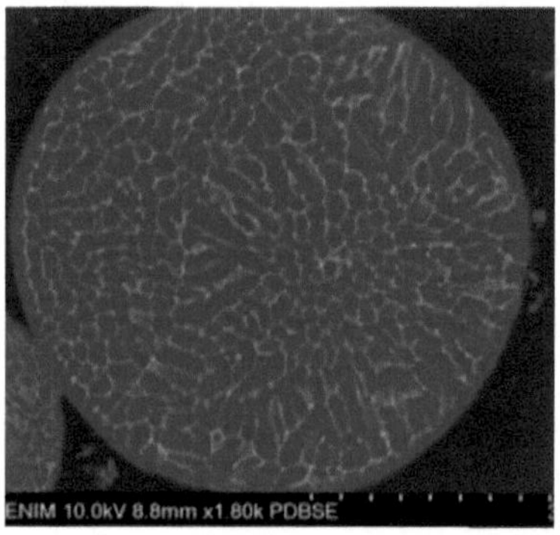


Figure 2 – Scanning Electron Microscope examination of BAS rapid solidified powder (>100micorns diameter) produced by rapid solidification in Argon atmosphere. In the picture two distinct finely distributed phases are clearly visible.

Figure 3 – Precursor material produced by cold compaction of rapid solidified powders made of BAS materials.

b) Once BAS precursor billets are ready, they are introduced into a lab-scale extrusion machinery available at Cenim laboratories. Results of 250°C extrusion of PMT are visible in figure 3 which shows some extruded bars (Semi-fished Sample Bars, or SSB) produced with different section shapes depending on the following validation mechanical and rolling tests to perform.

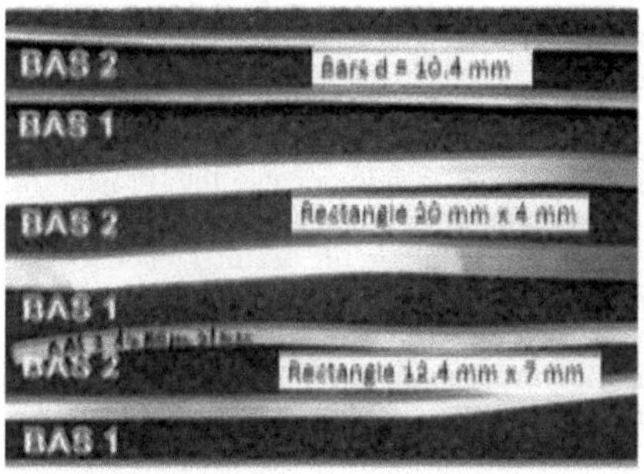


Figure 4 – Extruded bars with three different section shapes: round for tensile mechanical validation tests, square for fracture mechanics tests and flat section for secondary rolling tests.

c) Validation tests are conducted in order to check both microstructure and final mechanical properties obtained on SSB. Quality of the material processed is monitored by performance indicators listed in Table I compared to those targets defined by the project.

Table I. Performance indicators and project targets defined in the Green Metallurgy project .

| Monitoring indicators | Target |
|---|---|
| SSB grain size | <1 micron |
| UTS | >150% of conventional Mg alloys (casting, wrought Mg alloys) |
| YS | >150% of conventional Mg alloys (casting, wrought Mg alloys) |
| E% | >15% |
| KIc | >15 MPa √m |

Some metallographic samples cut from the semi-finished bars have been analyzed by Transmission Electron Microscope at Cenim to identify average grain size obtained after extrusion The average grain size obtained for SSBs, round and flat sections, respectively the lowest and the highest friction extrusion condition, that greatly affect further refining of grain size are shown in Table II.

Table II. Average grain size measured by TEM analyses on round and flat SSBs.

| Material | Average grain size (nm) for round SSB | Average grain size (nm) for square (20x4) SSB |
|---|---|---|
| BAS 1 | 437 ± 24 | 525 ± 38 |
| BAS 2 | 377 ± 39 | 502 ± 25 |

Validation of mechanical properties performed[3] of tensile tests on samples directly cut and machined from the semi-finished bars. Room temperature tensile tests were performed on samples in the as-received condition and after a heat treated condition that were defined during preliminary rolling tests conducted with laboratory rolling machinery. This particular heat treatment (1hour at 360°C) was initially set as optimal pre-heating and in-process heating treatment necessary to achieve best rolling conditions for flat samples cut directly from flat SSBs.
The data results from validation tensile tests are reported in Table III.

Table III. Static mechanical properties after validation test (Level I tests) conducted on round specimens (according to ASTM E8/E8M-09).

| Material | Test condition | YS (MPa) | UTS (MPa) | E% |
|---|---|---|---|---|
| BAS1 | Room temperature; as-extruded | 376 | 394 | 12 |
| BAS2 | Room temperature: as-extruded | 412 | 432 | 14 |
| BAS1 | Room temperature; Heat treated (1h@360°C) | 238 | 315 | 20 |
| BAS2 | Room temperature; Heat treated (1h@360°C) | 253 | 324 | 24 |

An extensive rolling test campaign was conducted using varying temperatures, reduction of ratio per rolling step (10%, 15% and 20% tested) and different rolling cylinder speed velocities (both testing asymmetric and symmetric conditions). As shown in figure 5, low temperature and/or high reduction ratio resulted in 45° angled fracture under rolling action; safer conditions were obtained by increasing pre-heating and rolling temperature (>300°C) at medium rolling speed and limiting reduction ratio per rolling step at 10%. As shown in figure 6, more satisfying results were obtained at 360°C (pre-heating), 300°C rolling temperature, medium rolling speed rate (symmetric rolling) and realizing a 30% of final reduction accumulating just 10% reduction ratio per each of three rolling steps that were intermediated by 360°C reheating treatment. Without considering border ruptures due to inhomogeneous friction caused by employing a not guided lab-scale rolling machinery, rolled sample in Figure 4 exhibited good surface finishing and no internal defects.
On the other hand, such satisfying results for a laboratory scale are not suitable for the next industrialization phase. It would not be economical or technologically viable to realize sheets starting from extruded SSB of BAS: at the least a 30% reduction is viable by an overly complex process lay-out of accumulated rolling single steps at high temperatures. While in industrialized lay-out the accumulative rolling step could be conducted under pre-heated roller cylinders, realistically limiting the total reduction ratio by 30%.

[3] When this manuscript is being written, non-standard mechanics fracture tests are in the designing phase. The fracture mechanics tests will not be conducted in standard condition because of limitations on sample geometry due to original dimensions of the SSB samples.

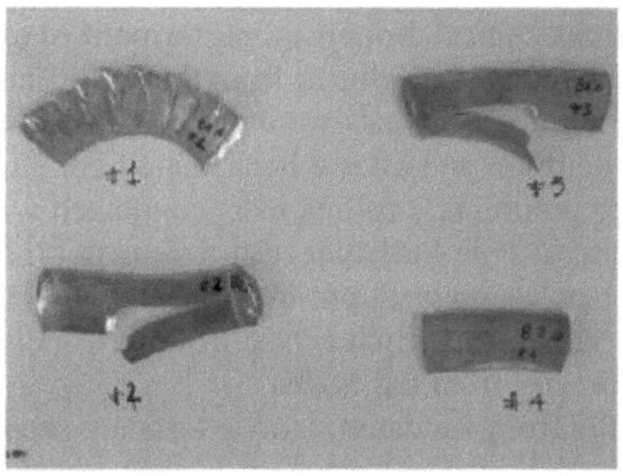

Figure 5 - Different cases from rolling tests: sample#1 ruptured 45° angle, because of low rolling temperature and high reduction ratio; sample#2 and sample#3) longitudinal crack caused by good temperature with too much high reduction ratio; sample#4)n safe rolling condition at higher rolling temperature and 10% of reduction ratio.

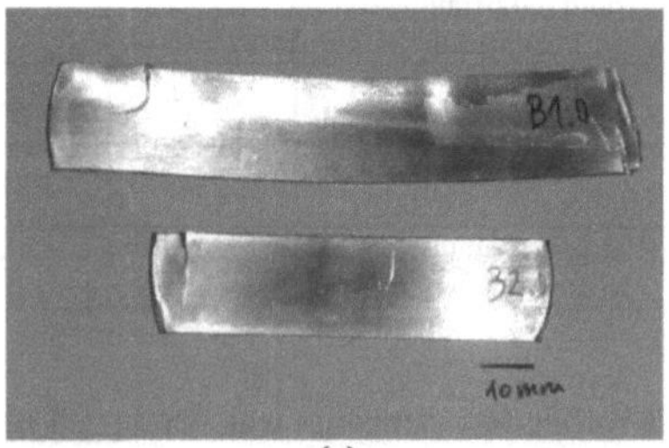

(a)

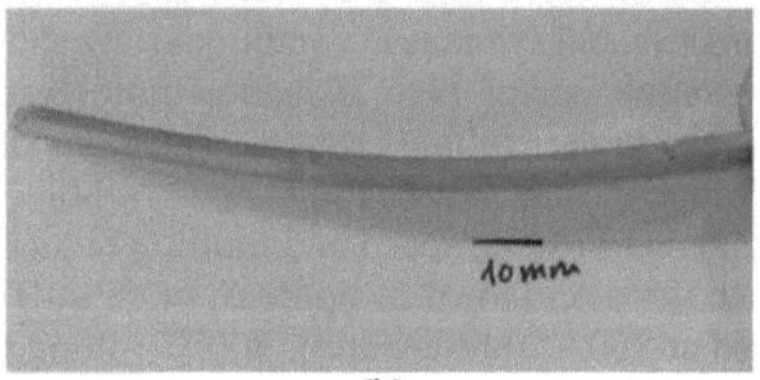

(b)

Figure 6 - Particular of safe rolling test: a) front view for 30% (accumulative rolling, 10% each step) reduction ratio of rolling test conducted at 300°C (360°C preheating temperature).

**Eco-magnesium® selected a unique Alternative Alloy System for full-recycling plate fabrication route**

Below is a brief description of the advancements in magnesium technology.

- Korean Institute of Technology (KITECH), one of the co-authors of this work, recently developed an environmentally friend magnesium alloy system called Eco-Magnesium®; these unique magnesium alloys are based on commercial chemical compositions (i.e. AZ91D, AZ31, AM60, etc.) but with the addition of CaO. KITECH discovered that CaO modified magnesium alloy systems can achieve a dramatic increase in oxidation resistance in the melt phase and avoid protective greenhouse gases, as $SF_6$ usually used in casting operations, which can lead to refinement of grain structure and diffusion strength and inhibiting oxidation in solid state [11-17]. Such magnesium CaO modified alloys are called Eco-Magnesium™ (Eco-Mg) since it is a less expensive "cleaner" to cast, with no oxidation or cover gas required, and potentially easy to recycle as it is more resistant to oxidation.

- Consolidated results of magnesium recyclability [18-20] obtained in the past by Japanese researchers shows that machined magnesium alloy chips can be extruded in a solid-state directly from recycled chips.

Basing on the above developments, the Project Management Board recently entered into an agreement of cooperation with KITECH for support during the testing phase of various Eco-AZ31 systems. This will allow them to check the capability of material to be processed by the Green Metallurgy full-recycling route, thus fully recycling machined chips that could be used successfully as raw material for the Green Metallurgy process route.

**Conclusions**

The first completed mechanical test campaign (level 1, static mechanical properties) through to test toughness behavior (level 2, fracture mechanics test campaign) on Basic Alloy System - two MgZn(Y) systems - produced by CENIM, very impressive mechanical properties were achieved by room temperature tensile tests conducted on the as-extruded materials.
BAS materials exhibited up to 420 MPa UTS and 380 MPa YS with minimum 9% of elongation, so that UTS and YS resulted 200% higher than conventional magnesium alloys, thus exceeding the planned target objectives.

Furthermore, when YS and UTS are heat-treated by a very short heating cycle, reduction is reduced by 30% compared to as-extruded conditions, but elongation increased dramatically, up to 22%. A closer investigation of the optimized microstructure modification by heat treatment is planned. BAS materials offer a more extensive combination of mechanical properties that can lead to optimization opportunities of engineering structural applications. These range from high performance applications that prefer the highest specific resistance with limited ductility or safer lightweight components favoring elongation at the UTS rupture.
Furthermore, previous experience [5] show that secondary heat treatments could ultimately be substituted by increasing the extrusion temperature: this would lead to further simplification of the optimized lay-out for single-step production of semi-finished bars
On the other hand, BAS showed not to be suitable for secondary rolling operations intended for flat parte generation. This does not mean that BAS cannot be rolled, but it is an overly complex and inefficient secondary operation that is not particularly suitable to be moved up to an industrial application.
For these reason, the Green Metallurgy project recently entered into a collaborative agreement to test Eco-AZ31 as a potential candidate as a low-impact and inexpensive material to apply in a fully-recycling route for flat components.

**Acknowledgements**

This work makes use of results produced by the Green Metallurgy project (http://www.green-metallurgy.eu), co-funded by the European Commission within the 2009 LIFE+ Programme.

Authors thank support provided by European Commission for supporting project dissemination activities. This work reflects only the authors' views. The Community is not liable for any use that may be made of the information contained therein.

**References**


1. L. Ridge, EUCAR - Automotive LCA Guidelines - Phase 2, In: Total Life Cycle Conference and Exposition; Graz, SAE paper 982185, 193-204 (1998).

2. A.Tharumarajah, and P.Koltun, "Is There an Environmental Advantage of Using Magnesium Components for Light-Weighting Cars?" *Journal of Cleaner Production*, 15 (2007), 1007–1013.

3. H.Friedrich, S. Schumann, "The second age of magnesium" (Paper presented at the Second Israeli International Conference, Sdom, Israel, February 2000).

4. H. Friedrich, S. Schumann, "Research for a new age of magnesium in the automotive industry", Journal of Material Process Technology, 117 (2001), 276–281.

5. G.Garces, M. Maeso, P. Pérez and P. Adeva, "Effect of Extrusion Temperature on Superplasticity of PM-WE54", Materials Science and Engineering A, 462(1-2), (2007), 127-131.

6. P.Pérez, S.Gonzales, G.Garces, G. Caruana, and P.Adeva, "High-strength Extruded Mg96Ni2Y1RE1 Alloy Exhibiting Superplastic Behaviour", Materials Science and Engineering A, 485(1-2), (2008) 194-199.

7. E.Mora, G.Garces, E.Onõrbe, P.Perez and P.Adeva, "High-strength Mg–Zn–Y Alloys Produced by Powder Metallurgy", *Scripta Materialia,* Volume 60, 2009, 776–779.

8. G. Garcés, A. Müller, E. Oñorbe, P. Pérez and P. Adeva, "Effect of Hot Forging on the Microstructure and Mechanical Properties of Mg-Zn-Y alloy", Journal of Materials Processing Technology, 206 (1-3), (2008), 99-105.

9. E.Mora, G. Garcés, E. Oñorbe, P.Pérez, and P.Adeva, "High-strength Mg-Zn-Y alloys Produced by Powder Metallurgy", Scripta Materialia, 60 (9), (2009), 776-779.

10. P. Pérez, G. Garcés, and P.Adeva, "Influence of Texture on the Mechanical Properties of Commercially Pure Magnesium Prepared by Powder Metallurgy", Journal of Materials Science, 42(11), (2007), 3969-3976.

11. S. K. Kim et al., "The behavior of CaO in Mg alloys," Proc. Magnesium Technology 2005, (2005), 285-289.

12. J. K. Lee et al., "Development of CaO added wrought Mg alloy for cleaner production," Proc. Magnesium Technology 2006, (2006), 185-189.

13. J. K. Lee et al., "Effect of CaO addition on ignition behavior in molten AZ31 and AZ91D Magnesium alloys." Rare Metals, 25 (2006) 155-159.

14. S. H. Ha et al., "The behavior of CaO and Ca in pure Mg", Rare Matals, 25 (2006) 150-154.

15. S. H. Ha et al., "Effect of CaO on oxidation resistance and microstructure of pure Mg", Materials Transactions, JIM, 49 (2008), 1081-1083.

16. D. I. Jang et al., "Effect of CaO on AZ31 Mg Strip Castings", Materials Transactions, JIM, 49 (2008), 976-979.

17. J. K. Lee et al., "Development of Novel Environment-friendly magnesium alloys", Advanced Materials Research, 47-50 (2008), 940-943.

18. M.Mabuchi,K.Kubota and K.Higashi, "New Recycling Process by Extrusion for Machined Chips of AZ91 Magnesium and Mechanical Properties of extruded Bars", Mater. Trans, JIM,36(1995), 1249-1254.

19. Y.Chono and M.Mabuchi,"Recycling of Magnesium Alloys Using Plastic Forming Process", Journal of the Japan Society for Technology of Plasticity, 44 (2003), 15-19.

20. K. Matsuzaki, Y. Murakoshi and T. Shimizu, "Microstructure and Mechanical Properties of Solid State Recycled Magnesium Alloy Chips", Magnesium Technology 2011, Proceedings of TMS (The Minerals, Metals & Materials Society), April 2011, 485-489.

**Magnesium Technology 2012**
*Edited by: Suveen N. Mathaudhu, Wim H. Sillekens, Neale R. Neelameggham, and Norbert Hort*
*TMS (The Minerals, Metals & Materials Society), 2012*

# Effect of Extrusion Conditions on Microstructure and Texture of Mg-1% Mn and Mg-1% Mn-1.6% Sr Alloys

Hemant Borkar, Mihriban Pekguleryuz

McGill University, Mining and Materials Engineering, 3610 University Street,
Montreal, QC, Canada H3A 2B2

Keywords: magnesium, extrusion, texture

## Abstract

Mg-1% Mn (M1) and Mg-1% Mn-1.6% Sr alloys have been subjected to hot extrusion and the effect of extrusion conditions on the microstructure and texture has been investigated. Addition of Sr to M1 alloy refines the extruded microstructure. Two different ram speeds of 4 mm/s and 8 mm/s and two different temperatures of 300°C and 350°C have been used in the study. The extent of recrystallization increases with increasing ram speed. The texture of extruded M1 weakens with Sr addition. The mechanism of texture weakening is suggested to be PSN with formation of new grains having orientations different than that of parent grains.

## Introduction

Despite their low density, wrought Mg alloys have limited applications due to low ductility and poor formability at room temperature. This is due to the limited number of slip systems in the hexagonal close packed (HCP) Mg alloys [1]. Another problem associated with Mg alloys is the strong fiber texture which is developed after extrusion in which the c-axis aligns perpendicular to the extrusion direction [2]. This leads to anisotropic behavior of deformed materials thus limiting their widespread applications.

Several methods have been attempted to improve the performance of wrought alloys such as the reduction of grain size [3], the activation of non basal slip by alloying (e.g Li addition) [4] and randomization or weakening of texture with RE elements (e.g Ce, Nd addition) [5-7]. Strontium has proven to be good grain refiner in Mg alloys thereby improving mechanical properties [8-10]. The Mg-Sr precipitates may also affect recrystallization during hot extrusion of Mg alloys containing Sr.

The deformation behaviour and the formation of new grains during recrystallisation is of importance for the control of microstructure and texture and therefore the properties of extruded profiles. The recrystallization behavior is closely related to the temperature and the strain induced during extrusion.

The current paper investigates the effect of extrusion conditions and effect of 1.6% Sr addition to Mg-1%Mn (M1) alloy by studying the microstructure and texture of extruded samples.

## Experimental Procedure

Mg-1%Mn (M1) and M1-1.6%Sr alloys were cast using 20kW, 5kHz induction furnace under protective gas mixture of $SF_6/CO_2$. Casting was done at 740 °C in preheated cylindrical shaped steel dies. 5 cm long, 2.5 cm dia extrusion samples were machined and subjected to extrusion at 300°C and 350°C and at ram speeds of 4mm/s and 8mm/s in lab-scale 100 Ton hydraulic press. Extrusion ratio of 1:8 was used in the experiments.

The samples for optical microscopy were obtained in such a way that the extrusion direction is parallel to the plane of the samples. Nikon Epishot 200 optical microscope equipped with Clemex optical image analyser was used for microstructure investigation. The samples for texture studies were obtained from cross sections of extrusions in such a way that the extrusion direction is perpendicular to the plane of the samples. The texture of these samples was measured with X-Ray diffraction using Bruker Materials Analyser texture goniometer equipped with Cu Ka radiation.

## Results and Discussion

### Microstructure

The stereoscopic images of M1 and M1-1.6Sr extruded at 350°C at ram speeds of 4 mm/s and 8 mm/s are shown in Figure 1. These figures depict the material flow during extrusion. Grain refining is observed with Sr addition to M1 in the samples extruded at both the speeds.

Microstructures of M1 and M1-1.6Sr extruded at 350°C at 4 mm/s speed are shown in Figure 2. Recrystallized grains originating at the grain boundaries of elongated parent grains can be observed in the microstructure of M1(Figure 2a). Deformation is accompanied with twins which can be spotted in the microstructure. Presence of twins can be attributed to the coarse grain structure of M1 as larger grain size promote increased twinning activity. M1-1.6Sr shows refined microstructure with presence of Mg-Sr intermetallic stringers elongated in the extrusion direction (Figure 2b). Twins are not observed in this microstructure. The recrystallized grains are seen to be associated with stringers.

At higher ram speed, the M1 (Figure 3a) shows similar microstructure to that of M1 at lower speed. M1-1.6Sr (Figure 3b), on the other hand, shows extensive recrystallization associated with Mg-Sr stringers. New grains are seen to be originating from the stringer and grow until they reach another stringer or grain boundary. This kind of recrystallization is called particle stimulated nucleation (PSN) of recrystallization. It has been observed in aluminium[11,12] and magnesium alloys[13] in which particle size of around 1.5 to 2 μm are sufficient to induce PSN.

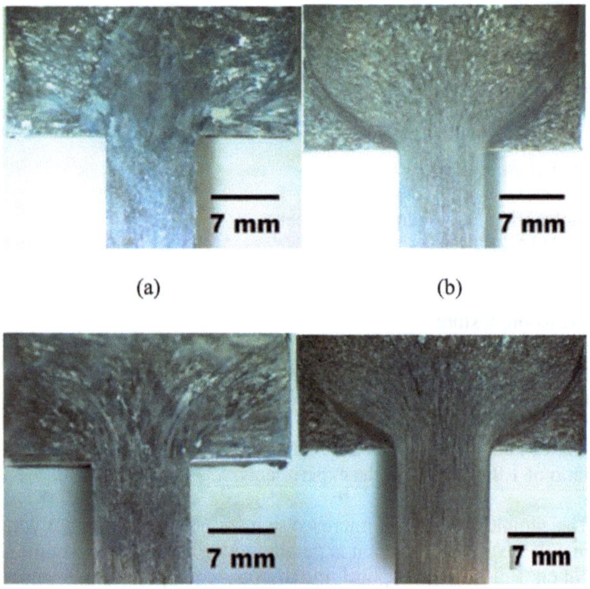


(a) (b)

(c) (d)

Figure 1. Stereoscopic images of extruded samples at 350°C
(a) M1, 4 mm/s speed (b) M1-1.6Sr , 4 mm/s speed
(c) M1,8 mm/s speed (d) M1-1.6Sr, 8 mm/s speed

In the process of PSN, deformation zones occur at the particles and can act as nucleation sites for recrystallization since they have high dislocation density and fine sub-grains. Higher ram speed in extrusion imposes higher strain and hence degree of recrystallization is higher in the samples extruded at higher speeds.

An increase in ram speed leads to an increase in the surface temperature of the billet. This increases homogeneity of deformation. The increase of ram speed from 4mm/s to 8 mm/s was found to improve properties (YS) slightly in ZK30 alloy but a further increase in speed to 16 mm/s led to the occurrence of grain growth and secondary recrystallisation, which lowers the properties[14]. In the current studies, both M1 and M1-1.6Sr alloys show improved surface quality after extrusion at higher ram speed of 8 mm/s.

Texture

The inverse pole figures of M1 and M1-1.6Sr extruded at 300°C and 350°C and at constant ram speed of 8 mm/s are shown in Figure 4 and Figure 5. The inverse pole figures are in extrusion direction which means that the axis of the figure is parallel to the extrusion direction. In M1 alloy extruded at 300°C (Figure 4a), the texture is mainly concentrated near $<10\bar{1}0>$ and $<11\bar{2}0>$ poles which means that grains are oriented in such a way that prismatic planes are perpendicular to the extrusion direction. It also means that basal planes are parallel to the extrusion direction. However, the texture is seen to be distributed up to 50° from extrusion direction and the maximum texture intensity is 1.5 times the random intensity. This shows that M1 extruded at this temperature does not show strong texture. M1.1.6Sr extruded at 300°C (Figure 4b) still shows relatively weaker texture with texture distribution up to 60° and some additional peaks developing in the pole figure. The maximum intensity in this case is 1.4 times the random intensity.

M1 alloy extruded at 350°C (Figure 5a) develops the texture which is closer to the typical fiber texture developed in magnesium alloys after extrusion. The texture is mainly concentrated near $<10\bar{1}0>$ and $<11\bar{2}0>$ poles and the distribution is within 30°C from the extrusion direction. The maximum texture intensity is 2.4 times the random intensity which is higher than that for M1 at 300°C. This probably means that the typical fiber texture in M1 alloy is more pronounced at higher extrusion temperature. M1-1.6Sr alloy extruded at 350°C (Figure 5b) produces relatively weaker textures than that of M1 with lower value of maximum intensity (2 times random intensity) and wider texture distribution (up to 50°). This shows that the texture weakening with Sr addition occurs more prominently at higher temperatures. As observed in Figure 3b, PSN grains are associated with stringers and extensive recrystallization occurs in M1-16.6Sr alloy extruded at 350°C. The important characteristic of PSN is that it produces new grains with different orientations than that of parent grains. As higher extrusion temperatures create

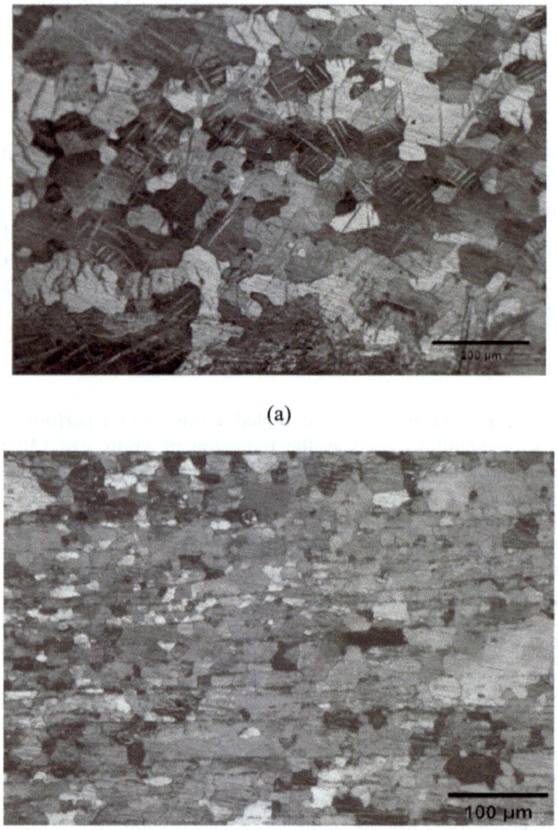


(a)

(b)

Figure 2. Microstructure of extruded samples at 350 °C and 4 mm/s ram speed (a) M1 (b) M1-1.6Sr

higher driving force for the recrystallization, large numbers of PSN grains with different orientations are available in the microstructure which can reduce the overall texture.

(a)

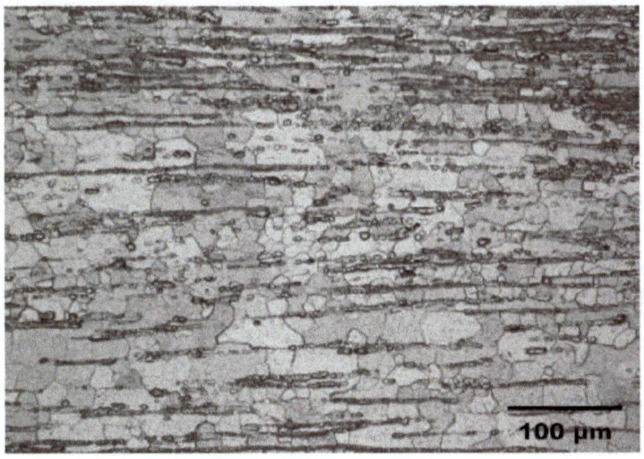


(b)

Figure 3. Microstructure of extruded samples at 350°C and 8 mm/s ram speed (a) M1 (b) M1-1.6Sr

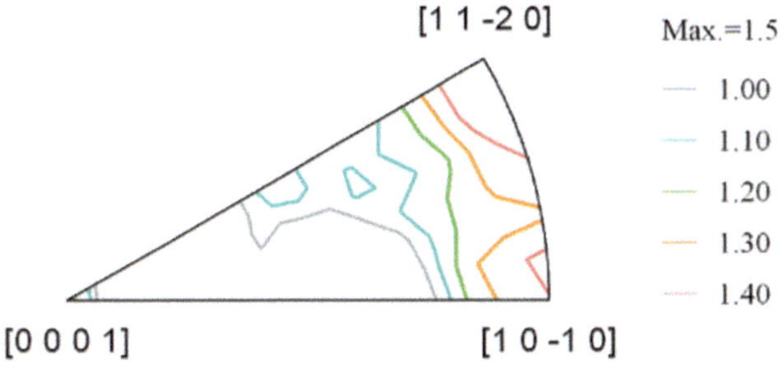


(a)

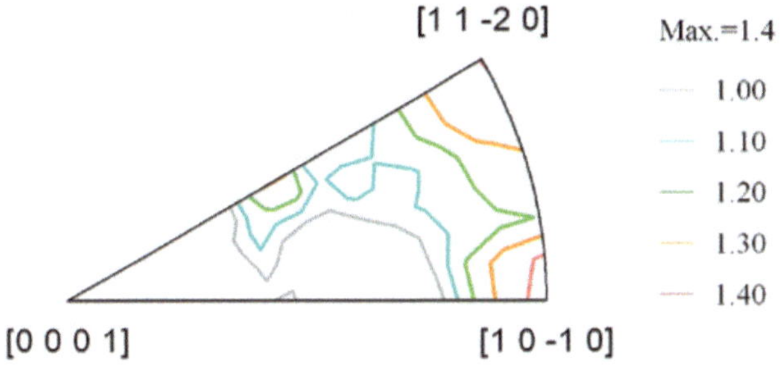


(b)

Figure 4. Inverse pole figures of samples extruded at 300°C and 8 mm/s ram speed (a) M1 and (b) M1-1.6Sr

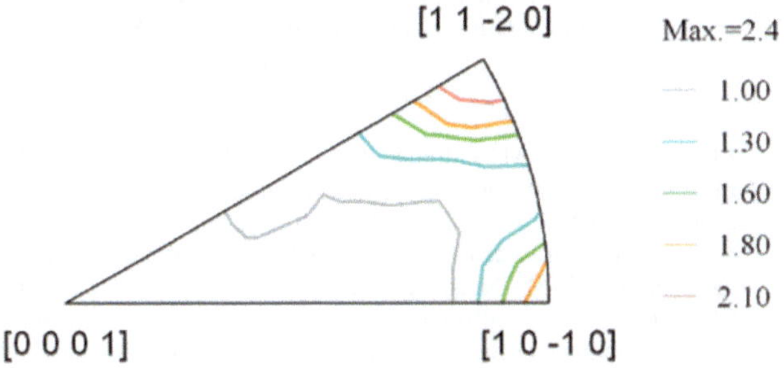


(a)

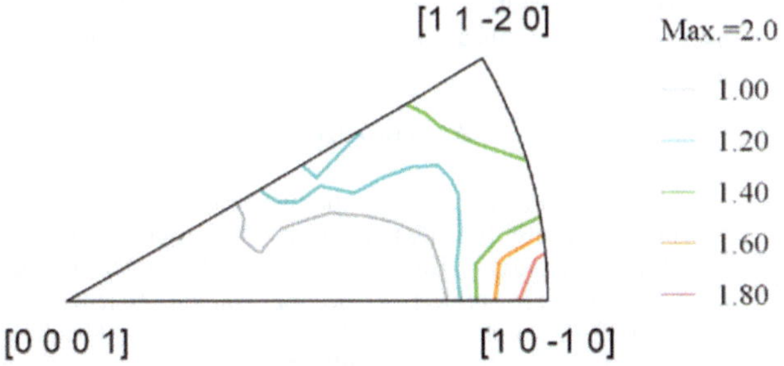


(b)

Figure 5. Inverse pole figures of samples extruded at 350°C and 8 mm/s ram speed (a) M1 and (b) M1-1.6Sr

Figure 6 shows the inverse pole figures of M1 and M1-1.6Sr extruded at 350°C and at ram speed of 4mm/s. Comparison of Figure 5a and 6a shows that M1 develops stronger texture at higher ram speed. However, higher ram speed produces higher strain which helps promoting PSN in M1-1.6Sr and hence the overall extruded texture is weaker in this alloy at higher ram speed than that at lower ram speed (Figure 5b and 6b). It can also be concluded that the effect of PSN is more pronounced at higher ram speed.

In the past, texture weakening of wrought Mg alloys with rare earth elements has been reported [5-7] in which rare earth addition produces a distinct peak in the inverse pole figure. However, with Sr addition, texture weakening occurs with random distribution and without typical distinct peak in the inverse pole figure.

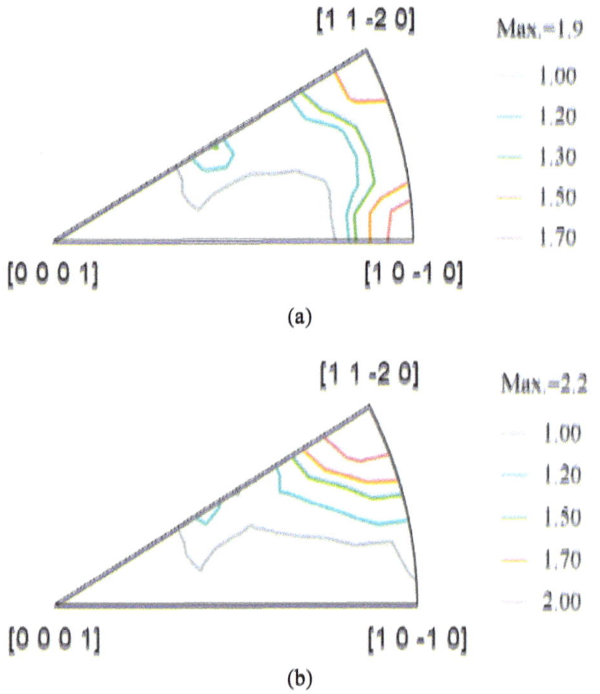


Figure 6. Inverse pole figures of samples extruded at 350°C and 4 mm/s ram speed (a) M1 and (b) M1-1.6Sr

### Conclusions

Microstructure of M1 extruded at 350°C and at lower and higher ram speeds shows presence of twins and recrystallized grains around parent grains. M1-1.6Sr extruded at these conditions shows refined microstructure in which recrystallized grains are associated with Mg-Sr stringers elongated in extrusion direction. The extent of recrystallization increases with increasing ram speed owing to higher amount of strain. The texture of extruded M1 weakens with Sr addition and this effect is more pronounced at higher extrusion temperatures and higher ram speeds. The mechanism of texture weakening is suggested to be PSN in which new grains are formed with orientations different than that of parent grains.

### Acknowledgements

This project is a collaborative research work between McGill University (Canada), Applied Magnesium (USA) and CANMET (Canada). The authors acknowledge the financial support (Strategic project grant) of Natural Sciences and Engineering Research Council (NSERC) of Canada. Thanks are due to Pierre Vermette (McGill) for his assistance to the experimental work.

### References

[1] B.L. Mordike, T. Ebert, Mater. Sci. Eng. A 302 (2001) 37-45.

[2] N. Stanford, Mater. Sci. Eng. A, 527(2010), 2669-2677.

[3] M.R. Barnett, Z. Keshavarz , A.G.Beer, D.Atwell, Acta Mater 52 (2004) 5093-5103.

[4] S.R Agnew, J.W.Senn, JOM 58 (May 2006), 62-69.

[5] J. Bohlen, M.R. Nurnberg, J.W. Senn, D. Letzig, S.R.Agnew, Acta Mater. 55 (2007) 2101.

[6] S.R.Agnew, O.Duygulu, Int.J.Plasticity 21(2005), 1161-1193.

[7] M. Masoumi, M. Hoseini, M. Pekguleryuz, Mater. Sci. Eng. A 528 (2011) 3122-3129.

[8] J.E.Gruzleski and C.A.Aliravci,U.S.Patent No 5,143,564,1992.

[9] M. Yang, F. Pan, R. Cheng, A. Tang, Mater. Sci. Eng. A, 491(2008) 440-445.

[10] X.Q. Zeng, Y.X.Wang,W.J. Ding, A. Luo, A.K. Sachdev, Metall. Mater. Trans. A 37 (2006) 1333.

[11] Z. Li, J. Dong, X.Q. Zeng, C. Lu, W.J. Ding, Mater. Sci. Eng. A, 466 (2007) 134-139.

[12] I.J. Polmear, Light Alloys: Metallurgy of the Light Metals, Arnold Publications, 1995.

[13] A. Sadeghi, M. Pekguleryuz, Mater. Sci. Eng. A 528 (2011) 1678-1685.

[14] Govind, K. S. Nair, M. C. Mittal, R. Sikand, A. K. Gupta, Materials Science and Technology, 2008, Vol 24,NO 4,399-405

**Magnesium Technology 2012**
*Edited by: Suveen N. Mathaudhu, Wim H. Sillekens, Neale R. Neelameggham, and Norbert Hort*
*TMS (The Minerals, Metals & Materials Society), 2012*

# MICROSTRUCTURAL CHARACTERIZATION OF HOMOGENISED AND AGED Mg-Gd-Nd-Zn-Y-Zr ALLOYS

S. Khawaled, M. Bamberger, A. Katsman
Technion – Israel Institute of Technology, Haifa 3200, Israel

Keywords: Mg-Gd_Nd alloys, Precipitation, Microstructure, Mechanical properties

## Abstract

The microstructure of the Mg—6%Gd-3.7%Nd-0.3%Zn-0.18%Y-0.15%Zr (%wt) alloy has been investigated after solution treatment at 540°C for 24hr followed by isothermal aging at 175°C up to 32 days by using of Vickers hardness, optical microscopy, scanning electron microscopy equipped with EDS, X-ray diffraction and transmission electron microscopy. It was observed that the homogenized alloy contained primary α-Mg solid solution, eutectic structures, cuboid shaped phases and Zr-rich clusters. The eutectic structures were the products of a 'quasibinary eutectic reaction' L→α-Mg+$\beta$-$Mg_5RE$. The eutectic phase was characterized to be of $Mg_5Gd$ prototype with the composition $Mg_5(Gd_xNd_{1-x}$, x≈0.2). The cuboid shaped phases, with the composition $Gd_4(Nd_xY_{1-x}$, x≈0.5), grew during aging and reached ~3μm average size. Precipitation of β″ and β′ phases during aging was observed. The maximum microhardness was achieved after 16 days of aging.

## Introduction

Magnesium alloys containing heavy rare earth metals are very attractive candidates for the automotive industry including racing cars and aerospace applications [1-8], due to their high strength properties combined with low density. Due to the large solubility of RE element in Mg matrix at high temperature and its rapid decrease with lowering temperature, the Mg–RE alloys show remarkable age-hardening response during isothermal aging at 175°C, and thus high strength is to be expected.

Microstructural characterization of the Mg–RE alloys performed by many researchers [1-11] has confirmed a four staged precipitation sequence: [α-Mg supersaturated solid solution (S.S.S.S.) → $\beta''(DO_{19})$ → $\beta'$(bco) → $\beta_1$(fcc) → β(fcc)] [2,3,9,10], where the later precipitate can coexist with the former. Among the four precipitated phases, the coherent β″and β′ phases are considered to be the primary strengthening phases [2,4,8,10]. The yield strength or hardness usually reaches a maximum, as the materials form a microstructure with fine β′ precipitates during aging [1-4].

In the present work, the microstructural evolution of a Mg-Gd-Nd-based alloy during isothermal aging at 175°C. This specific composition was selected in order to achieve maximum age hardening response, improved castability, grain refinement and strength. It was investigated by a combination of optical microscopy (OM), scanning electron microscopy (SEM), transmission electron microscopy (TEM), energy dispersive X-ray spectrometry (EDS) and X-ray diffraction (XRD). Microhardness measurements were performed to reveal the effect of precipitation on the mechanical properties of the alloy.

## Experimental

The chemical composition of the investigated alloy was Mg-6Gd-3.7Nd-0.3Zn-0.18Y-0.15Zr (wt.%).
Specimens cut from the cast ingot were solution treated (ST) at 540°C for 24hr, quenched into hot water of about 70°C, and then aged at 175°C in an oil bath for various periods of time up to 32days followed by water quenching.

The microhardness of aged specimens was measured by a Vickers hardness tester under a load of 50gr and holding time of 15sec. 20-40 indentations were conducted on each specimen. The specimens were polished with diamond paste and etched for 20sec in Ethanol 0.95%wt and $HNO_3$ 0.05%wt (Nital) solution.

A combination of OM, SEM, TEM, EDS and XRD was carried out to characterize as- cast and homogenized micro-structure of the alloy. Metallographic specimens were polished and etched in an Acetic Glycohol enchant (100ml ethanol, 20ml water, 6gr picric acid and 5ml acetic acid). Thin foils for TEM characterization were prepared by ion beam thinning, and were examined at the acceleration voltage of 200 kV. The compositions of the matrix and of secondary phases in the TEM specimens were analyzed by EDS instrument attached to the SEM, operating at 10kV. The constituent phases of the alloy in different conditions were identified by XRD with Cu K$\alpha$ radiation.

## Results

As-cast microstructure

Fig.1 shows the as-cast alloy consists mainly of the α-Mg phase as a matrix, eutectic-like structures, cuboid-like phases and Zr-rich clusters.

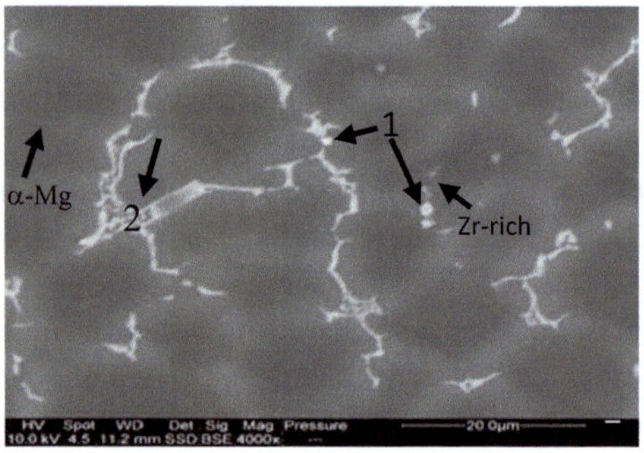


Fig. 1. Microstructure of the as-cast sample: SEM image shows the cuboid-like phase (1) and eutectic-like structure (2).

The cuboid-like particles in micrometer scale are frequently observed in the association with Zr-rich clusters and eutectic- like structures as shown in Fig.1. The cuboid-like particles are Y-rich phases containing Gd and Nd. Composition of these particles $Y_2(Gd_xNd_{1-x}$, $x\approx0.9$) was determined by EDS SEM using TEM samples.

ST microstructure

The purpose of homogenization for precipitate hardened Mg-RE alloys is to dissolve soluble as cast phases into α-Mg matrix and obtain supersaturated solid solution by quenching. The microstructure of the alloy was investigated in the ST and aged (175°C) conditions. It was observed that the homogenized alloy contained primary α-Mg solid solution, eutectic structures, cuboid shaped phases and Zr-rich clusters (Fig. 2).

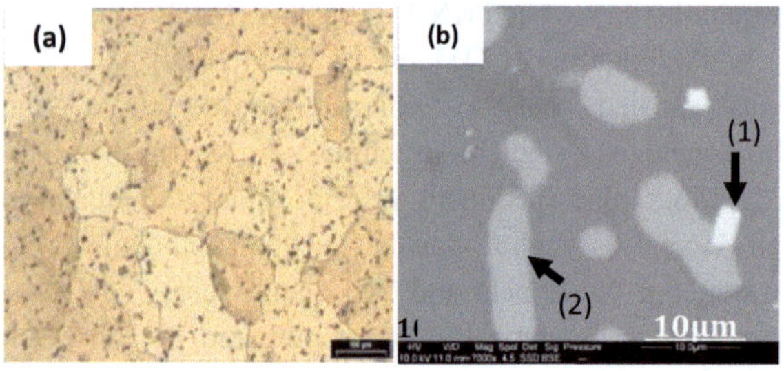


Fig. 2. Microstructure of the ST specimens: (a) Optical image showing the grain size and (b) SEM image showing the cuboid shaped phase (1) and eutectic structure (2).

The compositions of the phases were analyzed by EDS. The composition of the eutectic structures was characterized to be of α-Mg+β-$Mg_5(Gd_xNd_{1-x}$, $x\approx0.2$). The cuboid shaped phases have the composition $Gd_4(Nd_xY_{1-x}$, $x\approx0.5$). TEM micrographs (Fig.3) revealed the 1-2 μm cuboid one-phase particles (Fig.3a) and the two-phase skeleton-like eutectic structures (Fig.3b).

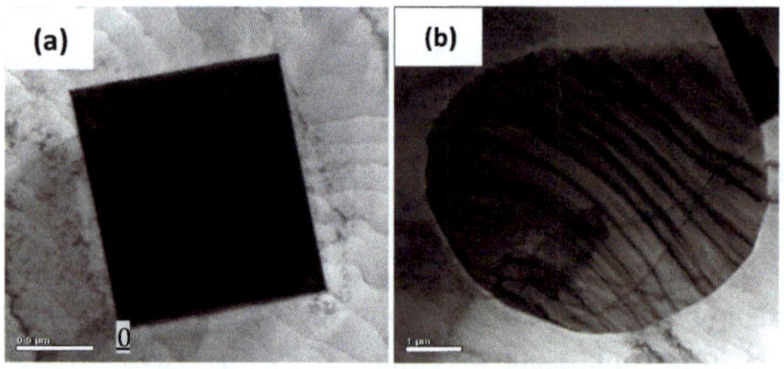


Fig. 3. TEM observations of **(a)** cuboid shaped phase and **(b)** eutectic-like structure.

Microstructure of aged samples

Microstructure of the aged samples consists of the eutectic structure, cuboid shaped particles and small precipitates uniformly distributed in the matrix. The cuboid shaped particles grow during aging up to ~3 μm in size; SEM+EDS analysis revealed that their composition did not change.

Crystal structure and composition of precipitates within the α-Mg matrix were investigated by TEM (Figs. 4-7); there were diffuse spots between the α-Mg matrix spots in the selected area electron diffraction (SAED) patterns, which were brighter than the matrix spots (Fig. 4). TEM images and corresponding SAED patterns of precipitates in a specimen aged for 16 days (peak-aged specimen) are shown in Fig.5. It was observed that the microstructure contains a high number density of fine plate shape $\beta''$ precipitates and a little number of globular particles $\beta'$ uniformly distributed within the α-Mg matrix (Fig. 6).

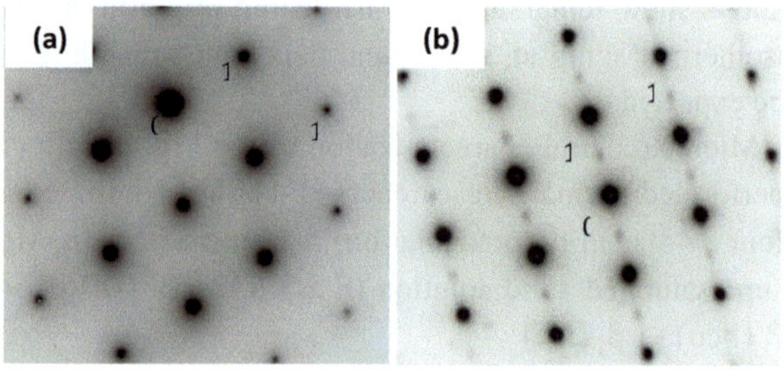


Fig. 4. SAED patterns of the specimen aged for (a) ST and (b) 32days with Z.A $[11\bar{2}3]_\alpha$ .

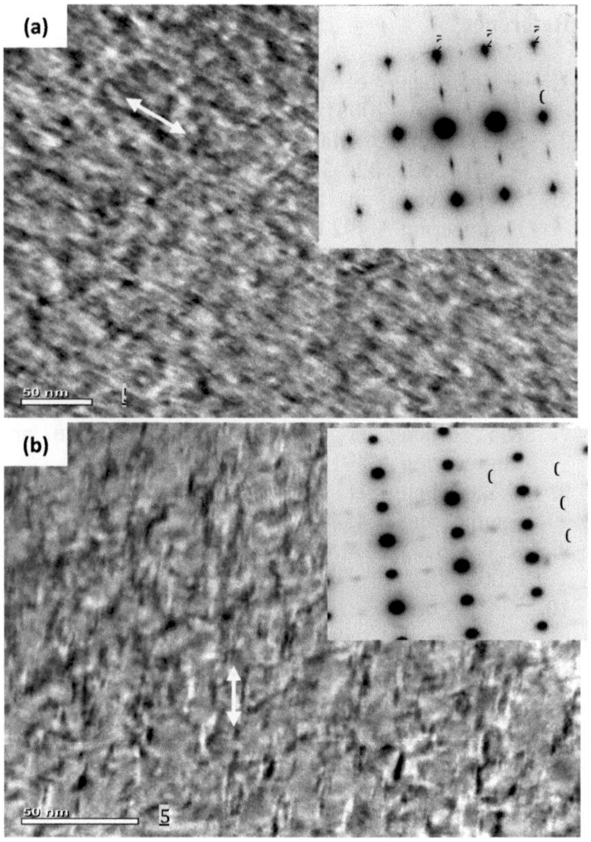


Fig. 5. TEM images and corresponding SAED patterns of the specimen aged for 16days at 175C: Z.A $[01\bar{1}0]_\alpha$ in (a) and Z.A $[2\bar{1}\bar{1}0]_\alpha$ in (b).

Combination of reflections from the two precipitates, $\beta''$and $\beta'$, appears in the corresponding SAED pattern (Fig. 5). The $\beta''$ precipitates with $DO_{19}$ structure $\left(a{\sim}2a_{\alpha-Mg} = 0.64\text{nm}, c{\sim}c_{\alpha-Mg} = 0.52\text{nm}\right)$ correspond to diffuse spots at 1/2 distance of $\{01\bar{1}0\}_\alpha$ or $\{2\bar{1}\bar{1}0\}_\alpha$ reflections, with the following orientation relationships (OR): $[0001]_{\beta''}//[0001]_\alpha$, $\{01\bar{1}0\}_{\beta''}//\{01\bar{1}0\}_\alpha$; and the $\beta'$ phase with Base Centered Orthorhombic (BCO) structure ($a{\sim}2a_{\alpha-Mg} = 0.64$nm, $b{\sim}2.2$nm, $c{\sim}c_{\alpha-Mg} = 0.52$nm) gives rise to the additional spots at 1/4 $\{01\bar{1}0\}_\alpha$, 1/2 $\{01\bar{1}0\}_\alpha$ and 3/4 $\{01\bar{1}0\}_\alpha$, with the following OR: $[001]_{\beta'}//[0001]_\alpha$, $\{100\}_{\beta'}//\{2\bar{1}\bar{1}0\}_\alpha$.

The microstructure of the sample aged for 32 days contains a little number of plate-shaped FCC $\beta_1$ precipitates ( $a \approx 0.74$nm) (Fig. 7) along with $\beta''$and $\beta'$ small precipitates.

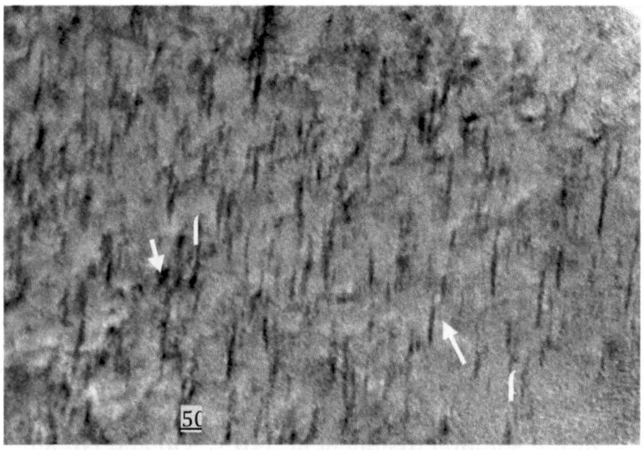

Fig. 6. TEM image of the specimen aged for 16days at 175C.

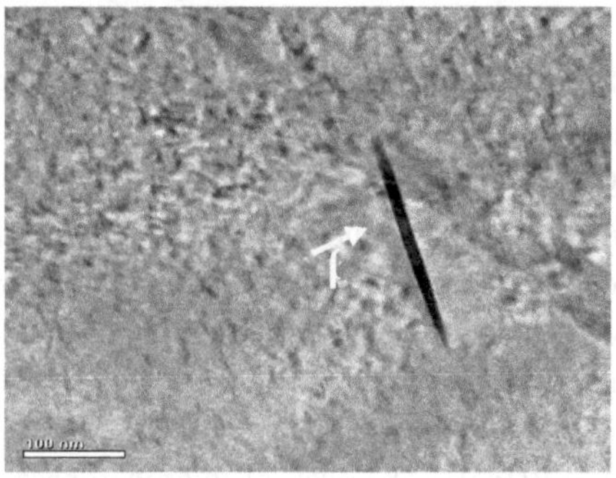

Fig. 7. TEM image of the specimen aged for 32days at 175C.

Microhardness

The microhardness evolution during isothermal aging at 175°C up to 32days is shown in Fig. 8. Microhardness increased and reached a maximum value of about 100 HV after 16 days of aging. Further aging led to decrease in the microhardness.

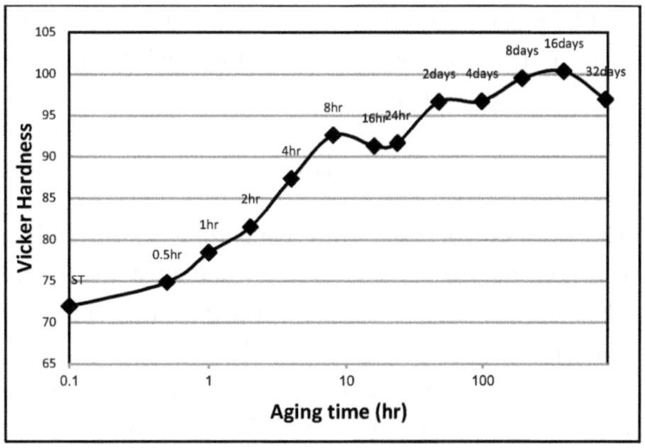


Fig. 8. Age hardening curve of the alloy aged at 175°C

## Discussion

The as-cast microstructure of the Mg-Gd-Nd-Zn-Y-Zr alloys consists of primary α-Mg solid solution, skeleton-like eutectic structure [1], small Y-rich cuboid-like phase and Zr-rich clusters. The eutectic structures were the products of a 'quasibinary eutectic reaction' L→α-Mg+β-$Mg_5RE$. The eutectic phase was characterized to have $Mg_5Gd$ type FCC crystal structure with a ≈ 2.2nm and a composition of $Mg_5(Gd_xNd_{1-x}$, x≈0.2).

During solution treatment at 540°C, Gd and Nd atoms are transferred from eutectic-like structure to cuboid-like phase to form the cuboid shaped phase with the stable composition $Gd_4(Nd_xY_{1-x}$, x≈0.5). The cuboid shaped particles grow during aging and reach ~3μm average size.

Aging of solution treated and quenched alloy results in precipitation of β″ and β′ phases. Generally, the precipitation sequence in the investigated alloy during aging is identical to that reported for many Mg-RE alloy systems [2,3,9,10], i.e. the four staged precipitation sequence:$[\alpha - \mathrm{Mg\ S.S.S.S.} \rightarrow \beta''(\mathrm{D0_{19}}) \rightarrow \beta'(\mathrm{bco}) \rightarrow \beta_1(\mathrm{fcc}) \rightarrow \beta(\mathrm{fcc})]$. The peak of microhardness after 16 days of aging corresponds to the combination of β″and β′ precipitates (Fig.5). The plate shape coherent β″ precipitates in $\alpha - \mathrm{Mg}$ matrix have OR: $[0001]_{\beta''}//[0001]_{\alpha}$, $\{01\bar{1}0\}_{\beta''}//\{01\bar{1}0\}_{\alpha}$. The globular precipitates β′ have OR: $[001]_{\beta'}//[0001]_{\alpha}$, $\{100\}_{\beta'}//\{2\bar{1}\bar{1}0\}_{\alpha}$. These results are similar to those reported by K.Y. Zheng et al. [10] for Mg-11Gd-2Nd-0.5Zr. The further aging up to 32 days results in coarsening of β″ and β′ precipitates and their transformation to $\beta_1$ phase (Fig.7). These processes are accompanied by decrease of microhardness (Fig.8).

## Conclusion

The Mg-Gd-Nd based alloy containing Y, Zn and Zr was investigated in the as-cast, ST and aged conditions. Skeleton-like eutectic structure α-Mg+β-$Mg_5(Gd_xNd_{1-x}$, x≈0.2) and cuboid shaped particles were found in all conditions. Y-rich cuboid-like particles found in the as-cast alloy transformed to cuboid shaped $Gd_4(Nd_xY_{1-x}$, x≈0.5) during ST and then grow during aging. Precipitation of plate shape β″ and globular β′ particles during aging resulted in maximum of microhardness increase after 16 days. The further aging results in coarsening of β″ and β′ precipitates and their transformation of $\beta_1$ phase.

## Reference


1. K.Y. Zheng et al. / Materials Science and Technology (2008) VOL24 NO3.
2. S.M. He et al. / Materials Science and Engineering A 431 (2006) 322-327.
3. K.Y. Zheng et al. / Materials Science and Engineering A 492 (2008) 185-190.
4. L. Gao et al. / J Materials Science (2009) 44:4443-4454.
5. Zheng Kai-yun, et al. / Trans. Nonferrous Met. Soc. China 17 (2007).
6. D. Li et al. /Journal of Alloys and Compounds 439 (2007) 254-257.
7. K.Y. Zheng et al. / Materials Science and Engineering A 491 (2008) 103-109.
8. K.Y. Zheng et al. / Materials Science and Engineering A 454-455 (2007) 314-321.
9. K.Y. Zheng et al. / Materials Characterization 59 (2008) 857-862.
10. K.Y. Zheng et al. / Materials Science and Engineering A 489 (2008) 44-54.
11. G. Sha et al. / Materials Science and Engineering A 527 (2010) 5092-5099.

Magnesium Technology 2012
Edited by: Suveen N. Mathaudhu, Wim H. Sillekens, Neale R. Neelameggham, and Norbert Hort
TMS (The Minerals, Metals & Materials Society), 2012

# On the Deformed Microstructure of Rolled Mg-2.9Y

S. A. Farzadfar[1], M. Sanjari[1], I.-H. Jung[1], E. Essadiqi[2] and S. Yue[1]

[1]McGill University, Department of Materials Engineering, Montreal, QC, Canada
[2]CANMET- Materials Technology Laboratory, Hamilton, ON, Canada

Keywords: Magnesium, Yttrium, Rolling, Texture

## Abstract

The hot and cold rolled microstructures of a Mg-2.9Y alloy were examined by means of metallography, X-ray texture measurement and EBSD technique. It is found that Y in solid solution suppresses dynamic recrystallization even at 450 °C. After rolling, a heterogeneous microstructure containing twins and bands is obtained. The bands were found to originate form double and compression twins, and are locations of high-stored energy. The bulk texture of the rolled material consists of the typical basal component, and also a component from the ND towards the TD. The latter component results from the presence of parent grains with *c*-axis along the TD. The strain is accommodated in such parent grains by extension twinning and possibly prismatic slip.

## Introduction

Available Mg sheets do not currently exhibit the room-temperature formability required in automotive applications. This is the main reason which retards the industrial application of this lightest structural material [1]. In rolled Mg alloy sheets, a strong basal texture develops, in which the basal planes are mainly distributed parallel to the RD-TD plane, where the RD and TD are rolling and transverse directions [2-4]. The basal texture develops mainly because of the dominant role played by basal slip and {10.2} extension twinning during deformation. These two mechanisms align the *c*-axis of grains with the direction of compressive strain, i.e., in rolled sheets, *c*-axes align with the ND (normal direction).

The presence of basal texture limits the room-temperature formability due to the restricted activities of <**a**> type slip systems under loading in the sheet normal and planar directions. Furthermore, the polarity of mechanical twinning in Mg results in the asymmetry of tension-compression strength in a textured material. It is therefore necessary to provide a way of weakening the texture in wrought Mg alloys.

It has recently been found that the addition of rare earth (RE) elements, such as Y, leads to significantly weaker textures [5-8]. One approach to study the effect of RE elements is by selecting binary Mg-RE alloys with the pure RE element in solid solution. Among the RE elements, Y is of particular interest due to its large solubility in Mg solid solution (2.2 wt% at 200 °C, and maximum solubility of 12 wt% at 567 °C) [9]. Very little data is available on the rolled microstructure of solid solution Mg-Y alloys [6, 10-12]. Some of these studies have been performed only on cold-rolled material that was pre-deformed by hot rolling [10, 12]. In other works [6, 11], the temperatures of pre/post-deformation annealing are dissimilar from that of deformation, which could result in precipitation (or dissolution of precipitates) and therefore variation in solute amount. A solid solution alloy, both before and during deformation, is therefore required to investigate the effect of Y solute.

The objective of this study is to characterize the as-rolled microstructure of a solid solution Mg-Y alloy at room temperature, 350 °C and 450 °C. More emphasis will be put on the microstructure features after hot rolling.

## Experimental Procedure

To avoid precipitation during hot rolling, a Mg-Y solid solution alloy with 2.9 wt% of Y was chosen. This composition was selected to ensure that solubility limit at the lowest hot rolling temperature (≈ 5.4 wt% at 350 °C) is not reached. The Mg-2.9Y alloy was prepared from high purity elements 99.9% Mg, 99.99% Zn, and 99.9% Y (all concentrations are given in wt%) by induction melting under $SF_6$ and $CO_2$ gas mixture. The molten alloys were cast into plates 100 mm in width and 6 mm in thickness in a steel mould pre-heated at 400 °C. Inductively coupled plasma atomic emission spectroscopy (ICP-AES) was used to determine the composition of the cast material, which was close to the nominal one ( ~ 0.12 wt% maximum difference). The major impurity levels were 0.03 wt% for Al and Fe, and 0.004 wt% for Si and Mn. To obtain solid solution alloys prior to rolling, the cast plates were heat treated at 430 °C for 24 h [13]. The average grain size in heat-treated cast material, measured by linear intercept method (ASTM standard E112), is 163(±10) μm.

Hot rolling at two temperatures, 350 °C and 450 °C, and cold rolling at room temperature were carried out in this work. The hot rolling was performed to 50% reduction in one pass to avoid inter-pass heating and possible meta-dynamic or static recrystallization. This is important since this study aims at clarifying the role of dynamic recrystallization on texture, and therefore any form other recrystallization types must be avoided. For rolling at room temperature, 30% reduction was successfully applied in 5 passes without major cracking. Minor edge cracking occurred at all rolling temperatures. Rolling was performed using laboratory rolling mills with a diameter of 150 mm and a rolling speed of 25 rpm. For hot rolling, the plates were pre-heated for 15 min prior to deformation, and quenched into room-temperature water within 1 s after rolling.

All microstructural and texture examinations were performed in the plane containing the RD and ND. The samples were ground down to 1200 grit SiC paper, and polished with 3 μm and 1 μm diamond suspension followed by 0.04 μm colloidal silica (Struers OP-S) polishing. In preparation for optical microscopy and electron back-scattered diffraction (EBSD), the polished samples were electropolished in 10% Nital solution at 20 V and temperatures ranging from -30 °C to -20 °C. For metallography, the samples were then etched with acetic picral (30 ml acetic acid, 15 ml $H_2O$, 6 g picric acid, 100 ml ethanol) for 3 to 5 s.

The macro-texture was evaluated by XRD technique using a Bruker D8 diffractometer with a Co Kα source. The orientation distribution function (ODF) was constructed from the incomplete pole figures of {10.0}, {00.2}, and {10.1}. Recalculated pole figures were then derived from the ODFs. Micro-texture was studied via EBSD technique in a Hitachi S 3000 FE-SEM fitted with a TSL EBSD camera operating at 20 kV, 70° tilt angle and a step size of 0.3 µm. The color format of EBSD maps is available on the CD version of the paper.

## Results and Discussion

### Macro-texture

The as-rolled macro-textures at different temperatures are shown in Figure 1 by (00.2) and (10.0) pole figures (PFs).

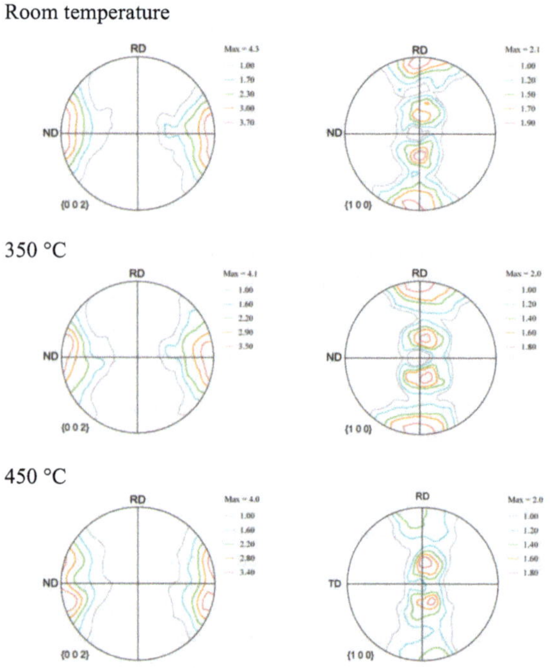


Figure 1. X-ray bulk texture after rolling at room temperature, 350 °C and 450 °C given by (00.2) and (10.0) PFs.

The as-rolled texture contains the typical strong basal component with the same value of maximum intensity in basal PFs at all temperatures. It should be mentioned that the average error in the maximum intensity of basal PFs is ±12%.

Two off-basal texture components, visible in the basal PFs, are also present: (i) ND-RD component in which *c*-axes are rotated towards the RD, and (ii) ND-TD component in which *c*-axes are tilted towards the TD. The intensity of the latter component ranges from 1.6 to 1.8 MROD (multiples of a random orientation distribution). It is noteworthy to mention that despite lower reduction at room temperature, the general features of the texture as well as the maximum intensity of basal PFs is similar to those at 350 °C. At 450 °C, the basal PF exhibits a double-peak shape, which could result from the possible activity of <**c** + **a**> slip system [2].

### Microstructure

The as-rolled microstructure of the Mg-2.9Y alloy at 450 °C is shown in Figure 2.

The interesting feature of the microstructure is the absence of dynamic recrystallization (DRX) even at 450 °C. This was found after close examination of high-stored energy locations such as twins, twin intersections, and bands. The suppression of DRX in the presence of Y has recently been observed [8].

(a)

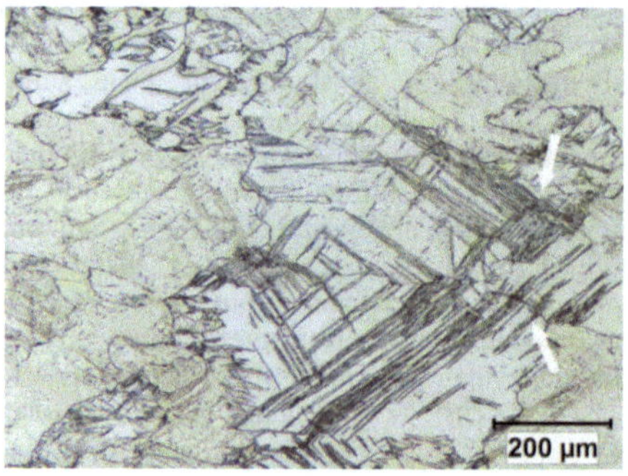


(b)

Figure 2. Optical micrographs of the as-rolled Mg-2.9Y alloy at 450 °C in RD-ND plane (RD is horizontal) from the center of sheets.

The role of Y in DRX suppression is not yet known, but is believed to be related to a change in the atomic potential. This change could possibly alter the lattice friction and consequently the critical resolved shear stress (CRSS) of slip systems. Another possibility is the pinning effect of Y on boundaries (low angle and high angle), which prevents recrystallization.

Twins with both thin and thick morphologies have evolved during deformation (Figure 2(b)). The thin twins, which are most probably contraction and/or double twins [14], pass sometimes through several grains, as is highlighted by arrows in Figure 2(a). The microstructure features at 350 °C and room temperature are similar to those at 450 °C.

### Micro-texture

To enhance the EBSD indexing, the as-rolled material was annealed for 3 min at 400 °C, and the obtained microstructures are depicted in Figure 3. High-angle boundaries (HABs with $\theta > 15°$) are shown by black lines.

(a)

(b)

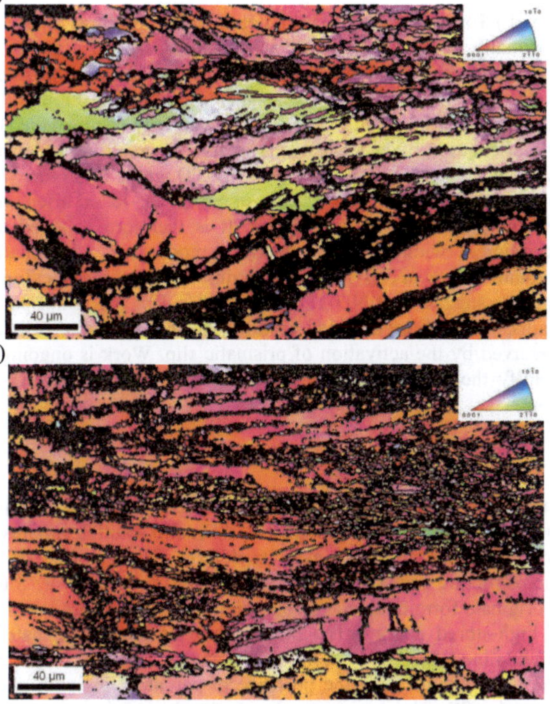


Figure 3. EBSD map (referring to the ND) after rolling at 350 °C (a) and 450 °C (b) followed by annealing at 400 °C for 3 min, showing band-like structures. The maps show the RD-ND plane with the RD being horizontal.

Un-indexed regions appearing in band-like structure are observable, which are most probably zones of deformation localization. Note that the short annealing treatment initiated formation of new grains at the bands. The bands, observed at all rolling temperatures, appear either at 30-40° from the RD or horizontally along the RD. The microstructure examination revealed that the horizontal bands are majorly limited to one basal parent grain, and do not pass through several grains. An example of horizontal bands is shown in Figure 4.

Figure 4. EBSD map (referring to the ND) after rolling at 450 °C (followed by annealing at 400 °C for 3 min) showing horizontal bands in basal parent grains. The maps show the RD-ND plane with the RD being horizontal.

The bands contain mainly compression and double twins as is evidenced by different types of twin boundary shown in Figure 5. The image quality (IQ) map of the microstructure region shown in Figure 3(a) is given in Figure 5 along with the length fraction of each twin boundary.

| Twin type | Angle/Axis | Length fraction |
|---|---|---|
| Contraction {10.1} | 56°@<11.0> | 0.17 |
| Contraction {10.3} | 64°@<11.0> | 0.19 |
| Extension {10.2} | 86°@<11.0> | 0.15 |
| Double {10.1}-{10.2} | 38°@<11.0> | 0.23 |
| Double {10.3}-{10.2} | 22°@<11.0> | 0.26 |

Figure 5. IQ map of the microstructure region illustrated in Figure 3(a). The boundaries corresponding to different twin types are also shown.

As can be seen in Fig. 5, two types of contraction twinning, {10.1} and {10.3}, extension twinning {10.2}, and two types of double twinning, {10.1}-{10.2} and {10.3}-{10.2}, are present. The bands, and noticeably the large one at the bottom of the map, contain mainly boundaries corresponding to double twinning. This is evidenced by their larger fraction compared to that of contraction and extension twins, i.e. 0.23 in the case of {10.1}-{10.2} twinning and 0.26 in the case of {10.3}-{10.2} twinning. The contraction (red and yellow boundaries) and extension (orange boundaries) twins, detected on the IQ map, are parts of the above-mentioned double twins. This is illustrated in Figure 6 by a detailed description of different twin boundaries for the rolled material at 450 °C.

A white arrow on Figure 6(b) highlights a white boundary pertaining to a {10.3}-{10.2} double twin. The upper boundary of this twin has formed by {10.3} contraction twinning (yellow boundary) and {10.2} extension twinning (orange boundary). This example illustrates the connection between double twinning and the constituent contraction and extension twinning. In other words, double twinning occurs when the matrix twins along either {10.1} or {10.3} plane, and then the twinned region twins along {10.2} plane. Therefore, when double twins are detected, the constituent primary twins are also present in the microstructure.

As can be seen in Figure 6(a), un-indexed parts exist inside the twins near the white arrow, which are indicative of deformation localization at these twins, and most probably the mechanism by which the bands form with straining. As a result of double twinning, unfavorably oriented basal planes reorient to a more favorable orientation (38° and 22° reorientation after {10.1}-{10.2} and {10.3}-{10.2} double twinning, respectively), which could accommodate further strain.

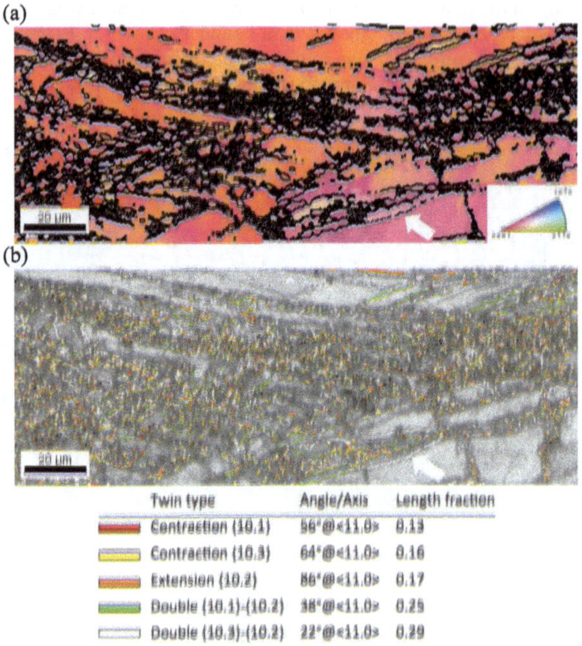


Figure 6. EBSD map (referring to the ND) of the rolled material at 450 °C, and the corresponding IQ map with different twin boundaries, showing the presence of contraction and double twins.

At all rolling temperatures, the parent grains were found to possess two different orientations: (i) the typical basal orientation (colored in red in Figure 7), or (ii) an orientation with *c*-axis along the TD (colored in green in Figure 7). The latter orientation pertains to the ND-TD component of the bulk texture displayed in Figure 1, and the parent grains possessing such orientation are hereafter termed ND-TD (parent) grains.

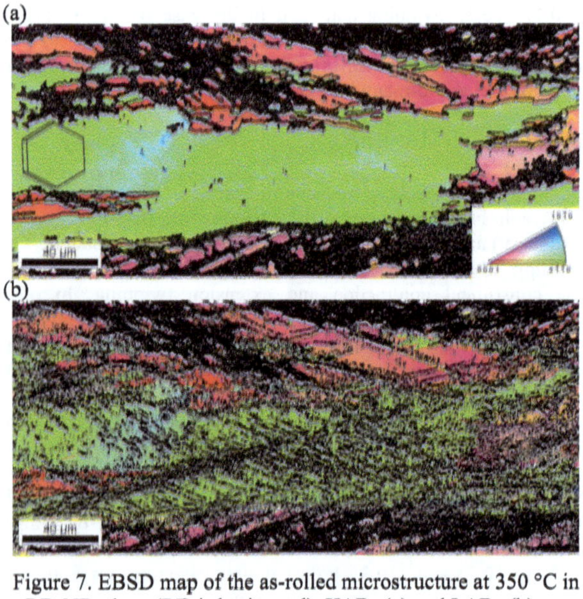


Figure 7. EBSD map of the as-rolled microstructure at 350 °C in RD-ND plane (RD is horizontal), HABs (a) and LABs (b) are shown separately on the EBSD map for clarity.

The LABs within ND-TD parent grains (Figure 7(b)) are essentially straight lines that intersect at an angle of approximately 120°, and coincide with the trace of prismatic planes as is illustrated by the hexagonal unit cell in Figure 7(a). These observations indicate the possible activity of prismatic slip. The orientation of ND-TD grains is also favorable for extension twinning. The upper area of the ND-TD grain in Figure 7(a) seems to be twinned parts because of both (near-)basal orientation and morphology of these regions. It is therefore believed that extension twinning is another mode of deformation accommodation in ND-TD parent grains. All the above-mentioned findings suggest that ND-TD grains are in fact the initial parent grains, which formed during casting, and that their orientation is preserved by the activation of prismatic slip. Work is ongoing to identify the origin and type of slip system activated in ND-TD grains.

## Summary

The as-rolled microstructure of a solid solution Mg-2.9Y alloy was characterized. It was found that Y in solid solution suppresses dynamic recrystallization even at 450 °C. The unrecrystallized microstructure consists of twins and bands. The bands were found to originate form double and compression twins, and are locations of high-stored energy. The bulk texture of the rolled material exhibits the typical basal component, and an ND-TD component. The latter results from the presence of parent grains with *c*-axis along the TD. The strain is accommodated in such parent grains by extension twinning and possibly prismatic slip.

## References

[1] K.U. Kainer, *Magnesium Alloys and Technologies* (Weinheim: Wiley-VCH, 2003).

[2] S.R. Agnew, M.H. Yoo, C.N. Tomé, "Application of Texture Simulation to Understanding Mechanical Behavior of Mg and Solid Solution Alloys Containing Li or Y," *Acta Mater.*, 49 (2001) 4277-4289.

[3] M.R. Barnett, M.D. Nave, C.J. Bettles, "Deformation Microstructures and Textures of Some Cold Rolled Mg Alloys," *Mater. Sci. Eng. A*, 386 (2004) 205-211.

[4] A. Jager, P. Lukac, V. Gartnerova, J. Haloda, M. Dopita, "Influence of Annealing on the Microstructure of Commercial Mg Alloy AZ31 after Mechanical Forming," *Mater. Sci. Eng. A*, 432 (2006) 20-25.

[5] J. Bohlen, M.R. Nürnberg, J.W. Senn, D. Letzig, S.R. Agnew, "The Texture and Anisotropy of Magnesium–Zinc–Rare Earth Alloy Sheets," *Acta Mater.*, 55 (2007) 2101-2112.

[6] K. Hantzsche, J. Bohlen, J. Wendt, K.U. Kainer, S.B. Yi, D. Letzig, "Effect of Rare Earth Additions on Microstructure and Texture Development of Magnesium Alloy Sheets," *Scripta Mater.*, 63 (2010) 725-730.

[7] N. Stanford, "Micro-alloying Mg with Y, Ce, Gd and La for Texture Modification—A Comparative Study," *Mater. Sci. Eng. A*, 527 (2010) 2669-2677.

[8] S. Yi, D. Letzig, K. Hantzsche, R.G. Martinez, J. Bohlen, I. Schestakow, S. Zaefferer, "Improvement of magnesium sheet formability by alloying addition of rare earth elements," *Mater. Sci. Forum*, 638-642 (2010) 1506-1511.

[9] L.L. Rokhlin, *Magnesium Alloys Containing Rare Earth Metals* (London: Taylor & Francis, 2003).

[10] R. Cottam, J. Robson, G. Lorimer, B. Davis, "Texture Evolution During the Static Recrystallization of Three Binary Mg-Y Alloys," *Ceram. Trans.*, 200 (2008) 501-508.
[11] R. Cottam, J. Robson, G. Lorimer, B. Davis, "Dynamic Recrystallization of Mg and Mg–Y Alloys: Crystallographic Texture Development," *Mater. Sci. Eng. A*, 485 (2008) 375-382.
[12] S. Sandlöbes, S. Zaefferer, I. Schestakow, S. Yi, R. Gonzalez-Martinez, "On the Role of Non-basal Deformation Mechanisms for the Ductility of Mg and Mg–Y Alloys," *Acta Mater.*, 59 (2011) 429-439.
[13] S.A. Farzadfar, M. Sanjari, I.-H. Jung, E. Essadiqi, S. Yue, "Role of Yttrium in the Microstructure and Texture Evolution of Mg," *Mater. Sci. Eng. A*, 528 (2011) 6742-6753.
[14] S.E. Ion, F.J. Humphreys, S.H. White, "Dynamic Recrystallisation and the Development of Microstructure During the High Temperaure Deformation of Magnesium," *Acta Metall.*, 30 (1982) 1909-1919.

# Magnesium Technology 2012

## Energy and Biomedical / Primary Production

### Thursday PM

*Session Chair:*

**Wim Sillekens**
**(TNO, Netherlands)**

**Neale Neelameggham**
**(IND LLC, USA)**

**Magnesium Technology 2012**
*Edited by: Suveen N. Mathaudhu, Wim H. Sillekens, Neale R. Neelameggham, and Norbert Hort*
***TMS (The Minerals, Metals & Materials Society), 2012***

# *IN-VITRO* CORROSION STUDIES OF BIOABSORBABLE ALLOYS

P. Gill, N. Munroe
Department of Mechanical and Materials Engineering, Florida International University, Miami, FL, USA

Keywords: magnesium, biodegradable, corrosion, SEM, XRD

## Abstract

Magnesium alloys have inspired a significant amount of attention from researchers all over the world for cardiovascular and orthopedic applications due to their light weight, mechanical integrity and degradation behavior. In this investigation, cast manufactured binary, ternary and quaternary magnesium alloys were studied for their degradation behavior by potentiodynamic polarization tests in phosphate buffer saline solution (PBS) and PBS containing amino acids (cysteine, C and tryptophan, W) at 37 °C. Electrochemical impedance spectroscopy (EIS) tests were performed to determine the charge transfer resistance and immersion tests were performed to assess corrosion rate and hydrogen evolution from the alloys. Furthermore, the surface morphology and surface chemistry of the alloys were observed by scanning electron microscopy (SEM) and X-ray diffraction (XRD).

## Introduction

Biodegradable alloys are a group of materials, which once implanted, negate further surgical procedures. The demand for better and improved implants has led researchers to explore different types of materials such as: polymers, ceramics, metals and composites. Recently, interest in biodegradable magnesium alloys has increased due to their light weight (low density, close to human bone), mechanical integrity (high strength to weight ratio) and biocompatibility (fourth most abundant cation in human body) [1, 2, 3]. However, these alloys face major challenges with regard to non-uniform corrosion, possibility of premature failure and evolved hydrogen [4]. Alloying is one method of improving the mechanical properties and corrosion resistance of Mg [4]. For bioabsorbable applications, the alloying element must be non-toxic and biocompatible [4, 5], so that the concentration of the dissolved ions once implanted, would not substantially exceed the concentration normally found in human blood [6]. The degradation behavior of magnesium alloys is mainly affected by both intrinsic (composition, microstructure, energy, etc.) and extrinsic factors (temperature, pH, amino acids, chloride ions, etc.) [7].

It has been reported that the corrosion susceptibility of implants may vary with the type of amino acid to which they are exposed [8]. The general structure of amino acids is R–CH($NH_2$)–COOH [4, 9], which have a zwitterionic or ampholyte character that is important for the adsorption of molecular species [9]. No significant difference in corrosion susceptibility was observed on exposing Nitinol to amino acids [8]. However, some titanium alloys displayed a slight increase in corrosion susceptibility [9]. There has been little or no significant corrosion studies reported on biodegradable magnesium alloys in amino acids. Chang et al. performed corrosion studies on Mg-Li with amino acids and reported it to have higher corrosion resistance [10]. Yamamoto and Hiromoto studied the effect of amino acids on pure magnesium and reported that amino acids reduced the barrier effect of insoluble salt layer against dissolution of magnesium, which leads to faster dissolution of Mg [11].

In this paper, cast MgZn, MgZnCa and MgZnCaGd alloys were studied for *in-vitro* corrosion properties. Potentiodynamic polarization studies were performed in PBS and PBS containing amino acids, in an attempt to explore their effect on the corrosion behavior of the alloys. Besides, the alloys were also tested for its hydrogen evolution in PBS and surface morphology.

## Materials and methods

Cast Mg1Zn, Mg1Zn1Ca and Mg1Zn1Ca1Gd (wt%) alloys were prepared using high purity alloying elements. The alloys were manufactured by melting the elements at 1000 °C under an inert atmosphere (Argon) and casting in a water cooled copper mould. The ramp-up time from room temperature to melting temperature was 1 minute and alloys were further heat treated at 350 °C and water quenched. The ingots were cut into cubes of dimensions 0.414x0.414x0.08 (inch), which were then mechanically polished to achieve a roughness close to 0.05 microns using abrasives and abrasive papers (Buehler) and lubricants (Sigma-Aldrich). Samples were polished in four steps as shown in Table 1. The usage of water based solutions was avoided during sample preparation, in order to prevent the hydrolysis of the alloy. PBS (Sigma Aldrich) and two amino acids (Acros Organics), cysteine (0.25 mM) and tryptophan (0.042 mM) were used as electrolytes for corrosion studies, with the latter at concentrations typically found in human blood [4].

Table 1. Lubricants and abrasives used during sample preparation.

| Surface | Lubricant | Abrasive |
|---|---|---|
| Carbimet® | Ethanol | SiC, 400 grit |
| Texmet® | Ethanol-Ethylene Glycol (3:1) | Metadi® Paste, 9 μm |
| Texmet® | Ethanol-Ethylene Glycol (3:1) | Metadi® Paste, 3 μm |
| Chemomet® | Ethanol-Ethylene Glycol (3:1) | Masterpolish®, 0.05 μm |

### Electrochemical testing

A typical three electrode (working, counter and reference) corrosion cell interphased with a potentiostat was used, in which the implant material (working electrode) is placed in deaerated simulated physiological solution, PBS and PBS containing amino acids (PBS+C and PBS+W). The potentiodynamic polarization tests were performed in accordance with ASTM: G 102-89 [12] in PBS (pH = 7.2) at 37 °C employing a GAMRY® potentiostat (G-750). All tests were conducted at a scan rate of 1.0 mV/s. The electrolyte was purged with high purity nitrogen for 30 minutes prior to immersion of the sample, as well as continuously during the corrosion test. The corrosion rate (CR) of the alloys was determined by:

$$CR = (I_{corr}.K.EW)/(\rho.A)$$

where, $I_{corr}$ is the corrosion current (amps), K is a constant for the corrosion rate (3272 mm/amp.cm.year), EW is the equivalent weight in grams/equivalent, A is the sample area (0.28 $cm^2$) and ρ is the density ($g/cm^3$) of the alloys calculated by Archimedes principal (MgZn, 1.75; MgZnCa, 1.76; and MgZnCaGd, 1.85).

Electrochemical impedance spectroscopy (EIS) tests were also performed to determine the effect of alloying elements on the charge transfer resistance at 37 °C in PBS. All tests were conducted in the frequency range from 1.0E-02 Hz to 1.0E+05 Hz with 10 points per decade.

### Immersion test

Immersion tests were performed in accordance with ASTM G31-72 [13]. Figure 1 show a schematic representation of the immersion test, where samples were placed at the bottom of a beaker filled with PBS, with a funnel placed above the sample. A solution filled measuring cylinder was placed over the funnel to collect and measure the hydrogen gas evolved as the solution was displaced. The samples were soaked in 300 mL of PBS for 192 hours (9 days) at 37 °C. The hydrogen evolution (per square cm surface area), weight gain and corrosion rate were evaluated during the experiment, by taking the average of three measurements. The samples were weighed before and after 24, 48, 72, 96, 120, 144, 168 and 192 hours. For each time interval, the sample was removed from the PBS, rinsed in distilled water, air dried and then weighed. Once the immersion test was completed, the samples were cleaned by immersing in 180 g/l of chromic acid for 20 min. The weight gain (wt) in grams was calculated as follows [14]:

*wt gain = (wt after immersion - wt before immersion)/surface area*

The corrosion rates were also calculated from the quantity of hydrogen evolved and weight change during the immersion test. The corrosion rate (CR, mm/year) was calculated as follows:

$$CR = (tW.10)/(\rho AT)$$

where, t is a constant (8760 h), W is mass loss, g (mass before immersion - mass after cleaning)/surface area, A is the surface area before immersion ($cm^2$), T is the immersion time (h) and ρ is the sample density ($g/cm^3$).

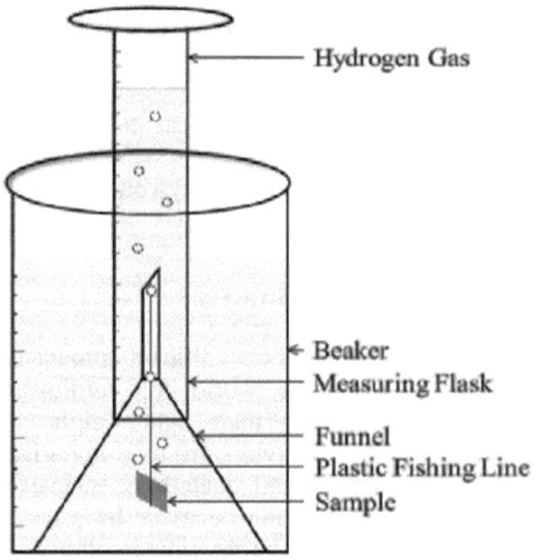


Figure 1. Schematic representation of immersion test setup.

The surface morphology of the mechanically polished alloys before and after corrosion tests was characterized by scanning electron microscopy (SEM, JEOL JSN 6330F). The microstructure phases were determined using XRD (Siemens 5000 D), where Cu-$K_\alpha$ radiation (wavelength, λ = 1.54 A°) operating at 35 mA and 40 kV with a scan rate of 0.01 °/sec over a 2 theta from 20-90° were used.

### Results and Discussion

Figure 2 shows the XRD patterns of mechanically polished alloys. The formation of binary phase precipitates was observed in all alloys. The phases: $Mg_2Ca$, $MgZn_2$, $Mg_5Gd$ [15], were instrumental for precipitation hardening. Zhou et al. reported that MgZnCa alloys synthesized by twin-roll rapid solidification technique produced a large number of precipitates, which contributed to precipitation hardening [16]. Peng et al. manufactured Mg-Gd based alloys and reported superior mechanical properties due to a microstructure composed of fine metastable precipitates dispersed in the matrix [17].

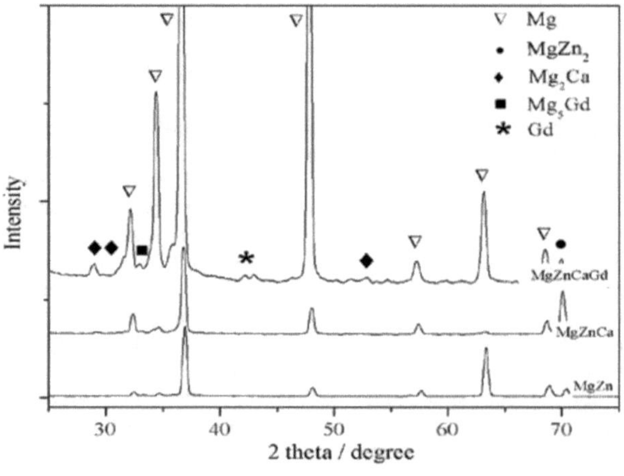


Figure 2. X-ray diffraction pattern of MgZn, MgZnCa and MgZnCaGd.

The morphologies of the alloys before (inset) and after potentiodynamic corrosion tests are compared in Figure 3. Stress induced cracks were observed on the surface of all alloys subjected to potentiodynamic polarization tests in PBS and PBS with amino acids except for MgZn in PBS. MgZnCaGd developed a significant amount of oxides in PBS with cysteine (Figure 2h). Similar cracked corrosion products were reported by various researchers with various Mg alloys [18, 19]. The adsorption of chloride ions on the surface oxides of the alloy lead to the formation of $Mg(OH)_2$ via the hydrolysis of $MgCl_2$ [20]. Dissolution of $MgCl_2$ subsequently exposing the underlying metal surface, which leads to further dissolution of magnesium. At anodic sites, $Mg^{++}$ ions are produced; $Cl^-$ ions diffuse to such sites in order to establish electrical neutrality. $MgCl_2$ is formed in the pits and hydrolyzed according to equation (1) producing hydrochloric acid and further dissolution of Mg.

$$MgCl_2 + 2H_2O \rightarrow Mg(OH)_2 + 2HCl \tag{1}$$

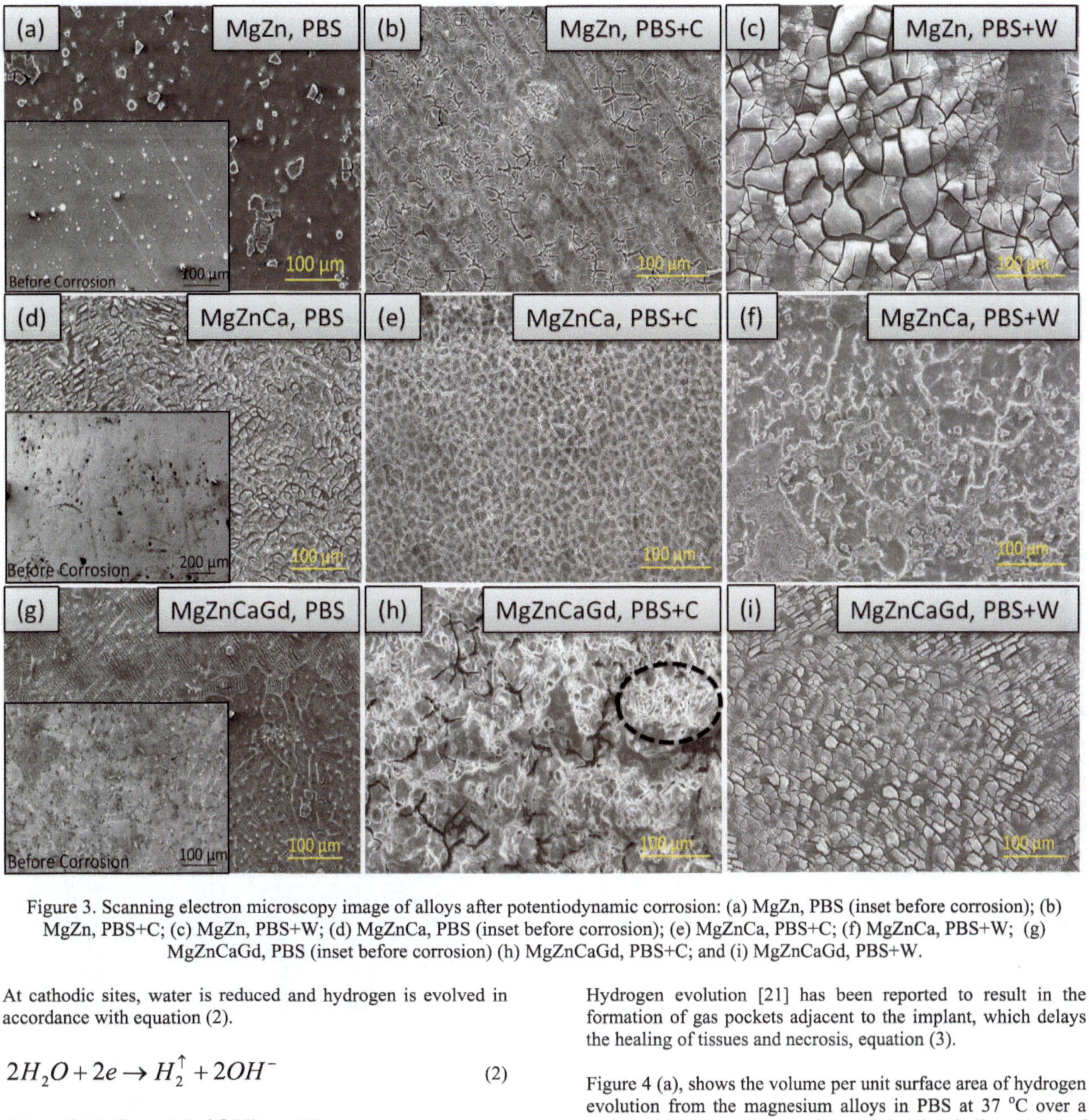


Figure 3. Scanning electron microscopy image of alloys after potentiodynamic corrosion: (a) MgZn, PBS (inset before corrosion); (b) MgZn, PBS+C; (c) MgZn, PBS+W; (d) MgZnCa, PBS (inset before corrosion); (e) MgZnCa, PBS+C; (f) MgZnCa, PBS+W; (g) MgZnCaGd, PBS (inset before corrosion) (h) MgZnCaGd, PBS+C; and (i) MgZnCaGd, PBS+W.

At cathodic sites, water is reduced and hydrogen is evolved in accordance with equation (2).

$$2H_2O + 2e \rightarrow H_2^{\uparrow} + 2OH^- \tag{2}$$

$$Mg + 2H_2O \rightarrow Mg(OH)_2 + H_2 \text{ (overall reaction)} \tag{3}$$

Hydrogen evolution [21] has been reported to result in the formation of gas pockets adjacent to the implant, which delays the healing of tissues and necrosis, equation (3).

Figure 4 (a), shows the volume per unit surface area of hydrogen evolution from the magnesium alloys in PBS at 37 °C over a period of 216 hours. It was observed that the similar volume of hydrogen was released from all of the alloys during the first 24 hours (Figure 4 (a) inset). After 24 hours, more hydrogen was released from MgZnCaGd, followed by MgZnCa and MgZn. This may be attributed to the formation of four oxides with MgZnCaGd, two of which are those of light metals Mg and Ca (porous oxides of smaller volume) and those of heavy metals, Zn and Gd (non-porous oxides of greater volume) [22]. Mg and Ca oxidize at a rate nearly constant with time (linear equation) whereas, Gd and Zn oxidize proportionally to the logarithm of the time (parabolic equation) [22].

$$y = K_1 t + A_1 \qquad \text{(Linear equation)}$$

$$y_2 = K_2 t + A_2 \qquad \text{(Parabolic equation)}$$

The different oxidation rates of the four oxides induce stresses, which result in the formation of surface cracks seen in the photomicrographs of Figure 3.

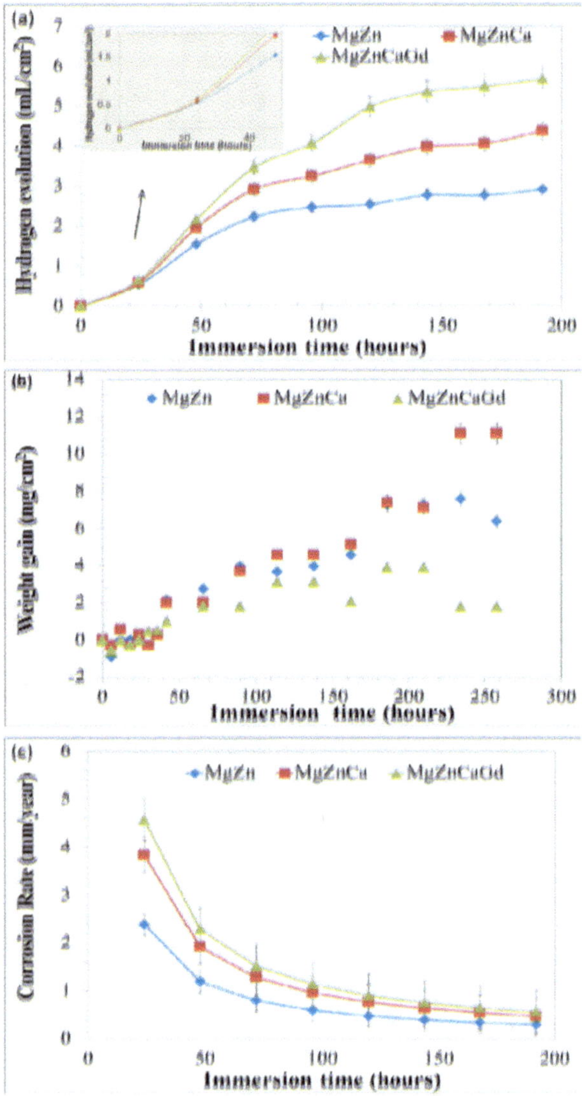


Figure 4. Immersion test of as-cast mechanically polished Mg alloys immersed in PBS for 192 hours at 37 °C: (a) hydrogen evolution volumes; (b) weight gain and (c) corrosion rate.

Figure 4 (b) shows the weight gain of the alloys as function of immersion time. MgZnCa and MgZn exhibited the highest weight gain followed by MgZnCaGd, which may be attributed to the stress induced delamination of the oxide (passivating layer) from the latter, as previously described. The highest weight gain observed in the MgZnCa may be due to the formation three oxides as compared with two for MgZn.

Figure 4 (c) shows the corrosion rate of the alloys as a function of immersion time in PBS at 37 °C. MgZnCaGd exhibited the highest corrosion rate followed by MgZnCa and MgZn. This may be attributed to the relative oxide porosity, cracking and delamination of the oxides in case of the former, which expose the metal to the electrolyte. The rates of corrosion can be inferred from SEM photomicrographs of Figure 4 (c), where there is evidence of uniform corrosion on MgZnCaGd as compared with localized corrosion on MgZn and MgZnCa.

Figure 5 shows the SEM photomicrographs after immersion test for 192 hours in PBS at 37 °C. It was observed that localized corrosion initially occurred at grain boundary sites and then propagated along the grain boundaries (arrows depict pits and circles depict corrosion propagation).

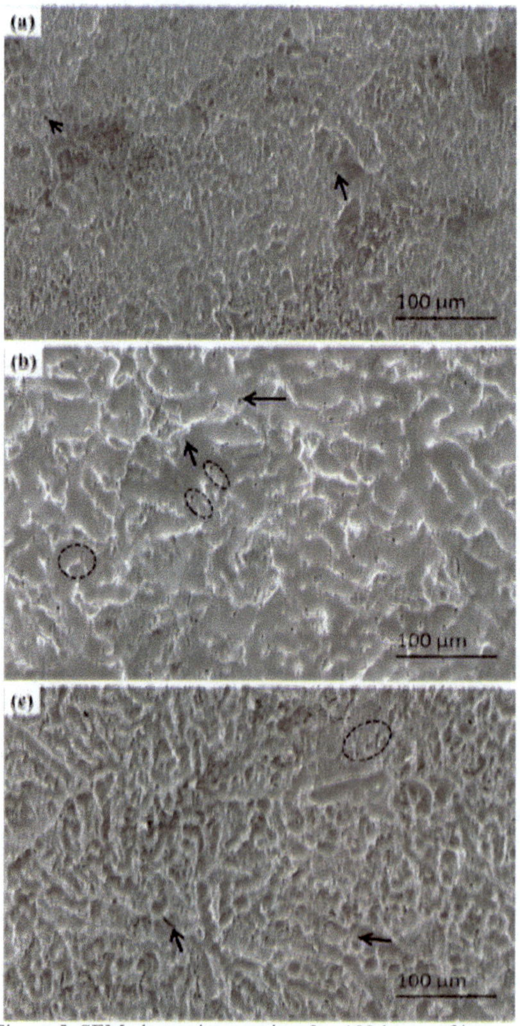


Figure 5. SEM photomicrographs after 192 hours of immersion in PBS: (a) MgZn; (b) MgZnCa and (c) MgZnCaGd.

EDS analysis on the surface of the alloys after immersion for 192 hours are shown in Table 2. The relative concentration of Mg, O, P, K, Zn, Ca and Gd on the surface of the alloys corresponds to the formation of a mixture of apatite, zinc phosphate and calcium phosphate.

Table 2. EDS analysis on the surface of the alloys after immersion in PBS for 192 hours.

| Alloys | Element percent, wt% | | | | | | |
|---|---|---|---|---|---|---|---|
| | Mg | O | P | K | Zn | Ca | Gd |
| MgZn | 74.0 | 18.4 | 4.6 | 0.2 | 2.7 | - | - |
| MgZnCa | 84.8 | 10.2 | 1.7 | 0.3 | 2.0 | 1.1 | - |
| MgZnCaGd | 61.1 | 13.4 | 3.5 | 0.5 | 2.4 | 0.9 | 18.2 |

Figure 6, Nyquist plots revealed an increase in corrosion resistance with the addition of Ca and Gd to MgZn. This could be attributed to the formation of various oxides, which act as an electric barrier. In the case of biodegradable alloys, initial slower corrosion rates followed by uniform degradation are desirable.

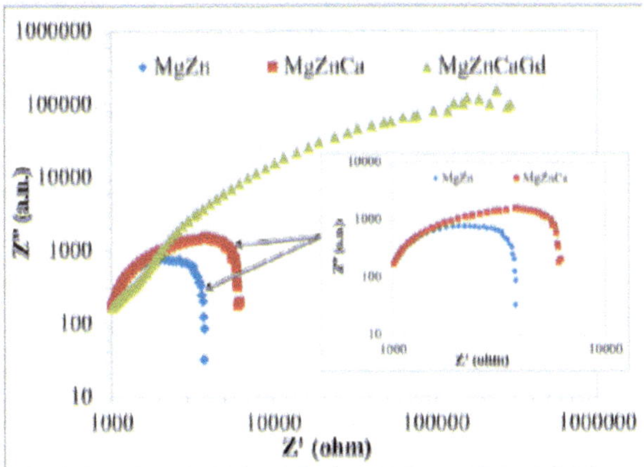


Figure 6. Nyquist plots (log-log representation) for MgZn, MgZnCa and MgZnCaGd in PBS at 37 °C.

Figure 7 shows typical polarization curves for MgZn, MgZnCa and MgZnCaGd alloys, where passivation control occurred between 0.001 - 0.002 A/cm$^2$. The rough appearance of the anodic curves resulted from continuous breakdown (pitting corrosion) and repassivation [18]. The corrosion parameters of the alloys in PBS and PBS with amino acids are shown in Table 3. MgZn had the lowest corrosion rate in PBS (23.17 mm/year) and exhibited the highest corrosion rate in PBS+W (130.11 mm/year). A similar trend was observed with MgZnCaGd, where the corrosion rate increased in PBS with amino acids. This could be attributed to the acid-base/polarity nature of amino acids. Thus, further studies are required; with titanium alloys, it was reported that amino acids are readily adsorbed onto the surface of $TiO_2$ via their carboxyl groups by replacing the basic hydroxyl group on a Ti site [9, 23].

MgZn had the highest corrosion potential, $E_{corr}$ (-1.4 V) in PBS, followed by MgZnCaGd (-1.6 V) and MgZnCa (-1.7 V), indicating that MgZn is least susceptible to corrosion followed by MgZnCaGd and MgZnCa.

Table 3. Average potentiodynamic polarization data from three test values in PBS and PBS containing amino acids.

| Alloys | $E_{corr}$, V | $I_{corr}$, μA | CR, mm/year |
|---|---|---|---|
| MgZn, PBS | -1.4 | 2.8E-4 | 23 |
| MgZnCa, PBS | -1.7 | 6.3E-4 | 51 |
| MgZnCaGd, PBS | -1.6 | 5.3E-4 | 44 |
| MgZn, PBS+C | -1.7 | 7.9E-4 | 64 |
| MgZnCa, PBS+C | -1.7 | 4.7E-4 | 38 |
| MgZnCaGd, PBS+C | -1.6 | 19.7E-4 | 164 |
| MgZn, PBS+W | -1.7 | 16.0E-4 | 130 |
| MgZnCa, PBS+W | -1.7 | 3.5E-4 | 28 |
| MgZnCaGd, PBS+W | -1.6 | 18.7E-4 | 156 |

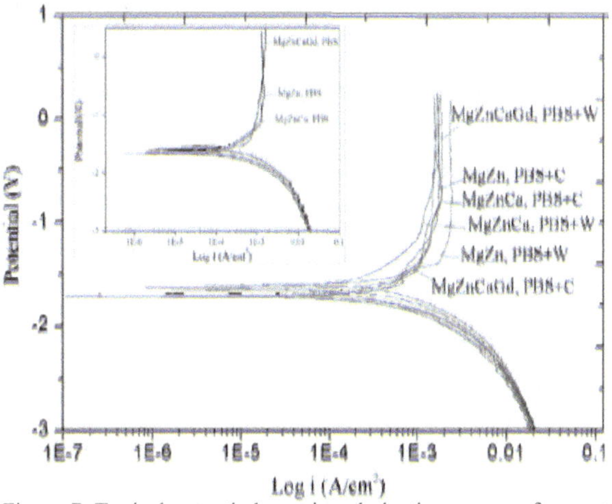


Figure 7. Typical potentiodynamic polarization curves of as-cast magnesium alloys in PBS (inset) and PBS containing amino acids at 37 °C.

## Conclusion

XRD depicted the presence of various binary phases, which may improve the mechanical properties of the alloys. Stress induced cracks developed on the surface of all alloys subjected to accelerated corrosion test except for MgZn. Potentiodynamic corrosion tests in PBS revealed that MgZnCa was the most susceptible to corrosion followed by MgZnCaGd and MgZn. In general, the corrosion rates of the magnesium alloys increased when amino acids were added to PBS. Immersion corrosion tests in PBS revealed that MgZnCaGd exhibited the highest corrosion rate followed by MgZnCa and MgZn. However, it should be noted that there was no significant difference in the corrosion rates of MgZnCa and MgZnCaGd in PBS for either accelerated or immersion corrosion tests. Photomicrographs of alloys immersed in PBS revealed that localized corrosion initially occurred at grain boundary sites, then, propagated along the grain boundaries. Nyquist plots depicted an increase in corrosion resistance with the addition of alloying elements, due to the formation of various oxides.

## Acknowledgement

P.G. and N.M. acknowledge Dr. H. T. Lin and Amit Datye from Oak Ridge National Laboratory (ORNL) for their helpful discussions. P.G. acknowledges the Dissertation Evidence Acquisition Fellowship from University Graduate School Florida International University.

N.M. would like to acknowledge the financial support from National Institute of General Medical Sciences Award Number SC3GM084816.


## References


1. R.K. Rude, "Magnesium deficiency: A cause of heterogeneous disease in humans", J Bone Miner Res, 13 (1998), 749-58.
2. M.E. Shils, Magnesium. In: Shils ME, Olson JA, Shike M, editors (Modern nutrition in health and disease, 8th ed, Philadelphia: Lea & Febiger (1994), 164–84.
3. G.L. Song, A. Atrens, "Corrosion mechanisms of magnesium alloys", Advanced Engineering Materials, 1 (1999), 11-33.
4. M.P. Staiger, A.M. Pietak, J. Huadmai, G. Dias, "Magnesium and its alloys as orthopedic biomaterials: A review", Biomaterials, 27 (2006), 1728–1734.
5. B. Heublein, R. Rohde, V. Kaese, M. Niemeyer, W. Hartung, A. Haverich, "Biocorrosion of magnesium alloys: a new principle in cardiovascular implant technology?", Heart, 89 (2003), 651–656.
6. H.X. Wang, S.K. Guan, X. Wang, C.X. Ren, L.G. Wang, "In vitro degradation and mechanical integrity of Mg–Zn–Ca alloy coated with Ca-deficient hydroxyapatite by the pulse electrodeposition process", Acta Biomaterialia, 6 (2010), 1743–1748.
7. W. Haider, "Enhanced Biocompatibility of NiTi (NITINOL) Via Surface Treatment and Alloying" Ph.D. thesis, Florida International University (2010), 177.
8. B.G. Pound, "Corrosion behavior of nitinol in blood serum and PBS containing amino acids", J Biomed Mater Res B Appl Biomater, 2 (2010), 287-295.
9. P. Gill, N. Munroe, C. Pulletikurthi, S. Pandya, W. Haider, "Effect of Manufacturing Process on the Biocompatibility and Mechanical Properties of Ti-30Ta Alloy", J Mater Eng Perform, 20 (2011), 819-823.
10. L. Chang, P. Wang, W. Liu, "Effect of amino acids on the structure and corrosion resistance of Mg-Li alloy anodic oxide film", Advanced Materials Research, 146-147 (2010), 785-788.
11. A. Yamamoto, S. Hiromoto, "Effect of inorganic salts, amino acids and proteins on the degradation of pure magnesium in vitro", Materials Science and Engineering C, 29 (2009), 1559–1568.
12. ASTM G102-89 (1999), Standard Practice for Calculation of Corrosion Rates and Related Information from Electrochemical Measurements. DOI: 10.1520/G0102-89R99.
13. ASTM G31-72 (2004), Standard Practice for Laboratory Immersion Corrosion Testing of Metals. DOI: 10.1520/G0031-72R04.
14. L. Xu, E. Zhang, D. Yin, S. Zeng, K. Yang, "In vitro corrosion behaviour of Mg alloys in a phosphate buffered solution for bone implant application", J Mater Sci: Mater Med, 19 (2008), 1017-1025.
15. Z. Li, X. Gu, S. Lou, Y. Zheng, "The development of binary Mg–Ca alloys for use as biodegradable materials within bone", Biomaterials, 29 (2008), 1329–1344.
16. T. Zhou, D. Chen, Z. Chen, "Microstructures and properties of rapidly solidified Mg-Zn-Ca alloys", Trans. Nonferrous Met. Soc. China, 18 (2008), 101-106.
17. Q. Peng, X. Hou, L. Wang, Y. Wu, Z. Cao, L. Wang, "Microstructure and mechanical properties of high performance Mg–Gd based alloys", Materials and Design, 30 (2009), 292–296.
18. Y. Wang, C.S. Lim, C.V. Lim, M.S. Yong, E.K. Teo, L.N. Moh, "In vitro degradation behavior of M1A magnesium alloy in protein-containing simulated body fluid", Materials Science Engineering C, 31 (2011), 579-587.
19. Y. Song, D. Shan, R. Chen, F. Zhang, E.H. Han, "Biodegradable behaviors of AZ31 magnesium alloy in simulated body fluid", Materials Science and Engineering C, 29 (2009), 1039–1045.
20. R. Ambat, N.N. Aung, W. Zhou, "Studies on the influence of chloride ion and pH on the corrosion and electrochemical behaviour of AZ91D magnesium alloy", Journal Applied Electrochemistry, 30 (2000), 865-874.
21. G. Song, "Control of biodegradation of biocompatable magnesium alloys", Corrosion Science, 49 (2007), 1696-1701.
22. H.H. Uhlig, The corrosion handbook (John Wiley & Sons, Inc. (1948), 15.
23. M. Schmidt, S.G. Steinemann, "XPS studies of amino acids adsorbed on titanium dioxide surfaces", J. Anal. Chem, 314 (1991), 412–415.

**Magnesium Technology 2012**
*Edited by: Suveen N. Mathaudhu, Wim H. Sillekens, Neale R. Neelameggham, and Norbert Hort*
***TMS (The Minerals, Metals & Materials Society), 2012***

# HIGH-CAPACITY HYDROGEN-BASED GREEN-ENERGY STORAGE SOLUTIONS FOR THE GRID BALANCING

F.D'Errico[1], A.Screnci[2]
[1]Politecnico di Milano, Department of Mechanical Engineering, Via La Masa 34, 20156 Milan, Italy
[2]Mc Phy Energy , Zone d'activité 26190 La Motte-Fanjas (France)

Keywords: Magnesium, nanostructure, hydrogen storage, sustainability, renewable sources

## Abstract

One of the current main challenges in green-power storage and smart grids is the lack of effective solutions for accommodating the unbalance between renewable energy sources, that offer intermittent electricity supply, and a variable electricity demand. Energy management systems have to be foreseen for the near future, while they still represent a major challenge. Integrating intermittent renewable energy sources, by safe and cost-effective energy storage systems based on solid state hydrogen is today achievable thanks to recently some technology breakthroughs. Optimized solid storage method made of magnesium-based hydrides guarantees a very rapid absorption and desorption kinetics. Coupled with electrolyzer technology, high-capacity storage of green-hydrogen is therefore practicable. Besides these aspects, magnesium has been emerging as environmentally friend energy storage method to sustain integration, monitoring and control of large quantity of GWh from high capacity renewable generation in the EU.

## Introduction

Renewable Energy Sources (RES), such as Wind and Solar Energy, biomass, and hydropower have been relatively highly costly, but advances over the past few decades have made some renewable resources much more price-competitive. While the Organisation for Economic Co-operation and Development (OECD) countries' global primary energy demand is expected to show only a slightly grow with respect to other countries worldwide (with an overall 36% increase by 2035), their demand for RES will show a relatively large increase, thus calling for appropriate solutions with a systematic approach.
The problems of climate change, air pollution and energy insecurity requires a large scale conversion to clean, perpetual, and reliable energy at low cost together with an increase in energy efficiency. That is why, for the next future, RES will enter more and more the current mainstream, also consistently with EU strategic policies and initiatives in the energy segment (such as the European Electricity Grid Initiative (EEGI), the European Strategic Energy Technology (SET) Plan, etc.).
To this purpose the European Strategic Energy Technology (SET) Plan has identified fuel cells and hydrogen among the technologies needed for Europe to achieve the targets for 2020 as well as to achieve the long-term vision for 2050 towards decarbonisation.
A possible scheme for the future evolution of the energy distribution usually refers to the so-called SMART GRID that consists widely in the intelligent (i.e. dual side communication) integration of systems for distributed generation, centralized renewable power generation, non-renewable but cleaner centralized power generation for RES peaks stabilizing and information from the users' side to efficiently manage - in real-time - electricity peripheral flows in order to avoid mismatching between demand side (users) and supply side (power production).
All these features seek to build IT framework to overcome constraints due to "passive" architecture of current grid. Peaks loads efficient management is crucial for give chance to further increase for share of power generation via-renewable sources.
One other important issue the International Energy Agency (IEA) identified as key is energy storage.
Energy storage is considered one of the promising means both to feed the grid by RES and also to balance the grid energy flow according to the demand needs. It is also clear that advanced storage solutions, of course coupled with other infrastructures and technology initiatives (such as the ones above mentioned on Smart Grids) can power off-grid utilities, such as a green mobility system with electric vehicles, that are powered by stored electricity once the grid can feed exclusively the requested demand.
Energy storage is therefore one of the most promising solution to support the development and usage of RES, as well as introduce other benefits as it enables:

- to capture surplus renewable energy, that cannot be utilized by the grid due to low load demand or transmission constraints;
- to provide supplementary or backup power in correspondence of RES reduction or absence;
- to effectively alleviate power quality issues (e.g. interferences arising from the connection of wind turbines to the grid), thus improving the electrical quality of the RES power output.

## Existing technologies for energy storage

Electricity storage systems can be categorized as mechanical, electromagnetic and electrochemical storage devices. Various technologies are at varying stages of technological maturity and they usually are exploited by their storage capacity (stored energy per mass or volume), or energy density, power (energy output per time), storage period (how long the energy should be stored) and size. Many of these key parameters actually can vary over a huge scale: from latent heat storage to prevent laptops from getting too hot (stored energy in the range of a few Wh) to the heat and cold thermal underground storage system underneath the German Reichstag in Berlin (stored energy in the range of some GWh).
Generally, energy density, power and energy ratings are essential criteria for any surveying process and synthetic selection maps have been developed to support selection. In Figure 1, currently electricity storage technologies are mapped over power and energy dimensioning basic features.
Together with above energy storage solutions today available, the "hydrogen battery" method is based on hydrogen energy storage (HES) route that consists in storing electricity into form of hydrogen (a high density energy carrier) produced by no-emission water electrolysis process powered by excess of electricity from RES. In Table I, some basic features of storage technologies are reported, as alternative solutions to HES systems.

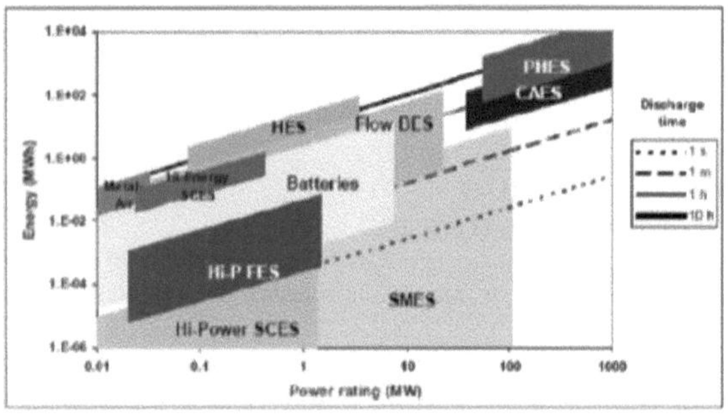


Figure 1 - Map of current available electricity storage technologies on energy vs. power comparison diagram (Source: Gonzales, 2004).

Table I – Main features, principles and drawbacks of the storage methods mapped in Figure 1.9, with the exception of HES.

| Technologies | Main features, advantages and drawbacks |
|---|---|
| **Compressed air energy storage (CAES)** | Utilizes off-peak electricity to compress air that is then stored in an airtight reservoir (typically an underground geological formation such as a salt or limestone cavern) thus releasing it from the reservoir to run it through a gas-fired combustion turbine. Main advantage of this method is a high efficiency around 70% (Robyns, 2005), technology maturity and high power and energy capacity[1], but low energy density, in the order of 12 kWh/m$^3$ (Multon, 2003). |
| **Pumped hydro energy storage (PHES)** | Utilizes off-peak electricity to pump water from a lower reservoir into a higher reservoir then producing electricity by releasing the potential energy stored in elevated water through turbines (in the same manner as conventional hydro facilities). As CAES, PHES have a high conversion efficiency of about 65–80% (Multon, 2003), meanwhile, as for the large energy storage capacity of CAES, the main shortcoming of this technology is the need for a site with different water elevations. |
| **Flywheel energy storage (FES)** | Store kinetic energy in a rotating mass with minimized friction losses to improve efficiency. They have a great cycling capacity (a few 10,000 to a few 100,000 cycles) . While the instantaneous efficiency is around of 85%, the main drawback is an overall efficiency that would drop to 78% by progressively release of kinetic energy to after 5 h, and 45% after one day. For a long-term storage, this type of apparatus is therefore not foreseeable. |
| **Batteries** | Conventional chemical storage achieved through accumulators (lead–acid, nickel–cadmium, etc.) are high energy densities (up to 150 and 2000Wh/kg for lithium), while they have relatively low durability for large-amplitude cycling (a few 100 to a few 1000 cycles). |
| **Flow batteries (Flow BES)** | A two-electrolyte system in which the chemical compounds used for energy storage are in liquid state, in solution with the electrolyte. They have a relative high efficiency about 75% (Couffin, 2004–2005) and want to overcome the limitations of standard battery system in which the electrochemical reactions create solid compounds that are stored directly on the electrodes on which they form. They are at development stage; in 2003 Regenesys Technologies (UK) developed a system with a storage capacity of 15MW /120MWh. |
| **Superconducting magnetic energy storage (SMES)** | Stores energy in the magnetic field created by the flow of direct current in a coil of cryogenically cooled, superconducting material. Great instantaneous efficiency, near 95% (Anzano 1989) and a fast response time (under 100 ms). Major shortcomings represented by the refrigeration system and necessity for massive storage (5000–10,000MWh) of very large coils (several 100m in diameter) to generate enormous electromagnetic forces. Underground installation is foresee to limit infrastructure costs. |
| **Super-Capacitors Energy Storage (SCES)** | Store energy in the form of an electric field between two electrodes. Generally very durable (8–10 years), with very high efficiency around 95%, drawbacks are represented by a very high self-discharge around 5% per day and the low energy density 3–5 W·h/kg, against 30-40 W·h/kg for a lead acid battery. |

[1] The first storage station using an underground compressed air reservoir has been in operation since November 1978 in Huntorf, near Bremen, Germany, while a larger installation has been set set in 1991 in MacIntosh, Alabama, (http://www.caes.net/mcintosh.html) began to deliver 100MW of power for 226 h, compressing ambient air and storing it at a pressure between 40 and 70 bars in a 2,555,000-m3 cavern, 700m deep in the ground.

Water electrolysis provide water decomposition when a direct current is passed between two electrodes immersed in water separated by a non-electrical conducting aqueous or solid electrolyte to transport ions and completing the circuit. Ideally, 39 kWh of electricity and 8.9 liters of water are required to produce 1 kg of hydrogen at 25 °C and 1 atmosphere pressure.

The energy efficiency of commercial water electrolyzer widely ranges from 50 to 70%, and it strongly depends also on how much system is efficient as external current density that feeds cell stack varies due to the variability and intermittency of RES power production.

As efficiency of the hydrogen generation system is one major drawback if compared with conventional battery, a very important key-point is that water electrolysis systems are cleaner in energy generation, modular as they can be easily grouped to produce hydrogen over wide scales ranging from few kilowatts to hundreds of megawatts and furthermore today commercial electrolyzers are capable of quick response to fluctuant electricity that is typically produced by intermittent and fluctuant wind turbines.

Regarding with modularity capabilities, power scale can be thought as unlimited, as a capillary distribution of these systems connected to the grid can build an infinite power capacity (Figure 2).

| Storage System | Typical Output Rating (Megawatt) | Typical Storage Rating (Megawatt hours) | Physical Linkage Reqd. |
|---|---|---|---|
| Pumped Hydro | 100 | 500- 10000 | Yes |
| CAES | 25 | 200 | No |
| Battery | 10 | 100 | Yes |
| Hydrogen | 10 | Unlimited | No |

Figure 2 – Summary of power and energy storage rating of HES compared with great capacity alternative energy storage technology (Price, 2000).

### Behavior of water electrolyzer based HES in presence of high intermittent power source

Experiences have been performed by 2004 to provide information about potential operation of alkaline commercial water electrolyzer HySTAT® by Hydrogenics under emulated intermittent conditions generated from actual wind speed data [1]. In this test research a period of 14 h of a typical day with good wind conditions in the Sierra del Perdon's wind farm (Navarra region, Spain) was selected.

The evolution with time of the wind speed during this period and the electric power output expected the specific wind turbine used according to its power curve are reported in the Figure 2.

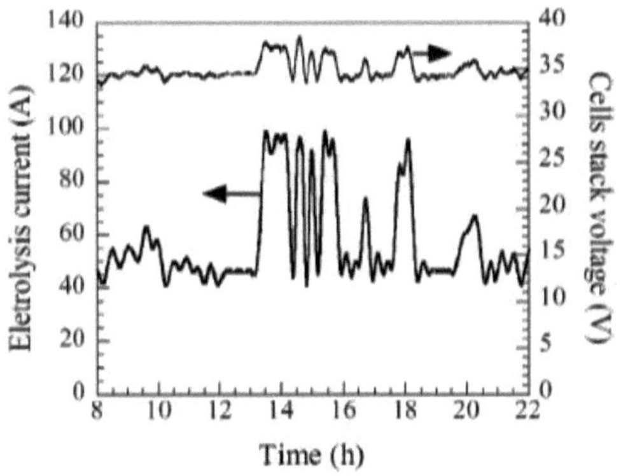


Figure 2- Evolution with time of the electrolysis current in a Hydrogenic HySTAT: total voltage established in the electrolyzer cell stack as a result of the passage of the electrolysis current (source: [1]).

The fast response of the stack voltage shown in Figure 2 demonstrated that electrolysis cell stack voltage readily matches a highly varying electric current-time profile, suggesting an essentially ohmic behavior governed by the internal resistances associated to the electrolyte and cells components. During test campaign electrolyzer worked reliably and at high efficiency, as it ranged 74-83% basing on the stack voltage. The hydrogen produced was at high purity as oxygen contents was always below 0.2 vol %.

The hydrogen produced by electrolyzer can be therefore stored to be released and used "on-demand" by various (also combined) user systems:

- an advanced modified turbine for power generation, advanced hydrogen fuelled turbines for power generation;
- a fuel cell stack to produce directly green-electricity, that functions in a "reverse mode" of the water electrolyzer, thus recombining hydrogen with oxygen to produce electricity and waste heat and pure water;
- as a green-fuel for either Fuel Cell vehicles or modified internal combustion engine vehicles;
- introducing this green-hydrogen by hydrogen merchants, thus providing an alternative to conventional hydrogen, normally produced by Steam Methane Reforming, SMR, at high carbon footprint.

## Electricity storage technologies for the RES integrating purposes

Conventional network operation of distribution network, at least in some regions of Europe, is actually based on a "fit and forget" approach. This means that verifications are performed during the connection phase and the connection schemas adopted ensures any reasonable behavior of network users is allowed.

From a technical point of view, however, it would be possible to increase the amount of distributed generation points connected to the grid where RES plants could contribute to voltage regulation and to provide ancillary services.

An appropriately sized storage device would reduce uncertainties in the capability of the system of self-regulating, thus allowing a further increase of RES units connectable to the grid. As stated, the various storage technologies have a number of features, such as efficiency, size, dynamic behavior, etc. which are all relevant in the selection of the most suitable system for a particular application or set of applications.

Hydrogen-based energy storage systems deploy the highest energy density; thanks to high modularity, they are of unlimited storage capacity, but, conversely, when the storage pathway is completed by a fuel cell system for the back conversion of the electricity the round-trip efficiency[2] of the system is low (around 30%).

On the other hand, thanking to varying deploying architectures today possible to develop, these energy storage systems could be really promising in near-term.

Economic viability of large storage system has to be evaluated taking into account these major requirements:

- working in combination with RES, especially fluctuant RES
- serving for secure balancing of the grid
- reducing curtailing and allowing progressive and potentially unlimited installation of RES.

Accordingly, the optimal choice should be driven mainly by these following important driving key-factors:

- The charge and discharge response to fluctuant electricity;
- The capital cost of the facility, including those ones of the RES plant (charging equipment, storage equipment, and discharge equipment);
- The O&M costs for the facility, including those ones of the RES plant;
- The transmission power costs;
- The power and energy rate;
- The power quality;
- Safety and security for the grid;
- The round-trip system efficiency;
- Reversibility of charge/discharge cycle;
- Modularity and easiness of upscale;
- The dispatchability;
- The flexibility and adaptability to future trends in the electric sector;
- Environmental sustainability and safe operation
- Possible synergies

At present time, very few storage systems can solve power peaks curtailment over a longer term (10-12 hours) of functioning.

## Magnesium as key-solution for safe and viable hydrogen solid storage for saving power curtailing

Both commercial water electrolyzers and fuel cell stacks have the major advantages of an instantaneous response: a) any electricity variability that flows into the water electrolyzer is used for producing hydrogen; b) any request of electricity by downstream users (i.e. demand) is matched by flowing variable hydrogen flux into the fuel cell stack.

On the other hand the major drawbacks of the supply-chain water electrolyzer→hydrogen storage→fuel cell stack is represented by:

[2] Round-trip efficiency is calculated throughout the first electricity conversion in hydrogen and the downstream back-conversion into electricity of hydrogen by use of fuel cell system.

- lost energy used during the storage phase, as hydrogen can be usually stored either as pressurized gas or liquefied cryogenically cooled to achieve adequate energy density;
- the low-efficiency of the fuel cell, that obviously affects the overall round-trip efficiency of the system: 70% for electrolyzer efficiency and not more than 50% for the fuel cell result generally in low round-trip efficiency of 35%.

On the other hand, one major advantage of using hydrogen is its high energy content, namely 39.4 kWh per kg of hydrogen[3]. But the actual energy density strongly depends on the storage methods used. Viability of the hydrogen storage systems are strictly related to the real capability of deploying a not expensive long-term safe storage of hydrogen.
The hydrogen produced can be stored in various forms: liquid (by cryogenic process, liquid hydrogen can be stored in tanks at 21.2 K at ambient pressure), compressed gas (high pressure gas cylinders usually operate at a maximum pressure of 20 MPa) or via solid–state methods. In any case the major objective of industrial hydrogen storage methods is reaching the highest volumetric density with full reversibility of hydrogen uptake and release. At ambient temperature and atmospheric pressure, 1 kg (equivalent to nominal 39 kWh) of the gas has a volume of 11 $m^3$.
To increase hydrogen density in conventional storage methods it is necessary to spend energy either for compressing gas or lowering temperature for liquefying. With an un-conventional solid-state method, energy must be used to contrast repulsion deriving from the interaction of hydrogen with another material.
Generally speaking a solid-state hydrogen storage (HSS) method is intrinsically safe (no pressurized reactive hydrogen gas is present) and a long-term storage, since no boil-off problems can occur.
HSS into metals physically consists of diffusion of hydrogen into metal (absorption phase) that is finally bonded with host material forming stable metal hydrides. Therefore, hydrogen storage capability of metals is strongly influenced by the way this two-phase process develops. Hydrogen atoms[4] penetrate and diffuse into metal by preferential routes: they flow throughout grain boundaries, i.e. preferential pathways due to higher concentration of vacations. The host metal initially dissolves some hydrogen as a solid phase. As the hydrogen pressure together with the concentration of hydrogen in the metal is increased, interactions between hydrogen atoms become locally important, and nucleation and growth of the hydride phase can start. Therefore the "storing" phase develops by the formation of intermetallic compounds (interstitial hydrides, or complex metal hydrides) at any sites (interstitial sites present in crystal lattice host) available for hydrogen locating and bonding to metal element.
Further constraints that work against absorption phase are: a) the presence of oxide layers that spontaneously formed in air on metals and, b) such a "barrier effect" that impedes hydrogen atoms to more deeply penetrate inside bulk because top layers saturate. These represents actually physical (oxide layers) and chemical (saturated layers) barriers that can prevent a fully hydrogen absorption inside bulk material; these barriers are generally responsible of lower sorption kinetics and practical absorption rate (weight percentage of hydrogen absorbed into host metal) far from theoretical values.

Among metals that are more attractive for hydrogen storage, magnesium is considered a key material since that it is the eighth most frequent element on the earth, it has a good reversible storage capacity, achieving theoretically the 7.6 wt.%, it is relatively inexpensive and it has very low density. On the other hand, Mg major drawback is its high desorption temperature of over 300 °C; the high stability of magnesium-hydrogen bond is the reason of such a high enthalpy of hydride formation that is responsible of high desorption temperatures and efficient heat management systems. During the absorption phase magnesium hydrides form with heat release, while the desorption phase host material needs to be heated up for the hydrogen realease. An efficient heat management systems is therefore required for industrial systems.
One possible route to produce Mg hydride reactions with lowered reaction enthalpy (i.e. low temperatures for uptake and release phases) is to add elements which exhibit lower enthalpies of formation. $Mg_2Ni$ is an example of this class of materials, as the enthalpies of formation for the ternary hydride $Mg_2NiH_4$ and the $Mg_2Ni$ is of the order of -67 $kJmol^{-1}$ per $H_2$, therefore lower than the -78 $kJmol^{-1}$ per $H_2$ that is enthalpy for the $MgH_2$ formation. Unfortunately this route reduces drastically the storage capacity and increases the density of the modified Mg-based material.
In last two decades further breakthroughs in hydrogen sorption kinetics have been achieved by pre-treating magnesium by high-energy ball milling operations (Schulz, 1995; Zaluska 1999). Ball milling applies severe plastic deformation to material for a high refining microstructure (i.e. dramatic increase of grain boundaries) that finally becomes more reactive to hydrogen diffusion.
During the last decade advancements in kinetics sorption of hydrogen in magnesium-based host materials have been done by selection of additives and nano-structuring techniques for increasing grain boundary density [2-10].
Exploiting economically viable solid-state hydrogen by magnesium-based storing solutions is possible if hydrogen kinetics sorption is rapid, the storage capacity can reach the theoretical value of 7.6 wt.% of hydrogen (corresponding at around 400 $kWh/m^3$) and heavy and expensive additives are limited.

A major breakthrough in magnesium storage is represented by proprietary patented manufacturing process of nanostructured magnesium-based material for hydrogen storage created by Mc Phy Energy (Mc Phy), a French company based in Grenoble. Cheaper magnesium-based nanostructured materials produced by Mc Phy are capable of hydrogen kinetics absorption that are industrially exploitable. The modularity concept exploited by Mc Phy is schematically represented in the Figure 4.
Basing on solid-state technology, the Mc Phy solution does not require hydrogen from the water electrolyzer unit to be compressed, as the pressure for starting hydrogen absorption is of the same level of the hydrogen pressure at the electrolyzer outlet. And pressure of hydrogen desorbed by the tank is similar to the level (around 2 bars) required for fuel cell stack, a costly and energy dissipative compressor is therefore eliminated. These solid storage system are highly efficient, since about 3-5% are the losses in absorption-desorption cycles due to that part of energy used to heat-up tank material for achieving desorption. Storage modules require less energy than liquid or compressed hydrogen

[3] Energy content of 1 kg hydrogen (equivalent to 141.9 MJ) at Higher Heating Value (HHV).

[4] Molecular hydrogen is dissociated at the surface before absorption; two H atoms recombine to $H_2$ in the desorption process.

technologies, which generally use up to 33% of the hydrogen energy content for storage phase.

These storing systems are produced in modular sub-units that can be installed and grouped in parallel to increase system capacity. As hydrogen is stored in solid state, storage modules are furthermore safe to be transported as any non-hazardous goods. Raw materials used in manufacturing do not generate any hazardous waste at the end of their life cycle, and they are fully recyclable.

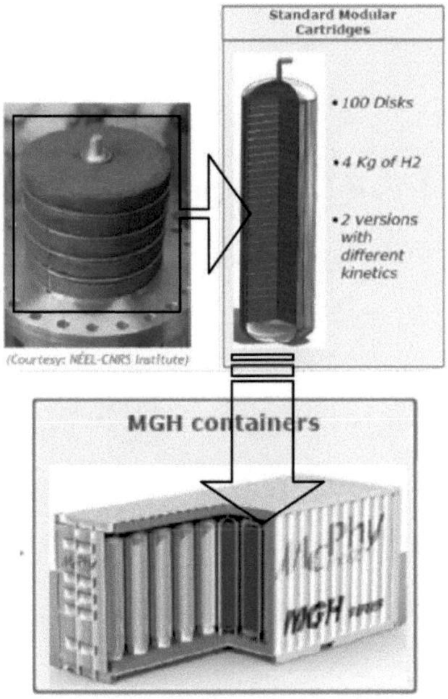


Figure 4 – Modularity concept realized by Mc Phy technology (courtesy of: Mc Phy Energy SA).

**Dynamic behavior of the WE coupled with Hydrogen Solid Storage system**

HSS needs to be optimally coupled with the WE, the upstream unit to assure high dynamic storage capability in transient and dynamic phase. This fact implies that HSS has to be capable to accept (absorb) in real time hydrogen at any rate that is produced by water electrolyzer that can function at varying conditions depending on external fluctuant electricity produced by RES.
For that, dynamic behavior of the HSS module has been extensively tested coupled with WE module that had to respond to an intermittent power source.
At CEA (Commissariat à l'Energie Atomique et aux énergies alternatives) facilities in Grenoble, a complete test campaign has been performed on a fully instrumented system constituted by a Mc Phy solid-state hydrogen storage tank, upstream coupled with a water electrolyzer and downstream connected to a fuel cell (figure 5). Extensive test campaign have done in order to analyze and optimize response of the system when connected with varying and fluctuant renewable source. Specifically the WE was directly linked to solar panel source emulator. Storage capability (i.e. sizing capacity for absorbing variable quantity of hydrogen) has been pointed out (figure 6).

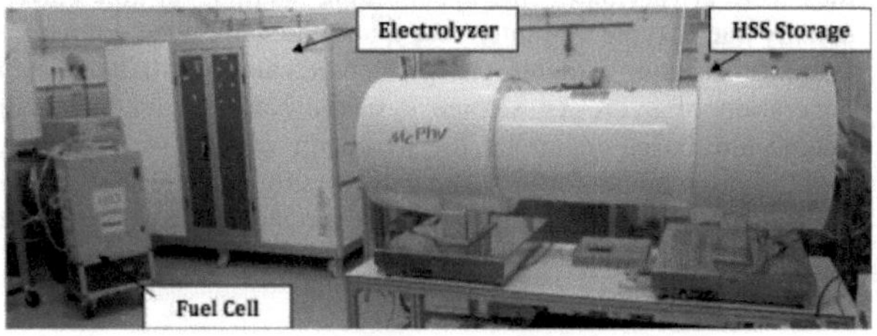


Figure 5- Test bench facility set at CEA premises.

A load power profile had been modeled in order to feed electrolyzer simulating electricity production of a common solar farm over 8 hours of functioning, sunny in the morning and cloudy in the afternoon. A total 200 grams of hydrogen have been produced by the electrolyzer and constantly absorbed by McPhy tank in 8 hours.

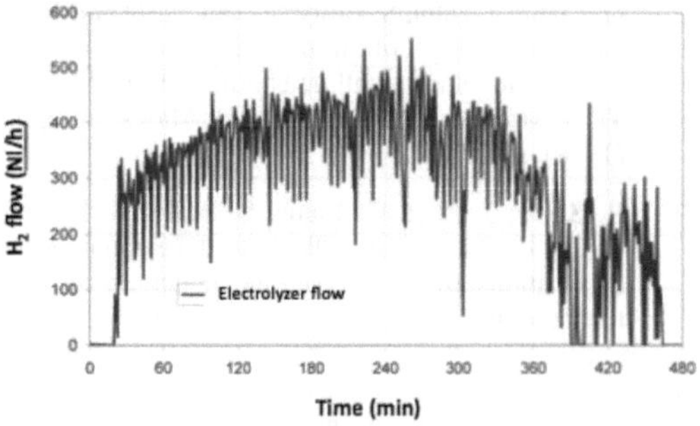


(a)

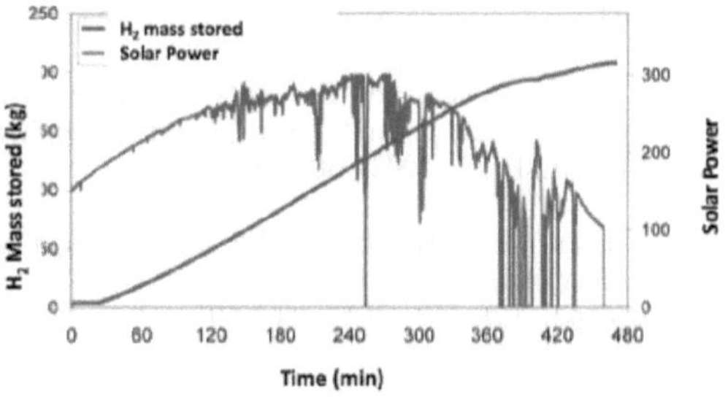


(b)

Figure 6 – Dynamic behavior diagrams showing: a) hydrogen flow produced by water electrolyzer fed by electricity produced by solar panels; b) hydrogen produced and stored in HSS module .

test results, water electrolyzer shown fully capability of accepting power variations from RES. Most important conclusions from test campaign were: a) electricity variations from power source feeding electrolyzer had no effects on hydrogen production; b) HSS module was well dimensioned to absorb all the hydrogen produced by water electrolyzer and it serves water electrolyzer unit in a fully stable mode.

**Conclusions**

Hydrogen storage via solid state (hydrides) is becoming a new challenge for magnesium based materials. Modular products commercially available are capable of sorption kinetics suitable to be coupled with up to date water electrolyzer technology.

Advancements have been done in the water electrolyzer technology specifically related to efficiency of such systems when they work at varying operating conditions, as high variable current density that is typical of intermittent and fluctuant wind sources. Behind this improvements, energy storage can be realized by new generation of hydrogen storage systems. New advancements in magnesium-based materials for hydrogen storage is leading to rapid improvements for high capacity "hydrogen batteries". Novel hydrogen solid storage modular units are suitable to be coupled with rapid and efficient electrolyzers to serve highly intermittent RES.

## References


1. L.M. Gandía, R. Oroz, A. Ursúa, P. Sanchis and P.M. Diéguez, "Renewable hydrogen production: performance of an alkaline water electrolyzer working under emulated wind conditions", Energy Fuels, 21 (2007), 1699–1706.

2. G. Liang, J. Huot, S. Boily, A. van Neste and R. Schulz, Catalytic effect of transition metals on hydrogen sorption in nanocrystalline ball milled MgH2–Tm(Tm = Ti, V,Mn,Fe and Ni) systems. J Alloys Compd, 292 (1999), p. 247-252

3. R. Shulz, J. Huot, G. Liang, S. Boily, G. Lalande and M.C. Denis, et al. "Recent developments in the applications of nanocrystalline materials to hydrogen technologies. Mater Sci Eng, A267 (1999), 240-245

1. A. Zaluska, L. Zaluski and J.O. Stroem-Olsen, "Structure, catalysis and atomic reactions on the nanoscale: a systematic approach to metal hydrides for hydrogen storage", Appl Phys, A72 (2001), 157-165

2. R.C. Bowman and B. Fultz, Metallic hydrides: hydrogen storage and other gas-phase applications. MRS Bull, 27 (2002), 688-693

3. J. Huot, M. L. Tremblay and R. Schulz, "Synthesis of nanocrystalline hydrogen storage materials", J Alloys Compd, 356–357 (2003), 603-607

Magnesium Technology 2012
*Edited by: Suveen N. Mathaudhu, Wim H. Sillekens, Neale R. Neelameggham, and Norbert Hort*
*TMS (The Minerals, Metals & Materials Society), 2012*

# Reaction Sintering of $Mg_2Si$ Thermoelectric Materials by Microwave Irradiation

ZHOU Shu-cai[1]; BAI Chen-guang[2]; FU Chun-lin[1]

[1]School of Metallurgical and Materials Engineering, Chongqing University of Science and Technology, Chongqing 401331,China;
[2]College of Material and science Engineering, Chongqing University, Chongqing, 400044,China
Email:zhoushucai71@126.com

***ABSTRACT:*** *In order to reduce the oxidizing and volatilizing caused by Mg element in the traditional methods for synthesizing $Mg_2Si$ compounds, solid state phase reaction at low temperature was introduced by microwave field. XRD was used to characterize the powders. At the same time, the influences of parameters during the synthesis processing were discussed. The results suggest that the heating profile is also dependent on the initial green density and higher green density provides lower heating rate while power setting are fixed and the oxidation of Mg can be rest rained by changing microwave heating programs. It was found that high purity $Mg_2Si$ intermetallic compound can be obtained with excessive content of 8at% Mg from the stoichiometric $Mg_2Si$, 853K and 30min.*

***Keywords*:** *$Mg_2Si$; thermoelectric materials, solid state reaction method, microwave synthesis*

## 1. Introduction

Magnesium silicide ($Mg_2Si$), having a face-centered-cubic $CaF_2$ type of structure, has been identified for a promising advanced thermoelectric materials in the temperature range from 500K to 800 K[1~4], because of their advantages of an environmental-friendly material, such as the abundance of its constituent elements in the earth's crust and the non-toxicity of its processing by-products[5,6].However, the phase purity and microstructure of the product $Mg_2Si$ are difficult to control by conventional melting processes such as ingot metallurgy (IM) because of the easy volatilization and oxidation of Mg and the great discrepancy of melting point between Mg and Si. Here to fore, many preparation methods have been developed to produce $Mg_2Si$ and improve its thermoelectric properties, such as vacuum melting[4], solid state reaction[7], mechanical alloying[8-10] and spark plasma sintering (SPS)[3,11], which are quite time consuming and complex. However, the above issues have not solved radically yet.

These problems mentioned above motivated the development of a new synthetic method that overcomes these problems and facilitates product formation. The microwave-assisted synthesis method is presented. This method is a very promising preparation method for many materials because it is fast, clean, energy efficient and does not suffer from the disadvantages of the classical preparation technique. Many materials have been synthesized by the microwave method at considerably lower temperature and in shorter time than the conventional methods[12,13]. Until recently, microwave processing was mostly restricted to ceramics, cemented carbides and ferrites[14]. Applicability of microwave sintering to metals was ignored due to the fact that they reflect microwaves. Roy et al.[12] reported that particulate metals can be heated rapidly in microwaves. This has led to the use of microwaves to consolidate a range of particulate metals and alloys[15].

In this paper, $Mg_2Si$ thermoelectric materials were prepared by microwave-assisted activation synthesis, which has not been reported before. It is a rapid process and results in a purity phase material. The results obtained seem to be very interesting and stimulate further research in this direction.

## 2. Experimental Procedure

Figure.1 shows a schematic illustration of the experimental equipment, which was used for the production of activated Mg and Si particles. It mainly consists of a microwave oven and a heat preservation system.

Due to the volatility of Mg during the synthesis of this compound, the excessive content of Mg is indispensable. In this study, samples with nominal formula $Mg_{2.08}Si$ were prepared. The samples were prepared as follows: according to the stoichiometric proportion of $Mg_2Si$, 2.08:1 atomic ratio of Mg (>99.5mass% purity, 200mesh) and Si (>99.5mass% purity, 300mesh) were mixed by an ultrasonic homogenizer in ethanol for 45min. After being completely dried in the cabinet, they were compressed to form pellet 10mm in thickness and 20mm in diameter under appropriate pressure. The samples were then placed in alumina crucible in the center of the microwave oven and the solid-state reaction was carried out under the protection of high purity Ar (99.9%). The microwave oven used in the present work consists of a 2.45GHz microwave generator with continuous adjustable power output 0–15kW. An infrared pyrometer was used (Raytek, Marathon series) for temperature measurement. Microwave heating used in the present work was carried out at the power level of 2.0kW–3.5kW and heated to a designed temperature (580°C–600°C) and kept at this temperature for 20-40min. The samples were cooled to room temperature for further control experiment raw materials. Phases of the fabricated materials were detected by X-ray diffraction (D/Max-IIIA) with Cu Kα radiation (λ=0.15406 nm).

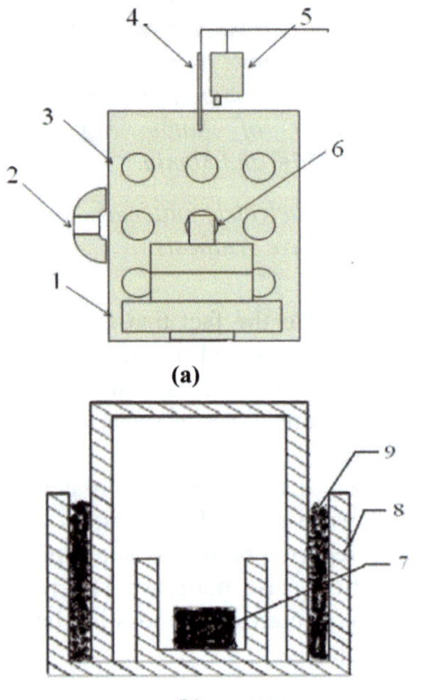


Figure1.Schematic view of microwave system setup(a) and Schematic view of the crucible (b)

1—rotating table;2—peep hole;3—magnetron tube; 4—infrared thermometer; 5—thermocouple; 6—crucible; 7—Mg、Si samples; 8—alumina; 9—Mg power

## 3. Results and Discussion

### 3.1 Heating Behavior of the Mg and Si fixed power compacts

Figure.2. shows the experimental temperature rise with time in Mg and Si fixed power compacts with 40%, 50% and 60% theoretical density, respectively, for a specific particle size while keeping constant the power setting and total exposure time. It is interesting to note that the porous metal compacts couple with microwaves and get heated up rapidly and the heating rate increases as the green density decreases (porosity increases). Higher is the porosity, higher is the heating rate. The temperature rise is restricted to a certain level.

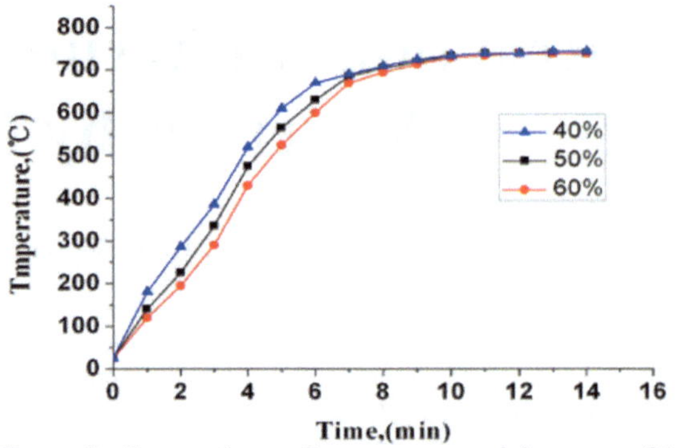


Figure 2. Comparison of temperature-rising rate of Mg and Si compacts as a function of initial porosity in microwave furnace.

It is well known that an incident wave on a conductive metallic material is mainly reflected, and microwaves for electrically conducting materials do not penetrate a bulk sample beyond the skin depth. The skin depth "δ" (m) is defined as the depth into the conductor from the surface at which the current density is 1/e of its value at the surface, given by the well-known expression (1)[16]

$$\delta = \sqrt{\frac{2}{\omega\sigma\mu}} \qquad (1)$$

Here, μ is the real part of the permeability taken to be $4 \times 10^{-7}$T m/A for nonmagnetic materials. σ and ω are the DC electrical conductivity and angular frequency, respectively. At 2.45 GHz, the skin depth of bulk magnesium is about 2.18μm. But as in case of porous material electrical conductivity decreases with an increase in porosity so from the above equation effective skin depth should also increase. Johnson and co-workers[15] measured the values of σ the conductivity of green copper powder. These measurement results show the sample conductivity is far below that of bulk copper. However, the initial skin depth of porous copper powder metal compacts is much larger—on the order of 0.05m (computed using Eq.(1)and the measured values of σ). For a lossy conductor, the electric currents may not be zero so the electric field **E** is not zero because of finite conductivity according to Ohm's law **J=σE.** which shows that at microwave frequencies the magnetic induction is much larger than the electric field, so the microwave heating of powdered metals mainly as the result of magnetic induction heating. Roy,et[11] al also attributed the heating of metals to eddy-current losses from the electric field. Thus an induction loss caused by eddy currents is one mechanism for microwave sintering of $Mg_2Si$ compound. A similar phenomenon was also found by Booske, el al[16].

### 3.2 Effect of the power of microwave on

## $Mg_2Si$ compounds

In order to find out the complete reaction temperature of Mg, Si to form $Mg_2Si$ in microwave chamber, the raw mixed powder were sintered by microwave at temperatures from 573 K to 903 K. When the input power was kept constant at 2.5kw and temperature is at 853K, $Mg_2Si$ single phase was formed, and no obvious Mg, Si, MgO and $SiO_2$ peaks are detected from XRD, as is shown in Figure.3. Phases of microwave samples at 903K (microwave power=3.5kW) show that some MgO and $SiO_2$ peak appear, as is shown in Figure.4. It can be attributed to the easier oxidation of Mg at higher temperature. So, the optimum temperature of solid-state synthesis by microwave is about 853K, which Mg and Si react completely and nearly no obvious oxidation is detected.

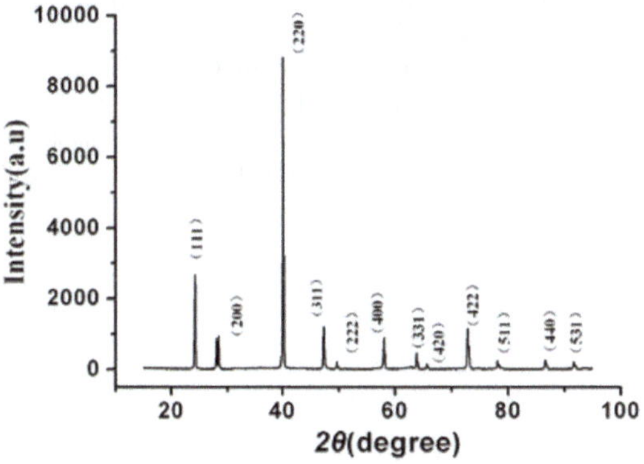


Figure 3. XRD patterns of $Mg_2Si$ samples (power=2.5kw)

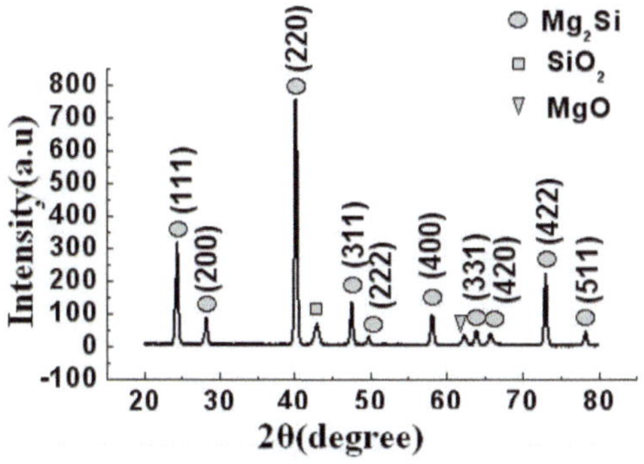


Figure 4. XRD patterns of $Mg_2Si$ samples (power=3.5kw)

### 3.3 Effect of isothermal holding time on $Mg_2Si$ compounds

Figure.5 shows XRD analysis of various holding-times of Mg, Si powder of stoichiometric proportion at 853K in microwave oven. The phase analysis of the XRD patterns reveals that $Mg_2Si$ is the major phase in the sample. It shows that complete reaction has been carried out when the holding-time is only 30 minutes. The peaks are very sharp at this situation and no obvious MgO peak appears. However, MgO and $SiO_2$ impurity peaks appears when the holding-time is 40 minutes as is shown in Figureure.5 (b). When the holding-time is only 20 minutes, there are a small amount of Mg peaks of raw powder, as is shown in Figureure.5(a). So, the optimum holding-time is 30 min and the excessive content of Mg is essential to compensate the volatilization of Mg during the processing.

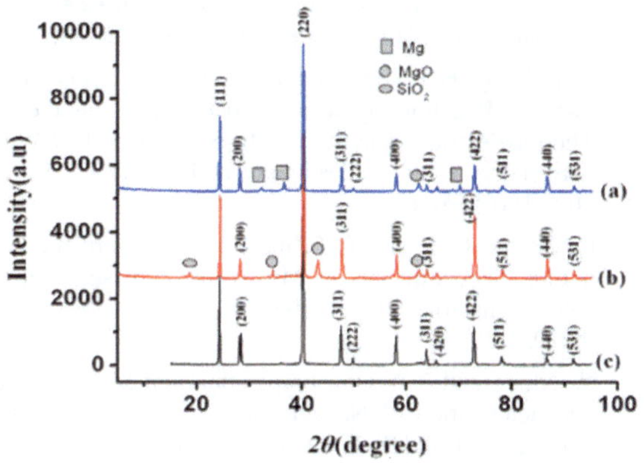


Figure 5. XRD patterns of $Mg_2Si$ samples of different holding-times during synthesis processing (a)—20min; (b)—40min; (c)—30min

## 4 Conclusions

a) The temperature-rising rate is also dependent on the initial green density. Higher green density provides lower heating rate while power setting are fixed. Irrespective of the green density, all the compacts attain the same steady-state temperature.

b) The optimum parameters by microwave heating to synthesize $Mg_2Si$ from Mg and Si raw powders are 853 K and the holding-time of 30 minutes with the excessive content of 8at% Mg from the stoichiometric ratio.

## 5 Acknowledgements

This work was supported by the Natural Science Foundation Project of CQCSTC (No.2009BB4228) and Research Foundation of Chongqing University of Science & Technology(No. CK2010Z09).

## References

[1] M Yoshinaga and T Iida. Bulk Crysta Growth of $Mg_2Si$ by the Vertical Bridgmam Method [J]. Thin Solid Film, 2004, 461:86-89.

[2] R Morris. G, R D Redin and G C Daniel. Semiconducting properties of $Mg_2Ge$ single Crystals[J].Physical. Review. 1958,109:1916-1920

[3] T Jun-ichi and K Hiroyasu. Thermoelectric Properties of Sb-doped $Mg_2Si$ Semiconductors [J]. Intermetallics, 2007, 15:1202-1207

[4] V. K Zaitsev, M. I Fedorov and E. A Gurieva. Highly effective $Mg_2Si_{1-x}Sn_x$ thermoelectrics[J]. Physical Review B , 2006, 74: 045207-1- 045207-5.

[5] T Caillat, A Borshchevsky and J P Fleurial. Properties of single crystalline semiconducting $CoSb_3$ [J]. Journal of Applied Physics. 1996,80:4442-4449.

[6] S Bose , H N Acharya and H D Banerjee. Electrocal, thermal, thermoelectric and related properties of magnesium silicide semiconductor prepared from rice husk[J]. Journal of Materials Science. 1993,28:5461-5468.

[7] H.Y Jiang, H.S Long and L.M Zhang. Effects of Solid-State Reaction Synthesis Processing Parameters on Thermoelectric Properties of $Mg_2Si$[J].Journal of Wuhan University of Technology-Materials Science, 2004,19(6):55-56.

[8] Q. Zhang, J. He and T. J Zhu. High figures of merit and natural nanostructures in $Mg_2Si_{0.4}Sn_{0.6}$ based thermoelectric materials[J]. Applied Physics Letters, 2008, 93:102109.

[9] W Xiong and X Y Qin. Preparation and Microstructural Characterization of Nanocrystalline $M_{g2}Si$ Intermetallic Compound Bulk[J]. The Chinese Journal of Nonferrous Metals, 2005,15:380-384.(in Chinese)

[10] X P Niu and L Li. Formation of Magnesium Silicide by Mechanical Alloying[J]. Advanced Performance Materials, 1997, 3:275-283.

[11] L Han，M Yang, Q Shen and L Zhang. Reaction sintering of magnesium sillicide thermoelectric material by the spark plasma sintering technique[J]. Journal of the Chinese Ceramic Society, 2008, 36(8):337-340.(in Chinese)

[12] R. Roy, D. Agrawal,and J.P Cheng. Full sintering of powdered metal bodics in a microwave field [J ] .Nature ,1999 ,399(17): 668-670.

[13] J W Lekse, T J Stagger and J A Aitken. Microwave Metallurgy: Synthesis of Intermetallic Compounds via Microwave Irradiation [J]. Chemistry of Materials, 2007;19:3601–3603.

[14] D E Clark and W H Sutton. Microwave Processing of Materials [J].Annual review of materials science, 1996,26 :299–331.

[15] S S Panda, V Singh, A Upadhyaya and D Agrawal. Sintering response of austenitic (316) and ferritic (434L) stainless steel consolidated in conventional and microwave furnaces [J].Scripta Mater, 2006,54:2179–2183.

[16] J H Booske, R.F Cooper, and S. A Freeman. Microwave ponderomotive forces in solid-state ionic plasmas[J]. Phys Plasmas. 1998,5(5):1664-1670.

# Charge-Discharge Mechanism of MgC powders and Mg-Li Alloy Thin Film Materials

Yen-Ting Chen[1,2], Fei-Yi Hung[1,2*], Truan-Sheng Lui[2], Ren-Syuan Xiao[1], Yi-Wei Tseng[1], Chih-Hsien Wang[2]

[1] Institute of Nanotechnology and Microsystems Engineering, Center for Micro/Nano Science and Technology, National Cheng Kung University, Tainan, TAIWAN 701.

[2] Department of Materials Science and Engineering, National Cheng Kung University, Tainan, TAIWAN 701.

Keywords: Mg-C, Mg-Li, Lithium battery, Charge-discharge

## Abstract

Magnesium-carbon powders and Magnesium-lithium powders were used as the anode materials for lithium ion batteries to investigate the structure and electrochemical behavior in room temperature. The composition of Mg-C powders contained 1:1 and 9:1. The powders and the thermal evaporated films of Mg-10Li were compared with Mg-C systems. In addition, Mg-10Li thermal evaporated film was used as the experimental materials to process the annealing treatment. The results show that Mg-C powders system had the interface effect of a Cu foil to reduce the electrochemical reaction. With increasing the carbon powder content, the charge-discharge characteristics of Mg-C powders was raised. Notably, the Mg-10Li specimen had better cycling properties than that of Mg-50C (1:1). After annealing at 200 $^0$C for 1hr, Mg-10Li alloy film not only increased the capacity, but also improved the charge-discharge cyclability.

## Introduction

Adding more active element (lithium) into the magnesium matrix will form the lightest structural materials, magnesium-lithium alloys [1-2]. It is notable that magnesium-lithium alloys possess higher theoretic voltage, theoretic capacities and electrochemical activity relative to other magnesium alloys [3-4]. Many researches [5-7] indicate that the anodes of magnesium-lithium alloys have feasibility for the fuel batteries. Other researches [7-8] indicate that the electrochemical mechanisms of the Mg-Li anode materials have higher capacities.

With the development of nanotechnology, the magnesium alloys whose nanostructures can enhance the performance of magnesium battery. Nowadays the nano or micro magnesium powders getting by ball-milling techniques have applied for magnesium-oxygen batteries [8-10]. (current density: 5mA/cm$^2$; energy density: 565Wh/Kg), but the problems of oxidation still affect the properties of battery powders. Furthermore, the life cycles of fuel battery of AZ magnesium alloys will be improved through the chemical coating of surface [11]. By buffering of the second phases, the magnesium-carbon thin film can also improve the performance of batteries [12-13].

According to our previous studies [14], the electrodes of Mg-Li alloy by vapor deposition had favorable charge-discharge characteristics and higher capacities. Since magnesium alloys apply for air batteries and fuel batteries mostly, the studies focusing on material characteristic for lithium batteries are few. Notably, the electrodes of magnesium matrix possess potential for application. To avoid oxidation and surface problems, we use the mixture of magnesium-carbon powders and the evaporating thin film to study the basic properties of Mg-C and Mg-10Li thin film. The present charge-discharge characteristics lead the Mg-Li alloy to be applied for lithium batteries.

## Experimental Procedures

Mg-C powders were prepared from mixture of magnesium powders and carbon powders. The weight ration of these powders were Mg:C=1:1 and Mg:C=9:1, respectively. Mg-C powders were stirred with carbon black (10 wt.%) uniformly and coated on Cu foils. The electrodes were cut for charge-discharge test (the same method with following thin film test). In addition, the Mg-10Li target was made by vacuum melting furnace and then performed the process of evaporating on 10μm Cu foil. The surface of the evaporating specimens were kept as deposited, part of specimens were annealed at 200$^0$C for 1hour under vacuum. Moreover, the thickness of Mg-10Li film was 1.2 μm (named as MgLi film). Every specimen was cut and then the analysis of the charge-discharge characteristic was carried out via the lithium half-cell. In addition, $LiPF_6$, ethylene carbonate (EC) and diethyl carbonate(DEC)(EC：DEC=1：1 vol.) were as the electrolyte.

The surface characteristics and microstructure of both the electrode powders and thin film were observed by SEM and FIB. In addition, the identification of the element was detected using X-Ray, the incident angle was 1°, scanning speed was 2°/min and scanning range was from 20° to 80°. The charge-discharge experiments were performed ten times in fixed current. The magnitude of current was set as 0.1C to theoretic capacity. Termination voltage was 0.01V and 1.5V, respectively. All of data were the average of 3~5 times.

## Results and Discussion

The appearances of magnesium powders and carbon powders are shown in Figure 1, the pure magnesium powders present the irregular patterns, belonged to the mechanical refinement powders and the particle size was approximately 150~220 μm (Fig.1 (a)). The average grain size of sphere carbon powders was in 185μm and the structural compactness was higher (Fig.1 (b)).

Figure 2 shows the XRD diffraction of pure magnesium powders and carbon powders. Figure 2 (a) confirms no oxidation of pure magnesium powders and the powders possessed an excellent crystallization. Form the intensity of low angle diffraction, the degree of graphitization for carbon powders was good and no impurity phases in the matrix (Figure 2 (b)).

The magnesium powders and carbon powders were mixed with 9:1 and 1:1 (carbon black: 10 wt.% and binder PVDF: 5wt.%) to perform the charge-discharge testing. Figure 3 (a) shows the surface morphology of Mg: C = 9:1 powders coating on the copper foil. The spherical carbon powders were distributed on the matrix and the surface of carbon powders possessed coating characteristics. When the increase in the proportion of carbon powders, the Mg: C = 1:1 powders had formed. For surface observation of coating layer, many carbon powders were found to have an aggregation (**Figure 3 (a)**). Regardless of mixing ratio, the magnesium particles could not easily be observed on the coating layers, and the matrix was the fine pore structure. The porosity was closely related with the charge-discharge characteristics.

(a)

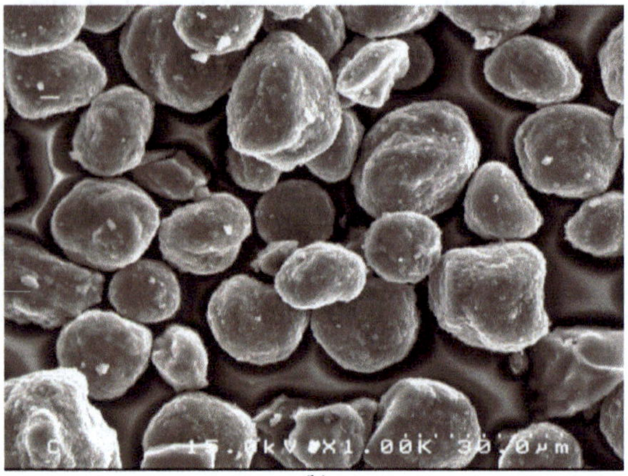


(b)

Fig. 1 Appearance of powders: (a) pure Mg powders (b) carbon powders.

Both charge-discharge cycle curves of Mg: C=9:1 electrode and Mg: C=1:1 electrode were shown in **Figure 4**. It was found that the initial curves of the two electrodes were increased with increasing the number of cycles, and then the trend of curves had stabilized. For Mg: C=9:1 curve (**Figure 4 (a)**), the charge-discharge cycle curve had two plateau areas could explain the non-stable state of ion reaction [15]. At the end of horizontal area, the ions of electrodes still had higher activation energy to make the charge-discharge capacity upgrade. In **Figure 4 (b)**, increasing the content of carbon powders, Mg: C=1:1 electrode had higher capacity than that of Mg: C=9:1 electrode in the same period of charge-discharge cycle. Notably, Mg: C=1:1 curve has a flat area that confirmed the balance mechanism at lithium ions inserted.

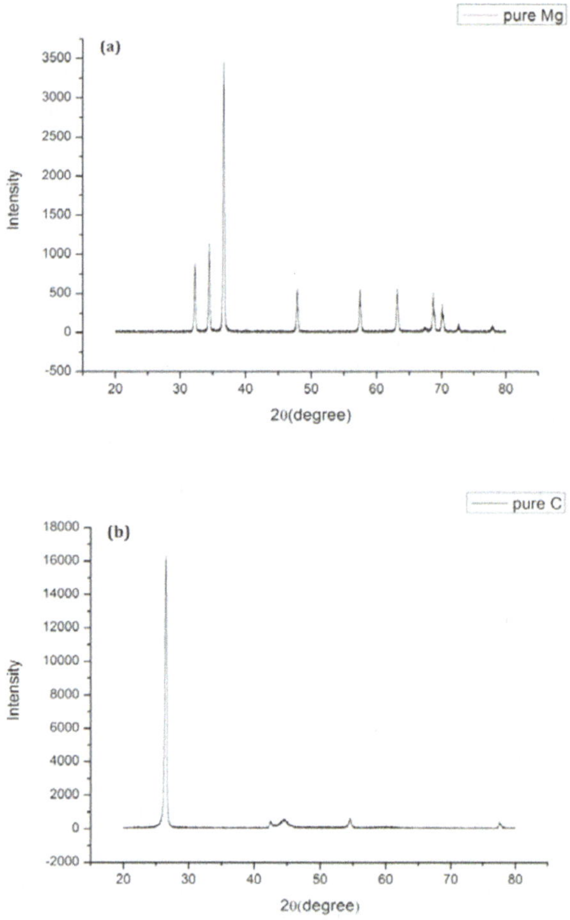


Fig. 2 XRD of powders: (a) pure Mg powders (b) carbon powders.

Compared with related literature of metal based [16], we found that the initial charge-discharge capacity of Mg-based electrodes increased, result from the magnesium ions and lithium-ions induced the activation gradually. In other words, the initial reaction duration, the lithium ion cannot reach the inside lattice of magnesium electrodes. Through the coating cracks in the interface between the carbon sphere and the matrix, the Li ions had contacted the copper foil easily to have lower initial capacitance. Increasing the reaction time, the depth of lithium ions into the magnesium based matrix increased to improve the charge-discharge capacity. This is why the charge-discharge cycle curve of the system increased with increasing the number of cycles. Due to lithium ions insert at magnesium matrix step by step, we can confirm that the phase structure of magnesium matrix had a closely related with the insertion lithium ions. Base on this fact, Mg-10Li alloy powder electrode was prepared in order to compare with related electrical data.

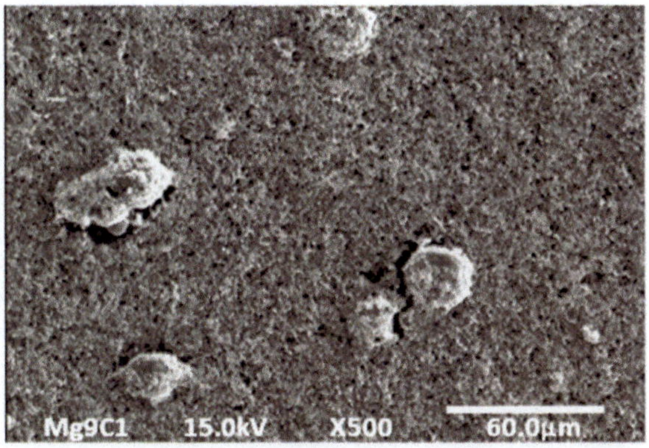


(a)

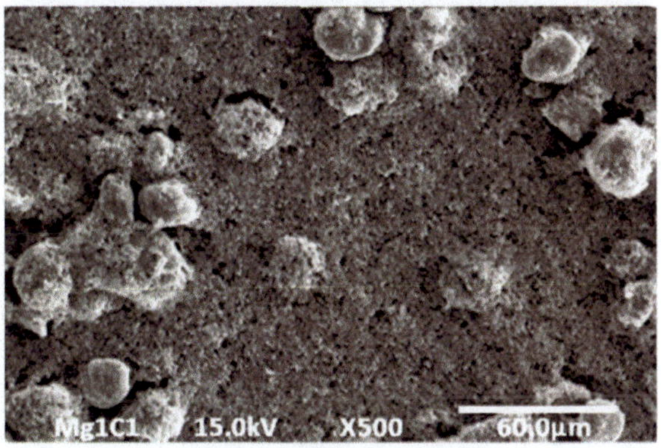


(b)

Fig. 3 Surface characteristics of Mg-C mixed powders: (a) Mg:C =9:1 powders (b) Mg:C =1:1 powders.

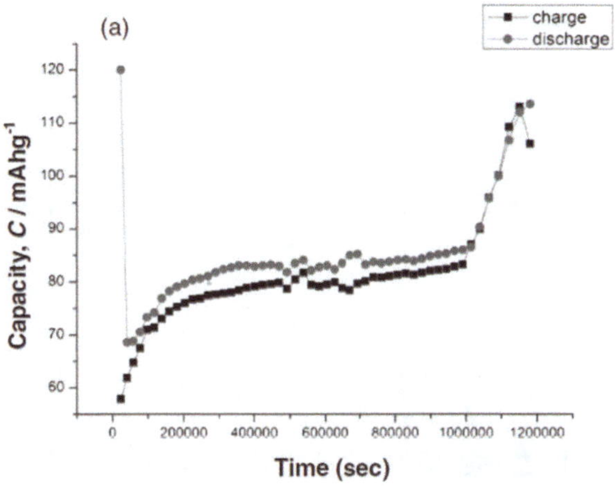


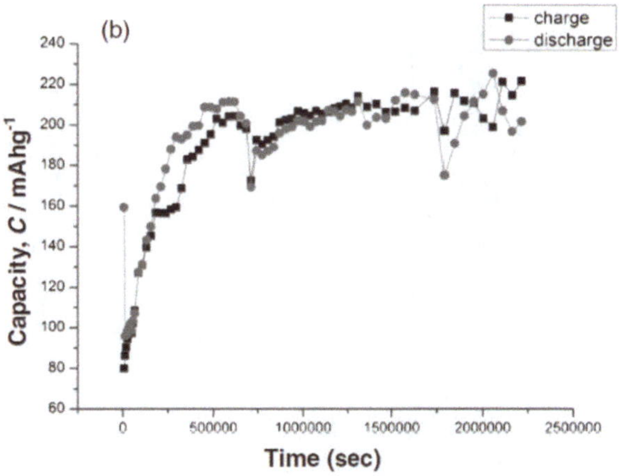


Fig. 4 Cycle performance of Mg-C mixed powders at 0.1 C-rate: (a) Mg:C =9:1 powders (b) Mg:C =1:1 powders.

Fig. 5(a) shows the appearance of Mg-10Li alloy powders, the particle size distribution was from 40um to 300um. Fig. 5(b) shows the surface characteristics of Mg-10Li alloy powders coated on Cu foils, it displayed a great decreasing of crack amount between Mg-10Li alloy powders and Cu foil. Notably, Mg-10Li alloy powders doesn't present a good charge-discharge property (was lower than 100mAh/g), it indicated that the matrix of Mg-10Li powders belonged to eutectic-like hcp structure [6], it was easy to form lithium oxide and hindered the insertion of lithium ions in the pulverization process, resulted in affecting the electrochemical characteristics of Mg-10Li powder electrode.

To avoid oxidation and keep the eutectic-like structure of Mg-10Li matrix, the thin film electrode of Mg-10Li using thermal evaporation was studied. Fig. 6(a) shows the surface characteristics of evaporating Mg-10Li alloy film, it was found that there was rolling textures at the surface of thin film. After annealing at $200^0$C in vacuum, the roughness of Mg-10Li thin film was improved and a small amount of Li-rich phase precipitated (Fig. 6(b): dark zone). In addition, the cross-section using FIB was observed (Fig. 7), the thickness of Mg-10Li thin film was about 1.2 um and the interface of Mg-10Li/Cu was not continuous (had interface cracks) after heat treatment.

Previous references [15] showed the interface characteristics (thin film /Cu foil) of electrodes using mechanical polishing were intact but it was not true. It was confirmed that there were not complete metallurgy bonding at interface between Mg-10Li thin film and Cu foil, and the interface existed some cracks (Fig. 7). Compared with the other thin films [17], the matrix of the evaporating Mg-10Li thin film was uniform and had no voids. According to electrodes references [18-19], the annealing would increase the uniformity of thin film, so the charge-discharge experiment of annealed Mg-10Li thin film was performed at room temperature.

Figure 8 shows the curve of charge-discharge capacity for annealed Mg-10Li alloy film, which takes the first cycle as the basic value (380 mAh/g). The result illustrated that Mg-10Li alloy film had the highest capacity at the first cycle and rapidly decreased with increasing of the cycle number. After 10 cycles, the discharge capacity of the electrodes still maintained about 70%. The decline mechanism had a relation with the phase change of electrode matrix.

(a)

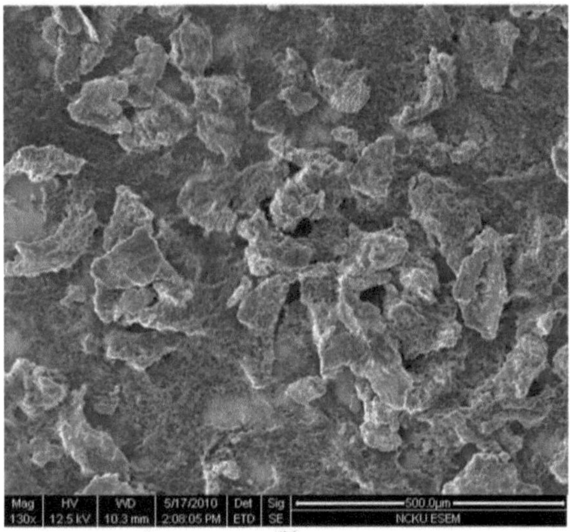

(b)

Fig. 5 SEM photograph of alloy powders: (a) MgLi powders (b) MgLi powders coated on the Cu foil.

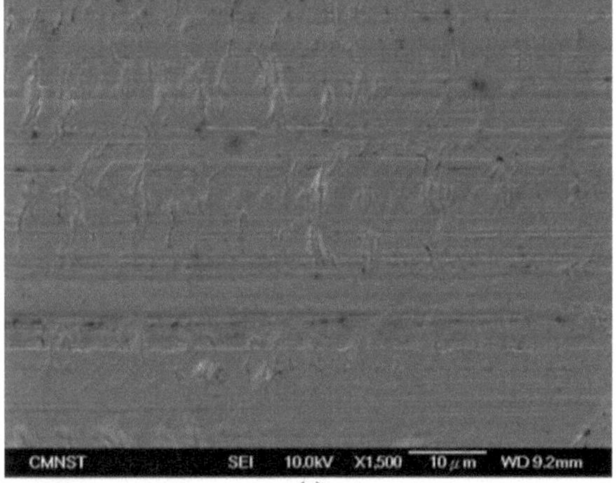

(a)

(b)

Fig. 6 SEM photograph of thermal coated films: (a) MgLi alloy film (b) annealed MgLi alloy film.

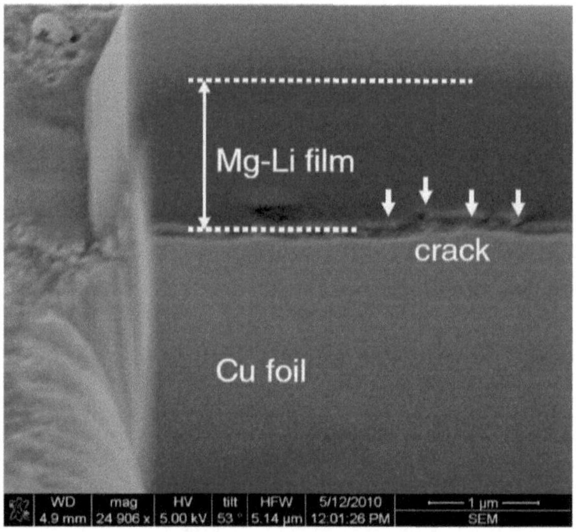


Fig. 7 Cross-section observations of annealed MgLi alloy films using FIB.

In fact, during the thermal evaporation and annealing process, Li-rich phases were observed in the matrix of Mg-10Li film, this phase (Li-rich) had higher activity and was not easy to form passivation layers. But it had dissolution effect for electrolyte and then hindered the insertion of ions. The magnesium matrix with solid solution of Li atom contributed to relieve the dissolution effect and electrochemical balance, so it could reduce the slope of charge-discharge curve. Therefore, the evaporating Mg-10Li thin film actually possesses the feasibility for the application of lithium ion batteries.

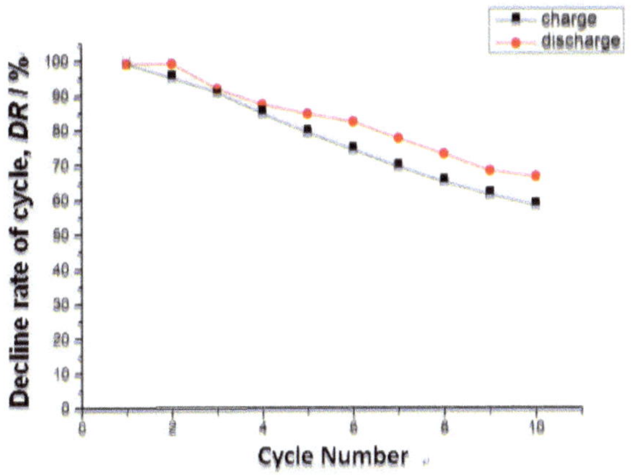


Fig. 8 Decline rate of charge-discharge capacities as a function of cycle number for annealed Mg-10Li alloy film. (Taking the first cycle as the base value)

## Conclusion

Both magnesium-carbon (Mg-C) powders and the evaporating Mg-10Li thin film possessed the feasibility for the application of lithium ion batteries. To avoid the oxidation and the generation of coating cracks, increasing the carbon content raised the charge-discharge characteristic of magnesium-carbon powder electrode. In evaporating Mg-10Li thin films, the heat treatment stabilized phase structure to improve the charge-discharge cycle life.

## Acknowledgements

The authors are grateful to National Cheng Kung University, the Center for Micro / Nano Science and Technology (D99-2700) and NSC 99-2221-E-006-132; NSC99-2622-E-006-033-CC3 for the financial support.

## References

1) S. Yi, J. Bohlen, F. Heinemann, D. Letzig: Acta Materialia 58 (2010) 592-605.
2) A. Askariani, M. H. Pishbin, R. Mahmudi: Materials and Design 31 (2010) 3618-3623.
3) N. S. Venkata Narayanan, B. V. Ashok Raj, S. Sampath; Journal of Power Sources 195 (2010) 4356-4364.
4) N. Yoshimotoa, M. Matsumotoa, M. Egashiaa, M. Moritaa; Journal of Power Sources 195 (2010) 2096-2098.
5) S. Ghosh, A. D. Sharma, A. K. Mukhopadhyay, P. Kundu, R. N. Basu; International Journal of hydrogen energy 35 (2010) 272-283.
6) C. P. Liang, H. R. Gong, Phase stability: Journal of Alloys and Compounds 489 (2010) 130-135.
7) L. Gao, C. Zhang, M. Zhang, X. Huang, X.Jiang: Journal of Alloys and Compounds 485 (2009) 789-793.
8) J. F. R. Castroa, S. F. Santos, F. R. Nikkuni, T. T. Ishikawac, E. A. Ticianelli; Journal of Alloys and Compounds 498 (2010) 57-61.
9) J. G. Zhang, D. Wang, W. Xu, J. Xiao, R. E. Williford: Journal of Power Sources 195 (2010) 4332-4337.
10) J. S. Oh, J. M. Ko, D. W. Kim: Electrochimica Acta 50 (2004) 903-906.
11) R. Renuka: Journal of Power Sources 87 (2000) 4-11.
12) S. Rousselot, A. Gazeau, D. l. Guay, L. Roue; Electrochimica Acta 55 (2010) 611-619.
13) A. S. Ingason, A. K. Eriksson, S. Olafsson; Journal of Alloys and Compounds 446-447 (2007) 530-533.
14) F. Y. Hung, T. S. Lu, R. S. Xiao: Annual Proceedings of Magnesium Alloy in Taiwan (2010) 60-66.
15) F. Y. Hung, H. C. Liao, T. S. Lui, L. H. Chen: Materials Transactions, vol.47 (11) (2006) 2759-2764.
16) B. Chen, J. Y. Xie, H. C. Yu, T. H. Wang: Electrochimica Acta, 53 (2008) 8149-8153.
17) S. Chen, F. Ke, M. Zhou, Y. Bai: Acta Materialia, 55 (2007) 3169-3175.
18) F. Y. Hung, T. S. Lui, L. H. Chen, K. J. Zhuang: Materials Transactions, Vol.49 (11) (2008) 2591-2596,
19) C. H. Wu, F. Y. Hung, T. S. Lui, L. H. Chen: Materials Transactions, Vol.50 (2) (2009) 381-387.

**Magnesium Technology 2012**
*Edited by: Suveen N. Mathaudhu, Wim H. Sillekens, Neale R. Neelameggham, and Norbert Hort*
***TMS (The Minerals, Metals & Materials Society), 2012***

# CONTROL OF YTTRIUM DIFFUSION OUT OF YTTRIA STABILIZED ZIRCONIA DURING SOM ELECTROLYSIS FOR MAGNESIUM PRODUCTION

E. Gratz[1], S. Pati[3], J. Milshtein[2], A. Powell[3] and U. Pal[1,2]
[1] Boston University, Division of Materials Science and Engineering, 15 Saint Mary's St. Boston, MA, 02215
[2]Boston University, Department of Mechanical Engineering, 101Cumminton St. Rm 101 Boston MA 02215
[3]Metal Oxygen Separation Technologies, Inc.11 Michigan Dr. Natick, MA 01760-1334

Keywords: Magnesium, SOM, YSZ, Membrane Stability, Diffusion

## Abstract

The solid oxide membrane (SOM) process has been used to produce magnesium by direct electrolysis of its oxide. In this process MgO is dissolved in a molten $CaF_2$-$MgF_2$ flux and an yttria-stabilized zirconia (YSZ) SOM membrane separates the cathode and the flux from the anode. YSZ membrane stability limits the operating life of the SOM electrolyzer. The YSZ membrane is known to degrade due to diffusion of yttrium into the flux. Yttrium diffusion can however be decreased by adding $YF_3$ to the flux. This study investigates the long-term stability of the YSZ membrane. Yttrium composition profiles in the YSZ membrane were determined using WDS as a function of immersion time and $YF_3$ content in the flux. An analytic solution to the diffusion equation was used to model the diffusion process. This study allows the determination of the optimum $YF_3$ content needed in the flux to minimize yttrium diffusion and increase membrane stability.

## Introduction

Magnesium is the third most abundant metal in the earth's crust. Since magnesium is the lightest of all metals used for construction alloys, magnesium is the leading candidate to replace steel and aluminum in automobiles. Automobile manufacturers would like to increase magnesium content from 4-5kg per vehicle to 40-100kg per vehicle [1]. However, in order for automobile manufacturers to replace steel, it is generally believed the magnesium to galvanized steel price ratio must be 4.33:1. For magnesium to replace aluminum in automobiles, the magnesium to aluminum price ratio must be 1.8:1 [2]. In 2008 this ratio was 8:1 for magnesium/steel and 2.5:1 for magnesium/aluminum.

The solid oxide membrane (SOM) process for magnesium production has been successfully employed for relatively short periods of time in a laboratory setting [3]. These results have been extrapolated and the SOM process has been proposed as a method to meet the demand for low cost primary magnesium [2].

In the SOM process the metal oxide is dissolved in an ionic molten fluoride flux, and a potential is applied to separate the oxygen anion from the metal cation. The metal cation is reduced at the cathode, and an oxygen-ion-conducting yttria stabilized zirconia (YSZ) tubular membrane allows vacancy diffusion of the oxygen anions to the anode where they lose electrons to produce oxygen gas. It is believed that the YSZ membrane stability limits the operating life of the SOM process.

Previous work has shown the YSZ membrane to be stable in the molten ionic $MgF_2$-$CaF_2$ based flux for short times, less than 50 hours. Additionally, it was proposed that one mechanism of membrane degradation is yttrium diffusion from the membrane into the flux [4]. However, no previous work has been reported investigating the long term stability of YSZ in molten fluoride flux, nor has the yttrium diffusion out of the YSZ been modeled in any detail.

This paper deals with the primary mechanism of YSZ membrane degradation in ionic based fluoride fluxes, which is believed to be yttrium diffusion from the membrane to the flux. It is also shown that the yttrium diffusion from the membrane is not dependent on the applied potential during magnesium electrolysis. Furthermore, it is believed that diffusion of yttrium from the membrane into the flux can be prevented by adding yttrium fluoride to the flux. The $YF_3$ increases the activity of yttrium in the flux and decreases [or reverses] the chemical potential gradient of yttrium at the flux-YSZ membrane interface. Therefore with appropriate additions of yttrium fluoride in the flux, it is possible to decrease or eliminate yttrium diffusion from the YSZ membrane into the flux and by reversing the chemical potential gradient with higher amounts of $YF_3$ in the flux it is possible for yttrium to diffuse into the membrane from the flux.

The diffusion process is modeled based on the semi-infinite solution to Fick's second law. The model correlates YSZ membrane stability to the concentration of yttrium fluoride in the flux and shows that the YSZ membrane is stable in the flux across a range of flux compositions without undergoing a phase transition or large changes in ionic conductivity. The range of flux compositions provides a basis for determining the parameters with which a large scale SOM process for magnesium production can run for long times, thousands of hours.

## Experimental

<u>*Experimental details:*</u> Yttrium diffusion profiles in the YSZ membranes were determined after performing static membrane stability tests. A static membrane stability test involved immersing a section of a YSZ tube in the $MgF_2$-$CaF_2$ based flux at 1190°C for a given time interval, with no applied potential. The YSZ tube was then raised out of the flux. After cooling, a section of the YSZ membrane was mounted and polished. The composition profile in the membrane was then determined by wavelength dispersive x-ray spectroscopy (WDS). All YSZ tubes were 0.75" outer diameter and contained 10.5w% yttria.

The compositions of flux tested in the static membrane stability tests are shown in Table 1. Magnesium fluoride hydrate was purchased from Alfa-Aesar and dried at 250°C for 12 hours. 96% pure magnesium oxide, 99.5% pure calcium fluoride and 99.99% pure yttrium fluoride were also purchased from Alfa-Aesar and dried for 12 hours at 250°C. After drying, the powders were mixed for 2 hours on a ball mill, at 200 rpm. Once mixed the powders were poured into a one end closed stainless steel tube.

The powders were then melted by heating to 1190°C in 95% argon, 5% hydrogen atmosphere. After equilibrating at

1190°C for 1 hour a 2 inch section of as received YSZ tube was lowered into the flux. The YSZ tubes were immersed in the flux for 16, 24, 32, or 100 hours before being raised out of the flux and then cooled to room temperature at 4°C/min.

The YSZ tubes were then cut with a diamond saw and mounted in Buehler Epothin Epoxy Resin (20-8140-032) with Epothin Epoxy Hardener (20-8142-016). Once mounted the samples were initially ground with a 45 micron diamond wheel, and polished (0.25 micron diamond polishing). A carbon coating was applied to the surface via sputtering and WDS spectroscopy was performed in order to determine the yttria concentration in the membranes. Compositional analysis was done every 5 microns for the first 40 microns from the membrane flux interface and every 10 microns thereafter.

| %YF3 | %CaF2 | %MgF2 | %MgO |
|---|---|---|---|
| 0 | 40.5 | 49.5 | 10.0 |
| 1.6 | 39.8 | 48.6 | 10.0 |
| 2.0 | 39.6 | 48.4 | 10.0 |
| 5.0 | 38.3 | 46.8 | 10.0 |

Table 1: Flux compositions tested in static membrane stability tests

*Electrolysis experiment:* SOM electrolysis was performed with flux containing no $YF_3$ in order to evaluate the effect of applied electric potential on yttrium diffusion out of the YSZ tube. Figure 1 shows the schematic of the SOM electrolyzer used in the electrolysis experiment. It consists of an upper electrolysis chamber, heated to 1190°C and a lower condensation chamber kept at a temperature range of 1050°C-300°C. The set-up is fabricated using SS-304 (Grade 304 stainless steel). It was heated to the aforementioned temperatures in a reducing atmosphere (5%$H_2$-Ar). The crucible wall of the electrolysis chamber served as the cathode. Silver inside the YSZ tube served as the anode. An alumina spacer was used to insulate the YSZ membrane from the steel which acted as the cathode. A 0.25" molybdenum tube served as the anode current collector. Before electrolysis was performed, the cell was purged with Ultra High Purity Argon at 300cc/min into the cathodic chamber through the annulus between the inner diameter of the stainless steel tube and the outer diameter of the YSZ tube to lower the magnesium partial pressure. The argon gas is required to keep the magnesium vapor produced from reducing the YSZ, as well as act as a carrier gas to transport the magnesium vapor to the condenser. The magnesium vapor produced on the cathodic side is transported to the condensation chamber through the two stainless steel tubes. These two tubes extend above the flux to prevent any molten flux from entering the condenser. On the anodic side, hydrogen was bubbled in the silver through the molybdenum tube at rate of 60cc/min to prevent oxidation of the molybdenum current collector.

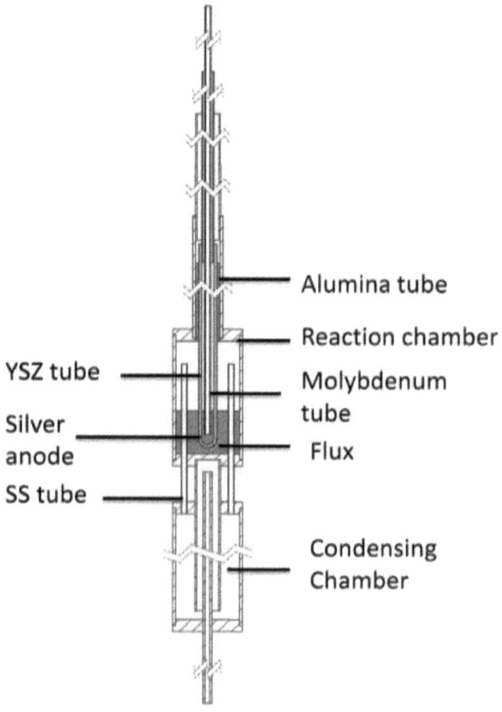


Figure 1: Design of experimental setup for SOM process.

An alumina tube attached to the YSZ tube (with Aremco 552 paste) was used to carry the $H_2O_{(g)}$ out of the anodic chamber and also prevent the $H_2O_{(g)}$ from reacting with the magnesium vapor produced on the cathodic side.

After the experiment the YSZ tube was sectioned and polished. The yttrium concentration profile in the YSZ tube was measured using Energy-dispersive X-ray spectroscopy (EDAX) and compared with the results of the static experiments with no $YF_3$ in the flux.

**Results and Discussion**

A static membrane stability test was preformed with no yttrium fluoride in the flux for 32 hours. The first 50 microns from the membrane flux interface contained no yttria and high porosity, 50-70 microns from the membrane flux interface the yttria concentration gradually increased and after 70 microns yttria concentration was uniform and matched the bulk concentration, Figure 2.

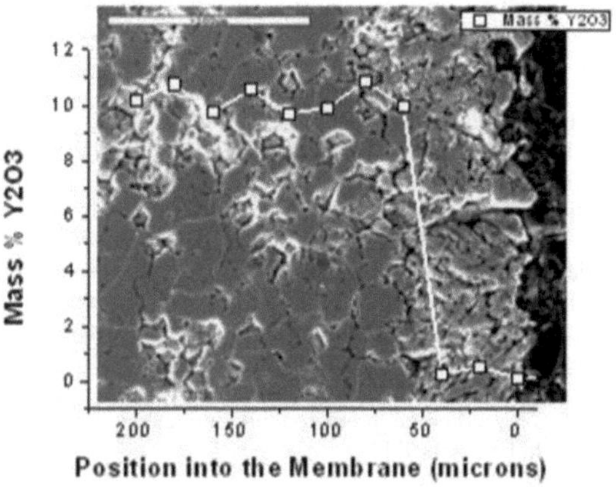


A

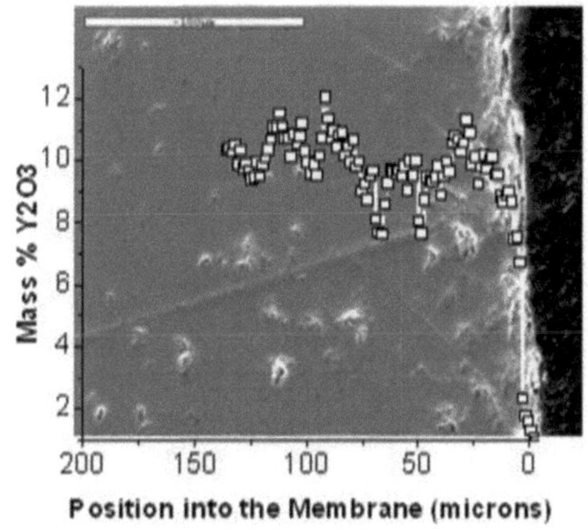


B

Figure 2: A. Yttria concentration profile overlaid on top of SEM image of YSZ exposed to the flux for 32hrs. B. As received YSZ.

It is believed that the decrease in yttria concentration at the membrane flux interface is due to yttrium diffusing from the membrane into the flux. If the mechanism for yttrium depletion in the membrane is chemical diffusion, then by controlling the activity of yttrium in the flux it should be possible to control the yttrium diffusion out of the membrane. It was found when the activity of yttrium in the flux is higher than that of the yttrium in the membrane, yttrium diffused from the flux into the membrane, increasing the yttria concentration in the membrane; this happened when 5w% $YF_3$ was added to the flux, Figure 3.

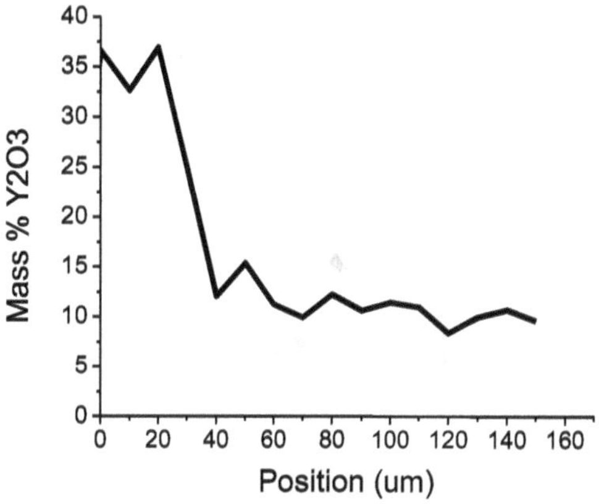


Figure 3. Yttria profile in the YSZ membrane exposed to flux with 5 w% $YF_3$ for 32hrs.

It was seen that when YSZ is exposed to a flux containing 2w% $YF_3$ for 100 hours the membrane has limited porosity, Figure 4, and concentration at the membrane flux interface increases to 15w%, Figure 5. $MgF_2$-$CaF_2$ flux containing 2 w% $YF_3$ appears to be a composition in which YSZ is stable for long time periods.

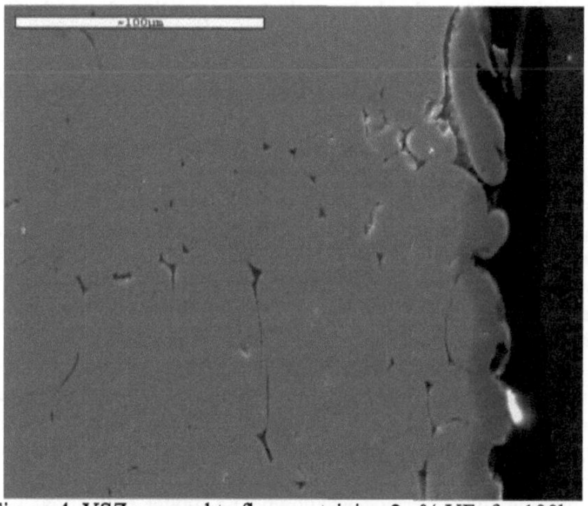

Figure 4. YSZ exposed to flux containing 2w% $YF_3$ for 100hrs.

The yttria concentration profile in YSZ exposed to flux containing yttrium fluoride can be modeled using the semi-infinite solution to Fick's second law, equation 1.

$$C = C_O + (C_\infty - C_O) * \text{erf}\left(\frac{x}{2\sqrt{Dt}}\right) \quad (1)$$

In equation 1 $C_0$ is the steady state concentration of yttria at the membrane flux interface, and $C_\infty$ is the bulk concentration in the membrane. The experimental results from the static membrane tests were used to fit for the diffusivity, using the least squares approximation. The results of this model are overlaid on top of the experimentally determined results from the static membrane stability tests in Figure 5.

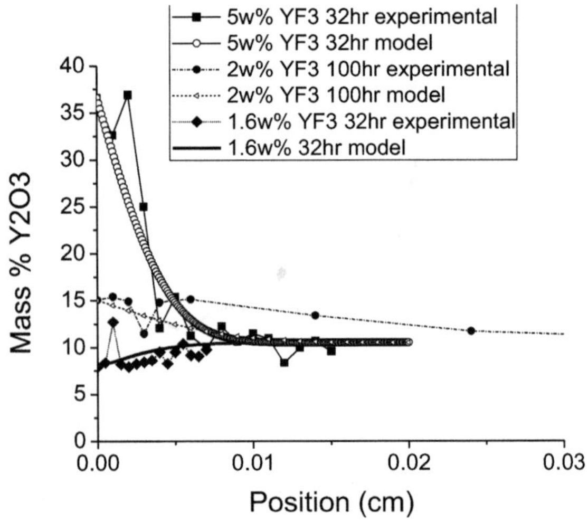


Figure 5: Experimental yttrium diffusion profiles and modeled results.

From the model, the diffusivity of yttrium in the membrane was determined to be $5.6x10^{-11}cm^2/sec$. According to the model the yttria concentration throughout the membrane will never reach the $C_o$ value during the operation of the SOM process, because the diffusivity is very low. However, it is important to determine the $C_0$ value because the corresponding change in $Y_2O_3$ concentration in the diffusion zone of the YSZ membrane may lead to changes in ionic conductivity or crystal structure.

A key factor in modeling the diffusion of yttrium in the membrane is knowing the $C_o$ value for a specific flux composition. Static membrane stability tests with 2w% $YF_3$ in the flux where performed for 32 and 100 hours. The $C_0$ value was measured to be 15.4 w% $Y_2O_3$ for the 100hr case and 16.1 w% $Y_2O_3$ for the 32 hour case. The difference in $C_0$ values for 32 and 100 hour studies falls within the experimental error for compositional analysis using WDS. It is therefore believed that the 32 hour static membrane stability tests are sufficiently long to determine the $C_0$ value for a given flux composition.

The experimentally determined values of $C_0$ were obtained using 32 hour experimental holds. Once more data points are acquired, a polynomial fit using a least squares approximation will be used to determine the $C_0$ value for a given $YF_3$ flux concentration. As long as the steady state value at the membrane flux interface is known, the model will allow us to determine the yttrium concentration profile in the YSZ membrane from the YSZ/flux interface.

10.5 w% $Y_2O_3$ stabilized zirconia after aging at 1200°C has both tetragonal and cubic phases present and high ionic conductivity [5]. Both the tetragonal and cubic phases are metastable from 5-16.5 w% $Y_2O_3$ [6]. From a mechanical stand point the YSZ membrane is most stable when both the tetragonal and cubic phases are present. Thus the $Y_2O_3$ stabilization in zirconia plays an important role in determining the phase, conductivity and mechanical properties of the membrane. Therefore, it is necessary to determine the amount of yttria in the YSZ during the entire SOM process in a commercial scale operation.

The equation acquired from the polynomial fit, will be used to determine the yttria concentration at the flux-YSZ interface as a function of the $YF_3$ content in the flux. This equation will provide a range of flux compostions where the SOM electrolyzer can successfully operate and will not undergo a phase change. Further analysis of the mechanical properties of the YSZ tubes is not a major concern for the SOM process since the tubes are suspended and stationary.

A YSZ tube was exposed to the flux for 23 hours and electrolysis of magnesium oxide was performed. A separate static membrane stability test was performed for 24hrs. In both cases no yttrium fluoride was present in the flux. The resulting yttrium composition profiles, measured with EDAX, are nearly identical, Figure 6. This shows that the diffusion process is not affected by the applied potential during electrolysis. Therefore it is believed the yttrium concentration in the YSZ membrane is dependent only on chemical diffusion and migration does not play a major role in determining the yttrium concentration profile. It is interesting to note earlier calculations on application of electrical potential across a YSZ membrane predicted no noticeable demixing of yttrium would be detected [7].

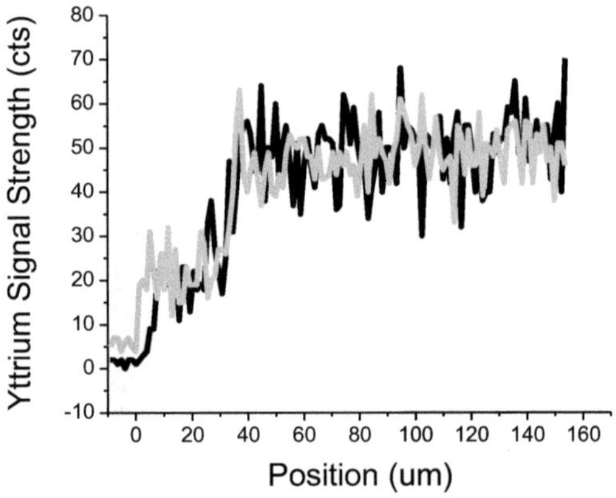


Figure 6: Yttrium concentration from the membrane flux interface. The red line is after 24 hours in the flux with no applied potential, black line is after a 23 hour electrolysis hold.

The diffusion model does not apply when there are trace amounts or no yttrium fluoride in the flux, because, it causes yttria concentration in the membrane to fall below 5 w% which leads to extensive porosity in the membrane. The porosity allows the flux to enter the membrane and increases the surface area of YSZ exposed to the flux; a complete diffusion profile is not observed and the yttrium concentration undergoes a step change. Figure 2(a).

## Conclusion

Yttrium diffuses from the YSZ membrane into molten $MgF_2$-$CaF_2$ flux over relatively short time scales. The diffusion of yttrium is related to the activity difference between yttrium in the membrane and yttrium in the molten flux. By increasing the activity of yttrium in the molten flux it is possible to prevent yttrium from diffusing into the flux. It was also shown that yttrium will diffuse into the YSZ membrane from the flux when the activity of yttrium in the flux is sufficiently high. Experimentally, YSZ was shown to be stable in flux containing 2w% yttrium fluoride. The diffusion of yttrium in the membrane was modeled using a semi-infinite solution to flicks second law. More data points are needed to correlate the concentration of yttrium fluoride in the flux to the concentration of yttrium in the membrane during the SOM electrolysis.

## References


1. Pollock, T.M., *Weight Loss with Magnesium Alloys.* Science, 2010. **328**(5981): p. 986-987.
2. Das, S., *Primary magnesium production costs for automotive applications.* JOM Journal of the Minerals, Metals and Materials Society, 2008. **60**(11): p. 63-69.
3. Krishnan, A., X.G. Lu, and U.B. Pal, *Solid oxide membrane process for magnesium production directly from magnesium oxide.* Metallurgical and Materials Transactions B-Process Metallurgy and Materials Processing Science, 2005. **36**(4): p. 463-473.
4. Krishnan, A., *Solid oxide membrane process for the direct reduction of magnesium from magnesium oxide.* ProQuest, 2006.
5. Kondoh, J., et al., *Effect of Aging on Yttria-Stabilized Zirconia.* Journal of The Electrochemical Society, 1998. **145**(5): p. 1527-1536.
6. Yashima, M., M. Kakihana, and M. Yoshimura, *Metastable-stable phase diagrams in the zirconia-containing systems utilized in solid-oxide fuel cell application.* Solid State Ionics, 1996. **86-88**(Part 2): p. 1131-1149.
7. Monceau, D., et al., *Kinetic demixing of ceramics in an electrical field.* Solid State Ionics, 1994. **73**(3-4): p. 221-225.


## ACKNOWLEDGEMENT


This material is based upon work supported by the National Science Foundation under Grant No. 1026639.

**Magnesium Technology 2012**
*Edited by: Suveen N. Mathaudhu, Wim H. Sillekens, Neale R. Neelameggham, and Norbert Hort*
***TMS (The Minerals, Metals & Materials Society), 2012***

# Study on the Thermodynamic and Experimental Carbothermic Reduction of Garnierite

Tao Qu * Yang Tian Bin Yang Da-chun Liu
Bao-qiang Xu Yong-nian Dai

(1 National Engineering Laboratory for Vacuum Metallurgy, Kunming 650093,China;

2 Key Laboratory of Nonferrous Metals Vacuum Metallurgy of Yunnan Province, Kunming 650093,China;

3 Faculty of Metallurgical and Energy Engineering, Kunming University of Science and Technology, Kunming 650093, China

4 State Key Laboratory Breeding Base of Complex Nonferrous Metal Resources Clear Utilization in Yunnan Province, Kunming 650093, China)

***Correspondence:** Tel: +8613518715200; E-mail: qutao_82@126.com

**Keywords:** garnierite; carbothermic reduction; magnesium; vacuum metallurgy; thermodynamics

**Abstract:**The overall utilization of magnesium and other metals should be systematically considered during the exploration of deficient garnierite.In this paper the thermodynamic analysis of the carbothermic reduction process for extracting metal magnesium from garnierite in vacuum was carried out to investigate its feasibility.The calculation results indicate that it is feasible technically that the carbothermic reduction process for extracting metal magnesium from garnierite in vacuum. Under the temperature of 1500℃ and vacuum degree is less than 300Pa, metal magnesium was obtained. The Nickel content in residue is more than twice as garnierite ore.

## 1. Introduction

Garnierite is not a single mineral but a mixture of the Ni-Mg-hydrosilicates serpentine, talc, sepiolite, chlorite and smectite. Garnierite is a hydrous nickel silicate, with the serpentine group of minerals, mostly a mixture of various Ni- and Ni-bearing magnesium layer silicates[1-2]. It is an alteration of olivine, serpentine rocks, and can be associated with olivine, serpentine, chromite and talc. Garnierite is mined from laterite deposits, formed by weathering and secondary mineralization of igneous rocks. The nickel is contained in the blue-green bands of the ore. The specific gravity of this mineral ranges from 2.3 to 2.8.

The Pyrometallurgical techniques are suited to treat garnierite, with processes typically involving drying, calcining, reduction and electric furnace smelting to produce a ferronickel[3-7]. The hydrometallurgical process is mainly used ammonia–ammonium carbonate leaching [8] and hydrochloric acid leaching [9]. All of these methods have a disadvantages is that the magnesium is not utilized well.

## 2. Thermodynamic Analysis

According to the test purposes, the main consideration in the nickel laterite NiO, $Fe_2O_3$, MgO and $SiO_2$ are that reduction of four major oxides and the possible reactions. First, consider the following a few basic oxidation reaction (phase state does not indicate that there is phase transition), according to Gass's law to determine possible reaction and starting temperature.

$$2C_{(s)} + O_{2(g)} = 2CO_{(g)} \quad \cdots\cdots (1)$$

$$2Ni + O_{2(g)} = 2NiO_{(s)} \quad \cdots\cdots (2)$$

$$\frac{4}{3}Fe_{(s)} + O_{2(g)} = \frac{2}{3}Fe_2O_{3(s)} \quad \cdots\cdots (3)$$

$$2Mg + O_{2(g)} = 2MgO_{(s)} \quad \cdots\cdots (4)$$

$$Si + O_{2(g)} = SiO_2 \ldots\ldots\ldots\ldots\ldots\ldots\ldots(5)$$

$$\frac{2}{3}SiC_{(s)} + O_{2(g)} = \frac{2}{3}SiO_2 + \frac{2}{3}CO_{(g)} \ldots(6)$$

In this paper, the Gibbs free energy of material function was studied, the Gibbs free energy changes of the reaction temperature at various points were calculated by equation (2-1), (2-2), (2-3), (2-4) and thermodynamic data of inorganic reference manual[10], the calculation process is omitted.

$$\Delta\Phi_T' = \sum(n_i\Delta\Phi_{i,T}')_{product} - \sum(n_i\Delta\Phi_{i,T}')_{reactant} \quad (2\text{-}1)$$

$$\Delta H_{298}^{\theta} = \sum V_i \Delta H_{298(product)}^{\theta} - \sum V_i \Delta H_{298(reactant)}^{\theta} \quad (2\text{-}2)$$

$$\Delta G_T^{\theta} = \Delta H_{298}^{\theta} - T\Delta\Phi_T' \quad (2\text{-}3)$$

$$\Delta G_T = \Delta G_T^{\theta} + RT\ln J^{\theta}(T) \quad (2\text{-}4)$$

Fig.1 shows the relationship between the Gibbs free energy changes of the main oxidate and temperature. The reaction and conditions of the lines 1-12 were shown below Fig.1, $P_{(O_2)} = P^{\theta} = 10^5 Pa$. According to Gass's law, the Gibbs free energy of the reaction between material and 1mol $O_2$ is becoming negative, the substance out of the more difficult to be reduced under a certain temperature, instead the substance can be reduced which $\Delta G_T$ larger than it. The lowest initial reaction temperature point is the intersection point of the two lines.

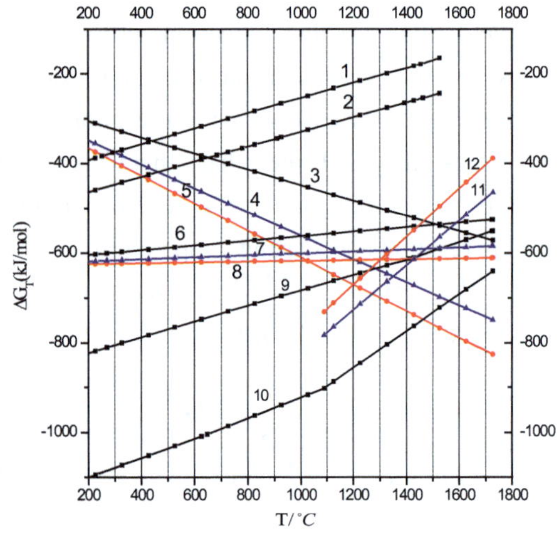


Fig.1 The relationship between the Gibbs free energy changes of the main oxidate and temperature

1- $2Ni + O_{2(g)} = 2NiO_{(s)}$

2- $\frac{4}{3}Fe_{(s)} + O_{2(g)} = \frac{2}{3}Fe_2O_{3(s)}$

3- $2C_{(s)} + O_{2(g)} = 2CO_{(g)}\ P_{(CO)} = P^{\theta}$

4- $2C_{(s)} + O_{2(g)} = 2CO_{(g)}\ P_{(CO)} = 500Pa$

5- $2C_{(s)} + O_{2(g)} = 2CO_{(g)}\ P_{(CO)} = 50Pa$

6- $\frac{2}{3}SiC_{(s)} + O_{2(g)} = \frac{2}{3}SiO_2 + \frac{2}{3}CO_{(g)}\ P_{(CO)} = P^{\theta}$

7- $\frac{2}{3}SiC_{(s)} + O_{2(g)} = \frac{2}{3}SiO_2 + \frac{2}{3}CO_{(g)}\ P_{(CO)} = 500Pa$

8- $\frac{2}{3}SiC_{(s)} + O_{2(g)} = \frac{2}{3}SiO_2 + \frac{2}{3}CO_{(g)}\ P_{(CO)} = 50Pa$

9-

10- $2Mg + O_{2(g)} = 2MgO_{(s)}\ P_{(Mg)} = P^{\theta}$

11-

12- $2Mg + O_{2(g)} = 2MgO_{(s)}\ P_{(Mg)} = 50Pa$

It can be seen from Fig.1 that NiO and $Fe_2O_3$ are poor stability, it can be reduced by C or Si at low temperatures Under atmospheric pressure above 450 ℃ the NiO can be reduced by C; $Fe_2O_3$ at 650 ℃ was reduced by C too, Lower initial reaction temperature is in vacuum; $SiO_2$, and MgO are stable, it can be reduced under higher temperature. Consider that the total pressure is 30-500Pa in furnace, the reduction temperature of $SiO_2$ and MgO decreased significantly.

Fig.2 shows that Fig.1 was magnified and remove the other reaction (2), (3), (5), (6) in order to facilitate observation, it can be seen that A:$P_{(CO)} = P_{(Mg)}$ = 50Pa, MgO can be reduced by C under 1473K ; C: $P_{(CO)} = P_{(Mg)}$ = 500Pa, MgO can be reduced by C under 1623K; At atmospheric pressure the reaction temperature is 1832K. However, we can see reaction (1) and (4) in the region will intersect within ABCD, that means between 1473K and 1623K, the carbothermic reduction magnesia is easily occurred.

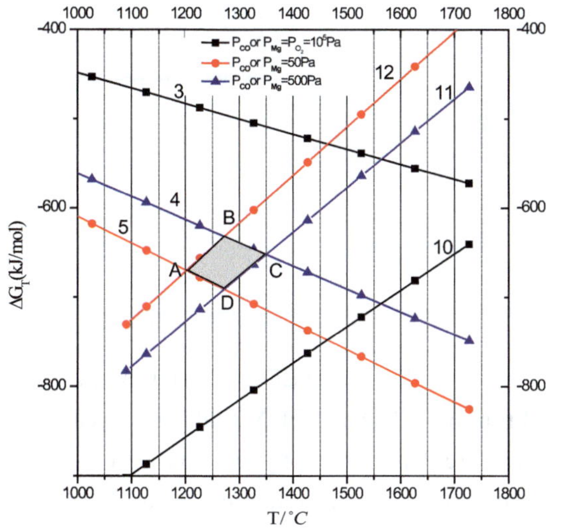


Fig.2 The relationship between the Gibbs free energy changes of MgO, CO and temperature

The above analysis shows that the carbothermic reduction magnesia is easily occurred under vacuum conditions. But the reverse reaction $Mg_{(g)}+CO_{(g)}=MgO_{(s)}+C_{(s)}$ [11] is also occurred.

3. Experimental

3.1 Materials

The materials used in this study were garnierite and coal with $CaF_2$ ($CaF_2 \geq 98\%$) as catalyst. The chemical composition of garnierite ores and coal were given in Table.1 and Table.2. And the XRD pattern of garnierite is shown in fig.3.

Tab.1 Chemical composition of garnierite

| composition | Ni | MgO | $Fe_2O_3$ | $SiO_2$ | $Al_2O_3$ |
|---|---|---|---|---|---|
| wt% | 1.18 | 27.74 | 11.10 | 36.91 | 2.73 |

Tab.2 composition of coal

| composition | $M_{ad}$ | $V_{daf}$ | $A_d$ | $FC_d$ |
|---|---|---|---|---|
| wt% | 0.78 | 16.43 | 19.21 | 63.58 |

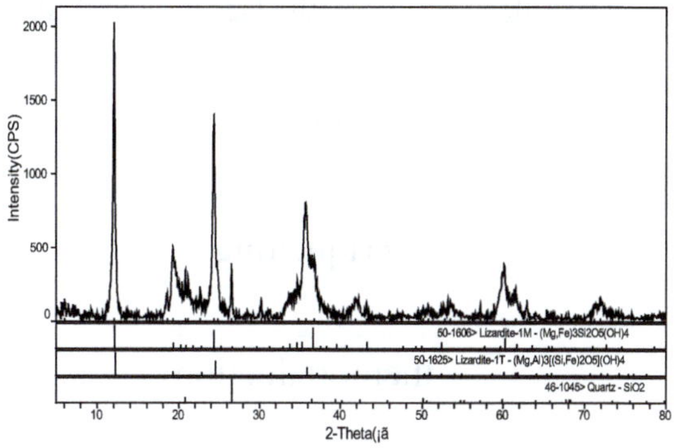


Fig.3 the XRD pattern of garnierite

3.2 Experimental device

The experimental device for vacuum thermal reduction is shown in Fig.4, which is compose of reactor, temperature control system, vacuum system and condenser system.1

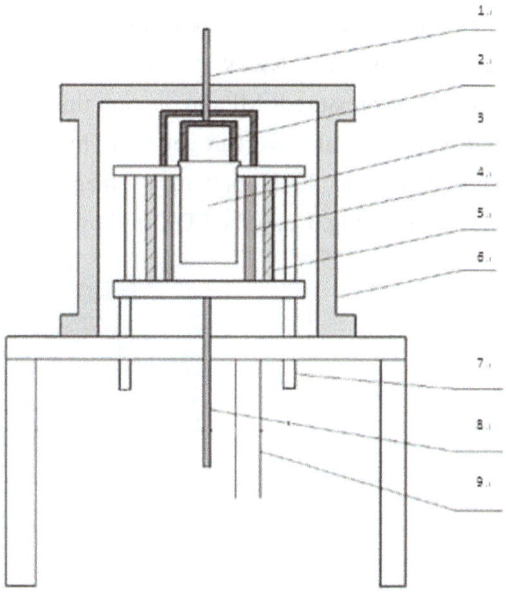


1. thermocouple; 2. condenser; 3. couple; 4. heating unit;

5. thermal insulating layer; 6. furnace; 7. electrode;

8. thermocouple; 9. exhaust pipe

Fig.4 Experimental device

3.3 Experimental method

In the vacuum and high temperature conditions, coal can reduced magnesium silicate and get magnesium vapor, and condensate it for getting magnesium metal. The technological flow sheet is shown in Fig.5.

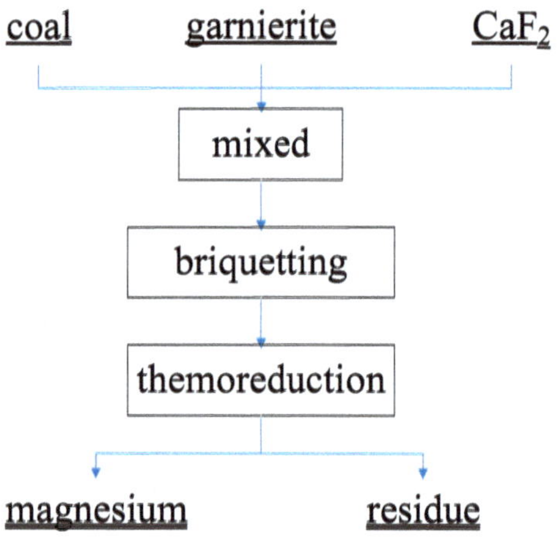


Fig. 5 Technical flow sheet of extracted magnesium from the garnierite

4 Results and Discussion

4.1 The magnesium extracted from the garnierite

On the condition of reduction temperature 1500℃, vacuum degree is less than 300Pa, briquetting pressure 8 MPa, reduction time 90min, and with 5% $CaF_2$, we can get some magnesium. The photo and XRD pattern of magnesium is shown in the Fig.6, and the EDX is shown in Fig.7.

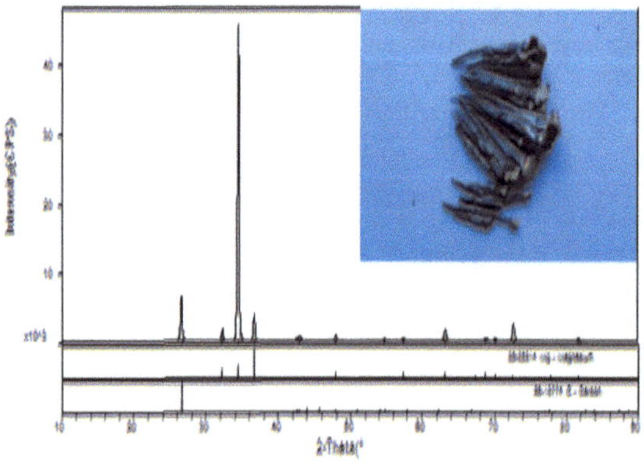

Fig.6 The photo and XRD pattern of magnesium

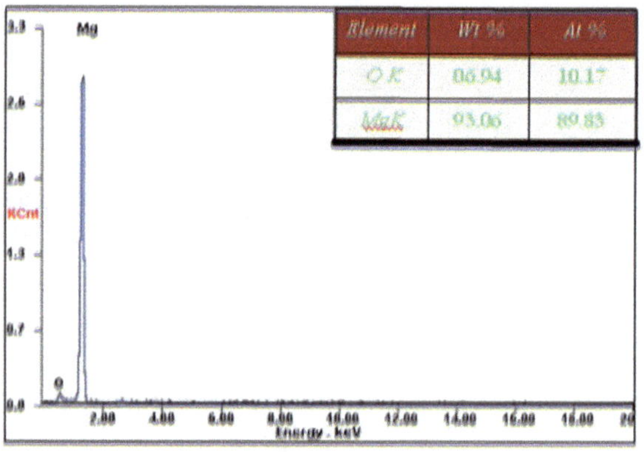


Fig.7 The EDX of magnesium

From the XRD and EDX analysis, we can know that there has some oxygen and carbon in the magnesium. We believe the oxygen and carbon get from the reverse reaction of reduction, the reaction formula is as following:

$$Mg(g)+CO(g)\rightarrow MgO(s)+C(s)$$

4.2 The XRD analysis of residue

On the condition of reduction temperature 1500℃, vacuum degree is less than 300Pa, briquetting pressure 8 MPa, reduction time 90min, and with 5% $CaF_2$, the XRD pattern of residue is shown in the fig.8.

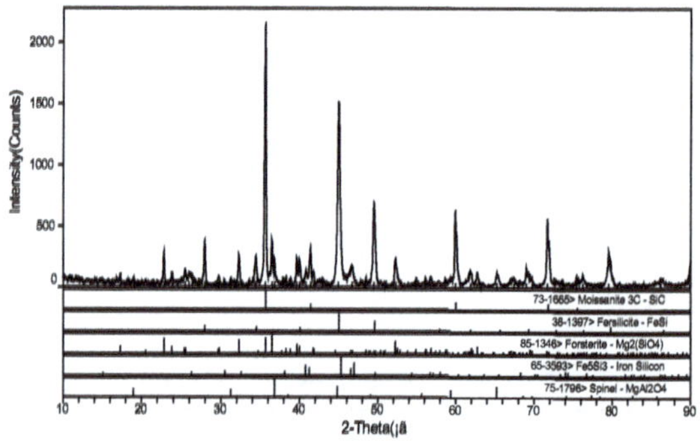


Fig.8 The XRD pattern of residue

It can be seen that there mainly are the phases of SiC and FeSi, and also contained some $Mg_2SiO_4$、$MgAl_2O_4$. It's shown that the MgO and $SiO_2$ still no reaction completely, and Si generated the SiC(main), FeSi and $Fe_5Si_3$, and it's not found the Iron and silicon elemental. Nickel does not be detected, mainly because of its content is low.

4.3 Effects of reduction time on reduction rate of magnesium

On the condition of reduction temperature 1500℃, vacuum degree is less than 300Pa, briquetting pressure 8 MPa, and with 5% $CaF_2$, the reduction rate of magnesium by coal at varying time conditions are shown in Fig.9. In these experiments, the reduction time is that the temperature of reactor keeps the 1500℃ after it is reach. As is shown from the chart, with the extension of reaction time, reduction rate of magnesium is increased. But when the reduction time is reached 90min, the curve has become flat, the reduction rate of magnesium with the increase in reaction time is not obvious.

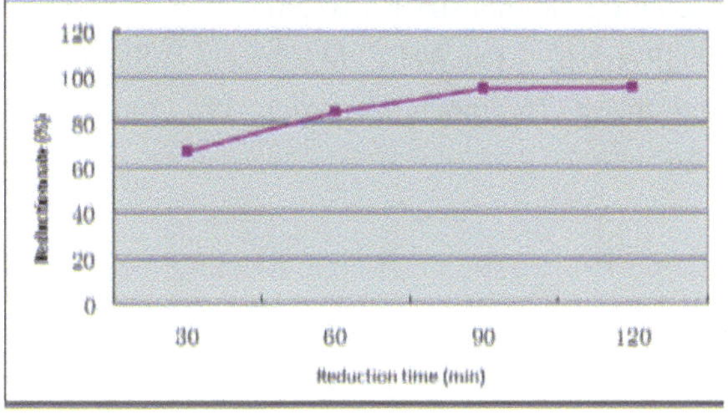


Fig.9 Relationship between reduction rate and time

4.4 Effects of reduction time on nickel content of residue

On the condition of reduction temperature 1500℃, vacuum degree is less than 300Pa, briquetting pressure 8 MPa, and with 5% $CaF_2$, the nickel content of residue at varying time conditions are shown in Fig.10. As is shown from the chart, with the extension of reaction time, the nickel content is increased. When the reduction time is reached 90min, the Nickel content is more than twice as garnierite ore.

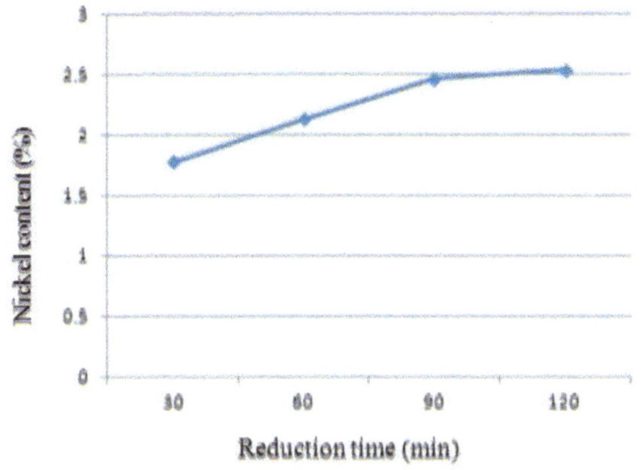


Fig.10 Relationship between nickel content and reduction time

5. Conclusion

In this paper, the coal was used to reduce the garnierite to prepare metal magnesium in vacuum. We can get the following conclusion:

1. The thermodynamic calculation results indicated the $P_{(CO)} = P_{(Mg)} = 50Pa$, that reduction temperature is down to 1473K from 1832K at atmospheric pressure.

2. This method can prepare the magnesium from garnierite, but has small amount oxygen and carbon, it should be got from the reverse reaction of reduction.

3. SiC and FeSi were obtained in residues by carbothermic reduction garnierite under a temperature of 1500℃, and it's not found the Iron and silicon elemental.

4. The reduction rate of magnesia is enhance with the extensive reduction time.

5. The nickel content is enhance with the extensive reduction time, when the reduction time is 90min, the Nickel content in residue is more than twice as garnierite ore

**Acknowledgements**

This work was financially supported by Science Research Foundation of Education Department of Yunnan Province, China (No. 2010C257)

## Reference

[1] Andrew B, Jeffrey H, Angus J M, et al, Development of a model for serpentine quantification in nickel laterite minerals by infrared spectroscopy, Minerals Engineering, 2010, vol23: 407-412;

[2] Sun Chuanyao, Yin Wanzhong, flotation principle of silicate minerals, Beijing, China: Science Press, 2001;

[3] Lagend ijk H, Schouken S. A. F. S, Smith P, et al. The production of ferronickel from nickel-containing laterite, South African Patent, EP19940306256, 2004.05.19;

[4] Nayak J. C, Production of Ferro-nickel from Sukinda Laterites in Rotary Kiln-electric

Furnace, Transaction of Indian Metallurgy, 1985, 38(3): 241-247

[5] Watanabe T, Ono S, Arai H, et al, Direct Reduction of Gamierite Ore for Production of Ferronickel with a Rotary Kiln at Nippon Yakin Co. Ltd. Oheyama Works, International Seminar on Laterites, MMIJ, Tokyo,1985: 103-115

[6] Tanaka K, Sakamoto K, Sato K, Production of ferronickel by blast furnace at Nippon Mining's Saganosaki smelter and refinery, International Seminar on Laterites, MMIJ, Tokyo,1985: 127-137;

[7] Ishii K, Development of Ferro-Nickel Smelting from Laterites in Japan, International Seminar on Laterites, MMIJ, Tokyo(1985): 13-24;

[8] Power L. F，Geiger G. H, The application of the reduction roast-ammoniacal ammonium carbonate leach to nickel laterites, Minerals Science, 1997, 9(1): 32-50

[9] Dutrizac, J.E. Chen, T.T. White, C.W. Fundamentals of serpentine leaching in hydrochloric acid media [J]. TMS Annu Meet. 2000:41～51;

[10] Dai Yong-nian, Yang Bin Vacuum metallurgy of nonferrous materials. Beijing, China: Metallurgical industry press.2003

[11]Rong-ti, PanWei, Masamichi Sano, Li Jian-qiang, Thermochimica Acta, 398 (2003) : 265 - 267

**Magnesium Technology 2012**
*Edited by: Suveen N. Mathaudhu, Wim H. Sillekens, Neale R. Neelameggham, and Norbert Hort*
***TMS (The Minerals, Metals & Materials Society), 2012***

# Mechanism of Carbothermic Reduction of Magnesia and Reversion Reaction

Yang Tian [1,2,3] Tao Qu [1,2,3] Bin Yang [1,2,3] Hong-xiang Liu [1,2,3] Cheng-bo Yang [1,2,3] Yong-nian Dai [1,2,3]

1.National Engineering Laboratory for Vacuum Metallurgy,Kunming University of Science and Technology,Kunming Yunnan,China,650093,

2.Vacuum Metallurgy Key Laboratory for Nonferrous Metal of Yunnan Province, Kunming University of Science and Technology, Kunming, Yunnan, China,650093,

3.Faculty of Metallurgy Engineering, Kunming University of Science and Technology, Kunming, Yunnan, China,650093

Keywords: Mechanism; Carbothermic reduction; Reversion reaction; Vacuum distillation; Magnesium

## Abstract

In this study, the mechanism of the carbothermic reduction process to extract magnesium from magnesia and the reversion reaction in vacuum were investigated. The carbon monoxide (CO) content of the gas, phases of the condensing product, surface morphology of the reduction slag and phase of the distillation product were measured by means of gas chromatography (GC), XRD, and SEM. The experimental results indicated that Mg was generated by magnesia and carbon at 1623K and 30~100Pa in the carbothermic reduction process.The main gas in carbothermic reduction process is carbon monoxide, no carbon dioxide occurred at any reaction time, the reduction reaction is $MgO_{(s)}+C_{(s)}=Mg_{(g)}+CO_{(g)}$. The gas-phase reversion will commence as soon as the saturated gas mixture is cooled. The reversion reactions are favored below 1373K.The distillation product by vacuum distillation process produced high purity metal magnesium product, it can be deduced in reversion reaction which occured at low temperature and 30~100Pa during carbothermic reduction. The reversion reaction was calculated, which gave the peak value $\gamma$ less than 9%.

## Introduction

Magnesium is the eighth most abundant element in the Earth's crust (approximately 2.5% by weight)and does not occur uncombined[1]. It is also the lightest of all metals used for structural alloys[2]. Magnesium alloys have attracted increasing interest for mass reduction in aerospace, automobile industry and transportation[3].The requirement have triggered renewed interest in magnesium[4].Now, the pigeon process is the main method of thermal reduction of magnesium metal production, However, since the low reactivity of ferrosilicon alloy, the high temperature and long reaction time were needed[5].

Carbothermic reduction is an alternative to both silicothermic and electrolytic processes for the production of magnesium[6-7]. But the reversion reaction between Mg vapor and CO are major technical challenges. In order to avoid the reversion reaction, Hori points that thermal control of the product gases is important throughout the process from the reaction chamber to the product collection point via the nozzle[8-9]. Now Tassios's invention involves the manner of heated nozzle. The invention heat is supplied to the nozzle over-and-above any heat that is supplied to the nozzle by gas flow[10]. But all of the invention of reversion reaction between Mg vapor and CO were only considered as apparatus and method, the mechanism of reaction was not studied.

Recently our research group[11-12] focused on the mechanism of the reduction process and reversion reaction. The effects of carbothermic reduction process, reversion reaction ratio, and condensing product of vacuum distillation were discussed. In order to investigate the mechanism of the carbothermic reduction process in vacuum, GC ,XRD and SEM techniques are also proposed.

## Experimental

### Raw material

Carbothermic reduction process: analytical grade of magnesia, carbon and $CaF_2$ were used as the raw materials in experiments.

Vacuum distillation process: condensing product of carbothermic reduction.

### Schematic diagram of vacuum furnace

Experiments were carried out in a self-made vacuum furnace shown in Fig.1.

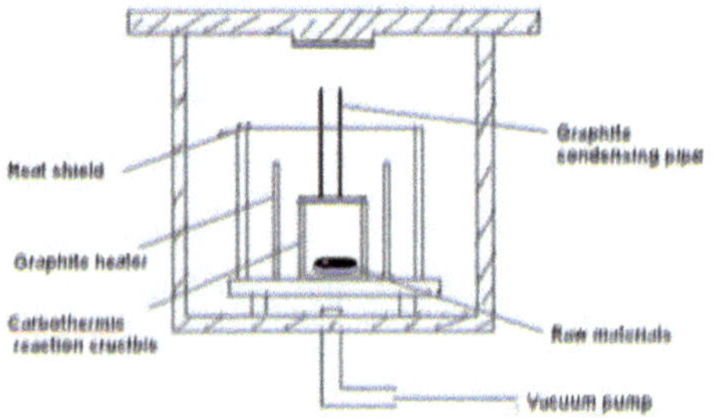


**Fig.1 Schematic diagram of vacuum furnace:**

### Experimental procedure

Carbothermic reduction process: magnesia and carbon with a certain molar ratio (MgO/C=1/2) were mixed, and cast into pellets of Φ30mm×10mm under 2-10MPa as shown in the Fig. 2. The heating rate of system temperature was about 15K•min$^{-1}$. In the magnesia carbothermic reduction process, the prepared pellets were heated from room temperature to 1073K and held at this temperature for 20 minutes to remove the adsorbed gases and water. Then, the temperature was raised to 1473K-1773K. After the reaction time(2h to 5h), the reaction product was cooled to room temperature and took out. The reduction reaction in this process was $MgO_{(s)}+C_{(s)}=Mg_{(g)}+CO_{(g)}$, and the reversion reaction in this process was $Mg_{(g)}+CO_{(g)}=MgO_{(s)}+C_{(s)}$.

Vacuum distillation process: used the condensing product of carbothermic reduction as distillation materials directly as shown in the Fig.3, the heating rate of system temperature was about 15K•min$^{-1}$. In the process, the condensing product was heated from room temperature to 1373K. After the distillation time(1h or 2h), the distillation product was cooled to room temperature and took out. The reaction take place during distillation process was $Mg_{(l)}=Mg_{(g)}$.

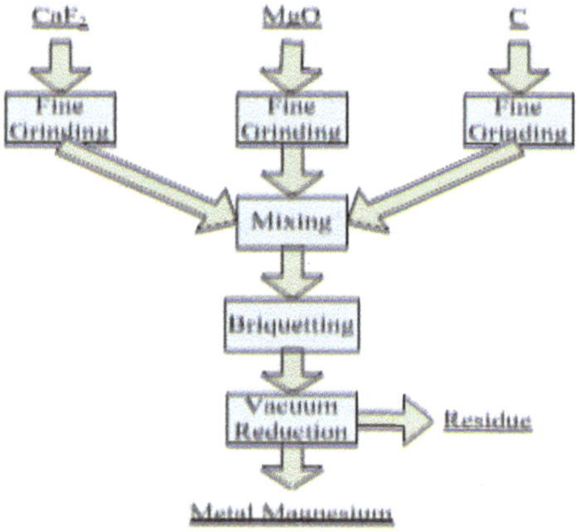


**Fig. 2 Reduction experimental flow chart**

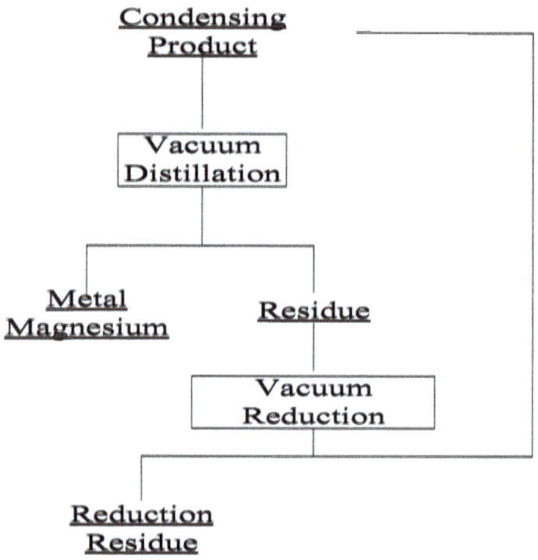


**Fig. 3 Distillation experimental flow chart**

### Analysis methods

The crystalline phase of the products was identified by X-ray diffraction (XRD) instrument (D/max-3B) using Cu $K_\alpha$ radiation with a scanning rate of 2 (°)/min which was made by Rigaku Corporation of Japan. Surface morphology of the condensing product was characterized by scanning electron microscopy (XL30ESEM-TMP, Phillips, Holland). The content of carbon monoxide was determined with gas chromatography (6890N, Agilent Technologies, Taiwan).

### Results and Discussion

**3.1Analysis of magnesia carbothermic reduction process**

The reactions of MgO and Carbon could occur at about 1625K and 10~100Pa[13], meanwhile, CO was generated continuously. Gas in the magnesia carbothermic reduction process was collected using the air-pocket at 60 minutes intervals, and the gas volume measured by drainage. The content of carbon monoxide was determined by GC. The volume of carbon monoxide per hour and total volume of carbon monoxide was calculated, and the average production rate of carbon monoxide per 60min determined.

**Tab.1 Gas chromatography analysis of carbothemic reduction**

| Gas Compositions / Reaction Temperature | CO | $CO_2$ |
|---|---|---|
| | Content/% | Content/% |
| 1723K | 39.59 | 0.00 |
| 1723K (after 1h) | 27.91 | 0.00 |
| 1723K (after 2h) | 58.21 | 0.00 |
| 1723K (after 3h) | 61.11 | 0.00 |
| 1723K (after 4h) | 54.29 | 0.00 |

Tab.1 shows the relationship between magnesia carbothermic reduction time at 1723K, the content of total volume of carbon dioxide and total volume of carbon monoxide. As shown in Tab.1, it can be seen that the main gas in carbothermic reduction process was carbon monoxide, no carbon dioxide occurred at any reaction time. Accordingly, the thermodynamic analysis of carbothermic process was assumed to generate gas approximation for 100%CO.The main reduction was $MgO_{(s)}+C_{(s)}=Mg_{(g)}+CO_{(g)}$.(1) The indirect reduction $MgO_{(s)}+CO_{(g)}=Mg_{(g)}+CO_{2(g)}$ (2) did not occur during the process.

Though the thermogravimetric reduction experiment, the mass loss of the sample was monitored as a function of time. The reduction ratio($\alpha$) at given instance was defined as $\alpha = \dfrac{M_0 - M_1}{M_0} \times 100\%$ (3)

where $M_0$ represents the initial mass of magnesium in the sample, and $M_1$ is the residual in the sample.

Fig.4 shows the relationships between the typical mass-loss percentage and temperature curve of reduction of magnesia by carbon. There was no massive loss detected when the temperature was below 1373K,indicating no reaction happens. When the temperature was higher than 1373K,the mass loss was less than 20% of the total mass, however, the temperature was above 1623K,the sample weight changed sharply with the increase of temperature. So the initial reduction temperature was 1623K, it was accorded with the theoretical calculation.

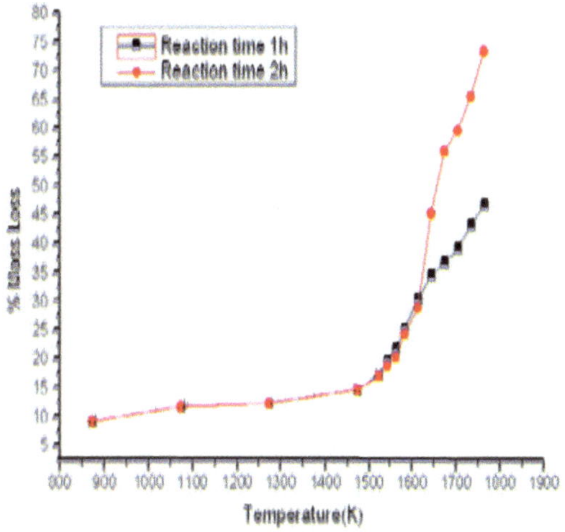


**Fig. 4 Relationships of the Reduction temperature and $\alpha$**

Fig.5 shows the SEM micrographs of pellets of magnesia and carbon before and after experiments. In SEM micrographs, the white cluster structured particles were MgO, the black granuiar-structure particles were carbon. From the micrograph(Fig.5(a)), it can be seen that the particle size of MgO was big and the mixing condition between magnesia and carbon particles was good before experiment. Then it was found that the particle size of carbon decrease, and few MgO detected(Fig.5(b)). All these factors result in the decrease of the contacting area between magnesia and carbon particles.

**(a)**

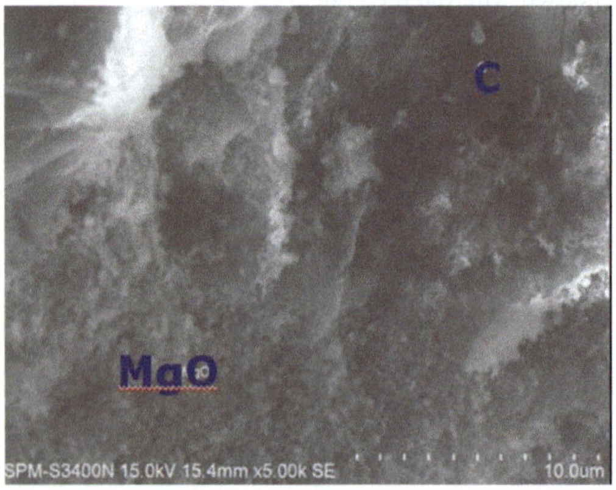


**(b)**

**Fig.5 The SEM micrographs of raw materials and reduction product samples: (a)the pellet of magnesia and carbon; (b)the residue obtained after carbothermic reduction at 1723K for 2h.**

### 3.2 Analysis of reversion reaction

A number of possible reversion reactions can take place during cooling of an Mg/CO/inert gas mixture:

gas phase reversion reaction:

$$Mg_{(g)}+CO_{(g)}=MgO_{(s)}+C_{(s)} \quad (4)$$

liquid-phase reversion reaction:

$$Mg_{(l)}+CO_{(g)}=MgO_{(s)}+C_{(s)} \quad (5)$$

solid-phase reversion reaction:

$$Mg_{(s)}+CO_{(g)}=MgO_{(s)}+C_{(s)} \quad (6)$$

The Gibbs free energy of reaction (4),(5),(6) at different temperatures and pressures were evaluated and the results reported in Fig.6.Thermodynamically, these reversion reactions were exothermic reaction, the gas-phase reversion will commence as soon as a saturated gas mixture was cooled. The magnesium condensation and liquid-phase reversion reaction were favored below 1373K, and magnesium solidification and liquid-phase reversion reactions were favored below 973K. It can be seen that these reversion reactions will be favored over the metal-forming reactions during cooling, meaning that the relative kinetics of each reaction will be critical to any process designed to produce metal. Fig.7 shows the XRD patterns of the condensing product in carbothermic reduction, it can be seen that metal magnesium was obtained. Meanwhile, magnesia was also found.

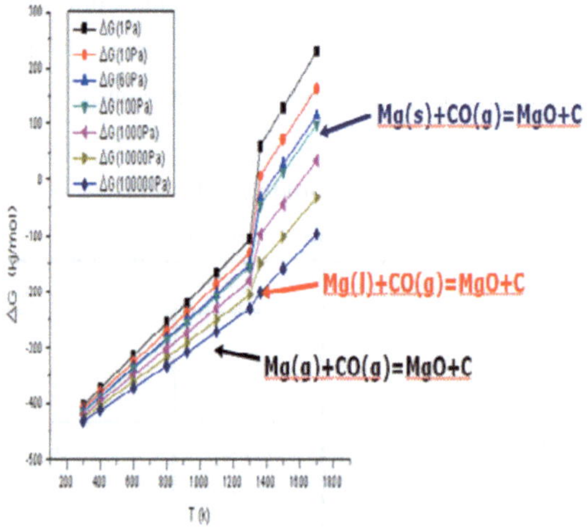


**Fig. 6 Free energy changes related to the temperature of reaction(4),(5),(6) under different pressure**

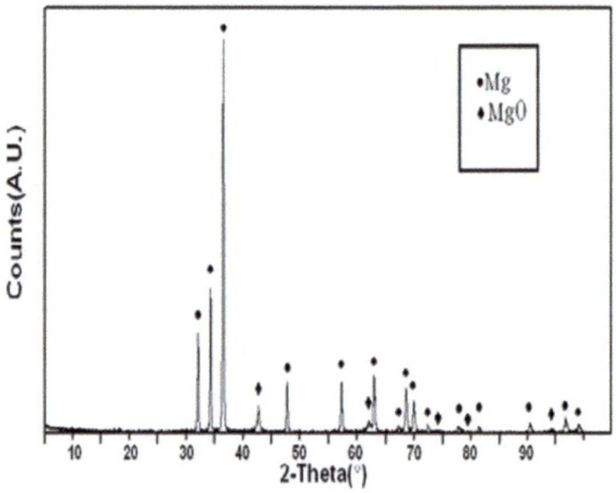


**Fig.7 XRD patterns of the condensing product in carbothermic reduction**

### 3.3 Analysis of vacuum distillation

After 1h(or 2h) of the vacuum distillation process ($Mg_{(l)}=Mg_{(g)}$), magnesia and carbon were loaded into the reaction crucible; metal magnesium was produced in the condensing towers. As can be seen from the XRD patterns of vacuum distillation product shown in Fig.8, it demonstrates that all peaks were sharp and well-defined, which indicated the high purity of the metal magnesium. Comparing the results of condensing product before and after vacuum distillation, it can be deduced that reversion reaction occured at low temperature and 30~100Pa during carbothermic reduction.

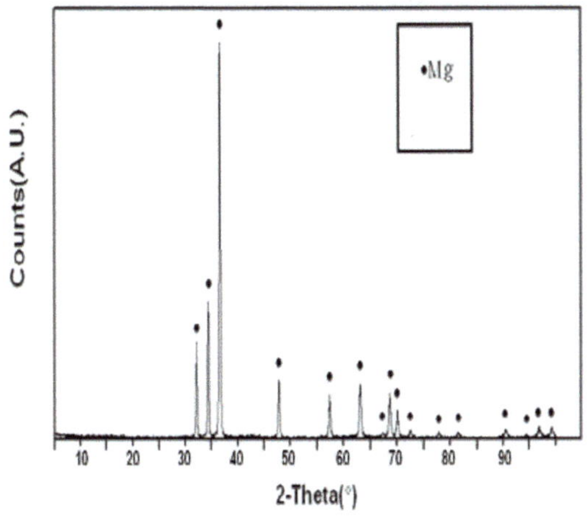


**Fig.8 XRD patterns of distillation product**

The reversion reaction ratio( $\gamma$ ) is defined as

$$\gamma = \frac{W_1}{W_0} \times 100\% \qquad (7)$$

where $W_0$ represents the total mass of magnesium and magnesia in the condensing product, and $W_1$ is the mass of magnesia in distillation residual.

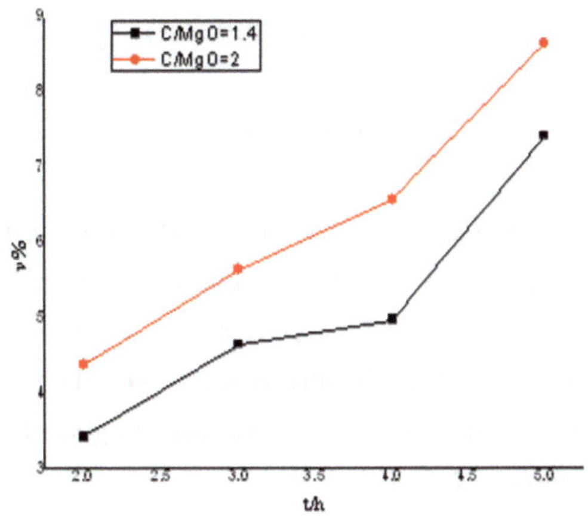


**Fig.9 Relationships between reversion ratio and reaction time in different molar ratio**

As shown in Fig.9, it was clear that the reversion reaction ratio changed slowly with the incremental molar ratio, the reversion reaction ratio increased with increasing reaction time. However, the peak value was less than 9%, it can be deduced that most magnesium vapor transforms into metal magnesium, the condensing condition of the vacuum equipment was appropriate for the carbothermic reduction.

## Conclusions

1) In the carbothermic reduction process, Mg was generated by magnesia and carbon at 1623K and 30~100Pa. The main gas in carbothermic reduction process was carbon monoxide, no carbon dioxide occurred at any reaction time, the reduction reaction was $MgO_{(s)}+C_{(s)}=Mg_{(g)}+CO_{(g)}$.

2)Thermodynamically, the three reversion reactions are exothermic reaction, the gas-phase reversion will commence as soon as a saturated gas mixture was cooled. The reversion reactions were favored below 1373K.

3) The distillation product by vacuum distillation process produced a high purity metal magnesium product. Through the calculation of reversion reaction, the peak value of $\gamma$ was less than 9%.

## Acknowledgments

This work was financially supported by Science

Research Fundation of Education Department of Yunnan Province, China (No. 2010C257).

## References

[1]YU Qing-chun, YANG Bin, XU Bao-qiang, LIU Da-chun, LI Zhi-hua, DAI Yong-nian. Study on the thermodynamic and experimental carbothermic reduction magnesia[C].**Proceedings of the 9th Vacuum Metallurgy and Surface Engineering Conference**, 2009:428～433.

[2]Li Rongti. Pan Wei, Masamichi Sano, Jianqiang Li.Catalytic reduction of magnesia by carbon[J].**Thermochimica Acta**,398(2003):265-267.

[3]ZHANG Ding-fei, QI Fu-gang, LAN Wei, SHI Guo-liang, ZHAO Xia-bing. Effects of Ce addition on microstructure and mechanical properties of Mg-6Zn-1Mn alloy[J].**Transactions of Nonferrous Metals Society of China**,2011,21(4):703-709.

[4]B. L. Mordike and T. Ebert: **Materials Science& Engineering A**, 2001, vol. 302, pp. 37-45.

[5] Hu Wen-xin,Liu Jian, Feng Nai-xiang, Peng Jian-ping.Theoretical analysis and experimental study on thermal reduction of calcined dolomite with Al-Si-Fe ternary alloy[C]. **Proceedings of the 9th Vacuum Metallurgy and Surface Engineering Conference**,2009:27～31.

[6] Geoffrey Brooks, Simon Trang, Peter Witt, M.N.H.Khan, Michael Nagle. **Journal of Materials**, 58(2006):51-55.

[7] Li Zhi-hua, Xue Huai-sheng, Dai Yong-nian. **Energy for Metallurgical Industry**.23(2004):20-22.

[8] Hori Fumio. Method for obtaining Mg and Ca through carbon reduction [P]. **US: 4147534.** 1979-04-03.

[9] Hori Fumio. Apparatus for obtaining Mg and Ca through carbon reduction [P].**US: 4200264.** 1980-04-29.

[10] Tassios Steven. Production Process[P].**WO: 2020012042A1**.2010-02-04.

[11] Tian Yang, Liu Hong-xiang, Yang Bin, Qu Tao, Dai Yong-nian.Process of magnesium production by magnesia carbothermic reduction in vacuum[J]. **Chinese Journal of Vacuum Science and Technology**,2012(3).

[12] Yu Qing-chun, Yang Bin, Ma Wen-hui, Li Zhi-hua, Dai Yong-nian.Study of carbothermic reduction of magnesia in vacuum[J]. **Chinese Journal of Vacuum Science and Technology**,2009, 5(29): 68-71.

[13] Li Zhi-hua, Dai Yong-nian, Xue Huai-sheng. Thermodynamical analysis and experimental test of magnesia vacuum carbothermic reduction[J]. **Nonferrous Metals**, 2005,2(57):56-59.

# Magnesium Technology 2012

## Poster Session

*Session Chair:*

**Eric A. Nyberg**
**(Pacific Northwest National Laboratory, USA)**

**Magnesium Technology 2012**
*Edited by: Suveen N. Mathaudhu, Wim H. Sillekens, Neale R. Neelameggham, and Norbert Hort*
***TMS (The Minerals, Metals & Materials Society), 2012***

# COMBINATION OF COOLING CURVE AND MICRO-CHEMICAL PHASE ANALYSIS OF RAPIDLY QUENCHED MAGNESIUM AM60B ALLOY

P.C. Marchwica[1], A.J. Gesing[2], J.H. Sokolowski[1], C. Blawert[3], J. Jekl[4], R. Berkmortel[4]

[1]University of Windsor
[2]Gesing Consultants Inc.
[3]Helmholtz-Zentrum Geesthacht Zentrum für Material- und Küstenforschung GmbH
[4]Meridian Lightweight Technologies Inc.

Key words: AM60, magnesium, thermal analysis, micro-chemical analysis, rapid quenching

## Abstract

Macro test samples of magnesium alloy AM60B were melted and quenched at maximum instantaneous cooling rates ranging from -5°C/s to -500°C/s and the resultant cooling curves were analyzed. Characteristic reactions on these curves corresponding to formation of individual phases were identified with the aid of literature data as well as metallographic and micro-chemical analysis. The results indicate that these phases, their size and location in the microstructure, their chemistry and their relative proportions all change in response to the increase in the cooling rate. These rapid cooling rates are typical of real industrial solidification processes such as die casting. These findings can be used to improve future computer models of casting solidification processes for magnesium and for other alloys.

## Introduction

With low density and good mechanical properties, magnesium alloys are increasingly being adopted for use in the automotive industry. As a lightweight structural material they feature excellent ductility, castability and strength and are currently being used for such components as instrument panels, seat frames and steering wheels [1-3]. Many magnesium alloys, such as AM60B, have been developed in order to obtain specific ranges of properties. AM60B, with additions of aluminum and manganese, has good energy-absorption characteristics and ductility [4].

Die casting is the predominant forming method for Mg-based alloys. Large markets for such die-cast alloys include portable consumer electronic housings and automotive/aerospace components. These applications involve use at ambient temperature. Other uses such as engine components expose the alloy to elevated temperature and thus require higher temperature strength and creep resistance. AM60B is a general purpose die-casting alloy containing ~6% Al and >0.3% Mn (wt %). Intended for ambient temperature applications, AM60B is designed for low cost, low density, good castability, and room temperature strength and improved ductility over the industry workhorse AZ91D alloy.

In AM60B, the addition of Al improves the castability by providing the $Mg_{17}Al_{12}$ intermetallic phase, which preserves fluidity during casting until a low temperature (~400°C) eutectic reaction with Mg. Mn has high affinity for impurities such as Fe and Ni (which lead to electrochemical corrosion) and traps them in $Al_8Mn_5$ intermetallic precipitates. Under thermodynamic equilibrium, these precipitates form in liquid Mg just above its melting point of 650 °C and can be removed with additional melt treatment. Without melt treatment, these precipitates can provide nucleation sites for the solidification of primary αMg grains. Such nucleation promotes reduction of grain size and thus contributes to the improvement in strength and ductility.

## Experimental

The Universal Metallurgical Simulator and Analyzer (UMSA) Technology Platform [5] is a desktop thermal analysis system capable of simulating various foundry processes in a quick and cost-effective manner. The basic platform function is heating samples using an induction coil and recording in-situ temperature changes using a thermocouple connected to a data acquisition system. The platform is also applicable to testing other process parameters including cooling/heat treatment cycles and the effects of vacuum/non-atmospheric environments. Recent developments of the platform have focused on improving the quenching capabilities in order to simulate higher cooling rates. While the previous platform was capable of reaching instantaneous cooling rates of around -30°C/s while blowing cold gas at the outer surface of a 18-mm solid cylindrical sample, the latest changes increase simulation capabilities to -500°C/s for macro test samples. The primary reason behind this improvement is the use of hollow test samples that accommodate a central cooling channel and increase the surface-area-to-volume ratio and allow the use of liquid coolants.

The AM60B alloy used in these experiments was provided by Meridian Technologies Inc. in the form of ingots. Nominally pure magnesium was provided in the form of bar stock. Hollow test samples were cut from the raw material with dimensions 12x18x23 mm (IDxODxL) with an additional hole placed within the wall to accommodate the thermocouple. K-type thermocouples (~0.5mm OD) were used in the experiments. Thermocouple calibration was performed using RTD probes. All samples were coated in colloidal graphite and placed in low-thermal-mass stainless steel foil crucibles. The thermocouple was also coated in graphite prior to insertion into the sample. Experimental conditions and process parameters were maintained constant throughout the course of the study. These include holding time, holding temperature, heating rate and atmosphere (argon).

In order to investigate capabilities of the modified UMSA platform, experiments were performed using various combinations of alloy and cooling methods. Each sample in the study was assigned to a particular cooling method and subjected to three repeated heating/cooling cycles. In each cycle, a sample would be heated past its melting point to 750°C and then quenched using its assigned method. Three trials were performed in order to ensure the repeatability of the methods as well as to maximize thermal contact between the thermocouple and the sample.

Using the modified system, gaseous coolants were found to cool samples at rates up to ~-30°C/s while liquid coolants cooled at rates of up to ~-500°C/s. Images from such rapidly cooled micro-structures are analyzed in this paper. However, rapid cooling induces significant turbulence in liquid coolants, which has a detrimental effect on signal quality. Because of this, thermal analysis in this paper is focused on gas-cooled thermal traces in the -15°C/s to -20°C/s range.

Micro-Chemical and Microstructural Analysis

Tables I shows the alloy compositions as well as the phase compositions and distributions calculated from SEM-EDX measurements. Directional solidification in radial and outward patterns allowed for identification of the sequence of changes in the alloy and phase composition during solidification.

Table I focuses on the AM60B samples solidified at -18°C/s and -280°C/s. Phase composition and distributions calculated in this table are later used in calculations of elemental mass balances for cooling trace analysis. For now, it is significant that at -18°C/s and slower cooling rates we observed equilibrium phases in the sample microstructure: αMg, $Mg_{17}Al_{12}$, $Al_8Mn_5$ and $MgSi_2$. At a rapid solidification rate of -280°C/s the quantity of intermetallic phases increases by 5 wt% and their composition changes to a metastable structure containing a large excess of Mg.

Table I. Comparison of Phase Compositions and Distributions for Rapidly Solidified AM60B Alloy

| | | Cooling Rate: -18°C/s to -15°C/s | | | |
|---|---|---|---|---|---|
| | **AM60B Alloy** | **αMg** | **$Mg_{17}Al_{12}$-αMg Eutectic** | **$Al_8Mn_5$** | **$Mg_2Si$** |
| **Phase Distribution** | **100.00** | **86.29** | **12.92** | **0.40** | **0.40** |
| Composition | wt% | wt% | wt% | wt% | wt% |
| Mg | 93.16 | 97.52 | 67.10 | 0.12 | 63.38 |
| Al | 6.26 | 2.27 | 32.27 | 33.58 | |
| Si | 0.27 | 0.07 | 0.52 | 0.27 | 36.62 |
| Mn | 0.38 | 0.14 | 0.11 | 60.73 | |
| Fe+Cr+Ni | 0.02 | | | 5.30 | |
| Total | 100 | 100 | 100% | 100% | 100% |

| | | Cooling Rate: -280°C/s | | | |
|---|---|---|---|---|---|
| | **AM60B Alloy** | **αMg** | **MgAl *IM*** | **MgAlMn *IM*** | **$Mg_2Si$** |
| **Phase Distribution** | **100.00** | **81.23** | **16.68** | **1.39** | **0.70** |
| Composition | wt% | wt% | wt% | wt% | wt% |
| Mg | 93.16 | 97.65 | 74.26 | 67.45 | 63.38 |
| Al | 6.26 | 2.10 | 25.56 | 21.12 | |
| Si | 0.27 | 0.01 | 0.04 | 0.30 | 36.62 |
| Mn | 0.38 | 0.24 | 0.14 | 11.13 | |
| Fe+Cr+Ni | 0.02 | | | | |
| Total | 100 | 100 | 100 | 100 | 100 |

* *IM* - Intermetallic

Figures 1-4 illustrate the AM60B alloy microstructures and their dependence on the solidification rates. The microstructures are all composed of equiaxed αMg grains, and their grain size is only slightly dependent on the cooling rate up to -240°C/s. Only at -280°C/s is there significant reduction in grain size. This is correlated with the following findings:

1. no $Al_8Mn_5$ inside αMg grains
2. metastable Mg-rich MgAlMn intermetallics in between αMg grains.

Apparently at very high cooling rates the nucleation of $Al_8Mn_5$ is suppressed and αMg grains nucleate at more abundant nucleation sites, probably small oxide inclusions.

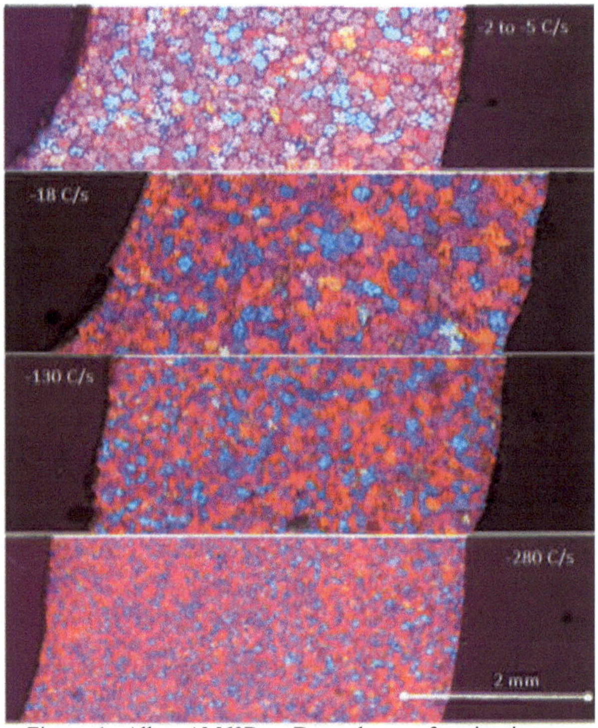


Figure 1. Alloy AM60B – Dependence of grain size on solidification rate. At >-20°C/s, the grain structure (equiaxed) and direction (radially outward) are not noticeably affected by solidification rate. At more rapid rates, grain size decreases and there is a slight size gradient in the solidification direction with grains growing larger later.

The peak cooling rate is recorded at the start of solidification. As the cooling proceeds, the baseline cooling rate slows down. Figure 2 illustrates how this affects the grain structure in a directionally solidified sample starting at -280°C/s. There is noticeable grain coarsening between the start and end regions of the sample (at inner and outer walls).

Figure 3 illustrates how interdendritic melt solidification at the end differs with the cooling rate. At slow cooling rates, equilibrium αMg-$Mg_{17}Al_{12}$ eutectic forms as a spongy two-phase structure. In the -240°C/s and -280°C/s samples, the equilibrium phases are replaced by the metastable Mg-rich intermetallics, $Mg_xAl_y$ and $Mg_xAl_yMn_z$.

Figure 4 illustrates that as the cooling rate increases αMg solidification becomes more dendritic and the equiaxed grains become equally equiaxed dendritic rosettes expanding radially out from the central nucleation site.

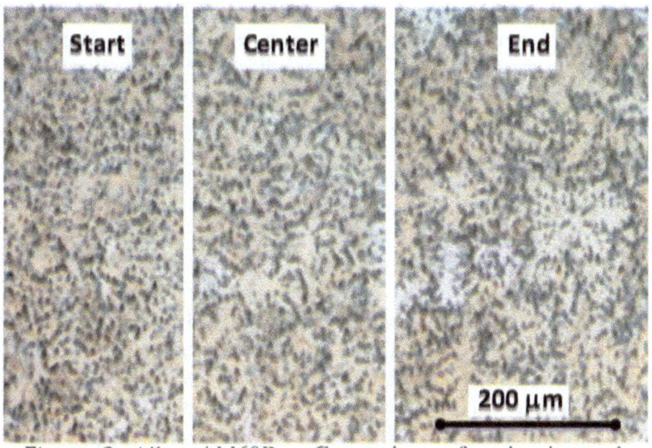


Figure 2. Alloy AM60B – Comparison of grain size and morphology in start, center and end regions for a sample solidified at -280°C/s. Images were taken from a polished and etched section accentuating the grain boundaries and intermetallics which are found both inside and between αMg grains.

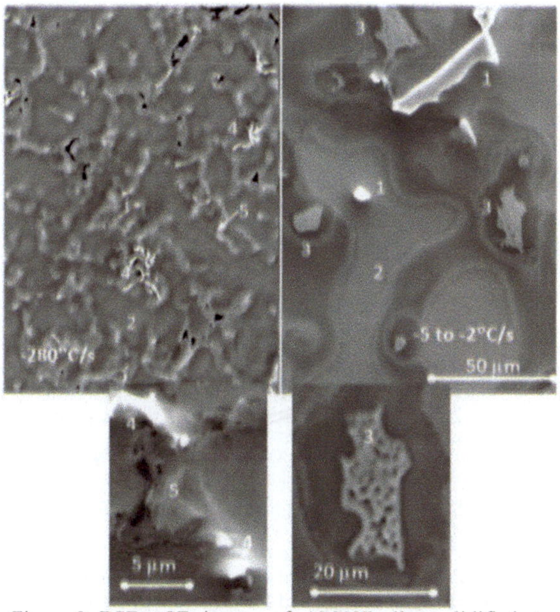


Figure 3. BSE + SE images of AM60B alloy solidified at -280°C/s (left) and -5 to -2 °C/s (right) showing changes in the micro-structure with cooling rate. At -5 to -2 °C/s we see: 1. blocky $Al_8Mn_5$ inside, 2. αMg grains and, 3. $\beta Mg_{17}Al_{12}$–αMg eutectic in-between, while at -280°C/s there is 4. dendritic $Mg_xAl_yMn_z$ and 5. single phase $Mg_xAl_y$ in between αMg grains.

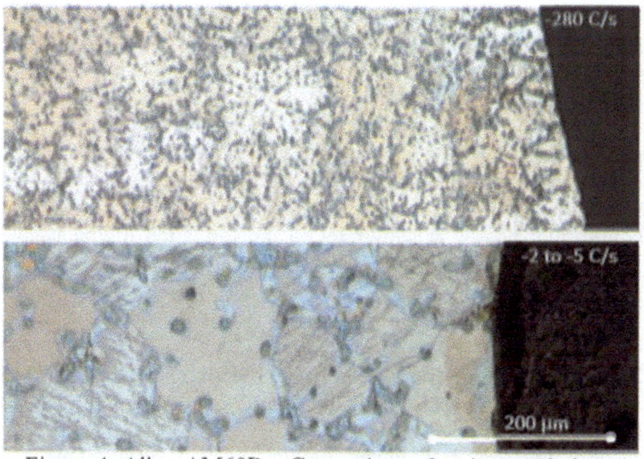


Figure 4. Alloy AM60B – Comparison of grain morphology in the start region between samples solidified at -280°C/s and at -2°C/s to -5°C/s. Enhanced tendency for dendritic growth is shown at higher solidification rates. Grains nucleate at solid particles and grow outward in all directions regardless of the direction of heat extraction. (Sample is cooled from the center.)

The AM60B alloy microstructures tell a story. There is a smaller but still significant variation in alloy microstructures attributable to the nearly two orders of magnitude change in the solidification rates.

In all cases the structure is dominated by a nearly equiaxed structure of pretty uniformly sized αMg grains. At slow cooling rates one tends to find small dendritic $Al_8Mn_5$ structures inside the αMg grains. There are also occasional small equiaxed $Mg_2Si$ crystals located near the grain boundaries but still typically surrounded with αMg phase. Between the αMg grains there is spongy eutectic structure of $\beta Mg_{17}Al_{12}$ with αMg inclusions. At rapid solidification rates between the αMg grains, there is a layer of Mg-rich MgAl intermetallic. The grain sizes decreased slightly and the equiaxed αMg structures changed from grains to small dendritic rosettes growing out from central $Al_8Mn_5$ intermetallics. At the fastest cooling rate of nearly -300°C/s, there was significant further reduction in the αMg grain size, and the $Al_8Mn_5$ intermetallics were replaced by Mg-rich MgAlMn crystals forming in the MgAl phase found at the αMg grain boundaries.

The sequence that would account for these observations at all but the fastest cooling rates involves starting with homogeneous nucleation of $Al_8Mn_5$, which starts growing and exhausts the supply of the Mn alloying element and reduces the concentration of other transition elements in the melt. Each $Al_8Mn_5$ structure appears to be an effective nucleation site for the αMg grains. As αMg grains grow they concentrate Al and Si in the melt. Eventually, at sufficient Si concentration, $Mg_2Si$ co-precipitates with the αMg. The equilibrium phase diagram predicts formation of αMg- $\beta Mg_{17}Al_{12}$ eutectic structure to finish the solidification sequence, which we observed experimentally at slow cooling rates. At rapid solidification rates the last phase to solidify was single-phase Mg-rich MgAl intermetallic in between the αMg grains and dendritic rosettes. Significantly this correlated to a loss of the exothermic peak from the cooling curve that we associated at the slow cooling curves with $\beta Mg_{17}Al_{12}$-αMg eutectic formation.

At the fastest cooling rate the undercooling is sufficient to delay $Al_8Mn_5$ nucleation and permit homogeneous nucleation of αMg throughout the melt volume, resulting in the observed extra fine grain size. Mn concentrates in the residual melt and finally comes out near the end as MgAlMn crystal in MgAl phase found between the αMg grains.

In the AM60B alloy the microstructure changed quite significantly with the increase in the solidification rate. This microstructural change is associated in the change in both the sequence of phase formation and in the shift from the formation of equilibrium phases at slow rates to the formation of metastable phases during rapid cooling.

## Cooling Curve Analysis

The cooling curve is derived from the time-temperature record of the sample being cycled through repeated melting and solidification cycles. We calculate the slope of a linear fit to several adjacent data points to smooth and differentiate the temperature data with respect to time. The cooling rate becomes less negative as the sample temperature drops, reducing the temperature difference between the sample and the coolant. The area of the peaks superposed on this sloped baseline represents the enthalpy of solidification of the various solid phases crystalizing out of the melt. To calculate the peak areas one must first calculate the portion of the baseline in the semi-solid region.

### Baseline Determination

For the magnesium parent metal, the heat capacities of liquid and solid are different, which results in a shift in the baseline between the fully solid and fully liquid regions. At any temperature:

$$B_s/B_l = c_{pl}/c_{ps} \quad (1)$$

where $B_s$ and $B_l$ are baselines determined in liquid and solid respectively, and $c_{pl}$ and $c_{ps}$ are heat capacities of the liquid and solid. Further, in the semi-solid region the baseline value for the semisolid mixture, B, can be calculated from the respective weight fractions of liquid and solid, $f_l$ and $f_s$ as follows:

$$1/B = (1-f_s)/B_l + f_s/B_s \quad (2)$$

Fraction solid at any time is estimated from the fraction of the total thermal solidification peak area that was integrated to that time. Since the peak area depends on the baseline, an iterative solution is required. The procedure is found to converge to a solution after a few iterations. The problem is simplified by ignoring the contribution of alloying elements to the baseline value. Figure 9 shows an example of the cooling curve as a function of temperature with the calculated baseline that accounts for the liquid and solid contributions in the semi-solid region. At the solidus (end of solidification) temperature the position of the $B_l$ is calculated from the measured $B_s$ value. This point is linearly interpolated to the measured $B_l$ curve at the liquidus (start of solidification) temperature. The $B_s$ curve between the liquidus and the solidus is then calculated from the $B_l$ interpolated values according to equation 1. The baseline B for the semisolid mix is then calculated from $B_s$ and $B_l$ by iterative solution of Equation 2.

### Enthalpy Calibration

The area of the baseline subtracted peak plotted against solidification time is proportional to the enthalpy of formation of the solid phase. The proportionality factor was determined by measuring the peak area for commercially pure Mg which has a latent heat of solidification of 8.48 KJ/mole reported for pure Mg by NIST at the melting point. The calibration is complicated by two facts. First, the measured area of the solidification peak depends on the cooling rate. The area, increases as the cooling rate goes more negative. Second, the enthalpy of solidification decreases as solidification temperature decreases. This decrease is significant for alloys with a wide solidification temperature range. The calibration measurements in commercially pure Mg spanned a range of -2°C/s to -20°C/s for a baseline value in during actual solidification. Linear interpolation fit to this data was used to calculate the conversion factor from the area measured in degrees centigrade to energy measured in joules:

$$C2J = \Delta H_m(T)/AP(dT/dt) \quad (3)$$

Here $\Delta H_m(T)$ is the latent heat of solidification of Mg at temperature T, and AP(dT/dt) is the measured area of the dT/dt peak with baseline subtracted and integrated with respect to time, at a specific baseline cooling rate. Individual C2J calibration factors were calculated for each data point and applied to the baseline subtracted dT/dt peak profile data before area integration. This allowed us to account for variation of the calibration factor with both temperature and the baseline cooling rate during the solidification event.

Figures 5, 6 and 7 show examples of the cooling curve baseline subtraction and calibration results for the alloy AM60B.

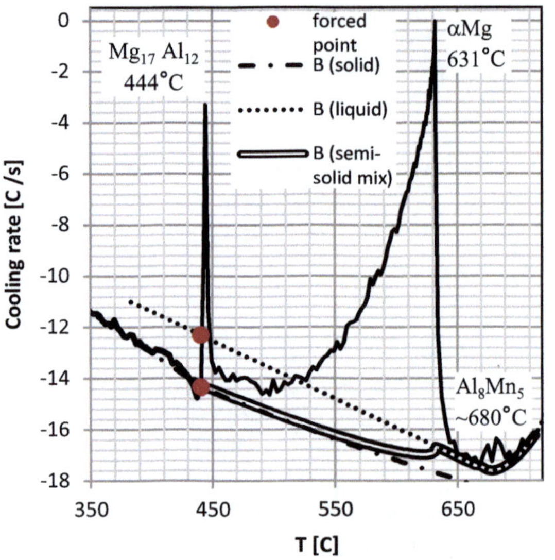


Figure 5. AM60B alloy cooling curve with calculated baseline. Liquid baseline is extrapolated linearly to a point located at solidus temperature and a value given by Equation 1. The solid baseline is extrapolated to the liquidus temperature based on Equation 1 and the interpolated value of liquid baseline. Semi-solid mix baseline is calculated from the solid and liquid baselines according to Equation 2.

The correlation between the cooling curve peaks, the phases observed in the microstructure and their identification is based on the phase quantity and location in the structure and the literature results on phase thermochemical and crystallographic properties and the alloy equilibrium phase diagram [7].

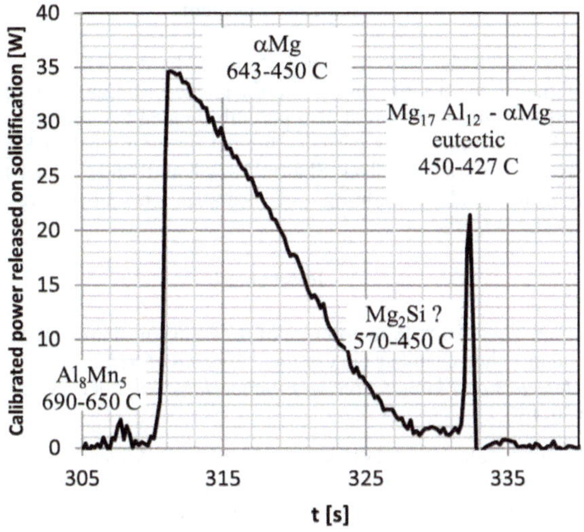


Figure 6. Calibrated values of the power released during solidification of a 5.7-g AM60B alloy sample at -18°C/s to -15°C/s. Integration of the individual peak areas yields experimental values for enthalpy of formation.

Phase Formation

The phase distribution information calculated from the microstructural and micro-chemical observations is combined with the evolution of fraction solid of each individual phase during the solidification event to generate a plot demonstrating the formation of the constituent solid phases and consumption of the melt as a function of the temperature. An example of such a plot for AM60B solidified at -18°C/s to -15°C/s is shown in Figure 7.

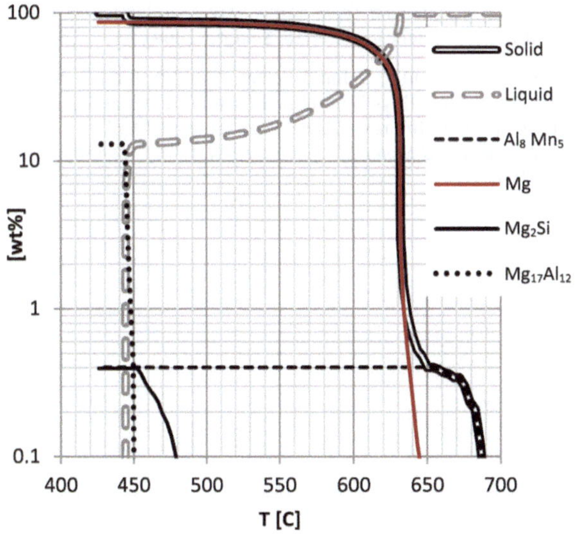


Figure 7. AM60B alloy phase evolution during solidification at -18°C/s to -14°C/s. The solid forms as 0.3% of $Al_8Mn_5$ between 687°C and 650°C; 85% as αMg between 643°C and 450°C and finally as 12% of $Mg_{17}Al_{12}$ -- αMg eutectic between 450°C and 428°C.

Table II: Enthalpy of formation of intermetallic phases from AM60B melt solidifying at -18°C/s

| | Alloy latent heat | αMg | $Mg_{17}Al_{12}$-αMg eutectic | $Al_8Mn_5$ | $Mg_2Si$ |
|---|---|---|---|---|---|
| $\Delta H_f$ [J/g] | -289 | -317 | -79 | -535 | -605 |
| $T_{start}$ [C] | 697 | 647 | 452 | 697 | 491 |
| $T_{max}$ [C] | 632 | 632 | 445 | 683 | 482 |
| $T_{end}$ [C] | 441 | 452 | 441 | 654 | 452 |

Table II summarises the key cooling curve information from solidification of AM60B at -18°C/s. Integration of peaks in Figure 10 associated with individual phases provides the information for this table. As expected the latent heat of solidification of the alloy is lower than that for pure Mg. Forming αMg in this alloy yields -317 J/g as compared with -348 J/g for formation of the same phase from pure Mg melt. Formation of the eutectic yields relatively little heat, whereas the formation of $Al_8Mn_5$ and $Mg_2Si$ are highly exothermic.

Solidification Path: Evolution of Melt Composition

Information illustrated in Figure 7 combined with the composition of the individual phases allow the calculation of the residual liquid composition during the solidification event. Knowledge of such solidification path is key to understanding the phase formation during solidification. The solidification path is plotted as concentration of the individual alloying elements as a function of temperature in Figure 8.

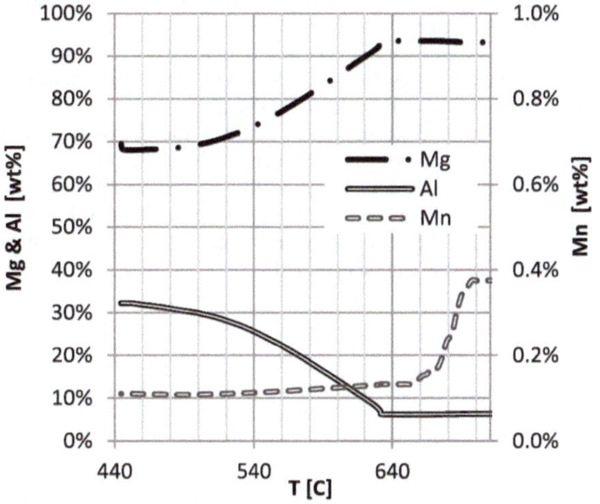


Figure 8. AM60B alloy melt composition variation during solidification at -18°C/s to -14°C/s

At very rapid cooling rates interpretation of the cooling curves becomes more difficult. Total solidification time reduces to a few seconds. During that time there is the initial transient of introducing the coolant to the molten sample and setting up a stable boundary layer and heat transfer coefficient between the coolant and the sample. When liquid coolant is used, it is common to observe vapor or air bubbles exiting the cooling tube. This suggests that an insulating gas film may form on the hot cooling tube surface and impede the cooling process. This can lead to irreproducible, apparent thermal peaks that have nothing to do with the enthalpy of alloy solidification. In this paper we limited the reported cooling curve data to >-25°C/s, which is a rate that we can consistently reach using compressed gas coolant. This eliminates the possibility of insulating vapor film formation on the cooling tube surface.

Further, there is an implicit assumption in the analysis that the sample is isothermal. This is valid at slow cooling rates because of very high thermal conductivity of the metal sample. However, as cooling rates reach hundreds of degrees per second, the heat flux required to achieve such a rate also requires a temperature gradient in the sample. The effect of the gradient causes the apparent exotherm to start at a higher temperature than recorded by the centrally located thermocouple and continues to a lower temperature after solidification is complete at the thermocouple location. This effect is reproducible and we calibrated it out for enthalpy measurements by performing the calibration measurements on pure Mg over the cooling rate range spanning the rates of interest for gas cooling. Extrapolation of these calibration values to more rapid liquid coolant rates was not successful. The temperature gradient in the sample spreads the exotherm over a range of temperatures. This can cause small and overlapping exotherms to be smeared out and lost in the baseline noise.

## Conclusions and Future Work

The UMSA technology platform has been modified to enable solidification of alloy melts at cooling rates up to -500°C/s.

A combination of cooling curve analysis with microstructural and micro-chemical results yields a wealth of information on the phases formed, sequence of events, the enthalpies of formation of component phases and evolution of residual melt composition during solidification.

Reliable and reproducible cooling curves were obtained with sufficient detail to resolve the component phase contributions at cooling rates up to -20°C/s for AM60B alloy.

At very rapid solidification rates detail is lost from the cooling traces due to a combination of actual change in the type and quantity of phase formed, and as well to experimental limitations. Future work will focus on eliminating the experimental limitations and extending the cooling rate range over which reliable, quantifiable results can be obtained.


## Acknowledgements

This research was partially funded by AUTO21, a member of the Networks of Centres of Excellence of Canada program. NSERC Discovery Grant also partially funded this research. The GCI contribution was partially funded by the Canadian SR&ED program. Assistance from Sharon Lackie and Andy Jenner in the analysis and machining of samples is also gratefully acknowledged.


## References


1. A. Kiełbus, "Microstructure of AE44 magnesium alloy before and after hot-chamber die casting," *Journal of Achievements in Materials and Manufacturing Engineering*, Vol. 20(1-2), (2007), 459-462.

2. T. Rzychoń, A. Kiełbus, J. Cwajna, and J. Mizera, "Microstructural stability and creep properties of die casting Mg–4Al-4RE magnesium alloy," *Materials Characterization,* Vol. 60, (2009), 1107-1113.

3. J.P. Weiler, J.T. Wood, R.J. Klassen, R. Berkmortel, and G. Wang, "Variability of skin thickness in an AM60B magnesium alloy die-casting," *Materials Science and Engineering*, Vol. 419, (2006), 297-305.

4. W. Huanga, B. Hou, Y. Pang, and Z. Zhou, "Fretting wear behavior of AZ91D and AM60B magnesium alloys," *Wear*, Vol. 260, (2006), 1173-1178.

5. A.E.W. Jarfors, K.-U. Kainer, Ming-Jen Tan, and J. Yong, "Recent Developments in the Manufacturing of Components from Aluminium-, Magnesium- and Titanium-Based Alloys," *COSMOS,* 5(1), 2009), 23-58.

6. A.J. Gesing, N.D. Reade, J.H. Sokolowski, and C. Blawert, "Solidification Behavior of Recyclable Mg Alloys - AZ91 and AZC1231," *Magnesium Technology 2009*, (San Francisco, CA: The Minerals, Metals & Materials Society, 2009), 117-128.

7. D. Mirković and R. Schmid-Fetzer, "Solidification Curves for Commercial Mg Alloys Determined from Differential Scanning Calorimetry with Improved Heat-Transfer Modeling," *Metallurgical and Materials Transactions A*, 38A (October 2007), 2575-2592.

**Magnesium Technology 2012**
*Edited by: Suveen N. Mathaudhu, Wim H. Sillekens, Neale R. Neelameggham, and Norbert Hort*
***TMS (The Minerals, Metals & Materials Society), 2012***

# Effects of Friction Stir Process on the Tensile Properties of AZ61 Magnesium Alloy at Room Temperature to 200 °C

Hsiang-Ching Chen [1], Truan-Sheng Lui[1], Li-Hui Chen[1], and Fei-Yi Hung[2]

[1] Department of Materials Science and Engineering, National Cheng Kung University, Tainan, TAIWAN 701.

[2] Institute of Nanotechnology and Microsystems Engineering, Center for Micro/Nano Science and Technology, National Cheng Kung University, Tainan, TAIWAN 701.

Keywords: AZ61 Mg alloy, Friction stir process , tensile mechanical property

## 1. Introduction

Friction stir welding (FSW) possesses more advantages than other traditional welding processes because it is a solid-state joint process [1]. In addition, friction stir processing is adopted as a new surface modification approach. It is acceptable that the dynamic recrystallization phenomenon during FSW/FSP results in the generation of fine and equiaxed grains in the stir zone [2-3].

Magnesium alloys are regarded as poor formability due to their limited available slip systems at room temperature. Increasing deformation temperature can improve the ductility of magnesium alloys. In addition, the distribution of (0002) basal planes dominates the deformation behavior of magnesium alloys. Woo et al. [4] and Lee et al. [5], discussed the tensile properties of friction stir processed AZ31 alloy, and showed that its flow stress and elongation were influenced by the texture variation after FSP.

In this study, the main aim is to discuss the deformation behavior of full-annealed and friction stir processed AZ61 Mg alloy, especially in their tensile ductility. Meanwhile, the influence of deformation temperatures on the tensile ductility is investigated as well.

## 2. Experimental procedures

A commercial AZ61 billet was hot extruded to a plate of 55 mm in width and 3 mm in thickness. Its nominal chemical composition in weight percent was 6.3% Al, 0.48% Zn, 0.37% Mn and balance Mg. The as-extruded plate was subjected to annealing at 345 °C for 12 hours. The microstructure of the full-annealed (AZ61-O) sample was shown in Fig.1. The full-annealed plate was also adopted as the substrate of the friction stir processing as shown in Fig. 2. The FSPed samples were designated as AZ61-FSP. The FSP was performed using these parameters: tool rotating speeds were fixed at 1243 rpm; tool moving speeds were fixed at 46 mm/min with 1.5° of tool angle; the downward push force was controlled at 14.7 MPa. The tool size used in FSP was 16 mm in shoulder diameter, 7 mm in pin diameter and 1.5 mm in pin height.

The AZ61-O and AZ61-FSP specimens were machined to tensile specimens shown in Fig. 3, and their tensile direction was parallel to extrusion direction (ED) or process direction (PD). In this study, tensile tests of AZ61-O and AZ61-FSP specimens were performed at an initial strain rate of $1.67 \times 10^{-3}$ $s^{-1}$ from room temperature to 200 °C.

Microstructure samples for optical microscopy were polished and then etched with a solution of 4.2 g picric acid, 70 mL ethanol, 5 mL acetic acid and 5 mL water. The information of texture and precipitation of the particles was identified by X-ray diffraction spectra.

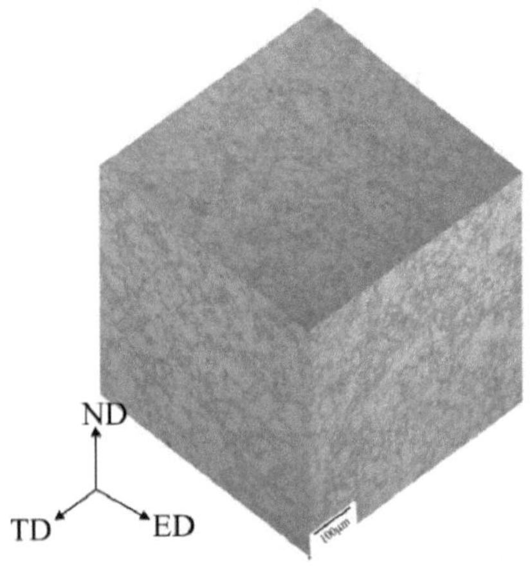

Fig.1. Microstructures of AZ61-O. The orientations are referred to extrusion direction (ED), normal direction (ND) and transverse direction (TD).

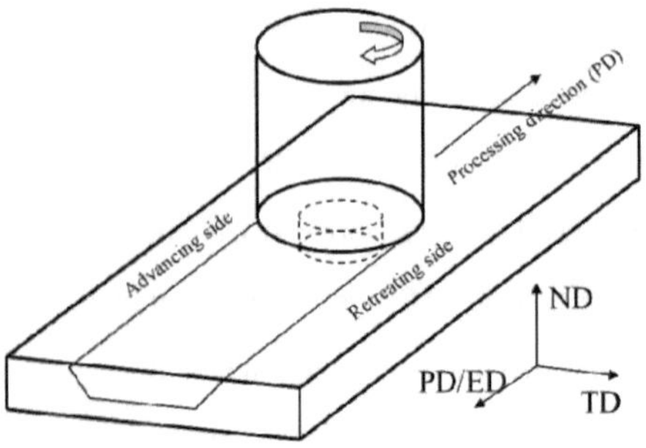

Fig.2. Schematic illustration of friction stir processing. The process direction was parallel to extrusion direction.

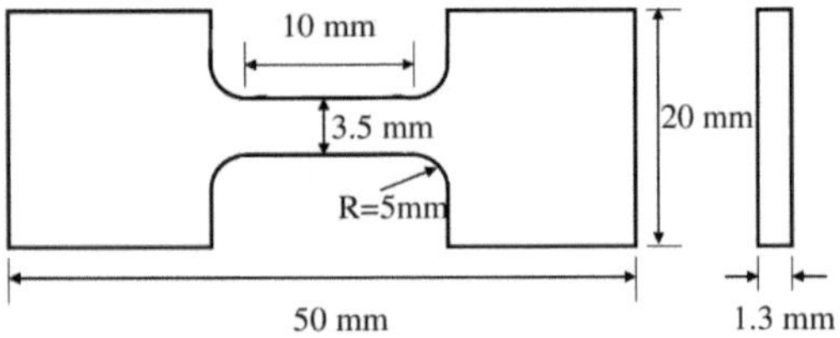

Fig.3. Schematic illustration of the dimensions of tensile specimens.

### 3. Results and discussion

The microstructure of AZ61-O is shown in Fig. 1, and its average grain size is 13 μm. On the other hand, the metal flow resulted from the friction stir processing can be observed in Fig. 4. The microstructure of AZ61-FSP consists of fine and coarse grains, and the grain size of fine grains is similar to that of AZ61-O. The macrostructure of metal flow is similar to the rotation trace of the stir pin because the inhomogeneous friction heat is introduced into the AZ61 specimen during friction stir processing [6]. The microstructures of AZ61-FSP specimens (Fig. 5) also reveal an inhomogeneous structure on the normal plane in the center of stir zone. Fine grains were fully recrystallized during FSP and the coarse grains are partially recrystallized. In Fig. 6, the schematic illustration of the grains structure is adopted to depict the microstructure evolution in the stir zone of the AZ61-FSP specimen [6]. On the other hand, comparing the X-ray diffraction patterns of AZ61-O and AZ61-FSP, the decrease in the peaks of γ-phase reveals the dissolution of the γ-phase $Mg_{17}Al_{12}$ during FSP (see in Fig. 7).

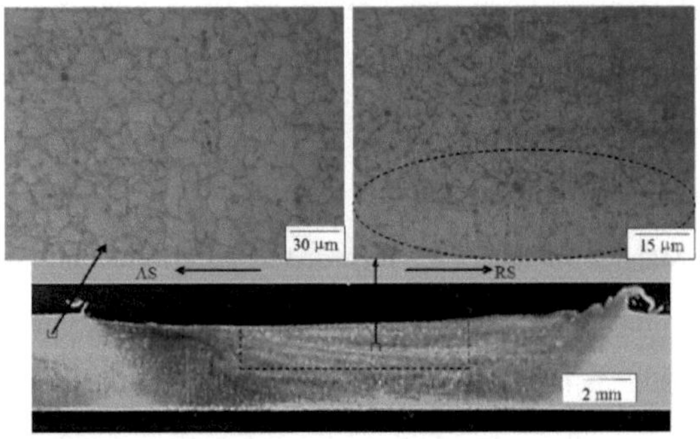

Fig.4. Microstructures of inhomogeneous structure of the base metal and the stir zone. Dot line refers to the coarse grains in the stir zone.

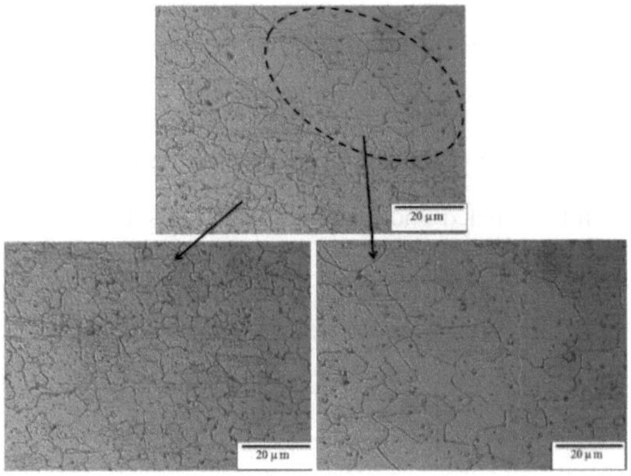

Fig.5. Microstructures of inhomogeneous structure in the center of stir zone, which is observed on the normal plane. The coarse grains are highlighted (dot line).

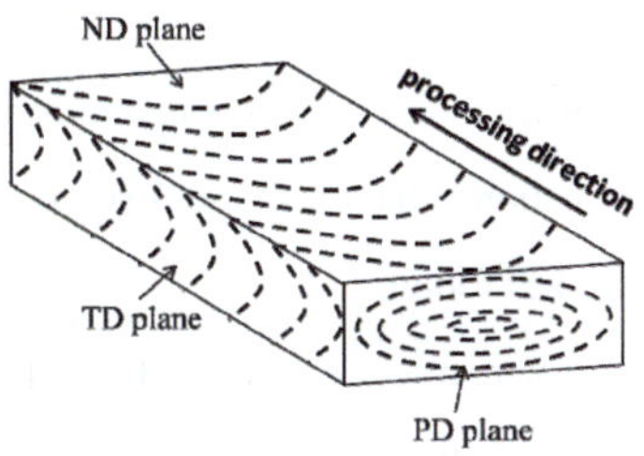

Fig.6. Schematic illustration of metal flow in the stir zone [6].

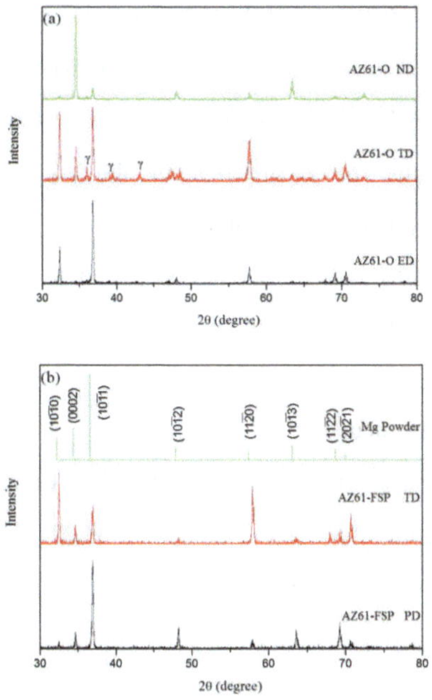

Fig.7. X-ray diffraction spectra: (a) AZ61-O and (b) AZ61-FSP detected on different planes. The samples of AZ61-FSP were cut from the center area of stir zone.

Fig. 8 and Fig. 9 show the tensile properties of AZ61-FSP and AZ61-O. It should be noted that the tensile flow stress of the AZ61-FSP specimen significantly decreased, which reveals the texture effect on the deformation behavior of AZ61-FSP and AZ61-O. As (0002) basal planes were rotated and rearranged during friction stir processing, (0002) basal planes were surrounded with the stir pin surface [4]. This texture is also identical with the metal flow illustrated in Fig. 6. The X-ray diffraction spectra of AZ61-FSP in Fig. 7 revealed that the texture feature of AZ61-FSP is beneficial to basal slip or twinning as tensile direction is parallel to process direction [5]. The microstructures of the fractured AZ61-FSP specimens have large numbers of deformation twins formed at room temperature to 100 °C (Fig. 10). Increasing test temperature, the flow stress of AZ61-O decreases more rapidly than that of AZ61-FSP.

As illustrated in Fig. 9, it should be noted that the AZ61-FSP specimens actually possess a better uniform elongation than AZ61-O at all test temperatures. Compared with AZ61-O specimens, the total elongation of the AZ61-FSP specimens is better from room temperature to 50 °C. It is pointed that the texture dependence is not only related to tensile flow stress but also affects the tensile elongation at room temperature [5]. Increasing the test temperature, the total elongation of the AZ61-O specimens increases more obviously than that of the AZ61-FSP samples. In the AZ61-O specimen tested at 100 °C, partially dynamic recrystallization phenomenon is observed on the grain boundaries as shown in Fig. 10(c). The dynamic recrystallization phenomenon occurrs in the high-temperature tensile test and the new grains initially nucleate on the original grain boundaries [7]. In Fig. 10(d), the AZ61-O specimen shows that the dynamic recrystallization produced the lots of new fine grains at 200 °C. The dynamic recrystallization occurring during tensile deformation decreases the flow stress and markedly raises the ductility at elevated temperatures [8]. It could be suggested that in the high-temperature tensile test, AZ61-O possessed a better elongation as a result of dynamic recrystallization.

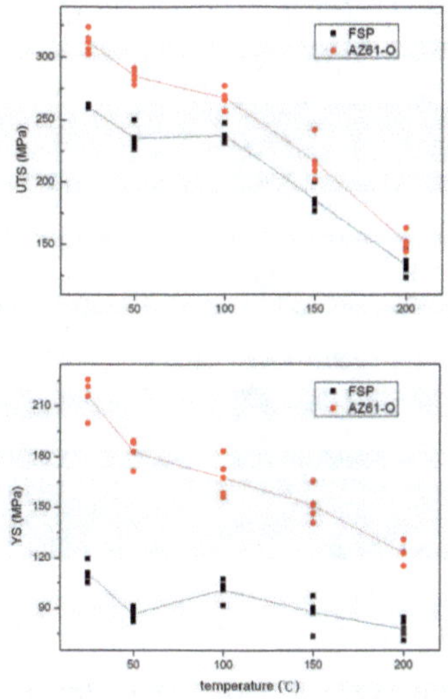

Fig.8. Flow stress of AZ61-O and AZ61-FSP against test temperatures: (a) ultimate tensile stress and (b) yield stress.

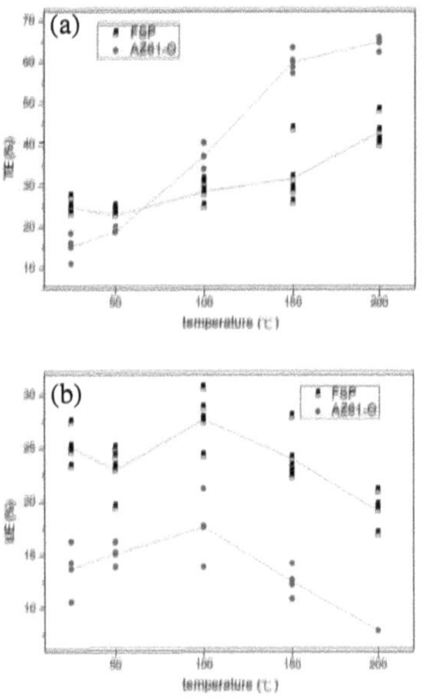


Fig.9. Elongation of AZ61-O and AZ61-FSP against test temperatures: (a) total elongation and (b) uniform elongation.

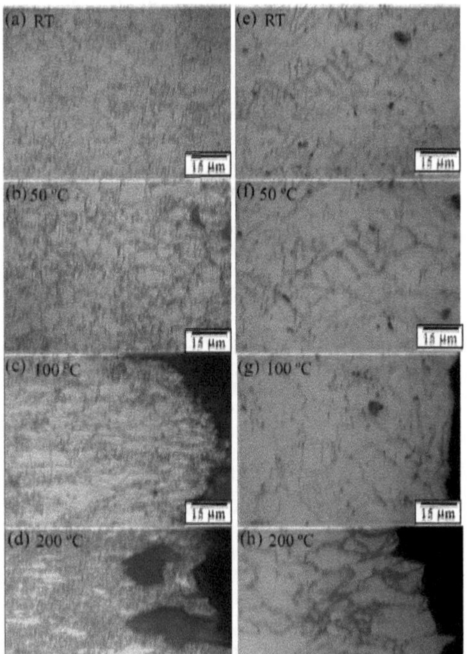


Fig.10. Subsurface of specimens deformed to fracture: (a)(b)(c)(d) AZ61-O tested from RT to 200 °C and (e)(f)(g)(h) AZ61-FSP tested from RT to 200 °C.

**5. Conclusion**

The effects of friction stir processing on the microstructure and tensile deformation behavior of the AZ61 magnesium alloy can be summarized as follows:

- Inhomogeneous structure resulted from the friction heat can be observed after friction stir processing.
- The AZ61-FSP specimen has a lower tensile flow stress than AZ61-O due to the texture generated by FSP.
- The total elongation of AZ61-FSP specimens is better than AZ61-O from room temperature to 50 °C.
- Dynamic recrystallization of the AZ61-O specimen during tensile test resulted in the new fine grains and markedly raised the total elongation of AZ61-O at the elevated temperatures.

**Acknowledgements**

This paper has been supported by the Chinese National Science Council (contract: No. NSC 100-2221-E-006 -093 -), for which the authors are grateful.

**References**

1. M.W. Mahoney, C.G. Rhodes, J.G. Flintoff, R.A. Spurling, and W.H. Bingel, "Properties of Friction-stur-welded 7075 T651 Aluminum," *Metallurgical and Materials Transactions A*, 29A (1998), 1955-1964.
2. B. London, M. Mahoney, B. Bingel, M. Calabrese, and D. Waldron, "Investigations on Residual Stresses in Friction Stir Welds" (Paper presented the 3th International Symposium on Friction Stir Welding, Kobe, Japan, 27 September 2001),27-28.
3. T.U. Seidel and A.P. Reynolds, "Visualization of material flow in AA2195 friction-stir welds using a marker insert technique," *Metallurgical and Materials Transactions A*, 32A (2001), 2879-2884.
4. W. Woo, H. Choo, D.W. Brown, P.K. Liaw, and Z. Feng, "Texture variation and its influence on the tensile behavior of a friction-stir processed magnesium alloy,"

*Scripta Materialia*, 54 (2006), 1859-1864.

5. H. W. Lee, T. S. Lui, and L. H. Chen, "Studies on the improvement of tensile ductility of hot-extrusion AZ31 alloy by subsequent friction stir process," *Journal of Alloys and Compounds*, 475 (2009), 139-144.
6. C. J. Lee, J. C. Huang, and X. H. Du, "Using Multiple FSP Passes to Cure Onion Splitting of Mg Alloys Deformed at Elevated Temperatures," *Materials Transactions*, 48 (4) (2007), 780-786.
7. S.M. Fatemi-Varzaneh, A. Zarei-Hanzaki, and H. Beladi, "Dynamic recrystallization in AZ31 magnesium alloy," *Materials Science and Engineering A*, 456 (2007), 52-57.
8. H.J. McQueena, and C.A.C. Imbert, "Dynamic recrystallization: plasticity enhancing structural development," *Journal of Alloys and Compounds*, 378 (2004), 35–43.

Magnesium Technology 2012
*Edited by: Suveen N. Mathaudhu, Wim H. Sillekens, Neale R. Neelameggham, and Norbert Hort*
***TMS (The Minerals, Metals & Materials Society), 2012***

# MAGNESIUM RECYCLING OF PARTIALLY OXIDIZED, MIXED MAGNESIUM-ALUMINUM SCRAP THROUGH COMBINED REFINING AND SOLID OXIDE MEMBRANE (SOM) ELECTROLYSIS PROCESSES

Xiaofei Guan[1], Peter Zink[1,2], Uday Pal[1,2]
[1]Boston University, Division of Materials Science and Engineering; 15 Saint Mary's St.; Brookline, MA, 02446
[2]Boston University, Department of Mechanical Engineering; 110 Cummington St.; Boston, MA, 02215

Keywords: Magnesium, Refining, SOM Electrolysis

## Abstract

Pure magnesium (Mg) is recycled from 19g of partially oxidized 50.5wt.%Mg-Aluminum (Al) alloy. During the refining process, potentiodynamic scans (PDS) were performed to determine the electrorefining potential for magnesium. The PDS show that the electrorefining potential increases over time as the Mg content inside the Mg-Al scrap decreases. Up to 100% percent of magnesium is refined from the Mg-Al scrap by a novel refining process of dissolving magnesium and its oxide into a flux followed by vapor phase removal of dissolved magnesium and subsequently condensing the magnesium vapors in a separate condenser. The solid oxide membrane (SOM) electrolysis process is employed in the refining system to enable additional recycling of magnesium from magnesium oxide (MgO) in the partially oxidized Mg-Al scrap. The combination of the refining and SOM processes yields 7.4g of pure magnesium; could not collect and weigh all of the magnesium recovered.

## Introduction

Magnesium is the least dense engineering metal, with an excellent stiffness-to-weight ratio. For this reason, U.S. auto makers would like to replace 630 lb. of steel and aluminum parts per vehicle with 340 lb. of magnesium alloy by the year 2020 [1]. This would reduce vehicle weight by 290 lbs and improve average fuel efficiency by 1.5-2 miles per gallon, reducing petroleum imports by up to 1 million barrels per day. Additionally, the fact that magnesium has good electromagnetic shielding makes it very attractive to the audio and electronic industry [2]. As magnesium consumption increases in the automotive, aerospace and electronic industries, magnesium recycling will become ever more important for both economic and environmental reasons.

Several recycling processes are currently in use. Some are based on the use of refining fluxes, and others are flux-free methods. Hydro Magnesium [3, 4] has a system for continuous melting with flux. Magnesium alloy is melted in one end of a large multi-chamber furnace and pumped out or cast from the other end. This refining process uses sufficient flux to absorb all the oxides present in the molten metal, but it cannot separate pure magnesium from aluminum and trace impurities such as zinc, iron and nickel; impurities like nickel and iron decrease corrosion resistance of magnesium [5]. Zhu et al [6] demonstrated that vacuum distillation for magnesium recycling. However, the vacuum distillation will leave 20-25% of the original scrap as the residue with a magnesium weight content as high as 60%. It is understood that further magnesium distillation is hindered by the magnesium oxide and aluminum oxide ($Al_2O_3$) on the surface of the residue.

In this paper, an innovative process for recycling pure magnesium metal from impure magnesium scrap alloy at high efficiency and low cost is proposed. As the typical magnesium alloys in automotive components are AZ91, AM50, AM60 and so on [7], all of which are predominantly Mg-Al alloys with small amount of impurities (Zn, Mn, Fe and Ni, etc). This paper reports proof of concept experiments of our process with partially oxidized 50.5wt%Mg-Al alloy. The refining efficiency is almost 100% by dissolving magnesium and its oxide from the scrap into a flux followed by vapor phase removal of dissolved magnesium and subsequently condensing the magnesium vapors in a separate condensing chamber. After refining, the SOM electrolysis process is employed in the system to enable additional recycling of magnesium from magnesium oxide in the scrap. The combination of the refining and SOM electrolysis processes proves to be a feasible method for recycling magnesium from partially-oxidized impure magnesium scrap alloy. Future work will involve recovery of magnesium from partially oxidized magnesium scrap containing many more impurities which make it unsuitable for reuse.

## Experimental

Figure 1 shows the design of the setup used in this experiment. The setup consists of an upper reaction chamber, heated to 1175°C and a lower condensing chamber with a temperature gradient of 1100-200°C. The setup was fabricated using grade 304 stainless steel (SS-304) and heated as mentioned above in an argon (Ar) atmosphere.

The Mg-Al alloy was prepared by melting a 9.6g piece of magnesium (Mg>99.8%) and another 9.4g piece of aluminum (Al>97.9%) together inside a small SS-304 crucible. The mixture of magnesium and aluminum was stirred with a SS-304 rod for 15 minutes at 800°C in an argon atmosphere, then held at the same temperature without stirring for 15 minutes and subsequently quenched. The alloy was homogeneous and the composition was 50.5wt%Mg-Al.

Since there is a finite solubility of iron in Mg-Al alloy, 2g of iron powder was added on the top of the Mg-Al alloy to decrease dissolution of iron from the crucible during the run. Also, the iron powder addition increases the density of the alloy (alloy sinks to the flux bottom in the alloy crucible) and lowers the vapor pressure of magnesium. A powdered flux ($MgF_2$-$CaF_2$-10wt.%MgO-2wt.%$YF_3$) containing a eutectic mixture of 45wt.% $MgF_2$-55wt.%$CaF_2$ with a melting point of 974°C [8] was used as the electrolyte and was packed both inside and outside the inverted crucible shown in Figure 1. In order to prevent the distillation of magnesium from the Mg-Al alloy during heating to 974°C, a layer of flux is melted on the surface of the crucible to hermetically seal the opening of the alloy crucible before heating. Thus, the alloy is trapped inside the alloy crucibles, and the direct distillation of magnesium is avoided. A result of melting the flux at the top of alloy crucible is that some of magnesium in the alloy becomes oxidized. This magnesium oxide is later reduced with the SOM electrolysis process.

During the refining process, the alloy crucible and inverted crucible served as the anode, and the reaction chamber wall and bubbling tube served as the cathode. The SOM tube was held above the flux. An alumina spacer was used to insulate the rod connecting the inverted crucible and the reaction chamber. Potentiodynamic scans were performed to determine the electrorefining potential for magnesium, as the refining of magnesium proceeded.

During the SOM electrolysis process, an yttria stabilized zirconia (YSZ) tube was used for recycling magnesium from magnesium oxide. The stainless steel wall of the reaction chamber still served as the cathode, but silver inside the YSZ tube served as the anode, and a carbon rod acted as the anodic current collector. When the applied electric potential between the cathode and anode exceeds the dissociation potential of magnesium oxide, magnesium vapor is produced at the cathode, and carbon reacts with oxygen to generate carbon monoxide [9, 10]. The overall cell reaction is given as:

$$MgO + C = Mg(g) + CO(g) \qquad (1)$$

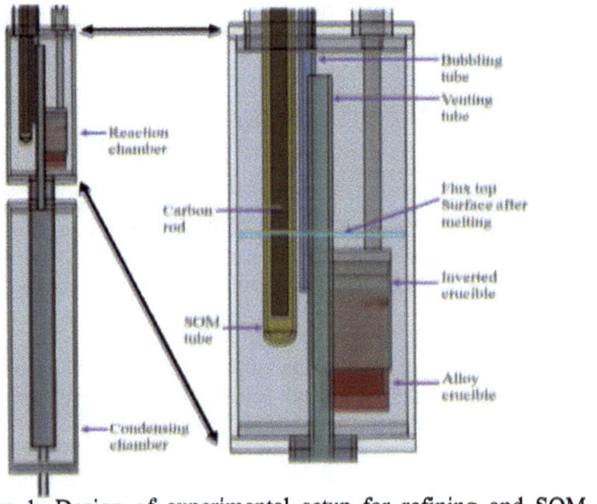


Figure 1. Design of experimental setup for refining and SOM electrolysis processes.

It should be noted that during both the refining and electrolysis processes, the reaction chamber is continually purged with 95%Ar-$H_2$ at 15cc/min through a bubbling tube, and at 30cc/min through the two annuli at the top of the reaction chamber. This is done to lower the partial pressure of magnesium vapor over the flux and to carry the magnesium vapor to the condensing chamber. The inlet of the venting tube is well above the flux surface, to prevent any molten flux from entering the condenser.

Electrochemical measurements were performed. A Solartron SI 1280B potentiostat was used for potentiodynamic scans and impedance spectroscopy during the refining process; an Agilent Technologies N5743A power supply was used for potentiodynamic scans and electrolysis.

## Results and Discussion

### 1. Refining process

Once the reaction chamber was at the desired temperature, potentiodynamic scans were performed as shown in Figure 2.The initial melted flux is entirely ionic and acts as an electronic insulator between the anode and the cathode. The current increases linearly as the applied potential increases, except at the two current jumps corresponding to magnesium dissolution and magnesium vapor formation, as can be seen in each potentiodynamic scan. The current-voltage curve is shifting in a positive direction over time from PDS1 to PDS5 due to decreasing magnesium concentration in the alloy as explained later in the paper. The ohmic resistance of the system was measured to be as low as 0.066Ω and 0.09Ω using impedance spectroscopy shown in Figure 3. One potential reason for the low resistance or the electronic conductivity is the dissolution of metallic magnesium in the flux. It has been shown that magnesium metal has some solubility in chloride based ionic salts [11]. The magnesium solubility inside the flux after the experiment was found to be 0.03wt% as measured with a manometer. Dilute acid was added to powdered flux in a closed container and the volume of gas produced due to the hydrogen evolution was measured. The flux was powdered in a glove box to avoid the oxidation of magnesium in the flux. The magnesium solubility inside a flux of similar composition (55%$MgF_2$-45%$CaF_2$-10%MgO) was reported to be 0.02-0.05wt.% [12], consistent with the experimental results .

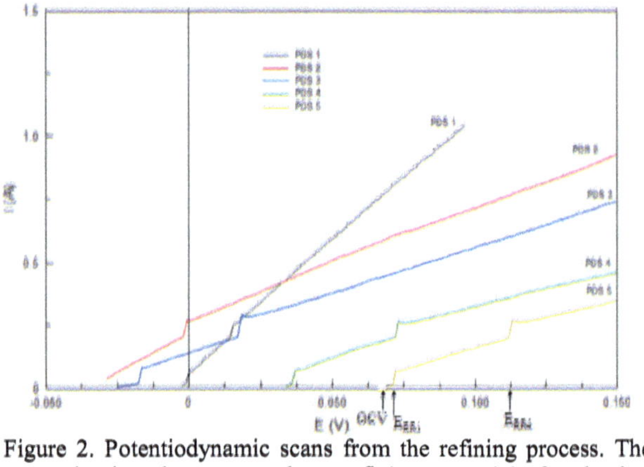


Figure 2. Potentiodynamic scans from the refining process. The open circuit voltage, two electrorefining potentials for the last potentiodynamic scan are indicated by arrows (OCV, $E_{ER1}$, $E_{ER2}$).

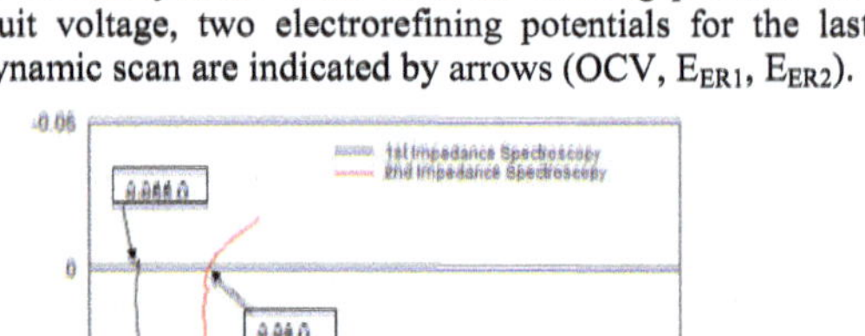


Figure 3. Impedance spectroscopy results during refining.

For each potentiodynamic scan, there are two current jumps which correspond to two different electrochemical reactions (Figure 2). The current jump at the first potential, close to the open circuit voltage is due to the following electrochemical reaction:

Anode: $Mg(alloy) \rightarrow Mg^{2+} + 2e^-$ (2)
Cathode: $Mg^{2+} + 2e^- \rightarrow Mg(flux)$ (3)

where the overall reaction is

$$Mg(alloy) \rightarrow Mg(flux) \quad (4)$$

This reaction is different from direct dissolution of magnesium from Mg-Al alloy into flux. Theoretically, the first electrorefining potential corresponding to the first current jump should be equal to the open circuit voltage, and its expression is given by

$$E_{ER1}=E_{OCV}=E_{anode}-E_{cathode}=\frac{RT}{2F}\ln\frac{a_{Mg(flux)}}{a_{Mg(alloy)}}=\frac{RT}{2F}\ln\frac{P_{Mg(flux)}/P^0_{Mg}}{\gamma x_{Mg(alloy)}}$$
$$=\frac{RT}{2F}\ln\frac{P_{Mg(flux)}}{\gamma x_{Mg(alloy)}P^0_{Mg}} \quad (5)$$

where $\gamma$ is the activity coefficient of magnesium in Mg-Al alloy; $x_{Mg(alloy)}$ is the magnesium molar content in Mg-Al alloy; $P^0_{Mg}$ =1.9atm is the magnesium vapor pressure at T=1175°C; and $P_{Mg(flux)}$ is the partial pressure of magnesium vapor in equilibrium with liquid magnesium dissolved in flux at the cathode, where $P_{Mg(flux)}<1$atm.

According to Equation (5), the first electrorefining potential and the open circuit voltage depend on the ratio of activity of magnesium dissolved in the flux and the activity of magnesium in the alloy. There is a minor difference between the open circuit voltage and the first electrorefining potential in the experimental measurement caused by the fluctuating value of $P_{Mg(flux)}$.

The second current jump at the higher potential is due to the following reaction:

Anode: $Mg(alloy) \rightarrow Mg^{2+} + 2e^-$ (6)
Cathode: $Mg^{2+} + 2e^- \rightarrow Mg(g)$ (7)

where the overall reaction is:

$$Mg(alloy) \rightarrow Mg(g) \quad (8)$$

The theoretical expression of electrorefining potential for Equation (8) is given by

$$E_{ER2}=E_{anode}-E_{cathode}=\frac{RT}{2F}\ln\frac{a_{Mg(g)}}{a_{Mg(alloy)}}=\frac{RT}{2F}\ln\frac{P_{Mg(g)}}{\gamma x_{Mg(alloy)}P^0_{Mg}} \quad (9)$$

and is plotted in Figure 4, where $\gamma$ is the activity coefficient of magnesium in Mg-Al alloy; $x_{Mg(alloy)}$ is the magnesium molar content in Mg-Al alloy; $P_{Mg}^0$ =1.9atm is the magnesium saturated vapor pressure at T=1175°C; and $P_{Mg(g)}$=1atm is the magnesium vapor partial pressure at the cathode. The electrorefining potential depends on the ratio of magnesium vapor activity at the cathode and magnesium activity in the alloy.

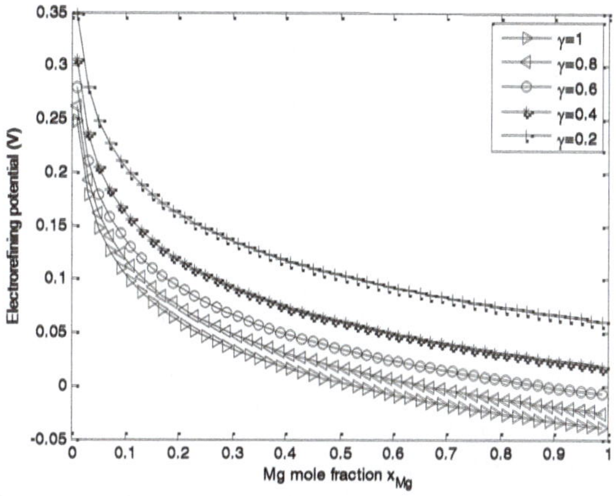

Figure 4. Theoretical values of the second electrorefining potential (based on Equation 9) of magnesium from Mg-Al scrap for different magnesium activity coefficients.

From Figure 4, as the magnesium content in the scrap goes down, the second electrorefining potential increases. The experimental result of the dependence of the second electrorefining potential and the open circuit voltage is shown in Figure 5. The second electrorefining potential and open circuit voltage increase with time, which indicates that the magnesium content in the scrap is decreasing as refining proceeds. The reason why the magnesium content in the scrap decreases is due to the finite solubility of magnesium in the flux. Magnesium dissolution in the flux is followed by vapor phase removal of the dissolved magnesium from the flux. The refining process can be expressed as: Mg(alloy) → Mg(flux) → Mg(g). Overall, pure magnesium is refined from magnesium scrap.

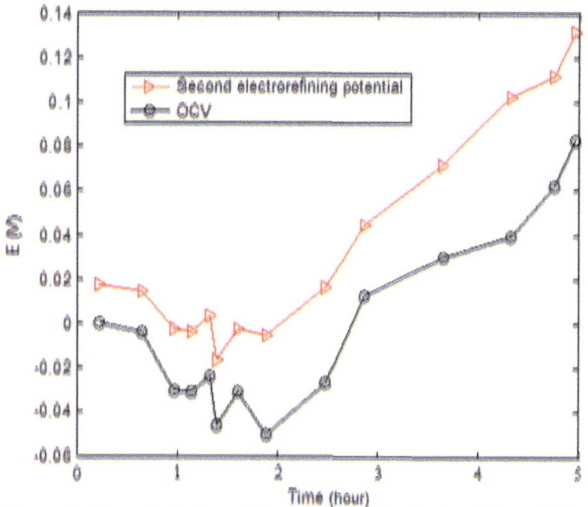

Figure 5. Second electrorefining potential (red triangle) and OCV (black circle) during refining process.

## 2. SOM electrolysis process

Once the second electrorefining potential reached 0.13V, the anode was switched from the scrap anode to the liquid silver and carbon rod inside the YSZ tube, and a new potential was applied.

When the potential reaches the dissociation potential of magnesium oxide, oxygen ions are pumped out of the flux through the YSZ membrane and are oxidized at the liquid silver anode by the carbon rod; meanwhile, magnesium ions at the cathode are reduced to magnesium metal vapor, which is collected in the condenser.

Electrolysis was performed for a total of two hours at a potential of 3V. The blue curve in Figure 6 shows the initial potentiodynamic scan before any SOM electrolysis was performed, and shows an electronic current of 0.35A due to the dissolution of magnesium from the scrap into the flux. The blue curve in Figure 7 shows the current-time relationship during the first hour of electrolysis. Current efficiency is defined as the ratio of Faradic current to total current, and was calculated to be approximately 41% for the first hour of electrolysis. This is done by measuring the flow rate of carbon monoxide produced from the reaction of the carbon rod and oxygen on the anode side of the YSZ membrane. Based on the volume of carbon monoxide generated, magnesium reduced at the cathode was calculated to be 0.18g during the first hour of SOM electrolysis.

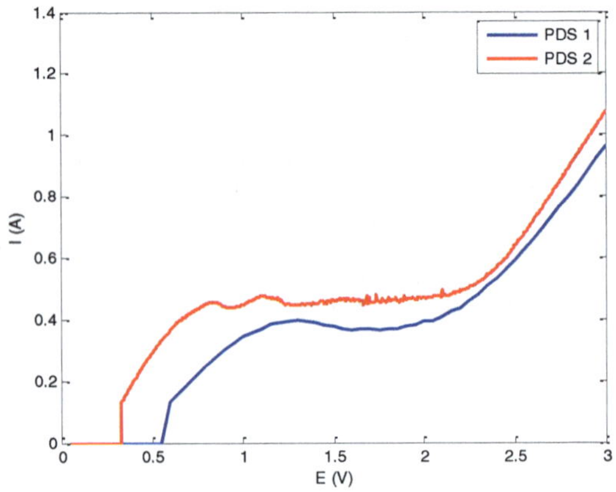


Figure 6. Potentiodynamic scan before the first hour of SOM electrolysis (blue curve) and before the second hour of SOM electrolysis (red curve).

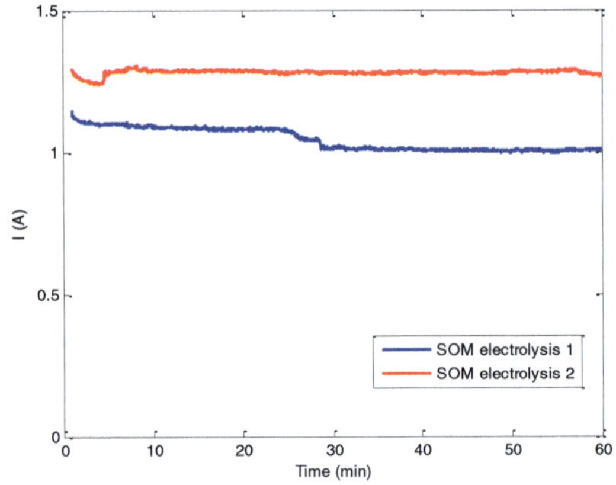


Figure 7. Current-time relationships for the first hour of electrolysis (blue curve) and the second hour of electrolysis (red curve). Both electrolyses were performed at 3V.

The red curve in Figure 6 shows the potentiodynamic scan before the second hour of SOM electrolysis, which shows the leakage current has increased due to a higher concentration of magnesium dissolved in the flux. The red curve in Figure 7 shows the current-time relationship during the second hour of SOM electrolysis. For the second hour of electrolysis, the current efficiency was 39%, and the amount of magnesium reduced was 0.22g. The total amount of magnesium reduced in the entire two hours of SOM electrolysis process was 0.4g.

The combination of refining and SOM electrolysis processes are shown in the schematic in Figure 8.

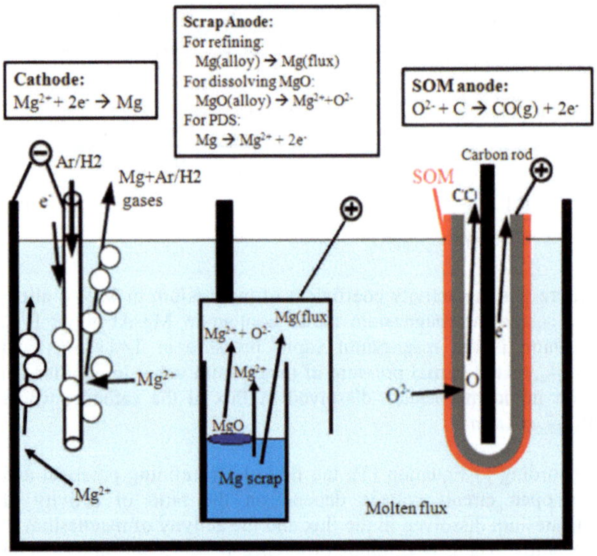


Figure 8. Schematic for refining and SOM electrolysis processes.

The total magnesium collected in the condenser from both the refining and SOM processes was weighed to be 7.4g. The Energy-dispersive X-ray spectroscopy (EDS) results for the collected magnesium are shown in Figure 9. The purity of the collected magnesium was determined to be 99.6wt%.

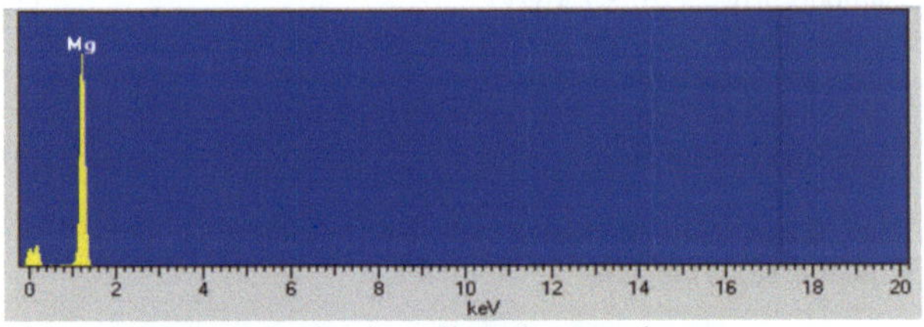


Figure 9. EDS results for the collected magnesium.

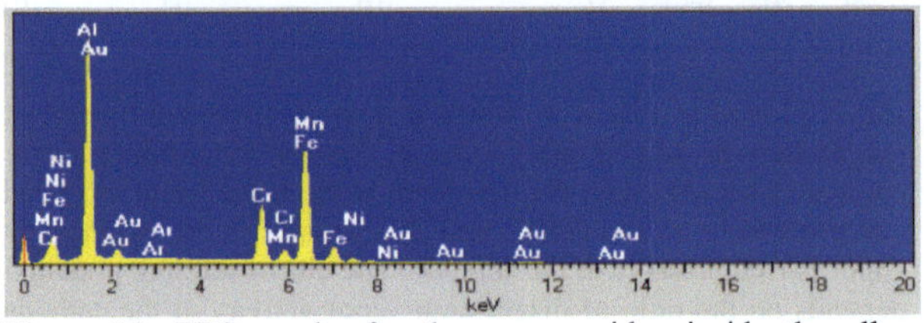


Figure 10. EDS results for the scrap residue inside the alloy crucible.

The scrap residue remained at the bottom of the alloy crucible. The iron powder and iron and chromium from the SS 304 crucibles alloyed with the scrap, which has a density higher than the flux. EDS was performed for the magnesium residue, and the results are shown in Figure 10. Only trace magnesium remains in the alloy, thus approximately 100% of the magnesium has been refined. Future experiments are also planned wherein the stainless steel parts will be replaced by low carbon steel to minimize additional dissolution of elements (Cr, Ni and Mn).

## Conclusions

The combination of the novel refining process and the SOM electrolysis process for recycling magnesium from magnesium scrap was carried out, and yielded 7.4g of pure magnesium metal. A small amount of magnesium condensed on the walls of the condenser and dissolved in the 680g of flux, and could not be quantified. No measurable amount of magnesium remained in the magnesium scrap, which implies that the refining efficiency is almost 100% using this novel refining process of dissolving magnesium and its oxide from the scrap into the flux, followed by vapor phase removal of dissolved magnesium. After the refining process, the SOM electrolysis process was employed in the refining system to enable the recycling of an additional 0.4g magnesium from magnesium oxide in both the scrap and the flux. This paper presented the technical feasibility of recycling magnesium from partially oxidized Mg-Al alloy by the combination of refining and SOM electrolysis processes. In the future, this approach will be tested using larger amounts of oxidized scrap with additional impurities and less initial oxide content in the flux.

## Acknowledgements

This material is based upon work supported by the Department of Energy under Award No. DE-EE0003454. The authors thank Adam C. Powell IV (Figure 8), Eric Gratz, and Soobhankar Pati for helpful discussion.

## References

[1] United States Automotive Materials Partnership (USAMP), *Magnesium Vision 2020: A North American Automotive Strategic Vision for Magnesium,* Technical Report, USCAR, November 2006.

[2] J. F. Kurgan, "Electromagnetic shield for electrical circuit," *US patent 5053924,* 1 October 1991.

[3] W. O, R. JB, "Method and apparatus for remelting and refining of magnesium and magnesium alloys," *US Patent 5167700*, 1 December 1992.

[4] H. E. Friedrich, B. L. Mordike, *Magnesium Technology: metallurgy, design data, applications* (Springer, 2006), 638.

[5] S. D. Cramer and B. S. Covino, *ASM Handbook, Volume 13A: Corrosion: Fundamentals, Testing, and Protection* (ASM international, 2003), 693.

[6] T. Zhu et al., "Innovative vacuum distillation for magnesium recycling", *Magnesium Technology 2001* (TMS, 2001), 55-60.

[7] G. Hanko, H. Antrekowitsch and P. Ebner, "Recycling automotive magnesium scrap," *Journal of the Minerals, Metals and Materials society*, 54 (2) (2002) 51-54.

[8] P. Chartrand and A. D. Pelton, "Thermodynamic evaluation and optimization of the LiF-NaF-KF-$MgF_2$-$CaF_2$ system using the modified quasi-chemical model," *Metallurgical and Materials Transactions A*, 32 (6) (2001) 1385-1396.

[9] A. Krishnan, U. B. Pal and X. G. Lu, "Solid Oxide Membrane Process for Magnesium Production Directly from Magnesium Oxide," *Metallurgical and Materials Transaction B*, 36 (4) (2005), 463-473.

[10] U. B. Pal, A. C. Powell, "The Use of Solid-oxide-membrane Technology for Electrometallurgy," *Journal of the Minerals, Metals and Materials Society*, 59 (5) (2007), 44-49.

[11] J. Wypartowicz, T. Ostvold and H. Oye, "The Solubility of Magnesium Metal and the Recombination Reaction in the Industrial Magnesium Electrolysis," *Electrochimica ACTA*, 25 (1980), 151-156.

[12] E. Gratz et al., "Efficiency and Stability of Solid Oxide Membrane Electrolyzers for Magnesium Production," *Magnesium Technology 2011* (Wiley-TMS, 2011), 39-42.

**Magnesium Technology 2012**
*Edited by: Suveen N. Mathaudhu, Wim H. Sillekens, Neale R. Neelameggham, and Norbert Hort*
***TMS (The Minerals, Metals & Materials Society), 2012***

# Microstructure and mechanical properties of Mg-5Sn-5Zn-xCa alloys

QIU Ke-qiang[1], LIU Bin[1], YOU Jun-hua[1] , REN Ying-lei[1]

[1](School of Materials Science and Engineering, Shenyang University of Technology, No.111, Shenliao West Road, Economic & Technological Development Zone; Shenyang, Liaoning 110870, P.R.CHINA)

Key words: Magnesium alloy, Microstructure, Ultimate tensile strength, Ca addition

## Abstract

Mg-5Sn-5Zn-xCa(x=0.5, 1, 2) alloys were melted by the vacuum melting furnace. The microstructures and phase compositions were analyzed by the scanning electron microscope (SEM) and X-ray diffraction (XRD). The mechanical properties were tested by the electronic universal test machine. The fracture surface were observed by scanning electron microscope (SEM). The results indicate that the microstructures of Mg-5Zn-5Sn-xCa alloys are composed of CaMgSn phases, layer $MgZn_2$ phases and Mg2Sn phases. Matrix precipitates onset $Mg_2Ca$ phase, when the content of Ca is 2 wt%. With the increasing of the Ca content, the needle CaMgSn phases become rods gradually, the layer $MgZn_2$ phases and plate $Mg_2Sn$ phases become continuous, ultimate tensile strength decrease. The ultimate tensile strength obtained by Mg-5Zn-5Sn-0.5Ca alloy is 184MPa.

## 1 Introduction

Magnesium alloys have greatly potential applications as the lightest structural material due to their inherent properties such as acceptable mechanical properties, low density, superior damping capacity and castability [1]. Nowdays, all of the researcher devote to improve the strength of the Mg alloys. The ultimate tensile strength (UTS) of Mg-RE alloys can reach 300MPa to 400MPa, but the high cost restricts these alloys from being extensively used. It is necessary to develop a new type Magnesium alloy [2]. Compared to Mg-Al and Mg-Zn alloys, the intermetallic phase Mg2Sn in Mg-Sn alloys has a much higher melting (770℃) than the $Mg_{17}Al_{12}$ phase (462℃) of Mg-Al alloys and MgZn phase (347℃) of Mg-Zn alloys. The solubility of Sn in the α-Mg solid solution drops sharply with the temperature decreasing. It provides a fundamental basis for improving the mechanical properties of these alloys through aging [3-4]. According to Liu, the microstructure of as-cast Mg-Sn alloys consisted of dendrite a-Mg and second Mg2Sn phases and the secondary dendrite arm spacing (DAS) of the a-Mg phase was decreased with the increase of tin content. The micro-hardness of the alloys increased when tin content rised, while the great tensile 123MPa were obtained by Mg-5 wt% Sn [5]. As one of the two major alloy element, Zn has strong strength effect. Solid solution strengthening and compounds to strengthing are two main way of Zn reinforced Mg alloys. Add 1wt% Zn to Mg-Al alloys will improve the comprehensive properties of the Mg alloys. Under the condition of tin, Zn solution reinforcement will be strengthened [6]. Ca is an important alloying element of Mg alloys; it is well know that Ca can behave as a grain refiner of magnesium alloys and form a stable $Mg_2Ca$ compound with Mg [7]. In addition, it also form intermetallic stable phase CaMgSn with content of tin [8-9]. Therefore, research on Mg-Sn-Zn-Ca alloys has important significance. This paper mainly investigate that the effect of 0.5 wt%, 1.0 wt%, 2.0 wt% Ca on as-cast microstructure and mechanical properties of Mg-5Sn-5Zn alloy.

## 2 Experimental

Mg-5Sn-5Zn-xCa(x=0.5,1,2) alloys were prepared using pure magnesium (99.99 wt%), tin (99.99 wt%), zinc (99.99 wt%), and calcium added in the form of Mg-30 wt% Ca master alloy. The experimental alloys were melted in a crucible under the protection of the nitrogen gas at 700℃. (The melting point of all metals is less than 700℃ except to Mg-Ca alloy. The melting point of intermediate phase of Mg-Ca alloy is 715℃, but it will react with other metals in the smelting process) The melt was homogenized by mechanical stirring and held at 700℃ for 30min, then it was poured into a preheated permanent mould in order to obtain a casting. The slices (shown in Fig.1) were cut by an electrodes charge wire-cut machine for tensile tests and structural characterization. The specimens were studied by scanning electron microscopy (SEM) and optical microscopy to examine the as-cast microscopy. The specimens were polished with chromium oxide solution. Etching was carried out using a 2% nitri acid and 98% ethyl alcohol solution at room temperature. The X-ray diffraction (XRD) and energy-dispersive X-ray (EDS) analysis were carried out on selected samples to identify the phases.

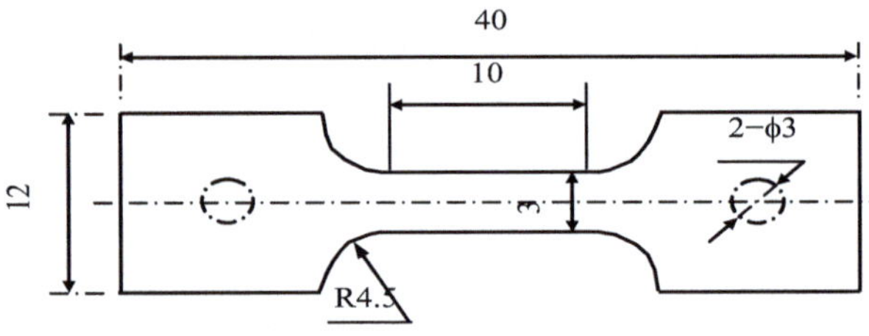


Fig.1 The size of tensile test specimen（mm）

## 3 Results and discussions

### 3.1 Microstructure observation

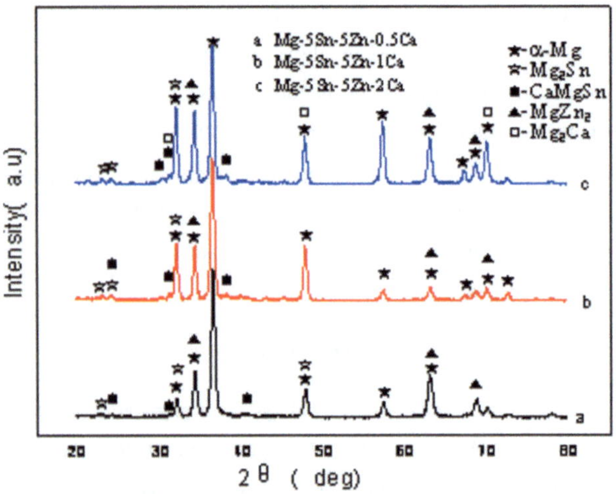


Fig.2 XRD patterns of Mg-Sn-Zn-Ca alloys

Fig.2 shows the XRD pattern of the as-cast experimental alloys. It can be observed from the Fig.2 that the as-cast experimental alloys are mainly composed of α-Mg, $MgZn_2$, $Mg_2Sn$ and CaMgSn phases with the addition of 0.5 wt% and 1.0 wt% Ca. With the Ca amount increasing, the intensity of the $Mg_2Sn$ and CaMgSn peaks become higher. The $Mg_2Ca$ strong peak appearing in the pattern of 2.0wt% Ca alloy indicates that there is a large number of the $Mg_2Ca$ phase in Mg-5Sn-5Zn-2Ca alloy. It is found in Fig.1 that the addition of 0.5 wt%-1.0 wt% Ca to the Mg-5Sn-5Zn alloy results in the formation of CaMgSn phase rather than $Mg_2Ca$ phase. In general, the difficulty level in forming the intermetallics can be estimated by the electronegative difference between different elements. The higher the electronegative difference is, the more easily the intermetallics forms. The electronegative values of Mg, Sn, Zn and Ca are 1.31, 1.65, 1.96 and 1.00, respectively. Obviously, the CaMgSn intermetallics is more easily formed than $Mg_2Ca$ [10]. When Ca content is 2.0wt%, the remaining amount of Ca formed $Mg_2Ca$ phase after the formation of CaMgSn phase.

Fig.3 show the optical microstructures of the as-cast alloys. Fig.3（a）is the microstructures of Mg-5Sn-5Zn-0.5Ca alloy. It can be observed from Fig.3 (a) that the phase precipitates along the grain boundary in the alloy with wide dendrite，and have needle-shaped precipitates in grain. Fig.3（b）is the microstructures of Mg-5Sn-5Zn-1Ca alloy. CoMPared with Fig.3 (a), it has discontinuous dendrite, and a part of needle-shaped precipitates change into rod-shaped. Fig.3（c）is the microstructures of Mg-5Sn-5Zn-2Ca alloy. It has the finest dendrite in some area. The phase precipitated in grain become almost all rod-shaped. The research indicates that it has two effects with the increase of Ca content: Firstly, The increase of Ca content (2%) can effectively refine the as-cast microstructure of the alloy. Secondly, the phase precipitated in grain change to rod-shaped from needle-shaped with the increase of Ca content. This will have important effect on mechanical properties of Mg alloys.

Fig.4 show the SEM microstructures of the as-cast alloys. The EDS results of the as-cast alloys are given in the table 1. Fig.4 (a) is the SEM microstructures of Mg-5Sn-5Zn-0.5Ca alloy. It can be observed from the Fig.4 (a) that the as-cast Mg-5Sn-5Zn-0.5Ca experimental alloys are mainly composed of three phases except α-Mg phase. According to the XRD and EDS results (Table 1), the plate phase in the grain boundary is $Mg_2Sn$, and the lamellar tissue is $MgZn_2$, and needle-shaped phase is identified as CaMgSn phase. According to the morphology and distribution, there are two eutectic reaction in the formation process: firstly, L → α-Mg+CaMgSn. secondly, L→α-Mg+$Mg_2Sn$+$MgZn_2$.

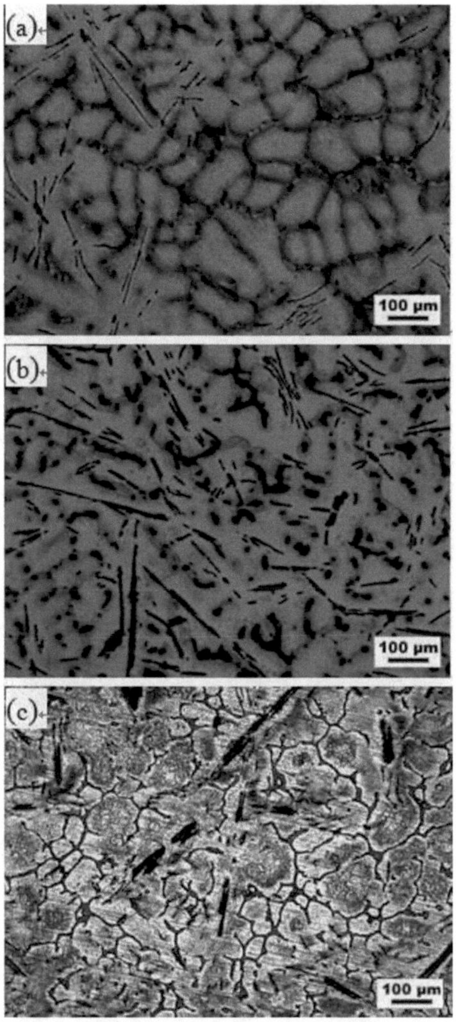


a) Mg-5Sn-5Zn-0.5Ca b) Mg-5Sn-5Zn-1Ca

c) Mg-5Sn-5Zn-2Ca

Fig.3 Microstructures of the as-cast Mg-Sn-Zn-Ca alloys

Fig.4 (b) is the SEM microstructures of Mg-5Sn-5Zn-1Ca alloy. CoMPared with Fig.4 (a), the $Mg_2Sn$ phase and $MgZn_2$ are not independent, but attached with each other. The rod-shaped CaMgSn precipitated in grain. According to the morphology and distribution, there are four eutectic reaction in the formation process: firstly, L → CaMgSn. Secondly, L+CaMgSn → α-Mg+CaMgSn. Thirdly, L→α-Mg+$Mg_2Sn$. fourth, L→α-Mg+ $MgZn_2$. Fig.4 (c) is the SEM microstructures of Mg-5Sn-5Zn-2Ca alloy. It differs from Fig.4 (a) and Fig.4 (b), there exist $Mg_2Ca$ phase (arrow D) in the grain boundary, it forming a network with $Mg_2Sn$ phase and $MgZn_2$ phase. According to the morphology and distribution, there are four eutectic reaction in the formation process: Firstly, L → CaMgSn. Secondly, L+CaMgSn → α-Mg+CaMgSn. Thirdly, L → α-Mg+$Mg_2Sn$. Fourth, L → α-Mg+CaMgSn+$Mg_2Ca$.

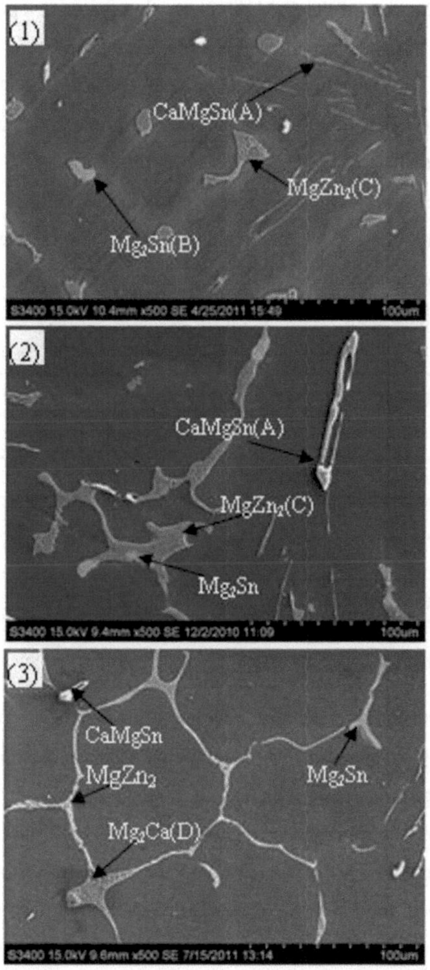


1) Mg-5Sn-5Zn-0.5Ca 2) Mg-5Sn-5Zn-1Ca

3) Mg-5Sn-5Zn-2Ca

Fig.4 SEM Microstructures of the as-cast Mg-Sn-Zn-Ca alloys

According to Mg-Sn binary diagram[11], the saturation sliod solubility of tin in magnesium is relatively large with 14.85 wt% at the temperature of 561.2℃, which decreases sharply with the temperature falling with 0.45 wt% at 200℃ and almost zero at room temperature. The fact that the saturation solid solubility declines fast with the decreasing of temperature case the tin atoms to precipitate, in the form of the $Mg_2Sn$ during the solidification process. According to the XRD and EDS result (Table 1), the layer phase(arrow C)precipitated on the grain boundary is $MgZn_2$ phase and another layer phase (arrow D) is $Mg_2Ca$ phase. Based on the calculated vertical phase diagram section of the Mg-Sn-Ca alloys with a constant Ca composition of 2.0 wt%, during the solidification of the CaMgSn phase first forms as a primary phase when passing through L+CaMgSn region, and then at the large stage of solidification a pseudo binary eutectic reaction (L+CaMgSn→α-Mg+CaMgSn) and a ternary eutectic reaction (L→α-Mg+CaMgSn+$Mg_2Ca$) occur at about 638℃ and 514℃, respectively [12]. It is found from the Fig.3 that $Mg_2Sn$ and $MgZn_2$ phases in the grain boundary are refined with Ca increasing and grain size become smaller. This study is consistent with Yang [10].

Table 1 EDS results of as-cast alloys (molar fraction,%)

| element | A | B | C | D |
|---|---|---|---|---|
| Mg | 83.22 | 67.15 | 27.15 | 74.31 |
| Sn | 9.72 | 32.85 | 0 | 19.53 |
| Zn | 0 | 0 | 72.85 | 0 |
| Ca | 6.96 | 0 | 0 | 6.61 |
| total amount | 100 | 100 | 100 | 100 |

### 3.2 Mechanical properties

The tensile properties including ultimate tensile strength (UTS), 0.2% yield strength（YS）, elongation (Elong) are listed in Table 2. It is observed that with the Ca amount increasing, the ultimate tensile strength and elongation increase, and the 0.2% yield strength（YS）first increase and then decrease. This is possibly ascribed to that the excess Ca cases the formation of coarse primary CaMgSn phase. CaMgSn phases possibly promotes the initiation and propagation of cracks and leads to an adverse effect on the ultimate tensile strength. The maximum ultimate tensile strength and elongation in the Mg-5Sn-5Zn alloy with addition of 0.5 wt% Ca are achieved. Fig.5 shows SEM tensile fractographs of as-cast experimental alloys. It can be observed that fractograph a which are composed of a large number of little dimples is typical ductile fracture, and it has excellent toughness. As shown in the Fig.5, fractograph b and fractograph c including a number of cleavage planes (A and B) and tearing ridge indicates that the tensile fracture surfaces of the experimental alloys exhibit mixed characteristics of cleavage and quasi-cleavage fractures. The crack caused by coarse primary CaMgSn particle can been seen in the fractograph C.

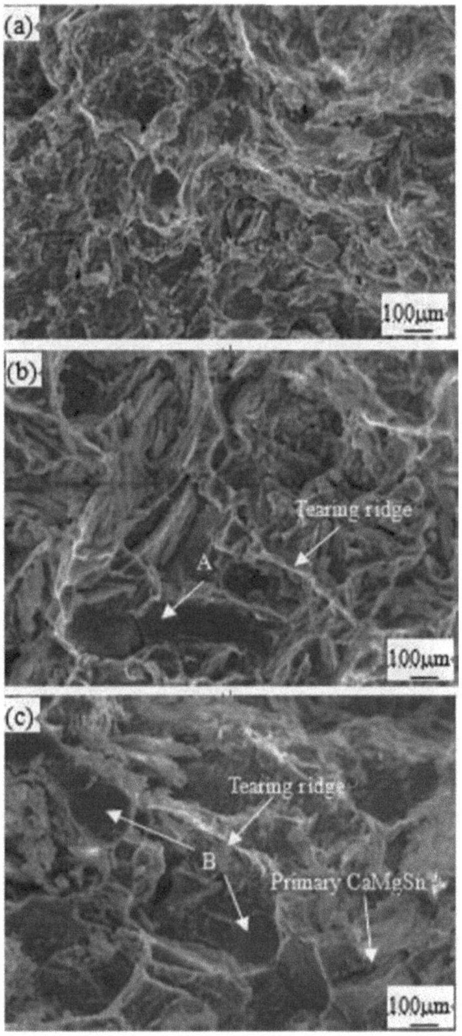


a)Mg-5Sn-5Zn-0.5Ca b)Mg-5Sn-5Zn-1Ca c)Mg-5Sn-5Zn-2Ca

Fig 5. SEM tensile fractographs of as-cast experimental alloys

Table 2 Tensile and properties of as-cast experimental alloys

| Alloy | Tensile property | | |
|---|---|---|---|
| | Uts/MPa | YS/MPa | Elong/% |
| Mg-5Sn-5Zn-0.5Ca | 184.72 | 85.71 | 3.29 |

| | | | |
|---|---|---|---|
| Mg-5Sn-5Zn-1.0Ca | 152.34 | 87.14 | 2.89 |
| Mg-5Sn-5Zn-1.0Ca | 123.28 | 52.13 | 1.05 |

## 4 Conclusions

1. The microstructures of Mg-5Zn-5Sn-xCa alloys are composed of CaMgSn phases, layer $MgZn_2$ phases and $Mg_2Sn$ phases. Matrix precipitates onset $Mg_2Ca$ phase, when the content of Ca is 2 wt%.

2. With the increasing of the Ca content, the needle CaMgSn phases gradually become rods, the layer $MgZn_2$ phases and plate $Mg_2Sn$ phases become continuous.

3. With the increasing of the Ca content, ultimate tensile strength and elongation decrease.

## References


[1]. G.Nayyeri, R.mahmudi, Enhanced creep properties of a cast Mg-5Sn alloy subjected to aging treatment [J]. Materials Science and Engineering A,2010:1-23.

[2]. REN Y L, RE Y, LIU Y Q. The research of high strength Mg-Zn-Ca-Al alloy [J]. Foundry, 2009, 58, (2): 101-103.

[3]. LIU H M, CHEN Y G, TANG Y B. The microstructure and mechanical properties of the as-cast Mg-Sn binary alloy [J]. Journal of Sichuan University, 2006, 38(2): 90-94.

[4]. ZHANG X P, CHEN Y G, XIAO S F. The microstructure and structure of the as-cast Mg-5Sn-xLa alloys[J]. Rare Metal Materials and Engineering, 2009,38:148-150.

[5]. LIU H M, CHEN Y G, TANG Y B, The microstructure, tensile properties, and creep behavior of as-cast Mg-(1-10)%Sn alloys[J]. Journal of Alloys and Compounds, 2007,440:122-126.

[6]. ZHAO H F, WANG L, YAN K. Effects of Zn addition on microstructure and mechanical properties of Mg-Sn alloy [J]. Rare Metal Materials and Engineering, 2010,39: 86-88.

[7]. LIU S F, FANG X M, WANG Z F. Effect of calcium on casting magnesium alloy [J]. Foundry, 2003, 52(4):246-248.

[8]. KOZLOV A, OHNO M, ABULELL T. Phase equilibria, thermodynamics and solidification microstructures of Mg-Sn-Ca alloys, Part 2: Prediction of phase formation in Mg-rich Mg-Sn-Ca cast alloys [J]. Intermetallics, 2008, 16: 316-321.

[9]. KIM DH, LEE J Y, LIM H K. The effect of microstruture evolution on the elevated temperature mechanical properties in Mg-Sn-Ca system [J]. Mater Trans, 2008, 49 (10): 2405-2413.

[10]. YANG M B, CHENG L, PAN F S. Effects of calcium addition on as-cast microstructure and mechanical properties of Mg-5Zn-5Sn alloy [J]. ScienceDirect, 2009, 20:769-775.

[11]. Bowles A L, Dieringa H, Blawert C, et al. Microstructural investigations of the Mg-Sn and Mg-Sn-Al alloy ststems [J]. Magnesiun Technology,2004:307-310.

[12]. YANG M B, PAN F S. Effects of Y addition on as-cast microstructure and mechanical properties of Mg–3Sn–2Ca (wt.%) magnesium alloy. Materials Science and Engineering A, 2009, 525:112-120.

# Phase dissolution of $\gamma$-$Mg_{17}Al_{12}$ during homogenization of as-cast AZ80 Magnesium alloy and its effect on room temperature mechanical properties

Rahul R. Kulkarni[1], Nityanand Prabhu[1], Peter D. Hodgson[2], Bhagwati P. Kashyap[1]
[1]Department of Metallurgical Engineering and Materials Science, Indian Institute of Technology Bombay, Mumbai, 400076, India
[2]Institute of Technology Research and Innovation, Deakin University, Waurn Ponds, Victoria 3217, Australia

Keywords: As-cast AZ80 Magnesium alloy, Homogenization treatment, Microstructure evolution, Mechanical properties

## Abstract

As-cast AZ80 Mg alloy contains α-Mg, partially divorce eutectic of α and γ ($Mg_{17}Al_{12}$), fully divorce eutectic of α and γ, and lamellar eutectic of α and γ phases. During homogenization, second phase (γ-$Mg_{17}Al_{12}$) gets dissolved can change the mechanical properties. Therefore, the aim of the present work is to bring out the kinetics of dissolution of γ phase and evaluate its effect on mechanical properties. Microstructure evolution during homogenization was investigated as a function of time for 0.5 to 100 h and at the temperatures of 400° and 439°C. In as-cast state, this material was found to contain 70% α-Mg and 30% eutectic phase. With increasing homogenization time, dissolution of lamellar eutectic occurs first which is followed by dissolution of fully divorce eutectic and partially divorce eutectic. The dissolution kinetics of γ phase was analyzed based on the decrease in its volume fraction as a function of time. The time exponent for dissolution was found to be 0.38 and the activation energy for the dissolution of γ phase was found to be 84.1 kJ/mol. This dissolution of γ phase leads to decrease in hardness and tensile strength with increase in homogenization time.

## 1. Introduction

Magnesium alloys are attractive material for the automotive, medical and electronic industries due their light weight, high specific strength, high stiffness and high damping capability. In addition, magnesium and its alloys are suitable candidates for temporary medical implants due to their biodegradable property [1]. Among all the magnesium alloys, Mg-Al-Zn alloys are the most widely used wrought alloys at present due to their low cost [2]. Earlier studies on as-cast Mg-Al-Zn alloy show the presence of cellular product of α-Mg and γ-$Mg_{17}Al_{12}$ with partially and fully divorce eutectic of the same [3]. The cellular structure is characterized by lamellar morphology; the partially divorce eutectic is characterized by 'island' of α-Mg within γ-$Mg_{17}Al_{12}$ and the fully divorce eutectic is characterized by the single γ-$Mg_{17}Al_{12}$ phase/grain surrounded by α-Mg. These morphologies develop as a result of differential cooling rate during casting. The γ-$Mg_{17}Al_{12}$ is an intermetallic compound whereas the α-Mg is a solid solution of Al in Mg, with the former being harder phase than the latter [7]. The study of homogenization heat treatment of as-cast AZ91 alloy shows the dissolution of the second phase i.e. $Mg_{17}Al_{12}$ [4], which influences the mechanical properties of this alloy.

In the present work, microstructure evolution during homogenization of as-cast AZ80 Mg alloy is studied with an attempt to bring out the kinetics and mechanism of dissolution of γ phase. The progress in homogenization is evaluated in terms of change in mechanical properties of this alloy at room temperature.

## 2. Experimental details

2.1 Material:

The material used for study is as-cast AZ80 Mg alloy with chemical composition given in table I.

**Table I Chemical Composition of AZ80 Mg alloy**

| Element | Al | Zn | Mn | Si | Mg |
|---|---|---|---|---|---|
| wt% | 7.7 | 0.4 | 0.15 | 0.2 | Balance |

2.2 Microstructure study:

The metallographic specimen of size 10 mm x 10 mm x 2 mm was cold mounted and mechanically polished with the diamond paste of 0.25 μm at the final stage. After polishing, the sample was cleaned with ethanol using ultrasonic cleaning. Etching was done with 10 ml HF and 90 ml water (which makes α-Mg as white and γ-$Mg_{17}Al_{12}$ as dark) and the same was examined under optical microscope at 1000X and in Hitachi S-3400N scanning electron microscope (SEM). To confirm the phases, energy dispersive spectrometry (EDS) attached with SEM was carried out.

2.3 X-ray diffraction study:

X-ray diffraction study was carried out on the sample of size 2 x 2 x 1 mm$^3$. Cu-K$_\alpha$ source was used in X'pert model Philips made XRD machine over the scanning range of 2θ equal to 5 to 90°.

2.4 Homogenization heat treatment:

Phase diagram of Mg-Al (see Figure 5 later) shows eutectic transformation at 437°C with aluminum composition around 33% in Mg. To know the transformation temperature range, differential scanning calorimetry (DSC) was used. To avoid the partial melting of segregated solute, homogenization heat treatment was carried out at 400° and 439°C over the time period of 0.5 hr to 100 hr. Samples were heated to these temperatures for particular time period and then quenched in water. Each sample was then polished and etched for metallography, and the phase analysis was done using Olympus-image analyzer at 100X. The volume fraction of γ-$Mg_{17}Al_{12}$ phase was measured over 10 fields within the sample. The fraction transformed was calculated using the statistical analysis for error bar estimated at 95% confidence level.

2.5 Mechanical Testing:

Tensile specimens of Φ 4.5 mm x 16mm gauge length were prepared. These were homogenized at 400° and 439°C for 3 hr. Tensile test was carried out at the constant initial strain rate of $10^{-4}$ $s^{-1}$ at room temperature using an Instron Universal testing machine with 5 kN capacity.

Macro-hardness was measured in HRt scale with 15 kg load and dwell time of 15 sec at five different places in each of the homogenized samples. An average of five readings for each sample was taken for analysis. The Vickers's micro-hardness was measured in the individual phases to know the relatively harder and softer phases. During this hardness measurement, the load was kept at 100 gm for 15 sec. Total 10 readings were taken in each phase. The mean microhardness (VHN) values are presented with error bar better than 4% at the confidence level of 95%.

## 3. Results and Discussion

3.1 Microstructure study:

As cast microstructure consists of mainly three parts α-Mg (both primary and eutectic), γ-eutectic particles and lamellar γ-precipitates (cooling rate of casting being sufficiently low, discontinuous precipitation can occur) which are shown in Figure 1 and Figure 2

When an alloy is solidified, it follows the primary solidification metallurgy. α-Mg will grow dendritically by rejecting the Al and Zn into the inter-dentritical liquid. When this region accumulates the Al, which is having eutectic composition, γ-$Mg_{17}Al_{12}$ will grow with further under-cooling of the alloy. This is known as divorce eutectic solidification [5]. This eutectic growth morphology has been studied by Nave et.al [3]. According to them, the morphologies were developed because of the differential cooling rate during solidification.

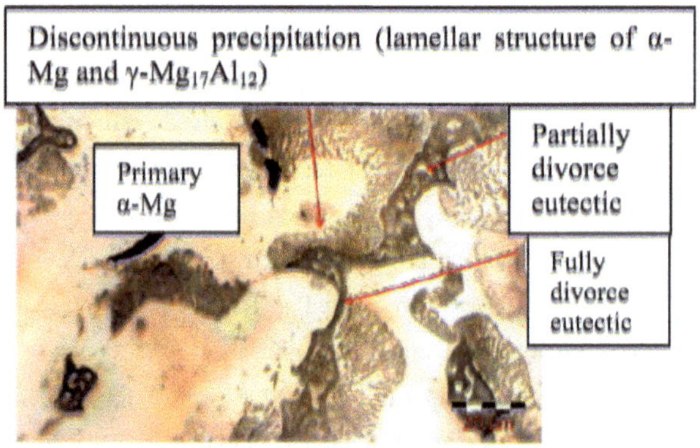


**Figure 1. Optical micrograph showing the as-cast structure of AZ80 Mg alloy**

The chart below shows that the eutectic growth in as-cast structure consists of two parts (i) fully divorce eutectic, (ii) partially divorce eutectic.

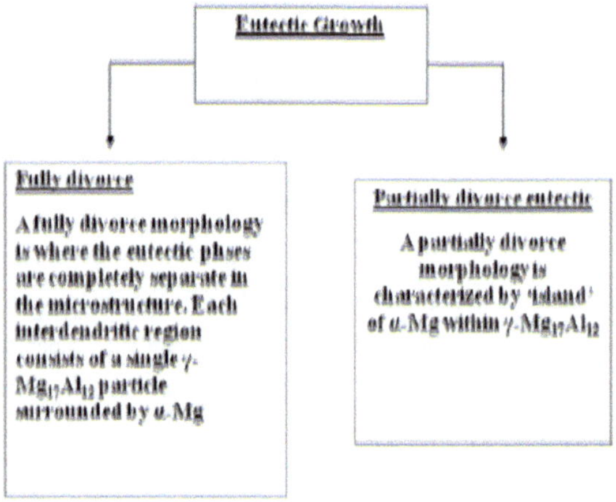

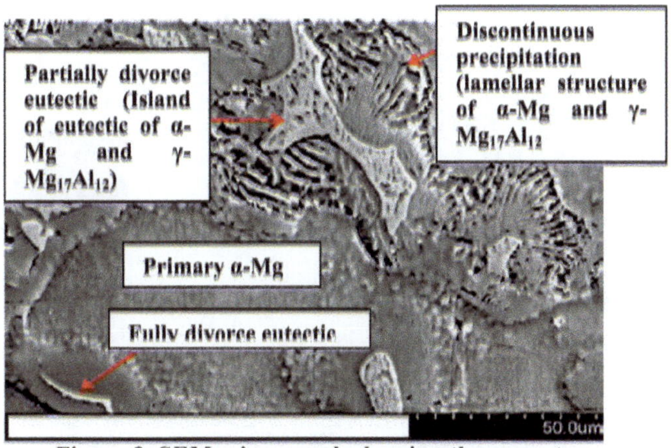


**Figure 2. SEM micrograph showing the as-cast structure of AZ80 Mg alloy**

EDS analysis (figure 3) confirms the phases α-Mg and $Mg_{17}Al_{12}$ in the form of divorce eutectic with different morphologies. Atomic % of elements in each of these phases are given in table 2. Actual ratio of Mg:Al = 60.39: 37.57 = 1.61 nearly matches with standard ratio of Mg:Al = 17:12 = 1.42, which confirms the intermetallic compound to be $Mg_{17}Al_{12}$.

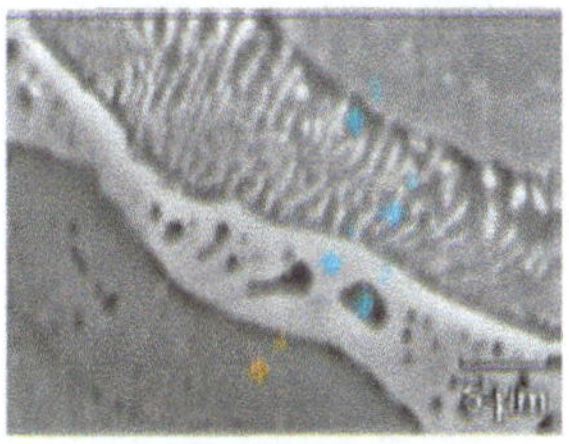

**Figure 3. SEM-EDS point analysis with composition stated in table II**

**Table II Elements in atomic % (SEM-EDS)**

| | Mg-K | Al-K | Si-K | Mn-K |
|---|---|---|---|---|
| pt1 | 60.39 | 37.57 | 0.03 | 0.08 |
| pt3 | 81.79 | 11.57 | 0.11 | 0.00 |

3.2 X-ray diffraction study:

Peaks in the figure 4 shows that $Mg_{17}Al_{12}$ is present in proportion less than that of Mg.

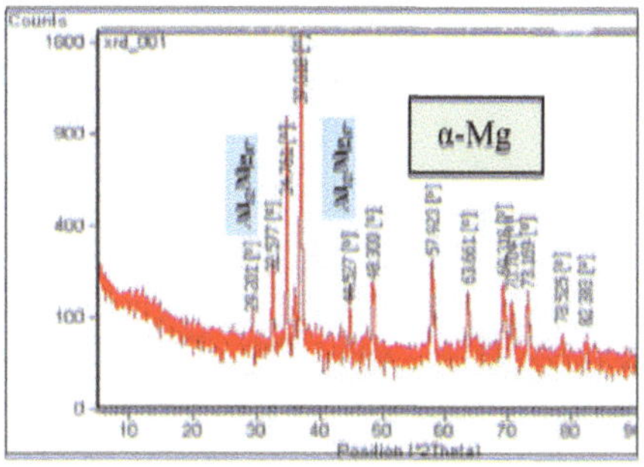


**Figure 4. XRD pattern for as-cast AZ80 Mg alloy**

3.3 Effect of Homogenization heat treatment:

The as-cast material is inhomogeneous in terms of phase size, phase morphology and their distributions. Also, from the phase diagram (Figure 5), it is observed that above the solvus line, the eutectic (a mixture of α-Mg and β-$Mg_{17}Al_{12}$) converts into the single phase α-Mg having maximum solubility of Al in Mg. So, precipitation behavior was studied at the temperatures below the transformation temperature and at the transformation temperature.

The DSC curve was obtained at the heating rate of 10°C/min (Figure 6). The endothermic peak in the DSC curve shows transformation from two phases α and γ to single phase α-Mg at 439°C

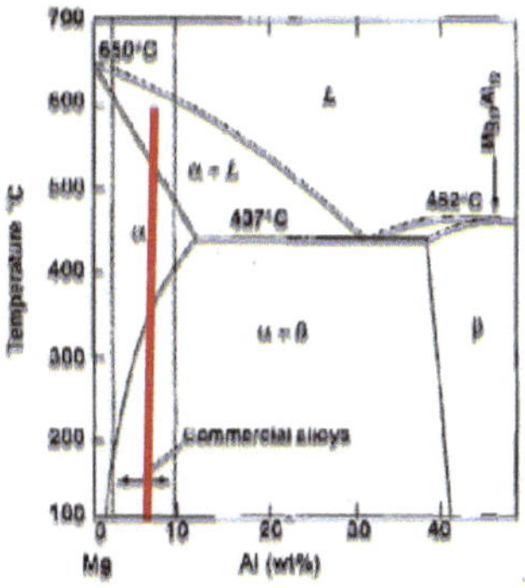


**Figure 5. Magnesium rich side of Mg-Al phase diagram with the Al content in AZ80 marked by bold solid line**

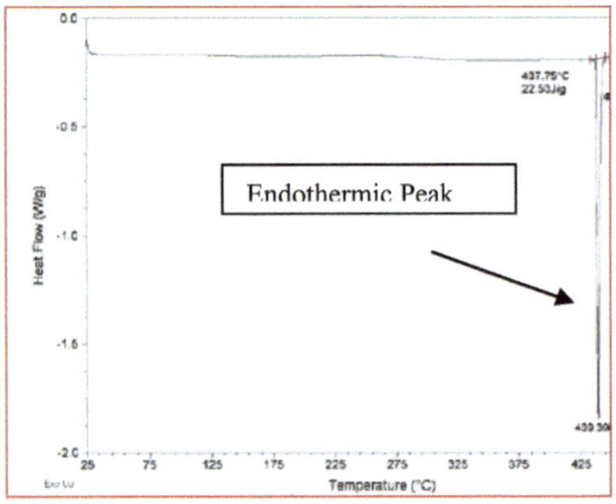


**Figure 6. DSC curve shows the endothermic peak at 439°C**

Figure 7 shows different microstructures of homogenized samples in terms of fraction transformed plotted as a function of time at 400°C. It can be seen from the microstructures that as the time period increases the more and more precipitates get dissolved; after 100 hr it is almost fully dissolved. The data points are found to fit into the sigmoidal curve with $R^2 = 0.93$. The same analysis was done at 439°C over the homogenization periods of 0.5, 1, 3, 24, 72, and 100 hr. These data points also fit well into similar sigmoidal curve (Figure 8). It can be seen that dissolution time required is less at 439°C as compared to that at 400°C which is due to the fact that diffusion for dissolution becomes faster at higher temperature.

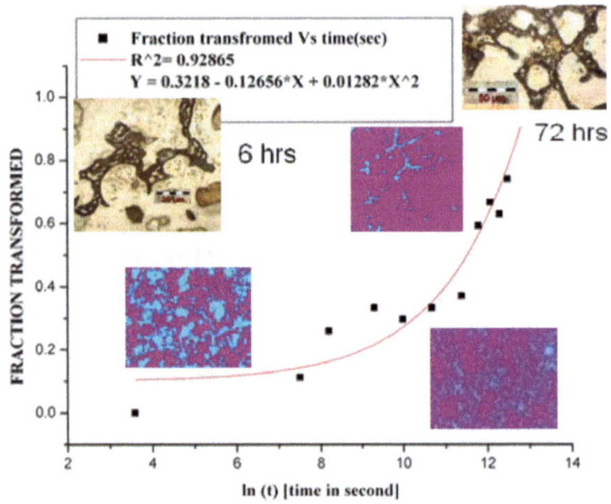


**Figure 7 Change in microstructure of homogenized samples plotted in terms of fraction transformed as a function of time at 400°C**

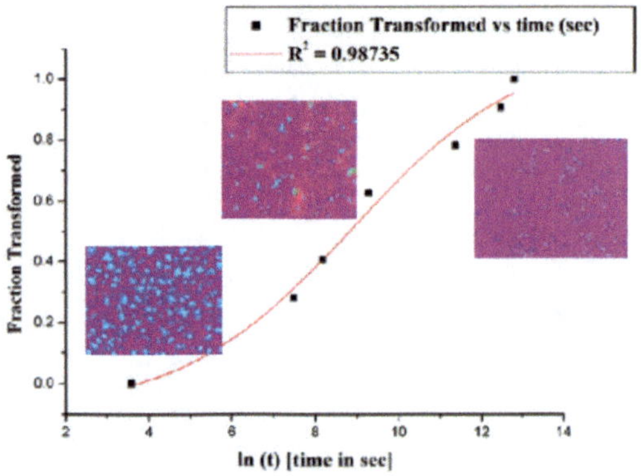


**Figure 8 Change in microstructure of homogenized samples plotted in terms of fraction transformed as a function of time at 439°C**

Further, it is observed that with increasing homogenization time, dissolution of lamellar eutectic occurs first, which is followed by dissolution of fully divorce eutectic and partially divorce eutectic (figure 7). The dissolution process is associated with diffusion. The concentration gradient is not uniform in partially divorce eutectic which may be a possible reason that the partially divorce eutectic takes more time for dissolution.

To study the kinetics of dissolution, Johnson-Mehl-Avrami- Kolmogorov (JMAK) equation for phase transformation [6] as given below was used-

$$X(t) = 1 - \exp\{-((k(T)t)^n\}$$

Here, $k(T)$ represents temperature dependent isothermal rate constant and n represents the time exponent. To calculate these parameters, log log (1/1-X) vs. log t was plotted from the data at 400°C and 439°C. From the plots presented in Figure 11, the value of k (T) is found to be 4.6 x $10^{-6}$ $s^{-1}$ and 5.68 x $10^{-7}$ $s^{-1}$, respectively. The value of n is determined to be 0.37 and 0.39. These values of n lie between 0 and 1, which is the indication that the rate of transformation decreases with increasing time.

$k(T)$ is taken to follow Arrhenius type relationship

$$k(T) = k_0 \exp\left(\frac{-Q_{eff}}{RT}\right)$$

where $Q_{eff}$ is the effective activation energy for transformation.

The value of $Q_{eff}$ is obtained to be 84.1 kJ/mol from vs. 1/T in figure 12. This value of $Q_{eff}$ is closer to the activation energy grain boundary diffusion in Mg (92 kJ/mole) [8]

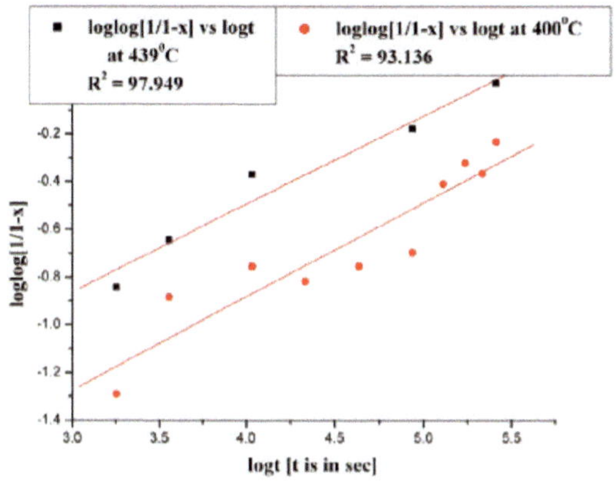


**Figure 11 plot of loglog (1/1-x) vs. log t at 400 and 439°C**

3.4 Variation in Hardness and Tensile Properties:

Micro-hardness test of the phases shows that the γ-$Mg_{17}Al_{12}$ (VHN = 89.77+/-3.59) is harder than α-Mg (VHN = 67+/-3.72). There is a lot of variation in macro-hardness over the homogenization period at both the temperatures of 400 and 439°C (Figure 12), which can be attributed to the variation in the distribution of second phase.

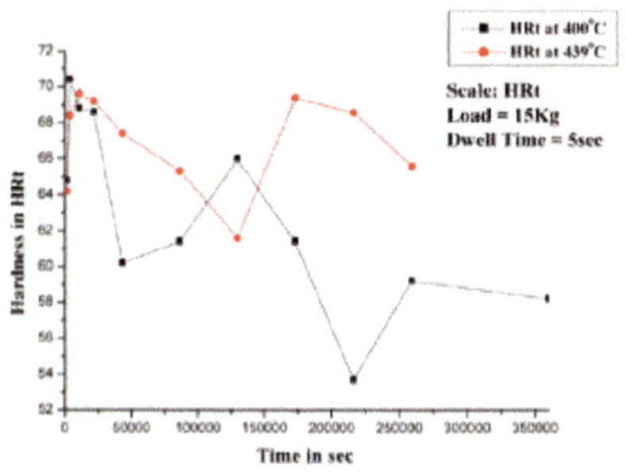


**Figure 42 Variation in hardness as a function of dissolution time at 400 and 439°C**

Tensile tests of samples subjected to isochronal annealing of 3 hr at 400 and 439°C show that the yield and ultimate tensile strengths are higher at 400°C than that at 439°C. This is possibly due to the decrease in the phase proportion of second phase as the temperature is increased. The proportion of second phaseγ-$Mg_{17}Al_{12}$ at 400°C (24%) is greater than at 439°C (13%). However, the tensile ductility is found to be less in case of the specimen which is homogenized at 439°C for 3 hrs (14%) than that at 400°C for 3 hrs (7%). This may be due to the distribution of γ-$Mg_{17}Al_{12}$ second phase which is more uniform for the specimen homogenized at 400°C than that of 439°C.

**Table 3 Tension Test Data**

| Tensile properties | 0.2% proof stress (MPa) | Ultimate Tensile strength (MPa) | Braking stress (MPa) | Elongation % |
|---|---|---|---|---|
| 400°C (3hr) | 109 | 179 | 178 | 14 |
| 439°C (3hr) | 55 | 78 | 75 | 7 |

In the present study, the solutionizing of γ-$Mg_{17}Al_{12}$ phase was done for the period of 0.5 to 100 hrs only at 400° and 439°C. The dissolution behavior at both the temperatures are seen to be similar with n = 0.37 to 0.39. However, the understanding of the mechanism for dissolution requires the extension of this study to a range of temperatures, in which case a precise value of the activation energy could bring some insight, instead of that based on the limited data only at the above two temperatures.

## 4. Conclusions

1. As cast AZ80 alloy consists of α-Mg, partially and fully divorce eutectic of α and γ ($Mg_{17}Al_{12}$), and lamellar eutectic of α and γ phases.
2. Phase proportion of γ - $Mg_{17}Al_{12}$ decreases with increase in time at both the temperatures of 400 and 439°C. The decrease of γ phase is associated with diffusion process.
3. With increasing homogenization time, dissolution of lamellar eutectic occurs first which is followed by dissolution of fully divorce eutectic and partially divorce eutectic.
4. The kinetics of transformation follows sigmoidal curve, which is an indication of time dependent phenomenon. The values of time exponent n are found to be 0.37 and 0.39 for homogenization temperature 400 and 439°C respectively. These values lying between 0 and 1, which suggest the rate of transformation decreases with increase in time. The effective activation energy for transformation was obtained to be 84.1 kJ/mol.

5. Homogenization at 439°C shows lower yield strength of 55 MPa and ultimate tensile strength of 78 MPa as compared to the corresponding values of 109 and 179 MPa obtained at 400°C. The room temperature elongations to failure after the above treatment were 7% and 14% respectively.

## 5. References


1. Hamid Reza Bakhsheshi Rad, Mohd Hasbullah Idris , Mohammed Rafiq Abdul Kadir, Saeed Farahany, *"Microstructure analysis and corrosion behaviour of biodegradable Mg-Ca implant alloys"*, Mat. Design, (2011) 1-30.
2. H.T. Zhou, Q.B. Li, Z. K. Zhao, Z.C. Liu, S.F.Wen, Q.D.Wang, *"Hot workability characteristics of magnesium alloy AZ80—A study using processing map"*, Mater. Sci. Engg. A, 527 (2010) 2022–2026.
3. M. D. Nave, A. K. Dahle and D.H. StJohn, *"Eutectic growth morphology in Magnesium-Aluminium alloys"*, Magnesium Technology, the minerals, metals materials society, 2000, 233-242.
4. Tianping Zhu, Zhan W. Chen, and Wei Gao, *"Dissolution of Eutectic β-$Mg_{17}Al_{12}$ Phase in Magnesium AZ91 Cast Alloy at Temperatures Close to Eutectic Temperature"*, Journal of Mater. Engg and Performance, 19 (2010), 860-867.
5. Tianping Zhua, , Zhan W. Chenb, Wei Gaoa, *"Microstructure formation in partially melted zone* during gas tungsten arc welding of AZ91 Mg cast alloy", mater. Charact., 59 (2008) 1550 – 1558.
6. A. K. Jena, M. C. Chaturvedi, "*Phase transformation in Materials*", prentice Hall, new Jersey, 1992.
7. I.A. Yakubtsova, B.J. Diaka, , C.A. Sagera, B. Bhattacharyab, W.D.MacDonaldc,1, M. Niewczasb, *"Effects of heat treatment on microstructure and tensile deformation of Mg AZ80 alloy at room temperature"*, Mater. Sci. and Engg. A 496 (2008) 247–255
8. B.H. Lee1, K.S. Shin and C.S. Lee1, *"High Temperature Deformation Behavior of AZ31 Mg Alloy"*, Mater. Sci. Forum 475-479 (2005) pp. 2927-2930

**Magnesium Technology 2012**
*Edited by: Suveen N. Mathaudhu, Wim H. Sillekens, Neale R. Neelameggham, and Norbert Hort*
***TMS (The Minerals, Metals & Materials Society), 2012***

# PRECIPITATE FORMATION AND GRAIN REFINEMENT OF MG-AL-SN ALLOY DURING HOT DEFORMATION

Abu Syed Humaun Kabir, Jing Su, Phuong Vo, In-Ho Jung and Stephen Yue

Department of Mining and Materials Engineering, McGill University, 3610 University Street, Montreal, Canada H3A 2B2

Keywords: Hot deformation, Precipitation, Dynamic recrystallization, Grain refinement

## Abstract

Magnesium alloys are very popular in the automobile industry due to its high strength to weight ratio. However, the use of commercial magnesium alloys as sheet is limited by room temperature ductility. One way to improve the ductility may be to form precipitates during hot rolling. These may delay dynamic recrystallization, possibly leading to grain refinement, which is known to improve ductility. Equilibrium diagrams obtained from thermodynamic modeling software FactSage were used to design Mg-3Al-2Sn alloy, which should form $Mg_2Sn$ precipitates during hot rolling temperatures around 300 °C. To investigate this prediction, the alloy was cast in a copper mould and precipitates formation characteristics were studied by using optical microscope (OM), scanning electron microscope (SEM) equipped with an energy-dispersive X-ray spectroscopy (EDS).

## Introduction

The reduction of energy consumption and environmental pollution are driving forces in the automotive industry to improve fuel efficiency through light weight automobiles. Magnesium, being the eighth most abundant element in the earth crust and third most dissolved in seawater [1], is the lightest structural metal with the highest specific strength and a potential candidate to reduce the most weight of the vehicle. Although 36 % and 78 % lighter than aluminum and steel [2], respectively, very limited wrought magnesium alloys are commercially available in comparison to aluminum and steel [3].

To be used as a structural component in the automobile, alloys should have sufficient room temperature ductility and specific strength. However, magnesium shows brittle fracture with less than 10 % elongation at room temperature due to its HCP structure [4]. For improved formability at room temperature, a refined grain magnesium alloy not only shows better strength and ductility but also a lower ductile brittle transition temperature (DBTT) [5]. It is well established that strength of polycrystalline metals at room temperature can be improved by grain refinement according to the Hall-Petch relationship [6] as the higher grain boundary area impedes dislocation movement [5].

In Mg alloys, grain refinement has been accomplished by thermomechanical processing. For example, Mohri *et al.* [7] reported improved strength and ductility of Mg-4Y-3RE alloy due to grain refinement (as small as 1.5 µm) by hot extrusion and, Kumar *et al.* [6] produced a grain size less than 5 µm for AZ91 alloy by dynamic recrystallization (DRX) during high temperature extrusion. Recrystallization is an important mechanism by which grain refinement can be achieved in metals with low or medium stacking fault energy, such as magnesium [8]. Recrystallization is a restoration and usually thermally activated process by which relatively strain free grains are formed within the deformed structure consuming the deformed grains [9]. During the hot deformation process the term 'dynamic' is used to distinguish from the 'static' recrystallization process, which takes place after the deformation or during the post deformation annealing process [10]. When a critical deformation condition, i.e. critical dislocation density, is reached, dynamic recrystallization is initiated by the bulging of pre-existing grain boundaries at low strain rate [11]. Recrystallization can be retarded by the pinning of precipitates at the grain and sub-boundaries and also individual dislocations. For grain boundary pinning to occur, it is necessary to have sufficient amount of precipitates and the precipitates size should be smaller than a critical value [12].

In this study, an Mg-3Al-2Sn alloy was thermomechanically processed to generate precipitates during hot deformation with a view to affecting dynamic recrystallization (DRX). Aluminum is the most widely used alloying element with magnesium as it improves the strength, hardness, corrosion resistance and castability of pure magnesium [13]. Tin with a small amount of aluminum increases the ductility of the magnesium alloy, and Sn in any magnesium alloy reduces the risk of hot tearing during the extrusion and forging process [3]. Finally, the addition of Sn in magnesium alloy produces thermally stable $Mg_2Sn$ precipitates [3].

## Experimental Procedures

Thermodynamics software FactSage [14] was used to design candidate Mg alloys with the alloying elements Al and Sn. Commercially pure Mg, Al and Sn were melted using an induction furnace (20 kW, 5 kHz) in a graphite crucible and five ingots were cast in a copper mold preheated to 250 °C. The alloy addition and casting were done at 690 °C and 730 °C respectively with a mixture of $CO_2$ (99.5 %) and $SF_6$ (0.5 %) gases used to protect the liquid metal. The cast ingot was homogenized to dissolve all the precipitates. The aim is to form the precipitates again during the hot deformation process. The temperature for homogenization was selected using the FactSage generated equilibrium diagram.

Compression samples of 11.4 mm in height and 7.6 mm in diameter (h/d ratio of 1.5) were machined from the homogenized bar. Isothermal compression testing was done at 300 °C with a strain rate of 0.01 $s^{-1}$ to strains of 0.2, 0.3, 0.4, 0.5, 0.7, and 0.9. Uniaxial compression test was performed using an MTS 810 coupled with a radiant furnace. The specimens were held at 300 °C for 10 minutes prior to deformation and water quenched immediately after the deformation.

Specimens for metallurgical examination were cut along the compression axis. Polished specimens were etched in a fresh solution of 10 ml acetic acid, 4.2 gm picric acid, 10 ml water and 70 ml alcohol for 3-8 seconds. Microstructural examination was carried out using a Nikon Epiphot 200 optical microscope with Clemex image analysis software. Grain size has been measured by linear intercept method. A Philips XL30 FEG SEM with energy dispersive X-ray spectroscopy (EDS) was used to examine the microstructure and characterization of the second phases. Microindentation hardness was measured by Clark CM-100AT machine with a load of 50 gram and a dwell time of 10 seconds.

## Results and Discussion

### Alloy design

The Mg-Al-Sn alloy was designed to target the homogenization and deformation temperatures, which are around 400 °C and 300 °C, respectively. Figure 1 shows the equilibrium phase fraction obtained from FactSage indicating the types and amount of precipitates formed during the equilibrium cooling of these two alloys. According to this diagram, a suitable homogenization (solution treatment) temperature is around 400 °C with secondary particles $Mg_2Sn$ and $Mg_{17}Al_{12}$ starting to precipitate out from the Mg matrix below 340 °C and 210 °C, respectively. However, the exponential dependence of nucleation rate on undercooling indicates that the precipitation phenomena usually requires a minimum amount of undercooling to produce the necessary driving force to nucleate a particle [15]. Consequently, the deformation temperature was selected at 300 °C to allow about 40-50 °C of undercooling. More uniform and fine precipitates are formed with plastic deformation, which can generate high energy sites like dislocations, twins and deformation bands. These defects can accelerate the diffusion of solute atoms enhancing the subsequent aging kinetics and are the preferential location for the solute atoms to gather due to the minimum of energy needed. Second phases are preferentially formed at the grain boundaries during the thermomechanical processing but may also form inside the grain when annealing treatment is associated with the thermomechanical processing [16].

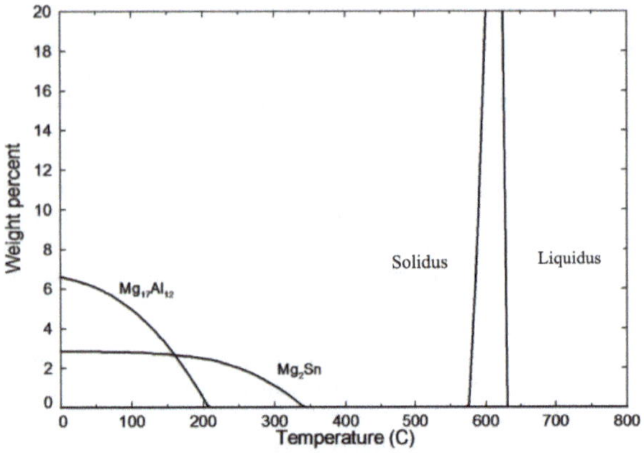


Figure 1: Equilibrium phase fraction of Mg-3Al-2Sn alloy predicted by FactSage.

### Microstructure

Figures 2 and 3 show the as-cast optical and SEM microstructures of the alloy and EDS analysis of precipitates. The microstructure consists of primary α-Mg, and secondary particles between dendritic arms. The average grain size was measured as 78 μm. The precipitates are coarse having the size around 2-5 μm. EDS analysis confirms that the precipitates are mainly composed of Mg, Al, and Sn.

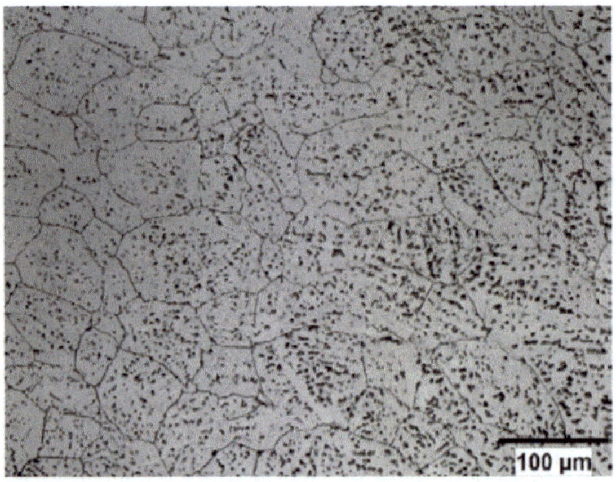


Figure 2: As-cast optical microstructure of Mg-3Al-2Sn alloy

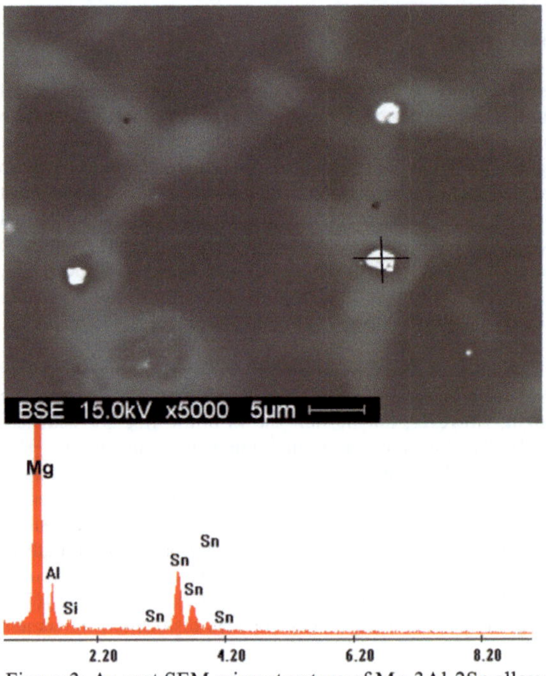


Figure 3: As-cast SEM microstructure of Mg-3Al-2Sn alloy and EDS quantitative analysis of the precipitates.

Although thermodynamically the precipitates were supposed to be dissolved above 340 °C, the dissolution of the precipitates in the matrix at this temperature is slow. In order to promote the homogenization, the temperature was selected as 420 °C. The samples were homogenized for 2, 4, 6, 8, 10 and 12 h to observe the dissolution behavior of the precipitates. No precipitate has been found after homogenization at 420 °C and for 10 h, as shown by the SEM microstructure in Figure 4-a. The average grain size of the homogenized sample is measured as 80 μm, which displayed almost no grain growth during the homogenization treatment. Before compression, the homogenized samples were held at the deformation temperature of 300 °C for 10 minutes to ensure a uniform temperature. Figure 4-b shows the SEM microstructure of the sample quenched immediately after the 10 minute hold and before compression. It is seen that the precipitates have started to form at the grain boundaries.

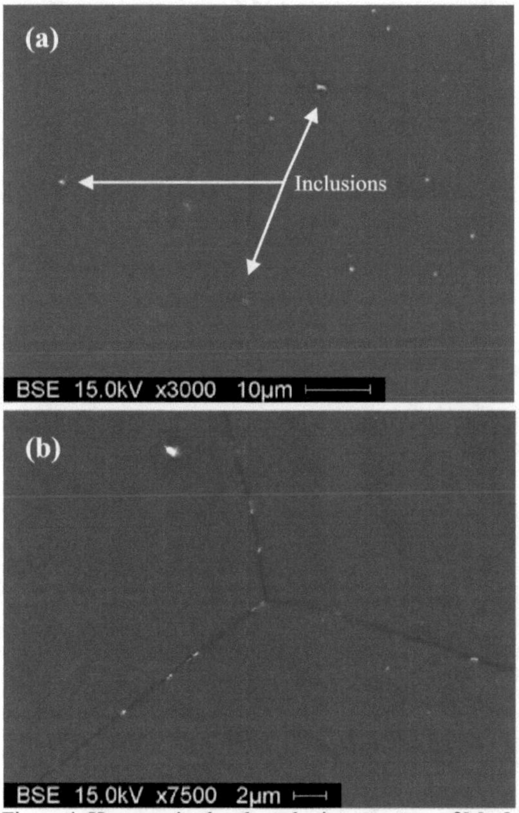


Figure 4: Homogenized and aged microstructure of Mg-3Al-2Sn alloy (a) Homogenized (b) Aging at 300 °C for 10 min and quenched

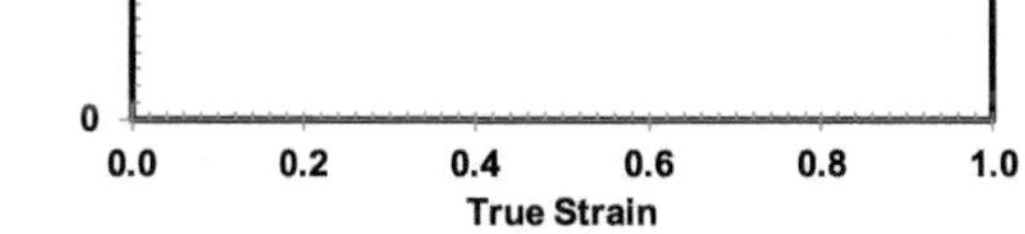


Figure 5: Flow behavior of Mg-3Al-2Sn alloy at 300 °C and strain rate of 0.01 $s^{-1}$.

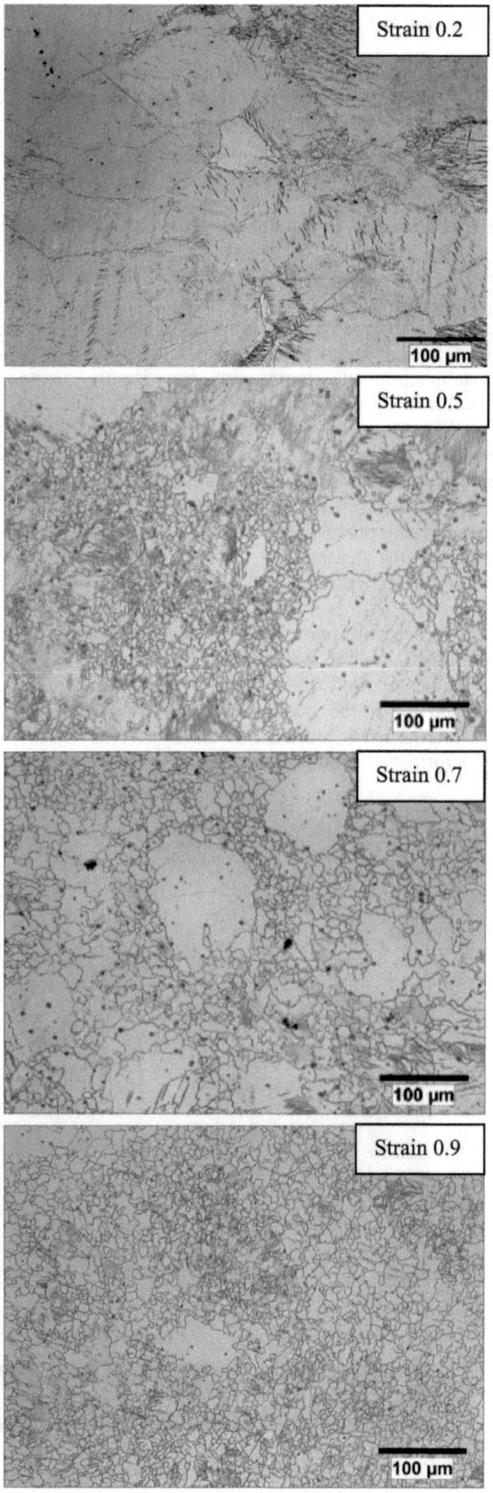


Figure 6: Microstructure evolution with different amount of deformation

## Deformation behavior

The typical true stress and true strain curve obtained at a strain of 0.9 is shown in Figure 5. At 300 °C, the material exhibits work hardening, a single peak at low strain, followed by flow softening. The softening behavior after peak stress is attributed to dynamic recrystallization phenomena [10].

## Microstructure evolution during deformation

Figure 6 shows the microstructures of hot deformed alloys subjected to various strains. A necklace type structure was observed with recrystallized grains distributed along pre-existing boundaries. The microstructures are characterized by dynamic recrystallization and grains are significantly refined. The recrystallized grains evolved dynamically as the static recrystallization was prevented by quenching the samples immediately after deformation. Dynamic recrystallization is attributed to magnesium alloy because of its low stacking fault energy (~ 78 mJ $m^{-2}$) and limited slip systems [8]. In general, for the hot working process, the dynamic recrystallization grain size depends on the deformation amount (strain), strain rate, and deformation temperature [6, 17, 18].

As can be seen in Figure 7, the DRX grain size increases with increasing strain up to strain of 0.7. The result is consistent with the results reported by Beer [19] for AZ31 alloy. Beer mentioned an initial increase in grain size with strain but reported to be fairly constant for strains more than 0.6 for AZ31 alloy deformed at a temperature of 400 °C and a strain rate of 0.1 $s^{-1}$. Figure 7 also shows that the % DRX increases with the increase in strain which indicates not only the occurrence of dynamic recrystallization but also the development of dynamic recrystallization with increasing the strain.

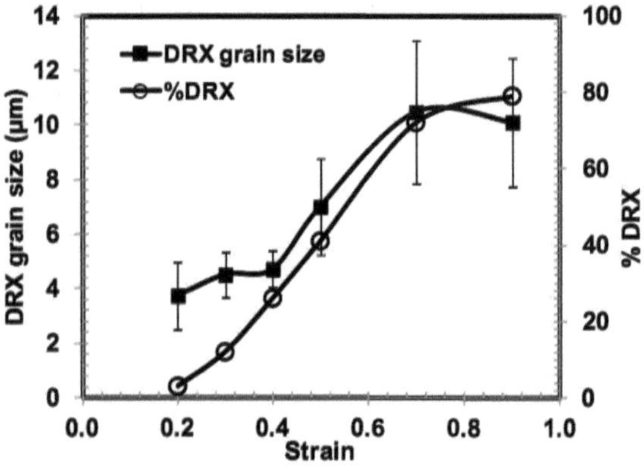


Figure 7: Variation of % DRX and DRX grain size with different strain.

Figure 8 shows the distribution of precipitates in the deformed alloy with precipitates 100-300 nm in size and mainly observed along the grain boundaries. The fine dispersion of precipitates may restrain grain growth by pinning the boundaries and dislocations.

Figure 8: Grain boundary pinning effect by precipitates (300 °C, strain rate 0.01$s^{-1}$, strain 0.9).

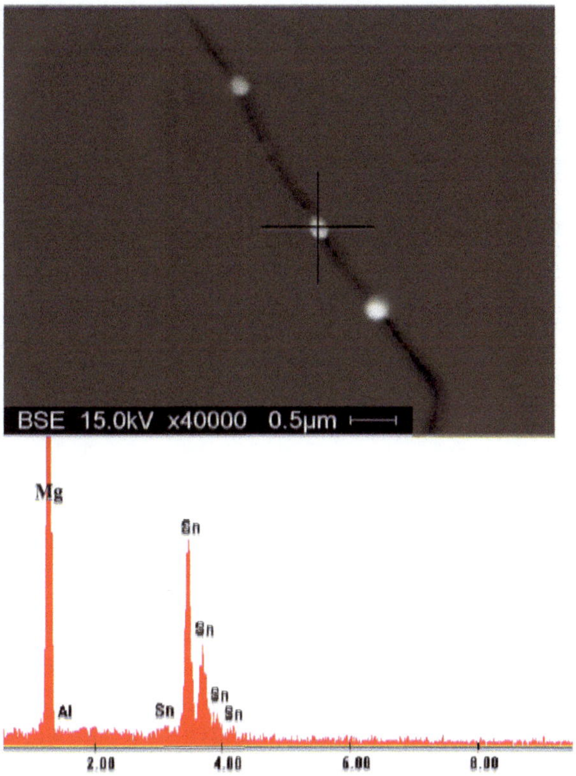

Figure 9: Precipitate formation at DRX grain boundary and EDS qualitative analysis of the precipitate.

The EDS study (Figure 9) confirms that the precipitates are mostly composed of Mg and Sn, which is consistent with the $Mg_2Sn$ precipitate predicted by the thermodynamic calculation results from FactSage at 300 °C. The onset of dynamic recrystallization occurs during deformation when the strain exceeds a critical value, referred to as the critical strain for dynamic recrystallization [20], which is slightly lower than the peak strain on the flow diagram [21]. The formation of precipitates at the DRX grain boundary begins at ~0.4 strain in this study which is higher than the critical strain for the recrystallization (~0.15-0.18). In order to start the precipitate formation at the critical strain for DRX, future work includes deformation at a lower temperature such as 250 °C. At this temperature, the weight fraction of precipitate is expected to be double than that obtained at 300 °C (Figure 1).

Figure 10 shows the variation of room temperature hardness with strains. Yang *et al.* [22] reported the same trend of hardness change with strain for AZ31 alloy but deformed at different temperature and strain rate. The increase in hardness is due to the increase in vol % of DRX grains i.e. more fine grains development due to DRX, according to the Hall-Petch relationship.

**Summary**

1. The experimental result supports the FactSage calculation in terms of homogenization temperature and precipitation prediction during deformation at 300 °C.

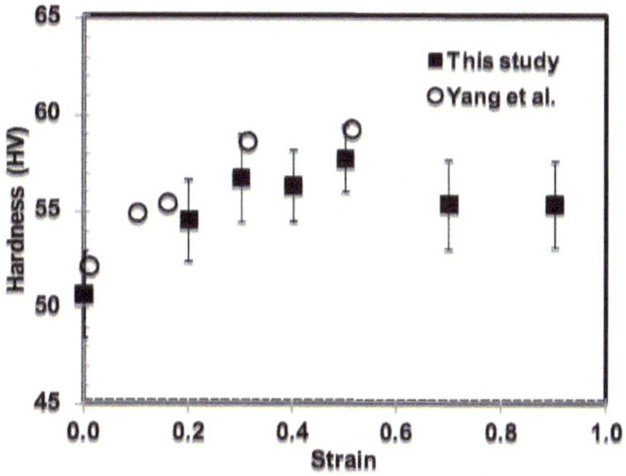

Figure 10: Variation of room temperature hardness with strain.

2. Amount of (volume percent) dynamic increases with increasing amount of deformation i.e. strain.
3. DRX grain size increases up to a certain strain then remain fairly constant with increasing strain.
4. Precipitation formation along the DRX grain may be associated with the grain refinement.
5. DRX grain boundary can be pinned by $Mg_2Sn$ precipitates formed during deformation and dynamic recrystallization.
6. Microindentation hardness increases with the increase in deformation.

**Acknowledgement**

The authors would like to thank General Motors (GM) and NSERC, Canada for the financial support.

**References**

1. K. Ozturk, Y. Zhong, Z.-K. Liu, and A. Luo, "Creep resistant Mg-Al-Ca alloys: Computational thermodynamics and experimental investigation", *JOM Journal of the Minerals, Metals and Materials Society*, 55(11) (2003), 40-44.
2. H. Watari, K. Davey, M. T. Rasgado, T. Haga, and S. Izawa, "Semi-solid manufacturing process of magnesium alloys by twin-roll casting", *Journal of Materials Processing Technology*, 155-156, (2004), 1662-1667.
3. H. E. Friedrich and B. L. Mordike, "Magnesium Technology: Metallurgy, Design Data, Applications" (2006), Springer.
4. T. Mukai, M. Yamanoi, H. Watanabe, and K. Higashi, "Ductility enhancement in AZ31 magnesium alloy by controlling its grain structure", *Scripta Materialia*, 45(1) (2001), 89-94.
5. T.-C. Chang, J.-Y. Wang, C.-M. O, and S. Lee, "Grain refining of magnesium alloy AZ31 by rolling", *Journal of Materials Processing Technology*, 140(1-3) (2003), 588-591.
6. N. V. Ravi Kumar, J. J. Blandin, C. Desrayaud, F. Montheillet, and M. Suéry, "Grain refinement in AZ91 magnesium alloy during thermomechanical processing", *Materials Science and Engineering A*, 359(1-2) (2003), 150-157.
7. T. Mohri, M. Mabuchi, N. Saito, and M. Nakamura, "Microstructure and mechanical properties of a Mg-4Y-3RE

alloy processed by thermo-mechanical treatment", *Materials Science and Engineering: A*, 257(2) (1998), 287-294.
8. H. Watanabe, H. Tsutsui, T. Mukai, K. Ishikawa, Y. Okanda, M. Kohzu, and H. Kenji, "Grain size control of commercial wrought Mg-Al-Zn alloys utilizing dynamic recrystallization", *Materials transactions*, 42(7) (2001), 1200-1205.
9. M. E. Kassner and S. R. Barrabes, "New developments in geometric dynamic recrystallization", *Materials Science and Engineering: A*, 410-411 (2005), 152-155.
10. F. J. Humphreys and M. Hatherly, "Recrystallization and related annealing phenomena" (2004), Elsevier.
11. G. J. Richardson, C. M. Sellars, and W. J. M. Tegart, "Recrystallization during creep of nickel", *Acta Metallurgica*, 14(10) (1966), 1225-1236.
12. T. Gladman, "On the Theory of the Effect of Precipitate Particles on Grain Growth in Metals", *Proceedings of the Royal Society A,* 294(1438) (1966), 298-309
13. M. M. Avedesian and H. Baker, "ASM Specialty Handbook: Magnesium and Magnesium Alloys" (1999), Materials Park, OH, USA, ASM International.
14. C. W. Bale, P. Chartrand, S. A. Degterov, G. Eriksson, K. Hack, R. B. Mahfoud, J. Melancon, A. D. Pelton, and S. Petersen, "FactSage thermochemical software and databases", *Calphad*, 26(2) (2002), 189-228.
15. D. A. Porter, K. E. Easterling, and M. Y. Sherif, "Phase transformations in metals and alloys" (2009), CRC press.
16. N. Hort, Y. Huang, and K. U. Kainer, "Intermetallics in Magnesium Alloys", *Advanced Engineering Materials*, 8(4) (2006), 235-240.
17. A. Galiyev, R. Kaibyshev, and G. Gottstein, "Correlation of plastic deformation and dynamic recrystallization in magnesium alloy ZK60", *Acta Materialia*, 49(7) (2001), 1199-1207.
18. S. M. Fatemi-Varzaneh, A. Zarei-Hanzaki, and H. Beladi, "Dynamic recrystallization in AZ31 magnesium alloy", *Materials Science and Engineering: A*, 456(1-2) (2007), 52-57.
19. A. G. Beer, "The role of strain on the recrystallization behaviour of hot worked and annealed magnesium alloy Mg-3Al-1Zn", Magnesium Technology 2010 : The Minerals, Metals & Materials Society (TMS), Seattle, WA (2010), 239-244.
20. S. Soheil, "Determination of critical strain for initiation of dynamic recrystallization", *Materials & Design*, 31(3), (2010) 1360-1364.
21. P. Bocher, "Dynamic recrystallization in a 304 stainless steel: microtexture measurement and models", (PhD thesis, McGill University, 1998).
22. X.-Y. Yang, Z.-S. Ji, H. Miura, and T. Sakai, "Dynamic recrystallization and texture development during hot deformation of magnesium alloy AZ31", *Transactions of Nonferrous Metals Society of China*, 19(1) (2009), 55-60.

**Magnesium Technology 2012**
*Edited by: Suveen N. Mathaudhu, Wim H. Sillekens, Neale R. Neelameggham, and Norbert Hort*
***TMS (The Minerals, Metals & Materials Society), 2012***

# The effect of precipitation on the mechanical properties of extruded AZ80

Ran Liu[1], De Liang Yin[2], Jing Tao Wang[3]
School of Materials Science and Engineering, Nanjing University of Science and Technology, Nanjing, 210094, P.R. China
[1]*liuran840211@163.com* [2]*dlyin@mail.njust.edu.cn* [3]*jtwang@mail.njust.edu.cn*

Key words: Ageing; Precipitated phase; Twinning; Mechanical properties

**Abstract**

The effect of ageing temperature and time on precipitates in AZ80 magnesium alloy were studied by optical microscopy to investigate the influence on mechanical properties. The results show the precipitation is distributed as bands along the extrusion direction, with no twinning. After ageing, the precipitation becomes dispersed uniformly. When ageing at 423 K, discontinuous precipitated phase increases the yield and tensile strength but reduces the ductility of the material. For ageing at 573 K, with increasing the ageing time, static ductility first increases and then decreases. This means that recrystallization induced by continuous precipitation is beneficial to both the strength and ductility. However, the effect of age hardening under 573 K is not better than 423 K. And $\{10\bar{1}2\} < \bar{1}011 >$ twinning was also activated in compression much more than in tensile samples, which is consistent with the behaviour of AZ31.

## Introduction

Magnesium alloys are considered as potential candidates for numerous applications, especially in transportation vehicles or lightweight enclosures for computer products owing to their excellent properties, such as low density, high strength-to-weigh ratios and high recycle-ability[1-2]. The Mg-Al-Zn is the most widely used magnesium alloy at present. Some studies on AZ80 have been reported on its microstructure and mechanical properties at different ageing temperatures [3-7]. However, the AZ80 alloy still has some drawbacks, such as poor ductility, low thermal conductivity and resistivity for example. The Mg-Al-Zn alloy usually contains large amounts of a coarse and network $\beta$-$Mg_{17}Al_{12}$ as a divorced eutectic distributed along the grain boundaries [8]. The network distributing of β will restrict the deformation and interfere with slip transfer. According to the Orowan theory, the existing of β phase which block the movement of dislocation can change the critical resolved shear stress (CRSS) of the slip systems. In the literature [9], it is considered that continuous precipitation is responsible for most of the age-hardening in Mg-Al-based alloys which is not the same as in this study.

So to further understand this alloy, the objective of this work is to investigate the effect of precipitation on the mechanical properties of AZ80 and the microstructure present during the yield stage to obtain a better understanding of how twinning affects deformation.

## Experimental

The material used in the present study was a commercial AZ80 alloy with the measured composition (wt. %): 8.3Al, 0.6Zn, 0.5 Mn, 0.002 Cu, Mg (balance). The as-cast ingot was homogenized at 683 K for 12h and extruded at 593 K with an extrusion ratio of 8 and an extrusion rate 4.2 $s^{-1}$ then air cooled. To clarify the precipitation behavior of this alloy, the material was then aged at either 423 K or 573 K for varying times. During ageing, the β phase precipitated out in two forms: discontinuous and continuous precipitations. Mechanical tests were conducted at room temperature with an initial strain rate of $2.67\times10^{-3}$ $s^{-1}$ on a CSS-4410 testing machine. Optical microstructural characterization was conducted using an OLYMPUS-BXD60M microscope. Specimens for microstructure observation were grinded and electrolytic polishing before etching for 5 s in a solution of 5 ml acetic acid, 1.5 g picric acid, 10 ml distilled water and 25ml ethanol.

## Results and discussion

Figure 1 shows the as-extruded microstructure with no twinning from a plane parallel to the extrusion direction (ED). During deformation, the original grains are refined, and their boundaries become serrated and in addition subgrains are formed.

Recrystallization is complete and the average grain size after extrusion is 27.04 μm. For magnesium alloys, which have low stacking fault energy, the activation of dislocation cross slip and climbing is not easy. Grain refinement could be attributed to continuous dynamic recrystallization which involves a progressive increase in grain boundary misorientation and changes of low angle boundaries into high angle boundaries [10]. For extrusion temperatures are high enough that non-basal slip is active then twinning is not observed. The as-cast alloy contains a microstructure consisting of α-Mg matrix, and eutectic β-$Mg_{17}Al_{12}$. The precipitated phase was broken up and distributed as bands along the ED on the grain boundaries. Because the atomic mismatch and more imperfections than those of the grain interior, and the grain boundary energy is high, the β-$Mg_{17}Al_{12}$ phase is apt to nucleate and grow along the grain boundaries.

After ageing, no matter discontinuous at 423 K or continuous at 573 K precipitation disperse but still on the boundaries. The morphology of precipitate at 573 K is not the same as samples ageing after solid solution. Duly [11] states that at high temperatures, discontinuous precipitation disappears both because volume diffusion prevents its growth and because its nucleation becomes difficult. Clark [12] had reported that the continuous precipitation consists of relatively large plates on the basal of the matrix. They have certain orientation relationship in literature [13-15]. However, in Wei Tang's [3] work, they clarified the discontinuous precipitation with a lamella structure which has different shape and distributing have the same orientation with the matrix comparing to the continuous precipitation. That means the orientation between the matrix and precipitate has no relationship with the latter mechanical behaviors.

Figure 2 shows the yield and ultimate strength of AZ80 after ageing at 423 K which can be compared to that after ageing at 573 K. The results show a very distinct trend and almost the same for the yield stress at the two temperatures with time. When ageing at 423 K, the yield stresses for either tension or compression both increases with increasing time. However, ageing at 573 K, the yield stress didn't increase significantly in tension, but decreased dramatically in compression. We also observed that ageing didn't affect the ultimate strength at either ageing temperature.

When ageing at higher temperature, the yield stress increases slightly whereas the ultimate strength decreases. One reason is that under high temperature, supersaturation of Al element is low which affects the volume fraction of β. Another main reason for the poor ageing hardness response of AZ80 alloy has been attributed to the orientation and coarseness of the continuous precipitation that makes them inefficient obstacles to dislocation movement. Slip in magnesium, like most hcp metals, occurs predominantly on the close-packed basal planes (0001). Since the continuous precipitation consists of coarsely dispersed plates lying parallel to the basal plane, there are many opportunities for dislocation to glide between the precipitates [9]. On the other hand, discontinuous precipitation can effectively prevent the activation of twinning and the movement of dislocations, this is responsible for the increase in strength.

Figure 4 and Figure 5 shows that the ageing temperature has a different impact on the static toughness and uniform strain. They all exhibited different trends under different temperatures. The static toughness and uniform strain for ageing at 423 K decrease continuously while they decreased initially and then increased for ageing at 573 K. With further ageing, the volume fraction of β-$Mg_{17}Al_{12}$ phase on the grain boundaries increases, which is an obstacle to the movement of dislocation. This leads to dislocations pile-ups and this enhances the stress concentration resulting in microcracking.

Microstructures after ageing at 423 K for 24h having been strained 2% in either tension or compressive testing are showed in Figure 6. It can be seen that twinning occurred in the mechanical testing of aged samples. For processing by traditional extrusion, the texture can be roughly considered as having a fiber texture with the basal planes parallel to the extrusion direction. It is well known that Mg and its alloys are low-symmetry materials with hexagonal-close packed (hcp) structures that have axial ratios (c/a) of around 1.633 which is less than $\sqrt{3}$. This pattern of texture does not allow any twinning under tensile loading but maximum twinning under compressive loading. Therefore, this starting texture could activate the $\{10\bar{1}2\}$ twinning at a low degree of deformation. The other reason maybe twinning and precipitation both have a certain relationship with the α-Mg matrix, they two must have interaction on each other. The pattern of twinning is $\{10\bar{1}2\}<\bar{1}011>$ which occurred much more in compression than in tension, this is the same result as observed in AZ31. Chengling

Lv [16] also found the same phenomenon. Twinning nucleated from the interface of the precipitation. From observations of the same region before and after deformation, the edge of precipitated phase is broken which is the spreading of the crack. That's because the matrix and the precipitated phase which have different structure are not sensitive resemblancely to the applied stress.

**Conclusions**

Ageing treatment with different times at either 423 K or 573 K have been applied to the extruded AZ80 alloy, aimed at investigating the effect of heat treatment on the mechanical properties and the presence of twinning during tensile and compressive testing at room temperature. The following conclusions have been made:

(1) The ageing temperature plays an important role in the precipitating of β such that continuous precipitation occurs in 573 K, and discontinuous precipitation occurs at 423 K.

(2) The material exhibits different trends of mechanical properties. And precipitation strengthening doesn't work in this study. Discontinuous precipitation is responsible for the age-hardening in this study which is different to that observed in former published work.

(3) Twinning is activated during compression at room temperature of hot extruded AZ80 alloy after ageing at 423 K since the existing sharp fiber texture favored the activation of $\{10\bar{1}2\}$ twinning which is the same result as in AZ31 alloy.

**Acknowledgement**

This work was supported by NUST Funding, under Grant No.2011XQTR04.

**References**

1. B.L. Mordike, T. Ebert, "Magnesium: Properties – appications - potential," Mater Sci Eng A, 302 (2001), 37-45.

2. Y. Kojima, "Platform science and technology for advanced magnesium alloys," Mater. Sci. Forum, 350-351 (2000), 3-18.

3. D. Duly et al., "Microstructure and local crystallographic evolution in an Al-1wt% Mg alloy deformed at intermediate temperature and high strain-rate," Acta Mater., 44 (1996), 2947-2962.

4. W. Tang et al., "Effect of heat treatment on microstructure and properties of AZ80 magnesium alloy," Acta Metallurica Sinica, 41 (2005), 1199-1206.

5. I.A. Yakubtsov et al., "Effects of heat treatment on microstructure and tensile deformation of Mg AZ80 alloy at room temperature," Mater Sci Eng A, 496 (2008), 247-255.

6. Y. Li, Z. Zhang and Y. Xue, "Influence of ageing on microstructure and mechanical properties of AZ80 and ZK60 magnesium alloys," Trans. Nonferrous Met. Coc. China, 21 (2011), 739-744.

7. Y .Li, B. Li and C. Yin, "Influence of ageing treatment on microstructure and mechanical properties of AZ80," Material & Heat Treatment, 38 (2009), 171-174.

8. A.K. Dahle, Journal of light metals. 2001, 1(1):61-72.

9. S. Celotto, "TEM study of continuous precipitation in Mg–9 wt%Al–1 wt%Zn alloy, " Acta Mater., 48 (2000), 1775-1787.

10. J.C. Tan, M.J. Tan, "Dynamic continuous recrystallization characteristics in two stage deformation of Mg–3Al–1Zn alloy sheet," Mater Sci Eng A, 339 (2003), 124-132.

11. D. Duly, J.P. Simon and Y. Berchet, "On the competition between continuous and discontinuous precipitations in binary Mg-Al alloys," Acta Metall. Mater., 43 (1995), 101-106.

12. J.B. Clark, "Age hardening in a Mg-9 wt.% Al alloy," Acta Metall., 16 (1968), 141-152.

13. A.F. Crawley, K.S.Milliken, "Precipitate morphology and orientation relationships in an aged Mg-9% Al-1% Zn-0.3% Mn alloy," Acta Metall., 22 (1974), 557-562.

14. A.F. Clawley, B. Lagowski, "Effect of two-step ageing on the precipitate structure in magnesium alloy AZ91," Metall Trans., 5 (1974), 949-951.

15. C.L. Xiao, C.P. Luo, and J.F. Nie, "Morphology and crystallography of β-(Mg17Al12) precipitate in an AZ91," Acta Metallurica Sinica, 37 (2001), 1-7.

16. C. Lv et al., "Effect of heat treatment on tension–compression yield asymmetry of AZ80 magnesium alloy," Materials & Design, 33 (2012), 529-533.

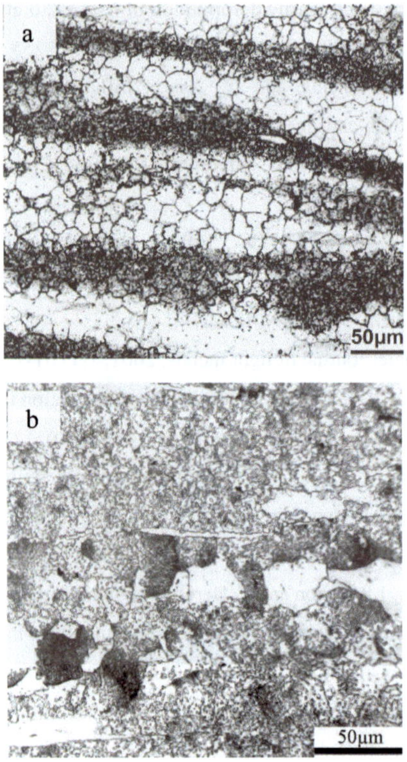


Figure 1. Optical microstructures of the cross section of AZ80 alloy after extrusion

(a) after extrusion (b) morphology of the continuous of the precipitated phase after extrusion

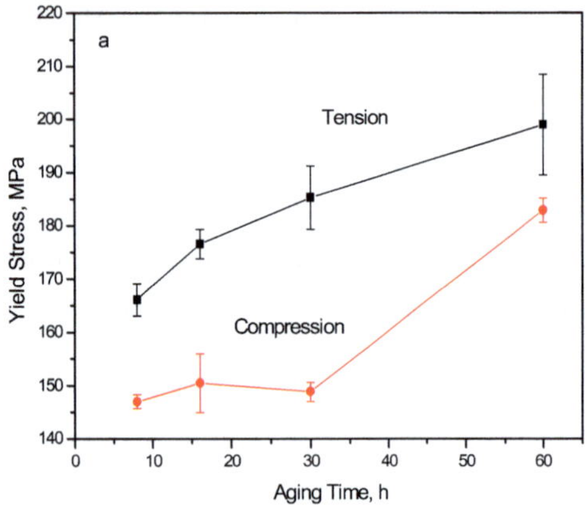


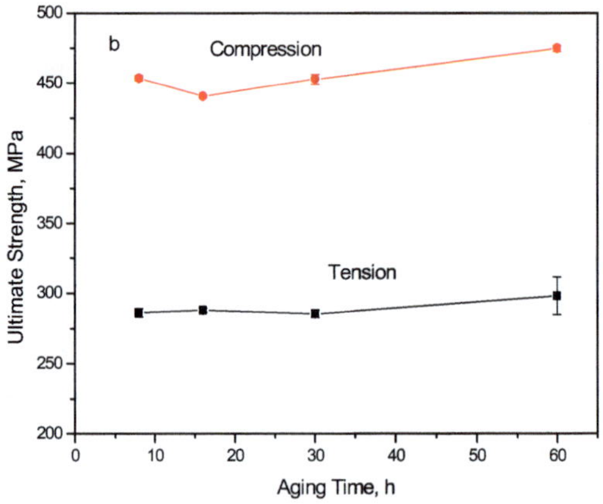


Figure 2. Yield (a) and ultimate strength (b) of AZ80 alloy ageing at 423 K

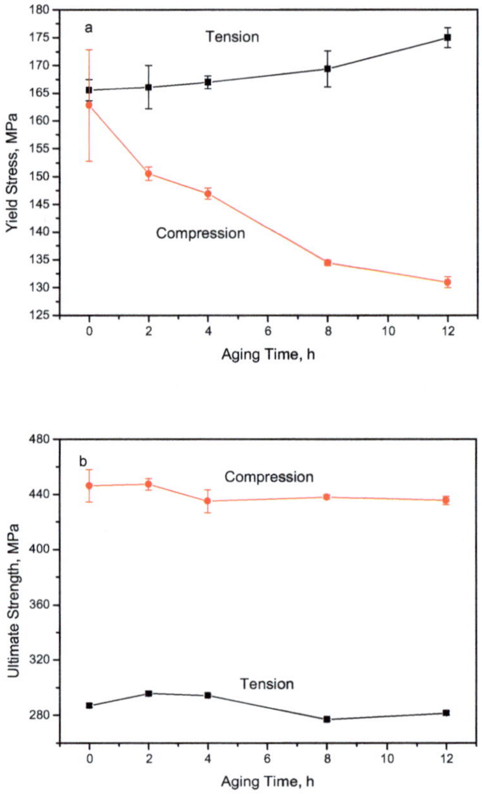


Figure 3. Yield stress (a) and ultimate strength (b) of AZ80 alloy ageing at 573 K

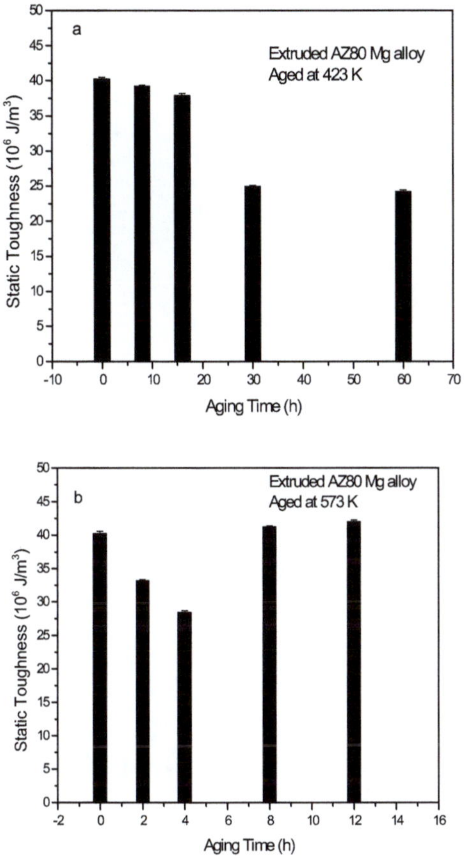


Figure 4. Static toughness of AZ80 alloy ageing at 423K (a) and 573 K (b)

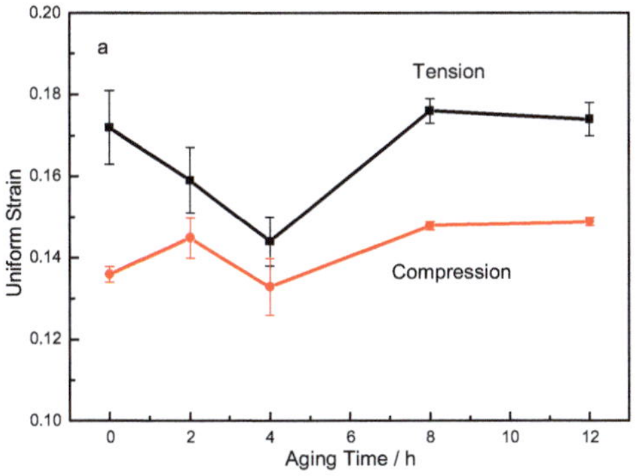


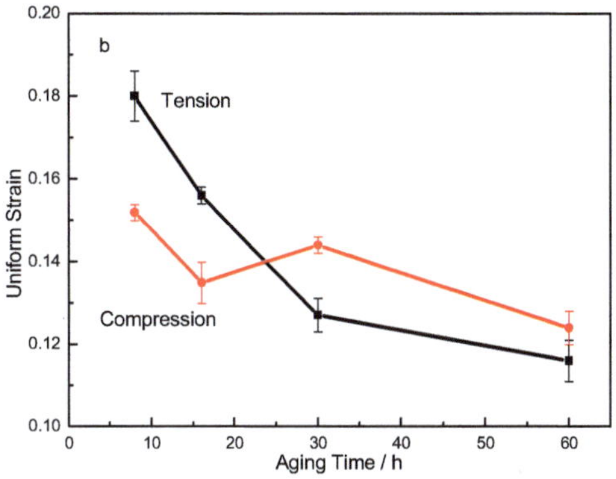


Figure 5. Uniform plastic deformation at 423 K (a) and 573 K (b) with different ageing time on extruded AZ80 alloy

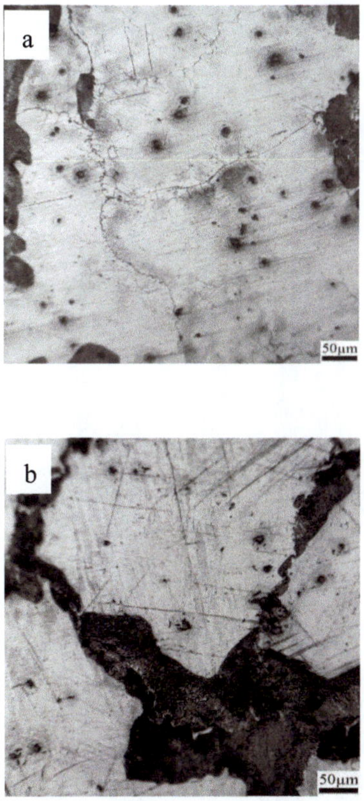


Figure 6. Microstructure of extruded AZ80 alloy with a strain of 2% for tension (a) and compressive (b)

**Magnesium Technology 2012**
*Edited by: Suveen N. Mathaudhu, Wim H. Sillekens, Neale R. Neelameggham, and Norbert Hort*
***TMS (The Minerals, Metals & Materials Society), 2012***

# INFLUENCE OF SECTION THICKNESS ON MICROSTRUCTURE AND MECHANICAL PROPERTIES OF SQUEEZE CAST MAGNESIUM ALLOY AM60

Xuezhi Zhang, Meng Wang, Zhizhong Sun, Henry Hu

Department of Mechanical, Automotive and Materials Engineering
University of Windsor, 401 Sunset Ave.
Windsor, Ontario, Canada N9B 3P4
E-mail: zhang11w@uwindsor.ca, wang112j@uwindsor.ca, sun11@uwindsor.ca, huh@uwindsor.ca

Keywords: Squeeze casting, Magnesium alloy AM60, Section thickness, Tensile properties

## Abstract

Squeeze cast light alloys has been approved for advanced engineering design of light integrity automotive applications. An understanding of the effect of section thicknesses on mechanical properties of squeeze cast magnesium alloys is essential for proper design of different applications. The present work studied the microstructure and tensile properties of magnesium alloy AM60 with different section thickness of 6, 10 and 20mm squeeze cast under an applied pressure of 30MPa. The results of tensile testing indicate that the yield strength (YS), ultimate tensile strength (UTS) and elongation ($E_f$) increase with a decreasing in section thicknesses of squeeze cast AM60. The microstructure analysis shows that the improvement in the tensile properties of squeeze cast AM60 is mainly attributed to the low level of gas porosity and the high content of eutectic phases and fine grain structure which resulted from high solidification rates taking place in the thin section.

## Introduction

Weight reduction and increasing on fuel efficiency have become a trend in automotive development. Magnesium is one-third lighter than aluminum, three-fourths lighter than zinc, and four-fifths lighter than steel [1,2]. Moreover, the combination of high specific strength and stiffness, and excellent castablity, high die casting rates and high dimensional accuracy qualify this interesting lightweight metal in the automotive industry. Potential applications of magnesium alloys on automotive could involve cross sections with different wall thicknesses and complex shapes. Squeeze casting is designed for production of relatively thick-walled parts and fine microstructure by means of slow filling velocity, semi-solid processing and solidification under high pressure [2].

Squeeze casting is termed to describe a process that involves the solidification of a molten metal in closed die under an imposed high pressure. This process has not been widely applied in the production of magnesium component in automotive industry. Thus, magnesium fabrication techniques must be varied and to extend the limits imposed by the current and traditional gravity and die casting techniques. The development of squeeze casting technique for magnesium alloys will enhance the competitiveness of magnesium components in the growing automotive market. In this work, magnesium alloy AM60 with various section thicknesses was squeeze cast under applied pressure of 30Mpa. The microstructure and tensile behaviour of alloy AM60 were studied. And, their relations with section thickness are presented. The mechanisms responsible for the resulted tensile properties are discussed based on the optical microstructural characterization [1-4].

## Experimental Procedure

The base magnesium alloy selected for this study was the conventional magnesium alloy AM60. This alloy was used to produce step castings without any addition. The chemical composition of this alloy is shown in Table 1. A step mold made of tool steel was used to fabricate step squeeze castings. The thicknesses of each step were 6mm, 10mm and 20mm as shown in figure 1. The step casting was then cut for density measurement, porosity evaluation, tensile testing and microstructure analysis.

Table 1: Chemical composition of the investigated alloy (wt. %)

| Alloy | Al | Mn | Si | Fe | Mg |
|---|---|---|---|---|---|
| AM60 | 5.93 | 0.18 | <0.02 | 0.013 | Bal. |

During casting process, protective gas mixture of sulfur hexafluoride, $SF_6$ and carbon dioxide, $CO_2$ in which $CO_2$ acted as the carrier gas, was applied to protect the melt from any excessive oxidation or possible burning. All tools for melt processing were preheated to 150 °C on the top of a box furnace for at least 20 minutes. Both top and lower die were preheated to 300 °C before pouring.

Specimens for tensile test, density measurement, porosity evaluation and microstructure analysis were obtained by sectioning the steps of squeeze castings.

### Porosity Measurement

Porosity was evaluated via density measurement. Following the measurement of specimen weight in air and distilled water, the actual density ($D_a$) of each specimen was determined using Archimedes principle based on ASTM Standard D3800 [5]:

$$D_a=W_aD_w/(W_a-W_w) \quad (1)$$

where $W_a$ and $W_w$ are the weight of the specimen in air and in water, respectively, and $D_w$ the density of water. The porosity of each specimen was calculated by the following equation (ASTM Standard C948):

$$\%Porosity=[(D_t-D_a)/D_t]\times 100\% \quad (2)$$

where $D_t$ is the theoretical density of the alloy AM60,which is 1.77 $g/cm^3$ [5].

Tensile Testing

The mechanical properties of the squeeze cast AM60 was evaluated by tensile testing. Subsize rectangular specimens were prepared according to ASTM standard B557M [6]. The gauge length and the width of the specimens were 25mm and 6mm, respectively. The thicknesses of specimens were kept the same as those of each step. The cross-section areas were measured after each specimen surface was polished in order to avoid stress concentration.

After preparation, the specimens were tested at ambient temperature on an Instron 8562 universal testing machine equipped with a computer data acquisition system. The outputted date (displacement and tensile load) was analyzed. The tensile properties, including ultimate tensile strength (UTS), 0.2% yield strength (YS), and elongation to failure ($E_f$) were obtained for each step thickness.

Figure 1: Schematic diagram showing a step squeeze casting.

Metallography

Specimens were prepared following the procedure which including: sectioning, mounting, grinding, polishing, T4 solution treatment and etching. Specimens were grounded using 240, 320, 400 and 600 grit silicon carbide papers, and ten polished using 1μm $Al_2O_3$ suspension. The etchant used was nitride (1ml $HNO_3$, 48ml ethanol). Etching was performed by merging the sample into the etchant for 30 seconds and ten washed the specimen surface with ethanol and running water. A Buehler optical image analyzer 2002 system was used to observe the grain structure.

**Results and Discussion**

Porosity Evaluation

Figure 2 shows the density and porosity measurement of squeeze cast AM60 with section thicknesses of 6, 10 and 20mm. It evident that the density of squeeze cast AM60 alloy samples increases and the porosity level decreases with an increase in section thickness. The porosity level of 6 mm specimen is significantly low (0.2%). The considerably low porosity level of the 6 mm sample may results from the high cooling rate. The numerical simulation of solidification of the step casting suggests the long solidification time should be responsible for the high level of porosity in the squeeze casting with thick cross-sections.

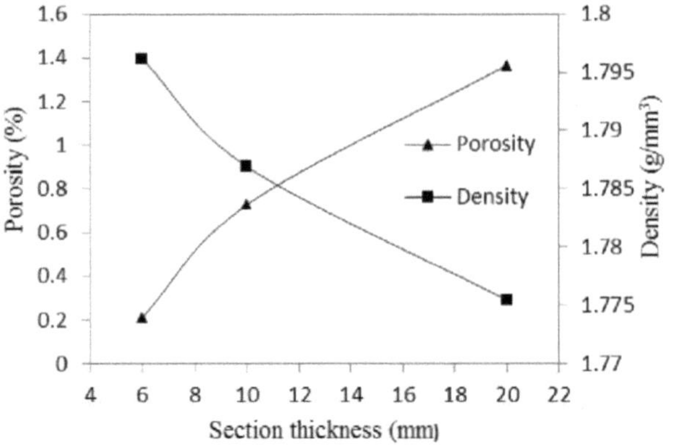


Figure 2: Porosity and density level of squeeze cast AM60 with thicknesses of 6, 10, and 20mm.

Tensile Properties

Figure 3 presents stress and strain curve for each step thickness of the squeeze cast AM60 alloy. For all three section thicknesses of specimens, the curves show that the alloy deforms elastically first under tensile loading. After the yield pointreaches, the alloy starts to deform plastically. It is obvious that the specimen with the thinner section (6 mm) has higher ultimate tensile strength, yield strength and elongation than those thicker specimens (10 and 20 mm). The variation of tensile properties with section thicknesses is summarized in table 2.

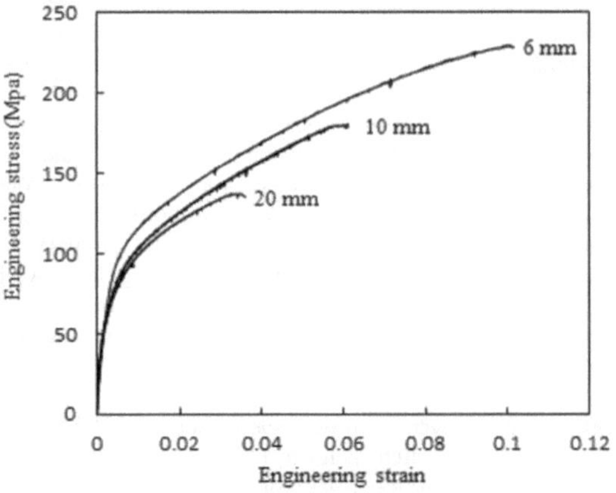


Figure 3: Typical engineering stress-strain curve of squeeze cast AM60 alloy.

Table 1: Effect of section thicknesses on UTS, YS and elongation.

| Thickness (mm) | YS (MPa) | UTS (MPa) | $E_f$ (%) |
|---|---|---|---|
| 6 | 96 | 229.89 | 10.19 |
| 10 | 74 | 181.04 | 6.52 |
| 20 | 67 | 137.68 | 3.69 |

Figure 4 shows the effect of section thickness on UTS, YS and $E_f$. it shows that a decease in section thicknesses enhances the elongation with an 176% increment from 3.69% (20 mm) to 10.19% (6 mm). Also, there are 43% and 67% increases in YS and UTS over the 20 mm thickness, respectively. The improvement in the tensile properties should be attributed to the low porosity level and fine cell structure of thin specimen. In other words, the low strength and poor elongation of the thick specimen should be resulted from coarse microstructure and high porosity content.

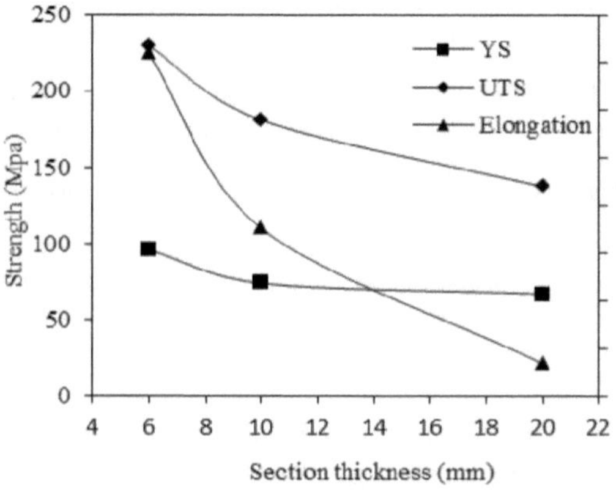


Figure 4: UTS, YS and elongation vs. section thickness.

Microstructure

Figure 5 parts presents optical micrographs generated at a magnification of 500X for squeeze cast AM60 specimens with the section thicknesses of 6 mm, 10 mm and 20 mm, respectively. Average grain size escalates from 16 μm for 6 mm specimen to 80 μm for 20 mm specimen. The greater total thermal energy in the thicker section of liquid metal requires more time for removal during solidification process if the thermal conductivity of the mold is the same for all sections [7-9]. As a result, the longer the time spent at elevated temperatures, the lager the grains grow. In other words, the thicker section in the same mold experienced a slower cooling rate and results in coarser microstructure.

(a)

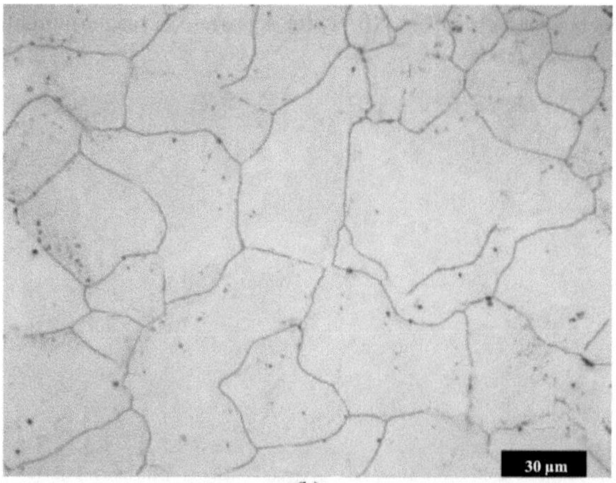


(b)

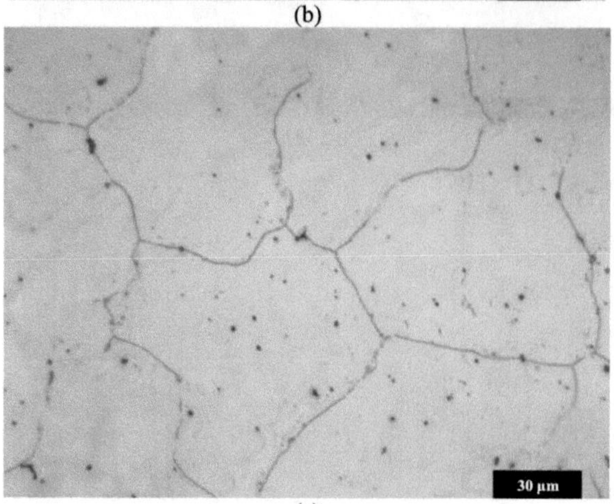


(c)

Figure 5: Optical micrographs showing grain structure of a) 6 mm b) 10mm and c) 20mm specimens.

The variation on grain size from one section to another has a large influence on the mechanical properties. It is especially true for the most prominent of the mechanical property degradation: elongation. As the grain size increases, the elongation drops significantly. This is in agreement with the results of tensile test data (table 2).

A severe embrittlement of the squeeze cast AM60 with increasing section thickness was observed. Fractographs in figure 6 (a) and (b) exemplify this range in fracture behaviour between the 6 mm and 20 mm thick specimens. Figure 6 (a) illustrate a ductile fracture surface containing dimples with dramatic height variation resulting from the elongated nature of the surface. In contrast, brittle fracture is shown in figure 6 (b) with the dominating presence of cleavage fracture, flat facets. In mechanism, the specimens fail by microvoid coalescence under tensile stress. The microvoid nucleate at the area of localized high plastic deformation which associate with second phase particles and grain boundaries. Eventually, continuous fracture surface forms as the microvoid grow. A considerable amount of energy is consumed of the formation of microvoids and finally leading to creation of cracks. Such embrittlement phenomena have been well documented elsewhere [10,11] for squeeze cast magnesium alloy including AM60.

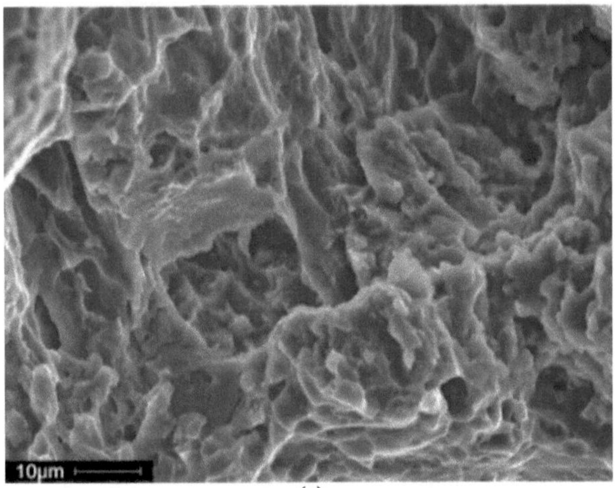


(a)

(b)

Figure 6: SEM fractograph of a) 6 mm and b) 20mm squeeze cast AM60 specimens.

## Conclusions

The influence of section thicknesses was investigated on tensile properties and microstructure of squeeze cast AM60 alloy. The significant increases in elongation (176%), UTS (67%) and YS(43%) of the 6 mm over the 20 mm section were achieved. The dependence of tensile properties on the section thickness should be attributed to the variation of solidification rates and the resulted microstructure of the squeeze cast AM60 alloy. The microstructure features of the 6, 10 and 20mm specimens were studied via optical metallography. The section thickness has significant influence on the grain size and porosity level of the squeeze cast samples. As the section thickness increased, the grain size porosity level increase and consequently reduced the tensile properties. The observation via SEM fractography illustrates that the fracture behaviour affected by the section thickness. The fracture of AM60 tends to transit from ductile to brittle as the section thickness increased.

## Acknowledgements

The authors would like to thank the Natural Sciences and Engineering Research Council of Canada, and University of Windsor for supporting this work.

## References

[1] G.S. Cole, *Summary of Magnesium Vision 2020:A North American Strategic Vision for Magnesium*, Magnesium Technology 2007, Editor(s) - Randy S. Beals, TMS, TMS, 2007 TMS Annual Meeting & Exhibition, 2007, 35-40.

[2] H. Hu, Alfred Yu, Naiyi Li and John E. Allison, *Potential Magnesium Alloys for High Temperature Die Cast Automotive Applications: A Review*, Material and manufactureing process (2003), Vol. 18, No.5, pp 687-717 .

[3] H. T. Gjestland S. Sannes, H. Westengen, D. Albright, *Effect of CastingTemperature, Section Thickness and Die Filling Sequence on Microstructure and Mechanical Properties of High Pressure Die Castings*, NADCA Transactions, Indianapolis, 2003, T03-036.

[4] M. Zhou, N. Li and H. Hu, *Effect of Section Thicknesses on Tensile Behavior and Microstructure of High Pressure Die Cast Magnesium Alloy AM50* , Materials Science Forum Vols. 475-479 (2005) pp 463-468.

[5] ASTM Standard D3800, 1999 (2004), *Standard Test Method for Density of High-Modulus Fibers*, ASTM International, 2004, DOI 10.1520/D3800-99R04

[6] *Standard Test Methods of Tension Testing Wrought and Cast Aluminum – and Magnesium – Alloy*, B557M, ASTM Standards, ASTM International, 2007, DOI 10. 1520/B0557M-07E01.

[7] M.S. Yong, A.J. Clegg, *Process Optimization for a Squeeze Cast Magnesium Alloy*, J Mater Process Technology (2004) 145:134–141.

[8] A.Yu, S. Wang, N. Li, H. Hu, *Pressurized Solidification of Magnesium Alloy AM50A*, Journal of Materials Processing Technology, 191, 247-250, 2007.

[9] Z. Sun, H. Hu, X. Niu, *Determination of Heat Transfer Coefficients by Extrapolation and Numerical Inverse Methods in Squeeze Casting of Magnesium Alloy AM60*, Journal of Materials Processing Technology, (2011) 211 1432-1440.

[10] M. Zhou, H. Hu, N. Li and J. Lo, *Microstructure and Tensile Properties of Squeeze Cast Magnesium Alloy AM50, Journal of Materials Engineering & Performance*, Vol. 14 No.4, 539-545, 2005.

[11] Z. Sun, M. Zhou, H. Hu and N. Li, *Strain-Hardening and Fracture Behavior of Die Cast Magnesium Alloy AM50*, Research Letters in Materials Science, Volume 2007, 1-5.

# AUTHOR INDEX
# Magnesium Technology 2012

# J



# K



# L



# M



# N



# O



# P

# SUBJECT INDEX
# Magnesium Technology 2012

## T



## U



## V



## W



## Y



## Z

MIX
Papier aus verantwortungsvollen Quellen
Paper from responsible sources
FSC® C105338

If you have any concerns about our products,
you can contact us on
**ProductSafety@springernature.com**

In case Publisher is established outside the EU,
the EU authorized representative is:
**Springer Nature Customer Service Center GmbH**
**Europaplatz 3, 69115 Heidelberg, Germany**

Printed by Libri Plureos GmbH
in Hamburg, Germany